OCCUPATIONAL NEUROLOGY

HANDBOOK OF CLINICAL NEUROLOGY

Series Editors

MICHAEL J. AMINOFF, FRANÇOIS BOLLER, AND DICK F. SWAAB

VOLUME 131

ELSEVIER

EDINBURGH LONDON NEW YORK OXFORD PHILADELPHIA
ST LOUIS SYDNEY TORONTO 2015

OCCUPATIONAL NEUROLOGY

Series Editors

MICHAEL J. AMINOFF, FRANÇOIS BOLLER, AND DICK F. SWAAB

Volume Editors

MARCELLO LOTTI AND MARGIT L. BLEECKER

VOLUME 131

3rd Series

ELSEVIER

EDINBURGH LONDON NEW YORK OXFORD PHILADELPHIA
ST LOUIS SYDNEY TORONTO 2015

ELSEVIER B.V.
Radarweg 29, PO Box 211, 1000 AE Amsterdam, Netherlands
The Boulevard, Langford Lane, Kidlington, Oxford OX5 1GB, UK
225 Wyman Street, Waltham, MA 02451, USA

ISBN: 9780444626271

British Library Cataloguing in Publication Data
A catalogue record for this book is available from the British Library

Library of Congress Cataloging in Publication Data
A catalog record for this book is available from the Library of Congress

Commissioning Editor: Mara E. Conner
Development Editor: Michael Parkinson
Project Manager: Sujatha Thirugnana Sambandam
Designer/Design Direction: Alan Studholme

The Publisher's policy is to use **paper manufactured from sustainable forests**

Handbook of Clinical Neurology 3rd Series

Available titles

Vol. 79, The human hypothalamus: basic and clinical aspects, Part I, D.F. Swaab ISBN 9780444513571
Vol. 80, The human hypothalamus: basic and clinical aspects, Part II, D.F. Swaab ISBN 9780444514905
Vol. 81, Pain, F. Cervero and T.S. Jensen, eds. ISBN 9780444519016
Vol. 82, Motor neurone disorders and related diseases, A.A. Eisen and P.J. Shaw, eds. ISBN 9780444518941
Vol. 83, Parkinson's disease and related disorders, Part I, W.C. Koller and E. Melamed, eds. ISBN 9780444519009
Vol. 84, Parkinson's disease and related disorders, Part II, W.C. Koller and E. Melamed, eds. ISBN 9780444528933
Vol. 85, HIV/AIDS and the nervous system, P. Portegies and J. Berger, eds. ISBN 9780444520104
Vol. 86, Myopathies, F.L. Mastaglia and D. Hilton Jones, eds. ISBN 9780444518996
Vol. 87, Malformations of the nervous system, H.B. Sarnat and P. Curatolo, eds. ISBN 9780444518965
Vol. 88, Neuropsychology and behavioural neurology, G. Goldenberg and B.C. Miller, eds. ISBN 9780444518972
Vol. 89, Dementias, C. Duyckaerts and I. Litvan, eds. ISBN 9780444518989
Vol. 90, Disorders of consciousness, G.B. Young and E.F.M. Wijdicks, eds. ISBN 9780444518958
Vol. 91, Neuromuscular junction disorders, A.G. Engel, ed. ISBN 9780444520081
Vol. 92, Stroke – Part I: Basic and epidemiological aspects, M. Fisher, ed. ISBN 9780444520036
Vol. 93, Stroke – Part II: Clinical manifestations and pathogenesis, M. Fisher, ed. ISBN 9780444520043
Vol. 94, Stroke – Part III: Investigations and management, M. Fisher, ed. ISBN 9780444520050
Vol. 95, History of neurology, S. Finger, F. Boller and K.L. Tyler, eds. ISBN 9780444520081
Vol. 96, Bacterial infections of the central nervous system, K.L. Roos and A.R. Tunkel, eds. ISBN 9780444520159
Vol. 97, Headache, G. Nappi and M.A. Moskowitz, eds. ISBN 9780444521392
Vol. 98, Sleep disorders, Part I, P. Montagna and S. Chokroverty, eds. ISBN 9780444520067
Vol. 99, Sleep disorders, Part II, P. Montagna and S. Chokroverty, eds. ISBN 9780444520074
Vol. 100, Hyperkinetic movement disorders, W.J. Weiner and E. Tolosa, eds. ISBN 9780444520142
Vol. 101, Muscular dystrophies, A. Amato and R.C. Griggs, eds. ISBN 9780080450315
Vol. 102, Neuro-ophthalmology, C. Kennard and R.J. Leigh, eds. ISBN 9780444529039
Vol. 103, Ataxic disorders, S.H. Subramony and A. Durr, eds. ISBN 9780444518927
Vol. 104, Neuro-oncology, Part I, W. Grisold and R. Sofietti, eds. ISBN 9780444521385
Vol. 105, Neuro-oncology, Part II, W. Grisold and R. Sofietti, eds. ISBN 9780444535023
Vol. 106, Neurobiology of psychiatric disorders, T. Schlaepfer and C.B. Nemeroff, eds. ISBN 9780444520029
Vol. 107, Epilepsy Part I, H. Stefan and W.H. Theodore, eds. ISBN 9780444528988
Vol. 108, Epilepsy Part II, H. Stefan and W.H. Theodore, eds. ISBN 9780444528995
Vol. 109, Spinal cord injury, J. Verhaagen and J.W. McDonald III, eds. ISBN 9780444521378
Vol. 110, Neurological rehabilitation, M. Barnes and D.C. Good, eds. ISBN 9780444529015
Vol. 111, Pediatric neurology Part I, O. Dulac, M. Lassonde and H.B. Sarnat, eds. ISBN 9780444528919
Vol. 112, Pediatric neurology Part II, O. Dulac, M. Lassonde and H.B. Sarnat, eds. ISBN 9780444529107
Vol. 113, Pediatric neurology Part III, O. Dulac, M. Lassonde and H.B. Sarnat, eds. ISBN 9780444595652
Vol. 114, Neuroparasitology and tropical neurology, H.H. Garcia, H.B. Tanowitz and O.H. Del Brutto, eds. ISBN 9780444534903
Vol. 115, Peripheral nerve disorders, G. Said and C. Krarup, eds. ISBN 9780444529022
Vol. 116, Brain stimulation, A.M. Lozano and M. Hallett, eds. ISBN 9780444534972
Vol. 117, Autonomic nervous system, R.M. Buijs and D.F. Swaab, eds. ISBN 9780444534910
Vol. 118, Ethical and legal issues in neurology, J.L. Bernat and H.R. Beresford, eds. ISBN 9780444535016
Vol. 119, Neurologic aspects of systemic disease Part I, J. Biller and J.M. Ferro, eds. ISBN 9780702040863
Vol. 120, Neurologic aspects of systemic disease Part II, J. Biller and J.M. Ferro, eds. ISBN 9780702040870
Vol. 121, Neurologic aspects of systemic disease Part III, J. Biller and J.M. Ferro, eds. ISBN 9780702040887
Vol. 122, Multiple sclerosis and related disorders, D.S. Goodin, ed. ISBN 9780444520012
Vol. 123, Neurovirology, A.C. Tselis and J. Booss, eds. ISBN 9780444534880

Foreword

We have long believed that there was a need for a volume on occupational neurology in the *Handbook of Clinical Neurology* series. Such a volume, we felt, would need to summarize the various chemical and physical agents that can damage the central or peripheral nervous systems or the sensory receptors, detailing both the individual agents and their underlying pathophysiologic mechanisms. It would also need to summarize the clinical, laboratory, and neuropsychologic evaluation of patients with neurologic disorders related to the workplace; the means of assessing the likely risks and level of neurotoxic exposure; and the clinical disorders that may result from occupational hazards, such as tremor, parkinsonism, encephalopathies, disorders of the special senses, pain syndromes, poly- and mononeuropathies, and abnormalities of the musculoskeletal system. These and other aspects are covered in the 28 chapters of the present volume, together with an account of the effects of shift work, considerations for returning to work after traumatic brain injury, and the accommodations needed for workers with chronic neurologic disorders.

The encyclopedic coverage provided in the present volume has been guided by the experience and skills of the two volume editors. Marcello Lotti is Professor of Environmental Medicine and Public Health at the University of Padua, Italy, and Margit Bleecker is affiliated with the Johns Hopkins Hospital and works at the Center for Occupational and Environmental Neurology in Baltimore, Maryland, USA. Both are widely recognized experts in the field and together they have made an outstanding team to develop and edit this new volume. They have gathered together a group of international contributors and crafted the text to ensure that the various chapters fit seamlessly together as a valuable synthesis of the literature. We are grateful to them and to all the contributors. We have reviewed and commented on each of the chapters in our capacity as series editors and believe that both clinicians and basic scientists will find much to appeal to them between the covers of this volume. The availability of the volume electronically on Elsevier's Science Direct site should ensure its ready accessibility and facilitate searches for specific information.

As always, it is a pleasure to thank Elsevier, our publishers – and in particular Mica Haley, Michael Parkinson, and Kristi Anderson – for their assistance in the development and production of this volume.

Michael J. Aminoff
François Boller
Dick F. Swaab

Preface

Velut in confinio duorum populorum constitutus ac simul ante retroque prospiciens.
[It is as if I were standing at the border between two peoples and were looking both forward and backward at the same time.] (Francesco Petrarca: Rerum Memorandarum Libri)

The neurotoxic effects of certain occupational exposures came to be appreciated initially by clinical observations. Later, with the advent of epidemiologic methods, studies of larger groups of workers led to the identification of neurotoxic effects that were clinically subtle or subclinical. More recently, certain work-related musculoskeletal disorders, previously disregarded, have come to be included in the practice of occupational neurology.

It may be asked why a new book on occupational neurology is necessary when important books on work-related neurologic disorders have been published over the years. Is it really necessary to report on certain diseases that have almost disappeared because of improvements in hygienic conditions in the workplace? Are there new criteria to identify and prevent these diseases? The many answers to such questions lead only to further questions.

New neurologic disorders have been associated with occupational exposure and the body of knowledge therefore requires updating, but the criteria governing the inclusion of new entries must be clarified, particularly when disorders are ill defined. Hygienic conditions are not the same worldwide and a given disease may be more prevalent in one country than another, where exposures are lower and disease uncommon. Nevertheless, less severe exposures and more subtle effects may occur even in well-developed countries where exposures are limited by law. New tools have been developed to detect neurologic disorders associated with occupational exposure. However, obtaining a meticulous occupational history is still of paramount importance and often the only means of establishing causality, and must not be overlooked. The effectiveness of available methods for monitoring workers at risk is often unclear and merits discussion.

It is for such reasons that we find ourselves, like the poet, looking backwards to what has been known for years and forwards to new developments. The aim of this volume of the *Handbook* is to condense all this knowledge in order to offer a comprehensive view of occupational neurology for neurologists, occupational medicine physicians, general practitioners, neuroscientists, industrial hygienists, and public health officials. We hope we have been successful in our endeavor.

Marcello Lotti
Margit L. Bleecker

Contributors

J.W. Albers
Department of Neurology, University of Michigan Health System, Ann Arbor, MI, USA

M.J. Aminoff
Department of Neurology, School of Medicine, University of California San Francisco, San Francisco, CA, USA

M. Aschner
Division of Clinical Pharmacology and Pediatric Toxicology; The Kennedy Center for Research on Human Development; Center in Molecular Toxicology and Center for Molecular Neuroscience, Vanderbilt University Medical Center, Nashville, TN, USA

K. Baigi
Montefiore Headache Center, Bronx, NY, USA

S. Bao
SHARP Program, Washington State Department of Labor and Industries, Olympia, WA, USA

S.K. Barnes
Center for Occupational and Environmental Neurology, Baltimore, MD, USA

M.L. Bleecker
Center for Occupational and Environmental Neurology, Baltimore, MD, USA

R. Bonfiglioli
Occupational Health Unit, Department of Medical and Surgical Sciences, University of Bologna, Bologna, Italy

R.M. Bowler
Department of Psychology, San Francisco State University, San Francisco, CA, USA

S. Caito
Division of Clinical Pharmacology and Pediatric Toxicology and The Kennedy Center for Research on Human Development, Vanderbilt University Medical Center, Nashville, TN, USA

W.M. Caudle
Department of Environmental Health, Rollins School of Public Health, Emory University, Atlanta, GA, USA

A.J. Cook
Neuroscience Institute, Texas A&M Health Science Center College of Medicine and Central Texas Veterans Healthcare System, Temple, TX, USA

G. Costa
Department of Clinical Sciences and Community Health, University of Milan and IRCCS "Ca' Granda Ospedale Maggiore Policlinico" Foundation, Milan, Italy

L.G. Costa
Department of Environmental and Occupational Health Sciences, University of Washington, Seattle, WA, USA and Department of Neuroscience, University of Parma, Parma, Italy

D. Dorman
Department of Molecular Biomedical Sciences, College of Veterinary Medicine, North Carolina State University, Raleigh, NC, USA

R.L. Doty
Smell and Taste Center, Department of Otorhinolaryngology; Head and Neck Surgery, Perelman School of Medicine, University of Pennsylvania, Philadelphia, PA, USA

C.M. Filley
Departments of Neurology and Psychiatry, University of Colorado School of Medicine and the Denver Veterans Affairs Medical Center, Aurora and Denver, CO, USA

D.A. Fox
Departments of Vision Sciences, Biology and Biochemistry, Pharmacology, and Health and Human Performance, University of Houston, Houston, TX, USA

T.L. Guidotti
Medical Advisory Services, Rockville, MD, USA

D. Hashim
Department of Preventive Medicine, Icahn School of Medicine at Mount Sinai, New York, NY, USA

S.H. Horowitz
Department of Neurology, Massachusetts General Hospital and Department of Neurology, School of Medicine, Tufts University, Boston, MA, USA

J.W. Klocek
Department of Psychology and Neuroscience, Baylor University, Waco, TX, USA

M.D. Lezak
Department of Neurology, Oregon Health and Sciences University, Portland, OR, USA

A.A. Little
Department of Neurology, University of Michigan Health System, Ann Arbor, MI, USA

D.M. Little
Baylor Scott and White Healthcare and Neuroscience Institute, Texas A&M Health Science Center College of Medicine, Temple, TX, USA

M. Lotti
Department of Cardiology, Thoracic and Vascular Sciences, School of Medicine, University of Padua, Padua, Italy

R.G. Lucchini
Department of Preventive Medicine, Icahn School of Medicine at Mount Sinai, New York, NY, USA and Unit of Occupational Medicine, Department of Medical Surgical Specialties, Radiological Sciences and Public Health, University of Brescia, Brescia, Italy

S. Mattioli
Occupational Health Unit, Department of Medical and Surgical Sciences, University of Bologna, Bologna, Italy

L.L. Maurer
Nicholas School of the Environment, Duke University, Durham, NC, USA

A. Moretto
Department of Biomedical and Clinical Sciences, University of Milan and International Centre for Pesticides and Health Risks Prevention (ICPS), Luigi Sacco Hospital, Milan, Italy

S.B. Morissette
Neuroscience Institute, Texas A&M Health Science Center College of Medicine and Central Texas Veterans Healthcare System, Temple, TX, USA

A.L. Oaklander
Department of Neurology and Department of Pathology (Neuropathology), Massachusetts General Hospital, Boston, MA, USA

M.A. Philbert
Toxicology Program, University of Michigan School of Public Health, Ann Arbor, MI, USA

M.A. Sainio Sr.
Development of Work and Organizations, The Finnish Institute of Occupational Health, Helsinki, Finland

M. Sliwinska-Kowalska
Department of Audiology and Phoniatrics, Nofer Institute of Occupational Medicine, Lodz, Poland

W.F. Stewart
Sutter Health, Walnut Creek, CA, USA

A. Vale
National Poisons Information Service (Birmingham Unit) and West Midlands Poisons Unit, City Hospital, Birmingham, UK; Honorary Professor, School of Biosciences, University of Birmingham, UK

F.S. Violante
Occupational Health Unit, Department of Medical and Surgical Sciences, University of Bologna, Bologna, Italy

Contents

SECTION 3 Movement disorders (parkinsonism or other movement disorders)

SECTION 4 Toxic neuropathies

SECTION 5 Special sensory organ systems

SECTION 6 Musculoskeletal disorders

SECTION 7 Other neurologic disorders and work restriction/accommodation

Section 1

Introduction

Handbook of Clinical Neurology, Vol. 131 (3rd series)
Occupational Neurology
M. Lotti and M.L. Bleecker, Editors

Chapter 1

Principles and practice of occupational neurology: an overview

MARCELLO LOTTI[1]* AND MARGIT L. BLEECKER[2]

[1]*Department of Cardiology, Thoracic and Vascular Sciences, School of Medicine, University of Padua, Padua, Italy*

[2]*Center for Occupational and Environmental Neurology, Baltimore, MD, USA*

INTRODUCTION

The presence of chemical and physical hazards in the workplace may result in several neurologic and psychiatric diseases. This variety of disorders reflects both the nature and the size of hazards and the complex web of the nervous system's organizations, functions, and targets.

Many chemicals have been shown to be neurotoxic, affecting the central nervous system (CNS), the peripheral nervous system (PNS), or both. Consideration should be provided to the chemistry of a given neurotoxicant because it may dictate the resulting syndrome, as in the case of mercury, because its elemental, inorganic, and organic forms cause quite different toxicities, or in the case of acrylamide, that is neurotoxic in its monomeric form only. The same chemical may have different targets in the nervous system. Certain organophosphorus esters cause both central/peripheral cholinergic overstimulation and delayed peripheral neuropathy and other chemicals can produce different effects depending on the dose, such as peripheral neuropathy or cerebellar syndrome after exposure to either relatively low or higher doses of acrylamide monomer, respectively. Acute and chronic exposures to the same chemical may result in different syndromes as, for example, in the case of *n*-hexane causing encephalopathy after extremely high single exposures and peripheral neuropathy after repeated lower exposures. In contrast, exposure to high levels of the industrial solvent trichloroethylene produces an encephalopathy but no peripheral neuropathy at lower doses. Neurologic syndromes may also derive from working conditions related to particular occupations and tasks. Depending on the body part that is involved in heavy work, posture, required force, repetition, or vibration, then different syndromes will arise.

The cells of the nervous system are affected in different ways: chemicals may cause neuronal death, axonal degeneration, or both, whereas most entrapment neuropathies are characterized by segmental demyelination. The outcome of exposure also depends on neuronal organization. Chemicals may selectively target cells of local neuronal circuits, such as in the case of the neurons of basal ganglia, when manganese causes an extrapyramidal syndrome. In this and other cases it is important to appreciate that the CNS displays high plasticity and redundancy of functions so that an initial damage could be efficiently compensated for over long periods of time. In these circumstances the neurotoxic effects may appear clinically after several years when the toxin-induced damage is superimposed by age-related physiologic neuronal loss. Also the toxin-induced nerve damage may be enhanced when the substrate, the peripheral nerve, has been previously altered by an entrapment neuropathy. However, this is not the case where neurons are hierarchically organized, as in the motor and sensory systems, where a damage at any level of these two neuron systems cannot be compensated for. Thus, chemicals such as organophosphates may cause axonal degeneration of long motor axons both in CNS and PNS with a loss of function that is shown as either flaccid paralysis, when the second or both motor neurons are affected or, later in the course of disease, as spastic ataxia, when regeneration of the second motor neuron has taken place. In addition, many forms of work-related neurologic disorders have a morphologic correlate such as the axonal degeneration in most toxic neuropathies; others do not, such as the acute neurotoxicities of organophosphates and pyrethroids, whereas in still other cases it is unknown, such as in low-back pain.

*Correspondence to: Marcello Lotti, MD, Professor of Occupational Medicine, Department of Cardiology, Thoracic and Vascular Sciences, School of Medicine, University of Padua, Via Giustiniani 2, 35128 Padua, Italy. E-mail: marcello.lotti@unipd.it

Occupational neurology encompasses prevention, etiologic diagnosis, and treatment of neurologic disorders that may occur at work. Prevention involves practices such as neuroepidemiology studies aimed at the detection of hazards and risks, biomonitoring of exposure devoted to the assessment of compliance with hygiene standards and medical surveillance directed at unmasking unnoticed or early signs of disease. Neuroepidemiologic studies may determine a functional domain that is disproportionately affected by a neurotoxic exposure and therefore useful for medical surveillance, namely visuospatial abilities and perchloroethylene exposure (White et al., 2014). Particular problems that the preventive aspects of occupational neurology should address also include accommodation of workers with preexisting neurologic disorders to avoid possible synergy with occupational exposure and to prevent the possible consequences of sleep deprivation due to shift work. In the clinic, the recognition of a neurologic disorder linked to occupational exposure should rule out common disorders and other etiologies, also considering the blurring of boundaries between occupational exposure and other environmental risk factors. Physicians must use all information from the above preventive practices, if available, and integrate it in the analysis of the individual patient. The treatment of work-related neurologic disorders is not different from those provided for the same common disease, except for some toxin-induced parkinsonism that clinically appears similar but whose underlying pathology is different, and therefore the standard drug treatment frequently has little benefit. Also neurotoxic effects of certain heavy metals are treated with chelating drugs such as sodium-EDTA and penicillamine, while acute organophosphate intoxication requires antidotes such as atropine and oximes.

OUTCOME OF EXPOSURE

Acute neurologic syndromes due to chemical and physical hazards are nowadays less common and are usually easily recognized because they are less variable in presentation. These include the narcotic effects of several solvents, the cholinergic overstimulation caused by organophosphate insecticides, and musculoskeletal disorders after acute overexertion events. However, most neurologic syndromes linked to work activities occur after subacute or chronic exposure, they may involve both CNS and PNS, and are usually insidious in onset. These disorders encompass encephalopathy, extrapyramidal syndrome, toxic and entrapment neuropathies, and alterations of sensory organs.

Whereas several work-related neurologic diseases have firmly been established and will be discussed in depth in the following chapters, others are still a matter of debate. In order to highlight challenges that may be encountered in occupational neurology, we report here some examples of either clear or controversial associations between exposure at work and certain neurologic disorders.

Persistent alterations of CNS functions, generally characterized prominently by cognitive impairment, are not well characterized and there is no consensus as to whether a chronic neurotoxic encephalopathy is a diagnostic entity (Albers and Berent, 2000; Schaumburg, 2000; Parry, 2008). This is the case with "chronic painters' syndrome," or "psycho-organic syndrome," described in workers with occupational exposure to solvents who had vague and mild symptoms of neurobehavioral impairment. This syndrome has variously been viewed with support or skepticism and the debate continues. However, other prolonged exposure such as to ethylene oxide may cause chronic toxic encephalopathy characterized by a steady decline in cognitive function (Estrin et al., 1987).

Another example of controversy is the risk of extrapyramidal disorders in welders that has been associated with low levels of exposure to manganese. Whereas the causal relationship between elevated levels of manganese exposure and parkinsonism has been well established (Racette, 2014), conflicting results have been reported about this risk in welders. An early study (Racette et al., 2005), followed by others, reported a high prevalence of disease among these workers but a subsequent meta-analysis did not (Mortimer et al., 2012). However, in a recent longitudinal cohort study on welders with no previous exposure, in which individuals served as their own controls, T1-weighted changes were detected in the basal ganglia at very low levels of manganese exposure with no manifest clinical symptoms (Baker et al., 2015).

Several peripheral neuropathies occurring at work have been well characterized and include *n*-hexane, acrylamide, some heavy metals, and some organophosphates. Minor abnormalities of nerve conduction can be demonstrated in individuals who are chronically exposed to other chemicals, although it is unlikely that these changes will progress to clinical neuropathy (Parry, 2008). However, these populations have not been followed for several years to determine if these subclinical effects may appear clinically when superimposed with age-related changes.

Occupational exposure to a variety of chemicals has been associated with neuropsychologic alterations. However, the causal correlation between these exposures and mild subclinical behavioral signs is often difficult to establish, because of confounding variables, including age, alcohol and drug intake, stress, assessment of exposure, choice of controls, and educational background, that are not always taken into account. Not surprisingly, results are often inconsistent, such as those from studies on altered neurobehavioral test performance reported in workers exposed to chronic low levels of organophosphorus pesticides (Rohlman et al., 2011; Starks et al., 2012). The overall evidence of these reported effects is difficult to interpret, especially because in most studies the absence of cholinesterase measurements hampers an accurate assessment of exposure. However, many epidemiologic studies with well-quantitated low-level neurotoxic exposure have demonstrated dose effects with neurobehavioral performance in the normal range. These "subclinical" effects are caused by the exposure and may be associated with functional complaints (White et al., 2014).

Work-related musculoskeletal disorders are among the most common occupational diseases in both the European Union and the USA (BLS, 2008; Schneider and Irastorza, 2010). Nerve compression – entrapment syndromes – occurring especially in the wrist and forearm is associated with repetitive tasks and vibration and the most frequent cause of lengthy disability is carpal tunnel syndrome (Daniell et al., 2009). Another example is the low-back pain that is associated with intensity and frequency of heavy lifting at work (Coenen et al., 2014). However, a number of risk factors unrelated to work are also involved in this disorder, including psychosocial ones that may interact, increasing the risk (Widanarko et al., 2015) and making etiologic distinction difficult.

ESTABLISHING CAUSATION

The notion of causality has received the longstanding attention of physicians, scientists, and philosophers, resulting in various interpretations in connection with phenomena of different kinds and scales. In the context of occupational diseases a distinction should be made between the criteria of causality established in epidemiology that are aimed at identification of hazards and risks in populations and those used in the clinical setting aimed at the etiologic diagnosis in individuals. Indeed, clinicians rely heavily up epidemiologic observations: hydrogen and oxygen are tightly bound in water, though they still remain distinct elements. Epidemiologic and clinical causalities are simply different.

In both circumstances, however, the key information required to establish causation is the assessment of exposure(s) to noxious agents, that is unfortunately almost always approximate. In epidemiologic surveys quantitative measurements of exposure are rarely available, particularly when exposures have been remote in time. Consequently, their intensity and duration are evaluated by means of surrogates such as questionnaires and job-exposure matrixes. In most patients with neurotoxic disorders it is impossible to assess exposure by detecting the offending chemicals in body fluids because of their short half-life and almost inevitably patients arrive at medical attention long after the exposure has ended. It is the short half-life of carboxyhemoglogin with carbon monoxide exposure that has led to examination of markers of biochemical effect, such as serum S-100B protein and neuron-specific enolase, biochemical markers of hypoxic brain damage, to better predict prognosis (Yardan et al., 2009; Park et al., 2012). Similarly, retrospective quantitative assessment of physical risk factors may be extremely difficult and is mostly based upon the length of exposure to a given risk only. Therefore, central to the assessment of possible offending exposures to both chemical and physical risks in the clinic is the taking of a very detailed patient's occupational history.

Neuroepidemiology

Epidemiology is the study of the relationships of the various factors determining the frequency and distribution of diseases in a human community. Epidemiology has developed its own criteria of causality that include strength of association, consistency of results, specificity of the risk factor, temporality between exposure and disease onset, biologic plausibility of the association, and analogy with other risk factors (Hill, 1965). Consequently, epidemiologic associations and the inferences derived from these studies cannot necessarily be applied to individual patients in clinical medicine (Angell and Kassirer, 1994).

Occupational neuroepidemiology involves systematic studies of large groups of workers that usually may present with more chronic and less obvious neurologic effects of exposure, although certain exposures seem to play a role in more specific disorders. However, the interpretation of findings has been often controversial, such as in the case of the chronic effects of solvent exposure, where symptoms are vague, mild, and nonspecific, there are no physical findings, and adequate means of ruling out confounding factors have not been taken. However when occupational exposure to low levels of solvents is quantitated for the individual worker over years of employment and confounding factors are taken into account, a dose–effect relationship with subclinical neuropsychiatric and neurobehavioral effects can be established (Bolla et al., 1990; Bleecker et al., 1991). Nonetheless, the exigency of public health problems demands action and despite imperfect knowledge, causal inferences must be made even in the absence of robust exposure data (Rothman, 1995).

Another example of difficult interpretation was that from epidemiologic studies on the association between pesticide exposures and risk of the neurodegenerative disorder, Parkinson's disease, due to inadequate documentation of exposure. This was reinforced by a recent meta-analysis of exposure to pesticides or solvents and risk of Parkinson's disease (Pezzoli and Cereda, 2013). However, Kamel (2013) reports on strong evidence for paraquat and rotenone based upon occupational neuroepidemiologic studies that were further supported by animal studies demonstrating features of Parkinson's disease after chronic low exposure to these pesticides. Also the organochlorine insecticides, which were higher in the brains of Parkinson's disease patients, are known to increase oxidative stress and inflammation, whch are important in the development of Parkinson's disease. From these studies important details of exposure are still missing, such as which are the critical pesticides that are associated with the disease, what life stage and time frame of exposure are involved, and whether intensity or duration of exposure, or both, is relevant (Kamel, 2013).

In other cases, and there are many, such as for example that musculoskeletal disorders are associated with physical risk factors, the strength of the association between exposure and specific effects is now better established (Bernard, 1997).

In conclusion, new clues suggesting ultimate causes or the triggering mechanisms for some neurologic disorders may be obtained from occupational neuroepidemiology, providing that all possible confounding factors are taken into account and more attention is given to quantitating exposure or finding markers of biochemical effect that better reflect brain pathology so that the results will be reproducible.

Clinical diagnosis

Neurologic disorders associated with occupational exposures rarely display pathognomonic signs and because of the nowadays common shortcomings of the multifactorial causality of disease, the physician has more or less to sneak into this network and try to assess the relevance of occupational components of this all-encompassing approach. To do so, clinical skill in clinical neurology and in particular history taking represents key elements (Nicholl and Appleton, 2015) because the true physician remains, as always, the person who takes the patient's history. Some factors in this respect include the ruling out or the relevance of other known risk factors such as drugs, nutritional supplements, recreational habits and hobbies, and hereditary degenerative diseases. The identification of occupational exposures and the assessment of their intensity and duration may prove difficult but it is essential, because in many circumstances this information represents the only criterion of causality. Therefore, questioning about occupation is challenging and requires some knowledge of toxicology, industrial processes, and industrial hygiene. Questioning should also be directed as to whether other co-workers have been affected. Whereas the usual techniques used in the clinic, including physical examination, electrodiagnostic testing, imaging, and neuropsychological evaluation, are essential in defining diagnosis, none is specific to establish etiology. Biopsies are not a standard part of the clinical evaluation but to document the presence of a small-fiber neuropathy, as found in complex regional pain syndrome, punch skin biopsies that quantitate the density of epidermal nerve fibers are now available as a routine test in the clinic.

For work-related disorders there is usually a relatively short temporal relationship between exposure and onset of symptoms, because they occur concurrently with exposure or shortly thereafter. Sometimes with low-level occupational solvent exposure, such as with *n*-hexane, it may take years before toxic neuropathy develops (Spencer et al., 1980). But when there is exposure to a combination of solvents, such as methyl ethyl ketone and *n*-hexane, potentiation of *n*-hexane metabolism may result in a toxic neuropathy in a matter of months (Altenkirch et al., 1982). Usually, neurotoxic diseases are self-limiting after exposure cessation, although sometimes a mild progression continues, a phenomenon called coasting. However, the delayed appearance of certain effects after exposure cessation should also be considered. This may occur when signs secondary to age-associated neuronal loss are superimposed upon earlier toxic degenerative changes that have been for a certain period of time compensated. This was reported for exposures to the drug of abuse MPTP and Parkinson's disease and for inorganic mercury exposure and sensory loss (Albers et al., 1988).

Occasionally, the concurrent presence of other signs at physical examination may smooth the path of etiologic diagnosis, as in the case of bullous dermatitis on hands and acrylamide neuropathy, Mees' lines on the fingernails and arsenic neuropathy, and neurogenic bladder and N,N' dimethylaminoproprionitrile neuropathy.

Finally, the clinical findings should obviously display a consistent pattern that corresponds to previously reported cases.

In conclusion, the proper application of knowledge to establish causation in an individual patient, as for instance that derived from epidemiologic associations, toxicology, and industrial hygiene, demands an accurate appreciation of its clinical conditions: a kind of particular judgment that a physician brings to the traditional single doctor/single patient situation.

MONITORING WORKERS AT RISK

The purposes of monitoring workers at risk of neurologic disorders are manifold and include compliance with hygienic standards, the detection of subclinical effects to prevent further exposure, the identification of susceptible individuals, and, in the case of unexpected outbreaks of neurotoxicity, to evaluate fellow workers who are apparently not affected.

However, no specific procedures are available to monitor the risks of developing musculoskeletal disorders. But workers in jobs at risk for neck and upper-limb musculoskeletal disorders have benefited from tailored workplace exercise programs (Rasotto et al., 2015). Indeed, instructions on manual handling should be given and the use of ergonomic aids implemented, although, for instance, there is little evidence for the effectiveness of most primary prevention initiatives for low-back pain. In this regard the European Guidelines reccommend that prevention efforts should be directed at minimizing its consequences (Burton et al., 2006). This may be achieved by stratifying patients through screening biopsychosocial risk factors and then applying interventions designed to target their specific problems and risk profile (Hill et al., 2011).

Biomarkers of exposure and effect that are available for neurotoxicants are relatively few when using the strict definition of biologic/biochemical/molecular markers. Exposure to neurotoxic chemicals can be monitored with the use of biomarkers in blood, urine, and exhaled air by measuring either parent compounds or their metabolites. Examples of the former include the concentrations of lead in blood and of styrene and toluene in exhaled air. Examples of the latter include the excretion of 2,5 hexanedione in workers exposed to *n*-hexane, and that of mandelic acid in those exposed to styrene. However, these tests only reflect recent exposure but, when monitored on a regular basis over years of employment, a cumulative dose may be calculated (Ford et al., 1991). In cases of chemicals that form covalent adducts to macromeolecules, such as acrylamide, the levels of such adducts, depending on the half-life of the chosen macromolecule, may reflect cumulative exposure over weeks to months (Calleman et al., 1994).

In very few circumstances the biochemical effects of exposures could be assessed, as for instance the inhibition of red blood cell acetylcholinesterases in workers exposed to organophosphorus insecticides and carbamates, that of lymphocyte neuropathy target esterase in workers exposed to certain organophosphates and that of red blood cell delta-aminolevulinic acid dehydratase in workers exposed to lead.

Outbreaks of neurologic disorders may occur with chemicals not previously known to be neurotoxic. For example, the neurotoxicity of methyl-*n*-butyl ketone was discovered years ago after an epidemic of neuropathy whose origin was identified in a factory in Ohio. The combination of industrial hygienic survey, clinical and electrophysiologic examinations and experimental studies confirmed the neurotoxicity of this chemical (Schaumburg, 2000). Another outbreak of neurologic disorders related to bladder and sexual dysfunction and peripheral neuropathy occurred in a foam rubber plant that used N,N' dimethylaminoproprionitrile as a catalyst. The causal association was established through epidemiologic studies, extensive clinical evaluation, and animal studies (Pestronk, 2000).

References

Albers JW, Berent S (2000). Controversies in neurotoxicology: current status. Neurol Clin 18: 741–764.

Albers JW, Kallenbach LR, Fine LJ et al. (1988). Neurological abnormalities associated with remote occupational mercury exposure. Ann Neurol 24: 651–659.

Altenkirch H, Wagner HM, Stoltenburg-Didinger G et al. (1982). Potentiation of hexacarbon-neurotoxicity by methyl-ethyl-ketone (MEK) and other substances: clinical and experimental aspects. Neurobehav Toxicol Teratol 4: 623–627.

Angell M, Kassirer JP (1994). Clinical research – what should the public believe? N Engl J Med 331: 189–190.

Baker MG, Criswell SR, Racette BA et al. (2015). Neurological outcomes associated with low level manganese exposure in an inception cohort of asymptomatic welding trainees. Scand J Work Environ Health 41: 94–101.

Bernard BP (1997). Musculoskeletal disorders and workplace factors. NIOSH, Cincinnati, OH, NIOSH Publication No. 97B141.

Bleecker ML, Bolla KI, Agnew J et al. (1991). Dose-related subclinical neurobehavioral effects of chronic exposure to low levels of organic solvents. Am J Ind Med 19: 715–728.

BLS (2008). Bureau of Labor Statistics. The Economics Daily, Musculoskeletal disorders and days away from work in 2007. U.S. Department of Labor, Washington, DC.

Bolla KI, Schwartz BS, Agnew J et al. (1990). Subclinical neuropsychiatric effects of chronic low-level solvent exposure in US paint manufacturers. J Occup Med 32 (8): 671–677.

Burton AK, Balagué F, Cardon G et al. (2006). Chapter 2, European guidelines for prevention in low back pain. Eur Spine J 15 (Suppl 2): s136–s168.

Calleman CJ, Wu Y, He F et al. (1994). Relationship between biomarkers of exposure and neurological effects in a group of workers exposed to acrylamide. Toxicol Appl Pharmacol 126: 361–371.

Coenen P, Gouttebarge V, van der Burght AS et al. (2014). The effect of lifting during work on low back pain: a health impact assessment based on a meta-analysis. Occup Environ Med 71: 871–877.

Daniell WE, Fulton-Kehoe D, Franklin GM (2009). Work-related carpal tunnel syndrome in Washington State worker's compensation: utilization of surgery and the duration of lost work. Am J Ind Med 52: 931–942.

Estrin WJ, Cavalieri SA, Wald P et al. (1987). Evidence of neurologic dysfunction related to long-term ethylene oxyde exposure. Arch Neurol 44: 1283–1286.

Ford DP, Schwartz BS, Powell S et al. (1991). A quantitative approach to the characterization of cumulative and average solvent exposure in paint manufacturing plants. Am Ind Hyg Assoc J 52 (6): 226–234.

Hill AB (1965). The environment and disease: association or causality? Proc R Soc Med 58: 295–300.

Hill JC, Whitehurst DG, Lewis M et al. (2011). Comparison of stratified primary care management for low back pain with current best practice (STarT Back): a randomized controlled trial. Lancet 378: 1560–1571.

Kamel F (2013). Paths from pesticides to Parkinson's. Science 341: 722–723.

Mortimer JA, Borenstein AR, Nelson LM (2012). Associations of welding and manganese exposure with Parkinson disease. Review and meta-analysis. Neurology 79: 1174–1180.

Nicholl DJ, Appleton JB (2015). Clinical neurology: why this still matters in the 21st century. J Neurol Neurosurg Psychiatry 86: 229–233.

Park E, Ahn J, Min YG et al. (2012). The usefulness of the serum s100b protein for predicting delayed neurological sequelae in acute carbon monoxide poisoning. Clin Toxicol (Phila) 50: 183–188.

Parry GJ (2008). Neurological complications of toxin exposure in the workplace. In: MJ Aminoff (Ed.), Neurology and general medicine. Chuchill Livingstone, Philadelphia, pp. 749–768.

Pestronk A (2000). N, N'-Dimethylaminopropionitrile. In: PS Spencer, HH Schaumburg, AC Ludolph (Eds.), Experimental and clinical neurotoxicology. Oxford University Press, New York, pp. 499–502.

Pezzoli G, Cereda E (2013). Exposure to pesticides or solvents and risk of Parkinson disease. Neurology 80: 2035–2041.

Racette BA (2014). Manganism in the 21st century: the Hanninen lecture. Neurotoxicology 45: 201–207.

Racette BA, Tabbal SD, Jennings D et al. (2005). Prevalence of parkinsonism in relationship to exposure in a large sample of Alabama welders. Neurology 64: 230–235.

Rasotto C, Bergamin M, Sieverdes JC et al. (2015). A tailored exercise program for women at risk for neck and upper limb musculoskeletal disorders. A randomized controlled trial. J Occup Environ Med 57: 178–183.

Rohlman Ds, Anger WK, Lein PJ (2011). Correlating neurobehavioral performance with biomarkers of organophosphorus pesticide exposure. Neurotoxicology 32: 268–276.

Rothman KJ (1995). Causes 1976. Am J Epidemiol 141: 90–95.

Schaumburg HH (2000). Human neurotoxic disease. In: PS Spencer, HH Schaumburg, AC Ludolph (Eds.), Experimental and clinical neurotoxicologypp. Oxford University Press, New York, pp. 55–82.

Schneider M, Irastorza X (2010). Work-related musculoskeletal disorders in the EU- Facts and figures. European Agency for Safety and Health at Work, Luxembourg.

Spencer PS, Couri D, Schaumberg HH (1980). n-Hexane and Methyl n-Butyl Ketone. In: PS Spencer, HH Schaumburg (Eds.), Experimental and Clinical Neurotoxicology. Williams & Wilkins, Baltimore, MD, pp. 456–475.

Starks SE, Gerr K, Kamel F et al. (2012). Neurobehavioral function and organophosphate insecticide use among pesticide applicators in the Agricultural Health Study. Neurotoxicol Teratol 34: 168–176.

White RF, Krengel M, Grashow R (2014). Neurotoxicology. In: MW Parsons, TA Hammeke (Eds.), Clinical Neuropsychology, A Pocket Handbook for Assessment, 3rd Edn, American Psychological Association, Washington, DC, pp. 338–362.

Widanarko B, Legg S, Devereux J et al. (2015). Interaction between physical and psychosocial risk factors on the presence of neck/schoulder symptoms and its consequences. Ergonomics 27: 1–12.

Yardan T, Cervik Y, Donderici O et al. (2009). Elevated serum S100B protein and neuron-specific enolase levels in carbon monoxide poisoning. Am J Emerg Med 27: 838–842.

Handbook of Clinical Neurology, Vol. 131 (3rd series)
Occupational Neurology
M. Lotti and M.L. Bleecker, Editors

Chapter 2

Evaluating suspected work-related neurologic disorders (clinical diagnosis)

MARCELLO LOTTI[1]* AND MICHAEL J. AMINOFF[2]

[1]*Department of Cardiology, Thoracic and Vascular Sciences, School of Medicine, University of Padua, Padua, Italy*

[2]*Department of Neurology, School of Medicine, University of California San Francisco, San Francisco, CA, USA*

INTRODUCTION

Occupationally related disease is responsible for considerable morbidity and mortality worldwide. Different organ systems may be affected, depending on the underlying cause. Neurotoxins are substances that interfere with the function of nervous tissue, and they may cause irreversible cell damage and cell death. Depending on the toxin, specific sites and elements of the nervous system are affected, and this is reflected in the resulting clinical disorder.

Occupationally related neurotoxic disorders have been known since antiquity and continue to occur. In the first century AD, for example, Pliny is said to have noted weakness in workers exposed to lead dust (Hunter, 1978), and in the last 50 years, outbreaks of peripheral neuropathy have followed occupational exposure to methyl-*n*-butyl ketone and 2-*t*-butylazo-2-hydroxy-5-methyl hexane (Allen et al., 1975; Horan et al., 1985). The prevalence of such work-related neurotoxic disorders is unknown, but it is probably higher than suspected because symptoms are often nonspecific and disorders may not be reported to public health officials.

Many agents in the work-related environment have neurotoxic effects reflecting involvement of the central or peripheral nervous systems (CNS or PNS), or both. Central involvement is typically manifested by neurobehavioral changes, convulsions, impaired arousal, or sleep disturbances, whereas either central or peripheral deficits may be characterized by motor, sensory, or autonomic abnormalities. Various combinations of symptoms and signs comprise syndromes associated with exposure to specific neurotoxins.

Occupational exposure to neurotoxic agents may occur in workers employed in the manufacture of potentially neurotoxic substances or working with such substances, and those subject to industrial, agricultural, horticultural, or military exposure. Indeed, neurotoxic disorders are an important cause of work-related disease. Early detection is important to limit exposure and thereby reduce both the severity of any neurotoxic syndrome and the number of workers affected. In particular, it may lead to early removal of an affected individual from the workplace, thereby limiting further exposure to the toxin and perhaps preventing irreversible neurologic damage. Furthermore, recognition of the earliest affected cases in a particular setting may lead to the recognition of subclinical cases and to changes that protect unaffected workers. Occupationally related injuries are also important but are beyond the scope of this chapter, as also are the neurotoxic effects of medications and substance abuse.

Certain basic criteria must be satisfied before it can be concluded that a clinical or subclinical disturbance of function has a neurotoxic basis. In the first instance, any manifestations of neurologic dysfunction must be related temporally to the toxin exposure, and the course and progression of the clinical disorder should be in accord with the biologic effects and mechanisms of action of the presumed toxin. Depending on the toxin and intensity of exposure, onset of symptoms may occur with little or no latent period, or after a delay of some weeks. Cessation of exposure to the presumed toxin should stop progression of the disorder (sometimes after an interval during which mild progression continues, a phenomenon called coasting). Symptoms may improve at weekends or during vacations, when subjects are away from the work environment, but a similar pattern may

*Correspondence to: Marcello Lotti, MD, Professor of Occupational Medicine, Department of Cardiology, Thoracic and Vascular Sciences, School of Medicine, University of Padua, Via Giustiniani 2, 35128 Padua, Italy. E-mail: marcello.lotti@unipd.it

relate to other factors, such as anxiety or disputes with coworkers. Second, any alternative explanation for the neurologic dysfunction must be excluded. Third, the clinical disorder should be biologically plausible, based on the known toxicity or effects of the toxin on biologic systems. And fourth, the presumed neurotoxic disorder should be similar in character to the disorder occurring in other cases of exposure to the same neurotoxin. In addition, it may be possible to detect the presence of the neurotoxin or its metabolites in the tissues of an affected individual, and it may be possible to reproduce the presumed neurotoxic disorder experimentally in animals, although a failure to reproduce it does not exclude the possibility of a neurotoxic etiology.

It is often difficult to document the precise level of exposure of workers to neurotoxic compounds, but in many instances it is clear that exposure has occurred to levels having neurotoxic effects. It is more problematic to determine the clinical significance of exposure for short periods to high concentrations or for prolonged periods to low concentrations of substances that are known to be neurotoxic.

A suspected neurotoxin typically will cause neurologic dysfunction of a consistent nature in exposed subjects. The severity of dysfunction may vary between individuals but, depending on exposure level, all exposed workers should have some manifestations of neurotoxicity. Neurologic manifestations in only a single worker suggests that the disorder is not due to a toxin, toxin exposure did not occur at work, or that the exposure was related to an individual work habit, such as not using a protective mask to limit exposure. Many disorders presumed to be occupationally related consist of common, nonspecific symptoms that, in fact, are of uncertain origin. The diagnostic difficulties are compounded by the fact that, when an outbreak of a possible neurotoxic disorder is publicized, identical symptoms may occur in suggestible subjects with less or no exposure to the suspected toxin (Parry, 2008). The risk of a neurotoxic disorder resulting from exposure in the workplace can be reduced by substituting less hazardous substances for hazardous materials when possible and by adopting preventive measures such as monitoring and limiting exposure. Thus, exposure to lead in the workplace can be reduced if premises are well ventilated; workers wear protective equipment such as goggles, boots, and special clothing, and have proper respiratory protection; if they shower after work; and if they undergo routine determination of blood lead level to monitor exposure, with exposure being terminated in those with an increased level before neuropathy develops.

In some instances, exposure to neurotoxic agents may render the peripheral nerves more susceptible to compression or entrapment injuries. The designation entrapment neuropathy refers to the compression, stretch, or angulation of an individual peripheral nerve by neighboring anatomic structures such that dysfunction occurs. The clinical consequences depend on the affected nerve, but are often superimposed on signs of a more widespread polyneuropathy. The nerve affected may depend on an individual worker's practices, such as leaning on the elbows or sitting with crossed legs.

The diagnosis of an occupationally related neurotoxic disorder is often difficult, then, because the time to symptom onset may be variable, there are usually no specific signs and symptoms, and clinical manifestations are frequently indistinguishable from those of other neurologic disorders. Moreover, information is lacking about the toxicity of many chemical substances, exposure is often to multiple substances, and workers are often uninformed about the nature and hazards of the materials to which they are exposed. Ancillary investigations may help to detect and quantify neurotoxic disorders. For example, it may be difficult to detect mild or subclinical neurobehavioral disturbances, especially after chronic low-dose exposure to a neurotoxin, but structured psychometric tests may reveal functional impairment in various cortical regions when possible confounding factors during testing, such as fatigue or exposure to psychotropic agents, are excluded. Electrophysiologic evaluation of function in the CNS and the major afferent pathways by electroencephalography (EEG), evoked potential studies, and recording event-related evoked potentials is generally of uncertain value and is discussed in a later section of this chapter, as also are motor and sensory nerve conduction studies, which are of considerable importance in evaluating the PNS. Quantitative tests of muscle strength and of coordination, body sway, and balance may also help in evaluating motor function. Quantitative sensory testing involves determination of vibration and discriminative tactile sensibility, cold and warm thermal thresholds, and heat pain thresholds, while quantitative tests of autonomic function mainly evaluate cardiovascular function.

Epidemiologic studies have been widely used in detecting occupation-related neurotoxic disorders. Case–control observations, however, require particular care in the selection of appropriate control subjects. For neurobehavioral observations, for example, subjects must be matched for age, sex, ethnic group, and educational and cultural background. Failure to rigorously control such studies has led to conflicting reports in the literature.

Reports on outbreaks of neurologic disease have often come from the clinical sector and then have led to further study of suspected neurotoxins. Unfortunately, most neurologists are poorly informed about neurotoxic disorders, and accordingly future training practices should be such as to increase their awareness of the possibility that patients may have diseases induced by toxic exposure in the workplace. Neurologists should also be aware of the implications in making such diagnoses, appreciate their responsibilities within the framework of any local or national workers' compensation system, and understand any requirements to alert health or regulatory agencies of suspected occupationally related neurotoxic disorders.

HISTORY

The history is as important in evaluating patients with suspected work-related neurotoxic disorders as it is in other clinical contexts. The location, quality, and time-course of symptoms, and any precipitating or alleviating factors must be determined, as must also the presence of any associated symptoms. Such associated symptoms are often the key to diagnosis. For example, chronic low-level exposure to arsenic may result from industrial contamination of ground and water or occur in workers involved in the manufacture of paints, fungicides, wood preservatives, and semiconductors; in certain mining and smelting operations; from combustion of certain arsenic-containing compounds; and in pesticide spraying. It may cause a neuropathy, but associated symptoms also occur, such as hyperkeratosis and desquamation of the skin, particularly in the palms and soles; redness and swelling of the hands and feet; patchy areas of skin discoloration; and gastrointestinal disturbances. Acute poisoning (as by ingestion of contaminated food or water or attempted homicide) is associated with transverse gray lines in the nails (Mees' lines).

Depending on the nature of symptoms, it may be important to obtain an account of their character and evolution from a spouse or other close relative. In addition to focusing on the patient's symptoms, the work environment must be explored. Job titles may be misleading and details should be obtained about the precise nature of an occupation and what it entails. A detailed list of chemicals used in the workplace should be obtained.

The diagnosis of a neurotoxic disorder is essentially one of exclusion, because symptoms are nonspecific. For example, the symptoms and signs of a toxic encephalopathy may resemble various metabolic, inflammatory, neoplastic, and even degenerative diseases of the nervous system. Symptoms may develop gradually over many years, making recognition of their etiology difficult, and confounding factors may add to the difficulty of determining whether the illness is work-related (for example, the patient may be diabetic). Even when a definite neurologic disorder, such as a peripheral neuropathy, is diagnosed, it may have arisen from causes other than neurotoxicity. Neurotoxicity should be suspected when symptoms are consistent with neurotoxic exposure and there is a history of such exposure, either in the workplace or elsewhere. A detailed occupational history focusing on exposure to different chemicals is required. When there is a long latent period between the onset of exposure and development of symptoms, however, it may become difficult to obtain detailed occupational and exposure histories.

Occupational exposure occurs most commonly by inhalation and through the skin. The work environment should be characterized with this in mind. Details of any measures to limit chemical exposure must be obtained, such as whether the workplace is well ventilated and the nature of any protective measures that are taken (such as a requirement to wear special clothing and gloves, and the use of masks, goggles, and other devices). Compliance with occupational exposure limits may not necessarily protect all workers from adverse effects, especially in the long term. Enquiry should be made as to whether coworkers have similar symptoms to those of a subject with suspected neurotoxicity. This may involve questioning coworkers and even examining those working in the same or a similar environment. When asymptomatic but exposed coworkers are screened, questioning may be focused, based on the clinical features in recognized cases; self-administered, standardized symptom questionnaires are often especially helpful (Parry, 2008). If several cases come to light, it may be possible to demonstrate a dose–response relationship by relating the likely exposure dose to the development or severity of the disorder.

Neurotoxin exposure can result from environmental factors (for example, a subject's residence or its location close to a source of industrial pollution) and to individual factors (such as hobbies or dietary peculiarities) rather than to the workplace. The history must therefore exclude these possibilities.

The history can be guided by knowledge of the likely syndromes produced by neurotoxins. Neurotoxins affecting the CNS may cause an acute or chronic encephalopathy. Symptoms are nonspecific and may consist solely of headache and malaise, but with more severe involvement, confusion, irritability, impaired concentration and judgment, drowsiness, unsteadiness, vertigo, tinnitus, sensory complaints, "weakness," or other complaints also occur. Disability may be out of proportion to the severity of symptoms. Neurologic examination is usually normal, but ataxia and nystagmus are sometimes encountered. When the PNS is affected, symptom onset is usually insidious with sensory impairment that progresses gradually. Numbness begins in the feet, spreads proximally, and eventually involves the hands. Distal motor disturbances may occur but commonly follow sensory changes; in severe cases, foot drop and weakness of the intrinsic hand muscles develop. Orthostatic symptoms are sometimes prominent. With intense exposure, progression is more rapid and symptoms are more severe. Examination shows any combination of distal sensory deficits, reduced or absent muscle stretch reflexes, and distal weakness. When a nerve entrapment syndrome is present, pain or sensory complaints in a restricted distribution is common and in severe cases weakness and wasting may occur in muscles innervated by the affected nerve.

In some instances, the history is less typical but then more revealing. For example, a history of facial numbness, analgesia, or dysesthesias with the later development of masticatory weakness is very suggestive of a distal axonopathy from exposure to trichloroethylene, a degreaser that involves the cranial nerves selectively, initially affecting the trigeminal nerve. The lower cranial nerves may be involved in some patients; and an optic neuropathy has sometimes occurred. Recovery, often incomplete, occurs with discontinuation of exposure, but patients may be left with residual patchy facial sensory loss. The neuropathy in this instance is attributed to dichloroacetylene, formed by decomposition of trichloroethylene.

An important aspect of the history is to distinguish neurotoxic from pseudoneurotoxic disease. In particular, it is important to determine whether a symptomatic individual exposed to a possible neurotoxin has a genuine neurotoxic disorder or has a pseudoneurotoxic disorder relating to psychogenic illness, the coincidental onset of a naturally occurring neurologic disorder, worsening of a preexisting disorder, or some combination of these factors (Schaumburg and Albers, 2005). Pseudoneurotoxic disorders are sometimes accompanied or followed by the so-called multiple chemical sensitivity syndrome. A variety of often vague and nonspecific symptoms involving several different organ systems and of stress occur on exposure to even low levels of various chemicals found in the workplace and elsewhere. Women are affected more often than men. Mood disturbances characterized by depression, anxiety, malaise, ease of fatigue, and poor concentration are common, as also is social withdrawal. The pathophysiology is unclear and there is no reliable treatment, but the mood disturbance should be treated.

Clinical neurologic examination is of limited utility for screening patients with suspected neurotoxin-related disease. The main aim is to exclude other conditions that might be the cause of symptoms. Most neurotoxic disorders are characterized by generalized rather than focal neurologic abnormalities but focal findings – such as parkinsonism from manganese exposure – are sometimes conspicuous.

PHYSICAL EXAMINATION

The aim of physical examination in occupational medicine does not differ from that in clinical neurology (Aminoff et al., 2005) and includes exclusion of naturally occurring conditions unrelated to neurotoxicity, and determining the consistency of the observed signs with the known effects of a given exposure. In addition, a more thorough examination should be directed to the target of toxicity/entrapment based on information obtained from the work history of the individual patient. Physical examination is less useful and often impractical for screening for neurotoxicity in large-worker populations (Parry, 2008), where other procedures such as quantitative sensory testing and tremor analysis are used (Dyck and O'Brien, 1999).

Whereas most neurotoxic diseases resemble various common diseases of the nervous system and rarely display pathognomonic signs, in some instances more widespread, coexisting features suggest a toxic origin. Therefore, physical examination should not be restricted to the nervous system. For instance, in cases of peripheral neuropathy, the presence of sweating and bullous dermatitis of the hands suggests acrylamide intoxication (He et al., 1989), and scalp alopecia may indicate thallium toxicity (Hoffman, 2003). Similarly, *N,N'*dimethylaminopropionitrile and 1-bromopropane peripheral neuropathies are associated with bladder dysfunction (Baker et al., 1981) and diarrhea (Ichihara et al., 2002), respectively.

Standard neurologic examination begins with the evaluation of mental status. Level of consciousness is depressed to a variable degree in several acute encephalopathies that follow excessive exposure to certain neurotoxins. In most cases the clinical features of these encephalopathies are not specific. Recovery may be rapid and complete or more protracted and incomplete, depending on the severity and duration of exposure. Cognitive function examination – orientation, judgment, concentration, language, mood, memory – may also reveal dysfunction. In the most severe instances, concomitant systemic effects, including pulmonary edema and circulatory collapse, may cause severe neurologic dysfunction independent of the underlying neurotoxicity. Examples of depressed consciousness and altered cognitive function include the consequences of high exposures to a variety of organic solvents (Dick, 1988) and other chemicals such as methyl bromide (de Souza et al., 2013), methylene chloride (Centers for Disease Control and Prevention, 2012), organic lead (Le Quesne, 1981), ethylene oxide (Ohnishi, 2000), and carbon monoxide (Hampson et al., 2012).

In the acute intoxication that occurs with exposure to organophosphorus esters, excessive CNS cholinergic stimulation leads to impaired consciousness and cognitive function, but this is accompanied by peripheral neurologic signs, such as muscle fasciculation and miosis, suggesting the diagnosis (Lotti, 2010).

Chronic toxic encephalopathy is now a rare event. It is characterized by a steady decline in cognitive function without alteration of consciousness, as exemplified by the encephalopathy resulting from chronic exposure to ethylene oxide (Crystal et al., 1988). The existence of a chronic solvent-induced encephalopathy is controversial (Albers and Berent, 2000; van Valen et al., 2012).

Clues to the nature of the underlying disease may derive from analysis of gait. A steppage gait characterized by foot drop and high step suggests distal symmetric polyneuropathy, as observed in most toxic neuropathies. Manganese-induced parkinsonism shows a specific dystonic gait (coq au pied) that is different from the gait observed in typical Parkinson's disease, which is characterized by difficulty in initiation, short steps, and tendency to festination (Olanow, 2004). Cerebellar wide-based gait may be observed in acrylamide poisoning (Myers and Macun, 1991).

Examination of cranial nerves may reveal impaired function of the olfactory nerve leading to anosmia/hyposmia in workers exposed to chromate salts and cadmium (Gobba, 2006), of the optic nerve causing defective color vision as from exposures to methyl bromide or organic solvents (Chavez et al., 1985; Paramei et al., 2004), and of the auditory nerve after acute or chronic exposures to noise or in association with solvent ones (Chang et al., 2003).

Tremor at rest is characteristic of intoxication by metallic mercury (Clarkson et al., 2003) and chlordecone (Taylor et al., 1978). Motor coordination abnormalities such as dysmetria on finger-to-nose testing and dysdiadochokinesia may occur, for instance, in patients with acrylamide poisoning and in manganism, reflecting cerebellar and extrapyramidal disorders. In manganese-induced parkinsonism, balance impairment occurs earlier than in classic Parkinson's disease, the neurologic disorder is symmetric, and there is a relative absence of tremor (Olanow, 2004). However, the described clinical differences between manganism and Parkinson's disease are not observed in parkinsonism occurring in welders exposed to manganese, where clinical signs overlap with those seen in the early stages of Parkinson's disease (Racette, 2013).

The overall pattern and distribution of motor and sensory disturbances suggest the diagnosis. Thus, focal impairment suggests entrapment, and bilateral and symmetric length-dependent distribution that commences in the legs is compatible with a toxic neuropathy. Motor examination may reveal muscle atrophy (wasting) in severe toxic neuropathies. In cases of entrapment neuropathies, for example in carpal tunnel syndrome that is associated with repetitive and strenuous processing work, pinch strength and thumb opposition and abduction may be altered (Marquardt et al., 2014). The muscle stretch (tendon) reflexes in the legs may be absent or reduced either symmetrically (particularly at the ankles) as in toxic polyneuropathies or in a lateralized distribution reflecting isolated peripheral nerve or radicular involvement such as, for instance, those resulting from prolapsed lumbosacral disks (L3–L4; S1–S2).

Sensory examination in toxic distal neuropathies reveals in most cases selective effects on large fibers. Loss of vibration sense is often an early sign in toxic neuropathies. Other findings include loss of joint position, impairment of touch pressure sensibility, sensory ataxia, and sometimes pseudoathetosis. The pattern of sensory loss in severe toxic neuropathy has a stocking-glove distribution. Small fibers are rarely involved, but pain and temperature loss are seen in a few toxic neuropathies such as in thallium poisoning.

LABORATORY INVESTIGATIONS

Laboratory toxicologic tests aimed at the diagnosis of occupational neurotoxic diseases depend on the clinical circumstances. In cases of acute poisoning, testing may only provide qualitative information for the differential diagnosis because correlations between concentration of the offending chemical in body fluids and clinical signs are often unknown. In a few cases of acute poisoning, some tests may predict clinical developments. However, in most cases, tests aimed at identifying the putative responsible chemical(s) are of limited value because these measurements usually reflect recent or previous exposures. By contrast, laboratory tests are widely used in the biomonitoring of occupational exposures to neurotoxicants (see Chapter 4).

Examples of laboratory investigations in acute poisoning

Virtually any chemical can be identified in the biologic fluids of a poisoned patient, providing that a skilled analytic laboratory is available. Depending on the chemical, it can be measured in the blood, as in the case of heavy metals; in the breath, as for solvents; or in the urine either as such (e.g., metals) or as specific metabolites (e.g., solvents). In cases of acute cholinergic overstimulation, such as occurs in organophosphate and carbamate poisoning, the measurement of acetylcholinesterase inhibition in red blood cells confirms the diagnosis (see Chapter 11). An elevated percentage of carboxyhemoglobin may confirm the diagnosis of coma caused by carbon monoxide (Hampson et al., 2012). In a few circumstances, certain laboratory tests may predict the delayed clinical consequences of an acute poisoning. For instance, early elevated concentrations in cerebrospinal fluid of myelin basic protein and interleukin-6 predict delayed encephalopathy from carbon monoxide poisoning (Ide and Kamijo, 2008, 2009). Another example of a predictive test is the measurement of neuropathy target esterase in circulating lymphocytes after poisoning by certain organophosphates. Inhibition soon after poisoning of this enzyme predicts the onset of polyneuropathy 2–3 weeks later (Lotti et al., 1986).

Examples of laboratory investigations for neurotoxic occupational diseases

Workers affected by neurotoxic diseases usually seek medical attention when exposure has already ceased. Since most neurotoxic chemicals remain for only a short period in the body, laboratory tests on body fluids have limited application. Some neurotoxic chemicals (e.g., acrylamide, carbon disulfide, and *n*-hexane) form covalent adducts to blood macromolecules such as albumin and hemoglobin, and the measurement of these adducts reflects past exposure over the previous weeks to months, depending on the half-life of the chosen macromolecule (Valentine et al., 1993; Calleman et al., 1994; Graham et al., 1995). Past exposures to metals may be revealed through the administration of chelators to mobilize the metal from body storage sites. Thus, lead is removed from bone by injecting ethylenediaminetetraacetic acid and its urinary excretion can be measured (Emmerson, 1963). For other metals, such as arsenic and mercury, measurements of their concentration in hair reflect prior and/or cumulative exposure rather than the recent exposure provided by blood measurements (Costa, 1996).

ELECTROPHYSIOLOGY

Electrophysiologic studies are important in helping to confirm the organic basis of symptoms, especially when there is little to find on clinical examination, and to detect subclinical or early neurologic involvement so that further exposure can be limited. They help to define the nature of any clinical disturbance and the site at which it arises. The severity and progression of the lesion can also be monitored (Aminoff and Albers, 2012). Electrophysiologic approaches are especially helpful in evaluating the PNS. When used to evaluate patients with suspected CNS dysfunction following neurotoxic exposure, the findings have been of limited utility and uncertain clinical relevance.

Electroencephalography

The EEG has been used to evaluate patients with neurobehavioral disturbances, but its variability in normal subjects limits its utility. A number of factors (such as age and level of arousal) affect it and may confound interpretation and comparison between subjects. Encephalopathies of diverse etiology lead to nonspecific slowing, and the EEG cannot distinguish between them reliably. Paroxysmal epileptiform activity may occur but is occasionally found in normal subjects and those taking certain medications or with certain metabolic disorders, as well as with mercury or lead poisoning, exposure to chlorinated hydrocarbons (as in the manufacture of DDT), or organophosphate poisoning. The limited practical utility of the EEG can be exemplified by reference to the encephalopathy induced by carbon disulfide. One study of 10 patients showed that none had abnormal EEGs (Huang et al., 1996), another revealed that the EEG findings were inconsistent but mostly normal (Aaserud et al., 1990), and a third study showed frequency changes of dubious relevance (Sińczuk-Walczak and Szymczak, 1997).

Quantitative analysis of the EEG of workers with possible exposure to neurotoxic agents has generally been disappointing. Most such studies have involved multiple comparisons of different aspects of the EEG, and false-positive results are therefore likely simply on the basis of chance. Methodologic ambiguity renders many studies invalid, and departures from normality are of uncertain relevance. Alterations in the EEG may be attributed erroneously to toxin exposure when they relate instead to a change in the level of arousal. Other artifacts must also be excluded before computerized analysis of the EEG.

Some of these issues can be considered by reference to the study of Matikainen et al. (1993), in which quantitative EEG analysis was performed on workers from several plastics factories. Various quantitative abnormalities were identified and some 798 comparisons were made: about 40 significant results would thus be expected by chance alone, with a significance criterion of 0.05 (Rebert and Hall, 1994). The number of comparisons was reduced to 192 by eliminating redundant parameters and restricting the channel count, but 10 significant findings would still be expected by chance alone. In fact, the number of significant comparisons was only 1, and this was of uncertain biologic meaning because there was no adequate control group. Neurometric discriminant analysis was used for classifying subjects in terms of EEG abnormalities, but the reference normative data were not an appropriate standard. Finally, even if abnormalities had been detected in this study of apparently healthy workers, there would have been no justification for concluding that the EEG changes reflect subclinical dysfunction related to neurotoxic exposure.

At the present time, then, it is hard to justify the use of expensive, quantitative techniques of EEG analysis to establish the presence of an encephalopathy that might have an occupational or toxic basis (Nuwer, 1997).

Sensory evoked potentials

Evoked potential studies of certain afferent systems provide quantitative data that permit objective evaluation and facilitate intra- or intersubject comparisons. The response of the CNS to visual, auditory, or somatosensory stimuli is recorded using established protocols. Depending on the sensory modality, the latency, amplitude, and intercomponent latency of the response are determined and compared with values in age- and sex-matched normal subjects studied under the same conditions. Evoked potential studies can help in determining the organic basis of equivocal symptoms and in detecting and localizing lesions of the afferent system under test. They may sometimes detect toxin-related changes at a subclinical stage so that further damage is avoided by limiting exposure, but the findings are not specific to any individual disorder.

The possible utility of evoked potential studies is exemplified by the work of Rosenberg and coworkers (1988), who defined the brainstem auditory evoked potential (BAEP) findings in 11 chronic toluene abusers. BAEPs were abnormal in five of them, and analysis of the group showed a prolongation of the absolute latency of wave V and of the I–V and III–V interpeak latencies compared to control subjects. Two of these five patients had normal clinical findings and no abnormality on magnetic resonance imaging (MRI) of the brain. These findings and those of others (Vrca et al., 1996) suggest that BAEPs may be useful in the early detection of CNS involvement from toluene inhalation.

Endogenous potentials

Endogenous (event-related) potentials are recorded over the vertex of the scalp in response to an infrequent stimulus to which the subject directs attention to distinguish it from other, more frequent stimuli. These potentials depend on the context of the target stimulus rather than on its physical characteristics. They may be delayed in patients with cognitive impairment compared to age-matched control subjects and have been used to distinguish between dementia and depression. Their utility in evaluating patients with neurobehavioral disturbances related to neurotoxic exposure is uncertain.

Electromyography and nerve conduction studies

These techniques are best considered as extensions of the neurologic examination and can be very helpful in evaluating patients with suspected neurotoxic disorders. They help to localize abnormalities to specific levels of the peripheral neuromuscular and sensory systems (such as to axons, dorsal root ganglion cells, anterior horn cells, neuromuscular junctions, or muscle) and to a specific site along the course of nerves, and suggest the most likely pathophysiology (such as axonal loss, demyelination, or conduction block), thereby assisting in differential diagnosis. The pattern of involvement can distinguish between polyneuropathy, mononeuropathy simplex or multiplex, and polyradiculopathy, and can reveal selective involvement of motor or sensory fibers. Studies can identify subclinical lesions and can be performed repeatedly to document changes over time. They allow the severity of a disorder to be defined quantitatively. In some instances, additional features characteristic of a particular neurotoxin may be encountered, as in patients with acute organophosphate toxicity (discussed later).

Electromyography

The electric activity of muscle is studied for diagnostic purposes by inserting a needle recording electrode directly into it. Bioelectric potentials are amplified, displayed on a cathode ray oscilloscope for visual analysis, and fed through a loudspeaker system for acoustic monitoring. Commercially available equipment includes analog-to-digital converters that permit the easy storage of data, advanced signal processing and analysis, and alteration in the display characteristics (e.g., time base and sensitivity) of stored potentials. Needle electromyography (EMG) is a sensitive indicator of partial denervation or muscle fiber necrosis. It is generally less important than nerve conduction studies in the evaluation of neurotoxic disorders, except in the evaluation of toxic myopathy. The findings in themselves are never pathognomonic of specific diseases and therefore do not provide a definitive diagnosis.

When the nerve supply to a muscle is interrupted (or parts of muscle fibers are isolated from their endplates, so that they are functionally denervated), abnormal spontaneous activity appears in the muscle within weeks and is characterized by positive waves and fibrillation potentials. These potentials represent involuntary muscle fiber action potentials and are associated with denervation hypersensitivity resulting from the proliferation of acetylcholine receptors on the muscle fiber surface. Other types of abnormal spontaneous activity may also occur. In addition, changes occur in the number and configuration of motor unit potentials seen with voluntary activation of the muscle. In myopathic disorders (with their reduced numbers of muscle fibers per unit), motor unit action potentials are typically small, short, and abundant for the degree of voluntary activity, and there is an increased number of polyphasic potentials. By contrast, in neuropathic disorders, the number of functional motor units is reduced, but the average number of muscle fibers per unit may actually be increased if reinnervation has occurred by collateral branching from surviving nerve fibers. The motor unit action potentials recorded in such circumstances are of longer duration than normal, are often large, and may be polyphasic. Recruitment of motor units is reduced. In axonal neuropathies, there may be EMG signs of chronic partial denervation in distal muscles (with or without accompanying evidence of reinnervation), whereas in the demyelinative neuropathies the only or most conspicuous abnormality on the needle examination is reduced recruitment of motor units.

Nerve conduction studies

Nerve conduction studies are performed using standardized techniques (Daube and Rubin, 2012). Motor nerve conduction studies are used to evaluate motor axons by recording the muscle responses (compound muscle action potentials) to nerve stimulation, whereas sensory nerve conduction studies assess sensory axons by stimulating or recording from nerves with predominantly sensory axons (sensory nerve action potentials). Normal values depend on various factors, including technique, patient age and size, temperature, and even occupation. Response amplitude depends on the number and size of functioning nerve or muscle fibers. Disorders with a marked loss of axons or muscle fibers (e.g., a toxic neuropathy or myopathy) have attenuated responses. Distal latency reflects conduction along the terminal portion of the nerve, whereas maximal nerve conduction velocity is the conduction velocity in the fastest-conducting (large myelinated) fibers. Late-response (F-wave or H-reflex) studies are sometimes helpful in detecting proximal pathology.

Blink reflex studies record reflex responses to stimulation of the trigeminal nerve. These reflexes involve peripheral and brainstem connections, and have occasional application in the evaluation of neurotoxic disorders.

The results of electrophysiologic studies are used to classify peripheral disorders, thereby reducing the number of disorders to be considered in the differential diagnosis. The electrophysiologic features of neurotoxic disorders vary with the causal agent, severity of involvement, and the time of testing in relation to the clinical course.

Some of these points can be illustrated by reference to organophosphates. During acute organophosphate intoxication, repetitive compound muscle action potentials characteristically occur after a single stimulus to a motor nerve, presumably from recurrent postsynaptic depolarization by persistent acetylcholine (Senanayake and Karalliedde, 1987). With exposure to certain of these insecticides at levels producing cholinergic toxicity, clinical or subclinical neuropathy may occur 2–4 weeks later (Lotti, 2002; Albers et al., 2007). The electrodiagnostic findings then are consistent with axonal degeneration of motor and, to a lesser extent, sensory fibers. Conduction velocity is usually normal, but compound muscle and sensory nerve action potentials are reduced in amplitude, and needle EMG shows evidence of severe denervation, with diffuse fibrillation potentials.

Tests of neuromuscular transmission

Electrophysiologic tests of neuromuscular transmission typically are performed at the time of standard nerve conduction studies and involve repetitive motor nerve stimulation or single-fiber electromyography. Details are provided in standard textbooks, and further comment here is unnecessary because the neuromuscular junction is an uncommon target for work-related neurotoxic agents.

Autonomic function

Dysautonomic symptoms occur in certain toxic neuropathies (e.g., related to use of or following exposure to organic solvents or acrylamide). Even when they are absent, autonomic function tests may be worthwhile because they assess the function of small-diameter axons that otherwise can only be evaluated physiologically by quantitative sensory testing. Adrenergic involvement is tested by assessing peripheral vasomotor function under various circumstances, such as an inspiratory gasp or a startle; sympathetic cholinergic (sudomotor) function by tests of sweating, as by the sympathetic skin response; and parasympathetic (vagal) impairment by the heart rate responses to various maneuvers such as deep inspiration, postural change, or the Valsalva maneuver. Sympathetic skin responses are recorded from the skin, between areas of high and low sweat gland density. They occur spontaneously or in response to a variety of stimuli, including electric stimulation or startle. Preserved responses argue against a significant abnormality of autonomic sudomotor function.

IMAGING

In general, imaging studies with computed tomography and MRI are of limited utility in the evaluation of a suspected neurotoxic disorder, except for helping to exclude other (structural) causes of the patient's clinical state and for confirming the site and extent of pathologic involvement. For example, slaughterhouse workers have been reported to develop a sensory-predominant polyradiculoneuropathy, sometimes with additional features of a transverse myelitis, meningoencephalitis, or aseptic meningitis, that probably has an autoimmune origin related to exposure to multiple aerosolized porcine brain tissue antigens (Lachance et al., 2010). The affected workers typically complained of painful paresthesias, distal sensory loss, and weakness developing 4 weeks or more after initial exposure. Examination also revealed diminished or absent muscle stretch reflexes. Electrophysiologic studies revealed abnormalities of the most proximal and distal nerve segments, involving both large- and small-diameter fibers. MRI findings confirmed proximal involvement at nerve roots or ganglia, showing ganglia enlargement, T2 hyperintensity, or enhancement in dorsal roots and enlarged (mostly ventral) roots in the cauda equina.

MRI is particularly helpful in indicating involvement of cerebral white matter, as this may be clinically inapparent and is not seen as well or at all by other imaging techniques. Indeed, MRI may reveal the specific vulnerability of the white matter to certain toxins, as is discussed in Chapter 7 of this volume.

Imaging may also be helpful in providing nonclinical markers of disease or treatment effects (Pogge and Slikker, 2004). Magnetic resonance spectroscopy (MRS) reveals information of physiologic or metabolic relevance; single-photon emission computed tomography (SPECT), functional MRI and positron emission tomography (PET) provide information on function. The utility of these techniques can be illustrated with regard to manganese neurotoxicity, and especially with reference to [^{18}F]-dopa PET, which is a noninvasive means of assessing the integrity of nigrostriatal dopaminergic neurons, i.e., of presynaptic dopamine terminals.

Occupational exposure to manganese dust or vapor occurs during manganese mining or smelting, and in manganese alloy producers, cell battery factory workers, and farmers using manganese-containing fungicides. In recent years, high levels of manganese also have occurred in recreational drug users who have injected certain home-made psychostimulant preparations (ephedron, also called methcathinone: Guilarte, 2013). Toxicity may result, manifesting initially with neurobehavioral and cognitive disturbances. Extrapyramidal changes, with dystonic posturing of the extremities or parkinsonism, may eventually develop. Neuroimaging has been used to study manganese-induced parkinsonism and has helped to distinguish it from classic Parkinson's disease (Kim, 2006). On T1-weighted MRI, increased signal intensity is present in the globus pallidus and midbrain bilaterally in exposed workers with signs of manganese intoxication, as well as in some who are symptomatic. [^{18}F]-dopa PET (which reflects nigrostriatal function, probably best correlating with dopamine content) and SPECT (with specific radioligands reflecting the availability of presynaptic dopamine transporters) have been reported as normal, indicating the integrity of the nigrostriatal dopaminergic system, as also has [^{11}C]-raclopride PET, reflecting postsynaptic dopamine receptors (Shinotoh et al., 1997). Further, neuroimaging studies performed in ephedrone users have shown normal dopamine terminals, based on dopamine transporter levels, in the striatum using SPECT imaging despite the presence of clinical hypokinetic parkinsonism (Selikhova et al., 2008; Sikk et al., 2013). These combined findings suggest that, in manganese toxicity, the pathologic process is downstream of the nigrostriatal system, accounting for the poor response to levodopa in such patients (Lu et al., 1994).

Further, imaging studies can provide a biomarker of manganese exposure. Criswell et al. (2011), using [^{18}F]-dopa PET, studied 20 asymptomatic welders exposed to manganese fumes, 20 subjects with idiopathic Parkinson's disease, and 20 normal controls. In Parkinson's disease, uptake was decreased in the caudate and, to a greater extent, the putamen, confirming the findings of others (Gallagher et al., 2011). The welders had a small (10%) but significantly lower level of [^{18}F]-dopa uptake in the caudate nucleus than controls but no change in the putamen. The caudate reduction may represent an early (asymptomatic) marker of manganese neurotoxicity.

Another possible biomarker is thalamic levels of the inhibitory neurotransmitter gamma-aminobutyric acid (GABA), which is involved in basal ganglia function. Animal studies have indicated that striatal GABA levels are increased after manganese exposure, and this has been confirmed in a study of humans exposed to manganese. Using MRS, Long et al. (2014) found that mean thalamic GABA level was significantly higher in smelter workers than controls; multiple linear regression analysis revealed an association between GABA level and the duration of exposure and an inverse association with fine motor performance. It was therefore suggested that increased thalamic GABA levels may be a biomarker for subtle deficits in motor control (Racette et al., 2012).

NEUROBEHAVIORAL TESTING

The mental status examination is used to elicit and quantify the behavioral changes, largely cognitive, that are found in patients with brain disease or dysfunction. Changes in neuropsychologic outcomes are detected and measured with a variety of structured psychometric test batteries, some known to be related to brain damage regardless of location, and others to damage of specific brain areas. A few examples of the uses of these tests in occupational neurology are given here but their description is beyond the scope of this chapter. More information on this issue may be found elsewhere (section 2). Work-related neurologic disorders resulting in diffuse and discrete cortical dysfunction may be suspected and detected in different circumstances. Massive exposures to chemicals in the workplace, resulting in confusion or a depressed level of consciousness, may occur accidentally. This may involve chemical exposure over a relatively short period of time to almost any solvent, carbon monoxide, pesticides (organophosphates, organochlorine, methyl bromide), cyanide, and others. In such cases the etiologic diagnosis is usually straightforward and based upon the history and sometimes on specific laboratory tests. Neuropsychologic evaluation is seldom performed because recovery is usually rapid and complete or, in some cases, because concurrent signs indicate the diagnosis. Also less frequent nowadays is the need to assess cognitive function, as for instance by using the Mini-Mental score (Simmons et al., 2011) to evaluate suspected mild dementia, possibly associated with prolonged high exposures to neurotoxins in the workplace. In fact, occupational exposures known to be associated with the development of dementia, as in the case of carbon disulfide exposure during the production of certain artificial fibers, are unlikely to have occurred in the last decades.

Improved conditions and safer practices in the workplace have led to a gradual shift in application of the neuropsychologic evaluation from the assessment of severe neurotoxic damage to the evaluation of mild subclinical disturbances, which may lead to disabling cumulative damage after several years of low-level exposure. Consequently, these neuropsychologic tests are now used extensively in screening workers exposed to neurotoxicants for the prevention of further intoxication and in an attempt to reduce further exposures. Thus, neuropsychologic test sensitivity rather than specificity is more important in screening for neurotoxicity in large populations.

Occupational exposures to certain neurotoxicants have shown a causal relationship with behavioral disturbances when associated with other signs of neurotoxicity. An example is that of the electrophysiologic and psychologic changes occurring in workers exposed to inhalation of sparsely soluble aluminum particles in welding fumes. These workers showed disturbances of cognitive processes, memory, concentration, and mood that were associated with EEG changes (Riihimäki and Aitio, 2012). Another example is given by the progressive cognitive changes that have been associated with a reduction of brain volumes in workers with past cumulative exposures to both inorganic and organic lead. (Schwartz et al., 2007, 2010).

Exposures to neurotoxicants may also represent a modifiable risk factor of behavioral changes associated with neurodegeneration. Among patients with Parkinson's disease, higher tibial lead concentrations were associated with worse cognition assessed with an analog of the Mini-Mental State Examination, suggesting that cumulative lead exposure may have exacerbated the disease (Weuve et al., 2013).

Occupational exposures to a variety of chemicals have been associated with neuropsychologic alterations. However, the causal correlation between these exposures and mild subclinical behavioral signs is often difficult to establish, because results may be mimicked or hidden by numerous confounding variables. Different epidemiologic reports cannot easily be compared and interpreted because the variables taken into account often differ. Such variables include age, alcohol and drug intake, stress, assessment of exposure, choice of controls, and educational background. The latter, for instance, was shown to be a relevant confounding factor in studies on lead- and solvent-exposed workers because those with a higher educational background seemed to compensate for the effects of these exposures on cognition (Bleecker et al., 2002; Sabbath et al., 2012).

An example of difficult interpretation of epidemiologic behavioral studies in exposed workers is represented by the early reports on chronic solvent-induced encephalopathy (Grasso, 1988) that lead to the dubious conclusion about the association of cognitive effects with exposures to solvent mixtures (Aminoff and Lotti, 2010). However, a recent evaluation of studies published in the period 1985–2012 has led to a consensus about the neuropsychologic characteristics, assessment, and diagnosis of this condition. Alteration of attention, speed of information processing, memory, and motor performance encompass the characteristic of this condition (van Valen et al., 2012) that appears to be reversible at least in workers who have been exposed only moderately (Sabbath et al., 2014). Nevertheless, the link between these changes and chronic low level of exposure to a variety of very different chemicals having in common solvent properties only is somehow counterintuitive because a shared underlying mechanism appears unlikely.

Epidemiologic studies in this field are often inconsistent. One example is that of occupational exposures to manganese, which in the past have been associated with a complex neurologic syndrome comprising cognitive deficits, several neuropsychologic disturbances, and parkinsonism (Guilarte, 2013). Although certain neurobehavioral changes have been observed in several studies of manganese workers without signs of parkinsonism (Bowler et al., 1999) and shown to persist for several years after the end of exposure (Bouchard et al., 2007), other studies on manganese miners and smelters were negative (Myers et al., 2003a, b).

Similarly inconsistent are the altered neurobehavioral test performances reported in workers exposed to chronic low levels of organophosphorus pesticides (Rohlman et al., 2011; Starks et al., 2012). Thus, the overall evidence of these reported effects is difficult to interpret, especially because in most studies the absence of cholinesterase measurements hampers the understanding of the underlying mechanism.

In conclusion, behavioral testing appears in most cases of limited uses for the etiologic diagnosis of neurologic diseases, whereas it may be helpful in both preventive medicine and research, providing that all possible confounding factors are taken into account so that the results will be reproducible.

References

Aaserud O, Hommeren OJ, Tvedt B et al. (1990). Carbon disulfide exposure and neurotoxic sequelae among viscose rayon workers. Am J Ind Med 18: 25–37.

Albers JW, Berent S (2000). Controversies in neurotoxicology: current status. Neurol Clin 18: 741–764.

Albers JW, Garabrant DH, Mattsson JL et al. (2007). Dose-effect analyses of occupational chlorpyrifos exposure and peripheral nerve electrophysiology. Toxicol Sci 97: 196–204.

Allen N, Mendell JR, Billmaier DJ et al. (1975). Toxic polyneuropathy due to methyl n-butyl ketone. Arch Neurol 32: 209–218.

Aminoff MJ, Albers JW (2012). Electrophysiologic techniques in the evaluation of patients with suspected neurotoxic disorders. In: MJ Aminoff (Ed.), Aminoff's Electrodiagnosis in Clinical Neurology, 6th Ed. Elsevier, Oxford, pp. 813–830.

Aminoff MJ, Lotti M (2010). Neurotoxic effects of workplace exposure. In: PJ Baxter, T-C Aw, A Cockroft et al. (Eds.), Hunter's Disease of Occupations, 10th Ed. Hodder Arnold, London, pp. 1151–1170.

Aminoff MJ, Greenberg DA, Simon RP (2005). Clinical Neurology, 6th Ed. McGraw-Hill, New York.

Baker EL, Christiani DC, Wegman DH et al. (1981). Follow-up studies of workers with bladder neuropathy caused by exposure to dimethylaminopropionitrile. Scand J Work Environ Health 7 (Suppl 4): 54–59.

Bleecker ML, Lindgren KN, Ford DP et al. (2002). The interaction of education and cumulative lead exposure on the mini-mental state examination. J Occup Environ Med 44: 574–578.

Bouchard M, Mergler D, Baldwin M et al. (2007). Neurobehavioral functioning after cessation of manganese exposure: a follow-up after 14 years. Am J Ind Med 50: 831–840.

Bowler RM, Mergler D, Sassine MP et al. (1999). Neuropsychiatric effects of manganese on mood. Neurotoxicology 20: 367–378.

Calleman CJ, Wu Y, He F et al. (1994). Relationship between biomarkers of exposure and neurological effects in a group of workers exposed to acrylamide. Toxicol Appl Pharmacol 126: 361–371.

Centers for Disease Control and Prevention (2012). Fatal exposure to methylene chloride among bathtub refinishers – United States, 2000-2011. MMWR Morb Mortal Wkly Rep 61: 119–122.

Chang SJ, Shih TS, Chou TC et al. (2003). Hearing loss in workers exposed to carbon disulfide and noise. Environ Health Perspect 111: 1620–1624.

Chavez CT, Hepler RS, Straatsma BR (1985). Methyl bromide optic atrophy. Am J Ophthalmol 99: 715–719.

Clarkson TW, Magos L, Myers GY (2003). The toxicology of mercury – current exposures and clinical manifestations. N Engl J Med 349: 1731–1737.

Costa LG (1996). Biomarkers research in neurotoxicology: the role of mechanistic studies to bridge the gap between the laboratory and epidemiological investigations. Environ Health Perspect 104: 55–67.

Criswell SR, Perlmutter JS, Videen TO et al. (2011). Reduced uptake of [^{18}F]DOPA PET in asymptomatic welders with occupational manganese exposure. Neurology 76: 1296–1301.

Crystal HA, Schaumburg HH, Grober E et al. (1988). Cognitive impairment and sensory loss associated with chronic low-level ethylene oxide exposure. Neurology 38: 567–569.

Daube JR, Rubin DL (2012). Nerve conduction studies. In: MJ Aminoff (Ed.), Aminoff's Electrodiagnosis in Clinical Neurology, 6th Ed. Elsevier, Oxford, pp. 289–325.

De Souza A, Narvencar KPS, Sindhoora KV (2013). The neurological effects of methyl bromide intoxication. J Neurol Sci 335: 36–41.

Dick RB (1988). Short duration exposures to organic solvents: The relationship between neurobehavioral test results and other indicators. Neurotoxicol Teratol 10: 39–50.

Dyck PJ, O'Brien PC (1999). Quantitative sensation testing in epidemiological and therapeutic studies of peripheral neuropathy. Muscle Nerve 22: 659–662.

Emmerson BT (1963). Chronic lead neuropathy: the diagnostic use of calcium EDTA and the association with gout. Australas. Ann Med 12: 310–319.

Gallagher CL, Oakes TR, Johnson SC et al. (2011). Rate of 6-[18F]-fluorodopa uptake decline in striatal subregions in Parkinson's disease. Mov Disord 26: 614–620.

Gobba F (2006). Olfactory toxicity: long term effects of occupational exposures. Int Arch Occup Environ Health 79: 322–331.

Graham DG, Amarnath V, Valentine WM et al. (1995). Pathogenetic studies of hexane and carbon disulfide neurotoxicity. Crit Rev Toxicol 25: 91–112.

Grasso P (1988). Neurotoxic and neurobehavioral effects of occupational exposures to organic solvents. Occup Med 3: 525–539.

Guilarte TR (2013). Manganese neurotoxicity: new perspectives from behavioral and neuropathological studies in humans and non-human primates. Front Aging Neurosci 5: 23.

Hampson NB, Plantadosi CA, Thom SR et al. (2012). Practice recommendation in the diagnosis, management, and prevention of carbon monoxide poisoning. Am J Respir Crit Care Med 186: 1095–1101.

He FS, Zhang SL, Wang HL et al. (1989). Neurological and electroneuromyographic assessment of the adverse effects of acrylamide on occupationally exposed workers. Scand J Work Environ Health 15: 125–129.

Hoffman RS (2003). Thallium toxicity and the role of Prussian blue in therapy. Toxicol Rev 22: 29–40.

Horan JM, Kurt TL, Landrigan PJ et al. (1985). Neurologic dysfunction from exposures to 2-t-butylazo-2-hydroxy-5-methylhexane (BHMH): a new occupational neuropathy. Am J Public Health 75: 513–517.

Huang C-C, Chu C-C, Chen R-S et al. (1996). Chronic carbon disulfide encephalopathy. Eur Neurol 36: 364–368.

Hunter D (1978). The Diseases of Occupations, Hodder & Stoughton, London.

Ichihara G, Miller JK, Ziolkowska A et al. (2002). Neurological disorders in three workers exposed to 1-bromopropane. J Occup Health 44: 1–7.

Ide T, Kamijo Y (2008). Myelin basic protein in cerebrospinal fluid: a predictive marker of delayed encephalopathy from carbon monoxide poisoning. Am J Emerg Med 26: 908–912.

Ide T, Kamijo Y (2009). The early elevation of interleukin-6 concentration in cerebrospinal fluid and delayed encephalopathy of carbon monoxide poisoning. Am J Emerg Med 27: 992–996.

Kim Y (2006). Neuroimaging in manganism. Neurotoxicology 27: 369–372.

Lachance DH, Lennon VA, Pittock SJ et al. (2010). An outbreak of neurological autoimmunity with polyradiculoneuropathy in workers exposed to aerosolized porcine neural tissue: a descriptive study. Lancet Neurol 9: 55–66.

Le Quesne P (1981). Toxic substances and the nervous system: the role of clinical observation. J Neurol Neurosurg Psychiatry 44: 1–8.

Long Z, Li XR, Xu J et al. (2014). Thalamic GABA predicts fine motor pperformance in manganese-exposed smelter workers. PLoS One 9 (2): e88220, 2014 Feb 4.

Lotti M (2002). Low-level exposures to organophosphorus esters and peripheral nerve function. Muscle Nerve 25: 492–504.

Lotti M (2010). Clinical toxicology of anticholinesterase agents in humans. In: R Krieger (Ed.), Hayes' Handbook of Pesticide Toxicology, 3rd Ed. Elsevier, San Diego, pp. 1543–1589.

Lotti M, Moretto A, Zoppellari R et al. (1986). Inhibition of lymphocytic neuropathy target esterase predicts the development of organophosphate-induced delayed neuropathy. Arch Toxicol 59: 176–179.

Lu C, Huang C, Chu N et al. (1994). Levodopa failure in chronic manganism. Neurology 44: 1600–1602.

Marquardt TL, Nataraj R, Evans PJ et al. (2014). Carpal tunnel sysndrome impairs thumb opposition and circumduction motion. Clin Orthop Relat Res 472: 2526–2533.

Matikainen E, Forsman-Gronholm L, Pfaffli P et al. (1993). Nervous system effects of occupational exposure to styrene: a clinical and neurophysiological study. Environ Res 61: 84–92.

Myers JE, Macun I (1991). Acrylamide neuropathy in a South African factory: an epidemiological investigation. Am J Ind Med 19: 487–493.

Myers JE, teWaterNaude J, Fourie M et al. (2003a). Nervous system effects of occupational manganese exposure on South African manganese mineworkers. Neurotoxicology 24: 649–656.

Myers JE, Thompson ML, Ramushu S et al. (2003b). The nervous system effects of occupational exposure on workers in a South African Manganese smelter. Neurotoxicology 24: 885–894.

Nuwer M (1997). Assessment of digital EEG, quantitative EEG, and EEG brain mapping: report of the American Academy of Neurology and the American Clinical Neurophysiology Society. Neurology 49: 277–292.

Ohnishi A (2000). Ethylene oxide. In: PS Spencer, HH Schaumburg (Eds.), Experimental and Clinical Neurotoxicology, 2nd Ed. Oxford University Press, New York, pp. 563–566.

Olanow CW (2004). Manganese-induced parkinsonism and Parkinson's disease. Ann N Y Acad Sci 1012: 209–223.

Paramei GV, Meyer-Baron M, Seeber A (2004). Impairment of color vision induced by organic solvents: a meta-analysis study. Neurotoxicology 25: 803–816.

Parry GJ (2008). Neurological complications of toxin exposure in the workplace. In: MJ Aminoff (Ed.), Neurology and General Medicine, 4th Ed. Elsevier, Philadelphia, pp. 749–768.

Pogge A, Slikker Jr W (2004). Neuroimaging: new approaches for neurotoxicology. Neurotoxicology 25: 525–531.

Racette BA (2013). Manganism in the 21st century: The Henninen lecture. Neurotoxicology 45C: 201–207.

Racette BA, Aschner M, Guilarte TR et al. (2012). Pathophysiology of manganese-associated neurotoxicity. Neurotoxicology 33: 881–886.

Rebert CS, Hall TA (1994). The neuroepidemiology of styrene: a critical review of representative literature. Crit Rev Toxicol 24: S57–S106.

Riihimäki V, Aitio A (2012). Occupational exposure to aluminum and its biomonitoring in perspective. Crit Rev Toxicol 42: 827–863.

Rohlman Ds, Anger WK, Lein PJ (2011). Correlating neurobehavioral performance with biomarkers of organophosphorous pesticide exposure. Neurotoxicology 32: 268–276.

Rosenberg NL, Spitz MC, Filley CM et al. (1988). Central nervous system effects of chronic toluene abuse – clinical, brainstem evoked response and magnetic resonance imaging studies. Neurotoxicology 10: 489–495.

Sabbath EL, Glymour MM, Berr C et al. (2012). Occupational solvent exposure and cognition: does the association vary with the level of education? Neurology 78: 1754–1780.

Sabbath EL, Gutierrez LA, Okechukwu CA et al. (2014). Time may not fully attenuate solvent-associated cognitive deficits in highly exposed workers. Neurology 82: 1716–1723.

Schaumburg HH, Albers JW (2005). Pseudoneurotoxic disease. Neurology 65: 22–26.

Schwartz BS, Chen S, Caffo B et al. (2007). Relations of brain volumes with cognitive function in males 45 years and older with past lead exposure. Neuroimage 37: 633–641.

Schwartz BS, Caffo B, Stewart WF et al. (2010). Evaluation of cumulative lead dose and longitudinal changes in structural magnetic resonance imaging in former organolead workers. J Occup Environ Med 52: 407–414.

Selikhova M, Fedoryshyn L, Matviyenko Y et al. (2008). Parkinsonism and dystonia caused by illicit use of ephedrone – a longitudinal study. Mov Disord 23: 2224–2231.

Senanayake N, Karalliedde L (1987). Neurotoxic effects of organophosphorus insecticides. An intermediate syndrome. N Engl J Med 316: 761–763.

Shinotoh H, Snow BJ, Chu NS et al. (1997). Presynaptic and postsynaptic striatal dopaminergic function in patients with manganese intoxication: a positron emission tomography study. Neurology 48: 1053–1056.

Sikk K, Haldre S, Aquilonius S-M et al. (2013). Manganese-induced parkinsonism in methcathinone abusers: bio-markers of exposure and follow up. Eur J Neurol 20: 915–920.

Simmons BB, Hartmann B, Deioseph D (2011). Evaluation of suspected dementia. Am Fam Physician 84: 895–902.

Sińczuk-Walczak H, Szymczak M (1997). Rhythm patterns of basic brain bioelectric activity in workers chronically exposed to carbon disulfide. Int J Occup Med Environ Health 10: 429–440.

Starks SE, Gerr K, Kamel F et al. (2012). Neurobehavioral function and organophosphate insecticide use among pesticide applicators in the Agricultural Health Study. Neurotoxicol Teratol 34: 168–176.

Taylor JR, Selhorst JB, Houff SA et al. (1978). Chlordecone intoxication in man I. Clinical observations. Neurology 28: 626–630.

Valentine WM, Graham DG, Antony DC (1993). Covalent cross-linking of erythrocyte spectrin by carbon disulfide in vivo. Toxicol Appl Pharmacol 121: 71–77.

Van Valen E, van Thriel C, Akila R et al. (2012). Chronic solvent-induced encephalopathy: European consensus of neurophysiological characteristic, assessment, and guidelines for diagnostics. Neurotoxicology 33: 710–726.

Vrca A, Karacić V, Bozicević D et al. (1996). Brainstem auditory evoked potentials in individuals exposed to long-term low concentrations of toluene. Am J Ind Med 30: 62–66.

Weuve J, Press DZ, Grodstein F et al. (2013). Cumulative exposure to lead and cognition in persons with Parkinson's disease. Mov Disord 28: 176–182.

Handbook of Clinical Neurology, Vol. 131 (3rd series)
Occupational Neurology
M. Lotti and M.L. Bleecker, Editors

Chapter 3

Neuropsychologic evaluation and exposure to neurotoxicants

ROSEMARIE M. BOWLER[1]* AND MURIEL D. LEZAK[2]

[1]*Department of Psychology, San Francisco State University, San Francisco, CA, USA*

[2]*Department of Neurology, Oregon Health and Sciences University, Portland, OR, USA*

BRIEF HISTORY OF NEUROPSYCHOLOGY

Neuropsychology is the study of brain–behavior relationships. Its history may be traced to an assertion by Hippocrates (460–370 BC) relating brain and mind and then by Galen (AD 130–200), who first described brain anatomy. The practice of neuropsychology, as we know it today, owes much to the work of Alexander Luria (1902–1977), who, in conjunction with his mentor, Lev Vygotsky (1896–1934), postulated that each area in the central nervous system served one of three basic functions: (1) the brainstem was involved in regulating arousal; (2) the posterior areas of the cortex were involved in reception, integration, and sensory analysis; and (3) the frontal and prefrontal lobes were involved in planning, executing, and carrying out the behavior (Luria, 1969, 1973). Luria conceptualized brain–behavior relationships in terms of functional systems representing interactive patterns of brain operations between defined areas of the brain.

Mental ability and sensory/perceptual tests were used by Kurt Goldstein (1939) in World War I and then by Hans-Lukas Teuber (1969) and Freda Newcombe (1969), among others, in World War II to evaluate brain-injured soldiers. The findings from these examinations spurred brain research as wounds due to penetrating missile shrapnel allowed examiners to relate small areas of brain damage to specific behavioral and cognitive deficits.

In the USA, the neurosurgeon Ward Halstead (1947) put together a battery of tests developed by T. G. Thurstone, C. E. Seashore, and others that he thought would identify patients with frontal-lobe dysfunction. Ralph Reitan, then a psychologist working with Halstead, added to Halstead's battery a test of communication abilities developed by J. Wepman and one (Trail Making) taken from the US Army battery of tests. Relying on Halstead's sample of 29 cases (actually, 28, as one person took the tests twice), Reitan advanced the notion that proper analysis of this battery's set of cutting scores alone could answer most questions concerning patients' abnormal behavior when brain dysfunction was suspected (Halstead, 1947; Reitan and Wolfson, 1985). As clinical psychology in the USA was changing from the "intuitive" approach of psychoanalytic theoreticians to more rigorous actuarial (statistical probability) techniques (Meehl, 1954), such a battery was readily accepted by many psychologists as a quick and easy answer to the thorny problems presented by psychiatrically and neurologically impaired patients.

From the early 1960s on, other psychologists were developing tests important for neuropsychology in the USA and elsewhere. Arthur Benton, at the University of Iowa, devised many relatively brief tests of specific cognitive functions, as did Elizabeth Warrington and her colleagues (McCarthy and Warrington, 1990) in England and Brenda Milner (1968, 1972) in Canada. In Australia, Kevin Walsh (1985) wrote about the necessity of going beyond test scores to include the qualitative aspects of patients' examination responses if a patient's condition is to be truly understood, an examination approach evolved independently by Edith Kaplan and her co-workers in Boston (Kaplan, 1988).

Before the days of sophisticated neuroimaging, neuropsychology was mostly used for the identification and localization of a cerebral lesion. The first systematic applications of neuropsychologic assessment thus dealt with diagnosis. Over time, neuropsychologists not only

*Correspondence to: Rosemarie M. Bowler, Ph.D., M.P.H., Licensed Psychologist and Qualified Medical Evaluator, 8371 Kent Drive, El Cerrito CA 94530, USA. Tel: +1-510-236-5599, Fax: +1-510-236-3370, E-mail: rbowl@sfsu.edu

became more skilled at localizing brain lesions on the basis of neuropsychologic test data, but developed specific criteria for predicting the nature of behavioral changes and functional loss.

In their efforts to separate psychiatric patients from those who were neurologically impaired, some clinicians in the 1940s and 1950s created single tests or test scoring systems purporting to achieve this goal, i.e., identify "organicity." Since the 1970s and 1980s, the clinical assessment of brain–behavior relationships has advanced from the use of such single tests and scoring systems to a complex, multifaceted process of assessment.

Neuropsychologic assessment procedures have developed rapidly during the last three decades, reflecting the growing appreciation of its value for neurodiagnostic questions and for the care and treatment of neurologically impaired patients, and for patients needing cognitive and behavioral rehabilitation (Lezak et al., 2012). Moreover, neuropsychologic assessment has made significant contributions to basic science research and to clinical knowledge in the neurologic sciences, psychiatry, and in clinical and cognitive psychology (Frackowiak et al., 1997; Gazzaniga, 2000).

The current integrated approach to neuropsychologic assessment is a complex undertaking requiring basic clinical psychology knowledge and training, knowledge of brain function and integration and how it relates to common and critical patterns of behavior, and an understanding of the tests and assessment techniques used to assess brain dysfunction. As noted by Vanderploeg (2000),

> *a competent practitioner must also have interviewing and counseling skills, the appreciation of social and cultural practices, and the psychodiagnostic knowledge of a clinical psychologist, but also possess the statistical knowledge and test familiarity of a psychometrician; in addition to a fairly comprehensive understanding of the human nervous system and its pathologies . . . Regardless of the structure of the evaluation, the process neither begins nor ends with giving tests alone (Vanderploeg, 2000; see Milberg et al., 1996; Lezak et al., 2012).*

CLINICAL ASSESSMENT OF NEUROTOXICANT-AFFECTED BRAIN–BEHAVIOR RELATIONSHIPS

A neurotoxicant is a nervous system poison (White et al., 1995; Bowler et al., 1998b; Bowler, 1999a). Most neurotoxicants found in occupational settings are solvents, heavy metals, carbon monoxide, and other toxic gases, or pesticides.

Most recent scientific studies on neurotoxicant-exposed workers have included neuropsychologic assessment methods. Consequently, in the last few decades, the literature on adverse occupational neurobehavioral effects has increased exponentially (Meyer-Baron et al., 2013). Early occupational health studies mainly investigated the neurotoxic effects of solvents and heavy metals. More recent work in this field has also dealt with neuropsychologic impairments associated with carbon monoxide and pesticides.

Helena Hänninen, at the Institute of Occupational Health in Helsinki, Finland, was one of the pioneers in recognizing and describing the clinical effects of carbon disulfide solvent exposure (Hänninen, 1971, 1974), exposure to mixtures of organic solvents (Hänninen et al., 1976), and exposure to low levels of lead (Hänninen et al., 1978). These early investigations of Scandinavian workers were often disregarded; their findings were called "the Scandinavian disease." This reflected a controversy between industry, labor, and health scientists about accepting deficits from overexposure to solvents.

The advent of reliable and comprehensive neuropsychologic evaluations led to a greater acceptance of neuropsychologic data and the conclusions drawn from them. Well-designed clinical neuropsychologic studies and evaluations have established usefulness in identifying workers' cognitive deficits and emotional disturbances that might be attributable to neurotoxicant exposure.

The incidence – or at least the awareness – of neurotoxic exposures from chemical substances has also increased in recent decades. It is now known that thousands of workers in various occupations have experienced high exposures to the categories described in Table 3.1 (Arlien-Søborg, 1991; Anger et al., 2000; Alessio et al., 2007). When workers were exposed without adequate personal protective equipment (PPE) and with poor, insufficient, or no ventilation, many developed cognitive, sensorimotor, and/or emotional dysfunction.

Occupational exposures, which were considerably higher in the past, have decreased in many industries. In the USA this has been due to the Occupational Safety and Health Administration (OSHA), which requires workers to have PPE provided by their employer and ventilated workspaces as needed (OSHA, 2012). Protective procedures were not always conducted in the past and may not always be present in small facilities. Environmental neurotoxic exposures in the vicinity of the polluting industrial sources are also of concern (Mergler et al., 1999; Rodriguez-Agudelo et al., 2006; Lucchini et al., 2009; Roels et al., 2012).

Patterns of data from neuropsychologic assessments can describe neurobehavioral consequences of specific neurotoxicant exposures (see Toxicological Profiles,

Table 3.1

Neurotoxicant type and occupational exposure examples

Neurotoxicant type	Neurotoxicant	Occupational exposure examples	References
Carbon monoxide and other gases	Carbon monoxide Carbon dioxide Carbon disulfide Hydrogen cyanide Hydrogen sulfide Nitrous oxide	Metal industry (welding, melting), breweries, mining, drivers, mechanics, firefighters, agricultural industry, sewage workers, chemical industry, power stations, anesthesia, rayon facilities	ATSDR (2012a) ATSDR (1996) ATSDR (2015) Bowler et al. (2003) Feldman et al. (1980) Arlien-Søborg and Simonsen (2011)
Metals	Aluminum Arsenic Copper Iron Lead Manganese Mercury Thallium Zinc	Metal work (melting, soldering, grinding, welding, etc.), mining, glazing, paper and pulp industry, steel and aluminum production, battery production, chemical industry, lead-based paint, electric equipment	ATSDR (2007) ATSDR (2012b) Bowler et al. (2006a) Bowler et al. (2007b) Feldman et al. (1980) Roels et al. (1999) Stewart and Schwartz (2007) Arlien-Søborg and Simonsen (2011)
Pesticides	Carbamates Methyl bromide Organochlorine Organophosphorous compounds Pyrethroids Triethyltin hydroxide	Farming, pesticide manufacturing, crop dusting, textile industry, agriculture and forestry industries, paper and pulp factories, warehouses, greenhouses, handling treated wood, fumigating, refrigerator manufacturing, gardening	Beseler et al. (2008) Mackenzie Ross et al. (2010) McCauley et al. (2006) Reidy et al. (1992) Roldán-Tapia et al. (2006) Arlien-Søborg and Simonsen (2011)
Solvents	1-Bromopropane Fluorinated hydrocarbons Isopropanol Methyl butyl ketone (MBK) Methyl ethyl ketone (MEK) Methylene chloride Mineral spirits *n*-Hexane Stoddard solvent Styrene Toluene Trichloroethylene Xylene Solvent mixtures	Dry cleaning, printing, painting, tile fixing, mechanic work, fiberglass production, degreasing, metal industry, refrigerator production, rubber and plastic industry, gluing, chemical industry, salon workers	Akila et al. (2006) Bowler et al. (1991) Bowler et al. (2001) Dick (2006) Feldman et al. (1980) Gregersen et al. (1984) Meyer-Baron et al. (2008) Tröster and Ruff (1990) van Hout et al. (2006) van Valen et al. (2012) White et al. (1995) Anonymous (1999) Arlien-Søborg and Simonsen (2011)

which are available on the Agency for Toxic Substances and Disease Registry (ATSDR) website: http://www.atsdr.cdc.gov/toxprofiles/index.asp). The predominant categories of neurotoxicants in occupational exposures are summarized in Table 3.1.

The American Conference of Governmental Industrial Hygienists (ACGIH) provides an annual review reporting what exposure levels are considered safe for humans (ACGIH, 2014). Additionally, the OSHA can set and update permissible exposure limits (PELs) for specific neurotoxicants, such as lead, solvents, aromatic amines (e.g., for *n*-hexane, 500 ppm or 1800 $\mu g/m^3$ over 8 hours/5 days per week). Most such PELs have not been based on quantitative risk assessments and many were adopted from preexisting industry consensus standards. Regulatory and recommended occupational exposure limits are available at https://www.osha.gov/dsg/annotated-pels/tablez-1.html.

Quantitative risk assessment

Few occupational exposure limits have been based on a quantitative risk assessment. Most were proposed following informal reviews of generally sparse information, usually from toxicologic or epidemiologic studies (Castleman and Ziem, 1988; Ziem and Castleman, 1989). Ultimately, estimates of exposure–response relationships are required for deriving scientifically sound health risk-based occupational exposure limits based on analyses of worker populations with adequate retrospective exposure assessments and measured outcomes. Typically a preferred statistical model is chosen to drive the final estimate. Understanding the modeling process and the uncertainties introduced is essential when developing an occupational exposure limit.

The benchmark dose procedure is one method for deriving an estimate of exposure corresponding to a specified level of excess prevalence for an adverse outcome, which is measured either as a discrete (Crump, 1984) or continuous variable (Crump, 1995; Bailer et al., 1997; Budtz-Jorgensen et al., 2001). It requires a specified exposure response function as well as the maximum acceptable excess risk and, in the case of a continuous outcome, a definition of what response level constitutes impairment. For neurobehavioral or neuropsychologic outcomes, clinical norms and definitions can be used in benchmark dose procedures (Park et al., 2006).

Not many quantitative risk assessments for neurologic-related impairments have been performed for occupational exposures. There have been several related to manganese (Mn)-associated parkinsonism (McClure and Odin, 1998; Clewell et al., 2003; Park et al., 2006, 2009).

NEUROTOXIC EXPOSURES IN OCCUPATIONAL SETTINGS

In the occupational settings listed in Table 3.1, varying levels of protection are offered to workers in the form of engineering controls; historically few forms of PPE were available. In the past, the ventilation was often insufficient in areas where workers performed tasks involving high air concentrations of neurotoxicants. Over the years, better enforcement of workplace regulations has lessened the impact of such overexposures. Earlier studies reported a number of common neurotoxic symptoms (Table 3.2).

Sequelae of exposure to neurotoxicants may vary depending on the source and length of exposure and if the exposure is acute or chronic. Some common neuropsychologic findings include dysfunction in attention/concentration, verbal fluency, visuospatial function, working memory, immediate and recent memory, learning, executive functions, motor abilities, tremor, color vision, olfaction, and mood (Feldman et al., 1980; Gregersen et al., 1984; Tröster and Ruff, 1990; Mergler et al., 1999; Bowler et al., 2001, 2007a; Roldán-Tapia et al., 2005; Dick, 2006; Bouchard et al., 2007; Mackenzie Ross et al., 2010).

Neuroimaging studies of some neurotoxicants have been useful; they can be used as a "biomarker" of exposure when behavioral abnormalities are found. The most widely used imaging techniques in neurotoxicant evaluations are magnetic resonance imaging (MRI) and spectroscopy (MRS) to identify, for example, accumulation of Mn or lead in specific brain regions. For Mn, increased T_1 intensity, indicated by shorter T_1 relaxation time, is used as an indicator of Mn accumulation in specified brain regions (Long et al., 2014). Long and co-workers reported that the T_1-weighted images are associated with gliosis and neurodegeneration. Increased signal intensity with Mn exposure was seen in the globus pallidus (Nelson et al., 1993; Kim et al., 1999; Josephs et al., 2005), caudate, putamen (Criswell et al., 2011), and the olfactory bulb (Dorman et al., 2006). With Mn, it is also important to measure the iron transporter –ferritin (Fe) concentrations (Kim et al., 2015) in blood as iron competes with Mn in crossing the blood–brain barrier.

The majority of neurotoxicants have a very short half-life and therefore cannot be imaged after traces of the chemical substance are no longer present in the brain, even though damage may have resulted. Of the classes of neurotoxicants, solvents have a particularly short half-life (0.5–2.7 days), while pesticides (i.e., organophosphates or carbamates) displace oxygen entering the brain and have a very long half-life of up to 120 days. Levels of organophosphates in the body are better assessed by measuring the cholinesterase activity and levels, but these data are generally no longer available by the time a study is undertaken. On the other hand, levels of some metals, such as lead or mercury, may remain elevated for years.

The lack of "hard data," such as seeing pathophysiologic changes in the brain with MRI or MRS, complicates legal challenges of impaired workers who had high exposures but who were not imaged immediately and for whom biomarkers in blood or urine were not collected soon enough. Heavy metals, however, with a somewhat longer half-life (e.g., the half-life of Mn in blood is over 30 days) afford an increased possibility for collecting biomarkers to confirm the exposure.

REFERRAL QUESTIONS TO THE NEUROPSYCHOLOGIST

Occupational medicine physicians or specialists in neurology may be familiar with lead poisoning and refer

Table 3.2

Symptom domains of neurotoxic exposures to solvents, carbon monoxide, heavy metals, and pesticides

	Symptom domain (examples)	Solvents	Carbon monoxide and other gases	Heavy metals	Pesticides
Physical and sensory	Dermatologic (skin rashes, cracking skin, etc.)	X			X
	Gastrointestinal (nausea, stomach cramps, diarrhea, etc.)	X	X		X
	Headaches/chemical sensitivity (headaches 2×/week or more, migraines, etc.)	X	X		X
	Olfactory (loss of sense of smell, etc.)	X		X	
	Respiratory/cardiac (shortness of breath, heart palpitations, etc.)	X	X	X	X
	Sensory/lower alcohol tolerance (metallic taste in mouth, etc.)	X		X	
	Sexual dysfunction (difficulty maintaining an erection, incontinence, loss of libido, etc.)	X	X	X	
	Sleep (sleeping more, difficulty falling asleep, etc.)	X	X	X	
	Visual (dark vision, blurred vision, etc.)	X	X		X
Neurologic and neuropsychologic	Cognitive disorder (attention/concentration, memory, etc.)	X	X	X	X
	Executive function (abstract reasoning, planning, etc.)	X	X	X	X
	Mood (irritability, depression, anxiety, etc.)	X	X	X	X
	Neurologic (muscle weakness, dizziness, etc.)	X	X	X	X
	Neuromotor/movement (loss of balance, slowness of movement, slow initiation of movement, postural sway, etc.)	X	X	X	X
	Tremor (shakiness, changes in handwriting, dropping items, difficulties in dressing, etc.)		X	X	

X = Symptoms are commonly reported in the various domains of function.

patients with complaints of mood or memory problems and confusion for neuropsychologic assessment. These same practitioners may be less knowledgeable about neuropsychologic complaints associated with other neurotoxicants and thus less likely to refer workers who complain of general malaise for a neuropsychologic evaluation. More commonly, general practitioners typically do not take an adequate work history or probe for specific exposure scenarios. When patients come with vague symptom complaints, many physicians will first search for psychologic causes such as life changes, stress, and distressing events and disruptions. Such symptoms and especially mood disturbances are typically attributed to family problems or social stresses. Neurologists who are more informed about the neurologic signs and symptoms of specific neurotoxicants will likely appreciate the importance of referring these patients to a neuropsychologist.

In considering referral for a clinical neuropsychologic evaluation when a work-related neurotoxic condition is suspected, the following questions should be explored:

1. Does the worker have neuropsychologic impairments that could be attributable to a workplace neurotoxicant exposure or a flare-up of a preexisting condition?

2. Do the worker's exposure conditions and time course support a work-related causation of the neuropsychologic impairments? (For example, was a new process or material introduced prior to the development of symptoms? Are there related symptoms of high exposure, such as eye, nose, and/or throat irritation, respiratory problems? Do symptoms ameliorate during vacation or other absences?)
3. Is the pattern of deficits consistent with the reported effects of neurotoxicant exposures that were plausibly present in the workplace?
4. Can other exposures for the dysfunction be definitively ruled out (e.g., hobbies requiring glue, grease, or pesticides at home, etc.)?
5. Is the worker aware of co-workers who have reported similar effects?

THE NEUROPSYCHOLOGIC ASSESSMENT PROCEDURE

Applications of neuropsychologic assessment

Although imaging can in some cases substantiate suspected brain neurotoxicity, it cannot document the psychologic expression of a worker's brain that has been overly exposed to toxins. How well or how poorly a worker functions cognitively, practically, and emotionally can be evaluated by neuropsychologic assessment (Bowler, 1999b). Occasionally the observable neuropsychologic findings of long-term solvent exposure may resemble early-onset dementia. The neuropsychologic examination can provide information for the differential diagnosis of early-onset dementias from exposure symptoms, and in differentiating depression from dementia in workers who may have had long-term neurotoxic exposure. Moreover, after an exposure to neurotoxic substances, neuropsychologic deficits may be the only residual evidence indicating the presence of cerebral damage (Lezak et al., 2012).

Today, neuropsychologic assessment is most often called upon for detailed behavioral descriptions necessary for the care of exposed workers, including effective treatment, appropriate rehabilitation, and vocational/educational placement. Neuropsychologic evaluations, when planned to be relevant to the patient's complaints, history, and situation, can guide medical and rehabilitative treatment. Repeated neuropsychologic examinations are necessary for predicting the degree and progression and/or remission of acute toxic encephalopathy or for documenting neurobehavioral decline requiring differentiation from other deteriorating conditions, e.g., dementias, mood or personality disorders.

Descriptive neuropsychologic data can also provide information needed for a worker's family members to make practical decisions in response to a disabled worker's neurologic, cognitive, or mood disorder. These decisions may concern the worker's legal status or financial options, whether the worker returns to his or her job, when a return to work is feasible, or domestic issues such as whether the worker needs help with household chores if living alone. Often without more data about their patients' cognitive and emotional status, physicians are at a loss for how to most effectively treat the disabled worker.

Neuropsychologic evaluations have proven invaluable in workers' compensation and forensic settings. Lawyers, judges, and juries have learned how cerebral dysfunction due to acute or chronic neurotoxic exposures may disable workers, compromising their ability to return to employment or to resume a normal home life.

Neuropsychologic assessment assists in formulating psychiatric or neurologic diagnoses. However valuable are the contributions of neuropsychologic assessment to an individual case, it should be kept in mind that testing procedures alone are neither sufficient nor the only part of an assessment protocol. Clinical interviews, clinical observation, and access to third-party information are also important in determining an examinee's neuropsychologic status.

In the USA, many standardized neuropsychologic tests have been statistically normed using only white populations of European origin. This significantly limits their usefulness in other cultural contexts. Of note, however, is that tests vary in their cultural sensitivity (i.e., some are more, some are less influenced by the cultural background and educational level of a geographic/ethnic area) (Mitrushina et al., 2005; Lezak et al., 2012). The wise examiner will take the worker's sociocultural background and language competencies into account when planning what tests to use. For some workers, translators may be required, but this produces validity and reliability considerations that only the examiner can resolve for each individual case (Artiola i Fortuny et al., 1998; Ardila et al., 2002).

Record review

A prior review (as available) of the worker's educational, occupational, psychosocial, and medical histories helps guide the neuropsychologist when planning the examination. Ideally, this will include a detailed work description and history of symptom onset in relation to neurotoxicant exposure. As in a neurologic examination, a history of alcohol, tobacco, and drug use will be taken, as well as a review of current medications.

Adverse reactions to long-term, low-level chronic neurotoxic exposures may have a slow onset and be hardly noticeable at first. Cumulatively, symptom evolution in response to long-term exposures will appear as increasingly impaired emotional, cognitive, and practical behavior with unfortunate repercussions in the worker's home life. Some clinicians generate a timeline of injuries, listing the worker's detailed prior work, medical and mental health history by date of onset.

Clinical interview

An interview typically precedes neuropsychologic testing. This enables the examiner to establish rapport with the worker to ensure cooperation and trust. A good working relationship between examiner and examinee is necessary for obtaining the worker's best performance.

The interview can provide descriptive information of the worker's physical, emotional, and cognitive changes since the neurotoxic exposure. Behavioral observations, typically considered in the course of a mental status examination, round out interpretation of the test and questionnaire data.

Clinical observations include both direct and indirect observations: observations of the worker by the examiner and those reported to the examiner by other observers, such as family members and workplace colleagues or co-workers. Reports on the worker's home life by a knowledgeable informant are most helpful. Questionnaires for significant others or close observers of the patient are possibly less biased than informant's verbal reports but limit the kind of information available to the examiner.

For example, Prigatano et al. (1986) developed a 30-item Patient Competency Rating Scale (PCRS) with a five-point Likert response scale (in the public domain), to assess patients' abilities to perform everyday tasks, as well as patients' self-awareness of their abilities. Workers' responses to the PCRS can be compared to the responses of a close observer/relative who completes a PCRS with identical items. The competency domains assessed in this rating scale are personal care, independent survival skills, social functioning, memory, and emotional control. The responses give useful practical information on the worker's competencies and also highlight any differences between the observer/relative or friend's ratings and the worker's self-rating.

Another useful questionnaire, also in the public domain, is the Behavioral Risk Factor Surveillance System (BRFSS), with questions on general health and quality of life, and a response to how many bad physical and mental health days a person has out of 30 and for how many days poor physical or mental health kept the worker from doing his or her usual activities. It has a very large state-by-state database for each US state and is available on the Centers for Disease Control and Prevention (CDC) website: http://www.cdc.gov/brfss/questionnaires/pdf-ques/2014_BRFSS.pdf (Centers for Disease Control and Prevention, 2014).

Use of questionnaires and inventories

Several different types of paper-and-pencil questionnaires may be included in a neuropsychologic examination. Questionnaires are more likely to be used in epidemiologic studies in which time for close questioning and observation of the worker is limited. In clinical practice, it is less likely that the examiner will look to questionnaires for information that can be readily obtained in the interview.

Sociodemographic information and lifestyle information

General information questionnaires are used to collect relevant sociodemographic information and lifestyle information. Topics typically covered in a general information questionnaire include:

- sociodemographics (age, sex, education, income, etc.)
- lifestyle questions and health habits (smoking, drinking, recreational drug use, etc.)
- major life events in the past 12 months and whether they affected physical and mental health.

Work/exposure history questionnaire

Work and exposure history questionnaires are important for understanding the exposure characteristics in the worker's occupational setting. In occupational exposures, detailed work and protective equipment use histories should be included. Thus, topics in these questionnaires deal with the following workplace specifics:

- current job title and year when employed, typical duties, percent of time spent doing each duty
- can the worker provide material safety data sheets from the workplace (required to be available under OSHA regulations). Has the employer provided confidential proprietary-ingredient information that is not disclosed on the material safety data sheets but is required upon request from a medical provider?
- does the employer have air-sampling data, in which case it must be provided upon request?

- use of PPE – type of PPE, frequency and duration of use
- symptoms experienced prior to work shift, during work shift, and after work shift
- previous jobs and years when employed – typical duties, hours spent doing each duty, if relevant
- symptoms developing during previous employment.

Physical/medical health questionnaire

This section of the questionnaire inquires about past and present medical history and dates of onset. It typically asks about:

- physician-diagnosed illnesses and date of diagnosis
- medications, both prescribed and over-the-counter, frequency of use, and dosage
- family history of major illnesses (parents, siblings).

Symptom questionnaire

Symptom questionnaires are useful tools in the assessment of neurotoxicity in both clinical and research evaluations. Symptom severity and date of onset are also important. Symptom questionnaires can be customized to correspond to the symptom profiles associated with a particular neurotoxicant. See Table 3.2, which shows symptom domains found to be affected by exposures to solvents, carbon monoxide and other gases, heavy metals, or pesticides.

It is good practice to evaluate the same patient over time. The symptom questionnaire, coupled with a follow-up interview, permits an update of the patient's neurobehavioral status and, as relevant, an assessment of the efficacy of the treatment/rehabilitation programs. Follow-up neuropsychologic testing will be useful if 1 year or more has elapsed after the initial test administration in order to avoid confounding the effects of test familiarity as well as giving time for changes to manifest clearly.

Follow-up questionnaires

Brief follow-up questionnaires can be used for repeated examinations to obtain updated exposure, symptoms, and illness history. They should include the following:

- update on lifestyle factors (smoking, etc.)
- update on major life events in the past 12 months
- update on occupational history (job change, reason for disability if disabled, etc.)
- changes in use of PPE
- update on physician-diagnosed illnesses and date of diagnosis
- update on medications – prescribed and over-the-counter, frequency, and dosage
- update on symptoms – current severity and date of onset.

Test selection and administration

Tests used in neuropsychologic assessment provide information on cognitive and motor abilities, mood, and personality factors, as well as effort. How the worker responds to the examination, and in particular, responds when urged or when a task is boring or difficult, is important information that is integral to the examination. Most tests generate scores and many non-standardized or not-normed assessment techniques elicit responses that can be coded with a three-point scale in which satisfactory performances are considered to be within normal limits, with the less than satisfactory responses considered either borderline or impaired.

In the USA, where neuropsychology is well established, neuropsychologic examiners can select from a vast array of standardized tests and examination techniques. To choose those tests that seem most appropriate for each individual worker, the neuropsychologist takes into account the worker's age and educational level. Additionally, the background and presenting problems, cultural relevance of the tests and the referral questions, which resulted in a request for a neuropsychologic examination, must be considered. Careful matching for age, gender, ethnicity, social, cultural, and educational background is necessary for group comparisons.

Neuropsychologists are required to keep informed of new test developments so changes in the culture (e.g., internet use) can be accounted for and used when evaluating the worker's knowledge level and intellectual capacity. Unawareness of current trends (e.g., news and sports events) can be evidence of cognitive impairment associated with the neurotoxic exposure. Ideally, the examination and tests selected for a worker will result in a specially tailored battery of tests, which may differ in some respects from that given to other workers, as the needs, competencies, and circumstances of each worker are unique. However, many standardized tests are administered to most patients because of their almost universal applicability and relevance (e.g., digit span, word naming, or tests of verbal learning, tests of memory, techniques for examining psychomotor efficiency, and mood questionnaires).

The interview and testing procedures typically produce two types of information: (1) substantive information – historical data or descriptions of current mood from the interview; the content of test item responses; and test scores from the tests and assessment techniques that were administered; and (2) qualitative information about how the worker responds – carefully, hesitantly, sluggishly, distractedly, impulsively; whether the worker observes errors in either interview or test responses and, if not, what the worker's response is to this failure; whether the worker cooperates intelligently, in a docile unquestioning manner, or only erratically. These qualitative observations may be as important for understanding a patient's neuropsychologic status as test scores (Lezak et al., 2012). These clinical observations and interactions with the worker are limited when using computerized tests. Visuospatial difficulties and motor slowing may confound the results of computerized tests.

For many examination purposes, the examiner identifies or makes a reasonable estimate of the patient's premorbid levels of cognitive, occupational, and social functioning. Premorbid test data of the individual patients are usually not available. Historical data, reports by patients, their family members, and other informants who knew the worker before his/her brain injury, as well as premorbid performance records (e.g., prior school and occupational achievements, occupational data, including past and present work performance ratings) provide reliable information about the patient's premorbid neuropsychologic status. The examiner also looks for abilities that are likely to be spared (i.e., remain essentially unimpaired, such as word knowledge) to estimate the level of premorbid functioning. This helps establish a level of premorbid functioning against which the worker's present neuropsychologic functioning can be compared.

Neuropsychologic assessment procedures must be performed by qualified neuropsychologists (Figs 3.1 and 3.2). Although, in certain cases, some tests may be administered by well-trained psychometricians or technicians, the integration and interpretation of all data remain in the domain of the neuropsychologist, as it requires highly specialized knowledge and experience. Testing conditions require a private, quiet, and comfortable setting where the worker and examiner have minimal distractions. Sensible adherence to the published directions of each test and attention to precise timing are required in order to have accurate test scores.

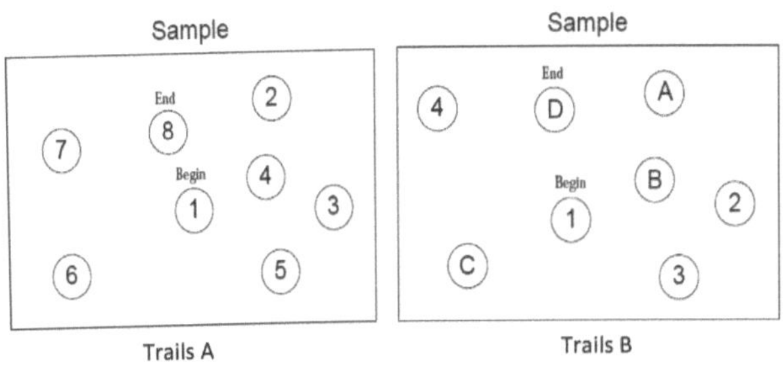

Fig. 3.1. Trail-making tests A (left) and B (right).

The lack of trained neuropsychologists in international neurotoxicological research can be a serious problem. It is strongly urged that local clinical psychologists and neuropsychologists (often few outside of major urban areas) take responsibility for administering the tests or reviewing data from psychometrician/technician administrations. Additionally, advanced-level health staff may be trained to administer some of the neuropsychologic assessment procedures and tests. Findings, however, should only be interpreted with the close collaboration of trained neuropsychologists and clinical psychologists familiar with the tests and methods described here.

TESTING THE MOST RELEVANT NEUROPSYCHOLOGIC FUNCTIONS IN NEUROTOXICOLOGY RESEARCH

Although specific neurotoxins differ in how they affect the central nervous system, there are several functional areas of cognitive deficits that are associated with many, if not most, kinds of neurotoxins. Attention, memory, learning, and mental processing speed are the commonly vulnerable cognitive functions. Some neurotoxins, typically metals, also involve motor system dysfunction. Thus, when planning an assessment battery to study neurotoxic effects, it is tests of the relevant functions that the examiner will want to include.

However, in order to document deficits, it becomes important to identify – in so far as is possible – the worker's premorbid levels of functioning. The test battery, therefore, will include not only tests of vulnerable cognitive functions but also several tests that are unlikely to be affected by neurotoxic exposure. Tests of these more sturdy functions typically involve overlearned information and skills, e.g., verbal abilities, which are examined by vocabulary or reading tests, or perceptual functions measured with simple visual recognition or organizing tasks. When selecting tests for neurotoxic research, beyond a test's relevance for the neurotoxin(s) being studied, several other issues must be considered. Not least of these is time – the test battery cannot take so long to complete that the injured worker's stamina or time costs become an issue. Test cost can

Fig. 3.2. Cancel "H" test.

become another issue, as some tests are expensive. When funding for these neuropsychologic studies is limited, costs can be minimized by using tests in the public domain or, in large-scale studies, the principal investigator (PI) may bargain with the test publisher.

Many of the tests used in neurotoxic research come from standardized test batteries. The more recent versions of the Wechsler Adult Intelligence Scales (WAIS-IV: Wechsler, 2009a) and Wechsler Memory Scales (WMS-IV: Wechsler, 2009b) include both tests sensitive to the cognitive effects of neurotoxic exposure and some that are relatively resistant to such effects. Some batteries, as well as other frequently used individual tests, were devised to examine specific functions that, over time, have acquired good normative samples (e.g., Auditory-Verbal Learning Test; Complex Figure Test; Trail Making Test: see Mitrushina et al., 2005). Readers interested in learning about specific tests and test batteries can find comprehensive information in Mitrushina et al. (2005); Strauss et al. (2006); Gurd et al. (2010), and Lezak et al. (2012).

Attention and concentration

Reduced attentional capacity is frequently the most usual indicator of most kinds of brain dysfunction. It becomes particularly apparent in response to most neurotoxic overexposures, both acute and chronic. Attentional deficits are closely tied to slowed mental processing speed.

Attentional deficits may account for many of the complaints brought by brain-damaged patients. They may wrongly believe they have impaired memory, mental dullness, confusion, and fatigability when they have problems sustaining concentration (Howieson and Lezak, 1994). Thus, the examination of attention and concentration is a central feature of every neuropsychologic examination (Lezak et al., 2012). Clarifying the nature of attentional problems involves both observations of patients' overall behavior, as well as their performance on tests assessing attentional functions.

Attention per se is measured by the amount of information recalled immediately after hearing or seeing the stimulus, which is typically a string of numbers or words, but can be a sentence or a string of pictured items. Some tasks require the exact sequence of a stimulus string for credit (e.g., digit span, sentence recall); others, which present more items than can normally be recalled, give credit for the quantity of items recalled regardless of sequence (e.g., word list or picture item). Although some clinicians may refer to this attentional function as "immediate memory," deficits in attentional capacity are closely associated with other aspects of attention and can be unrelated to memory and learning abilities.

Slowed mental processing speed may be the most common problem for neurotoxicant-exposed workers. It can be indirectly examined with timed tests in which the score depends upon response speed, such as symbol or number substitution tasks or tests requiring visual/mental tracking. Both visuographic and oral response formats are available for these tests.

Computerized reaction time tests are commonly used to assess processing speed and to help clarify the nature of associated attentional deficits (Wild et al., 2008; Bilder, 2011). Simple reaction time slowing is often present in brain-injured patients and with significant increases in slowing with greater task complexity. A number of testing programs have been developed, most for purchase.

Memory involves a complex array of systems dealing with registration, storage, retention, and retrieval of information. Deficits in any of these functions can impair a worker's ability to work and quality of life in general.

Memory

Working memory

Working memory refers to the capacity to hold information briefly in memory while performing other mental operations on the information (Mirsky et al., 1995). It is as much an attentional function as a memory ability. Deficits in working memory are so frequently associated with executive function disorders that some neuropsychologists classify working memory with the executive functions.

Tasks involving reversed operations, such as repeating a set of digits or spelling a word in reverse, examine working memory. Other working-memory tests ask the subject to give one kind of response – such as naming animals or counting backwards – while holding in memory three letters or a name.

Learning and retrieval

Research and clinical experience have distinguished verbal memory from memory for nonverbal information. Memory deficits due to neurotoxicity, which typically involves widespread brain areas, are generally unlikely to affect verbal and nonverbal information storage or retrieval differentially. Other kinds of memory have also been identified, such as autobiographic memory, procedural memory, and perceptual learning (Lezak et al., 2012). The latter two tend to be hardy functions of little interest for neurotoxicologists; autobiographic memory is routinely covered in a well-planned interview.

Memory testing requires that test sequences be spaced to allow for time between stimulus presentation (the "immediate trial," which, when an immediate response is called for, becomes an attentional test) and a "delay trial." This delayed response measures both how much information is retained (i.e., learned) and how well-retained information is retrieved. A low-level delayed response may reflect an impaired learning ability or it may be due to defective retrieval in persons whose learning ability is intact. Defective retrieval can be as handicapping as impaired learning as the behavioral outcome is the same although the underlying processes differ. Moreover, the rehabilitation potential for these two conditions varies, with brighter prospects in those who learn adequately but have difficulty pulling out information spontaneously.

It is important to distinguish these two kinds of memory impairment, which can be readily done by appropriate testing. When the amount of information spontaneously retrieved is below normal limits, a "recognition trial" is given or, in some test procedures, the patient hears or sees associated cues. Since it is much easier to recognize learned information or have recall prompted by cues than it is to produce it spontaneously, a notable difference between the quantity and accuracy of spontaneous responses and responses prompted by cues or a recognition format provides information about whether the low-level original response was due to an impaired learning ability or to impaired retrieval.

Typically, persons who learn well but have pronounced retrieval problems will recall meaningful material – such as sentences or stories – better than the unrelated items in list-learning tests. Including both kinds of learning tests in the test battery helps clarify the nature of a memory complaint; for a neurotoxic research battery, including both is essential.

Many memory tests assess only one aspect of memory. Some batteries purport to contain tests that assess all levels of the memory system. Both stand-alone tests that were developed to examine specific aspects of memory and tests from standardized batteries have been incorporated into testing programs for neurotoxicity research (Bowler et al., 1998a, 2006b; White and Proctor, 1992; White et al., 1992).

Visuoperception, visuomotor, and construction

These abilities get to the heart of many activities in workplaces where neurotoxins are used. Fortunately, they are not very vulnerable to neurotoxic exposure. A well-balanced test battery designed for neurotoxicity research will include one or two tests of visuoperception and construction. Visuoperceptual tests may ask the subject to make a spatial judgment, solve visually presented

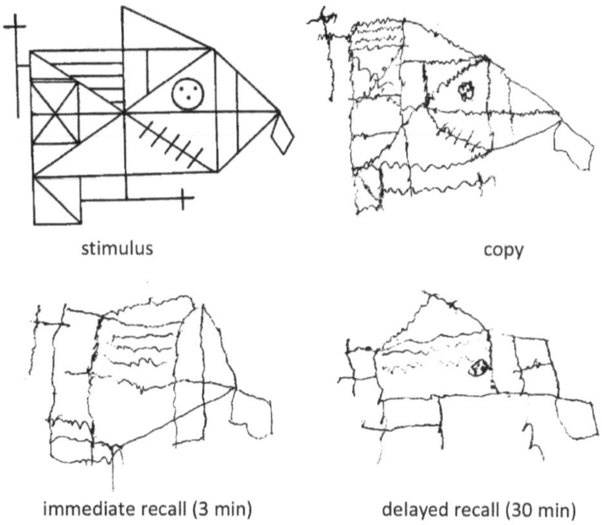

Fig. 3.3. Rey–Osterrieth complex figure stimulus and copy, immediate, and delayed-recall productions by a 53-year-old African American welder with 12 years of education and 27 years of welding experience.

puzzles (e.g., cut-up pictures) or make fine discriminations from complex visual material. Visuomotor functions are typically not examined in themselves but impairments will show up in inaccuracies in drawing, tracing, or tracking with a pencil.

When the constructional test is a drawing task requiring the subject to copy a model it can also serve as the stimulus for testing visual learning (Fig. 3.3). Standardized scoring systems with age norms are available for most of these tests; for many of them, reliable scoring requires examiner expertise. Free drawings, such as drawing a bicycle or a clock, provide rich information regarding perception, graphomotor competency, constructional ability, plus executive functioning. Scoring systems for these tasks call upon experienced examiner judgment more than for many other kinds of tests. Although numeric scores on construction tests are informative, a qualitative analysis of how different subjects approach and work through the task can tell much about how they deal with real-life problems and situations (see Chapter 11, Lezak, 2012).

Language and verbal fluency

Word knowledge, typically measured by vocabulary tests, is often quite resistant to brain insult, which makes these useful for estimating premorbid ability levels. Neurotoxicant-exposed patients' language functions are rarely impaired.

However, verbal fluency, measured as speed of producing words meeting specific requirements (e.g., words beginning with a specific letter; category – phonetic fluency; names of animal or vegetable or furniture – category fluency). Response times are frequently slowed and range of responses can become impoverished in workers who were overexposed to neurotoxicants. Naming speed can be affected by adverse mood states and anxiety. Verbal fluency test performances can give information on the subject's capacity for planning and organizing ongoing activity as well as the ability to adhere to, understand, and maintain instructions, all necessary for effective executive functioning (Lezak et al., 2012). Thus, like most tests used in neuropsychology, verbal fluency assesses a complex of behaviors, each of which can be analyzed separately.

Executive functions, reasoning, and concept formation

Executive functions and the higher-level thought processes are frequently impaired following neurotoxic exposures. The construct of executive function includes initiation and self-regulation of behaviors necessary to begin and carry out tasks purposefully and effectively. It includes the abilities to formulate reasonable goals and to anticipate and design the steps necessary to achieve that result (planfulness), to evaluate one's ongoing behavior, and correct errors as needed (Lezak et al., 2012). Executive function disorders typically do not show up as cognitive deficits per se, but affect how (reasonably, appropriately, timely, etc.) activities – cognitive and otherwise – are performed.

Workers with executive function disorders due to neurotoxicant exposure often have damage in the frontal and prefrontal cortex and its subcortical connections (Gazzaley and D'Esposito, 2007). Significant deficits in executive functioning have also been associated with subcortical lesions linked to the frontal cortex through frontal-subcortical circuits (Miller and Cummings, 2007). Damage elsewhere in the brain, to limbic and subcortical structures, can occur with anoxic events or with heavy-metal exposure to the basal ganglia or to thalamic nuclei.

Exposed workers with compromised executive functioning tend to have ineffective and disordered interactions with their environment. Executive dysfunctions can affect all aspects of a person's behavior, whether work-related, in the home, socializing with friends, or making purchases. Tests that require examinees to plan their responses, structure the responses independently, and allow for subtle errors can elicit evidence of executive dysfunction. Drawing tests, such as "draw a bicycle," or "draw a house, a tree, and a person," are examples of tests sensitive to executive disorders.

Effective planning, for example, requires aspects of executive functioning that enable a person to develop and use effective strategies for accomplishing a task; to identify, gather, and organize necessary resources; to create sequences and hierarchies to guide efforts in a logical order; and to control and regulate the response process. Maze tests have been the classic tests for planning (Porteus, 1959). Successful solutions to maze problems depend upon many of these aspects of executive functioning. Many other tests of planning are available (Luria, 1969, 1973; Strauss et al., 2006; Lezak et al., 2012).

There are no tests for evaluating the ability to carry out completion of goal-directed activities or for the important executive functions of self-monitoring, self-regulation, and self-correction. Rather, these functions are inherent in the qualitative aspects of the examination – this includes how the worker responds to the testing situation and goes about doing the tests as well as the test responses themselves.

The assessment of effort

In neuropsychologic assessment, it can be important to differentiate poor performance and low effort due to depression or disinclination to do well from true cognitive deficits. This differentiation can be most relevant when assessing attentional functions or memory and learning, as these abilities may be compromised by anxiety, mood disorders, and other emotional disturbances. Formalized guidelines are available for criteria of malingering determination (Slick et al., 1999), but close observation and a knowledgeable evaluation of test performance differences may be sufficient when evaluating questionable performances.

Unexpectedly poor performances raise the question of whether complaints are valid or the worker is withholding a better performance. Workers who have a reason – financial or personal – to feign or exaggerate impairment may do poorly, often on memory tests, especially if they are already experiencing diminished memory efficiency. These workers with genuine deficits have been known to perform below even their already reduced level in efforts – whether consciously or not – to make sure their problems become apparent to the examiner. Tests of effort and malingering may help the examiner determine whether memory or other cognitive complaints are believable. When evaluating what may seem to be a questionable performance, it is important to look for evidence of good effort on tests other than those with low scores. It is also important to remember that, for most workers, their dignity and sense of self-worth cannot be bought.

Psychomotor efficiency, tremor, and reaction time

The examination of possibly injured workers will typically include tests of simple motor speed and motor control, such as finger-tapping speed or grip strength (Figs 3.4 and 3.5). As tremor is a serious issue with some neurotoxicants, such as Mn and lead, special techniques have been devised to test for this problem (Bowler et al., 2001). Typically, only test protocols for examining workers exposed to tremor-producing neurotoxins will include tremor tests which measure performance bilaterally.

The presence and severity of motor and movement symptoms have also been assessed with the Unified Parkinson's Disease Rating Scale (UPDRS: Fig. 3.6) (Movement Disorder Society Task Force on Rating Scales for Parkinson's Disease, 2003). The UPDRS motor scale has been used in a number of neurotoxicant studies (Ballard et al., 1997; Beuter et al., 1999; Koller et al., 2004; Kim et al., 2011). When Mn or lead is the element of exposure, the results of the studies have been sensitive to qualitative ratings of impairment.

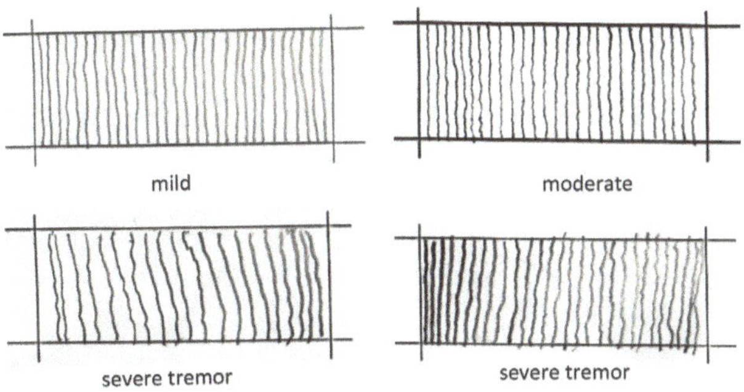

Fig. 3.4. Parallel-lines test.

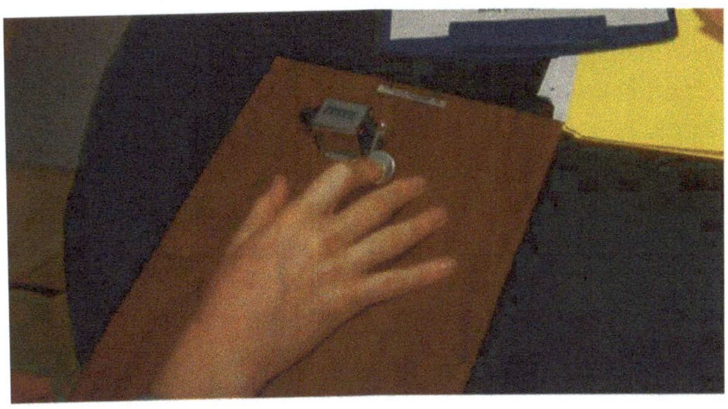

Fig. 3.5. Finger-tapping test.

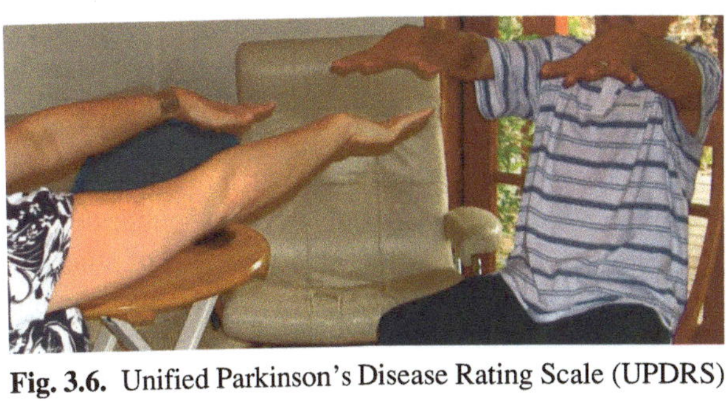

Fig. 3.6. Unified Parkinson's Disease Rating Scale (UPDRS) tremor test.

Sensory function

Sensory disorders are not uncommon with neurotoxicant exposures. Both vision and smell have been shown to be affected in neurotoxin exposures. The University of Pennsylvania Smell Identification Test (UPSIT: Doty, 1995), VISTECH Contrast Sensitivity Test (Kennedy and Dunlap, 1990), and, for color, the Lanthony D-15 (Lanthony, 1978) are common tests to assess sensory function. Testing techniques and apparatus have been developed for examining all the various aspects of sensory competency (Tate, 2010).

Emotional status

The interactive relationship between affective (i.e., emotional) processes and cognitive/executive functioning is both subtle and complex, requiring careful evaluation. Following a traumatic event or a newly acquired brain injury, a person's usual emotional status is likely to be disturbed. Because of the complexity of the interactions between the experience of impairment and the emotional response to it, it is virtually impossible to delineate the direction and strength of any causal relationships between behavioral impairment and affective states. When the event or the brain injury is fairly severe, absence of emotional disturbance suggests that the person's emotional capacity is impaired.

Tests of personality and mood are commonly employed to assess an individual's emotional condition. Such tests supplement the less formally structured face-to-face interview.

The tests specific to personality function and mood most frequently included in a neuropsychologic evaluation battery are self-report questionnaires, which are relatively brief, easy to administer and score, and more or less sensitive to the limitations of an injured patient. Some tests were designed to measure current emotional mood and mood changes; others assess more enduring personality characteristics. The assessment of mood and personality of neurotoxicant-exposed workers may include tests measuring levels of anxiety and depression, tests with questions inquiring about common psychiatric symptoms, and personality inventories designed to provide a multidimensional description of the subject's personality. Which dimensions are measured depends upon the personality theory on which the inventory was based. For traumatic events in the workplace, it may be useful to assess the development of posttraumatic stress disorder (PTSD) symptoms, either immediately after a trauma which may be perceived as life-threatening, or after a significant delay of time (Bowler et al., 2011). One of the most frequently used tests to assess PTSD is the brief stressor-specific PTSD Checklist – Civilian Version (PCL-17), a scale which covers the PTSD symptom clusters from the Diagnostic and Statistical Manual IV (Blanchard et al., 1996; American Psychiatric Association, 2000).

In addition to these paper-and-pencil quantitative measures of mood and personality, a clinical neuropsychologic evaluation includes a qualitative assessment of the worker's emotional status. Observation of the worker's appearance, manner of approach to the testing, mental status, and responsivity provides important information about his or her affective functioning. Especially useful is the clinician's appraisal of the patient's interactions with the examiner. When undergoing a psychologic examination, subjects' responses to the examiner and to this often stressful situation may tell a lot about their personality, self-regard, and attitudes toward perceived authority: e.g., are the examinee's responses to the examiner friendly, suspicious, hostile, cooperative, passive, challenging, anxious?

Workers exposed to neurotoxicants often are angry once they realize that their exposure and its consequent problems could have been prevented. Some may internalize these feelings and become depressed or externalize their anger as seemingly random hostility. Moreover, some neurotoxicants can exacerbate anger and psychotic reactions in affected workers (e.g., with Mn exposure, anger, depression, and anxiety scales on mood tests are frequently elevated).

Brief neuropsychologic screening test battery suited for repeated administrations

In case of a neurotoxic spill or other release in the workplace, a brief screening test battery, which is safe, acceptable, easy to perform, affordable, precise, and reliable, could be quickly administered to a large group of workers who might have had an acute toxic exposure(s). One outstanding example of an instrument that can be used for any neurotoxicant was developed in Quebec, Canada, following identification of a cluster of Mn-overexposed workers (Commission de la Santé et de la Securité du Travail (CSST), Workers Compensation Board in Quebec, Canada). In 2005, the Institut de Recherche Robert-Sauvé en Santé et en Sécurité du Travail (IRSST), an arm of the CSST, formed an expert committee and created a multidisciplinary and multicenter prevention network to deal with workplace problems of Mn-exposed workers (Ostiguy et al., 2006). This committee consisted of industrial hygiene, occupational, neurologic, and neuropsychologic specialists who recommended a medical review of the exposed workers, in addition to the administration of a battery of neuropsychologic screening tests that could be readily administered under the supervision of a psychologist or neuropsychologist (Ostiguy et al., 2005). Although this battery is a response to continuous-exposure effects and recommended for the annual screening of workers, the test battery can also be used for workers with acute high exposures to neurotoxicants.

Possible sources of overexposure in workers, which may result in adverse health effects, include malfunctioning ventilation systems, defective process design, accidental spills, releases, and/or inadequate PPE. After a workplace has undergone industrial hygiene improvements, readministration of the same or an alternate neuropsychologic test battery can provide some assurance of the efficacy of the workplace exposure reduction. Both industrial hygiene measurements in the workplace and a comparison of each worker's prior and follow-up test results, when analyzed as a group, can help assure the company and workers that the adverse health risks of workers are diminished and possibly low enough to no longer be of concern. In the absence of baseline measurements, in usual circumstances, a cross-sectional analysis that can be compared to population norms may be helpful.

The initial test battery constructed for the IRSST (Ostiguy et al., 2005) was designed and recommended as a rapid problem-solving response to limit potentially critical health effects in exposed workers. Time considerations in using individual neuropsychologic screening tests will always be a major concern. Test selection will depend upon knowledge of what neuropsychologic domains are likely to be affected by the neurotoxicant under consideration. A literature review can help the PI to identify which tests in the relevant domain(s) have shown sensitivity and specificity with respect to the neurotoxicant to be studied.

Qualified technicians are essential for effective data gathering; appropriate training and supervision can best assure the validity of the data. A brief mental status questionnaire, designed for use by the technician to complete after the testing, can aid in identifying behavioral problems or limitations that might compromise individual test results. Additionally, double scoring and double data entry are desirable, with the PI resolving any conflicting scores.

Cost considerations are practical limitations. Fortunately, there are many excellent tests in the public domain that have good track records for these kinds of studies.

The battery of tests listed in Table 3.3 will require initial training in test administration, scoring, and interpretation with consultation by a psychologist or neuropsychologist who is knowledgeable of the tests.

Table 3.3

Sample brief screening test battery for large numbers of workers

Domains	Tests	Measures	Admin. time
Attention, concentration, working memory	Reliable Digit Span Test [P] forward and backward	Attention, concentration, and auditory working memory	3 minutes
Visuospatial	Trail Making Test A [P]	Scanning, tracking, speed	3 minutes
	Trail Making Test B [P]	Including symbol set shifting	4 minutes
Verbal fluency	Category Naming [P] (animals, fruits, transportation)	Verbal fluency, word generation	2 minutes
Cognitive flexibility	Stroop Color Word: Word, Color, Color-Word	Cognitive flexibility, set shifting, reading speed, color naming	4 minutes
Psychomotor efficiency	Finger Tapping Test (FTT) [P]	Index finger motor speed Bilateral – three trials	4 minutes
	Grooved Pegboard Test [P]	Finger and hand fine dexterity	3 minutes
	Hand Dynamometer Test [P]	Bilateral grip strength	2 minutes
Effort	Rey 15 Item Test	Symptom validity test	2 minutes
Affect and mood (optional)	Brief Symptom Inventory (BSI)*	Symptoms, depression, anxiety, psychoticism	8 minutes
	Behavioral Risk Factors (BRFSS)*	Physical and mental health	2 minutes
	PTSD Checklist (PCL)	Posttraumatic stress symptoms	5 minutes
		Total examiner time (optional tests)	27 minutes (15 minutes)

Note. [P] indicates a public domain test; scoring sheets are not copyrighted.
PTSD, posttraumatic stress disorder.
*Self-rating mood in the BSI and the BRFFS of bad days/month of physical and mental health does not require examiner test administration time.

Sample brief screening test battery to be used with large groups of workers

The neuropsychologic domains and tests recommended for the brief screening of workers are shown above with time of administration given in Table 3.3.

TYPES OF TEST SCORES, RELIABILITY, VALIDITY, AND CUT-OFF SCORES

Most of the tests used for neuropsychologic assessments come with normative data derived from healthy population samples. In so far as possible, the demographic characteristics of the normative sample should be similar to the demographics of the population from which the worker is drawn. When normative data for a similar group are not available, an unexposed "healthy" control group, matching the demographic characteristics of the exposed sample, can provide the needed comparison data.

Standard scores

Raw test scores obtained from the assessment are generally converted to standard scores, scores statistically converted into one scale with identical units. Standard scores may, in turn, be converted into age-adjusted "scaled scores," e.g., WAIS-IV (Wechsler, 2009a); WMS-IV (Wechsler, 2009b). Many other tests and batteries use age-graded standard scores (White and Stern, 2001), or some other score-naming system, such as IQ score. All standard scores can be converted into a percentile score (e.g., 5th percentile, 84th percentile; numbers more familiar to the general public). Percentile scores may be used to identify impairment levels: scores below the 5th percentile indicate severe impairment; persons scoring between the 6th and 16th percentiles are moderately impaired; scores in the 17th to 24th percentile indicate mild impairment; scores at the 25th percentile and above are within normal limits and better (Fig. 3.7).

Whether in a standard score format or in percentiles, these transformed scores can then be compared with each other. This will result in a profile of scores. A healthy person's profile is likely to be fairly flat, without big dips, but may have some elevations in areas of specialized interests, whereas persons with brain damage are more likely to produce score profiles with big dip patterns that reflect the damaged area(s). The examiner looks for statistically significant differences between scores, and, particularly, sets of scores on tests of related functions, to identify the presence of brain dysfunction. For example, if a person has been exposed to lead or Mn,

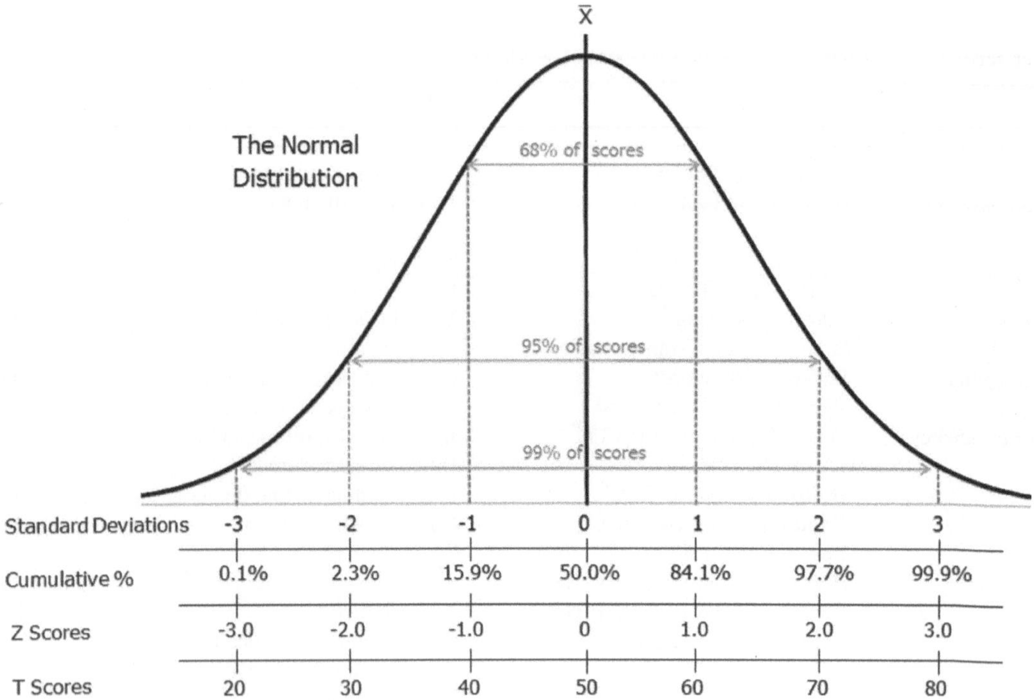

Fig. 3.7. Normal distribution with standardized scores and percentiles.

metals known to affect the basal ganglia region of the brain, that person's scores on tests of psychomotor function will likely be among the lower, if not the lowest, scores in the test battery as the worker with psychomotor slowing may still score well in other areas, such as language tests. Profiles of scores can be plotted on charts and graphs, facilitating ready interpretation of the test data.

In addition, the concept of cognitive reserve needs to be considered. Brain reserve capacity (BRC) is described by Lezak et al. (2012) as "the mental capacity construct of BRC," which may be enhanced by level of education and career achievements. It may result in "later and slower onset" of conditions such as Alzheimer's disease and may present advantages for higher-educated or gifted persons with a greater store of knowledge and competency.

Reliability

Reliability is the regularity with which a test generates the same score under similar retest conditions; or the regularity with which different items of a test produce similar findings. Test reliabilities are usually reported in the test publishers' manual. See McCaffrey et al. (1993) for test–retest data for many independently developed single tests.

Validity

Different validities serve different purposes. "Content validity" represents the extent to which a test measures what it purports to measure. "Construct validity" is the extent to which a test measures the specific construct it purports to measure. "Predictive validity" is determined by comparing test scores with relevant behavioral data. "Face validity" refers to whether the test appears to examine what it is supposed to examine. "Ecologic validity" is the degree to which the test data relate to everyday activities and abilities.

Cut-off scores

"Cut-off" or "cutting" scores separate the "normal" or "unimpaired" from the "abnormal" or "impaired" ends of a continuum of scores. They can be useful for identifying the level of impairment in a set of scores. The "cut" is usually the score at the 5th percentile; all test scores at that level or below will be considered as indicating an impaired performance. Cut-off scores should be used with caution when evaluating test performances of individual workers as demographic variations from the normative sample need to be considered. A related technique for evaluating the likelihood of an individual's score being within or below the normal range is the odds ratio, based on statistical analyses of data from the large-scale group study in which that individual participated.

Treatment and rehabilitation

Appropriate early treatment and rehabilitation may ameliorate both the heavy financial costs to society in loss of earning capacity and social dependency, and the financial, social, and emotional burden on these patients and their families. Unfortunately, in the USA, the worker's compensation system is an adversarial one, which places the burden of proof on claimant workers, resulting in drawn-out legal proceedings with varying levels of success. If workers and/or their health providers or evaluators can prove that their disability is at minimum directly associated with their occupational exposures, comprehensive rehabilitation, including physical occupational therapy, speech/cognitive therapy, psychotherapy, and vocational rehabilitation may be provided.

Cognitive exercises, including computer-assisted strategies, have been used to improve specific neuropsychologic processes, mostly focused on attention, memory, and executive skills. Compensatory devices, such as memory books and electronic paging systems, are used both to bystep intractable deficits, such as learning problems, and to compensate for specific deficits. The efficacy of many of these interventions has been demonstrated (Hardy and Scanlon, 2009; Posit Science, 2015).

Counseling and emotional/behavioral education are important components of most comprehensive rehabilitation programs. Many neurotoxicant-poisoned workers need such treatment as they often suffer from depression and loss of self-esteem associated with cognitive dysfunction and their diminished social status as unemployed or, even worse, unemployable. Their family members, too, will often need guidance and support in understanding and dealing with the difficult behaviors resulting from patients' brain dysfunction and distress.

Counseling and reeducation can help these patients restore self-esteem, enable acceptance of their limitations, and improve their ability for satisfying family and community relations. A key question addresses the costs and benefits of rehabilitation. Given the possibility of tremors, balance or visual problems, rehabilitation may involve more than cognitive deficits. If resources are expended to improve the functioning of a disabled worker to the level of assisted employment (a job with consistent support from a trainer), this once-skilled worker may be satisfied with a lower-level job than held premorbidly or be able to return to the prior occupation. Without vocational rehabilitation or supported employment programs, many disabled workers are unable to keep their jobs or regain employment. Yet, even with such assistance, there will be workers who will never return to employment. However, any improvement, afforded by a rehabilitation experience, and the worker's awareness and enjoyment of such improvement, will be welcome to workers and family alike.

DESIGN OF NEUROPSYCHOLOGIC/ NEUROTOXICANT RESEARCH STUDIES

1. The study should begin with a literature review of recent research on the neurotoxicant of interest and on current neuropsychologic practice and research results relevant to examining the neurotoxin. A multidisciplinary team should be selected. Funding for the health study should be explored.
2. Company managers in an industry using the neurotoxin of interest need to be approached for inclusion in a proposed health study. The request will emphasize that minimizing health risks can benefit both workers and the employer.
3. Involvement of the employers' local and federal health agencies and community leaders can be enlisted in support of such a study. The study PI will serve as mediator between these groups as needed. Exposure measurements determining potential risk from overexposure to the neurotoxicant can demonstrate whether such a study is needed. Additional measurements and potential biomarkers of the neurotoxin(s) under study may need to be identified.
4. For an "in-house" study: When the proposed research has been approved by the company's administration, departments with occupational neurotoxicant exposure and unexposed departments that could serve as controls should be evaluated to determine similarity in vocational proficiencies and competencies. It may be important to engage the interest of managers in these departments to ensure their willingness to support a study involving their workers. Feedback and close collaboration with the relevant occupational management teams should be built into the study design. Close collaboration will provide ongoing feedback and serve to ensure cooperation during and after the study. It is important to present the study plan to workers being recruited for the study.
5. Design of study and Institutional Review Board certification from both the researcher's institution and the funder require formal written approval. Ethics and confidentiality training of all staff in administering tests and collecting data are necessary conditions. Financial incentives for participation or compensation for possibly missed wages should be provided to the workers.
6. Choice of questionnaires and tests used in occupational health studies can be tailored to the particular industry and site; examination materials from other sources may be included.
7. The importance of conducting the examinations in privacy, in a quiet comfortable setting, and at a neutral location if possible requires careful selection of the study location.

8. In addition to test administration, clinical observations should be noted for future coding. The psychometrician and technician examiners need to be trained so that their examinations are standardized and they know how to make clinical observations and produce scorable research data. Confidentiality should be reiterated, as described in the consent form. All participants should be reminded of the anonymity of their personal information. Only group data, without any identifying data of the participants, can be provided to management.

RESOURCES OF STATE AND FEDERAL AGENCIES THAT ADDRESS NEUROTOXICANT EXPOSURES, HEALTH EFFECTS, AND ENFORCEMENT OF OCCUPATIONAL STANDARDS

US Environmental Protection Agency's (EPA's) Integrated Risk Information System (IRIS), according to its website: http://www.epa.gov/iris/

The EPA's IRIS is a human health assessment program that evaluates information on health effects that may result from exposure to environmental contaminants. Through the IRIS program, the EPA provides high-quality science-based human health assessments to support the agency's regulatory activities. The IRIS database is web-accessible and contains information on more than 550 chemical substances.

Agency for Toxic Substances and Disease Registry (ATSDR), according to its website: http://www.atsdr.cdc.gov/

The ATSDR, based in Atlanta, Georgia, is a federal public health agency of the US Department of Health and Human Services. ATSDR serves the public by applying science, taking responsive public health actions, and providing trusted health information to prevent harmful exposures and diseases related to toxic substances.

The complete list of substances with toxicologic profiles developed by the ATSDR can be accessed here: http://www.atsdr.cdc.gov/toxprofiles/index.asp.

California EPA's Office of Environmental Health Hazard Assessment (OEHHA) as per its website: http://www.oehha.ca.gov/about/description.html

The OEHHA is the lead California-state agency for the assessment of health risks posed by environmental contaminants. OEHHA's mission is to protect human health and the environment through scientific evaluation of risks posed by hazardous substances. The OEHHA is one of five state departments within the California EPA.

OEHHA implements the Safe Drinking Water and Toxic Enforcement Act of 1986, commonly known as Proposition 65, and compiles the state's list of substances that cause cancer or reproductive harm. The OEHHA also develops health-protective exposure levels for contaminants in air, water, and soil as guidance for regulatory agencies and the public. These include public health goals for contaminants in drinking water and both cancer potency factors and noncancer reference exposure levels for the Air Toxics Hot Spots Program.

National Institute for Occupational Safety and Health (NIOSH)

"Providing National and World Leadership to Prevent Workplace Illnesses and Injuries"

From http://www.cdc.gov/niosh/about.html :

The National Institute for Occupational Safety and Health (NIOSH) is the U.S. federal agency that conducts research and makes recommendations to prevent worker injury and illness.

NIOSH research is key to these efforts. It provides practical solutions to identified problems. The Institute's work protects the safety and health of the nation's 155 million workers. NIOSH provides the only dedicated federal investment for research needed to prevent the societal cost of work-related fatalities, injuries and illnesses in the United States – estimated in 2007 at $250 billion in medical costs and productivity losses alone. Safety and health risks take huge tolls on workers, their families, businesses, communities, and the nation's economy; NIOSH works to promote a healthy, safe and capable workforce that can rise to the challenges of the 21st century.

Origin of NIOSH and OSHA

The Occupational Safety and Health Act of 1970 established the NIOSH, which is located within the CDC and the OSHA, part of the US Department of Labor. NIOSH provides technical support to OSHA, which develops and enforces workplace safety and health regulations. Please see https://www.osha.gov/ for more information on OSHA.

A link to milestones in NIOSH history can be found at http://www.cdc.gov/niosh/timeline.html.

The National Personal Protective Technology Laboratory (NPPTL)

NPPTL was established at the NIOSH Pittsburgh site to provide leadership for the prevention of injury and illness among workers who must rely on PPE, including respirators, gloves, and hard hats. NPPTL's strategic research program will ensure that the development of new PPE will meet real needs as work settings, technologies, and worker populations' change and new threats emerge.

American Conference of Governmental Industrial Hygienists (ACGIH): http://www.acgih.org/

Threshold limit values are published by the ACGIH, a private consortium. Very few threshold limit values have been based on quantitative risk assessments. For many substances, limits were proposed following informal reviews of generally sparse information, usually from toxicologic or epidemiologic studies (Castleman and Ziem, 1988; Ziem and Castleman, 1989).

Additional CDC/NIOSH and OSHA resources

CDC: NIOSH Pocket Guide to Chemical Hazards: www.cdc.gov/niosh/npg/

The *NIOSH Pocket Guide to Chemical Hazards* provides information on hundreds of chemicals used in occupational settings.

CDC: Chemical safety: www.cdc.gov/niosh/topics/chemical-safety/

A comprehensive database of chemical information and health effects.

CalOSHA: Department of Industrial Relations: http://www.dir.ca.gov/dosh/

With a link and also the Washington State OSHA-delegated agency: http://www.lni.wa.gov/Safety/Rules/Find/.

(Under the OSHA Act, states can assume regulatory authority subject to some conditions.)

CDC: Respirators: www.cdc.gov/niosh/topics/respirators/default.html

A "Respirator Trusted-Source Information Page," providing information on NIOSH-approved respirators and applications.

CDC: Safety and prevention: www.cdc.gov/niosh/topics/safety.html

An overall index of workplace safety and prevention materials (NIOSH).

CDC: Traumatic incident stress: www.cdc.gov/niosh/topics/traumaticincident/default.html

Issues in traumatic incident stress in emergency first-responders to natural or manmade disasters, describing physical, cognitive, emotional, or behavioral symptoms.

CDC: Work schedules: shift work and long work hours: www.cdc.gov/niosh/topics/workschedules/default.html

According to 2004 data from the Bureau of Labor Statistics, almost 15 million Americans work full time on evening shift, night shift, rotating shifts, or other employer arranged irregular schedules. The International Labor Office in 2003 reports that working hours in the United States exceed Japan and most of Western Europe.

Lack of sleep results in functional impairments.

CDC: Carbon monoxide hazards from small gasoline-powered engines: www.cdc.gov/niosh/topics/co/default.html

Guidelines, recommendations, and publications for workers who use gasoline-powered tools or other small gasoline-powered engines.

FUTURE DEVELOPMENTS IN NEUROPSYCHOLOGY

The advent of computerized test batteries (e.g., Baker et al., 1985) held much promise for occupational health use as it no longer seemed to require psychologists or neuropsychologists to conduct studies in the field. Its goal was to quantify cognitive, intellectual, and affective functioning in working populations. It was heralded as a major advance. However, within a decade, questions arose about the comparability of test responses on a computer to test data as traditionally administered

An early investigation into the comparability of the computerized and traditional versions of the same tests of visuomotor and psychomotor functions was found to have low to very modest correlations ($r = 0.23$ to $r = 0.47$) (Bowler et al., 1991). Comparisons of tests of word knowledge had higher correlations ($r = 0.72$). Computerized testing, however, is showing promise for attention assessments when test responses are measured in milliseconds.

Computerized testing will likely never replace a live neuropsychologist carefully observing and testing a worker, especially in individual evaluations or as supervisors of technicians in occupational research studies. However, it is increasingly used for tasks that require only the push of a button, such as answering questionnaires or on reaction time tasks.

An important aid to understanding the neuropsychologic consequences of neurotoxicant exposure will be the integration of neuroimaging techniques into these studies. Long et al. (2014) illustrated the usefulness of neuroimaging in assessing as MRI and/or MRS. Kochunov et al. (2010) correlated processing speed with cerebral health markers in the frontal lobes and documented frontal-lobe dysfunction in currently working welders.

Neurotransmitter changes have also been documented in Mn-exposed workers. For example, abnormal brain γ-aminobutyric acid concentration in smelter workers was reported by Dydak et al. (2011). Studies such as this open up new techniques for studying the neuropsychologic implications of occupational toxic exposures.

References

ACGIH, American Conference of Governmental Industrial Hygienists (2014). TLVs and BEIs: Based on the documentation of the threshold limit values for chemical substances and physical agents and biological exposure indices. ACGIH Signature Publications. Cincinnati.

Akila R, Müller K, Kaukiainen A et al. (2006). Memory performance profile in occupational chronic solvent encephalopathy suggests working memory dysfunction. J Clin Exp Neuropsychol 28: 1307–1326.

Alessio L, Campagna M, Lucchini R (2007). From lead to manganese through mercury: mythology, science, and lessons for prevention. Am J Ind Med 50: 779–787.

American Psychiatric Association, APA (2000). Diagnostic and Statistical Manual of Mental Disorders. 4th ed, Washington, D.C., text rev.

Anger WK, Liang Y, Nell V et al. (2000). Lessons learned – 15 years of the WHO-NCTB: a review. Neurotoxicology 21: 837–846.

Anonymous (1999). Neuropsychological assessment of adult neurotoxic exposure. Continuum: Lifelong in Neurology 5: 51–69.

Ardila A, Rodriguez-Menendez G, Rosselli M (2002). Current issues in neuropsychological assessment with Hispanic/Latinos. In: Fr Ferraro (Ed.), Minority and cross-cultural aspects of neuropsychological assessment. Studies on neuropsychology, development, and cognition. Swets & Zeitlinger, Lisse.

Arlien-Søborg P (1991). Solvent Neurotoxicity. CRC Press, Boca Raton.

Arlien-Søborg P, Simonsen L (2011). Chemical neurotoxic agents. Switzerland, International Labor Organization, Geneva.

Artiola i Fortuny L, Heaton R, Hermosillo D (1998). Neuropsychological comparisons of Spanish-speaking participants from the U.S.–Mexico border region versus Spain. J Int Neuropsychol Soc 4: 363–379.

ATSDR, (Agency for Toxic Substances and Disease Registry) (1996). Toxicological Profile for Carbon Disulfide, U.S. Department of Health and Human Services, Public Health Service, Atlanta, GA.

ATSDR, (Agency for Toxic Substances and Disease Registry) (2007). Toxicological Profile for Lead, U.S. Department of Health and Human Services, Public Health Service, Atlanta, GA.

ATSDR, (Agency for Toxic Substances and Disease Registry) (2012a). Toxicological Profile for Carbon Monoxide, U.S. Department of Health and Human Services, Public Health Service, Atlanta, GA.

ATSDR, (Agency for Toxic Substances and Disease Registry) (2012b). Toxicological Profile for Manganese, U.S. Department of Health and Human Services, Public Health Service, Atlanta, GA.

ATSDR, (Agency for Toxic Substances and Disease Registry) (2015). Toxicological Profile for Hydrogen Sulfide/Carbonyl Sulfide. U.S. Department of Health and Human Services, Public Health Service, Atlanta, GA.

Bailer AJ, Stayner LT, Smith RJ et al. (1997). Estimating benchmark concentrations and other noncancer endpoints in epidemiology studies. Risk Anal 17: 771–779.

Baker E, Letz R, Fidler A (1985). A computer-administered neurobehavioral evaluation system for occupational and environmental epidemiology: Rationale, methodology, and pilot study results. J Occup Med 27: 206–212.

Ballard C, McKeith I, Burn D et al. (1997). The UPDRS scale as a means of identifying extrapyramidal signs in patients suffering with dementia with Lewy bodies. Acta Neurol Scand 96: 366–371.

Beseler C, Stallones L, Hoppin J et al. (2008). Depression and pesticide exposures among private pesticide applicators enrolled in the Agricultural Health Study. Environ Health Perspect 116: 1713–1719.

Beuter A, Edwards R, de Geoffroy A et al. (1999). Quantification of neuromotor function for detection of the effects of manganese. Neurotoxicology 20: 355–366.

Bilder R (2011). Neuropsychology 3.0: Evidence-based science and practice. J Int Neuropsychol Soc 17: 7–13.

Blanchard EB, Jones-Alexander J, Buckley TC et al. (1996). Psychometric properties of the PTSD Checklist (PCL). Behav Res Ther 34: 669–673.

Bouchard M, Mergler D, Baldwin M et al. (2007). Neuropsychiatric symptoms and past manganese exposure in a ferro-alloy plant. Neurotoxicology 28: 290–297.

Bowler R (1999a). Occupational and Environmental Neuropsychology. Occupational and Environmental Medicine Secrets, Hanley and Belfus, New York.

Bowler RM (1999b). Occupational and Environmental Neuropsychology. In: RM Bowler, JE Cone (Eds.), Occupational Medicine Secrets, Hanley & Belfus, Philadelphia.

Bowler R, Mergler D, Huel G et al. (1991). Neuropsychological impairment among former microelectronics workers. Neurotoxicology 12: 87–103.

Bowler R, Hartney C, Ngo L (1998a). Amnestic disturbance and PTSD in the aftermath of a chemical release. Arch Clin Neuropsychol 13: 455–471.

Bowler R, Mergler D, Huel G et al. (1998b). Relationship of affective and neuropsychological function among solvent-exposed microelectronics workers. In: R Schwarzer (Ed.), Advances in Health Psychology Research. Freie Universitaet, Berlin, Germany.

Bowler R, Lezak MD, Booty A et al. (2001). Neuropsychological dysfunction, mood disturbance, and emotional status of munitions workers. Applications of Neuropsychology 8: 74–90.

Bowler RM, Gysens S, Hartney C (2003). Neuropsychological effects of ethylene dichloride exposure. Neurotoxicology 24: 553–562.

Bowler RM, Gysens E, Diamond S et al. (2006a). Manganese exposure: Neuropsychological and neurological symptoms and effects in welders. Neurotoxicology 27: 315–326.

Bowler RM, Gysens S, Diamond E et al. (2006b). Manganese exposure: neuropsychological and neurological symptoms and effects in welders. Neurotoxicology 27: 315–326.

Bowler RM, Nakagawa S, Drezgic M et al. (2007a). Sequelae of fume exposure in confined space welding: A neurological and neuropsychological case series. Neurotoxicology 28: 298–311.

Bowler RM, Roels HA, Nakagawa S et al. (2007b). Dose–effect relationships between manganese exposure and neurological, neuropsychological and pulmonary function in confined space bridge welders. Occup Environ Med 64: 167–177.

Bowler RM, Harris M, Li J et al. (2011). Longitudinal mental health impact among police responders to the 9/11 terrorist attack. Am J Ind Med 55: 297–312.

Budtz-Jorgensen E, Keiding N, Grandjean P (2001). Benchmark dose calculation from epidemiological data. Biometrics 57: 698–706.

Castleman BI, Ziem GE (1988). Corporate influence on threshold limit values. Am. J. Ind. Med. 13: 531–559.

Centers for Disease Control and Prevention, CDC (2014). Behavioral Risk Factor Surveillance System [Online], National Center for Chronic Disease Prevention and Health Promotion: Division of Population Health, Atlanta, GA, Available: http://www.cdc.gov/brfss/questionnaires/pdf-ques/2014_BRFSS.pdf (accessed January 22 2015).

Clewell HJ, Lawrence GA, Calne DB et al. (2003). Determination of an occupational exposure guideline for manganese using the benchmark method. Risk Anal 23: 1031–1046.

Criswell SR, Perlmutter JS, Videen TO et al. (2011). Reduced uptake of [18 F]FDOPA PET in asymptomatic welders with occupational manganese exposure. Neurology 76: 1296–1301.

Crump KS (1984). A new method for determining allowable daily intakes. Fundam Appl Toxicol 4: 854–871.

Crump KS (1995). Calculation of benchmark dose from continuous data. Risk Anal 15: 79–89.

Dick F (2006). Solvent neurotoxicity. Occup Environ Med 63: 221–226.

Dorman DC, Struve MF, Wong BA et al. (2006). Correlation of brain magnetic resonance imaging changes with pallidal manganese concentrations in rhesus monkeys following subchronic manganese inhalation. Toxicol Sci 92: 219–227.

Doty RL (1995). The Smell Identification Test Administration Manual. 3rd edition. Haddon Heights, NJ, Sensonics.

Dydak U, Jiang Y-M, Long L-L et al. (2011). In vivo measurement of brain GABA concentrations by magnetic resonance spectroscopy in smelters occupationally exposed to manganese. Environ Health Perspect 119: 219–224.

Feldman R, Ricks N, Baker E (1980). Neuropsychological effects of industrial toxins: A review. Am J Ind Med 1: 211–227.

Frackowiak R, Friston K, Fristh C (1997). Human Brain Function. Academic Press, San Diego.

Gazzaley A, D'Esposito M (2007). Unifying prefrontal cortex function: executive control, neural networks, and top-down modulation. In: B Miller, J Cummings (Eds.), The Human Frontal Lobes: Functions and Disorders, 2nd edn. The Guilford Press, New York.

Gazzaniga MS (2000). The Mind's Past. University of California Press, Berkeley, CA.

Goldstein K (1939). The organism: a holistic approach to biology derived from pathological data in man. American Book Publishing, Salt Lake City.

Gregersen P, Angleso B, Nielsen T et al. (1984). Neurotoxic effects of organic solvents in exposed workers: an occupational, neuropsychological, and neurological investigation. Am J Ind Med 5: 201–225.

Gurd J, Kischka U, Marshall J (2010). Handbook of Clinical Neuropsychology, 2nd ed. Oxford University Press, Oxford, UK.

Halstead WC (1947). Brain and Intelligence, University of Chicago Press, Chicago, IL.

Hänninen H (1971). Psychological picture of manifest and latent carbon disulfide poisoning. Br J Ind Med 28: 374–381.

Hänninen H (1974). Behavioral study of the effects of carbon disulfide. U.S. Department of Health, Education and Welfare, Washington, D.C.

Hänninen H, Eskelinen L, Husman K et al. (1976). Behavioral effects of long-term exposure to a mixture of organic solvents. Scand J Work Environ Health 2: 240–255.

Hänninen H, Hernberg S, Mantere P et al. (1978). Psychological performance of subjects with low exposure to lead. J Occup Med 20: 683–689.

Hardy J, Scanlon M (2009). The science behind Lumosity, Lumos Labs, San Francisco.

Howieson DB, Lezak MD (1994). Separating memory from other cognitive problems. In: A Baddeley, BA Wilson

(Eds.), Handbook of memory disorders, John Wiley, Chichester, Sussex, England.

Josephs KA, Ahlskog JE, Klos KJ et al. (2005). Neurologic manifestations in welders with pallidal MRI T1 hyperintensity. Neurology 64: 2033–2039.

Kaplan E (1988). A process approach to neuropsychological assessment. In: T Boll, BK Bryant (Eds.), Clinical neuropsychology and brain function: Research, measurement, and practice. American Psychological Association, Washington, D.C.

Kennedy RS, Dunlap WP (1990). Assessment of the Vistech contrast sensitivity test for repeated measures applications. Optom Vis Sci 67: 248–251.

Kim KS, Kim Y, Jin Y et al. (1999). Factors associated with psychoneurobehavioral outcomes in workers exposed to manganese. Korean Journal of Occupational and Environmental Medicine 11: 213–228.

Kim Y, Bowler R, Abdelouahab N et al. (2011). Motor function in adults of an Ohio community with environmental manganese exposure. Neurotoxicology 32: 606–614.

Kim Y, Lobdell D, Wright C et al. (2015). Blood metal concentrations of manganese, lead, and cadmium in relation to serum ferritin levels in Ohio residents. Biol Trace Elem Res 1–9.

Kochunov P, Coyle T, Lancaster J et al. (2010). Processing speed is correlated with cerebral health markers in the frontal lobes as quantified by neuroimaging. Neuroimage 49: 1190–1199.

Koller W, Lyons K, Truly W (2004). Effect of levodopa treatment for parkinsonism in welders: a double-blind study. Neurology 62: 730–733.

Lanthony P (1978). The desaturated panel D-15. Doc Ophthalmol 46: 185–189.

Lezak MD (2012). Chapter 11: Memory I: Tests. In: MD Lezak, D Howieson, E Bigler et al. (Eds.), Neuropsychological Assessment, 5th edn. Oxford University Press, Oxford.

Lezak MD, Howieson D, Bigler E et al. (2012). Neuropsychological Assessment. 5th edn. Oxford University Press, Oxford.

Long Z, Jiang Y-M, Li X-R et al. (2014). Vulnerability of welders to maganese exposure – A neuroimaging study. Neurotoxicology 45: 285–292.

Lucchini RG, Martin CJ, Doney BC (2009). From manganism to manganese-induced parkinsonism: a conceptual model based on the evolution of exposure. Neuromolecular Med 11: 311–321.

Luria AR (1969). The frontal syndrome. In: PJ Vinken, GW Bruyn (Eds.), Handbook of Clinical Neurology, 2. North Holland Publishing, Amsterdam.

Luria AR (1973). The Working Brain: An Introduction to Neuropsychology, Basic Books, New York, NY.

Mackenzie Ross S, Brewin C, Curran H et al. (2010). Neuropsychological and psychiatric functioning in sheep farmers exposed to low levels of organophosphate pesticides. Neurotoxicol Teratol 32: 452–459.

McCaffrey RJ, Ortega A, Haase R (1993). Effects of repeated neuropsychological assessments. Arch Clin Neuropsychol 8: 519–524.

McCarthy R, Warrington E (1990). Cognitive neuropsychology: A clinical introduction. Academic Press, San Diego, CA.

McCauley LA, Anger KW, Keifer M et al. (2006). Studying health outcomes in farmworker populations exposed to pesticides. Environ Health Perspect 114: 2006.

McClure P, Odin M (1998). Exposure–response relationships for subclinical neurological effects in mangnanese-exposed workers. OSHA Contact J-9-F 5-0051. Syracuse Research Corporation, New York.

Meehl PE (1954). Clinical versus statistical prediction: A theoretical analysis and a review of the evidence, Available online at: https://www.psych.umn.edu/people/meehlp/032ClinstixBook.pdf (accessed 6 April 2015).

Mergler D, Baldwin M, Bélanger S et al. (1999). Manganese neurotoxicity, a continuum of dysfunction: results from a community based study. Neurotoxicology 20: 327–342.

Meyer-Baron M, Blaszkewicz M, Henke H et al. (2008). The impact of solvent mixtures on neurobehavioral performance – conclusions from epidemiological data. Neurotoxicology 29: 349–360.

Meyer-Baron M, Schaper M, Knapp G et al. (2013). The neurobehavioral impact of manganese: results and challenges obtained by a meta-analysis of individual participant data. Neurotoxicology 36: 1–9.

Milberg WP, Hebben N, Kaplan E (1996). The Boston Process approach to neuropsychological assessment. In: I Grant, K Adams (Eds.), Neuropsychological assessment of neuropsychiatric disorders, 2nd edn. Oxford University Press, New York, NY.

Miller B, Cummings J (2007). The Human Frontal Lobes: Functions and Disorders, Guilford Press, New York.

Milner B (1968). Visual recognition and recall after right temporal-lobe excision in man. Neuropsychologia 6: 191–209.

Milner B (1972). Disorders of learning and memory after temporal lobe lesions in man. Clinical Neuropsychology 19: 421–446.

Mirsky A, Fantie B, Tatman J (1995). Assessment of attention across the lifespan. In: RL Mapou, J Spector (Eds.), Clinical Neuropsychological Assessment: A Cognitive Approach. Plenum, New York, NY.

Mitrushina M, Boone K, Razani J et al. (2005). Handbook of normative data for neuropsychological assessment. Oxford University Press, Oxford.

Movement Disorder Society Task Force on Rating Scales for Parkinson's Disease (2003). The Unified Parkinson's Disease Rating Scale (UPDRS): status and recommendations. Mov Disord 18: 738–750.

Nelson K, Golnick J, Korn T et al. (1993). Manganese encephalopathy: Utility of early magnetic resonance imaging. Br J Ind Med 50: 510–513.

Newcombe F (1969). Missile wounds of the brain. Oxford University Press, London.

OSHA, Occupational Safety and Health Administration (2012). Occupational Safety and Health Standards, Toxic and Hazardous Substances. Code of Federal Regulations 29, U.S. Department of Labor, Washington, D.C.

Ostiguy C, Asselin P, Malo S et al. (2005). Management of Occupational Manganism: Consensus of an Expert's Panel. Studies and Research Projects Report. IRSST, Montreal, Quebec.

Ostiguy C, Asselin P, Malo S (2006). The emergence of manganese-related health problems in Quebec: an integrated approach to evaluation, diagnosis, management and control. Neurotoxicology 27: 350–356.

Park RM, Bowler RM, Eggerth DE et al. (2006). Issues in neurological risk assessment for occupational exposures: The Bay Bridge Welders. Neurotoxicology 27: 373–384.

Park RM, Bowler RM, Roels HA (2009). Exposure–response relationship and risk assessment for cognitive deficits in early welding-induced manganism. J Occup Environ Med 51: 1125–1136.

Porteus SD (1959). The Maze Test and clinical psychology. Pacific Books, Palo Alto.

Posit Science (2015). Brain HQ [Online]. Available: http://www.brainhq.com/ (accessed February 11, 2015).

Prigatano GP, Fordyce D, Zeiner H et al. (1986). Neuropsychological rehabilitation after brain injury. Johns Hopkins University Press, Baltimore.

Reidy T, Bowler R, Rauch S et al. (1992). Pesticide exposure and neuropsychological impairment in migrant farm workers. Arch Clin Neuropsychol 7: 85–95.

Reitan RM, Wolfson D (1985). The Halstead-Reitan Neuropsychological Test Battery. Neuropsychology Press, Tuscon, AZ.

Rodriguez-Agudelo Y, Riojas-Rodriguez H, Rios C et al. (2006). Motor alterations associated with exposure to manganese in the environment in Mexico. Sci Total Environ 368: 542–556.

Roels HA, Ortega Eslava MI, Ceulemans E et al. (1999). Prospective study on the reversibility of neurobehavioral effects in workers exposed to manganese dioxide. Neurotoxicology 20: 255–272.

Roels H, Bowler R, Kim Y et al. (2012). Manganese exposure and cognitive deficits: A growing concern for manganese neurotoxicity. Neurotoxicology 33: 872–880.

Roldán-Tapia L, Parrón T, Sánchez-Santed F (2005). Neuropsychological effects of long-term exposure to organophosphate pesticides. Neurotoxicol Teratol 27: 259–266.

Roldán-Tapia L, Nieto-Escamez F, del Aguila E et al. (2006). Neuropsychological sequelae from acute poisoning and long-term exposure to carbamate and organophosphate pesticides. Neurotoxicol Teratol 28: 694–703.

Slick DJ, Sherman E, Iverson GL (1999). Diagnostic criteria for malingered neurocognitive dysfunction: proposed standards for clinical practice and research. Clin Neuropsychol 13: 545–561.

Stewart W, Schwartz B (2007). Effects of lead on the adult brain: A 15-year exploration. Am J Ind Med 50: 729–739.

Strauss E, Sherman E, Spreen O (2006). A compendium of neuropsychological tests: Administration, norms, and commentary, 3rd ed, Oxford University Press, New York.

Tate RL (2010). A compendium of tests, scales and questionnaires: The practitioner's guide to measuring outcomes after acquired brain impairment. Psychology Press, Hove, U.K.

Teuber H-L (1969). Neglected aspects of the post-traumatic syndrome. In: A Walker (Ed.), The late effects of head injury. Thomas, Springfield, IL.

Tröster A, Ruff R (1990). Neuropsychological sequelae of exposure to the chlorinated hydrocarbon solvents trichloroethylene and trichloroethane. Arch Clin Neuropsychol 5: 31–47.

van Hout M, Schmand B, Wekking E et al. (2006). Cognitive functioning in patients with suspected chronic toxic encephalopathy: evidence for neuropsychological disturbances after controlling for insufficient effort. J Neurol Neurosurg Psychiatry 77: 296–303.

van Valen E, van Thriel C, Akila R et al. (2012). Chronic solvent-induced encephalopathy: European consensus of neuropsychological characteristics, assessment, and guidelines for diagnostics. Neurotoxicology 33: 710–726.

Vanderploeg RD (2000). Clinician's guide to neuropsychological assessment. Lawrence Erlbaum Associates, Mahwah, NJ.

Walsh K (1985). Understanding brain damage. Churchill-Livingstone, Edinburgh.

Wechsler D (2009a). WAIS-IV Manual, Psychological Corporation, San Antonio, TX.

Wechsler D (2009b). WMS-IV Manual, Psychological Corporation, San Antonio, TX.

White R, Proctor S (1992). Research and clinical criteria for development of neurobehavioral test batteries. J Occup Med 34: 140–148.

White T, Stern R (2001). NAB Neuropsychological Assessment Battery, Psychological Assessment Resources, Lutz, FL.

White R, Feldman R, Proctor S (1992). Neurobehavioral effects of toxic exposure. In: R White (Ed.), Clinical syndromes in adult neuropsychology: The practitioner's handbook. Elsevier, New York.

White R, Proctor S, Echeverria D et al. (1995). Neurobehavioral effects of acute and chronic mixed-solvent exposure in the screen printing industry. Am J Ind Med 28: 221–231.

Wild K, Howieson D, Webbe F et al. (2008). Status of computerized cognitive testing in aging: A systematic review. Alzheimers Dement 4: 428–437.

Ziem GE, Castleman BI (1989). Threshold limit values: Historical perspectives and current practice. J Occup Environ Med 31: 910–918.

Handbook of Clinical Neurology, Vol. 131 (3rd series)
Occupational Neurology
M. Lotti and M.L. Bleecker, Editors

Chapter 4

Exposure assessment for chemical and physical agents

ANGELO MORETTO*

Department of Biomedical and Clinical Sciences, University of Milan and International Centre for Pesticides and Health Risks Prevention (ICPS), Luigi Sacco Hospital, Milan, Italy

INTRODUCTION

Exposure assessment to either chemical or physical agents, as applicable, is a necessary step for both etiologic diagnosis of occupational disease and health surveillance of workers. In fact, paraphrasing Paracelsus (Philippus Aureolus Theophrastus Bombastus von Hohenheim, 1493–1541), all agents are poisons, and only the dose makes them nonpoisonous. Exposure assessment is the tool to measure or estimate the dose of an agent to which a worker or a group of workers is or has been exposed. This information is essential to conclude on the possible relationship between a disease and an occupational exposure. On the other hand, measuring or estimating exposure is also one of the fundamental pillars for health surveillance of workers. In particular, identification of unacceptable exposure levels of a working population will lead to appropriate intervention in order to prevent future illness. Therefore, exposure assessment necessarily includes the type, extent and magnitude, source, and duration of the contact between the worker and the chemical or physical agent(s) in question.

Regarding chemicals, it is evident that assessing the exposure is a measure or an estimate of the contact with a substance, which can lead to absorption of the chemical into the organism. It is, in fact, the amount absorbed that is relevant for a toxicologic effect, unless the chemical has a local effect at the point of contact (e.g., the skin, the eye). No local neurotoxic effects are known following occupational exposure.

Substances may be present in the workplace as airborne materials (e.g., gases, vapors, or inhalable dusts) and be absorbed via the respiratory tract, this being the most frequent and relevant route of absorption in occupational settings. Substances might be present as gases, vapors, dusts, or fibers that have different rates of absorption depending on their chemical and physical characteristics. Other substances can also deposit on the skin and be dermally absorbed; this occurs with materials that enter in direct contact with the hands, or less frequently, with other parts of the skin. Some of the airborne substances may also deposit on the skin. For other substances, particularly dusty materials, there might be significant gastrointestinal absorption due to unwanted hand-to-mouth contact. Workers may also be exposed by direct injection following puncture of the skin by a sharp object.

Exposure assessment is not simple and straightforward: on the one hand airborne concentrations in the same workplace do fluctuate over time and space, whereas on the other hand there are no direct ways to quantify individual exposure via dermal and gastrointestinal routes. In the former case, strategies have been developed to reduce uncertainties on the estimation that include ambient air sampling and personal sampling (see below). In the latter case, where available, measurements in biologic fluids of the chemicals or of their metabolites can be used to estimate the so-called "aggregate" dose that derives from inhalation, dermal and gastrointestinal absorption: this activity is called biologic monitoring (see below).

Figure 4.1 is a graphical representation of the relationship between an exposure to an agent (hazard), the assessment of the exposure, of the possible biochemical and physiologic effects, and the disease.

This chapter will describe how we can assess workers' exposure, and how we define, interpret, and use exposure and biologic monitoring data either to control working conditions or to attribute a disease to an exposure.

Certain physical agents may also cause damage to the nervous system. Particularly relevant in the workplace

*Correspondence to: Angelo Moretto, M.D., Ph.D., International Centre for Pesticides and Health Risks Prevention (ICPS), Luigi Sacco Hospital, Via GB Grassi 74, 20157 Milan, Italy. Tel: +39-02-3568-661, E-mail: angelo.moretto@unimi.it

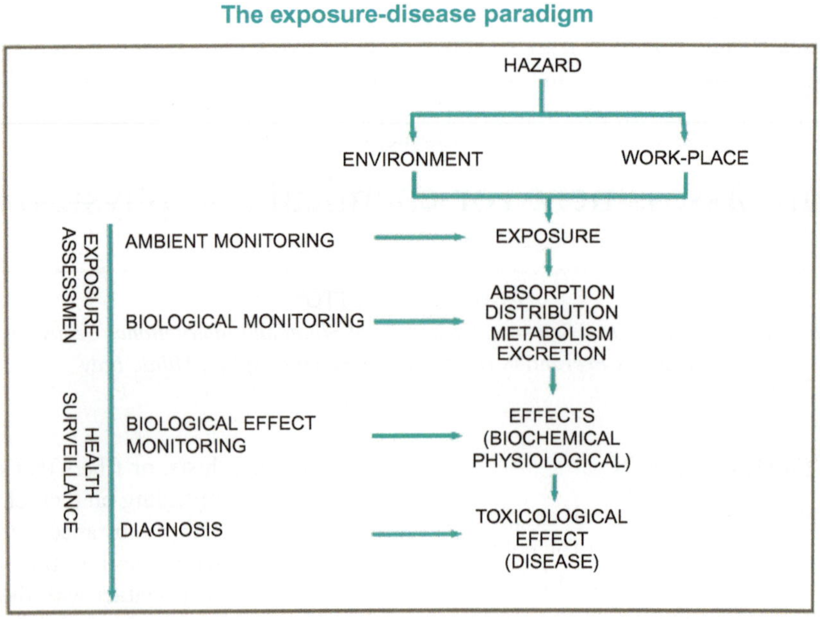

Fig. 4.1. The exposure–disease paradigm.

are vibrations, both whole-body and hand–arm, and noise. These are discussed in separate chapters and therefore are not presented here. It should only be noted that the general considerations presented in this chapter regarding the use of limit values also apply to limit values for physical agents.

GENERAL CRITERIA OF RISK ASSESSMENT: DEFINITION OF EXPOSURE LIMITS

All substances have the potential to cause a toxic effect, which might, in fact, occur in specific circumstances of exposure. In order to protect workers' health it is necessary to define the conditions in which the exposure is such that no adverse effects are expected. The process, called "risk assessment and management," is graphically represented in Figure 4.2. Detailed descriptions of this process can be found in a number of texts, e.g., van Leeuwen (2007), Greim and Snyder (2008), WHO (2009), and Faustman and Omenn (2013). Based on available toxicologic data, the conditions in which no effects are expected in humans are determined and consequently exposure limits are defined for humans and expressed as concentration of substance(s) in different media (air, soil, water, food) or as levels of a compound or its metabolite(s) in accessible fluids, or as acceptable modifications of physiologic/biochemical parameters following exposure (i.e., biologic monitoring).

Although strictly connected, risk assessment and risk management are different processes that try to answer different questions. Risk assessment is the attempt to define the conditions and associated probabilities for an adverse effect to occur and its severity. Risk management instead deals with all regulatory, legal, and technical issues related to actions to be taken in order to ensure, obtain, and control conditions in which the risk is acceptable.

The first step in risk assessment entails the collection and evaluation of toxicologic data to identify the

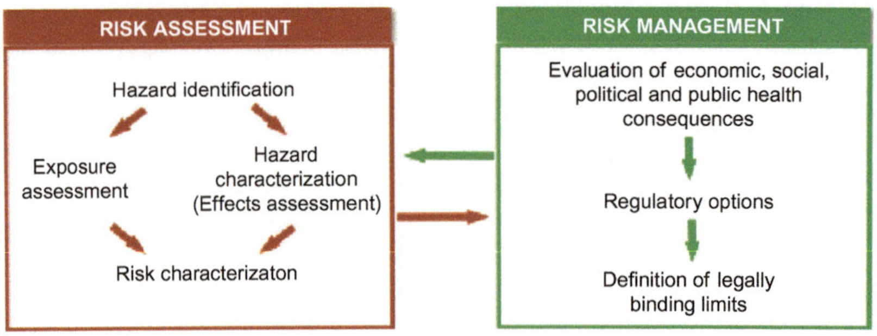

Fig. 4.2. Risk assessment and risk management. Specific activities and interactions.

fect(s) (hazard) on health caused by the substance. The ources of information can be different and include *vitro* or *in vivo* experimental data in animals, poisong case reports, epidemiologic studies, and studies on xposed workers. Once the hazard has been identified, e dose–response relationship needs to be characterid to define the severity of an effect in an individual r its frequency in the population. The critical effect generally considered the toxicologic effect that has e lowest threshold (i.e., that occurs at the lowest ose) among all the effects caused by the substance. he goal for the purpose of risk assessment is to idenify the no-observable-(adverse)-effect-level (NO(A) L) for this effect. The data may be available as either xternal dose (e.g., mg/m^3 or ppm for inhalation expoure), or internal dose (mg/kg body weight for oral nd dermal absorption, i.e., exposure corrected for he rate of absorption via the specific route). Most requently this information is only available in experinental animals and it needs to be extrapolated to humans. Therefore, correction factors are applied to the numeric value of the NOAEL. These correction factors are called "uncertainty," "safety," "extrapolation," "protection," or "assessment" factors by different bodies, and are factors meant to overcome the uncertainties in the extrapolation process to a generic working population of the available data. Usually, for the working population these factors are lower than those used for the general population, where the accepted values are 10 for extrapolation from human NOAEL, to account for the variability of sensitivity within the human population, and 100 from animal NOAEL, to account also for a possible higher sensitivity of humans versus experimental animals. Additional factors are sometimes added to account for an incomplete data set, availability of only the lowest observed adverse effect level (LOAEL), and other considerations (WHO, 2005, 2009; van Leeuwen, 2007; Greim and Snyder, 2008).

There is no generally agreed approach to the application of these factors for the working population but, based on three main considerations, these factors are lower, and sometimes significantly lower than those used for the general population exposed to environmental and consumer chemicals. In fact, it is considered that: (1) the working population is more homogeneous than the general population; (2) occupational exposure is not lifetime, but is conventionally considered to be 8 hours/day, 5 days/week/, 240 days/year for up to 45 years; and (3) it is considered that the health of workers is expected to be controlled by periodic health surveillance and monitoring programs.

In relation to the type of exposure, characteristics of the compound, and the exposed population, different limits are set. Table 4.1 describes in detail the most important types of limits used in occupational settings. Limits applicable to the general population include the tolerable/acceptable daily intake for repeated oral exposures (e.g., for substances present in food or in drinking water), and the acute reference dose for acute oral exposures. These limits are expressed in mg/kg body weight. Other limits are referred to the source of exposure. In case of inhalatory exposures, the no observed adverse effect concentration is used; metrics changes (e.g., from mg/kg body weight to mg/m^3 air), but the general concepts do not change. Similarly, the limits for water or food contaminants will be expressed as concentrations (mg/L or mg/kg, respectively). For genotoxic carcinogenic compounds a different approach is taken, since it is believed that these substances have no threshold. However, since this does not apply to neurotoxic compounds it will not be addressed.

It should be underlined that some of these occupational limits can be derived from studies where exposure was via the oral or dermal route, and extrapolated to ambient air concentration based on physiologic and kinetic parameters. In any case, these threshold limit values (TLVs)/occupational exposure limits (OEL) do not take into quantitative account exposures by routes other than inhalation, and do not consider the presence of the individual factors that may influence either the exposure or the sensitivity to adverse effects like genetic differences and variants, or the person's work rate and associated breath rate that affects the quantity of chemical inhaled. Therefore, while these values are widely accepted as a way of judging whether occupational chemical exposure is acceptable, their interpretation needs careful and expert judgment both in the industrial health and in the clinical settings.

EXPOSURE ROUTES

Inhalation

Exposure by inhalation is conventionally expressed as the concentration of the substance in ambient air over a fixed duration. This is called time-weighted average (TWA), when applied to an 8-hour period (or similar length of time), and short-term exposure level (STEL) when it refers to shorter periods (e.g., 15 minutes). For those compounds causing acute effects, the concentration is expressed as a peak, instantaneous or over a few minutes (ceiling). The concentration in air is expressed as mg/m^3 or in parts per million (ppm); see Figure 4.3 for the conversion of ppm to mg/m^3. Extent of inhalation exposure is determined not only by the concentration in air but also by the breathing rate that changes significantly from resting to heavy physical activity. Therefore, it is expected that in the same environment, effort will affect the intake of individuals,

Table 4.1

Occupational limits: ambient air exposure limits and biological limits

Name and acronym	Organization	Explanation
Threshold limit value (TLV) Occupational exposure level (OEL)	Limit values recommended by American Conference of Governmental Industrial Hygienists (ACGIH) Limit values recommended by the Scientific Committee on Occupational Exposure Limits (SCOEL) in the European Union	TLV/OEL-TWA (time-weighted average) Limit expressed as time-weighted average concentration of the substance. This value is intended to protect the health of "most" workers exposed for 8 hours/day, 5 days/week, for the entire working TLV/OEL-STEL (short-term exposure limit) Highest allowed value for short-term exposures: generally for 15 minutes, not more than four times in a workshift, at intervals of at least 60 minutes. It is the concentration to which it is believed that workers can be exposed for a short period of time without the occurrence of irritation, chronic or irreversible damage, dose-related toxic effects, narcosis such that it increases the probability of accidents or to interfere with the ability to counteract adverse situations, or decrease work efficiency. This limit is applied to substances that have systemic toxicity deriving from short high exposures. The TLV-STEL should be taken in conjunction with the TLV-TWA, where the latter indicates the long-term limits to prevent chronic effects, whereas the former represents the maximum allowed fluctuation above the TLV-TWA when the substance has also recognized effects that can derive from an acute, short-term exposure TLV/OEL-Ceiling Threshold limit value that should always be respected any time during the workshift. It is a limit value that is applied to very short exposures. This limit is generally applied to substances with significant acute effects such as irritation or corrosivity
Biologic exposure index (BEI) Biologic limit value (BLV)	Limit values recommended by ACGIH (BEI) or by SCOEL (BLV). Other organizations may propose or set similar indexes	BEI/BLV It represents the mean value of the biologic index associated with an inhalatory exposure at the TLV/OEL-TWA. Certain BEIs/BLVs are based on initial biochemical effects (e.g., inhibition of an enzyme)

Actual limit values can be found at:
ACGIH: http://www.acgih.org/tlv/
NIOSH: http://www.cdc.gov/niosh/npg/default.html
OSHA: https://www.osha.gov/dts/chemicalsampling/toc/chmn_A.html
SCOEL: http://ec.europa.eu/social/BlobServlet?docId=3803&langId=en
WHO International Programme on Chemical Safety: http://www.who.int/ipcs/en/

Transformation from ppm to mg/m³

$$\text{ppm} = \text{mg/m}^{-3} \times 22.41/\text{MW}$$

ppm: parts per million
MW = Molecular weight

Fig. 4.3. How to transform ppm into mg/m^3.

introducing differences that are related to breathing rate. This should be remembered when considering data of ambient air concentration for the assessment of actual systemic intake of a chemical.

Dermal

Chemicals that come into contact with the skin may be absorbed. To do so, they must be deposited on the skin, cross the stratum corneum and other skin layers, and diffuse until they reach the peripheral blood supply. Dermal exposure should be ideally expressed as the amount deposited on the skin corrected by the absorption rate, that is, the percentage of the amount deposited that reaches the blood. However, it is not easy to measure the concentration of chemicals on the skin and, as an alternative, the mass that comes into contact with the skin during a period of time is generally used as a surrogate. This is defined as mass per unit area over a defined part of the body, e.g., mg/cm^2, or as the mass on a specific body part, e.g., mg on both hands. In any case, a correction still needs to be done for the effective absorption of the compound. Because of this complexity, there are no exposure limits established for dermal exposure. However, a semiquantitative evaluation of dermal exposure can be done based on dermal absorption data or on chemicophysical characteristics of the substance. The latter include liposolubility, ionization, and molecular weight. When dermal absorption is considered significant, a warning ("skin" notation) is usually associated with the TLV/OEL to indicate this fact.

Oral

In the occupational setting, oral exposure is generally accidental or associated with poor hygienic practices during work, such as smoking or eating in the workplace, or any other hand-to-mouth contact. In the occupational setting there are obviously no exposure limits of oral intake. In addition, there is no consensus on how exposure by inadvertent ingestion should be assessed. Other important factors to be considered for this route of exposure are the extent of hand contamination, the type of work, the use of protective gloves, and personal traits of individual workers, e.g., nail biting.

Injection

This particular route of exposure is associated with accidental injuries caused by needles or other sharp objects that might be contaminated by chemicals. Exposure assessment in this condition is almost impossible and an estimation can only be made by calculating the number of injuries, e.g., per day.

ASSESSING THE LEVEL OF EXPOSURE

Measurements or estimates of exposure are a surrogate for the dose. This in turn will depend on concentration, duration, and frequency of exposure. There are several methods that have been developed in the past 50 years, to assess, during a given period of time, the amount of contact between a worker and a chemical. Methods of exposure assessment can provide detailed and precise real-time (e.g., using direct-reading instruments) or delayed (sampling followed by analysis in the laboratory) quantification. These methods apply to exposure to airborne material such as gases, fumes, vapor, and dust. The concentration is generally expressed in milligrams per cubic metre of air (mg/m^3). The values in parts per million (ppm) are also given for gases and vapours (see Fig. 4.3 for the conversion from mg/m^3 to ppm). In certain situations only broad categorization into qualitative bands such as "low," "medium," and "high" can be obtained and might even be sufficient for the purpose of assessing the safety of exposures in the working place.

Data can be collected in the general ambience or with personal sampler. In the former case, collecting devices are positioned in the workplace based on previous analysis of the environment with the aim of defining the "average" ambient concentration, or they might be placed near the sources of the contaminant identified as most probable or most important. When the collecting device is positioned near the nose/mouth of the worker, a "personal" monitoring can be obtained that reflects the individual's exposure while performing the different job tasks that may be associated with different levels of exposure.

For certain compounds/agents there are direct-reading instruments that can assist when there is a need to have either immediate results or data about changes in exposure during a short period of time. Those methods that require collection and subsequent analysis in the laboratory only provide information on the average concentration over the duration of the sampling and fluctuation in time of the concentration and possible unacceptable high peaks of exposure of short duration are missed. See Figure 4.4 for examples on the timing issues of sampling strategy. For compounds that do not present

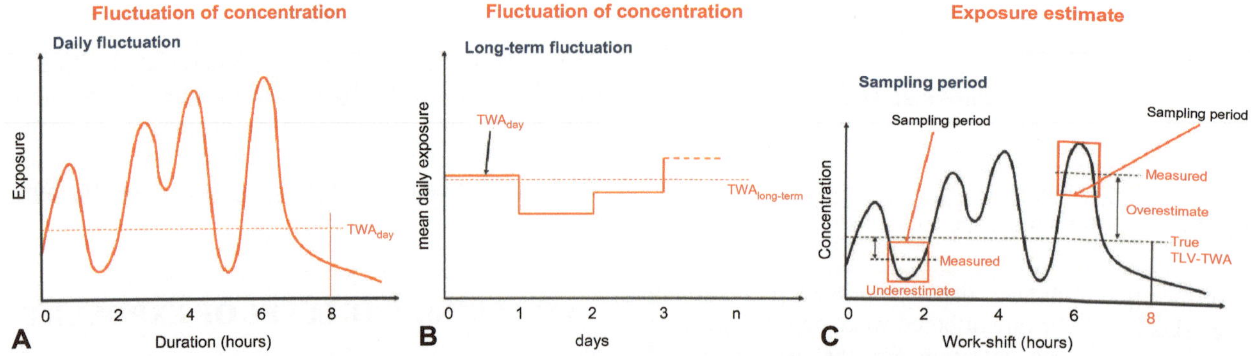

Fig. 4.4. Fluctuations of environmental concentrations in the workplace and issues related to sampling. (**A**) concentrations may vary during the workshift. If sampling is carried out during the whole shift, the resulting concentration will be time-weighted average (TWA, dotted horizontal line) and any information on the within-day variability is lost. (**B**) Concentration (TWA) may vary between workdays. Sampling should be carried out during several days if working conditions are not usually consistent. (**C**) Examples of over/underestimation of concentration (TWA). If there is large within-day variability, reduced sampling times may cause underestimation (left box) or overestimation (right box) of the TWA. TLV, threshold limit value.

significant acute effects, relatively high fluctuations of the concentration during a working day or even week or month are not relevant for the workers exposed. However, if the sampling time only covers part of the work day, there can be significant under/overestimation of the exposure (Fig. 4.4C). Also, given the possibility of day-to-day fluctuations (Fig. 4.4B), the data collected over one working day cannot be taken as the definitive value for that working place. For this reason, sampling strategies of different working days might be required if significant variability is expected (Thorne, 2013).

Workers in charge of the same job may have different exposure levels because of differences in their work tasks, their behavior and work practices, and their location in the plant. Moreover, on average some workers appear to have higher exposure than others (Kromhout et al., 1993; Cherrie and Semple, 2011). In certain workplaces, a number of measurements needs to be done before meaningful conclusions could be drawn on the extent of exposure of a group of workers in that particular workplace.

The availability of direct-reading instruments, which provide an (almost) instantaneous reading of the concentration, represents a great advantage in these situations.

In recent years, as a substitute for the generally expensive and time-consuming monitoring activity, a number of exposure models have been developed making use of the wealth of existing data and workplace information. There is a growing attention to these models and a push to use them. Some of these models are available online where data can be entered and the output is generally provided as an estimated distribution of exposure concentrations generated for the scenario provided. See, e.g. Stoffenmanager (www.stoffenmanager.nl), TRA (www.ecetoc.org/tra), EASE (www.hse.goc.uk/research/rrpdf/rr136.pdf), and the advanced REACH tool (McNally et al., 2014).

EXPOSURE LIMITS

Evaluation of the risk of a worker or of a group of workers is generally based on the comparison between the measured or estimated exposure and the appropriate OEL. As previously described, OELs are generally based on toxicologic data obtained from either experimental studies or studies in humans. OELs are derived by a number of scientific and/or technical bodies and typically have legal status in the country where they were derived. Most industrialized (old and recent) countries have their own regulatory systems, although OELs from different countries are often broadly comparable, especially in the case of old industrialized countries. In the European Union (EU), there are OELs that are legally enforceable and are set by the European Commission upon advice by the Scientific Committee on Occupational Exposure Limits, and implemented in each EU country by national legislation. These limits are believed to "protect workers from chemical risk." In the USA, the Occupational Safety and Health Administration (OSHA) under the Department of Labor sets permissible exposure limits (PELs) that are also legally binding and that employers should comply with. These limits are set on the basis of the best scientific literature in such a way that they assure "to the extent feasible…that no employee will suffer material impairment of health or functional capacity" from exposure "for the period of his working life." The National Institute for Occupational Safety and Health (NIOSH) provides OSHA with recommended exposure limits (RELs) that might be used to set the PELs. RELs are more frequently updated than PELs and are generally more stringent. However, very influential worldwide is the list of limits proposed by the American Conference of Governmental Industrial Hygienists (ACGIH), a nonprofit scientific association. These limits are frequently revised and are developed as guidelines that have no legal value. ACGIH was the first organization to publish a list of limits, known as threshold limit values (TLVs), in the 1940s. Currently the ACGIH provides TLVs for almost 700 substances (www.acgih.org/TLV: data available for a fee), with supporting documentation that was used to set those limits.

Metrics of inhalation exposure

Depending on the type of adverse health effects, different metrics can be used to set the exposure limits (Table 4.1). For compounds that do not cause significant acute effects, following short-term exposures, the OEL/TLV is established in relation to a reference period, that typically is as described above, based on an 8-hour workshift. These limits are defined as TWAs, meaning that it is acceptable for concentrations to fluctuate above and below the limit provided that they average over the workshift. There are, however, substances for which an 8-hour TWA limit provides insufficient protection because, e.g., some effect can occur after a short high exposure or the set TWA limit is not very much lower than exposures at which there might be a risk of short-term effects occurring. In such cases STELs may also be set, usually involving a 15-minute reference period. Such periods are, generally and unless stated differently, allowed four times in a workshift, with a 60-minute interval. The aim of a STEL is therefore to prevent adverse health effects and other unwanted effects, such as impaired alertness or ability to self-rescue, central nervous system depression, irritation, and nuisance, due to peaks in exposure that are not controlled by the application of the 8-hour TWA limit. For substances that would necessitate a STEL over a very short period of time – shorter than 15 minutes – a TLV/OEL-C (ceiling) is set. It should be noted that such C-limits are only meaningful if there are methodologies that provide instantaneous measurements of the compound concentration, since C-limits are intended not to be exceeded any time during the workshift.

Dermal absorption has a significant relative impact on total body burden of certain chemicals, especially when exposure by the inhalation route must be kept at a relatively low level (i.e., for substances that have low OEL/TLV) or when the rate of dermal absorption is high. While this event must be taken into account and the rate of absorption of a substance can be measured or estimated with sufficient approximation, it is not easy, and generally is not done, to measure the extent of skin deposition of a substance. Therefore, for such substances an alert in the form of the so-called "skin notation" is provided, e.g., OEL-TWA S(kin), but no quantification can be provided. This notation is meant to warn of the need to adopt measures to prevent or reduce skin contact. Dermal absorption of gases and vapors is generally of minor importance in relation to inhalation absorption. Similarly, direct skin contact with highly volatile liquids (i.e., those with low boiling point) is not likely to result in appreciable dermal absorption, as the liquid evaporates rapidly from the skin. Solids and liquids with higher boiling temperature and low vapor pressure may cause skin exposure by both direct contact and by impingement of aerosols on the skin. When available, biologic monitoring (see below) is recommended for those substances that have significant dermal exposure and absorption.

BIOLOGIC MONITORING

The analysis of a chemical present in the workplace and/or of its metabolite(s) in a biologic fluid represents the basis for biologic monitoring (Lauwerys and Hoet, 2001; Thorne, 2013). As such, biologic monitoring is not an assessment of exposure in the sense described above, but it is a tool to assess systemic exposure, i.e., to quantify the absorption of the chemical by all possible routes, mainly inhalation, ingestion, and through the skin. An extensive use of the term "biologic monitoring" includes also the measurement of early biochemical changes or other early consequences of systemic exposure, and this is sometimes called "biologic effect monitoring" (Lauwerys and Hoet, 2001; Thorne, 2013): an example of effect monitoring is the measurement of erythrocyte acetycholinesterase activity in workers exposed to organophosphorous compounds (see Chapter 11). While the measurement of the chemical or its metabolite(s) provides assessment that spans from lower, ineffective dose to higher, toxic doses, measurements of early effects only provide information on doses that cause effects, whereas the absence of effect does not provide any quantitative estimation of the internal systemic dose. It only provides some reassurance that the exposure levels are or were below those causing effects, but does not provide any quantification of how much lower they are. In this respect, biologic effect monitoring is more a useful complement of health surveillance of workers than a proper assessment of exposure.

Biologic monitoring can help to assess actual systemic exposure, and is particularly useful when air monitoring alone cannot provide a comprehensive and reliable measurement or estimation of exposure. For instance, this may occur because other routes of exposure, e.g., dermal (e.g., "skin" notation) or oral, are relevant. On the other hand, route-specific exposure data cannot be provided by biologic monitoring. For instance, the relevance of the oral route may be deduced from the work practices and the hygienic conditions of the worker and of the workplace.

Biologic monitoring may also provide an indication of the efficacy and efficiency of personal respiratory protective equipment used in certain conditions where the primary source of pollution and, hence, the air concentration cannot be controlled. In this case, measurements of ambient air concentrations of the substance are not very informative and biologic monitoring may also prove useful to provide feedback to the worker on the proper work practices and adequate use of protective equipment. Biologic monitoring is generally more expensive than environmental monitoring techniques.

When to collect the sample for biologic (effect) monitoring

The toxicokinetic characteristics of the substance determine the optimum time for sample collection; in particular, the half-life of elimination. On the basis of this toxicokinetic characteristic, a value of a specific biologic parameter in a sample collected in specified time during or after exposure can reflect different periods of exposures, from minutes/hours to months/years. Therefore, the ideal sampling time varies according to the compound involved (Table 4.2). During health surveillance of workers, sampling times are often compromised or dictated by the practicalities in the workplace, so that sample collection occurs at the "end of exposure," or at the "end of shift," or prior to the start of the shift ("pre-shift"). In the clinical setting, time of measurement of the biologic monitoring parameters is dictated by the access of the patient to the specialist physician usually driven by the appearance of clinical signs. Table 4.3 reports a summary of the information and relevance for time and duration of exposure for biologic and biologic effect monitoring of the most important neurotoxicants.

Another important determinant for the appropriate sampling time is the route of exposure, particularly for those compounds that have a short half-life. For instance, because of the lag phase and the slower passage through the skin, the apparent half-life of a substance will generally differ if it is inhaled or dermally absorbed.

Interfering factors

The results of biologic monitoring can be influenced by factors not related to occupational exposure. There may be compounds to which workers can be exposed for nonprofessional reasons; the compound can be a ubiquitous contaminant (e.g., mercury and lead), it can be used in domestic settings (e.g., solvents) or because it is related to specific hobbies (e.g., solvents and lead). The compound that is measured for biologic monitoring may be a metabolite of another nonoccupational substance, or even a substance present in the diet or any other common source of exposure. Obviously, this interference is important in the occupational setting because a distinction will need to be made between the contribution of occupational exposure and nonoccupational exposure. On the other hand, in the clinical setting the source of

Table 4.2

Relationship between elimination half-lives, time of sampling, and exposure duration that the sample reflects

Elimination half-life (hours)	Sampling time	Exposure time-frame reflected by the sample
<2	End of exposure. The time between exposure and urine voiding is relevant	Large uncertainty due to variable retention time of urine in bladder
2–10	End of exposure, end of shift	For shorter half-life, the sample mostly reflect contribution of exposure of the last few hours before sampling. For longer half-life, the sample mostly reflects the contribution of the whole workshift
10–100	End of exposure at the end of the work week or after several exposures	The contribution of exposure during the previous days is higher with increasing half-life, up to more than 50% for half-life around 100 hours
>100	Sampling time not critical	For longer half-life, the contribution of recent exposure becomes irrelevant while the contribution of exposure during the previous weeks or even months/years becomes predominant

Useful parameters for biologic and biologic effect monitoring of compounds associated with neurologic diseases

Compound	Parameter		Normal values (NV) or BEI/BLV	Timeframe	Notes
	Biologic monitoring	Biologic effect monitoring			
Acrylamide	*N*-(2-carbamoyl ethyl) valine (AAVal)		NV (95th percentile of European population): 80 pmol/g globin (SCOEL)	It provides information on cumulative exposure over several weeks	Non-smokers. Smokers are expected to have 3–4 times higher values. Acrylamide air levels higher than ~10–15 μg/m^3 have been shown to result in AAVal hemoglobin adduct levels higher than background levels in normal nonsmoking populations
Arsenic	Arsenic plus methylated arsenic in urine		BEI 35 μg/L (ACGIH)	End of shift	
Carbon disulfide	2-thio-thiazolidine-4-carboxylic acid (TTCA) in urine		BLV 1.5 mg/g creatinine (SCOEL) BEI 0.5 mg/g creatinine (ACGIH)	End of shift	
Carbon monoxide		Carboxyhemoglobin	NV 3% (SCOEL) BEI 3.5% (ACGIH)	End of exposure End of shift	Smokers have higher levels. Short half-life
	Exhaled air carbon monoxide		20 ppm (ACGIH)	End of shift	
n-Hexane	2,5-hexanedione in urine		BEI 0.4 mg/L (ACGIH)	End of shift at end of work week	
Lead	Lead in blood		NV <100 μg/L BEI 400 μg/L (ACGIH); BLV 300 μg/L (SCOEL)	Within weeks of last exposure	Bound to erythrocytes, best related to somewhat recent exposure
	Lead in urine		NV <50 μg/g creatinine		Not very informative
	Lead in urine after chelation		NV <150 μg/24 h	Any time	Amount and correlation with blood levels depend on length of exposure
		Free erythrocyte protoporphyrins/zinc protoporphyrin	NV <800/400 μg/L	Any time	Correlates with blood level; increases from blood level >350 μg/L

Continued

Table 4.3

Continued

Compound	Parameter		Normal values (NV) or BEI/BLV	Timeframe	Notes
	Biologic monitoring	Biologic effect monitoring			
		Δ-aminolevulinic acid dehydratase in erythrocytes			Inhibited at <100 μg/L blood lead
Manganese	Manganese in blood and urine		NV: Blood 4–15 μg/L Urine 1–8 μg/L		Poor correlation between urinary/ blood manganese levels and toxic effects*
Mercury (elemental and inorganic divalent compounds)	Mercury in blood		BLV 10 μg/L (SCOEL)	–	
			BEI 15 μg/L (ACGIH)	End of shift at end of work week	
	Mercury in urine		BLV 30 μg/g creatinine (SCOEL)	–	
			BEI 35 μg/g creatinine (ACGIH)	Prior to shift	
Methanol	Methanol in urine		BEI 15 mg/L (ACGIH)	End of shift	
Organophosphorous compounds		Acetylcholinesterase activity in erythrocytes	BEI 70% of baseline (ACGIH)	End of shift	
Phenol	Phenol in urine		BLV 120 mg/g creatinine (SCOEL)	End of shift	Nonoccupational exposure, dietary intake
			BEI 250 mg/g creatinine (ACGIH)	End of shift	Short half-life
Styrene	Mandelic acid plus phenylglyoxylic acid in urine		BEI 400 mg/g creatinine (ACGIH)	End of shift prior to last shift of work week	
Toluene	Toluene in blood		BEI 0.02 mg/L (ACGIH)	End of shift	
	Toluene in urine		BEI 0.03 mg/L (ACGIH)	End of shift	
	o-cresol in urine		BEI 0.3 mg/g creatinine (ACGIH)	End of shift	

BEI, biologic exposure index; BLV, biologic limit value; SCOEL, Scientific Committee on Occupational Exposure Limits; ACGIH, American Conference of Governmental Industrial Hygienists.
*See, e.g., Lander et al. (1999).

exposure is irrelevant except for the possible social and legal consequences of identifying an occupational exposure as responsible for a disease.

It should also be noted that, when a metabolite is measured, it might be the inactive form of the parent: in this case, its presence due to another source and different compound may not be associated with the toxic effect caused by the parent. For this reason, the collection of additional information with the subject by a colloquium, typically in the clinical setting, or a questionnaire, sometimes possible in the workplace, is necessary. For instance, total urinary mercury and total arsenic levels may increase after consumption of seafood and occupational exposures; more complex chemical analysis with speciation of the metals contributing to total urinary excretion can help attribute exposure to occupational and dietary sources, or their relative contribution. A compound used as a biomarker may also be formed endogenously: in this case preexposure or baseline samples need to be collected, and information on circadian rhythm should also be known.

Physiologic and personal conditions can also modify the excretion of the biomarker: for instance, acute intake of alcohol or of some drugs can be associated with competition for metabolism of many solvents, resulting in a prolongation of their half-life of excretion. On the other hand, enzymes potentially involved in metabolism of some solvents and chemicals are induced after chronic intake of alcohol or of some drugs, and this also modifies the relationship between excretion of the biomarker and exposure. Among the physiologic factors, work rate, with associated increased respiration, leads to increased absorption and consequently increased systemic dose, e.g., moderately hard work may cause up to three times the increase in respiration rate and, hence, of urinary metabolite levels compared to a worker at rest with the same airborne levels of the chemical.

Creatinine correction

Renal excretion of compounds is related to plasma concentration of the compound and urine concentrations depends on the volume of urine excreted, which varies because of differences in fluid intake and insensible losses, such as sweating, which can alter urinary flow rate over a wide physiologic range. For practical reasons, most biologic monitoring samples are collected as a spot either right before or after the workshift and therefore adjustments to the analytical results are needed to compensate for short-term concentration or dilution effects. Commonly, adjustments are based either on urine specific gravity or creatinine concentration, both of which provide a good approximation for the rate of urine production. This is not always true; for instance, when a chemical, e.g., methanol, simply equilibrates between intra- and extracellular fluids, including urine. Also, corrections for extreme urine dilution or concentration, e.g., creatinine concentration is less than 3 mmol/L or more than 30 mmol/L, are subject to considerable error and it is recommended not to use these measurements.

Biologic effect monitoring

Measurements of parameters that reflect the interaction between the chemical, or its active metabolite(s), and the organisms do not generally provide an estimation of the absorbed dose; this process is called biologic effect monitoring. Table 4.3 reports some of these parameters that can be used in the occupational setting or for clinical evaluation. When a marker of biologic effect appears or is increased above background it provides an indication of an increased risk, or early warning of an incipient specific poor health outcome in an individual. Also for these parameters the data should be evaluated against the whole toxicologic database if they are to be meaningfully used in the clinical setting.

In health surveillance in the workplace, biologic effect measurements can be useful because they can provide a grading of health risks, and in certain instances they can be applied to the individual worker, allowing the identification of incorrect individual work practices or physiologic, genetic, or other predisposing factors. Various international organizations propose biologic monitoring guidance values (Table 4.3).

Biologic limits

Similarly to exposure limits, some limits have also been set for biologic monitoring parameters, called biological exposure indices (BEI) in the USA, and biological limit values (BLVs) in EU. These are guideline values for biologic monitoring and, according to a widely accepted ACGIH definition, they represent the level "most likely to be observed in specimens collected from healthy workers who have been exposed to chemicals to the same extent, as workers with inhalation exposure at the TLV." As for acceptable OELs/OELs, they correspond to exposure conditions considered to pose acceptable risk to workers.

How to use and interpret the results

The biologic limits are most frequently derived from the relationship between the biologic parameter and the airborne concentration limit that is itself "health-based." In a few other instances, they are directly derived from the dose–response relationship between (the absence of) health effects and the biologic parameter being measured. Rarely, there is not enough information to allow the derivation of such health-based values, and therefore values are proposed that are based on or derived from good occupational hygiene practice.

Given the modality of derivation of such guidance values, measurements that exceed those values cannot be directly linked to or predictive of ill health. In the health surveillance setting in the workplace, such exceedances indicate the need for an investigation of exposure control systems that probably need to be improved. In the clinical setting careful consideration should be given to the entire database available for the substance in order to understand the real toxicologic meaning of those measurements. As indicated by the ACGIH and many other bodies, these limits are not fine lines between health and disease. Besides available data from regulatory organizations, there may be data in the literature that indicate a possible relation between certain biologic monitoring parameters. These have not (yet) been incorporated into BEI/TLVs for various reasons, but they can be evaluated in the specific situation and expert judgment can be applied to extrapolate information useful for exposure assessment in an attempt to find an etiologic attribution.

EXPOSURE ASSESSMENT IN THE CLINICAL SETTING

Assessment of exposure is a fundamental step for both proper health surveillance in the workplace and etiologic attribution of disease or exclusion of any causative role of occupation. Most activities previously described refer to groups of workers and in only a few cases the worker is individually addressed. However, the tools that are available for exposure assessment can guide the search for the etiologic agent of a disease. A number of points should be underlined in this respect when collecting the work history of a subject. A careful occupational history should identify relevant exposures in association with specific jobs or tasks performed by the individual. Also the source of the substance(s) needs to be identified as clearly as possible, with indication of worker's interaction, the general preventive measures to control the emissions, the time spent near or at distance from the source, the local controls in use, the work environment (i.e., size of the room and general ventilation), the use of personal protective equipment (PPE). Other important information will be that on accidental spillage or accidental high exposures. All this information will be useful to put the quantitative data on exposure assessment in the right frame.

As described, exposure assessment can have different levels of precision. When direct information on the worksite is not available, a surrogate assessment can be derived by a detailed occupational history combined with information that might be available for comparable worksites. There are methods to reconstruct past exposures by, e.g., mathematical modeling based on certain characteristics of the worksite using descriptions and analogous data. A better situation is where actual monitoring data from the worksite are available. In this case, the limitations of those measurements should not be underestimated to avoid undue conclusions on etiology. Measured concentrations in the air should be evaluated regarding their representativeness of the working situation. It should be considered whether there are enough data, repeated over time, to provide a reliable estimate of the "average" ambient concentration. Moreover, it should be understood if the measurements were aimed at identifying sources of release of the substance rather than determining the general conditions of the workplace.

Consideration should also be given to the use of PPE. Proper use of PPE reduces, sometimes to a significant extent, the actual exposure of workers. On the other hand it is also known that the use of PPE tends to reduce the care in respecting proper working practices and workers may behave less carefully, thereby reducing the real effect of PPE. If available, biologic monitoring data often provide more accurate data on exposure; in fact, the levels can be lower than expected based on ambient air concentration. On the other hand, as described above, biologic monitoring integrates all routes of exposure, such as oral and dermal, that cannot be properly assessed as external exposure.

When interpreting both ambient and biologic monitoring, and comparing the data with existing limit values, it is important to remember that the primary purpose of those limits, as pointed out by the ACGIH in the introduction to the list of TLV and BEI, is "intended for use in the practice of industrial hygiene as guidelines or recommendations in the control of potential workplace health hazards and for no other use e.g. … as proof or disproof of an existing disease … these limits are not fine lines between safe and dangerous concentrations." In fact, because of the way they are derived, values just above the limits do not mean that a disease or an adverse health effect will ensue. Generally, the margin between the set limit value and the dose/concentration at which health effects are expected is wide enough that they cannot be exceeded. Therefore, in the clinical setting the limits will serve as a guide for further evaluations; values below the limit can be generally regarded as safe and hence not indicative of a causal role in a disease. Values above the limits should be considered in conjunction with all other information to judge on the likelihood that the disease under investigation may have been caused by exposure to the substance. This will depend on, e.g., the nature of the toxic effect, the amount by which the exposure exceeded the limit, and the characteristics of the individual worker. A partial exception is represented by the comparatively few biologic effect limit values that, in fact, indicate an interaction of the substance with the organism.

Moreover, exposure does not necessarily stop at the workplace. Exposure can be "taken home" when contaminated clothing is brought back home; in such instances exposure continues for the worker and may also involve other family members. Cases have been described of spouses and children of workers dealing with asbestos who suffered long-term health effects even if never directly occupationally exposed. This may conceivably occur also with other agents besides asbestos, such as metals, pesticides, dust, and animal allergens. In addition, workers may be exposed when performing the same job/task while engaged in those activities as self-employed work, e.g., during evenings or weekends. The physician should always investigate the possibility that such exposures are higher than those experienced at the workplace. Diet can be a significant source of intake of compounds that are also present in the workplace. For instance, old domestic lead piping can cause high lead intake, comparable or even higher than those occurring in a well-controlled workplace. Also, offshore workers may have a high dietary consumption of shellfish that are known to contain mercury. In this case, measurement of urinary mercury excretion may provide little information

CONCLUSIONS

Defining a causal exposure in an individual is important for both the clinical management of the individual and for social and legal reasons (e.g., there might be compensation issues or civil or criminal court proceedings). Exposure assessment is fundamental in this definition, because it allows determination of the dose. This definition should be quantitative to the extent possible, but tools are also available for a semiquantitative estimation. In any case, expert judgment will be needed for careful interpretation of the data. A dialog between the neurologist and the expert in occupational medicine and hygiene is necessary.

References

Cherrie JW, Semple S (2011). Occupational exposure to hazardous substances. In: PJ Baxter, T-C Aw, A Cockcroft et al. (Eds.), Hunter's Diseases of Occupations, CRC Press, Boca Raton, pp. 43–55.

Faustman EM, Omenn GS (2013). Risk assessment. In: CD Klaassen (Ed.), Casarett and Doull's Toxicology. The Basic Science of Poisons, McGraw Hill, New York, pp. 123–149.

Greim H, Snyder R (2008). Introduction to the discipline of toxicology. In: H Greim, R Snyder (Eds.), Toxicology and Risk Assessment. A Comprehensive Introduction, Wiley, Chichester, UK, pp. 1–18.

Kromhout H, Symanski E, Rappaport SM (1993). A comprehensive evaluation of within- and between-worker components of occupational exposure to chemicals. Ann Occup Hyg 37: 253–270.

Lander F, Kristiansen J, Lauritsen JM (1999). Manganese exposure in foundry furnacemen and scrap recycling workers. Int Arch Occup Environ Health 72: 546–550.

Lauwerys RR, Hoet P (2001). Industrial Chemical Exposure. Guidelines for Biological Monitoring. Lewis Publishers, Boca Raton, pp. 1–19.

McNally K, Warren N, Fransman W et al. (2014). Advanced REACH Tool: A Bayesian Model for Occupational Exposure Assessment. Ann Occup Hyg 1–15.

Thorne PS (2013). Occupational toxicology. In: CD Klaassen (Ed.), Casarett and Doull's Toxicology. The Basic Science of Poisons, McGraw Hill, New York, pp. 1391–1411.

Van Leeuwen CJ (2007). General introduction. In: CJ Van Leeuwen, TG Vermeire (Eds.), Risk Assessment of Chemicals: An Introduction. Springer, Dordrecht, Netherlands, pp. 1–36.

WHO (2005). Chemical-specific adjustment factors for interspecies differences and human variability: Guidance document for use of data in dose/concentration–response assessment. WHO, Geneva.

WHO (2009). Principles and Methods for the Risk Assessment of Chemicals in Food. WHO, Geneva.

Handbook of Clinical Neurology, Vol. 131 (3rd series)
Occupational Neurology
M. Lotti and M.L. Bleecker, Editors

Chapter 5

The mechanisms of neurotoxicity and the selective vulnerability of nervous system sites

LAURA L. MAURER[1] AND MARTIN A. PHILBERT[2]*
[1]*Nicholas School of the Environment, Duke University, Durham, NC, USA*
[2]*Toxicology Program, University of Michigan School of Public Health, Ann Arbor, MI, USA*

INTRODUCTION

The nervous system functions as the coordinating center of mammals. It relies on the taking in of sensory information from highly specialized organs in peripheral tissues, filtering, processing, and relaying of inputs through convergent and divergent, excitatory and inhibitory/modulatory pathways, and finally prioritizing and coordinating appropriate responses to the initial stimulus. This elegant set of functions is achieved, in part, through an ordered and varied array of a wide variety of types and subtypes of neurons, neurotransmitters/neuromodulators, receptors, vasculature, and neuroanatomic relationships that differ substantially by location and function. It is thus that neurotoxicants that affect specific conserved functions express their toxicity by virtue of regional and functional distribution, neurodevelopmental window, or decline in function with age. In all cases, the clinical signs of intoxication align with the vulnerable region.

One of the distinguishing features of neurons is that they are postmitotic. However, the loss of a neuron does not result in loss of a function per se due to the relative functional reserve of the system and the ability of adjacent neurons to reestablish synaptic contacts with surviving neurons through mechanisms of neuronal plasticity within the central nervous system (CNS). They are also made vulnerable by the requirement for large amounts of adenosine triphosphate (ATP) for repolarization of normally functioning axons, the transport of organelles over considerable distances from the neuronal soma to the synaptic terminus, recycling of neurotransmitters and synaptic machinery, and the routine biosynthetic processes that maintain cellular structure. Tonic firing of neuronal subpopulations also makes them singularly sensitive to the effects of energy deprivation whether it is the product of reductions in vascular flow or the direct result of metabolic poisons. Similarly, the destruction of insulating myelin sheaths produced by oligodendrocytes in the CNS and Schwann cells in the PNS results in the unregulated depolarization of internodal segments of axons and the futile use of ATP in repolarizing the denuded axonal membrane. However, repair of CNS injury comes at the cost of diverting energy from the routine maintenance of nerve impulse conduction to the fueling of inflammatory responses (a blunt instrument that itself can induce additional injury) and the formation of a stabilizing, but impenetrable, glial scar. The inherent gross anatomic, regional, microscopic, and biochemical/neurochemical heterogeneity of the nervous system is the substrate on which neurotoxicants produce targeted neuronal and/or glial toxicity confined to specific regions or functionalities.

Localized variability in energy demand in the brain lends itself to regional susceptibility to injury by neurotoxicant exposures that perturb glycolysis or mitochondrial production of ATP. The heavy reliance of neurons on aerobic respiration is directly related to the maintenance of ion gradients to continue nerve impulse conductance. Thus, basal levels of metabolism depend on, and fluctuate in response to, localized neuronal activity. The reliance on constant energy sources renders neurons vulnerable in the absence of metabolic reserves. The brain relies heavily on aerobic glycolysis, does not store glycogen, and is acutely sensitive to even small or temporary interruptions in the supply of glucose and oxygen. Not surprisingly, chemicals causing mitochondrial dysfunction in the brain can severely impair

*Correspondence to: Martin A. Philbert, Ph.D., Professor and Dean, University of Michigan School of Public Health, Toxicology Program, 1415 Washington Heights, Ann Arbor, MI 48109-2029, USA. Tel: +1-734-763-5454, E-mail: philbert@umich.edu

localized neuronal function. Astrocytes function to biochemically and structurally support adjacent neurons, and are equipped to sacrifice their own energy reserves to maintain neuronal function in the event of neuronal energetic deficit as a result of chemical exposure. Because astrocytes have proliferative capacity, this is a desirable response to neurotoxicity; astrocytes can be replaced, whereas neurons are postmitotic and, therefore, are irrevocably lost. There exists much current debate regarding the ability of multipotent subependymal progenitor cells to penetrate the parenchyma and repopulate regions in which chemical injury has resulted in loss of neurons (Sokolowski et al., 2013; Rossman et al., 2014). These rodent models are highly suggestive of latent mechanisms that may ameliorate focal injury. However, the ability of neural progenitor cells to successfully traverse the larger distances in the brains of higher mammals, coupled with the fact that chemicals (especially those that have potency as developmental neurotoxicants) are likely to exert deleterious effects on the progenitor cell itself, makes this a fertile area for basic research.

Outward clinical signs and symptoms of intoxication typically reflect localization of lesions. However, lesion distribution does not always correspond with spatial distribution of the parent compound. Neurotoxicant exposure can result in distribution of the chemical throughout the brain (including regions not presenting with lesions) and detoxification of the compound via reduction mechanisms, as seen in the example of 1,3-dinitrobenzene (DNB) exposure (Romero et al., 1995; Hu et al., 1997, 1999; Reeve et al., 2002). Despite global distribution of the parent compound, DNB exposure results in region-specific lesions (symmetric spongiform lesions in the auditory and vestibular complex) and corresponding clinical signs of intoxication (ataxia, dizziness) (Philbert et al., 1987; Ray et al., 1992). The ability of some chemicals to exert selective regional and/or cellular neurotoxicity therefore depends upon factors beyond mere variations in absorption and distribution throughout the tissue, such as localized temporal and spatial fluctuations in energy requirements, availability of energy substrates, resting metabolic state, and blood flow.

METABOLIC SUBSTRATES FOR VARIABILITY

Significant sex-dependent (Klenke et al., 2014), age- and anatomically related heterogeneities in the cellular and regional distribution of several important metabolic and functional pathways in the nervous system exist (Noda et al., 2002; Schweizer et al., 2013; Yuan et al., 2013; Coumailleau and Kah, 2014; Tsukada et al., 2014), and may contribute to regional susceptibility to neurotoxicant-induced damage. Long-term treatment with the anticonvulsant phenytoin provides at least one relevant example of region-specific neurotoxicity which correlates well with the heterogeneous distribution of metabolic enzymes. Chronic treatment with phenytoin results in ataxia (Botez et al., 1985; Robertson et al., 2013) accompanied by cerebellar degeneration (McLain et al., 1980; Lindvall and Nilsson, 1984). Within the cerebellum, Purkinje cell damage is observed (Volk and Kirchgassner, 1985; Kiefer et al., 1989); recent *in vitro* evidence suggests that phenytoin induces regionally restricted and cell-specific expression of cytochrome P450 2E1 (Boussadia et al., 2014), likely contributing to regional differences in phenytoin metabolism. Differences in the distribution of key enzymes in the nervous system have been proposed as the substrate for selective responses to neurotoxicants (Cleary et al., 2013) and have been used to model the disposition and fate of xenobiotics in a number of neural and nonneural tissues (reviewed by Stepensky, 2013). Several lines of evidence suggest significant cellular compartmentalization and regulation of phase I metabolic enzymes such as cytochrome P450 (reviewed by Xu et al., 2013) in brain. Intuitively, the heterogeneous distribution of xenobiotic activating systems may render specific populations of cells vulnerable to attack. However, similar heterogeneity in the regulation and distribution of phase II and antioxidant systems may compensate for any local increased production of reactive intermediates. Recent evidence points to the interplay between hepatic and cerebral xenobiotic metabolizing systems, further compounding the inherent complexity of regional and cellular metabolism and laying waste to early, oversimplified models of cellular vulnerability merely as a function of the presence or absence of phase I and/or phase II enzyme systems (Montelli et al., 2012; Bonfanti et al., 2014).

CENTRAL NERVOUS SYSTEM

The CNS includes the brain and spinal cord, and is protected by the blood–brain barrier. Neurotoxicant exposures can breach the blood–brain barrier by causing astrocytic foot process swelling, endothelial tight junction compromise, or damage to pericytes, or can diffuse through the membranes if small enough. Local breakdown of the blood–brain barrier is seen in examples of neurotoxicant-induced neurotoxicity and often contributes to regional susceptibility. Metabolic cross-talk between astrocytes and neurons modulates tightness of the blood–brain barrier: mechanisms of neurospecific toxicity cause adjacent astrocytic responses (e.g., astocytic buffering of excess ions in the extracellular space leading to cellular swelling). Conversely, exposures causing mechanisms of toxicity originating in the astrocyte can lead to secondary mechanisms of neurotoxicity through production of toxic metabolites in the astrocyte that are shuttled into the extracellular space, with the potential to exert direct effects on surrounding neurons. Regions of the brain with a higher resting metabolic rate are sometimes more susceptible to neurotoxicants which target mitochondria in either neurons or surrounding glia.

The overwhelming majority of clinical signs related to neurotoxicity are nausea, vomiting, blurred vision, vertigo, mental confusion, and aphasia/dysphasia with attendant loss in sensory function and/or ataxia. The number of chemicals producing these signs of intoxication suggests that there are common manifestations of intoxication with a multiplicity of cellular and biochemical mechanisms. These signs and symptoms are frequently useful in identifying the regional component of injury, but, may have initiating mechanisms in the macroglia, microglia, neuronal somata, axons, synapse, and/or dysfunction of the blood–brain barrier. What follows is a brief description of the major targets of neurotoxic chemicals in the CNS.

Differential regional susceptibility to axonopathy

Oligodendrocytes within the CNS prevent neurite outgrowth, thus preventing axonal repair, by secreting inhibitory factors. This differentiates the CNS response to axonal degeneration from that of the peripheral nervous system (PNS). In the PNS, Schwann cells promote axonal regeneration by forming bands of Bungner and facilitating remyelination. Axonal degeneration in the CNS is irreversible, due in part to astrocyte scarring (Qiu et al., 2000); however, when this detrimental glial effect is removed by transplantation of CNS neurons within the PNS, the neurons from the CNS gain the ability to regenerate neurites.

Axonopathies are those neurotoxic disorders in which the axon is the primary target of neurotoxicity, with involvement of the surrounding myelin and the neuron cell body remaining undamaged. As a function of their length, longer axons have more surface area than shorter axons, and are thus more prone to chemical toxicity than shorter axons. Since lesions arising in axonopathies bear a resemblance to axonal transection, axonal transport is normally disrupted in neurotoxicant-induced axonopathy. As axonal degeneration progresses, the resulting clinical manifestation is peripheral neuropathy, with impaired sensory awareness and motor strength first impaired in the most distal part of the axon, in the extremities. If left untreated, the deficit develops to include areas proximal to the extremities and toward the spinal cord. In cases of axon damage due to neurotoxicant exposure, if the exposure is removed in a timely and complete manner, the potential for axonal regeneration exists. While originally believed to be a PNS-specific phenomenon, the CNS retains the ability to grow processes from existing axons (termed "sprouting") and is a repair mechanism employed after stroke (reviewed by Overman and Carmichael, 2014) and following exposure to neurotoxicants *in vitro* (Leong et al., 2001). The process of sprouting in the CNS can be dampened in aged individuals (reviewed by Crutcher, 2002), implying a reduced ability to effectively recover from exposure to denervating agents in the elderly.

An important intracellular component of axons is the cytoskeleton. Organelles, such as mitochondria, utilize microtubules and neurofilament as primary means of intracellular transport within axons. Microtubules are composed of α-tubulin and β-tubulin heterodimers whose association is exceptionally responsive to phosphorylation, which is the molecular switch for microtubule dynamics (polymerization and depolymerization of the tubulin heterodimer). Neurofilament transport of mitochondria depends on mitochondrial membrane potential and phosphorylation of the neurofilament (Wagner et al., 2003). Exposure to the organophosphate leptophos concurrently with industrial solvents toluene and *n*-hexane results in central peripheral distal axonopathy through a mechanism involving aberrant phosphorylation of neurofilament and microtubule components (Abou-Donia et al., 1988). Administration of the neurotoxic metabolite 2,5-hexanedione (a product of *n*-hexane) to male Wistar rats results in a significant decrease in neurofilament content associated with the eventual development of severe abnormal gait (Song et al., 2007). Early decreases in neurofilament light-chain degradation in the spinal cord may be a precipitating event in 2,5-hexanedione neurotoxicity (Song et al., 2012).

Regional specificity of toxicants targeting myelin in the CNS

Toxicant exposure can initiate damage to myelin within the CNS. Myelin, acting as an insulator for electrical signaling in neurons, can be damaged by toxicant exposure in two ways: intramyelinic edema (separation of myelin) and through demyelination. Though initially separate mechanistic processes, intramyelinic edema can progress to demyelination. If damage to myelin is sufficient or remyelination processes are inadequate, it can ultimately disrupt electric conductance through neuron processes. Because remyelination capabilities within the CNS are limited in comparison to that of the PNS, demyelination of neurons within the CNS can be especially debilitating. Chemical exposures which lead to differential regional demyelination correspond with functional neurologic outcomes which parallel specific sites of myelin loss.

Human exposure to 1-bromopropane, a solvent used to replace ozone-depleting chemicals after regulations were put in place in 1996 in developed countries, causes deficits in motor function and memory (Ichihara et al., 2004; Majersik et al., 2007). Recent mechanistic studies on the hippocampus and cerebellum of rats exposed to 1-bromopropane suggest that memory impairment is caused in part by a decrease in myelin basic protein, and loss of controlled motor function is caused in part by loss of oligodendrocytes with a resultant loss of granular cells and Purkinje cells (Mohideen et al., 2013). The mechanism of the observed neuronal loss is postulated to be due in part to induction of oxidative stress by depletion of glutathione and vulnerability of sulfhydryl groups to oxidation (Wang et al., 2002).

Recent studies implicate emerging roles of region-specific astrocyte dynamics that may contribute to progression of myelinopathies within the CNS. A recently proposed mechanism underlying selective vulnerability to central pontine myelinolysis, an idiopathic condition in which the myelin within the pons is destroyed, includes differential astrocytic thresholds for sodium overload leading to apoptosis (Takeda et al., 2014). Cuprizone, a copper chelator, causes demyelination primarily in corpus callosum of mice, with oligodendrocyte loss most pronounced in the dorsal corpus callosum (Steelman et al., 2012), and could be linked to glutamate excitotoxicity mediated in part by astrocytes (Azami Tameh et al., 2013). Glial-specific mechanisms in cuprizone-induced neurotoxicity likely involve a combination of demyelination and incomplete or insufficient remyelination. Despite myelin existing throughout the CNS and PNS, a range of neurotoxicants produce regionally specific myelinopathies resulting in functional neurologic outcomes.

Regional specificity of toxicants targeting astrocytes in the CNS prior to neuronal involvement

Relatively recently, astrocytes have been recognized for providing a wide range of physiologic functions in the CNS. These physiologic roles are in addition to their canonic function of providing physical support to adjacent neurons. Astrocytic functional perturbations often result in aberrations in neuronal physiology, despite an accompanying change in neuronal morphology. By serving as a spatial buffering system for ions in the extracellular space and actively metabolizing both endogenous molecules and xenobiotics, astrocytes serve as a means of defense during neurotoxicant exposure. Despite the presence of astrocytes throughout the CNS, neurotoxicant exposure primarily affecting astrocytic function does not often result in homogeneous distribution of astrocytic damage and resultant neuronal dysfunction, nor are the clinical manifestations diffuse.

The dinitrobenzenes are industrial intermediates in the production of dyes, plastics, and explosives. Exposure to DNB results in gliovascular lesions with astrocytes as a primary cellular target in the brainstem and deep cerebellar roof nuclei (Philbert et al., 1987). While the precise molecular mechanisms by which the regional sensitivity of astrocytes to DNB exposure is not yet elucidated, it may be due at least in part to bioactivation of DNB by NADPH-dependent cytochrome c reductase (Romero et al., 1991; Hu et al., 1997), with resultant oxidative stress proposed as part of the mechanism of toxicity (Ray et al., 1992, 1994; Romero et al., 1995; Hu et al., 1999). Brainstem nuclei in which lesions appear have a high demand for glucose, but this demand is parallel to unaffected regions, such as forebrain and mesencephalic structures (Bagley et al., 1989; Mastrogiacomo et al., 1993; Calingasan et al., 1994). The selective vulnerability of brainstem astrocytes to DNB is not fully elucidated, but a growing body of evidence suggests region-specific differences in mitochondrial-specific mechanisms (respiratory capacity, induction of the mitochondrial permeability transition pore (mtPTP)). *In vitro* studies suggest that DNB induces the mtPTP in C6 glioma cells (Tjalkens et al., 2000). Metronidazole, a nitro compound used for its antimicrobial properties, can exert neurotoxic effects. Prolonged use of metronidazole as a treatment for infections is associated with peripheral neuropathy characterized by paresthesias, dysesthesias, and pruritus, in addition to other somatosensory disorders. Sensorimotor deficits caused by prolonged use of metronidazole can be irreversible (Kapoor et al., 1999), and can result in irreversible encephalopathy, leading to death (Groothoff et al., 2010); higher doses of metronidazole for extended periods of time cause epileptiform seizures, hallucinations, and attendant encephalopathy. Metronidazole metabolism to its toxic hydroxylamine intermediate mediates its mechanism of toxicity, as the hydroxylamine intermediate readily binds to a wide range of cellular macromolecules (Coxon and Pallis, 1976). Lesions occur, similar in pattern to those induced by DNB toxicity, but with equal susceptibility of both neurons and astrocytes (Schentag et al., 1982), as opposed to the observance of primary involvement of astrocytes in DNB exposure. Regionally specific development of lesions by metronidazole is due at least in part to the structural resemblance of the compound itself and its reactive metabolite with that of thiamine. Thiamine triphosphate, also known as vitamin B_1, is required for mitochondria to metabolize α-ketoglutarate and pyruvate. Thus, the primary mechanism of neurotoxicity exerted by metronidazole and pyrithiamine is likely due to a common antimetabolite activity based on similarity of lesions (Evans et al., 1975; Watanabe and Kanabe, 1978; Kapoor et al., 1999).

Astrocytes within areas other than the brainstem are also uniquely susceptible to certain neurotoxicant exposures. Exposure to methionine sulfoximine results in irreversible inhibition of glutamine synthetase. Methionine sulfoximine causes region-specific changes to both neurons and astrocytes in the hippocampal fascia dentata and pyramidal cell layer, in the lower layers of the cerebral cortex, and in cerebellar Purkinje cells. Increased astrocytic glycogen levels (Phelps, 1975) and swollen and damaged astrocytic mitochondria (Hevor, 1994) accompany neuronal loss in these regions. Inhibition of glutamine synthetase in astrocytes may be a mechanistic link between regionally specific lesions and the generation of seizures due to methionine sulfoximine exposure; however, this has yet to be definitively proven. It is postulated that, because glutamine is a precursor for both glutamate and γ-aminobutyric acid, the irreversible inhibition of glutamine synthetase may cause a severe enough depletion of glutamine to cause seizures.

Microglial activation in response to neurotoxicants

Microglia are the resident macrophages in the brain and spinal cord. Microglia have multiple functions, ranging from phagocytosis to release of cytokines; the balance of normal phagocytic responses and release of proinflammatory cytokines can have differential spatial consequences for neuronal survival in chemical exposures. And while microglia often work in concert with astrocytes to maintain the local neuronal environment, differential astrocyte and microglial responses to toxicants can determine regional susceptibility. For example, injection of 1-methyl-4-phenyl-1,2,3,6-tetrahydropyridine (MPTP) to rhesus monkeys resulted in microglial-mediated inflammation in midbrain ipsilateral to the injection, whereas there was no correlation with development of dopaminergic neuron degeneration and astrocytic response; it should be noted that these responses also correlated with the age of the animal (Kanaan et al., 2008).

Also governing the tolerance of the aged animal to detrimental effects of neurotoxicants are dietary measures, including caloric restriction. Caloric restriction has been shown to extend the lifespan of laboratory animals (reviewed by Walsh et al., 2014), with the potential for translation to multiple benefits in humans (reviewed by Robertson and Mitchell, 2013). Recent evidence suggests that caloric restriction reduces microglial activation in response to lipopolysaccharide, and that it was highly region-specific: microglial activation was significantly reduced in arcuate nucleus, ventromedial nucleus of the hypothalamus, and subfornical organ in animals that underwent caloric restriction (Radler et al., 2014). Thus, regionally differential mechanisms governing microglial activation and related responses (proinflammatory, anti-inflammatory) may also underlie regional susceptibility to neurotoxicant exposure.

Blood–brain barrier susceptibility to neurotoxicants

While it is generally understood that the presence of a blood–brain barrier presents a physical barrier to any hydrophilic moiety that does not have a channel, receptor, or transporter, it also presents a formidable biochemical barrier between the blood and CNS parenchyma (reviewed by Palmer and Alavijeh, 2013). Similarly, the choroid plexus provides both physical and biochemical barriers to hydrophilic moieties that might otherwise gain access to the cerebrospinal fluid (Janssen et al., 2013). The advent of nanomedicine and the widespread ability to create monodisperse preparation of micelles and other nanostructures has opened up new strategies for the targeting of poorly soluble drugs and hormones to the CNS, circumventing both the BBB and choroid plexus (Zheng et al., 2014). While it is unlikely that ambient or engineered nanomaterials will gain access to the systemic circulation in sufficient quantities to allow for acute neurotoxicity, the hijacking of natural pathways for the transport of small particles (micropinocytosis and endocytosis, for example) poses interesting possibilities for the chronic accumulation of materials in the CNS (Vilella et al., 2014). The use of preexisting portals for entry of nanomaterials into epithelial cells has been shown in other organ systems (Latunde-Dada et al., 2014; Pereira et al., 2014). Nevertheless, in brain, the presence of anatomic and biochemical barriers to access, the heavy reliance of neural tissues on energy, considerations of heterogeneity in neurotransmitter systems and metabolic pathways all conspire to yield a picture of vulnerability that will likely require a systems approach to untangle in such a way as to yield predictive models of regional/cellular susceptibility.

Despite the blood–brain barrier comprising a biochemical and physical barrier to neurotoxicants, a number of mechanisms of toxicity involve permeabilization of the blood–brain barrier. Regional permeability of the blood–brain barrier is highly dependent upon the integrity of local populations of endothelial cells, pericytes, and astrocytes. Certain neurotoxicants, like methylmercury, exert neurotoxic effects in part by accumulating in the dorsal root ganglion. The cell body region of the dorsal root ganglion contains relatively dense (Jimenez-Andrade et al., 2008) and permeable capillaries (Hirakawa et al., 2004; Abram et al., 2006), which partially explains sensory clinical signs of intoxication caused by methylmercury toxicity. Recent evidence suggests that astrocytic laminin is necessary for the expression of claudin-5 and occludin (proteins essential for the maintenance of endothelial tight junctions) in capillaries and prevents pericyte differentiation involved in increased blood–brain barrier permeability using a conditional laminin knockout mouse model (Yao et al., 2014).

PERIPHERAL NERVOUS SYSTEM

The PNS consists of nerves outside the brain and spinal cord, including the 12 cranial nerves and motor nerves, providing sensory information to the CNS from the limbs and extremities. There are fundamental differences in structure and function between the CNS and PNS, which renders the PNS uniquely susceptible to chemical and traumatic injury, but also uniquely equipped to cope with such injury. Specialized cell types within the PNS perform tasks specifically to maintain neuronal function in instances of neurotoxicant exposure, such as Schwann cells.

Axonal repair in the PNS

Schwann cells in the PNS are responsible for remediating axonal loss by decreasing myelin lipid synthesis, downregulating myelin protein-encoding genes, and reverting back to a premyelinating mitotic Schwann cell phenotype (Stoll and Muller, 1999). Proliferating Schwann cells provide physical support for regenerating axons by forming bands of Bungner in the original basal lamina. The band of Bungner also provides trophic support from multiple growth factors (brain-derived nerve growth factor, nerve growth factor, insulin-like growth factor). Myelin breakdown also triggers recruitment of macrophages (both resident and hematogenous), aiding in the dissolution of damaged myelin through expression of complement receptor 3, which recognizes complement 3 on the surface of myelin sheaths. Schwann cells secrete nerve growth factor in response to macrophage-specific release of interleukin-1, further supporting the reestablishment of healthy myelin. However, if axonal contact is not reestablished, Schwann cells eventually degenerate and collagen production from fibroblasts results in insufficient regeneration processes. Like the CNS, the PNS retains the ability to grow new processes from existing axons following axonal injury. Exposure to neurotoxicants, such as acrylamide (DeGrandchamp et al., 1990), 2,5-hexanedione (Simonati and Cavanagh, 1984), and botulinum toxin (Angaut-Petit et al., 1990), induces terminal sprouting at the neuromuscular junction. Age influences the ability of the PNS to undergo sprouting, not unlike the CNS (reviewed by Verdu et al., 2000). In both the CNS and PNS, it appears that aging causes a diminished ability to undergo sprouting, not necessarily that it is a completely inactive process.

Recent evidence suggests a role for extracellular-regulated kinases 1 and 2 in governing axonal regeneration by determining the appropriate thickness of myelin in axon development; although this mechanism is active in oligodendrocytes in the CNS (Ishii et al., 2012; Fyffe-Maricich et al., 2013), it is just as integral a developmental myelination mechanism by Schwann cells in the PNS (Ishii et al., 2013). This is a potentially important mechanism for regeneration of myelin in the event of exposure to a demyelinating agent targeting the PNS, as ERK signaling reverts Schwann cells back to a progenitor state in a reversible model of demyelination *in vivo* (Napoli et al., 2012). While ERK signaling itself has multiple functions, it appears to be relatively conserved between the CNS and PNS with regard to remyelination. However, the entire mechanism may be fundamentally different between the two regions of the nervous system, as neuramidinase-3 is required for ERK signaling in axons within the PNS, but not the CNS (Kappagantula et al., 2014). Thus, calcium-signaling mechanisms active in PNS and CNS that are important for myelin repair, but governed by different enzymes, may underlie regional differences in myelin recovery and regeneration.

Chemically induced axonopathies in the PNS

Axonopathies in the PNS can affect either sensory or motor neuron function (or both). Sensory and motor neuron axonopathies result from exposure to a range of compounds, including 3,3'-iminodipropionitrile (Soler-Martin et al., 2012), 4-hydroxy-2-nonenal (Akude et al., 2010), and 1,2-diacetylbenzene (Sabri et al., 2007). 3,3'-Iminodipropionitrile (IDPN) exposure results in lesions and symptoms mirroring that observed in cases of amyotrophic lateral sclerosis (Carpenter, 1968). The chief manifestation of toxicity in the development of IDPN-induced proximal axonopathy is axonal atrophy, caused by accumulation of neurofilaments in swollen proximal axons associated with decreased neurofilament within the neuromuscular junction (Soler-Martin et al., 2012). Accumulation of neurofilaments in the motor neuron axon could be due to impaired energy production, as IDPN exposure decreases ATP synthesis in isolated brain mitochondria (Sickles et al., 1990). Exposure to 4-hydroxy-2-nonenal mechanisms of toxicity may also be related to aberrant responses to energy demands, as it causes disturbances to axonal mitochondrial transport and the development of protein adducts on mitochondria and the cytoskeleton (Akude et al., 2010). 1,2-Diacetylbenzene causes differential expression of proteins involved in energy metabolism and protein folding (Tshala-Katumbay et al., 2008), likely contributing to the resultant pathology (neurofilament disorganization, accumulation of organelles).

Acrylamide exposure can occur occupationally during mining and in paper manufacturing. In examples of high acute occupational exposures, peripheral neuropathies result (Garland and Patterson, 1967; Myers and Macun, 1991). While repeated, low-dose exposures of acrylamide in rats mimic the effects seen in high levels of acute occupational exposures in humans (Crofton et al., 1996), chronic low-dose occupational exposures in humans have been associated with nonneural outcomes, including pancreatic cancer (Swaen et al., 2007). Similar to other neurotoxicants that cause peripheral axonopathy, acrylamide exposure disrupts cytoskeletal components of the axon (Gupta and Abou-Donia, 1997).

CONCLUDING REMARKS

By virtue of its complex cellular composition and dynamic biochemical mechanisms to perform the requisite neural functions, the nervous system (namely the brain) is uniquely susceptible to the action of a range of diverse chemicals. Despite neuroprotective mechanisms provided by supporting cell types and structures such as the blood–brain barrier, chemicals can disrupt neural function. Neurotoxicity is not always distributed uniformly throughout the tissue, and often results in targeted lesions that correspond to outward clinical signs of intoxication. Mechanisms underlying regional susceptibility to chemically induced neurotoxicity can sometimes be explained by differences in the distribution or amount of mixed-cell populations, some lesions affect neuronal populations that are functionally different from unaffected regions, and some neurotoxicants target regionally specific glial populations (i.e., Schwann cells vs astrocytes). However, not all neurotoxic exposures can be explained wholly by cellular makeup and the resultant "division of labor" between and among cell types, or by region. There are multiple neurotoxicant exposures that result in damage by complex molecular mechanisms that may even involve erroneous glial or neuronal responses to such damage, whether it is incorrect, insufficient, or a combination of the two.

References

Abou-Donia MB, Lapadula DM, Suwita E (1988). Cytoskeletal proteins as targets for organophosphorus compound and aliphatic hexacarbon-induced neurotoxicity. Toxicology 49: 469–477.

Abram SE, Yi J, Fuchs A et al. (2006). Permeability of injured and intact peripheral nerves and dorsal root ganglia. Anesthesiology 105: 146–153.

Akude E, Zherebitskaya E, Roy Chowdhury SK et al. (2010). 4-Hydroxy-2-nonenal induces mitochondrial dysfunction and aberrant axonal outgrowth in adult sensory neurons that mimics features of diabetic neuropathy. Neurotox Res 17: 28–38.

Angaut-Petit D, Molgo J, Comella JX et al. (1990). Terminal sprouting in mouse neuromuscular junctions poisoned with botulinum type A toxin: morphological and electrophysiological features. Neuroscience 37: 799–808.

Azami Tameh A, Clarner T, Beyer C et al. (2013). Regional regulation of glutamate signaling during cuprizone-induced demyelination in the brain. Ann Anat 195: 415–423.

Bagley PR, Tucker SP, Nolan C et al. (1989). Anatomical mapping of glucose transporter protein and pyruvate dehydrogenase in rat brain: an immunogold study. Brain Res 499: 214–224.

Bonfanti P, Comelli F, Assi L et al. (2014). Responsiveness of hepatic and cerebral cytochrome P450 in rat offspring prenatally and lactationally exposed to a reconstituted PCB mixture. Environ Toxicol 29: 856–866.

Botez MI, Gravel J, Attig E et al. (1985). Reversible chronic cerebellar ataxia after phenytoin intoxication: possible role of cerebellum in cognitive thought. Neurology 35: 1152–1157.

Boussadia B, Ghosh C, Plaud C et al. (2014). Effect of status epilepticus and antiepileptic drugs on CYP2E1 brain expression. Neuroscience 281C: 124–134.

Calingasan NY, Baker H, Sheu KF et al. (1994). Selective enrichment of cholinergic neurons with the alpha-ketoglutarate dehydrogenase complex in rat brain. Neurosci Lett 168: 209–212.

Carpenter S (1968). Proximal axonal enlargement in motor neuron disease. Neurology 18: 841–851.

Cleary M, Trefz F, Muntau AC et al. (2013). Fluctuations in phenylalanine concentrations in phenylketonuria: a review of possible relationships with outcomes. Mol Genet Metab 110: 418–423.

Coumailleau P, Kah O (2014). Cyp19a1 (aromatase) expression in the *Xenopus* brain at different developmental stages. J Neuroendocrinol 26: 226–236.

Coxon A, Pallis CA (1976). Metronidazole neuropathy. J Neurol Neurosurg Psychiatry 39: 403–405.

Crofton KM, Padilla S, Tilson HA et al. (1996). The impact of dose rate on the neurotoxicity of acrylamide: the interaction of administered dose, target tissue concentrations, tissue damage, and functional effects. Toxicol Appl Pharmacol 139: 163–176.

Crutcher KA (2002). Aging and neuronal plasticity: lessons from a model. Auton Neurosci 96: 25–32.

Degrandchamp RL, Reuhl KR, Lowndes HE (1990). Synaptic terminal degeneration and remodeling at the rat neuromuscular junction resulting from a single exposure to acrylamide. Toxicol Appl Pharmacol 105: 422–433.

Evans WC, Evans IA, Humphreys DJ et al. (1975). Induction of thiamine deficiency in sheep, with lesions similar to those of cerebrocortical necrosis. J Comp Pathol 85: 253–267.

Fyffe-Maricich SL, Schott A, Karl M et al. (2013). Signaling through ERK1/2 controls myelin thickness during myelin repair in the adult central nervous system. J Neurosci 33: 18402–18408.

Garland TO, Patterson MW (1967). Six cases of acrylamide poisoning. Br Med J 4: 134–138.

Groothoff MV, Hofmeijer J, Sikma MA et al. (2010). Irreversible encephalopathy after treatment with high-dose intravenous metronidazole. Clin Ther 32: 60–64.

Gupta RP, Abou-Donia MB (1997). Acrylamide and carbon disulfide treatments increase the rate of rat brain tubulin polymerization. Mol Chem Neuropathol 30: 223–237.

Hevor TK (1994). Some aspects of carbohydrate metabolism in the brain. Biochimie 76: 111–120.

Hirakawa H, Okajima S, Nagaoka T et al. (2004). Regional differences in blood-nerve barrier function and tight-junction protein expression within the rat dorsal root ganglion. Neuroreport 15: 405–408.

Hu HL, Bennett N, Lamb JH et al. (1997). Capacity of rat brain to metabolize m-dinitrobenzene: an in vitro study. Neurotoxicology 18: 363–370.

Hu HL, Bennett N, Holton JL (1999). Glutathione depletion increases brain susceptibility to m-dinitrobenzene neurotoxicity. Neurotoxicology 20: 83–90.

Ichihara G, Li W, Shibata E et al. (2004). Neurologic abnormalities in workers of a 1-bromopropane factory. Environ Health Perspect 112: 1319–1325.

Ishii A, Fyffe-Maricich SL, Furusho M et al. (2012). ERK1/ERK2 MAPK signaling is required to increase myelin thickness independent of oligodendrocyte differentiation and initiation of myelination. J Neurosci 32: 8855–8864.

Ishii A, Furusho M, Bansal R (2013). Sustained activation of ERK1/2 MAPK in oligodendrocytes and Schwann cells enhances myelin growth and stimulates oligodendrocyte progenitor expansion. J Neurosci 33: 175–186.

Janssen SF, Van Der Spek SJ, Ten Brink JB et al. (2013). Gene expression and functional annotation of the human and mouse choroid plexus epithelium. PLoS One 8: e83345.

Jimenez-Andrade JM, Herrera MB, Ghilardi JR et al. (2008). Vascularization of the dorsal root ganglia and peripheral nerve of the mouse: implications for chemical-induced peripheral sensory neuropathies. Mol Pain 4: 10.

Kanaan NM, Kordower JH, Collier TJ (2008). Age and region-specific responses of microglia, but not astrocytes, suggest a role in selective vulnerability of dopamine neurons after 1-methyl-4-phenyl-1,2,3,6-tetrahydropyridine exposure in monkeys. Glia 56: 1199–1214.

Kapoor K, Chandra M, Nag D et al. (1999). Evaluation of metronidazole toxicity: a prospective study. Int J Clin Pharmacol Res 19: 83–88.

Kappagantula S, Andrews MR, Cheah M et al. (2014). Neu3 sialidase-mediated ganglioside conversion is necessary for axon regeneration and is blocked in CNS axons. J Neurosci 34: 2477–2492.

Kiefer R, Knoth R, Anagnostopoulos J et al. (1989). Cerebellar injury due to phenytoin. Identification and evolution of Purkinje cell axonal swellings in deep cerebellar nuclei of mice. Acta Neuropathol 77: 289–298.

Klenke U, Taylor-Burds C, Wray S (2014). Metabolic influences on reproduction: adiponectin attenuates GnRH neuronal activity in female mice. Endocrinology 155: 1851–1863.

Latunde-Dada GO, Pereira D, Tempest B et al. (2014). A nanoparticulate ferritin-core mimetic is well taken up by HuTu 80 duodenal cells and its absorption in mice is regulated by body iron. J Nutr 144: 1896–1902.

Leong CC, Syed NI, Lorscheider FL (2001). Retrograde degeneration of neurite membrane structural integrity of nerve growth cones following in vitro exposure to mercury. Neuroreport 12: 733–737.

Lindvall O, Nilsson B (1984). Cerebellar atrophy following phenytoin intoxication. Ann Neurol 16: 258–260.

Majersik JJ, Caravati EM, Steffens JD (2007). Severe neurotoxicity associated with exposure to the solvent 1-bromopropane (n-propyl bromide). Clin Toxicol (Phila.) 45: 270–276.

Mastrogiacomo F, Bergeron C, Kish SJ (1993). Brain alpha-ketoglutarate dehydrogenase complex activity in Alzheimer's disease. J Neurochem 61: 2007–2014.

Mclain Jr LW, Martin JT, Allen JH (1980). Cerebellar degeneration due to chronic phenytoin therapy. Ann Neurol 7: 18–23.

Mohideen SS, Ichihara S, Subramanian K et al. (2013). Effects of exposure to 1-bromopropane on astrocytes and oligodendrocytes in rat brain. J Occup Health 55: 29–38.

Montelli S, Peruffo A, Zambenedetti P et al. (2012). Expression of aromatase P450(AROM) in the human fetal and early postnatal cerebral cortex. Brain Res 1475: 11–18.

Myers JE, Macun I (1991). Acrylamide neuropathy in a South African factory: an epidemiologic investigation. Am J Ind Med 19: 487–493.

Napoli I, Noon LA, Ribeiro S et al. (2012). A central role for the ERK-signaling pathway in controlling Schwann cell plasticity and peripheral nerve regeneration in vivo. Neuron 73: 729–742.

Noda A, Ohba H, Kakiuchi T et al. (2002). Age-related changes in cerebral blood flow and glucose metabolism in conscious rhesus monkeys. Brain Res 936: 76–81.

Overman JJ, Carmichael ST (2014). Plasticity in the injured brain: more than molecules matter. Neuroscientist 20: 15–28.

Palmer AM, Alavijeh MS (2013). Overview of experimental models of the blood–brain barrier in CNS drug discovery. Curr Protoc Pharmacol 62, Unit 7 15.

Pereira DI, Bruggraber SF, Faria N et al. (2014). Nanoparticulate iron(III) oxo-hydroxide delivers safe iron that is well absorbed and utilised in humans. Nanomedicine 10: 1877–1886.

Phelps CH (1975). An ultrastructural study of methionine sulphoximine-induced glycogen accumulation in astrocytes of the mouse cerebral cortex. J Neurocytol 4: 479–490.

Philbert MA, Nolan CC, Cremer JE (1987). 1,3-Dinitrobenzene-induced encephalopathy in rats. Neuropathol Appl Neurobiol 13: 371–389.

Qiu J, Cai D, Filbin MT (2000). Glial inhibition of nerve regeneration in the mature mammalian CNS. Glia 29: 166–174.

Radler ME, Hale MW, Kent S (2014). Calorie restriction attenuates lipopolysaccharide (LPS)-induced microglial activation in discrete regions of the hypothalamus and the subfornical organ. Brain Behav Immun 38: 13–24.

Ray DE, Brown AW, Cavanagh JB et al. (1992). Functional/metabolic modulation of the brain stem lesions caused by 1,3-dinitrobenzene in the rat. Neurotoxicology 13: 379–388.

Ray DE, Abbott NJ, Chan MW et al. (1994). Increased oxidative metabolism and oxidative stress in m-dinitrobenzene neurotoxicity. Biochem Soc Trans 22: 407S.

Reeve IT, Voss JC, Miller MG (2002). 1,3-Dinitrobenzene metabolism and GSH depletion. Chem Res Toxicol 15: 361–366.

Robertson LT, Mitchell JR (2013). Benefits of short-term dietary restriction in mammals. Exp Gerontol 48: 1043–1048.

Robertson K, Von Stempel CB, Arnold I (2013). When less is more: a case of phenytoin toxicity. BMJ 2013, bcr2012008023.

Romero I, Brown AW, Cavanagh JB et al. (1991). Vascular factors in the neurotoxic damage caused by 1,3-dinitrobenzene in the rat. Neuropathol Appl Neurobiol 17: 495–508.

Romero IA, Lister T, Richards HK et al. (1995). Early metabolic changes during m-dinitrobenzene neurotoxicity and the possible role of oxidative stress. Free Radic Biol Med 18: 311–319.

Rossman IT, Lin L, Morgan KM et al. (2014). Engrailed2 modulates cerebellar granule neuron precursor proliferation, differentiation and insulin-like growth factor 1 signaling during postnatal development. Mol Autism 5: 9.

Sabri MI, Hashemi SB, Lasarev MR et al. (2007). Axonopathy-inducing 1,2-diacetylbenzene forms adducts with motor and cytoskeletal proteins required for axonal transport. Neurochem Res 32: 2152–2159.

Schentag JJ, Ziemniak JA, Greco JM et al. (1982). Mental confusion in a patient treated with metronidazole – a concentration-related effect? Pharmacotherapy 2: 384–387.

Schweizer M, Markmann S, Braulke T et al. (2013). Ultrastructural analysis of neuronal and non-neuronal lysosomal storage in mucolipidosis type II knock-in mice. Ultrastruct Pathol 37: 366–372.

Sickles DW, Fowler SR, Testino AR (1990). Effects of neurofilamentous axonopathy-producing neurotoxicants on in vitro production of ATP by brain mitochondria. Brain Res 528: 25–31.

Simonati A, Cavanagh JB (1984). Changes in terminal sprout formation in rat sternocostalis muscle during chronic intoxication with 2,5 hexanedione. Muscle Nerve 7: 355–361.

Sokolowski K, Obiorah M, Robinson K et al. (2013). Neural stem cell apoptosis after low-methylmercury exposures in postnatal hippocampus produce persistent cell loss and adolescent memory deficits. Dev Neurobiol 73: 936–949.

Soler-Martin C, Vilardosa U, Saldana-Ruiz S et al. (2012). Loss of neurofilaments in the neuromuscular junction in a rat model of proximal axonopathy. Neuropathol Appl Neurobiol 38: 61–71.

Song F, Zhang C, Yu S et al. (2007). Time-dependent alteration of cytoskeletal proteins in cerebral cortex of rat during 2,5-hexanedione-induced neuropathy. Neurochem Res 32: 1407–1414.

Song F, Zhang Q, Kou R et al. (2012). 2,5-hexanedione altered the degradation of low-molecular-weight neurofilament in rat nerve tissues. Food Chem Toxicol 50: 4277–4284.

Steelman AJ, Thompson JP, Li J (2012). Demyelination and remyelination in anatomically distinct regions of the corpus callosum following cuprizone intoxication. Neurosci Res 72: 32–42.

Stepensky D (2013). Prediction of drug disposition on the basis of its chemical structure. Clin Pharmacokinet 52: 415–431.

Stoll G, Muller HW (1999). Nerve injury, axonal degeneration and neural regeneration: basic insights. Brain Pathol 9: 313–325.

Swaen GM, Haidar S, Burns CJ et al. (2007). Mortality study update of acrylamide workers. Occup Environ Med 64: 396–401.

Takeda T, Makinodan M, Fukami S et al. (2014). Primary cerebral and cerebellar astrocytes display differential sensitivity to extracellular sodium with significant effects on apoptosis. Cell Biochem Funct 32: 395–400.

Tjalkens RB, Ewing MM, Philbert MA (2000). Differential cellular regulation of the mitochondrial permeability transition in an in vitro model of 1,3-dinitrobenzene-induced encephalopathy. Brain Res 874: 165–177.

Tshala-Katumbay D, Monterroso V, Kayton R et al. (2008). Probing mechanisms of axonopathy. Part I: Protein targets of 1,2-diacetylbenzene, the neurotoxic metabolite of aromatic solvent 1,2-diethylbenzene. Toxicol Sci 105: 134–141.

Tsukada H, Ohba H, Nishiyama S et al. (2014). PET imaging of ischemia-induced impairment of mitochondrial complex I function in monkey brain. J Cereb Blood Flow Metab 34: 708–714.

Verdu E, Ceballos D, Vilches JJ et al. (2000). Influence of aging on peripheral nerve function and regeneration. J Peripher Nerv Syst 5: 191–208.

Vilella A, Tosi G, Grabrucker AM et al. (2014). Insight on the fate of CNS-targeted nanoparticles. Part I: Rab5-dependent cell-specific uptake and distribution. J Control Release 174: 195–201.

Volk B, Kirchgassner N (1985). Damage of Purkinje cell axons following chronic phenytoin administration: an animal model of distal axonopathy. Acta Neuropathol 67: 67–74.

Wagner OI, Lifshitz J, Janmey PA et al. (2003). Mechanisms of mitochondria-neurofilament interactions. J Neurosci 23: 9046–9058.

Walsh ME, Shi Y, Van Remmen H (2014). The effects of dietary restriction on oxidative stress in rodents. Free Radic Biol Med 66: 88–99.

Wang H, Ichihara G, Ito H et al. (2002). Biochemical changes in the central nervous system of rats exposed to 1-bromopropane for seven days. Toxicol Sci 67: 114–120.

Watanabe I, Kanabe S (1978). Early edematous lesion of pyrithiamine induced acute thiamine deficient encephalopathy in the mouse. J Neuropathol Exp Neurol 37: 401–413.

Xu M, Ju W, Hao H et al. (2013). Cytochrome P450 2J2: distribution, function, regulation, genetic polymorphisms and clinical significance. Drug Metab Rev 45: 311–352.

Yao Y, Chen ZL, Norris EH et al. (2014). Astrocytic laminin regulates pericyte differentiation and maintains blood brain barrier integrity. Nat Commun 5: 3413.

Yuan H, Frank JE, Hong Y et al. (2013). Spatiotemporal uptake characteristics of ^{18}F-2-fluoro-2-deoxy-D-glucose in a rat middle cerebral artery occlusion model. Stroke 44: 2292–2299.

Zheng S, Xie Y, Li Y et al. (2014). Development of high drug-loading nanomicelles targeting steroids to the brain. Int J Nanomedicine 9: 55–66.

Section 2

Disorders associated with exposure to neurointoxicants

Handbook of Clinical Neurology, Vol. 131 (3rd series)
Occupational Neurology
M. Lotti and M.L. Bleecker, Editors

Chapter 6

Occupation and the risk of chronic toxic leukoencephalopathy

CHRISTOPHER M. FILLEY*

Departments of Neurology and Psychiatry, University of Colorado School of Medicine and the Denver Veterans Affairs Medical Center, Aurora and Denver, CO, USA

INTRODUCTION

Occupational settings can pose a wide variety of risks to workers, among which neurotoxins have attracted much attention in recent decades. Exposure to neurotoxins can occur in a range of work settings, including manufacturing, construction, agriculture, military service, and many others. Many chemicals, metals, and other agents can be responsible, and exposure most often occurs via inhalation, ingestion, or skin contact (Harris and Blain, 2004). The nervous system is vulnerable to many hundreds of toxins, and both the peripheral nervous system (PNS) and central nervous system (CNS) can be injured (Spencer and Schaumburg, 2000). The sheer number of toxins within the field of neurotoxicology is imposing, presenting many problems in the understanding of nervous system effects, and new neurotoxins are being discovered on a regular basis. The complexity of nervous tissue, with many potential sites of toxicity, also poses significant problems in understanding neurotoxic syndromes, and injury to the brain is particularly difficult to investigate because clinical syndromes may involve not only motor and sensory disturbances but also several forms of neurobehavioral dysfunction (Rosenberg, 1995; Spencer and Schaumburg, 2000; Harris and Blain, 2004). In the workplace, some of the most vexing clinical problems have involved cognitive and emotional symptoms in exposed individuals, which can be inherently challenging to quantify with reliable clinical-pathologic data. Indeed, toxic encephalopathy, implying direct brain damage as a result of toxic injury, has been regarded as one of the most common of all suspected neurotoxic injuries that occur in the workplace (Rosenberg, 1995).

Among the syndromes producing the clinical manifestations of toxic encephalopathy is toxic leukoencephalopathy, a disorder of brain white matter that became apparent and has been much studied since the advent of magnetic resonance imaging (MRI) in the 1980s enabled detailed *in vivo* depiction of white-matter injury (Filley and Kleinschmidt-DeMasters, 2001). Whereas acute toxic leukoencephalopathy (ATL) can be observed with high-level exposure to certain toxins, far more relevant in the occupational setting is the issue of chronic toxic leukoencephalopathy (CTL). The number of workers at risk for CTL is considerable, and although a figure for those actually affected is presently not possible to estimate, the establishment of evidence-based preventive measures is clearly an important goal. This review will consider what is known about this poorly understood syndrome, particularly with respect to workers engaged in occupational settings where exposure may occur, and suggest how a better understanding can be achieved with further investigation.

WHITE MATTER AND NEUROTOXICOLOGY

The majority of neurotoxins affecting the brain have traditionally been assumed to be toxic primarily by virtue of their effects on gray matter, where cell bodies, dendrites, and synapses offer many potential sites for toxic injury (Schaumburg and Spencer, 2000; Dobbs, 2011). The white matter of the brain, often relatively neglected in the cognitive neurosciences, has only recently been recognized as a potential target of neurotoxins. This development was inaugurated and greatly propelled by MRI, which offers the unprecedented capacity to enable the *in vivo* visualization of white matter. Leukotoxic disorders were recognized in the past, but with the introduction of MRI, major steps forward were made in the characterization of leukotoxic injury and the identification of previously unrecognized leukotoxins. Almost

*Correspondence to: Christopher M. Filley, MD, Behavioral Neurology Section, 12631 East 17th Avenue, B-185, Aurora CO 80045, USA. Tel: +1-303-724-2187, Fax: +1-303-724-2202, E-mail: christopher.filley@ucdenver.edu

immediately, the importance of neurobehavioral dysfunction was appreciated in these disorders. This observation was somewhat unexpected in view of the traditional neuroscientific view that the gray matter of the brain, particularly the cerebral cortex, is responsible for the higher functions. Nevertheless, the entity of toxic leukoencephalopathy came to be recognized as a neurotoxic syndrome caused by a variety of white-matter toxins and typically associated with prominent neurobehavioral disturbance (Filley and Kleinschmidt-DeMasters, 2001). In this section, the neurobiology of white matter will be discussed in light of its neurobehavioral affiliations and susceptibility to certain toxins.

Anatomy

White matter makes up about half the adult brain volume (Filley, 2012). Coursing in large tracts, bundles, and fasciculi that are located within and between the hemispheres, white-matter pathways are traditionally divided into the three categories of projection, commissural, and association tracts (Nolte, 2002). Projection fibers are those that travel to and from the cerebral cortex to convey elemental information involving afferent input and efferent output; examples of afferent tracts are the thalamocortical sensory relays and the geniculocalcarine tracts, and representative efferent tracts are the corticospinal and corticobulbar tracts. Of more importance for neurobehavioral function are the commissural and association tracts, which convey cognitive and emotional information between and within the hemispheres, and also to and from the cerebellum and brainstem. By far the largest commissural tract is the corpus callosum, a massive midline collection of myelinated fibers that connect all cerebral cortical regions in a homologous fashion, The association fibers are divided into short and long association tracts; the former, also known as U or arcuate fibers, serve to link adjacent cortical gyri, and the latter constitute a major portion of brain volume by virtue of their connectivity across all lobes of the cerebrum. Whereas classic neuroanatomy has provided an essential basis of knowledge that has proven enormously helpful for clinical and research purposes, much remains to be learned, and modern neuroimaging of white-matter tracts has been a most instructive complementary source information in recent years (Schmahmann and Pandya, 2006).

Physiology

The main function of white matter is the enhancement of conduction velocity in axons that are invested with myelin (Filley, 2012). Myelin is a complex mixture comprised of 70% lipids and 30% protein, and is laid down by oligodendrocytes in a manner that leaves small segments of the axon unmyelinated. The nodes of Ranvier left free of myelin allow for electric activity in restricted areas of the axon and permit the phenomenon of saltatory conduction, a key evolutionary development that greatly enhances neuronal efficiency. Myelinated axons conduct action potentials as much as 100 times faster than unmyelinated axons, thus vastly facilitating the transfer of information between widely dispersed cortical and subcortical gray-matter regions. As further evidence of the special role myelinated systems have assumed in brain function, white matter has actually expanded more than gray matter over the course of mammalian evolution, by a ratio of 4:3 (Zhang and Sejnowski, 2000). As much as gray-matter expansion has been crucial for enabling the singular cognitive capacities of *Homo sapiens*, the expansion of white matter has been even more impressive, implying that human intelligence owes at least as much to white matter as to the cortical and subcortical gray matter.

White matter and behavioral neurology

The dominant perspective of behavioral neurology and related disciplines in recent decades posits the existence of multiple distributed neural networks subserving all aspects of higher function (Geschwind, 1965; Mesulam, 1990), and white matter figures prominently in every one of these networks (Filley, 2010, 2012). Gray-matter structures are best conceptualized as mediating the information processing of the brain, while matter subserves the equally vital role of information transfer (Filley, 2012). Despite the continuing bias of what has been called "corticocentrism" in cognitive neuroscience (Parvizi, 2009) – implying that only cortical gray matter merits consideration in the study of higher functions – it is impossible to imagine that normal cognition and emotion can be mediated by a brain in which either gray or white matter is dysfunctional. The neurobehavioral operations of white matter are in general more difficult to study, as functional techniques such as electroencephalography, magnetoencephalography, positron emission tomography (PET), single-photon emission tomography (SPECT), and functional MRI (fMRI) are all designed to assess the cerebral cortex (Parvizi, 2009), but the functions of myelinated tracts are no less important than those of gray-matter gyri and nuclei.

Susceptibiliity to neurotoxins

Neurotoxic injury can occasionally occur from sources external to the body, such as cranial irradiation for the treatment of cancer, but most neurotoxic injury is sustained as a result of toxins reaching the nervous system via transport through the vasculature. After being inhaled, ingested, or absorbed transdermally, these toxins enter the blood stream, and hepatic metabolism is often sufficient to render them nontoxic (Harris and Blain, 2004). For toxins that remain, it is of course necessary that they enter the brain if they are to be neurotoxic and produce neurobehavioral dysfunction. To do so, these substances must traverse the blood–brain barrier (BBB), a protective anatomic arrangement that functions to prevent many toxins and pathogens from entering the CNS. The BBB has unique cellular features – capillary endothelial cells with tight junctions and astrocytic foot processes – that act to restrict the CNS entry of many toxins that may damage other organs without the advantage of such protection (Rosenberg, 1992). Neurotoxins more likely to penetrate the BBB are smaller molecules and those that are lipid-soluble and nonpolar (Rosenberg, 1992). Once neurotoxins enter the CNS, they may exert their effects at a number of cellular and subcellular sites associated with neurons, synapses, glial cells, and myelin (Schaumburg and Spencer, 2000). Whereas it is difficult to assess the relative vulnerability of white matter to gray matter, and there is doubtless neuropathologic overlap in the toxicity of many neurotoxins, it is clear that white matter is specifically targeted by some of these agents (Rosenberg, 1995; Filley and Kleinschmidt-DeMasters, 2001). As will be discussed, the chemical agents that can injure white matter are highly diverse, but all of them can be grouped together as leukotoxins.

One general principle that in part explains the action of leukotoxins is that neurons with long processes offer many potential sites of attack (Harris and Blain, 2004), and indeed myelinated axons are lengthy components of neurons that may extend over long intracerebral distances. Myelin is to date the most widely appreciated target of leukotoxins, a point made clear by the lipophilicity of organic solvents, most notably toluene, that are known to produce myelin damage with heavy and prolonged exposure (Filley, 2013). Study of other leukotoxins, however, has disclosed that injury can occur not only in the myelin sheath, but also the axon itself, adjacent oligodendrocytes, nearby astrocytes, and in the vasculature perfusing white matter (Filley and Kleinschmidt-DeMasters, 2001). While much more remains to be learned of the pathogenesis of toxic leukoencephalopathy, the white matter represents a complex tissue compartment of the brain with a variety of possible leukotoxic sites.

EVALUATION OF LEUKOTOXIC SYNDROMES

In a worker or any other individual with possible leukotoxic injury of the brain, the key to accurate diagnosis is a thorough clinical evaluation coupled with additional diagnostic information when necessary. Although neurotoxic syndromes are occasionally straightforward to identify, such as parkinsonism after 1-methyl-4-phenyl-1,2,3,6-tetrahydropyridine exposure or tardive dyskinesia from prolonged neuroleptic drug use, most are not so obvious, and this adage holds true for leukotoxic disorders as well. Moreover, few cases can be usefully illuminated by adequate laboratory testing, as specific blood or urine tests for most leukotoxins are not readily available (Schaumburg and Spencer, 1987; Dobbs, 2011). Thus the assessment of leukotoxicity rests mainly on careful clinical evaluation.

History

A detailed history is the cornerstone of diagnosis. The history taking should begin with a thorough review of the exposure event(s), including the onset of symptoms, the setting, the presumed toxin(s), the length of exposure, and quantification if available. Details of the work environment relevant to safety are often very helpful, such as use of personal protective equipment, the adequacy of ventilation, and the proper handling of materials. In keeping with the basic principles of neurotoxicology (Schaumburg and Spencer, 1987), a leukotoxic disorder is supported by a strong dose–response relationship and improvement of symptoms when the toxin is removed. Another useful feature of toxic exposure is that other persons in the work environment typically have similar symptoms. Lasting symptoms of the type typical of CTL may indeed suggest chronic toxic exposure, but in such cases the possibility of diseases within other neuropathologic categories – metabolic, demyelinative, and degenerative – and psychiatric illness should be considered. In addition, normal aging may produce some cognitive symptoms than can mimic those of CTL (Filley, 2011).

Neurologic examination

The neurologic exam is a crucial companion of the history, and should be thorough while still concentrating on components of the exam that are specific for the patient's complaints. Symptoms referable to the PNS should prompt a focused examination of motor and sensory function in the extremities, for example, whereas cognitive complaints warrant a special focus on the mental status. A useful general principle of the neurologic examination in neurotoxic cases is the typical nonfocality of findings (Schaumburg and Spencer, 1987), and this is equally true of leukotoxic syndromes. For the toxic leukoencephalopathies, the mental status examination is the most important component of the neurologic exam, although one that is often overlooked. Many approaches to the mental status examination can be adopted, but a standard survey of cognitive domains includes the assessment of attention, memory, language, visuospatial skills, executive function, personality, and emotion (Filley, 2011). Standardized short cognitive tests such as the Mini-Mental State Examination (MMSE: Folstein et al., 1975) and the Montreal Cognitive Assessment (MoCA: Nasreddine et al., 2005) are also available. The MMSE and the MoCA yield a score of 0–30, and can offer a convenient and quantifiable measure of cognitive function. A measure especially suited for detecting frontal-lobe dysfunction, typical of CTL, is the Frontal Assessment Battery (FAB: Dubois et al., 2000). None of these tests alone can be used to establish the presence of CTL, but they can be useful in measuring and tracking cognitive impairment in these patients. In particular, the MoCA (Griebe et al., 2011) and the FAB (Kanno et al., 2011) have been shown to be sensitive to white-matter dysfunction in other diseases.

Magnetic resonance imaging

MRI of the brain is essential in the evaluation of potential leukoencephalopathy, since it is the only noninvasive test that can reliably disclose white-matter pathology. The first studies of toxic leukoencephalopathy were done with magnet field strengths of 0.35 T, and now magnets with field strengths up 3.0 T and higher are becoming routinely available, thus steadily improving the resolution of brain structures and detection of white-matter lesions. The use of axial, sagittal, and coronal images – especially T2-weighted, proton density, and fluid attenuated inversion recovery (FLAIR) – is very helpful in identifying macrostructural white-matter lesions of any type. The pattern of involvement in CTL is usually diffuse, and both the cerebral and cerebellar white matter can be affected. There is often loss of gray–white-matter differentiation in the cerebral hemispheres. With more severe involvement of white matter, ventricular enlargement can develop, most visibly of the lateral ventricles.

Neuropsychology

The neuropsychologic evaluation is an important component of the assessment of potential leukotoxic disorders. Often considered an extension of the mental status examination, a neuropsychologic assessment offers special advantages. This evaluation can provide an extensive set of data on cognitive and emotional function, gathered under standardized conditions, that permits the comparison of an individual's performance with that of many others who have been similarly tested. Both the level of impairment, if any, and the pattern of cognitive performance are useful. In the toxic leukoencephalopathies, a typical profile of deficits is often seen (Filley and Kleinschmidt-DeMasters, 2001), consistent with deficits encountered with other white-matter disorders (Filley et al., 1988; Filley, 2012). A key point, however, is that whereas neuropsychologic testing is highly sensitive to cognitive dysfunction, it is considerably less specific to the actual disease process that might be responsible. Thus the neuropsychologic data must be interpreted in light of all the assembled clinical information. This point underscores the value of an experienced neuropsychologist who can interpret test results with attention to all relevant details. Computerized test batteries, which have at times been promoted for conducting cognitive testing, lack the level of sophistication conferred by a neuropsychologist who can assess the entire clinical picture, including not only test scores but also the process underlying how each individual displays his or her unique cognitive performance (Kaplan, 1990).

Autopsy

The postmortem examination is rarely performed for the assessment of occupational leukoencephalopathy because individuals who have had this degree of exposure are likely to come to autopsy for some other reason. The value of autopsy in the leukoencephalopathies has been most apparent in cases of severe white-matter involvement, and substance abusers have provided some of the most striking examples (Kornfeld et al., 1994; Rosenberg et al., 1998b; Filley et al., 2004; Al-Hajri and Del Bigio, 2010). Postmortem data of individuals with occupational exposure to potential leukotoxins have been reported, however, and the brains of 103 workers who had been occupationally exposed to organic solvents over several years were found to have no specific histologic abnormalities (Klinken and Arlien-Søborg, 1993). These findings have several possible explanations: (1) brain white-matter injury is not sustained with low-level exposure to leukotoxins; (2) the damage is reversible after cessation of toxic exposure; and (3) current neuropathologic techniques are insufficiently sensitive to detect subtle white-matter abnormalities.

CHRONIC TOXIC LEUKOENCEPHALOPATHY

CTL is a leukotoxic syndrome characterized by clinically significant neurobehavioral morbidity in association with demonstrable white-matter injury in the brain. A key distinction to be made at the outset of this discussion is that CTL differs from ATL, a condition related to abrupt or recent exposure to leukotoxins that is typically self-limited if the toxic insult is eliminated. ATL most often presents with a syndrome dominated by acute confusion and headache, and diffuse cerebral white matter changes are evident on MRI, particularly FLAIR and diffusion-weighted imaging sequences (McKinney et al., 2009). ATL is not a trivial problem, as it often implies substantial morbidity and even death if the toxic exposure has irreversible effects (McKinney et al., 2009). For purposes of this review, however, a more relevant problem is that of CTL, which by definition implies a protracted, if not permanent, period of neurobehavioral dysfunction that may involve a range of neurobehavioral syndromes, including dementia. It is apparent that CTL may be highly relevant to the workplace, as workers in many occupations may be exposed to a host of potential leukotoxins. However, very little is known about the risk of CTL in workers because toxic exposure in the workplace is typically low and it is unclear if leukotoxicity has in fact occurred. The vast majority of data on CTL comes from study of obviously impaired individuals who have been exposed through substance abuse to organic solvents at high concentration and for prolonged periods. Whereas much useful information has been gathered from study of these unfortunate individuals, illuminating the effects of extreme solvent exposure levels on white-matter structure and function, a far more difficult problem is determining the relevance of these observations to persons who have had low-level exposure to solvents and other leukotoxins. Low-level exposure may be sufficient to produce syndromes similar to, but more subtle than, those of high-level exposure, but it is likely many exposures occur that are innocuous because a threshold for leukotoxicity has not been reached. In this section, the available data on high level exposure will be summarized as a background for considering the possible impact of low-level exposure in the workplace.

Background

Toxic injury to the white matter of the brain has long been appreciated because of the propensity of cancer treatments – both cranial irradiation and a variety of chemotherapeutic agents – to produce clinically significant, often disabling, and sometimes fatal leukoencephalopathy (Filley and Kleinschmidt-DeMasters, 2001). These injuries were recognized before the MRI era, most notably from the neuropathologic study of the effects of therapeutic radiation on cerebral white matter (Sheline et al., 1980), but as MRI became available, *in vivo* diagnosis of toxic white-matter injury became increasingly common. As investigation proceeded, it became clear that other agents can produce such a syndrome, including other medications, environmental toxins, and drugs of abuse (Filley and Kleinschmidt-DeMasters, 2001). But the problem of CTL was most usefully illuminated by the unexpected, but highly informative, discovery that inhalant abuse could produce a dramatic clinical syndrome of leukoencephalopathy. This widespread but relatively underappreciated problem involves millions of mostly young adults in the USA and around the world, and is the least-studied form of substance abuse (Howard et al., 2011). The opportunity to study the neurologic and neurobehavioral.effects of solvents at high exposure levels with the powerful technique of MRI was unprecedented, and has led to many advances in the understanding not only of the leukotoxicity of organic solvents, but also the importance of white matter in higher function.

Clinical features

CTL is a neurotoxic syndrome, with its most prominent effects exerted on neurobehavioral competence. Whereas the diversity of leukotoxins studied might suggest that few commonalities exist between them, generalizations can in fact be made about their clinical impact. Focal neurologic deficits can occur, involving the cerebellum, motor system, cranial nerves, and brainstem, but are less clinically significant than the neurobehavioral dysfunction caused by the typically diffuse cerebral damage of leukotoxins (Rosenberg, 1995; Filley and Kleinschmidt-DeMasters, 2001). As is true of behavioral neurology in general, the location rather than the type of brain damage is the major determinant of clinical impact (Filley, 2011), and diffuse white-matter involvement resulting from leukotoxins produces a clinical picture that closely resembles that of other white-matter disorders (Filley, 2012). Major features of this syndrome include a pervasive slowing of information-processing speed, executive dysfunction, attentional disturbances, and the relative preservation of language.

CTL as a consequence of inhalant abuse, now known to result primarily from toluene, is the most illustrative clinical model of this syndrome (Filley, 2013). Inhalants may contain a variety of solvents, but toluene – methylbenzene – is very common and has provided the bulk of information on severe CTL from organic solvents. Toluene is favored by inhalant abusers because of its euphorigenic properties, ready availability, and low cost. Fumes of the solvent are inhaled from soaked rags ("huffing") or from plastic bags ("bagging"), and rapidly reaches the blood stream en route to the brain. Euphoria follows when a sufficient brain level of toluene is reached, and, similar to other drugs of abuse, likely results from activation of the brain's reward system mediated by dopaminergic transmission in the ventral tegmental area and nucleus accumbens (Lubman et al., 2008). The first report of persistent neurologic injury related to chronic toluene inhalation appeared in 1961 and emphasized cerebellar degeneration (Grabski,

1961). The same individual reported at the time was again presented after 8 more years of toluene inhalation, and then the case report called attention to permanent encephalopathy characterized by bizarre and inappropriate behavior, and emotional lability; diffuse cerebral atrophy was noted on pneumoencephalography, the only available brain imaging technique (Knox and Nelson, 1966).

The major studies on the neurologic sequelae of toluene abuse were conducted in the 1980s, when detailed clinical, MRI, neuropsychologic, and autopsy findings were reported in well-defined series of patients. The clinical picture of abstinent toluene abusers disclosed multifocal brain involvement leading to cerebellar, corticospinal, cranial nerve, brainstem, and, most importantly, neurobehavioral dysfunction (Hormes et al., 1986; Rosenberg et al., 1988a). An important strength of these studies is that the toluene abusers were studied during abstinence, so that the acute effects of toluene intoxication were not a confounding issue. Dementia was striking in abstinent toluene abusers, and was characterized by impaired memory, apathy, and preserved language (Hormes et al., 1986). Brain MRI was remarkable in disclosing loss of gray–white-matter differentiation in the cerebral hemispheres and diffuse white-matter hyperintensity (Rosenberg et al., 1988a). Comprehensive neuropsychologic assessment in these subjects disclosed that cognitive deficits correlated with the duration of toluene abuse and with the severity of MRI white-matter involvement (Filley et al., 1990). Patients began coming to autopsy after being evaluated in life, and diffuse white-matter pathology was confirmed, not only showing severe loss of myelin in the cerebrum and cerebellum but also sparing of gray matter and even the axons within white-matter tracts (Rosenberg et al., 1988b; Kornfeld et al., 1994).

These observations were so dramatic that they helped stimulate a new concept in behavioral neurology, the notion of white-matter dementia (Filley et al., 1988). This idea was introduced and developed in an effort to highlight the potentially disabling cognitive impairment that can occur with cerebral white-matter disorders, expanding the locus of brain–behavior relationships from the cerebral cortex, which has long held a dominant position in the cognitive neurosciences (Parvizi, 2009), to the half of the brain occupied by white matter. White-matter dementia has been steadily investigated, and with collected findings from a vast literature on white-matter disorders and several key studies of white matter–behavior relationships (Filley et al., 1989 1990; Lafosse et al., 2007), a remarkably consistent pattern of deficits and strengths has been observed (Filley, 2012). This profile – slowed processing speed, executive dysfunction, sustained attention deficit, memory retrieval deficit, visuospatial dyfunction, psychiatric disturbance, combined with sparing of language, extrapyramidal function, and procedural memory – is specific for white-matter involvement, and differs from the profiles of both the cortical and classic subcortical dementias. Given that white-matter involvement is often subtle and causes cognitive impairment that is measurable but does not qualify as dementia, the entity of mild cognitive dysfunction has also been introduced as a proposed precursor syndrome to white-matter dementia (Kozora and Filley, 2011). Mild cognitive dysfunction may in fact describe many individuals with leukotoxic exposure that has injured cerebral white matter but not yet led to the onset of dementia.

Magnetic resonance imaging

The initial work with toluene abuse was conducted with MRI, using very early technology, and the field strength of the magnets was 0.35 T. As awareness increased and experience grew with the study of toluene abusers, MRI using higher field strength was employed, and evidence accumulated that white matter was indeed the primary target of this solvent (Caldemeyer et al., 1993; Xiong et al., 1993; Yamanouchi et al., 1997; Kamran and Bakshi, 1998; Rosenberg et al., 2002). Diffuse involvement of the cerebral and cerebellar white matter and concomitant cognitive impairment was repeatedly confirmed. In Figure 6.1, severe cerebral leukoencephalopathy is evident on the MRI scan of a chronic toluene abuser. Some studies did find hypointensity in the

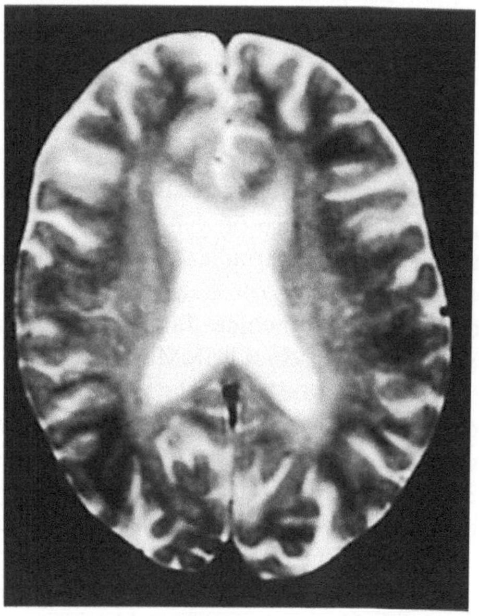

Fig. 6.1. Axial brain magnetic resonance imaging scan of a chronic toluene abuser showing severe leukoencephalopathy in the cerebral hemispheres and lateral ventricular enlargement. (Reproduced from Filley et al., 2004.)

thalamus and basal ganglia (Caldemeyer et al., 1993; Unger et al., 1994; Kamran and Bakshi, 1998; Aydin et al., 2002), or the cerebral cortex of toluene abusers (Kamran and Bakshi, 1998). These observations may be explained by toluene located within the sparse white matter of these structures, or possibly by iron deposition, but in either case, gray-matter involvement does not appear to influence cognition. A 2008 review of 30 studies concluded that evidence supports the conclusion that toluene is a specific CNS leukotoxin with the potential to cause significant cognitive dysfunction (Yücel et al., 2008).

Neuropsychology

The detailed testing of inhalant abusers is helpful for documenting and characterizing deficits, and for correlation of these deficits with MRI changes (Filley et al., 1990). However, cognitive assessment in this setting is inherently limited by the fact that performance on a given test may be determined by many factors in addition to toxic exposure. Even the most sophisticated and expert neuropsychologic assessment is by its very nature more sensitive than specific. Deficits consistent with exposure injury can be found in control subjects as well as exposed persons (Rosenberg et al., 2002), and caution is advisable in the attribution of observed deficits to a toxic exposure. However, in the appropriate clinical setting, neuropsychologic testing can be very helpful in the evaluation of inhalant abusers by identifying characteristic deficits in processing speed, sustained attention, memory retrieval, and executive function (Yücel et al., 2008) that are entirely consistent with the concept of white-matter dementia (Filley et al., 1988; Filley, 2012).

Neuropathology

Although few in number, other autopsy cases of toluene abuse came to light, and consistently showed selective white-matter involvement in CTL (Filley et al., 2004), which could be evident even years after the achievement of abstinence (Fornazzari et al., 2003). More substantial numbers of autopsy cases were not readily available because the follow-up of toluene abusers is inconsistent or not possible. Among the most convincing studies, however, was a large autopsy series of solvent abusers from Canada showing that 22 of 75 affected individuals had selective myelin loss with sparing of gray-matter structures and axons within the injured white matter; toluene was considered the major solvent, and the 22 affected brains were generally from people with longer periods of abuse (Al-Hajri and Del Bigio, 2010). A study from Colombia examined the brains of young adults who had abused glue, another vehicle for the

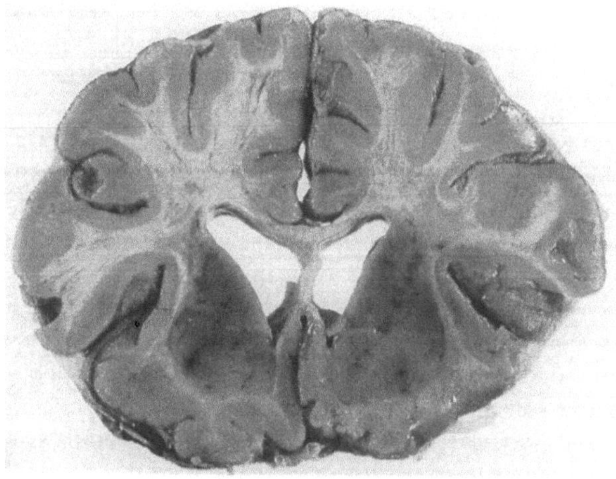

Fig. 6.2. Autopsied brain in coronal section of a chronic solvent abuser showing diffuse discoloration of the cerebral white matter and thinning of the corpus callosum. (Reproduced from Kornfeld et al., 1994.)

intentional inhalation of high levels of toluene, and leukoencephalopathy was again the major finding, with myelin damage and preservation of axons supporting a selective myelinotoxic effect (Marulanda and Colegial, 2005). All of these studies found support in experimental work documenting the preferential distribution of inhaled toluene within lipid-rich regions of the brain (Gospe and Calaban, 1988) and sparing the cortex and hippocampus (Ameno et al., 1992). Figure 6.2 shows diffuse white-matter involvement in the autopsied brain of a chronic solvent abuser. In the aggregate, postmortem studies of toluene abuse have consistently supported the early findings of diffuse cerebral myelin loss as the origin of neurobehavioral dysfunction.

OCCUPATIONAL LEUKOTOXINS

Neurotoxicology has generally not classified toxic syndromes based on a distinction between gray- and white-matter injury, but as a general rule, the major CNS target of most members of the vast range of neurotoxins is considered to be the gray matter, either cortical or subcortical. Secondary involvement of white matter is known to occur in some cases (Dobbs, 2011), but prominent effects on nucleic acid and protein synthesis, mitochondrial metabolism, and neurotransmitter function (Schaumburg and Spencer, 2000) all suggest major effects on the cell bodies of neurons located in gray matter. The main exception to this generalization is the organic solvents, which by virtue of their chemical structure are highly lipophilic, explaining their ready penetration of the BBB and their clear predilection for the lipid-rich myelin sheath. Organic solvents are widely employed in industry, and among the more commonly used are the aromatic hydrocarbons toluene, benzene, xylene, and styrene, the aliphatic hydrocarbons methyl-*n*-butyl ketone and *n*-hexane, the chlorinated hydrocarbons trichloroethylene, perchloroethylene, methylene chloride, trichloroethane, and carbon tetrachloride, and others, including methanol, hydrogen sulfide, and carbon disulfide. Millions of tons of these solvents are produced yearly in the USA alone, with exposure occurring routinely in millions of workers (Rosenberg, 1995). Some of these compounds, such as the aliphatic hydrocarbons, have prominent effects on the PNS and no apparent central neurotoxicity, but many solvents have been found to have neurobehavioral sequelae, and white matter in the brain has emerged as the most important target. In this section, the occupational leukotoxicity of solvents will be considered, along with other white-matter toxins that can be problematic in the workplace.

An important point to be emphasized is that occupational exposure to any neurotoxic substance is typically far less than that associated with exposure related to drug abuse. Again, the example of toluene proves most illustrative. In the USA, exposure to solvents is regulated by reference to the concept of time-weighted average (TWA), the ambient level to which workers can be exposed for a normal 8-hour day over 40 hours per week without ill effects. For toluene, the TWA established by the Occupational Safety and Health Administration (OSHA) is 200 ppm and for the National Institute for Occupational Safety and Health the TWA is 100 ppm (Rosenberg, 1995). Headache, fatigue, and slowed reflexes can appear at levels slightly above 200 ppm (Benignus, 1981), but acute confusion does not develop until around 600 ppm, and euphoria is not reached until 800 ppm and above (Rosenberg, 1995). Laboratory animals can develop cognitive deficits and high-frequency hearing loss after 5 weeks of exposure to 1200–1400 ppm toluene, but these are reversible after removal from exposure (Pryor et al., 1983). In contrast, inhalant abusers can be chronically exposed to toluene fumes that may reach many times legally permissible levels. Given that high-level exposure to certain leukotoxins has been securely linked with severe CTL, it is now appropriate to consider the range of leukotoxic agents for which some evidence of low-level exposure leading to less severe CTL has been obtained. These agents have been studied to a variable extent with regard to workplace toxicity, and overall conclusions should only be drawn with caution.

Solvents

Organic solvents are hydrocarbons that are widely prevalent and used in many industries as well as in the household. Occupational exposure can occur among workers in many settings who use or come into contact with these compounds. An extensive literature from Denmark and other Scandinavian countries first appeared in the 1970s claiming that painters exposed to solvents could develop encephalopathy as a result of their workplace exposure. This condition, known as chronic painter's syndrome (Axelson et al., 1976; Hänninen et al., 1976; Hane et al., 1977; Arlien-Søborg et al., 1979), and sometimes referred to as painter's encephalopathy, chronic toxic encephalopathy, the neurasthenic syndrome, or the psycho-organic syndrome, reportedly involved a variety of neurobehavioral features such as memory loss, personality change, apathy, depression, and even progressive dementia. These data were of sufficient concern that in 1985 the World Health Organization (WHO) developed a classification of toxic encephalopathy from the effect of solvents involving four syndromes: type 1, consisting of symptoms only and potentially reversible; type 2A, sustained personality or mood change that is not reversible; type 2B, impairment of intellectual function that is objectively documented but mild and potentially reversible; and type 3, marked intellectual deterioration consistent with irreversible dementia.

These early reports, and the WHO classification, were then challenged by critical reviews with close attention to methodologic issues, and the quality of the research in this area was questioned (Errebo-Knudsen and Olsen, 1986; Rosenberg, 1995; Albers and Berent, 2000). Shortcomings of the early studies included little attention to control groups, exposure measurements, dose–response relationships, and comorbid problems such as high alcohol consumption (Errebo-Knudsen and Olsen, 1986; Rosenberg, 1995; Albers and Berent, 2000). A reanalysis of some of the initial subjects some years after the first reports in fact disclosed that cognitive impairment could not be confirmed when the influence of age, education, and intelligence was considered (Gade et al., 1988). Later study found that cognitive performance in exposed workers remained stable (Edling et al., 1990), indicating that progressive dementia could not be documented.

More rigorous investigation examining ambient levels of solvents experienced by workers in US industries found no dose–response relationships between solvent exposure and behavioral dysfunction (Maizlish et al., 1985) or neurologic signs (Maizlish et al., 1987). Well-controlled studies of workers at two US paint-manufacturing plants also could not document an association between low-level solvent exposure and neuropsychiatric symptoms of the psycho-organic syndrome (Bolla et al., 1990). Thus the neurotoxicity of industrial solvents at low levels that was accepted by some (Baker et al., 1985) came into serious question with further study (Rosenberg, 1995; Albers and Berent, 2000). Perhaps the most telling criticism of this body of work is that competing explanations for putative neurotoxic effects were not excluded by the reports alleging this kind of injury (Rosenberg, 1995; Albers and Berent, 2000).

None of this initial work, however, involved MRI to any significant extent, as the technique was still in its infancy. A report from 1977 described cognitive impairment in an optician exposed to high levels of toluene at work that resolved within a month after exposure was eliminated; computed tomography (CT) showed cerebral and cerebellar atrophy but white matter could not be adequately assessed with this technique (Boor and Hurtig, 1977). Later, a large, controlled CT study of painters exposed to solvents in occupational settings found no brain atrophy in excess of that expected for age (Triebig and Lang, 1993), confirming the emerging recognition that CT is not sensitive to white-matter pathology. In light of the data on toluene abuse that first appeared in the 1980s, it might be expected that the use of MRI would disclose subtle degrees of white-matter involvement in people with lesser degrees of solvent exposure. A 1996 report of workers exposed to industrial solvents found that 11 of 32 subjects had diffuse white-matter hyperintensity and loss of gray–white differentiation (Thuomas et al., 1996), but clinical details were limited, and the consensus at the time was that MRI could at best disclose nonspecific abnormalities that require clinical correlation (Frey, 2000). A thorough review in 2003 concluded that the existing evidence, by then including MRI, was insufficient to support a neurotoxic effect of occupational exposure to organic solvents (Ridgway et al., 2003) A similar conclusion was reached some years later, when an MRI study of 71 painters diagnosed with chronic solvent encephalopathy found that, whereas cerebral atrophy correlated with duration of exposure, white-matter findings were considered nonspecific and not helpful in diagnosis (Keski-Säntti et al., 2009).

Scattered case reports have appeared, however, raising the possibility that CTL may occur in workers exposed to solvents in the workplace. The majority of these reports implicate toluene. Jung and colleagues (2007) described a 29-year-old man with occupational exposure to toluene in a chemical manufacturing plant who had severe leukoencephalopathy on MRI, and after death showed diffuse loss of myelin; the coexistence of myelodysplastic syndrome causing white-matter vasculopathy was thought to exacerbate the leukotoxicity of toluene. A case of reversible toluene leukoencephalopathy was described in a solvent salesman who was accidentally exposed to toluene when he left several toluene-containing products in the back seat of his car for 6–8 months; recovery was complete 6 months after correction of the problem (Qureshi et al., 2009). In 2013, a report appeared of two migrant workers exposed to solvents in enclosed spaces and without protective equipment who developed toxic leukoencephalopathy related to toluene and 1,2,3-tricholopropane; after removal from the toxin a return to normal life was possible for both men in 6 months (Mi et al., 2013). Kao and colleagues (2014) described a 49-year-old man who died after a lacquer thinner fire causing intense exposure to toluene; the case was unusual in that loss of myelin was not diffuse, but rather most evident in the left parieto-occipital white matter. All of these cases represent exceptional and highly unusual exposure events.

A literature has also appeared on the leukotoxic effects of carbon disulfide, to which workers in the viscous rayon and rubber industries, and in agriculture, may be exposed. Carbon disulfide has damaging effects on peripheral nerves and the basal ganglia, leading to peripheral neuropathy and parkinsonism, but MRI studies of exposed workers have also disclosed white-matter changes (Cha et al., 2002; Cho et al., 2002; Ku et al., 2003; Chuang et al., 2007). These studies have found a combination of diffuse and focal white-matter lesions, suggesting that demyelination, lacunar infarction, or both may be present. Endothelial proliferation of cerebral arterioles has been observed neuropathologically (Rosenberg, 1995), and diffuse microangiopathy has been suggested from studies of cerebral blood flow showing decreased perfusion of the total brain parenchyma (Chuang et al., 2007).

Metals

As has been pointed out above, most neurotoxins are thought to exert their noxious effects on gray matter, and this generalization is true for heavy-metal poisoning as well. A literature has recently been developed regarding lead, a well-known neurotoxin that is being found capable of leukotoxicity with the application of MRI in exposed individuals. Lead has often been observed to cause axonal peripheral neuropathy in adults, and recently a case was reported with neuropathy and cognitive dysfunction thought secondary to extensive cerebral and cerebellar white-matter lesions on MRI (de Souza et al., 2013). The metal is best known, however, for producing encephalopathy in children (Sanders et al., 2009; Dobbs, 2011). Lead has many toxic effects, but one of the most important aspects of lead poisoning (plumbism) in children is lead encephalopathy. This problem became notorious for causing neurosychologic impairment among inner-city children who, because of an eating disorder known as pica, had a propensity for ingesting leaded paint from dilapidated old houses (Needleman et al., 1990). Lead was banned in the USA from gasoline and household paint in the late 1970s, mitigating lead exposure in the general population, but exposure to this metal can still occur, and no safe exposure threshold level has been identified (Sanders et al., 2009). One case report described lead encephalopathy in an infant with accidental heavy exposure, and symmetric cerebral dysmyelination sparing the cerebral cortex and basal ganglia was striking (Sahu et al., 2010).

A specific effect on white matter in occupational settings has also been observed, and good evidence for a leukotoxic effect of lead in humans has come from studies of chemical manufacturing workers exposed to both organic and inorganic lead many years before (Stewart et al., 1999, 2006). Lead was measured in the tibia of these individuals, and higher tibia lead was significantly associated with quantified memory and executive function deficits (Stewart et al., 1999), and with increased white-matter lesion burden and decreased total brain volume, frontal and total gray-matter volume, and parietal white-matter volume (Stewart et al., 2006). Other studies have shown that lead exposure predicts impaired psychomotor speed and dexterity (Bleecker et al., 1997), at least in part explained by MRI white-matter changes (Bleecker et al., 2007).

Another metal known to be leukotoxic is arsenic, which has rarely been noted to cause leukoencephalopathy from occupational exposure. Arsenic neurotoxicity was common in the pre-antibiotic era when the use of arsphenamine for neurosyphilis produced hemorrhagic encephalitis that presented as acute encephalopathy (Beckett et al., 1986). Arsenic poisoning is rare today, although it can be seen with exposure to insecticides, rodenticides, certain medicinals, and in certain occupations (Freeman and Couch, 1978). The most familiar neurotoxicity of this metal is peripheral neuropathy (Freeman and Couch, 1978), but acute (Beckett et al., 1986) and prolonged (Freeman and Couch, 1978) encephalopathy may occur. The neuropathology involves multiple regions of necrosis and hemorrhage in the white matter of the cerebral hemispheres (Cole et al., 1966).

Pesticides

Pesticides comprise a large group of chemical substances used to control insects, weeds, and fungi (Kamel and Hoppin, 2004). Pesticides in common use include organophosphates (such as tabun, sarin, cyclosarin, and soman), organochlorines, carbamates, and pyrethroids, and farmers and farmworkers are the individuals most likely to be occupationally exposed to these agents (Kamel and Hoppin, 2004). As with other common neurotoxins, pesticides accumulated over the course of chronic exposure are considered most likely to injure the brain by attacking gray matter (Dobbs, 2011). The emerging work on Parkinson's disease risk from exposure to pesticides exemplifies this tendency, as Parkinson's disease characteristically involves subcortical gray-matter pathology, well known to be most prominent in the substantia nigra (Allen and Levy, 2013). Unfortunately, certain pesticides also find use in wartime as chemical weapons, and nerve agents are among the deadliest in the arsenal of chemical warfare (Newmark, 2007). These drugs act via peripheral acetylcholinesterase inhibition, and can produce a dramatic cholinergic crisis with rapidly evolving cerebral hypoxia and death from respiratory failure (Newmark, 2007). The CNS effects of these agents have been less clear.

In this regard, a recent clinical problem in combat veterans that has brought the issue of pesticides to widespread attention as an occupational hazard of military service is the Gulf War syndrome, a controversial condition putatively linked with exposure to pesticides, nerve gas, and pyridostigmine in the Gulf War of 1990–1991 (U.S. Department of Veterans Affairs, 2008). Almost 700 000 American troops served in this conflict, and the Gulf War syndrome has evolved as a proposed clinical entity among many of its veterans, consisting of a wide range of complaints such as headache, fatigue, memory loss, musculoskeletal pain, and other symptoms (US. Department of Veterans Affairs, 2008). The nonspecific nature of these complaints, suggesting multisystem involvement, and lack of reliable diagnostic biomarker has led to doubt about whether any neurotoxic disorder can be substantiated (Albers and Berent, 2000), but controversy persists. The syndrome has indeed been challenging to assess, but MRI studies have shown white-matter changes that could be a result of exposure to the organophosphates sarin and cyclcosarin. Sarin (*o*-isopropyl methylphosphonoflouridate) and cyclosarin (cyclohexyl methylphosphonoflouridate) are potent organophosphate nerve agents to which some 100 000 troops were potentially exposed in the destruction of a munitions storage complex in Khamisiyah, Iraq, in 1991 (Heaton et al., 2007). Volumetric MRI of veterans of this incident disclosed reduction of cerebral white-matter volume that was associated with higher levels of sarin and cyclosarin exposure (Heaton et al., 2007). Later, an association was found in further studies between total white-matter volume and measures of executive function and visuospatial abilities in these individuals (Chao et al., 2010). Using a 4.0 T magnet, MRI investigation of exposed veterans then found reduced volumes of cerebral white matter, and also gray matter, both of which were associated with typical symptoms of the Gulf War illness (Chao et al., 2011). In support of the notion that organophosphates may be leukotoxic is a report of sarin causing similar MRI white-matter changes in victims of the Toyko subway attack of 1995 (Yamasue et al., 2007).

Hypobaria

A recently recognized and novel form of leukotoxic injury is that caused by hypobaria, the exposure to low barometric pressure that can occur in pilots flying at extremely high altitude. The US Air Force, for example, operates U-2 high-altitude reconnaissance aircraft, and neurologic decompression sickness (NDCS) is a known occupational risk of pilots flying at extremely high altitudes and therefore exposed to low ambient pressure (McGuire et al., 2012, 2013). One pilot reported with near-fatal NDCS was found to have permanent cognitive deficits in association with multiple bilateral frontal and parietal white-matter lesions on MRI (Jersey et al., 2010). This problem was then more extensively investigated, and large series of pilots with occupational exposure to hypobaria disclosed a significant increase in MRI white-matter hyperintensity volume and number (McGuire et al., 2012, 2013). The mechanism of this injury is not fully established, but is believed to be related to microembolic gas bubbles occluding multiple arterioles in the cerebral white matter (McGuire et al., 2012, 2013). The white-matter changes of hypobaria were unexpected but clinically significant, indicating once again how much MRI has informed the study of toxic leukoencephalopathy.

WHAT IS THE RISK OF CTL IN WORKERS?

Overall, the risk of CTL in workers is in all likelihood quite low. A more precise statement cannot be made because no surveillance information in exposed populations is available, and it is far from clear what leukotoxins – recognized and still awaiting recognition – should be included in such an analysis. Individuals exposed to very high levels of known leukotoxins may develop severe CTL, but the great majority of exposure events involve low-level exposure that is likely of no consequence. Severe CTL developing in workers exposed to leukotoxins is exceedingly rare, and even well-documented examples of less severe CTL are uncommon. The exposure levels necessary to produce the severe forms of CTL seen, for example, in toluene abusers are not routinely present in occupational settings, and the development of milder CTL in workers with ordinary exposure, while possible, is difficult to establish (Klinken and Arlien-Søborg, 1993; Rosenberg, 1995). The diagnosis of CTL in workers must thus be considered in the context of many factors that can confound the establishment of a secure causal relationship between workplace exposure and a neurobehavioral syndrome.

Uncertain exposure

One of the most obvious problems in determining a firm link between exposure and a clinical syndrome is the uncertain identity of the presumed offending agent, and the extent to which a worker is exposed (Rosenberg, 1995). It is a simple matter to diagnose a toxic syndrome when a clear exposure history is available and a characteristic clinical picture can be observed – as in a case of CTL from long-standing toluene abuse – but quite another when exposure is uncertain and no suggestive clinical profile can be found. Laboratory testing for many toxins is unavailable, and even if the toxin is identified, workplace exposure levels are not typically quantified. Many are the patients whose complaints of memory loss, fatigue, and dysphoria after putative toxic exposure are nearly impossible to judge clinically because of the paucity or complete absence of information as to what exposure may have occurred.

Combined exposure

Another common issue is the problem of multiple concurrent exposures. The literature on solvents features several reports of workers exposed to "mixed solvents," which immediately introduces problems in interpreting the effects of any one solvent in isolation. White-matter changes may be seen on MRI and cognitive impairment may be documented (Feldman et al., 1999), but in such cases it is not possible to determine whether one solvent, more than one solvent, or none of them is responsible. Many workers are exposed to more than one potential toxin, and the combined effects of these agents, even if it were possible to measure them all, is difficult to imagine. Whether and how solvents may interact to potentiate or mitigate neurotoxicity are questions for which no information is available.

Uncertain cause-and-effect relationship

Even if a neurotoxic exposure is well documented, difficulty can still arise as to the relationship between the exposure incident and the presumed effect. Association is not causation, and the fact that a cluster of symptoms follows neurotoxic exposure does not necessarily mean that the symptoms were caused by the exposure. This distinction brings up the logical fallacy of *post hoc, ergo propter hoc* ("after the fact, therefore because of the fact"), a common error that leads to the often vigorous but unsupported contention that an exposure event produced illness (Rosenberg, 1995). Neurotoxicology does not usually have the advantage enjoyed by a field such as infectious disease, for example, where the application of Koch's postulates can lead to definitive and convincing evidence that a pathogen unequivocally produced a disease state, and no other explanation is in any way plausible (Rosenberg, 1995). Rather, much uncertainty attends the attribution of a clinical sequel to the presumed neurotoxic exposure.

Comorbid medical or neurologic disorders

A difficult confound often present in the analysis of potential leukotoxicity is the related problem of comorbid medical and neurologic disorders. A worker exposed to a putative leukotoxin may also have concurrent cerebrovascular disease, alcoholism, degenerative dementia, traumatic brain injury, autoimmune disease, multiple sclerosis, or any combination of these and many other disorders that can manifest in a manner similar to what might be predicted with mild CTL. Ischemic cerebral white-matter disease, typically a result of hypertension, diabetes, obesity, hyperlipidemia, the metabolic syndrome, and cigarette smoking, may produce cognitive dysfunction very similar to that of workers exposed to leukotoxins (Filley, 2012). Another common issue is that of alcohol and alcohol abuse, problems well known to clinicians that can interfere with the determination of causality in neurotoxic exposures (Rosenberg, 1995). In the autopsy study of workers exposed to solvents discussed above (Klinken and Arlien-Søborg, 1993), chronic alcoholism was correlated with reduced brain weight, implying that these workers may have sustained damage from alcohol rather than solvents An apt analogy can be made with the assessment of occupational respiratory disease: a worker with putative brain disease who has inhaled fumes but also consumes alcohol is very similar to the worker with putative lung disease who is exposed to inhaled toxins and also smokes cigarettes. A particularly confusing aspect of alcohol and the brain is that evidence exists for a direct toxic effect of alcohol on white matter (Filley, 2012), complicating the use of MRI because the leukotoxic neuropathology of alcohol may be quite similar to that of the solvent(s) in question. Medication effects must also be considered, as many centrally active medications – antidepressants, neuroleptics, benzodiazepines, anticonvulsants, and a variety of drugs used for the treatment of medical diseases – can produce cognitive side effects (Meador, 1998).

Psychiatric disorders

Another very common clinical problem in individuals presenting with possible neurotoxic injury is psychiatric disease. This issue implicates a wide range of possible disorders, including anxiety, depression, personality disorders, and psychosis. In the workplace, a predisposing problem is occupational stress, which disproportionally affects blue-collar employees such as assembly-line workers and machine operators rather than those with more stimulating and flexible occupations (Rosenberg, 1995; Lundberg, 1999). The psychiatric diseases in general are frequently encountered in neurologic and other medical settings, and are well known to produce neurobehavioral manifestations that mimic neurologic diseases such as Alzheimer's disease (AD), frontotemporal dementia, and neurotoxic states (Kenfield et al., 2010). A useful approach to dealing with the complex challenges raised by possible psychiatric comorbidity has been devised by Schaumburg and Albers (2005), and postulates three varieties of "pseudoneurotoxic" disease that can be distinguished from genuine neurotoxic disease. Pseudoneurotoxic disease can be type 1, a neurologic illness that begins after toxic exposure but is unrelated; type 2, a psychologic illness beginning after toxic exposure; or type 3, a preexisting neurologic or psychologic illness that is interpreted by the patient as being worsened by toxic exposure (Schaumburg and Albers, 2005). The distinction between these variants of pseudoneurotoxic disease can be challenging, but thorough clinical evaluation in light of the principles of recognizing neurotoxic disease (Schaumburg and Spencer, 1987) can be most helpful. Such principles include a clear association of exposure with symptoms, the appearance of symptoms – in most cases – immediately after exposure and without a delay, a strong dose–response relationship, neurologic signs reflecting nonfocal involvement, and clinical improvement after removal of the toxin (Schaumburg and Spencer, 1987).

Litigation

A final problem attending the assessment of many patients with suspected CTL from workplace exposure is the role of litigation. Workers may often be entitled to generous disability payments if a neurotoxic syndrome can be substantiated in the workplace, and clinicians in all areas of medicine are well aware of the potential for intentional exaggeration or fabrication of illness to enhance the prospects for receiving such benefits. Malingering is the term most often used for such behavior, and often the clinician can observe limited effort on routine cognitive testing. Since the assessment of possible chronic CTL typically involves neuropsychologic evaluation, malingering is referred to alternatively as "poor effort" on testing, such that the neuropsychologist cannot reasonably judge the level and the pattern of the individual's cognitive dysfunction. Malingering in persons reporting exposure to occupational and environmental substance has been reported to occur with a prevalence of 40% (Greve et al., 2006), although others uspect a figure lower than this (Schaumburg and Albers, 2005). The issue is particularly difficult because the symptoms of leukotoxic exposure are so nonspecific that intentional fabrication or elaboration of complaints is relatively straightforward. Careful clinical assessment may offer the most useful information on this often difficult issue, considering the putative toxic exposure in light of the entirety of the worker's past history and present motivations.

OCCUPATIONAL TOXIN EXPOSURE AND ALZHEIMER'S DISEASE

AD is the most common degenerative dementia, and, as discussed above, the presence of AD or related neurodegenerative diseases may complicate the assessment of possible CTL in workers. AD occurs mostly in people over 65, but the possibility of early-onset disease frequently becomes an issue in the assessment of cognitive complaints. Although AD is widely regarded as a disease of the cerebral cortex, the etiopathogenesis of the disease remains obscure in most cases, and white-matter pathology developing before the onset of dementia and cortical neuropathology is receiving considerable attention (Bartzokis, 2011). In this respect, leukotoxicity from environmental agents becomes a consideration as one of the early white-matter insults that could lead to AD in later years. Occupational risk factors have been studied to a limited extent in AD, using epidemiologic methods to focus mainly on the categories of solvents, metals, and pesticides. Recent reviews of this literature have concluded that, while solvents and metals are not unlikely to be implicated, exposure to pesticides has some support as a risk factor for AD (Santibáñez et al., 2007; Zaganas et al., 2013). Most recently, elevated serum levels of dichlorodiphenyldichloroethylene (DDE), the metabolite of the organochlorine pesticide dichlorodiphenyltrichloroethane (DDT), have been associated with an increase in AD risk (Richardson et al., 2014). DDT was commonly used in the USA from the 1940s until its ban in 1972, and many older adults may be at risk from exposure to this pesticide during those years. The mechanism underlying the possible neurotoxic effect of DDT may be related to the cortical neuropathology of AD, as DDT increases amyloid precursor protein levels (Richardson et al., 2014). However, despite the possibility that pesticides may increase AD risk, the data supporting this claim are far from conclusive. Even if AD is found to have a firm association with pesticides, it appears that the neurotoxicity of these chemicals will involve gray-matter regions. A primary leukotoxic injury of pesticides leading to AD is conceivable, but evidence for this pathogenetic mechanism is lacking.

SUMMARY

CTL is rare in workers, as occupational exposure to potential leukotoxins is typically low, but the syndrome is more likely to occur in individuals exposed to higher and more hazardous levels of leukotoxins. This summary statement implies that, while much has been learned about this field, much more needs to be investigated. Extrapolating from neurobehavioral syndromes involving extreme leukotoxic exposure to the low-level exposures typical of the workplace is fraught with difficulty, and thresholds for leukotoxic injury can only be roughly estimated. In considering the entire range of leukotoxins that have come to be recognized in recent decades, it is clear that toxic white-matter injury is a potential hazard of many settings, clinical and otherwise (Filley and Kleinschmidt-DeMasters, 2001). It is equally apparent that white-matter toxicity supports much evidence that myelinated systems are critical in the organization of distributed neural networks subserving cognition and emotion (Schmahmann et al., 2008; Filley, 2012). Human beings, however, are exposed to an extraordinary range of toxins – affecting the nervous system and many others – on a regular basis, and it is highly implausible that low-level exposures to the great majority of these agents is sufficient to cause tissue injury by overwhelming the body's natural defense mechanisms. For workers and others in society who may be exposed to low levels of leukotoxins, the most important question is what levels of these toxins can be considered safe. Only a general answer to this question can be made with current knowledge. However, the means by which these challenging issues can be productively addressed are becoming ever more sophisticated, and there is ample reason for looking forward to more satisfying answers to previously intractable problems.

FUTURE DIRECTIONS

Neurotoxicology is a challenging field, and many of the reasons for this statement are evident in the complex and nascent area of toxic white-matter disease, in which neurobehavioral function is so critically involved. Not only is the study of human behavior fraught with inherent issues of subjectivity and imprecision, but the variety of diverse leukotoxins, the difficulty in measuring their effects, and the multiplicity of other factors that can impact behavior are all major confounds. CTL, however, differs from other areas of neurotoxicology in that an emerging diagnostic technique exists that promises to improve the assessment of target tissue damage in an objective and reliable manner. Neuroimaging, which has already in essence defined this field of study, has the additional potential to clarify with ever-increasing precision the toxic effects of many known and suspected agents on white matter. The MRI era has made possible the investigation of white-matter lesions visible to the naked eye, and advances now available include a wide array of techniques that permit the visualization of white matter that appears, but may not in fact be, normal. The next epoch in the study of toxic leukoencephalopathy will very likely take advantage of the opportunity to evaluate the normal-appearing white matter (NAWM) of the brain with the goal of detecting tissue injury more subtle than that which is visible on conventional MRI. The microstructural white-matter damage that may be uncovered by such investigation will need to be carefully interpreted in light of clinical, neuropsychologic, and autopsy data when available, but the potential for elucidating previously mysterious details of low-level exposure is promising indeed. Not only will this work disclose more about threshold effects for leukotoxicity in a wide range of settings, thus enabling more informed decisions about workplace exposure levels and prevention of leukotoxic injury, it will also allow a more complete understanding of the role of white matter in neurobehavioral function.

NEUROIMAGING OF WHITE-MATTER MICROSTRUCTURE

As discussed above, MRI has proven transformative not only in neurotoxicology but in neurology as a whole, clearly surpassing CT for the clinical evaluation of cognitive disorders (Tanridag and Kirshner, 1987). In particular, MRI provides excellent depiction of white-matter lesions that can be readily seen on standard T2-weighted, proton density, and FLAIR sequences. As time passed, however, advances in neuroimaging technologies were made, and the macrostructural white-matter lesions seen on conventional MRI were soon recognized as representing only a portion of the neuropathology in many disorders. The notion of the NAWM was thus proposed, implying that even the white matter that looks to be normal on standard MRI may harbor early neuropathology of the type that may later develop into macrostructural lesions (Filley, 2012). Changes in white-matter microstructure may involve myelin, axons, glial cells, or any combination of these elements, and have been observed in all varieties of white-matter disorder as well as several psychiatric diseases and normal aging (Filley, 2012). Whereas it can be reasonably asked how much these neuroimaging changes reflect true structural damage within the white matter, there can be no doubt that new research avenues are made possible by these technologies. Given that white matter is not as readily accessible to detailed study as gray matter (Parvizi, 2009), any success in drilling down to the microscopic level of myelinated systems with noninvasive neuroimaging is welcome, and may well prove as transformative as was MRI three decades ago.

Before discussing these new methods, it is relevant to consider the development of the major functional neuroimaging techniques that have flourished in recent decades. These modalities – PET, SPECT, and fMRI – all offer the prospect of imaging the brain in action because they are designed to capture brain activity rather than details of its anatomy. Whereas these techniques offer impressive views of brain function, which can be at rest or task-based, they are less useful for CTL and other white-matter disorders because they only assess gray-matter regions where metabolic activity in association with brain function is most apparent. Functional neuroimaging may in time elucidate the gray-matter correlates of white-matter tract dysfunction, but at this point the detailed description of white-matter structure is the crucial priority of research in this area.

The three new MRI techniques that have led to the interest in visualizing the microstructure of white matter are magnetization transfer imaging (Van Buchem, 1999), magnetic resonance spectroscopy (MRS: Rudkin and Arnold, 1999), and diffusion tensor imaging (DTI: Horsfield and Jones, 2002). Details of these techniques can be found elsewhere (Filley, 2012), but for the purposes of this review, three measurements are most important. MRS generates a spectrum of neurometabolites, of which *N*-acetyl-aspartate (NAA), an indicator of axonal integrity, and choline, a marker of demyelination and inflammation, are most useful. DTI yields several measures, but the most helpful is fractional anisotropy (FA), a general indicator of large-tract function (Filley, 2012). The key point is that each of these measures can reveal changes in the NAWM that may be clinically significant and helpful in understanding the etiopathogenesis of disease. In the toxic leukoencephalopathies, MRS and DTI have thus been most usefully employed.

To begin with studies of the higher exposures associated with toluene abuse, both MRS and DTI observations have been informative. Findings of reduced cerebral white-matter NAA on MRS (Aydin et al., 2003) and lower FA on DTI (Yücel et al., 2008) in toluene abuse indicate that early white-matter involvement may be detectable before conventional MRI becomes abnormal. Interestingly, similar but less severe DTI white-matter changes have been found in cannabis use (Yücel et al., 2010; Zalesky et al., 2012), and also in opiate (Bora et al., 2010) and methamphetamine addiction (Kim et al., 2009), other forms of substance abuse in which leukotoxicity is likely (Filley, 2012). With regard to occupational toluene exposure, an MRS study of shoemakers found increased choline in the parietal white matter, and in the thalamus and basal ganglia (Alkan et al., 2004).

There has been additional work on white-matter microstructural neuroimaging in lead exposure. An MRS study of adults with childhood lead exposure found decreased choline in the frontal white matter and decreased NAA in the basal ganglia and cerebellum (Cecil et al., 2011). Similarly, DTI studies of adults with childhood lead exposure have found significant FA decreases in the cerebral hemispheric white matter, including the corpus callosum (Brubaker et al., 2009). Most relevant from an occupational medicine perspective was a controlled DTI study of factory workers exposed to lead revealing that lead exposure was associated with significantly decreased FA in the frontal, temporal, parietal, and occipital lobes of both hemispheres (Hsieh et al., 2009).

These studies are clearly preliminary, and far more work is needed before any conclusions can be drawn about the utility of advanced neuroimaging in the detection, quantitation, and clinical significance of occupational white-matter toxic injury. While the early results are intriguing, and in many respects consistent with what is known of more intense exposure to leukotoxins, the existing data raise at least as many questions as they answer. One of the crucial issues, for example, is whether the finding of leukotoxicity in a given population means that only white matter is affected. Indeed, recent studies of lead toxicity also highlight the adverse effects of this metal on gray matter, as cortical volume loss has been associated with lead exposure (Brubaker et al., 2010). Conversely, neurotoxic exposures that have been thought to involve only gray-matter toxicity may be found to have unexpected white-matter involvement. In this regard, a DTI study of welders with manganese exposure disclosed reductions of FA in the frontal white matter and the corpus callosum (Kim et al., 2011), an unexpected finding in view of much evidence that manganese is toxic to the basal ganglia (Olanow, 2004). Another question concerns the reproducibility of MRS and DTI results. Important differences in equipment capability and technical expertise may exist between centers, rendering study comparisons suspect. Because these techniques are rapidly evolving, differences in technical sophistication may develop at such a pace that an innovation introduced to one research group is not readily available to another.

Nevertheless, the promise of microstructural white-matter neuroimaging is undeniable. Not only in neurotoxicology but in behavioral neurology and neurology broadly considered, the new techniques of advanced neuroimaging that enable the detailed assessment of cerebral white matter are highly appealing in their potential utility as a means of defining subtle but clinically important white-matter neuropathology at an early stage when prevention and treatment are more likely to succeed. Combined with careful clinical evaluation, and sensitive neuropsychologic testing appropriately interpreted, advanced neuroimaging of white-matter microstructure may herald a new era in neurotoxicology. The benefits of these exciting new technologies will not only add to our knowledge of leukotoxic syndromes, but also a wide range of other white-matter disorders that have so often been dismissed as irrelevant for the complex operations of cognition and emotion. Toxic leukoencephalopathy, particularly its chronic form, has helped to develop an innovative approach that considers white-matter–behavior relationships as an integral component of how the brain mediates higher function.

References

Albers JW, Berent S (2000). Controversies in neurotoxicology: current status. Neurol Clin 18: 741–764.

Al-Hajri Z, Del Bigio MR (2010). Brain damage in a large cohort of solvent abusers. Acta Neuropathol 119: 435–445.

Alkan A, Kutlu R, Hallac T et al. (2004). Occupational prolonged organic solvent exposure in shoemakers: brain MR spectroscopy findings. Magn Reson Imaging 22: 707–713.

Allen MT, Levy LS (2013). Parkinson's disease and pesticide exposure – a new assessment. Crit Rev Toxicol 43: 515–534.

Ameno K, Kiriu T, Fuke C et al. (1992). Regional brain distribution of toluene in rats and in a human autopsy. Arch Toxicol 66: 153–156.

Arlien-Søborg P, Bruhn P, Gyldensted C et al. (1979). Chronic painters' syndrome. Chronic toxic encephalopathy in house painters. Acta Neurol Scand 60: 149–156.

Axelson O, Hane M, Hogstedt C (1976). A case-referent study on neuropsychiatric disorders among workers exposed to solvents. Scand J Work Environ Health 2: 14–20.

Aydin K, Sencer S, Demir T et al. (2002). Cranial MR findings in chronic toluene abuse by inhalation. AJNR Am J Neuroradiol 23: 1173–1179.

Aydin K, Sencer S, Ogel K et al. (2003). Single-voxel proton MR spectroscopy in toluene abuse. Magn Reson Imaging 21: 777–785.

Baker Jr EL, Smith TJ, Landrigan PJ (1985). The neurotoxicity of industrial solvents: a review of the literature. Am J Ind Med 8: 207–217.

Bartzokis G (2011). Alzheimer's disease as homeostatic responses to age-related myelin breakdown. Neurobiol Aging 32: 1341–1371.

Beckett WS, Moore JL, Keough JP et al. (1986). Acute encephalopathy due to occupational exposure to arsenic. Br J Ind Med 43: 66–67.

Benignus VA (1981). Health effects of toluene: a review. Neurotoxicology 2: 567–588.

Bleecker ML, Lindgren KN, Ford DP (1997). Differential contribution of current and cumulative indices of lead dose to neuropsychological performance by age. Neurology 48: 639–645.

Bleecker ML, Ford DP, Vaughan CG et al. (2007). The association of lead exposure and motor performance mediated by cerebral white matter change. Neurotoxicology 28: 318–323.

Bolla KI, Schwartz BS, Agnew J et al. (1990). Subclinical neuropsychiatric effects of chronic low-level solvent exposure in US paint manufacturers. J Occup Med 32: 671–677.

Boor JW, Hurtig HI (1977). Persistent cerebellar ataxia after exposure to toluene. Ann Neurol 2: 440–442.

Bora E, Yücel M, Fornito A et al. (2010). White matter microstructure in opiate addiction. Addict Biol 17: 141–148.

Brubaker CJ, Schmithorst VJ, Haynes EN et al. (2009). Altered myelination and axonal integrity in adults with childhood lead exposure: a diffusion tensor imaging study. Neurotoxicology 30: 867–875.

Brubaker CJ, Dietrich KN, Lanphear BP et al. (2010). The influence of age of lead exposure on adult gray matter volume. Neurotoxicology 31: 259–266.

Caldemeyer KS, Pascuzzi RM, Moran CC et al. (1993). Toluene abuse causing reduced MR signal intensity in the brain. AJR Am J Roentgenol 161: 1259–1261.

Cecil KM, Dietrich KN, Altaye M et al. (2011). Proton magnetic resonance spectroscopy in adults with childhood lead exposure. Environ Health Perspect 119: 403–408.

Cha JH, Kim SS, Han H et al. (2002). Brain MRI findings of carbon disulfide poisoning. Korean J Radiol 3: 158–162.

Chao LL, Rothlind JC, Cardenas VA et al. (2010). Effects of low-level exposure to sarin and cyclosarin during the 1991 Gulf War on brain function and brain structure in US veterans. Neurotoxicology 31: 493–501.

Chao LL, Abadjian L, Hlavin J et al. (2011). Effects of low-level sarin and cyclosarin exposure and Gulf War illness on brain structure and function: a study at 4T. Neurotoxicology 32: 814–822.

Cho SK, Kim RH, Yim SH et al. (2002). Long-term neuropsychological effects and MRI findings in patients with CS2 poisoning. Acta Neurol Scand 106: 269–275.

Chuang WL, Huang CC, Chen CJ et al. (2007). Carbon disulfide encephalopathy: cerebral microangiopathy. Neurotoxicology 28: 387–393.

Cole M, Scheulein M, Kerwin D (1966). Arsenical encephalopathy due to use of Milibis. Arch Intern Med 17: 706–711.

de Souza A, Narvencar KP, Desai PK et al. (2013). Adult lead encephalopathy. Neurol Res 35: 54–58.

Dobbs MR (2011). Toxic encephalopathy. Semin Neurol 31: 184–193.

Dubois B, Slachevsky A, Litvan I et al. (2000). The FAB: a Frontal Assessment Battery at bedside. Neurology 55: 1621–1626.

Edling C, Ekberg K, Ahlborg Jr G et al. (1990). Long-term follow up of workers exposed to solvents. Br J Ind Med 47: 75–82.

Errebo-Knudsen EO, Olsen F (1986). Organic solvents and presenile dementia (the painters' syndrome). A critical review of the Danish literature. Sci Total Environ 48: 45–67.

Feldman RG, Ratner MH, Ptak T (1999). Chronic toxic encephalopathy in a painter exposed to mixed solvents. Environ Health Perspect 107: 417–422.

Filley CM (2010). White matter: organization and functional relevance. Neuropsychol Rev 20: 158–173.

Filley CM (2011). Neurobehavioral anatomy, 3rd ed. University of Colorado Press, Boulder.

Filley CM (2012). The behavioral neurology of white matter, 2nd ed. Oxford University Press, New York.

Filley CM (2013). Toluene abuse and white matter: a model of toxic leukoencephalopathy. Psychiatr Clin North Am 36: 293–302.

Filley CM, Kleinschmidt-DeMasters BK (2001). Toxic leukoencephalopathy. N Engl J Med 345: 425–432.

Filley CM, Franklin GM, Heaton RK et al. (1988). White matter dementia: clinical disorders and implications. Neuropsychiatry Neuropsychol Behav Neurol 1: 239–254.

Filley CM, Heaton RK, Nelson LM et al. (1989). A comparison of dementia in Alzheimer's disease and multiple sclerosis. Arch Neurol 46: 157–161.

Filley CM, Rosenberg NL, Heaton RK (1990). White matter dementia in chronic toluene abuse. Neurology 40: 532–534.

Filley CM, Halliday W, Kleinschmidt-DeMasters BK (2004). The effects of toluene on the central nervous system. J Neuropathol Exp Neurol 63: 1–12.

Folstein MF, Folstein SE, McHugh PR (1975). Mini-mental state. A practical method for grading the cognitive state of patients for the clinician. J Psychiatr Res 12: 189–198.

Fornazzari L, Pollanen MS, Myers V et al. (2003). Solvent abuse-related toluene leukoencephalopathy. J Clin Forensic Med 10: 93–95.

Freeman JW, Couch JR (1978). Prolonged encephalopathy with arsenic poisoning. Neurology 28: 853–855.

Frey KA (2000). Neuroimaging in neurotoxicology. Neurol Clin 18: 615–629.

Gade A, Mortensen EL, Bruhn P (1988). "Chronic painter's syndrome". A reanalysis of psychological test data in a group of diagnosed cases, based on comparisons with matched controls. Acta Neurol Scand 77: 293–306.

Geschwind N (1965). Disconnexion syndromes in animals and man. Brain 88 (237–294): 585–644.

Gospe Jr SM, Calaban MJ (1988). Central nervous system distribution of inhaled toluene. Fundam Appl Toxicol 11: 540–545.

Grabski DA (1961). Toluene sniffing producing cerebellar degeneration. Am J Psychiatry 118: 461–462.

Greve KW, Bianchini KJ, Black FW et al. (2006). The prevalence of cognitive malingering in persons reporting exposure to occupational and environmental substances. Neurotoxicology 27: 940–950.

Griebe M, Förster A, Wessa M et al. (2011). Loss of callosal fibre integrity in healthy elderly with age-related white matter changes. J Neurol 258: 1451–1459.

Hane M, Axelson O, Blume J et al. (1977). Psychological function changes among house painters. Scand J Work Environ Health 3: 91–99.

Hänninen H, Eskelinen L, Husman K et al. (1976). Behavioral effects of long-term exposure to a mixture of organic solvents. Scand J Work Environ Health 2: 240–255.

Harris JB, Blain PG (2004). Neurotoxicology: what the neurologist needs to know. J Neurol Neurosurg Psychiatry 75 (Suppl 3): iii29–iii34.

Heaton KJ, Palumbo CL, Proctor SP et al. (2007). Quantitative magnetic resonance brain imaging in US army veterans of the 1991 Gulf War potentially exposed to sarin and cyclosarin. Neurotoxicology 28: 761–769.

Hormes JT, Filley CM, Rosenberg NL (1986). Neurologic sequelae of chronic solvent vapor abuse. Neurology 36: 698–702.

Horsfield MA, Jones DK (2002). Applications of diffusion-weighted and diffusion tensor MRI to white matter diseases – a review. NMR Biomed 15: 570–577.

Howard MO, Bowen SE, Garland EL et al. (2011). Inhalant use and inhalant use disorders in the United States. Addict Sci Clin Pract 6: 18–31.

Hsieh TJ, Chuang HY, Chen YC et al. (2009). Subclinical white matter integrity in subjects with cumulative lead exposure. Radiology 252: 509–517.

Jersey SL, Baril RT, McCarty RD et al. (2010). Severe neurological decompression sickness in a U-2 pilot. Aviat Space Environ Med 81: 64–68.

Jung KH, Chu K, Kim YA et al. (2007). Rapidly progressive toxic leukoencephalomyelopathy with myelodysplastic syndrome: a clinicopathological correlation. J Clin Neurol 3: 45–49.

Kamel F, Hoppin JA (2004). Association of pesticide exposure with neurologic dysfunction and disease. Environ Health Perspect 112: 950–958.

Kamran S, Bakshi R (1998). MRI in chronic toluene abuse: low signal in the cerebral cortex on T2-weighted images. Neuroradiology 40: 519–521.

Kanno S, Abe N, Saito M et al. (2011). White matter involvement in idiopathic normal pressure hydrocephalus: a voxel-based diffusion tensor imaging study. J Neurol 258: 1949–1957.

Kao HW, Pare L, Kim R et al. (2014). Toxic leukoencephalopathy with atypical MRI features following a lacquer thinner fire. J Clin Neurosci 21: 878–880.

Kaplan E (1990). The process approach to neuropsychological assessment of psychiatric patients. J Neuropsychiatry Clin Neurosci 2: 72–87.

Kenfield MC, Arciniegas DB, Anderson CA et al. (2010). When cognitive evaluation does not disclose a neurologic disorder: experience of a university behavioral neurology clinic. Cogn Behav Neurol 23: 112–118.

Keski-Säntti P, Mäntylä R, Lamminen A et al. (2009). Magnetic resonance imaging in occupational chronic solvent encephalopathy. Int Arch Occup Environ Health 82: 595–602.

Kim IS, Kim YT, Song HJ et al. (2009). Reduced corpus callosum white matter microstructural integrity revealed by diffusion tensor eigenvalues in abstinent methamphetamine addicts. Neurotoxicology 30: 209–213.

Kim Y, Jeong KS, Song HJ et al. (2011). Altered white matter microstructural integrity revealed by voxel-wise analysis of diffusion tensor imaging in welders with manganese exposure. Neurotoxicology 32: 100–109.

Klinken L, Arlien-Søborg P (1993). Brain autopsy in organic solvent syndrome. Acta Neurol Scand 87: 371–375.

Knox JW, Nelson JR (1966). Permanent encephalopathy from toluene inhalation. N Engl J Med 275: 1494–1496.

Kornfeld M, Moser AB, Moser HW et al. (1994). Solvent vapor abuse leukoencephalopathy. Comparison to adrenoleukodystrophy. J Neuropathol Exp Neurol 53: 389–398.

Kozora E, Filley CM (2011). Cognitive dysfunction and white matter abnormalities in systemic lupus erythematosus. J Int Neuropsychol Soc 22: 1–8.

Ku MC, Huang CC, Kuo HC et al. (2003). Diffuse white matter lesions in carbon disulfide intoxication: microangiopathy or demyelination. Eur Neurol 50: 220–224.

Lafosse JM, Corboy JR, Leehey MA et al. (2007). MS vs. HD: can white matter and subcortical gray matter pathology be distinguished neuropsychologically? J Clin Exp Neuropsychol 29: 142–154.

Lubman DI, Yücel M, Lawrence AJ (2008). Inhalant abuse among adolescents: neurobiological considerations. Br J Pharmacol 154: 316–326.

Lundberg U (1999). Stress responses in low-status jobs and their relationship to health risks: musculoskeletal disorders. Ann N Y Acad Sci 896: 162–172.

Maizlish NA, Langolf GD, Whitehead LW et al. (1985). Behavioural evaluation of workers exposed to mixtures of organic solvents. Br J Ind Med 42: 579–590.

Maizlish NA, Fine LJ, Albers JW et al. (1987). A neurological evaluation of workers exposed to mixtures of organic solvents. Br J Ind Med 44: 14–25.

Marulanda N, Colegial C (2005). Neurotoxicity of solvents in brain of glue abusers. Environ Toxicol Pharmacol 19: 671–675.

McGuire SA, Sherman PM, Brown AC et al. (2012). Hyperintense white matter lesions in 50 high-altitude pilots with neurologic decompression sickness. Aviat Space Environ Med 83: 1117–1122.

McGuire S, Sherman P, Profenna L et al. (2013). White matter hyperintensities on MRI in high-altitude U-2 pilots. Neurology 81: 729–735.

McKinney AM, Kieffer SA, Paylor RT et al. (2009). Acute toxic leukoencephalopathy: potential for reversibility clinically and on MRI with diffusion-weighted and FLAIR imaging. AJR Am J Roentgenol 193: 192–206.

Meador KJ (1998). Cognitive side effects of medications. Neurol Clin 16: 141–155.

Mesulam M-M (1990). Large-scale neurocognitive networks and distributed processing for attention, language, and memory. Ann Neurol 28: 597–613.

Mi T, Han C, Wang Y et al. (2013). Acute toxic leukoencephalopathy in migrant workers exposed to organic solvents in construction materials. Occup Environ Med 70: 435–436.

Nasreddine ZS, Phillips NA, Bédirian V et al. (2005). The Montreal Cognitive Assessment, MoCA: a brief screening

tool for mild cognitive impairment. J Am Geriatr Soc 53: 695–699.
Needleman HL, Schell A, Bellinger D et al. (1990). The long-term effects of exposure to low doses of lead in childhood. An 11-year follow-up report. N Engl J Med 322: 83–88.
Newmark J (2007). Nerve agents. Neurologist 13: 20–32.
Nolte J (2002). The human brain, 5th ed. Mosby, St. Louis, 2002.
Olanow CW (2004). Manganese-induced parkinsonism and Parkinson's disease. Ann N Y Acad Sci 1012: 209–223.
Parvizi J (2009). Corticocentric myopia: old bias in new cognitive sciences. Trends Cogn Sci 13: 354–359.
Pryor GT, Dickinson J, Howd RA et al. (1983). Transient cognitive deficits and high-frequency hearing loss in weanling rats exposed to toluene. Neurobehav Toxicol Teratol 15: 53–57.
Qureshi SU, Blanchette AR, Jawaid A et al. (2009). Reversible leukoencephalopathy due to chronic unintentional exposure to toluene. Can J Neurol Sci 36: 388–389.
Richardson JR, Roy A, Shalat SL et al. (2014). Elevated serum pesticide levels and risk for Alzheimer Disease. JAMA Neurol 71: 284–290.
Ridgway P, Nixon TE, Leach JP (2003). Occupational exposure to organic solvents and long-term nervous system damage detectable by brain imaging, neurophysiology or histopathology. Food Chem Toxicol 41: 153–187.
Rosenberg NL (1992). Neurotoxicology. In: JB Sullivan, GR Kreiger (Eds.), Hazardous materials toxicology, Williams and Wilkins, Baltimore, pp. 143–153.
Rosenberg NL (1995). Occupational and environmental neurology, Butterworth-Heinemann, Boston.
Rosenberg NL, Spitz MC, Filley CM et al. (1988a). Central nervous system effects of chronic toluene abuse – clinical, brainstem evoked response and magnetic resonance imaging studies. Neurotoxicol Teratol 10: 489–495.
Rosenberg NL, Kleinschmidt-DeMasters BK, Davis KA et al. (1988b). Toluene abuse causes diffuse central nervous system white matter changes. Ann Neurol 23: 611–614.
Rosenberg NL, Grigsby J, Dreisbach J et al. (2002). Neuropsychologic impairment and MRI abnormalities associated with chronic solvent abuse. J Toxicol Clin Toxicol 40: 21–34.
Rudkin TM, Arnold DL (1999). Proton magnetic resonance spectroscopy for the diagnosis and management of cerebral disorders. Arch Neurol 56: 919–926.
Sahu JK, Sharma S, Kamate M et al. (2010). Lead encephalopathy in an infant mimicking a neurometabolic disorder. J Child Neurol 25: 390–392.
Sanders T, Liu Y, Buchner V et al. (2009). Neurotoxic effects and biomarkers of lead exposure: a review. Rev Environ Health 24: 15–45.
Santibáñez M, Bolumar F, García AM (2007). Occupational risk factors in Alzheimer's disease: a review assessing the quality of published epidemiological studies. Occup Environ Med 64: 723–732.
Schaumburg HH, Albers JW (2005). Pseudoneurotoxic disease. Neurology 65: 22–26.
Schaumburg HH, Spencer PS (1987). Recognizing neurotoxic disease. Neurology 37: 276–278.
Schaumburg HH, Spencer PS (2000). Classification of neurotoxic responses based on vulnerability of cellular sites. Neurol Clin 18: 517–524.
Schmahmann JD, Pandya DN (2006). Fiber pathways of the brain. Oxford University Press, New York.
Schmahmann JD, Smith EE, Eichler FS et al. (2008). Cerebral white matter: neuroanatomy, clinical neurology, and neurobehavioral correlates. Ann N Y Acad Sci 1142: 266–309.
Sheline GE, Wara WM, Smith V (1980). Therapeutic irradiation and brain injury. Int J Radiat Oncol Biol Phys 6: 1215–1228.
Spencer PS, Schaumburg HH (Eds.), (2000). Experimental and Clinical Neurotoxicology, 2nd ed. New York, Oxford University Press.
Stewart WF, Schwartz BS, Simon D et al. (1999). Neurobehavioral function and tibial and chelatable lead levels in 543 former organolead workers. Neurology 52: 1610–1617.
Stewart WF, Schwartz BS, Davatzikos C et al. (2006). Past adult lead exposure is linked to neurodegeneration measured by brain MRI. Neurology 66: 1476–1484.
Tanridag O, Kirshner HS (1987). Magnetic resonance imaging and CT scanning in neurobehavioral syndromes. Comparative neuroradiologic findings. Psychosomatics 28: 517–528.
Thuomas KA, Möller C, Odkvist LM et al. (1996). MR imaging in solvent-induced chronic toxic encephalopathy. Acta Radiol 37: 177–179.
Triebig G, Lang C (1993). Brain imaging techniques applied to chronically solvent-exposed workers: current results and clinical evaluation. Environ Res 61: 239–250.
U.S. Department of Veterans Affairs (2008). Research Advisory Committee on Gulf War Veterans' Illnesses, Washington, D.C.
Unger E, Alexander A, Fritz T et al. (1994). Toluene abuse: physical basis for hypointensity of the basal ganglia on T2-weighted MR images. Radiology 193: 473–476.
Van Buchem MA (1999). Magnetization transfer: applications in neuroradiology. J Comput Assist Tomogr 23 (suppl 1): S9–S18.
World Health Organization (1985). Chronic Effects of Organic Solvents on the Central Nervous System and Diagnostic Criteria, World Health Organization and Nordic Council of Ministers, Copenhagen.
Xiong L, Matthes JD, Li J et al. (1993). MR imaging of "spray heads": toluene abuse via aerosol paint inhalation. AJNR Am J Neuroradiol 14: 1195–1199.
Yamanouchi N, Okada S, Kodama K et al. (1997). Effects of MRI abnormalities on WAIS-R performance in solvent abusers. Acta Neurol Scand 96: 34–39.
Yamasue H, Abe O, Kasai K et al. (2007). Human brain structural change related to acute single exposure to sarin. Ann Neurol 61: 37–46.
Yücel M, Takagi M, Walterfang M et al. (2008). Toluene misuse and long-term harms: a systematic review of the

neuropsychological and neuroimaging literature. Neurosci Biobehav Rev 32: 910–926.

Yücel M, Zalesky A, Takagi MJ et al. (2010). White-matter abnormalities in adolescents with long-term inhalant and cannabis use: a diffusion magnetic resonance imaging study. J Psychiatry Neurosci 35: 409–412.

Zaganas I, Kapetanaki S, Mastorodemos V et al. (2013). Linking pesticide exposure and dementia: what is the evidence? Toxicology 307: 3–11.

Zalesky A, Solowij N, Yücel M et al. (2012). Effect of long-term cannabis use on axonal fibre connectivity. Brain 135: 2245–2255.

Zhang K, Sejnowski TJ (2000). A universal scaling law between gray matter and white matter of cerebral cortex. Proc Natl Acad Sci U S A 97: 5621–5626.

Handbook of Clinical Neurology, Vol. 131 (3rd series)
Occupational Neurology
M. Lotti and M.L. Bleecker, Editors

Chapter 7

Neurotoxicity of solvents

MARKKU ALARIK SAINIO SR.*

Development of Work and Organizations, The Finnish Institute of Occupational Health, Helsinki, Finland

ORGANIC SOLVENTS

Organic solvents is a common designation for a large group of more than 200 chemical compounds capable of dissolving nonwater-soluble materials such as fats, oils, waxes, resins, rubber, asphalt, cellulose filaments, and plastic materials. Compounds from many different chemical groups can serve as solvents. Organic refers to compounds that contain carbon bonds and in which at least one carbon atom is covalently linked to an atom of another type (commonly hydrogen, oxygen, or nitrogen). Organic solvents contain at least one carbon and one hydrogen molecule. Aliphatic compounds form a chain, whereas aromatic compounds form a six-carbon ring. The hydrogen group may be substituted by some other element such as a hydroxyl group in alcohols or a carbonyl group in ketones and esters. Halogenated hydrocarbons contain a substituted halogen element, often a chloride. As a group, they share few physical features and even fewer chemical properties, but a common feature and the potential explanation to neurotoxicity is their lipophilicity and thereby affinity for the nervous system.

Organic solvents include aliphatic, aromatic, or chlorinated hydrocarbons and alcohols, ethers, esters, ketones, carbon disulfide, and fuels (Table 7.1). Solvent products are often mixtures of chemical substances and solvents, e.g., white spirit with varying concentrations of aliphatic and aromatic hydrocarbons. They are usually fluids at room temperature and evaporate easily. They enter the body mainly by inhalation and most solvents penetrate through the skin. Solvents are metabolized mainly in the liver and they accumulate to lipid-rich tissues such as the brain. A chemical is considered neurotoxic if it is capable of inducing a consistent pattern of neuroanatomic change or neural dysfunction that causes neurochemical, neurophysiologic, or behavioral effects (Harry et al., 2001). Aromatic and halogenated hydrocarbons are considered more neurotoxic than other hydrocarbons and alcohols.

Selected chemicals are presented here in more detail due to their widespread use, neurotoxic potential in the central nervous system, and ability to cause occupational diseases. This selection does not exclude the lack of neurotoxic potential in other solvents. The adverse solvent effects to the nervous system may not be directly derived based on the neurotoxic properties of a specific solvent because solvents are mostly used in mixtures with varying components and concentrations. Hydrocarbons serve as a typical representative of the clinically relevant neurotoxic solvents. The clinical picture of chronic solvent encephalopathy due to different solvents seems to be very similar, which is rather surprising considering their few physical and chemical properties (Keski-Säntti et al., 2010; Spee et al., 2012). Thus, the focus is on solvent encephalopathy seen in occupational settings due to hydrocarbons and not the more rarely seen effects of certain solvents such carbon disulfide-related cerebral vasculopathy.

Hereon, the term "solvent" is used instead of organic solvents, CNS for central nervous system, and CSE for solvent-induced or chronic solvent or toxic encephalopathies. CSE cases are workers which have been diagnosed to have encephalopathy related to occupational solvent exposure.

OCCUPATIONAL USE OF SOLVENTS

Millions of workers are regularly occupationally exposed to organic solvents that are considered neurotoxic (Caldwell et al., 2000). Large groups of workers in a great variety of professions are exposed daily, or at least frequently. Solvents may be components in paints, varnishes, lacquers, adhesives, glues, and

*Correspondence to: Markku Alarik Sainio Sr., M.D., Ph.D., Adjunct Professor, Specialist in Neurology, The Finnish Institute of Occupational Health, Topliuksenkatu 41 a A, Helsinki 00250, Finland. Tel: +35-8405693186, E-mail: markku.sainio@ttl.fi

Table 7.1

Occupational use of hydrocarbon solvents with neurotoxicity

Chemical group	Compound (example)	Sources of exposure and main occupational use (example)
Aliphatic hydrocarbons	*n*-Hexane	In glues, paints, lacquers, and printing inks; rubber and shoe industry
Aromatic hydrocarbons	Toluene	In paints, fuel oil, cleaning agents, lacquers, paint thinners; photogravure, printing, car and spray painting, linoleum laying
	Xylene	In paints, lacquers, adhesives, inks, varnishes, dyes, glues; polyester production, photogravure, spray painting, textile and rubber industry
	Styrene	In polymerized plastic, rubber, aircraft fuel; boat building, reinforced plastic industry, fiberglass production, synthesis and manufacture of polymers, copolymers, polyester resins
Chlorinated hydrocarbons	Trichloroethylene	In adhesives, paint removers; degreasing of metal components, dry cleaning, textile and leather industry
	Perchloroethylene	In metal degreasers; dry cleaning, textile industry
	Trichloroethane	In adhesives, inks, glues, paints; cold and dip cleaning, degreasing, metal work, printing, dry cleaning, leather work
	Methylene chloride	In metal degreasers, paint and varnish removers; degreasing, metal painting
Alcohols and glycols	Methanol, ethanol, propanol, butanol	In solvents and detergents
	Ethylene glycol	In antifreezes, a polymer precursor
Ketones	Methyl ethyl ketone	In adhesives, paints, dyes; cleaning, coating, paint stripping, in chemical and textile industry
Esters	Ethyl acetate	In paints, coatings, inks, adhesives; painting, manufacture of chemicals

degreasing or cleaning agents, for example and used in the production of dyes, polymers, plastics, textiles, printing inks, agricultural products, and pharmaceuticals (Table 7.1). Workers in several occupations and work tasks, such as construction, industrial, metal, and car painting, printing-trade work, floor layering and lacquering, reinforced plastic lamination, and wooden surface finishing are typically exposed. Usually, solvents are used in mixtures to achieve optimal dissolving properties. A typical example is white spirit, also called mineral spirit, Stoddard solvent, mineral turpentine, petroleum spirit, or mineral naphtha, which is a mixture of aliphatic hydrocarbons with a content of aromatic hydrocarbons ranging from $<1\%$ to 25%. It is widely used in cleaning and degreasing and in paints, lacquers, varnishes, and wood preservatives. Single-solvent exposure occurs mainly to styrene in boat building and lamination, to toluene in printing, and perchloroethylene in dry cleaning. The use of some solvents with evident toxicity has been forbidden or restricted by national or international bans. These include halogenated hydrocarbons such as benzene and carbon tetrachloride.

In developed countries, improvements in workplace hygiene and the wider use of personal protection and substitution to less harmful products have reduced occupational exposure levels (Caldwell et al., 2000; Dick, 2006; Kim et al., 2010). However, solvents are still a major hygienic problem at many workplaces. High exposure is found in confined surroundings with poor ventilation and high temperature, and with skin contact in addition to vapor inhalation. Despite improvements, there are still work tasks where workers are exposed excessively to solvents, even in countries with advanced work safety. In Norway, 11% of male and 7% of female workers are exposed to solvents daily (Leira et al., 2006). In Finland, more than every fifth employee meets solvents in their working environment and every 10th experiences inconvenience due to them (Tuomi and Ahonen, 2010).

Work tasks with high risk of CSE

The risk of CSE is still high in certain work tasks, where the incidence of CSE cases shows the risk of workers in the field (Keski-Säntti et al., 2010) (Table 7.2). In Finland, the highest risk of CSE is among wooden surface finishers, in industrial, metal, or car painters, in floor layers and lacquerers, in reinforced plastic laminators, and in construction painters (Keski-Säntti et al., 2010). The analyses of exposure of CSE cases by Keski-Säntti and coworkers (2010) revealed that exposure to solvents with known neurotoxic properties had been extensive (median duration 31.5 years, range 9–60 years) and above 50% of the occupational exposure limits given in 1981 to prevent neurologic diseases and corresponding to the threshold limit values of the

Table 7.2

The solvent work tasks that led to chronic solvent encephalopathy (CSE) in 128 cases diagnosed in 1995–2007 in Finland. The incidence of CSE in the work task in relation to the number of employed workforce in that field (Keski-Säntti et al., 2010)

	n (%)*	Incidence[†]
Industrial, metal, or car painting	48 (38)	0.8
Construction painting	26 (20)	0.2
Printing-trade work	12 (9)	0.3
Floor laying and lacquering	11 (9)	0.7
Reinforced plastic lamination	11 (9)	0.3
Wooden surface finishing	9 (7)	0.9
Gluing	9 (7)	NA
Motor vehicle cleaning	8 (6)	NA
Other industrial work	4 (3)	NA
Degreasing	4 (3)	NA
Working with rubber products	2 (2)	0.1
Working with footwear	2 (2)	<0.1
Gasoline handling	2 (2)	<0.1
Mechanic work with vehicles and machines	2 (2)	NA
Maintenance and repair work in industry	2 (2)	NA
Production of paint, lacquer, and printing inks	2 (2)	NA
Laundry worker (dry cleaning)	1 (1)	0.1
Paint removing	1 (1)	NA
Laboratory work	0 (0)	–
Cleaning work	0 (0)	–

*Some cases had more than one main work task.
[†]Average incidence 1995–2007: CSE cases/1000 workers at the work task under consideration according to the National Workforce Survey. NA, not available.

American Conference of Governmental Industrial Hygienists (Riala et al., 1984; Keski-Säntti et al., 2010).

In South Korea, with a population of 48.9 million, only six cases of occupational CSE were reported by 2009, and five of them were painters and one degreaser (Kim et al., 2010). In the Netherlands, between 2003 and 2010, the most common occupations of the CSE cases were painters, paint sprayers, and printers (www.occupationaldiseases.nl), which may be partly due to active screening of painters. In New Zealand, the most common occupations where CSE was detected were spray painters, printers, and boat builders (Dryson and Ogden, 1998).

Solvent pharmacokinetics and neurotoxicity

In work-related exposure, the main route of uptake is the inhalation of solvent vapors and absorption through the skin. High solvent volatility and large surfaces of evaporation with significant air concentrations of vapor, lack of appropriate ventilation systems, and relatively high temperature of the work environment all contribute to increased inhalation (Xiao and Levin, 2000). Lipophilic solvents are readily absorbed in direct skin contact, which is dependent on skin integrity, chemical properties, and concentration of the solvent, duration of exposure, and the surface area of skin exposed. Increase in physical workload adds to solvent absorption by elevating alveolar ventilation and pulmonary perfusion. Uptake depends on the specific air/blood partition coefficients of each individual solvent, determined by the alveolocapillary membrane permeability and the solubility of the solvent in blood (Lof and Johanson, 1998). Substances with low water solubility, e.g., 1,1,1-trichloroethane, reach saturation levels at relatively low blood concentrations and associate with less CNS disturbances, whereas styrene, with higher blood solubility and a potential for progressively rising blood concentrations, is associated with more severe CNS effects (Baker et al., 1985). However, neurotoxicity may not be predicted from the water solubility or lipophilicity factors. The absorbed solvent is metabolized in the liver and removed as metabolites in the urine or, to a lesser extent, in the bile. Metabolism of solvents in the liver occurs in two phases. Phase I, controlled by the cytochrome P450 (CYP) enzyme complex, yields reactive metabolites, further detoxified in phase II (Thier et al., 2003). The toxicity of solvents depends on the toxicity of the solvent itself or the reactive intermediate metabolites.

Interactions between solvents, often used as mixtures, are complicated and incompletely characterized. Solvents may inhibit or induce each other's metabolism, and the effects may be additive, synergistic, or potentiated (Lof and Johanson, 1998). Enzyme competition and induction may occur also between solvents and alcohol or drugs. Blood levels of toluene, xylene, and styrene-7,8-oxide, a highly toxic metabolite of styrene, increase acutely with concomitant alcohol ingestion, whereas workers chronically ingesting alcohol have lower blood levels of toluene and xylene due to metabolizing enzyme induction (Savolainen, 1980; Waldron et al., 1983).

Solvents and their metabolites are distributed to target tissues according to their blood supply and the lipid content of the organ system. The low molecular weight and lipophilicity of solvents enable rapid penetration of the blood–brain barrier and an accumulation in the brain, especially in myelin, with a lipid content of approximately 70% (Jin et al., 2004).

Solvent metabolism-related neurotoxicity

Detoxification enzymes of phase II, such as glutathione S-transferase M1 (GSTM1), T1 (GSTT1), P1 (GSTP1),

CYP2E1, and microsomal epoxide hydrolase (mEH), show genetic polymorphisms affecting the neurotoxicity of solvents (Lof and Johanson, 1998; Viaene et al., 2001; Bolt et al., 2003; Kezic et al., 2006; Godderis et al., 2010). The more slow solvent detoxifying GSTM1 null genotype (GSTM1*0) is more prevalent in CSE patients than in solvent-exposed non-CSE cases (Söderkvist et al., 1996; Ahmadi et al., 2002) and more common in solvent-exposed workers with psychiatric, neurologic, or cognitive symptoms (especially deficits in sustained attention and short-term memory) (Godderis et al., 2010). GSTM1*0 is also associated with an increased risk for Parkinson disease in heavily solvent-exposed individuals (Dick et al., 2007). Solvent workers and chronic toxic encephalopathy patients showed GSTP1 genotypic polymorphism which associated also to DNA hypomethylation with increasing solvent exposure (Godderis et al., 2012).

Solvent-related effects on the neurotransmitter systems

Cellular mechanism have been studied *in vitro*, in cell cultures and animal studies, which indicate that solvents induce changes in the lipid structure of cell membranes which interfere with synaptic membrane transport mechanisms and disturb intercellular communication (Dodd et al., 2000; Jin et al., 2004). Solvents also interact with the lipophilic portions of cell proteins which disrupt, for example, fast axonal transport of proteins such as neurofilaments (Jin et al., 2004). The neurotoxic mechanisms may differ from one solvent to another, but a common factor is the ability to promote the synthesis of reactive oxygen species (Corsi et al., 2007). This may lead to free-radical-induced lipid peroxidation, mitochondrial and nucleic acid damage, and oxidative stress, all of which play a role in the early phases of neuronal apoptosis (Vettori et al., 2005).

The effects of solvents involve complex interactions with neurotransmitters and ion channel systems. Inhalation of some solvents, such as toluene, benzene, xylene, and 1,1,1-trichloroethane, and ingested ethanol acutely inhibit the function of one of the glutamate receptors, the *N*-methyl-D-aspartic acid (NMDA) receptor (Bowen et al., 2006; Chastain, 2006). Chronic exposure causes adaptive upregulation in the sensitivity of NMDA receptors, which results in an increased vulnerability to glutamate-induced excitotoxicity and production of reactive oxygen species (Dodd et al., 2000; Bowen et al., 2006).

Animal studies have revealed chronic solvent effects on monoamine (dopamine, serotonin, and norepinephrine) and cholinergic neurotransmitter systems (Arlien-Søborg, 1992). Rats exposed to toluene or white spirit show increase in nerve-specific marker proteins (e.g., neuron-specific enolase, creatine kinase-B, beta-S100, glial fibrillary acidic protein), suggesting activation of astrocytes and gliosis in the white matter (Huang et al., 1993; Gotohda et al., 2000; Nielsen et al., 2006). The best evidence of solvent effects on neurotransmitters exists for the dopaminergic (DA) system. Rats exposed to solvents showed an increase in dopamine synthesis in the brain without affecting the number of presynaptic terminals or postsynaptic dopamine receptors which may result from solvent-enhanced catalytic activity of the dopadecarboxylase enzyme or as a response to reduced dopamine D2-receptor affinity (von Euler et al., 1987). Although in styrene-exposed workers, an increase in prolactin secretion has been reported, suggesting a possible dysfunction of DA neurons in the hypothalamus which controls pituitary gland prolactin secretion (Mutti et al., 1984; Ferroni et al., 1992; Mutti and Smargiassi, 1998), an increase of secretion in the CNS could not be shown in rats (Jarry et al., 2002). Rats exposed to styrene showed cell loss and DA depletion in retinal amacrine cells (Vettori et al., 2000). The vulnerability of the DA neurons may be due to their characteristics: slow-conducting, small-diameter axon fibers with a low conduction safety factor and a large projection area. In the hypothalamus and ventral tegmental area they are in direct contact with the walls of capillary vessels, with greater risk for exposure to exogenous substances present in blood (Gralewicz and Dyzma, 2005).

SOLVENT ENCEPHALOPATHY

Acute high solvent exposure causes symptoms such as dizziness, disequilibrium, disorientation, euphoria, and a feeling of drunkenness. Exposure to very high solvent concentrations may lead to unconsciousness, convulsions, and death. The acute and transient effects of solvents due to pharmacologic effects and neurochemical changes in the CNS do not necessarily indicate neurotoxicity (Harry et al., 2001). Acute intoxication to very high solvent concentrations (Welch et al., 1991) and long-term low-level exposure (Gregersen et al., 1987; Mikkelsen, 1997; Dick et al., 2000; Visser et al., 2008, Keski-Säntti et al., 2010; Spee et al., 2012; van Valen et al., 2012) may lead to permanent neurotoxic effects in the brain, i.e. CSE.

Evidence for chronic adverse effects in the brain due to occupational solvent exposure began to emerge in the early 1960s after the Finnish psychologist Helena Hänninen described cases with carbon disulfide intoxication in rubber-manufacturing workers (Hänninen, 1964). She proposed the "psycho-organic syndrome" or "organic solvent syndrome" as a new occupational disease. In the 1970s, there was growing awareness, beginning from the Nordic countries, of a pattern of symptoms and signs related to long-term solvent exposure. However, there has been uncertainty of the nature and characteristics of this condition even between experts (Ridgway et al., 2003). Since then, a body of literature has arisen to characterize the brain effects of occupational solvent exposure and CSE, recognised as an occupational disease in many countries.

In the current literature, to avoid confusion with reversible effects, CSE is usually defined as permanent neurotoxic effects to the CNS ranging from mild cognitive signs to neuropsychiatric dysfunction with loss of work ability (van Valen et al., 2012). At the occupational exposure levels of the developed countries, the main target of long-term solvent exposure seems to be the CNS without clinical peripheral neuropathy (Keski-Säntti et al., 2010; www.occupationaldiseases.nl).

Symptoms and clinical signs

The emergence of permanent solvent-effects in the CNS is insidious. Age of onset varies depending on the exposure. In working conditions of a developed country, Finland, mean age at diagnosis was 52.8 years (range 35–73 years) in a series of 129 CSE cases (Keski-Säntti et al., 2010). The cases were mainly males (90%), most likely due to male preponderance in exposed workers (Keski-Säntti et al., 2010). Of these CSE cases only one had an acute onset due to massive exposure intoxication (Keski-Säntti et al., 2010). Manifestations of CSE include a wide variety of nonspecific cognitive, neurologic, somatic, and mood-related symptoms. Symptoms may be categorized as: (1) impairment of memory and other cognitive functions (attention, concentration, abstraction, thinking, planning, etc.) and (2) impaired emotional control and motivation (e.g., emotional lability, irritability, and reduced initiative and energy). The hallmarks of CSE are the cognitive symptoms: impaired memory and difficulties in concentration (Juntunen, 1978, 1993; Arlien-Søborg et al., 1979; Edling et al., 1990; Kaukiainen et al., 2009a, b). Symptom awareness is typical and it has been used in screening and characterisation of CSE (Hooisma et al., 1994; Karlson et al., 2000; Carter et al., 2002; Kaukiainen et al., 2009a, b; Furu et al., 2012; Spee et al., 2012). Poor symptom awareness and clinical progression suggest non-occupational etiology (Kaukiainen et al., 2009a; Keski-Säntti et al., 2010).

Symptoms are nonspecific and can thus be due to, or coexist with, non-occupational factors, e.g., poor state of health, psychiatric disorders, high alcohol consumption, and unemployment (Friis et al., 1997; Kaukiainen et al., 2004a, 2009a, b; Furu et al., 2012). The symptom severity in CSE may not reflect the degree of neuropsychologic findings (Bast-Pettersen, 2009). On the other hand, the cognitive dysfunction often explains the experienced cognitive complaints.

The clinical neurologic examination is usually normal or may reveal slight nonspecific abnormalities, such as disequilibrium, impaired fine motor control, and dyscordination (Arlien-Søborg et al., 1979; Gregersen et al., 1987; Antti-Poika et al., 1989; Juntunen, 1993). Clinical polyneuropathy is usually not present in the absence of intoxication to *n*-hexane or methyl *n*-butyl ketone (Juntunen and Haltia, 1982; Keski-Säntti et al., 2010).

Neurotoxic symptom questionnaires

Symptom questionnaires have been used for epidemiologic studies, screening, early detection, and follow-up of workers with CSE. Neurotoxic questionnaires share very similar features and have been shown to detect CSE: Euroquest (EQ) (Chouanière et al., 1997), Questionnaire 16 or 18 (Q16 or Q18) (Lundberg et al., 1997; Nasterlack et al., 1999; Ihrig et al., 2001; Nordling Nilson et al., 2007), Psychological–Neurological Questionnaire PNF (Kiesswetter et al., 1997) and the Neurotoxicity Symptom Checklist (NSC-60) (Hooisma et al., 1994; Viaene et al., 2001).

EQ has been well validated and shown to be cost-effective and efficient to screen and evaluate CSE (Chouanière et al., 1997; Karlson et al., 2000; Deschamps et al., 2001; Carter et al., 2002; Rouch et al., 2003; Kaukiainen et al., 2004a, 2009a, b; Furu et al., 2012, 2014). The core EQ, with 59 questions, is directed to chronic effects of neurotoxicity. The six symptom domains of core EQ are neurologic, psychosomatic, mood, memory and concentration, fatigue, and sleep disturbance. The symptom domain of memory and concentration (items 38–47) has the best power to discriminate workers with CSE from unexposed referents (Carter et al., 2002; Kaukiainen et al., 2009a). Based on symptom analyses of well-studied CSE cases, a cut-off level of at least three memory and concentration symptoms occurring often or very often has been proposed for screening and diagnostic purposes (Kaukiainen et al., 2009a, b).

Structural findings in the brain

An autopsy study of workers with CSE revealed no neuropathologic abnormalities in the brain (Klinken and Arlien-Soborg, 1993). Another postmortem examination of a patient with parkinsonism induced by prolonged exposure to a mixture of aliphatic hydrocarbons disclosed severe and widespread DA neuronal loss, associated with severe gliosis in the substantia nigra and almost complete loss of tyrosine hydroxylase immunostaining in the striatum (Pezzoli et al., 1996).

Neuroanatomic findings have been studied by neuroimaging. Computed tomography of the brain of workers with CSE has been found either to be normal or to show slight atrophy (Arlien-Søborg et al., 1979; Ørbæk et al., 1987; Leira et al., 1992; Ellingsen et al., 1993; Mikkelsen et al., 1998). Magnetic resonance imaging (MRI), a more sensitive method to assess brain atrophy and white-matter changes, has been found to be normal or to show central or cortical atrophy in both solvent-exposed workers and workers with CSE (Bernsen et al., 1992; Leira et al., 1992; Kelafant et al., 1994; Lundberg et al., 1995; Thuomas et al., 1996; Aaserud et al., 2000; Alkan et al., 2004; Visser et al., 2008; Keski-Säntti et al., 2009). Some of the MRI studies have revealed changes similar to those seen in association with toluene and alcohol abuse, i.e., reduced corpus callosum volume, loss of gray/white-matter discrimination, periventricular white-matter hyperintensities, and hypointensity in the basal ganglia in T2-weighted images, suggesting solvent effects on white matter (Thuomas et al., 1996; Alkan et al., 2004; Haut et al., 2006; Yücel et al., 2008; Harper, 2009). Keski-Säntti and co-workers (2009) found that 38% of the MRI findings of 71 workers with CSE could be classified as abnormal, i.e., presence of atrophy and/or abnormal white-matter hyperintensities (Keski-Säntti et al., 2009). In the same study, brain atrophy in any brain area was found in 24%, and apart from vermian atrophy, which correlated with alcohol consumption, atrophy associated with the duration of solvent exposure (Keski-Säntti et al., 2009). The only diffusion tensor imaging study on workers with CSE failed to reveal abnormalities in the white-matter tracts, but the fractional anisotropy in the thalamus and caudate nucleus predicted reduced performance of attention and psychomotor speed (Visser et al., 2008).

The moderate and nonspecific structural findings weaken their contribution to clinical diagnostics, but they are useful in the differential diagnosis of nonsolvent-related pathology. Altogether in CSE, there is evidence of neuronal loss, axonopathy, and demyelination with frontoparietal preponderance, which is supported by functional imaging studies (see below).

Findings by functional brain imaging

Single-photon emission computed tomography (SPECT), a measure of blood flow in the brain and, indirectly, brain metabolism, has revealed in workers with CSE decreased blood flow mainly in temporal, frontal, and parietal areas (Arlien-Sobørg et al., 1982; Hagstadius et al., 1989; Callender et al., 1993; Heuser et al., 1994; Heuser and Mena, 1998) or higher occipital blood flow (Deschamps et al., 1993). Dopamine transporter binding measured by ^{123}I-ioflupane SPECT was significantly lower both in the striatum and the putamen in solvent-exposed patients with Parkinson's disease than in nonexposed patients with Parkinson's disease (Canesi et al., 2007).

Positron emission tomography (PET) with fluorodeoxyglucose, a direct measure of glucose metabolism, has in two CSE case reports, one with long-term solvent exposure and one with acute tetrabromoethane intoxication, revealed frontoparietal and subcortical (basal ganglia, amygdala, hippocampus) hypometabolism (Morrow et al., 1990; Varney et al., 1998).

In CSE cases and solvent-exposed workers, some findings suggest a dysfunction of the DA neurons in the basal ganglia and an increase in the risk for parkinsonism and Parkinson's disease. Long-term exposure to organic solvents has been shown to increase the rate of dopamine synthesis in the brain without affecting the number of presynaptic terminals or postsynaptic dopamine receptors (Edling et al., 1997). A more recent study has shown reduced striatal D2-binding in CSE patients and also in asymptomatic solvent-exposed workers, indicating a more specific postsynaptic damage (Visser et al., 2008). PET imaging study of a worker with acute parkinsonism due to solvent abuse showed normal fluorodopa uptake but reduced raclopride binding, indicating an unusual striatal D2 dopamine receptor binding (Uitti et al., 1994). Case reports involving PET imaging of patients with signs of parkinsonism and long-term occupational exposure to hydrocarbons have revealed similar but more severe and widespread DA dysfunction and loss of DA nerve terminals than in Parkinson's disease, and also a reduction in dopamine D2-binding sites in the caudate nucleus, findings not present in idiopathic Parkinson's disease (Pezzoli et al., 1995; Hageman et al., 1999). Magnetic resonance spectroscopy (MRS), a noninvasive method to study the metabolism of the brain tissue, has shown in patients with Parkinson's disease with high-degree solvent exposure, normal *N*-acetyl aspartate (NAA) content in brain white matter and occipital lobe, but low NAA levels in the lentiform nucleus of the basal ganglia, suggesting neuronal damage (Rango et al., 2006).

MRS in solvent-exposed workers has shown increased choline levels in the thalamus, basal ganglia, and parietal white matter, indicating demyelination (Alkan et al., 2004). There was a positive correlation between choline/creatinine ratio in basal ganglia and duration of exposure (Alkan et al., 2004). In contrast, MRS of workers with CSE, as well as solvent-exposed workers, showed only a trend for reduced choline, but reductions in frontal gray-matter metabolites NAA and *N*-acetylaspartyl-glutamate (Visser et al., 2008). This finding suggests abnormality in neuronal viability and axonal density, although after post hoc analyses, this effect was significant only in exposed control subjects, but not in the CSE group (Visser et al., 2008).

Functional imaging studies during cognitive tasks have revealed dysfunction in relevant brain areas. In solvent-exposed workers with cognitive impairment, PET during a working-memory task has revealed activation of the frontal cortex in atypical areas, suggesting neural compensation (Haut et al., 2000). Reduced striatal D2 binding has been shown in CSE patients and asymptomatic solvent-exposed workers which was associated with impairment in neuropsychologic tests requiring abstraction and mental flexibility, attention, and response inhibition, i.e., frontal brain functions (Visser et al., 2008). This is in agreement with a functional MRI study of solvent-exposed workers by Tang and co-workers (2011). During the N-Back task, an activation of the anterior cingulate, prefrontal, and parietal cortices – areas serving working-memory function and attention – was significantly lower in the solvent-exposed workers relative to controls (Tang et al., 2011). The finding correlated with neurobehavioral deficits and lifetime solvent exposure (Tang et al., 2011).

In the near future, functional imaging may reveal the different neural basis of conditions with similarities in cognitive deficits, i.e., CSE, alcoholism (Sullivan et al., 2010), Parkinson's disease (Owen, 2004), and aging (Backman et al., 2000; Salthouse, 2000; McEvoy et al., 2001), all suggesting frontotemporoparietal dysfunction and disturbances in the frontostriatothalamic circuitry.

Dysfunction in CNS neurophysiology

Electrophysiologic methods, i.e., electroencephalography (EEG) and evoked potentials, have been used to study solvent effects in the CNS. EEG studies on solvent-exposed workers and workers with CSE have shown either normal activity or increased beta activity and mainly diffuse but, in some cases, focal slow-wave abnormalities (Elofsson et al., 1980; Seppäläinen et al., 1980; Seppäläinen and Antti-Poika, 1983; Ørbæk et al., 1985; Gregersen et al., 1987; Seppäläinen, 1988; Antti-Poika et al., 1989; Callender et al., 1993). Due to large individual variations in EEG and subjective visual interpretation of the recordings, this method is less sensitive to reveal slight brain involvement. Quantitatively analyzed EEG (QEEG) provides a more sensitive and objective method to analyze electric activity of the brain. QEEG has revealed increased total power and differences in the anteroposterior distribution, i.e., increased power over anterior and decreased power over posterior areas in solvent-exposed workers (Elofsson et al., 1980; Ørbæk et al., 1985; Jonkman et al., 1992; Lundberg et al., 1995; Indulski et al., 1996) and in workers with CSE (Ørbæk et al., 1988; Lindgren et al., 1997). A more recent study on quantitative EEG in CSE cases revealed also an increase in absolute frontal theta power, indicating neuronal dysfunction in the frontal areas (Keski-Säntti et al., 2008). Some similarity is thus shared with aging-related slowing of EEG, decreased amplitude of alpha activity, and increased theta and delta power (Rossini et al., 2007), and chronic alcoholism, with the most frequent finding of an increase in beta power, but also increased theta activity, especially in central and parietal areas (Coutin-Churchman et al., 2006; Campanella et al., 2009).

Findings in visual evoked potential and somatosensory evoked potential on solvent-exposed workers and CSE patients have been inconsistent, ranging from normal to prolonged latency, and have shown both decreased and increased amplitudes (Elofsson et al., 1980; Antti-Poika et al., 1989; Massioui et al., 1990; Urban and Lukas, 1990; Deschamps et al., 1993; Lille et al., 1993; Stetkarova et al., 1993; Vrca et al., 1995, 1997; Indulski et al., 1996; Gong et al., 2003; Verberk et al., 2004).

In solvent-exposed workers and workers with CSE, auditory event-related potentials during a classic oddball paradigm have been either normal (Massioui et al., 1990) or shown prolonged P300 latency, a positive wave at about 300 ms after the stimulus (Massioui et al., 1990; Morrow et al., 1992a; Lundberg et al., 1995; Steinhauer et al., 1997), decreased amplitude (Lindgren et al., 1997), or both (Morrow et al., 1996). In CSE cases, auditory P300 amplitudes were decreased at Pz location (Keski-Säntti et al., 2007, 2012). In two follow-up studies of 1–2 years, either no change (Keski-Säntti et al., 2007) or improvement in P300 latency was seen in association to the length without solvent exposure (Morrow et al., 1998). The findings in dual-task event-related potential paradigms are promising. In the dual task, CSE cases had either decreased P300 amplitudes (Lindgren et al., 1997) or P300 was more frequently unrecognizable (Keski-Säntti et al., 2012). The latter may indicate that CSE patients present with slower performance speed and difficulties in allocation of attention, suggesting disturbed brain activity in posterior aspects of frontoparietal continuity (Keski-Säntti et al., 2012).

In exposed workers, findings in evoked potentials have correlated to the extent of exposure history and longer time from last solvent exposure in some studies (Vrca et al., 1997; Morrow et al., 1998), but in another study, no correlation was found between exposure indices of CSE cases and findings in P300 or quantitative EEG (Keski-Säntti et al., 2007, 2008).

Altogether, the neurophysiologic findings are nonspecific and have been of little use in the diagnostics of CSE. Changes in P300 latency and amplitude may occur in many conditions affecting cognitive processing, i.e., aging, depression, alcoholism, Parkinson's disease and Alzheimer's disease (Polich, 1997; Polich and Herbst, 2000; Roschke and Wagner, 2003; Campanella et al., 2009), and P300 amplitude is influenced by many factors, i.e., smoking, medication, vigilance, fatigue, circadian rhythm, personality, and genetic factors (Polich and Herbst, 2000; Polich, 2007).

Cognitive dysfunction

Cognitive dysfunction is the hallmark of solvent-related adverse effects in the CNS. The findings in clinical neuropsychologic assessment of workers with CSE show mild to moderate deficits in memory, especially working memory, and learning, attention, and allocation of attentional resources, visuospatial functions, abstract reasoning, and speed of information processing (Morrow et al., 1992b; Song-Yen et al., 1997; Chouaniére et al., 2002; Akila et al., 2006; van Hout et al., 2006; Meyer-Baron et al., 2008; Bast-Pettersen, 2009; van Valen et al., 2012). The working-memory dysfunction is similar to that of moderate to severe Parkinson's disease but different from Alzheimer's disease (Akila et al., 2006; Poewe, 2008). In affected individuals, the problems in these cognitive processes explain their everyday cognitive complaints, predominantly difficulties in short-memory, encoding of new information, and retrieving information from memory storage. Similar disturbances, although more severe, and even dementia, are seen in toluene abuse (Rosenberg et al., 2002; Filley et al., 2004; Yücel et al., 2008). It has been hypothesized that neurotoxic solvent effects lead to premature aging (Nilson et al., 2002; Ridgway et al., 2003), which has been suggested also to occur in alcoholism (Boutros et al., 2000). In neuropsychologic testing, the cognitive performance of CSE patients corresponds to the normative performance of 20-year-older healthy subjects (Nilson et al., 1999, 2002).

van Valen and the consensus group (2012) conclude that the most common neuropsychologic impairments in CSE patients are within the domains of attention and memory and, to a lesser extent, in motor performance. Impairments within the domains of attention and motor performance were most often observed in the subdomains of information-processing speed and motor speed. The influence of long-term solvent exposure on memory processes appears to involve immediate recall more than other memory processes and generally influences memory for verbal, visual, and visuospatial information. Regardless of the memory processes that were studied, poorer performance in CSE cases was found in comparison to normal controls. Although the neural basis of this type of cognitive impairment is not yet understood, both impaired processing speed at several levels and limitations in the capacity of working memory may be the underlying mechanisms (van Valen et al., 2012).

This is supported by the findings in functional imaging. Solvent-related cognitive impairment, with reduced-working memory performance, indicated altered function in the striatum, anterior cingulate, prefrontal, and parietal cortices (Haut et al., 2000; Visser et al., 2008; Tang et al., 2011).

Clinical course of CSE

Typically, emergence of CSE due to more than a decade of low-level exposure is insidious over a period of weeks and months. At first, dysfunction may be reversible after prolonged nonexposure, but later becomes irreversible. The cognitive symptoms of memory and concentration are among the earliest symptoms (Chouaniére et al., 2002; Furu et al., 2012). After the cessation of exposure, the symptoms of CSE usually do not progress and may be alleviated slightly, although encephalopathy, i.e., difficulties of memory and concentration, persists even when exposure has ceased (Edling et al., 1990; Baker, 1994, Nordling Nilson et al., 2007; Kaukiainen et al., 2009a, b; van Valen et al., 2009). The partial alleviation in noncognitive symptoms seen at 1-year follow-up after CSE diagnosis suggests a coexistence of functional symptoms (Kaukiainen et al., 2009a, b). A systematic review (1970–2008) on the course of CSE showed that symptoms and complaints seem to be stable or improve slightly with time, but do not disappear (van Valen et al., 2009). This is supported by the studies of a cohort of retired workers, where a high risk for cognitive impairment associated with a history of high solvent exposure at work, which had ended 12–30 years before (Sabbath et al., 2014). In the same prospective cohorts, solvent exposure was associated with poor cognition only among less-educated individuals, which might be explained by a higher cognitive reserve in the more-educated group (Sabbath et al., 2012).

CSE does not seem to appear after a latency in solvent exposure (Keski-Säntti et al., 2010). Some evidence supports the hypothesis that exposure to solvents may increase with normal aging. In 18-year and 20-year follow-up studies, formerly highly exposed floor layers and printers reported more symptoms and exacerbation of cognitive impairments (Nilson et al., 2002; Nordling Nilson et al., 2007, 2010).

There is some evidence that occupational solvent exposure is associated with parkinsonism and Parkinson's disease. Exposure to some solvents, such as carbon disulfide, *n*-hexane, methyl alcohol, toluene, methanol, trichloroethylene, and several solvent mixtures, has been associated with parkinsonism (Pezzoli et al., 1989, 1995, 1996; Guehl et al., 1999; Hageman et al., 1999; Vanacore et al., 2000; Gralewicz and Dyzma, 2005). Moreover, two case-control studies have shown that patients with Parkinson's disease have been exposed to solvents more frequently than healthy referents (Seidler et al., 1996; Pals et al., 2003). Occupational exposure to hydrocarbon solvents has been found as a risk factor for earlier onset of Parkinson's disease symptoms and a more severe disease course (Smargiassi et al., 1998; Pezzoli et al., 2000). There is less evidence of an increased risk of Alzheimer's disease in association with occupational exposure to solvents (Santibáñez et al., 2007).

CHRONIC SOLVENT ENCEPHALOPATHY: RECOGNITION OF AN OCCUPATIONAL DISEASE

CSE with insidious increase of cognitive and mental effects due to long-term low-level solvent exposure is difficult to diagnose objectively. This has hindered its recognition and widespread acceptance as an occupational disease. During the last decades, many countries have accepted CSE as an occupational disease (Triebig and Hallermann, 2001). Since 1990 CSE was recognized as an occupational disease in the European List of Occupational Diseases, with a supporting document as a guide for diagnosis (European Commission, 2009). It is also included in the 2009 International Labour Organization List of Occupational Diseases.

Diagnostic criteria of occupational CSE

In the Nordic countries, clinical diagnostic criteria for occupational CSE have been used since the 1970s (Juntunen, 1978). In 1985, a working group of the World Health Organization (WHO) presented diagnostic criteria and a classification for solvent-induced chronic toxic encephalopathy (Table 7.3) (WHO, 1985). In Raleigh, North Carolina, USA, the Workshop on Neurobehavioral Effects of Solvents introduced a somewhat different classification, which divided patients of WHO class II into those with sustained personality or mood change (type 2A) and those with impairment of cognitive functions (type 2B) (Baker and Seppäläinen, 1986; Cranmer and Golberg, 1987).

In the 10th edition of *The International Classification of Diseases* (ICD-10), CSE is classified as toxic encephalopathy (G92) with a defined causal agent (*T52.x; organic solvents) (WHO, 2007). In *The Diagnostic and Statistical Manual for Mental Disorders*, fourth edition (DSM-IV), CSE is classified as a substance-induced persistent dementia (292.82), amnestic (292.83), or other cognitive disorder (294.9) (American Psychiatric Association 1994).

The diagnosis of CSE relies on: (1) verification of substantial long-term exposure to neurotoxic solvents; (2) a characteristic clinical picture of nervous system damage with typical subjective symptoms and objective findings in clinical examinations; and (3) exclusion of other organic brain disorders and primary psychiatric diseases (WHO, 1985; Cranmer and Golberg, 1987; Juntunen, 1993; Van der Hoek et al., 2000). The current concept is that CSE is defined by irreversible cognitive disorder and not only personality or mood changes, and thus corresponds to the type II stage in the WHO classification and type 2B in Raleigh classification (van Valen et al., 2012). Initial symptoms of impairment usually manifest before cessation of exposure (van Valen et al., 2009; Keski-Säntti et al., 2010). However, the exact time of symptom onset may be difficult to establish, since they develop gradually. In 2009, an expert group convened by the European Commission suggested 10 years as the minimum duration of daily exposure, and a latency between symptom onset and cessation of exposure of no more than a few months. Less exposure duration may be accepted in cases with particularly high concentrations.

The presence of neuropsychologic impairment is the hallmark of CSE and thus its valid assessment with clinical neuropsychologic procedures is crucial for the diagnosis. An important step forward was the European consensus document on neuropsychologic assessment of CSE providing the neuropsychologic characteristics and guidelines for the neuropsychologic assessment in the diagnostic process (van Valen et al., 2012).

There is coexistence of mood disorders in CSE patients and solvent-exposed workers, mainly depression and anxiety (Condray et al., 2000; Morrow et al., 2000, 2001; Visser et al., 2011). In a large cohort of CSE cases, an increased risk was detected for major depressive disorder (relative risk (RR) = 7.4), dysthymia (RR = 6.0), panic disorders (RR = 7.1), agoraphobia (RR = 5.5), and generalized anxiety disorder (RR = 15.8), but reduced risks for alcohol-related disorders (RR = 0.3) (Visser et al., 2011). Currently, the evidence on psychiatric symptoms and findings of CSE is insufficient to discriminate neurotoxic etiology from psychologic reactions secondary to the decrease in functionality and quality of life or nonexposure-related factors (Morrow et al., 1993, 2001; van Hout et al., 2003; Keski-Säntti et al., 2010; Visser et al., 2011).

Sleep-related problems are frequent in CSE, e.g., diminished sleep quality, difficulties falling asleep, frequent awakenings, and trouble falling asleep again,

Table 7.3

Classification systems for chronic solvent encephalopathy according to World Health Organization (WHO) and Raleigh workshops

WHO criteria		Raleigh criteria	
Type I	**Organic affective syndrome**	**Type 1**	**Symptoms only**
Pathophysiology	Unclear		The patient complains of nonspecific symptoms such as fatigability, memory impairment, difficulty in concentration, and loss of initiative. These symptoms are reversible if exposure is discontinued, and there is no objective evidence of neuropsychiatric dysfunction
Course	Days to weeks, no sequelae		
Clinical manifestations	Depression, irritability, loss of interest in daily activities		
Reduced CNS function	–		
Type II	**Mild chronic toxic encephalopathy**	**Type 2 A**	**Sustained personality or mood change**
Pathophysiology	Unclear		There is a marked and sustained change in personality involving fatigue, emotional lability, impulse control, and general mood and motivation
Course	Insidious onset, duration weeks to months, reversibility variable		
Clinical manifestations	Fatigue, mood disturbances, memory complaints, attentional complaints		
Reduced CNS function	Psychomotor function (speed, attention, dexterity); short-term memory; other abnormalities common	**Type 2B**	**Impairment in intellectual function** There is difficulty in concentration, impairment of memory, and a decrease in learning capacity. These symptoms are accompanied by objective evidence of impairment. There may also be minor neurologic signs. The complete reversibility of type 2B is questionable
Type III	**Severe chronic toxic encephalopathy**	**Type 3**	**Dementia**
Pathophysiology	Unclear, often associated with structural CNS damage		In this condition, marked global deterioration in intellect and memory is often accompanied by neurologic signs and neuroradiologic findings. This condition is, at best, poorly reversible, but is generally nonprogressive, once exposure has ceased
Course	Insidious onset, duration indefinite, usually irreversible		
Clinical manifestations	Loss of intellectual abilities of sufficient severity to interfere with social or occupational functioning; impairment of abstract thinking; impaired judgment; other disturbances of cortical function; personality change		
Reduced CNS function	Types of abnormality similar to mild toxic encephalopathy; more pronounced and pervasive functional deficits; some neurophysiologic and neuroradiologic test abnormalities		

CNS, central nervous system.

waking up too early, and feeling tired upon awakening (Kaukiainen et al., 2009a, b). These symptoms are prevalent also in other conditions such as mood disorders (Viaene et al., 2009). It has been argued that solvent-exposed workers with sleep apnea may have been misdiagnosed as CSE due to the similarity in symptoms of sleep deprivation, i.e., fatigue, forgetfulness, and concentration problems (Monstad et al., 1992; Viaene et al., 2009). On the other hand, acute solvent exposure may induce or at least aggravate sleep apnea (Monstad et al., 1992; Edling et al., 1993; Laire et al., 1997; Viaene et al., 2009).

Diagnostic procedure

Recognition of adverse effects on the CNS due to long-term occupational solvent exposure has been shown to be difficult in clinical practice (Bleecker et al., 1991;

Table 7.4

Evaluation steps of central nervous neurotoxic effects

I	Relevant exposure level, length, and type
II	Appropriate insidiously increasing central nervous system symptoms along with ongoing exposure
III	Signs and objective findings of dysfunction. Central nervous system dysfunction examination by neurologic and neuropsychologic evaluations
IV	Other causes of nervous system dysfunction excluded by differential diagnosis. Treatable conditions should be treated and reevaluation done after follow-up

Juntunen, 1993; Ridgway et al., 2003; Spurgeon, 2006). The steps listed in Table 7.4 increase the accuracy of the diagnostic process (Kaukiainen et al., 2009a).

In specialized clinics, diagnostic procedures have been set up to diagnose CSE due to occupational exposure to solvents. Examinations include evaluation of all previous medical records and available documents relating to exposure, interviews and clinical examinations by specialists in occupational medicine and neurology, brain imaging with MRI, differential diagnostic laboratory testing, and at least one – usually several – comprehensive clinical neuropsychologic assessments. If other possible causes of cognitive dysfunction – most often psychiatric or sleep disorders – are suspected, further studies are carried out. Examinations, such as EEG, evoked potentials, cerebrospinal fluid, ophthalmic and visual function examinations, and laboratory tests, have been used, but their diagnostic value has remained low (Kaukiainen et al., 2004b; Näsänen et al., 2005; Ojanpää et al., 2006; Keski-Säntti et al., 2007, 2008, 2009, 2012; Päällysaho et al., 2007). Cerebrospinal fluid has been normal or shown nonspecific abnormalities, such as a slight lymphoid reaction or an increase in protein concentration, suggesting protein leak through the blood–brain barrier or a nonspecific immunologic activation (Juntunen et al., 1982; Wikkelsø et al., 1984; Barregård et al., 1990; Moen et al., 1990).

Misdiagnosis of CSE can be substantially reduced with thorough differential diagnostics, and follow-up of at least 1 year, and when necessary up to 3 years, to exclude reversible or progressive symptomatology (Kaukiainen et al., 2009a; Furu et al., 2012; van Valen et al., 2012). During follow-up, solvent exposure is halted or minimized, and other possible medical disorders are treated. Finally, clinical and neuropsychologic reevaluations are carried out, and the CSE diagnosis is verified by a multidisciplinary team (Juntunen, 1993; European Commission, 2009; Kaukiainen et al., 2009a; Keski-Säntti et al., 2010; Spee et al., 2012).

Differential diagnoses

Symptoms and findings in CSE are nonspecific, and common in the general population. Thus, the differential diagnostics of CSE is the key to reliable diagnostics. Nonoccupational CNS disorders such as cerebrovascular and neurodegenerative diseases (e.g., Alzheimer's and Parkinson's disease), infections and inflammatory brain diseases (e.g., multiple sclerosis), neoplasms and paraneoplastic syndromes, and traumatic brain injury should be excluded. Due to the nonspecificity of the symptoms and findings of CSE, many conditions affecting cognition should be taken into account, e.g., metabolic (e.g., avitaminosis, thyroid disorders), psychologic and psychiatric causes, chronic pain, sleep disorders, and alcohol or substance abuse (European Commission, 2009; van Valen et al., 2012).

Neurobehavioral test performance is influenced by many factors, such as primary intellectual capacity, developmental learning difficulties, education, motivation, alertness, and vigilance in the test situation, pain, alcohol use, and medication, which may add to interpretation difficulties in individual cases (Fidler et al., 1987; Williamson, 1996). Furthermore, patients with suspected CSE may be involved in litigation or financial compensation procedures and may show insufficient effort and suboptimal performance in test situations (van Hout et al., 2003, 2006). However, insufficient neuropsychologic test effort was not associated with DSM-IV disorders in subjects suspected of CSE (Visser et al., 2011). Also, repeated assessment might strengthen patient beliefs in a medical etiology and may lead to iatrogenic effects (van Hout et al., 2014). Thus, an open discussion of medical, psychologic, and social factors that may influence symptoms is important from the first interview.

Among solvent-exposed workers, there seems to be an increased risk for excessive alcohol consumption and alcoholism (Juntunen et al., 1982; Antti-Poika et al., 1985; Lundberg et al., 1992). In advanced cases of alcohol encephalopathy, it may not be possible to identify CSE-related adverse effects, but typical alcohol-related manifestations such as ataxia and polyneuropathy are not characteristic of CSE (Arlien-Søborg et al., 1979; Juntunen and Haltia, 1982; Gregersen et al., 1987; Berstad et al., 1989; Juntunen, 1993; Keski-Säntti et al., 2010).

PREVENTION OF SOLVENT-RELATED ADVERSE EFFECTS AND CSE

Legislative action, improvement of workplace hygiene, and substitution of chemicals to less harmful ones has enabled the prevention of solvent-related adverse effects

(Dick, 2006; Spurgeon, 2006). In recent decades, the number and proportion of solvent-exposed workers and the levels of solvent exposure in many occupations have decreased (Caldwell et al., 2000). Nowadays, the majority of paints and lacquers in construction painting are water-soluble. The active replacement of chemicals by less harmful ones is reflected in the decreasing proportion of workers exposed to aliphatic, aromatic, and halogenated hydrocarbons at levels above 50% of occupational exposure limits (Keski-Säntti et al., 2010).

In developed countries, the incidence of CSE is declining and cases are encountered with less prominent manifestations (Keski-Säntti et al., 2010; Furu et al., 2012; Spee et al., 2012). In Finland, from 1995 to 2007, the age of onset of CSE had increased yearly by 0.6 years and the duration of exposure by 0.8 years, suggesting less neurotoxicity of work-related solvent exposure (Keski-Säntti et al., 2010).

Most likely, the occurrence of CSE is higher in developing countries, where solvents are used widely, occupational hygiene is often inadequate, and CSE is poorly recognized, although described in case reports (Myers et al., 1999; Purvis et al., 2001; Saddik et al., 2003; Misra and Kalita, 2009; Zhan et al., 2010). Surprisingly, CSE has recently been shown to be underdiagnosed also in countries with advanced occupational health services (Kaukiainen et al., 2009a; Furu et al., 2012, 2014).

CSE often leads to early retirement and loss of ability to work, even though the degree of cognitive dysfunction is regarded as moderate (Gregersen et al., 1987; van Valen et al., 2009; Keski-Säntti et al., 2010). The majority of CSE cases end up outside the labor market due to disability pensions or long-term sick leave (Gregersen et al., 1987; Mikkelsen, 1997; van Valen et al., 2009). From consecutive CSE patients at the Finnish Institute of Occupational Health during 2002–2004, 74% were considered unable to work due to CSE, and 95% were on disability pension 1 year later (Keski-Säntti et al., 2010). Employment of workers with residual work ability often fails, due to restrictions set on their solvent exposure, and cognitive disturbances preventing occupational rehabilitation. Interestingly, the retirement rate was diminished (6.7% vs 74%) in CSE cases found with proactive questionnaire screening followed by clinical examinations of symptomatic workers in fields at high risk of CSE (Furu et al., 2012, 2014). New cases of CSE at an earlier stage of disability were also detected by screening in The Netherlands (Spee et al., 2012). The cases found by active screening seem to be less severely affected, their work ability is less strongly affected, and the CSE symptoms – difficulties in learning, memory, and concentration – may not limit the possibilities to continue working, retrain, and become reemployed (Spee et al., 2012; Furu et al., 2014).

References

Aaserud O, Nakstad PH, Bakke SJ et al. (2000). Failure to confirm neurotoxic impairment using cerebral magnetic resonance imaging on solvent-exposed workers. Scand J Work Environ Health 26: 346–352.

Ahmadi A, Jonsson P, Flodin U et al. (2002). Interaction between smoking and glutathione S-transferase polymorphisms in solvent-induced chronic toxic encephalopathy. Toxicol Ind Health 18: 289–296.

Akila R, Müller K, Kaukiainen A et al. (2006). Memory performance profile in occupational chronic solvent encephalopathy suggests working memory dysfunction. J Clin Exp Neuropsychol 28: 1307–1326.

Alkan A, Kutlu R, Hallac T et al. (2004). Occupational prolonged organic solvent exposure in shoemakers: brain MR spectroscopy findings. Magn Reson Imaging 22: 707–713.

American Psychiatric Association (1994). DSM-IV. Diagnostic and statistical manual of mental disorders, 4th edition. American Psychiatric Association, Washington DC.

Antti-Poika M, Juntunen J, Matikainen E et al. (1985). Occupational exposure to toluene: neurotoxic effects with special emphasis on drinking habits. Int Arch Occup Environ Health 56: 31–40.

Antti-Poika M, Ojala M, Matikainen E et al. (1989). Occupational exposure to solvents and cerebellar, brainstem and vestibular functions. Int Arch Occup Environ Health 61: 397–401.

Arlien-Søborg P (1992). Solvent Neurotoxicity, CRC, Florida.

Arlien-Søborg P, Bruhn P, Gyldensted C et al. (1979). Chronic painters' syndrome. Chronic toxic encephalopathy in house painters. Acta Neurol Scand 60: 149–156.

Arlien-Søborg P, Henriksen L, Gade A et al. (1982). Cerebral blood flow in chronic toxic encephalopathy in house painters exposed to organic solvents. Acta Neurol Scand 66: 34–41.

Backman L, Ginovart N, Dixon RA et al. (2000). Age-related cognitive deficits mediated by changes in the striatal dopamine system. Am J Psychiatry 157: 635–637.

Baker EL (1994). A review of recent research on health effects of human occupational exposure to organic solvents. A critical review. J Occup Med 36: 1079–1092.

Baker EL, Seppäläinen AM (1986). Human aspects of neurobehavioral effects. Report of the workshop session on clinical and epidemiological topics. Neurotoxicology 7: 45–56.

Baker Jr EL, Smith TJ, Landrigan PJ (1985). The neurotoxicity of industrial solvents: a review of the literature. Am J Ind Med 8: 207–217.

Barregård L, Wikkelsø C, Rosengren LE et al. (1990). Cerebrospinal fluid proteins in men with chronic encephalopathy after exposure to organic solvents. Scand J Work Environ Health 16: 423–427.

Bast-Pettersen R (2009). The neuropsychological diagnosis of chronic solvent induced encephalopathy (CSE) – a reanalysis of neuropsychological test results in a group of CSE patients diagnosed 20 years ago, based on comparisons with matched controls. Neurotoxicology 30: 1195–1201.

Bernsen HJ, Verhagen WI, de Bijl MA et al. (1992). Magnetic resonance studies on brain dysfunction induced by organic solvents. Acta Neurol Belg 92: 207–214.

Berstad J, Flekkoy K, Pedersen ON (1989). Encephalopathy and polyneuropathy induced by organic solvents. J Oslo City Hosp 39: 81–86.

Bleecker ML, Bolla KI, Agnew J et al. (1991). Dose-related subclinical neurobehavioral effects of chronic exposure to low levels of organic solvents. Am J Ind Med 19: 715–728.

Bolt HM, Roos PH, Thier R (2003). The cytochrome P-450 isoenzyme CYP2E1 in the biological processing of industrial chemicals: consequences for occupational and environmental medicine. Int Arch Occup Environ Health 76: 174–185.

Boutros NN, Reid MC, Petrakis I et al. (2000). Similarities in the disturbances in cortical information processing in alcoholism and aging: a pilot evoked potential study. Int Psychogeriatr 12: 513–525.

Bowen SE, Batis JC, Paez-Martinez N et al. (2006). The last decade of solvent research in animal models of abuse: mechanistic and behavioral studies. Neurotoxicol Teratol 28: 636–647.

Caldwell DJ, Armstrong TW, Barone NJ et al. (2000). Hydrocarbon solvent exposure data: compilation and analysis of the literature. AIHAJ 2000 (61): 881–894.

Callender TJ, Morrow L, Subramanian K et al. (1993). Three-dimensional brain metabolic imaging in patients with toxic encephalopathy. Environ Res 60: 295–319.

Campanella S, Petit G, Maurage P et al. (2009). Chronic alcoholism: insights from neurophysiology. Neurophysiol Clin 39: 191–207.

Canesi M, Benti R, Marotta G et al. (2007). Striatal dopamine transporter binding in patients with Parkinson's disease and severe occupational hydrocarbon exposure. Eur J Neurol 14: 297–299.

Carter N, Iregren A, Söderman E et al. (2002). EUROQUEST – A questionnaire for solvent related symptoms: factor structure, item analysis and predictive validity. Neurotoxicology 23: 711–717.

Chastain G (2006). Alcohol, neurotransmitter systems, and behavior. J Gen Psychol 133: 329–335.

Chouanière D, Cassitto MG, Spurgeon A et al. (1997). An international questionnaire to explore neurotoxic symptoms. Environ Res 73: 70–72.

Chouanière D, Wild P, Fontana JM et al. (2002). Neurobehavioral disturbances arising from occupational toluene exposure. Am J Ind Med 41: 77–88.

Condray R, Morrow LA, Steinhauer SR et al. (2000). Mood and behavioral symptoms in individuals with chronic solvent exposure. Psychiatry Res 97: 191–206.

Corsi P, D'Aprile A, Nico B et al. (2007). Synaptic contacts impaired by styrene-7,8-oxide toxicity. Toxicol Appl Pharmacol 224: 49–59.

Coutin-Churchman P, Moreno R et al. (2006). Clinical correlates of quantitative EEG alterations in alcoholic patients. Clin Neurophysiol 117: 740–751.

Cranmer JM, Golberg L (1987). Human aspects of solvent neurobehavioral effects. Report of the workshop session on clinical and epidemiological topics. Proceedings of the work-shop on neuro-behavioral effect of solvents., 7. 45–56.

Deschamps D, Garnier R, Lille F et al. (1993). Evoked potentials and cerebral blood flow in solvent induced psycho-organic syndrome. Br J Ind Med 50 (4): 325–330, Apr.

Deschamps D, Geraud C, Dally S (2001). Cognitive functions in workers exposed to toluene: evaluation at least 48 hours after removal from exposure. Int Arch Occup Environ Health 74: 285–288.

Dick FD (2006). Solvent neurotoxicity. Occup Environ Med 63: 221–226.

Dick F, Semple S, Chen R et al. (2000). Neurological deficits in solvent-exposed painters: a syndrome including impaired colour vision, cognitive defects, tremor and loss of vibration sensation. QJM 2000 (93): 655–661.

Dick FD, De Palma G, Ahmadi A et al. (2007). Gene–environment interactions in parkinsonism and Parkinson's disease: the Geoparkinson study. Occup Environ Med 64: 673–680.

Dodd PR, Beckmann AM, Davidson MS et al. (2000). Glutamate-mediated transmission, alcohol, and alcoholism. Neurochem Int 37: 509–533.

Dryson EW, Ogden JA (1998). Chronic solvent neurotoxicity in New Zealand: notified cases between 1993 and 1997. N Z Med J 111: 425–427.

Edling C, Ekberg K, Ahlborg Jr G et al. (1990). Long-term follow up of workers exposed to solvents. Br J Ind Med 47: 75–82.

Edling C, Lindberg A, Ulfberg J (1993). Occupational exposure to organic solvents as a cause of sleep apnoea. Br J Ind Med 50: 276–279.

Edling C, Hellman B, Arvidson B et al. (1997). Do organic solvents induce changes in the dopaminergic system? Positron emission tomography studies of occupationally exposed subjects. Int Arch Occup Environ Health 70: 180–186.

Ellingsen DG, Bekken M, Kolsaker L et al. (1993). Patients with suspected solvent-induced encephalopathy examined with cerebral computed tomography. J Occup Med 35: 155–160.

Elofsson SA, Gamberale F, Hindmarsh T et al. (1980). Exposure to organic solvents. A cross-sectional epidemiologic investigation on occupationally exposed care and industrial spray painters with special reference to the nervous system. Scand J Work Environ Health 6: 239–273.

European Commission (2009). Information notices on occupational diseases: a guide to diagnosis. Office for Official Publications of the European Communities, Luxembourg. http://ec.europa.eu/social/main.jsp?catId=716andlangId=enandintPageId=229 (Accessed 5 March 2015).

Ferroni C, Selis L, Mutti A et al. (1992). Neurobehavioral and neuroendocrine effects of occupational exposure to perchloroethylene. Neurotoxicology 13: 243–247.

Fidler AT, Baker EL, Letz RE (1987). Estimation of long term exposure to mixed solvents from questionnaire data: a tool for epidemiological investigations. Br J Ind Med 44: 133–141.

Filley CM, Halliday W, Kleinschmidt-DeMasters BK (2004). The effects of toluene on the central nervous system. J Neuropathol Exp Neurol 63: 1–12.

Friis L, Norbäck D, Edling C (1997). Occurrence of neuropsychiatric symptoms at low levels of occupational exposure to organic solvents and relationships to health, lifestyle, and stress. Int J Occup Environ Health 3: 184–189.

Furu H, Sainio M, Hyvärinen HK et al. (2012). Detecting chronic solvent encephalopathy in occupations at risk. Neurotoxicology 33: 734–741.

Furu H, Sainio M, Ahonen G et al. (2014). Cost of detecting a chronic solvent encephalopathy case by screening. Neurotoxicology 45: 253–259.

Godderis L, Maertens N, de Gelder V et al. (2010). Genetic susceptibility in solvent induced neurobehavioral effects. Neurotox Res 17: 268–278.

Godderis L, De Raedt K, Tabish AM et al. (2012). Epigenetic changes in lymphocytes of solvent-exposed individuals. Epigenomics 4: 269–277.

Gong Y, Kishi R, Kasai S et al. (2003). Visual dysfunction in workers exposed to a mixture of organic solvents. Neurotoxicology 24: 703–710.

Gotohda T, Tokunaga I, Kubo S et al. (2000). Effect of toluene inhalation on astrocytes and neurotrophic factor in rat brain. Forensic Sci Int 113: 233–238.

Gralewicz S, Dyzma M (2005). Organic solvents and the dopaminergic system. Int J Occup Med Environ Health 18: 103–113.

Gregersen P, Klausen H, Elsnab C (1987). Chronic toxic encephalopathy in solvent-exposed painters in Denmark 1976-1980: clinical cases and social consequences after a 5-year follow-up. Am J Ind Med 11: 399–417.

Guehl D, Bezard E, Dovero S et al. (1999). Trichloroethylene and parkinsonism: a human and experimental observation. Eur J Neurol 6: 609–611.

Hageman G, van der Hoek J, van Hout M et al. (1999). Parkinsonism, pyramidal signs, polyneuropathy, and cognitive decline after long-term occupational solvent exposure. J Neurol 246: 198–206.

Hagstadius S, Ørbæk P, Risberg J et al. (1989). Regional cerebral blood flow at the time of diagnosis of chronic toxic encephalopathy induced by organic-solvent exposure and after the cessation of exposure. Scand J Work Environ Health 15: 130–135.

Hänninen H (1964). Psychische symptome bei schwefelkohlen-stoffvergiftung. XIVth International Congress on Occupational Health, Excerpta Medica Foundation, Amsterdam, pp. 894–897, International Congress Series, No. 62.

Harper C (2009). The neuropathology of alcohol-related brain damage. Alcohol Alcohol 44: 136–140.

Harry J, Kulig B, Lotti M et al. (2001). Neurotoxicity risk assessment for human health: principles and approaches. In: Environmental Health Criteria 223, World Health Organization, Geneva, Switzerland. http://www.inchem.org/documents/ehc/ehc/ehc223.htm (Accessed 5 March 2015).

Haut MW, Leach S, Kuwabara H et al. (2000). Verbal working memory and solvent exposure: a positron emission tomography study. Neuropsychology 14: 551–558.

Haut MW, Kuwabara H, Ducatman AM et al. (2006). Corpus callosum volume in railroad workers with chronic exposure to solvents. J. Occup Environ Med 48: 615–624.

Heuser G, Mena I (1998). Neurospect in neurotoxic chemical exposure demonstration of long-term functional abnormalities. Toxicol Ind Health 14: 813–827.

Heuser G, Mena I, Alamos F (1994). NeuroSPECT findings in patients exposed to neurotoxic chemicals. Toxicol Ind Health 10: 561–571.

Hooisma J, Hänninen H, Emmen H et al. (1994). Symptoms indicative of the effects of organic solvent exposure in Dutch painters. Neurotoxicol Teratol 16: 613–622.

Huang J, Kato K, Shibata E et al. (1993). Nerve-specific marker proteins as indicators of organic solvent neurotoxicity. Environ Res 63: 82–87.

Ihrig A, Triebig G, Dietz MC (2001). Evaluation of a modified German version of the Q16 questionnaire for neurotoxic symptoms in workers exposed to solvents. Occup Environ Med 58: 19–23.

Indulski JA, Sinczuk-Walczak H, Szymczak M et al. (1996). Neurological and neurophysiological examinations of workers occupationally exposed to organic solvent mixtures used in the paint and varnish production. Int J Occup Med Environ Health 9: 235–244.

Jarry H, Metten M, Gamer AO et al. (2002). Effects of 5-day styrene inhalation on serum prolactin and dopamine levels and on hypothalamic and striatal catecholamine concentrations in male rats. Arch Toxicol 76: 657–663.

Jin CF, Haut M, Ducatman A (2004). Industrial solvents and psychological effects. Clin Occup Environ Med 4: 597–620.

Jonkman EJ, De Weerd AW, Poortvliet DC et al. (1992). Electroencephalographic studies in workers exposed to solvents or pesticides. Electroencephalogr Clin Neurophysiol 82: 438–444.

Juntunen J (1978). Neurotoxic Syndromes in Man. Third International Course in Industrial Toxicology. Institute of Occupational Health, Helsinki, 164–168.

Juntunen J (1993). Neurotoxic syndromes and occupational exposure to solvents. Environ Res 60: 98–111.

Juntunen J, Haltia M (1982). Polyneuropathies in occupational neurology: pathogenetic and clinical aspects. Acta Neurol Scand Suppl 92: 59–73.

Juntunen J, Taskinen E, Luisto M et al. (1982). Cerebrospinal fluid cells and proteins in patients occupationally exposed to organic solvents. J Neurol Sci 54: 413–425.

Karlson B, Österberg K, Ørbæk P (2000). Euroquest: the validity of a new symptom questionnaire. Neurotoxicology 21: 783–789.

Kaukiainen A, Riala R, Martikainen R et al. (2004a). Solvent-related health effects among construction painters with decreasing exposure. Am J Ind Med 45: 627–636.

Kaukiainen A, Vehmas T, Rantala K et al. (2004b). Results of common laboratory tests in solvent-exposed workers. Int Arch Occup Environ Health 77: 39–46.

Kaukiainen A, Akila R, Martikainen R et al. (2009a). Symptom screening in detection of occupational solvent-related encephalopathy. Int Arch Occup Environ Health 82: 343–355.

Kaukiainen A, Hyvärinen HK, Akila R et al. (2009b). Symptoms of chronic solvent encephalopathy: Euroquest questionnaire study. Neurotoxicology 30: 1187–1194.

Kelafant GA, Berg RA, Schleenbaker R (1994). Toxic encephalopathy due to 1,1,1-trichloroethane exposure. Am J Ind Med 25: 439–446.

Keski-Säntti P, Holm A, Akila R et al. (2007). P300 of auditory event related potentials in occupational chronic solvent encephalopathy. Neurotoxicology 28: 1230–1236.

Keski-Säntti P, Kovala T, Holm A et al. (2008). Quantitative EEG in occupational chronic solvent encephalopathy. Hum Exp Toxicol 27: 315–320.

Keski-Säntti P, Mäntylä R, Lamminen A et al. (2009). Magnetic resonance imaging in occupational chronic solvent encephalopathy. Int Arch Occup Environ Health 82: 595–602.

Keski-Säntti P, Kaukiainen A, Hyvärinen HK et al. (2010). Occupational chronic solvent encephalopathy in Finland 1995–2007: incidence and exposure. Int Arch Occup Environ Health 83: 703–712.

Keski-Säntti P, Palmu K, Pitkonen M et al. (2012). Multimodal event-related potentials in occupational chronic solvent encephalopathy. Neurotoxicology 33: 703–709.

Kezic S, Calkoen F, Wenker MA et al. (2006). Genetic polymorphism of metabolic enzymes modifies the risk of chronic solvent-induced encephalopathy. Toxicol Ind Health 22: 281–289.

Kiesswetter E, Sietmann B, Seeber A (1997). Standardization of a questionnaire for neurotoxic symptoms. Environ Res 73: 73–80.

Kim Y, Jeong K, Yog-Hun Y et al. (2010). Occupational neurologic disorders in Korea. J Clin Neurol 6: 64–72.

Klinken L, Arlien-Soborg P (1993). Brain autopsy in organic solvent syndrome. Acta Neurol Scand 87: 371–375.

Laire G, Viaene MK, Veulemans H et al. (1997). Nocturnal oxygen desaturation, as assessed by home oximetry, in long-term solvent-exposed workers. Am J Ind Med 32: 656–664.

Leira HL, Myhr G, Nilsen G et al. (1992). Cerebral magnetic resonance imaging and cerebral computerized tomography for patients with solvent-induced encephalopathy. Scand J Work Environ Health 18: 68–70.

Leira H, Austrheim H, Wannag A (2006). Løysemiddelskader meldt arbeidstilsynet 1985–2005. Ramazzini 4: 4–5.

Lille F, Margules S, Mallet A et al. (1993). Evoked potentials in workers occupationally exposed to organic solvents. Electromyogr Clin Neurophysiol 33: 279–283.

Lindgren M, Österberg K, Ørbæk P et al. (1997). Solvent-induced toxic encephalopathy: electrophysiological data in relation to neuropsychological findings. J Clin Exp Neuropsychol 19: 772–783.

Lof A, Johanson G (1998). Toxicokinetics of organic solvents: a review of modifying factors. Crit Rev Toxicol 28: 571–650.

Lundberg I, Gustavsson A, Högberg M et al. (1992). Diagnoses of alcohol abuse and other neuropsychiatric disorders among house painters compared with house carpenters. Br J Ind Med 49: 409–415.

Lundberg I, Michelsen H, Nise G et al. (1995). Neuropsychiatric function of housepainters with previous long-term heavy exposure to organic solvents. Scand J Work Environ Health 21: 1–44.

Lundberg I, Högberg M, Michélsen H et al. (1997). Evaluation of the Q16 questionnaire on neurotoxic symptoms and a review of its use. Occup Environ Med 54: 343–350.

Massioui FE, Lille F, Lesevre N et al. (1990). Sensory and cognitive event-related potentials in workers chronically exposed to solvents. J Toxicol Clin Toxicol 28: 203–219.

McEvoy LK, Pellouchoud E, Smith ME et al. (2001). Neurophysiological signals of working memory in normal aging. Brain Res Cogn Brain Res 11: 363–376.

Meyer-Baron M, Blaszkewicz M, Henke H et al. (2008). The impact of solvent mixtures on neurobehavioral performance: conclusions from epidemiological data. Neurotoxicology 29: 349–360.

Mikkelsen S (1997). Epidemiological update on solvent neurotoxicity. Environ Res 73: 101–112.

Mikkelsen S, Jorgensen M, Browne E et al. (1998). Mixed solvent exposure and organic brain damage. A study of painters. Acta Neurol Scand Suppl 118: 1–143.

Misra U, Kalita J (2009). Toxic neuropathies. Indian perspective. Neurol India 57: 697–705.

Moen BE, Kyvik KR, Engelsen BA et al. (1990). Cerebrospinal fluid proteins and free amino acids in patients with solvent induced chronic toxic encephalopathy and healthy controls. Br J Ind Med 47: 277–280.

Monstad P, Mellgren SI, Sulg IA (1992). The clinical significance of sleep apnoea in workers exposed to organic solvents: implications for the diagnosis of organic solvent encephalopathy. J Neurol 239: 195–198.

Morrow LA, Callender T, Lottenberg S et al. (1990). PET and neurobehavioral evidence of tetrabromoethane encephalopathy. J Neuropsychiatry Clin Neurosci 2: 431–435.

Morrow LA, Steinhauer SR, Hodgson MJ (1992a). Delay in P300 latency in patients with organic solvent exposure. Arch Neurol 49: 315–320.

Morrow LA, Robin N, Hodgson MJ et al. (1992b). Assessment of attention and memory efficiency in persons with solvent neurotoxicity. Neuropsychologia 30: 911–922.

Morrow LA, Kamis H, Hodgson MJ (1993). Psychiatric symptomatology in persons with organic solvent exposure. J Consult Clin Psychol 61: 171–174.

Morrow LA, Steinhauer SR, Condray R (1996). Differential associations of P300 amplitude and latency with cognitive and psychiatric function in solvent-exposed adults. J Neuropsychiatry Clin Neurosci 8: 446–449.

Morrow LA, Steinhauer SR, Condray R (1998). Predictors of improvement in P300 latency in solvent-exposed adults. Neuropsychiatry Neuropsychol Behav Neurol 11: 146–150.

Morrow LA, Gibson C, Bagovich GR et al. (2000). Increased incidence of anxiety and depressive disorders in persons with organic solvent exposure. Psychosom Med 62: 746–750.

Morrow LA, Stein L, Bagovich GR et al. (2001). Neuropsychological assessment, depression, and past exposure to organic solvents. Appl Neuropsychol 8: 65–73.

Mutti A, Smargiassi A (1998). Selective vulnerability of dopaminergic systems to industrial chemicals: risk assessment of related neuroendocrine changes. Toxicol Ind Health 14: 311–323.

Mutti A, Vescovi PP, Falzoi M et al. (1984). Neuroendocrine effects of styrene on occupationally exposed workers. Scand J Work Environ Health 10: 225–228.

Myers J, Nell V, Colvin M et al. (1999). Neuropsychological function in solvent-exposed South African paint makers. J Occup Environ Med 41: 1011–1018.

Näsänen R, Kaukiainen A, Hero V et al. (2005). Effects of long-term occupational solvent exposure on contrast sensitivity and performance in visual search. Environ Toxicol Pharmacol 19: 497–504.

Nasterlack M, Dietz MC, Frank KH et al. (1999). A multidisciplinary cross-sectional study on solvent-related health effects in painters compared with construction workers. Int Arch Occup Environ Health 72: 205–214.

Nielsen GD, Lund SP, Ladefoged O (2006). Neurological effects of white spirit: contribution of animal studies during a 30-year period. Basic Clin Pharmacol Toxicol 98: 115–123.

Nilson L, Barregard L, Backman L (1999). Trail making test in chronic toxic encephalopathy: performance and discriminative potential. Clin Neuropsychol 13: 314–327.

Nilson LN, Sällsten G, Hagberg S et al. (2002). Influence of solvent exposure and aging on cognitive functioning: an 18 year follow up of formerly exposed floor layers and their controls. Occup Environ Med 59: 49–57.

Nordling Nilson L, Barregård L, Sällsten G et al. (2007). Self-reported symptoms and their effects on cognitive functioning in workers with past exposure to solvent-based glues: an 18-year follow-up. Int Arch Occup Environ Health 81: 69–79.

Nordling Nilson L, Karlson B, Nise G et al. (2010). Delayed manifestations of CNS effects in formerly exposed printers – a 20-year follow-up. Neurotoxicol Teratol 32: 620–626.

Ojanpää H, Näsänen R, Päällysaho J et al. (2006). Visual search and eye movements in patients with chronic solvent-induced toxic encephalopathy. Neurotoxicology 27: 1013–1023.

Ørbæk P, Risberg J, Rosen I et al. (1985). Effects of long-term exposure to solvents in the paint industry. A cross-sectional epidemiologic study with clinical and laboratory methods. Scand J Work Environ Health 11: 1–28.

Ørbæk P, Lindgren M, Olivecrona H et al. (1987). Computed tomography and psychometric test performances in patients with solvent induced chronic toxic encephalopathy and healthy controls. Br J Ind Med 44: 175–179.

Ørbæk P, Rosen I, Svensson K (1988). Power spectrum analysis of EEG at diagnosis and follow up of patients with solvent induced chronic toxic encephalopathy. Br J Ind Med 45: 476–482.

Owen AM (2004). Cognitive dysfunction in Parkinson's disease: the role of frontostriatal circuitry. Neuroscientist 10: 525–537.

Päällysaho J, Näsänen R, Mäntyjärvi M et al. (2007). Colour vision defects in occupational chronic solvent encephalopathy. Hum Exp Toxicol 26: 375–383.

Pals P, Van Everbroeck B, Grubben B et al. (2003). Case-control study of environmental risk factors for Parkinson's disease in Belgium. Eur J Epidemiol 18: 1133–1142.

Pezzoli G, Barbieri S, Ferrante C et al. (1989). Parkinsonism due to n-hexane exposure. Lancet 2: 874.

Pezzoli G, Antonini A, Barbieri S et al. (1995). *n*-Hexane-induced parkinsonism: pathogenetic hypotheses. Mov Disord 10: 279–282.

Pezzoli G, Strada O, Silani V et al. (1996). Clinical and pathological features in hydrocarbon-induced parkinsonism. Ann Neurol 40: 922–925.

Pezzoli G, Canesi M, Antonini A et al. (2000). Hydrocarbon exposure and Parkinson's disease. Neurology 55: 667–673.

Poewe W (2008). Non-motor symptoms in Parkinson's disease. Eur J Neurol 15: 14–20.

Polich J (1997). EEG and ERP assessment of normal aging. Electroencephalogr Clin Neurophysiol 104: 244–256.

Polich J (2007). Updating P300: an integrative theory of P3a and P3b. Clin Neurophysiol 118: 2128–2148.

Polich J, Herbst KL (2000). P300 as a clinical assay: rationale, evaluation, and findings. Int J Psychophysiol 38: 3–19.

Purvis K, Jumba I, Wandiga S et al. (2001). Worker exposure and health risks from volatile organic compounds utilized in the paint manufacturing industry in Kenya. Appl Occup Environ Hyg 16: 1035–1042.

Rango M, Canesi M, Ghione I et al. (2006). Parkinson's disease, chronic hydrocarbon exposure and striatal neuronal damage: a 1-H MRS study. Neurotoxicology 27: 164–168.

Riala R, Kalliokoski P, Pyy L et al. (1984). Solvent exposure in construction and maintenance painting. Scand J Work Environ Health 10: 263–266.

Ridgway P, Nixon TE, Leach JP (2003). Occupational exposure to organic solvents and long-term nervous system damage detectable by brain imaging, neurophysiology or histopathology. Food Chem Toxicol 41: 153–187.

Roschke J, Wagner P (2003). A confirmatory study on the mechanisms behind reduced P300 waves in depression. Neuropsychopharmacology 28: S9–S12.

Rosenberg NL, Grigsby J, Dreisbach J et al. (2002). Neuropsychologic impairment and MRI abnormalities associated with chronic solvent abuse. J Toxicol Clin Toxicol 40: 21–34.

Rossini PM, Rossi S, Babiloni C et al. (2007). Clinical neurophysiology of aging brain: from normal aging to neurodegeneration. Prog Neurobiol 83: 375–400.

Rouch I, Wild P, Fontana JM et al. (2003). Evaluation of the French version of EUROQUEST: a questionnaire for neurotoxic symptoms. Neurotoxicology 24: 541–546.

Sabbath EL, Glymour MM, Berr C et al. (2012). Occupational solvent exposure and cognition: does the association vary by level of education? Neurology 78: 1754–1760.

Sabbath EL, Gutierrez LA, Okechukwu CA et al. (2014). Time may not fully attenuate solvent-associated cognitive deficits in highly exposed workers. Neurology 82: 1716–1723.

Saddik B, Nuwayhid I, Williamson A et al. (2003). Evidence of Neurotoxicity in working children in Lebanon. Neurotoxicology 24: 733–739.

Salthouse TA (2000). Aging and measures of processing speed. Biol Psychol 54: 35–54.

Santibáñez M, Bolumar F, García AM (2007). Occupational risk factors in Alzheimer's disease: a review assessing the quality of published epidemiological studies. Occup Environ Med 64: 723–732.

Savolainen K (1980). Combined effects of xylene and alcohol on the central nervous system. Acta Pharmacol Toxicol (Copenh) 46: 366–372.

Seidler A, Hellenbrand W, Robra BP et al. (1996). Possible environmental, occupational, and other etiologic factors for Parkinson's disease: a case-control study in Germany. Neurology 46: 1275–1284.

Seppäläinen AM (1988). Neurophysiological approaches to the detection of early neurotoxicity in humans. Crit Rev Toxicol 18: 245–298.

Seppäläinen AM, Antti-Poika M (1983). Time course of electrophysiological findings for patients with solvent poisoning. A descriptive study. Scand J Work Environ Health 9: 15–24.

Seppäläinen AM, Lindström K, Martelin T (1980). Neurophysiological and psychological picture of solvent poisoning. Am J Ind Med 1: 31–42.

Smargiassi A, Mutti A, De Rosa A et al. (1998). A case-control study of occupational and environmental risk factors for Parkinson's disease in the Emilia-Romagna region of Italy. Neurotoxicology 19: 709–712.

Söderkvist P, Ahmadi A, Akerbäck A et al. (1996). Glutathione S-transferase M1 null genotype as a risk modifier for solvent-induced chronic toxic encephalopathy. Scand J Work Environ Health 22: 360–363.

Song-Yen T, Jong-Dar C, Wen-ying C et al. (1997). Neurobehavioral effects of occupational exposure to low-level organic solvents among Taiwanese workers in paint factories. Environ Res 73: 146–155.

Spee T, van Valen E, van Duivenbooden C et al. (2012). A screening programme on chronic solvent-induced encephalopathy among Dutch painters. Neurotoxicology 33: 727–733.

Spurgeon A (2006). Watching paint dry: organic solvent syndrome in late-twentieth-century Britain. Med Hist 50: 167–188.

Steinhauer SR, Morrow LA, Condray R et al. (1997). Event-related potentials in workers with ongoing occupational exposure. Biol Psychiatry 42: 854–858.

Stetkarova I, Urban P, Prochazka B et al. (1993). Somatosensory evoked potentials in workers exposed to toluene and styrene. Br J Ind Med 50: 520–527.

Sullivan EV, Harris RA, Pfefferbaum A (2010). Alcohol's effects on brain and behavior. Alcohol Res Health 33: 127–143.

Tang CY, Carpenter DM, Eaves EL et al. (2011). Occupational solvent exposure and brain function: an fMRI study. Environ Health Perspect 119: 908–913.

Thier R, Bruning T, Roos PH et al. (2003). Markers of genetic susceptibility in human environmental hygiene and toxicology: the role of selected CYP, NAT and GST genes. Int J Hyg Environ Health 206: 149–171.

Thuomas KA, Moller C, Ödkvist LM et al. (1996). MR imaging in solvent-induced chronic toxic encephalopathy. Acta Radiol 37: 177–179.

Triebig G, Hallermann J (2001). Survey of solvent related chronic encephalopathy as an occupational disease in European countries. Occup Environ Med 58: 575–581.

Tuomi T, Ahonen I (2010). Kemialliset tekijät (Chemical factors). In: T Kauppinen, R Hanhela, I Kandolin et al. (Eds.), Työ ja terveys Suomessa (Work and Health in Finland), Finnish Institute of Occupational Health, Helsinki, pp. 52–59 (In Finnish).

Uitti RJ, Snow BJ, Shinotoh H et al. (1994). Parkinsonism induced by solvent abuse. Ann Neurol 35: 616–619.

Urban P, Lukas E (1990). Visual evoked potentials in rotogravure printers exposed to toluene. Br J Ind Med 47: 819–823.

van der Hoek JA, Verberk MM, Hageman G (2000). Criteria for solvent-induced chronic toxic encephalopathy: a systematic review. Int Arch Occup Environ Health 73: 362–368.

van Hout MS, Schmand B, Wekking EM et al. (2003). Suboptimal performance on neuropsychological tests in patients with suspected chronic toxic encephalopathy. Neurotoxicology 24: 547–551.

van Hout MS, Schmand B, Wekking EM et al. (2006). Cognitive functioning in patients with suspected chronic toxic encephalopathy: evidence for neuropsychological disturbances after controlling for insufficient effort. J Neurol Neurosurg Psychiatry 77: 296–303.

van Hout M, Hageman G, van Valen E (2014). Pitfalls in clinical assessment of neurotoxic diseases: negative effects of repeated diagnostic evaluation, illustrated by a clinical case. Neurotoxicology 45: 247–252.

van Valen E, Wekking E, van der Laan G et al. (2009). The course of chronic solvent induced encephalopathy: a systematic review. Neurotoxicology 30: 1172–11786.

van Valen E, van Thriel C, Akila R et al. (2012). Chronic solvent-induced encephalopathy: European consensus of neuropsychological characteristics, assessment, and guidelines for diagnostics. Neurotoxicology 33: 710–726.

Vanacore N, Gasparini M, Brusa L et al. (2000). A possible association between exposure to *n*-hexane and parkinsonism. Neurol Sci 21: 49–52.

Varney NR, Morrow LA, Pinkston JB et al. (1998). PET scan findings in a patient with a remote history of exposure to organic solvents. Appl Neuropsychol 5: 100–106.

Verberk MM, Brons JT, Salle HJ (2004). Visual evoked potentials in workers with chronic solvent encephalopathy. Int Arch Occup Environ Health 77: 328–334.

Vettori MV, Corradi D, Coccini T et al. (2000). Styrene-induced changes in amacrine retinal cells: an experimental study in the rat. Neurotoxicology 21: 607–614.

Vettori MV, Caglieri A, Goldoni M et al. (2005). Analysis of oxidative stress in SK-N-MC neurons exposed to styrene-7,8-oxide. Toxicol In Vitro 19: 11–20.

Viaene MK, Pauwels W, Veulemans H et al. (2001). Neurobehavioural changes and persistence of complaints in workers exposed to styrene in a polyester boat building plant: influence of exposure characteristics and microsomal epoxide hydrolase phenotype. Occup Environ Med 58: 103–112.

Viaene M, Vermeir G, Godderis L (2009). Sleep disturbances and occupational exposure to solvents. Sleep Med Rev 13: 235–243.

Visser I, Lavini C, Booij J et al. (2008). Cerebral impairment in chronic solvent-induced encephalopathy. Ann Neurol 63: 572–580.

Visser I, Wekking E, de Boer A et al. (2011). Prevalence of psychiatric disorders in patients with chronic solvent induced encephalopathy (CSE). Neurotoxicology 32: 916–922.

von Euler G, Fuxe K, Agnati LF et al. (1987). Ganglioside GM1 treatment prevents the effects of subacute exposure to toluene on N-[3H]propylnorapomorphine binding characteristics in rat striatal membranes. Neurosci Lett 82: 181–184.

Vrca A, Bozicevic D, Karacic V et al. (1995). Visual evoked potentials in individuals exposed to long-term low concentrations of toluene. Arch Toxicol 69: 337–340.

Vrca A, Bozicevic D, Bozikov V et al. (1997). Brain stem evoked potentials and visual evoked potentials in relation to the length of occupational exposure to low levels of toluene. Acta Med Croatica 51: 215–219.

Waldron HA, Cherry N, Johnston JD (1983). The effects of ethanol on blood toluene concentrations. Int Arch Occup Environ Health 51: 365–369.

Welch L, Kirshner H, Heath A et al. (1991). Chronic neuropsychological and neurological impairment following acute exposure to a solvent mixture of toluene and methyl ethyl ketone (MEK). J Toxicol Clin Toxicol 29: 435–445.

Wikkelsø C, Ekberg K, Lillienberg L et al. (1984). Cerebrospinal fluid proteins and cells in men subjected to long-term exposure to organic solvents. Acta Neurol Scand Suppl 100: 113–119.

Williamson AM (1996). The role of confounding variables in the assessment of neurobehavioural effects of chronic solvent exposure. Neurotoxicology 17: 761–767.

World Health Organization (1985). Chronic effects of organic solvents on the central nervous system and diagnostic criteria, World Health Organization, Copenhagen.

World Health Organization. International Statistical Classification of Diseases and Related Health Problems 10th Revision Version for 2007. http://www.who.int/classifications/icd/en/ (accessed 5 March 2015).

Xiao JQ, Levin SM (2000). The diagnosis and management of solvent-related disorders. Am J Ind Med 37: 44–61.

Yücel M, Takagi M, Walterfang M et al. (2008). Toluene misuse and long-term harms: a systematic review of the neuropsychological and neuroimaging literature. Neurosci Biobehav Rev 32: 910–926.

Zhan F, Zheng W, Liu L et al. (2010). Diagnosis and prognosis evaluation of 1,2-dichloroethane encephalopathy – Magnetic resonance imaging combined with diffusion tensor imagining and magnetic resonance spectroscopy study. Neurol India 59: 108–110.

Handbook of Clinical Neurology, Vol. 131 (3rd series)
Occupational Neurology
M. Lotti and M.L. Bleecker, Editors

Chapter 8

Hydrogen sulfide intoxication

TEE L. GUIDOTTI*
Medical Advisory Services, Rockville, MD, USA

Hydrogen sulfide is a relatively common inhalation hazard for its high level of toxicity, encountered in settings as diverse as sewers, the oil and gas industry, fishing and storage, aquaculture, and (mostly successful) suicides (Agency for Toxic Substances and Disease Registry, 2006). Occupational exposure to hydrogen sulfide was described by Ramazzini, elegantly and accurately, in 1700.

The clinical picture of acute toxicity from hydrogen sulfide has intrigued numerous investigators because of the unusual suddenness of its most characteristic feature, immediate loss of consciousness. It has been written of hydrogen sulfide that "death may come on like a stroke of lightning" (Oesterhelweg and Puschel, 2008). Toxicity from this gas ranges from mild and quickly reversible, in the cases of sudden loss of consciousness called "knockdown," to instantly lethal (Woodall et al., 2005). A more complete understanding of the toxicology of hydrogen sulfide has been impeded by confusion in the literature, which often and unfortunately perpetuates demonstrably incorrect information. One of them is that hydrogen sulfide exerts its major effects by the same mechanism of action as cyanide, which is clearly not the case.

As with most uncommon problems in clinical toxicology, much of the literature on hydrogen sulfide toxicity consists of case reports of occupational exposure, both individual and multiple (Stine et al., 1976; Peters, 1981; Hoidal et al., 1986; Mack, 1987; Knight and Presnell, 2005; Ago et al., 2008; Oesterhelweg and Puschel, 2008; Lee et al., 2009; Poli et al., 2010; Fujita et al., 2011; Nogue et al., 2011; Lancia et al., 2013), and, even more tragically, case series from typically multiple casualities arising from the same incident where a rescue has been attempted, with tragic results (Demaret and Fialaire, 1974; Gunn and Wong, 2001; Gallay et al., 2002; US Chemical Safety and Hazard Investigation Board, 2003; Kage et al., 2004; Poli et al., 2010; Bassindale and Hosking, 2011; Nogue et al., 2011; Zaba et al., 2011). Fortunately, there have been very few pediatric cases (Claudet et al., 2012; Sastre et al., 2013). However, this may change with the growing popularity of the form of suicide involving its use.

Acute hydrogen sulfide toxicity cases are sporadic, unpredictable, and relatively uncommon in absolute terms. However, they do occur more often in the oil and gas sector where there is an opportunity for such cases to be observed and documented.

When clinical information must come from individual case reports, it is inevitable that exposure cannot be quantified, that some outcome measures will not be available in a timely fashion, and that the baseline preexposure state will not be documented, especially neurobehavioral function. Confounding factors are often not mentioned in case reports, although it is important to exclude them, such as head trauma, prolonged hypoxemia due to hypoventilation, and the possibility of already oxygen-deficient atmospheres in a confined space. Even so, case reports from many dissimilar situations are remarkably consistent.

The emphasis in this chapter is on clinical neurology and the pathophysiology of neurologicmanifestations of toxicity. Reproductive, developmental, and carcinogenicity issues are not considered in this review, because of limited relevance to human exposure. Pulmonary toxicity is discussed in some detail, both because it is an important practical issue in management and because hypoxemia may complicate the picture of neurotoxicity.

CONTEXT AND INTRODUCTION

Hydrogen sulfide is the second most common cause of fatal gas inhalation exposures in the workplace (at 7.7%, second to carbon monoxide at 36%) and has unique

*Correspondence to: Tee L. Guidotti, Medical Advisory Services, P.O. Box 7479, Gaithersburg MD 20898-7479 USA. E-mail: tee@teeguidotti.com

features that make hydrogen sulfide cases toxicologically unusual and stereotypical. Due to the situations in which it occurs, acute hydrogen sulfide toxicity is reported almost exclusively among fit, previously healthy working-age men.

As in most such incidents, casualties usually occur in multiples as would-be rescuers rush to save their co-workers and, in their heroic but misguided haste, neglect to protect themselves with self-contained breathing apparatus. Intensive training and ready availability of personal protective equipment are required to prevent such situations (Mattarano 1999). Largely because of recent accelerated development that is now settling into maintenance and behavior-dependent routine operations, there have been a disproportionate number of cases reported recently from industries in China, particularly in the oil and gas sector (Jianwen et al., 2014). However, sporadic cases can occur anywhere because of the disseminated nature of sewer and agricultural operations.

Chronic, low-concentration exposure on a community-wide basis may occur where there is a common or point source, such as an industrial facility or a geologic source, but is sometimes alleged to occur from persistent sewer or drain malfunction with resulting gas exposure. Misunderstandings, and sometimes misrepresentations, of the subacute and chronic toxicity of hydrogen sulfide have led to improbable accusations and even false claims. This has led to litigation and often contentious community issues. The problem of sorting out true toxicity from the symptoms of apprehension is complicated by the odor of the gas, which even at very low concentrations may arouse awareness, concern, and anxiety when it is perceived as a threat.

Exposure scenarios and sources

Hydrogen sulfide toxicity is generated wherever sulfur-containing compounds decompose under reducing conditions, either chemical or by microbial action. The substrate may be petroleum and natural gas (which are formed from long-decomposed archaic biologic material (petroleum), sewage, animal waste, rotting animal matter, pulp and paper mill effluent, tanneries, asphalt, sulfide ore smelting operations, or geologic sources of sulfur such as volcanoes.

In industry, the agent is probably most recognized as a hazard in the oil and gas industry, where it is a major hazard in upstream "sour" (high-sulfur-containing) crude oil and natural gas production, in pumping stations and gas plants, and in refineries (where sulfur is scrubbed from the gas stream in sulfur recovery units and may be concentrated). In industrial settings such as refineries, incidents typically occur in confined spaces. In the field, they often occur in depressions in the land or pits, where the gas settles because it is heavier than air. Unpublished information on hydrogen sulfide-related knockdowns exists in the form of incident reports in the oil and gas industry but this is difficult or impossible for outsiders to access, although efforts have been made to establish a registry (Guidotti, 1996a). (Privileged opportunities granted to the author to review some of these unpublished reports have revealed no deviations from what is reported in the literature.) Sour gas, natural gas containing hydrogen sulfide, contains more than hydrocarbons and hydrogen sulfide, including some small amounts of carbon disulfide, carbonyl sulfide, methyl mercaptans, and trace metals, but these other components are toxicologically insignificant in practice and can be ignored. The hazard is managed mainly by monitoring technology, which has a reputation for "false" alarms, although some of these undoubtedly reflect transient passage of higher concentrations of the gas through the area. In the past, sour gas was often flared (ignited) in order to oxidize hydrogen sulfide to less acutely toxic but irritating sulfur dioxide; this is still done as an emergency measure (Fig. 8.1).

Occasionally there is confusion over the odor of processed natural gas, for example, arising from gas leaks or suicide attempts in houses. Mercaptans are added to natural gas as a safety measure to ensure detection of leaks. They have a sweeter odor than hydrogen sulfide and are essentially nontoxic at the levels used.

Very rarely, liquid hydrogen sulfide is encountered as a special-use reagent or chemical feedstock. (There is only one producer in the USA.) The principles of

Fig. 8.1. A sour-gas plant in the foothills of the Rocky Mountains, in Alberta. This plant receives from many wells in the vicinity a stream of natural gas rich in hydrogen sulfide, which it removes by absorption in a concentrated amine solution and which is eventually converted to sulfur. It also strips out small-molecular-weight "natural-gas liquids" for separate sale and dries the natural gas stream for entry into a pipeline. Note that there is no visible flare for venting and oxidizing hydrogen sulfide, which used to be a common sight. The industry has succeeded in reducing flaring to a minimum by better control over gas flow and desulfurization.

management are the same as for incidents involving the gas, except that clothing on the victim may have to be removed for the protection of first responders and the person decontaminated before transport.

Wastewater treatment plants present a hazard, even in the open, away from confined spaces (Nogue et al., 2011). Sewer gas is a combination of hydrogen sulfide and methane but the latter has very low toxicity and so contributes nothing to the clinical presentation, although it is often remarked upon as if it did. Outgassing from sewage can, and occasionally does, generate lethal exposures, particularly when a stagnant pool is disturbed. This is a known hazard in workers who maintain sewerage, and clean out septic tanks and cesspools.

Similarly, concentrations of animal waste also produce hydrogen sulfide in agriculture. Sporadic lethal exposures occur most often in swine confinement or liquid manure operations. Such events do not always occur in confined spaces (Costigan, 2003; Knubben-Schweizer et al., 2011). Lethal exposures occur most often when a worker is mucking or pumping manure, because the disturbance releases the gas.

Decomposition during storage of organic material in the holds of fishing boats or in failed refrigeration units may yield significant concentrations of hydrogen sulfide in confined spaces (Tvedt et al., 1991; Gerasimon et al., 2007). Most recently, aquaculture has been associated with a growing number of fatalities worldwide (Myers, 2010).

Volcanic activity releases enormous quantities of hydrogen sulfide to the atmosphere, potentially causing local deaths, and has been implicated by some geoscientists as a possible cause of the Permian extinction, which followed the eruption of several volcano chains (Lamarque and Orlando, 2007). Likewise, hydrogen sulfide seeps from vents in the deep ocean, which are fortunately beyond human exposure, and sustains life evolved to sulfur metabolism in the anaerobic deep ocean. As a practical matter, geologic sources are a hazard in tapping geothermal wells for energy.

A small number of cases in the broader community occur from exposure associated with natural water sources, principally geysers, natural hot springs, and pools (Daldal et al., 2010). The geothermal hot pools of Rotorua, New Zealand, have sadly had a disproportionate share of these incidents, sometimes involving tourists (Bassindale and Hosking, 2011).

Hydrogen sulfide recently became an instrument of choice for suicide among the younger generation in Japan, peaking in 2008 as a result of social media communicating the means to generate the gas from common household chemicals. As a means of suicide, it was very effective, with a fatality rate of 95% (Morii et al., 2010; Fujita et al., 2011). The method unfortunately works well to achieve fatally toxic concentrations in confined spaces and so is most often used in locked cars. Often the suicidant is wearing tight-fitting goggles to avoid painful eye irritation. However tragic for the individual involved, this method of suicide also presents grave risks to others, with the potential for multiple secondary casualties to the community, especially would-be rescuers, first responders, family members, friends arriving at the scene, and even, when performed in closets and apartments, entire neighborhoods (Truscott, 2008; Ishikawa, 2010). Knowing this, suicidants often leave signs visible through the car window warning first responders and others of hazardous materials and toxic gas. This method of suicide has been reported rarely but increasingly in the USA, Australia, and Canada (Lamarque and Orlando, 2007; Goode, 2011; Chemical Suicides, 2012; Bott and Dodd, 2013; Sams et al., 2013).

Physiologic role

Sulfide is produced endogenously and also by commensal bacterial in the gut (giving odor to flatus, which is mostly methane) and mouth (contributing to halitosis). Sulfide is present normally in tissues of the body, especially the gut. Given the toxicity of the gas, the exceptionally high normal concentrations in brain tissue puzzled investigators until the physiologic role of the gas was discovered (Reiffenstein et al., 1992).

The physiologic role of hydrogen sulfide, as opposed to its toxicologic role, is only now emerging (Wang, 2012; Wang et al., 2014). A detailed review of the physiologic role of hydrogen sulfide is premature and beyond the scope of this chapter, except to indicate that this knowledge is likely to change and refine toxicologic models in the future. There are early indications that hydrogen sulfide has a role in the response to stroke, brain and spinal cord injury, and dementia, but the fragmentary evidence is too early to be coherent.

There is a close similarity and interaction among the gas transmitters hydrogen sulfide, carbon monoxide, and nitric oxide (which is much less toxic). Perturbation in their physiologic role is likely to have consequences in both exacerbating and limiting injury, but was undiscovered and therefore completely outside the thinking of toxicology until very recently.

Hydrogen sulfide is a paracrine mediator, a diffusible vasorelaxing agent, immune modulator, and neuromodulator (Popov, 2013). It may have a role in insulin sensitivity and glucose metabolism (Zhang et al., 2013). Hydrogen sulfide may act following acute injury (for example, by stroke or trauma) to maintain perfusion, preserve the integrity of the microcirculation, dampen glutamate-mediated excitotoxicity, preventing overexuberant apoptosis, and to quench oxidative stress.

On the other hand, it may aggravate the situation by elevating calcium influx and so contributing to calcium overload in secondary neuronal injury (Olson, 2011; Wang et al., 2014). Toxic exposure that interferes with these functions may involve direct effects on the nervous system, perhaps mediating cell responses in stroke (Qu et al., 2006) as well as in other organ systems. The state of "suspended animation" induced during knockdown shows promise as a therapeutic intervention (Olson, 2011; Li et al., 2012).

PROPERTIES

Hydrogen sulfide (H_2S, molecular weight 34.08) is a flammable, colorless gas that is offensively odorous (smell of rotten eggs or cabbage), soluble in lipid, and poorly soluble in water (vapor pressure in water 18.75 10^2 kPa), and heavier than air (density 1.19). In aqueous media it dissociates readily into hydrosulfide, but not to sulfide as follows. Undissociated hydrogen sulfide and the hydrosulfide ion are predominant at a 1:2 ratio at pH 7.4 (Reiffenstein et al., 1992; Agency for Toxic Substances and Disease Registry, 2006):

$$H_2S \leftrightarrow H^+ + HS^- \leftrightarrow 2H^+ + S^{2-} pK_{a1} = 7.04. pK_{a2} = 11.96$$

Biologic sources of hydrogen sulfide tend to release intermittent peaks or pulses of the gas, usually together with methane (which is for all practical purposes nontoxic). Once released, the dense gas may persist until displaced by wind or ventilation in low-lying ground, depressions such as trenches or pits, or in confined spaces. Hydrogen sulfide in water, for example in sewage, swamps, and the deep ocean, tends to accumulate to saturation and then is released spontaneously or with perturbation, such as heating or physical disturbance. This results in "burping" of the gas and a bolus effect, which can then be trapped in confined spaces or pits (Nogue et al., 2011).

Sustained release tends to be more common from geologic sources, as in the "oilpatch." Even there, volcanic activity, tapped pockets of gas, and the vagaries of oil and gas extraction may result in a variable exposure profile over short periods.

CLINICAL TOXIDROME

A "toxidrome" in toxicology is a syndrome of symptoms and signs characteristic of the effects of a toxic agent. Hydrogen sulfide is unique in its toxidrome for acute and subacute exposure. Whether a toxic effect exists at low exposure levels remains controversial.

Acute

Acute hydrogen sulfide toxicity is one of the most unusual and reliable toxidromes in medical toxicology, and features several bizarre signs that allow clinical differentiation to be made from other toxic gases (Yant, 1930; Burnett and King, 1978; Smith and Gosselin, 1979; Beauchamp et al., 1984; Arnold et al., 1985; Wang, 1989; Reiffenstein et al., 1992; Milby and Baselt, 1999; Woodall et al., 2005; Guidotti, 2010). These will be discussed in order of clinical significance in surviving victims.

The toxidrome of hydrogen sulfide includes the following progressive symptoms at increasing exposure levels, such that individual cases, unless occurring suddenly, typically include all the effects that precede them:

1. odor perception
2. conjunctivitis
3. olfactory paralysis, not to be confused with olfactory fatigue
4. "knockdown" (acute central neurotoxicity, reversible)
5. pulmonary edema (uncommon)
6. apnea (not observed).

Cases of toxicity from hydrogen sulfide always show these features, singly or in combination, almost always in the sequence numbered, regardless of the susceptibility or characteristics of the victim or subject. Atypical presentations are so uncommon as to call the diagnosis into question, even in highly unusual cases. When they are not reported, the history is often incomplete or the victim was overcome too quickly for the progression to be observed or to develop fully.

Apnea, immediate paralysis of breathing, is a lethal risk at very high concentrations, and may be the cause of death in victims who are already unconscious from knockdown. Clinically it is not observed separately because the victim invariably is already dead. Apnea is an important effect used in animal studies, however, because it is reproducible of neurotoxicity in the rat, which is more susceptible than human beings. Approximate thresholds for each cardinal effect are given in Table 8.1.

The stereotypical toxidrome assumes, of course, that there has not been trauma to the head from falling or a period of anoxia or hypoperfusion, as might occur with near-lethal exposures. It also assumes that there is no premorbid condition or susceptibility state that would substantially modify the presentation; to date, none has been demonstrated.

Knockdown

In the oilpatch, "knockdown" specifically means sudden loss of consciousness due to hydrogen sulfide exposure. (The term has other meanings in other industries and

Table 8.1

Health effects of hydrogen sulfide at various approximate exposure levels (Guidotti, 1994, 1996b, 2007, 2010)

Concentration (ppm)	Effects
0.01–0.3	Odor threshold (highly variable)
1–5	Moderate offensive odor, may be associated with nausea, tearing of the eyes, headaches, or loss of sleep with prolonged exposure; healthy young male subjects experience no decline in maximal physical work capacity
5	Aerobic metabolism inhibited in skeletal muscle (lowest observed effect level, probably close to threshold), no observable signs or symptoms; 8-hour occupational exposure limit in Alberta
10	Ceiling level (15 minutes) in Alberta; OSHA PEL 8-hour time-weighted average; anaerobic metabolism threshold inhibited during exercise; eye irritation
15	15-minute occupational exposure limit in Alberta; short-term exposure limit OSHA PEL
20	Evacuation level in Alberta; ceiling (15-minute time-weighted average) exposure limit OSHA, odor very strong; conjunctivitis may occur
20–50	Conjunctivitis (eye irritation) and lung irritation. Possible eye damage after several days of exposure; may cause digestive upset and loss of appetite; 50 ppm is the peak (never to be exceeded) OSHA PEL
100	"Gas eye" and lung irritation; olfactory paralysis, sensation of odor disappears
150–200	Olfactory paralysis; severe eye and lung irritation
250–500	Pulmonary edema may occur, especially if exposure is prolonged
500	Serious damage to eyes within 30 minutes; severe lung irritation; "knockdown" (sudden unconsciousness) and death (especially if exposure prolonged by hours); amnesia for period of exposure
1000	Breathing may stop within one or two breaths (apnea); immediate collapse

OSHA, Occupational Safety and Health Administration; PEL, permissible exposure limit.

working life, and the phenomenon does not have a special name in other industries where it occurs.) Hydrogen sulfide-induced acute central neurotoxicity leads to abrupt and reversible loss of consciousness and collapse, often described by those who witness it as like turning off a switch. A knockdown may easily be fatal if exposure at high concentration (roughly 500–1000 ppm) is prolonged, but if exposure is transient, as it often is in an oilfield due to air movement, it may also be quickly reversible. The victim typically falls down as if letting the strings loose on a marionette. If exposure is transient, as usually happens in the oilpatch, recovery may be equally rapid and apparently complete in functional terms. Survivors have even reported to the author that the experience is not unpleasant and may be mildly euphoric, assuming they have not injured themselves in the fall.

Given long experience in the "oilpatch" (areas where oil and gas production are prevalent), the author has interviewed many oil and gas workers who have had near-misses or actual knockdowns associated with hydrogen sulfide, and has been involved in fatality investigations. The histories given by patients or observers told after the fact are remarkably consistent with each other and with the relevant literature (Burnett and King, 1978). Observers who witness a knockdown consistently describe it as immediate and complete collapse, often using the metaphor that it is as if a switch had been turned off or a puppeteer had dropped the strings on a marionette. Recovery from a knockdown is described as being as quick, almost instantaneous, with the person suddenly realizing that he or she is lying on the ground but experiencing no confusion or postictal effects. Oilfield workers, at least above a certain age, usually immediately recognize what happened and get up easily. In the past it was quite common for them to return to work without reporting the incident. Those who observe recovery from a knockdown often describe it using the metaphor of someone turning a switch back on. Those who actually experience a knockdown and recover from it have sometimes said to the author that it was not an unpleasant sensation and that they woke up immediately without mental confusion.

Pulmonary edema and irritant toxicity

The significance of pulmonary edema or apnea (another acute, high-level effect) to clinical neurology is obvious: prolonged hypoxemia resulting in anoxic brain damage. However, this appears to be uncommon.

Pulmonary edema is associated with prolonged exposure while the victim is unconscious (Richardson, 1995; Tanaka et al., 1999; Daldal et al., 2010). Hydrogen sulfide is irritating to mucous membranes. This effect mostly involves the deep lung and, as discussed in the next subsection, the epithelium (outer lining) of the eye. Because gaseous hydrogen sulfide is poorly soluble in water, it tends to be carried deeply into the lung, where diffuse alveolar damage may occur, resulting in toxic inhalation and a risk of pulmonary edema (Guidotti, 2001, 2006).

Older studies suggest that 20% of cases of hydrogen sulfide toxicity reaching the emergency department in a setting where most cases came from the oil and gas sector showed at least some evidence of pulmonary edema (Roth et al., 1997). Although this seems implausibly high, it may reflect the residual cases between those who did not seek care (oil companies now demand it) and early fatalities.

Compared to other toxic gases, death by pulmonary edema seems to be quite unusual in hydrogen sulfide toxicity. In one series of 152 cases from China, an overall mortality of only 5% was reported, far below the 50% usually cited for mortality from noncardiogenic pulmonary edema (Wang, 1989). First, concentrations high enough to induce pulmonary edema would in most cases already have induced central effects (at levels well above the threshold for knockdown) that would already be lethal, before progression to alveolar edema. The modulatory effect of hydrogen sulfide on inflammation may also spare the lung from early injury and protect against acute endothelial disruption and diffuse alveolar damage. Also, hydrogen sulfide may be much less toxic to lung than other tissues and may even have profound protective effects. For example, it has a potent effect in reducing experimentally induced oxidative injury in the lung (Cao et al., 2014).

Experimental studies have shown that hydrogen sulfide is only moderately cytotoxic for pulmonary cells, it does not initiate severe irreversible changes despite extensive pulmonary edema, and does not seem to disrupt the basement membrane of the alveolar endothelium (Lopez et al., 1988; Cao et al., 2014). Thus, the ultimate prognosis for recovery and lung remodeling is likely to be good if the patient can be supported through the acute episode. There are several possible reasons for this. This may be why there are very few reports of chronic lung disease, either airways or interstitial (Duong et al., 2001) associated with recovery from hydrogen sulfide inhalation, and those in the literature more often suggest relatively mild obstructive rather than restrictive effects (Richardson, 1995).

Gas eye (irritative conjunctivitis)

"Gas eye" is a superficial inflammation of the cornea and conjunctiva due to the irritant effect of hydrogen sulfide (Tansy et al., 1981; Lambert et al., 2006). It is often recurrent in workers who are exposed for prolonged periods to relatively low occupational concentrations, 20 ppm and possibly lower. Some workers have said to the author that they relied on eye irritation as a second warning sign of exposure over the permissible occupational exposure limit, because it occurs at a lower level than loss of olfaction. (This is a dangerous practice.)

Gas eye is characterized by microvacuolization and edema in the corneal epithelium, leading to mild refractory and chromatic distortion in the form of a colored halo surrounding objects and other visual changes. Blepharospasm, an acute spasm of the orbicularis oculi muscle, may result in tearing, periorbital pain, and photophobia (Tansy et al., 1981; Guidotti, 1994; Milby and Baselt, 1999; Lambert et al., 2006). This form of blepharospasm is acute and not to be confused with eyelid fasciculation.

In Rotorua, New Zealand, there was a statistically significant doubling of risk for admission to hospital due to eye disorders, including conjunctivitis, which is an uncommon reason for hospitalization. Given ambient levels, this is a plausible effect of exposure (Bates et al., 1997).

Olfactory effects

The odor threshold and olfactory effects of hydrogen sulfide are critical in occupational health because odor provides the primary warning signal to workers that the gas is present.

Hydrogen sulfide is very odorous, with a low olfactory threshold, from less than 0.01 to 0.3 ppm depending on individual sensitivity. By 1–5 ppm, the odor is very offensive, aptly described as rotten eggs. Thus, at lower levels the gas has excellent warning properties. However, at high exposure levels the warning properties disappear because at relatively high concentrations (about 100 ppm) hydrogen sulfide paralyzes the olfactory mechanism, preventing perception of any smell. This effect removes the primary warning sign of hydrogen sulfide exposure and the principal warning to, say, oilfield workers suddenly caught in a plume (Turner and Fairhurst, 1990; Smith and Gosselin, 1979; Milby and Baselt, 1999; Guidotti, 2010).

Two distinct mechanisms are involved in loss of olfactory acuity to hydrogen sulfide. As with most strong odors, workers may experience olfactory fatigue at lower levels of exposure and may become accustomed to them in the short term. Olfactory fatigue is common to the sensory processing of many strong odors and leads to behavioral tolerance of the odor. There is also a mechanism specific and almost unique to hydrogen sulfide, known as olfactory paralysis. Olfactory fatigue is a sensory adaptation. Olfactory paralysis is neurotoxicity of the olfactory nerve and bulb. Unfortunately, the two phenomena are frequently confused in the literature.

Olfactory paralysis due to both neurotoxicity affecting the olfactory bulb and fibers may be followed by hyposmia or anosmia, a permanent loss of the ability to perceive odor. This pattern of toxicity was demonstrated experimentally to be attributable to selective toxicity to olfactory mucosa in the nasal passages and to olfactory neuron loss after subchronic inhalation (Turner and Fairhurst, 1990; Brenneman et al., 2000; Roberts et al., 2006). Hyposmia has been found to be present in most men who recovered from severe, potentially lethal hydrogen sulfide toxicity (Hirsch, 2002).

Human subjects exposed to transient high levels of hydrogen sulfide have been reported to show deficits on standardized tests of smell and taste years later (Hirsch, 2002). This is not invariable, however, because tolerance can be induced in the short term and the olfactory mucosa can recover if exposure does not persist too long (Roberts et al., 2006).

SECONDARY SYMPTOMS AND SIGNS

As with all toxic conditions, there are secondary symptoms and signs of hydrogen sulfide toxicity that vary and in some cases are secondary rather than direct effects, and sometimes side effects of treatment (such as hypotension from excessive nitrite). The literature documenting the prevalence and profile of these secondary effects is much weaker than that reporting the cardinal effects, not surprisingly.

Acute respiratory effects

Bronchospasm occurs in many, but not all, cases but, remarkably, is not reported to result in bronchitis, chronic irritant-induced asthma, or reactive airways dysfunction syndrome (Shivanthan et al., 2013). This may be an example of the lung-sparing effect of hydrogen sulfide as a physiologic modulator of inflammation.

Shortness of breath is to be expected for a respiratory irritant and may be a sign of the onset of pulmonary edema. However, in at least one case, dyspnea and cough were relieved during hyperbaric oxygen treatment following hydrogen sulfide exposure, suggesting that shortness of breath may reflect hypoxia (Gallay et al., 2002).

Exposure in the short term at 10 ppm does not appear to be associated with reduced lung function or increased airways reactivity (Bhambhani et al., 1996a). However, other studies have suggested that an isolated reduction in residual volume, occurring in 23% of cases exposed in an industrial setting, in the absence of any radiologic or clinical signs, constitutes a subclinical effect to acute exposure (Buick et al., 2000). If confirmed, this unusual effect may be related to inhibitory effects on pulmonary receptors and reflex activity, rather than an effect on lung mechanics, as assumed by the author of the study (Klentz and Fedde, 1978; Almeida and Guidotti, 1999).

Further emphasizing the possible significance of a modulatory effect of hydrogen sulfide on the lung is the observation that there is an apparent exposure-related reduction in asthma risk (incidence and prevalence) in Rotorua, where ambient levels of hydrogen sulfide are high.

Acute cardiac effects

Given the purported mechanism of sudden cytochrome inhibition and disruption in energetic metabolically active cells, one would expect the heart to be a frequent target organ with a toxic cardiomyopathy resulting on recovery. However, there have been surprisingly few validated reports of cardiotoxicity (Christia-Lotter et al., 2007; Amino et al., 2009). Two case reports suggest myocardial ischemia resulting in the absence of any sign of coronary artery disease in otherwise healthy young male workers. One in Korea initially went into acute heart failure with left ventricular dilation but responded well to the usual treatment of acute congestive failure. After recovery, he was significantly impaired with symptoms of dyspnea on exertion and showed a reduced ejection fraction and reduced ventricular chamber size. He showed no neurologic sequelae (Lee et al., 2009). The other, in Japan, was the second victim in a stereotypical unsuccessful rescue attempt and showed evidence of skeletal rhabdomyolysis but delayed myocardial injury, from which he eventually recovered completely (Hirakawa et al., 2013).

Nonspecific symptoms and signs

Unfortunately, the secondary literature usually does not distinguish among secondary symptoms that occur in exposed persons who do not experience knockdown or conjunctivitis (two benchmark clinical symptoms that establish exposure levels) and those that do, and almost never is there an exposure assessment refined enough to establish an exposure–response relationship.

Headache and short-term cognitive changes, such as short-term memory loss, are common but very nonspecific, as they may occur in many other situations and as a response to distraction and posttraumatic stress. Headache occurred in about a third of hydrogen sulfide exposures but was a transient phenomenon lasting a day or so (Audeau et al., 1985; Sjaasted and Bakketeig, 2006). One exception involved recovery from a prolonged knockdown and cardiac arrest in which chronic headache followed treatment by heroic measures (Asif and Exline, 2012).

Seizure disorders are only reported for 2% of cases, historically, but most such cases have probably been associated with anoxic brain damage in near-lethal cases

(Arnold et al., 1985). In a case series involving a release in an oil refinery in Sri Lanka, about which exposure details were unavailable, a worker with neurofibromatosis (confirmed after the incident) was found unconscious and cyanotic, and developed decerebrate posturing with repeated generalized tonic-clonic seizures while unconscious. After initially responding to ventilation with 100% oxygen and benzodiazepines, the patient went back into status epilepticus after being started on sodium nitrite (hypotension not reported). Astonishingly, he recovered with initial neurologic, predominantly cerebellar, signs that improved and was left with only residual dysarthria and mild ataxia (Shivanthan et al., 2013). In at least one case in which a detailed clinical description is available, an apparently new-onset seizure occurred during a knockdown and resolved without evolving into a chronic seizure disorder (Audeau et al., 1985). One may speculate that the nerve tissue-sparing effect of hydrogen sulfide may play a role in the favorable outcomes.

Gastrointestinal symptoms (including jaundice and diarrhea) were reported in the older literature but not in more recent, systematic case series (Arnold et al., 1985). Nausea and vomiting are to be expected given the malodorous exposure, which most people find disgusting (Guidotti, 2010).

CHRONIC EFFECTS

There are many unanswered questions about neurotoxicity from hydrogen sulfide.

Neurotoxicity

A central issue in studying neurotoxicity associated with hydrogen sulfide has been to differentiate between the primary effects of toxicity in the vicinity of 500 ppm, which cause knockdown as a neurologic event, and the effects of anoxia which can occur as a result or because the atmosphere was low in oxygen to begin with (as in confined spaces and sewers). Hydrogen sulfide toxicity may be the result of several effects, including interference with oxygen uptake and metabolism and also hypoxia from oxygen deprivation due to apnea or respiratory insufficiency. These mechanisms would produce the constellation of findings identified in anoxic brain injury. There may also be other, direct mechanisms of neurotoxicity.

Occupational exposure to hydrogen sulfide in four workers did result in loss of consciousness during a release, associated with abnormal evoked response potentials (assessing cortical function) and subclinical neuropsychiatric clinical findings without overt disease. The subjects had cardinal symptoms (eye irritation), confirming that exposure had occurred, possibly repeatedly, at exposure levels that must have exceeded 50 ppm much of the time and may have reached 243 ppm, as noted in other incidents at this refinery. The pattern of symptoms and impairment was not uniform among subjects. However, these same subjects also had evidence of posttraumatic stress disorder, which can influence neurobehavioral testing, and were not representative of the exposed population, having been selected precisely because they demonstrated abnormalities (Hirsch, 2002).

It is now accepted that neurologic sequelae can follow knockdown in the context of acute high-level exposure. Evidence remains weak for effects associated with chronic, low-level exposure, discussed below. It would be expected that, if these effects occur, they would be observed first and most consistently at higher exposures. However, in the case of hydrogen sulfide, exposures that result in knockdown may easily be confounded by head trauma during falls, which is common (Arnold et al., 1985; Wang, 1989; Gabbay and Perrone, 2001), and hypoxemia from apnea or seizure activity (Wang, 1989; Tvedt et al., 1991; Snyder et al., 1995; Tanaka et al., 1999). Even so, in the past, when these events were far more common, oilfield workers often went back to work after a knockdown.

Peripheral neuropathy

Despite the observation that demyelination can occur in experimental models (Solnyshkova, 2003), peripheral neuropathy is not often or consistently reported in cases of hydrogen sulfide toxicity. Some well-documented cases describe a tingling paresthesia which does not appear to be associated with chronic neuropathy (Wasch et al., 1989). The finding does not necessarily imply selective neurotoxicity because a paresthesia of this type could also be central or mediated by vascular changes.

Chronic respiratory effects following recovery

It is not clear whether prolonged or repeated hydrogen sulfide exposure is associated with chronic respiratory impairment. A single case report has suggested that interstitial fibrosis followed exposure to hydrogen sulfide, but the case has many other unusual features and may not be representative (Duong et al., 2001).

Cross-sectional studies of sewer workers, who are exposed to hydrogen sulfide but also have other risk factors, such as exposure to bioaerosols and endotoxin, suggest that lung function is significantly reduced after accounting for smoking and may show an accelerated rate of decline with age. The relationship between age and lung function was skewed in this sample, however, and showed an exposure–response relationship only at the extremes, and did not take outmigration from the reference group (which may have been exposed to chlorine) into consideration (Richardson, 1995). Longitudinal studies required to confirm this finding are not available.

Another cross-sectional study of oilfield workers did not find such an effect. Subjects who had experienced a knockdown did not show a difference from those who had not in this population. However, this study had profound methodologic issues, not least of which was that respiratory symptoms were part of the exclusion criteria for their selection (Hessel et al., 1997). (The subjects had been recruited to be a reference group for another study, a point which was not made clear in the article.)

It is a paradox of the hydrogen sulfide literature that exposure is not known to induce airways reactivity or the condition known as "reactive airways dysfunction syndrome," although this might be expected based on the gas's known irritant effect. One may speculate that this could be a direct consequence of the physiologic anti-inflammatory role of hydrogen sulfide.

CHRONIC EFFECTS OF LOWER-LEVEL EXPOSURE (<10 PPM)

Whether effects exist after exposure at lower levels representative of routine occupational exposure or a community incident, has been controversial.

General

At 5 and 10 ppm, hydrogen sulfide exposure appears to induce an earlier transition from aerobic to anaerobic metabolism in resting and exercising adults, consistent with partial inhibition of aerobic metabolism, but this is not associated with subjective symptoms or clinically evident effects (Bhambhani et al., 1994, 1996a, b, 1997).

Living with much lower levels of exposure, but certainly elevated compared to most communities, is the city of Rotorua, New Zealand, a cultural center for the Maori people, where there are also many sulfur springs and geothermal sources. In Rotorua, ambient hydrogen sulfide levels are high, at an average 5 ppm, sometimes peaking at 100 ppm. Ecologic studies of residents of Rotorua suggest some excess morbidity in both central and peripheral nervous system disorders, with an elevation in peripheral disorders, including mononeuritis (Bates et al., 2002).

In Rotorua, there was an elevated risk of cardiovascular disease observed but a lower risk for stroke and evidence for an elevated prevalence of high blood pressure, which is inconsistent, unless somehow reflecting the agent's physiologic mechanisms. The study did not take into account a variety of potential confounders, including mercury (which was not measured) and other sulfides (not reported) (Bates et al., 1997). Ecologic studies of this type are intrinsically a weak study design for etiologic investigation and should not be considered strong evidence of causation.

Odor

Hydrogen sulfide provokes complicated psychologic and physiologic responses because of its foul odor and the concerns this raises (Woodall et al., 2005). At 5 ppm, which is highly noxious, experimental exposure results in reported anxiety (Fiedler et al., 2008). The odor of hydrogen sulfide at that level is perceived by most people as disgusting and may therefore provoke both aversion and physiologic responses (nausea and increase in vagal tone). These physiologic and anxiety-generating effects may be perceived as toxic by people experiencing them.

Exposures on the order of 5–10 ppm are sometimes encountered in community settings, although they are more common as background in workplaces such as oil refineries. Levels above 1 ppm are very uncommon in the community, where concentrations above the odor threshold (generally above 0.05 ppm for most people) are normally unacceptable and prosecuted as a nuisance. It has been alleged that health effects may be observed at exposures < 1 ppm, but the evidence is weak.

Rotorua, New Zealand (North Island), famously has hot springs and vents that release hydrogen sulfide continuously and with sporadic elevations. The city is a population and cultural center for the Maori, who are historically socioeconomically disadvantaged, and so there are many confounding issues that affect the interpretation of population studies (Bates et al., 1997, 1998, 2002). Maori culture and arts are also widely practiced there, both for aboriginal authenticity and for tourists and therefore certain occupational exposures, such as wood dust from carving (a risk for sinusitis and sinonasal cancers), tend to be colocated with exposure sources. For these reasons, acute cases from Rotorua are plausibly causal as related to hydrogen sulfide, but chronic disease associations, such as cancer risk, require careful interpretation because of confounding. Interpretation must also take into account misclassification of ethnicity and differential utilization of healthcare between Maori and non-Maori residents.

The Medical Diagnostic Review, a population-based prevalence survey and clinical screening conducted in southern Alberta, Canada, in the mid-1980s, found no evidence of an excess in prevalence of neurologic signs and findings, both soft and hard, or other nonrespiratory outcomes in populations living downwind of sour-gas facilities (Spitzer, 1986). The study initially reported to the community that there was no elevation in respiratory outcomes (Spitzer et al., 1989), then later in the literature reported an increased prevalence in respiratory symptoms among children, which they attributed to parental concern (Dales et al., 1989).

Neurologic effects

Well-designed experimental studies have not shown evidence of neurotoxic sensory or cognitive effects associated with acute exposure to hydrogen sulfide at 5 ppm and below, notwithstanding the apparent physiologic shift to anaerobic metabolism at 10 ppm (Bhambhani et al., 1994, 1996a, b; Bhambhani and Singh, 1991). This, again, suggests that the mechanisms are not the same.

Sewer workers have been a popular group to study for evidence of neurologic impairment due to hydrogen sulfide at levels near 10 ppm. Egyptian sewer workers in Cairo exposed to levels estimated to be near 10 ppm were found to have evidence of "cognitive impairment" compared to a control group, for which no description was provided. However, the symptom profile reported was very nonspecific (headache, memory deficits, and lack of concentration) and commonly seen as secondary effects of psychologic factors (stress, job dissatisfaction, sleep disorder, low socioeconomic status or education). As may be imagined, sewer workers in developing countries are likely to have low socioeconomic status and other issues that confound performance on standardized psychometric tests. More persuasive was the observation of a difference in P300 latency, suggesting slowing of cognitive function (Sahar and Nirmeen, 2010). This converges with a similar finding following higher exposure levels (Hirsch, 2002). Even so, there are confounding factors in all such studies that require replication and validation.

Controversies in assessing toxicity

There are a number of studies in the literature that attempt to document neurologic effects (beyond odor responses) at levels relevant to community exposures.

The most difficult studies to interpret come from one investigator (Kilburn, 1993, 1997, 2001, 2003, 2010; Kilburn and Warshaw, 1995; Kilburn et al., 2010). They require evaluation as a group and close attention because of issues regarding quality assurance and testing methodology, some of which were revealed at a major conference in 2000 (Woodall et al., 2005) and in legal proceedings since then (Kilburn, 2004). Kilburn uses a number of tests of his own adaptation and interpretation is often confused because he reports the results in terms of percentage predicted (on analogy to pulmonary function testing) rather than raw scores.

The intellectual problem with this body of work is that, despite questions with respect to validity, it cannot be summarily dismissed. Findings of neurobehavioral impairment that suggest impairment in highly exposed workers who had experienced loss of consciousness are plausible (Kilburn, 2003). On the other hand, Kilburn's findings of similar or more severe impairment at much lower concentrations in the context of prolonged community exposure have not been reproduced when the same population was studied by other, appropriately trained and experienced investigators (Campagna et al., 2004). He has, for example, suggested that knockdown and death may occur at exposure levels of 50 ppm, for which there is no evidence (Kilburn and Warshaw, 1995). (This was not a typographic error for 500 ppm, as it is repeated in the discussion.)

Kilburn and Warshaw (1995) studied 13 former workers at a refinery in California and 22 neighbors who lived downwind, all of them parties to a lawsuit. Referents were "friends and relatives nominated by the exposed subjects" who had not worked at the refinery or lived downwind of it. Ambient exposure levels of 10–100 ppb were recorded near the residents' homes and occupational exposure by environmental monitoring suggested exposures of 0–8.8 ppm in the area of concern, a desulfurization plant. In this group, blink reflex latencies were on the order of 14 ms. The most consistent and largest differences between the "exposed group," which combined workers and residents, and the referent group was in the profile of mood states, where the exposed group showed much higher scores for anger, depression, confusion, and fatigue, all of which are also known confounding factors in neurobehavioral tests and plausibly associated with litigation. All of 33 health complaints were more frequent in the exposed group (prevalence was not reported), most at high levels of statistical significance exceeding $p < 0.00001$, and, for some symptoms like "exhilaration" and "cough with blood," by implausibly large margins. (Obviously, coughing up blood should be very rare in any community-derived population: Kilburn and Warshaw, 1995.)

In 1997, Kilburn published a case series of 16 subjects, 12 of them parties to lawsuits, five of whom had experienced knockdowns. In his opinion and by his methods, some of which are not standard, all showed neurologic abnormalities. Unlike the earlier study, referents were drawn from three communities elsewhere in the USA who had been recruited at random for other studies from voter registration rolls. Similar to the earlier study, health complaints were universal save for two ("loss of consciousness" and "exhilaration" were not elevated in frequency) at extraordinary levels of statistical significance, although by less in absolute magnitude. Among neurobehavioral tests, blink reflex latencies were reported to be on the order of 15–20 ms and symmetric for unexposed individuals and on the order of 85–100 and asymmetric among exposed subjects when presented as group data, a highly implausible finding with no explanation for the discrepancy (Kilburn, 1997).

Subsequently, in order to address the issue of knockdown as a risk factor, Kilburn published a case series of 19 new subjects alleged to have been exposed to hydrogen sulfide and to have experienced adverse effects but not knockdown, studied from months to 10 years after the exposure. No information was provided on how the subjects were exposed or whether they had been referred for evaluation for medicolegal purposes, as had his other cases. No exposure assessment or estimate was available and the opportunity for exposure varied enormously, from a purported plumbing communication with sewers to living in a house across a road from an open waste dump that received gypsum board. (Small quantities of hydrogen sulfide may be released by sulfur-reducing bacteria from defective construction materials made from gypsum containing pyrite, in the presence of water.) Subjects completed a battery of health questions and neurophysiologic and neurobehavioral tests. Referents were a group of subjects from Wickenburg, Arizona, USA, recruited for a previous study. The results were unusual. Despite showing wide and overlapping confidence intervals, they were reported to be different and highly significant statistically. There were many anomalies, such as blink reflex latency times, on the order of 100 ms for both exposed and referent subjects, that were asymmetric in the referent group, and asymmetric color vision scores in the exposed group (Kilburn, 2003).

Kilburn has also published data suggesting neurobehavioral effects observed in populations exposed to low levels of hydrogen sulfide, including single exposures below 1.0 or 0.5 ppm. The subjects consisted primarily of litigants, involved in a lawsuit for compensation for damages arising out of their exposure. The studies describe a pattern of poor recall, delayed reaction time, unilateral (not bilateral) changes in visual performance, and abnormal balance (requiring special technology to detect). The group has described similarly nonspecific patterns for other potential hazards (Kilburn, 1997).

The effects of litigation, anxiety, posttraumatic stress, anticipation of compensation, and test bias cannot help but render this and similar studies uninterpretable, untrustworthy, and biased toward false positive results. For example, in one table purporting to show the findings on retest of 8 subjects exposed to hydrogen sulfide in an oil refinery, the stated interval is 15 months but the average age of the same group between first test and retest changed by only 5.8 months; still, regardless of the true interval, the relatively few findings that show significant differences are difficult to interpret, such as the "balance sway speed" (Kilburn's proprietary coordination test), which shows deterioration when the subject's eyes are open but not when closed.

Neurobehavioral testing is also easily confounded by effects of learning, practice, motivation, sleep, education, distraction, depressed mood, and stress. The wide range of "normal" performance on these tests renders them rather insensitive and nonspecific in identifying neurotoxic outcomes, limitations freely acknowledged in the mainstream neurobehavioral literature and the subject of much discussion. Above all, however, the tests have to be performed correctly to be valid.

During a symposium in 2000 (Woodall et al., 2005), Dr. Kilburn admitted publicly that he conducted many of these tests in a manner that did not conform to standardized instructions, used methods of his own devising, and did not always conform to standard practice in neuroscience. A further problem with Kilburn's work is the frequent reporting of implausible outcomes, such as lengthened blink reflex latency in one eye but not another, and the implausibly universal patterns of responses in symptom surveys.

Kilburn's subjects are usually engaged in litigation or are seeking compensation at the time of testing and much of the work is usually supported by legal fees and treatment services. He rejects, as an epidemiologic proposition, that this recruitment factor biases his work. His referent groups have often been church members (Mormon, with implications for lifestyle) and others who are not likely to be representative of their own communities. He does not compare his exposed populations with referent cases drawn from nearby or similarly traumatized or psychologically affected communities.

Kilburn has admitted under oath in a deposition (Ahmed *v.* Legacy Homes, 2004: Kilburn, 2004) that he could not replicate his earlier findings in his control groups when subjects were retested. His explanation was that a rapid decline in neurologic capacity in the USA as a whole reflected itself as a decline in performance in his referent groups. He also admitted that his "community" control groups often deviated in important ways from his subject groups (for example, being drawn from rural communities, which is associated with a markedly different distribution of neurologic lifestyle factors, such as caffeine and nicotine intake).

Kilburn's work, if taken at face value, also suggests a lack of specificity in the toxic effects of chemical agents, since it suggests that very similar deficits are reliably found for agents with very different mechanisms of action and neurotoxic potential, such as hydrogen sulfide, polychlorinated biphenyls, arsenic, chlorine, chlorpyrifos, nickel carbonyl, and ammonia (Kilburn, 1997, 2001). The similarity of outcomes and potency described for such a disparate group of chemicals must lead one to at least consider that what is being detected as an abnormality is actually a methodologic artifact. Kilburn (2001) rejects this explanation.

Confusingly, some of Kilburn's findings cannot be summarily dismissed. Findings of neurobehavioral impairment that suggest impairment in highly exposed workers who had been rendered unconscious are in fact plausible (Kilburn, 1993). On the other hand, the findings become less plausible as studies address lower concentrations. One of Kilburn's earlier contributions purported to document headache, nausea, vomiting, depression, and personality changes, nosebleeds, and shortness of breath among refinery workers exposed to hydrogen sulfide; some of these outcomes are plausible, such as headache, others are unlikely (it seems improbable that refinery workers who could not adapt to nausea from the odor would continue to work there), and others are implausible (such as nosebleeds: Kilburn and Warshaw, 1995). Furthermore, Kilburn's findings of similar or more severe impairment at much lower concentrations in the context of prolonged, documented community exposure (Inserra et al., 2002) were not replicated when the same population was studied by experienced investigators using conventional means (Campagna et al., 2004; Inserra et al., 2004).

In a similar vein, a brace of paired community symptom surveys were conducted by Legator et al. (2001) between two communities exposed to hydrogen sulfide emissions, from wastewater and geothermal sources, and local reference communities without such sources of exposure. Again, symptom prevalence was uniformly (except for the category of "endocrine") and massively different, with point estimates of odds ratios reaching as high as 12.7 and 11.9 in point estimate, and including such alarming symptoms as "coughing up blood." The study groups were not drawn from a representative sample; most were plaintiffs in a legal case who had an incentive to overreport symptoms. Legator et al. (2001) also were rather casual in their manipulation of the comparison groups, excluding one or the other as it suited them, on grounds that should have been apparent at the beginning as criteria for exclusion from the study. There were also marked differences as well in the age distribution and racial composition (important in these specific locations as a marker for both socioeconomic status and language fluency). Other than tracking smoking rates, the study did not account for other confounding variables, such as other industrial activity and exposure to agricultural pesticides in the regions where the study was conducted. These problems are reflected in the bizarre outcomes, in which not only plausible symptoms but dental complaints and "spleen problems" all achieved the same high level of statistical significance. The authors incorporated an insightful discussion of the limitations of their study. However, this study probably should not be relied upon.

TREATMENT

The key to survival in virtually all hydrogen sulfide incidents is rescue at site and prehospital management. Table 8.2 is a draft protocol, prepared for a company with a recurrent hydrogen sulfide exposure problem, outlining appropriate procedures in stepwise fashion.

Decontamination is not required for the protection of health providers and bystanders, except in the rare instance in which liquid hydrogen sulfide is involved in a spill. However, the foul odor of hydrogen sulfide may complicate management and cause anxiety among other patients. Removal of clothing and simple skin washing with soap and water should remove the odor from the patient. Airing clothing outdoors should greatly reduce the odor within minutes.

Despite considerable advances in understanding the physiology of hydrogen sulfide and incremental clinical advances in treating acute toxicity, the management of hydrogen sulfide poisoning is still empiric and poorly grounded in evidence derived from research (Yalamanchili and Smith, 2008). Treatment of transient exposures, and of knockdowns in which the worker has regained consciousness, is not specific, hinges on rapid removal from exposure, and is clearly unsatisfactory given the current state of the art (Hoffman, 1997; Guidotti, 2007; Policastro and Otten, 2007).

Overwhelming exposures and prolonged knockdowns may require aggressive treatment but the circumstances in which these exposures occur often lead to delays in transport and treatment. There is a modest literature on antidotes, which are largely experimental or anecdotal, except for nitrite administration and, more recently, hyperbaric oxygen. Human experience is acquired one case at a time, because hydrogen sulfide toxicity cases, like most occupational toxicity situations, are infrequent and therefore difficult to prepare for and to study by protocol.

Hyperbaric oxygen therapy is an attractive and logical option for treating hydrogen sulfide intoxication, made practical by the increasingly widespread availability of hyperbaric chambers and familiarity gained by treating carbon monoxide exposure. Uncontrolled clinical evidence suggests that it may be effective. Rapid improvement in oxygenation and acid–base balance, and an incidental reduction in methemoglobinemia, suggests that hyperbaric oxygen treatment improved physiologic effects in ways that could be critical to short-term survival, that may reduce the risk of brain injury from hypoxemia, that may reduce the risk of cardiac complications from acidosis, and that would not interfere with longer-term

Table 8.2

Management of sour-gas and hydrogen sulfide emergencies

1. Purpose
To provide guidance for the appropriate management of cases known or suspected to be knockdown due to hydrogen sulfide
2. Policies
2.1 Presume that exposure to hydrogen sulfide has occurred in any "man down" situation in any of the following settings:
2.1.1 Facilities involving sour natural gas or crude oil
2.1.2 Facilities involving sewer lines
2.1.3 Confined spaces in which organic matter has been held
2.2 Self-contained, positive-pressure breathing apparatus (SCBA) must be available on site and close (specify practical distance) in any operation in which hydrogen sulfide could be encountered and in any entry into confined spaces
2.3 Rescue attempts undertaken without SCBA are forbidden: breath holding is not an alternative to SCBA and could be lethal to both attempted rescuer and the initial victim, by delaying successful rescue and transport
2.4 Any and all personnel engaged in the rescue effort must be adequately protected:
2.4.1 SCBA is mandatory for rescue involving entry into confined spaces or when emissions from a source persists (e.g., well blowout). (Never use air-purifying respirators, such as a half-mask respirator with charcoal filter cartridges. These are ineffective against hydrogen sulfide and the mistake may be lethal)
2.4.2 A protective suit is not required
2.4.3 Clinicians providing emergency services who are not familiar with hydrogen sulfide toxicity should consult with a toxicologist, if available, or a specialist in pulmonary or internal medicine or an emergency medicine specialist
3. Initial management
3.1 Call (Emergency Response, Security, Fire and Rescue, as appropriate) immediately
3.2 Rescue and extrication must only be attempted by trained personnel wearing SCBA and following safety precautions
3.2.1 After donning self-contained breathing apparatus, immediately remove the victim from the area and take to a safe area upwind and away from the source of exposure
3.2.2 If the victim is unconscious or shows evidence of trauma, stabilize the cervical spine
3.3 Document the presence of hydrogen sulfide (this should be done by personnel not directly engaged in the rescue effort)
3.3.1 Odor of rotten eggs will be present around the area
3.3.2 History from workers that they smelled rotten eggs and that the odor then disappeared (olfactory paralysis)
3.3.3 History of sudden eye irritation
4. Prehospital management
4.1 Provide basic life support and trauma stabilization on site if required
4.1.1 If there is reason to suspect a fall, stabilize the neck
4.1.2 Provide oxygen
4.1.3 Provide positive-pressure ventilation if the victim is not alert and breathing spontaneously
4.1.4 Irrigate the eyes with saline if the victim reports eye irritation
4.1.5 Decontamination of person and clothes is not necessary
4.2 The emergency medical technician conveying the patient by ambulance should:
4.2.1 Manage the airway as appropriate and provide positive-pressure ventilation with 100% oxygen if the victim is not alert and breathing spontaneously
4.2.2 Follow usual procedures for basic life support
4.3 Stabilize and transport to hospital for definitive care
5. Emergency care
5.1 The physician or nurse attending the patient should:
5.1.1 Evaluate critical status immediately on arrival
5.1.1.1 Assess baseline respiratory status, cardiovascular, neurologic status on arrival
5.1.1.2 Perform usual measures if patient is in cardiopulmonary arrest or compromised
5.1.1.3 Monitor closely thereafter
5.1.2 Send blood for carbon monoxide and cyanide determinations if there is a potential for mixed exposure
5.1.3 Take the history of exposure from the victim and document
5.1.4 Decontamination is not required
5.2 The examining physician providing emergency care should:
5.2.1 Take the medical history from the victim or others, with care to identify other reasons for loss of consciousness and comorbidities
5.2.2 Conduct a thorough examination emphasizing the following:
5.2.2.1 Neurologic and mental status
5.2.2.1.1 Assess and monitor level of consciousness using Glasgow coma scale, if applicable

Continued

Table 8.2

Continued

5.2.2.1.2 Evidence of head trauma

5.2.2.2 Evidence of hypoxia. Signs of hypoxemia such as cyanosis and low arterial blood oxygen tension may be absent: do not rely on them. Oxygen saturation is usually low but this is subject to error because it is a calculated number, not a direct measurement

5.2.3 With consultation and under specialist care, consider antidotal therapy

5.2.3.1 Hyperbaric oxygen therapy is supported by strong but anecdotal evidence, requires rapid transport to a hyperbaric chamber, and is only effective for immediate treatment

5.2.3.2 Sodium nitrite is used for the induction of methemoglobinemia; sulfide has a higher affinity for methemoglobin than hemoglobin. The effectiveness of induced methemoglobinemia has not been established and the treatment is known to be useless if initiated more than an hour after exposure. Treatment with nitrite creates a risk of hypotension and cardiovascular complications. Therefore sodium nitrite should not be used for the treatment of hydrogen sulfide-exposed patients with a history of high probability of heart disease or a cause of hypotension. Exceptions may be acceptable if the patient is moribund and the treatment can be administered promptly after exposure and monitored closely

5.2.3.3 Hydroxocobalamin, as for cyanide, has not been systematically evaluated and is supported only by anecdotal evidence but has low risk and a rationale for treatment. Like sodium nitrite, it is unlikely to be helpful unless given very soon after exposure

6. Definitive care

6.1 The attending caregiver providing definitive care should:

6.1.1 Consult with a physician experienced in hydrogen sulfide toxicity (preferably an emergency medicine specialist, toxicologist, pulmonary specialist, internist, or intensivist)

6.1.2 Provide eye care (irrigation, patch) if patient is experiencing visible eye irritation or discomfort

6.1.3 Monitor patient for respiratory, cardiac, and neurologic changes and treat as needed. Special emphasis on risk of pulmonary edema:

6.1.3.1 Follow with serial chest films for development of pulmonary edema

6.1.3.2 Noncardiogenic edema usually occurs within 24 hours but may occur as long as 72 hours after exposure

6.1.3.3 Treat as adult respiratory distress syndrome

6.2 The occupational medicine specialist should:

6.2.1 Advise the attending physician as required

6.2.2 Evaluate the patient for fitness to return to work

6.2.2.1 Report fitness for duty determination (fit, unfit, fit with accommodation)

6.2.2.2 Determine work accommodation, if such is required

6.2.2.3 Determine permanent impairment if applicable

6.2.2.4 Review medical standards periodically and revise as necessary in collaboration with toxicologists and emergency medical services

recovery (Lindenmann et al., 2010). The utility of oxygen is probably related to the oxygenation of marginally anoxic tissues and the displacement of sulfide from binding sites on cytochrome c oxidase, but this has not been worked out. The standard treatment, as for carbon monoxide poisoning, has been used successfully, especially in multiple casualties where individualization may not have been practical (Lindenmann et al., 2010). A series of three treatments at 2.5 atmospheres for 90 minutes each has been recommended (Whitcraft et al., 1985). Hyperbaric oxygen treatment is safe in skilled hands but transport to a facility should not delay supportive care, which is a more important factor in survival (Demaret and Fialaire, 1974; Peters, 1981; Turner and Fairhurst, 1990).

In one remarkable case, a young oil worker was knocked down during a failed rescue attempt of a co-worker, and was immobilized for 20 minutes in the cloud. By the time he could be rescued, he was unresponsive and in cardiac arrest but was resuscitated, intubated, and ventilated, then transported, stabilized hemodynamically (withholding nitrite), and was treated, over 9 hours after the event, with both hyperbaric oxygen and hypothermia, as if for head trauma. Five days later, he was discharged, initially with only the complaint of short-term memory loss. Over time, other cognitive deficits (not quantified in the report) and a mild expressive aphasia declared themselves, but the cognitive function appeared to be improving when he was lost to follow-up (Asif and Exline, 2012).

4-Dimethylaminophenol has been used as a specific antidote in analogy with its use in cyanide toxicity (Lindenmann et al., 2010). The rationale for regenerating cytochrome oxidase c that is the rationale for its use in cyanide may or may not work equally well for hydrogen sulfide. Since its use is combined with other interventions, its efficacy cannot be easily determined, but no obviously adverse consequences to its use have been observed (Lindenmann et al., 2010).

Nitrite treatment is based on the theory that methemoglobin generated by nitrite would displace the sulfide, as it does cyanide, and regenerate the active cytochrome oxidase (Smith, 1981). Although methemoglobin is more effective in binding sulfide than oxyhemoglobin, the complex between sulfide and methemoglobin does not last long enough to make a difference. Sulfide disappears quickly from the circulation under conditions of good oxygenation anyway. The evidence suggests that nitrite can only be effective within the first few minutes following exposure and may actually slow sulfide removal thereafter. There are also reasons to believe that the methemoglobin hypothesis is insufficient to explain the action of nitrite even in the first few minutes. A further practical problem with this approach is that it may add to the anoxic burden that already may exist from the cytochrome poisoning, respiratory paralysis due to central toxicity, and ventilation–perfusion mismatch associated with pulmonary edema. It may also induce hypotension and further complicate the anoxia with hypoperfusion. Increased oxygen tension may overcome some of the physiologic disadvantages of nitrite treatment.

Hydroxocobalamin, which is a standard treatment for cyanide, also complexes with hydrogen sulfide, has low toxicity, and has a rationale for therapy that suggests that it would be appropriate for a trial. However, it has been reportedly used only in extreme cases, which were probably not salvageable in the event, and so has really not been tested for efficacy (Fujita et al., 2011).

Other pharmacologic treatments have been proposed, but efficacy has not been demonstrated (Guidotti, 2007). Thiosulfite has been used as an antidote, on the presumption that treatment clears free sulfide, but the therapeutic rationale has been questioned, especially since thiosulfite is a metabolite of sulfide (Smith and Smith, 1977), and the treatment has been abandoned (Guidotti, 2001). Pretreatment with several agents has been associated with increased survival in animal models, including pyruvate (Dulaney, 1988), dithiothreitol (Warenycia et al., 1990), and α-ketocarboxylic acids (Hume, 1988). These have not shown therapeutic efficacy and will not be discussed further here.

There is a potentially promising lead involving two readily available agents, bicarbonate and glucose (Almeida et al., 2008). Almeida et al. investigated the toxicity of dimethyl sulfide (Me_2S), a less potent and more easily handled analog of hydrosulfide, in Sprague–Dawley rats. The incidence and duration of coma (the rat equivalent of knockdown) induced by this agent over a range of dose were determined and with this model the efficacy of purported antidotes for H_2S was determined. Using a standard dose of dimethyl sulfide that reliably induced the coma effect, three agents proposed to protect against lethality from hydrogen sulfide, nitrite, pyruvate, and dithiothreitol were found to have no significant effect in shortening duration of coma and preventing lethality, whether given at onset of coma or as pretreatment. Bicarbonate, on the other hand, reduced duration of coma by about 28% but did not affect lethality. Bicarbonate and glucose administered together, however, reduced coma duration 37% and mortality 50% if given at onset, and the effect was even greater when given as a pretreatment, reducing coma duration by 80% and preventing mortality altogether. Sulfide toxicity in the central nervous system may therefore be associated in some way with the metabolism of glucose. Perhaps cellular hypoglycemia and carbonic anhydrase activity play a role in the pathophysiology of sulfide toxicity, although earlier studies have suggested a hyperglycemic response (Hayden et al., 1990), perhaps due to stress and catecholamine release.

These observations demonstrate that a treatment protocol for both knockdown and death from H_2S may ultimately be possible, although the feasibility of administering bicarbonate and glucose in prehospital, field situations remains to be determined.

TOXICOKINETICS

Little is known of the toxicokinetics of H_2S and sulfide, because absorption, dissociation, and formation of the hydrosulfide ion, and distribution by the circulation occur so rapidly.

Absorption

The field of inhalation toxicology is based on certain principles and empiric observations that are different in practice from more familiar approaches in pharmacology, drug side effects, or toxicity by other routes. For an inhaled gas, concentration over time, yielding cumulative dose, is not necessarily the best metric of exposure. In the real world, the exposure profile is irregular as concentration changes with the evolution of the incident and especially with the passage of the "plume" of the gas over and around the victim in an open area. The potential of a gas to reach the deep lung is inversely proportional to its solubility in water, so that insoluble gases such as hydrogen sulfide and carbon monoxide arrive at a relatively higher concentrations in alveolar air than a soluble gas (such as cyanide, which is highly soluble). Diffusion across the alveolar–capillary barrier and dissociation occur almost instantaneously, so that a bolus enters the blood, not unlike an intravenous injection.

Distribution

The concentration of sulfide in blood rises in an abrupt step function so that a flux or front is carried to target tissues that, in the case of brain, are already high in sulfide. Because sulfide is cleared so quickly, the concentration in blood and then tissue drops to physiologic levels almost as quickly as it rose. As a consequence, nonlethal cases do not demonstrate elevated sulfide in blood or tissue for long enough to be useful in diagnosis. However, sulfide accumulation in tissue can be used as one marker, together with thiosulfate, of exposure postmortem because circulation stops. The sulfide ion binds to heme compounds but there is no evidence that this is an important depot site, toxicologically.

Metabolism

The metabolism of hydrogen sulfide is of importance for its toxicologic and forensic implications, because it occurs too quickly for treatment interventions. Once absorbed, H_2S dissociates in part into hydrosulfide ion, HS^-, then is spontaneously transformed into sulfide ($S^=$) and is enzymatically oxidized to thiosulfate ($S_2O_3^=$), which is oxidized to sulfate ($SO_4^=$) and enters general metabolism (Hildebrandt and Grieshaber, 2008; Olson et al., 2013). Some H_2S remains as free H_2S in blood and this fraction appears to interact with metalloproteins, disulfide-containing proteins, and thio-*S*-methyltransferase, forming methyl sulfides.

Excretion

Sulfate is then excreted by the kidney or recycled in intermediary metabolism (Haggard, 1925). Excretion of sulfide is not a limiting step in toxicity.

TOXICODYNAMICS

Concentration is much more important than duration of exposure in the toxicity of hydrogen sulfide. The exposure–response curve for lethality is extremely steep for hydrogen sulfide, among the steepest known in inhalation toxicology (Prior et al., 1988; Guidotti, 1996b; US Environmental Protection Agency, 2009). Many current models for risk assessment for hydrogen sulfide use a concentration–time constant for lethality of the general form $C^n \times t$, where n ranges from 1.43 to over 4.36, but the empiric evidence strongly favors the higher exponents (Guidotti, 1996b).

Mechanisms of action

The primary mechanism of hydrogen sulfide toxicity was long believed to be cytochrome oxidase c poisoning, with resulting shutdown of oxidative phosphorylation and energy to the cell, analogous to cyanide. However, this cannot explain the many observed anomalies and departures from the model of cyanide toxicity or the reversibility of many effects (Smith et al., 1976; Smith and Smith, 1977; Guidotti, 2007). If irreversibly or slowly reversible poisoning of cytochrome oxidase were the only or even primary mechanism of action in causing the cardinal symptoms of hydrogen sulfide toxicity that constitute the human toxidrome (except apnea), then almost all exposures would be fatal, anoxic brain injury would be universal in survivors, and cardiac manifestations, both arrhythmia and acute infarction with resulting myopathy and other dysfunction, would be a major clinical presentation. That most cases survive and these cyanide-like outcomes rarely occur suggests that poisoning of cytochrome oxidase c cannot be the major mechanism at usual exposure levels.

Because anoxia at the cellular level was assumed to be the cause of toxicity, hydrogen sulfide is classified toxicologically but unhelpfully as an "asphyxiant," together with cyanide, carbon monoxide, and azide. The asphyxiants, as a class, do not represent a coherent toxicologic category, because they are heterogeneous in chemistry, mode of action, and clinical presentation. This has perpetuated a simplistic view of the mechanism of action of the gas. Classification of hydrogen sulfide as an asphyxiant has probably impeded the achievement of more useful insights.

H_2S is, however, known to inhibit cytochrome c oxidase and possibly other heme-containing macromolecules by binding as aqueous sulfide (Smith and Smith, 1977; Smith and Gosselin, 1979; Roth et al., 1997). H_2S does interact with a number of enzymes and other macromolecules, especially hemoglobin and myoglobin. However the affinity of hydrogen sulfide for hemoglobin is weak, especially compared to carbon monoxide. It is stronger for cytochrome oxidase c but not as strong as cyanide.

There are clearly other mechanisms at work (Greer et al., 1995; Almeida and Guidotti, 1999; Truong et al., 2006; Almeida et al., 2008).

One lethal endpoint of H_2S exposure is respiratory paralysis, a dose-dependent reduction in ventilatory drive resulting in apnea, which follows an initial hyperpnea (Haggard, 1925; Klentz and Fedde, 1978). Infusion of the sulfide ion alone into the circulation mimics the systemic effects of H_2S inhalation. The apneic response to H_2S is not enhanced by intra-arterial injection of sulfide into the carotid artery, which delivers blood directly to the brain, compared to intravenous injection into the femoral artery, in which mixing occurs in the general circulation. The apneic response can also be blocked by paralysis of the vagal nerve by lidocaine or by infusion of sodium bicarbonate solution, in the absence of acidosis. These findings strongly suggest that a peripheral receptor in the lung is responsible for the apneic response (Almedia and Guidotti, 1999).

Hydrogen sulfide may also be a neuromodulator and so toxic exposure may involve direct functional effects as well as cell toxicity.

Because hydrosulfide binds to cytochrome c, it has always been assumed to act in the same way as cyanide, causing anoxia at the cellular level (Smith and Smith, 1977; Smith and Gosselin, 1979; Roth et al., 1997). Cellular anoxia probably does exert a devastating effect at high exposure levels, and there may be enough inhibition at lower levels to serve as a histologic biomarker (Dorman et al., 2002). However, this cannot be the only significant mechanism and may not even be the predominant toxic effect (Reiffenstein et al., 1992). Cytochrome oxidase is not upregulated, as would be expected, following repeated high but nonlethal exposure to hydrogen sulfide in rats despite many other changes in gene expression (Roberts and Dorman, 2008). As noted earlier, there is no consistent pattern of anoxic brain injury following knockdown, as there is with coma equivalents following cyanide and carbon monoxide.

Hydrogen sulfide clearly can poison cytochrome c *in vitro* at physiologically relevant concentrations and can inhibit aerobic metabolism even at relatively low levels. In studies designed to explore the use of hydrogen sulfide in inducing suspended animation on a mouse model, conducted at 80 ppm (comparable to much higher levels in humans), the gas was found to be a rapidly reversible inhibitor of heart rate, cardiac output, core body temperature, and respiratory rate while preserving stroke volume and blood pressure (Volpato et al., 2008). In healthy, fit human volunteers, hydrogen sulfide exposure as low as 5 ppm during exercise (30 minutes) is associated with an early shift from aerobic to anaerobic metabolism, as indicated by increasing blood lactate levels, but without symptoms (Bhambhani et al., 1996b). This observation precipitated a review of occupational exposure levels in the UK (Costigan, 2003).

There are clearly other mechanisms of toxicity that are likely to be at least as important in the toxidrome, however (Greer et al., 1995; Almeida and Guidotti, 1999).

It has been known for many years that H_2S has potent neurotoxic properties. Respiratory paralysis may occur if exposure is prolonged, presumably as a direct consequence of sulfide toxicity inhibiting brainstem respiratory nuclei. However, the studies cited on apnea, supported by studies suggesting a chemoreceptor response, suggest that there are other pathways that affect central activity. Measurable brain levels of sulfide may or may not change appreciably following inhalation of hydrogen sulfide (even as high as 400 ppm in the rat, which is equivalent to a much higher concentration in humans), as reports disagree (Goodwin et al., 1989; Dorman et al., 2002), and concentrations in other tissues increase only transiently (Dorman et al., 2002). Even so, electroencephalographic changes are detectable even at lower concentrations (25 ppm) in the rat and persist for weeks (Skrajny et al., 1996). In additional, peripheral reflex mechanisms may play an important role in the intact animal and in the observed clinical syndrome (Almeida and Guidotti, 1999; Qu et al., 2006; Almeida et al., 2008).

In the past there was a controversy over whether sulfhemoglobin occurs following high-level hydrogen sulfide exposure, but the weight of evidence is that it does not (Policastro and Otten, 2007; Smith and Gosselin, 1984).

FORENSICS

Reconstruction of hydrogen sulfide-related events is usually circumstantial because of the unavailability of forensic testing. Time of death is difficult to determine with these indicators, but sequence of death may be reconstructed in multiple fatalities by integrating inferences on exposure with reconstruction of the events of the incident (Poli et al., 2010).

Discoloration of skin and brain tissue appears to be associated with very high exposure levels (Ago et al., 2008).

Sulfide and thiosulfate testing are never available in a timely manner for diagnosis and as a practical matter are only used in fatal cases to establish the cause of death. Now that the broader physiologic role of hydrogen sulfide is apparent, there is likely to be more interest and opportunity for the development of biomonitoring indicators.

Sulfide is rapidly eliminated from the body and is unstable postmortem. To be useful, blood sulfide levels must be obtained no later than 2 hours after exposure (Ballerino-Regan and Longmire, 2010). Blood should be drawn in a 5 mL "red-top" container (heparin-free). As quickly as possible, blood and any urine samples that can be obtained should be frozen for analysis. "Detergent suicide" cases, on the other, hand, expire quickly and often attain much higher levels than accidental cases, with sulfide detectable even in cerebrospinal fluid and pleural fluid.

Thiosulfate, however, lasts longer and is stable after death, and therefore serves as a marker for hydrogen sulfide exposure. Interpretation is by comparison to reference values, as there is no clear threshold for toxic levels of the gas in blood or tissue (Bassindale and Hosking, 2011). Reference thiosulfate levels are rather high in normal tissue, approximately 0.3 μg/mL. Elevated levels must therefore be substantially raised above that to suggest toxicity, empirically by a multiple of 8–77 (Ballerino-Regan and Longmire, 2010; Daldal et al., 2010; Poli et al., 2010). However, detergent suicide cases show the opposite relationship, with sulfide higher than thiosulfate, presumably because there is so little time for oxidation (Maebashi et al., 2011).

Urinary thiosulfate levels are only elevated in subjects who live long enough to filter and excrete the ion in urine. The presence of thiosulfate in urine may therefore be used as an indication of whether a person was killed outright or survived some time after a knockdown, but this information would not be very useful as a practical matter in most cases (Ballerino-Regan and Longmire, 2010).

Tissue levels of thiosulfate have also been found to be highly elevated postmortem, including in brain (Poli et al., 2010).

A method for postmortem evaluation of alveolar air has been reported but it is difficult to see how it would work and replication has not been reported by other investigators (Sidlo et al., 2009).

Sulfhemoglobin is not used for documenting hydrogen sulfide exposure.

PREVENTION

Hydrogen sulfide acts instantaneously to produce a knockdown and provides little margin for safety. Its effects render escape impossible at that point and create a trap for unwary rescuers. Events often occur in situations where the worker is new on the job or poorly trained (Lancia et al., 2013). Protocols such as that laid out in Table 8.2 can prevent avoidable fatalities only if they are intensively, realistically, and frequently drilled, because, as numerous examples show, the impulse to rescue without following procedures inevitably causes multiple casualties.

Air-supply personal respiratory protection is required for a gas as toxic as hydrogen sulfide, in the form of self-contained breathing apparatus (SCBA). Air-purifying devices are not adequate and reliance on such devices (such as canister-type full- or half-mask respirators) can be a fatal mistake. In the past, tympanic membrane defects (perforated eardrums) excluded workers from certification for SCBA gear, on the grounds that H_2S could bypass the SCBA and enough could be inhaled through the ear to impede response, but this is not supported by the evidence and intact eardrums are no longer required (Ronk and White, 1985).

The first line of defense in protection against hydrogen sulfide, especially in the oil and gas industry, is process management, so that the hazard is contained, isolated from the worker, and if necessary, as in an oil or gas well upset, diverted and flared. In any situation in which there is a risk of encountering hydrogen sulfide, SCBA must be available and workers must be trained in its use. Unfortunately, this has not prevented the recurrence of multiple casualties resulting from impulsive and doomed rescue attempts, in which a co-worker rushes to the assistance of a "man down" without stopping to don SCBA. It is a noble but sad illusion that any responder (invariably male) can hold his breath long enough, especially under conditions of maximal exertion, as required to rescue a co-worker in a toxic atmosphere. This often results in two unnecessary deaths because, given the reversibility of much of the toxidrome, the first victim often could be saved if the would-be rescuer had followed procedures. Intensive training and ready availability of personal protective equipment are required to prevent such situations (Mattarano, 1999).

Confined spaces, such as tanks, manholes, and ship's holds, are particularly hazardous even under normal circumstances, and are prone to the same type of multiple casualty. Even before hydrogen sulfide enters the space, it may have low oxygen, may be difficult to escape from in the event of an emergency, and hold toxic atmospheres from other sources, for example, solvent or welding fumes. Work performed in confined spaces, such as in refineries, absolutely requires a permit system and testing prior to entry and procedures must be rigidly enforced.

Most standards for the protection of occupational and community populations are based on prevention of conjunctivitis (eye irritation), knockdown, and respiratory tract irritation (US Public Health Service, 1977; Costigan, 2003). Occupational exposure levels based on time-weighted averages, such as the 8-hour time-weighted average permissible exposure limit of the US Occupational Safety and Health Administration (OSHA), do not take the steep exposure–response curve into account, and therefore tend to perpetuate the idea that an 8-hour time-weighted average concentration over time has as much significance as, or more significance than, the peak exposure, when this is not the case (Prior et al., 1988; Guidotti, 1996b).

Hydrogen sulfide, having a very low olfactory threshold, has good warning properties at lower levels of exposure, but these valuable warning properties are lost at higher concentrations because of olfactory paralysis. More than one oilfield worker has observed to the author that "you're not really in trouble until you can't smell it any more." This is obviously very dangerous for the worker, because, while the appearance of a foul odor is appreciated, the disappearance of a familiar odor may not be noticed. Some nasal sensation may persist, mediated by other pathways, as in the perception of ammonia and solvents, which are not so much smelled as felt in the nose. Some workers, quite ill-advisedly, depend on conjunctivitis as a warning sign. It is probably fair to conclude that workers who rely on these secondary warning properties have denied themselves the margin of safety and opportunity for escape that could ensure their survival as an incident unfolds suddenly.

Hydrogen sulfide alarms, typically set to go off at 5 ppm, are a mainstay of protection in gas plants and other industrial operations, but not in agricultural operations. These devices have a reputation for false alarms, but it is probably more correct that they have a "hair trigger" and are going off with exposure that is transient due to air movement.

Current exposure standards and guidelines for hydrogen sulfide promulgated by various authorities are mostly based on conjunctivitis and, in the case of the UK Health and Safety Executive, the transition to anaerobic metabolism (Costigan, 2003; US Environmental Protection Agency, 2009; Guidotti, 2010). The OSHA permissible exposure limit and the National Institute for Occupational Safety and Health-recommended exposure limit for H_2S are both 10 ppm, 8-hour time-weighted average. OSHA also has a short-term standard of 15 ppm. Alberta, Canada, a province dominated by oil and gas development, with abundant sour-gas resources, has an evacuation standard for the general population of 5 ppm and much lower occupational exposure levels.

"Set-backs," or buffer zones from sources of hydrogen sulfide, are standard in the sour-gas industry to prevent community exposure risk. The encroachment of communities on industrial sites and oil- and gasfields that were once relatively isolated, together with economic incentives to site wells and facilities where deposits are known to exist, including urban locations, has resulted in many conflicts in land use and regulatory requirements for emergency planning to protect residents and create setbacks, between gas facilities and populated areas. The Alberta Energy Regulator (formerly the Energy Resources Conservation Board) has been particularly proactive in emergency planning based on projected concentrations generated by computer models in the event of an uncontrolled release.

ACKNOWLEDGMENTS

We thank Mr. Jeffrey Curran for assistance in the literature search and retrieval. Development of this chapter was internally supported by Medical Advisory Services, Rockville, Maryland.

References

Agency for Toxic Substances and Disease Registry (2006). Toxicological Profile for Hydrogen Sulfide, US Department of Health and Human Services, US Public Health Service, Washington, DC.

Ago M, Ago K, Ogata M (2008). Two fatalities by hydrogen sulfide poisoning: variation of pathological and toxicological findings. Leg Med 10: 148–152.

Almeida AF, Guidotti TL (1999). Differential sensitivity of lung and brain to sulfide exposure: a peripheral mechanism for apnea. Toxicol Sci 50: 287–293.

Almeida AF, Nation PN, Guidotti TL (2008). Mechanism and treatment of sulfide-induced coma: a rat model. Int J Toxicol 27: 287–293.

Amino M, Yoshioka K, Suzuki Y et al. (2009). Improvement in a patient suffering from cardiac injury due to severe hydrogen sulfide poisoning: a long-term examination of the process of recovery of heart failure by performing nuclear medicine study. Intern Med 48: 1745–1748.

Arnold IM, Dufresne RM, Alleyne BC et al. (1985). Health implication of occupational exposures to hydrogen sulfide. J Occup Med 27: 373–376.

Asif MJ, Exline MC (2012). Utilization of hyperbaric oxygen therapy and induced hypothermia after hydrogen sulfide exposure. Respir Care 57: 307–310.

Audeau FM, Gnanaharan C, Davey K (1985). Hydrogen sulphide poisoning: associated with pelt processing. N Z Med J 98: 145–147.

Ballerino-Regan D, Longmire AW (2010). Hydrogen sulfide exposure as a cause of sudden occupational death. Arch Pathol Lab Med 134: 1105.

Bassindale T, Hosking M (2011). Deaths in Rotorua's geothermal hot pools: hydrogen sulphide poisoning. Forensic Sci Int 207: e28–e29.

Bates MN, Garrett N, Graham B et al. (1997). Air pollution and mortality in the Rotorua geothermal area. Aust N Z J Public Health 21: 581–586.

Bates MN, Garrett N, Graham B et al. (1998). Cancer incidence, morbidity and geothermal air pollution in Rotorua, New Zealand. Int J Epidemiol 27: 10–14.

Bates MN, Garrett N, Shoemack P (2002). Investigation of health effects of hydrogen sulfide from a geothermal source. Arch Environ Health 57: 405–411.

Beauchamp Jr RO, Bus JS, Popp JA et al. (1984). A critical review of the literature on hydrogen sulfide toxicity. Crit Rev Toxicol 13: 25–97.

Bhambhani Y, Singh M (1991). Physiological effects of hydrogen sulfide inhalation during exercise in healthy men. J Appl Physiol 71: 1872–1877.

Bhambhani Y, Burnham R, Snydmiller G et al. (1994). Comparative physiological responses of exercising men and women to 5 ppm hydrogen sulfide exposure. Am Ind Hyg Assoc J 55: 1030–1035.

Bhambhani Y, Burnham R, Snydmiller G et al. (1996a). Effects of 10-ppm hydrogen sulfide inhalation on pulmonary function in healthy men and women. J Occup Environ Med 38: 1012–1017.

Bhambhani Y, Burnham R, Snydmiller G et al. (1996b). Effects of 5 ppm hydrogen sulfide inhalation on biochemical properties of skeletal muscle in exercising men and women. Am Ind Hyg Assoc J 57: 464–468.

Bhambhani Y, Burnham R, Snydmiller G et al. (1997). Effects of 10-ppm hydrogen sulfide inhalation in exercising men and women. Cardiovascular, metabolic, and biochemical responses. J Occup Environ Med 39: 122–129.

Bott E, Dodd M (2013). Suicide by hydrogen sulfide inhalation. Am J Forensic Med Pathol 34: 23–25.

Brenneman KA, James RA, Gross EA et al. (2000). Olfactory neuron loss in adult male CD rats following subchronic inhalation exposure to hydrogen sulfide. Toxicol Pathol 28: 326–333.

Buick JB, Lowry RC, Magee TR (2000). Is a reduction in residual volume a sub-clinical manifestation of hydrogen sulfide intoxication? Am J Ind Med 37: 296–299.

Burnett WW, King EG (1978). Hydrogen sulfide poisoning. Can Med Assoc J 118: 776.

Campagna D, Kathman SJ, Pierson R et al. (2004). Ambient hydrogen sulfide, total reduced sulfur, and hospital visits for respiratory diseases in northeast Nebraska, 1998-2000. J Expo Anal Environ Epidemiol 14: 180–187.

Cao HZX, Zhang J, Huang X et al. (2014). Hydrogen sulfide protects against bleomycin-induced pulmonary fibrosis in rats by inhibiting NF-KH expression and regulating Th1/Th2 balance. Toxicol Lett 224: 387–394.

Chemical Suicides (2012). Available online at: www.firetraining.ca/firetraining/images/Chemical20%Suicide.ppt.

Christia-Lotter A, Bartoli C, Piercecchi-Marti MD et al. (2007). Fatal occupational inhalation of hydrogen sulfide. Forensic Sci Int 169: 206–209.

Claudet I, Marcoux MO, Karsenty C et al. (2012). Severe hydrogen sulfide intoxication: a pediatric case of survival. Ann Fr Anesth Reanim 31: 255–258.

Costigan MG (2003). Hydrogen sulfide: UK occupational exposure limits. Occup Environ Med 60: 918–928.

Daldal H, Beder B, Serin S et al. (2010). Hydrogen sulfide toxicity in a thermal spring: a fatal outcome. Clin Toxicol (Phila) 48: 755–756.

Dales RE, Spitzer WO, Suissa S et al. (1989). Respiratory health of a population living downwind from natural gas refineries. Am Rev Respir Dis 139: 595–600.

Demaret D, Fialaire J (1974). Intoxication by hydrogen sulfide in a natural gas refinery. Eur J Toxicol Environ Hyg 7: 32–36.

Dorman DC, Moulin FJ-M, McManus BE, Mahle KC, James RA et al. (2002). Cytochrome oxidase inhibition induced by acute hydrogen sulfide inhalation: correlation with tissue sulfide concentrations in the rat brain, liver, lung and nasal epithelium. Toxicol Sci 65: 18–25.

Dulaney Jr M, Hume AS (1988). Pyruvic acid protects against the lethality of sulfide. Res Commun Chem Pathol Pharmacol 59: 133.

Duong TX, Suruda AJ, Maier LA (2001). Interstitial fibrosis following hydrogen sulfide exposure. Am J Ind Med 40: 221–224.

Fiedler N, Kipen H, Ohman-Strickland P et al. (2008). Sensory and cognitive effects of acute exposure to hydrogen sulfide. Environ Health Perspect 116: 78–85.

Fujita Y, Fujino Y, Onodera M et al. (2011). A fatal case of acute hydrogen sulfide poisoning caused by hydrogen sulfide: hydroxocobalamin therapy for acute hydrogen sulfide poisoning. J Anal Toxicol 35: 119–123.

Gabbay DS, De Roos F, Perrone J (2001). Twenty-foot fall averts fatality from massive hydrogen sulfide exposure. J Emerg Med 20: 141–144.

Gallay AVLF, Demarest S, Van der Heyden J et al. (2002). Belgian coca-cola-related outbreak: intoxication, mass sociogenic illness, or both? Am J Epidemiol 155: 140–147.

Greer JJ, Reiffenstein RJ, Almeida AF, Carter JE (1995). Sulfide-induced perturbations of the neuronal mechanisms controlling breathing in rats. J Appl Physiol (1985) 78: 433–440.

Gerasimon G, Bennett S, Musser J et al. (2007). Acute hydrogen sulfide poisoning in a dairy farmer. Clin Toxicol (Phila) 45: 420–423.

Goode E (2011). Chemical suicides, popular in Japan, are increasing in the U.S. N Y Times Mag, 18 June 2011.

Goodwin LR, Francom D, Dieken FP, Taylor JD et al. (1989). Determination of sulfide in brain tissue by gas dialysis/ion chromatography: post-mortem studies and two case reports. J Anal Toxicol 13: 105–109.

Guidotti TL (1994). Occupational exposure to hydrogen sulfide in the sour gas industry: some unresolved issues. Int Arch Occup Environ Health 66: 153–160.

Guidotti TL (1996a). An international registry for toxic inhalation and pulmonary edema: notes from work in progress. Int Arch Occup Environ Health 68: 380–386.

Guidotti TL (1996b). Hydrogen sulphide. Occup Med (Lond) 46: 367–371.

Guidotti TLRS (2001). Science on the Witness Stand: Evaluating Scientific Evidence in Law, Adjudication and Policy, OEM Press, Beverley Farms, MA.

Guidotti TL (2006). Respiratory tract irritants. In: WR Rom (Ed.), Environmental and Occupational Medicine, 4 ed. Lippincott, Philadelphia, pp. 543–555.

Guidotti TL (2010). Hydrogen sulfide: advances in understanding human toxicity. Int J Toxicol 29: 569–581.

Guidotti TL (2007). Hydrogen sulfide. In: BS Shannon MW, MJ Burns (Eds.), Haddad and Winchester's Clinical Management of Poisoning and Drug Overdose, 4 ed. Saunders/Elsevier, Philadelphia, pp. 1335–1342.

Gunn B, Wong R (2001). Noxious gas exposure in the outback: two cases of hydrogen sulfide toxicity. Emerg Med 13: 240–246.

Haggard HW (1925). The toxicology of hydrogen sulfide. J Industrial Hyg 7: 113–121.

Hayden LJ, Goeden H, Roth SH (1990). Growth and development in the rat during sub-chronic exposure to low levels of hydrogen sulfide. Toxicol Ind Health 6: 389–401.

Hessel PA, Herbert FA, Melenka LS et al. (1997). Lung health in relation to hydrogen sulfide exposure in oil and gas workers in Alberta, Canada. Am J Ind Med 31: 554–557.

Hildebrandt TM, Grieshaber MK (2008). Three enzymatic activities catalyze the oxidation of sulfide to thiosulfate in mammalian and invertebrate mitochondria. FEBS J 275: 3352–3361.

Hirakawa A, Takeyama N, Iwatsuki S et al. (2013). Delayed myocardial injury following acute hydrogen sulfide intoxication. Japanese journal of toxicology 26: 44–48.

Hirsch AR (2002). Hydrogen sulfide exposure without loss of consciousness: chronic effects in four cases. Toxicol Ind Health 18: 51–61.

Hoffman H, Guidotti T (1997). Natural gas. In: MI Greenberg (Ed.), Occupational, Industrial and Environmental Toxicology, Mosby, Philadelphia, pp. 359–366.

Hoidal CR, Hall AH, Robinson MD et al. (1986). Hydrogen sulfide poisoning from toxic inhalations of roofing asphalt fumes. Ann Emerg Med 15: 826–830.

Hume AS, Dulaney MD (1988). The effectiveness of various ketocarboxylic acids in preventing sulfide-induced lethality. Toxicologist 8: 28.

Inserra S, Phifer B, Pierson R et al. (2002). Community-based exposure estimate for hydrogen sulfide. J Expo Anal Environ Epidemiol 12: 124–129.

Inserra SG, Phifer BL, Anger WK et al. (2004). Neurobehavioral evaluation for a community with chronic exposure to hydrogen sulfide gas. Environ Res 95: 53–61.

Ishikawa Y (2010). Response of the Tokyo fire department to hydrogen sulfide poisoning incidents caused by detergents and other household chemicals. Japanese journal of toxicology 23: 206–211.

Jianwen Z, Da L, Wenxing F (2014). An approach for estimating toxic releases of H_2S-containing natural gas. J Hazard Mater 264: 350–362.

Kage S, Ikeda H, Ikeda N et al. (2004). Fatal hydrogen sulfide poisoning at a dye works. Leg Med 6: 182–186.

Kilburn KH (1993). Case report: profound neurobehavioral deficits in an oil field worker overcome by hydrogen sulfide. Am J Med Sci 306: 301–305.

Kilburn KH (1997). Exposure to reduced sulfur gases impairs neurobehavioral function. South Med J 90: 997–1006.

Kilburn KH (2001). Function testing for chemical brain damage: a review. Arch Environ Health 56: 132–137.

Kilburn KH (2003). Effects of hydrogen sulfide on neurobehavioral function. South Med J 96: 639–646.

Kilburn KH (2004). In: MD Kilburn (Ed.), Deposition of Kaye H, Jamil and Hasima Ahmed vs Legacy/Monterey Homes LP.

Kilburn KH (2012). Human impairment from living near confined animal (hog) feeding operations. J Environ Public Health 2012: 565690.

Kilburn KH, Warshaw RH (1995). Hydrogen sulfide and reduced-sulfur gases adversely affect neurophysiological functions. Toxicol Ind Health 11: 185–197.

Kilburn KH, Thrasher JD, Gray MR (2010). Low-level hydrogen sulfide and central nervous system dysfunction. Toxicol Ind Health 26: 387–405.

Klentz RD, Fedde M (1978). Hydrogen sulfide: effects on avian respiratory control and intrapulmonary CO_2 receptors. Respir Physiol 32: 355–367.

Knight LD, Presnell SE (2005). Death by sewer gas: case report of a double fatality and review of the literature. Am J Forensic Med Pathol 26: 181–185.

Knubben-Schweizer G, Brosinski K, Steiner B (2011). Hydrogen sulfide poisoning (manure gas poisoning) in a cattle facility with open stall design. Schweiz Arch Tierheilkd 153: 127–129.

Lamarque J-F, Kiehl JT, Orlando JJ (2007). Role of hydrogen sulfide in a Permian-Traiassic boundary ozone collapse. Geophys Res Lett 34.

Lambert TW, Goodwin VM, Stefani D et al. (2006). Hydrogen sulfide (H_2S) and sour gas effects on the eye. A historical perspective. Sci Total Environ 367: 1–22.

Lancia M, Panata L, Tondi V et al. (2013). A fatal work-related poisoning by hydrogen sulfide: report on a case. Am J Forensic Med Pathol 34: 315–317.

Lee EC, Kwan J, Leem JH et al. (2009). Hydrogen sulfide intoxication with dilated cardiomyopathy. J Occup Health 51: 522–525.

Legator MS, Singleton CR, Morris DL et al. (2001). Health effects from chronic low-level exposure to hydrogen sulfide. Arch Environ Health 56: 123–131.

Li RQ, McKinstry AR, Moore JT et al. (2012). Is hydrogen sulfide-induced suspended animation general anesthesia? J Pharmacol Exp Ther 341: 735–742.

Lindenmann J, Matzi V, Anegg U et al. (2010). Hyperbaric oxygen in the treatment of hydrogen sulphide intoxication. Acta Anaesthesiol Scand 54: 784–785.

Lopez A, Prior M, Lillie LE, Gulayets C, Atwal OS (1988). Histologic and ultrastructural alterations in lungs of rats exposed to sub-lethal concentrations of hydrogen sulfide. Vet Pathol 25: 376–384.

Mack RB (1987). World enough and time – hydrogen sulfide poisoning. N C Med J 48: 33–34.

Maebashi K, Iwadate K, Sakai K et al. (2011). Toxicological analysis of 17 autopsy cases of hydrogen sulfide poisoning resulting from the inhalation of intentionally generated hydrogen sulfide gas. Forensic Sci Int 207: 91–95.

Mattarano DA, Merinar T (1999). Respiratory protection on offshore drilling rigs. Appl Occup Environ Hyg 14: 141–148.

Milby TH, Baselt RC (1999). Hydrogen sulfide poisoning: clarification of some controversial issues. Am J Ind Med 35: 192–195.

Morii D, Miyagatani Y, Nakamae N et al. (2010). Japanese experience of hydrogen sulfide: the suicide craze in 2008. J Occup Med Toxicol 5: 28.

Myers ML (2010). Review of occupational hazards associated with aquaculture. J Agromedicine 15: 412–426.

Nogue S, Pou R, Fernandez J et al. (2011). Fatal hydrogen sulphide poisoning in unconfined spaces. Occup Med (Lond) 61: 212–214.

Oesterhelweg L, Puschel K (2008). "Death may come on like a stroke of lightening": phenomenological and morphological aspects of fatalities caused by manure gas. Int J Leg Med 122: 101–107.

Olson KR (2011). The therapeutic potential of hydrogen sulfide: separating hype from hope. Am J Physiol Regul Integr Comp Physiol 301: R297–R312.

Olson KR, Deleon ER, Gao Y et al. (2013). Thiosulfate: a readily accessible source of hydrogen sulfide in oxygen sensing. Am J Physiol Regul Integr Comp Physiol 305: R592–R603.

Peters JW (1981). Hydrogen sulfide poisoning in a hospital setting. JAMA 246: 1588–1589.

Poli D, Solarino B, Di Vella G et al. (2010). Occupational asphyxiation by unknown compound(s): environmental and toxicological approach. Forensic Sci Int 197: e19–e26.

Policastro MA, Otten E (2007). Case files of the University of Cincinnati Fellowship in Medical Toxicology: two patients with acute lethal occupational exposure to hydrogen sulfide. J Med Toxicol 3: 73–81.

Popov D (2013). An outlook on vascular hydrogen sulphide effects, signalling, and therapeutic potential. Arch Physiol Biochem 119: 189–194.

Prior MG, Sharma AK, Yong S, Lopez A (1988). Concentration-time interactions in hydrogen sulfide toxicity in rats. Can J Vet Res 52: 375–379.

Qu K, Chen CP, Halliwell B et al. (2006). Hydrogen sulfide is a mediator of cerebral ischemic damage. Stroke 37: 889–893.

Ramazzini B (1700). De morbis artificum diatriba, Modena (Italy), Corona.

Reiffenstein RJ, Hulbert WC, Roth SH (1992). Toxicology of hydrogen sulfide. Annu Rev Pharmacol Toxicol 32: 109–134.

Richardson DB (1995). Respiratory effects of chronic hydrogen sulfide exposure. Am J Ind Med 28: 99–108.

Roberts ES, Wong VA, McManus BE et al. (2006). Changes in intracellular pH play a secondary role in hydrogen sulfide-induced nasal cytotoxicity. Inhal Toxicol 18: 159–167.

Roberts ES, Thomas RS, Dorman DC (2008). Gene expression changes following acute hydrogen sulfide (H_2S)-induced nasal respiratory epithelial injury. Toxicol Pathol 36: 560–567.

Ronk R, White MK (1985). Hydrogen sulfide and the probabilities of 'inhalation' through a tympanic membrane defect. J Occup Med 27: 337–340.

Roth SH, Skrajny B, Bennington R et al. (1997). Neurotoxicity of hydrogen sulfide may result from inhibition of respiratory enzymes. Proc West Pharmacol Soc 40: 41–43.

Sahar Ali F, Nirmeen Adel K (2010). Cognitive functions changes among Egyptian sewage network workers. Toxicol Ind Health 26: 229–238.

Sams RN, Carver 2nd HW, Catanese C et al. (2013). Suicide with hydrogen sulfide. Am J Forensic Med Pathol 34: 81–82.

Sastre C, Baillif-Couniou V, Kintz P et al. (2013). Fatal accidental hydrogen sulfide poisoning: a domestic case. J Forensic Sci 58 (Suppl 1): S280–S284.

Shivanthan MC, Perera H, Jayasinghe S et al. (2013). Hydrogen sulphide inhalational toxicity at a petroleum refinery in Sri Lanka: a case series of seven survivors following an industrial accident and a brief review of medical literature. J Occup Med Toxicol 8: 9.

Sidlo J, Bauer M, Bauerova J et al. (2009). Diagnostics of fatal hydrogen sulfide poisonings. Soud Lek 54: 37–40.

Sjaasted O, Bakketeig LS (2006). Hydrogen sulphide headache and other rare, global headaches: Vågå study. Cephalalgia 26: 466–476.

Skrajny B, Reiffenstein RJ, Sainsbury RS et al. (1996). Effects of repeated exposures of hydrogen sulphide on rat hippocampal EEG. Toxicol Lett 84: 43–53.

Smith RP (1981). Nitrite treatment for hydrogen sulfide poisoning. Ann Intern Med 95: 782.

Smith RP, Gosselin RE (1979). Hydrogen sulfide poisoning. J Occup Med 21: 93–97.

Smith RP, Gosselin RE (1984). The influence of methemoglobinemia on the lethality of some toxic anions. II. Sulfide. Toxicol Appl Pharmacol 6: 564–592.

Smith RP, Kruszyna R, Kruszyna H (1976). Management of acute sulfide poisoning. Effects of oxygen, thiosulfate, and nitrite. Arch Environ Health 31: 166–169.

Smith L, Kruszyna H, Smith RP (1977). The effect of methemoglobin on the inhibition of cyctochrome c oxidase by cyanide, sulfide or azide. Biochem Pharmacol 26: 2247–2250.

Snyder JW, Safir EF, Summerville GP et al. (1995). Occupational fatality and persistent neurological sequelae after mass exposure to hydrogen sulfide. Am J Emerg Med 13: 199–203.

Solnyshkova TG (2003). Demyelination of nerve fibers in the central nervous system caused by chronic exposure to natural hydrogen sulfide-containing gas. Bull Exp Biol Med 136: 328–332.

Spitzer WO, MI-URG (June 1986). The Southwestern Alberta Medical Diagnostic Review: Summary. Montreal.

Spitzer WO, Dales RE, Schechter MT et al. (1989). Chronic exposure to sour gas emissions: meeting a community concern with epidemiologic evidence. CMAJ 141: 685–691.

Stine RJ, Slosberg B, Beacham BE (1976). Hydrogen sulfide intoxication. A case report and discussion of treatment. Ann Intern Med 85: 756–758.

Tanaka S, Fujimoto S, Tamagaki Y, Wakayama K et al. (1999). Bronchial injury and pulmonary edema caused by hudrogen sulfide poisoning. Am J Emerg Med 17: 427–429.

Tansy MF, Kendall FM, Fantasia J, Landin WE et al. (1981). Acute and subchronic toxicity studies of rats exposed to vapors of methyl mercaptan and other reduced-sulphur compounds. J Tox Env Health 8: 71–88.

Truong DH, Eghbal MA, Hindmarsh W et al. (2006). Molecular mechanisms of hydrogen sulfide toxicity. Drug Metab Rev 38: 733–744.

Truscott A (2008). Suicide fad threatens neighbours, rescuers. CMAJ 179: 312–313.

Turner RM, Fairhurst S (1990). Toxicology of substances in relation to major hazards: hydrogen sulfide, Health and Safety Executive, HMSO, London, pp. 1–14.

Tvedt B, Skyberg K, Aaserud O et al. (1991). Brain damage caused by hydrogen sulfide: a follow-up study of six patients. Am J Ind Med 20: 91–101.

US Chemical Safety and Hazard Investigation Board (2003). Hydrogen Sulfide Poisoning: Georgia-Pacific Naheola Mill, Pennington, Alabama, January 16, 2002. National Technical Information Service, US Chemical Safety and Hazard Investigation Board, Springfield VA.

US Environmental Protection Agency (2009). Graphical Arrays of Chemical-Specific Health Effect Reference Values for Inhalation Exposures (2009 Final Report). Washington, DC, US EPA, EPA/600/R-09/061, 2009. Available at http://cfpub.epa.gov/ncea/cfm/recordisplay.cfm?deid=211003. Accessed 11 August 2010.

US Public Health Service (May 1977). Criteria for a Recommended Standard: Occupational Exposure to hydrogen sulfide, U.S. Department of Health, Education and Welfare, Center for Disease Control.

Volpato GP, Searles R, Yu B et al. (2008). Inhaled hydrogen sulfide: a rapidly reversible inhibitor of cardiac and metabolic function in the mouse. Anesthesiology 108: 659–668.

Wang DX (1989). A review of 152 cases of acute poisoning of hydrogen sulfide. Chinese journal of preventive medicine 23: 330–332.

Wang R (2012). Physiological implications of hydrogen sulfide: a whiff exploration that blossomed. Physiol Rev 92: 791–896.

Wang JF, Li Y, Song JN et al. (2014). Role of hydrogen sulfide in secondary neuronal injury. Neurochem Int 64: 37–47.

Warenycia MW, GL, Francom DM, Dieken GP et al. (1990). Dithiothreitol liberates non-acid labile sulfide from brain tissue of H_2S-poisoned animals. Arch Toxicol 64: 650–655.

Wasch HH, Estrin WJ, Yip P et al. (1989). Prolongation of the P-300 latency associated with hydrogen sulfide exposure. Arch Neurol 46: 902–904.

Whitcraft 3rd DD, Bailey TD, Hart GB (1985). Hydrogen sulfide poisoning treated with hyperbaric oxygen. J Emerg Med 3: 23–25.

Woodall GM, Smith RL, Granville GC (2005). Proceedings of the Hydrogen Sulfide Health Research and Risk Assessment Symposium October 31–November 2, 2000. Inhal Toxicol 17: 593–639.

Yalamanchili C, Smith MD (2008). Acute hydrogen sulfide toxicity due to sewer gas exposure. Am J Emerg Med 26: 518, e5-7.

Yant WP (1930). Hydrogen sulphide in industry-occurrence, effects, and treatment. Am J Public Health Nations Health 20: 598–608.

Zaba C, Marcinkowski JT, Wojtyla A et al. (2011). Acute collective gas poisoning at work in a manure storage tank. Ann Agric Environ Med 18: 448–451.

Zhang L, Yang G, Untereiner A et al. (2013). Hydrogen sulfide impairs glucose utilization and increases gluconeogenesis in hepatocytes. Endocrinology 154: 114–126.

Handbook of Clinical Neurology, Vol. 131 (3rd series)
Occupational Neurology
M. Lotti and M.L. Bleecker, Editors

Chapter 9

The neurotoxicity of organochlorine and pyrethroid pesticides

LUCIO G. COSTA*

Department of Environmental and Occupational Health Sciences, University of Washington, Seattle, WA, USA and Department of Neuroscience, University of Parma, Parma, Italy

INTRODUCTION

Among pesticides, organochlorine and pyrethroid compounds are classified as insecticides, as their most relevant role is in the control of insect pests. Apart from natural compounds such as nicotine or rotenone, which have been used as insecticides for a few centuries, organochlorines represent the first important class of insecticides developed by the burgeoning chemical industry in the first half of the 20th century. Pyrethroids were developed in the 1970s, after two other major classes of insecticides, the organophosphates and the carbamates. Additional novel classes of insecticides, such as the neonicotinoids or the phenylpyrazoles, have been more recently introduced. In contrast to the organochlorine compounds, most of which have been banned in the past several years, the pyrethroids still demand a substantial share of the insecticide market.

All of the chemical insecticides are neurotoxicants, as they act by poisoning the nervous system of the target organisms (Costa et al., 2008; Casida, 2009; Table 9.1). The central nervous system of insects is highly developed and not unlike that of mammals, and the peripheral nervous system, though less complex, also presents striking similarities (Costa, 2013). Thus, insecticides are mostly not species-selective with regard to targets of toxicity, and mammals, including humans, are highly sensitive to their toxic effects. This chapter will review the major aspects of the neurotoxicity of organochlorine and pyrethroid insecticides with a particular emphasis on their effects in humans, and their potential contribution to nervous system disorders.

ORGANOCHLORINE COMPOUNDS

A classification of organochlorine insecticides

There are several organochlorine compounds that have been used as insecticides. They differ in their chemical structure, their mechanism of toxicity in target and non-target species, and in various other characteristics. The most widely known organochlorine insecticide is undoubtedly the chlorinated ethane derivative 1,1,1-trichloro-2, 2-*bis* (4-chlorophenyl) ethane (DDT), together with some of its analogs. While DDT is the universally accepted common name of the insecticide, technical-trade DDT is a mixture of several isomers, with *p,p'*-DDT being responsible for the insecticidal activity, and *o,p'*-DDT displaying endocrine-disrupting (estrogenic) properties (Fig. 9.1). A similar insecticide is methoxychlor (2,2-*bis* (*p*-methoxyphenyl)-1,1,1-trichloroethane), which is the *p,p'*-dimethoxy analog of *p,p'*-DDT.

A second important class of organochlorine insecticides is that of the cyclodienes, which include chlordane, dieldrin, aldrin (which is rapidly metabolized to dieldrin), heptachlor, and endrin. Benzene hexachloride (BHC), also known as 1,2,3,4,5,6- hexachlorocyclohexane, is composed of eight stereoisomers; the γ isomer, which was given the name lindane, is the one with insecticidal activity. Toxaphene, a complex mixture of chlorinated bornanes and camphenes, is also classified as an organochlorine insecticide. Finally, mirex and chlordecone (Kepone) have chlorinated caged structures.

*Correspondence to: Dr. Lucio G. Costa, Department of Environmental and Occupational Health Sciences, University of Washington, 4225 Roosevelt Way NE, Suite 100, Seattle, WA 98105, USA. Tel: +1-206-543-2831, Fax: +1-206-685-4696, E-mail: lgcosta@u.washington.edu

Table 9.1

Targets of insecticides in the nervous system

Insecticide	Target	Type of interaction
Pyrethroids (all)	Sodium channels	Activation
DDT and analogues	Sodium channels	Activation
Dihydropyrazoles	Sodium channels	Inhibition
Cyclodienes	GABA receptor-gated chloride channels	Inhibition
Hexachlocyclohexanes	GABA receptor-gated chloride channels	Inhibition
Pyrethroids (type II)	GABA receptor-gated chloride channels	Inhibition
Phenylpyrazoles	GABA receptor-gated chloride channels	Inhibition
Organophosphates	Acetylcholinesterase	Inhibition
Carbamates	Acetylcholinesterase	Inhibition
Nicotine	Acetylcholine nicotinic receptors	Activation
Neonicotinoids	Acetylcholine nicotinic receptors	Activation
Rotenone	Complex I in mitochondria	Inhibition
Avermectins	Glutamate-gated chloride channels*	Activation
Formamidines	Octopamine receptors†	Activation
Diamides	Ryanodine receptors	Activation

*Found only in insects. In mammals avermectins activate $GABA_A$ receptors.
†In mammals formamidines activate $alpha_2$-adrenergic receptors.
GABA, gamma-aminobutyric acid.

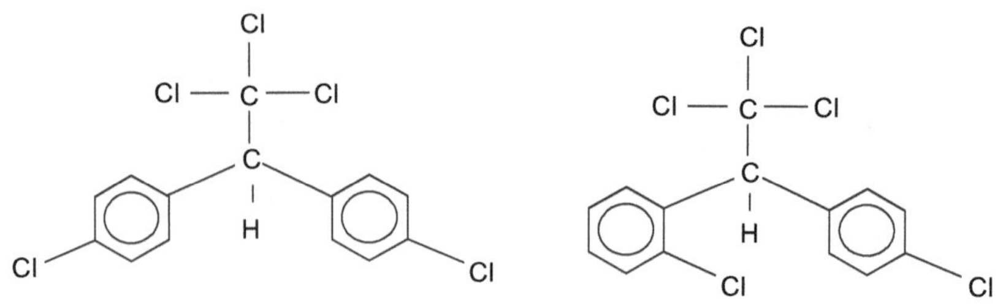

Fig. 9.1. Structure of the organochlorine insecticide *p,p'*-DDT and of its isomer *o,p'*-DDT. Note that, in technical-grade DDT, the *p,p'* isomer, which has insecticidal activity, is ~77%, while *o,p'*-DDT, which has endocrine-disrupting properties, is ~15%. Other isomers (e.g., *p,p'*- or *o,p'*- DDE and DDD) are also present in technical DDT.

With the exception of DDT and lindane, which still have limited uses, all organochlorine insecticides have been banned in most countries in the past 40 years, primarily because of ecologic considerations. Yet, because of their environmental persistence and high lipophilicity, exposure to these compounds continues, most notably through the diet; hence, a discussion of their neurotoxicity has more than historic value.

History and uses

The period between 1935 and 1950 was characterized by the development of DDT and of other chlorinated hydrocarbon insecticides. In the late 1930s, the Swiss firm J.R. Geigy was very active, researching a contact poison for flies, mosquitoes, and other insects. Though first synthesized by O. Zeidler in 1874, it was only in 1939 that Paul Müller discovered that DDT was a very powerful insecticide. The first patent was obtained in 1940, and from the beginning of 1942 the preparation appeared on the market under the names Gesarol and Neocid, among others (Holmstedt and Liljestrand, 1963). During World War II, the practical value of DDT was evidenced by its effectiveness in curtailing a severe epidemic of typhus in the city of Naples, Italy; on that occasion more than one million people were sprayed with DDT (Costa, 1987). Afterwards, one of the most successful uses of DDT was in malaria eradication programs, since before its development most attempts to control or eliminate malaria had been unsuccessful. For example, in the Italian province of Latina there were 175 new cases of malaria in 1944, but after a DDT spray control program was initiated, no new cases of malaria appeared by 1949 (Murphy, 1986).

A more recent example of the efficacy of DDT in malaria control comes from South Africa. In this country, DDT was only banned in 1996, and at the time $<10\ 000$ cases of malaria were registered. By the year 2000, cases of malaria had increased to 62 000, but with the reintroduction of DDT at the end of that year, cases were down to 12 500 (Maharaj et al., 2005). In addition to being effective against insects that transmit some of the world's most serious diseases, such as typhus, malaria, and yellow fever, DDT was also effective against a wide variety of agricultural pests. In the USA, annual use of DDT rose until 1959 to about 36 000 tons, after which it gradually declined until its ban in 1972 (Rogan and Chen, 2005).

Together with DDT, other chlorinated insecticides were developed in the 1940s and the 1950s. BHC may be considered as the oldest, having been first synthesized by Faraday in 1825. Some 50 years later Meunier discovered the α and β isomers, and in 1912 van der Linden reported the presence of the γ and δ isomers. Its insecticidal properties were discovered only in the early 1940s by scientists in England and in France, who recognized that the γ isomer of BHC, commonly known as lindane (after van der Linden), was a highly potent insecticide and introduced it to the market (Brooks, 1974; Ecobichon and Joy, 1982). The cyclodiene-type chlorinated hydrocarbons were introduced to the market starting in the mid-1940s. The insecticidal properties of chlordane were described in 1945, heptachlor was introduced for agricultural use in 1948, and in the next 5 years dieldrin and aldrin were also marketed. All of these compounds have experienced wide use before being banned in most countries due to their persistence and environmental and human health effects. Lindane's use as an insecticide in agriculture has been banned, though it remains available as a scabicide and pediculocide in lotions and shampoos, as a second line of intervention. The primary use of chlordane has been for termite control, while other compounds were primarily used in agriculture. Endrin was registered for use as a rodenticide to control voles in orchards. Mirex and chlordecone were introduced in the late 1950s for use against fire ants and leaf-eating insects, respectively, and have also since been banned.

The properties (low volatility, chemical stability, lipid solubility, slow rate of biotransformation and degradation) that made organochlorine compounds such effective insecticides also brought about their demise because of their persistence in the environment, bioconcentration and biomagnification in food chains, and the acquisition of biologically active body burdens at higher trophic levels (Costa, 2013). Sweden banned DDT in 1970, the USA in 1972, and the UK in 1986. Most other organochlorine insecticides were also banned in the USA during this period; for example, aldrin and dieldrin were banned in 1975, toxaphene in 1982, and chlordane in 1988. These bans occurred largely on the basis of ecologic considerations. Their extensive use and their environmental persistence made them widespread pollutants, with a tendency to bioaccumulate in the food chain. Adverse effects on bird reproduction (e.g., eggshell thinning) were among the first ecologic effects to be identified. Particularly compelling was the publication in 1962 of Rachel Carson's book, *Silent Spring*, an impassioned denouncement of the consequences of chemical contamination of the environment, with particular emphasis on the bioaccumulation of DDT and its effects on birds (Carson, 1962; Peakall, 1970). This book, and the discussions that it generated, played an important role in the decision to ban DDT and other organochlorines (U.S. Environmental Protection Agency, 1975). Monitoring of organochlorine insecticides in fish in the Great Lakes showed extremely high levels in 1970 and a progressive decline after the use of these compounds was restricted or banned (Hickey et al., 2006).

Neurotoxicity of organochlorine insecticides

DDT AND ITS ANALOGS

DDT has a moderate acute toxicity when given by the oral route, with an LD_{50} of about 250 mg/kg, and *p,p'*-DDT is at least 10-fold more toxic than *o,p'*-DDT (Smith, 2010; Table 9.2). Dermal absorption of DDT is very limited, resulting in dermal LD_{50} values of >1000 mg/kg. In humans, oral doses of 10–20 mg/kg produce illness, but doses as high as 285 mg/kg have been ingested accidentally without fatal results. Toxicity from dermal exposure in humans is also low, as evidenced by the lack of significant adverse health effects when thousands of people were liberally dusted with this compound. Upon absorption, DDT distributes in all tissues, and the highest concentrations are found in adipose tissue. DDT is also extensively but slowly metabolized, with DDE, DDD, and DDA (in each case both the *p,p'* and the *o,p'* isomers) being the primary metabolites in humans (Smith, 2010). DDE is also stored in adipose tissue. Excretion is through bile, urine, and milk.

Acute exposure to high doses of DDT causes motor unrest, increased frequency of spontaneous movements, abnormal susceptibility to fear, and hypersusceptibility to external stimuli (light, touch, sound). This is followed by the development of fine tremors, progressing to coarse tremors, and eventually tonic-clonic convulsions. Symptoms usually appear several hours after exposure, and death, usually due to respiratory failure, may follow after 24–72 hours (Ecobichon and Joy, 1982). Signs and symptoms of poisoning are similar in most animal species, with dogs displaying prominent convulsions. There

Table 9.2

Acute toxicity of organochlorine and pyrethroid insecticides

Organochlorine	LD_{50} (mg/kg)	Pyrethroid	LD_{50} (mg/kg)
DDT	250	Cismethrin (I)	63
Methoxychlor	5000–6000	Allethrin (I)	200
Endrin	3	Tetramethrin (I)	>4000
Heptachlor	40		
Dieldrin	46	Deltamethrin (II)	52
Aldrin	67	Fenvalerate (II)	450
Chlordane	283	Cypermethrin (II)	900
Mirex	365		
Chlordecone	125		

Adapted from Ray (2010).
Indicated values are of oral LD_{50} (mg/kg of body weight) in rats. Type I and type II pyrethroids are shown.

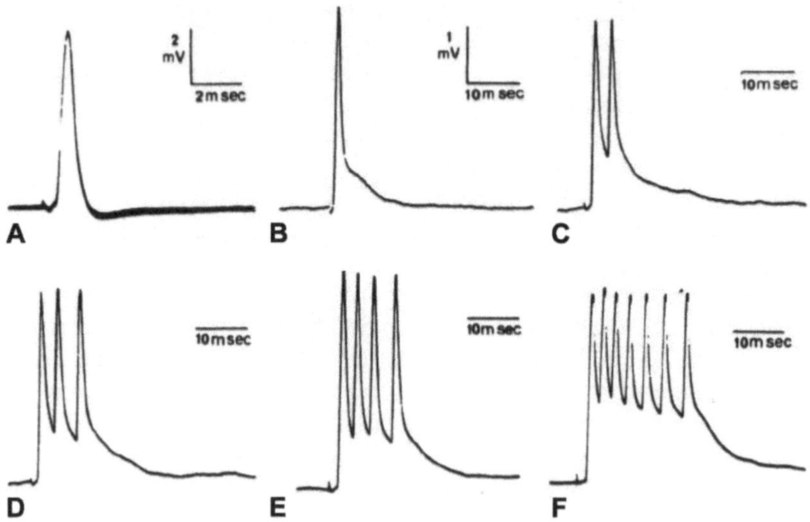

Fig. 9.2. Mechanism of action of DDT. Action potential of a single nerve fiber from the sciatic nerve of the toad *Xenopus laevis*, before (**A**), and 65, 85, 90, 105, and 135 minutes after (**B–F**) *in vitro* exposure to 100 μM DDT. Note the difference in timescale between **A** and **B**. The first observable effect is the development of a negative after-potential (**B**). With time of exposure, repetitive spike discharge is initiated and increased in proportion to the amplitude of the negative after-potential (**C–F**). (Reproduced from van den Bercken (1972), with permission.)

is no evidence that young animals may be more sensitive than adults to the acute toxicity of DDT, and they may actually be less sensitive (Smith, 2010). In humans, ingestion of DDT is symptomatic above 10 mg/kg of body weight (Ray, 2010). At higher doses, the earliest symptom of poisoning by DDT is hyperesthesia of the mouth and lower part of the face, followed by paresthesia of the same area and of the tongue. Dizziness, tremor of the extremities, hyperexcitability, confusion, and vomiting can follow, while convulsions occur only in severe poisoning.

Signs and symptoms of acute poisoning clearly point at the nervous system as the primary target for DDT toxicity (Woolley, 1982). Both in insects and in mammals, DDT interferes with the sodium channels in the axonal membrane (Vijverberg et al., 1982). DDT has little or no effect on the resting potential or the rising phase and peak amplitude of the action potential. However, it greatly prolongs the depolarizing (negative) afterpotential of the action potential, and this produces a period of increased neuronal excitability immediately after the spike phase. This, in turn, enhances the probability of repetitive firing, and the insurgence of a "train" of action potentials (Fig. 9.2). Voltage clamp studies have shown that the principal effect of DDT is to slow down the closing of sodium channels once they have opened, while having little or no effect on closed gates. DDT also inhibits Ca^{2+}-ATPase (an ecto-ATPase, located on the outside of the cell membrane), whose function is believed to be that of maintaining high external calcium concentrations; its inhibition by DDT would thus lower external calcium and contribute to membrane instability and repetitive firing (Matsumura and Ghiasuddin, 1979). Neurochemical studies have also shown that DDT exposure alters the levels of some neurotransmitters such as acetylcholine, norepinephrine, and serotonin, as well as of cyclic guanosine monophosphate, but these effects appear to be the results, rather than the cause, of DDT-induced neurotoxicity (Woolley, 1982). Treatment for DDT poisoning focuses on the nervous system. In animals, phenytoin and calcium gluconate have been found to reduce DDT-induced tremors and mortality, respectively. In humans, in addition to decontamination and supportive treatment, diazepam, or phenobarbital may be beneficial to control convulsions, if present.

Compared to DDT, its analog methoxychlor has lower acute toxicity (Table 9.2) and a shorter biologic half-life. Methoxychlor is rapidly metabolized, and does not accumulate in tissues. Though convulsions have been reported in dogs after high doses, in the rat depression of the central nervous system, mild tremors, diarrhea, and anorexia are observed (Smith, 2010).

Hexachlorocyclohexanes and cyclodienes

These two families of organochlorine insecticides comprise a large number of compounds that share a similar mechanism of neurotoxic action (Brooks, 2010). Lindane and cyclodienes have moderate to high acute oral toxicity (Table 9.2). However, in contrast to DDT, these compounds are readily absorbed through the skin. The primary target for their toxicity is also the central nervous system. Unlike DDT, tremor is essentially absent, but convulsions are a prominent aspect of poisoning. Convulsive seizures are often sudden and abrupt in onset, and may occur without warning, or can be preceded by hyperexcitability, myoclonic jerks, and general malaise (Ray, 2010). Convulsions are believed to be due to the ability of these compounds to interfere with γ-aminobutyric acid (GABA)-mediated neurotransmission. GABA is an important neurotransmitter in the mammalian and insect central nervous system and in the insect neuromuscular junction. $GABA_A$ receptors are members of the superfamily of ligand-gated ion channels that contain a chloride ionophore; by binding to these receptors, endogenous GABA causes the opening of chloride channels resulting in hyperpolarization of the membrane. Lindane and cyclodienes bind to a specific site (the picrotoxin site) on the chloride channel, thereby blocking its opening and thus antagonizing the "inhibitory" action of GABA (Cole and Casida, 1986; Eldefrawi and Eldefrawi, 1987; Narahashi et al., 2007). Treatment of acute poisoning is symptomatic; phenobarbital and diazepam can be used as anticonvulsants. Additional reported neurochemical effects of organochlorine insecticides include inhibition of Na^+- K^+, Ca^{2+}, and Mg^{2+}-ATPases, and changes in neurotransmitter levels (Costa, 2013).

Mirex and chlordecone

Of these two organochlorine insecticides, which have a similar cage-like structure, and moderate acute toxicity (Table 9.2), chlordecone (Kepone) has been the most studied because of an episode of chronic intoxication which involved 148 workers in a chlordecone-producing factory in Hopewell, Virginia, between 1973 and 1975 (Taylor, 1982). The primary manifestation of chlordecone toxicity was the presence of tremors, which are observed in animals as well as in humans (Guzelian, 1982). The hands were particularly involved, but fine tremors of the head and trembling of the entire body (called "Kepone shakes" by the workers) were also observed (Ray, 2010). In addition, ataxia, opsoclonus, slurred speech, mood changes, irritability, memory loss, and peripheral neuropathy were observed in exposed workers (Martinez et al., 1978; Taylor et al., 1978). Slow reversibility of neurotoxic signs was observed in most individuals, compatible with the long half-life of chlordecone in blood (4–6 months); however, some individuals complained of tremors and memory loss several years after cessation of exposure (Taylor, 1982). The exact mechanism of chlordecone neurotoxicity has not been elucidated, but it is believed to involve inhibition of ATPases (both Na^+-K^+ and Mg^{2+} ATPases), and ensuing inhibition of the uptake of catecholamines (Desaiah, 1982). In contrast to cyclodienes, chlordecone does not cause seizures.

Chlordecone may also affect the developing nervous system. Studies in rats have shown that neonatal exposure to chlordecone causes memory impairment and fine body tremors (Mactutus and Tilson, 1984, 1985). In humans, chlordecone is known to cross the placental barrier and is excreted in breast milk. A recent study in the French West Indies reported continuous exposure to chlordecone, despite its ban two decades ago. *In utero* and neonatal exposure to chlordecone was associated with negative effects on cognitive and motor development in 7-month-old infants (Dallaire et al., 2012).

Other adverse health effects of organochlorine insecticides and human body burden

Organochlorines also target other organs in addition to the nervous system, in particular the liver. Chronic exposure to most organochlorine insecticides (e.g., DDT,

DDE, cyclodienes, chlordecone) increases liver weight and causes hepatic cell hypertrophy and necrosis, and these compounds are potent inducers of cytochromes P450 (Smith, 1991, 2010). None of the organochlorine is genotoxic, but most have been reported to induce tumors in rodents. For example, DDT has been shown to be hepatocarcinogenic in mice and rats, but not in other species. DDT also increases incidence of lung tumors and adenomas in mice. Based on animal data, DDT is classified as a possible human carcinogen by the International Agency for Research on Cancer, though evidence of human carcinogenicity is still inconclusive (Beard, 2006). In contrast to DDT, its analog methoxychlor is a modest inducer of liver microsomal enzymes, but causes chronic nephritis, hypertrophy of kidneys, mammary glands, and uterus, testicular atrophy, and decreased spermatogenesis. Evidences of mutagenicity and carcinogenicity are inconclusive. Cyclodienes have been shown to act as tumor promoters and cause liver tumors in mice, effects also seen with chlordecone (Smith, 1991).

Because of their stability and high lipophilicity, organochlorines are present in adipose tissues of most individuals. For example, DDE, the breakdown product of DDT in the environment and the result of its biotransformation, is an extremely stable compound ($t_{1/2}$=7–11 years) which accumulates in fat, and high concentrations are usually found in human adipose tissue. Studies in populations from different countries in the 1960s showed that all adipose tissue samples contained DDT at concentrations ranging from 5 to 20 mg/kg (ppm), with DDE accounting for about 60% of total DDT (World Health Organization, 1979). Though concentrations of DDT in humans show a decreasing trend, those of DDE, which is ingested with food, particularly meat and fish, remain more constant or decrease only slightly. Organochlorine insecticides are also excreted in the milk. Though levels have decreased in most countries, elevated levels remain in several developing countries (Costa, 2013).

Organochlorines as endocrine disruptors

An endocrine, or hormone, disruptor can be defined as an exogenous agent that interferes with the synthesis, secretion, transport, binding, action, or elimination of natural hormones in the body that are responsible for the maintenance of homeostasis, reproduction, development, and behavior (Stoker and Kavlock, 2010). In addition to affecting the reproductive systems, endocrine disruption also impairs the functioning of the nervous system, particularly during development (Weiss, 2012).

Several organochlorine insecticides are endocrine disruptors. The *o,p*′-isomer of DDT, which comprises approximately 15% of the technical-grade product, has estrogenic properties, in that it can act as an agonist at estrogen receptors (ER) α and β. A metabolite of DDT, *p,p*′-DDE, on the other hand, inhibits androgen binding to androgen receptors (Kelce et al., 1995). Metabolites of the DDT analog methoxychlor are active as ERα agonists or ERβ antagonists, and are antiandrogenic. Chlordecone also has estrogenic properties (Guzelian, 1982). Other organochlorine compounds with weak estrogenic activity are dieldrin, endosulfan, toxaphene, lindane, and the β isomer of hexachlorocyclohexane. In all cases, the potency of the insecticides or their metabolites is several orders of magnitude less than those of the hormones. Experimental studies have shown that administration of these compounds to animals can cause hormonally mediated adverse health effects. For example, *p,p*′-DDE had clear antiandrogenic effect when given to fetal, pubertal, and adult male rats (Kelce et al., 1995). Several studies have shown that chlordecone and methoxychlor affect the male reproductive system (Guzelian, 1982; Kavlock, 2001). Low or absent sperm count was found in chlordecone-exposed workers (Guzelian, 1982; Taylor, 1982).

Exposure to methoxychlor during gestation (E8–E15) induces transgenerational defects in spermatogenic capacity and sperm viability, i.e., these abnormalities are also observed in F2 generation animals (Anway et al., 2005). It has been suggested that estrogen equivalents from phytoestrogens present in foods and beverages are several orders of magnitude higher than estrogen equivalents resulting from estrogenic organochlorine insecticide contaminants in food (Safe, 2000), and that the association between exposure to organochlorine compounds and possible related diseases in humans appeared to be weak (Golden et al., 1998; Safe, 2005). However, the Pine River Statement reviewed 494 studies published between 2003 and 2008 and found that "DDT and its breakdown product DDE may be associated with adverse health outcomes such as breast cancer, diabetes, decreased semen quality, spontaneous abortion, and impaired development in childhood" (Eskenazi et al., 2009). Early life exposure to *p,p*′-DDT has been shown to be associated with decreased triiodothyronine levels in preschool children (Alvarez-Pedrerol et al., 2008). Furthermore, of >20 epidemiologic studies published in 2009, more than half showed significant associations between DDT and/or DDE with type 2 diabetes, abnormal hormonal levels, birth mass, pancreatic duct carcinoma, and altered sperm parameters (Bouwman et al., 2011).

The debate on the continued use of DDT and lindane

The Stockholm Convention on Persistent Organic Pollutants, ratified in 2004 by 50 states, outlawed the use of

12 industrial chemicals (the "Dirty Dozen"), including DDT. Yet, an exemption clause allows malaria-endemic nations to continue utilizing DDT for indoor residual wall spraying. The United Nations Environment Program estimates that about 25 countries would use DDT under this exemption from its ban. While the environmental and human health effects of DDT are evident, one has to come to grips with the burden of mortality from malaria worldwide. Indeed, there are an estimated 250 million clinical cases of malaria, causing almost one million deaths each year, mostly in children <5 years of age, and mostly in sub-Saharan Africa (van den Berg, 2009; World Health Organization, 2010). Thus, use of DDT might reduce mortality from malaria and overall infant mortality. However, though indoor residual spraying might not cause the ecologic effects that caused the ban of DDT, it would expose humans to amounts of DDT that may cause adverse health effects. In this regard, reproductive outcomes are of most concern (Chen and Rogan, 2003; Rogan and Chen, 2005). Furthermore, DDT levels in breast milk often exceed the tolerable daily intake and would cause great concern in any developed country (Okankwo et al., 2008), and high serum levels of DDT have also been found in applicators when proper personal protective equipment is not utilized (Bouwman et al., 2011).

Lindane is the other organochlorine insecticide still in use today in most of the world, though with very limited applications. Most uses of lindane (e.g., in agriculture) have been banned in more than 50 countries starting with the Netherlands in 1979 to Canada in 2008. In 2007 it was recommended that lindane be added to the Stockholm Convention list of chemicals to be outlawed worldwide, but be granted an exception for use as a pharmaceutical. Lindane-based shampoos were once the main treatment for lice and scabies; however, the pharmaceutical use of lindane has been decreasing over the years (e.g., the number of prescriptions in the USA were 1.8 million in 1997 and only 200 000 in 2006) because of tighter regulations. For head lice the first line of treatment is currently represented by shampoos and lotion containing pyrethrins or the organophosphorus insecticide malathion, while lindane remains very effective against scabies (Lebwohl et al., 2007). However, in some US states (e.g., California) use of lindane has been fully banned, even as a pharmaceutical, primarily on the basis of concerns for water contamination (Humphreys et al., 2008).

PYRETHROIDS

Chemistry

All pyrethroid insecticides contain an acid moiety, a central ester bond, and an alcohol moiety (Fig. 9.3). The acid moiety contains two chiral carbons, thus pyrethroids typically exist as stereoisomeric compounds (*trans* and *cis*). Additionally, some pyrethroids also have a chiral carbon on the alcohol moiety, allowing for a total of eight different stereoenantiomers. These chemical considerations are relevant, as pyrethroids' effects on sodium channels, their insecticidal activity, and their mammalian toxicity, are stereospecific. The *cis* isomers are generally more toxic than the corresponding *trans* isomers (Casida et al., 1983).

Based on toxic signs in rats, pyrethroids have been divided into two types, causing a type I or a type II syndrome (see below). A key structural difference between type I and type II pyrethroids is the presence only in the latter of a cyano group at the α carbon of the alcohol moiety of the compound (Fig. 9.3). However, certain pyrethroids (e.g., cyphenothrin, flucythrinate) elude such classification, as they produce a combination of the two syndromes (Soderlund et al., 2002; Ray and Fry, 2006).

History and uses

Extracts from the flower heads of *Chrysanthemum cinerariaefolium* were appreciated as insecticides in ancient China and Persia to control body lice. The flowers were first cultivated for commercial use in Armenia in 1828. They were first commercialized in the former Yugoslavia in 1840, and then expanded to Asia and East Africa (Casida, 1980). Pyrethrum extracts contain six closely related insecticidal esters, collectively

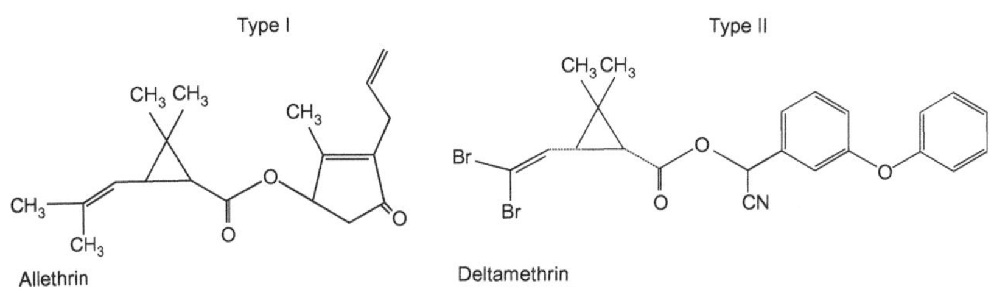

Fig. 9.3. Structure of type I (left) and type II (right) pyrethroid insecticides. Shown are allethrin and deltamethrin. Note that all type II pyrethroids have a cyano (CN) group. Type I pyrethroids cause the type I (tremor) sydrome, while type II pyrethroids cause the type II (choreoatethosis and salivation) syndrome.

referred to as pyrethrins, whose main structures were elucidated between 1910 and 1924 (Casida, 1980). However, because pyrethrins are decomposed rapidly by light, synthetic analogs, the pyrethroids, were developed. The first pyrethroid, allethrin, was followed by several others in the early 1970s, particularly because of the work of Michael Elliott at the Rothamsted Experimental Station in Harpenden, England, and of scientists at Sumitomo Chemical Company in Japan (Casida, 2010). Because of their high insecticidal potency, generally low mammalian toxicity, relatively low tendency to induce insect resistance, and lack of environmental persistence, pyrethroids have encountered much success in the past 30 years, and now account for almost 20% of the global insecticide market (Soderlund, 2012). Pyrethroids are used widely as insecticides both in the house and in agriculture, in medicine for the topical treatment of head lice, and in tropical countries for malaria control, both in soaked bed nets to prevent mosquito bites, as well as in indoor residual spraying.

Neurotoxicity of pyrethroids

The acute oral mammalian toxicity of pyrethroids is generally low (Table 9.2). The relevance of stereospecificity is underscored by the fact that the LD_{50} for 1R,*trans*-resmethrin is 8000 mg/kg, but that of 1R,*cis*-resmethrin is 100 mg/kg (Casida et al., 1983). The relatively low mammalian toxicity of pyrethroids is confirmed by the fact that, despite their extensive worldwide use, there are relatively few reports of human poisonings, and only a dozen deaths (Bradberry et al., 2005). Most deaths occurred following accidental or intentional exposure to pyrethroids. For example, a 45-year-old man died 3 hours after eating beans and cheese prepared using a 10% cypermethrin solution instead of oil (Poulos et al., 1982). He et al. (1989) reported on 573 cases of acute pyrethroid poisoning in China between 1983 and 1988, which resulted in seven deaths. The dermal toxicity of pyrethroids is even lower, because of limited absorption through the skin.

Pyrethroids induce two distinct intoxication syndromes in rats following oral or intravenous dosing. After testing 36 pyrethroids, Verschoyle and Aldridge (1980) noted that some produced signs of intoxication characterized by marked behavioral arousal, aggressive sparring, increased startle response, and fine body tremor progressing to whole-body tremor and prostration. This syndrome was designated as the T (tremor) syndrome. In contrast, a second group of pyrethroids produced a syndrome characterized by profuse salivation, coarse tremor progressing to choreoathetosis, and clonic seizures. This second syndrome was named CS (choreoathetosis with salivation) syndrome (Soderlund et al., 2002; Ray and Fry, 2006 Breckenridge et al., 2009; Soderlund, 2012). The main difference between these two groups of pyrethroids was the presence of an α-cyano-phenoxybenzyl moiety in compounds causing the CS syndrome. Subsequent studies on the intracerebral toxicity of 29 pyrethroids in mice confirmed the existence of two separate syndromes, which were named I (T) and II (CS) (Lawrence and Casida, 1982). Additional confirmation was provided by a study in which 12 pyrethroids were given to rats by the oral route (Weiner et al., 2009). However, the correspondence between structure and syndrome is not perfect, as some compounds with type II structure (i.e., with a cyano group) produce either the T syndrome or a mixed T and CS syndrome (Soderlund, 2012).

Mechanism of pyrethroid toxicity: interaction with the sodium channel

The insecticidal action of pyrethroids depends on their ability to bind and disrupt voltage-gated sodium channels, and the same mechanism is believed to be responsible for their neurotoxic effects in mammals (Narahashi, 1996; Soderlund, 2010, 2012). However, secondary targets have also been identified, such as calcium and chloride channels, which may contribute to the two distinct toxic syndromes (Breckenridge et al., 2009). Mammalian voltage-gated sodium channels are composed of one α and two β subunits. Ten different α subunits (Na_V1.1–1.9, Nax) and four different β subunits (β1–β4) have been identified, and they are expressed in a tissue-, region-, and time-specific manner (Catterall, 2000). Pyrethroids bind to the α subunit of the sodium channel and slow the activation (opening), as well as the rate of inactivation (closing) of the sodium channel, leading to a stable hyperexcitable state. Sodium channels then open at more hyperpolarized potentials, and are held open longer, allowing more sodium ions to cross and depolarize the neuronal membrane (Shafer et al., 2005). In general, type II compounds delay the inactivation of sodium channels substantially longer ($>$10 ms) than do type I compounds ($<$10 ms) (Ray and Fry, 2006). Type I compounds prolong channel opening only long enough to cause repetitive firing of action potential (repetitive discharge), analogously to DDT (Vijverberg et al., 1982). In contrast, type II compounds hold the channels open for such long periods that the membrane potential ultimately becomes depolarized, to the point at which generation of action potential is not possible (depolarization-dependent block) (Shafer et al., 2005). These differences in the time of opening of sodium channels are believed to contribute to the differences observed between the T and CS syndromes (Ray and Fry, 2006; Breckenridge et al., 2009).

The lower sensitivity of mammalian sodium channel isoforms compared to insect channels is believed to contribute to the selective toxicity of pyrethroids. Even the most sensitive rat sodium channel isoforms (e.g., $Na_V1.3$, $Na_V1.6$, or $Na_V1.8$) are at least one order of magnitude less sensitive to pyrethroids than insect channels. A key difference appears to be a single amino acid in a specific domain of the channel; the rat has an isoleucine while insects have a methionine. Replacing the isoleucine with a methionine in rats greatly increases their sensitivity to pyrethroids (Soderlund, 2012). Limited studies on human $Na_V1.3$ channels indicate that they are actually less sensitive to modification by pyrethroids than the same rat channels, though the molecular basis for such difference is unknown (Tan and Soderlund, 2009). In addition to the higher sensitivity of insect sodium channels, lower body temperature (as pyrethroids show a negative temperature coefficient of action), and slower biotransformation in insects contribute to the high selectivity of pyrethroids (Ray and Fry, 2006).

In humans, upon occupational or accidental exposure, the primary adverse effect resulting from dermal contact with pyrethroids is paresthesia (Flannigan et al., 1985; He et al., 1989). Symptoms include continuous tingling or pricking or, when more severe, burning. The condition reverses in about 24 hours, and topical application of vitamin E has been shown to be an effective treatment. Paresthesia is presumably due to abnormal pyrethroid-induced repetitive activity in skin nerve terminals (Ray and Fry, 2006), and is more pronounced with type II compounds. Cases of mild acute illness, with neurologic and respiratory symptoms, have been reported in flight attendants following aircraft disinfection with permethrin (Sutton et al., 2007), and upon the use of pyrethrin/pyrethroid to control bed bugs (MMWR, 2011). When ingested, main symptoms of pyrethroid poisoning are gastric in nature (epigastric pain, nausea, vomiting), and can be followed by dizziness, headache, fatigue, disturbances in consciousness, and convulsions (He et al., 1989). Given their widespread use, reports on incidents due to pyrethroid exposure have been increasing over the past decade, though such exposures resulted in minor or no symptoms in most cases (Power and Sudakin, 2007). A recent study identified a total of 4974 cases of acute pyrethrin/pyrethroid-related illnesses in the USA from 2000 to 2008, with a yearly increase over this period. Most cases (85%) were of low severity, and 34% were work-related. Respiratory effects were the most common symptoms reported (Hudson et al., 2014).

Additional targets for pyrethroids

Type II pyrethroids, but not type I compounds, also bind to and inhibit $GABA_A$-gated chloride channels (Lawrence and Casida, 1983), albeit at higher concentrations than those sufficient to affect sodium channels (10^{-7} M vs 10^{-10} M). This effect is believed to contribute to the seizures that accompany severe type II pyrethroid poisoning. However, drugs that enhance GABAergic transmission (e.g., diazepam), have modest effects toward deltamethrin-induced choreoatetosis or seizures (Soderlund, 2010). Type II pyrethroids, such as deltamethrin, also inhibit at low concentration (10^{-10} M) voltage-dependent chloride channels (Forshaw et al., 1993). Since agents that open these chloride channels, such as ivermectin and pentobarbital, antagonize pyrethroid-induced choreoatetosis and salivation, inhibition of maxi voltage-dependent chloride channels by pyrethroid may contribute to some of the signs associated with the type II poisoning syndrome (Forshaw et al., 2000; Breckenridge et al., 2009; Soderlund, 2010).

Other reported targets for pyrethroids include calcium ATPase and voltage-gated calcium channels (Shafer and Meyer, 2004), which are, however, affected at higher concentrations (10^{-5} M and 10^{-6} M, respectively). The latter interaction may be of interest, as pyrethroids have been shown to increase calcium-dependent neurotransmitter release (Soderlund, 2010). However, results of biochemical and electrophysiologic studies on calcium channels are not always consistent, as both stimulation and inhibition have been reported. Thus, the relevance of such interactions in the *in vivo* toxicity of pyrethroids remains unclear (Soderlund, 2012). Also of note is the observation that several pyrethroids could stimulate protein kinase C-dependent protein phosphorylation at very low concentrations (10^{-13} M) (Enan and Matsumura, 1993), but whether this interaction is involved in the modulation of sodium and/or chloride channels remains to be determined.

Potential developmental neurotoxicity of pyrethroids

Young animals are more sensitive to the acute toxicity of certain pyrethroids, such as deltramethrin and cypermethrin (Sheets, 2000), most likely because of a lesser capacity for metabolic detoxification (Anand et al., 2006); however, only minor age-related differences were found for other compounds (Sheets, 2000). Some studies have suggested that certain pyrethroids may cause developmental neurotoxicity. For example, behavioral and biochemical alterations have been reported in rats and mice upon pre- or postnatal exposure to pyrethroids (Eriksson and Fredriksson, 1991; Lazarini et al., 2001). $Na_V1.3\beta3$ channels, which predominate in the developing rat brain, are particularly susceptible to disruption by type II compounds (Meacham et al., 2008). However, overall evidence for a clear developmental neurotoxicity of pyrethroids has been judged inadequate (Shafer et al.,

2005), and developmental neurotoxicity studies carried out with some pyrethroids (e.g., deltamethrin) according to regulatory guidelines have provided negative results (European Food Safety Agency, 2009). Furthermore, levels of background pyrethroid exposure (presumably through residues in the diet) in children have been found to be orders of magnitude lower than the corresponding acceptable daily intake (Heudorf et al., 2004). Also, the use of deltamethrin-impregnated bed nets does not appear to pose any health risk in children and neonates, while substantially reducing infant mortality from malaria (Alonso et al., 1991; Barlow et al., 2001).

Organochlorine and pyrethroid insecticides and neurodegenerative diseases

Environmental factors are believed to play a role in the etiology of various neurodegenerative diseases. This is especially true for Parkinson's disease (PD), in which selected genes may only explain certain forms of familial or inherited disease, but not the majority of disease cases which are sporadic. PD is a movement disorder pathologically characterized by degeneration of dopaminergic cell bodies in the substantia nigra pars compacta with subsequent loss of dopamine in the nigrostriatal system. It is also characterized by the presence of intracytoplasmic inclusions known as "Lewy bodies," primarily composed of α-synuclein aggregates and ubiquitin. Clinical signs include rigidity, resting tremors, bradykinesia, and postural instability (Lees et al., 2009). Suggested mechanisms of PD pathogenesis include defective protein handling, mitochondrial dysfunction, oxidative stress, and neuroinflammation. Among environmental factors potentially associated with the etiology of PD, pesticides have been among those most studied. Dozens of epidemiologic studies (ecologic, cross-sectional, and prospective) have been reported, which have overall provided evidence of an association between pesticide exposure and PD (Priyadarshi et al., 2000; Brown et al., 2006; Freire and Koifman, 2012). Most interest has been devoted to two pesticides, the herbicide paraquat and the insecticide rotenone, because of their structure, mechanism of action, and robust animal data (Franco et al., 2010; Moretto and Colosio, 2011). However, there is also some evidence of an association between organochlorine insecticides and PD, while for pyrethroids the evidence is much weaker (Hatcher et al., 2008a).

An increased risk of PD has been associated with exposure to organochlorine insecticides in at least two epidemiologic studies (Hancock et al., 2008; Elbaz et al., 2009). Furthermore, a third study found that the association was stronger in homozygotes for the ABCB1 transporter, who had a diminished ability to clear xenobiotics from the brain (Dutheil et al., 2010). Two additional studies found a correlation between PD and serum levels of various organochlorines (e.g. dieldrin, *p,p'*-DDT, *p,p'*-DDE, β-hexachlorocyclohexane) (Weisskopf et al., 2010; Richardson et al., 2011). As there are also two reports of elevated levels of dieldrin in brain of postmortem PD patients compared to controls (Fleming et al., 1994; Corrigan et al., 2000), subsequent efforts were directed to elucidate the potential interactions between this insecticide and other organochlorine insecticides and dopaminergic neurons. Several *in vitro* and *in vivo* studies have revealed the ability of dieldrin and heptachlor to interfere with dopaminergic neurotransmission by altering the dopamine transporter and the vesicular monoamine transporter, leading to dopamine depletion (Miller et al., 1999; Kanthasamy et al., 2005). In contrast, DDT and DDE were found to affect the plasma membrane dopamine transporter and the vesicular monoamine transporter *in vitro*, but caused no nigrostriatal damage upon *in vivo* administration (Hatcher et al., 2008b). Dieldrin also causes oxidative stress and mitochondrial dysfunction, which leads to apoptotic death of dopaminergic neurons (Kanthasamy et al., 2005). Furthermore, dieldrin increases the fibrillation of α-synuclein (Uversky et al., 2002), and induces ubiquitin-proteasome dysfunction in dopaminergic cells (Sun et al., 2005). Of interest are also the results of a study in which developmental exposure of mice to low doses of dieldrin altered the synaptic and vesicular dopamine transporters, and enhanced susceptibility to the neurotoxicity of the known dopaminergic neurotoxin 1-methyl-4-phenyl-1,2,3,6-tetrahydropyridine (MPTP) (Richardson et al., 2006). Nevertheless, in contrast to rotenone and paraquat, loss of dopaminergic neurons and other characteristics of PD have not been observed with organochlorine insecticides *in vivo*.

With regard to pyrethroids, evidence of a potential association with PD is much more limited. Various pyrethroids (mostly type II compounds) have been shown to interfere with dopaminergic transport *in vivo*, but the underlying mechanisms have not been clarified (Bloomquist et al., 2002; Pittman et al., 2003; Elwan et al., 2006; Nasuti et al., 2007). Cypermethrin has been shown to cause nigrostriatal degeneration upon long-term exposure, and this has been attributed to its ability to induce oxidative stress (Singh et al., 2011, 2012). So far, epidemiologic studies have not reported associations between PD and exposure to pyrethroids (Moretto and Colosio, 2011).

ACKNOWLEDGMENTS

Research by the author has been supported by grants from the National Institutes of Health, Environmental Protection Agency, and National Science Foundation.

References

Alonso L, Lindsay SW, Armstrong JRM et al. (1991). The effect of insecticide-treated bed nets on mortality of Gambian children. Lancet 337: 1499–1502.

Alvarez-Pedrerol M, Ribas-Fito N, Torrent M et al. (2008). Effects of PCBs, p, p'-DDT, p, p'-DDE, HCB, and β-HCH on thyroid function in preschool children. Occup Environ Med 65: 452–457.

Anand SS, Bim KB, Padilla S et al. (2006). Ontogeny of hepatic and plasma metabolism of deltamethrin in vitro: role in age-dependent acute neurotoxicity. Drug Metab Dispos 34: 389–397.

Anway MD, Cupp AS, Uzumcu M et al. (2005). Epigenetic transgenerational actions of endocrine disruptors and male fertility. Science 306: 1466–1468.

Barlow SM, Sullivan FM, Lines J (2001). Risk assessment of the use of deltamethrin on bed nets for the prevention of malaria. Food Chem Toxicol 39: 407–422.

Beard J (2006). DDT and human health. Science Tot Environ 355: 78–89.

Bloomquist JR, Barlow ML, Gillette JS et al. (2002). Selective effects of insecticides on nigrostriatal dopaminergic nerve pathways. Neurotoxicology 23: 537–544.

Bouwman H, van de Berg H, Kylin H (2011). DDT and malaria prevention: addressing the paradox. Environ Health Perspect 119: 744–747.

Bradberry SM, Cage SA, Proudfoot AT et al. (2005). Poisoning due to pyrethroids. Toxicol Rev 24: 93–106.

Breckenridge CB, Holden L, Sturgess N et al. (2009). Evidence for separate mechanism of toxicity for type I and type II pyrethroid insecticides. Neurotoxicology 30S: S17–S31.

Brooks GT (1974). Chlorinated Insecticides. Vol I. Technology and Application; Vol. II. Biological and Environmental Aspects, CRC Press, Cleveland, pp. 1–249 and 1–197.

Brooks GT (2010). Interactions with the gamma-aminobutyric acid A-receptor: polychlorocycloalkanes and recent congeners and other ligands. In: R Krieger (Ed.), Hayes' Handbook of Pesticide Toxicology, Academic Press, San Diego, pp. 2065–2092.

Brown TP, Rumsby PC, Capleton AC et al. (2006). Pesticides and Parkinson's disease – is there a link? Environ Health Perspect 114: 117–121.

Carson R (1962). Silent Spring, Houghton Mifflin, Boston.

Casida JE (1980). Pyrethrum flowers and pyrethroid insecticides. Environ Health Perspect 34: 189–202.

Casida JE (2009). Pest toxicology: the primary mechanisms of pesticide actions. Chem Res Toxicol 22: 609–619.

Casida JE (2010). Michael Elliot's billion dollar crystals and other discoveries in insecticide chemistry. Pest Manag Sci 66: 1163–1170.

Casida JE, Gammon DW, Glickman AH et al. (1983). Mechanisms of selective action of pyrethroid insecticides. Annu Rev Pharmacol Toxicol 23: 413–438.

Catterall WA (2000). From ionic currents to molecular mechanisms: the structure and function of voltage-gated sodium channels. Neuron 26: 13–25.

Chen A, Rogan WJ (2003). Nonmalarial infant deaths and DDT use for malaria control. Emerg Infect Dis 9: 960–964.

Cole LM, Casida JE (1986). Polychlorocycloalkane insecticide-induced convulsions in mice in relation to disruption of the GABA-regulated chloride ionophore. Life Sci 39: 1855–1862.

Corrigan FM, Wienburg CL, Shore RF et al. (2000). Organochlorine insecticides in substantia nigra in Parkinson's disease. J Toxicol Environ Health A 59: 229–234.

Costa LG (1987). Toxicology of pesticides: a brief history. In: LG Costa, CL Galli, SD Murphy (Eds.), Toxicology of Pesticides: Experimental, Clinical and Regulatory Perspectives, Springer-Verlag, Heidelberg, Heidelberg, pp. 1–10.

Costa LG (2013). Toxic effects of pesticides. In: CD Klaassen (Ed.), Casarett and Doull's Toxicology. The Basic Science of Poisons, 8th Edition, McGraw-Hill, New York, pp. 933–980.

Costa LG, Giordano G, Guizzetti M et al. (2008). Neurotoxicity of pesticides: a brief review. Front Biosci 13: 1240–1249.

Dallaire R, Muckle G, Rouget F et al. (2012). Cognitive, visual, and motor development of 7-month-old Guadeloupean infants exposed to chlordecone. Environ Res 118: 79–85.

Desaiah D (1982). Biochemical mechanisms of chlordecone neurotoxicity: a review. Neurotoxicology 3: 103–110.

Dutheil F, Beaune P, Tzourio C et al. (2010). Interaction between ABCB1 and professional exposure to organochlorine insecticides in Parkinson's disease. Arch Neurol 67: 739–745.

Ecobichon DJ, Joy RM (1982). Pesticides and Neurological Diseases, CRC Press, Boca Raton, pp. 1–281.

Elbaz A, Clavel J, Rathouz PJ et al. (2009). Professional exposure to pesticides and Parkinson's disease. Ann Neurol 66: 494–504.

Eldefrawi AT, Eldefrawi ME (1987). Receptors for γ-aminobutyric acid and voltage-dependent chloride channels as targets for drugs and toxicants. FASEB J 1: 262–271.

Elwan MA, Richardson JR, Guillot TS et al. (2006). Pyrethroid pesticide-induced alteration in dopamine transporter function. Toxicol Appl Pharmacol 211: 188–197.

Enan E, Matsumura F (1993). Activation of phosphoinositide protein kinase C pathway in rat brain tissue by pyrethroids. Biochem Pharmacol 45: 703–710.

Eriksson P, Fredriksson A (1991). Neurotoxic effects of two different pyrethroids, bioallethrin and deltamethrin, on immature and adult mice: changes in behavioural and muscarinic receptors variables. Toxicol Appl Pharmacol 108: 78–85.

Eskenazi B, Chevrier J, Rosas LG et al. (2009). The Pine River Statement: human health consequences of DDT use. Environ Health Perspect 117: 1359–1367.

European Food Safety Agency (2009). Scientific opinion of the Panel on Plant Protection Products and their Residues (PPR) on a request from the European Commission on potential developmental neurotoxicity of deltamethrin. EFSA J 921: 1–34.

Flannigan SA, Tucker SB, Key MM et al. (1985). Synthetic pyrethroid insecticides: a dermatological evaluation. Br J Ind Med 42: 363–372.

Fleming L, Mann JB, Bean J et al. (1994). Parkinson's disease and brain levels of organochlorine pesticides. Ann Neurol 36: 100–103.

Forshaw PJ, Lister T, Ray DE (1993). Inhibition of a neuronal voltage-dependent chloride channel by the type II pyrethroid, deltamethrin. Neuropharmacology 32: 105–111.

Forshaw PJ, Lister T, Ray DE (2000). The role of voltage-gated chloride channels in type II pyrethroid insecticide poisoning. Toxicol Appl Pharmacol 163: 1–8.

Franco R, Li S, Rodriguez-Rocha H et al. (2010). Molecular mechanisms of pesticide-induced neurotoxicity: relevance to Parkinson's disease. Chem Biol Interact 188: 289–300.

Freire C, Koifman S (2012). Pesticide exposure and Parkinson's disease: epidemiological evidence of association. Neurotoxicology 33: 947–971.

Golden RJ, Noller KL, Titus-Enrstoff L et al. (1998). Environmental endocrine modulators and human health: an assessment of biological evidence. Crit Rev Toxicol 28: 109–227.

Guzelian PS (1982). Comparative toxicology of chlordecone (Kepone) in humans and experimental animals. Annu Rev Pharmacol Toxicol 22: 89–113.

Hancock DB, Martin ER, Mayhew GM et al. (2008). Pesticide exposure and risk of Parkinson's disease: a family-based case-control study. BMC Neurol 8: 6.

Hatcher JM, Pennell KD, Miller GW (2008a). Parkinson's disease and pesticides: a toxicological perspective. Trends Pharmacol Sci 29: 322–329.

Hatcher JM, Delea KC, Richardson JR et al. (2008b). Disruption of dopamine transport by DDT and its metabolites. Neurotoxicology 29: 682–690.

He F, Wang S, Liu L et al. (1989). Clinical manifestations and diagnosis of acute pyrethroid poisoning. Arch Toxicol 63: 54–58.

Heudorf U, Angerer J, Drexler H (2004). Current internal exposure to pesticides in children and adolescents in Germany: urinary levels of metabolites of pyrethroid and organophosphorus insecticides. Int Arch Occup Environ Health 77: 67–72.

Hickey JP, Batterman SA, Chernyak SM (2006). Trends of chlorinated organic contaminants in Great Lakes trout and walleye from 1970 to 1998. Arch Environ Contam Toxicol 50: 97–110.

Holmstedt B, Liljestrand G (1963). Readings in Pharmacology, Pergamon Press, Oxford, pp. 1–395.

Hudson NL, Kasner EJ, Beckman J et al. (2014). Characteristics and magnitude of acute pesticide-related illnesses and injuries associated with pyrethrin and pyrethroid exposures – 11 states, 2000–2008. Am J Ind Med 57: 15–30.

Humphreys EH, Janssen S, Heil A et al. (2008). Outcomes of the California ban on pharmaceutical lindane: clinical and ecologic impacts. Environ Health Perspect 116: 297–302.

Kanthasamy AG, Kitazawa M, Kanthasamy A et al. (2005). Dieldrin-induced neurotoxicity: relevance to Parkinson's disease pathogenesis. Neurotoxicology 26: 701–719.

Kavlock RJ (2001). Pesticides as endocrine-disrupting chemicals. In: R Krieger (Ed.), Handbook of Pesticide Toxicology, Academic Press, San Diego, pp. 727–746.

Kelce WR, Stone CR, Laws SC et al. (1995). Persistent DDT metabolite p, p'-DDE is a potent androgen receptor antagonist. Nature 375: 581–585.

Lawrence LJ, Casida JE (1982). Pyrethroid toxicology: mouse intracerebral structure–toxicity relationship. Pestic Biochem Physiol 18: 9–14.

Lawrence LJ, Casida JE (1983). Stereospecific action of pyrethroid insecticides on the γ-aminobutyric acid receptor–ionophore complex. Science 221: 1399–1401.

Lazarini CA, Florio JC, Lemonica IP et al. (2001). Effects of prenatal exposure to deltamethrin on forced swimming behavior, motor activity, and striatal dopamine levels in male and female rats. Neurotoxicol Teratol 23: 665–673.

Lebwohl M, Clark L, Levitt J (2007). Therapy for head lice based on life cycle, resistance, and safety considerations. Pediatrics 119: 965–974.

Lees AJ, Hardy J, Revesz T (2009). Parkinson's disease. Lancet 373: 2055–2066.

Mactutus CF, Tilson HA (1984). Neonatal chlordecone exposure impairs early learning and retention of active avoidance in the rat. Neurobehav Toxicol Teratol 6: 75–83.

Mactutus CF, Tilson HA (1985). Evaluation of long-term consequences in behavioural and/or neural function following neonatal chlordecone exposure. Teratology 31: 177–186.

Maharaj R, Mthembu DJ, Sharp BL (2005). Impact of DDT re-introduction on malaria transmission in KwaZulu-Natal. S Afr Med J 95: 871–874.

Martinez AJ, Taylor JR, Dyck PJ et al. (1978). Chlordecone intoxication in man. II. Ultrastructure of peripheral nerves and skeletal muscle. Neurology 28: 631–635.

Matsumura F, Ghiasuddin SM (1979). Characteristics of DDT-sensitive Ca-ATPase in the axonic membrane. In: T Narahashi (Ed.), Neurotoxicology of Insecticides and Pheromones, Plenum Press, New York, pp. 245–257.

Meacham CA, Brodfuehrer PD, Watkins JA et al. (2008). Developmentally-regulated sodium channel subunits are differentially sensitive to α-cyano containing pyrethroids. Toxicol Appl Pharmacol 231: 273–281.

Miller GW, Kirby ML, Levey AI et al. (1999). Heptachlor alters expression and function of dopamine transporters. Neurotoxicology 20: 631–637.

MMWR (2011). Acute illness associated with insecticide used to control bed bugs – seven stages, 2003–2010. MMWR Morb Mortal Wkly Rep 60: 1269–1274.

Moretto A, Colosio C (2011). Biochemical and toxicological evidence of neurological effects of pesticides: the example of Parkinson's disease. Neurotoxicology 32: 383–391.

Murphy SD (1986). Toxic effects of pesticides. In: CD Klaassen, MO Amdur, J Doull (Eds.), Casarett and Doull's Toxicology. The Basic Science of Poisons, Macmillan Publishing, New York, pp. 519–581.

Narahashi T (1996). Neuronal ion channels as the target sites of insecticides. Pharmacol Toxicol 78: 1–14.

Narahashi T, Zhao X, Ikeda T et al. (2007). Differential actions of insecticides on target sites: basis for selective toxicity. Hum Exp Toxicol 26: 361–366.

Nasuti C, Gabbianelli R, Falcioni ML et al. (2007). Dopaminergic system modulation, behavioral changes and oxidative stress after neonatal adminsitration of pyrethroids. Toxicology 229: 194–205.

Okankwo JO, Mutshatsi TN, Botha B et al. (2008). DDT, DDE and DDD in human milk from South Africa. Bull Environ Contam Toxicol 81: 348–354.

Peakall DB (1970). Pesticides and the reproduction of birds. Sci American 222: 72–78.

Pittman JT, Dodd CA, Klein BG (2003). Immunohistochemical changes in the mouse striatum induced by the pyrethroid insecticide permethrin. Int J Toxicol 22: 359–370.

Poulos L, Athanaselis S, Coutselinis A (1982). Acute intoxication with cypermethrin (NRDC 149). J Toxicol Clin Toxicol 19: 519–520.

Power LE, Sudakin DL (2007). Pyrethrin and pyrethroid exposure in the United States: a longitudinal analysis of incidents reported to poison centers. J Med Toxicol 3: 94–99.

Priyadarshi A, Khuder SA, Schaub EA et al. (2000). A meta-analysis of Parkinson's disease and exposure to pesticides. Neurotoxicology 21: 435–440.

Ray D (2010). Organochlorine and pyrethroid insecticides. In: M Philbert (Ed.), Nervous System and Behavioral Toxicology, Elsevier, New York, pp. 445–457.

Ray DE, Fry JR (2006). A reassessment of the neurotoxicity of pyrethroid insecticides. Pharmacol Ther 111: 174–193.

Richardson JR, Caudle WM, Wang M et al. (2006). Developmental exposure to the pesticide dieldrin alters the dopamine system and increases neurotoxicity in an animal model of Parkinson's disease. FASEB J 20: E976–E985.

Richardson JR, Roy A, Shalat SL et al. (2011). β-Hexachlorocyclohexane levels in serum and risk of Parkinson's disease. Neurotoxicology 32: 640–645.

Rogan WJ, Chen A (2005). Health risks and benefits of bis (4-chlorophenyl) -1, 1,1 –trichloroethane (DDT). Lancet 366: 763–773.

Safe S (2000). Endocrine disruptors and human health – is there a problem? An update. Environ Health Perspect 108: 487–493.

Safe S (2005). Clinical correlates of environmental endocrine disruptors. Trends Endocrinol Metab 16: 139–144.

Shafer TJ, Meyer DA (2004). Effects of pyrethroids on voltage-sensitive calcium channels: a critical evaluation of strengths, weaknesses, data needs, and relationship to assessment of cumulative neurotoxicity. Toxicol Appl Pharmacol 196: 303–318.

Shafer TJ, Meyer DA, Crofton KM (2005). Developmental neurotoxicity of pyrethroid insecticides: critical review and future research needs. Environ Health Perspect 113: 123–136.

Sheets LP (2000). A consideration of age dependent differences in susceptibility to organophosphorus and pyrethroid insecticides. Neurotoxicology 21: 57–64.

Singh AK, Tiwari MN, Dixit A et al. (2011). Nigrostriatal proteomics of cypermethrin-induced dopaminergic neurodegeneration: microglial activation dependent and independent regulation. Toxicol Sci 122: 526–538.

Singh AK, Tiwari MN, Prakash O et al. (2012). A current review of cypermethrin-induced neurotoxicity and nigrostriatal dopaminergic neurodegeneration. Curr Neuropharmacol 10: 64–71.

Smith AG (1991). Chlorinated hydrocarbon insecticides. In: WJ Hayes, ER Laws (Eds.), Handbook of Pesticide Toxicology, Academic Press, San Diego, pp. 731–915.

Smith AG (2010). Toxicology of DDT and some analogues. In: R Krieger (Ed.), Hayes's Handbook of Pesticide Toxicology, Academic Press, San Diego, pp. 1975–2032.

Soderlund DM (2010). Toxicology and mode of action of pyrethroid insecticides. In: R Krieger (Ed.), Hayes' Handbook of Pesticide Toxicology, Academic Press, San Diego, pp. 1665–1686.

Soderlund DM (2012). Molecular mechanisms of pyrethroid insecticide neurotoxicity: recent advances. Arch Toxicol 86: 165–181.

Soderlund DM, Clark JM, Sheets LP et al. (2002). Mechanisms of pyrethroid neurotoxicity: implications for cumulative risk assessment. Toxicology 171: 3–59.

Stoker TE, Kavlock RJ (2010). Pesticides as endocrine-disrupting chemicals. In: R Krieger (Ed.), Hayes' Handbook of Pesticide Toxicology, Academic Press, San Diego, pp. 551–569.

Sun F, Anantharam V, Latchoumycandane C et al. (2005). Dieldrin induces ubiquitin-proteasome dysfunction in α-synuclein overexpressing dopaminergic neuronal cells and enhances susceptibility to apoptotic cell death. J Pharmacol Exp Ther 315: 69–79.

Sutton PM, Vergara X, Beckman J et al. (2007). Pesticide illness among flight attendants due to aircraft disinsection. Am J Ind Med 50: 345–356.

Tan J, Soderlund DM (2009). Human and rat $Na_v1.3$ voltage-gated sodium channels differ in inactivation properties and sensitivity to the pyrethroid tefluthrin. Neurotoxicology 30: 81–89.

Taylor JR (1982). Neurological manifestations in humans exposed to chlordecone and follow-up studies. Neurotoxicology 3: 9–16.

Taylor JR, Selhorst JB, Houff SA et al. (1978). Chlordecone intoxication in man. I. Clinical observations. Neurology 28: 626–630.

U.S. Environmental Protection Agency (1975). DDT: A review of Scientific and Economic Aspects of the Decision to Ban its Use as a Pesticide, USEPA, Washington DC, pp. 1–300.

Uversky VN, Li J, Bower K et al. (2002). Synergistic effects of pesticides and metals on the fibrillation of α-synuclein: implications for Parkinson's disease. Neurotoxicology 23: 527–536.

Van den Bercken J (1972). The effect of DDT and dieldrin on myelinated nerve fibers. Eur J Pharmacol 20: 205–214.

Van den Berg H (2009). Global status of DDT and its alternatives for use in vector control to prevent disease. Environ Health Perspect 117: 1656–1663.

Verschoyle RD, Aldridge WN (1980). Structure-activity relationships of some pyrethroids in rats. Arch Toxicol 45: 325–329.

Vijverberg HPM, van der Zalm JM, van den Bercken J (1982). Similar mode of action of pyrethroids and DDT on sodium channel gating in myelinated nerves. Nature 295: 601–603.

Weiner ML, Nemec M, Sheets L et al. (2009). Comparative functional observational battery of twelve commercial pyrethroid insecticides in male rats following acute oral exposure. Neurotoxicology 30S: S1–S16.

Weiss B (2012). The intersection of neurotoxicology and endocrine disruption. Neurotoxicology 33: 1410–1419.

Weisskopf MG, Knekt P, O'Reilly EJ et al. (2010). Persistent organochlorine pesticides in serum and risk of Parkinson's disease. Neurology 74: 1055–1061.

Woolley DE (1982). Neurotoxicity of DDT and possible mechanisms of action. In: KN Prasad, A Vernadakis (Eds.), Mechanisms of Action of Neurotoxic Substances, Raven Press, New York, pp. 95–141.

World Health Organization (1979). DDT and its derivatives. Environmental Health Criteria 9. WHO, Geneva, pp. 1–194.

World Health Organization (2010). World Malaria report 2010, WHO, Geneva, pp. 1–238.

Handbook of Clinical Neurology, Vol. 131 (3rd series)
Occupational Neurology
M. Lotti and M.L. Bleecker, Editors

Chapter 10

Organophosphorus and carbamate insecticide poisoning

ALLISTER VALE[1]* AND MARCELLO LOTTI[2]

[1]*National Poisons Information Service (Birmingham Unit) and West Midlands Poisons Unit, City Hospital, Birmingham, UK; Honorary Professor, School of Biosciences, University of Birmingham, UK*

[2]*Department of Cardiology, Thoracic and Vascular Sciences, School of Medicine, University of Padua, Padua, Italy*

INTRODUCTION

Organophosphorus (OP) insecticides are used widely in agriculture, horticulture, and veterinary medicine. In addition, they are used domestically and in public hygiene to control vectors of disease. Some (e.g., malathion) are used to treat human infestation with scabies, head lice, and crab lice. OP nerve agents have also been used as warfare agents and in terrorist attacks.

Carbamate insecticides were developed in the 1950s in response to a search for insecticides with anticholinesterase activity but with greater specificity and lower mammalian toxicity than OP insecticides. Carbamates are used widely in many countries where agriculture is a major industry and pest control essential and are employed frequently as alternatives to OP insecticides. Certain carbamates, such as physostigmine and pyridostigmine, are drugs commonly used in clinical practice.

OP insecticides

OP insecticides are complex molecules and usually esters, amides, or thiol derivatives of phosphoric acid (Fig. 10.1). The R^1 and R^2 moieties are usually alkyl or aryl groups. When they are linked to phosphorus via oxygen, the resulting compound is termed a phosphate (Fig. 10.1A), while bonding via sulfur results in a phosphorothiolate or S-substituted phosphorothioate (Fig. 10.1B). Molecules in which phosphorus is linked by a double bond to sulfur (P=S) are known as phosphorothioates (Fig. 10.1C). Phosphoramidates (Fig. 10.1D) have a carbon atom linked to phosphorus through an NH group.

The X moiety of an OP can be one of a wide range of substituted or branched aliphatic, aromatic, or heterocyclic groups linked to the phosphorus atom via an -O- or -S- to make it more labile. During inhibition of the target enzymes, such as acetylcholinesterase (AChE), the phosphorus atom binds to an amino acid on the enzyme and X is eliminated; X is thus often referred to as the "leaving group." The lability of the linkage between X and the phosphorus atom is critical in determining the reactivity of the OP with AChE. The P=O-containing structure is often referred to as an oxon and the P=S structure as a thion. The term thion is often incorporated in the trivial name, e.g., parathion.

MECHANISMS OF TOXICITY

Phosphates (P=O) are biologically active, whereas phosphorothioates (P=S) need bioactivation to the corresponding metabolite (oxon) before becoming so.

The oxon phosphorylates the serine hydroxyl group located at the active site of AChE, thus inhibiting it. Inhibition occurs in blood, brain, and other tissues in a time-dependent manner. The four stages of interaction of OP oxons with AChE have been reviewed (Johnson et al., 2000). The extent of inhibition depends on the rate constant for the reaction and the time that the enzyme is exposed to the oxon. Inhibition of AChE results in accumulation of acetylcholine (ACh) at autonomic and some central synapses and at autonomic postganglionic and neuromuscular junctions. As a consequence, ACh binds to, and stimulates, muscarinic and nicotinic receptors. Phosphorylated enzyme may become "aged" by partial dealkylation of the serine group at the active site of AChE. The rate at which it occurs depends on the structure of the OP compound but results in an inactive

*Correspondence to: Professor Allister Vale, MD FRCP, Director, National Poisons Information Service (Birmingham Unit) and West Midlands Poisons Unit; City Hospital, Birmingham B18 7QH, UK. Tel: +44-121-507-4123, Fax: +44-121-507-5580, E-mail: allistervale@npis.org

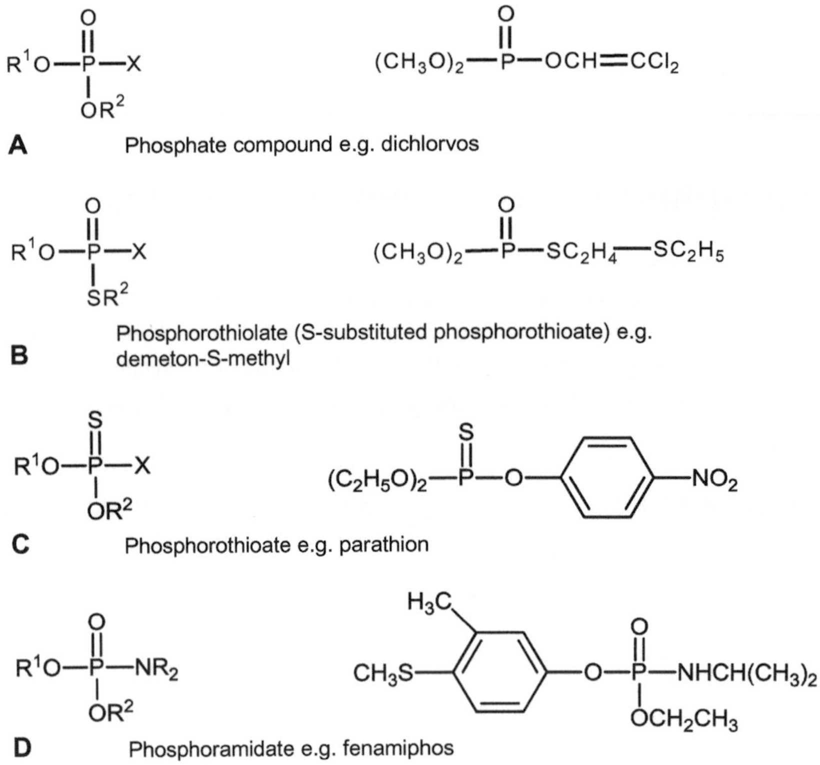

Fig. 10.1. Structures of some organophosphorus insecticides. (**A**) Phosphate compound, e.g., dichlorvos. (**B**) Phosphorothiolate (S-substituted phosphorothioate), e.g., demeton-*S*-methyl. (**C**) Phosphorothioate, e.g., parathion. (**D**) Phosphoramidate, e.g., fenamiphos.

enzyme which can no longer be reactivated. Recovery of AChE activity requires synthesis of new enzyme in the liver (Burgen and Hobbiger, 1951).

Because most OP insecticides are dialkylphosphates or dialkylphosphorothioates, a dialkoxyphosphoryl derivative is formed when they interact with AChE. The rates of spontaneous reactivation, aging, and oxime-induced reactivation of AChE depend solely on the nature of the alkoxy groups; the rates of inhibition also depend on the nature of the leaving group. OP pesticides are also mostly dimethyl (e.g., demeton-S-methyl, dichlorvos, dimethoate, malathion) or diethyl (e.g., chlorpyrifos, diazinon, parathion) compounds and either give rise to dimethoxyphosphoryl AChE or diethoxyphosphoryl AChE. The rate of spontaneous reactivation is faster for dimethoxyphosphoryl than diethoxyphosphoryl enzyme (Maroni et al., 1986; Wilson et al., 1992) and this also appears to be true of oxime-induced reactivation (Hansen and Wilson, 1999). Spontaneous reactivation of dimethoxy phosphorylated AChE proceeds quite rapidly so that the patient's condition should improve even without oxime therapy. Unless oximes are employed, however, there is no such expectation of rapid recovery for patients intoxicated with diethoxy phosphoryl insecticides.

The chemical structure of the OP insecticide and particularly the presence of other ingredients in the formulation may have an impact on the speed of onset of features of intoxication. In addition, the fact that many OP insecticides are lipophilic means that they are distributed to, and stored in, body fat and therefore elimination takes place slowly. Thus, the severity of intoxication may increase for 12–36 hours after exposure, intoxication may be prolonged, and relapse may occur after apparent recovery.

CLINICAL FEATURES

Acute exposure

The clinical presentation and severity of OP poisoning are dependent not only on the particular OP compound and the magnitude of exposure but on other factors, including the route of exposure, the age of the patient, whether exposure was a suicidal attempt (when a substantial ingestion is likely), and the solvent in the formulation. Not only may skin absorption of the OP be enhanced by the solvent but ingestion of solvent may induce vomiting with risk of aspiration, pneumonia, and depressed consciousness. There is increasing evidence that the solvents in formulations are responsible for the high morbidity and mortality (Eddleston et al., 2010, 2012).

Toxicity may occur following inhalation, ingestion, or skin contamination. Although dermal absorption of OP compounds tends to be slow, severe poisoning may still ensue if exposure is prolonged. Skin contact and subsequent absorption are the major route of exposure occupationally. Inhalation of OP insecticides, particularly during the manufacture of formulations, for example, because of inefficiently operating ventilation equipment (Vale and Scott, 1974) or during spraying or mixing, is a recognized occupational hazard. Ingestion is uncommon in the workplace but can occur accidentally in workers with poor personal hygiene, such as those who do not remove contaminated clothing or fail to wash their hands. Such exposure is not likely to lead to other than mild features. However, the deliberate ingestion of an OP insecticide is likely to result in more severe features of intoxication.

The onset and severity of toxicity depend on the speed and degree of depression of AChE activity. As some OP compounds require biotransformation to become biologically active, signs of intoxication may be delayed. Furthermore, extensive occupational misuse of OP compounds may cause progressive depletion of AChE activity until toxic effects occur. Symptoms can present within 5 minutes of massive ingestion and almost always occur within 12 hours. Muscarinic features (Table 10.1) generally appear first and characterize mild to moderate poisoning but are not always present. In one study, no single symptom was noted in more than 60% of cases (Hirschberg and Lerman, 1984). For example, miosis, although the most prevalent specific sign, was found in only 44% of cases (Hirschberg and Lerman, 1984). In other studies, miosis was observed in 82% of 61 cases (Bardin et al., 1987) and in 83% of 23 cases (Lee and Tai, 2001). Muscarinic symptoms rarely present more than 24 hours after ingestion.

The initial symptoms of poisoning are often a feeling of exhaustion and weakness. Vomiting, cramp-like abdominal pain, sweating, and hypersalivation may follow (Table 10.2). Miosis and tightness in the chest during inspiration may occur at an early stage but are not reliable indicators of the severity of systemic poisoning because they may be caused by local anticholinesterase effects of spray mist on the eye or bronchi.

In more severely poisoned patients, nicotinic features predominate (Table 10.1) at an early stage. Muscle twitching affects the eyelids, tongue, face, and calf muscles; respiratory muscles then become involved and generalized weakness ensues. There is a decrease in dynamic lung compliance, an increase in total pulmonary resistance, and an increased alveolar-arterial oxygen gradient. Convulsions may also occur (Lee and Tai, 2001). Sinus tachycardia is likely to be present (although some patients may be bradycardic when effects on heart muscarinic innervation prevail) and mydriasis may be observed (Wyckoff et al., 1968), particularly if atropine is given. Later effects may include diarrhea, tenesmus, incontinence, ataxia, and confusion. Cardiac effects include atrioventricular block, S-T changes, peaked T waves, and QT prolongation. Ventricular arrhythmias are a common cause of death (Saadeh et al., 1997) and tachyarrhythmias of the torsades de pointes type (Ludomirsky et al., 1982; Kiss and Fazekas, 1983) may progress to ventricular fibrillation and/or asystole. Nasal and bronchial hypersecretion (rhinorrhea and bronchorrhea) with bronchoconstriction, cyanosis, respiratory

Table 10.1

Systemic features of organophosphorus insecticide poisoning

Effect	Features
Muscarinic	Cough, wheeze, dyspnea, bronchorrhea, bronchoconstriction, pulmonary edema, cyanosis
	Rhinitis, rhinorrhea, increased salivation, lacrimation, sweating
	Urinary and fecal incontinence
	Nausea, vomiting, abdominal cramps, diarrhea, tenesmus
	Bradycardia, hypotension
	Blurred vision, miosis
Nicotinic	Muscle fasciculation, including diaphragm muscle weakness
	Tachycardia, pallor
	Mydriasis
	Hyperglycemia
Central nervous system	Headaches, anxiety, dizziness, restlessness, insomnia, nightmares, drowsiness, confusion, tremor, ataxia, dysarthria, dystonic reactions
	Hypotension, respiratory depression
	Convulsions, coma

Table 10.2

Common initial features of organophosphorus insecticide poisoning by route of exposure

Route of exposure	Features
Ingestion	Abdominal pain Nausea and vomiting Diarrhea Involuntary defecation
Inhalation	Chest tightness Rhinorrhea Increased salivation Bronchorrhea
Dermal	Localized sweating and muscle fasciculation
Ocular	Miosis Eye pain

Note: Systemic features may follow exposure by these routes.

depression, and coma supervene in severe cases and death may then follow from respiratory failure. A triad of muscarinic, nicotinic, and central nervous system (CNS) symptoms will be apparent in severe cases and was found in 17% of patients in one series (Hirschberg and Lerman, 1984), although in a study of severely poisoned patients requiring ventilatory support it occurred in 60% (Finkelstein et al., 1989).

Central nervous system features

CNS involvement ranging from headache, anxiety, confusion, sleep disturbances, and blurred vision to tremor and convulsions to coma, hypothermia, and central respiratory depression has been described. The failure of the central respiratory drive may be an important factor in causing respiratory failure (Grob, 1956; Tsao et al., 1990), although clinical discrimination of the selective effects of anticholinesterases on the CNS is difficult, particularly in assessing their relative importance in causing death (Lotti, 1992a), because toxicity to the CNS, in general, and of anticholinesterases, in particular (Glow and Rose, 1965), always holds a strong peripheral component.

Early CNS signs and symptoms may last for several days, in which only moderate additional cholinergic symptomatology may be detectable (Grob and Harvey, 1958). Electroencephalogram (EEG) abnormalities can be detected at the onset of symptoms and are characterized by irregularities in rhythm, variation and increase in potential, and intermittent bursts of abnormally slow waves of elevated voltage similar to those seen in epilepsy; these symptoms usually persist for about a week or longer (Grob et al., 1947).

Coma is usually due to direct CNS depression by OPs and solvents in the commercial formulation. In these circumstances the EEG will show profound depression of cortical activity (Lotti and Becker, 1982). OPs vary in their potency to induce seizures (Hoskins et al., 1986) and this manifestation may not be due entirely to AChE inhibition (Van Meter et al., 1978), as it is blocked by benzodiazepines, which are known to act via gamma-aminobutyric acidergic (GABAergic) mechanisms (Lipp, 1973).

Occasionally, during the acute phase of severe OP intoxication (3–40 hours after exposure), patients display opsoclonus (Hata et al., 1986; Pullicino and Aquilina, 1989; De Bleecker, 1992), an abnormal, rapid, involuntary, repetitive, chaotic, and conjugated ocular movement. Opsoclonus is most likely another sign of cholinergic overstimulation, which lasts for several hours to a few days and recovers spontaneously.

Patients may also present with choreoathetosis (Joubert et al., 1984; Joubert and Joubert, 1988b).

Intermediate syndrome

Relapse after apparent resolution of cholinergic symptoms has been reported, particularly in patients who have ingested highly lipophilic OP insecticides, and is termed the intermediate syndrome. The syndrome was first described by Senanayake and Karalliedde (1987), who reported the onset of muscle paralysis affecting particularly upper-limb muscles, neck flexors, and cranial nerves some 24–96 hours after OP exposure, though there are reports of paralysis occurring before 24 hours and even after 96 hours (He et al., 1998; Singh et al., 2000; Eddleston et al., 2006). It is often associated with the development of respiratory failure, a major contributor to the high mortality of OP poisoning. Several recent reviews have described the syndrome in detail (De Bleecker, 2006; Karalliedde et al., 2006; Jayawardane et al., 2009; Abdollahi and Karami-Mohajeri, 2012). Although the etiology of the syndrome is not clearly understood (Benson et al., 1992; De Bleecker et al., 1993; Karalliedde and Henry, 1993; Sedgwick and Senanayake, 1997; Avasthi and Singh, 2000; Dandapani et al., 2003; John et al., 2003), it is generally believed to result from a persistent excess of ACh at the neuromuscular junction (Besser et al., 1989a, b, 1990; De Bleecker et al., 1992, 1994a, b; Besser and Gutmann, 1995; Katz et al., 1997; Singh et al., 2000).

The intermediate syndrome can be regarded as a spectrum disorder affecting the neuromuscular junction: at one end of the spectrum, patients develop only the electrophysiological counterpart of neuromuscular transmission failure, while at the other end, patients develop weakness of 3/5 or less than 3/5 on the Medical Research Council (MRC) grading of all three muscle groups and are at risk of developing respiratory failure (Jayawardane et al., 2009). The selective involvement of certain muscles remains unexplained.

Jayawardane et al. (2009) have shown that in the intermediate syndrome subclinical electrophysiological abnormalities are common and progressive, and precede the onset of the clinical syndrome. Serial repetitive nerve stimulation studies have been most commonly used and are the most accessible for clinicians. Clinical and experimental studies demonstrate a progression through early initial decrement–increment patterns at high rates of stimulations, which correlate with moderate muscle weakness, to decrement–increment patterns at intermediate- and low-frequency stimulations. Progression to a combination of decrement–increment and repetitive fade patterns correlates with clinical deterioration; severe decrement pattern is usually observed immediately before the onset of respiratory failure. Although electrophysiological features closely parallel clinical severity during progression of the syndrome, the same is not true during recovery. Electrophysiological changes sometimes improve long before the patient recovers normal strength and respiratory function.

OP-induced delayed neuropathy

OP-induced delayed neuropathy is a rare complication of acute exposure to some OP insecticides (Lotti et al., 1984; Lotti, 1992b; Lotti and Moretto, 2005; Moretto and Lotti, 2006). It results from phosphorylation and subsequent aging of at least 70% of an esterase, neuropathy target esterase (NTE), in peripheral nerves that is distinct from AChE (Johnson, 1990); NTE is a 147-kDa membrane-bound protein of the patatin-like phospholipase family (Casida et al., 2008; Wijeyesakere and Richardson, 2010). After NTE becomes aged and as a result of loss of NTE activity, membrane phospholipid homeostasis and endoplasmic reticulum functions, including axonal transport and glial–axonal interactions, are disrupted (Glynn, 2006); axonal degeneration ensues and is followed by demyelination of nerve fibers (Johnson, 1969). Only a small number of marketed OP insecticides, for example methamidophos (Senanayake and Johnson, 1982), are capable of causing this syndrome. While there is compelling evidence of the involvement of NTE in the initiation of OP-induced delayed neuropathy, its role in axonal degradation remains obscure.

Lotti and Moretto (2005) have reviewed the features resulting from distal degeneration of some axons in both the peripheral nervous system and CNS occurring 1–4 weeks after single or short-term exposures. Cramping muscle pain in the lower limbs, distal numbness, and paresthesiae are followed by progressive weakness, depression of deep tendon reflexes in the lower limbs and, in severe cases, in the upper limbs. Signs include a high-stepping gait from bilateral foot drop and, in severe cases, quadriplegia with foot and wrist drop, as well as pyramidal signs. In time, there might be significant recovery of peripheral nerve function but, depending on the degree of pyramidal involvement, spastic ataxia may be permanent (Lotti and Moretto, 2005). Human and experimental data indicate that recovery is usually complete in the young.

Delayed neuropsychological sequelae after OP poisoning

Since acute poisoning with OP insecticides can cause major effects, such as convulsions, respiratory failure, and cardiac arrhythmias, all of which can result in cerebral anoxia, it would not be surprising if severe acute OP poisoning was associated with long-term neurological sequelae. Moreover, there is now evidence that such complications can occur (Sánchez-Santed et al., 2004).

Rosenstock et al. (1991) and McConnell et al. (1994) investigated the neuropsychological effects and, as an index of delayed neuropathy, measured vibrotactile thresholds in 36 male agricultural workers in Nicaragua who had been admitted to hospital between 10 and 34 months earlier with occupationally related acute OP poisoning: 21 of 36 workers had been poisoned with methamidophos, a recognized peripheral neurotoxin. In a battery of neuropsychological tests the exposed group performed significantly worse than the control group and a significant decrease in vibrotactile sensitivity was also observed.

Steenland et al. (1994) studied 128 men exposed to OP insecticides, the majority occupationally, who had sought medical attention and of whom 28% had been admitted to hospital in California for more than one night. The study group was divided into two subgroups:

"probable" cases (45/128) who had symptoms following OP exposure but in whom cholinesterase activity was not measured, and "definite" cases (83/128) that had both symptoms and a red cell or plasma cholinesterase activity $\leq 20\%$ of normal activity. Neurobehavioral tests, nerve conduction tests, and vibrotactile threshold responses were undertaken several years after exposure. The study group performed significantly worse in tests of sustained visual attention and mood and those in the "definitely" poisoned subgroup also showed significantly worse vibrotactile sensitivity of finger and toe, but had no neurological symptoms.

Savage et al. (1988) investigated the presence of chronic neurological or neuropsychological abnormalities in 100 subjects who had experienced at least one episode of acute OP poisoning, confirmed by a physician. The authors found significant deficits in a wide range of neuropsychological variables, including visuomotor, attention, and language function. Persistent abnormalities in affective behavior, especially anxiety, were also found, though no differences on EEG or neurological examination were identified.

Miranda et al. (2002b) studied a cohort of men who were exposed predominantly occupationally to an OP insecticide and who had been admitted for at least one day to two hospitals in Nicaragua for treatment of OP poisoning. Data on AChE activity were available only in some 38% of exposed individuals and in 28% of controls. Quantitative tactile vibration thresholds were measured in 56 subjects on the first occasion 1–24 days (median 7) after exposure and for the second time 24–128 days (median 50) after poisoning and compared to 39 nonexposed men. The study group had their symptoms classified as mild ($n=0$), moderate ($n=35$), or severe ($n=21$). Big-toe vibrotactile thresholds increased markedly from the first to the second examinations in those with severe intentional poisoning (ingestion) due to neuropathic OPs (methamidophos, chlorpyrifos). No significant impairment of vibrotactile thresholds was detected in association with occupational poisonings or with less severe intentional poisonings with neuropathic or nonneuropathic OPs.

In a further study (Miranda et al., 2002a) of the same Nicaraguan population ($n=59$), marked impairment of grip and pinch strength in the dominant hand was found, particularly among those with severe poisoning due either to neuropathic OPs or intentional ingestion. In a third study (Miranda et al., 2004), hand strength and vibration thresholds were performed in 48 men 2 years after admission for acute OP poisoning. In those exposed to nonneuropathic OPs, grip and strength had recovered and was not different from controls; pinch strength remained lower than for controls. Those only moderately poisoned with neuropathic OPs showed an evolution similar to those exposed to nonneuropathic OPs, whereas those severely poisoned (notably, as a result of intentional ingestion) still remained significantly weaker than controls. Index finger and toe vibration thresholds were slightly increased at the end of the 2-year period.

In a fourth study (Delgado et al., 2004) ($n=53$), changes in immediate verbal memory, visuomotor performance, and neuropsychiatric symptoms at the time of discharge, 7 weeks postpoisoning and at 2 years were evaluated. Immediate verbal learning showed deficits in the high-exposure group, particularly at the time of discharge. Visuomotor performance showed a deficit at 7 weeks but improved thereafter. A 2-year delayed excess of neuropsychiatric symptoms was present in those occupationally exposed. The authors concluded that visuomotor performance and possibly short-term verbal memory were affected early after severe acute OP poisoning and recovered, either truly or by some compensatory mechanism. Neuropsychiatric symptoms seemed to increase after a longer latent period, particularly in the less severely exposed subjects.

However, shortcomings in the methodology of these studies, including poor assessment of exposures and of the severity of poisoning, different timing of examination after poisoning, and choice of controls, do not allow a firm conclusion to be drawn that severe acute OP poisoning can produce subclinical damage to the CNS and peripheral nervous system.

Neuropsychological effects of low-level OP exposure

There remains controversy, however, as to whether low-level exposures to OP insecticides (without the features of acute poisoning or the cholinergic syndrome having occurred) can result in peripheral nerve dysfunction. We are still of the view (Lotti, 2002) that a single pattern of subclinical disturbances that relates low-level OP exposure to human peripheral neuropathy has not been identified. The clinical characteristics of the mild neuropathies described point to a condition that differs from OP-induced delayed polyneuropathy. The distribution of involved peripheral nerves is unusual for a toxic polyneuropathy, including OP-induced delayed neuropathy, because both upper and lower limbs seem to be affected at once. In some studies, defects seem irreversible, because they were detected several months after cessation of exposure. Although OPs are rapidly cleared from the body and peripheral nerves regenerate, it is not clear what mechanism would sustain a peripheral neuropathy for a long time after cessation of exposure. Therefore, in the absence of histopathology and follow-up studies, the significance of these changes remains unclear. These difficulties in interpretation are further amplified by the lack of an experimental model other than the classic OP-induced delayed polyneuropathy one. These mild, inconsistent, and unexplained findings provide no evidence to support the proposal that low-level exposures to OPs caused peripheral neuropathy in humans, even among possibly hypersusceptible individuals.

On the other hand, Mackenzie Ross et al. (2013), having reviewed the available evidence concerning the impact of low-level occupational exposure to OPs (by conducting a meta-analysis of 14 studies involving more than 1600 participants), found a significant association between "low-level exposure to OPs and impaired neurobehavioural function which is consistent, small to moderate in magnitude and concerned primarily with cognitive functions such as psychomotor speed, executive function, visuospatial ability, working and visual memory."

DIAGNOSIS AND INVESTIGATIONS

The diagnosis of OP poisoning is based on the patient's history, clinical presentation, and laboratory tests. In a patient with a positive history, a typical smell on the breath, characteristic symptoms, and depressed erythrocyte AChE activity, diagnosis is not difficult to make. Unfortunately, the history is often unobtainable and in one study was missing in 36% of all cases (Hayes et al., 1978). Moreover, the clinical features of OP poisoning may not be recognized as such if the patient presents, for example, with heart block, gastroenteritis, convulsions, or ketoacidosis. An awareness of this diversity of presentation is the first step to accurate diagnosis.

The activity of two enzymes may be measured to confirm a diagnosis of OP poisoning. These are red-cell AChE and plasma butyrylcholinesterase (plasma cholinesterase; plasma pseudocholinesterase: BuChE). Both are surrogates for AChE activity in the CNS and peripheral nervous system. The erythrocyte AChE activity is invariably more specific than plasma BuChE activity as a marker of OP insecticide exposure, though some OP insecticides (e.g., chlorpyrifos, demeton, malathion) depress plasma BuChE activity to a greater degree than erythrocyte cholinesterase activity.

Usually a good correlation exists between the levels of erythrocyte AChE inhibition and the severity of symptoms in OP poisoning (Lotti, 1995). Hence, it is not necessary to wait for confirmation from the laboratory for the diagnosis of acute OP poisoning to be confirmed; specific treatment should be started immediately.

MANAGEMENT

All cases of OP poisoning should be dealt with as an emergency and all patients with more than minor symptoms should be admitted to a critical care unit as quickly as possible.

Once the patient is stabilized, gastric lavage may be considered in all potentially serious cases if ingestion has occurred less than 1 hour previously, although its value is unproven (Li et al., 2009). If performed, it should be undertaken with care and with an endotracheal tube *in situ* if consciousness is depressed, as hydrocarbons are present in many OP insecticide formulations. Induced emesis should be avoided as it is dangerous in a patient whose level of consciousness might deteriorate; aspiration pneumonia is a well-recognized hazard and can also occur in conscious patients given emetics when hydrocarbons are present in the pesticide formulation.

The capacity of activated charcoal to adsorb most OP compounds has not yet been demonstrated. In a study in Sri Lanka involving OP and carbamate insecticide poisoning, no benefit from single- or multiple-dose charcoal was demonstrated (Eddleston et al., 2008).

If dermal exposure is the major route of exposure, contaminated clothes should be removed and stored safely, without carers themselves being contaminated. Contaminated skin should be washed thoroughly with alkaline soap and cold water.

Ventilation must be maintained. Rhinorrhea and bronchorrhea require prompt relief with intravenous atropine (see below) and supplemental oxygen should be given to maintain $Pa\text{O}_2 > 10$ kPa. If these measures fail, the patient should be intubated and mechanical ventilation instituted. Careful attention must be given to fluid and electrolyte balance and adjustments to infusion fluids made as necessary. Heart rate, blood pressure, electrocardiogram (ECG), and arterial blood gases should be monitored routinely.

Hypotension, cardiac arrhythmias, and convulsions should be treated conventionally. The management of convulsions and muscle fasciculation with benzodiazepines is discussed below.

Antidotes

The evidence base for atropine, oximes, and diazepam has been reviewed (Blain, 2011). While the therapeutic combination of oxime and atropine is well established in the treatment of OP pesticide poisoning, diazepam may also be of benefit by reducing anxiety, restlessness, and muscle fasciculation, suppressing seizures, and reducing morbidity and mortality when used in conjunction with an oxime and atropine.

ATROPINE

Atropine competes with ACh for a common binding site on the peripheral muscarinic receptor, thus effectively antagonizing its actions in producing increased tracheobronchial and salivary secretions, bronchoconstriction, and bradycardia. Atropine sulfate 2 mg (0.02–0.1 mg/kg in a child) intravenously should be given as soon as possible in moderately and severely poisoned adults and may be life-saving in those with rhinorrhea and bronchorrhea. If the initial dose produces only a partial response, it should be doubled and doubled again if there is only a limited clinical response (Eddleston et al., 2004a). The dose should be titrated to control rhinorrhea and bronchorrhea, to raise the pulse rate above 80 bpm, and restore systolic blood pressure to more than 80 mmHg. In severe cases, several hundred mg may be required, though in a study in Sri Lanka the mean dose was 23.4 mg (standard deviation (SD) 22.0, range 1–75 mg) (Eddleston et al., 2004a).

The peripheral antimuscarinic effects of atropine may not be the only antidotal property of the drug; it may also be of value in treating the acute dystonic reactions occasionally observed (Smith, 1977; Joubert et al., 1984; Joubert and Joubert, 1988a; Wedin, 1988).

Oximes

The fundamental action of the pyridinium oximes is to reactivate AChE inhibited by OPs, thus allowing ACh to be hydrolyzed in the usual way and resumption of normal cholinergic neurotransmission. It is generally held that the beneficial effects of oximes in OP poisoning are confined to peripheral nicotinic sites and that central nervous effects are clinically insignificant (Bismuth et al., 1992), although there is some evidence that pralidoxime crosses the blood–brain barrier (Sakurada et al., 2003). This means that the beneficial effects of oximes are mainly on neuromuscular transmission and that there is little action on parasympathetic effects such as bronchorrhea, bronchoconstriction, and rhinorrhea, or on CNS effects.

The reactivation of inhibited AChE and clinical improvement depends on:

- the chemical form of inhibited AChE. While spontaneous reactivation of dimethyl phosphorylated AChE proceeds rapidly, in the case of diethyl phosphorylated AChE it is very slow. Spontaneous reactivation of diethyl phosphorylated AChE can be accelerated by oximes
- the plasma concentration of the OP insecticide (Willems et al., 1993; Worek et al., 1997)
- aging which results from monodealkylation of dialkoxyphosphorylated AChE (Karalliedde and Szinicz, 2001); aged enzyme reactivates neither spontaneously nor under the influence of oximes so that recovery depends on synthesis *de novo* of more enzyme
- the plasma oxime concentration
- the duration of oxime therapy.

With the typical dimethoxy and diethoxy structure of pesticides, aging rates only become a problem when treatment is delayed and/or exposure prolonged. It is commonly, but erroneously, believed that one day after intoxication with an OP insecticide virtually all the phosphorylated enzyme will be aged so that oxime therapy will be ineffective by this time. However, this interpretation derives from *in vitro* studies in which AChE is rapidly inhibited and is maintained fully inhibited by an excess of inhibitor and in the absence of oxime throughout. These experiments do not represent *in vivo* circumstances and should not be a reason to abandon oxime therapy after 24 hours.

The role of oximes

Pawar et al. (2006) administered two pralidoxime-dosing schedules in 200 patients who had moderately severe OP insecticide poisoning. Patients ($n = 100$) who received the high-dose regimen (pralidoxime iodide 2 g loading dose over 30 minutes, then 1 g/h for 48 hours and then 1 g every 4 hours until weaned from the ventilator) had a lower mortality (1% vs 8%) and less need of intubation and ventilator support than the 100 patients receiving a lower-dose regimen (pralidoxime iodide 2 g loading dose, then 1 g every 4 hours until weaned). They also developed less muscle weakness and required less atropine (median 6 mg vs 30 mg) during the first day and fewer developed pneumonia.

In contrast, De Silva et al. (1992) concluded that nothing is to be gained by the addition of an oxime to the standard regimen of atropine plus mechanical ventilation based on a study in which 21 patients received atropine alone, and 24 atropine plus pralidoxime chloride in a median dose of 4 g (range 3–7 g) intravenously during the first 24 hours of treatment, then 1 g daily for up to 5 days (median 2–5 g, range 0–5 g). The mortality in both groups was 29%.

Eddleston et al. (2009) came to a similar conclusion after performing a double-blind, randomized, placebo-controlled trial comparing pralidoxime chloride (2 g loading dose, followed by a constant infusion of 0.5 g/h for up to 7 days) with saline. Despite clear reactivation of red-cell AChE activity in diethyl OP pesticide-poisoned patients, there was no evidence of improved survival or reduced need for intubation (Eddleston et al., 2009). The reason may lie in the increasing evidence that the solvents in formulations are primarily responsible for the high morbidity and mortality and against which oximes would have no impact (Eddleston et al., 2010).

The value of oximes has been challenged in a systematic review of clinical trials (Eddleston et al., 2002) and in a Cochrane review (Buckley et al., 2011), which concluded that:

> *Current evidence is insufficient to indicate whether oximes are harmful or beneficial in the management of acute organophosphorus pesticide poisoning. The World Health Organization recommended pralidoxime regimen (30 mg/kg pralidoxime chloride bolus followed by 8 mg/kg/h infusion) is not supported by outcomes to date. Lower doses have also shown worse outcomes in one RCT [randomized controlled trial]. Thus the published RCTs provide limited guidance for clinicians. Despite this there are consistent animal data supporting their effectiveness, when given early. Based on our understanding of the mechanism,* in vitro *and animal data, benefit–risk ratio is more likely to be favourable when they are given early, to patients with serious poisoning by diethyl OPs. However, the RCTs have not defined effective doses or sub-groups that are likely to benefit.*

Why might oximes be ineffective? In an important study performed in mini-pigs using orally administered clinically relevant doses of agricultural emulsifiable concentrate (EC) dimethoate, dimethoate active ingredient alone, or solvents, Eddleston et al. (2012) found that administration of agricultural dimethoate EC, but not saline, caused respiratory arrest within 30 minutes, severe distributive shock, and neuromuscular joint dysfunction, that was similar to human poisoning. Moderate toxicity resulted from poisoning with dimethoate active ingredient alone, or the major solvent cyclohexanone. Combining dimethoate with cyclohexanone reproduced severe poisoning characteristic of agricultural dimethoate EC poisoning. A formulation without cyclohexanone showed less mammalian toxicity. These results indicate that solvents play a crucial role in OP (and specifically dimethoate) toxicity, which could explain why oximes seem to be less effective clinically than in experimental studies where pure OP insecticide rather than marketed formulations are often employed.

Pralidoxime

Early experiments in anesthetized cats (possibly seven animals) given lethal doses of intravenous sarin and pralidoxime mesylate 10 mg/kg intramuscularly, but not atropine, suggested that plasma pralidoxime concentrations of above 4 mg/L were required to counteract neuromuscular block *in vitro*, and bradycardia, hypotension, and respiratory failure *in vivo* (Sundwall, 1961). Willems et al. (1993) reported that parathion-ethyl and parathion-methyl poisoning could be treated effectively with pralidoxime methylsulfate (plasma concentrations approximately 4 mg/L) and atropine provided that the pesticide concentrations were <30 μg/L. In more severe cases, even plasma pralidoxime concentrations of 14.6 mg/L did not produce improvement. In dimethoate poisoning, AChE could not be reactivated at concentrations of up to 20 mg/L. Data from a case report suggest that pralidoxime concentrations of >40 mg/L may be required to produce reactivation of inhibited AChE (Casey et al., 1995a, b).

These studies support the recommendation that plasma oxime concentrations of far more than 4 mg/L are required to produce AChE reactivation and substantial clinical improvement in severe OP poisoning. Based on clinical studies (Willems and Belpaire, 1992; Willems et al., 1993; Casey et al., 1995a), it has been recommended that pralidoxime chloride 30 mg/kg should be administered intravenously as soon as possible in any severe case. An intravenous infusion of 8–10 mg/kg/h should then be continued. As a rule, pralidoxime should be continued for as long as atropine is required, that is, until clear clinical improvement is achieved and maintained even after discontinuing the infusion, a process which may take many days while residual insecticide is cleared from body stores.

Obidoxime

Based on pharmacokinetic data of obidoxime in healthy human volunteers (Sidell et al., 1972), Thiermann et al. (2010) considered that a bolus dose of obidoxime 250 mg intravenously (IV), followed by a continuous infusion at 30 mg/h, was an appropriate therapeutic regimen, which they employed in single patients (Eyer, 1996; Thiermann et al., 1997). The efficacy and safety of this regimen were evaluated in an observational study (Eyer et al., 2009; Thiermann et al., 2009). These authors further investigated this regimen in 34 severely poisoned patients (Thiermann et al., 2010). The mean (± SD) total dose of obidoxime infused over 65 (±55) hours was 2269 (±1726) mg, which resulted in a mean steady-state plasma concentration of 5.2 (±2.6) mg/L. Reactivation of inhibited erythrocyte AChE was achieved in 23 of 34 patients, while in 7 patients aging was complete before the commencement of obidoxime. In 4 patients high OP insecticide load prevented significant reactivation and complete aging ensued in spite of this obidoxime regimen.

Thiermann et al. (1997) reported that, in parathion poisoning, obidoxime (250 mg IV as a bolus followed by 750 mg/day by infusion) was effective, but that in severe poisoning AChE reactivation did not occur until the concentration of inhibitor in the plasma had declined.

The same dosage regimen was ineffective with oxydemeton methyl when oxime therapy was delayed more than one day after poisoning. In another series, Thiermann et al. (1999) reported that, in parathion poisoning, reactivation was possible 7 days after poisoning whereas with oxydemeton methyl a response was only seen when obidoxime therapy was instituted soon after poisoning. Similarly, Zilker et al. (1997) found that obidoxime (750 mg/day by infusion) reduced the need for atropine in parathion poisoning, but that demeton-S-methyl poisoning only responded if obidoxime therapy was instituted shortly after intoxication.

HI-6

Kusić et al. (1991) reported a case series in which HI-6 was administered four times daily as a single intramuscular injection of 500 mg in conjunction with atropine and diazepam therapy. The oxime was started on admission and continued for a minimum of 48 hours and a maximum of 7 days. HI-6 rapidly reactivated human red blood cell AChE inhibited by diethoxy OPs (phorate, pyridaphenthion, and quinalphos), as well as that inhibited by dichlorvos. With the exception of dichlorvos, enzyme inhibited by dimethoxy OPs (dimethoate, phosphamidon) was generally resistant to treatment by HI-6.

Conclusions

In experimental *in vivo* studies where pure OP insecticide rather than a marketed formulation is usually administered, oximes administered in appropriate dose have been shown to have proven efficacy in reversing inhibition of erythrocyte AChE and producing clinical improvement. In contrast, in some clinical studies, no such improvement has been demonstrated, even when high-dose pralidoxime therapy has been administered. Studies by Eddleston et al. (2012) suggest that this is because the solvents in commercial formulations play a crucial role in enhancing toxicity. Despite this observation, oximes should still be administered in acute OP poisoning.

Benzodiazepines

A large number of anticonvulsants have been studied in animals or used in OP poisoning (Sellström, 1992), but diazepam is the one that has been investigated most for use in OP pesticide poisoning (Marrs, 2003). In a variety of pharmacodynamic studies in experimental animals, it was concluded that diazepam adds to the benefits of atropine and of the classic combination of atropine and oximes (Rump and Grudzinska, 1974; Rump et al., 1976; Kleinrok and Jagiello-Wojtowicz, 1977; Gupta, 1984; Krutak-Krol and Domino, 1985; Bokonjic et al., 1987; Kassa and Bajgar, 1994). Experimentally, diazepam prevents, stops, and/or ameliorates convulsions induced by OP insecticides and renders less severe or even prevents related structural changes in the brain. However, the effects on lethality are less clear (Marrs, 2003).

In most reports, the indication for the use of diazepam was convulsions, but it has also been used to control muscle fasciculation. In the case of convulsions, the adult dose is 10–20 mg IV, while that in children is 0.2–0.3 mg/kg IV. If large doses of diazepam are required to suppress seizure activity, phenytoin should be considered as an alternative (Johnson and Vale, 1992). Diazepam in doses of 5–10 mg IV has also been recommended in the absence of convulsions in cases where there is anxiety and restlessness (Johnson and Vale, 1992).

Prevention and treatment of intermediate syndrome

If intermediate syndrome is a consequence of ACh accumulation at the neuromuscular junction, oximes should protect against its development. It is noteworthy, and possibly relevant, that pralidoxime methylsulfate protected against development of muscle fiber necrosis (Cavaliere et al., 1998). Johnson et al. (1996) reported a prospective trial with two treatment regimens of pralidoxime in the treatment of patients with OP poisoning. Seventy-two adults with a history of consumption of these compounds and requiring intensive care were entered into the trial. They were then were randomized using a block randomization to receive either a single bolus of 1 g pralidoxime at admission followed by placebo infusion over the next 4 days or a single placebo bolus at admission followed by pralidoxime 12 g as a continuous infusion over the next 4 days. A higher prevalence of intermediate syndrome was observed in the latter group. Comparison of patients who received at least 1 g of pralidoxime within 12 hours of ingestion of the OP poison with those who received pralidoxime after 12 hours suggested that the time of administration of pralidoxime was important; the first group were less likely to develop the syndrome.

Other case reports have suggested that high dose of oxime may not avert intermediate syndrome. In a case of severe malathion self-poisoning, pralidoxime 1 g was administered IV approximately 7 hours after poisoning and at 12 hours an infusion of pralidoxime 400 mg/h was given. The patient developed intermediate syndrome on hospital day 3 (Sudakin et al., 2000).

Carbamate insecticides

They are derivatives of *N*-methyl carbamic acid and have the general formula $ROC(O)NHCH_3$ (Fig. 10.2). In most

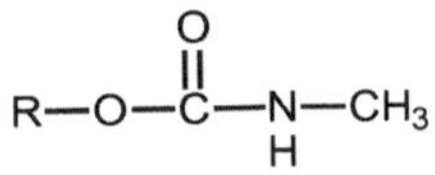

Fig. 10.2. General formula of *N*-methyl carbamates.

instances R is a substituted phenol or heterocycle, notable exceptions being aldicarb, which has an *N*-alkyl R group, and pirimicarb, which has two *N*-methyl groups. They are generally poorly soluble in water but highly soluble in polar organic solvents such as methanol, ethanol, and acetone.

MECHANISMS OF TOXICITY

Carbamates act by inhibiting AChE. This is achieved by carbamylation of a serine hydroxyl residue at the active site of the enzyme, a process that involves cleavage of the carbamate molecule, which in effect is treated by the enzyme as if it were ACh (Alvares, 1992). AChE inhibition results in accumulation of ACh at synapses in the CNS, at neuromuscular junctions and at parasympathetic effector sites. AChE activity is restored when spontaneous hydrolysis of the carbamylated enzyme occurs. This ensues very rapidly in comparison to reversal of the interaction of AChE with OP insecticides. Moreover, "aging" does not occur and many carbamates do not cross the blood–brain barrier readily, so the effect on cerebral AChE activity is considerably less than that caused by OP insecticides.

Poisoning with pesticides almost always involves commercial formulations, not isolated active ingredients, and the additional chemicals they contain not infrequently contribute to toxicity (Hu et al., 2010; Gil et al., 2012, 2013; Huan et al., 2013).

CLINICAL FEATURES

Acute exposure

Systemic toxicity may occur following ingestion, inhalation, or skin contamination and severe poisoning of a neonate by transplacental transfer has been reported with propoxur (Sarkar et al., 1994). Although carbamate insecticide poisoning is generally less severe than OP insecticide poisoning, the toxicity of individual members of the group varies and they should not be considered toxicologically benign; acute poisoning with them can be severe and fatal (Zwiener and Ginsburg, 1988; Tsao et al., 1990; Tay et al., 1998; Ameno et al., 2001; Pinakini and Kumar, 2006). A retrospective comparison of 17 methomyl poisonings with 42 World Health Organization class I OP insecticide poisonings (methidathion, dichlorvos, phosphamidon, parathion) (Lee et al., 2011) showed the mortality in each group to be comparable (17.6% and 9.5% respectively; $p=0.339$). However, the features and causes of death differed dramatically. Seven patients in the methomyl group presented in cardiac arrest rather than in cholinergic crisis and 3 died of subsequent multiorgan failure. In contrast, no patient poisoned with an OP insecticide presented in this manner; the 4 who died did so from respiratory complications such as acute respiratory distress syndrome and pneumonia (Lee et al., 2011). In another series of 52 carbamate-poisoned patients, 35 developed medical complications, principally respiratory failure, within 24 hours and 7 (13.5%) died. An initial Glasgow Coma Score of 13 or less was the most reliable predictor of complications (Moon et al., 2012).

INGESTION

As carbamate insecticides do not require bioactivation, toxicity manifests rapidly, usually within 1 hour after dosing in animals (Padilla et al., 2007). Poisoned humans also usually develop cholinergic symptoms within a few minutes (Miyazaki et al., 1989; Ragoucy-Sengler et al., 2000; Hoffmann et al., 2008) and in the most severe cases muscle twitching, profound weakness, profuse sweating, incontinence, mental confusion, and progressive cardiac and respiratory failure ensue (Liddle et al., 1979; Miyazaki et al., 1989; Tracqui et al., 2001). In less severe cases, cholinergic symptoms are usually evident within 2 hours and typically resolve within 24 hours (Lifshitz et al., 1997). There is some evidence that CNS depression in moderate carbamate poisoning is more common in young children than adults, possibly due to a greater blood–brain barrier permeability in infants (Lifshitz et al., 1997). Coma may persist for 18–24 hours (Miyazaki et al., 1989). Seizures (Hoffmann et al., 2008) are relatively uncommon as a primary complication in severe carbamate poisoning since carbamate penetration into the CNS is limited, but they may occur secondary to hypoxia.

The effect of carbamates on heart rate is unpredictable and tachycardia occurs as commonly as the bradycardia that one would predict from the mechanism of action. Arrhythmias, including torsades de pointes, are recognized (Saadeh et al., 1997). Complete heart block has been reported following carbaryl poisoning (Siegal et al., 2009). Acute respiratory distress syndrome developed in a child poisoned with propoxur (Kumar et al., 2014).

There has been only one report of the intermediate syndrome following carbamate (carbofuran) poisoning in a patient who abused alcohol chronically (Paul and Mannathukkaran, 2005). The diagnosis was based on the development of dyspnea (respiratory rate of 40/minute and fall in oxygen saturation in the absence of signs of pulmonary infection) and weakness of neck flexors. However, there was no laboratory confirmation and cholinesterase activity was not measured, so it cannot be certain that the patient was poisoned with a carbamate.

Of 37 adults with acute carbamate insecticide poisoning, 9 had glycosuria alone and 5 a combination of glycosuria and hyperglycemia (random blood glucose >200 mg/dL). Glycosuria and hyperglycemia persisted for a mean of 2.25 days and 2.75 days respectively (Sudhir et al., 2013). Acute pancreatitis complicates acute carbamate ingestion rarely (Petkovska et al., 2000), but has been described with aldicarb (Moritz et al., 1994), carbofuran (Rizos et al., 2004), methomyl (Brahmi et al., 2005; Makrides et al., 2005), and propoxur (Votanopoulos et al., 2007), and may contribute to hyperglycemia (Weizman and Sofer, 1992). Fulminant hepatic failure developed in one man who consumed methomyl (Agha et al., 2010).

Death has occurred within a few hours in untreated cases (Thierauf et al., 2009) and is usually due to respiratory failure (Liddle et al., 1979; Tsatsakis et al., 1996). Pulmonary edema and hypoxic brain damage are the main findings at autopsy (Liddle et al., 1979; Zwiener and Ginsburg, 1988; Miyazaki et al., 1989; Tsao et al., 1990; Al-Samarraie et al., 2009). Organs have been transplanted from patients who died from carbamate insecticide poisoning (Garcia et al., 2010; Martinez-Mier et al., 2010); one was hyperacutely rejected (Chan et al., 2012).

Inhalation

Exposures in pesticide applicators are typically inhalational, though the dermal route may also be involved. Life-threatening poisoning in these circumstances is rare, though cholinergic symptoms have occurred following equipment malfunction or failure to observe appropriate protective measures (Simpson and Bermingham, 1977; McConnell and Hruska, 1993; Testud et al., 1999). In these cases the rapid (often within 30 minutes) onset of mild cholinergic features (nausea, headache, sweating, lacrimation, salivation, chest tightness, coughing, bronchorrhea, and blurred vision caused by constriction of one or both pupils) serves to alert affected individuals to remove themselves from exposure (Simpson and Bermingham, 1977; Coleman et al., 1990; Testud et al., 1999). Resolution of symptoms occurs typically within hours (Coleman et al., 1990), often without the need for atropinization (Testud et al., 1999). If exposure continues, vomiting, diarrhea, abdominal pain, muscle fasciculation, and weakness develop (Testud et al., 1999). Hypotension, sinus tachycardia, or bradycardia and dyspnea may also occur (Testud et al., 1999). ECG T-wave changes, including inversion, are recognized (Saiyed et al., 1992; Lim et al., 2010). Stevens–Johnson syndrome has been reported (Lim et al., 2010).

Severe sequelae following occupational carbamate exposure are rare, have often involved methomyl, and include several cases resulting from pilots of crop-spraying aircraft being exposed to carbamate spray that gained entry to the cockpit (Richardson and Batteese, 1973; Driskell et al., 1991; Cable and Doherty, 1999). One of these exposures (Driskell et al., 1991) proved fatal as a result of the pilot crashing the plane.

Dermal exposure

Between January 2002 and August 2004, 13 farm workers were admitted to one emergency department with nausea, vomiting, headache, dizziness, weakness, and blurred vision after mixing seeds in liquid carbofuran. Their hands were red on presentation, three were agitated, one lethargic and the most common laboratory abnormality was hyperglycemia, which was found in six (Satar et al., 2005).

Delayed neurological sequelae

Carbamate poisoning is generally held not to be associated with delayed polyneuropathy such as is seen with some OP insecticides, but three cases have been reported involving ingestion of carbaryl (Dickoff et al., 1987), m-tolyl methylcarbamate (Umehara et al., 1991) and carbofuran (Yang et al., 2000). The first developed a sensorimotor neuropathy (Dickoff et al., 1987), while the second had loss of large myelinated fibers and axonal degeneration on sural nerve biopsy (Umehara et al., 1991). The third (Yang et al., 2000) developed arm and foot sensory symptoms and muscle weakness on the sixth day after hospitalization; the sensory component resolved rapidly but motor features accompanied by abnormal electrophysiological studies persisted for several months. Lotti and Moretto (2006) concluded that high doses of carbamate may cause a sensorimotor polyneuropathy in humans similar to OP insecticide-induced delayed polyneuropathy. Moreover, carbamates may exacerbate polyneuropathy initiated by prior dosing with an OP insecticide or by underlying subclinical axonal damage.

Peripheral neuropathy was also seen in 3 out of 36 patients aged 14–70 years poisoned with carbamates, 80% of which involved carbofuran (Winnik et al., 1997). Though it is not clear whether this finding was carbamate-induced or predated the poisoning, other common causes of neuropathy, especially ethanol, were carefully excluded, leaving carbamates as the probable explanation (Winnik et al., 1997).

A 41-year-old woman who ingested some 200 mL of a formulation containing methomyl and alphamethrin required cardiopulmonary resuscitation shortly after ingestion (Hu et al., 2010). Four days later she could barely identify objects within a foot (30 cm) of her. Magnetic resonance imaging (MRI) scanning revealed bilateral occipital-lobe and basal ganglia abnormalities that indicated her blindness was cortical in nature. Her visual acuity improved considerably over the next few months. To add to her problems, however, she developed neurological symptoms in both lower limbs 3 weeks after ingestion and electrophysiological studies showed changes consistent with an axonal neuropathy of both peroneal nerves.

Yang et al. (2000) also reported a patient who developed sensorimotor neuropathy following ingestion of carbofuran. Recovery commenced after 1 week and continued for 4 months. Pontine myelinolysis, confirmed by MRI, has been described in a child poisoned with carbaryl; it improved after 4 months and disappeared after 2 years (Santinelli et al., 2006).

DIAGNOSIS AND INVESTIGATIONS

Many laboratories are not acquainted with the special procedure required to assay red blood cell cholinesterase in the presence of carbamates. Samples must be kept on ice or frozen at –20 °C during transportation to the laboratory and prior to analysis. Spontaneous reversal of the inhibition is rapid and is accelerated by the time interval between sampling and analysis, dilution of the sample, the addition of substrate (usually ACh at high concentration, which competes successfully for the active site on the enzyme), and duration of the assay. The latter should, therefore, take less than 3 minutes and employ minimal dilution and minimal amounts of substrate (Ecobichon, 2001). Although red-cell AChE activity correlates well with the severity of intoxication, its measurement adds little to management of carbamate poisoning, as the majority of patients recover rapidly.

MANAGEMENT

All cases of carbamate poisoning should be dealt with as an emergency and admitted to hospital as quickly as possible. Severe cholinergic features necessitate supportive care and atropine administration, most appropriately in a critical care unit (Baban et al., 1998; Tay et al., 1998) if significant morbidity and mortality is to be prevented.

Bronchorrhea requires removal of secretions by suction and intravenous atropine should be given as below. Supplemental oxygen should be given to maintain arterial $Pao_2 > 10$ kPa (75 mmHg). If these measures fail the patient should be intubated and mechanical ventilation instituted. The use of succinylcholine might be best avoided; it caused life-threatening trismus in an 18-year-old woman poisoned with methomyl (Ouanes et al., 2009). Heart rate, blood pressure, ECG, and arterial oxygen saturation or blood gas tensions should be monitored. Convulsions should be treated conventionally with a benzodiazepine.

Gastric lavage may be considered in all potentially serious cases if ingestion has occurred less than 1 hour previously, although its value is unproven. If performed, it should be undertaken with care and with an endotracheal tube *in situ* if consciousness is depressed as hydrocarbon solvents are often present in the formulations. Alternatively, oral activated charcoal (50–100 g in an adult) may be given in potentially serious cases if ingestion has occurred less than 1 hour previously, though the capacity of activated charcoal to adsorb most carbamate compounds (and other chemicals in the carbamate formulation) has not yet been demonstrated. In a study in Sri Lanka involving carbamate and OP insecticide poisoning, no benefit from single- or multiple-dose charcoal was demonstrated (Eddleston et al., 2008).

If dermal exposure is the major route of exposure, contaminated clothes should be removed and stored safely, without carers themselves being contaminated. Contaminated skin should be washed thoroughly with soap and cold water.

Ventilation must be maintained. Rhinorrhea and bronchorrhea require prompt relief with intravenous atropine (see below) and supplemental oxygen should be given to maintain $Pao_2 > 10$ kPa. If these measures fail, the patient should be intubated and mechanical ventilation instituted. Careful attention must be given to fluid and electrolyte balance and adjustments to infusion fluids made as necessary. Heart rate, blood pressure, ECG, and arterial blood gases should be monitored routinely.

Hypotension, cardiac arrhythmias, and convulsions should be treated conventionally.

Antidotes

Atropine

Atropine competes with ACh for a common binding site on the muscarinic receptor, thus effectively antagonizing its actions in producing increased tracheobronchial and salivary secretions, bronchoconstriction, and bradycardia. A dose of atropine 2 mg IV (0.02 mg/kg body weight for a child) should be given as soon as possible in moderately or severely poisoned patients, particularly in those who develop rhinorrhea and bronchorrhea. If the initial dose produces only a partial response, it should be doubled every 5–10 minutes until clinical improvement is evident (Eddleston et al., 2004b). The dose should be titrated to control rhinorrhea and bronchorrhea, raise the pulse rate above 80 bpm, and restore systolic blood pressure to more than 80 mmHg.

Oximes

In vitro studies by Eckert et al. (2008) demonstrated that HI-6, and to a lesser extent obidoxime, could accelerate the rate of decarbamylation of human erythrocyte AChE inhibited by physostigmine or pyridostigmine. In some animal studies (Carpenter et al., 1961; Sanderson, 1961; Natoff and Reiff, 1973; Sterri et al., 1979; Harris et al., 1989) pralidoxime increased the toxicity of carbaryl while in others (Mercurio-Zappala et al., 2007) it protected against poisoning. More importantly, some human carbamate poisonings (Zwiener and Ginsburg, 1988; Burgess et al., 1994; Ekins and Geller, 1994; Goswamy et al., 1994; Nelson et al., 2001) have benefited from pralidoxime, though there has been at least one death despite its use (Farago, 1969). At present there is insufficient evidence to either recommend or prohibit the use of pralidoxime in severe poisoning with carbamate insecticides. It should seldom need to be considered in most cases since carbamates have a shorter duration of action than OP insecticides. However, if intoxication is life-threatening and unresponsive to atropine and supportive measures, pralidoxime chloride 30 mg/kg body weight can be given by intravenous injection over 5–10 minutes, followed by an infusion of 8–10 mg/kg/h; alternatively repeated bolus doses may be given 4–6-hourly. Patients poisoned with mixtures of carbamates and OP compounds should be treated as for OP poisoning.

PREVENTION OF CHOLINERGIC TOXICITY IN OCCUPATIONAL EXPOSURES TO INSECTICIDES

Occupational exposures to OPs are commonly monitored by measuring urinary excretion of alkylphosphates, activities of blood cholinesterases or both, whereas exposures to carbamates are monitored by blood cholinesterase activities and, rarely, by urinary excretion of metabolites (Maroni et al., 2000a, b).

Numerous field studies have assessed exposures to OPs by means of urinary alkylphosphates but their toxicological significance is unclear and they should be considered simply as an evidence of exposure. The reasons are manifold. Common agricultural practices lead to concurrent exposures to several OPs, each with different acute toxicity, though some of them are metabolized to the same alkylphosphate (Moretto et al., 1995). The pharmacokinetics of OPs as well as that of different alkylphosphates are quite variable and peak excretion of alkylphosphates in the urine might be reached at significantly different time points after the end of exposure (Morgan et al., 1977; Fenske and Elkner, 1990). Finally almost nothing is known about the correlation between urinary excretion of metabolites and inhibition of either erythrocyte AChE and/or plasma BuChE activity.

Erythrocyte AChE is the same enzyme that is involved in synaptic transmission and its inhibition is considered a mirror of that in the nervous system, even if the interpretation may pose some difficulties (Lotti, 1995). Biomonitoring exposures to AChE inhibitors requires measurement of erythrocyte AChE activity before the exposure in order to assess the "real" inhibition as the consequence of exposure. This is because there is a large inter- and intraindividual variability in the baseline activities of this enzyme (Hayes, 1982).

Based on clinical and toxicological data rough estimates of the activities of erythrocyte AChE inhibition requiring action after occupational exposures are as follows: up to 20–29% inhibition from preexposure values indicates evidence of exposure and requires improvements in hygienic conditions. Inhibition of 30–50% suggests hazard and, besides improvement in hygienic conditions, the individual should be removed from exposure. An individual with >50% inhibition is considered poisoned and must be admitted to the hospital (IPCS, 1986).

Inhibition of plasma BuChE should be interpreted as an index of exposure to OPs and carbamates, not of toxicity, because no physiologic functions are known for this enzyme.

Measurement of plasma paraoxonase, an enzyme involved in the detoxification of several OPs, has been suggested as a marker of susceptibility to OP toxicities given the wide polymorphism of this enzyme (Costa et al., 2003).

References

Abdollahi M, Karami-Mohajeri S (2012). A comprehensive review on experimental and clinical findings in intermediate syndrome caused by organophosphate poisoning. Toxicol Appl Pharmacol 258: 309–314.

Agha A, Dib S, Al-Hakami M et al. (2010). A fatal case of methomyl poisoning leading to fulminant hepatic failure. Int J Toxicol 7.

Al-Samarraie MSJ, Karinen R, Rognum T et al. (2009). Lethal poisoning with ethiofencarb and ethanol. J Anal Toxicol 33: 389–392.

Alvares AP (1992). Pharmacology and toxicology of carbamates, In: B Ballantyne, TC Marrs (Eds.), Clinical and experimental toxicology of organophosphates and carbamates, Butterworth-Heinemann, Oxford, pp. 40–46.

Ameno K, Lee S-K, In S-W et al. (2001). Blood carbofuran concentrations in suicidal ingestion cases. Forensic Sci Int 116: 59–61.

Avasthi G, Singh G (2000). Serial neuro-electrophysiological studies in acute organophosphate poisoning – correlation with clinical findings, serum cholinesterase levels and atropine dosages. J Assoc Physicians India 48: 794–799.

Baban NK, Nunley DL, Borges Jr AS et al. (1998). Human sequelae of severe carbamate poisoning. Tenn Med 91: 103–106.

Bardin PG, Van Eeden SF, Joubert JR (1987). Intensive care management of acute organophosphate poisoning. A 7-year experience in the western Cape. S Afr Med J 72: 593–597.

Benson B, Tolo D, McIntire M (1992). Is the intermediate syndrome in organophosphate poisoning the result of insufficient oxime therapy? J Toxicol Clin Toxicol 30: 347–349.

Besser R, Gutmann L (1995). A quantitative study of the pancuronium antagonism at the motor endplate in human organophosphorus intoxication. Muscle Nerve 18: 956–960.

Besser R, Gutman L, Dillman U et al. (1989a). End-plate dysfunction in acute organophosphate intoxication. Neurology 39: 561–567.

Besser R, Gutman L, Weilemann LS (1989b). Intoxication of end-plate acetylcholinesterase during the course of organophosphate intoxications. Arch Toxicol 63: 412–415.

Besser R, Vogt T, Gutmann L (1990). Pancuronium improves the neuromuscular transmission defect of human organophosphate intoxication. Neurology 40: 1275–1277.

Bismuth C, Inns RH, Marrs TC (1992). Efficacy, toxicity and clinical use of oximes in anticholinesterase poisoning. In: B Ballantyne, TC Marrs (Eds.), Clinical and experimental toxicology of organophosphates and carbamates, Butterworth-Heinemann, Oxford, pp. 555–577.

Blain PG (2011). Organophosphorus poisoning (acute). Clin Evid, 2011: pii: 2102.

Bokonjic D, Jovanovic D, Jokanovic M et al. (1987). Protective effects of oximes HI-6 and PAM-2 applied by osmotic minipumps in quinalphos-poisoned rats. Arch Int Pharmacodyn Ther 288: 309–318.

Brahmi N, Blel Y, Kouraichi N et al. (2005). Acute pancreatitis subsequent to voluntary methomyl and dichlorvos intoxication. Pancreas 31: 424–427.

Buckley NA, Eddleston M, Li Y et al. (2011). Oximes for acute organophosphate pesticide poisoning. Cochrane Database Syst Rev 1–36.

Burgen ASV, Hobbiger F (1951). The inhibition of cholinesterases by alkyl-phosphates and alkylphenolphosphates. Br J Pharmacol 6: 593–605.

Burgess JL, Bernstein JN, Hurlbut K (1994). Aldicarb poisoning: a case report with prolonged cholinesterase inhibition and improvement after pralidoxime therapy. Arch Intern Med 154: 221–224.

Cable GG, Doherty S (1999). Acute carbamate and organochlorine toxicity causing convulsions in an agricultural pilot: a case report. Aviat Space Environ Med 70: 68–72.

Carpenter CP, Weil CS, Palm PE et al. (1961). Mammalian toxicity of 1-naphthyl-*N*-methylcarbamate (Sevin insecticide). J Agric Food Chem 9: 30–39.

Casey PB, Blakey L, Bradberry SM et al. (1995a). Late reactivation of erythrocyte cholinesterase activity by pralidoxime in a case of chlorpyrifos poisoning. Przegl Lek 52: 206.

Casey PB, Gosden E, Blakey L et al. (1995b). Plasma pralidoxime concentrations following bolus injection and continuous infusion. Przegl Lek 52: 203–204.

Casida JE, Nomura DK, Vose SC et al. (2008). Organophosphate-sensitive lipases modulate brain lysophospholipids, ether lipids and endocannabinoids. Chem Biol Interact 175: 355–364.

Cavaliere MJ, Puga FR, Calore EE et al. (1998). Protective effect of pralidoxime on muscle fiber necrosis induced by organophosphate compounds. J Toxicol Clin Toxicol 36: 295–300.

Chan Y-H, Liew K-P, Sun CC-FS et al. (2012). Hyperacute rejection from a donor who died of carbamate intoxication – A case report. Am J Emerg Med 30: 1661.e1–1661.e4.

Coleman AME, Smith A, Watson L (1990). Occupational carbamate pesticide intoxication in three farm workers. Implications and significance for occupational health in Jamaica. West Indian Med J 39: 109–113.

Costa LG, Richter RJ, Li W-F et al. (2003). Paraoxonase (PON1) as a biomarker of susceptibility for organophosphate toxicity. Biomarkers 8: 1–12.

Dandapani M, Zachariah A, Kavitha MR et al. (2003). Oxidative damage in intermediate syndrome of acute organophosphorus poisoning. Indian J Med Res 117: 253–259.

De Bleecker J (1992). Transient opsoclonus in organophosphate poisoning. Acta Neurol Scand 86: 529–531.

De Bleecker JL (2006). Intermediate syndrome in organophosphate poisoning. In: RC Gupta (Ed.), Toxicology of organophosphate and carbamate compounds, Academic Press, Amsterdam, pp. 371–380.

De Bleecker J, Vogelaers D, Ceuterick C et al. (1992). Intermediate syndrome due to prolonged parathion poisoning. Acta Neurol Scand 86: 421–424.

De Bleecker J, van den Neucker K, Colardyn F (1993). Intermediate syndrome in organophosphorus poisoning: a prospective study. Crit Care Med 21: 1706–1711.

De Bleecker J, Lison D, Van Den Abeele K et al. (1994a). Acute and subacute organophosphate poisoning in the rat. Neurotoxicology 15: 341–348.

De Bleecker J, Van Den Abeele K, de Reuck J (1994b). Electromyography in relation to end-plate acetylcholinesterase in rats poisoned by different organophosphates. Neurotoxicology 15: 331–340.

De Silva HJ, Wijewickrema R, Senanayake N (1992). Does pralidoxime affect outcome of management in acute organophosphorus poisoning? Lancet 339: 1136–1138.

Delgado E, McConnell R, Miranda J et al. (2004). Central nervous system effects of acute organophosphate poisoning in a two-year follow-up. Scand J Work Environ Health 30: 362–370.

Dickoff DJ, Gerber O, Turovsky Z (1987). Delayed neurotoxicity after ingestion of carbamate pesticide. Neurology 37: 1229–1231.

Driskell WJ, Groce DF, Hill Jr RH et al. (1991). Methomyl in the blood of a pilot who crashed during aerial spraying. J Anal Toxicol 15: 339–340.

Eckert S, Eyer P, Melzer M et al. (2008). Effects of oximes on rate of decarbamylation of human red blood cell AChE measured with two different methods. Biochem Pharmacol 75: 1561–1566.

Ecobichon DJ (2001). Carbamate insecticides. In: RI Krieger (Ed.), Handbook of pesticide toxicology. Principles, 2 ed. Academic Press, San Diego, California, pp. 1087–1106.

Eddleston M, Szinicz L, Eyer P et al. (2002). Oximes in acute organophosphorus pesticide poisoning: a systematic review of clinical trials. Q J Med 95: 275–283.

Eddleston M, Buckley NA, Checketts H et al. (2004a). Speed of initial atropinisation in significant organophosphorus pesticide poisoning – a systematic comparison of recommended regimens. J Toxicol Clin Toxicol 42: 865–875.

Eddleston M, Dawson A, Karalliedde L et al. (2004b). Early management after self-poisoning with an organophosphorus or carbamate pesticide – a treatment protocol for junior doctors. Crit Care 8: R391–R397.

Eddleston M, Mohamed F, Davies JOJ et al. (2006). Respiratory failure in acute organophosphorus pesticide self-poisoning. Q J Med 99: 513–522.

Eddleston M, Juszczak E, Buckley NA et al. (2008). Multiple-dose activated charcoal in acute self-poisoning: a randomised controlled trial. Lancet 371: 579–587.

Eddleston M, Eyer P, Worek F et al. (2009). Pralidoxime in acute organophosphorus insecticide poisoning — A randomised controlled trial. PLoS Med 6, e1000104.

Eddleston M, Harris JB, Self I et al. (2010). The major solvent in agricultural dimethoate preparations is essential for toxicity in minipigs. Clin Toxicol 48: 297.

Eddleston M, Street JM, Self I et al. (2012). A role for solvents in the toxicity of agricultural organophosphorus pesticides. Toxicology 294: 94–103.

Ekins BR, Geller RJ (1994). Methomyl-induced carbamate poisoning treated with pralidoxime chloride. West J Med 161: 68–70.

Eyer P (1996). Use of obidoxime in OP poisoning (Abstract from workshop on organophosphate poisoning, Munich, 4–5 September 1995). Hum Exp Toxicol 15: 78.

Eyer F, Worek F, Eyer P et al. (2009). Obidoxime in acute organophosphate poisoning: 1 – clinical effectiveness. Clin Toxicol 47: 798–806.

Farago A (1969). Fatal suicidal case of Sevin (1-naphthyl-*N*-methyl-carbamate) poisoning. Arch Toxikol 24: 309–315.

Fenske RA, Elkner KP (1990). Multi-route exposure assessment and biological monitoring of urban pesticide applicators during structural control treatments with chlorpyrifos. Toxicol Ind Health 6: 349–371.

Finkelstein Y, Kushnir A, Raikhlin-Eisenkraft B et al. (1989). Antidotal therapy of severe acute organophosphate poisoning: a multihospital study. Neurotoxicol Teratol 11: 593–596.

Garcia JH, Coelho GR, Marques GA et al. (2010). Successful organ transplantation from donors poisoned with a carbamate insecticide. Am J Transplant 10: 1490–1492.

Gil H, Hong J, Song H et al. (2012). A case of methanol intoxication caused by methomyl pesticide ingestion. Hum Exp Toxicol 31: 1299–1302.

Gil H-W, Jeong M-H, Park JS et al. (2013). An outbreak of food borne illness due to methomyl pesticide intoxication in Korea. J Korean Med Sci 28: 1677–1681.

Glow PH, Rose S (1965). Effects of reduced acetylcholinesterase levels on extinction of a conditioned response. Nature 206: 475–477.

Glynn P (2006). A mechanism for organophosphate-induced delayed neuropathy. Toxicol Lett 162: 94–97.

Goswamy R, Chaudhuri A, Mahashur AA (1994). Study of respiratory failure in organophosphate and carbamate poisoning. Heart Lung 23: 466–472.

Grob D (1956). The manifestations and treatment of poisoning due to nerve gas and other organic phosphate anticholinesterase compounds. AMA Arch Intern Med 98: 221.

Grob D, Harvey JC (1958). Effects in man of the anticholinesterase compound sarin (isopropyl methyl phosphonofluoridate). J Clin Invest 37: 350–368.

Grob D, Harvey AM, Langworthy OR et al. (1947). The administration of di-isopropyl fluorophosphate (DFP) to man. III. Effect on the central nervous system with special reference to the electrical activity of the brain. Bull Johns Hopkins Hosp 81: 257–266.

Gupta RC (1984). Acute malathion toxicosis and related enzymatic alterations in *Bubalus bubalis*: antidotal treatment with atropine, 2-PAM, and diazepam. J Toxicol Environ Health 14: 291–303.

Hansen ME, Wilson BW (1999). Oxime reactivation of RBC acetylcholinesterases for biomonitoring. Arch Environ Contam Toxicol 37: 283–289.

Harris LW, Talbot BG, Lennox WJ et al. (1989). The relationship between oxime-induced reactivation of carbamylated acetylcholinesterase and antidotal efficacy against carbamate intoxication. Toxicol Appl Pharmacol 98: 128–133.

Hata S, Bernstein E, Davis LE (1986). Atypical ocular bobbing in acute organophosphate poisoning. Arch Neurol 43: 185–186.

Hayes Jr WJ (1982). Pesticides studied in man, Williams & Wilkins, Baltimore, Maryland.

Hayes MMM, van der Westhuizen NG, Gelfand M (1978). Organophosphate poisoning in Rhodesia. A study of the clinical features and management of 105 patients. S Afr Med J 54: 230–234.

He F, Xu H, Qin F et al. (1998). Intermediate myasthenic syndrome following acute organophosphates poisoning – an analysis of 21 cases. Hum Exp Toxicol 17: 40–45.

Hirschberg A, Lerman Y (1984). Clinical problems in organophosphate insecticide poisoning. The use of a computerised information system. Fundam Appl Toxicol 4: S209–S214.

Hoffmann U, Hecker U, Abel P (2008). Acute poisoning by pirimicarb: clinical and toxicological features. Clin Toxicol 46: 694–696.

Hoskins B, Fernando JC, Dulaney MD et al. (1986). Relationship between the neurotoxicities of Soman, Sarin and Tabun, and acetylcholinesterase inhibition. Toxicol Lett 30: 121–129.

Hu Y-H, Yang C-C, Deng J-F et al. (2010). Methomyl-alphamethrin poisoning presented with cholinergic crisis, cortical blindness, and delayed peripheral neuropathy. Clin Toxicol 48: 859–862.

Huan WM, Cheng CH, Yih WJ (2013). Carbamate poisoning mixed with methanol intoxication – Misfortunes never come singly: case report. Clin Toxicol 51: 356–357.

IPCS (1986). Environmental health criteria 63. Organophosphorus insecticides: a general introduction, World Health Organization, Geneva.

Jayawardane P, Senanayake N, Dawson A (2009). Electrophysiological correlates of intermediate syndrome following acute organophosphate poisoning. Clin Toxicol 47: 193–205.

John M, Oommen A, Zachariah A (2003). Muscle injury in organophosphorous poisoning and its role in the development of intermediate syndrome. Neurotoxicology 24: 43–53.

Johnson MK (1969). The delayed neurotoxic effect of some organophosphorus compounds. Identification of the phosphorylation site as an esterase. Biochem J 114: 711–717.

Johnson MK (1990). Organophosphates and delayed neuropathy – is NTE alive and well? Toxicol Appl Pharmacol 102: 385–399.

Johnson MK, Vale JA (1992). Clinical management of acute organophosphate poisoning: an overview. In: B Ballantyne, TC Marrs (Eds.), Clinical and experimental toxicology of organophosphates and carbamates, Butterworth-Heinemann, Oxford, pp. 528–535.

Johnson S, Peter JV, Thomas K et al. (1996). Evaluation of two treatment regimens of pralidoxime (1 gm single bolus dose vs 12 gm infusion) in the management of organophosphorus poisoning. J Assoc Physicians India 44: 529–531.

Johnson MK, Jacobsen D, Meredith TJ et al. (2000). Evaluation of antidotes for poisoning by organophosphorus pesticides. Emerg Med 12: 22–37.

Joubert J, Joubert PH (1988a). Chorea and psychiatric changes in organophosphate poisoning. S Afr Med J 74: 32–34.

Joubert J, Joubert PH (1988b). Chorea and psychiatric changes in organophosphate poisoning. A report of 2 further cases. S Afr Med J 74: 32–34.

Joubert J, Joubert PH, van der Spuy M et al. (1984). Acute organophosphate poisoning presenting with choreoathetosis. J Toxicol Clin Toxicol 22: 187–191.

Karalliedde L, Henry JA (1993). Effects of organophosphates on skeletal muscle. Hum Exp Toxicol 12: 289–296.

Karalliedde L, Szinicz L (2001). Management of organophosphorus compound poisoning. In: L Karalliedde, S Feldman, J Henry et al. (Eds.), Organophosphates and health. Imperial College Press, London, pp. 257–293.

Karalliedde L, Baker D, Marrs TC (2006). Organophosphate-induced intermediate syndrome: aetiology and relationships with myopathy. Toxicol Rev 25: 1–14.

Kassa J, Bajgar J (1994). Treatment of the stressogenic effect of dichlorvos. Sb Ved Pr Lek Fak Karlovy Univerzity Hradci Kralove 37: 29–32.

Katz EJ, Cortes VI, Eldefrawi ME et al. (1997). Chlorpyrifos, parathion, and their oxons bind to and desensitize a nicotinic acetylcholine receptor: relevance to their toxicities. Toxicol Appl Pharmacol 146: 227–236.

Kiss Z, Fazekas T (1983). Organophosphates and *torsade de pointes* ventricular tachycardia. J R Soc Med 76: 984–985.

Kleinrok Z, Jagiello-Wojtowicz E (1977). Wplyw wybranych leków przeciwpadaczkowych na toksycznosc ostra estru dwuizopropylowego kwasu fluorofosforowego u myszy. Ann Univ Mariae Curie Sklodowska [Med] 32: 155–160.

Krutak-Krol H, Domino EF (1985). Comparative effects of diazepam and midazolam on paraoxon toxicity in rats. Toxicol Appl Pharmacol 81: 545–550.

Kumar S, Mahajan G, Walia V et al. (2014). Baygon (propoxur) poisoning leading to acute respiratory distress syndrome (ARDS). Indian J Pediatr 51: 730–731.

Kusić R, Jovanovic D, Randjelovic S et al. (1991). HI-6 in man: efficacy of the oxime in poisoning by organophosphorus insecticides. Hum Exp Toxicol 10: 113–118.

Lee P, Tai DYH (2001). Clinical features of patients with acute organophosphate poisoning requiring intensive care. Intensive Care Med 27: 694–699.

Lee BK, Jeung KW, Lee HY et al. (2011). Mortality rate and pattern following carbamate methomyl poisoning. Comparison with organophosphate poisoning of comparable toxicity. Clin Toxicol 49: 828–833.

Li Y, Tse ML, Gawarammana I et al. (2009). Systematic review of controlled clinical trials of gastric lavage in acute organophosphorus pesticide poisoning. Clin Toxicol 47: 179–192.

Liddle JA, Kimbrough RD, Needham LL et al. (1979). A fatal episode of accidental methomyl poisoning. J Toxicol Clin Toxicol 15: 159–167.

Lifshitz M, Shahak E, Bolotin A et al. (1997). Carbamate poisoning in early childhood and in adults. J Toxicol Clin Toxicol 35: 25–27.

Lim JH, Kim HS, Kim HO et al. (2010). Stevens–Johnson syndrome following occupational exposure to carbamate insecticide. J Dermatol 37: 182–184.

Lipp JA (1973). Effect of benzodiazepine derivatives on soman-induced seizure activity and convulsions in the monkey. Arch Int Pharmacodyn Ther 202: 244–251.

Lotti M (1992a). Central neurotoxicity and behavioural effects of anticholinesterases. In: B Ballantyne, TC Marrs (Eds.), Clinical and experimental toxicology of organophosphates and carbamates, Butterworth-Heinemann, Oxford, pp. 75–83.

Lotti M (1992b). The pathogenesis of organophosphate polyneuropathy. Crit Rev Toxicol 21: 465–487.

Lotti M (1995). Cholinesterase inhibition: complexities in interpretation. Clin Chem 41: 1814–1818.

Lotti M (2002). Low-level exposures to organophosphorus esters and peripheral nerve function. Muscle Nerve 25: 492–504.

Lotti M, Becker CE (1982). Treatment of acute organophosphate poisoning: evidence of a direct effect on central nervous system by 2-PAM (pyridine-2-aldoxime methyl chloride). J Toxicol Clin Toxicol 19: 121–127.

Lotti M, Moretto A (2005). Organophosphate-induced delayed polyneuropathy. Toxicol Rev 24: 37–49.

Lotti M, Moretto A (2006). Do carbamates cause polyneuropathy? Muscle Nerve 34: 499–502.

Lotti M, Becker CE, Aminoff MJ (1984). Organophosphate polyneuropathy: pathogenesis and prevention. Neurology 34: 658–662.

Ludomirsky A, Klein HO, Sarelli P et al. (1982). Q-T prolongation and polymorphous ("torsade de pointes") ventricular arrhythmias associated with organophosphorus insecticide poisoning. Am J Cardiol 49: 1654–1658.

Mackenzie Ross S, McManus IC, Harrison V et al. (2013). Neurobehavioral problems following low-level exposure to organophosphate pesticides: a systematic and meta-analytic review. Crit Rev Toxicol 43: 21–44.

Makrides C, Koukouvas M, Achillews G et al. (2005). Methomyl-induced severe acute pancreatitis: possible etiological association. JOP 6: 166–171.

Maroni M, Järvisalo J, La Ferla F (1986). The WHO-UNDP epidemiological study on the health effects of exposure to organophosphorus pesticides. Toxicol Lett 33: 115–123.

Maroni M, Colosio C, Ferioli A et al. (2000a). Biological monitoring of pesticide exposure: a review. Carbamate pesticides. Toxicology 143: 39–46.

Maroni M, Colosio C, Ferioli A et al. (2000b). Biological monitoring of pesticide exposure: a review. Organophosphorus pesticides. Toxicology 143: 9–37.

Marrs TC (2003). Diazepam in the treatment of organophosphorus ester pesticide poisoning. Toxicol Rev 22: 75–81.

Martinez-Mier G, Avila-Pardo SF, Guraieb-Barragan E et al. (2010). Successful renal transplantation from a deceased donor with pesticide intoxication: a case report. Transplant Proc 42: 2397–2398.

McConnell R, Hruska AJ (1993). An epidemic of pesticide poisoning in Nicaragua: implications for prevention in developing countries. Am J Public Health 83: 1559–1562.

McConnell R, Keifer M, Rosenstock L (1994). Elevated quantitative vibrotactile threshold among workers previously poisoned with methamidophos and other organophosphate pesticides. Am J Ind Med 25: 325–334.

Mercurio-Zappala M, Hack JB, Salvador A et al. (2007). Pralidoxime in carbaryl poisoning: an animal model. Hum Exp Toxicol 26: 125–129.

Miranda J, Lundberg I, McConnell R et al. (2002a). Onset of grip- and pinch-strength impairment after acute poisonings with organophosphate insecticides. Int J Occup Environ Health 8: 19–26.

Miranda J, McConnell R, Delgado E et al. (2002b). Tactile vibration thresholds after acute poisonings with organophosphate insecticides. Int J Occup Environ Health 8: 212–219.

Miranda J, McConnell R, Wesseling C et al. (2004). Muscular strength and vibration thresholds during two years after acute poisoning with organophosphate insecticides. Occup Environ Med 61, e4.

Miyazaki T, Yashiki M, Kojima T et al. (1989). Fatal and nonfatal methomyl intoxication in an attempted double suicide. Forensic Sci Int 42: 263–270.

Moon JM, Chun BJ, Lee BK (2012). Glasgow coma scale score in the prognosis of acute carbamate insecticide intoxication. Clin Toxicol 50: 832–837.

Moretto A, Lotti M (2006). Peripheral nervous system effects and delayed neuropathy. In: RC Gupta (Ed.), Toxicology of organophosphate and carbamate compounds. Academic Press, Amsterdam, pp. 361–370.

Moretto A, Capodicasa E, Bertolazzi M et al. (1995). Biological monitoring of occupational exposures to organophosphorus insecticides. In: HH McDuffie, JA Dosman, KM Semchuk et al. (Eds.), Agricultural health and safety: workplace, environment, sustainability. CRC Press, Boca Raton, FL, pp. 217–221.

Morgan DP, Hetzler HL, Slach EF et al. (1977). Urinary excretion of paranitrophenol and alkyl phosphates following ingestion of methyl or ethyl parathion by human subjects. Arch Environ Contam Toxicol 6: 159–173.

Moritz F, Droy JM, Dutheil G et al. (1994). Acute pancreatitis after carbamate insecticide intoxication. Intensive Care Med 20: 49–50.

Natoff IL, Reiff B (1973). Effect of oximes on the acute toxicity of anticholinesterase carbamates. Toxicol Appl Pharmacol 25: 569–575.

Nelson LS, Perrone J, DeRoos F et al. (2001). Aldicarb poisoning by an illicit rodenticide imported into the United States: *Tres Pasitos*. J Toxicol Clin Toxicol 39: 447–452.

Ouanes I, Ouanes-Besbes L, Abroug F (2009). Major trismus associated with succinylcholine administration in carbamate insecticide poisoning. Can J Anaesth 56: 627–628.

Padilla S, Marshall RS, Hunter DL et al. (2007). Time course of cholinesterase inhibition in adult rats treated acutely with carbaryl, carbofuran, formetanate, methomyl, methiocarb, oxamyl or propoxur. Toxicol Appl Pharmacol 219: 202–209.

Paul N, Mannathukkaran TJ (2005). Intermediate syndrome following carbamate poisoning. Clin Toxicol 43: 867–868.

Pawar KS, Bhoite RR, Pillay CP et al. (2006). Continuous pralidoxime infusion versus repeated bolus injection to treat organophosphorus pesticide poisoning: a randomised controlled trial. Lancet 368: 2136–2141.

Petkovska L, Naumovski Dz, Pilovski Gj et al. (2000). Carbamate induced pancreatitis. Archives of Toxicology, Kinetics, and Xenobiotic Metabolism 8: 173–174.

Pinakini KS, Kumar TSM (2006). Serial cholinesterase estimation in carbamate poisoning. J Clin Forensic Med 13: 274–276.

Pullicino P, Aquilina J (1989). Opsoclonus in organophosphate poisoning. Arch Neurol 46: 704–705.

Ragoucy-Sengler C, Tracqui A, Chavonnet A et al. (2000). Aldicarb poisoning. Hum Exp Toxicol 19: 657–662.

Richardson EM, Batteese Jr RI (1973). An incident of Zectran poisoning. J Maine Med Assoc 64: 158–159.

Rizos E, Liberopoulos E, Kosta P et al. (2004). Carbofuran-induced acute pancreatitis. JOP 5: 44–47.

Rosenstock L, Keifer M, Daniell WE et al. (1991). Chronic central nervous system effects of acute organophosphate pesticide intoxication. Lancet 338: 223–227.

Rump S, Grudzinska E (1974). Investigations on the effects of diazepam in acute experimental intoxication with fluostigmine. Arch Toxikol 31: 223–232.

Rump S, Faff J, Szymanska T et al. (1976). Efficacy of repeated pharmacotherapy in experimental acute poisonings with fluostigmine. Arch Toxicol 35: 275–280.

Saadeh AM, Farsakh NA, Al-Ali MK (1997). Cardiac manifestations of acute carbamate and organophosphate poisoning. Heart 77: 461–464.

Saiyed HN, Sadhu HG, Bhatnagar VK et al. (1992). Cardiac toxicity following short-term exposure to methomyl in spraymen and rabbits. Hum Exp Toxicol 11: 93–97.

Sakurada K, Matsubara K, Shimizu K et al. (2003). Pralidoxime iodide (2-PAM) penetrates across the blood–brain barrier. Neurochem Res 28: 1401–1407.

Sánchez-Santed F, Cañadas F, Flores P et al. (2004). Long-term functional neurotoxicity of paraoxon and chlorpyrifos: behavioural and pharmacological evidence. Neurotoxicol Teratol 26: 305–317.

Sanderson DM (1961). Treatment of poisoning by anticholinesterase insecticides in the rat. J Pharm Pharmacol 13: 435–442.

Santinelli R, Tolone C, D'Avanzo A et al. (2006). Pontine myelinolysis in a child with carbamate poisoning. Clin Toxicol 44: 327–328.

Sarkar S, Narang A, Singh S (1994). Transplacentally acquired carbamate insecticide (Baygon) poisoning in a neonate. Indian Pediatr 31: 343–346.

Satar S, Satar S, Sebe A et al. (2005). Carbofuran poisoning among farm workers. Mt Sinai J Med 72: 389–392.

Savage EP, Keefe TJ, Mounce LM et al. (1988). Chronic neurological sequelae of acute organophosphate pesticide poisoning. Arch Environ Health 43: 38–45.

Sedgwick EM, Senanayake N (1997). Pathophysiology of the intermediate syndrome of organophosphorus poisoning. J Neurol Neurosurg Psychiatry 62: 201–202.

Sellström Å (1992). Anticonvulsants in anticholinesterase poisoning. In: B Ballantyne, TC Marrs (Eds.), Clinical and experimental toxicology of organophosphates and carbamates, Butterworth-Heinemann, Oxford, pp. 578–586.

Senanayake N, Johnson MK (1982). Acute polyneuropathy after poisoning by a new organophosphate insecticide. N Engl J Med 306: 155–157.

Senanayake N, Karalliedde L (1987). Neurotoxic effects of organophosphorus insecticides: an intermediate syndrome. N Engl J Med 316: 761–763.

Sidell FR, Groff WA, Kaminskis A (1972). Toxogonin and pralidoxime: kinetic comparison after intravenous administration to man. J Pharm Sci 61: 1765–1769.

Siegal D, Kotowycz MA, Methot M et al. (2009). Complete heart block following intentional carbamate ingestion. Can J Cardiol 25: e288–e290.

Simpson GR, Bermingham S (1977). Poisoning by carbamate pesticides. Med J Aust 2: 148–149.

Singh G, Sidhu UPS, Mahajan R et al. (2000). Phrenic nerve conduction studies in acute organophosphate poisoning. Muscle Nerve 23: 627–632.

Smith DM (1977). Organophosphorus poisoning from emergency use of a handsprayer. Practitioner 218: 877–883.

Steenland K, Jenkins B, Ames RG et al. (1994). Chronic neurological sequelae to organophosphate pesticide poisoning. Am J Public Health 84: 731–736.

Sterri SH, Rognerud B, Fiskum SE et al. (1979). Effect of toxogonin and P2S on the toxicity of carbamates and organophosphorus compounds. Acta Pharmacol Toxicol 45: 9–15.

Sudakin DL, Mullins ME, Horowitz BZ et al. (2000). Intermediate syndrome after malathion ingestion despite continuous infusion of pralidoxime. J Toxicol Clin Toxicol 38: 47–50.

Sudhir U, Chandrashekar Pai R et al. (2013). Glycemic changes in acute anticholinesterase insecticide poisoning. W Lond Med J 5: 27–33.

Sundwall A (1961). Minimum concentrations of N-methylpyridinium-2-aldoxime methane sulphonate (P2S) which reverse neuromuscular block. Biochem Pharmacol 8: 413–417.

Tay SY, Tai DYH, Seow E et al. (1998). Patients admitted to an intensive care unit for poisoning. Ann Acad Med Singapore 27: 347–352.

Testud F, De Larquier A, Descotes J (1999). Occupational acute poisonings with methomyl: role of dermal exposure and need for preventive measures. J Environ Med 1: 137–139.

Thierauf A, Gnann H, Bohnert M et al. (2009). Suicidal poisoning with mercaptodimethur – morphological findings and toxicological analysis. Int J Leg Med 123: 327–331.

Thiermann H, Mast U, Klimmek R et al. (1997). Cholinesterase status, pharmacokinetics and laboratory findings during obidoxime therapy in organophosphate poisoned patients. Hum Exp Toxicol 16: 473–480.

Thiermann H, Szinicz L, Eyer F et al. (1999). Modern strategies in therapy of organophosphate poisoning. Toxicol Lett 107: 233–239.

Thiermann H, Worek F, Eyer P et al. (2009). Obidoxime in acute organophosphate poisoning: 2: PK/PD relationships. Clin Toxicol 47: 807–813.

Thiermann H, Eyer F, Felgenhauer N et al. (2010). Pharmacokinetics of obidoxime in patients poisoned with organophosphorus compounds. Toxicol Lett 197: 236–242.

Tracqui A, Flesch F, Sauder P et al. (2001). Repeated measurements of aldicarb in blood and urine in a case of nonfatal poisoning. Hum Exp Toxicol 20: 657–660.

Tsao TCY, Juang Y-C, Lan R-S et al. (1990). Respiratory failure of acute organophosphate and carbamate poisoning. Chest 98: 631–636.

Tsatsakis AM, Tsakalof AK, Siatitsas Y et al. (1996). Acute poisoning with carbamate pesticides: the Cretan experience. Sci Justice 36: 35–39.

Umehara F, Izumo S, Arimura K et al. (1991). Polyneuropathy induced by *m*-tolyl methyl carbamate intoxication. J Neurol 238: 47–48.

Vale JA, Scott GW (1974). Organophosphorus poisoning. Guys Hosp Rep 123: 13–25.

Van Meter WG, Karczmar AG, Fiscus RR (1978). CNS effects of anticholinesterases in the presence of inhibited cholinesterases. Arch Int Pharmacodyn Ther 231: 249–260.

Votanopoulos KI, Lee TC, Dominguez EP et al. (2007). Propoxur induced pancreatitis after inhalation of Baygon pesticide. Pancreas 34: 379–380.

Wedin GP (1988). Comment: anticholinesterase insecticide reactions. Drug Intell Clin Pharm 22: 919–920.

Weizman Z, Sofer S (1992). Acute pancreatitis in children with anticholinesterase insecticide intoxication. Pediatrics 90: 204–206.

Wijeyesakere SJ, Richardson RJ (2010). Neuropathy target esterase. In: RI Krieger (Ed.), Hayes' handbook of pesticide toxicology, 3 ed. Academic Press, San Diego, California, pp. 1435–1455.

Willems JL, Belpaire FM (1992). Anticholinergic poisoning: An overview of pharmacotherapy. In: B Ballantyne, TC Marrs (Eds.), Clinical and experimental toxicology of organophosphates and carbamates, Butterworth-Heinemann, Oxford, pp. 536–542.

Willems JL, De Bisschop HC, Verstraete AG et al. (1993). Cholinesterase reactivation in organophosphorus poisoned patients depends on the plasma concentrations of the oxime pralidoxime methylsulphate and of the organophosphate. Arch Toxicol 67: 79–84.

Wilson BH, Hooper MJ, Hansen ME et al. (1992). Reactivation of organophosphorus inhibited AChE with oximes. In: JE Chambers, PE Levi (Eds.), Organophosphates – chemistry, fate and effects, Academic Press, San Diego, pp. 107–137.

Winnik L, Pach D, Gawlikowski T et al. (1997). Multiorgan damage in acute oral carbamates poisonings. Przegl Lek 54: 684–688.

Worek F, Bäcker M, Thiermann H et al. (1997). Reappraisal of indications and limitations of oxime therapy in organophosphate poisoning. Hum Exp Toxicol 16: 466–472.

Wyckoff DW, Davies JE, Barquet A et al. (1968). Diagnostic and therapeutic problems of parathion poisonings. Ann Intern Med 68: 875–882.

Yang P-Y, Tsao TCY, Lin J-L et al. (2000). Carbofuran-induced delayed neuropathy. J Toxicol Clin Toxicol 38: 43–46.

Zilker T, Felgenhauer N, Hibler A et al. (1997). Factors influencing the efficacy of obidoxime in organophosphate pesticides poisoning. Przegl Lek 54: 662–664.

Zwiener RJ, Ginsburg CM (1988). Organophosphate and carbamate poisoning in infants and children. Pediatrics 81: 121–126.

Handbook of Clinical Neurology, Vol. 131 (3rd series)
Occupational Neurology
M. Lotti and M.L. Bleecker, Editors

Chapter 11

Neurotoxicity of metals

SAMUEL CAITO[1,2]* AND MICHAEL ASCHNER[1,2,3,4]

[1]*Division of Clinical Pharmacology and Pediatric Toxicology, Vanderbilt University Medical Center, Nashville, TN, USA*

[2]*The Kennedy Center for Research on Human Development, Vanderbilt University Medical Center, Nashville, TN, USA*

[3]*Center in Molecular Toxicology, Vanderbilt University Medical Center, Nashville, TN, USA*

[4]*Center for Molecular Neuroscience, Vanderbilt University Medical Center, Nashville, TN, USA*

INTRODUCTION

Humans have long recognized the unique properties of metals as both a necessary dietary component for life as well as a resource for manufactured goods. Metals make up the bulk of the periodic table and are therefore abundant in our environment. Due to their unique chemistry and characteristics, such as reflectivity, malleability, ductility, and conductivity, metals are used for numerous purposes in nature and industry. Metalloids, such as arsenic (As), selenium (Se), antimony (Sb), and polonium (Po), share many of the properties of both metals and nonmetals, and for this sake they are included in this chapter. In our twenty-first-century world, metals and metalloids surround us, from jewelry and currency, to planes and automobiles, the wires, batteries, and microchips encased in plastic electronic devices, and pigments and dyes on painted surfaces. While metals are beneficial in our day-to-day lives, the opposite side of the coin reveals the highly toxic nature of these elements. Toxicity to metals can result from malicious poisonings, environmental exposures, and occupational exposures. The focus of this chapter is to investigate neurotoxicities that result largely from occupational exposures to metals, to delve into the pathologic and physiologic mechanisms behind these toxicities, and to discuss the measures that should be in place to prevent the development of these conditions.

Occupational exposures to metals, and subsequent toxicities, have been well documented throughout history. Key periods of human civilization are characterized by their use of metals, such as the bronze and iron ages. Pigments used in paints, dyes, and enamels have been used by humans since prehistory. As has been consistently used for its green pigmentation from the Egyptians through the Victorian era. Both chromium (Cr) and As have been phased out of the production of red and green dyes due to their toxicity. Cadmium (Cd), while carcinogenic, is still in use in the production of red, yellow, and orange pigments. Lead (Pb) has been used in manufacturing from antiquity for purposes as diverse as for pipes, ceramic glazes, paint additives, cosmetics, and even as a sweetener for wine. Perhaps the most recognizable occupational toxicity to the general public is that of mercury (Hg). While Hg had been used by physicians as a panacea for diverse conditions, from constipation to syphilis, it was during the industrial revolution when English hat factory workers were exposed to mercuric nitrate vapor during the felting process that occupational exposures to Hg entered the consciousness of the population. The behavioral instability of erethism (see below) is believed to be the inspiration for Lewis Carroll's Mad Hatter in *Alice in Wonderland*. While the exposure conditions that placed these metals in history may not be relevant today, many are still used and valued in industry and continue to pose concern for human health through modern sources of exposure. Occupational exposures to metals have been linked to a number of diseases, including learning deficits, carcinogenesis, and neurodegeneration.

Manufacturing and other industrial processes influence not only occupational exposures, but also environmental exposures to metals through generation and disposal of hazardous waste products, air and water pollution, and by accidental spills during transport of waste or raw materials. Metals are naturally found in soils and

*Correspondence to: Samuel Caito, PhD, Vanderbilt University Medical Center, Division of Pediatric Toxicology, 11415 MRB IV, 2215-B Garland Ave, Nashville, TN 37232-0414, USA. Tel: +1-615-343-6227, E-mail: samuel.w.caito@vanderbilt.edu

minerals, and natural sources of metals include waste runoff, eruption of volcanoes, and erosion of natural deposits. Mining and quarrying are responsible for the extraction of metals in the raw state for human use. Miners are exposed to high levels of metals, often higher than safety guidelines. Basu et al. (2011) have measured Cr and As levels in the urine of gold miners in Ghana, and found 52% and 34%, respectively, above health guidelines. The local environment of the mine will determine the metals that miners will be exposed to. Radiation exposure is an additional risk that miners face: for example, Kavasi et al. (2010) have measured significant radon (Ra) and thoron (a natural decay product of thorium (^{232}Th) in European manganese (Mn) mines. Exposure to metals during the extraction of minerals can have negative impacts on an individual's health, especially over time (see below). In addition, extraction of minerals often results in contamination of surrounding native plants, animals, water, and soil (Antonijevic et al., 2012; Geremias et al., 2012; McIntyre and Whiting, 2012; Rodgher et al., 2013). Metal dusts can travel through the air and increase concentrations of ambient metals thousands of miles away. Increased levels of Hg in Sweden were traced to coal burning in the US Midwest (Clarkson and Magos, 2006), providing further evidence for global effects of air pollution. Contamination of food and water sources is a major source of metal exposure.

Some metals can be converted from the inorganic state to an organic species by bacteria. A classic example is the formation of methylmercury (MeHg). During the 1950s in Minamata, Japan, industrial waste containing inorganic Hg was released into the environment by factories, where it was methylated in the bay by anaerobic microorganisms, forming MeHg. This resulted in bioaccumulation of MeHg in the fish, contaminating the local food supply. Women who consumed the contaminated fish during pregnancy gave birth to children with Minamata disease, characterized by paresthesia, sensory deficits, slurred speech, unsteady gait, muscle weakness, irritability, memory loss, depression, and sleeping disturbances (Farina et al., 2011). Similar symptoms were observed in Basra, Iraq after consumption of bread made from grain treated with ethylmercury intended for planting, not human consumption (Farina et al., 2011).

MeHg contamination of fish remains a problem. MeHg is efficiently absorbed (with an absorption rate >90%), but poorly eliminated in aquatic organisms. It therefore bioaccumulates and biomagnifies in the food chain, with species higher up the food chain containing the most MeHg. Ingestion of the high trophic level fish, such as tuna, swordfish, and shark, increases human exposure and risk for deleterious effects. While environmental exposures to metals represent a significant proportion of metal-induced neurotoxicities, they are beyond the scope of this book. Readers are recommended to read the following resources (Waksman and Brody, 2007; Myers et al., 2009; Guilarte, 2010; Winneke, 2011).

Metals exert their toxicity by disrupting highly regulated the physiologic metal balance that is essential for life. Metals are classified as either essential or nonessential for life depending on whether they are required for a physiologic role (Table 11.1). These metals play important roles as cofactors for enzymes or through incorporation into metalloproteins. Essential metals show a hormetic U-shaped dose–response curve, where both too little and too much intake of an essential metal is deleterious. Essential metals have a range in optimal intake levels where the effects of too little or too much of the metal are outweighed by the beneficial physiologic roles. As an example, Mn is required for all living organisms as a cofactor for multiple enzymes, such as oxidoreductases, transferases, hydrolases, lyases, isomerases, ligases, lectins, and integrins (Aschner and Aschner, 2005). Overexposure to Mn often occurs in occupation settings, leading to a condition termed manganism. This is a rare neurologic disorder with symptoms analogous to Parkinson's disease (PD). Nonessential metals do not show U-shaped dose–response curves and have no known biologic function. For xenobiotics in general, there are concentrations of exposures with no observable adverse effect level (NOAEL), as well as exposure concentrations at which adverse health effects are first noticed, the lowest observable adverse effect level (LOAEL). Exposure to metals will lead to toxic sequelae if the exposure is above the LOAEL concentration. Metals disrupt metabolic processes by altering antioxidant balance,

Table 11.1

Essential and nonessential metals and metalloids

	Nonessential metals	
Essential metals	Not radioactive	Radioactive
Calcium	Aluminum*	Neptunium
Cobalt*	Arsenic*	Polonium
Copper*	Cadmium*	Plutonium
Iron*	Chromium*	Thallium*
Magnesium	Lead*	Thorium
Manganese*	Mercury*	Uranium*
Molybdenum	Nickel	
Selenium*	Tin*	
Sodium	Titanium*	
Zinc*		

*Metal is characterized as neurotoxic.

binding to free sulfhydryl groups (-SH), substituting and competing with essential metals for binding sites on enzymes, receptors, and transport proteins. For this reason cells have developed highly sophisticated mechanisms to regulate metal homeostasis.

BASIC PRINCIPLES OF METAL TOXICITY

There are several occupations in which workers have the potential to be exposed to toxic levels of metals. While exposure to the metal may vary by the type of work, there are basic principles that govern the outcome of exposure that are independent of individual metal species. These factors are similar to any other xenobiotic exposure in the industrial setting, such as dose, duration, and route of exposure. Age and genetics of the individual can influence toxicity, as there have been gene variants that affect an individual's ability to respond to metals. Developmental and perinatal exposure to metals is often more toxic than exposure to adults due to differences in absorption and metabolism. As a result, occupational exposure is a major concern for pregnant women in the workplace.

Duration and route of exposure

The duration of exposure determines the physiologic effects of metal exposure. Acute metal toxicity is often caused by sudden exposure to a large dose of a metal. This type of exposure usually results from industrial accidents, accidental contamination of a food supply, suicides, and homicidal poisonings. Symptoms from acute exposures are usually severe and rapid in onset. They can include cramping, nausea, vomiting, pain, sweating, headaches, difficulty in breathing, impaired cognitive, motor and language skills, mania, radiation poisoning, convulsions, and death. Radiation poisoning can occur from exposure to radioactive metals, such as uranium (U), Th, neptunium (Np), plutonium (Pu), and Po. It can manifest as nausea and vomiting, dizziness, weakness, diarrhea, headache, fatigue, fever, seizures, lethargy, ataxia, tremor, hypotension, and hemorrhage. Chronic toxicity is caused by long-term exposure. This often occurs in both the occupational setting and in residential areas near industrial centers and in metal-polluted environments. Chronic exposures usually occur at lower doses than acute exposures, and thus the toxicity from the chronic exposure results from distribution and accumulation of metals in tissues over time. Symptoms of chronic toxicity may include irritability, impaired cognitive, motor, language skills, learning difficulties, nervousness and emotional instability, insomnia, nausea, lethargy, allergic reactions, and cancer. Many of these symptoms are subtle and not unique to metal exposure, such as irritability and insomnia, which can complicate diagnosis. Diagnosis can also be difficult since, by definition, chronic toxicity is late in onset and there are instances where an individual starts to develop noticeable signs of toxicity years after being removed from the occupational environment that initiated the development of symptoms.

Epidemiologic studies have associated chronic exposure to metals with an increased risk for developing several progressive neurodegenerative disorders, such as Alzheimer's disease (AD), PD, Huntington's disease (HD) and amyotrophic lateral sclerosis (ALS), and ataxias (Breydo and Uversky, 2011). Cancer may develop from exposure to radioactive metals or from carcinogenic metals, such as As or Cd, which cause DNA damage and genetic mutations upon failure to repair DNA lesions. As one can see, there are a variety of effects from toxic acute and chronic exposures to metals. While some of the possible symptoms overlap between the two types of exposure, it is not guaranteed that chronic and acute exposure to the same metal will display the same symptoms. For example, acute exposure to iron (Fe) leads to vomiting, hemorrhage of gastrointestinal tract, cardiac depression, and metabolic acidosis, while chronic exposures lead to hepatic cirrhosis (Weinberg, 2010).

In the occupational setting, exposure to metals can come from a variety of sources, including ingestion (oral), inhalation, dermal, and injection. Ingestion of metals can result from accidental ingestion of metals from spills and sprayed materials, as well as from consumption of contaminated food or water. Oral uptake of metals occurs through the gastrointestinal tract, and the metals are subjugated to first pass phase detoxification in the liver and kidney. Metals such as Hg, As, Mn, Pb, and Ni are absorbed through the gastrointestinal tract with differing absorption rates due to the differential affinities of membrane transport mechanisms for metals, as well as differing chemical properties of the metals, such as speciation. Chemical speciation can severely affect a metal's toxicity to humans; for example, Cr^{III} is nontoxic to humans, whereas Cr^{VI} is a class A (known) carcinogen (Lindsay et al., 2012). Inhalation of metals occurs when metals gain access to the respiratory tract. Occupations that have a high risk of metal inhalant exposure include welders, miners, smelters, smiths, and other professions involved in manufacturing, such as batteries, glass, ammunition, microchips and circuit boards, jet engines, and ceramic glazes. Additionally workers who use metal-containing tools and machines can also be at risk, especially dentists, plumbers and fitters, auto mechanics, construction workers, firing range instructors, and cooks who use uncoated or worn copper cookware. The use of tobacco products can contribute synergistically to inhaled metal exposure as there are trace levels of metals in cigarettes (Pappas, 2011).

While dermal exposure has been the least studied of all exposure pathways in regard to metals, it represents a significant source of exposure to humans. Occupational conditions which produce dust, such as mining, represent sources of exposure for both the workers as well as for inhabitants of surrounding areas (Moody and Chu, 1995; Kerin and Lin, 2010). Dermal exposures can lead to altered immunocompetence and allergic dermatitis (Forte et al., 2008; Kirshen and Pratt, 2012; Thyssen, 2012). Radiation exposure on the worksite occurs primarily through dermal contact, with the dose of exposure dependent on the type of radiation (α, β, or γ) and the surface area of exposure. Injection of metals is not the most common route of exposure; however accidental exposure conditions may allow for it. Metal dusts and other metal-containing compounds can gain direct entry into the blood stream through contact with wounded skin or the eyes. The route of exposure is a major determinant of toxicity to metals; absorption rates vary by route for each metal. For example, inorganic Hg is most toxic if inhaled or injected but is not well absorbed through the skin, while Ni is very toxic when inhaled, leading to altered immunocompetence. It is also a well-known skin irritant, causing allergic dermatitis. Wearing the proper protective gear can significantly decrease exposure and the development of toxic endpoints.

Absorption and transport of metals in the body

Upon absorption into the systemic circulation, regardless of point of entry, metals are transported and distributed to multiple organs and tissues. This is a physiologic process utilized by dietary essential metals to distribute to areas of need and is hijacked by toxic nonessential metals. Tissues that accumulate metals have an innate high demand for metals (brain and heart), are areas of metal storage (bone), or are involved in the detoxification and elimination of metals (liver and kidney). Transport and distribution of metals are under tight homeostatic control due to the cytotoxic consequences of both metal deficiency and overaccumulation of metals. Net accumulation of metals occurs by a large exposure, an elevation in cellular uptake, and/or a reduction in cellular efflux due to mutations in transporters. Accumulation of metals in the brain has been observed in neurodegenerative diseases, including AD and PD. This may arise through occupational exposures as well as by reduced transport of metals in the aged brain (Tarohda et al., 2004).

The divalent metal transporter (DMT1) is an important metal transporter shown to transport Mn, Cu, zinc (Zn), and Fe. In a mouse model of PD, increased DMT1 expression was observed in the ventral mesencephalon, and DMT1 polymorphisms have been associated with an increased risk for PD in a Han Chinese population (Salazar et al., 2008; He et al., 2011). It is important to consider that each metal behaves differently once inside the body and will distribute amongst organ systems based on the metal's chemistry and the location of metal transport proteins in different tissues.

Transporter proteins (transporters) allow for the movement of metals from the extracellular space through the cell membrane. These proteins are required to move essential anions, cations, and metals which cannot enter the cell through passive diffusion due to their size and charge. The transporters can facilitate movement of metals following concentration gradients or via energy-dependent pumps to facilitate active transport against a gradient. Transporters are usually functionally redundant; the same metal entering a cell has several related and nonrelated transporter proteins. Additionally a given transporter may be used by several metals. Essential metals, such as Cu, Zn, Mn, and Fe, have evolutionary conserved specialized transport systems to traverse the plasma membranes, such as DMT1 and copper transporter 1 (Martinez-Finley et al., 2012).

Just as there are specific transporters that regulate metal uptake, efflux of metals from the cell is equally important. Interestingly, there are some transporters which are capable of transporting metals in either direction depending on the protein's subcellular localization. Metals can also gain access to the inside of a cell by forming a complex with organic molecules and piggybacking on their transport. This has been observed with metal–amino acid complexes and metals bound to organic anions. Transport of nonessential metals occurs either through the functional redundancy of metal transporter cargos or through the process of molecular mimicry. Metal transporters lack high levels of specificity, therefore metals of similar size, valence, or chemical properties can be transported by the same protein. This redundancy is an evolutionary advantage under physiologic conditions because one transporter protein is able to move multiple cargos; however, this property also allows for the entrance of toxic metals. For example, Cd^{2+} enters cells through the ZIP family of transporters, whose function is to carry Zn^{2+} from the extracellular space to the cytoplasm (Liu et al., 2008). Both Cd^{2+} and Pb^{2+} are substrates for DMT1, a transporter that carries Fe^{2+} and Mn^{2+} under physiologic conditions (Picard et al., 2000; I Bannon et al., 2002). Some metals, such as MeHg and As, enter cell transporters utilized by essential metals.

MeHg enters cells through molecular mimicry, a process by which a metal or a metal–complex closely resembles an endogenous ligand for a transporter. MeHg is highly nucleophilic and has high affinity for cysteine residues, resulting in the formation of a MeHg–cysteine conjugate. This conjugate structurally resembles methionine, allowing it to be transported across the cell membrane by the large neutral amino acid transporter (LAT1) (Yin et al., 2008). A transporter for MeHg in an unconjugated cysteine has yet to be identified.

As^{III} has been shown to enter cells through multiple transporters, including glucose transporters (GLUT1/GLUT2/GLUT5, and SGLT), aquaporins (AQP3 and AQP10), and organic anion transporting polypeptides (OATPB), while As^{V} can enter cells through the phosphate transporter NaPiIIb (Liu et al., 2006b; Calatayud et al., 2012). It is not clear how these diverse proteins can transport As, but it is known that As enters through GLUT1 independently of glucose (Jiang et al., 2010). Further research is needed to elucidate these mechanisms.

Intracellular sequestration of metals and elimination

Once inside cells, metals are trafficked to various organelles for utilization, sequestration, storage, and elimination. While not entirely characterized, mechanisms have evolved to ensure proper delivery to metalloproteins while simultaneously preventing free radical formation and oxidative stress. This is an essential process, as nearly one-third of proteins require metals for optimal function (Waldron et al., 2009). Excess metals are sequestered by binding to sulfhydryl groups of glutathione (GSH) or metallothionines, removing them from the pool of free metals, and therefore they cannot be incorporated into metalloproteins. GSH is the major intracellular thiol responsible for maintaining the proper redox potential of the cell by reacting with nucleophilic compounds and reactive oxygen species (ROS) and aiding in their sequestration and elimination.

Metallothionines are proteins which contain several thiol-binding sites for metals and are transcriptionally upregulated by the presence of metals. Metal-binding chaperone proteins, such as heat shock proteins, are utilized to regulate the incorporation of the correct metal into specific proteins and to allow for proper protein folding. The complex interplay between metal-binding chaperones, metal sequestration, and metalloprotein incorporation is poorly understood.

Detoxification, excretion, and elimination are equally important to a metal's toxicity. Cells have developed mechanisms to remove free metals from their cytoplasm and transport the metal to extracellular compartments. GSH and metallothionines can bind metals to facilitate their transport out of the cell. As with import of metals, transporters move metals from the cell interior to outside the cell membrane. Once past the plasma membrane, metals bind circulating plasma proteins, such as ferritin, transferrin, albumin, and hemoglobin. Binding to these proteins can facilitate both distribution to tissues or detoxification and elimination in the liver and kidney. The liver is the major site of detoxification for xenobiotics. The liver metabolizes metals into less toxic and hydrophilic forms to be subsequently excreted in urine. For example, enzymes in the liver cleave long-chain alkyl Hg compounds, rendering them more readily excretable. Filtration and excretion in the urine avoid the enterohepatic circulation, preventing reabsorption of metals in the gastrointestinal tract. The enterohepatic circulation also is a site of metal elimination. Metals that enter the enterohepatic circulation and are not reabsorbed by the gastrointestinal tract are excreted in the bile. Due to the recycling of bile salts, metals can be reabsorbed by the gastrointestinal tract.

SENSITIVITY OF THE NERVOUS SYSTEM TO METALS

As previously mentioned, the nervous system readily accumulates metals, allowing for neurotoxicity. In addition to the exposure conditions experienced by neural tissue, the nervous system has unique characteristics that make it highly susceptible to metals. There are several characteristics that make the nervous system vulnerable to xenobiotics in general; these include complex structure and long period of development, interconnectivity to other organ systems, composition of differentiated postmitotic cells, selective transport into the central nervous system (CNS), high rate of metabolism, and myelination. While all factors are important to metal-induced neurotoxicity, the following sections will examine the contributions of the blood–brain barrier (BBB) and the metabolic activity of the brain.

Anatomic barriers

The CNS systemic circulation is largely separated from the rest of the body by the BBB. The BBB is comprised of an extensive network of specialized endothelial cells of the microvasculature of the brain, spinal cord, retina, and peripheral nervous system as well as glial cells. Endothelial cells of the BBB form tight junctions around the vascular lumen of the capillaries, creating a tighter seal between the blood and surrounding nervous tissue than the fenestrae present in systemic capillaries. The BBB glial cells are primarily astrocytes, which project "astocytic feet" to surround the endothelial cells to provide structure and regulation of the tightness of the barrier. There are a small number of sites in the brain, lining the ventricles, that are not protected by the BBB: the circumventricular organs, which include the pineal gland, posterior pituitary, and area postrema. The lack of complete BBB in these regions allows circumventricular organs to sense and respond to signals in the blood stream, such as osmolality, toxins, and peptide hormones. Small lipophilic substances and water can passively diffuse across the BBB; however, larger molecules must be actively transported in and out of the nervous system across the BBB.

Several transport systems are used to move essential anions, cations, and metals from the vascular lumen to the neuronal tissues. Toxic metals, whether essential metals in large doses or nonessential metals, must traverse the BBB. Essential metals pass through the BBB via transporters similar to those found in the gastrointestinal tract. Nonessential metals often pass through the BBB by molecular mimicry. MeHg gains access to the CNS through molecular mimicry as a cysteine conjugate that structurally resembles methionine, a substrate for LAT1 on the BBB endothelial cells (Yin et al., 2008).

The blood–cerebrospinal fluid (CSF) barrier is comprised of modified epithelial cells of the choroid plexus and surrounding capillaries called ependymal cells. There are four choroid plexi which are located in the third and fourth ventricles of the brain. The choroid plexus synthesizes CSF and acts as a filter to remove xenobiotics, excess neurotransmitters, and metabolic waste products from the CSF. The ependymal cells contain numerous villi and microvilli, which increase the total surface area available for filtration and transport. Just as the cells of the BBB are held tightly together, the ependymal cells of the choroid plexus contain tight junctions to create a barrier between the capillary lumen and the ventricles, although not as tight as the BBB. The choroid plexus contains transporters, which can be the target for entry of metals. Many of the metals that cross the BBB also enter the CNS through the blood–CSF barrier and accumulate in the CSF. Pb enters the brain through both the blood–CSF barrier and BBB. Pb exposure is known to cause the microvasculature of the brain to become leaky, resulting in increased permeability of the BBB, brain swelling, herniation, ventricular compression, petechial and cerebral hemorrhages, thrombosis, and arteriosclerosis (Zheng et al., 2003). The endothelial cells in the BBB microvasculature and the choroid plexus cells accumulate Pb. Pb can activate protein kinase C signaling due to Pb-induced disruption of intracellular Ca^{2+} levels. Activation of protein kinase C has been shown to increase endothelial permeability, which is observed during Pb toxicity (Markovac and Goldstein, 1988). Pb also inhibits the functioning of the choroid plexus epithelial cells, significantly decreasing transthyretin production (Zheng et al., 1996). Transthyretin, which is involved in thyroid hormone binding and delivery, is only produced by the choroid plexus. By disrupting transthyretin homeostasis, Pb exposure disrupts thyroid hormone signaling.

Energy dependence

Neurons are energetically costly cells that depend highly on aerobic respiration to maintain sufficient energy stores to generate action potentials and maintain electric conductivity. At rest the membrane potential is maintained by ion channels and energy-consuming ion pumps. Because of this high-energy need, a continual source of energy must be available to keep the neurons healthy. Mitochondrial production of adenosine triphosphate (ATP) involves several proteins working simultaneously to transfer electrons, produce ATP, and reduce ROS production. However the mitochondria are important targets for many metals.

Mitochondria produce the cell's supply of ATP through oxidative phosphorylation. The electron transport chain is composed of several metalloenzymes, such as cytochrome C oxidase (complex IV). These enzymes can be vulnerable to xenobiotic or excess physiologic metals, which can lead to inhibition of ATP production. This can be extremely damaging in tissues with high-energy demands, such as the brain. Pb has been shown to decrease ATP levels and reduce Na^+/K^+ ATPase activity in rats, while As is a known complex I, complex II, and complex IV inhibitor (Baranowska-Bosiacka et al., 2011; Dwivedi et al., 2011). Damage to the electron transport chain results in the formation of free radicals, which, in combination with the ROS produced during metal exposure, can lead to the activation of the intrinsic apoptotic pathway through the opening of a permeability transition pore in the mitochondrial inner membrane. This allows for cytochrome c release and caspase activation, the final steps of apoptosis. The permeability transition pore can be opened upon exposure to As, thallium, Cd, Hg, and Cu (Korotkov and Lapin, 2003; Wang et al., 2009b; Belyaeva et al., 2012).

TOXICITY OF SELECTED METALS

Metals are highly toxic systemic toxins which are routinely used for industrial purposes. Currently 18 metals and one metalloid (As) appear on the top 150 of National Priorities List from the Agency for Toxic Substances and Disease Registry (ATSDR). The majority of the metals are highly toxic, and exposure to them can result in severe consequences in the occupational setting. These include Pb, As, and Hg. There are metals on the ATSDR list that are essential metals for humans that can present toxic effects with overexposure; these include cobalt (Co), Zn, and Mn. Some of the metals on the registry are dangerous to humans, primarily due to their radioactive properties, such as U, Np, Pu, and Po. As radiation is the primary concern, these will not be further discussed.

In the following section, selected metals from the ATSDR that produce neurotoxic endpoints will be examined, and the physiologic effects, pathology, diagnosis of toxicity, and treatment of metal exposure will be discussed.

Accurate detection of metals in biologic samples is essential for diagnostic tests since clinical manifestations of metal exposure can produce general or subtle symptoms. Blood, urine, nails, and hair samples are common indicator tissues collected from patients for biologic monitoring for heavy-metal toxicity. Blood and urine levels correlate well with acute exposures and accurately reflect recent exposures. Hair and nail samples provide a more detailed overview of exposure due to the fact that both grow at constant rates and are very stable, and samples are easily collected, stored, and processed (Moon et al., 1988). However, metal concentrations from hair can be difficult to interpret due to contamination from shampoos and other hair-styling products. Direct measurement of metal concentrations in biologic samples can be performed by several analytic methods using spectrometry (atomic absorption spectrometry, graphite furnace atomic absorption spectrometry, atomic fluorescence spectrometry, atomic emission spectrometry, and inductively coupled plasma mass spectrometry). Metals can also be visualized in patients using magnetic resonance imaging (MRI) and positron emission tomography (PET). Direct measurement of metal concentrations at specific organs is not routinely performed in the clinical setting; however, bone marrow sampling may be performed in certain instances. Indirect estimates can be calculated from metabolic distribution models derived from examining samples postmortem and from animal studies. Additionally assessments, such as occupational and nonoccupational health histories, are required to accurately diagnose heavy-metal toxicity.

Proper safety is essential in preventing exposures to toxic metals and subsequent injury. Appropriate personal protective gear, labeling chemicals, workforce education, and proper handling and disposal protocols are important measures to prevent injury. In the USA, both the Occupational Safety and Health Administration (OSHA) and the National Institute for Occupational Safety and Health (NIOSH) are responsible for determining safe working conditions and setting limits and regulations for metal exposure in the workplace. A NIOSH-recommended exposure limits (RELs) is the level that is considered protective to worker safety and health over a working lifetime when used in combination with exposure and medical monitoring, worker training, and personal protective equipment. NIOSH RELs are not legally enforceable limits, but are used by the OSHA during the formation of legally enforceable permissible exposure limits (PELs). RELs and PELs are in place to protect the worker from the deleterious effects of exposure. For the following metals, NIOSH and OSHA safety guidelines are presented in Table 11.2.

Table 11.2

Regulatory standards in the USA for metal exposure in the workplace

Metal	OSHA PEL	NIOSH REL
Aluminum	8-hour time-weighted average 15 mg/m^3, total dust 8-hour time-weighted average 5 mg/m^3, respirable fraction	10-hour time-weighted average 5 mg/m^3
Arsenic	8-hour time-weighted average 0.01 mg/m^3 5 µg/m^3 Action level	Ceiling limit* of 0.002 mg/m^3
Lead (organic and inorganic)	8-hour time-weighted average 0.05 mg/m^3	10-hour time-weighted average 0.05 mg/m^3
Manganese	Ceiling limit of 5 mg/m^3	10-hour time-weighted average 1 mg/m^3 Ceiling limit of 3 mg/m^3
Mercury (inorganic)	8-hour time-weighted average 0.1 mg/m^3	Mercury vapor: 10-hour time-weighted average 0.05 mg/m^3 Other mercury: ceiling limit of 0.1 mg/m^3
Mercury (organic)	8-hour time-weighted average 0.01 mg/m^3 Ceiling limit of 0.04 mg/m^3	10-hour time-weighted average 0.01 mg/m^3 Ceiling limit of 0.03 mg/m^3
Trimethyltin	8-hour time-weighted average 0.1 mg/m^3	10-hour time-weighted average 0.1 mg/m^3

OSHA, Occupational Safety and Health Administration; PEL, permissible exposure limit; NIOSH, National Institute for Occupational Safety and Health; REL, recommended exposure limit.

Both PEL and REL are for "general industry."

*Ceiling limit defined as maximum exposure allowed in period of 15–30 minutes.

Aluminum

Aluminum (Al) is the third most common element on earth, found in several mineral deposits, such as feldspars, mica, cryolite, and bauxite. Pure Al is rarely found in the environment and is produced as a raw material from the refinement of bauxite to Al hydroxide and Al oxide, which is then electrolytically reduced to elemental Al. Recycling of scrap Al in a smelter is also a significant source of Al production. There are multiple uses of Al in industry. Al oxide spontaneously forms on Al surfaces, creating a protective coating, utilized for corrosion protection on steel. Al and Al alloys are used in vehicles, electronics, wiring, building materials, packaging, pigments, rocket propellants, explosives, fireworks, fire-resistant fibers, and abrasives (Krewski et al., 2007; Riihimaki and Aitio, 2012). Al salts are used in the drug industry for antacids, buffered aspirins, vaccine adjuvants, and antiperspirants. Al sulfate is used for water purification.

Al compounds primarily occur as airborne particles in the workplace; as a result, regulatory agencies set exposure limits based upon both total Al in the air and respirable fraction (Table 11.2). The amount of Al workers are exposed to will depend upon their profession and the industrial process involved. For example, during the production of Al sulfates used in water purification, exposure to airborne Al is low to moderate (0.02–4 mg/m^3), while workers exposed to Al powder in the fireworks and explosive industry work with higher concentrations (5–21 mg/m^3) (Simonsson et al., 1985; Letzel et al., 1996; Riihimaki et al., 2008). Exposures to Al from smelting are accompanied by exposure to particulate fluoride salts, ammonia, hydrofluoric and hydrochloric acid, Fe, Zn, Pb, Cu, Mn, and thermal degradation products of organic compounds, such as polycyclic aromatic hydrocarbons and vinyl chloride (Kilburn, 1998; Healy et al., 2001).

Pathology/physiology

There is no identified function for Al in the human body and exposure to Al can produce toxicity. Large exposures of Al that produce intoxication are rare, and usually manifest in persons with compromised kidney functioning, resulting in dialysis-associated bone diseases, such as osteitis fibrosa cystica. Al can also affect hematopoiesis in these patients, decreasing heme synthesis and increasing hemolysis, leading to anemia.

While the CNS is not a target for acute Al exposures, the brain readily accumulates Al and can be affected over time. There have been several studies examining the neurologic effects of chronic Al exposure in smelters and welders. The first study to examine the effects of Al on the brain of a ball-mill operator in an Al flake powder factory characterized a progressive loss of CNS functions, including forgetfulness, speech, myoclonic jerks, hemiparesis, convulsions, and dementia (McLaughlin et al., 1962). Postmortem analysis of the brain found Al levels 10–20 times higher than nonexposed controls.

More recent epidemiology studies have characterized a different set of symptoms: balance disorders, memory and concentration deficits, fatigue, emotional irritability, and depression (Hosovski et al., 1990; White et al., 1992; Riihimaki et al., 2000; Polizzi et al., 2002; Sinczuk-Walczak et al., 2003; Halatek et al., 2005). Although these symptoms can appear nonspecific, studies comparing Al welders and steel welders have shown a greater prevalence of concentration deficits and depression in the welders exposed to Al, providing evidence that Al exposure can elicit specific neurotoxic effects (Sjogren et al., 1990). Lower cognitive measures have been observed in Canadian miners who intentionally inhaled Al hydroxide as a preventive measure against lung silicosis than nonexposed miners (Rifat et al., 1990). Interestingly, the alveolar burden of Al in the miners corresponds to similar levels of Al deposition experienced by metal inert gas welders (1 mg/m^3) (Rifat et al., 1990; Riihimaki et al., 2008). Al neurotoxicity does not resolve after cessation of exposure. Polizzi et al. (2002) examined neurologic functioning in 64 secondary smelt workers who had been retired for at least 10 years following a mean of 31 years of employment against 32 age-, education-, and location-matched retirees. Former smelters showed significantly increased serum Al levels as compared to the referents, as well as inferior performance in cognitive and neuropsychologic tests (Polizzi et al., 2002).

Within the past two decades there has been much debate about whether Al is involved in AD. AD is the most common neurodegenerative disease in the USA, and is characterized by loss of cholinergic neurons of the cerebral cortex and the accumulation of amyloid plaques and neurofibrillary tangles. The majority of AD cases are not genetic, but rather sporadic in origin, giving support to the role of environmental exposures in neurodegenerative diseases. Levels of Al are increased in brains of patients with PD (Trapp et al., 1978). Studies from animal models have shown that Al distributes to all regions of the brain, with the highest concentrations in the hippocampus (Julka et al., 1995). This is due to Al's ability to cross the BBB complexed to transferrin or citrate (Roskams and Connor, 1990; Julka et al., 1995; Nagasawa et al., 2005). Al disturbs cholinergic neurotransmission by decreasing acetylcholinesterase and choline acetyltransferase enzymatic activity (Peng et al., 1992; Yellamma et al., 2010). Additionally, Al has been shown to induce oxidative stress and cause Fe accumulation (Kumar et al., 2008; Wu et al., 2012).

Furthermore, Al causes aggregation of amyloid-β and microtubule associated protein tau, hallmarks of AD (Langui et al., 1990; Shigematsu and McGeer, 1992; Kushkuley et al., 2010). As previously mentioned, occupational exposures have been shown to cause cognitive declines in individuals exposed to Al; however, no correlation with AD has been found. Epidemiologic studies have shown an increased risk of developing AD with living in areas where Al concentrations in drinking water are greater than 100 μg/L (McLachlan et al., 1996; Rondeau et al., 2000). This is puzzling, as no association between AD or dementia has been found with intake of food rich in Al or antacid use (Rogers and Simon, 1999; Krewski et al., 2007). Currently Al is thought to be a contributing factor for the development of AD.

Diagnosis and treatment

Accurate measurements of Al exposure are difficult to obtain. Due to the abundance of Al in the earth's crust and prevalence in industry, humans consume around 10 mg Al from food and less than 1 mg from water daily (Willhite et al., 2012). Levels of Al in serum and urine are very low and require great care in interpreting. Contamination of the specimen can occur from the sampling and analysis methods due to the ubiquitous nature of Al in our environment. Confounding interpretations, oral doses of Al in buffered aspirins and antacids can be as high as thousands of milligrams a day (Yokel and McNamara, 2001; Riihimaki and Aitio, 2012). If clinical symptoms indicate Al poisoning, or an individual is involved in an acute exposure incident, a deferoxamine infusion test may be performed to aid in the detection of Al from urine or serum. Deferoxamine is a metal-free ligand of an iron-chelating agent originally isolated from *Streptomyces pilosus*. It is used in chelation therapy for acute and chronic Fe toxicity and Al toxicity. As deferoxamine liberates Al from tissues, there is an increase in serum levels of Al, which can be compared to samples taken prior to infusion (McCarthy et al., 1990). In general, serum levels of Al greater than 50 μg/L correlate with toxicity. Additionally, since Al affects both bones and hematopoiesis, bone and bone marrow biopsies may be taken and aid in diagnosis. Upon diagnosis with Al poisoning, chelation therapy with deferoxamine is usually started and the patient is recommended to reduce Al intake from the diet, medications, and antiperspirants.

Arsenic

As is a metalloid that exists in multiple allotropes, with the most commonly used in industry being gray As. Gray As adopts a double-layer structure of interlocking rings, is soft and brittle, and is used for a semiconductor. As is in the same group as phosphorus in the periodic table, and both the yellow and black allotropes of As resemble tetraphosphorus and red phosphorus, respectively. Both yellow and black As are less stable than gray As, and have fewer uses in industry. As is found in multiple oxidation states in minerals: arsenides (–3), arsenites (+3), and As acid (+5). Furthermore As is found in alloys, such as gallium arsenide, indium arsenide, Al arsenide, and Cd arsenide (Magalhaes, 2002; Jomova et al., 2011).

As has been used since the bronze age for strengthening alloys of copper (Cu) and Pb, an important process still in use. There are several classic examples of As toxicity due to its incorporation into commercial products, such as in a green pigment (Paris green, emerald green), food coloring (Scheele's green), Victorian-era skin whitener, medical treatment for hemorrhoids, gingivitis, angina, leukemia, syphilis, and trypanosomiasis, and lewisite, a chemical weapon used during World War I. Current uses of As include semiconductors, automobile batteries, pyrotechnics, laser diodes and LEDs, feed additive in poultry (nitarsone), wood preservative, and pesticides (monosodium methyl arsenate and disodium methyl arsenate). Recently it has been determined that there is As, along with Pb and Hg, above safety levels found in 20% of Chinese and Tibetan traditional medicines (Martena et al., 2010). Manufacturing of As-containing products, mining, refining, and chemical synthesis of As compounds represent the major sources of occupational exposure. Park et al. (2010) have found that workers involved in maintenance of machinery and of the facilities that utilize As have a higher risk of exposure than workers in routine production work. This may in part be due to safety regulations being designed around production workers and not maintenance staff.

Pathology/physiology

The toxicity of As has been appreciated since ancient times. Due to being odorless, highly potent, and producing symptoms similar to common diseases, As was the poison of choice of European royalty, earning the titles "king of poisons" and "poison of kings." As poisoning was common in homicides until the invention of the Marsh test in the early 19th century. As affects multiple organ systems (liver, reproductive system, hematopoietic system) and is carcinogenic. Inorganic As can cause neurologic effects after either inhalation or oral exposure (Lagerkvist and Zetterlund, 1994; Uede and Furukawa, 2003).

Acute exposure to As above 2 mg As/kg/day can lead to As encephalopathy, characterized by headache, confusion, seizure, coma, and death. Chronic exposures to low levels of As (0.03–0.1 mg As/day) affect the peripheral nervous system. As peripheral neuropathy starts as a symmetric numbness in hands and feet, sensory changes, and muscle tenderness. It then progresses into muscle weakness and a painful pins-and-needles sensation. As neuropathy is also characterized by reduced nerve-conducting velocity, which is a typical hallmark of Wallerian axonal degeneration (Le Quesne and McLeod, 1977; Greenberg, 1996). Symptoms of peripheral neuropathy can occur 1–2 weeks after exposure, and can persist even after the individual is no longer exposed to As. The reduction in nerve-conducting velocity is only partially resolved over time.

The exact mechanisms underlying As-induced neuropathy are not well understood. As increases oxidative stress and has been shown to be substituted for phosphate in ATP (Vahidnia et al., 2007). This will have a profound impact on energy-dependent cells like neurons. Additionally, As has been shown to interact with cytoskeletal proteins important for neuronal health. Neurofilament light subunit (NF-L) is significantly decreased in rats exposed to As (Vahidnia et al., 2006). NF-L is responsible for organizing and co-assembling neurofilaments composed of NF-L, NF-H, and NF-M. Neurofilament proteins are important structural proteins that control axonal caliber, transport, and signal, and their disruption is implicated in neurodegeneration (Al-Chalabi and Miller, 2003).

Diagnosis and treatment

Detection and diagnosis of As exposure are easily performed. Chronic arsenicosis causes nonmalignant dermal manifestations, such as hyperpigmentation and hyperkeratosis (Fig. 11.1). Pigmentation changes may be blotchy or a finely freckled raindrop pattern on the trunk and extremities. However, these visual dermatologic signs will not appear following an acute exposure. If clinical symptoms suggest As exposure or an As-containing compound is involved in an accidental spill, urine analysis is performed. Rapid urine spot tests are available to diagnose elevated levels of As, but they do not distinguish between the organic and inorganic As species. A speciation test is required in all cases in which total urine As is elevated (usually over 50 μg/L). As levels can also be accurately measured in hair and nails; due to known rates of hair and nail growth, this can be used to estimate duration of exposure. Chelation therapy is the most effective treatment for As exposure, the most commonly used being meso-2,3-dimercaptosuccinic acid (DMSA) and sodium 2,3-dimercapto-1propane sulfonic acid (DMPS). Interestingly, British anti-lewisite, the first chelating agent used in medicine, was developed during World War II for treatment of exposure to As-containing gases. British anti-lewisite was found to bind to a number of metals, including Cd, Pb, Hg, Ni, gold, Co, and Cr; however, it is no longer used due to multiple toxic side effects. Hemodialysis is also used to remove As from the blood stream, although there is only a short time frame for effectiveness before the As is absorbed.

Lead

The toxic effects of Pb have been appreciated for centuries. While Pb occurs naturally in the environment as ores with other metals (e.g., cerussite, anglesite, and galena), the highest levels of Pb occur in nature due to human activities. Pb compounds have been utilized by humans due to their softness, malleability, ductility, poor conductivity, and resistance to corrosion. Pb contamination of polar regions in Greenland date back to 500 BC–300 AD, when it is estimated that around 400 tons of Pb were deposited in the environment by the ancient Greeks and Romans. With all the Pb in the environment, Pb poisoning was first characterized and documented by early scientists and physicians (Needleman, 2004). The Romans were the first to use Pb on a large scale for water supply pipes, kitchen utensils and tableware, ceramic pigment, and sweetener for wine. Historically, Pb has been used in paints, stained glass, antiknock agents, and book printing (Hernberg, 2000). Current uses for metallic inorganic Pb include electrodes for electrolysis and in automobile batteries, radiation shielding and reactor coolant, leaded glass, raw material for machinery, semiconductors,

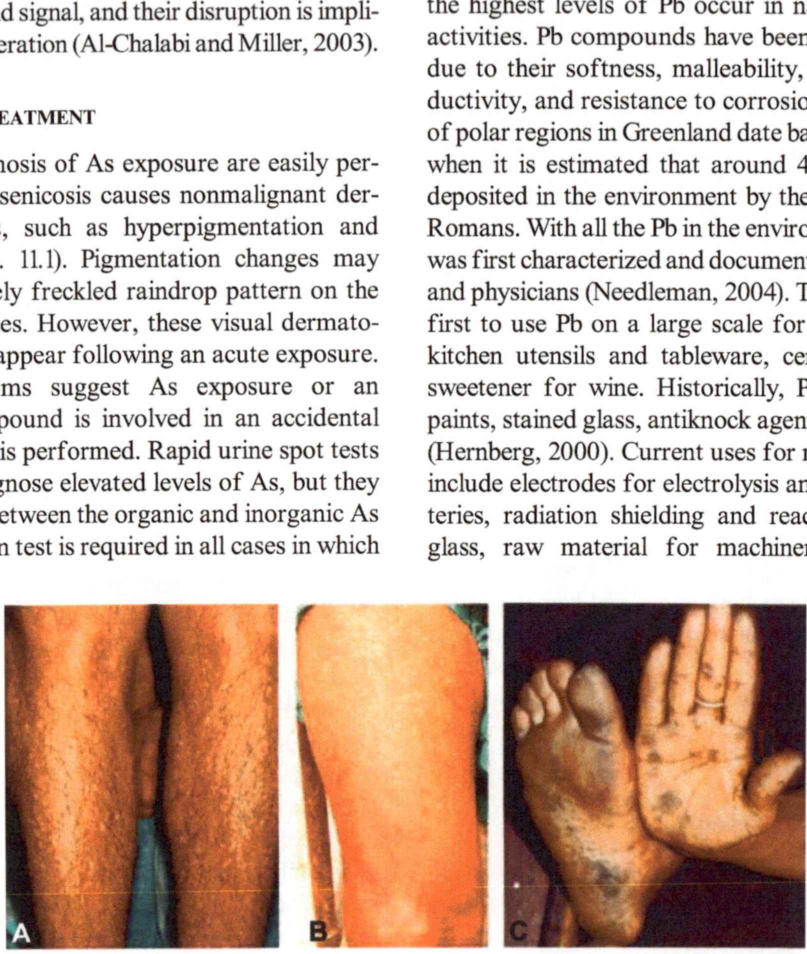

Fig. 11.1. Dermatologic symptoms of arsenic toxicity. (**A**) Raindrop pigmentation on both legs. (**B**) Raindrop pigmentation on right leg. (**C**) Hyperkeratosis on palm and sole. (Modified from Ghosh et al., 2007.)

ammunition, polyvinyl chloride (PVC) plastics, and sailing ballasts. Organolead compounds are in limited use due to their high toxicity and concerns for their ability to bioaccumulate in the environment. Tetraethyllead was widely used as an antiknock fuel additive for automobiles, but has been banned in the USA since 1996. Tetraethyllead and other organolead compounds still can be added to aviation fuels. Pb naphthenates and stearates are used in stabilizers for plastics and as Pb soaps.

Professions that have the highest risk for Pb exposure include manufacturing of Pb-containing products, recycling of batteries, car repair, Pb paint removal, demolition, refining, welding, and smelting. Regulations from OSHA have reduced the amount of Pb poisoning in the workplace; however the regulations are not strictly followed in some professions. Professions that may not be considered as high a risk for Pb exposure may not monitor Pb levels in their working conditions. For example, Sen et al. (2002) have found significant blood Pb levels in scaffolders who erected and dismantled scaffolding on Pb-painted structures. Both inorganic and organic Pb are absorbed through the lungs and gastrointestinal tract, while only organic Pb compounds can be absorbed through the skin. In occupational settings exposure is typically through inhalation. Once in the human body Pb distributes to brain, liver, kidney, and bone.

Pathology/physiology

While there are no physiologic functions for inorganic or organic Pb in the human body, the symptoms of Pb exposure significantly differ between the two.

Inorganic Pb affects several organ systems, including skeletal system, hematopoietic system, kidneys, reproductive system, and both the peripheral nervous system and CNS. Exposure to inorganic Pb has been associated with anemia, hypertension, sterility, osteoporosis, and spontaneous abortions (Gidlow, 2004). Neurologic symptoms characteristic of Pb exposure include muscle pain and weakness, constipation, colic, paresthesia, tremor, incoordination, asthenia, fatigue, headaches, insomnia, hearing deficits, nervous irritability, memory deficits, and deficits in attention and verbal comprehension. Severe chronic exposures to Pb lead to Pb encephalopathy, characterized by insomnia, poor attention, vomiting, convulsions, and coma. Chronic exposures to low levels of Pb have less severe neurologic and behavioral effects, such as decreased libido, mood changes (irritability, nervousness), headache, depression, decreased dexterity and reaction time, dizziness, fatigue, forgetfulness, lethargy, and impaired concentration. Demyelination of peripheral neurons severely affects conductance and leads to incoordination, muscle weakness, and pain.

Studies of Pb workers have shown that peripheral neuropathies significantly increase with blood Pb levels of 30 μg/100 mL and CNS dysfunction occurring with Pb levels over 60 μg/100 mL (Sappalainen et al., 1979; Triebig et al., 1984). The neuropathologic effects of chronic inorganic Pb exposure in the CNS include endothelial cell swelling and necrosis of the cerebral and cerebellar capillaries, capillary leakage and cerebral edema, loss of neuronal cells, cytoplasmic vacuolization, interstitial edema, and demyelination of nerve fibers. There are several molecular mechanisms implicated in Pb exposure, including generation of oxidative stress, disruption of Ca^{2+}-dependent cell signaling, inhibition of nitric oxide synthase, and alteration of glutamatergic signaling by antagonizing the *N*-methyl-D-aspartate receptor (Neal and Guilarte, 2010; Nava-Ruiz et al., 2012). Currently there is investigation on whether Pb is a risk factor for the development of AD. The Veterans Affairs normative aging study is a longitudinal cohort of men who were disease-free at time of recruited in 1963. High levels of bone and blood Pb were associated with decreased spatial copying skills, reduced pattern memory, and declines in cognitive functions (Payton et al., 1998; Schwartz et al., 2000). Animal studies also support this link. Pb exposure increased amyloid precursor protein mRNA and aggregated antibodies in rats, and increased amyloidogenesis and senile plaque deposition and upregulated amyloid precursor proteins in nonhuman primates (Basha et al., 2005a, b; Wu et al., 2008). Interestingly, there have not been studies in workers exposed to Pb correlating to development of AD.

While inorganic Pb can affect multiple organ systems, organic Pb is primarily a CNS toxin. The clinical symptoms of exposure to tetraethyllead and triethyllead include lethargy, inappetence, tremor, hypermotility, hyperexcitability, aggression, psychosis, hypothermia, convulsion, incoordination, ataxia, paralysis, and death. Pathologic studies in rabbits and mice have revealed that organolead compounds cause neuronal necrosis in the pyriform/entorhinal cortex, hippocampal formation, neocortex, and the amygdaloid nuclei. There is also neuronal swelling and necrosis in the brainstem and midbrain nuclei, pontine nuclei, basal nuclei, and anterior cervical spinal cord. Organolead compounds have been shown to alter glial cell populations in the brain and cause hypomethylation. Tetraethyllead is dealkylated to triethyllead by the liver, which has been shown to disturb Cl^- ionic flux, reversing γ-aminobutyric acid (GABA) neuron hyperpolarization, allowing the neurons to depolarize with exaggerated excitation (Chang, 1990). Triethyllead has been shown to increase the neuron utilization of glucose, allowing for accumulation of lactate and pyruvate, as well as mitochondrial degeneration (Chang, 1990).

Diagnosis and treatment

Diagnosis of Pb toxicity involves multiple measures to ascertain the extent of exposure. Levels of Pb in the blood and urine can be measured and are representative of current exposure. Chelatable Pb level is measured from the blood and urine following the administration of Ca-Na_2-ethylenediaminetetraacetic acid (EDTA). Pb levels in blood are in equilibrium with Pb levels in deposit tissues. Addition of a chelation agent removes diffusible Pb in plasma. The diffusible Pb in tissues then equilibrate with the plasma, which can be measured for additional insight into the Pb exposure (Alessio et al., 2007). Blood and urinary δ-aminolevulinic acid (ALA) levels and ALA dehydrogenase activity are often measured, as Pb is known to inhibit the enzyme, which is integral in the formation of hemoglobin. This measure can estimate recent Pb exposure as well as reflect extent of damage to the hematopoietic system. Bone Pb levels more accurately represent total body burden. As Pb^{2+} competes with Ca^{2+}, it is therefore incorporated into bones during normal osteoblast/osteoclast homeostatic activities. Storage in bone allows for a 5–19-year half-life for Pb (Rabinowitz, 1991). Levels of Pb in bone can be measured from a bone marrow biopsy or from X-ray fluorimetry (Skerfving et al., 1993).

Chelation therapy with EDTA, DMSA, and D-penicillamine is the primary therapy for Pb toxicity. Effects of Pb on the kidney and hematopoietic system are reversible following treatment; however, the effects on the CNS are only partially reversible (Yassi, 1980).

Manganese

Mn is the 12th most abundant element on earth and is found in a variety of ores, the most important for industry being pyrolusite (MnO_2). Mn can exist in 11 different oxidation states, the most common being +2, +3, +4, and +7. It is used in a wide range of industrial processes and commercial products. Mn is essential for steel and stainless-steel production and formation of Al alloys. Mn is used in the production of oxygen and chlorine, as a brown or pink pigment, as cathode in alkaline batteries, and as an antiknock agent in unleaded gasoline (methylcyclopentadienyl Mn tricarbonyl). Mn is also incorporated into fungicides, such as maneb and mancozeb. Occupational exposure to Mn is the primary cause of human Mn intoxication. Exposure to Mn in the occupational setting is predominantly inhalational, as many of the industrial processes utilizing Mn involve welding, smelting, or the creation of fine dusts.

Unlike many of the metals discussed in this chapter, Mn is an essential metal for humans, entering our diet in foods such as grains (wheat germ and bran), nuts (hazelnuts and pecans), chocolate, mussels, clams, soybeans, seeds (flax, sesame, pumpkin, sunflower, and pine nuts), and spices (chilli powder, cloves, and saffron). Mn is necessary for protein and energy metabolism, bone mineralization, metabolic regulation, and cellular protection from ROS, and acts as a cofactor for many enzymes, lectins, and integrins. While Mn is essential for life, too much Mn is associated with a parkinsonian-like disease termed manganism.

Pathology/physiology

Manganism results from excessive Mn levels in the brain. From its first description in 1837 by Couper of Glasgow, manganism has been compared to PD. Both diseases present with akinetic rigidity, dystonia, tremor, posture instability, and bradykinesia, and can develop dementia. However the neuropathology between the two diseases is strikingly different. Manganism is characterized by a loss of neurons in the globus pallidus and substantia nigra pars reticulata. Additionally, the putamen, caudate nucleus, and subthalamic nucleus can be damaged. In contrast, PD mainly affects the substantia nigra pars compacta and locus coeruleus; the dopaminergic neurons in these brain regions are typically spared in manganism (Olanow, 2004). Lewy body protein aggregates present in PD are not present in manganism (Perl and Olanow, 2007). The extrapyramidal symptoms arising in Mn-exposed individuals include headaches, muscular cramps, fatigue, psychosis, and hypertonia with cogwheel rigidity. Manganism progresses after the cessation of exposure, with a plateau for the neurologic symptoms occurring after the initial 5–10 years (Huang et al., 2007). Development of manganism can occur after chronic exposure to Mn, but in some cases it appears after a subchronic exposure of one or more months to large doses of Mn (Aschner et al., 2009).

It is unclear why Mn affects specific brain regions over others. Mn easily crosses the BBB and is transported by DMT1, ZIP8, transferrin, organic anion transporter, monocarboxylase transporter, and OATP. Mn alters several neurotransmitter systems (Aschner et al., 2009). Mn alters norepinephrine signaling, affecting norepinephrine levels and uptake, and decreasing α_2-adrenogeric receptor expression in the locus coeruleus and substantia nigra (Anderson et al., 2009). This may contribute to behavioral problems associated with manganism. Striatal GABA levels and uptake are decreased in rats following Mn exposure (Garcia et al., 2006; Anderson et al., 2007). Mn-exposed rats show decreased locomotor activity and motor deficits (Normandin et al., 2004), suggesting that alteration of GABA by Mn may contribute to locomotor effects, such as dystonia, observed in manganism. Levels of glutamate are increased in rats exposed to Mn, while the glutamate transporter and glutamine synthetase are decreased in astrocytes (Erikson et al., 2002, 2008; Xu et al., 2010), which may lead to excitotoxicity. Additionally, due to the similarities with PD, researchers are now focusing on how Mn affects the dopaminergic system. Along with the alteration in neurotransmission, there is inflammatory activation of astrocytes and microglia during manganism. Astrocytes accumulate more than 50-fold greater concentrations of Mn than neurons (Wedler et al., 1989), making them a target for Mn-induced toxicity. Studies of rat brains from animals exposed to Mn have shown increased number of glial cells (Alzheimer type II astrocytosis) in the globus pallidus (Hazell et al., 2006). Astrocytes and microglia may contribute to Mn toxicity through the release of proinflammatory mediators, such as prostaglandin E_2, nitric oxide, tumor necrosis factor-α, and interleukin-6. Levels of nitric oxide synthase are significantly increased in astrocytes surrounding areas of degeneration in the globus pallidus and striatum (Liu et al., 2006a), which can increase ROS and contribute to cell death.

The involvement of Mn in the development of PD is currently a debated topic. The pathologic differences between manganism and PD and the similarity of clinical symptoms suggest that there are shared mechanisms between the two conditions related to the homeostasis of Mn in tissue. Diagnosis of manganism vs PD is typically made based on exposure history; however, in areas with high levels of Mn, either naturally occurring or from industrial pollution, there is an increased PD prevalence in comparison to surrounding regions (Lucchini et al., 2007). This suggests a role for Mn in the development of PD in people who are not exposed to Mn occupationally.

Diagnosis and treatment

Blood and urine analysis is initially performed if clinical symptoms suggest manganism. As the globus pallidus is the main site of damage in manganism, MRI scans can be performed to examine its structure. Using the Pallidal Index, the ratio between the intensity of the T1 signal in the pallidum and that of the frontal white matter, one can semiquantitatively estimate Mn concentrations (Krieger et al., 1995). Fluorodopa PET scans can assess the integrity of the substantia nigra pars compacta dopaminergic system (Fig. 11.2). PD patients have a loss of cells in the substantia nigra pars compacta; however, fluorodopa PET scans have been shown to be normal in Mn-induced parkinsonism (Wolters et al., 1989; Shinotoh et al., 1995). It is important to differentiate between PD and manganism, as patients with manganism respond poorly to levodopa (Poewe and Wenning, 2002). Chelation therapy is the most common treatment for manganism. EDTA significantly increases the amount of Mn excreted in the urine and the blood Mn levels; however, the effects on clinical neurologic symptoms are debated (Calne et al., 1994; Ono et al., 2002; Crossgrove and Zheng, 2004).

Mercury

Hg is a highly toxic element which occurs as several different species – elemental Hg (Hg^0), inorganic Hg compounds (mercurous Hg_2^{+2} and mercuric Hg^{+2}), and organic Hg compounds (e.g., MeHg, ethylmercury, etc.). These species have different chemical properties utilized in industrial processes. Additionally the three species have different toxicologic effects following exposure (Clarkson and Magos, 2006). Hg^0 is rarely found in its pure state, but is extracted from cinnabar, corderoite, or livingstonite ores. Hg is present in the

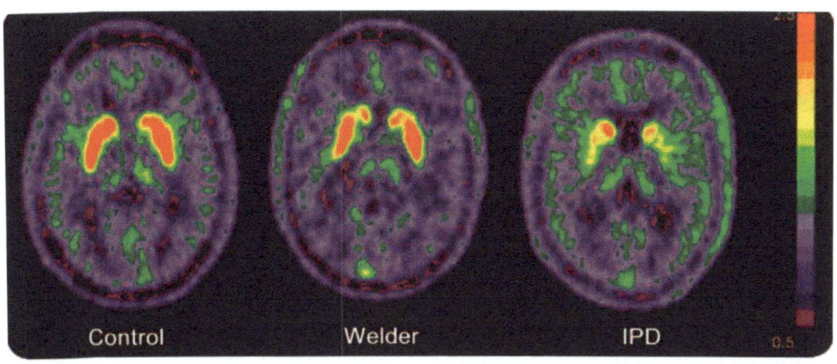

Fig. 11.2. Differential positron emission tomography scans from patients with Parkinson's disease and manganism. Composite images of decay-corrected counts from 24 to 94 minutes from a representative control, welder, and subject with idiopathic Parkinson disease (IPD) normalized to the reference region. [^{18}F] FDOPA uptake is reduced in the caudate of the welder in comparison to the control, while the posterior putamen is the most affected region in the subject with IPD. (Reproduced from Criswell et al., 2011.)

earth's crust and there are trace levels of Hg in the air and water. Volcanic activity and forest fires are responsible for natural deposition of Hg into our environment, while industrial processes (coal combustion, gold production, smelting, and cement production) and industrial waste are man-made sources. Release of Hg into the environment allows for the conversion of inorganic Hg into MeHg, a major environmental contaminant, by bacteria. MeHg bioaccumulates and biomagnifies up the food chain in aquatic environments, with the largest predatory fish (e.g., tuna, swordfish, and sharks) containing the highest levels of MeHg. Over time Hg can combine with sulfur, forming cinnabar. Collectively this is termed the Hg cycle.

Elemental Hg is a volatile liquid at room temperature and pressure. It was once commonly used in thermometers; however, due to concerns of human exposure and release of Hg into the environment, Hg is no longer used. Hg^0 is not as commonly used in industry as it once was, with the exception of manufacturing of chlorine-alkali, dental amalgams, electronic switches, and fluorescent lamps. Inorganic Hg compounds are used in a wide range of medical and cosmetic products: antiseptics, teething powders, and skin-lightening creams. Organomercury compounds consist of Hg complexed to alkyl or phenyl groups. Phenyl Hg compounds are used as preservatives in medicines, such as thimerosal, antiseptics, and disinfectants (nitromersol, merbromin, and phenylmercuric borate). Alkyl Hg compounds have been used as insecticides and fungicides; however, due to mass poisoning incidents involving Hg-treated grain seeds, it is no longer used. Organomercury compounds used in industry are synthesized from chemical reactions of inorganic Hg with diazonium salts, Girgnard reagents, organolithium compounds, alkenes, or arenes.

Pathology/physiology

Acute exposures to Hg^0 vapor do not generally affect the CNS but cause respiratory distress and dyspnea. Chronic exposures, however, can induce tremors, delusions, memory loss, erethism, anxiety, and cognition problems. These effects can eventually disappear after exposure ceases unless the individual is exposed to a large dose (Albers et al., 1988). Preparation of and release from dental amalgam fillings represents an area of active debate. Amalgam fillings consist of 50% Hg^0 with Cu, silver, and Zn which are mixed from a powder immediately before use. There is concern for an increased risk of cognitive and neuropsychologic symptoms for dental professionals as urine concentrations of Hg^0 exceed 500 nmol/L (normal levels: 10–20 μg/L) (Langworth et al., 1997; Rohling and Demakis, 2006; Hilt et al., 2009). There have been claims that people with amalgam fillings have symptoms similar to Hg vapor exposure or that amalgams induce AD; however no association has been found by several epidemiologic studies (Nitschke et al., 2000; Factor-Litvak et al., 2003; Watson et al., 2011).

Inorganic mercurous and mercuric compounds do not affect the CNS; their target organs are the gastrointestinal tract, immune system, and kidneys.

Organomercury compounds have been long recognized as neurotoxic, with the most well characterized being MeHg. The effects of chronic MeHg were made apparent following the Minamata incident in Japan, where around 200 000 people were exposed to MeHg over the course of 20 years. Adults developed a wide range of neurologic problems, such as paresthesia, blurred vision, hearing impairment, olfactory and gustatory disturbances, slurred speech, ataxic gait, clumsiness, muscle weakness, dysarthria, irritability, memory loss, depression, and sleeping disturbance. Analysis of the postmortem brain showed a loss of neurons in the granular layer of cerebellum as well as loss of granular cells in the somatosensory, visual and auditory cortical areas (Ekino et al., 2007). Similar effects were observed following consumption of contaminated grain in Iraq. Studies in the Iraqi cohort were critical in establishing a correlation between hair levels of Hg with physiologic symptoms (Bakir et al., 1973), allowing for more effective biomonitoring of Hg exposure.

Several mechanisms have been implicated in MeHg toxicity. In the nervous system, MeHg induces oxidative stress, disrupts calcium homeostasis, alters glutamate and GABA signaling, and alters cellular signaling and gene expression. MeHg is a soft electrophile and interacts with thiol (-SH) and selenol (-SeH) groups, forming stable complexes with defined stoichiometry. Thiol groups are important functional moieties in antioxidant molecules and enzymes, such as GSH, thioredoxins, and glutaredoxins. MeHg interacts with GSH to form an excretable glutamine synthetase (GS)–$HgCH_3$ complex (Ballatori and Clarkson, 1985), increasing the GSSG: GSH ratio and reducing the antioxidant capacity both in astrocytes and microglia (Wang et al., 2009a; Ni et al., 2010). Additionally, MeHg increases production of ROS through inhibition of glutathione peroxidases and of the electron transport chain of mitochondria (Mori et al., 2007, 2011). This results in lipid peroxidation, protein oxidation, and DNA oxidation.

MeHg directly inhibits proteins necessary for calcium homeostasis, glutamate transport, and GABA signaling. Glutamate is the major excitatory neurotransmitter in the nervous system, and accumulates in astrocytes following MeHg exposure (Aschner et al., 1993). This increase in glutamate can result in excitotoxicity. Mercuric chloride, which accumulates in the brain following

MeHg exposure, decreases GS activity and inhibits glutamine transport in cultured astrocytes (Allen et al., 2001). MeHg decreases GABA signaling, the most abundant inhibitory neurotransmitter, by modifying $GABA_A$ receptor conformation (Fonfria et al., 2001). Glutamate and GABA signaling are crucial for establishing equilibrium between excitation and inhibition; their imbalance may play a key role in MeHg neurotoxicity. Finally MeHg has been shown to alter several cell signaling pathways, including phospholipase C, calcium signaling, and phosphatidylinositol 3-kinases (PI3K/Akt), and nuclear factor kappa-light-chain-enhancer of activated B cells (NF-κB) (Dong et al., 2001; Chen et al., 2006; Kang et al., 2006; Chang, 2011).

Diagnosis and treatment

Hg levels are usually measured in urine and blood for diagnostic purposes following acute exposures, although levels are measured in hair and nails to estimate levels of Hg following chronic exposure. Radiography is also performed if Hg is ingested; the location of the metal will appear radiopaque on the radiograph.

Patients who exhibit Hg poisoning will immediately be put on chelation therapy with DMPS. DMPS has been shown to bind Hg and significantly increase its excretion from the human body. In one of the largest studies examining DMPS efficacy in Philippine gold miners and people living downstream of the gold mine whose diet included locally caught fish, DMPS increased urinary Hg up to 85% and there were significant improvements in tests for tremor, ataxia, and visual perception. Patients also reported subjective improvements in memory, sleeplessness, fatigue, anxiety, and paresthesias (Drasch et al., 2001; Bose-O'Reilly et al., 2003). Inorganic Hg is caustic; dilution with water or other fluids is often performed. For ingested Hg, gastric lavage and bowel irrigation are performed to remove as much Hg as possible before it enters the circulation.

Trimethyltin

There are several types of organotin compounds (also called stannanes) that are used in industry. Organotin compounds are commonly generated through reaction of tin (Sn) halides with a Grignard reagent. All stannanes are characterized by a central Sn atom complexed to multiple carbon groups. The most common compounds include tetrabutyltin, tributyltin oxide, triphenyltin acetate, hydroxide, or chloride, azocyclotin, cyhexatin, tetraethyltin, triethyltin, and trimethyltin (TMT) chloride. Organotin compounds are most often used in the chemical synthesis of polymers as PVC heat stabilizers, manufacturing of polyurethane and silicone, as well as for wood preservatives. While the tetraorganotins are very stable and demonstrate low toxicity, the triorganotins are significantly more toxic. This property has been exploited for the pesticide industry; triorganotins are used as bactericides, fungicides, miticides, and acaricides. Neurotoxicity resulting from exposure to organotins has been recognized, with one of the earliest reported large-scale incidents occurring in France in 1954 resulting from the contamination of Stalinon, an oral preparation of diethyltin diiodide for the treatment of boils, staphylococcal skin infections, osteomyelitis, anthrax, and acne, with triethyltin iodide being a major contaminant. Around 270 people were intoxicated, with the primary clinical signs being nausea, vertigo, visual disturbances, papilledema, paraplegia, and convulsions. Autopsies of deceased individuals revealed pronounced cerebral edema of the white matter (Gruner, 1958). Studies from human occupational exposures as well as rodent models have shown the propensity of organotin compounds to cause cerebral edema and neurodegeneration. The neurotoxicity of the most well-characterized compound, TMT, will be discussed below.

Pathology/physiology

Reports of TMT toxicity in humans have primarily focused on acute accidental exposures. Headaches, memory deficits, confabulation, poor concentration, forgetfulness, confusion, disorientation, aggressiveness, vision and hearing abnormalities, insomnia, fatigue, anorexia, ataxia, loss of vigilance, loss of libido, stuttering, hyperactivity, seizures, and death have been described in cases involving chemists and industrial workers accidently exposed to TMT from spills and explosions (Fortemps et al., 1978; Brown et al., 1979; Ross et al., 1981; Rey et al., 1984; Feldman et al., 1993; Yoo et al., 2007). Accurate measurements on the size of dose the individuals were exposed to were not well documented. Exposure of rodents to TMT recapitulates many of the symptoms of TMT poisoning, including tremors and memory loss (Chang et al., 1982; Mignini et al., 2012). Currently there are no reports on chronic exposure to TMT.

The changes in memory, mood, and behavior observed upon TMT exposure are consistent with damage to the limbic system. Histopathology from rodent studies shows a preferential vulnerability of the hippocampus to TMT, with changes occurring in the pyramidal cells of the CA2 and CA3 areas and in the granule cells of the dentate gyrus (Chang, 1986; Whittington et al., 1989). The structure of the amygdala, neocortex, pyriform cortex, entorhinal cortex, and olfactory bulb also is affected by TMT exposure. Neuronal death in these locations is induced by apoptosis, as several studies have characterized chromatin condensation, DNA fragmentation, and activated caspase-3 following TMT exposure (Fiedorowicz et al., 2001; Geloso et al., 2002). It is unknown what mechanism is responsible for the selective neurodegeneration in the hippocampus. This is an active area of research, as TMT-induced neurodegeneration has been used to model AD and to evaluate the effects of neuroprotective agents in neurodegeneration.

TMT exposure induces oxidative stress and activates pathways that can lead to cytotoxic responses, such as NF-κB, c-fos, and JNK. Additionally, TMT has been shown to activate astrocytes and microglia, which can lead to neuroinflammation through the release of inflammatory mediators, such as cytokines and chemokines, which can induce or exacerbate neuronal damage (Brabeck et al., 2002; Brock and O'Callaghan, 1987; Rohl and Sievers, 2005).

Diagnosis and treatment

Diagnosis of TMT poisoning has primarily relied on measurements of urine and blood Sn levels. Treatments for TMT poisoning include dilution with water or milk, administration of activated charcoal, and gastric lavage for oral exposures. Benzodiazepines, such as diazepam, may be given for seizures, as well as phenobarbital or propofol. Patients are usually hospitalized following exposure to monitor for hypotension, dysrhythmias, hypoglycemia, electrolyte disturbances, respiratory depression, hypoxia, and coma (Rumack, 2013). Unfortunately, TMT induces irreversible CNS changes, and as a result very few people can completely recover. Of the 270 cases of triethyltin poisonings from Stalinon, 100 people died shortly after exposure, while only 10 made a full recovery (Feldman et al., 1993).

SUMMARY

Occupational exposure to metals is an all too common occurrence in our industrialized society. The effects of metal exposure can be on a variety of organ systems, including the CNS. Neurotoxic effects of metals can be varied and will depend on the metal, dose, duration, and route of exposure. Reversibility of metal-induced neurotoxicity may not always be possible, but treatment upon diagnosis can prevent death and help lessen symptoms.

ACKNOWLEDGMENTS

This review was supported in part by grants from the National Institute of Environmental Health Sciences ESR01-10563, R01-07331, and ES-020852.

References

Albers JW, Kallenbach LR, Fine LJ et al. (1988). Neurological abnormalities associated with remote occupational elemental mercury exposure. Ann Neurol 24: 651–659.

Al-Chalabi A, Miller CC (2003). Neurofilaments and neurological disease. Bioessays 25: 346–355.

Alessio L, Campagna M, Lucchini R (2007). From lead to manganese through mercury: mythology, science, and lessons for prevention. Am J Ind Med 50: 779–787.

Allen JW, Mutkus LA, Aschner M (2001). Mercuric chloride, but not methylmercury, inhibits glutamine synthetase activity in primary cultures of cortical astrocytes. Brain Res 891: 148–157.

Anderson JG, Cooney PT, Erikson KM (2007). Brain manganese accumulation is inversely related to gamma-amino butyric acid uptake in male and female rats. Toxicol Sci 95: 188–195.

Anderson JG, Fordahl SC, Cooney PT et al. (2009). Extracellular norepinephrine, norepinephrine receptor and transporter protein and mRNA levels are differentially altered in the developing rat brain due to dietary iron deficiency and manganese exposure. Brain Res 1281: 1–14.

Antonijevic MM, Dimitrijevic MD, Milic SM et al. (2012). Metal concentrations in the soils and native plants surrounding the old flotation tailings pond of the copper mining and smelting complex Bor (Serbia). J Environ Monit 14: 866–877.

Aschner JL, Aschner M (2005). Nutritional aspects of manganese homeostasis. Mol Aspects Med 26: 353–362.

Aschner M, Du YL, Gannon M et al. (1993). Methylmercury-induced alterations in excitatory amino acid transport in rat primary astrocyte cultures. Brain Res 602: 181–186.

Aschner M, Erikson KM, Herrero Hernandez E et al. (2009). Manganese and its role in Parkinson's disease: from transport to neuropathology. Neuromolecular Med 11: 252–266.

Bakir F, Damluji SF, Amin-Zaki L et al. (1973). Methylmercury poisoning in Iraq. Science 181: 230–241.

Ballatori N, Clarkson TW (1985). Biliary secretion of glutathione and of glutathione-metal complexes. Fundam Appl Toxicol 5: 816–831.

Bannon DI, Portnoy ME, Olivi L et al. (2002). Uptake of lead and iron by divalent metal transporter 1 in yeast and mammalian cells. Biochem Biophys Res Commun 295: 978–984.

Baranowska-Bosiacka I, Gutowska I, Marchetti C et al. (2011). Altered energy status of primary cerebellar granule neuronal cultures from rats exposed to lead in the pre- and neonatal period. Toxicology 280: 24–32.

Basha MR, Murali M, Siddiqi HK et al. (2005a). Lead (Pb) exposure and its effect on APP proteolysis and Abeta aggregation. FASEB J 19: 2083–2084.

Basha MR, Wei W, Bakheet SA et al. (2005b). The fetal basis of amyloidogenesis: exposure to lead and latent overexpression of amyloid precursor protein and beta-amyloid in the aging brain. J Neurosci 25: 823–829.

Basu N, Nam DH, Kwansaa-Ansah E et al. (2011). Multiple metals exposure in a small-scale artisanal gold mining community. Environ Res 111: 463–467.

Belyaeva EA, Sokolova TV, Emelyanova LV et al. (2012). Mitochondrial electron transport chain in heavy metal-induced neurotoxicity: effects of cadmium, mercury, and copper. Scientific World Journal 2012: 136063.

Bose-O'Reilly S, Drasch G, Beinhoff C et al. (2003). The Mt. Diwata study on the Philippines 2000 – treatment of mercury intoxicated inhabitants of a gold mining area with DMPS (2,3-dimercapto-1-propane-sulfonic acid, Dimaval). Sci Total Environ 307: 71–82.

Brabeck C, Michetti F, Geloso MC et al. (2002). Expression of EMAP-II by activated monocytes/microglial cells in different regions of the rat hippocampus after trimethyltin-induced brain damage. Exp Neurol 177: 341–346.

Breydo L, Uversky VN (2011). Role of metal ions in aggregation of intrinsically disordered proteins in neurodegenerative diseases. Metallomics 3: 1163–1180.

Brock TO, O'Callaghan JP (1987). Quantitative changes in the synaptic vesicle proteins synapsin I and p38 and the astrocyte-specific protein glial fibrillary acidic protein are associated with chemical-induced injury to the rat central nervous system. J Neurosci 7: 931–942.

Brown AW, Aldridge WN, Street BW et al. (1979). The behavioral and neuropathologic sequelae of intoxication by trimethyltin compounds in the rat. Am J Pathol 97: 59–82.

Calatayud M, Barrios JA, Velez D et al. (2012). In vitro study of transporters involved in intestinal absorption of inorganic arsenic. Chem Res Toxicol 25: 446–453.

Calne DB, Chu NS, Huang CC et al. (1994). Manganism and idiopathic parkinsonism: similarities and differences. Neurology 44: 1583–1586.

Chang LW (1986). Neuropathology of trimethyltin: a proposed pathogenetic mechanism. Fundam Appl Toxicol 6: 217–232.

Chang LW (1990). The neurotoxicology and pathology of organomercury, organolead, and organotin. J Toxicol Sci 15 (Suppl 4): 125–151.

Chang JY (2011). Methylmercury-induced IL-6 release requires phospholipase C activities. Neurosci Lett 496: 152–156.

Chang LW, Tiemeyer TM, Wenger GR et al. (1982). Neuropathology of trimethyltin intoxication. II. Electron microscopic study on the hippocampus. Environ Res 29: 445–458.

Chen XL, Dodd G, Thomas S et al. (2006). Activation of Nrf2/ARE pathway protects endothelial cells from oxidant injury and inhibits inflammatory gene expression. Am J Physiol Heart Circ Physiol 290: H1862–H1870.

Clarkson TW, Magos L (2006). The toxicology of mercury and its chemical compounds. Crit Rev Toxicol 36: 609–662.

Criswell SR, Perlmutter JS, Videen TO et al. (2011). Reduced uptake of [^{18}F]FDOPA PET in asymptomatic welders with occupational manganese exposure. Neurology 76 (15): 1296–1301.

Crossgrove J, Zheng W (2004). Manganese toxicity upon overexposure. NMR Biomed 17: 544–553.

Dong L, Li Z, Bi X et al. (2001). Effects of methyl mercury chloride on nuclear factor-kappa B DNA binding activities of nuclear protein extracts from developing rat cerebra and cerebella. Wei Sheng Yan Jiu 30: 7–9.

Drasch G, Bose-O'Reilly S, Beinhoff C et al. (2001). The Mt. Diwata study on the Philippines 1999 – assessing mercury intoxication of the population by small scale gold mining. Sci Total Environ 267: 151–168.

Dwivedi N, Mehta A, Yadav A et al. (2011). MiADMSA reverses impaired mitochondrial energy metabolism and neuronal apoptotic cell death after arsenic exposure in rats. Toxicol Appl Pharmacol 256: 241–248.

Ekino S, Susa M, Ninomiya T et al. (2007). Minamata disease revisited: an update on the acute and chronic manifestations of methyl mercury poisoning. J Neurol Sci 262: 131–144.

Erikson KM, Suber RL, Aschner M (2002). Glutamate/aspartate transporter (GLAST), taurine transporter and metallothionein mRNA levels are differentially altered in astrocytes exposed to manganese chloride, manganese phosphate or manganese sulfate. Neurotoxicology 23: 281–288.

Erikson KM, Dorman DC, Lash LH et al. (2008). Duration of airborne-manganese exposure in rhesus monkeys is associated with brain regional changes in biomarkers of neurotoxicity. Neurotoxicology 29: 377–385.

Factor-Litvak P, Hasselgren G, Jacobs D et al. (2003). Mercury derived from dental amalgams and neuropsychologic function. Environ Health Perspect 111: 719–723.

Farina M, Rocha JB, Aschner M (2011). Mechanisms of methylmercury-induced neurotoxicity: evidence from experimental studies. Life Sci 89: 555–563.

Feldman RG, White RF, Eriator II (1993). Trimethyltin encephalopathy. Arch Neurol 50: 1320–1324.

Fiedorowicz A, Figiel I, Kaminska B et al. (2001). Dentate granule neuron apoptosis and glia activation in murine hippocampus induced by trimethyltin exposure. Brain Res 912: 116–127.

Fonfria E, Rodriguez-Farre E, Sunol C (2001). Mercury interaction with the GABA(A) receptor modulates the benzodiazepine binding site in primary cultures of mouse cerebellar granule cells. Neuropharmacology 41: 819–833.

Forte G, Petrucci F, Bocca B (2008). Metal allergens of growing significance: epidemiology, immunotoxicology, strategies for testing and prevention. Inflamm Allergy Drug Targets 7: 145–162.

Fortemps E, Amand G, Bomboir A et al. (1978). Trimethyltin poisoning. Report of two cases. Int Arch Occup Environ Health 41: 1–6.

Garcia SJ, Gellein K, Syversen T et al. (2006). A manganese-enhanced diet alters brain metals and transporters in the developing rat. Toxicol Sci 92: 516–525.

Geloso MC, Vercelli A, Corvino V et al. (2002). Cyclooxygenase-2 and caspase 3 expression in trimethyltin-induced apoptosis in the mouse hippocampus. Exp Neurol 175: 152–160.

Geremias R, Bortolotto T, Wilhelm-Filho D et al. (2012). Efficacy assessment of acid mine drainage treatment with coal mining waste using *Allium cepa* L. as a bioindicator. Ecotoxicol Environ Saf 79: 116–121.

Ghosh P, Banerjee M, De Chaudhuri S et al. (2007). Comparison of health effects between individuals with and without skin lesions in the population exposed to arsenic through drinking water in West Bengal, India. J Expo Anal Environ Epidemiol 17: 215–223.

Gidlow DA (2004). Lead toxicity. Occup Med (Lond) 54: 76–81.

Greenberg SA (1996). Acute demyelinating polyneuropathy with arsenic ingestion. Muscle Nerve 19: 1611–1613.

Gruner JE (1958). Cerebrospinal lesions following the ingestion of diethyl-tin diiodide (stalinon). Rev Neurol (Paris) 98: 109–116.

Guilarte TR (2010). Manganese and Parkinson's disease: a critical review and new findings. Environ Health Perspect 118: 1071–1080.

Halatek T, Sinczuk-Walczak H, Rydzynski K (2005). Prognostic significance of low serum levels of Clara cell phospholipid-binding protein in occupational aluminium neurotoxicity. J Inorg Biochem 99: 1904–1911.

Hazell AS, Normandin L, Norenberg MD et al. (2006). Alzheimer type II astrocytic changes following sub-acute exposure to manganese and its prevention by antioxidant treatment. Neurosci Lett 396: 167–171.

Healy J, Bradley SD, Northage C et al. (2001). Inhalation exposure in secondary aluminium smelting. Ann Occup Hyg 45: 217–225.

He Q, Du T, Yu X et al. (2011). DMT1 polymorphism and risk of Parkinson's disease. Neuroscience Letters 501: 128–131.

Hernberg S (2000). Lead poisoning in a historical perspective. Am J Ind Med 38: 244–254.

Hilt B, Svendsen K, Syversen T et al. (2009). Occurrence of cognitive symptoms in dental assistants with previous occupational exposure to metallic mercury. Neurotoxicology 30: 1202–1206.

Hosovski E, Mastelica Z, Sunderic D et al. (1990). Mental abilities of workers exposed to aluminium. Med Lav 81: 119–123.

Huang CC, Chu NS, Lu CS et al. (2007). The natural history of neurological manganism over 18 years. Parkinsonism Relat Disord 13: 143–145.

Jiang X, McDermott JR, Ajees AA et al. (2010). Trivalent arsenicals and glucose use different translocation pathways in mammalian GLUT1. Metallomics 2: 211–219.

Jomova K, Jenisova Z, Feszterova M et al. (2011). Arsenic: toxicity, oxidative stress and human disease. J Appl Toxicol 31: 95–107.

Julka D, Sandhir R, Gill KD (1995). Altered cholinergic metabolism in rat CNS following aluminum exposure: implications on learning performance. J Neurochem 65: 2157–2164.

Kang MS, Jeong JY, Seo JH et al. (2006). Methylmercury-induced toxicity is mediated by enhanced intracellular calcium through activation of phosphatidylcholine-specific phospholipase C. Toxicol Appl Pharmacol 216: 206–215.

Kavasi N, Vigh T, Sorimachi A et al. (2010). Effective dose of miners due to natural radioactivity in a manganese mine in Hungary. Radiat Prot Dosimetry 141: 432–435.

Kerin EJ, Lin HK (2010). Fugitive dust and human exposure to heavy metals around the Red Dog Mine. Rev Environ Contam Toxicol 206: 49–63.

Kilburn KH (1998). Neurobehavioral impairment and symptoms associated with aluminum remelting. Arch Environ Health 53: 329–335.

Kirshen C, Pratt M (2012). Dental allergic contact dermatitis: an interesting case series and review of the literature. Dermatitis 23: 222–226.

Korotkov SM, Lapin AV (2003). Thallium induces opening of the mitochondrial permeability transition pore in the inner membrane of rat liver mitochondria. Dokl Biochem Biophys 392: 247–252.

Krewski D, Yokel RA, Nieboer E et al. (2007). Human health risk assessment for aluminium, aluminium oxide, and aluminium hydroxide. J Toxicol Environ Health B Crit Rev 10 (Suppl 1): 1–269.

Krieger D, Krieger S, Jansen O et al. (1995). Manganese and chronic hepatic encephalopathy. Lancet 346: 270–274.

Kumar V, Bal A, Gill KD (2008). Impairment of mitochondrial energy metabolism in different regions of rat brain following chronic exposure to aluminium. Brain Res 1232: 94–103.

Kushkuley J, Metkar S, Chan WK et al. (2010). Aluminum induces neurofilament aggregation by stabilizing cross-bridging of phosphorylated c-terminal sidearms. Brain Res 1322: 118–123.

Lagerkvist BJ, Zetterlund B (1994). Assessment of exposure to arsenic among smelter workers: a five-year follow-up. Am J Ind Med 25: 477–488.

Langui D, Probst A, Anderton B et al. (1990). Aluminium-induced tangles in cultured rat neurones. Enhanced effect of aluminium by addition of maltol. Acta Neuropathol 80: 649–655.

Langworth S, Sallsten G, Barregard L et al. (1997). Exposure to mercury vapor and impact on health in the dental profession in Sweden. J Dent Res 76: 1397–1404.

Le Quesne PM, McLeod JG (1977). Peripheral neuropathy following a single exposure to arsenic. Clinical course in four patients with electrophysiological and histological studies. J Neurol Sci 32: 437–451.

Letzel S, Schaller K, Angerer J et al. (1996). Biological monitoring of occupational aluminum powder exposure. Ann Occup Hyg 3: 271–280.

Lindsay DR, Farley KJ, Carbonaro RF (2012). Oxidation of Cr(III) to Cr(VI) during chlorination of drinking water. J Environ Monit 14: 1789–1797.

Liu X, Sullivan KA, Madl JE et al. (2006a). Manganese-induced neurotoxicity: the role of astroglial-derived nitric oxide in striatal interneuron degeneration. Toxicol Sci 91: 521–531.

Liu Z, Sanchez MA, Jiang X et al. (2006b). Mammalian glucose permease GLUT1 facilitates transport of arsenic trioxide and methylarsonous acid. Biochem Biophys Res Commun 351: 424–430.

Liu N, Yue Q, Wang YQ et al. (2008). Coordination compounds of *bis*(5-tetrazolyl)amine with manganese(II), zinc(II) and cadmium(II): synthesis, structure and magnetic properties. Dalton Trans 4621–4629.

Lucchini RG, Albini E, Benedetti L et al. (2007). High prevalence of Parkinsonian disorders associated to manganese exposure in the vicinities of ferroalloy industries. Am J Ind Med 50: 788–800.

Magalhaes M (2002). Arsenic. An environmental problem limited by solubility. Pure Appl Chem 74: 1843–1850.

Markovac J, Goldstein GW (1988). Picomolar concentrations of lead stimulate brain protein kinase C. Nature 334: 71–73.

Martena MJ, Van Der Wielen JC, Rietjens IM et al. (2010). Monitoring of mercury, arsenic, and lead in traditional Asian herbal preparations on the Dutch market and estimation of associated risks. Food Addit Contam Part A Chem Anal Control Expo Risk Assess 27: 190–205.

Martinez-Finley EJ, Chakraborty S, Fretham SJ et al. (2012). Cellular transport and homeostasis of essential and nonessential metals. Metallomics 4: 593–605.

McCarthy JT, Milliner DS, Johnson WJ (1990). Clinical experience with desferrioxamine in dialysis patients with aluminium toxicity. Q J Med 74: 257–276.

McIntyre T, Whiting MJ (2012). Increased metal concentrations in giant sungazer lizards (*Smaug giganteus*) from mining areas in South Africa. Arch Environ Contam Toxicol 63: 574–585.

McLachlan DR, Bergeron C, Smith JE et al. (1996). Risk for neuropathologically confirmed Alzheimer's disease and residual aluminum in municipal drinking water employing weighted residential histories. Neurology 46: 401–405.

McLaughlin AI, Kazantzis G, King E et al. (1962). Pulmonary fibrosis and encephalopathy associated with the inhalation of aluminium dust. Br J Ind Med 19: 253–263.

Mignini F, Nasuti C, Artico M et al. (2012). Effects of trimethyltin on hippocampal dopaminergic markers and cognitive behaviour. Int J Immunopathol Pharmacol 25: 1107–1119.

Moody RP, Chu I (1995). Dermal exposure to environmental contaminants in the Great Lakes. Environ Health Perspect 103 (Suppl 9): 103–114.

Moon J, Davison AJ, Smith TJ et al. (1988). Correlation clusters in the accumulation of metals in human scalp hair: effects of age, community of residence, and abundances of metals in air and water supplies. Sci Total Environ 72: 87–112.

Mori N, Yasutake A, Hirayama K (2007). Comparative study of activities in reactive oxygen species production/defense system in mitochondria of rat brain and liver, and their susceptibility to methylmercury toxicity. Arch Toxicol 81: 769–776.

Mori N, Yasutake A, Marumoto M et al. (2011). Methylmercury inhibits electron transport chain activity and induces cytochrome c release in cerebellum mitochondria. J Toxicol Sci 36: 253–259.

Myers GJ, Thurston SW, Pearson AT et al. (2009). Postnatal exposure to methyl mercury from fish consumption: a review and new data from the Seychelles Child Development Study. Neurotoxicology 30: 338–349.

Nagasawa K, Ito S, Kakuda T et al. (2005). Transport mechanism for aluminum citrate at the blood–brain barrier: kinetic evidence implies involvement of system Xc- in immortalized rat brain endothelial cells. Toxicol Lett 155: 289–296.

Nava-Ruiz C, Mendez-Armenta M, Rios C (2012). Lead neurotoxicity: effects on brain nitric oxide synthase. J Mol Histol 43: 553–563.

Neal AP, Guilarte TR (2010). Molecular neurobiology of lead (Pb(2+)): effects on synaptic function. Mol Neurobiol 42: 151–160.

Needleman H (2004). Lead poisoning. Annu Rev Med 55: 209–222.

Ni M, Li X, Yin Z et al. (2010). Methylmercury induces acute oxidative stress, altering Nrf2 protein level in primary microglial cells. Toxicol Sci 116: 590–603.

Nitschke I, Muller F, Smith J et al. (2000). Amalgam fillings and cognitive abilities in a representative sample of the elderly population. Gerodontology 17: 39–44.

Normandin L, Ann Beaupre L, Salehi F et al. (2004). Manganese distribution in the brain and neurobehavioral changes following inhalation exposure of rats to three chemical forms of manganese. Neurotoxicology 25: 433–441.

Olanow CW (2004). Manganese-induced parkinsonism and Parkinson's disease. Ann N Y Acad Sci 1012: 209–223.

Ono K, Komai K, Yamada M (2002). Myoclonic involuntary movement associated with chronic manganese poisoning. J Neurol Sci 199: 93–96.

Pappas RS (2011). Toxic elements in tobacco and in cigarette smoke: inflammation and sensitization. Metallomics 3: 1181–1198.

Park D, Yang H, Jeong J et al. (2010). A comprehensive review of arsenic levels in the semiconductor manufacturing industry. Ann Occup Hyg 54: 869–879.

Payton M, Riggs KM, Spiro 3rd A et al. (1998). Relations of bone and blood lead to cognitive function: the VA Normative Aging Study. Neurotoxicol Teratol 20: 19–27.

Peng JH, Xu ZC, Xu ZX et al. (1992). Aluminum-induced acute cholinergic neurotoxicity in rat. Mol Chem Neuropathol 17: 79–89.

Perl DP, Olanow CW (2007). The neuropathology of manganese-induced Parkinsonism. J Neuropathol Exp Neurol 66: 675–682.

Picard V, Govoni G, Jabado N et al. (2000). Nramp 2 (DCT1/DMT1) expressed at the plasma membrane transports iron and other divalent cations into a calcein-accessible cytoplasmic pool. J Biol Chem 275: 35738–35745.

Poewe W, Wenning G (2002). The differential diagnosis of Parkinson's disease. Eur J Neurol 9 (Suppl 3): 23–30.

Polizzi S, Pira E, Ferrara M et al. (2002). Neurotoxic effects of aluminium among foundry workers and Alzheimer's disease. Neurotoxicology 23: 761–774.

Rabinowitz MB (1991). Toxicokinetics of bone lead. Environ Health Perspect 91: 33–37.

Rey C, Reinecke HJ, Besser R (1984). Methyltin intoxication in six men; toxicologic and clinical aspects. Vet Hum Toxicol 26: 121–122.

Rifat SL, Eastwood MR, McLachlan DR et al. (1990). Effect of exposure of miners to aluminium powder. Lancet 336: 1162–1165.

Riihimaki V, Aitio A (2012). Occupational exposure to aluminum and its biomonitoring in perspective. Crit Rev Toxicol 42: 827–853.

Riihimaki V, Hanninen H, Akila R et al. (2000). Body burden of aluminum in relation to central nervous system function among metal inert-gas welders. Scand J Work Environ Health 26: 118–130.

Riihimaki V, Valkonen S, Engstrom B et al. (2008). Behavior of aluminum in aluminum welders and manufacturers of aluminum sulfate – impact on biological monitoring. Scand J Work Environ Health 34: 451–462.

Rodgher S, de Azevedo H, Ferrari CR et al. (2013). Evaluation of surface water quality in aquatic bodies under the influence of uranium mining (MG, Brazil). Environ Monit Assess 185: 2395–2406.

Rogers MA, Simon DG (1999). A preliminary study of dietary aluminium intake and risk of Alzheimer's disease. Age Ageing 28: 205–209.

Rohl C, Sievers J (2005). Microglia is activated by astrocytes in trimethyltin intoxication. Toxicol Appl Pharmacol 204: 36–45.

Rohling ML, Demakis GJ (2006). A meta-analysis of the neuropsychological effects of occupational exposure to mercury. Clin Neuropsychol 20: 108–132.

Rondeau V, Commenges D, Jacqmin-Gadda H et al. (2000). Relation between aluminum concentrations in drinking water and Alzheimer's disease: an 8-year follow-up study. Am J Epidemiol 152: 59–66.

Roskams AJ, Connor JR (1990). Aluminum access to the brain: a role for transferrin and its receptor. Proc Natl Acad Sci U S A 87: 9024–9027.

Ross WD, Emmett EA, Steiner J et al. (1981). Neurotoxic effects of occupational exposure to organotins. Am J Psychiatry 138: 1092–1095.

Rumack B (2013). Trimethyltin Chloride. In: Hazardous Substances Data Bank (Information System Microdex).

Salazar J, Mena N, Hunot S et al. (2008). Divalent metal transporter 1 (DMT1) contributes to neurodegeneration in animal models of Parkinson's disease. PNAS 105: 18578–18583.

Sappalainen A, Hernber S, Kock B (1979). Relatioship between blood lead levels and nerve conduction velocities. Neurotoxicology 1: 313–332.

Schwartz BS, Stewart WF, Bolla KI et al. (2000). Past adult lead exposure is associated with longitudinal decline in cognitive function. Neurology 55: 1144–1150.

Sen D, Wolfson H, Dilworth M (2002). Lead exposure in scaffolders during refurbishment construction activity – an observational study. Occup Med (Lond) 52: 49–54.

Shigematsu K, McGeer PL (1992). Accumulation of amyloid precursor protein in damaged neuronal processes and microglia following intracerebral administration of aluminum salts. Brain Res 593: 117–123.

Shinotoh H, Snow BJ, Hewitt KA et al. (1995). MRI and PET studies of manganese-intoxicated monkeys. Neurology 45: 1199–1204.

Simonsson BG, Sjoberg A, Rolf C et al. (1985). Acute and long-term airway hyperreactivity in aluminium-salt exposed workers with nocturnal asthma. Eur J Respir Dis 66: 105–118.

Sinczuk-Walczak H, Szymczak M, Razniewska G et al. (2003). Effects of occupational exposure to aluminum on nervous system: clinical and electroencephalographic findings. Int J Occup Med Environ Health 16: 301–310.

Sjogren B, Gustavsson P, Hogstedt C (1990). Neuropsychiatric symptoms among welders exposed to neurotoxic metals. Br J Ind Med 47: 704–707.

Skerfving S, Nilsson U, Schutz A et al. (1993). Biological monitoring of inorganic lead. Scand J Work Environ Health 19 (Suppl 1): 59–64.

Tarohda T, Yamamoto M, Amamo R (2004). Regional distribution of manganese, iron, copper, and zinc in the rat brain during development. Anal Bioanal Chem 380: 240–246.

Thyssen JP (2012). Cobalt sensitization and dermatitis: considerations for the clinician. Dermatitis 23: 203–209.

Trapp GA, Miner GD, Zimmerman RL et al. (1978). Aluminum levels in brain in Alzheimer's disease. Biol Psychiatry 13: 709–718.

Triebig G, Weltle D, Valentin H (1984). Investigations on neurotoxicity of chemical substances at the workplace. V. Determination of the motor and sensory nerve conduction velocity in persons occupationally exposed to lead. Int Arch Occup Environ Health 53: 189–203.

Uede K, Furukawa F (2003). Skin manifestations in acute arsenic poisoning from the Wakayama curry-poisoning incident. Br J Dermatol 149: 757–762.

Vahidnia A, Romijn F, Tiller M et al. (2006). Arsenic-induced toxicity: effect on protein composition in sciatic nerve. Hum Exp Toxicol 25: 667–674.

Vahidnia A, van der Voet GB, de Wolff FA (2007). Arsenic neurotoxicity – a review. Hum Exp Toxicol 26: 823–832.

Waksman JC, Brody A (2007). Contrast sensitivity in occupational and environmental neurotoxicology: what does it really mean? Arch Environ Occup Health 62: 177–181.

Waldron KJ, Rutherford JC, Ford D et al. (2009). Metalloproteins and metal sensing. Nature 460: 823–830.

Wang L, Jiang H, Yin Z et al. (2009a). Methylmercury toxicity and Nrf2-dependent detoxification in astrocytes. Toxicol Sci 107: 135–143.

Wang Y, Xu Y, Wang H et al. (2009b). Arsenic induces mitochondria-dependent apoptosis by reactive oxygen species generation rather than glutathione depletion in Chang human hepatocytes. Arch Toxicol 83: 899–908.

Watson GE, Lynch M, Myers GJ et al. (2011). Prenatal exposure to dental amalgam: evidence from the Seychelles Child Development Study main cohort. J Am Dent Assoc 142: 1283–1294.

Wedler FC, Ley BW, Grippo AA (1989). Manganese(II) dynamics and distribution in glial cells cultured from chick cerebral cortex. Neurochem Res 14: 1129–1135.

Weinberg ED (2010). The hazards of iron loading. Metallomics 2: 732–740.

White DM, Longstreth Jr WT et al. (1992). Neurologic syndrome in 25 workers from an aluminum smelting plant. Arch Intern Med 152: 1443–1448.

Whittington DL, Woodruff ML, Baisden RH (1989). The time-course of trimethyltin-induced fiber and terminal degeneration in hippocampus. Neurotoxicol Teratol 11: 21–33.

Willhite CC, Ball GL, McLellan CJ (2012). Total allowable concentrations of monomeric inorganic aluminum and hydrated aluminum silicates in drinking water. Crit Rev Toxicol 42: 358–442.

Winneke G (2011). Developmental aspects of environmental neurotoxicology: lessons from lead and polychlorinated biphenyls. J Neurol Sci 308: 9–15.

Wolters EC, Huang CC, Clark C et al. (1989). Positron emission tomography in manganese intoxication. Ann Neurol 26: 647–651.

Wu J, Basha MR, Brock B et al. (2008). Alzheimer's disease (AD)-like pathology in aged monkeys after infantile exposure to environmental metal lead (Pb): evidence for a developmental origin and environmental link for AD. J Neurosci 28: 3–9.

Wu Z, Du Y, Xue H et al. (2012). Aluminum induces neurodegeneration and its toxicity arises from increased iron accumulation and reactive oxygen species (ROS) production. Neurobiol Aging 33: 199, e191-112.

Xu B, Xu ZF, Deng Y (2010). Protective effects of MK-801 on manganese-induced glutamate metabolism disorder in rat striatum. Exp Toxicol Pathol 62: 381–390.

Yassi A (1980). Chronic lead poisoning from industrial exposure: a review. Can Fam Physician 26: 1056–1062.

Yellamma K, Saraswathamma S, Kumari BN (2010). Cholinergic system under aluminium toxicity in rat brain. Toxicol Int 17: 106–112.

Yin Z, Jiang H, Syversen T et al. (2008). The methylmercury-L-cysteine conjugate is a substrate for the L-type large neutral amino acid transporter. J Neurochem 107: 1083–1090.

Yokel RA, McNamara PJ (2001). Aluminium toxicokinetics: an updated minireview. Pharmacol Toxicol 88: 159–167.

Yoo CI, Kim Y, Jeong KS et al. (2007). A case of acute organotin poisoning. J Occup Health 49: 305–310.

Zheng W, Shen H, Blaner WS et al. (1996). Chronic lead exposure alters transthyretin concentration in rat cerebrospinal fluid: the role of the choroid plexus. Toxicol Appl Pharmacol 139: 445–450.

Zheng W, Aschner M, Ghersi-Egea JF (2003). Brain barrier systems: a new frontier in metal neurotoxicological research. Toxicol Appl Pharmacol 192: 1–11.

Handbook of Clinical Neurology, Vol. 131 (3rd series)
Occupational Neurology
M. Lotti and M.L. Bleecker, Editors

Chapter 12

Carbon monoxide intoxication

MARGIT L. BLEECKER*
Center for Occupational and Environmental Neurology, Baltimore, MD, USA

INTRODUCTION

Carbon monoxide (CO) is a colorless, odorless, nonirritant gas absorbed through the lungs. The rate of CO uptake depends upon duration of exposure, concentration of CO in the environment, and minute ventilation (Forbes et al., 1945). Carboxyhemoglobin (COHb) is formed when CO binds with hemoglobin with an affinity 200 times greater than oxygen, thereby decreasing oxygen-carrying capacity and the release of oxygen to tissues, leading to tissue hypoxia. Since CO causes the remaining oxyhemoglobin to hold the oxygen more strongly, the oxygen dissociation curve is shifted to the left, making it more difficult to release the remaining oxygen in the blood to the tissues.

At low COHb levels (1–5%) intrinsic compensatory physiologic mechanisms prevent CO hypoxia through increased blood flow and increased oxygen extraction. At a COHb below 20% changes in oxygen consumption in the brain are negligible and therefore brain function should not be affected (Benignus et al., 1990). A 10% decrement in brain oxygen consumption does not occur until a COHb level of 27% is reached (Raub and Benignus, 2002). See Table 12.1 for progression of effects in humans as COHb levels increase. Once loss of consciousness accompanied by hypotension occurs with CO poisoning there is the possibility for the development of delayed neurologic sequelae (DNS). DNS may present after recovery with a lucid interval of 2–40 days during which diffuse demyelination develops in the brain with a constellation of signs and symptoms that is reversible in the majority of patients (Lee, 1978; Choi, 1983; Min, 1986; Crystal and Ginsberg, 2000; Lin et al., 2009).

SOURCES OF CO EXPOSURE

CO is a product of incomplete combustion of hydrocarbons, leading to elevated air levels, in poorly functioning heating systems and inadequate ventilation of flame-based heating sources. CO poisoning has occurred when gasoline-powered pressure washers were used indoors to clean livestock (CDC, 1993); in warehouses where propane-powered forklifts are needed (Fawcett et al., 1992; Wesley et al., 1995); with indoor burning of charcoal briquettes (Hampson et al., 1994); riding in the back of pick-up trucks (Hampson and Norkool, 1992); or in ice skating rinks using propane-powered resurfacing machines (CDC, 1996).

Data from the US National Poison Data System during 2000–2009 showed 68316 CO exposures, with 45.1% managed on site and 53.7% treated at a healthcare facility. Most occur during the winter months and are primarily located in the Midwest and Northeast, with 77.6% at residence site and 12% at the workplace. The most common clinical symptoms with CO exposure are headache, nausea, and dizziness (CDC, 2011).

From 2008 to 2010, 87 hyperbaric facilities reported information on 864 patients who received treatment for CO poisoning (Clower et al., 2012). The most common CO sources were furnace or boiler 32%, generator 20%, motor vehicle 13%, grill 6%, and space heater 5%, with 75% of the patients treated in the fall and winter. The most common fuel involved was gasoline 48%, natural gas/propane 36%, and diesel 2%.

CO poisoning occurs following hurricanes due to the use of portable generators and gasoline-powered appliances that are used improperly by inappropriate placement and ventilation. Following Hurricane Katrina during August 29 to September 24, 2005, 51 cases of CO poisoning were treated at hyperbaric facilities in Alabama, Louisiana, and Mississippi (CDC, 2005). Investigations in Alabama and Texas after hurricanes found nearly all were due to gasoline-powered generators that, even when outside, were placed close to window air conditioners and few homes had CO detectors (CDC, 2006).

*Correspondence to: Margit L. Bleecker, MD, PhD, Director, Center for Occupational and Environmental Neurology, 2 Hamill Road, Suite 225, Baltimore MD 21210, USA. Tel: +1-410-433-2077, Fax: +1-410-433-0622, E-mail: coen@msn.com

Table 12.1

Human response with varying levels of carboxyhemoglobin (COHb)

COHb (%)	Findings in humans
<1	Endogenous carbon monoxide (CO) from catabolism of hemoglobin; at these concentrations CO has many neuroprotective effects
1–9	Normal levels in tobacco smokers; chest pain and decreased exercise duration with ischemic heart disease
15–20	Headache; changes in visual evoked potentials
20–30	Severe headache, dizziness, nausea, fatigue; impaired manual dexterity
30–40	Severe headache, nausea, vomiting, increased heart and respiratory rate, loss of consciousness with hypotension
40–50	Coma and seizures, hypotension
60–70	Lethal if untreated

There also is CO in the atmosphere, with higher levels in urban areas. Outdoor levels of CO vary depending upon emissions from motor vehicles, trucks, and industrial processes generating combustion gases. Outside relatively enclosed spaces such as tunnels and parking garages have higher levels as automobile exhaust may emit 100 000 ppm CO (Widdop, 2002). When in a closed garage lethal concentrations of COHb may occur within 10 minutes (Stewart, 1975).

Explosives used in surface blasting may produce CO through detonation. This CO can travel through the soil and accumulate in confined spaces, creating a hazard for workers. Also homes in the vicinity of a blast site may acquire elevated CO levels in the inside air from leakage of CO in the soil water to floor drains (CDC, 1998).

Smoke-polluted environments, a common source of CO exposure for smokers, also places nonsmokers at risk. While smoking a cigarette a smoker is exposed to 400–500 ppm CO, resulting in a COHb level of 3–8%; in a heavy smoker the COHb may increase to 15% (Raub et al., 2000) as compared to COHb of 1–3% in nonsmokers. A popular activity is hookah smoking (waterpipe), where the exposure to CO from the combustion of charcoal is 10 times that of a cigarette smoker. Four young adults treated for CO poisoning from hookah smoking had loss of consciousness in three, with COHb ranging from 20.1% to 29.6% (von Rappard et al., 2014).

The World Health Organization recommends an ambient air level for CO of 9 ppm/8 hours (10 mg/m^3) that should protect susceptible populations, as the COHb would not rise above 2.5% (WHO, 1999). By contrast, to protect healthy workers during an 8-hour work day the occupational exposure to CO ranges from 25 to 50 ppm (30–60 mg/m^3) worldwide (CO 25 ppm = COHb 3.5%).

NEUROPROTECTIVE EFFECTS OF CO

Endogenous CO is generated by the catabolism of hemoglobin that accounts for COHb levels less than 1%. These physiologic levels of endogenous CO may serve as a neurotransmitter or modulate other neurotransmitters and hormones. CO's many functions at these concentrations include vasodilation, neurotransmission, inhibition of platelet aggregation, antiproliferation effects of smooth muscle, and anti-inflammatory agent (Verma et al., 1993; Hanafy et al., 2013). For this reason, some would argue that CO is not neurotoxic but neuroprotective. Animal studies found that CO improved the effects of ischemic stroke and intraparenchymal hemorrhage (Wang et al., 2011; Yabluchanskiy et al., 2012). An animal model of multiple sclerosis showed decreased cerebral inflammatory burden with CO treatment. Low levels of CO are therapeutic and there are ongoing clinical trials using CO treatment in idiopathic pulmonary fibrosis and acute respiratory distress syndrome, after kidney transplant and after colon resection to prevent paralytic ileus (Hanafy et al., 2013). CO in its neuroprotective role is involved with memory as a neurotransmitter, circadian rhythm and vasomotor tone (Hanafy et al., 2013).

MECHANISMS OF TOXICITY

CO toxicity is a combination of tissue hypoxia-ischemia by the formation of COHb and direct cellular damage by CO. Animal studies showed that the initial increase in cerebral blood flow with CO exposure remains until loss of consciousness, when cardiac compromise results in hypotension (Lo et al., 2007). Ischemia in the arterial border zones of the brain is the result of hypotension with CO poisoning (Okeda et al., 1981). Areas of the brain affected by hypoxia secondary to CO include the cortex, layers three, four, and five, the hippocampi, the white matter, the basal nuclei, and Purkinje cells in the cerebellum.

CO enters the blood through the lungs and 10–15% binds to heme-containing protein, particularly cytochrome c oxidase and myoglobin. Increased intracellular oxidative stress is the result of inflammation caused by CO through increased concentration of cytosolic heme and the heme oxygenase-1 protein. By binding to platelet heme protein, CO causes the release of nitric oxide that leads to impaired mitochondrial function and increased tissue hypoxia by the formation of peroxynitrile. CO exposure leads to increases in oxidative stress, lipid peroxidation, and apoptosis through platelet-neutrophil aggregation and neutrophil degranulation, release of myeloperoxidase, proteases, and reactive oxygen species. The proteases interact with xanthine dehydrogenase in endothelial cells to form xanthine oxygenase that inhibits mechanisms against oxidative stress. CO-induced hypoxia activates hypoxia-inducible factor-la, that may affect either protective or injurious gene regulation depending on the CO dose and factors in the exposed individual (Kao and Nanagas, 2006; Weaver, 2009).

A potential mechanism of neurotoxicity suggested by Park et al. (2014) attributes injury in acute CO poisoning to catecholamine crisis from increased sympathetic secretion in nerve terminals that leads to increased dopamine levels in the mesolimbic system, including the globus pallidus, resulting in acute injury. Sustained higher levels of catecholamine (norepinephrine/dopamine) in synapse or oligodendrocyte of deep white matter lead to delayed neurologic lesions in deep white matter. This is accompanied by ongoing inflammation in the brain between 1 week and 6 months after CO exposure.

Another proposed mechanism for ongoing inflammation in the brain causing delayed CO-mediated neuropathology is an immunologic response (Thom et al., 2004). CO poisoning triggers adduct formation between MBP and malonylaldehyde, a lipid peroxidation product. MBP is altered over days along with influx of macrophages and CD-4 lymphocytes and increase of activated microglia. If rats were rendered immunologically tolerant to MBP before CO exposure, the lymphocyte proliferation and microglia activation did not occur. Also with CO exposure these immunologically tolerant rats had no decrements in maze learning, as observed in rats with chemically modified MBP. The authors suggest that this immunologic response of CO poisoning, resulting in DNS, may be treatable with anti-inflammatory compounds.

BIOMARKERS OF CO EXPOSURE

The physiologic variables that determine COHb levels with CO exposure were published by Coburn, Forster, and Kane (CFK) in 1965 (Coburn et al., 1965) (Table 12.2). Subsequently the CFK equation was found to predict accurately COHb levels in men and women at rest and with exercise (Peterson and Stewart, 1970, 1975).

COHb has a short half-life of 4–6 hours and frequently CO-intoxicated patients receive oxygen treatment by emergency medical technician, decreasing the half-life to approximately 1 hour before a COHb level is ever obtained. If COHb level is determined 2 minutes after the cessation of CO exposure, levels of COHb could be accurately predicted using the CFK equation. However as this time interval increases, so do the errors in prediction (Benignus et al., 1994). Benignus et al. (1994) studied rapid CO uptake following transient exposure to 6683 ppm $C^{18}O$ and found that venous blood predictions of COHb were overestimated due to marked variations in level and amount of delay of appearance of COHb in venous blood among subjects. By contrast, change in arterial COHb was always underestimated based upon predictions of venous blood samples. The authors concluded that, during transient high-level CO exposure the heart and brain may be exposed to much higher levels of COHb than previously expected. This provides one explanation for the poor correlation between COHb concentration and clinical outcomes.

Many have looked for markers of CO toxicity that better reflect brain pathology. Because of the complexity of the numerous systems affected by CO, Thom et al. (2010) examined acute-phase reactants and proteins associated with inflammatory responses, including chemokines/cytokines and interleukins, growth factors, hormones, and an array of autoantibodies, and found elevations of all in CO-poisoned patients with no discrimination by severity of exposure, including patients with normal COHb and brief exposure. Early after CO exposure elevations of interleukin-6 concentration and S100B protein, a marker of neuronal damage after brain trauma, occurred in cerebrospinal fluid but not in serum

Table 12.2

Basic form of the CFK equations

$$\frac{A[\mathrm{HbCO}]_t - BV_{\mathrm{CO}} + \mathrm{PI}_{\mathrm{CO}}}{A[\mathrm{HbCO}]_0 - BV_{\mathrm{CO}} + \mathrm{PI}_{\mathrm{CO}}} = \exp - \mathrm{tAVbB}$$

where:

$A = P_{c,O2} / M\,[HbO_2]$

$B = 1 / DL_{CO} + PL / VA$

M = ratio of the affinity of blood for CO to that for O_2

$[HbO_2]$ = mL of O_2 per mL blood

$[HbCO]_t$ = mL of CO per mL blood at time t

$[HbCO]_0$ = mL of CO/mL blood at the beginning of the exposure interval

$P_{c,O2}$ = average partial pressure of O_2 in lung capillaries (mmHg)

V_{CO} = rate of endogenous CO production (mL/min)

DL_{CO} = diffusivity of the lung for CO (mL/min/mmHg)

PL = barometric pressure minus the vapor pressure of water at body temperature (mmHg)

V_b = blood volume (mL)

PI_{CO} = partial pressure of CO in inhaled air (mmHg)

V_A = alveolar ventilation rate (mL/min)

t = exposure duration (min)

exp = 2.7182, the base of natural logarithms, raised to the power of the bracketed expression and predicted the development of delayed encephalopathy associated with CO exposure (Ide and Kamijo, 2009; Ide et al., 2012). By contrast, serum S100B protein, was higher in CO-poisoned patients who developed delayed encephalopathy with a level of more than 0.165 µg/L, a strong predictor (Park et al., 2012). Compared to controls, CO-poisoned patients with loss of consciousness had elevated levels of serum S100B and neuron-specific enolase, another biochemical marker of hypoxic brain damage (Yardan et al., 2009). However, another study with multiple determinations of serum S100B and neuron-specific enolase over the course of 48 hours could not differentiate groups with CO exposure by level of consciousness and controls (Rasmussen et al., 2004). Acute brain dysfunction with CO poisoning found significant elevations in serum tau protein that was higher in those with loss of consciousness and correlated with COHb levels; in this study serum S100B protein did not differ significantly from controls (Gawlikowski et al., 2014). Blood lactate concentration in CO-poisoned patients is recommended for risk stratification to determine those requiring hospitalization (Cervellin et al., 2014).

CLINICAL PRESENTATION OF CO TOXICITY

Early studies with controlled CO exposure in volunteers studied effects with COHb ranging from 4% to 20% (Laties and Merigan, 1979). Minor changes in performance (visual perception, complex tasks of judgment and motor coordination) at the time of exposure were noted (Beard and Wertheim, 1967; Beard and Grandstaff, 1970; Putz, 1979; Gliner et al., 1983), but frequently these findings could not be replicated, especially when a double-blind exposure paradigm was employed (Mikulka et al., 1970; Stewart et al., 1970, 1973; McFarland, 1973; Benignus et al., 1977, 1987a; Mihevic et al., 1983). One of the early studies with controlled CO exposure (Stewart et al., 1970) that included double-blind exposure for half of the 25 experiments found changes only in the visual evoked response at a COHb of 20% that increased as COHb approached 30% and manual dexterity that was normal at COHb 22%, but very impaired at COHb of 28%. Benignus (1994) performed a meta-analysis of behavioral effects and CO exposure and determined that a COHb level of 18–25% was associated with a 10% decrement in behavior, while a COHb concentration of 12–18% produced a 5% decrement in behavior. However there is no evidence to support permanent harmful effects to the brain from CO exposure producing a COHb in the target range of 4–20% (Benignus et al., 1977; Townsend and Maynard, 2002).

The first consistent symptom with controlled CO exposure is a mild frontal headache when COHb was 15–20% (Stewart et al., 1970). As headache is the most common symptom with CO exposure, its characteristics were examined to determine if there were features that would alert the physician to obtain a COHb level. One hundred patients referred for hyperbaric oxygen (HBO) treatment for CO exposure answered questions concerning their headaches (Hampson and Hampson, 2002). The location was frontal for 66%, while 58% had more than one location. Pain was dull in 72%, sharp in 36%; constant in 74%, throbbing in 41%, and intermittent in 16%. Headache improved in 72% prior to hyperbaric treatment and resolved completely in 21%. Hyperbaric treatment resulted in improved headache pain in 97% and resolution in 44%. Unfortunately, there were no unique features of these headaches. Sometimes the physical location and when the headache occurs are more helpful, for example, if the exposure is only at work or only in the home and resolves when not in that environment.

Studies examining the onset of symptoms with CO exposure found they usually occur above a COHb of 20% (Schulte, 1963; Benignus et al., 1987b). However in patients with significant coronary vascular disease chest pain may occur at much lower COHb concentrations (1–9%) (Anderson et al., 1973; Aronow and Issell, 1973; Aronow, 1981). The Expert Panel on Air Quality (Department of the Environment, 1994) reported that patients with ischemic heart disease had decreased exercise duration due to increased chest pain when COHb levels were 2.7–5.1%. In fact, Stewart (1975) believed, "there is no concentration of CO that does not exert a significant and measurable untoward effect upon a diseased cardiovascular system." Even in young healthy men, maximal short-term exercise duration was decreased with COHb levels of 2.5–4% (Department of the Environment, 1994) and at 10% (Koike et al., 1991). Healthy young Swiss soldiers with no underlying coronary vascular disease developed chest pain and palpitations with CO exposure while sleeping (Henz and Maeder, 2005). Electrocardiogram changes with CO exposure may be nonspecific, such as dysrhythmias, or changes of myocardial ischemia. Troponin 1 and creatine phosphokinase, cardiac markers of ischemia, may be elevated (Holstege et al., 2004).

Once COHb reaches 21–35%, symptoms include headache, dizziness, nausea, fatigue, confusion, and memory problems (O'Donoghue, 1985).

Three hundred and forty-five patients with a mean COHb of 21% treated with normobaric or HBO returned to baseline immediately or within a matter of days (Raphael et al., 1989; Annane et al., 2001). Ninety-seven percent of these patients were able to return to work at

1 month. Persistent symptoms in 32% included fatigue, headache, memory and sleeping problems, decreased concentration, behavioral and visual disturbances. In another study 32 patients with a mean COHb of 30% at the time of exposure were treated and examined 1 month after exposure with a psychometric battery. The researchers found no significant difference for memory, attention, executive function, or reaction time compared to controls (Deschamps et al., 2003).

Some of the literature does not stratify CO-exposed patients by initial level of consciousness when analyzing outcome status following treatment (Gale et al., 1999; Weaver et al., 2002, 2007; Jasper et al., 2005). This makes it impossible to interpret the results, as DNS is combined with those that may have persistent symptoms but never had loss of consciousness or hypotension with exposure. This confusion is present in the Jasper et al. (2005) paper, where 51% had level of consciousness that includes both accidental and suicide attempt, but outcomes on various measures of depression are never stratified by this variable. Also the accidentally poisoned with no cognitive sequelae at 6 weeks were seven times more likely to have depression and anxiety at 12 months than at 6 weeks. This suggests that factors other than CO exposure may be responsible for symptoms present at 12 months that were not there at 6 weeks. Until these flaws in study design are addressed, there is no good evidence that any long-term sequelae occur in the group with no LOC at the time of CO exposure.

DELAYED NEUROLOGIC SEQUELAE

Loss of consciousness/coma occurs with a COHb level of 30–50% (Widdop, 2002). DNS may occur in those individuals rendered unconscious and hypotensive at the time of CO exposure (Kao and Nanagas, 2006). After a lucid interval of 2–40 days diffuse demyelination develops in the brain with a constellation of signs and symptoms, including lethargy, behavior changes, forgetfulness, memory loss, and parkinsonian features (Lee, 1978; Choi, 1983; Min, 1986; Crystal and Ginsberg, 2000; Lin et al., 2009). DNS develops in approximately 3–12% of those with acute or chronic CO exposure (Choi, 1983; Min, 1986). A small percentage (3%) of patients with DNS will develop parkinsonism when demyelination involves the globus pallidus (Choi, 1983). Recovery from DNS occurs in 75% of patients within 1 year (Choi, 1983).

In a second publication (Choi, 2002), focusing on parkinsonism with CO poisoning in 242 patients, 9.5% developed parkinsonism with a median latency of 4 weeks and mild to severe impairment of cognitive functions, gait disturbances, urinary incontinence, and mutism. There was no resting tremor but some had an intentional tremor. Other signs included small stride, hypokinesia, masked facies, rigidity, release reflexes, and retropulsion. Abnormalities in the white matter and globus pallidus seen on computed tomography (CT) were similar to CO-exposed patients who made a full recovery. Standard drug treatment for Parkinson's disease was not effective. Sixteen patients followed for 1 year had recovery at 6 months in 81%.

Another case that developed DNS 2 months after CO poisoning had bilateral globus pallidus lesions located in the inner segments compatible with focal hemorrhage. ^{14}C-N-methylspiperone positron emission tomography imaging showed increase dopamine D_2 receptor binding in the caudate and putamen that resolved after 5 months at the same time that the clinical condition of disorientation, abnormal behavior, bradykinesia, muscle rigidity, and gait disturbances improved. Of note, treatment with levodopa had no effect but bromocriptine, a postsynaptic D_2 agonist, resulted in dramatic improvement of the parkinsonian signs (Yoshii et al., 1998). Another similar case with no improvement with carbidopa/levodopa responded to trihexyphenidyl (Klawans et al., 1982).

Another feature of DNS/parkinsonism that developed in a patient obtunded with a COHb of 30% was depression with apathy. CT showed bilateral globus pallidus necrosis 4 weeks post CO exposure. With Sinemet treatment both parkinsonism and depression resolved despite persistence of the pallidal lesions (Jaeckle and Nasrallah, 1985).

Because of the similarity in appearance of Parkinson's disease and parkinsonism with DNS, the presence of Parkinson protein 2 gene (PARK2) in CO-poisoned patients was examined to determine whether it influenced susceptibility of DNS (Li et al., 2013). The presence of PARK2 polymorphism, namely the allelic variant of rs1784594, is a risk factor for DNS with CO poisoning, especially in women (Liang et al., 2013).

Onset of choreoathetosis 10–30 days after CO poisoning with involvement of the basal ganglia had spontaneous resolution after 14–90 days (Schwartz et al., 1985; Park and Choi, 2004). Gilles de la Tourette's syndrome presented in a previously healthy 56-year-old man who was rendered comatose for several weeks following CO poisoning. CT of the brain found ventricular enlargement, bitemporal tissue loss, and irregular, patchy low-density lesions in the basal ganglia (Pulst et al., 1983).

CO exposure for 40 minutes in rats resulted in increased extracellular dopamine in the striatum and a decrease in its metabolites while CO withdrawal enhanced oxidative metabolism of the increased dopamine. Therefore CO is participating in the neurotoxicity of dopamine via the formation of reactive compounds through dopamine oxidation that may be responsible for some of the clinical presentations of DNS (Hara et al., 2002).

NEUROPSYCHOLOGIC EFFECTS

A screening neuropsychologic battery developed for use in the emergency room to determine the level of cognitive impairment associated with CO exposure was flawed in its initial test phase (Messier and Myers, 1991). No adjustments were made for education that was significantly different between the exposed and control groups, with the control group rating higher. This could definitely contribute to the group differences that were present before and after treatment. No dose–effect between CO exposure and performance was examined to support that CO exposure was the source of the difference between the groups. The components of the CO neuropsychologic battery were general orientation, Trail Making, Digit Symbol, Aphasia, and Block Design. There was no test predictive of premorbid abilities that would be resistant to neurotoxic exposure, such as reading ability or vocabulary. In another study, Rottman et al. (1995) demonstrated that the CO battery was not able to differentiate CO-exposed individuals from controls of a similar age and educational level.

The CO battery was subsequently used to define a clinical condition of chronic CO exposure in 7 patients exposed from 1 to 18 months (Myers et al., 1998). All had motor slowing and memory problems on the neuropsychologic battery. Three patients had a conversion disorder on Minnesota Multiphasic Personality Inventory. There were few neurologic complaints. After multiple HBO treatments it was difficult to determine when maximum medical improvement had occurred due to the presence of litigation.

A frequently cited reference for neuropsychologic impairment following low-level CO exposure had many flaws (Amitai et al., 1998). The 45 CO-exposed university students had COHb $< 6\%$ levels, except for four individuals (COHb 7–11%) despite ambient air concentration of CO of 17–100 ppm, suggesting that the air measurement of CO did not reflect the personal breathing zone of the exposed individual. Smoking in 13% of the CO group and 17% of the control group was not considered in the analyses and no COHb was obtained for the control group. Exposed and control groups were matched for age and sex, but not for premorbid intelligence. This is critical if abnormal performance on neuropsychologic tests is attributed to CO exposure, when it could be premorbid intelligence. There were findings on the neuropsychologic battery for the CO group that were not biologically plausible, namely poorer performance on digit span forward, but not on digit span backward, a more difficult task. Also Trail-Making Test part A, a simple psychomotor task, was slower in the CO group while Trail-Making Test part B – a more difficult task that requires cognitive flexibility and psychomotor speed – had similar performance in both groups. In a test of verbal learning and memory, the CO-exposed group recalled more words (Bleecker and Lindgren, 1999). Therefore the difference in performance on the neuropsychologic battery was most likely related to pre-existing differences in the two groups.

With chronic, intermittent CO exposure (around 106 ppm), neuropsychologic testing can show significant cognitive impairment without abnormalities on magnetic resonance imaging (MRI) or magnetic resonance spectroscopy (MRS) (Prockop, 2005). The functional domains affected included executive function, nonverbal sequences, complex copying, visuospatial judgment, organization, psychomotor speed, and attention.

NEUROIMAGING STUDIES FOLLOWING CO EXPOSURE

Mechanisms of CO poisoning that contribute to findings on MRI include the hypoxic-hypotension process that produces ischemic changes in the arterial border zones of the brain. Brain hypoxia injures the cerebral cortex through increases in excitatory amino acids besides oxidative stress, cellular necrosis, apoptosis, and inflammation. Conventional MRI may show involvement of the globus pallidus with CO exposure, as it is at risk for damage due to the hypotension-hypoxia process because of poor collateral blood flow, or from CO binding to heme iron in the globus pallidus where the highest concentration of iron in the brain is located (Beppu, 2014).

Gray-matter changes in the hippocampus and cerebral cortex are more frequently reported in those patients with persistent symptoms from CO exposure (Bruno et al., 1993; Gale et al., 1999; Henke et al., 1999). As neurons have a higher metabolic demand than white matter, they would be expected to be more sensitive to hypoxia; however they can withstand pure hypoxia for a longer period than exposure to ischemia (Singhal et al., 2002).

Seven patients with impaired level of consciousness and COHb ranging from 39% to 68% were treated with HBO and four immediately had a normal examination, while two had pyramidal signs for a few days and another had an amnestic syndrome that persisted after 4 weeks. Positron emission tomography studies performed a few days after HBO found decreased blood flow in the frontal and temporal cortex in the whole group compared to controls. Prolonged ischemia was demonstrated by the uncoupling of oxygen consumption and blood flow supporting an unstable metabolic condition. The three patients with residuals following HBO had a greater decrease in blood flow and oxygen consumption, primarily in the thalamus and striatum (De Reuck et al., 1993).

When 73 consecutive patients with CO poisoning had MR, 12% had white-matter damage and only one patient had a basal ganglia lesion at 2 weeks and 6 months (Hopkins et al., 2006). In a more homogeneous group of 19 patients all rendered unconscious by CO exposure, 6 had persistent normal MR studies of the brain while, of the remaining 13 patients, 11 had abnormalities in the globus pallidus and 6 had white-matter lesions (O'Donnell et al., 2000). In 22 cases that eventually died from CO poisoning, the globus pallidus and white-matter lesions were equally present in 16/22 (Lapresle and Fardeau, 1967). The centrum semiovale and periventricular white matter are most commonly affected (Durak et al., 2005; Beppu, 2014). Reported less frequently is white-matter involvement of the temporal lobe, occipital lobe, parietal lobe, and corpus callosum (Beppu, 2014). When the cerebellum is involved the prognosis is poorer (O'Donnell et al., 2000). White-matter abnormalities in the centrum semiovale, fornix, and corpus callosum are associated with cognitive impairment following CO intoxication (Kesler et al., 2001; Parkinson et al., 2002; Porter et al., 2002; Chang et al., 2010). These white-matter lesions when followed over time may improve along with clinical presentation, suggesting that a reversible demyelinating process may be part of the pathology associated with CO intoxication (Chang et al., 1992).

The pathogenesis of white-matter damage with CO exposure was examined in animals and found to be correlated with the intensity of the blood pressure drop and not duration of exposure or COHb level. Okeda et al. (1981) concluded that selective white-matter damage was secondary to decrease in cerebral blood flow and the fact that the white matter is supplied by long arteries with thin medial layer and limited capillary beds.

MRI changes in white matter with CO exposure include hypointensity on T1-weighted images and hyperintensity on T2-weighted and fluid-attenuated inversion recovery (FLAIR) images, supporting a demyelinating process (Lo et al., 2007). Other sequences of MRI, such as diffusion weighted imaging (DWI), diffusion tensor imaging (DTI), and MRS, when used in the setting of CO poisoning may depict abnormalities earlier than conventional MRI. Hours after CO exposure DWI showed evidence of cytotoxic edema in the frontal and parietal white matter (Teksam et al., 2002; Sener, 2003). DWI findings with CO poisoning differed from cerebral infarction, as apparent diffusion coefficient (ADC) remained low for months with DNS, supporting the underlying pathology of progressive demyelination with cytotoxic edema (Kim et al., 2003). By the same token, increases in ADC accompanied improvement of DNS symptoms (Kim et al., 2002, 2003; Teksam et al., 2002).

To better quantitate demyelination with CO poisoning, DTI studies specifically focused on fractional anisotropy (FA) values in the centrum semiovale and found decreased values before the onset of DNS. Therefore, FA values could be used as a predictor for the development of DNS (Beppu et al., 2010). FA values in the centrum semiovale also correlated with other markers of demyelination, such as myelin basic protein (MBP) (Beppu et al., 2012). In patients with chronic symptoms months after CO exposure, DTI studies found the greatest damage in the deep white matter, the centrum semiovale (Fujiwara et al., 2012).

MRS provides data on gray-matter and white-matter integrity following CO poisoning. Decreased *N*-acetyl aspartate (NAA) indicates neuronal damage while increase in choline (Cho) supports the presence of demyelination. Lactate peaks point to a conversion of aerobic metabolism to anaerobic metabolism. Five patients in coma from CO exposure had MRI hyperintensities of white matter and globus pallidus. MRS showed increased Cho and decreased NAA in white matter and lactate peaks were present in 2 patients early on. The MRS remained abnormal even after the MRI had normalized in these 5 patients. Another 2 patients with chronic low CO exposure and no loss of consciousness had normal MRI and MRS scans (Kondziella et al., 2009).

The timeline for changes in Cho and NAA in patients with chronic symptoms following CO poisoning are first increase in Cho, which then decreases months later, while NAA begins to decrease at 3 weeks and lactate usually appears at 1–2 months (Murata et al., 2001; Prockop, 2005; Terajima et al., 2008). Cho levels in the centrum semiovale are higher in those patients with persistent chronic symptoms compared to patients with acute transient symptoms with CO exposure. Increased Cho reflects inflammation and ongoing demyelination (Beppu et al., 2011).

NEUROPHYSIOLOGIC ABNORMALITIES WITH CO POISONING

Electroencephalographic (EEG) abnormalities associated with CO poisoning are correlated with clinical presentation, namely duration of unconsciousness or level of recovery. Within 24–48 hours of acute CO poisoning, EEG may show diffuse, irregular slow waves and bifrontal slow waves to low-voltage and isoelectric tracings (Ogawa et al., 1973). As patients regain consciousness there is increased 8–12 Hz activity mixed with theta activity. Even when alpha returns to normal it is mixed with theta and beta activity (Bokonjie and Buchthal, 1961). In one case report of CO poisoning, seizures were believed to be due to necrosis that occurs at arterial border zones between anterior cerebral and middle cerebral arteries or between middle cerebral and posterior cerebral arteries (Neufeld et al., 1981). Gilbert and Glaser (1959) reported on a case with chronic CO poisoning with numerous episodes of dizziness, lethargy, ataxia, and loss of consciousness whose COHb was 20% after being in hospital for 30 hours. Examination found mental dullness and impaired concentration when EEG showed irregular 10–12 alpha and intermixed fast activity with frequent spike and sharp wave and spike wave. Anticonvulsant medication was discontinued when repeat EEG found no seizure activity.

Peripheral neuropathies occurring in the setting of CO poisoning show histologically demyelination with preservation of axons (Wilson and Winkelman, 1924). The mechanism is unknown but may be secondary to compression, ischemic injury, petechial hemorrhages, or possibly CO toxicity (McDonald, 1967). Snyder (1970) reported on a family with CO poisoning in which one member developed bilateral ulnar palsy and another had bilateral facial diplegia that Snyder did not believe supported compressive trauma or petechial hemorrhage as an underlying mechanism. Toxic or ischemic neuropathy appeared to be a more likely etiology.

Evaluation of peripheral nerve function in animals following CO exposure found block of conduction in the acute phase. This is followed by a second phase with conduction slowing and then loss of conduction that may be due to the effect of anoxia on axoplasmic flow. In the third phase, conduction velocity is markedly decreased due to changes affecting the nodes of Ranvier. The effects in the peripheral nervous system were more closely associated with ischemia due to lowering of the blood pressure by CO exposure than CO-induced hypoxia (Grunnet and Petajan, 1976; Petajan et al., 1976).

TREATMENT

As noted previously in this chapter under Biomarkers of CO Exposure, the half-life of COHb when breathing room air is 4–6 hours. The half-life decreases to 40–60 minutes when breathing 100% oxygen and further decreases to 15–30 minutes when HBO therapy is administered.

The cornerstone of treatment for CO poisoning is 100% normobaric oxygen using a tight-fitting mask for >6 hours to treat hypoxia, for blood pressure support, and to address any cardiac issues if present.

From 2008 to 2010, 87 hyperbaric facilities reported information on 864 patients who received treatment for CO poisoning (Clower et al., 2012), of which 353 required hospitalization (median COHb 25%, range 1–77%) and 475 were discharged after treatment (median COHb 21%, range 0.1–46%). The most common symptoms were headache (66%), dizziness (51%), nausea/vomiting (46%), loss of consciousness (44%), and confusion (30%). Twenty-one patients reported no symptoms. Thirty-eight percent had loss of consciousness for ≤ 10 minutes, while for 49% the duration was unknown.

Since the indications of when to use HBO therapy for CO poisoning have remained controversial (Raphael et al., 1989; Weaver et al., 1996), Ernst and Zibrak (1998) published guidelines in 1998. These included: (1) coma; (2) any period of unconsciousness; (3) any abnormal score on the CO Neuropsychological Screening Battery; (4) COHb > 40%; (5) pregnancy and COHb level > 15%; (6) signs of cardiac ischemia or arrhythmia; (7) history of ischemic heart disease and COHb level > 20%; (8) recurrent symptoms for up to 3 weeks; and (9) symptoms that do not resolve with normobaric oxygen after 4–6 hours. As discussed in the section on neuropsychologic effects, above, if the only finding was an abnormal score on one of the tests in the CO Neuropsychological Screening Battery, this should not warrant treatment with HBO.

The Undersea and Hyperbaric Medical Society recommends HBO treatment for CO-poisoned patients regardless of COHb levels when there is transient or prolonged unconsciousness, neurologic signs, cardiovascular dysfunction, or severe acidosis, or if their age is ≥36 years, or if the CO exposure duration interval is ≥24 hours. They agree that the role of neuropsychologic testing to determine need for HBO therapy is not clear. The optimal protocol for HBO treatment with CO poisoning has not been determined and whether clinical improvement or reduced rate of neurocognitive sequelae occurs when HBO is administered beyond 6 hours from poisoning is unknown (Weaver, 2008). These recommendations are made in the absence of well-designed randomized control studies or appropriate study design to allow for meaningful interpretation of the results.

The controversy over HBO treatment for CO poisoning continues as the American College of Emergency Physicians subcommittee could not reach a consensus on any variable for which HBO was indicated with CO poisoning (Wolf et al., 2008). Two Cochrane reviews critiqued multiple randomized controlled trials and concluded that the evidence did not support that HBO therapy for CO poisoning reduced the persistence of adverse neurologic outcomes (Juurlink et al., 2005; Buckley et al., 2011). In light of marked variability in patient selection and CO poisoning in these studies, a more conservative recommendation suggested one HBO treatment at 2 atmosphere within 12 hours of CO exposure for comatose patients with acute nonsuicidal CO poisoning and COHb greater than 25%.

Any pregnant woman with CO poisoning should receive HBO therapy (Guzman, 2012) because CO exposure in pregnant women is dangerous, as there may be a lag time for CO uptake, but eventually the COHb level in the fetus is higher than in the mother (Long and Hill, 1977). The fetal hemoglobin releases less oxygen to the tissues, resulting in significant hypoxia (Farrow et al., 1990).

To test the hypothesis that HBO therapy is needed to prevent the development of DNS, Thom et al. (1995) treated patients with CO poisoning with HBO: none developed DNS while 23% developed DNS after treatment with ambient-pressure oxygen. There was a problem with case definition for DNS. None of these patients had loss of consciousness and their level of function (going to work) with persistent symptoms of headache, dizziness, and difficulty concentrating is not commonly used to define DNS. In this study differences in performance on the neuropsychologic battery between the two treatment groups could be attributed to premorbid abilities as the ambient pressure oxygen group had fewer years of education and this was not taken into account in the analyses (Thom et al., 1995).

There is concern that advocates for HBO treatment are located at facilities that offer this treatment. Some believe further randomized controlled trials are unnecessary, as withholding HBO treatment in CO-poisoned patient is unethical, while others feel further trials are unethical given the absence of data showing the effectiveness of HBO treatment and the expense of transferring patients to such facilities (Kao and Nanagas, 2006).

HBO is not free of side-effects; some include painful barotrauma, decompression sickness, pulmonary edema and hemorrhage, seizures, and oxygen toxicity. The presence of a pneumothorax is an absolute contraindication for HBO treatment (Kao and Nanagas, 2006).

References

Amitai Y, Zlotogorski Z, Golan-Katzav V et al. (1998). Neuropsychological impairment from acute low-level exposure to carbon monoxide. Arch Neurol 55: 845–848.

Anderson EW, Andelman RJ, Strauch JM (1973). Effect of low-level carbon monoxide exposure on onset and duration of angina pectoris. A study in ten patients with ischemic heart disease. Ann Intern Med 79: 46–50.

Annane D, Chevret S, Jars-Guincestre MC et al. (2001). Prognostic factors in unintentional mild carbon monoxide poisoning. Intensive Care Med 27: 1776–1781.

Aronow WS (1981). Aggravation of angina pectoris by two percent carboxyhemoglobin. Am Heart J 101: 154.

Aronow WS, Issell MW (1973). Carbon monoxide effects on exercise-induced angina pectoris. Ann Intern Med 79: 392–395.

Beard RR, Grandstaff N (1970). Carbon monoxide exposure and cerebral function. Ann N Y Acad Sci 174: 385–395.

Beard RR, Wertheim GA (1967). Behavioral impairment associated with small doses of carbon monoxide. Am J Public Health 57 (11): 2012–2022, Nov.

Benignus VA (1994). Behavioral effects of carbon monoxide: meta analyses and extrapolations. J Appl Physiol 76 (3): 1310–1316.

Benignus VA, Otto DA, Prah JD et al. (1977). Lack of effects of carbon monoxide on human vigilance. Percept Mot Skills 45: 1007–1014.

Benignus VA, Muller KE, Barton CN et al. (1987a). Effect of low level carbon monoxide on compensatory tracking and event monitoring. Neurotoxicol Teratol 9: 227–234.

Benignus VA, Kafer ER, Muller KE et al. (1987b). Absence of symptoms with carboxyhemoglobin levels of 16–23%. Neurotoxicol Teratol 9: 345–348.

Benignus VA, Muller KE, Malott CM (1990). Dose-effects functions for carboxyhemoglobin and behavior. Neurotoxicol Teratol 12: 111–118.

Benignus VA, Hazucha MJ, Smith MV et al. (1994). Prediction of carboxyhemoglobin formation due to transient exposure to carbon monoxide. J Appl Physiol 76 (4): 1739–1745.

Beppu T (2014). The role of MR imaging in assessment of brain damage from carbon monoxide poisoning: a review of the literature. AJNR Am J Neuroradiol 35: 625–631.

Beppu T, Nishimoto H, Ishigaki D et al. (2010). Assessment of damage to cerebral white matter fiber in the subacute phase after carbon monoxide poisoning using fractional anisotropy in diffusion tensor imaging. Neuroradiology 52: 735–743.

Beppu T, Nishimoto H, Fujiwara S et al. (2011). ^{1}H-magnetic resonance spectroscopy indicates damage to cerebral white matter in the subacute phase after CO poisoning. J Neurol Neurosurg Psychiatry 82: 869–875.

Beppu T, Fujiwara S, Nishimoto H et al. (2012). Fractional anisotropy in the centrum semiovale as a quantitative indicator of cerebral white matter damage in the subacute phase in patients with carbon monoxide poisoning: correlation with the concentration of myelin basic protein in cerebrospinal fluid. J Neurol 259: 1689–1705.

Bleecker ML, Lindgren KN (1999). The mere presence of low levels of carboxy- hemoglobin is not causal proof for altered neuropsychological performance. Arch Neurol 56: 1299.

Bokonjic N, Buchthal F (1961). Postanoxic unconsciousness as related to clinical and EEG recovery in stagnant anoxia and carbon monoxide poisoning. In: H Gastaut, JS Meyer (Eds.), Cerebral anoxia and the electroencephalogram, pp. 118–127, Springfield, Ill, Charles C Thomas.

Bruno A, Wagner W, Orrison WW (1993). Clinical outcome and brain MRI four years after carbon monoxide intoxication. Acta Neurol Scand 87: 205–209.

Buckley NA, Juurlink D, Isbister G et al. (2011). Hyperbaric oxygen for carbon monoxide poisoning. Cochrane Database Syst Rev 4, CD002041.

CDC (1993). Unintentional carbon monoxide poisoning from indoor use of pressure washers – Iowa, January 1992 – January 1993. MMWR Morb Mortal Wkly Rep 42 (40): 777–779, 785.

CDC (1996). Carbon monoxide poisoning at an indoor ice arena and bingo hall – Seattle 1996. MMWR Morb Mortal Wkly Rep 45 (13): 265–267.

CDC (1998). Carbon monoxide poisoning and death after the use of explosives in a sewer construction project. DHHS (NIOSH) Publication number 98-122, National Institute for Occupational Safety and Health (NIOSH), Atlanta, GA.

CDC (2005). Carbon monoxide poisoning after Hurricane Katrina – Alabama, Louisiana, and Mississippi, August–September 2005. MMWR Morb Mortal Wkly Rep 54 (39): 996–998.

CDC (2006). Carbon monoxide poisoning after two major hurricanes – Alabama and Texas. August–October 2005. MMWR Morb Mortal Wkly Rep 55 (9): 236–239.

CDC (2011). Carbon monoxide exposures – United States, 2000–2009. MMWR Morb Mortal Wkly Rep 60 (30): 1014–1017.

Cervellin G, Comelli I, Rastelli G et al. (2014). Initial blood lactate correlates with carboxyhemoglobin and clinical severity in carbon monoxide poisoned patients. Clin Biochem 47 (18): 298–301.

Chang KH, Han MH, Kin HS et al. (1992). Delayed encephalopathy after acute carbon monoxide intoxication: imaging features and distribution of cerebral white matter lesions. Radiology 184: 117–122.

Chang CC, Chang WN, Lui CC et al. (2010). Longitudinal study of carbon monoxide intoxication by diffuse tensor imaging with neuropsychiatric correlation. J Psychiatry Neurosci 35: 115–125.

Choi IS (1983). Delayed neurological sequelae in carbon monoxide intoxication. Arch Neurol 40: 433–435.

Choi IS (2002). Parkinsonism after carbon monoxide poisoning. Eur Neurol 48: 30–33.

Clower JH, Hampson NB, Iqbal S et al. (2012). Recipients of hyperbaric oxygen treatment for carbon monoxide poisoning and exposure circumstances. Am J Emerg Med 30: 846–851.

Coburn RF, Forster RE, Kane PB (1965). Considerations of the psychological variables that determine the blood carboxyhemoglobin concentration in men. J Clin Invest 44: 1899–1910.

Crystal HA, Ginsberg MD (2000). Carbon monoxide. In: PS Spencer, HH Schaumburg (Eds.), Experimental and clinical neurotoxicology, 2nd edn. Oxford University Press, New York, pp. 318–329.

De Reuck J, Decoo D, Lemahieu I et al. (1993). A positron emission tomography study of patients with acute carbon monoxide poisoning treated by hyperbaric oxygen. J Neurol 240: 430–434.

Department of the Environment (1994). Expert panel on air quality standards. Carbon Monoxide. HMSO, London.

Deschamps D, Geraud C, Julien H et al. (2003). Memory one month after acute carbon monoxide intoxication: a prospective study. Occup Environ Med 60: 212–216.

Durak AC, Coskun A, Yikilmaz A et al. (2005). Magnetic resonance imaging findings in chronic carbon monoxide intoxication. Acta Radiol 46: 322–327.

Ernst A, Zibrak JD (1998). Carbon monoxide poisoning. N Engl J Med 339 (22): 1603–1608.

Farrow JR, Davis GJ, Weaver LK et al. (1990). Fetal death due to nonlethal maternal carbon monoxide poisoning. J Forensic Sci 35 (6): 1448–1452.

Fawcett TA, Moon RE, Fracica PJ et al. (1992). Warehouse workers' headache, carbon monoxide poisoning from propane-fueled forklifts. J Occup Med 34 (1): 12–15.

Forbes WH, Sargent F, Roughton FJW (1945). The rate of carbon monoxide uptake by normal men. Am J Physiol 143 (4): 594–608.

Fujiwara S, Beppu T, Nishimoto H et al. (2012). Detecting damaged regions of cerebral white matter in the subacute phase after carbon monoxide poisoning using voxel-based analysis with diffusion tensor imaging. Neuroradiology 54: 681–689.

Gale SD, Hopkins RO, Weaver LK et al. (1999). MRI, quantitative MRI, SPECT, and neuropsychological findings following carbon monoxide poisoning. Brain Inj 13 (4): 229–243.

Gawlikowski T, Golasik M, Gomolka E et al. (2014). Proteins as biomarkers of carbon monoxide neurotoxicity. Inhal Toxicol 26: 885–890.

Gilbert GJ, Glaser GH (1959). Neurologic manifestations of chronic carbon monoxide poisoning. N Engl J Med 10: 1217–1220.

Gliner JA, Horvath SM, Mihevic PM (1983). Carbon monoxide and human performance in a single and dual task methodology. Aviat Space Environ Med 54 (8): 714–717, Aug.

Grunnet ML, Petajan JH (1976). Carbon monoxide-induced neuropathy in the rat. Ultrastructural changes. Arch Neurol 33: 158–163.

Guzman JA (2012). Carbon monoxide poisoning. Crit Care Clin 28: 537–548.

Hampson NB, Hampson LA (2002). Characteristics of headache associated with acute carbon monoxide poisoning. Headache 42: 220–223.

Hampson NB, Norkool DM (1992). Carbon monoxide poisoning in children riding in the back of pickup trucks. J Amer Med Assoc 267 (4): 538–540.

Hampson NB, Kramer CC, Dunford RG et al. (1994). Norkool DM. Carbon monoxide poisoning from indoor burning of charcoal briquets. J Amer Med Assoc 271 (1): 52–53.

Hanafy KA, Oh J, Otterbein LE (2013). Carbon monoxide and the brain: time to rethink the dogma. Curr Pharm Des 19: 2771–2775.

Hara S, Mukai T, Kurosaki K et al. (2002). Modification of the striatal dopaminergic neuron system by carbon monoxide exposure in free-moving rats, as determined by in vivo brain microdialysis. Arch Toxicol 76: 596–605.

Henke K, Kroll NE, Behniea H et al. (1999). Memory lost and regained following bilateral hippocampal damage. J Cogn Neurosci 11: 682–697.

Henz S, Maeder M (2005). Prospective study of accidental carbon monoxide poisoning in 38 Swiss soldiers. Swiss Med Wkly 135: 398–406.

Holstege CP, Baer AB, Eldridge DL et al. (2004). Case series of elevated troponin I following carbon monoxide poisoning. J Toxicol Clin Toxicol 42 (5): 742–743.

Hopkins RO, Fearing MS, Weaver LK et al. (2006). Basal ganglia lesions following carbon monoxide poisoning. Brain Inj 20 (3): 273–281.

Ide T, Kamijo Y (2009). The early elevation of interleukin 6 concentration in cerebrospinal fluid and delayed encephalopathy of carbon monoxide poisoning. Am J Emerg Med 27: 992–996.

Ide T, Kamijo Y, Ide A, Yoshimura K et al. (2012). Elevated S100B level in cerebrospinal fluid could predict poor outcome of carbon monoxide poisoning. Am J Emerg Med 30: 222–225.

Jaeckle RS, Nasrallah HA (1985). Major depression and carbon monoxide-induced parkinsonism: diagnosis, computerized axial tomography, and response to L-dopa. J Nerve Ment Dis 173 (8): 503–508, Aug.

Jasper BW, Hopkins RO, Van Duker H et al. (2005). Affective outcome following carbon monoxide poisoning a prospective longitudinal study. Cogn Behav Neurol 18 (2): 127–134, June.

Juurlink DN, Buckley NA, Stanbrook MB et al. (2005). Hyperbaric oxygen for carbon monoxide poisoning. Cochrane Database Syst Rrev 1, CD002041.

Kao LW, Nanagas KA (2006). Toxicity associated with carbon monoxide. Clin Lab Med 26: 99–125.

Kesler SR, Hopkins RO, Blatter DD et al. (2001). Verbal memory deficits associated with fornix atrophy in carbon monoxide poisoning. J Int Neuropsychol Soc 7: 640–646.

Kim HY, Kim BJ, Moon SY et al. (2002). Serial diffusion-weighted MR imaging in delayed postanoxic encephalopathy. A case study. J Neuroradiol 29 (3): 211–215, Sept.

Kim JH, Chang KH, Song IC et al. (2003). Delayed encephalopathy of acute carbon monoxide intoxication: Diffusivity of cerebral white matter lesions. AJNR Am J Neuroradiol 24: 1592–1597.

Klawans HL, Stein RW, Tanner CM et al. (1982). A pure parkinsonian syndrome follow acute carbon monoxide intoxication. Arch Neurol 39: 302.

Koike A, Wasserman K, Armon Y et al. (1991). The work-rate dependent effect of carbon monoxide on ventilatory control during exercise. Respir Physiol 85: 169–183.

Kondziella D, Danielsen ER, Hansen K et al. (2009). 1H MR spectroscopy of gray and white matter in carbon monoxide poisoning. J Neurol 256: 970–979.

Lapresle J, Fardeau M (1967). The central nervous system and carbon monoxide poisoning. II. Anatomical study of brain lesions following intoxication with carbon monoxide (22 cases). Prog Brain Res 24: 31–74.

Laties VG, Merigan WH (1979). Behavioral effects of carbon monoxide on animals and man. Annu Rev Pharmacol Toxicol 19: 357–392.

Lee MH (1978). Clinical studies on delayed sequelae in carbon monoxide intoxication. J Korean Neuropsychiatr Assoc 17: 374–385.

Li W, Zhang Y, Gu R et al. (2013). DNA pooling base genome-wide association study identifies variants at NRXN3 associated with delayed encephalopathy after acute carbon monoxide poisoning. PLoS One 8 (11): e79159.

Liang F, Li W, Zhang P et al. (2013). A PARK2 polymorphism associated with delayed neuropsychological sequelae after carbon monoxide poisoning. BMC Med Genet 14: 99.

Lin WC, Lu CH, Lee YC et al. (2009). White matter damage in carbon monoxide intoxication assessed in vivo using diffusion tensor MR imaging. AJNR Am J Neuroradiol 30 (5): 1017–1021.

Lo CP, Chen SY, Lee KW et al. (2007). Brain injury after acute carbon monoxide poisoning: early and late complications. AJR Am J Roentgenol 189 (4): W205–W211.

Long LD, Hill EP (1977). Carbon monoxide uptake and elimination in fetal and maternal sheep. Am J Physiol 232: H324–H330.

McDonald WL (1967). Structural and functional changes in human and experimental neuropathy. In: Modern trends in neurology, Vol. 4. Butterworth, London, p. 145.

McFarland RA (1973). Low level exposure to carbon monoxide and driving performance. Arch Environ Health 27: 355–359, Dec.

Messier LD, Myers RA (1991). A neuropsychological screening battery for emergency assessment of carbon-monoxide-poisoned patients. J Clin Psychol 47: 675–684.

Mihevic PM, Gliner JA, Ho SM (1983). Carbon monoxide exposure and information processing during perceptual-motor performance. Int Arch Occup Environ Health 51: 355–363.

Mikulka P, O'Donnell, Heinig P et al. (1970). The effect of carbon monoxide on human performance. Ann N Y Acad Sci 174: 409–420.

Min SK (1986). A brain syndrome associated with delayed neuropsychiatric sequelae following acute carbon monoxide intoxication. Acta Psychiatr Scand 73: 80–86.

Murata T, Kimura H, Kado H et al. (2001). Neuronal damage in the interval form of CO poisoning determined by serial diffusion weighted magnetic resonance imaging plus 1 H-magnetic resonance spectroscopy. J Neurol Neurosurg Psychiatry 71: 250–253.

Myers RAM, DeFazio A, Kelly MP (1998). Chronic carbon monoxide exposure: a clinical syndrome detected by neuropsychological tests. J Clin Psychol 54: 555–567.

Neufeld MY, Swanson JW, Klass DW (1981). Localized EEG abnormalities in acute carbon monoxide poisoning. Arch Neurol 38: 524–527.

O'Donnell P, Buxton PJ, Pitkin A et al. (2000). The magnetic resonance imaging appearances of the brain in acute carbon monoxide poisoning. Clin Radiol 55: 273–280.

O'Donoghue JL (1985). Carbon monoxide, inorganic nitrogenous compounds, and phosphorus. In: JL O'Donoghue (Ed.), Neurotoxicity of industrial and commercial chemicals, 1. CRC Press, Florida, pp. 193–200.

Ogawa M, Minami T, Katsurada K et al. (1973). Early electroencephalographic change following acute carbon monoxide poisoning in relation to cerebral metabolism. Med J Osaka Univ 24: 85–90.

Okeda R, Funata N, Takano T et al. (1981). The pathogenesis of carbon monoxide encephalopathy in the acute phase – physiological and morphological correlation. Acta Neuropathol (Berl) 54: 1–10.

Park S, Choi IS (2004). Chorea following acute carbon monoxide poisoning. Yonsei Med J 45: 363–366.

Park E, Ahn J, Min YG et al. (2012). The usefulness of the serum s100b protein for predicting delayed neurological sequelae in acute carbon monoxide poisoning. Clin Toxicol (Phila) 50: 183–188.

Park EJ, Min YG, Kim GW et al. (2014). Pathophysiology of brain injuries in acute carbon monoxide poisoning: a novel hypothesis. Med Hypotheses 83: 186–189.

Parkinson RB, Hopkins RO, Cleavinger HB et al. (2002). White matter hyperintensities and neuropsychological outcome following carbon monoxide poisoning. Neurology 58: 1525–1532.

Petajan JH, Packham SC, Frens DB et al. (1976). Sequelae of carbon monoxide- induced hypoxia in the rat. Arch Neurol 33: 152–157.

Peterson JE, Stewart RD (1970). Absorption and elimination of carbon monoxide by inactive young men. Arch Environ Health 21: 165–171.

Peterson JE, Stewart RD (1975). Predicting the carboxyhemoglobin levels resulting from carbon monoxide exposures. J Appl Physiol 39: 633–638.

Porter SS, Hopkins RO, Weaver LK et al. (2002). Corpus callosum atrophy and neuropsychological outcome following carbon monoxide poisoning. Arch Clin Neuropsychol 17: 195–204.

Prockop LD (2005). Carbon monoxide brain toxicity: clinical, magnetic resonance imaging, magnetic resonance spectroscopy, and neuropsychological effects in 9 people. J Neuroimaging 15 (2): 144–149, Apr.

Pulst SM, Walshe TM, Romero JA (1983). Carbon monoxide poisoning with features of Gilles de la Tourette's syndrome. Arch Neurol 40: 443–444.

Putz VR (1979). The effects of carbon monoxide on dual-task performance. Hum Factors 21 (1): 13–24.

Raphael JC, Elkharrat D, Jars-Guincestre MC et al. (1989). Trial of normobaric and hyperbaric oxygen for acute carbon monoxide intoxication. Lancet 334: 414–419.

Rasmussen LS, Poulsen M, Christiansen M et al. (2004). Biochemical markers for brain damage after carbon monoxide poisoning. Acta Anaesthesiol Scand 48: 469–473.

Raub JA, Benignus VA (2002). Carbon monoxide and the nervous system. Neurosci Biobehav Rev 26: 925–940.

Raub J, Mathieu-Nolf M, Hampson N et al. (2000). Carbon monoxide poisoning – a public health perspective. Toxicology 145: 1–14.

Rottman SJ, Kaser-Boyd N, Cannis T et al. (1995). Low-level carbon-monoxide poisoning: inability of neuropsychological testing to identify patients who benefit from hyperbaric oxygen therapy. Prehosp Disaster Med 10: 276–282.

Schulte JH (1963). Effects of mild carbon monoxide intoxication. Arch Environ Health 7: 524–530.

Schwartz A, Hennerici M, Wegener OH (1985). Delayed choreoathetosis following acute carbon monoxide poisoning. Neurology 35: 98–99.

Sener RD (2003). Acute carbon monoxide poisoning: diffusion MR imaging findings. AJNR Am J Neuroradiol 24: 1475–1477.

Singhal AB, Topcuoglu MA, Koroshetz WJ (2002). Diffusion MRI in three types of anoxic encephalopathy. J Neurol Sci 196: 37–40.

Snyder RD (1970). Carbon monoxide intoxication with peripheral neuropathy. Neurology 20: 177–180.

Stewart RD (1975). The effect of carbon monoxide on humans. Annu Rev Pharmacol 15: 409–423.

Stewart RD, Peterson JE, Baretta ED et al. (1970). Experimental human exposure to carbon monoxide. Arch Environ Health 21: 154–164.

Stewart RD, Newton PE, Hosko M et al. (1973). Effects of carbon monoxide on time perception. Arch Environ Health 27: 155–160.

Teksam M, Casey SO, Michel E et al. (2002). Diffusion-weighted MR imaging findings in carbon monoxide poisoning. Neuroradiology 44: 109–113.

Terajima K, Igarashi H, Hirose M et al. (2008). Serial assessments of delayed encephalopathy after carbon monoxide poisoning using magnetic resonance spectroscopy and diffusion tensor imaging on 3.0 T system. Eur Neurol 59: 55–61.

Thom SR, Taber RL, Mendiguren II et al. (1995). Delayed neuropsychologic sequelae after carbon monoxide poisoning: prevention by treatment with hyperbaric oxygen. Ann Emerg Med 25: 474–480.

Thom SR, Bhopale VM, Fisher D et al. (2004). Delayed neuropathy after carbon monoxide poisoning is immune-mediated. Proc Natl Acad Sci U S A 101: 13660–13665.

Thom S, Bhopale V, Milovanova T et al. (2010). Plasma biomarkers in carbon monoxide poisoning. Clin Toxicol (Phila) 48 (1): 47–56.

Townsend CL, Maynard RL (2002). Effects on health of prolonged exposure to low concentrations of carbon monoxide. Occup Environ Med 59: 708–711.

Verma A, Hirsch DJ, Glatt CE et al. (1993). Carbon monoxide: a putative neural messenger. Science 259: 381–384.

von Rappard J, Schönenberger M, Bärlocher L (2014). Case report: carbon monoxide poisoning following use of a water pipe/hookah. Dtsch Arztebl Int 111: 674–679.

Wang B, Cao W, Biswal S (2011). Carbon monoxide-activated Nrf2 pathway leads to protection against permanent focal cerebral ischemia. Stroke 42: 2605–2610.

Weaver LK (2008). Carbon monoxide poisoning. In: LB Gesell (Ed.), Hyperbaric oxygen therapy indications, 12th edn. Undersea and Hyperbaric Medical Society, Durham, NC, pp. 19–28.

Weaver LK (2009). Carbon monoxide poisoning. N Engl J Med 360 (12): 1217–1225.

Weaver LK, Hopkins RO, Larson-Lohr V (1996). Neuropsychologic and functional recovery from severe carbon monoxide poisoning without hyperbaric oxygen therapy. Ann Emerg Med 27 (6), June.

Weaver LK, Hopkins RO, Chan KJ et al. (2002). Hyperbaric oxygen for acute carbon monoxide poisoning. N Engl J Med 347: 1057–1067.

Weaver LK, Valentine KJ, Hopkins RO (2007). Carbon monoxide poisoning risk factors for cognitive sequelae and the role of hyperbaric oxygen. Am J Respir Crit Care Med 176: 491–497.

WHO (1999). Environmental health criteria 213: Carbon monoxide. 2nd edn. World Health Organization, Finland.

Widdop B (2002). Analysis of carbon monoxide. Ann Clin Biochem 39: 378–391.

Wilson G, Winkleman NW (1924). Multiple neuritis following carbon monoxide poisoning. J Amer Med Assoc 82: 1407.

Wolf SJ, Lavonas EJ, Sloan EP et al. (2008). Clinical policy: critical issue in the management of adult patients presenting to the emergency department with acute carbon monoxide poisoning. Ann Emerg Med 51 (2): 138–152.

Yabluchanskiy A, Sawle P, Homer-Vanniasinkam S (2012). CORM-3, a carbon monoxide- releasing molecule, alters the inflammatory response and reduces brain damage in a rat model of hemorrhagic stroke. Crit Care Med 42: 544–552.

Yardan T, Cervik Y, Donderici O et al. (2009). Elevated serum S100B protein and neuron-specific enolase levels in carbon monoxide poisoning. Am J Emerg Med 27: 838–842.

Yoshii F, Kozuma R, Takahashi W et al. (1998). Magnetic resonance imaging and ^{11}C-*N*-methylsiperone/positron emission tomography studies in a patient with the interval form of carbon monoxide poisoning. J Neurol Sci 160: 87–91.

Section 3

Movement disorders (parkinsonism or other movement disorders)

Handbook of Clinical Neurology, Vol. 131 (3rd series)
Occupational Neurology
M. Lotti and M.L. Bleecker, Editors

Chapter 13

Extrapyramidal system neurotoxicity: animal models

DAVID DORMAN*
Department of Molecular Biomedical Sciences, College of Veterinary Medicine, North Carolina State University, Raleigh, NC, USA

INTRODUCTION

The pyramidal and extrapyramidal systems provide motor control to the muscles of the head, neck, and limbs. Some neuroanatomists describe the pyramidal system as being involved with initiation and control of "voluntary" movements, whereas the extrapyramidal system controls "involuntary" movements (Sengul and Watson, 2014). Extrapyramidal motor disease; therefore, is generally characterized by impaired motor control, which is usually the result of basal ganglionic dysfunction (Sanders and Gillig, 2012). The World Health Organization's *International Classification of Diseases* includes a classification for extrapyramidal and movement disorders (Jankovic, 1995). This World Health Organization category includes, among others, Parkinson's disease, secondary parkinsonism, other degenerative diseases of the basal ganglia, and clinical syndromes that result in dystonia, dyskinesia, essential tremor, and other forms of tremor and chorea.

This chapter briefly reviews the anatomy of the extrapyramidal system and discusses several naturally occurring and experimental chemical models that target the mammalian (nonhuman) extrapyramidal system. One goal of this chapter is to illustrate the similarity in responses seen in animals and people. Topics discussed include extrapyramidal syndromes associated with antipsychotic drugs, carbon monoxide, reserpine, cyanide, rotenone, paraquat, 1-methyl-4-phenyl-1,2,3,6-tetrahydropyridine (MPTP), and manganese. Of these, manganese neurotoxicity is one of the most important occupationally occurring causes of secondary parkinsonism affecting the human basal ganglia. No attempt has been made to provide an exhaustive review of each of the topics addressed, but wherever possible references to review articles are provided.

EXTRAPYRAMIDAL SYSTEM ANATOMY

An anatomic distinction between the pyramidal and extrapyramidal systems is purely artificial. The synaptic connections found within the extrapyramidal system are extremely complex and remain poorly understood. The major parts of the extrapyramidal system are the subcortical nuclei. This includes the caudate, putamen, and globus pallidus; which are collectively known as the basal ganglia. Other structures that are considered part of the extrapyramidal system include the substantia nigra, parts of the thalamus, subthalamic nucleus, and the red nucleus. These nuclei are synaptically interconnected with connections to the brainstem, cerebellum and the pyramidal system. The role of the basal ganglia (extrapyramidal system) as a major motor control system diminishes as brain development evolved from lower to higher vertebrates (Reeves and Swenson, 2008). Agents that target the basal ganglia (caudate, putamen, and globus pallidus) and the subthalamic nucleus will be the primary focus of this chapter.

ASSESSMENT OF ANIMAL MODELS FOR EXTRAPYRAMIDAL EFFECTS

Methods used in experimental animal studies to assess extrapyramidal effects have been recently reviewed (Tieu, 2011). Methods used can be broadly classified as:

1. Behavioral. Endpoints of interest can include:
   a. functional observation battery and other clinical assessments for the presence or absence of visible clinical signs;
   b. assessment of motor activity using automated test apparatuses, including rotarod tests, circling behaviors (rotation test), stride length

*Correspondence to: David Dorman, DVM, PhD, North Carolina State University, College of Veterinary Medicine, Department of Molecular Biomedical Sciences, 1060 William Moore Drive, Raleigh, NC 27607, USA. Tel: +1-919-513-6237, E-mail: david_dorman@ncsu.edu

and gait evaluations, spontaneous motor activity.

2. Neurochemical. Endpoints of interest can include:
   a. quantification of striatal dopamine and dopamine metabolite concentration using use reverse-phase high-performance liquid chromatography coupled with an electrochemical detector or other suitable analytic method;
   b. experimental evidence for oxidative stress in many animal models can include measurement of increased lipid peroxidation and altered ratios of oxidized/reduced glutathione.
3. Histologic evaluation of dopaminergic neuron populations. Endpoints of interest can include:
   a. assessment of tyrosine hydroxylase (TH) immunoreactivity. TH catalyzes the conversion of L-tyrosine to L-3,4-dihydroxyphenylalanine (L-DOPA) and is used as a histologic marker of dopaminergic cells;
   b. qualitative or quantitative morphologic assessments, that include evaluation of dopaminergic neuron number and distribution. Quantitative stereologic approaches are generally preferred. The estimated average number of dopaminergic neurons in an adult mouse nigra is commonly reported to be ~8000–14000. There is a significant difference in the density of dopaminergic neurons between the caudal and rostral regions. Sampling the population of dopaminergic neurons should occur at different levels throughout the entire nigra;
   c. evaluation of the topographic distribution of dopaminergic neurons in the nigral and striatal regions, including assessment of dopaminergic terminals in the striatum (e.g., density);
   d. evaluation of the number and distribution of dopamine transporters;
   e. detection of Lewy body-like aggregation. Several methods have been used, including assessment of fibrillary protein with thioflavin S or immunohistochemical evaluation of intracellular α-synuclein.
4. Response to L-DOPA and other therapeutic interventions.

SPONTANEOUS TOXIC SECONDARY PARKINSONISM IN ANIMALS

The following agents have been reported to produce secondary parkinsonism in animals under nonexperimental conditions. In some cases a toxic extrapyramidal syndrome has been identified in humans. Because of the importance of these human toxic syndromes, experimental animal models have also been developed and are briefly mentioned below.

Antipsychotic drugs

A common form of toxic secondary parkinsonism occurs in people taking medications for psychotic disorders and depression. One of the first drugs to be associated with a "drug-induced extrapyramidal reaction" was the phenothiazine neuroleptic, chlorpromazine (Burkhard, 2014). Today drug-induced parkinsonism is most commonly associated with the use of dopamine receptor-blocking agents (Morley et al., 2014). Drug-induced D_2 receptor blockade is associated with tremor, rigidity, akinesia, and bradykinesia – a series of effects collectively known as extrapyramidal symptoms.

A spectrum of responses is associated with this syndrome. For example, acute (within 1–5 days) dystonic reactions are associated with neuroleptic drugs and clinical manifestations are usually initially restricted to the head and neck, but may extend toward the upper and lower limbs as well as the trunk (Burkhard, 2014). Pathogenic mechanisms of acute dystonic reactions may involve dopamine receptor blockade and many cases can be effectively managed with anticholinergic drugs (Burkhard, 2014). A nondegenerative drug-induced parkinsonism syndrome has also been associated with more prolonged use of neuroleptics and calcium-channel blockers. Studies of patients have shown that, when compared with Parkinson's disease, drug-induced parkinsonism has more symmetric bilateral involvement with greater effects on the upper limbs. Drug-induced parkinsonism is associated with bradykinesia, tremor, and rigidity, whereas Parkinson's disease more commonly involves gait and posture (Hassin-Baer et al., 2001). Reduced olfactory function has also been associated with the appearance and severity of extrapyramidal symptoms associated with drug-induced parkinsonism (Kruger et al., 2008; Morley and Duda, 2014).

A variety of veterinary species also develop extrapyramidal signs following exposure to neuroleptic agents. Extrapyramidal signs have been seen in horses following the use of the phenothiazine neuroleptic fluphenazine (Kauffman et al., 1989; Brewer et al., 1990; Baird et al., 2006) and dogs given haloperidol (de Leeuw and Sangster, 1979). Extrapyramidal signs have also been reported in birds. For example, Starkey and coworkers (2008) reported dystonia, intermittent ataxia, and coarse-muscle tremors consistent with an extrapyramidal syndrome in a macaw bird treated with haloperidol and clomipramine. Experimental models for antipsychotic drug-induced parkinsonism have been developed (reviewed in Castagné et al., 2009). Traditional species include rodents (Messiha, 1991; Migler et al., 1993;

Crocker and Hemsley, 2001; Tada et al., 2004; Trevizol et al., 2011), cats (Tan et al., 1992), and nonhuman primates (Porsolt, 2013). Zebra fish, medium-throughput animal-based assay, have also been recently proposed (Farrell et al., 2011).

Carbon monoxide

Carbon monoxide is one of the most common causes of poisoning in the modern world, and the leading cause of death by poisoning in the USA (Prockop and Chichkova, 2007). Carbon monoxide poisoning commonly occurs following house fire smoke inhalation. Carbon monoxide competitively binds to hemoglobin, forming carboxyhemoglobin. Carbon monoxide has a 200–230-fold greater affinity for hemoglobin than oxygen. This results in a conformational change in the hemoglobin molecule, reduced oxygen-carrying capacity, hypoxemia, and hypoxia resulting in neurotoxicity and other effects. Hypoxic brain damage predominates in the cerebral cortex, cerebral white matter, and basal ganglia, especially in the globus pallidus.

Acute extrapyramidal lesions associated with carbon monoxide inhalation include hemorrhagic necrosis of the basal ganglion and acute damage to the pallidum and substantia nigra pars reticulata which is also referred to as the pallidoreticular system (Sohn et al., 2000; Choi, 2002; Hopkins et al., 2006; Subhaschandra et al., 2008; Chen et al., 2012). The exact pathogenesis of carbon monoxide neurotoxicity remains unclear: hypoxia, reperfusion injury, lipid peroxidation, and excitatory amino acid neurotoxicity have been proposed (Thom et al., 2004; Prockop and Chichkova, 2007).

A variety of veterinary species, including dogs, cats, and other companion animals, can be exposed to carbon monoxide (Drobatz et al., 1999a, b; Jackson and Drobatz, 2004). Neurologic signs of carbon monoxide toxicity reported in dogs include ataxia, absent menace response, dull sensorium, loss of consciousness, head tremors, twitching, seizures, and dementia (Mariani, 2003; Kent et al., 2010). Extrapyramidal lesions have been reported experimentally in monkeys, cats, and dogs (Ginsberg et al., 1974; Okeda et al., 1981, 1982; Song et al., 1983). The use of rodents and other animal models has also been reviewed (Penney, 1990).

Equine nigropallidal encephalomalacia

Russian knapweed (*Acroptilon repens*; previously classified as *Centaurea repens*) and yellow starthistle (*Centaurea solsitialis* L.) produce an unusual neurotoxicologic disease colloquially known as "chewing disease" (equine nigropallidal encephalomalacia) in horses. These plants are nonindigenous invasive weeds that were introduced from Europe into the USA in the1800s. Since that time these plants have continued to spread throughout Idaho, Oregon, Washington, and other western US states (Maddox et al., 1985; Juliá et al., 2007).

The clinical signs and lesions associated with these plants are well recognized amongst veterinarians as a toxic parkinsonism syndrome (Chang et al., 2012). Although neurologic signs often appear suddenly in horses, development of the syndrome requires prolonged (>30–80 days) exposure (Young et al., 1970a, b). Clinical signs associated with so-called "chewing disease" require chronic high-dose exposure to the plants and include hypertonicity and dystonia of the lips and tongue, inability to prehend food, central nervous system depression, and locomotor deficits. With time, the animal's ability to eat and drink becomes impaired and hypokinesia and death occur in later stages of the disease. Gross examination of the brain often reveals bilateral necrosis of the globus pallidus and the substantia nigra (Cordy, 1954; Fowler, 1965). Histologic studies have shown the presence of lesions within the substantia nigra pars reticulata, sparing the cell bodies of the dopaminergic neurons in the substantia nigra pars compacta, and in the rostral portion of the globus pallidus, with partial disruption of dopaminergic (TH-positive) fibers passing through the globus pallidus (Chang et al., 2012). Based on these lesions, the disease has also been named nigropallidal encephalomalacia. These neuropathologic features are different from Parkinson's disease, in which the substantia nigra pars reticulata and globus pallidus are often unaffected, whereas the number of pigmented dopaminergic neurons in the substantia nigra pars compacta may be reduced. Likewise, abnormal cytoplasmic inclusions (Lewy-like bodies) seen in humans with Parkinson's disease are absent in horses with nigropallidal encephalomalacia (Chang et al., 2012).

The exact toxicologic principle associated with this equine syndrome remains unknown. The sesquiterpene lactone, repin, has been identified as one possible toxin (Tukov et al., 2004a). Experimental repin administration to mice and rats produces ataxia, postural tremors, hypothermia, and decreased response to auditory and tactile stimuli that may be due to glutathione depletion and inhibition of neuronal dopamine release (Robles et al., 1998). Recent studies of repin in rodents and cultured PC12 neurons (derived from a rat adrenal pheochromocytoma) suggests that the toxin interferes with striatal dopamine release and glutathione metabolism, and promotes oxidative stress: these mechanisms may be involved in the pathogenesis of the nigropallidal lesions (Robles et al., 1998; Tukov et al., 2004b). Rodents exposed to repin did not develop histologic lesions in either the striatum or substantia nigra (Robles et al., 1998).

TOXIC SECONDARY PARKINSONISM IN EXPERIMENTAL ANIMALS

Toxicant-exposed animals have been used as experimental models for parkinsonism syndromes. In particular several models have explored the association between brain dopamine and the development of clinical signs or neuropathology. They have also been used to identify modes of actions of these neurotoxicants.

Reserpine

One of the first chemicals to be evaluated as a potential model for Parkinson's disease was reserpine (Carlsson et al., 1957). Reserpine interferes with vesicle storage of monoamines by irreversibly blocking monoamine transporters 1 and 2 (VMAT-1 and VMAT-2). This blockage results in the depletion of dopamine and other catecholamines in nerve terminals, resulting in akinesia that could be rescued by administration of L-DOPA (Cannon and Greenamyre, 2010). Reserpine is also associated with increased striatal oxidative stress.

Reserpine administration continues to be an important pharmacologic tool which, when coupled with appropriate animal models, is used to evaluate parkinsonism, akinesia, tardive dyskinesia, and other extrapyramidal disorders (Trevizol et al., 2011; Bisong et al., 2013). Rats given reserpine demonstrate impaired performance on a novel object recognition task, decreased striatal TH levels, and reduced number of TH-positive cells in substantia nigra pars compacta (Santos et al., 2013). Mice given reserpine (1 mg/kg), administered once every other day for 4 days, develop orofacial dyskinesia, tongue protrusion, and vacuous chewing, tremor, and catalepsy (Dutra et al., 2002). Cats given intramuscular injections of reserpine at 3–5 mg/kg/day for 1 or 5 days developed ataxia, rigidity, and tremor (Beresford et al., 1961). Similar clinical effects have also been reported in dogs given reserpine (Adams et al., 1971). The developmental toxicity of reserpine towards the developing putamen has also been evaluated in rabbits (Tennyson et al., 1983).

Cyanide

Acute high-dose human exposure to cyanide following accidental or deliberate exposure has been associated with clinical signs of parkinsonism (e.g., akinesia and rigidity), brain-imaging findings, and neurologic lesions suggestive of selective dopaminergic neuronal loss in the basal ganglia (Uitti et al., 1985; Rosenberg et al., 1989; Feldman and Feldman, 1990; Kasamo et al., 1993; Sarikaya et al., 2006). The primary action of cyanide is inhibition of cytochrome c oxidase in complex IV of the oxidative phosphorylation chain, thereby blocking intracellular oxygen utilization (histotoxic hypoxia) and reducing cellular adenosine triphosphate generation (Leavesley et al., 2008). The brain's limited anaerobic metabolic capacity and high dependence on mitochondrial energy production contribute to its increased susceptibility to cyanide.

Several animal models have been used to evaluate the mode of action of cyanide neurotoxicity. In the mouse, sublethal doses of cyanide produced bilateral neuronal death in the substantia nigra within 3 days of cyanide administration (Mills et al., 1999). Zhang and coworkers (2010) showed that cyanide-induced mitochondrial dysfunction and death of dopaminergic cells was mediated by up-regulation and localization of Bcl-2/adenovirus E1B 19-kDa interacting protein 3 to mitochondria and endoplasmic reticulum in a mouse model of cyanide neurotoxicity. Although brain lesions were observed in each study, clinical signs of cyanide parkinsonism went unreported. Kanthasamy and coworkers (1994) reported reduced numbers of TH-positive cells and dopamine concentrations in the substantia nigra and decreased motor activity of mice following repeated cyanide exposure.

TOXICANT-BASED MODELS FOR PARKINSON'S DISEASE

Toxicant-exposed animals are also widely used as experimental models for Parkinson's disease. They are used to investigate pathogenic mechanisms such as oxidative stress, mitochondrial defects, proteolytic stress, neuroinflammation, impaired ubiquitin proteasomal system, and autophagy as well as targeted therapies in these toxic disease syndromes (Le et al., 2014). Results of several decades of research effort have shown that some animal models incompletely replicate the classic findings associated with human Parkinson's disease, namely, progressive loss of dopaminergic neurons with deposition of Lewy body-like inclusions in the brain. Moreover, animal models often fail to demonstrate certain clinical features of Parkinson's disease, including response to L-DOPA treatment (Bezard et al., 1998; Uversky, 2004; Cannon and Greenamyre, 2010; Blandini and Armentero, 2012; Blesa et al., 2012a; Le et al., 2014). Importantly, clinical signs seen in people with Parkinson's disease (e.g., bradykinesia, tremor, postural abnormalities, and rigidity) are incompletely replicated in many animal studies, especially those using rodents and other lower vertebrates.

This chapter's discussion of primarily *in vivo* animal models focuses on traditional mammalian laboratory animal models. However, there have also been mechanistic studies conducted in *Drosophila* (reviewed in Guo, 2010, 2012), *Caenorhabditis elegans* (Schmidt et al., 2007; Harrington et al., 2010; Vistbakka et al., 2012), and zebra fish (Flinn et al., 2008; Sager et al., 2010), amongst others (Pienaar et al., 2010).

NEUROCHEMICAL CHANGES ASSOCIATED WITH PARKINSON'S DISEASE

Classic Parkinson's disease is characterized clinically by asymmetric resting tremor, bradykinesia, rigidity, and later postural instability and pathologically by the loss of neurons, particularly dopaminergic neurons, in the substantia nigra that project primarily to the striatum. Postmortem studies demonstrate severe reductions in dopamine concentration in the striatum of patients with Parkinson's disease, with greater reductions in the putamen than the caudate, particularly in the caudal putamen (Kish et al., 1988; Piggott et al., 1999). Thus, the fundamental neurochemical lesion in classic Parkinson's disease is deficiency of dopamine in the extrapyramidal system (caudate and putamen). In this brain region dopamine enhances the excitability of the motor areas of the cerebral cortex and increases movement (Reeves and Swenson, 2008). Once basal ganglionic dopaminergic neurons are lost, the motor and premotor cortex become less excitable and decrease spontaneous movement.

Several dopamine-related imaging agents have been developed to assess the integrity of dopaminergic neurons using single-photon emission computed tomography or positron emission tomography (PET). These radioligands image presynaptic targets (e.g., dopamine transporter, vesicular monoamine transporter or dihydroxyphenylalanine (DOPA) decarboxylase activity) or postsynaptic D2 dopamine receptors (Cummings et al., 2011). Dopaminergic imaging changes are usually evident in the preclinical stages in most patients with Parkinson's disease or another neurodegenerative striatonigral disease (Cummings et al., 2011). Conversely, their absence in a symptomatic patient is suggestive of a diagnosis other than neurodegenerative parkinsonism.

Pesticides

Associations between agricultural pesticide exposure and Parkinson's disease and other human neurodegenerative disease have long been suspected (Baltazar et al., 2014; Goldman, 2014). A complete review of this literature is beyond the scope of this chapter; instead, this discussion will focus on animal models for rotenone and paraquat, with special emphasis on clinical signs and neuropathology.

Rotenone

Rotenone is a naturally occurring plant-derived (e.g., *Derris elliptica*) nonselective pesticide and inhibitor of mitochondrial complex I, causing increased formation of reactive oxygen species (Li et al., 2003; Chen et al., 2007; Abdulwahid Arif and Ahmad Khan, 2010). Deficiency in complex I has been reported in the substantia nigra in patients with Parkinson's disease (Schapira et al., 1990; Papa and De Rasmo, 2013). Rotenone is highly lipophilic and readily crosses the blood–brain barrier (Caboni et al., 2004). Studies with rotenone have used a variety of exposure routes, including oral administration, subcutaneous injection, osmotic pumps, intraperitoneal (IP) injection, and intravenous (IV) administration (Höglinger et al., 2006). Subcutaneous injection of biodegradable rotenone microspheres has also been used successfully in rats (Ferris et al., 2013). In contrast, intranasal instillation in rodents was not associated with motor dysfunction or nigrostriatal system pathology (Rojo et al., 2007).

Experimental studies evaluating extrapyramidal effects of rotenone have their genesis with studies performed by Betarbet and coworkers (2000), who described selective degeneration of the nigrostriatal pathway with alpha-synuclein accumulation (so-called "Lewy neurites") in rats following repeated (5-week) exposure to rotenone. Since that time other investigators have shown that rotenone exposure in rats produced selective damage in the striatum and the globus pallidus, but the substantia nigra was spared (Ferrante et al., 1997; Sherer et al., 2003). Rotenone is also characterized by the development of a rodent behavioral phenotype characterized by hypokinesia, hunched posture, and rigidity (Schmidt and Alam, 2006). Adult ovariectomized Long–Evans rats given rotenone microspheres develop hypokinesia, loss of dopaminergic neurons, and the appearance of putative Lewy bodies in the substantia nigra (Ferris et al., 2013). Alterations in nerve conduction velocity and histologic changes consistent with peripheral nephropathy along with striatal lesions have occurred in rats exposed subchronically to rotenone (Binienda et al., 2013).

Several challenges exist with the rotenone model, including the variability in neuropathologic response and incidence seen with treated animals and excess morbidity and mortality due to the inherent toxicity of this pesticide (Lapointe et al., 2004; Höglinger et al., 2006; Ferris et al., 2013). In addition, there are no documented cases of rotenone-induced Parkinson's disease in humans.

Paraquat

Another toxic model of Parkinson's disease relies on animal administration of paraquat (*N,N'*-dimethyl-4-4-4'-bypiridinium). Paraquat is a nonselective bipyridyl contact herbicide, which shares certain structural similarities with a metabolite of MPTP (see below). When in its charged state, paraquat does not cross the blood–brain barrier. Recent studies suggest that paraquat is not a substrate for *p*-glycoproten (Lacher et al., 2014). However, brain transport in rodents apparently occurs via a neutral amino acid transporter (Shimizu et al., 2001; McCormack and Di Monte, 2003), where it reaches a concentration lower than in peripheral tissues (Prasad et al., 2009). Brain delivery demonstrates age dependency, with the highest levels detected in neonatal or aged rats (Corasaniti et al., 1991). Other studies using quantitative PET imaging have shown minimal blood–brain barrier penetration (during 90 minutes) of [^{11}C] paraquat in the rhesus monkey brain following a single low-dose (0.1–0.4 μg/kg, IV) exposure (Bartlett et al., 2009).

Paraquat neurotoxicity is assumed to be related to its redox cycling potential, leading to oxidative stress (Taetzsch and Block, 2013). Paraquat also demonstrates partial selectivity for nigrostriatal dopaminergic neurons. One mechanism that may account for this selective toxicity relates to interactions of paraquat with the dopamine transporter (Rappold et al., 2011). Most studies evaluating paraquat neurotoxicity have been performed in rodents. In mice, paraquat produces motor deficits and loss of nigral dopaminergic neurons in a dose- and age-dependent manner (Brooks et al., 1999; McCormack et al., 2002; Thiruchelvam et al., 2003; Breckenridge et al., 2013). Responses in mice following paraquat exposure have been variable. Several investigators have reported that IP exposure of mice causes a loss of TH-positive dopaminergic neurons in the substantia nigra pars compacta (Brooks et al., 1999; McCormack et al., 2002; Yin et al., 2011; Jiao et al., 2012). Other studies show that paraquat exposure in mice may reduce the number of TH-positive neurons; however, it does not alter the concentration of dopamine and related catecholamines, increase dopamine turnover in the striatum, or result in dopaminergic neuronal cell death (Breckenridge et al., 2013; Minnema et al., 2014).

Several investigators have developed a paraquat toxicity model that also includes simultaneous administration of the fungicide maneb (manganese ethylene-*bis*(dithiocarbamate)) in mice (Thiruchelvam et al., 2003; Patel et al., 2006; Desplats et al., 2012; Prakash et al., 2013) and rats (Cicchetti et al., 2005; Saint-Pierre et al., 2006; Xu et al., 2011; Bastías-Candida et al., 2013). When administered together, paraquat and maneb appear to induce synergistic effects, causing more severe changes in dopaminergic neurotransmission and increased neuronal cell loss in the substantia nigra par compacta (Tieu, 2011).

Like rotenone, animal models of paraquat neurotoxicity demonstrate variable responses (Berry et al., 2010; Tieu, 2011) that have led some to question its relevance as an animal model for Parkinson's disease (Miller, 2007). Some studies report brain delivery; however, this may be dose-, route-, or species-dependent (reviewed in Bartlett et al., 2009). Despite these limitations in the animal model, paraquat has proven useful in improving our understanding of mechanisms of nigrostriatal degeneration, especially the role of α-synuclein aggregation in neurodegeneration (Cory-Slechta et al., 2008). Knowledge concerning acute to chronic effects of paraquat on the human basal ganglia is limited. Wu and colleagues (2012) examined several acutely poisoned patients with magnetic resonance imaging (MRI) and found changes in the extrapyramidal ganglia that may be associated with excessive iron deposition. Acute symptoms in humans following high-dose (>30 mg/kg) paraquat exposure include pulmonary edema, liver damage, or death due to multiple organ failure within hours to days (Saravu et al., 2013).

1-Methyl-4-phenyl-1,2,3,6-tetrahydropyridine (MPTP)

A toxic parkinsonism syndrome associated with MPTP was first recognized among a small cohort of Californian IV meperidine users showing acute clinical signs compatible with Parkinson's disease (Davis et al., 1979; Langston et al., 1983). One affected patient died several years after the initial MPTP exposure and had substantia nigra lesions with Lewy body-like aggregations that were suggestive of Parkinson's disease (Langston et al., 1983). Since this outbreak MPTP has been the subject of numerous studies that have revealed the following information about this chemical:

- In humans, exposure to MPTP results in the appearance of extrapyramidal symptoms within several days.
- MPTP is highly lipophilic and after systemic administration rapidly crosses the blood–brain barrier.
- Conversion of MPTP to the toxic metabolite 1-methyl-4-phenylpyridinium (MPP+) is required for neurotoxicity. MPTP is converted to the toxic metabolite MPP+ by perisynaptic astrocytes in a two-step reaction that requires the enzyme monoamine oxidase B. Once released from the astrocytes into the extracellular space via the OCT-3 transporter, MPP+ is taken up into the neuron by the dopamine transporter and can be stored in vesicles via uptake by the vesicular monoamine transporter, VMAT2. MPP+ is an excellent substrate for the dopamine transporter, and is selectively transported into the dopaminergic neurons, where it accumulates in mitochondria.
- Mitochondrial MPP+ quenches the activity of reduced nicotinamide adenine dinucleotide dehydrogenase, the first enzyme in the electron transport chain, thereby inhibiting respiration at the level of complex I (Greenamyre et al., 2001).

- In humans, MPTP toxicity responds, at least initially, to L-DOPA and other dopamine agonists.
- In general, MPTP does not produce the Lewy body pathology seen with classic Parkinson's disease (Langston et al., 1999).

In vivo rodent models

One of the earliest attempts at replicating MPTP neurotoxicity in an animal model involved the injection of MPTP rats, but only transient parkinsonian-like symptoms were induced, since rats are exceptionally resistant to MPTP, as was later learned (Boyce et al., 1984; Chiueh et al., 1984). Boyce and coworkers (1984) gave MPTP (10 mg/kg/day IP) to rats for 16 days, which were then observed for a further 9–11 days. In this study rats given MPTP failed to develop changes in motor activity, striatal dopamine concentrations, or altered uptake of [^{3}H]-dopamine in the striatal region. Chiueh et al. (1984) observed clinical signs, including immobility and clonic movements following administration of MPTP at 5–10 mg/kg and increases (e.g., caudate nucleus) and decreases (e.g., substantia nigra) in dopamine concentration without neuronal loss at these sites.

Studies using direct MPP+ into the rat brain can result in degeneration of the ipsilateral nigrostriatal dopamine pathway (Sirinathsinghji et al., 1988; Staal and Sonsalla, 2000). Other studies have shown that the dose required to induce comparable dopaminergic degeneration in rats is associated with a high mortality rate (Giovanni et al., 1994). It has also been demonstrated that intranasal administration of MPTP to rats triggers changes in motor function and neurodegeneration in the substantia nigra par compacta (reviewed in Prediger et al., 2012) and mice (Rojo et al., 2007). The relative resistance of rats to MPTP poisoning has been attributed in part to lower MPP+ levels in the brains of rats and more rapid clearance of MPTP and its metabolites when compared with more susceptible monkey models (Johannessen et al., 1985; Riachi et al., 1988). Other studies have suggested that an inherent difference in brain sensitivity to MPTP may also account for some of the species differences seen between rats and mice (Giovanni et al., 1994).

The mouse is the most commonly used rodent model to evaluate MPTP neurotoxicity. Several studies have shown different responses among various mouse strains with respect to changes in striatal dopamine concentration, behavior responses, and substantia nigral neuropathology (reviewed in Meredith and Rademacher, 2011). The exact basis for this difference in response remains incompletely understood and likely reflects pharmacodynamics and pharmacokinetic differences (Meredith and Rademacher, 2011). Other factors have also been identified, including sex (female mice are generally more susceptible to systemic toxicity), age, and body weight. The extent to which brain lesions develop also depends on the route of MPTP administration, with subcutaneous injection resulting in greater neurotoxicity when compared with IP administration.

One of many examples of an acute regimen involves administration of four injections (20 mg/kg, q 2 hours) to C57/bl mice, resulting in an approximate 90% depletion of striatal dopamine and a 70% loss of dopaminergic neurons in the substantia nigra pars compacta (Jackson-Lewis et al., 1995). A wide range of dosing schedules has been developed that clearly show neurochemical and histologic changes similar to those seen in people following MPTP exposure. Some dosing protocols use male C57/bl mice given probenecid to potentiate the effects of MPTP, allowing for a more subchronic model of dopaminergic neurotoxicity (Meredith et al., 2008). One advantage of the mouse model is the availability of genetically modified animals to evaluate modes of actions and drug therapies (reviewed in Blandini and Armentero, 2012).

Nonhuman primate models

Nonhuman primate models were also evaluated early in the history of MPTP neurotoxicity (Burns et al., 1983). When administered to nonhuman primates through repeated bilateral intracarotid injections, MPTP produces a syndrome that replicates most of the clinical signs of MPTP neurotoxicity in humans, including clinical signs (e.g., akinesia, rigidity, postural tremor, flexed posture), dopamine depletion, responsiveness to L-DOPA treatment, and neuropathology (e.g., dopaminergic neuronal loss in the substantia nigra pars compacta). Species used have included cynomolgus monkeys (Przedborski et al., 1991), marmosets (reviewed in Eslamboli, 2005), and rhesus monkeys (Burns et al., 1983; Chiueh et al., 1985; Meissner et al., 2003; reviewed in Cork et al., 1987). Most experiments include acute exposures where monkeys are treated with high doses of MPTP for a short time; however, longer exposure models (e.g., 0.5 mg/kg every 2 weeks for several months) have also been developed (Blesa et al., 2012b).

One advantage of the monkey MPTP model is that it readily lends itself to MRI, PET, and other brain-imaging methods, and electrophysiologic studies of deep brain stimulation used with Parkinson's disease patients (Emborg, 2007; Ellens and Leventhal, 2013). Another advantage is the use of a unilateral intracarotid injection that results in unilateral lesions (Emborg et al., 2006; Ding et al., 2008). In addition, the nonhuman primate model is preferable over rodents as these animals show clear and lasting behavioral features after MPTP treatment, which reflect many aspects of human Parkinson's disease.

Other species

Rabbits are relatively insensitive to MPTP neurotoxicity (Lermontova et al., 1995). MPTP has also been used in other species, including domesticated dogs and cats (Wilson et al., 1987; Johannessen et al., 1991; Akunne et al., 1992; Choi et al., 2011) and cats (Schneider and Markham, 1986; Chen et al., 1993; Aznavour et al., 2012). Experiments in dogs and cats demonstrate dopamine depeletion, clinical signs, and neuropathology seen in mice and monkeys given MPTP.

ANIMAL MODELS FOR MANGANESE NEUROTOXICITY

Manganese is an essential nutrient that under certain exposure conditions can demonstrate regional neurotoxicity for the human extrapyramidal system. Manganese is an important occupational hazard. Inhalation exposure to manganese dust occurs primarily in mining, ore crushing, and metallurgic operations for iron, steel, ferrous and nonferrous alloys. Manufacturing of dry-cell batteries, anti-knock gasoline additives, pesticides (e.g., maneb), pigments, dyes, inks, and incendiary devices can also lead to occupational manganese exposure. Manganese fumes are also produced during metallurgic operations and several types of welding operation. The discussion in this chapter will focus on occupational manganese neurotoxicity rather than environmental exposure, including effects in children (reviewed in Zoni and Lucchini, 2013) or the potential role of manganese on the etiology of idiopathic Parkinson's disease (reviewed in Dusek et al., 2014).

An association between manganese inhalation and neurotoxicity was first recognized in the mid 19th century in workers at an ore-grinding plant where "black oxide of manganese" was processed (Couper, 1837). Most epidemiologic research on manganese conducted during the late 20th century focused on occupational exposures associated with inhalation. The result of these studies clearly established a causal association between chronic manganese inhalation and a clinical syndrome known as manganism that mimics Parkinson's disease (Lucchini et al., 2009). Hallmarks of manganism include behavioral changes, extrapyramidal motor dysfunction, and neurochemical and neuropathologic changes in the basal ganglia and globus pallidus (Roth, 2006; Roels et al., 2012). Manganese-induced cognitive deficits have also been reported in workers with exposure to manganese-based welding fumes (Bowler et al., 2006; Ellingsen et al., 2008; Park et al., 2009; Chang et al., 2010).

An ideal animal model of manganese neurotoxicity should replicate the following observations seen in occupationally exposed people under high-dose conditions:

- Manganese accumulates within the human striatum, globus pallidus, and substantia nigra (Calne et al., 1994; Nagatomo et al., 1999).
- Since manganese is paramagnetic, it increases the signal intensity seen with T1-weighted MRI. Changes in the T1-weighted image correlate with manganese tissue concentration (Dorman et al., 2006) and allow for visual inspection of the body for manganese accumulation. Brain MRI studies of highly exposed people reveal signal changes in the globus pallidus, striatum, and midbrain consistent with manganese accumulation at that site (Fitsanakis et al., 2006; Tuschl et al., 2013).
- Neurologic manifestations of manganism include progressive bradykinesia, dystonia, and gait abnormalities (Rivera-Mancía et al., 2011; reviewed in Guilarte, 2013). Our knowledge of manganese effects on cognition and other neurologic function is growing (Roels et al., 2012).
- The primary neuropathologic target is the globus pallidus (particularly the internal segment) with sparing of the substantia nigra pars compacta and an absence of Lewy bodies (reviewed in Yamada et al., 1986; Pal et al., 1999; Perl and Olanow, 2007).

Significant differences have been widely documented regarding the species-specific sensitivity to manganese neurotoxicity (Newland, 1999; Aschner et al., 2005). Rodents generally fail to develop a behavioral syndrome or neuropathologic lesion comparable to that seen in manganese-poisoned humans. In contrast, nonhuman primates best replicate the neurotoxic effects observed in humans. Therefore, this discussion will focus on studies performed in nonhuman primates. Results from these studies have shown that manganese-exposed monkeys develop:

- gait and other motor abnormalities that mimic those observed in affected humans (Olanow et al., 1996; Guilarte, 2013; Kim et al., 2013);
- MRI changes consistent with elevated brain manganese concentrations localized to the striatum, globus pallidus, and substantia nigra. Changes in the pallidal index (i.e., ratio of the signal intensity in the globus pallidus over the intensity in the frontal white matter using T1-weighted MRI) have also been used as a measure of manganese accumulation in the globus pallidus. Manganese-exposed monkeys develop a higher pallidal index than controls (Dorman et al., 2006; Guilarte et al., 2006b; Park et al., 2007; Sung et al., 2007). Han et al. (2008) also reported reversal

of pallidal MRI changes following cessation of manganese exposure – this recovery is also noted in welders following removal from the workplace (Han et al., 2014). Likewise, manganese-exposed monkeys demonstrate accumulation in the olfactory bulb following manganese inhalation (Dorman et al., 2006), a change also seen in exposed workers (Sen et al., 2011);

- reduced levels of striatal and pallidal dopamine and 3,4-dihydroxyphenylacetic acid in conjunction with loss of dopaminergic neurons analogous to those found in manganese-poisoned humans (Neff et al., 1969; Bird et al., 1984; Struve et al., 2007);
- deficits in spatial and nonspatial working memory as well as effects on visuospatial-paired associate learning (Schneider et al., 2006, 2009, 2013).

Chronic exposure did not produce the loss of dopamine terminals in the caudate and putamen based on [^{11}C]-methylphenidate PET imaging of exposed animals (Guilarte et al., 2006a, 2008).

Histologic assessment of the frontal cortex from manganese-exposed monkeys has also shown the presence of cells with apoptotic stigmata and astrocytosis in both the gray and white matter and α-synuclein aggregation in the frontal cortex gray and white matter (Verina et al., 2013). Neuropathologic description of lesions in the globus pallidus of manganese-exposed monkeys is largely lacking (Ulrich et al., 1979), although biomarkers consistent with oxidative stress are present at this location in monkeys following manganese inhalation (Erikson et al., 2007).

Multiple studies have shown that homeostatic control mechanisms play an important role in regulating intestinal absorption and biliary excretion of manganese (Aschner and Dorman, 2006). Several physiologically based pharmacokinetic models for manganese have been reported (Taylor et al., 2012). These models effectively simulate manganese tissue kinetics from inhaled, oral, and parenteral manganese intake (Andersen et al., 2010). This multidose route capability is achieved by incorporating homeostatic control processes, saturable tissue-binding capacities, and preferential fluxes in various tissue regions (see Taylor et al., 2012, for review).

Several recent studies have quantitatively evaluated manganese pharmacokinetics in order to assess the predictive value of animal studies for manganese neurotoxicity (Gwiazda et al., 2007; Schroeter et al., 2012). Gwiazda et al.'s (2007) review of subchronic to chronic rodent and nonhuman primate studies led them to conclude that most existing animal studies might be of limited relevance for the risk assessment of chronic low-level manganese exposure to humans. Schroeter and coworkers (2012) used a physiologically based pharmacokinetic model that accounts for the dose dependencies of manganese distribution to estimate increases in brain manganese concentrations in monkeys following manganese exposure. They evaluated exposures by inhalation, oral, IV, IP, and subcutaneous dose routes, and spanned durations ranging from several weeks to over 2 years. This analysis confirmed that the dose–response relationship for the neurotoxic effects of manganese in monkeys was independent of exposure route.

Rodent models have also been widely used to assess the toxicity and pharmacokinetics of manganese. One area of special interest has included the study of olfactory delivery of manganese and associated effects on the sense of smell. This area of research has taken on greater importance with our scientific understanding that inhaled manganese can undergo direct "nose-to-brain" transport and increased recognition that changes in the sense of smell occur following manganese exposure (Guarneros et al., 2013) as well as with Parkinson's disease and several other neurodegenerative disease syndromes (Dorman, 2010; Lucchini et al., 2012; Zoni et al., 2012; Morley and Duda, 2014).

Initial studies examining direct nose-to-brain transport of manganese relied on intranasal instillation of an aqueous solution of radiolabeled manganese chloride ($^{54}MnCl_2$) into one side of the nasal cavity of animals whereas the opposite portion of the nasal cavity would receive the aqueous vehicle control. One of the first studies was performed by Tjälve and Henriksson (1999) and used freshwater pike as the animal model. In these studies, gamma spectrometry and autoradiography were used to determine the fate of the intranasally instilled manganese. These experiments demonstrated the presence of ^{54}Mn in the olfactory bulb on the same (i.e., ipsilateral) side to the instilled nostril (Tjälve and Henriksson, 1999). Qualitatively similar intranasal instillation studies performed in the same laboratory that used rats were able to replicate these initial findings (Tjälve et al., 1996). Moreover, rats given an IP injection of $^{54}MnCl_2$ failed to accumulate elevated olfactory bulb manganese concentrations when compared with levels seen in the rest of the brain (Tjälve et al., 1996).

Brenneman and coworkers (2000) extended this general approach to an inhalation route of exposure. This group used a rat animal model in which one nostril was physically occluded using a plastic plug and tissue glue. In these studies, direct delivery along the olfactory route accounted for nearly all of the ^{54}Mn found in the olfactory bulb and tract of the rat brain following acute manganese inhalation.

Elder and coworkers (2006) used this nasal occlusion model to evaluate the distribution of inhaled manganese oxide nanoparticle aerosols. They used short-term exposures to demonstrate that manganese accumulated only in the olfactory bulb on the side of the head with a patent nostril. In general, smaller (≤1.5 μm) manganese particles more efficiently undergo olfactory transport when compared with larger (~10 μm) aerosol particles (Fechter et al., 2002; Elder et al., 2006). Olfactory transport is also favored when soluble manganese particles are used (Dorman et al., 2002; Leavens et al., 2007).

Cross et al. (2004) used unilateral intranasal $MnCl_2$ instillation in rats and brain MRI to show significant transport of manganese in the olfactory bulb, lateral olfactory tract, and olfactory tubercle ipsilateral to site of manganese administration. Dorman et al. (2006) also used brain MRI and measurement of tissue manganese concentrations to demonstrate presumed olfactory transport of inhaled manganese sulfate in rhesus monkeys. Visual evaluation of the manganese-enhanced neuronal tracts illustrated changes in the olfactory bulb but failed to demonstrate evidence for direct translocation of manganese from the olfactory bulb to the globus pallidus (Dorman et al., 2006). Sen and colleagues (2011) used brain MRI to evaluate region-specific manganese accumulation in seven welders without obvious neurologic deficits. When compared with age- and gender-matched controls, the manganese-exposed welders had significantly higher T1 relaxation rates in the olfactory bulb as well as evidence of manganese accumulation in the frontal white matter, globus pallidus, and putamen. Although the results of the Dorman and Sen studies are consistent with animal studies evaluating olfactory transport of manganese, these studies do not provide direct evidence that olfactory transport of manganese occurs in humans and other primates.

The toxicologic significance of olfactory transport of manganese remains incompletely understood. Although olfactory transport rapidly delivers manganese to brain structures in the olfactory pathway, it appears to be relatively slow (and perhaps inefficient) in delivering inhaled manganese to the rat striatum or primate globus pallidus and other more distant brain structures that are typically associated with manganism (Dorman et al., 2006; Leavens et al., 2007; Gutman et al., 2013).

CONCLUSIONS

Animal models have been widely used to study a variety of neurotoxicants that target the human extrapyramidal system. In these cases, the animal model and toxicant exposure are used to improve our understanding of toxicologic syndromes of importance. In other cases, these animal-toxicant models have also been used to study Parkinson's disease and other extrapyramidal syndromes have been developed to understand the pathogenesis and test potential therapeutics. Despite the limitations some animal-toxicant models have shown for human neurodegenerative disease, the toxicology community should not lose sight of the fact that they are still useful tools for the study of chemical neurotoxicity. Each model has advantages and disadvantages that need to be considered prior to the conduct of an experimental study. The choice of the animal model to be used rests on the research question(s) being evaluated.

References

Abdulwahid Arif I, Ahmad Khan H (2010). Environmental toxins and Parkinson's disease: putative roles of impaired electron transport chain and oxidative stress. Toxicol Ind Health 26: 121–128.

Adams HR, Smookler HH, Clarke DE et al. (1971). Clinicopathologic effects of chronic reserpine administration in mongrel dogs. J Pharm Sci 60: 1134–1138.

Akunne HC, Johannessen JN, de Costa BR et al. (1992). MPTP lesions of the nigrostriatal dopaminergic projection decrease [3H]1-[1-(2-thienyl)cyclohexyl]piperidine binding to PCP site 2: further evidence that PCP site 2 is associated with the biogenic amine reuptake complex. Neurochem Res 17: 261–264.

Andersen ME, Dorman DC, Clewell 3rd HJ et al. (2010). Multi-dose-route, multi-species pharmacokinetic models for manganese and their use in risk assessment. J Toxicol Environ Health A 73: 217–234.

Aschner M, Dorman DC (2006). Manganese: pharmacokinetics and molecular mechanisms of brain uptake. Toxicol Rev 25: 147–154.

Aschner M, Erikson KM, Dorman DC (2005). Manganese dosimetry: species differences and implications for neurotoxicity. Crit Rev Toxicol 35: 1–32.

Aznavour N, Cendres-Bozzi C, Lemoine L et al. (2012). MPTP animal model of Parkinsonism: dopamine cell death or only tyrosine hydroxylase impairment? A study using PET imaging, autoradiography, and immunohistochemistry in the cat. CNS Neurosci Ther 18: 934–941.

Baird JD, Arroyo LG, Vengust M et al. (2006). Adverse extrapyramidal effects in four horse given fluphenazine decanoate. J Am Vet Med Assoc 229: 104–110.

Baltazar MT, Dinis-Oliveira RJ, de Lourdes Bastos M et al. (2014). Pesticides exposure as etiological factors of Parkinson's disease and other neurodegenerative diseases – A mechanistic approach. Toxicol Lett 230: 85–103.

Bartlett RM, Holden JE, Nickles RJ et al. (2009). Paraquat is excluded by the blood brain barrier in rhesus macaque: an in vivo pet study. Brain Res 1259: 74–79.

Bastías-Candia S, Di Benedetto M, D'Addario C et al. (2013). Combined exposure to agriculture pesticides, paraquat and maneb, induces alterations in the N/OFQ-NOPr and PDYN/KOPr systems in rats: relevance to sporadic Parkinson's disease. Environ Toxicol Dec 23.

Beresford WA, Cole JC, Glees P (1961). The toxic and behavioural effects of reserpine on cats. Anim Behav 9: 111–112.

Berry C, La Vecchia C, Nicotera P (2010). Paraquat and Parkinson's disease. Cell Death Differ 17: 1115–1125.

Betarbet R, Sherer TB, MacKenzie G et al. (2000). Chronic systemic pesticide exposure reproduces features of Parkinson's disease. Nat Neurosci 3: 1301–1306.

Bezard E, Imbert C, Gross CE (1998). Experimental models of Parkinson's disease: from the static to the dynamic. Rev Neurosci 9: 71–90.

Binienda ZK, Sarkar S, Mohammed-Saeed L et al. (2013). Chronic exposure to rotenone, a dopaminergic toxin, results in peripheral neuropathy associated with dopaminergic damage. Neurosci Lett 541: 233–237.

Bird ED, Anton AH, Bullock B (1984). The effect of manganese inhalation on basal ganglia dopamine concentrations in rhesus monkey. Neurotoxicology 5: 59–65.

Bisong SA, Brown RE, Osim EE (2013). Comparative extrapyramidal effects of *Rauwolfia vomitoria*, chlorpromazine and reserpine in mice. J Nat Med 67: 107–112.

Blandini F, Armentero MT (2012). Animal models of Parkinson's disease. FEBS J 279: 1156–1166.

Blesa J, Phani S, Jackson-Lewis V et al. (2012a). Classic and new animal models of Parkinson's disease. J Biomed Biotechnol 2012: 845618.

Blesa J, Pifl C, Sánchez-González MA et al. (2012b). The nigrostriatal system in the presymptomatic and symptomatic stages in the MPTP monkey model: a PET, histological and biochemical study. Neurobiol Dis 48: 79–91.

Bowler RM, Gysens S, Diamond E et al. (2006). Manganese exposure: neuropsychological and neurological symptoms and effects in welders. Neurotoxicology 27: 315–326.

Boyce S, Kelly E, Reavill C et al. (1984). Repeated administration of N-methyl-4-phenyl 1,2,5,6-tetrahydropyridine to rats is not toxic to striatal dopamine neurones. Biochem Pharmacol 33: 1747–1752.

Breckenridge CB, Sturgess NC, Butt M et al. (2013). Pharmacokinetic, neurochemical, stereological and neuropathological studies on the potential effects of paraquat in the substantia nigra pars compacta and striatum of male C57BL/6 J mice. Neurotoxicology 37: 1–14.

Brenneman KA, Wong BA, Buccellato MA et al. (2000). Direct olfactory transport of inhaled manganese ($^{54}MnCl_2$) to the rat brain: toxicokinetic investigations in a unilateral nasal occlusion model. Toxicol Appl Pharmacol 169: 238–248.

Brewer BD, Hines MT, Stewart JT et al. (1990). Fluphenazine induced Parkinson-like syndrome in a horse. Equine Vet J 22: 136–137.

Brooks AI, Chadwick CA, Gelbard HA et al. (1999). Paraquat elicited neurobehavioral syndrome caused by dopaminergic neuron loss. Brain Res 823: 1–10.

Burkhard PR (2014). Acute and subacute drug-induced movement disorders. Parkinsonism Relat Disord 20 (Suppl 1): S108–S112.

Burns RS, Chiueh CC, Markey SP et al. (1983). A primate model of parkinsonism: selective destruction of dopaminergic neurons in the pars compacta of the substantia nigra by N-methyl-4-phenyl-1,2,3,6-tetrahydropyridine. Proc Natl Acad Sci U S A 80: 4546–4550.

Caboni P, Sherer TB, Zhang N et al. (2004). Rotenone, deguelin, their metabolites, and the rat model of Parkinson's disease. Chem Res Toxicol 17: 1540–1548.

Calne DB, Chu NS, Huang CC et al. (1994 Sep). Manganism and idiopathic parkinsonism: similarities and differences. Neurology 44 (9): 1583–1586.

Cannon JR, Greenamyre JT (2010). Neurotoxic in vivo models of Parkinson's disease recent advances. Prog Brain Res 184: 17–33.

Carlsson A, Lindquist M, Magnusson T (1957). 3,4-dihydroxyphenylalanine and 5-hydroxytryptophan as reserpine antagonists. Nature 180: 1200.

Castagné V, Moser PC, Porsolt RD (2009). Preclinical behavioral models for predicting antipsychotic activity. Adv Pharmacol 57: 381–418.

Chang Y, Lee JJ, Seo JH et al. (2010). Altered working memory process in the manganese-exposed brain. Neuroimage 53: 1279–1285.

Chang HT, Rumbeiha WK, Patterson JS et al. (2012). Toxic equine parkinsonism: an immunohistochemical study of 10 horses with nigropallidal encephalomalacia. Vet Pathol 49: 398–402.

Chen S, Zhou X, Xu D et al. (1993). Parkinsonism induced with 1-methyl-4-phenyl-1,2,3,6-tetrahydropyridine (MPTP) in cats: behavioral, biochemical and pathological studies. Chin Med Sci J 8: 223–226.

Chen Y, McMillan-Ward E, Kong J et al. (2007). Mitochondrial electron-transport-chain inhibitors of complexes I and II induce autophagic cell death mediated by reactive oxygen species. J Cell Sci 120: 4155–4166.

Chen NC, Lui CC, Huang SH et al. (2012). Pallidoreticular lesion in carbon monoxide intoxication by gradient echo: report of a case with parkinsonism features and review of the literature. Acta Neurol Taiwan 21: 44–48.

Chiueh CC, Markey SP, Burns RS et al. (1984). Neurochemical and behavioral effects of systemic and intranigral administration of N-methyl-4-phenyl-1,2,3,6-tetrahydropyridine in the rat. Eur J Pharmacol 100: 189–194.

Chiueh CC, Burns RS, Markey SP et al. (1985). Primate model of parkinsonism: selective lesion of nigrostriatal neurons by 1-methyl-4-phenyl-1,2,3,6-tetrahydropyridine produces an extrapyramidal syndrome in rhesus monkeys. Life Sci 36: 213–218.

Choi IS (2002). Parkinsonism after carbon monoxide poisoning. Eur Neurol 48: 30–33.

Choi CB, Kim SY, Lee SH et al. (2011). Assessment of metabolic changes in the striatum of a MPTP-intoxicated canine model: in vivo ^{1}H-MRS study of an animal model for Parkinson's disease. Magn Reson Imaging 29: 32–39.

Cicchetti F, Lapointe N, Roberge-Tremblay A et al. (2005). Systemic exposure to paraquat and maneb models early Parkinson's disease in young adult rats. Neurobiol Dis 20: 360–371.

Corasaniti MT, Defilippo R, Rodino P et al. (1991). Evidence that paraquat is able to cross the blood–brain barrier to a different extent in rats of various age. Funct Neurol 6: 385–391.

Cordy DR (1954). Nigropallidal encephalomalacia in horses associated with ingestion of yellow star thistle. J Neuropathol Exp Neurol 13: 330–342.

Cork LC, Kitt CA, Struble RG et al. (1987). Animal models of degenerative neurological disease. Prog Clin Biol Res 229: 241–269.

Cory-Slechta DA, Thiruchelvam M, Di Monte DA (2008). Letter regarding: "Paraquat: the red herring of Parkinson's disease research". Toxicol Sci 103: 215–216.

Couper J (1837). On the effects of black oxide of manganese when inhaled into the lungs. Br Ann Med Pharm Vital Stat Gen Sci 1: 41–42.

Crocker AD, Hemsley KM (2001). An animal model of extrapyramidal side effects induced by antipsychotic drugs: relationship with D2 dopamine receptor occupancy. Prog Neuropsychopharmacol Biol Psychiatry 25: 573–590.

Cross DJ, Minoshima S, Anzai Y et al. (2004). Statistical mapping of functional olfactory connections of the rat brain in vivo. Neuroimage 23: 1326–1335.

Cummings JL, Henchcliffe C, Schaier S et al. (2011). The role of dopaminergic imaging in patients with symptoms of dopaminergic system neurodegeneration. Brain 134: 3146–3166.

Davis GC, Williams AC, Markey SP et al. (1979). Chronic Parkinsonism secondary to intravenous injection of meperidine analogues. Psychiatry Res. 1: 249–254.

de Leeuw B, Sangster B (1979). Severe extrapyramidal syndrome in a dog caused by a haloperidol (Serenas®) intoxication. Vet Quart 1: 134–137.

Desplats P, Patel P, Kosberg K et al. (2012). Combined exposure to maneb and paraquat alters transcriptional regulation of neurogenesis-related genes in mice models of Parkinson's disease. Mol Neurodegener 7: 49.

Ding F, Luan L, Ai Y et al. (2008). Development of a stable, early stage unilateral model of Parkinson's disease in middle-aged rhesus monkeys. Exp Neurol 212: 431–439.

Dorman DC (2010). Olfactory system. In: CA McQueen (Ed.), *Comprehensive Toxicology, Volume 13*: Nervous System and Behavioral Toxicology. Academic Press, Oxford, pp. 263–276.

Dorman DC, Brenneman KA, McElveen AM et al. (2002). Olfactory transport: a direct route of delivery of inhaled manganese phosphate to the rat brain. J Toxicol Environ Health A 65: 1493–1511.

Dorman DC, Struve MF, Wong BA et al. (2006). Correlation of brain magnetic resonance imaging changes with pallidal manganese concentrations in rhesus monkeys following subchronic manganese inhalation. Toxicol Sci 92: 219–227.

Drobatz KJ, Walker LM, Hendricks JC (1999a). Smoke exposure in dogs: 27 cases (1988–1997). J Am Vet Med Assoc 215: 1306–1311.

Drobatz KJ, Walker LM, Hendricks JC (1999b). Smoke exposure in cats: 22 cases (1986–1997). J Am Vet Med Assoc 215: 1312–1316.

Dusek P, Roos PM, Litwin T et al. (2014). The neurotoxicity of iron, copper and manganese in Parkinson's and Wilson's diseases. J Trace Elem Med Biol [EPub ahead of Print].

Dutra RC, Andreazza AP, Andreatini R et al. (2002). Behavioral effects of MK-801 on reserpine-treated mice. Prog Neuropsychopharmacol Biol Psychiatry 26: 487–495.

Elder A, Gelein R, Silva V et al. (2006). Translocation of inhaled ultrafine manganese oxide particles to the central nervous system. Environ Health Perspect 114: 1172–1178.

Ellens DJ, Leventhal DK (2013). Review: electrophysiology of basal ganglia and cortex in models of Parkinson disease. J Parkinsons Dis 3: 241–254.

Ellingsen DG, Konstantinov R, Bast-Pettersen R et al. (2008). A neurobehavioral study of current and former welders exposed to manganese. Neurotoxicology 29: 48–59.

Emborg ME (2007). Nonhuman primate models of Parkinson's disease. ILAR J 48: 339–355.

Emborg ME, Moirano J, Schafernak KT et al. (2006). Basal ganglia lesions after MPTP administration in rhesus monkeys. Neurobiol Dis 23: 281–289.

Erikson KM, Dorman DC, Lash LH et al. (2007). Manganese inhalation by rhesus monkeys is associated with brain regional changes in biomarkers of neurotoxicity. Toxicol Sci 97: 459–466.

Eslamboli A (2005). Marmoset monkey models of Parkinson's disease: which model, when and why? Brain Res Bull 68: 140–149.

Farrell TC, Cario CL, Milanese C et al. (2011). Evaluation of spontaneous propulsive movement as a screening tool to detect rescue of Parkinsonism phenotypes in zebrafish models. Neurobiol Dis 44: 9–18.

Fechter LD, Johnson DL, Lynch RA (2002). The relationship of particle size to olfactory nerve uptake of a non-soluble form of manganese into brain. Neurotoxicology 23: 177–183.

Feldman JM, Feldman MD (1990). Sequelae of attempted suicide by cyanide ingestion: a case report. Int J Psychiatry Med 20: 173–179.

Ferrante RJ, Schulz JB, Kowall NW et al. (1997). Systemic administration of rotenone produces selective damage in the striatum and globus pallidus, but not in the substantia nigra. Brain Res 753: 157–162.

Ferris CF, Marella M, Smerkers B et al. (2013). A phenotypic model recapitulating the neuropathology of Parkinson's disease. Brain Behav Evol 3: 351–366.

Fitsanakis VA, Zhang N, Avison MJ et al. (2006). The use of magnetic resonance imaging (MRI) in the study of manganese neurotoxicity. Neurotoxicology 27: 798–806.

Flinn L, Bretaud S, Lo C et al. (2008). Zebrafish as a new animal model for movement disorders. J Neurochem 106: 1991–1997.

Fowler ME (1965). Nigropallidal encephalomalacia in the horse. J Am Vet Med Assoc 147: 607–616.

Ginsberg MD, Myers RE, McDonagh BF (1974). Experimental carbon monoxide encephalopathy in the primate. II. Clinical aspects, neuropathology, and physiologic correlation. Arch Neurol 30: 209–216.

Giovanni A, Sieber BA, Heikkila RE et al. (1994). Studies on species sensitivity to the dopaminergic neurotoxin 1-methyl-4-phenyl-1,2,3,6-tetrahydropyridine. Part 1:

Systemic administration. J Pharmacol Exp Ther 270: 1000–1007.

Goldman SM (2014). Environmental toxins and Parkinson's disease. Annu Rev Pharmacol Toxicol 54: 141–164.

Greenamyre JT, Sherer TB, Betarbet R et al. (2001 Sep–Nov). Complex I and Parkinson's disease. IUBMB Life 52 (3–5): 135–141.

Guarneros M, Ortiz-Romo N, Alcaraz-Zubeldia M et al. (2013). Nonoccupational environmental exposure to manganese is linked to deficits in peripheral and central olfactory function. Chem Senses 38: 783–791.

Guilarte TR (2013). Manganese neurotoxicity: new perspectives from behavioral, neuroimaging, and neuropathological studies in humans and non-human primates. Front Aging Neurosci 5: 23.

Guilarte TR, Chen MK, McGlothan JL et al. (2006a). Nigrostriatal dopamine system dysfunction and subtle motor deficits in manganese-exposed non-human primates. Exp Neurol 202: 381–390.

Guilarte TR, McGlothan JL, Degaonkar M et al. (2006b). Evidence for cortical dysfunction and widespread manganese accumulation in the nonhuman primate brain following chronic manganese exposure: a 1H-MRS and MRI study. Toxicol Sci 94: 351–358.

Guilarte TR, Burton NC, McGlothan JL et al. (2008). Impairment of nigrostriatal dopamine neurotransmission by manganese is mediated by pre-synaptic mechanism(s): implications to manganese-induced parkinsonism. J Neurochem 107: 1236–1247.

Guo M (2010). What have we learned from *Drosophila* models of Parkinson's disease? Prog Brain Res 184: 3–16.

Guo M (2012). *Drosophila* as a model to study mitochondrial dysfunction in Parkinson's disease. Cold Spring Harb Perspect Med 2 (11).

Gutman DA, Magnuson M, Majeed W et al. (2013 Mar). Mapping of the mouse olfactory system with manganese-enhanced magnetic resonance imaging and diffusion tensor imaging. Brain Struct Funct 218 (2): 527–537.

Gwiazda R, Lucchini R, Smith D (2007). Adequacy and consistency of animal studies to evaluate the neurotoxicity of chronic low-level manganese exposure in humans. J Toxicol Environ Health A 70: 594–605.

Han JH, Chung YH, Park JD et al. (2008). Recovery from welding-fume-exposure-induced MRI T1 signal intensities after cessation of welding-fume exposure in brains of cynomolgus monkeys. Inhal Toxicol 20: 1075–1083.

Han SH, Ahn SW, Youn YC et al. (2014). Reversal of pallidal magnetic resonance imaging T1 hyperintensity in a welder presenting as reversible parkinsonism. Neurol India 62: 117–118.

Harrington AJ, Hamamichi S, Caldwell GA et al. (2010). *C. elegans* as a model organism to investigate molecular pathways involved with Parkinson's disease. Dev Dyn 239: 1282–1295.

Hassin-Baer S, Sirota P, Korczyn AD et al. (2001). Clinical characteristics of neuroleptic-induced parkinsonism. J Neural Transm 108: 1299–1308.

Höglinger GU, Oertel WH, Hirsch EC (2006). The rotenone model of parkinsonism – the five years inspection. J Neural Transm Suppl 70: 269–272.

Hopkins RO, Fearing MA, Weaver LK et al. (2006). Basal ganglia lesions following carbon monoxide poisoning. Brain Inj 20: 273–281.

Jackson CB, Drobatz K (2004). Neurologic dysfunction associated with smoke exposure in dogs. J Vet Emerg Crit Care 12: 193.

Jackson-Lewis V, Jakowec M, Burke RE et al. (1995). Time course and morphology of dopaminergic neuronal death caused by the neurotoxin 1-methyl-4-phenyl-1,2,3,6-tetrahydropyridine. Neurodegeneration 4: 257–269.

Jankovic J (1995). International Classification of Diseases, tenth revision: neurological adaptation (ICD-10 NA): extrapyramidal and movement disorders. Mov Disord 10: 533–540.

Jiao Y, Lu L, Williams RW et al. (2012). Genetic dissection of strain dependent paraquat-induced neurodegeneration in the substantia nigra pars compacta. PLoS One 7 (1): e29447.

Johannessen JN, Chiueh CC, Burns RS et al. (1985). Differences in the metabolism of MPTP in the rodent and primate parallel differences in sensitivity to its neurotoxic effects. Life Sci 36: 219–224.

Johannessen JN, Sobotka TJ, Weise VK et al. (1991). Prolonged alterations in canine striatal dopamine metabolism following subtoxic doses of 1-methyl-4-phenyl-1,2,3,6-tetrahydropyridine (MPTP) and 4'-amino-MPTP are linked to the persistence of pyridinium metabolites. J Neurochem 57: 981–990.

Juliá R, Holland DW, Guenthner J (2007). Assessing the economic impact of invasive species: the case of yellow starthistle (*Centaurea solsitialis L.*) in the rangelands of Idaho, USA. J Environ Manage 85: 876–882.

Kanthasamy AG, Borowitz JL, Pavlakovic G et al. (1994). Dopaminergic neurotoxicity of cyanide: neurochemical, histological, and behavioral characterization. Toxicol Appl Pharmacol 126: 156–163.

Kasamo K, Okuhata Y, Satoh R et al. (1993). Chronological changes of MRI findings on striatal damage after acute cyanide intoxication: pathogenesis of the damage and its selectivity, and prevention for neurological sequelae: a case report. Eur Arch Psychiatry Clin Neurosci 243: 71–74.

Kauffman VG, Soma L, Divers TJ et al. (1989). Extrapyramidal side effects caused by fluphenazine decanoate in a horse. J Am Vet Med Assoc 195: 1128–1130.

Kent M, Creevy KE, Delahunta A (2010). Clinical and neuropathological findings of acute carbon monoxide toxicity in chihuahuas following smoke inhalation. J Am Anim Hosp Assoc 46: 259–264.

Kim CY, Sung JH, Chung YH et al. (2013). Home cage locomotor changes in non-human primates after prolonged welding-fume exposure. Inhal Toxicol 25: 794–801.

Kish SJ, Shannak K, Hirnykiewicz O (1988). Uneven pattern of dopamine loss in the striatum of patients with idiopathic Parkinson's disease. Pathophysiological and clinical implications. N Engl J Med 318: 876–880.

Kruger S, Haehner A, Thiem C et al. (2008). Neuroleptic-induced parkinsonism is associated with olfactory dysfunction. J Neurol 255: 1574–1579.

Lacher SE, Gremaud JN, Skagen K et al. (2014). Absence of P-glycoprotein transport in the pharmacokinetics and toxicity of the herbicide paraquat. J Pharmacol Exp Ther 348: 336–345.

Langston JW, Ballard P, Tetrud JW et al. (1983). Chronic Parkinsonism in humans due to a product of meperidine-analog synthesis. Science 219: 979–980.

Langston JW, Forno LS, Tetrud J et al. (1999 Oct). Evidence of active nerve cell degeneration in the substantia nigra of humans years after 1-methyl-4-phenyl-1,2,3,6-tetrahydropyridine exposure. Ann Neurol 46 (4): 598–605.

Lapointe N, St-Hilaire M, Martinoli MG et al. (2004). Rotenone induces non-specific central nervous system and systemic toxicity. FASEB J 18: 717–719.

Le W, Sayana P, Jankovic J (2014). Animal models of Parkinson's disease: a gateway to therapeutics? Neurotherapeutics 11: 92–110.

Leavens TL, Rao D, Andersen ME et al. (2007). Evaluating transport of manganese from olfactory mucosa to striatum by pharmacokinetic modeling. Toxicol Sci 97: 265–278.

Leavesley HB, Li L, Prabhakaran K et al. (2008). Interaction of cyanide and nitric oxide with cytochrome c oxidase: implications for acute cyanide toxicity. Toxicol Sci 101: 101–111.

Lermontova N, Drany O, Dobrota D et al. (1995). Relative resistance of rabbits to MPTP neurotoxicity. Mol Chem Neuropathol 25: 135–141.

Li N, Ragheb K, Lawler G et al. (2003). Mitochondrial complex I inhibitor rotenone induces apoptosis through enhancing mitochondrial reactive oxygen species production. J Biol Chem 278: 8516–8525.

Lucchini RG, Martin CJ, Doney BC (2009). From manganism to manganese-induced parkinsonism: a conceptual model based on the evolution of exposure. Neuromolecular Med 11: 311–321.

Lucchini RG, Dorman DC, Elder A et al. (2012). Neurological impacts from inhalation of pollutants and the nose-brain connection. Neurotoxicology 33: 838–841.

Maddox DM, Mayfield A, Poritz NH (1985). Distribution of yellow starthistle (*Centaurea solstitialis*) and Russian knapweed (*Centaurea repens*). Weed Sci 33: 315–327.

Mariani CL (2003). Full recovery following delayed neurologic signs after smoke inhalation in a dog. J Vet Emerg Crit Care 13: 235–239.

McCormack AL, Di Monte DA (2003). Effects of L-dopa and other amino acids against paraquat-induced nigrostriatal degeneration. J Neurochem 85: 82–86.

McCormack AL, Thiruchelvam M, Manning-Bog AB et al. (2002). Environmental risk factors and Parkinson's disease: Selective degeneration of nigral dopaminergic neurons caused by the herbicide paraquat. Neurobiol Dis 10: 119–127.

Meissner W, Prunier C, Guilloteau D et al. (2003). Time-course of nigrostriatal degeneration in a progressive MPTP-lesioned macaque model of Parkinson's disease. Mol Neurobiol 28: 209–218.

Meredith GE, Rademacher DJ (2011). MPTP mouse models of Parkinson's disease: an update. J Parkinsons Dis 1: 19–33.

Meredith GE, Totterdell S, Potashkin JA et al. (2008). Modeling PD pathogenesis in mice: advantages of a chronic MPTP protocol. Parkinsonism Relat Disord 14 (Suppl 2): S112–S115.

Messiha FS (1991). Neurotoxicity of chlorpromazine and modulation by amantadine as a function of mouse strain. Neurotoxicology 12: 571–581.

Migler BM, Warawa EJ, Malick JB (1993). Seroquel: behavioral effects in conventional and novel tests for atypical antipsychotic drug. Psychopharmacology (Berl) 112: 299–307.

Miller GW (2007). Paraquat: the red herring of Parkinson's disease research. Toxicol Sci 100: 1–2.

Mills EM, Gunasekar PG, Li L et al. (1999). Differential susceptibility of brain areas to cyanide involves different modes of cell death. Toxicol Appl Pharmacol 156: 6–16.

Minnema DJ, Travis KZ, Breckenridge CB et al. (2014). Dietary administration of paraquat for 13 weeks does not result in a loss of dopaminergic neurons in the substantia nigra of C57BL/6 J mice. Regul Toxicol Pharmacol 68: 250–258.

Morley JF, Duda JE (2014). Use of hyposmia and other non-motor symptoms to distinguish between drug-induced parkinsonism and Parkinson's disease. J Parkinsons Dis 4: 169–173.

Morley JF, Pawlowski SM, Kesari A et al. (2014). Motor and non-motor features of Parkinson's disease that predict persistent drug-induced Parkinsonism. Parkinsonism Relat Disord 20: 738–742.

Nagatomo S, Umehara F, Hanada K et al. (1999 Jan 1). Manganese intoxication during total parenteral nutrition: report of two cases and review of the literature. J Neurol Sci 162 (1): 102–105.

Neff NH, Barrett RE, Costa E (1969). Selective depletion of caudate nucleus dopamine and serotonin during chronic manganese dioxide administration to squirrel monkeys. Experientia 25: 1140–1141.

Newland MC (1999). Animal models of manganese's neurotoxicity. Neurotoxicology 20: 415–432.

Okeda R, Funata N, Takano T et al. (1981). The pathogenesis of carbon monoxide encephalopathy in the acute phase – physiological and morphological correlation. Acta Neuropathol (Berl) 54: 1–10.

Okeda R, Funata N, Song SJ et al. (1982). Comparative study on pathogenesis of selective cerebral lesions in carbon monoxide poisoning and nitrogen hypoxia in cats. Acta Neuropathol 56: 265–272.

Olanow CW, Good PF, Shinotoh H et al. (1996). Manganese intoxication in the rhesus monkey: a clinical, imaging, pathologic, and biochemical study. Neurology 46: 492–498.

Pal PK, Samil A, Calne DB (1999). Manganese neurotoxicity: a review of clinical features, imaging and pathology. Neurotoxicology 20: 227–238.

Papa S, De Rasmo D (2013). Complex I deficiencies in neurological disorders. Trends Mol Med 19: 61–69.

Park JD, Chung YH, Kim CY et al. (2007). Comparison of high MRI T1 signals with manganese concentration in brains of cynomolgus monkeys after 8 months of stainless steel welding-fume exposure. Inhal Toxicol 19: 965–971.

Park RM, Bowler RM, Roels HA (2009). Exposure–response relationship and risk assessment for cognitive deficits in early welding-induced manganism. J Occup Environ Med 51: 1125–1136.

Patel S, Singh V, Kumar A et al. (2006). Status of antioxidant defense system and expression of toxicant responsive genes in striatum of maneb- and paraquat-induced Parkinson's disease phenotype in mouse: mechanism of neurodegeneration. Brain Res 1081: 9–18.

Penney DG (1990). Acute carbon monoxide poisoning: animal models: a review. Toxicology 62: 123–160.

Perl DP, Olanow CW (2007). The neuropathology of manganese-induced Parkinsonism. J Neuropathol Exp Neurol 66: 675–682.

Pienaar IS, Götz J, Feany MB (2010). Parkinson's disease: insights from non-traditional model organisms. Prog Neurobiol 92: 558–571.

Piggott MA, Marshall EF, Thomas N et al. (1999). Striatal dopaminergic markers in dementia with Lewy bodies, Alzheimer's and Parkinson's diseases: rostrocaudal distribution. Brain 122: 1449–1468.

Porsolt RD (2013). The usefulness of non-human primates in central nervous system safety pharmacology. J Pharmacol Toxicol Methods 68: 23–29.

Prakash J, Yadav SK, Chouhan S et al. (2013). Neuroprotective role of *Withania somnifera* root extract in maneb-paraquat induced mouse model of parkinsonism. Neurochem Res 38: 972–980.

Prasad K, Tarasewicz E, Mathew J et al. (2009). Toxicokinetics and toxicodynamics of paraquat accumulation in mouse brain. Exp Neurol 215: 358–367.

Prediger RD, Aguiar Jr AS, Matheus FC et al. (2012). Intranasal administration of neurotoxicants in animals: support for the olfactory vector hypothesis of Parkinson's disease. Neurotox Res 21: 90–116.

Prockop LD, Chichkova RI (2007). Carbon monoxide intoxication: an updated review. J Neurol Sci 262: 122–130.

Przedborski S, Jackson-Lewis V, Popilskis S et al. (1991). Unilateral MPTP-induced parkinsonism in monkeys. A quantitative autoradiographic study of dopamine D1 and D2 receptors and re-uptake sites. Neurochirurgie 37: 377–382.

Rappold PM, Cui M, Chesser AS et al. (2011). Paraquat neurotoxicity is mediated by the dopamine transporter and organic cation transporter-3. Proc Natl Acad Sci U S A 108: 20766–20771.

Reeves AG, Swenson RS (2008). Disorders of basal ganglia function. In: Disorders of the Nervous System. A Primer, Available at:https://www.dartmouth.edu/~dons/part_3/chapter_26.html [Accessed October 1, 2014].

Riachi NJ, Harik SI, Kalaria RN et al. (1988). On the mechanisms underlying 1-methyl-4-phenyl-1,2,3,6-tetrahydropyridine neurotoxicity. II. Susceptibility among mammalian species correlates with the toxin's metabolic patterns in brain microvessels and liver. J Pharmacol Exp Ther 244: 443–448.

Rivera-Mancía S, Ríos C, Montes S (2011). Manganese accumulation in the CNS and associated pathologies. Biometals 24: 811–825.

Robles M, Choi BH, Han B et al. (1998). Repin-induced neurotoxicity in rodents. Exp Neurol 152: 129–136.

Roels HA, Bowler RM, Kim Y et al. (2012). Manganese exposure and cognitive deficits: a growing concern for manganese neurotoxicity. Neurotoxicology 33: 872–880.

Rojo AI, Cavada C, de Sagarra MR et al. (2007). Chronic inhalation of rotenone or paraquat does not induce Parkinson's disease symptoms in mice or rats. Exp Neurol 208: 120–126.

Rosenberg NL, Myers JA, Martin WR (1989). Cyanide-induced parkinsonism: clinical, MRI, and 6-fluorodopa PET studies. Neurology 39: 142–144.

Roth JA (2006). Homeostatic and toxic mechanisms regulating manganese uptake, retention, and elimination. Biol Res 39 (1): 45–57.

Sager JJ, Bai Q, Burton EA (2010). Transgenic zebrafish models of neurodegenerative diseases. Brain Struct Funct 214: 285–302.

Saint-Pierre M, Tremblay ME, Sik A et al. (2006). Temporal effects of paraquat/maneb on microglial activation and dopamine neuronal loss in older rats. J Neurochem 98: 760–772.

Sanders RD, Gillig PM (2012). Extrapyramidal examinations in psychiatry. Innov Clin Neurosci 9: 10–16.

Santos JR, Cunha JA, Dierschnabel AL et al. (2013). Cognitive, motor and tyrosine hydroxylase temporal impairment in a model of parkinsonism induced by reserpine. Behav Brain Res 253: 68–77.

Saravu K, Sekhar S, Pai A et al. (2013). Paraquat - A deadly poison: Report of a case and review. Indian J Crit Care Med 17: 182–184.

Sarikaya I, Apaydin H, Topal U et al. (2006). Cyanide-induced Parkinsonism and F-18 FDG PET/CT findings. Clin Nucl Med 31: 363–364.

Schapira AHV, Cooper JM, Dexter D et al. (1990). Mitochondrial complex I deficiency in Parkinson's disease. J Neurochem 54: 823–827.

Schmidt WJ, Alam M (2006). Controversies on new animal models of Parkinson's disease pro and con: the rotenone model of Parkinson's disease (PD). J Neural Transm Suppl 70: 273–276.

Schmidt E, Seifert M, Baumeister R (2007). *Caenorhabditis elegans* as a model system for Parkinson's disease. Neurodegener Dis 4: 199–217.

Schneider JS, Markham CH (1986). Neurotoxic effects of N-methyl-4-phenyl-1,2,3,6-tetrahydropyridine (MPTP) in the cat. Tyrosine hydroxylase immunohistochemistry. Brain Res 373: 258–267.

Schneider JS, Decamp E, Koser AJ et al. (2006). Effects of chronic manganese exposure on cognitive and motor functioning in non-human primates. Brain Res 1118: 222–231.

Schneider JS, Decamp E, Clark K et al. (2009). Effects of chronic manganese exposure on working memory in non-human primates. Brain Res 1258: 86–95.

Schneider JS, Williams C, Ault M et al. (2013). Chronic manganese exposure impairs visuospatial associative learning in non-human primates. Toxicol Lett 221: 146–151.

Schroeter JD, Dorman DC, Yoon M et al. (2012). Application of a multi-route physiologically based pharmacokinetic model for manganese to evaluate dose-dependent neurological effects in monkeys. Toxicol Sci 129: 432–446.

Sen S, Flynn MR, Du G et al. (2011). Manganese accumulation in the olfactory bulbs and other brain regions of "asymptomatic" welders. Toxicol Sci 121: 160–167.

Sengul G, Watson C (2014). Ascending and descending pathways in the spinal cord. In: G Paxinos (Ed.), The Rat Nervous System, 4th edn. Academic Press, Waltham, MA p. 122.

Sherer TB, Kim JH, Betarbet R et al. (2003). Subcutaneous rotenone exposure causes highly selective dopaminergic degeneration and alpha-synuclein aggregation. Exp Neurol 179: 9–16.

Shimizu K, Ohtaki K, Matsubara K et al. (2001). Carrier-mediated processes in blood–brain barrier penetration and neural uptake of paraquat. Brain Res 906: 135–142.

Sirinathsinghji DJ, Heavens RP, Richards SJ et al. (1988). Experimental hemiparkinsonism in the rat following chronic unilateral infusion of MPP+ into the nigrostriatal dopamine pathway. I. Behavioural, neurochemical and histological characterization of the lesion. Neuroscience 27: 117–128.

Sohn YH, Jeong Y, Kim HS et al. (2000). The brain lesion responsible for parkinsonism after carbon monoxide poisoning. Arch Neurol 57: 1214–1218.

Song SY, Okeda R, Funata N et al. (1983). An experimental study of the pathogenesis of the selective lesion of the globus pallidus in acute carbon monoxide poisoning in cats. With special reference to the chronologic change in the cerebral local blood flow. Acta Neuropathol 61: 232–238.

Staal RG, Sonsalla PK (2000). Inhibition of brain vesicular monoamine transporter (VMAT2) enhances 1-methyl-4-phenylpyridinium neurotoxicity in vivo in rat striata. J Pharmacol Exp Ther 293: 336–342.

Starkey SR, Morrisey JK, Hickam HD et al. (2008). Extrapyramidal side effects in a blue and gold macaw (*Ara ararauna*) treated with haloperidol and clomipramine. J Avian Med Surg 22: 234–239.

Struve MF, McManus BE, Wong BA et al. (2007). Basal ganglia neurotransmitter concentrations in rhesus monkeys following subchronic manganese sulfate inhalation. Am J Ind Med 50: 772–778.

Subhaschandra S, Jatishwor W, Suraj T (2008). Isolated symmetrical bilateral basal ganglia T2 hyperintensity in carbon monoxide poisoning. Ann Indian Acad Neurol 11: 251–253.

Sung JH, Kim CY, Yang SO et al. (2007). Changes in blood manganese concentration and MRI t1 relaxation time during 180 days of stainless steel welding-fume exposure in cynomolgus monkeys. Inhal Toxicol 19: 47–55.

Tada M, Shirakawa K, Matsuoka N et al. (2004). Combined treatment of quetiapine with haloperidol in animal models of antipsychotic effect and extrapyramidal side effects: comparison with risperidone and chlorpromazine. Psychopharmacology (Berl) 176: 94–100.

Taetzsch T, Block ML (2013). Pesticides, microglial NOX2, and Parkinson's disease. J Biochem Mol Toxicol 27: 137–149.

Tan U, Gepdiremen A, Kutlu N et al. (1992). Parkinsonian-like tremor induced by a combined application of chlorpromazine and pentobarbitone in cats with left-paw preference. Int J Neurosci 63: 157–162.

Taylor MD, Clewell 3rd HJ, Andersen ME et al. (2012). Update on a pharmacokinetic-centric alternative tier II program for MMT-Part II: Physiologically based pharmacokinetic modeling and manganese risk assessment. J Toxicol Clin Exp 2012: 791431.

Tennyson VM, Gershon P, Budininkas-Schoenebeck M et al. (1983). Effects of extended periods of reserpine and α-methyl-p-tyrosine treatment on the development of the putamen in fetal rabbits. Int J Dev Neurosci 1: 305–318.

Thiruchelvam M, McCormack A, Richfield EK et al. (2003). Age-related irreversible progressive nigrostriatal dopaminergic neurotoxicity in the paraquat and maneb model of the Parkinson's disease phenotype. Eur J Neurosci 18: 589–600.

Thom SR, Fisher D, Zhang J et al. (2004). Neuronal nitric oxide synthase and N-methyl-D-aspartate neurons in experimental carbon monoxide poisoning. Toxicol Appl Pharmacol 194: 280–295.

Tieu K (2011). A guide to neurotoxic animal models of Parkinson's disease. Cold Spring Harb Perspect Med 1: a009316.

Tjälve H, Henriksson J (1999). Uptake of metals in the brain via olfactory pathways. Neurotoxicology 20: 181–195.

Tjälve H, Henriksson J, Tallkvist J et al. (1996). Uptake of manganese and cadmium from the nasal mucosa into the central nervous system via olfactory pathways in rats. Pharmacol Toxicol 79: 347–356.

Trevizol F, Benvegnú DM, Barcelos RC et al. (2011). Comparative study between two animal models of extrapyramidal movement disorders: prevention and reversion by pecan nut shell aqueous extract. Behav Brain Res 221: 13–18.

Tukov FF, Anand S, Gadepalli RS et al. (2004a). Inactivation of the cytotoxic activity of repin, a sesquiterpene lactone from *Centaurea repens*. Chem Res Toxicol 17: 1170–1176.

Tukov FF, Rimoldi JM, Matthews JC (2004b). Characterization of the role of glutathione in repin-induced mitochondrial dysfunction, oxidative stress and dopaminergic neurotoxicity in rat pheochromocytoma (PC12) cells. Neurotoxicology 25: 989–999.

Tuschl K, Mills PB, Clayton PT (2013). Manganese and the brain. Int Rev Neurobiol 110: 277–312.

Uitti RJ, Rajput AH, Ashenhurst EM et al. (1985). Cyanide-induced parkinsonism: a clinicopathologic report. Neurology 35: 921–925.

Ulrich CE, Rinehart W, Busey W et al. (1979). Evaluation of the chronic inhalation toxocity of a manganese oxide aerosol. II. Clinical observations, hematology, clinical chemistry and histopathology. Am Ind Hyg Assoc J 40: 322–329.

Uversky VN (2004). Neurotoxicant-induced animal models of Parkinson's disease: understanding the role of rotenone, maneb and paraquat in neurodegeneration. Cell Tissue Res 318: 225–241.

Verina T, Schneider JS, Guilarte TR (2013). Manganese induces α-synuclein aggregation in the frontal cortex of non-human primates. Toxicol Lett 217: 177–183.

Vistbakka J, VanDuyn N, Wong G et al. (2012). C. elegans as a genetic model system to identify Parkinson's disease-associated therapeutic targets. CNS Neurol Disord Drug Targets 11: 957–964.

Wilson JS, Turner BH, Morrow GD et al. (1987). MPTP produces a mosaic-like pattern of terminal degeneration in the caudate nucleus of dog. Brain Res 423: 329–332.

Wu B, Song B, Tian S et al. (2012). Central nervous system damage due to acute paraquat poisoning: a neuroimaging study with 3.0 T MRI. Neurotoxicology 33: 1330–1337.

Xu H, Chen R, Cai X et al. (2011). Differential effects of activating D1 and D2 receptors on electrophysiology of neostriatal neurons in a rat model of Parkinson's disease induced by paraquat and maneb. Neurosci Res 71: 411–420.

Yamada M, Ohno S, Okayasu I et al. (1986). Chronic manganese poisoning: a neuropathological study with determination of manganese distribution in the brain. Acta Neuropathol (Berl) 70: 273–278.

Yin L, Lu L, Prasad K et al. (2011). Genetic-based, differential susceptibility to paraquat neurotoxicity in mice. Neurotoxicol Teratol 33: 415–421.

Young S, Brown WW, Klinger B (1970a). Nigropallidal encephalomalacia in horses fed Russian knapweed: *Centaurea repens L.* Am J Vet Res 31: 1393–1404.

Young S, Brown WW, Klinger B (1970b). Nigropallidal encephalomalacia in horses caused by ingestion of weeds of the genus Centaurea. J Am Vet Med Assoc 157: 1602–1605.

Zhang L, Li L, Leavesley HW et al. (2010). Cyanide-induced apoptosis of dopaminergic cells is promoted by BNIP3 and Bax modulation of endoplasmic reticulum-mitochondrial $Ca2+$ levels. J Pharmacol Exp Ther 332: 97–105.

Zoni S, Lucchini RG (2013). Manganese exposure: cognitive, motor and behavioral effects on children: a review of recent findings. Curr Opin Pediatr 25: 255–260.

Zoni S, Bonetti G, Lucchini R (2012). Olfactory functions at the intersection between environmental exposure to manganese and Parkinsonism. J Trace Elem Med Biol 26: 179–182.

Handbook of Clinical Neurology, Vol. 131 (3rd series)
Occupational Neurology
M. Lotti and M.L. Bleecker, Editors

Chapter 14

Occupational exposures and parkinsonism

W. MICHAEL CAUDLE*

Department of Environmental Health, Rollins School of Public Health, Emory University, Atlanta, GA, USA

The term parkinsonism refers to a suite of neurologic disorders, such as Parkinson's disease (PD), multiple systems atrophy (MSA), and progressive supranuclear palsy (PSP), that share a diverse constellation of clinical features that include bradykinesisa, postural instability, rigidity, and tremor (Dickson, 2012). This heterogeneity is further observed in the pathologic manifestations of these disorders. For instance, while the presence of accumulations of the cytoplasmic protein α-synuclein is a common pathologic feature of PD and MSA, PSP is more aptly pathologically defined by inclusions of the micotubule protein, tau. However, a common unifying pathologic feature of many parkinsonian disorders is the loss of dopaminergic neurons in the substantia nigra that occurs following loss of dopaminergic projections to the basal ganglia, which seem to underlie many of the major clinical features associated with parkinsonism (Dickson, 2012). It can be appreciated that this spectrum of parkinsonian clinicopathologic presentations can significantly complicate the differential diagnosis of PD and other parkinsonian disorders.

Parkinsonian disorders have a varied etiology and may arise from defined genetic alterations or can be induced following exposure to a host of different chemicals, drugs, or toxicants that selectively damage the nervous system. The contribution of inherited and sporadic forms of parkinsonism has been best elucidated in PD, which has shown a clear influence of genetic mutations as well as exogenous factors involved in disease etiology. Indeed, mutations to several genes have been identified as genetic risk factors for the disease, yet these genetic alterations are only able to account for 5–10% of the cases of PD (Farrer, 2006). This would suggest that there are exogenous or environmental factors that influence the risk of development of PD, that either work independently or in conjunction with genetic predisposition to facilitate the onset of the disease (Gao and Hong, 2011). Epidemiologic evidence further supports the contribution of environmental factors in PD, showing the incidence of PD in monozygotic and dizygotic twins was almost identical, suggesting the heritability of PD was low (Tanner et al., 1999; Wirdefeldt et al., 2008, 2011b). More importantly, it provided evidence that an exogenous factor was significantly influencing the relative risk of PD.

Additional epidemiologic and laboratory work over the last several decades has provided extensive support for the idea that exposure to different environmental factors, including pesticides, metals, and solvents, as well as other toxicants found in industrial and commercial uses, could be a significant risk factor for the development of PD as well as other parkinsonian syndromes (Table 14.1) (Wirdefeldt et al., 2011a). The majority of these exposures occur in an occupational setting to individuals who are most likely to have direct interaction with the compound. Indeed, the highest level of pesticide exposure appears to occur with those who are using the chemicals, such as pesticide applicators, farmers, and gardeners. A similar concern is seen with workers highly exposed to solvents during metal manufacturing, as well as individuals exposed to metal dust and fumes during mining, smelting, or welding activities. Each of these settings, coupled with the potential for less than optimal personal protective equipment, leaves the individual vulnerable to prolonged, high-level exposure scenarios that can lead to significant damage of the nervous system. In the following sections I will provide a discussion of several classes of compounds and specific chemicals that have been shown in both epidemiologic and laboratory work to increase the risk of PD and parkinsonism.

*Correspondence to: W. Michael Caudle, PhD, Assistant Professor, Department of Environmental Health, Rollins School of Public Health, Claudia Nance Rollins Building, Room 2033, Mailstop: 1518-002-1AA, Emory University, 1518 Clifton Rd NE, Atlanta, GA 30322, USA. Tel: +1-404-712-8432, Fax: +1-404-727-8744, E-mail: william.m.caudle@emory.edu

Table 14.1

Environmental toxicants associated with parkinsonism

Chemical	Exposure setting(s)	Mechanism(s) of action	Reference
Dieldrin	Occupational, diet	Dopamine mishandling, generation of reactive oxygen species (ROS)	Corrigan et al. (1996, 2000); Kitazawa et al. (2001); Kamel et al. (2007); Hatcher et al. (2008)
Chlorpyrifos	Occupational	Alteration in dopamine signaling	Dhillon et al. (2008); Gatto et al. (2009); Torres-Altoro et al. (2011)
Rotenone	Occupational	Inhibition of mitochondrial complex I	Betarbet et al. (2000); Dhillon et al. (2008); Tanner et al. (2011)
Paraquat	Occupational	Generation of ROS	Thiruchelvam et al. (2000b); Manning-Bog et al. (2002); McCormack et al. (2002); Kamel et al. (2007); Tanner et al. (2011)
Polychlorinated biphenyls (PCBs)	Occupational, diet	Dopamine mishandling	Seegal et al. (1994); Corrigan et al. (1998, 2000); Caudle et al. (2006); Steenland et al. (2006); Hatcher-Martin et al. (2012)
Iron	Occupational, diet	Generation of ROS	Gorell et al. (1997); Gerlach et al. (2006); Kaur et al. (2007)
Manganese	Occupational	Generation of ROS, NMDA-mediated excitotoxicity, mitochondrial inhibition	Gavin et al. (1999); Guilarte et al. (2006); Racette et al. (2012)
Trichloroethylene (TCE)	Occupational	Generation of ROS, inhibition of mitochondrial complex I	Gash et al. (2008); Liu et al. (2010); Goldman et al. (2012)

NMDA, *N*-methyl-D-aspartate.

PESTICIDES

Pesticides are a broad class of compounds used in both agricultural as well as household settings to control or eliminate various nuisance plants, insects, or animals that could impact human and crop health. Specifically, pesticides include compounds formulated to manage rodents (rodenticides), fungus (fumigants), plants (herbicides), and insects (insecticides) through various mechanisms of action that may be specifically formulated to target the central nervous system of pests, such as insecticides, or may affect the function of other cellular components, leading to demise. The most prominent route for human exposure to these pesticides is through their direct application by an individual, again, either in a large commercial agricultural setting, like a farm, or even on a smaller scale, such as a home garden or in an indoor setting. Less direct exposures can also occur to individuals who are working or living in close proximity to the area where these compounds are being applied as a consequence of off-target effects or by handling food that has been sprayed with pesticides. Over the last several decades a wealth of studies have been conducted that demonstrate a strong association between exposure to pesticides and risk or incidence of PD. However, because multiple pesticides are usually used it has been difficult to delineate specific classes of pesticides and even more cumbersome to identify specific compounds in each relevant class that that may underlie the risk for PD. In more recent years the use of novel techniques and technologies as well as better-defined study populations have allowed for a more focused investigation of the impact of individual pesticides on PD, providing a better understanding of the chemicals and potential mechanisms of action that contribute to PD pathogenesis. Although there are currently dozens of individual pesticides being used, I will narrow the following discussion in order to appraise the classes and individual chemicals that have demonstrated the strongest epidemiologic and laboratory support for an association between human exposure and PD risk.

Insecticides

By far the most widely used pesticides in the USA and abroad are insecticides. Chemicals in this class of pesticide can be further stratified into different groups that are defined by their chemical structure and their specific mechanisms of action. In general, insecticides are formulated to directly target the central nervous system of insects, leading to neurologic dysfunction and death. The most prominent classes of insecticides are organochlorines, organophosphates, pyrethroids, and rotenoids. Of these compounds, the majority of epidemiologic studies have focused on the risk of PD associated with exposure to organochlorines. However, more recent work has also begun to uncover the contribution of exposure to organophosphates and rotenoids as possible risk factors. While several laboratory studies have demonstrated the ability of pyrethroids to alter the function of the nigrostriatal dopamine system (Bloomquist et al., 2002; Elwan et al., 2006), these compounds have not been explicitly shown to impact the risk of PD in the human population.

Organochlorines

Organochlorine insecticides, including DDT, dieldrin, heptachlor, lindane, and β-hexachlorocyclohexane (β-HCH), were extensively used, beginning in the 1940s and continuing through the 1970s, when their manufacture and use were discontinued due to environmental and human health concerns. While the vast majority of these compounds have been phased out, many of these compounds are still used in other developing countries, while others, such as endosulfan, have only recently seen their applications and use reduced. Organochlorine insecticides have varied targets and mechanisms of action within the central nervous system that affect the function of sodium and calcium channels and transporters as well as interfering with γ-aminobutyric acid (GABA) neurotransmission by blocking specific GABA receptors, contributing to their neurotoxic effects (Narahashi et al., 1995). These compounds are of further concern as their physical and chemical properties confer a high degree of lipophilicity and resistance to breakdown and metabolism, both in the environment as well as the human body, which allows them to persist and bioaccumulate in these settings. Thus, although the use of many of these compounds has been discontinued for many years, their presence is still detectable in the environment and in human tissue, even in populations that never had a direct exposure, creating a continued health concern for the human population.

The impact of occupational exposure to organochlorine compounds on the nigrostriatal dopamine system and the risk of PD has been demonstrated in several studies in the last 15 years (Corrigan et al., 1996, 2000; Kamel and Hoppin, 2004; Kamel et al., 2007; Elbaz et al., 2009; Ross et al., 2012). Of these compounds, occupational exposure to dieldrin and β-HCH has received the most attention as possible risk factors for PD (Kamel et al., 2007; Weisskopf et al., 2010; Kamel, 2013; Rhodes et al., 2013). One of the first studies to show an association between exposure to dieldrin and PD was that by Corrigan et al. (1996, 2000), which showed an elevation of dieldrin in the caudate nucleus and substantia nigra of patients with PD compared with controls. Similar elevations in serum levels of β-HCH have been found in PD patients compared with both control as well as individuals diagnosed with Alzheimer's disease (Richardson et al., 2009, 2011), suggesting a link between exposure to these compounds and risk of PD. Several laboratory investigations utilizing cell- and animal-based models further support these findings, especially for dieldrin. Work by Hatcher et al. (2008) demonstrated a significant alteration in dopamine handling as well as an increase in oxidative stress and α-synuclein in mice exposed to dieldrin. Selective damage to dopamine neurons and elevations in oxidative stress, alterations to mitochondrial function and elevated expression of α-synuclein have also been shown in primary cultured and clonal cell models of dieldrin exposure (Sanchez-Ramos et al., 1998; Kitazawa et al., 2001). Furthermore, studies that utilized a developmental exposure to dieldrin also showed alterations in the expression of both the dopamine transporter (DAT) and the vesicular monoamine transporter 2 (VMAT2) in offspring. These same animals also demonstrated an increased vulnerability to dopaminergic damage following subsequent exposure to 1-methyl-4-phenyl-1,2,3,6-tetrahydropyridine (MPTP) (Richardson et al., 2006).

Organophosphates

Although developed in the 1940s, the use and application of organophosphate insecticides for agricultural and household use did not escalate until the late 1960s as a response to the restrictions placed on the use of organochlorine compounds. Similar to organochlorines, organophosphate insecticides have been formulated to target specific cellular components of the central nervous system. In contrast to the organochlorine compounds, organophosphates exert their neurotoxicity by interaction and inhibition of the acetylcholinesterase enzyme, which functions to degrade the neurotransmitter, acetylcholine, into choline and acetyl coenzyme A, thus terminating the action of acetylcholine at the postsynaptic receptors. In the presence of an organophosphate compound this degradation is inhibited and acetylcholine is allowed to remain in the synaptic cleft and continue to interact with and activate acetylcholine receptors, resulting in a suite of neurologic symptoms that can lead to death. The most common organophosphate compounds found in the environment are chlorpyrifos, malathion, and methyl parathion. Like organochlorines, the manufacture and use of organophosphate insecticides have been greatly restricted or completely discontinued due to their general toxicity to the human population; however some compounds are still in use, both in the USA as well as abroad.

Although their use has been widespread, only a few studies have established a link between occupational exposure to organophosphate insecticides and the risk or incidence of PD. However, work in the last 5 years has begun to demonstrate significant associations between exposure to specific organophosphates and PD. Most notably, Hancock et al. (2008) showed an increased incidence of PD in patients occupationally exposed to organophosphate insecticides, while Dhillon et al. (2008) and Gatto et al. (2009) both specifically identified exposure to chlorpyrifos as a major contributor to PD risk. These findings are given further credence through laboratory investigations that suggest exposure to chlorpyrifos can significantly alter the normal functioning of the nigrostriatal dopamine pathway/circuit in rats (Torres-Altoro et al., 2011). One potential explanation for the lack of evidence relating exposure to organophosphates and PD could lie in the fact that, unlike their organochlorine predecessors, organophosphate compounds do not accumulate and persist in the environment to a similar extent as that seen with organochlorines. For example, the half-life of chlorpyrifos in soil is 10-fold less than that seen for an organochlorine insecticide such as dieldrin (Hatcher et al., 2008).

Interestingly, recent work has demonstrated an increased vulnerability to organophosphate neurotoxicity and risk for PD in individuals that carry one of several different mutations in their paraoxonase 1 (PON1) gene. PON1 serves an important function in the detoxication of specific organophosphate compounds, including chlorpyrifos and diazinon. Thus, alteration in the expression or function of PON1 would reduce the breakdown and elimination of these compounds, allowing them to persist and continue to affect acetylcholinesterase (Akhmedova et al., 2001; Manthripragada et al., 2010; Lee et al., 2013).

Rotenone

Rotenone is a naturally occurring compound extracted from the roots, leaves, and seeds of plants in the pea family and used for centuries by native Indians as a tool for hunting fish in rivers and streams. More recently, rotenone has been employed as a means to clear or manage nuisance fish in larger bodies of water as well as a small-scale pesticide on botanicals (Isman, 2006). As rotenone is highly hydrophobic, it is able to easily cross the cell membrane without the assistance of specific transport mechanisms. Unlike organochlorines and organophosphates, rotenone does not explicitly target the central nervous system. Rather, as a potent mitochondrial complex I inhibitor, it exerts its toxic and neurotoxic actions through disruption of adenosine triphosphate synthesis in the mitochondria and generation of oxidative stress (Degli Esposti, 1998). However, its reported specificity for damage to the nigrostriatal dopamine system may reside in the inherent vulnerability of the dopamine neurons to alteration in mitochondria function and oxidative stress and damage. Indeed, several reports have identified mitochondrial deficits in patients with PD (Schapira, 2008). While the interest in rotenone as a potential environmental risk factor for PD was initially bolstered by evidence demonstrating alterations in complex I in PD patients, it was previously believed that the short half-life in the environment and limited commercial use of rotenone reduced the potential exposure scenarios for the human population. However, recent epidemiologic findings suggest that, indeed, exposure to rotenone is a significant risk factor for development of PD in specific populations (Dhillon et al., 2008; Tanner et al., 2011).

Experimental support for these findings has been well established with findings that exposure of mice and rats to rotenone is capable of recapitulating many of the key pathologic features seen in human PD. The most prominent study, by Betarbet et al. (2000), found that treatment of rats with rotenone caused a significant reduction in dopamine in the striatum, loss of dopaminergic neurons and terminals in the substantia nigra pars compacta (SNpc) and striatum, respectively, in addition to elevated expression of oxidative stress and α-synuclein. Most interestingly, the rotenone-induced pathology was preferential for the dopamine system, sparing the GABA and glutamatergic neurotransmitter circuits in the basal ganglia. Since these studies, these findings have been repeated and expanded using various other model systems (Cannon and Greenamyre, 2010).

Herbicides

Herbicides are a broad class of pesticides that are used to remove nuisance plants, such as grasses and weeds, that may compromise the growth and yield of desired crops that are in close proximity. Herbicidal compounds are represented by numerous different formulations that target diverse cellular functions of plant cells, leading to their demise (Duke, 1990). Interestingly, many of these compounds are formulated to affect specific cellular components found only in plants and not in mammals, thus attempting to limit or reduce the potential for toxicity to the human population. For example, the widely used herbicide, 2,4-dichlorophenoxyacetic acid (2,4-D) targets auxin, which is a plant growth hormone. Although 2,4-D was synthesized to confer these specific mechanisms of action, there is still great concern regarding their toxic effects on the human population. Of the herbicides currently used, glyphosate, atrazine, 2,4-D, and paraquat are the most recognizable and have received the most attention related to their potential for toxic effects following occupational exposure. While each compound has been extensively studied, only paraquat has been identified as a risk factor for PD.

Paraquat

Although the toxicologic properties of paraquat have been well known since the 1930s, paraquat has only been used as a herbicide since the 1960s. Like other herbicides, its primary function is to kill and control unwanted plants, weeds, and grasses both in commercial and private agricultural settings. Paraquat's primary mechanism of toxicity is through its ability to redox cycle within a cell, causing a massive production of reactive oxygen species, including superoxide anion (O_2^-) and hydroxyl radicals, which can subsequently damage the expression and function of multiple cellular components imperative for normal cell function (Bus and Gibson, 1984). Indeed, of the general class of herbicides, paraquat is deemed as one of the more toxic compounds.

Interest in the possible role of paraquat toxicity on the nigrostriatal dopamine system was stimulated in the 1980s when the chemical structure of paraquat was found to resemble that of the selective dopaminergic neurotoxin, 1-methyl-4-phenylpyridinium (MPP+), which is taken up into the dopamine neurons by DAT (Snyder and D'Amato, 1985). Based upon these similarities it was suggested that paraquat may preferentially damage dopamine neurons and, like MPP+, may also target the mitochondria as a mitochondrial complex I inhibitor. However, laboratory studies focused on these concerns have found that paraquat does not enter dopamine neurons via the DAT and is not a potent complex I inhibitor, as compared with rotenone or MPP+ (Richardson et al., 2005). These findings suggest that paraquat's effects on dopamine neurons, as discussed below, may be mediated by the inherent vulnerability of the dopamine neuron to alteration in oxidative stress.

Epidemiologic support for the association between exposure to paraquat and the incidence or risk of PD is limited, but significant links have been identified in studies of occupationally exposed individuals (Kamel et al., 2007; Tanner et al., 2011; Wang et al., 2011; Freire and Koifman, 2012). Interestingly, the effect of paraquat on PD risk is often evaluated in the presence of additional pesticidal compounds, including fungicides, such as ziram and maneb. Indeed, in many studies, the risk of PD is elevated in individuals who reported exposures to a mixture of paraquat and maneb and/or ziram, suggesting a potential synergistic pathology between the two classes of pesticide (Costello et al., 2009).

While the epidemiologic support for paraquat as a risk factor for PD has gained ground in more recent years, the role for paraquat as a dopaminergic neurotoxin and its effects on the nigrostriatal dopamine circuit have been extensively studied in the laboratory. Consistent findings demonstrate that exposure of mice to paraquat routinely causes a selective loss of dopamine neurons in the SNpc in addition to a significant influx in reactive oxygen species and aggregation of α-synuclein protein (Thiruchelvam et al., 2000b; Manning-Bog et al., 2002; McCormack et al., 2002). These effects appear to be mediated by the redox cycling capability of paraquat in the dopamine neurons, as mice overexpressing superoxide dismutase, a scavenger of superoxide, are protected from the dopaminergic effects of paraquat (Peng et al., 2005). As discussed above, experimental data appear to mimic the epidemiologic findings, suggesting that exposure to a mixture of paraquat and maneb exacerbates many of the dopaminergic effects seen by paraquat alone (Thiruchelvam et al., 2000a).

Fungicides

Fungicidal compounds are used to combat fungal diseases in plants, either prior to the incidence of infection, or they can be applied to treat plants already infected with fungi or mold. Fungicides are comprised of a broad group of compounds, characterized by their diverse chemical structures and mechanisms of action. The most widely used class of fungicides is dithiocarbamates, which include compounds such as maneb, ziram, and mancozeb. In addition, other fungicidal compounds, like captan, folpet, and benomyl, which have a similar structure to the dithiocarbamates, have been extensively used. A significant amount of data has shown generalized, yet severe, toxicity of these compounds following exposure, including tumor development, dermal irritation, endocrine disruption, and some effects on the central nervous system. Of these compounds, the neurotoxic effects of the dithiocarbamate compounds, maneb and ziram, have been extensively studied, either as singular exposures or in conjunction with other pesticides, such as paraquat. As benomyl contains a carbamate moiety, similar to that seen with maneb and ziram, these compounds will be discussed as a group.

Carbamate fungicides

Although the dopaminergic neurotoxicity of exposure to carbamate fungicides has been well documented, the specific mechanisms of action are still being elucidated. Reports using cell-based models suggest that carbamate fungicides may exert their neurotoxicity on the dopamine system through several different pathways. Specifically, maneb has been shown to cause mitochondrial dysfunction through inhibition of the mitochondrial complex III, while ziram appears to affect dopamine neurons by inhibiting the function of the ubiquitin-proteasomal system

(UPS) (Zhang et al., 2003; Chou et al., 2008). Furthermore, in addition to affecting the UPS, benomyl has been shown to damage microtubule formation and inhibit the enzyme aldehyde dehydrogenase, which is involved in the metabolism of dopamine to nontoxic metabolites (Fitzmaurice et al., 2013). The majority of interest involving exposure to fungicides and PD has focused on the ability of maneb to augment the neurotoxic effects of paraquat on the nigrostriatal dopamine system. As discussed in the section on paraquat, maneb alone does not appear to cause overt dopaminergic neurotoxicity. Only when administered simultaneously with paraquat are its potential neurotoxic effects observed. These findings are further supported in epidemiologic studies that have identified a co-exposure to paraquat and maneb as a risk factor for PD, while a single exposure to only maneb does not appear to affect PD incidence (Costello et al., 2009). Similar epidemiologic studies have also found that concomitant occupational exposure to ziram and paraquat is a significant risk factor for PD, while exposure to ziram on its own is not (Wang et al., 2011). More recently, occupational exposure to benomyl was also suggested to be a risk factor for PD (Fitzmaurice et al., 2013).

Halogenated industrial compounds

Halogenated compounds are a class of chemicals characterized by varying degrees of halogen substitution on carbon atoms. Of these, substitutions with chlorine, bromine, or fluorine are the most commonly occurring and produce organochlorine, organobromine, and organofluorine compounds, respectively. The utility of these compounds in industrial as well as commercial settings is varied and can range from electric insulating coatings, flame-retardant oils, and in adhesives and plastics. The same physical and chemical properties that have made these compounds so attractive for use, including being highly stable and resistant to degradation, have also made them extremely dangerous to the environment and the human population, as they tend to have very long half-lives, thus allowing them to persist in the environment and in human tissue. Of the numerous compounds that can be categorized as organohalogens, our discussion will be focused on polychlorinated biphenyls (PCBs) and polybrominated diphenyl ethers (PBDEs), as these have received the most attention given their effects on the nigrostriatal dopamine system and as a risk factor for PD.

Polychlorinated biphenyls

PCBs are composed of biphenyl rings with varying amounts and positions of chlorine substitutions located around them, allowing for the formation of 209 different chemical conformations, also known as congeners. Their use in industrial and commercial settings began in the 1930s in the USA, where they were manufactured as mixtures of various congeners and sold under the trade name Aroclor. Of these mixtures, Aroclor 1242, 1254, and 1260, which contain 42%, 54%, and 60% chlorine, respectively, were the major mixtures produced. Due to their thermal stability, PCBs have been extensively utilized as lubricants and coolants to attenuate the generation of heat in transformers and capacitors, in hydraulic fluids, as well as other electric equipment (Erickson, 1986). Health concerns related to PCB exposure resulted in a discontinuation of their manufacturing and use in 1977. Even almost 40 years following this ban, significant levels of PCBs are still present in our environment and continue to impact human exposure and health (Safe, 1993; Kamrin et al., 1994).

Coupled with the persistence of PCBs in the environment, they also display a high degree of lipophilicity, which allows them to deposit in lipid-rich areas of the body, including the brain. Just as in the environment, degradation or metabolism of PCBs and elimination from the body is also difficult, presenting a situation for PCBs to accumulate within the body and create a toxic setting of constant exposure to these compounds over several years. This type of exposure is of serious concern as PCBs have been shown to be detrimental to the central nervous system (Faroon et al., 2001). While exposure to PCBs has been linked with several neurobehavioral deficits (Park et al., 2010), exposure to and deposition of PCBs in the brain have been associated with the incidence of PD. Work by Corrigan et al. (1998, 2000) found significantly elevated concentrations of PCBs, as well as specific PCB congeners in the brains of PD patients compared with controls. An additional study by Hatcher-Martin et al. (2012) also identified a greater deposition of PCBs and congeners in the brains of PD patients versus controls. The association between PCB exposure and risk for PD was also demonstrated in a cohort who had been occupationally exposed to high levels of PCBs while working in a factory that manufactured electric capacitors. Interestingly, while an increased incidence of PD was observed, it was only seen in female employees and not in males (Steenland et al., 2006). A cohort of these workers was further studied by $[^{123}\beta]$ CIT single-photon emission computed tomography imaging to determine if occupational exposure to PCBs altered DAT in the striatum. As before, female workers were found to show an inverse relationship between serum PCB levels and DAT densities in the striatum, suggesting that PCB exposure can specifically alter the nigrostriatal dopamine system (Seegal et al., 2010). Interestingly, these reductions in DAT are similar to those observed in animal studies (discussed below). Furthermore, the relevance of gender in the above findings warrants additional research as it may provide important information regarding the cellular and molecular mechanisms involved in PCB-mediated neurologic disease.

The effects of PCBs on the nigrostriatal dopamine system in laboratory animals have been well recognized for many years. Indeed, numerous studies have used various experimental models, including dopaminergic cell lines or brain slices, as well as mice, rats, and nonhuman primates to demonstrate PCB-mediated reductions in dopamine and dopamine neurotransmission (Seegal et al., 1986, 1989, 1990, 1991, 1994, 1998; Shain et al., 1991; Lee and Opanashuk, 2004; Richardson and Miller, 2004). The mechanisms leading to the reduction in dopamine levels are varied as exposure has been demonstrated to inhibit tyrosine hydroxylase and aromatic acid decarboylase, which are both enzymes involved in the synthesis of dopamine. Additionally, PCBs have been shown to alter the expression and function of the plasmalemmal DAT as well as VMAT2 to sequester dopamine, both of which are known to tightly regulate dopamine levels in the neuron (Giros et al., 1996; Richardson and Miller, 2004; Bemis and Seegal, 2004; Caudle et al., 2006, 2008; Fonnum et al., 2006; Lee et al., 2012). The implication of alterations to dopamine handling via disruption of DAT and VMAT2 as contributors to damage to the nigrostriatal dopamine system by PCBs is still under investigation. However, it should be noted that deficits in the ability of VMAT2 to sequester cytosolic dopamine can result in elevated levels of oxidative stress in the striatum, subsequent loss of striatal dopamine and DAT, degeneration of dopamine neurons in the SNpc as well as behavioral alterations reminiscent of those found in PD (Caudle et al., 2007; Taylor et al., 2009). It has been postulated that this dopaminergic degeneration is a result of the accumulation of dopamine in the cytosol and its subsequent oxidation to neurotoxic species and the formation of reactive oxygen and nitrogen species (Caudle et al., 2008).

Polybrominated diphenyl ethers

Structurally similar to PCBs, PBDEs are composed of two aromatic rings, with each ring decorated with varying degrees of bromine substitutions, allowing for 209 different congeners. PBDEs were introduced and quickly utilized in a similar fashion to PCBs in the 1970s following their phase-out. PBDEs are additive flame retardants primarily used in the manufacture and insulation of electronic equipment, as well as reduction of flammability in polyurethane foam used in carpeting, furniture cushions, and home insulation (Darnerud et al., 2001; de Wit, 2002). The physical and chemical properties of PBDEs that also make them highly lipophilic, resistant to breakdown, persistent in the environment, as well as able to bioaccumulate and biomagnify in the environment and the human body make them a relevant neurologic concern, especially given the effects on the nigrostriatal dopamine system seen with PCBs (Norstrom et al., 2002).

While data concerning the effects of PBDEs on the nigrostriatal dopamine system are minimal, recent work in the last few years has begun to highlight some concerning findings. Initial work with these compounds found the PBDE mixture DE-71 appeared to target the DAT and VMAT2 proteins, causing a significant inhibition of their ability to sequester dopamine (Mariussen and Fonnum, 2003). More recently, these findings have been elaborated in an *in vivo* model of PBDE exposure. Mice treated with DE-71 demonstrated a loss of DAT and VMAT2 expression in the striatum as well as a reduction in striatal dopamine and a subsequent decrease in locomotor activity (Bradner et al., 2013). Interestingly, when PBDEs were administered to mice with a 90% reduction in VMAT2 expression and function, many of these dopaminergic findings were exacerbated (Bradner et al., 2013).

METALS

Since the beginnings of cellular life, metals, including copper, zinc, iron, and manganese, have been a critical part of numerous biologic processes, especially enzymes. The use of these metals as enzymes arises from their efficient catalytic properties as they significantly increase the rates of enzymatic reactions owing to their ability to transfer electrons. In addition to their biologic use, metals have been used for centuries in commerce and industry. As a result, metals are recognized as some of the earliest toxicants in history, with reports of metal toxicity by Hippocrates around 370 BC. While the toxic effects of metal exposure on the human body have been well documented, the last 20–30 years have seen an increase in the recognition and appreciation for the neurotoxicologic impacts metal exposure can have on the human population. Indeed, the brain appears to be exquisitely vulnerable to metal toxicity as metals target various aspects of the neuron through generation of oxidative stress. These oxidants are extremely reactive and can readily interact with and modify various intraneuronal components, including proteins, lipids, and DNA, causing disruption in their normal function and expression and leading to more widespread alterations in neuronal processes. These effects are especially detrimental when they occur in the basal ganglia. As with other environmental factors, metal exposure has been suggested to be a risk factor for the development of PD and other parkinsonian-related movement disorders. For instance, manganese, iron, copper, and lead have all been shown to alter the function of the dopamine system and increase the risk for parkinsonism. This section will present recent findings concerning the potential role and mechanisms of action of metal toxicity in PD and parkinsonism.

Iron

Iron is an essential metal for proper functioning of many biologic species through its extensive use by mitochondrial enzymes involved in the electron transport chain and oxygen transport by hemoglobin. Our exposure to iron occurs via the diet, especially meat, poultry, and fish, as well as iron supplementation. In addition, occupational exposure to iron, predominantly from metal fumes or metal dust generated during welding and in iron and steel production, can contribute to our body burdens of iron. Once inside the body, iron is transported across cell membranes by the divalent metal transporter 1 (DMT1) prior to being bound by proteins, such as transferrin, which traffics iron into tissues, and ferritin (H- and L-ferritin), which serves as a storage depot for free iron. These mechanisms are critical for regulating and maintaining iron homeostasis in the cell. As a consequence, alterations to these functions or overwhelming of their capacities can cause an elevation in free iron, which is highly susceptible to free radical generation via the Fenton reaction, which catalyzes the conversion of hydrogen peroxide to the highly reactive hydroxyl radical. These hydroxyl radicals easily react with various components of the cell, including DNA, membrane lipids, and protein, leading to their dysfunction. Dopamine neurons within the SNpc are highly susceptible to disruptions in iron homeostasis as a readily available supply of hydrogen peroxide is present as a normal byproduct of the production and metabolism of dopamine within the neuron. This suggests that a decrement in iron homeostasis could facilitate the further production of reactive species that could subsequently damage various aspects of the dopamine neuron. Indeed, a genetic form of iron mishandling in pantothenate kinase-associated neurodegeneration leads to iron accumulation in the basal ganglia, resulting in neuronal damage and parkinsonian symptomatology (Thomas et al., 2004).

In addition to these two disorders, several epidemiologic studies have established exposure to iron, most notably through occupational settings, as a risk factor for parkinsonism (Gorell et al., 1997, 1999a; Rybicki et al., 1999). Related to these findings, an increased accumulation of iron has been found in the serum and SNpc of PD patients compared with controls (Sofic et al., 1988; Gerlach et al., 2006; Oakley et al., 2007; Zhao et al., 2013). While the cause for this accumulation is suggested to be due to excessive exposure in an occupational setting, there is evidence that other molecular factors within the nigrostriatal dopamine system may exacerbate the neurotoxic effects of iron. For example, studies have shown elevated levels of the lactoferrin receptors and DMT1 on dopaminergic neurons in the SNpc of patients with PD, which could facilitate the transport of iron into the dopamine neurons. In a similar study, Friedman et al. (2011) found a decrease in the amount of L-ferritin, which serves to sequester and store free cytosolic iron, in PD patients versus controls. Taken in concert, coupling occupational exposure to iron with an increase in iron transport mechanisms as well as a reduction in the ability of the dopamine neurons to store free iron could provide an advantageous environment for the generation of oxidative stress.

The deleterious effects of iron mishandling have also been demonstrated in several animal models of PD. In a study by Kaur et al. (2007), exposure of neonatal mice to levels of iron routinely found in baby formula resulted in an age-dependent reduction in TH+ neurons in the SNpc and a loss of dopamine in the striatum of older animals. Additionally, unilateral injection of $FeCl_3$ into the SNpc of adult rats resulted in a significant reduction in striatal dopamine and other dopaminergic markers and a concomitant deficit in dopamine-related behaviors (Sengstock et al., 1994; Junxia et al., 2003). Given the broad and far-reaching cellular effects of iron-mediated oxidative stress, it is difficult to assign a precise cause or neuronal cascade that is affected by these events. However, studies have identified α-synuclein as a key target. In addition to acting directly on α-synuclein, the formation of oxidative species has been shown to facilitate the fibrilization and aggregation of this protein. Indeed, it is this aggregated and fibrillar conformation that is considered to the predominant toxic form of α-synuclein and involved in the degenerative process of dopaminergic neurons (Giasson et al., 2000; Uversky et al., 2001).

Copper

Like iron, copper is an essential metal that serves several important biologic functions in the body, including an important component of cytochrome c oxidase in mitochondrial respiration, as well as being an integral part of key enzymes, such as superoxide dismutase and dopamine ß-hydroxylase (Harris, 2000). Similar to iron, copper is very reactive and can exploit the Fenton reaction in order to generate hydroxyl radicals from hydrogen peroxide in the dopamine neurons. Thus, regulation and maintenance of manageable levels of copper are imperative to normal neuronal function. The primary routes of occupational exposure to copper occur through the inhalation of fumes generated from welding, as well as metal mining and smelting activities. Copper in the blood is usually bound to ceruloplasmin prior to being transported into cells via the copper transporter ATPase (ATP7A), which serves to maintain copper homeostasis in the cell (Hellman and Gitlin, 2002). Indeed, mutation of this gene, as seen in Wilson's disease, results in accumulation of copper in several body tissues and damage to the basal ganglia and substantia nigra (Barbeau and Friesen, 1970; Hitoshi et al., 1991; Oder et al., 1993; Barthel et al., 2003). Further studies by Gorell et al. (1999b) have identified occupational exposures to copper as a risk factor for PD (Gorell et al., 1997, 1999a). These studies showed an almost 2.5-fold increase in risk for PD following a 20-year occupational exposure to copper. More recently, further evidence for copper exposure as a risk factor for PD has been demonstrated following intranigral injection of copper into rats. Investigators noted a significant reduction in several indices of dopaminergic neurodegeneration, including reduced striatal dopamine and loss of tyrosine hydroxylase projections and neurons in the striatum and substantia nigra, respectively (Yu et al., 2008).

The purported mechanisms of action that underlie copper-induced neuropathology are still tentative and somewhat contrasting, given our current understanding of copper levels in PD brains. It could be assumed that a persistent occupational exposure would result in an elevated deposition of copper in the brain and the ability of copper to generate neurotoxic reactive species, facilitating the neurodegenerative process, similar to iron. Related to this, recent reports have suggested that mitochondria are preferentially damaged following exposure to copper (Paris et al., 2009). Alteration to mitochondria function has long been appreciated as a potential pathogenic cascade in PD, most notably, through the aberrant generation of oxidative species following damage (Schapira, 2008). Additionally, like iron, copper has been shown to accelerate the oligomerization of α-synuclein monomers into neurotoxic fibrils, which can impair mitochondrial function, as well as other aspects of the dopamine neuron (Uversky et al., 2001). However, these speculated cascades would appear to be in contrast to previous studies that have reported unchanged or even reduced levels of copper in the SN of PD patients versus control (Davies et al., 2014). Given these findings, it has been suggested that dopaminergic pathology may arise from a reduction in the activity of the superoxide dismutase 1, a key cellular antioxidant in dopamine neurons, that results from the reduction in neuronal copper levels. Whether or not this interaction underlies dopaminergic loss is still unclear.

Manganese

Manganese is another essential metal with important functions in multiple biologic processes, including serving as cofactors for enzymes, including superoxide dismutase, as well as participating in neurotransmitter metabolism (Schroeder et al., 1966; Hurley et al., 1984; Golub et al., 2005). Although manganese exposure can occur through several different forms, including ingestion of food and exposure to manganese-containing products, exposure in an occupational setting such as mining, smelting, and welding appears to be a major contributor to these toxicants. The majority of these exposures are via inhalation of manganese dust, as in mining, or volatilization and inhalation of manganese fumes from manganese substrates during smelting and welding (Huang et al., 1989; Hudnell, 1999). Our current understanding suggests that inhalation of particulate manganese is able to bypass the blood–brain barrier where it is taken up directly by presynaptic nerve endings in the olfactory bulb. After being taken up, manganese is retrogradely transported to the cell body, where it can be released into the interstitial space (Tjalve and Henriksson, 1999; Vitarella et al., 2000; Fechter et al., 2002; Normandin et al., 2004). However, these pathways have been suggested based upon rodent studies and their relevance to human exposure is still being investigated (Brenneman et al., 2000; Aschner et al., 2005). Exposure to high levels of airborne manganese has been associated with several neurologic symptoms, including reduced neurobehavioral performance and neuropsychologic impairment (Huang et al., 1993; Gibbs et al., 1999). However, the most significant consequence of manganese toxicity appears to be impaired motor function, which manifests as a constellation of parkinsonian features, including muscular rigidity, tremor, reduced motor movement, and a "cock-walk" gait. Given these findings, a significant amount of attention has been focused on the neurologic effects of elevated occupational exposure to manganese, as they share several characteristic features of PD.

The association between elevated occupational exposure to manganese and neurobehavioral and motor dysfunction has been understood since the 1830s, when James Couper first described a neurologic dysfunction that shared many similarities to PD in workers exposed to manganese ore. Interestingly, at this time, Couper was able to delineate the symptomology related to manganese exposure that was distinct from typical PD. Since these findings, reports relating manganese exposure with parkinsonism have been mixed, with some studies identifying a clear association with occupational exposure (Gorell et al., 1999a, b; Racette et al., 2012), while others have not identified a similar link (Semchuk et al., 1993; Seidler et al., 1996; Marsh and Gula, 2006). The reasons for this controversy are varied, suggesting that this is an area that needs further investigation.

Regardless of the controversy, several reports have confirmed the effects of manganese exposure on the brain, particularly nuclei of the basal ganglia. Of these nuclei, Criswell et al. (2011, 2012) found the pallidal nuclei to be more susceptible to manganese accumulation compared with the caudate and putamen in welders with high occupational exposure to manganese. These data are further aligned with pathologic findings that symptoms related to elevated manganese exposure are attributed primarily to damage to neurons in the globus pallidus (internal and external segments), while largely sparing the caudate, putamen, and SNpc. Further studies using imaging techniques have identified significant reductions in the dopaminergic D2 receptor in the basal ganglia. However, while most studies report the preservation of dopaminergic regions, such as the caudate, putamen, and substantia nigra (Shinotoh et al., 1995; Olanow et al., 1996; Pal et al., 1999; Olanow, 2004), others suggest a mild damage to these regions as well as a general alteration in their function following manganese exposure (Suzuki et al., 1975; Eriksson et al., 1992; Kim et al., 2002; Chen et al., 2006; Wright et al., 2004; Guilarte et al., 2006; Criswell et al., 2011).

Similar findings have been observed in laboratory studies using nonhuman primates. Using an exposure paradigm that closely mimics that seen in humans, pathology appeared to be preferential for the globus pallidus, while leaving the dopaminergic inputs relatively untouched. In general, no change in DAT expression, striatal dopamine levels, with minimal reductions in the dopamine D2 receptor expression, and nonresponse to L-DOPA treatment was seen (Guilarte, 2013).

The mechanisms by which manganese mediates neurotoxicity are not wholly understood. It is known that manganese is predominantly transported into neurons by DMT1. Manganese, when in excess, can inhibit mitochondrial function, reduce glutathione levels, increase *N*-methyl-D-aspartate-mediated neurotoxicity and alter calcium homeostasis, all of which culminate in cellular dysfunction (Maynard and Cotzias, 1955; Brouillet et al., 1993; Gavin et al., 1999). Furthermore, manganese is a potent mediator of pro-oxidant activities, through its production of reactive species, particularly superoxides, peroxide, and hydroxyl radical (Graham et al., 1978; Cohen, 1984). More recently, the role of astrocytes in manganese-mediated neurotoxicity has been investigated. Manganese has been found to accumulate in significantly higher concentrations in astrocytes compared with other neuronal cells, and alter the antioxidant functions of astrocytes by altering glutathione synthesis (Dukhande et al., 2006; Erikson et al., 2008), which may further contribute to neuronal demise.

SOLVENTS

Solvents represent a broad range of chemicals with the common utility of dissolving one substance into another. The most common solvents in use today include trichloroethylene (TCE), toluene, acetone, hexane, and carbon disulfide, which all serve multiple purposes in industrial and home uses. Exposure to solvents can occur through several different routes, with the most prevalent route being occupational exposure via inhalation or dermal exposure. However, nonoccupational exposure can also occur through ingestion of contaminated water as well as inhalation exposure, both following improper waste disposal or accidental release of solvents. In general, solvents are lipophilic, allowing them to be amenable to quick and easy absorption into the body and to target organs following exposure. Over the years, several neurologic deficits have been associated with solvent exposure. Yet, the strongest support for solvent exposure as a risk factor for parkinsonism is found with TCE.

Trichloroethylene

TCE is a chlorinated hydrocarbon that has proved to be very versatile in its uses and applications, including its early use as an anesthetic given its properties as a central nervous system depressant. However, TCE is also extensively used in other industrial settings, including as a solvent in printing inks, paints, lacquers, varnishes, adhesives, and paint strippers. While the use of TCE in many of these applications has been reduced or replaced with other compounds, TCE is still widely used in the degreasing of metals and as an intermediate for hydrofluorocarbon production, highlighting its continued status as a potential health risk to those highly exposed.

The toxicity of TCE can be attributed to several mechanisms, including damage to cellular membrane integrity, formation of oxidative free radicals resulting in lipid peroxidation, as well as disruption of calcium transport. This membrane disruption leads to demyelination, with TCE exposure having been shown to result in loss of sensory nerve function (Huber, 1969; Feldman et al., 1970, 1985). As the potential neurotoxic effects of TCE exposure have become better recognized, an increased interest in the possible link between chronic exposure to TCE and parkinsonism has been explored. Indeed, over the last several years, multiple case reports have identified populations highly exposed to TCE exhibiting a potential increased risk incidence of PD and PD pathology (Bringmann et al., 1992; Guehl et al., 1999; Kochen et al., 2003; Gash et al., 2008; Goldman, 2010). Most recently a study of twins discordant for PD identified exposure to TCE as having a sixfold increased risk for PD (Goldman et al., 2012). These epidemiologic findings are largely supported by the experimental literature, utilizing animal models of TCE exposure. Specifically, Gash et al. (2008) found TCE exposure to cause significant damage to the nigrostriatal dopamine system, as shown by loss of dopamine neurons in the SNpc as well as a subsequent loss of dopamine in the striatum and midbrain. While the exact mechanisms of action of TCE that led to dopaminergic neurotoxicity are not known, the authors were able to determine that TCE inhibits mitochondrial complex I, similar to other parkinsonian mimetics, such as MPTP and rotenone. Similar dopaminergic damage was also seen by Liu et al. (2010), as they found treatment of rats with TCE resulted in dopamine neuron degeneration in the SNpc as well as accumulation of α-synuclein and an increase in reactive oxygen species. Interestingly, neuronal damage appeared to be preferential to the dopamine neurons in the SNpc as no loss of GABAergic or cholinergic neurons was observed in this area and the dopaminergic neurons of the ventral tegmental area were similarly spared, as seen in PD.

CONCLUSIONS

Appreciation of the role of the environment in neurologic disease has made great strides in the last several decades, providing an enriched understanding of the relative contributions and interplay between environmental and genetic factors in parkinsonism. As described in this chapter, the nigrostriatal dopamine system appears to be uniquely vulnerable to toxic insults, containing a multitude of targets that can be taken advantage of to elicit a neurotoxic cascade. Identification of specific chemicals and compounds has served to elaborate our general understanding of which chemicals are impacting the central nervous system (Table 14.1). However, these compounds may simply be "scratching the surface" of the chemicals that can harm the dopamine system and that have not been identified yet. Thus, a rigorous approach, which seems to already be in place, needs to continue in order to delineate environmental factors that may be involved in parkinsonism risk. Beyond this identification, investigations should also be focused on elucidating the specific cellular and molecular targets and mechanisms of these chemicals that may underlie their neurotoxic properties. In doing so, not only will our understanding grow of how these toxicants are damaging the dopamine system, but we will achieve a greater appreciation of the cellular processes of the dopamine circuit and the aspects that may make it especially vulnerable to toxic insult. Finally, this knowledge could be utilized to identify molecular targets of the dopamine neuron and pathway that could be exploited for therapeutic interventions aimed at attenuating or abrogating the neurotoxic consequences.

REFERENCES

Akhmedova SN, Yakimovsky AK, Schwartz EI (2001). Paraoxonase 1 Met–Leu 54 polymorphism is associated with Parkinson's disease. J Neurol Sci 184: 179–182.

Aschner M, Erikson KM, Dorman DC (2005). Manganese dosimetry: species differences and implications for neurotoxicity. Crit Rev Toxicol 35: 1–32.

Barbeau A, Friesen H (1970). Treatment of Wilson's disease with L-dopa after failure with penicillamine. Lancet 1: 1180–1181.

Barthel H, Hermann W, Kluge R et al. (2003). Concordant pre- and postsynaptic deficits of dopaminergic neurotransmission in neurologic Wilson disease. AJNR Am J Neuroradiol 24: 234–238.

Bemis JC, Seegal RF (2004). PCB-induced inhibition of the vesicular monoamine transporter predicts reductions in synaptosomal dopamine content. Toxicol Sci 80: 288–295.

Betarbet R, Sherer TB, MacKenzie G et al. (2000). Chronic systemic pesticide exposure reproduces features of Parkinson's disease. Nat Neurosci 3: 1301–1306.

Bloomquist JR, Barlow RL, Gillette JS et al. (2002). Selective effects of insecticides on nigrostriatal dopaminergic nerve pathways. Neurotoxicology 23: 537–544.

Bradner JM, Suragh TA, Wilson WW et al. (2013). Exposure to the polybrominated diphenyl ether mixture DE-71 damages the nigrostriatal dopamine system: role of dopamine handling in neurotoxicity. Exp Neurol 241: 138–147.

Brenneman KA, Wong BA, Buccellato MA et al. (2000). Direct olfactory transport of inhaled manganese ((54)MnCl(2)) to the rat brain: toxicokinetic investigations in a unilateral nasal occlusion model. Toxicol Appl Pharmacol 169: 238–248.

Bringmann G, Friedrich H, Feineis D (1992). Trichloroharmanes as potential endogenously formed inducers of Morbus Parkinson: synthesis, analytics, and first in vivo-investigations. J Neural Transm Suppl 38: 15–26.

Brouillet EP, Shinobu L, McGarvey U et al. (1993). Manganese injection into the rat striatum produces excitotoxic lesions by impairing energy metabolism. Exp Neurol 120: 89–94.

Bus JS, Gibson JE (1984). Paraquat: model for oxidant-initiated toxicity. Environ Health Perspect 55: 37–46.

Cannon JR, Greenamyre JT (2010). Neurotoxic in vivo models of Parkinson's disease recent advances. Prog Brain Res 184: 17–33.

Caudle WM, Richardson JR, Delea KC et al. (2006). Polychlorinated biphenyl-induced reduction of dopamine transporter expression as a precursor to Parkinson's disease-associated dopamine toxicity. Toxicol Sci 92: 490–499.

Caudle WM, Richardson JR, Wang MZ et al. (2007). Reduced vesicular storage of dopamine causes progressive nigrostriatal neurodegeneration. J Neurosci 27: 8138–8148.

Caudle WM, Colebrooke RE, Emson PC et al. (2008). Altered vesicular dopamine storage in Parkinson's disease: a premature demise. Trends Neurosci 31: 303–308.

Chen MK, Lee JS, McGlothan JL et al. (2006). Acute manganese administration alters dopamine transporter levels in the non-human primate striatum. Neurotoxicology 27: 229–236.

Chou AP, Maidment N, Klintenberg R et al. (2008). Ziram causes dopaminergic cell damage by inhibiting E1 ligase of the proteasome. J Biol Chem 283: 34696–34703.

Cohen G (1984). Oxy-radical toxicity in catecholamine neurons. Neurotoxicology 5: 77–82.

Corrigan FM, French M, Murray L (1996). Organochlorine compounds in human brain. Hum Exp Toxicol 15: 262–264.

Corrigan FM, Murray L, Wyatt CL et al. (1998). Diorthosubstituted polychlorinated biphenyls in caudate nucleus in Parkinson's disease. Exp Neurol 150: 339–342.

Corrigan FM, Wienburg CL, Shore RF et al. (2000). Organochlorine insecticides in substantia nigra in Parkinson's disease. J Toxicol Environ Health A 59: 229–234.

Costello S, Cockburn M, Bronstein J et al. (2009). Parkinson's disease and residential exposure to maneb and paraquat from agricultural applications in the central valley of California. Am J Epidemiol 169: 919–926.

Criswell SR, Perlmutter JS, Videen TO et al. (2011). Reduced uptake of [(1)(8)F]FDOPA PET in asymptomatic welders with occupational manganese exposure. Neurology 76: 1296–1301.

Criswell SR, Perlmutter JS, Huang JL et al. (2012). Basal ganglia intensity indices and diffusion weighted imaging in manganese-exposed welders. Occup Environ Med 69: 437–443.

Darnerud PO, Eriksen GS, Johannesson T et al. (2001). Polybrominated diphenyl ethers: occurrence, dietary exposure, and toxicology. Environ Health Perspect 109 (Suppl 1): 49–68.

Davies KM, Bohic S, Carmona A et al. (2014). Copper pathology in vulnerable brain regions in Parkinson's disease. Neurobiol Aging 35: 858–866.

de Wit CA (2002). An overview of brominated flame retardants in the environment. Chemosphere 46: 583–624.

Degli Esposti M (1998). Inhibitors of NADH-ubiquinone reductase: an overview. Biochim Biophys Acta 1364: 222–235.

Dhillon AS, Tarbutton GL, Levin JL et al. (2008). Pesticide/environmental exposures and Parkinson's disease in East Texas. J Agromedicine 13: 37–48.

Dickson DW (2012). Parkinson's disease and parkinsonism: neuropathology. Cold Spring Harb Perspect Med 2.

Duke SO (1990). Overview of herbicide mechanisms of action. Environ Health Perspect 87: 263–271.

Dukhande VV, Malthankar-Phatak GH, Hugus JJ et al. (2006). Manganese-induced neurotoxicity is differentially enhanced by glutathione depletion in astrocytoma and neuroblastoma cells. Neurochem Res 31: 1349–1357.

Elbaz A, Clavel J, Rathouz PJ et al. (2009). Professional exposure to pesticides and Parkinson disease. Ann Neurol 66: 494–504.

Elwan MA, Richardson JR, Guillot TS et al. (2006). Pyrethroid pesticide-induced alterations in dopamine transporter function. Toxicol Appl Pharmacol 211: 188–197.

Erickson BD (1986). Analytical Chemistry of PCBs. Butterworth, Boston.

Erikson KM, Dorman DC, Lash LH et al. (2008). Duration of airborne-manganese exposure in rhesus monkeys is associated with brain regional changes in biomarkers of neurotoxicity. Neurotoxicology 29: 377–385.

Eriksson H, Tedroff J, Thuomas KA et al. (1992). Manganese induced brain lesions in *Macaca fascicularis* as revealed by positron emission tomography and magnetic resonance imaging. Arch Toxicol 66: 403–407.

Faroon O, Jones D, de Rosa C (2001). Effects of polychlorinated biphenyls on the nervous system. Toxicol Ind Health 16: 305–333.

Farrer MJ (2006). Genetics of Parkinson disease: paradigm shifts and future prospects. Nat Rev Genet 7: 306–318.

Fechter LD, Johnson DL, Lynch RA (2002). The relationship of particle size to olfactory nerve uptake of a non-soluble form of manganese into brain. Neurotoxicology 23: 177–183.

Feldman RG, Mayer R, Taub A (1970). Evidence for peripheral neurotoxic effect of trichloroethylene. Neurology 20: 599–606.

Feldman RG, White RF, Currie JN et al. (1985). Long-term follow-up after single toxic exposure to trichloroethylene. Am J Ind Med 8: 119–126.

Fitzmaurice AG, Rhodes SL, Lulla A et al. (2013). Aldehyde dehydrogenase inhibition as a pathogenic mechanism in Parkinson disease. Proc Natl Acad Sci U S A 110: 636–641.

Fonnum F, Mariussen E, Reistad T (2006). Molecular mechanisms involved in the toxic effects of polychlorinated biphenyls (PCBs) and brominated flame retardants (BFRs). J Toxicol Environ Health A 69: 21–35.

Freire C, Koifman S (2012). Pesticide exposure and Parkinson's disease: epidemiological evidence of association. Neurotoxicology 33: 947–971.

Friedman A, Arosio P, Finazzi D et al. (2011). Ferritin as an important player in neurodegeneration. Parkinsonism Relat Disord 17: 423–430.

Gao HM, Hong JS (2011). Gene–environment interactions: key to unraveling the mystery of Parkinson's disease. Prog Neurobiol 94: 1–19.

Gash DM, Rutland K, Hudson NL et al. (2008). Trichloroethylene: Parkinsonism and complex 1 mitochondrial neurotoxicity. Ann Neurol 63: 184–192.

Gatto NM, Cockburn M, Bronstein J et al. (2009). Well-water consumption and Parkinson's disease in rural California. Environ Health Perspect 117: 1912–1918.

Gavin CE, Gunter KK, Gunter TE (1999). Manganese and calcium transport in mitochondria: implications for manganese toxicity. Neurotoxicology 20: 445–453.

Gerlach M, Double KL, Youdim MB et al. (2006). Potential sources of increased iron in the substantia nigra of parkinsonian patients. J Neural Transm Suppl 133–142.

Giasson BI, Duda JE, Murray IV et al. (2000). Oxidative damage linked to neurodegeneration by selective alpha-synuclein nitration in synucleinopathy lesions. Science 290: 985–989.

Gibbs JP, Crump KS, Houck DP et al. (1999). Focused medical surveillance: a search for subclinical movement disorders in a cohort of U.S. workers exposed to low levels of manganese dust. Neurotoxicology 20: 299–313.

Giros B, Jaber M, Jones SR et al. (1996). Hyperlocomotion and indifference to cocaine and amphetamine in mice lacking the dopamine transporter. Nature 379: 606–612.

Goldman SM (2010). Trichloroethylene and Parkinson's disease: dissolving the puzzle. Expert Rev Neurother 10: 835–837.

Goldman SM, Quinlan PJ, Ross GW et al. (2012). Solvent exposures and parkinson disease risk in twins. Ann Neurol 71: 776–784.

Golub MS, Hogrefe CE, Germann SL et al. (2005). Neurobehavioral evaluation of rhesus monkey infants fed cow's milk formula, soy formula, or soy formula with added manganese. Neurotoxicol Teratol 27: 615–627.

Gorell JM, Johnson CC, Rybicki BA et al. (1997). Occupational exposures to metals as risk factors for Parkinson's disease. Neurology 48: 650–658.

Gorell JM, Johnson CC, Rybicki BA et al. (1999a). Occupational exposure to manganese, copper, lead, iron, mercury and zinc and the risk of Parkinson's disease. Neurotoxicology 20: 239–247.

Gorell JM, Rybicki BA, Cole Johnson C et al. (1999b). Occupational metal exposures and the risk of Parkinson's disease. Neuroepidemiology 18: 303–308.

Graham DG, Tiffany SM, Bell Jr WR et al. (1978). Autoxidation versus covalent binding of quinones as the mechanism of toxicity of dopamine, 6-hydroxydopamine, and related compounds toward C1300 neuroblastoma cells in vitro. Mol Pharmacol 14: 644–653.

Guehl D, Bezard E, Dovero S et al. (1999). Trichloroethylene and parkinsonism: a human and experimental observation. Eur J Neurol 6: 609–611.

Guilarte TR (2013). Manganese neurotoxicity: new perspectives from behavioral, neuroimaging, and neuropathological studies in humans and non-human primates. Front Aging Neurosci 5: 23.

Guilarte TR, Chen MK, McGlothan JL et al. (2006). Nigrostriatal dopamine system dysfunction and subtle motor deficits in manganese-exposed non-human primates. Exp Neurol 202: 381–390.

Hancock DB, Martin ER, Mayhew GM et al. (2008). Pesticide exposure and risk of Parkinson's disease: a family-based case-control study. BMC Neurol 8: 6.

Harris ED (2000). Cellular copper transport and metabolism. Annu Rev Nutr 20: 291–310.

Hatcher JM, Pennell KD, Miller GW (2008). Parkinson's disease and pesticides: a toxicological perspective. Trends Pharmacol Sci 29: 322–329.

Hatcher-Martin JM, Gearing M, Steenland K et al. (2012). Association between polychlorinated biphenyls and Parkinson's disease neuropathology. Neurotoxicology 33: 1298–1304.

Hellman NE, Gitlin JD (2002). Ceruloplasmin metabolism and function. Annu Rev Nutr 22: 439–458.

Hitoshi S, Iwata M, Yoshikawa K (1991). Mid-brain pathology of Wilson's disease: MRI analysis of three cases. J Neurol Neurosurg Psychiatry 54: 624–626.

Huang CC, Chu NS, Lu CS et al. (1989). Chronic manganese intoxication. Arch Neurol 46: 1104–1106.

Huang CC, Lu CS, Chu NS et al. (1993). Progression after chronic manganese exposure. Neurology 43: 1479–1483.

Huber F (1969). Clinical aspects and neuropathology of trichloroethylene poisoning. Z Unfallmed Berufskr 62: 226–267.

Hudnell HK (1999). Effects from environmental Mn exposures: a review of the evidence from non-occupational exposure studies. Neurotoxicology 20: 379–397.

Hurley LS, Keen CL, Baly DL (1984). Manganese deficiency and toxicity: effects on carbohydrate metabolism in the rat. Neurotoxicology 5: 97–104.

Isman MB (2006). Botanical insecticides, deterrents, and repellents in modern agriculture and an increasingly regulated world. Annu Rev Entomol 51: 45–66.

Junxia X, Hong J, Wenfang C et al. (2003). Dopamine release rather than content in the caudate putamen is associated with behavioral changes in the iron rat model of Parkinson's disease. Exp Neurol 182: 483–489.

Kamel F (2013). Epidemiology. Paths from pesticides to Parkinson's. Science 341: 722–723.

Kamel F, Hoppin JA (2004). Association of pesticide exposure with neurologic dysfunction and disease. Environ Health Perspect 112: 950–958.

Kamel F, Tanner C, Umbach D et al. (2007). Pesticide exposure and self-reported Parkinson's disease in the agricultural health study. Am J Epidemiol 165: 364–374.

Kamrin MA, Fischer LJ, Suk WA et al. (1994). Assessment of human exposure to chemicals from Superfund sites. Environ Health Perspect 102 (Suppl 1): 221–228.

Kaur D, Peng J, Chinta SJ et al. (2007). Increased murine neonatal iron intake results in Parkinson-like neurodegeneration with age. Neurobiol Aging 28: 907–913.

Kim Y, Kim JM, Kim JW et al. (2002). Dopamine transporter density is decreased in parkinsonian patients with a history of manganese exposure: what does it mean? Mov Disord 17: 568–575.

Kitazawa M, Anantharam V, Kanthasamy AG (2001). Dieldrin-induced oxidative stress and neurochemical changes contribute to apoptopic cell death in dopaminergic cells. Free Radic Biol Med 31: 1473–1485.

Kochen W, Kohlmuller D, De Biasi P et al. (2003). The endogeneous formation of highly chlorinated tetrahydro-beta-carbolines as a possible causative mechanism in idiopathic Parkinson's disease. Adv Exp Med Biol 527: 253–263.

Lee DW, Opanashuk LA (2004). Polychlorinated biphenyl mixture aroclor 1254-induced oxidative stress plays a role in dopaminergic cell injury. Neurotoxicology 25: 925–939.

Lee DW, Notter SA, Thiruchelvam M et al. (2012). Subchronic polychorinated biphenyl (Aroclor 1254) exposure produces oxidative damage and neuronal death of ventral midbrain dopaminergic systems. Toxicol Sci 125: 496–508.

Lee PC, Rhodes SL, Sinsheimer JS et al. (2013). Functional paraoxonase 1 variants modify the risk of Parkinson's disease due to organophosphate exposure. Environ Int 56: 42–47.

Liu M, Choi D-Y, Hunter RL et al. (2010). Trichloroethylene induces dopaminergic neurodegeneration in Fisher 344 rats. J Neurochem 112: 773–783.

Manning-Bog AB, McCormack AL, Li J et al. (2002). The herbicide paraquat causes up-regulation and aggregation of alpha-synuclein in mice: paraquat and alpha-synuclein. J Biol Chem 277: 1641–1644.

Manthripragada AD, Costello S, Cockburn MG et al. (2010). Paraoxonase 1, agricultural organophosphate exposure, and Parkinson disease. Epidemiology 21: 87–94.

Mariussen E, Fonnum F (2003). The effect of brominated flame retardants on neurotransmitter uptake into rat brain synaptosomes and vesicles. Neurochem Int 43: 533–542.

Marsh GM, Gula MJ (2006). Employment as a welder and Parkinson disease among heavy equipment manufacturing workers. J Occup Environ Med 48: 1031–1046.

Maynard LS, Cotzias GC (1955). The partition of manganese among organs and intracellular organelles of the rat. J Biol Chem 214: 489–495.

McCormack AL, Thiruchelvam M, Manning-Bog AB et al. (2002). Environmental risk factors and Parkinson's disease: selective degeneration of nigral dopaminergic neurons caused by the herbicide paraquat. Neurobiol Dis 10: 119–127.

Narahashi T, Carter DB, Frey J et al. (1995). Sodium channels and GABAA receptor-channel complex as targets of environmental toxicants. Toxicol Lett 82–83: 239–245.

Normandin L, Ann Beaupre L, Salehi F et al. (2004). Manganese distribution in the brain and neurobehavioral changes following inhalation exposure of rats to three chemical forms of manganese. Neurotoxicology 25: 433–441.

Norstrom RJ, Simon M, Moisey J et al. (2002). Geographical distribution (2000) and temporal trends (1981-2000) of brominated diphenyl ethers in Great Lakes hewing gull eggs. Environ Sci Technol 36: 4783–4789.

Oakley AE, Collingwood JF, Dobson J et al. (2007). Individual dopaminergic neurons show raised iron levels in Parkinson disease. Neurology 68: 1820–1825.

Oder W, Prayer L, Grimm G et al. (1993). Wilson's disease: evidence of subgroups derived from clinical findings and brain lesions. Neurology 43: 120–124.

Olanow CW (2004). Manganese-induced parkinsonism and Parkinson's disease. Ann N Y Acad Sci 1012: 209–223.

Olanow CW, Good PF, Shinotoh H et al. (1996). Manganese intoxication in the rhesus monkey: a clinical, imaging, pathologic, and biochemical study. Neurology 46: 492–498.

Pal PK, Samii A, Calne DB (1999). Manganese neurotoxicity: a review of clinical features, imaging and pathology. Neurotoxicology 20: 227–238.

Paris I, Perez-Pastene C, Couve E et al. (2009). Copper dopamine complex induces mitochondrial autophagy preceding caspase-independent apoptotic cell death. J Biol Chem 284: 13306–13315.

Park HY, Hertz-Picciotto I, Sovcikova E et al. (2010). Neurodevelopmental toxicity of prenatal polychlorinated biphenyls (PCBs) by chemical structure and activity: a birth cohort study. Environ Health 9: 51.

Peng J, Stevenson FF, Doctrow SR et al. (2005). Superoxide dismutase/catalase mimetics are neuroprotective against selective paraquat-mediated dopaminergic neuron death in the substantial nigra: implications for Parkinson disease. J Biol Chem 280: 29194–29198.

Racette BA, Criswell SR, Lundin JI et al. (2012). Increased risk of parkinsonism associated with welding exposure. Neurotoxicology 33: 1356–1361.

Rhodes SL, Fitzmaurice AG, Cockburn M et al. (2013). Pesticides that inhibit the ubiquitin-proteasome system: effect measure modification by genetic variation in SKP1 in Parkinson's disease. Environ Res 126: 1–8.

Richardson JR, Miller GW (2004). Acute exposure to aroclor 1016 or 1260 differentially affects dopamine transporter and vesicular monoamine transporter 2 levels. Toxicol Lett 148: 29–40.

Richardson JR, Quan Y, Sherer TB et al. (2005). Paraquat neurotoxicity is distinct from that of MPTP and rotenone. Toxicol Sci 88: 193–201.

Richardson JR, Caudle WM, Wang M et al. (2006). Developmental exposure to the pesticide dieldrin alters the dopamine system and increases neurotoxicity in an animal model of Parkinson's disease. FASEB J 20: 1695–1697.

Richardson JR, Shalat SL, Buckley B et al. (2009). Elevated serum pesticide levels and risk of Parkinson disease. Arch Neurol 66: 870–875.

Richardson JR, Roy A, Shalat SL et al. (2011). beta-Hexachlorocyclohexane levels in serum and risk of Parkinson's disease. Neurotoxicology 32: 640–645.

Ross GW, Duda JE, Abbott RD et al. (2012). Brain organochlorines and Lewy pathology: the Honolulu-Asia Aging Study. Mov Disord 27: 1418–1424.

Rybicki BA, Johnson CC, Peterson EL et al. (1999). A family history of Parkinson's disease and its effect on other PD risk factors. Neuroepidemiology 18: 270–278.

Safe S (1993). Toxicology, structure-function relationship, and human and environmental health impacts of polychlorinated biphenyls: progress and problems. Environ Health Perspect 100: 259–268.

Sanchez-Ramos J, Facca A, Basit A et al. (1998). Toxicity of dieldrin for dopaminergic neurons in mesencephalic cultures. Exp Neurol 150: 263–271.

Schapira AH (2008). Mitochondria in the aetiology and pathogenesis of Parkinson's disease. Lancet Neurol 7: 97–109.

Schroeder HA, Balassa JJ, Tipton IH (1966). Essential trace metals in man: manganese. A study in homeostasis. J Chronic Dis 19: 545–571.

Seegal RF, Brosch KO, Bush B (1986). Polychlorinated biphenyls produce regional alterations of dopamine metabolism in rat brain. Toxicol Lett 30: 197–202.

Seegal RF, Brosch K, Bush B et al. (1989). Effects of Aroclor 1254 on dopamine and norepinephrine concentrations in pheochromocytoma (PC-12) cells. Neurotoxicology 10: 757–764.

Seegal RF, Bush B, Shain W (1990). Lightly chlorinated ortho-substituted PCB congeners decrease dopamine in nonhuman primate brain and in tissue culture. Toxicol Appl Pharmacol 106: 136–144.

Seegal RF, Bush B, Brosch KO (1991). Sub-chronic exposure of the adult rat to Aroclor 1254 yields regionally-specific changes in central dopaminergic function. Neurotoxicology 12: 55–65.

Seegal RF, Bush B, Brosch KO (1994). Decreases in dopamine concentrations in adult, non-human primate brain persist following removal from polychlorinated biphenyls. Toxicology 86: 71–87.

Seegal RF, Pappas BA, Park GA (1998). Neurochemical effects of consumption of Great Lakes salmon by rats. Regul Toxicol Pharmacol 27: S68–S75.

Seegal RF, Marek KL, Seibyl JP et al. (2010). Occupational exposure to PCBs reduces striatal dopamine transporter densities only in women: a beta-CIT imaging study. Neurobiol Dis 38: 219–225.

Seidler A, Hellenbrand W, Robra BP et al. (1996). Possible environmental, occupational, and other etiologic factors for Parkinson's disease: a case-control study in Germany. Neurology 46: 1275–1284.

Semchuk KM, Love EJ, Lee RG (1993). Parkinson's disease: a test of the multifactorial etiologic hypothesis. Neurology 43: 1173–1180.

Sengstock GJ, Olanow CW, Dunn AJ et al. (1994). Progressive changes in striatal dopaminergic markers, nigral volume, and rotational behavior following iron infusion into the rat substantia nigra. Exp Neurol 130: 82–94.

Shain W, Bush B, Seegal R (1991). Neurotoxicity of polychlorinated biphenyls: structure–activity relationship of individual congeners. Toxicol Appl Pharmacol 111: 33–42.

Shinotoh H, Snow BJ, Hewitt KA et al. (1995). MRI and PET studies of manganese-intoxicated monkeys. Neurology 45: 1199–1204.

Snyder SH, D'Amato RJ (1985). Predicting Parkinson's disease. Nature 317: 198–199.

Sofic E, Riederer P, Heinsen H et al. (1988). Increased iron (III) and total iron content in post mortem substantia nigra of parkinsonian brain. J Neural Transm 74: 199–205.

Steenland K, Hein MJ, Cassinelli 2nd RT et al. (2006). Polychlorinated biphenyls and neurodegenerative disease mortality in an occupational cohort. Epidemiology 17: 8–13.

Suzuki Y, Mouri T, Suzuki Y et al. (1975). Study of subacute toxicity of manganese dioxide in monkeys. Tokushima J Exp Med 22: 5–10.

Tanner CM, Ottman R, Goldman SM et al. (1999). Parkinson disease in twins: an etiologic study. JAMA 281: 341–346.

Tanner CM, Kamel F, Ross GW et al. (2011). Rotenone, paraquat, and Parkinson's disease. Environ Health Perspect 119: 866–872.

Taylor TN, Caudle WM, Shepherd KR et al. (2009). Nonmotor symptoms of Parkinson's disease revealed in an animal model with reduced monoamine storage capacity. J Neurosci 29: 8103–8113.

Thiruchelvam M, Brockel BJ, Richfield EK et al. (2000a). Potentiated and preferential effects of combined paraquat and maneb on nigrostriatal dopamine systems: environmental risk factors for Parkinson's disease? Brain Res 873: 225–234.

Thiruchelvam M, Richfield EK, Baggs RB et al. (2000b). The nigrostriatal dopaminergic system as a preferential target of repeated exposures to combined paraquat and maneb: implications for Parkinson's disease. J Neurosci 20: 9207–9214.

Thomas M, Hayflick SJ, Jankovic J (2004). Clinical heterogeneity of neurodegeneration with brain iron accumulation (Hallervorden-Spatz syndrome) and pantothenate kinase-associated neurodegeneration. Mov Disord 19: 36–42.

Tjalve H, Henriksson J (1999). Uptake of metals in the brain via olfactory pathways. Neurotoxicology 20: 181–195.

Torres-Altoro MI, Mathur BN, Drerup JM et al. (2011). Organophosphates dysregulate dopamine signaling, glutamatergic neurotransmission, and induce neuronal injury markers in striatum. J Neurochem 119: 303–313.

Uversky VN, Li J, Fink AL (2001). Metal-triggered structural transformations, aggregation, and fibrillation of human alpha-synuclein. A possible molecular NK between Parkinson's disease and heavy metal exposure. J Biol Chem 276: 44284–44296.

Vitarella D, Wong BA, Moss OR et al. (2000). Pharmacokinetics of inhaled manganese phosphate in male Sprague-Dawley rats following subacute (14-day) exposure. Toxicol Appl Pharmacol 163: 279–285.

Wang A, Costello S, Cockburn M et al. (2011). Parkinson's disease risk from ambient exposure to pesticides. Eur J Epidemiol 26: 547–555.

Weisskopf MG, Knekt P, O'Reilly EJ et al. (2010). Persistent organochlorine pesticides in serum and risk of Parkinson disease. Neurology 74: 1055–1061.

Wirdefeldt K, Gatz M, Bakaysa SL et al. (2008). Complete ascertainment of Parkinson disease in the Swedish Twin Registry. Neurobiol Aging 29: 1765–1773.

Wirdefeldt K, Adami HO, Cole P et al. (2011a). Epidemiology and etiology of Parkinson's disease: a review of the evidence. Eur J Epidemiol 26 (Suppl 1): S1–S58.

Wirdefeldt K, Gatz M, Reynolds CA et al. (2011b). Heritability of Parkinson disease in Swedish twins: a longitudinal study. Neurobiol Aging 32 (1923): e1921–e1928.

Wright AK, Atherton JF, Norrie L et al. (2004). Death of dopaminergic neurones in the rat substantia nigra can be induced by damage to globus pallidus. Eur J Neurosci 20: 1737–1744.

Yu WR, Jiang H, Wang J et al. (2008). Copper (Cu^{2+}) induces degeneration of dopaminergic neurons in the nigrostriatal system of rats. Neurosci Bull 24: 73–78.

Zhang J, Fitsanakis VA, Gu G et al. (2003). Manganese ethylene-bis-dithiocarbamate and selective dopaminergic neurodegeneration in rat: a link through mitochondrial dysfunction. J Neurochem 84: 336–346.

Zhao HW, Lin J, Wang XB et al. (2013). Assessing plasma levels of selenium, copper, iron and zinc in patients of Parkinson's disease. PLoS One 8: e83060.

Handbook of Clinical Neurology, Vol. 131 (3rd series)
Occupational Neurology
M. Lotti and M.L. Bleecker, Editors

Chapter 15

Tremor secondary to neurotoxic exposure: mercury, lead, solvents, pesticides

ROBERTO G. LUCCHINI[1,2]* AND DANA HASHIM[1]

[1]*Department of Preventive Medicine, Icahn School of Medicine at Mount Sinai, New York, NY, USA*

[2]*Unit of Occupational Medicine, Department of Medical Surgical Specialties, Radiological Sciences and Public Health, University of Brescia, Brescia, Italy*

TREMORS

A tremor is an unintentional, oscillatory rhythmic movement of one or more parts of the body that is often a symptom of an underlying disorder that affects the nervous system. Tremors may manifest as an adjunct effect of Parkinson's disease, degenerative neurologic diseases (e.g., multiple sclerosis), traumatic injury, environmental exposure, or recreational drug use. Moreover, there are several etiologies of tremor and patients typically present with tremor as one complaint among other symptoms.

The most common tremor type is essential tremor (ET), which has a wide variability of characteristic neurologic deficit associations (Jankovic, 2002). Estimated worldwide prevalence is 5%, with a higher prevalence in males (Louis et al., 1998; Dogu et al., 2003; Blond and Netterstrom, 2007; Louis and Ferreira, 2010). ET may be idiopathic or, when demonstrated, may be familial, with differences in age of onset and tremor severity within members of the same family (Louis et al., 2001). Due to this variability, environmental factors are suspected to be involved in the modifying ET susceptibility, although those mechanisms have yet to be fully understood (Louis, 2008; Jimenez-Jimenez et al., 2013).

There are a number of toxic agents that are known to cause fine, rapid tremors (8–12 Hz) reminiscent of ET, including alcohol, lead, mercury, and drug compounds (caffeine, bromides, barbiturates, cocaine, amphetamine, ephedrine) (Plumb and Bains, 2006). Particularly, lead, mercury, solvents, and pesticides have been among the most studied toxic agents and are also associated with occupational exposures. Long-term occupational exposure to these agents in conjunction with an increase in tremor severity commonly observed with advancing age may affect the ability to perform certain tasks, occupational or otherwise (NINDS, 2012). Additionally, the presence of tremors may lead to psychologic or physical stress, two risk factors that are also known to increase tremor severity (NINDS, 2012). The effects of ET can be devastating for workers' health, safety, and employment, and can result in embarrassment, difficulty performing daily tasks, job change, or early retirement (Bain et al., 1994).

LEAD AND TREMOR

Lead exposure has been demonstrated to result in an acute or chronic progressive neurologic disorder in which fine tremors are a prominent feature (Young et al., 1977; Seshia et al., 1978; Valpey et al., 1978; Goldings and Stewart, 1982; Coulehan et al., 1983). Both high and low blood lead concentrations have been shown in replicated studies to be significantly associated with ET (Louis et al., 2003, 2005; Dogu et al., 2007).

Nowadays, the major cause of elevated adult blood lead concentrations comes from workplace exposures (Abadin et al., 2007). Blood lead concentrations have been shown to increase with age and decrease with education level, and male workers have almost twice the geometric mean blood concentration as female workers (Yassin et al., 2004).

Workers may be exposed to organic and inorganic forms of lead through a number of sources (Winegar

*Correspondence to: Roberto G. Lucchini, MD, Professor and Vice Chair of Preventive Medicine, Director, Division of Occupational Medicine, Icahn School of Medicine at Mount Sinai, One Gustave L. Levy Place, Box 1057, New York, NY 10029-6574, USA. Tel: +1-212-824-7052, E-mail: roberto.lucchini@mssm.edu

et al., 1977; Coulehan et al., 1983). The major routes of lead exposure are inhalation and ingestion of lead-containing dusts and fumes (Abadin et al., 2007). The highest concentrations of lead exposure may occur in the following industries: lead-smelting and refining industries, battery-manufacturing plants, steel-welding or cutting operations, construction, rubber products and plastics industries, printing industries, firing ranges, radiator repair shops and other industries requiring flame soldering of lead solder, paint, car radiators, cable and wires, batteries, pigments, some cosmetics, ceramic ware with lead glazes, and tin cans. The lowest concentrations of lead exposure have been observed for workers involved in professional labor categories such as managerial positions and healthcare professionals, where contact with lead-bearing fumes may be indirect or limited (Abadin et al., 2007).

Once absorbed through inhalation or digestion, lead is distributed to the blood, soft tissues, and bones. In the blood, 99% of lead is bound to erythrocytes and 1% remains in the plasma to exchange with soft tissues (Rabinowitz, 1991). Lead affects many biologic systems once absorbed, including the kidney, brain, liver, and bone marrow. The mean half-life of blood-bound lead is about 30 days before lead is cleared through renal excretion (Rabinowitz, 1991). However, clearance can take longer in workers and other individuals with a long history of lead exposure and large bone stores (Hryhorczuk et al., 1985). In the bones, the half-life of lead is much longer at 5–19 years (Rabinowitz, 1991) and the bones carry 95% of the lead body burden (Rabinowitz, 1991). Females are especially affected by elevated bone lead reservoirs as lead is typically released during times of bone turnover, such as in menopause, pregnancy (Riess and Halm, 2007), breast feeding (Gulson et al., 1998), or in cases of hyperthyroidism (Goldman et al., 1994).

The pathologic mechanism for lead toxicity and ET remains unclear. In rat studies, however, it has been demonstrated that lead exposure produces a destructive cerebellar pathology due to degenerative changes in Purkinje cell morphology in animal studies (McConnell and Berry, 1979; Walsh and Tilson, 1984). In humans with known chronic lead exposure, similar findings have been revealed by postmortem examinations (Valpey et al., 1978), positron emission tomography (Wills et al., 1995), functional magnetic resonance imaging (Bucher et al., 1997), magnetic resonance spectroscopic imaging (Louis et al., 2002), clinical studies, and electrophysiologic studies (Deuschl et al., 2000; Stolze et al., 2001; Gironell et al., 2002).

Chronic occupational lead exposures have also been associated with tremors at rest in case series reports and epidemiologic studies (Winkel et al., 1995; Gorell et al., 1997; Kuhn et al., 1998; Coon et al., 2006). Additional presenting signs in these workers included hallmarks of Parkinson's disease such as rigidity, short-term memory impairment, ataxia, and akinesia (Kuhn et al., 1998; Manto, 2012). Although nonoccupational associations between lead exposure and parkinsonian symptoms have been found (Coon et al., 2006; Weisskopf et al., 2010; Weuve et al., 2013), the mechanisms by which these occur are still unclear (Winkel et al., 1995; Knowles and Donaldson, 1996; Kuhn et al., 1998). Regardless, patients presenting with a history of occupational lead exposure and resting tremors have been reported to improve following parkinsonian pharmacotherapy that included levodopa.

Lead toxicity susceptibility may be conferred by the ALAD-2 allele coded by the δ-amino-levulinic acid dehydratase (ALAD) gene in certain individuals (Louis et al., 2005). According to a case–control study, more ET cases had an ALAD-2 allele than controls (Louis et al., 2005). Additionally, this study found a significant interaction between blood lead concentration and ALAD allele status, with the highest likelihood of ET risk in individuals with both an ALAD-2 allele and elevated blood lead concentration.

ET may manifest as an early symptom of lead poisoning within weeks of initial exposure. Symptoms that may occur concurrently include altered sensations, irritability, poor attention span, headache, memory loss, and hallucinations (Kumar et al., 1987). Occupational history should include inquiries of job task description, potential lead exposure sources, engineering controls, the use and maintenance of personal protective equipment (including correct type of respirator with proper fit), housekeeping practices, smoking, and food consumption at work (which can result in hand-to-mouth exposures). Nonoccupational exposures that may also contribute to lead poisoning and should be taken into consideration include activities such as distillation of "moonshine" alcohol, use of guns and firing ranges (National Research Council of the National Academies, 2012), home repair work, and use of psychiatric medications that cause tremors, such as tricyclic antidepressants, phenothiazine, and valproate, and possible lead-bearing imported or dietary supplements (Agency for Toxic Substances and Disease Registry, 2007). Symptom timing, changes in frequency or intensity with work duration, and the relationship to exposure conditions are also helpful diagnostic information. It is also important to note past history of lead exposure on previous jobs or childhood lead poisoning.

In addition to occupational and environmental history, blood lead concentration remains the mainstay of lead toxicity diagnosis because it is the most accurate lead exposure indicator of previous few weeks as well as an indicator of endogenous lead release from the bone and soft-tissue stores. The Centers for Disease Control and Prevention (CDC) considers elevated blood lead concentrations in adults ≥30 μg/dL (Centers for Disease Control 2005). For occupational lead exposure, the Agency for Toxic Substances and Disease Registry recommends limiting blood lead concentrations to 60 μg/dL (Patrick, 2006; Agency for Toxic Substances and Disease Registry, 2007; Kosnett et al., 2007). However, cumulative lead exposure from chronic low-level exposures may also have detrimental long-term effects on worker health (Hu et al., 1998). Therefore, keeping a record of repeated measures of blood lead, from which a cumulative blood lead index (CBLI) can be calculated, is advised (Hu et al., 2007). CBLI has been shown to closely parallel bone lead concentrations measured by X-ray fluorescence (Hu et al., 2007). X-ray fluorescence is a rapid and noninvasive measurement of bone lead concentration that is a more accurate assessment of long-term lead exposure (Todd and Chettle, 1994) and is becoming increasingly standardized. If blood lead concentration is elevated, auxiliary tests such as complete blood count with peripheral smear for basophilic stippling (Paglia et al., 1975), serum creatinine, and urinalysis should be performed to assess blood and renal toxicity.

MERCURY AND TREMOR

Elemental, organic, and inorganic mercury have long been established as potent neurotoxicants (Bakir et al., 1973; Davis et al., 1974; Grandjean, 1997; Davidson et al., 1998) capable of inducing both resting and intentional tremors (Vroom and Greer, 1972; Levine et al., 1982; Albers et al., 1988; Liu et al., 2011). Traces of mercury exposure may occur from food and amalgam fillings; however occupational exposure to mercury typically occurs in higher dosage and workers are exposed to mercury through both inhalation and oral routes (World Health Organization International Agency for Research on Cancer, 1997). The majority of reported fine tremor cases related to mercury have been observed in elemental mercury (Hg^0) exposure only (Lucchini et al., 2002; Iwata et al., 2007; Liu et al., 2011). Elemental mercury is also a major source of occupationally related exposure (EPA, 1993).

Occupations in which the highest mercury vapor exposures occur include mercury mining, work in chloralkali and alkaline battery plants, and production of devices for measuring temperature and pressure (e.g., thermometers, barometers) (World Health Organization International Agency for Research on Cancer, 1997). Occupational exposure can also occur in glass cutting, fluorescent lamp manufacturing, and manufacturing of scientific instruments in dental personnel following release of Hg^0 from amalgam tooth fillings. Products containing inorganic mercury have been banned since the 1990s in the USA; however some exposure may occur through the use of old latex paint. The efficiency of absorption, elimination route, and tissue deposition of mercury depend upon the route of exposure and the chemical form of the metal (Bridges and Zalups, 2010). For the inhalation route, about 70–80% of elemental mercury vapor is absorbed while 15% of this same form is absorbed in the gastrointestinal tract (WHO, 1991). Alkyl mercurials, such as methyl mercury, are highly absorbed from the gastrointestinal tract at 95% when digested and later dealkylated. Mercury clearance half-life varies depending on physiologic compartment, with a total body biologic half-life of 58 days (Hursh et al., 1976). The kidney, hair, and central nervous system are major sites of deposition for all mercury forms (WHO, 1991).

Tremors and irritability are the most prominent symptoms of inhaled elemental mercury (Bidstrup et al., 1951; Fawer et al., 1983; Piikivi et al., 1984; Albers et al., 1988). Similar to lead toxicity, the tremors can occur after both acute high-concentration mercury exposures and chronic low-concentration mercury exposures. Acute high-exposure mercury toxicity presents with concurrent respiratory symptoms (cough, dyspnea, and chest pain), vomiting, and metallic taste in the mouth. Tremors, muscle fasciculations, myoclonus, and muscle pains have also been reported after acute toxicity (McFarland and Reigel, 1978; Bluhm et al., 1992). First clues of chronic occupational mercury exposure, however, are gingivitis and central nervous system-related symptoms, specifically intention and resting tremors, cognitive difficulties, and erethism (a characteristic feature of mercury toxicity that includes excessive shyness intermixed with nervousness and irritability) (Fawer et al., 1983; Agency for Toxic Substances and Disease Registry, 1999). Other clinical features of toxicity include psychologic disturbances such as memory loss, increased excitability, insomnia, and personality changes (Al-Batanony et al., 2013).

The most accurate assessment of both acute and chronic elemental mercury exposure is urine mercury concentration level, which is also correlated with tremors and hand–eye coordination (Miller et al., 1975; Williamson et al., 1982; Verberk et al., 1986). The CDC considers mercury concentrations greater than 4 μg/L to be above the background range (CDC, 2001–2002). Mercury blood concentration accurately assesses exposure within the first 3 days due to the short blood mercury half-life, while mercury hair deposition analysis measures primarily organic methyl mercury and is accurate for assessment of chronic exposures (Rafati-Rahimzadeh et al., 2014).

SOLVENTS AND TREMOR

Organic solvents are volatile chemical compounds that share a common structure of at least one carbon and one hydrogen atom, a low molecular weight, and lipophilicity. For practical purposes, solvents as related to occupational exposure-related tremors are defined as "a liquid that has the ability to dissolve, suspend or extract other materials, without chemical change to the material or solvent" (HSE, 1998).

Organic solvents are widely used in the production and manufacturing of common items, including paints, pharmaceuticals, degreasants, adhesives, printing inks, pesticides, cosmetics, dry cleaning, and household cleaners (HSE, 1998). The highest occupational exposure occurs in dry cleaning, screen printing, glass manufacturing, and painting. Lowest occupational exposure occurs in petrol pump attendants, carpenters, chemical process operators, lab technicians, and cleaners using polish (Dick, 2006). The most common occupational exposure route is inhalation; however, dermal exposure may also occur due to the highly lipophilic nature of solvents (Semple, 2004).

Although solvent usage has recently decreased in some occupations in North America and Europe, inhalation sources persist, particularly in confined-space painting (Dick et al., 2000). Tremors may be the first symptom reported as a result of solvent exposure (Iwata et al., 2005; Dick, 2006; Papageorgiou et al., 2009). Solvents that are still widely used and are commonly associated with tremor include carbon disulfide (Chapman et al., 1991; Frumkin, 1998; Takebayashi et al., 1998), toluene (Deleu and Hanssens, 2000), xylene (Iwata et al., 2005), *n*-hexane, trichloroethylene, and perchloroethylene (CDC, 1987; Dick, 2006; NIOSH, 2013). In addition to tremors, acute solvent toxicity may include cognitive difficulties such as difficulty concentrating and general malaise such as light-headedness, headache, nausea, and vomiting. Chronic effects of solvent exposure are characterized by cognitive performance decline and psychologic changes (Berr et al., 2010; Sabbath et al., 2012, 2014) and are a risk factor for motor neuron disease (Gunnarsson et al., 1992; Schulte et al., 1996). Attention impairment, slow information processing, short-term memory loss, including fatigue, irritability, depression, and insomnia have also been reported (van der Hoek et al., 2001; van Valen et al., 2012). Signs of solvent toxicity vary depending on the specific solvent exposure and patient symptoms can be nonspecific and subjective.

The mechanism underlying the neurotoxic influence of organic solvents on the central nervous system is not yet fully understood (Ridgway et al., 2003), although studies on severe solvent abuse (e.g., sniffing) have resulted in structural damage to periventricular brain regions and cortical atrophy (Yucel et al., 2008; Keski-Santti et al., 2009). Studies specifically examining chronic occupational solvent exposure effects have yet to elucidate mechanisms behind the tremor symptoms.

Diagnosing tremor secondary to solvent toxicity is challenging for physicians and healthcare workers because a normative criteria inclusion/exclusion list for solvent toxicity has not yet been well established. Unlike toxic metals, for which cumulative exposure measures are available, solvents do not leave a cumulative trace. Therefore, questioning patients on occupation, including duration, intensity, and frequency of solvent exposure, is crucial to diagnosing tremor secondary to solvent exposures (Ridgway et al., 2003). Similarly to diagnosing chronic solvent-induced encephalopathy, however, the World Health Organization and European researchers recommend that solvent toxicity should be a diagnosis of exclusion (van der Laan and Sainio, 2012), after ruling out other more common causes of tremor such as stroke, medication side effects, and degenerative neurologic conditions. A specialist team that includes an occupational physician, neurologist, occupational hygienist, and psychiatrist may aid in reaching a definitive diagnosis.

PESTICIDES AND TREMOR

Common tremor-inducing pesticides that have been used for the past 50 years include organophosphates and carbamate agents (Rotenberg et al., 1995). Signs and symptoms of toxicity by these agents are similar. Worldwide, an estimated three million people are exposed to organophosphate or carbamate agents each year, with up to 300 000 fatalities (Eyer, 2003; Eddleston and Phillips, 2004). In the USA, there were more than 8000 reported exposures to these agents in 2008 (Bronstein et al., 2009). Thus, the exposure to pesticides is widespread.

In the occupational setting, exposures to pesticides are almost entirely by inhalation, although ingestion may occur, and pesticides are highly absorbed by the lungs. Organophosphates bind to acetylcholinesterase, interfering with acetylcholinesterase muscarinic and nicotinic receptors. Acute cholinergic crisis develops within hours of exposure that includes salivation, miosis, and lacrimation. Nicotinic effects may also occur as a result of acute organophosphate exposure, including tachycardia and weakness in addition to tremor-like muscle fasciculation (Jett, 2011). Although referred to in studies interchangeably, muscle fasciculation is distinct from tremors and involves a single motor unit, thus being smaller in scale. Detecting muscle fasciculation may require electromyogram testing for visualization. Additionally, muscle fasciculations are not aggravated by psychologic or physical factors as typical of tremors, and have a nonrhythmic, "random" sensation when felt.

Pesticides can also produce both rest and action tremors in humans (Cannon et al., 1978; Taylor et al., 1978; Taylor, 1982; Gorell et al., 2004; Louis et al., 2006; Jett, 2011). Early age of ET onset has also been shown to be associated with brief pesticide exposure most strongly in individuals with a family history of ET (Jimenez-Jimenez et al., 2007). Tremor may also occur following chronic exposure, relating to central effects on dopamine and resulting in a parkinsonian tremor (Baltazar et al., 2014; Liew et al., 2014).

DIAGNOSING TREMOR

Tremors rarely present as an isolated sign or symptom. Rather, tremor is a manifestation of another underlying disorder that must be identified. For lead, mercury, and solvent toxicity, identification of the cause of tremor is a process of elimination. Physical examination, blood tests, and neuroimaging may be necessary to rule out secondary causes.

Since the etiologies of ET may not be easily recognizable, they may be misdiagnosed as idiopathic parkinsonian tremor (Marshall et al., 2009; Newman et al., 2009). ET is clearly distinguished by not occurring at rest and has been classically illustrated by difficulty in drawing a two-dimensional spiral, in which impairment of fine motor control is evident. ET tremors can vary from low-amplitude and high-frequency (4–12 Hz or 4–12 shakes of the arm per second) postural hand tremors to much larger amplitudes involving other body parts, including the head or vocal cords (Louis, 2005). ET can also affect the face, legs, and gait and postural and kinetic type of action tremors are characteristic. Postural tremors occur when the affected body part is held in position against gravity and kinetic tremors occur during voluntary movement while kinetic tremors occur during movement. Other distinguishing ET characteristics include early age of onset (less than 60 years old, although ET can also occur over age 60) and worsening during psychologic or physical stress. Additionally, ET can be treated with propranolol, other beta-blockers, or primidone while levodopa is only effective in parkinsonian tremors.

For chronic tremor assessment, physicians can look at old handwriting samples to compare apparent fine motor control while writing. In some patients with ET, an intention tremor is also present. This is apparent when the hand makes a purposeful movement towards an object under visual guidance and manifests as an increase in tremor as the hand approaches the object. The self-conscious embarrassment and reluctance to expose the shaking in public can have a negative effect on worker quality of life (Plumb and Bains, 2006).

TREATMENT OF TREMOR

Once the exposure cause of the tremor is identified, removal or reduction from the source is the first step in ameliorating tremor symptoms. After the source of toxicity is removed, actions can be taken at work to prevent further excessive exposures, such as implementing work practices and control measures, including engineering controls or personal protection (respirators, clothing), hand washing, and other improved clean-up practices that generate less exposure to the toxic agent.

For lead poisoning, removal or job transfer away from lead exposure when blood lead levels rise to 50 μg/dL (2.42 μmol/L) is mandated under the standards of the US Occupational Safety Health Administration. In most cases of mildly elevated lead levels (>20 μg/dL), removal of the patient from the exposure may be the only therapy indicated. Chelation should be considered for patients with blood lead levels >80 μg/dL (3.86 μmol/L) or for symptomatic patients with blood levels 50–80 μg/dL (2.42–3.86 μmol/L). Primary care clinicians may find it advisable to refer patients with elevated blood lead levels, particularly above 20 μg/dL (0.97 μmol/L), to occupational medicine clinicians.

For mercury inhalation toxicity, patients should receive supportive care and be monitored for development of acute pneumonitis and pulmonary edema in a hospital setting. For severe symptoms and highly elevated circulating levels of elemental mercury, chelation may be required. Agents such as dimercaptosuccinic acid contain sulfhydryl groups, which bind mercury ions and facilitate their excretion through urine and feces. The efficacy of chelation therapy should be monitored by repeated clinical assessment and measurement of mercury in blood and urine. The majority of studies suggest that motor system disturbances are reversible upon exposure cessation of mercury exposure, while cognitive impairments, primarily memory deficits, may be permanent (Chaffin et al., 1973; Miller et al., 1975; Hanninen, 1982).

For solvent toxicity tremor symptoms have been shown to abate after chronic exposure cessation (Orti-Pareja et al., 1996). However, for solvent toxicity that progresses to chronic solvent-induced encephalopathy, cognitive sequelae may persist (Bruhn et al., 1981; Kaukiainen et al., 2009).

SUMMARY AND CONCLUSION

Compared to many degenerative neurologic disorders, the mechanisms of ET onset are not fully understood.

Furthermore, most published articles on ET are case studies rather than case–control or prospective studies. Nevertheless, tremor remains one of the most commonly encountered neurologic symptoms. The traditional view of ET as having no underlying etiology is gradually being replaced by the notion that many of these tremors may be caused by chronic environmental exposures in addition to genetic susceptibility. A better understanding of the disease etiology and mechanisms will ultimately lead to advances in the development of treatments.

Since tremors are a nonspecific symptom, a preventive emphasis on known neurotoxic exposure minimization is crucial. The costs and consequences of acute and chronic metals, solvents, and pesticide exposures can be ameliorated by early recognition of toxicity, decreasing sources of exposure through the implementation of recommended exposure limits, protective protocol, and safety measures facilitated by a team of occupational medicine specialists.

Although legislation efforts have been made to reduce occupational exposures to toxic metals, solvents, and pesticides, the threat of acute and chronic toxicity remains. Over 4000 facilities reported the release of over 12 million lb of lead into the environment in 2004 and this problem continues to affect workers (Dooyema et al., 2012). Organochlorine pesticides have also created an enduring environmental problem, are resistant to degradation, and will be detectible in soil for 30 years after their application (Nunes and Tajara, 1998; Carr et al., 1999).

There is a demand for further research into lead and solvent toxicity mechanisms in particular. The effects of low-level chronic exposure are still not certain. In addition, an environment and genetic susceptibility link is required to initiate curative treatment for common neurotoxin sequelae such as tremors.

References

Abadin H, Ashizawa A, Stevens YW et al. (2007). Toxicological Profile for Lead. Potential for Human Exposure, Agency for Toxic Substances and Disease Registry (US), Atlanta (GA).

Agency for Toxic Substances and Disease Registry (1999). Medical Management Guidelines for Mercury, Agency for Toxic Substances and Disease Registry, Atlanta, GA.

Agency for Toxic Substances and Disease Registry (2007). Toxicological profile for Lead, Agency for Toxic Substances and Disease Registry, Atlanta, GA.

Al-Batanony MA, Abdel-Rasul GM, Abu-Salem MA et al. (2013). Occupational exposure to mercury among workers in a fluorescent lamp factory, Quisna Industrial Zone, Egypt. Int J Occup Environ Med 4: 149–156.

Albers JW, Kallenbach LR, Fine LJ et al. (1988). Neurological abnormalities associated with remote occupational elemental mercury exposure. Ann Neurol 24: 651–659.

Bain PG, Findley LJ, Thompson PD et al. (1994). A study of hereditary essential tremor. Brain 117 (Pt 4): 805–824.

Bakir F, Damluji SF, Amin Zaki I et al. (1973). Methylmercury poisoning in Iraq: an interuniversity report. Science 181: 230–241.

Baltazar MT, Dinis-Oliveira RJ, De Lourdes Bastos M et al. (2014). Pesticides exposure as etiological factors of Parkinson's disease and other neurodegenerative diseases – A mechanistic approach. Toxicol Lett 230: 85–103.

Berr C, Vercambre MN, Bonenfant S et al. (2010). Occupational exposure to solvents and cognitive performance in the GAZEL cohort: preliminary results. Dement Geriatr Cogn Disord 30: 12–19.

Bidstrup PL, Bonnell JA, Harvey DG et al. (1951). Chronic mercury poisoning in men repairing direct-current meters. Lancet 2: 856–861.

Blond M, Netterstrom B (2007). Neuromotor function in a cohort of Danish steel workers. Neurotoxicology 28: 336–344.

Bluhm RE, Breyer JA, Bobbitt RG et al. (1992). Elemental mercury vapour toxicity, treatment, and prognosis after acute, intensive exposure in chloralkali plant workers. Part II: Hyperchloraemia and genitourinary symptoms. Hum Exp Toxicol 11: 211–215.

Bridges CC, Zalups RK (2010). Transport of inorganic mercury and methylmercury in target tissues and organs. J Toxicol Environ Health B Crit Rev 13: 385–410.

Bronstein AC, Spyker DA, Cantilena Jr LR et al. (2009). 2008 Annual Report of the American Association of Poison Control Centers' National Poison Data System (NPDS): 26th Annual Report. Clin Toxicol (Phila) 47: 911–1084.

Bruhn P, Arlien-Soborg P, Gyldensted C et al. (1981). Prognosis in chronic toxic encephalopathy. A two-year follow-up study in 26 house painters with occupational encephalopathy. Acta Neurol Scand 64: 259–272.

Bucher SF, Seelos KC, Dodel RC et al. (1997). Activation mapping in essential tremor with functional magnetic resonance imaging. Ann Neurol 41: 32–40.

Cannon SB, Veazey Jr JM, Jackson RS et al. (1978). Epidemic kepone poisoning in chemical workers. Am J Epidemiol 107: 529–537.

Carr RL, Couch TA, Liu J et al. (1999). The interaction of chlorinated alicyclic insecticides with brain GABA(A) receptors in channel catfish (*Ictalurus punctatus*). J Toxicol Environ Health 56: 543–553.

CDC (1987). Organic Solvent Neurotoxicity. Current Intelligence Bulletin, Available online at: http://www.cdc.gov/niosh/docs/87-104/.

CDC (2001–2002). National Health and Nutrition Examination Survey (NHANES. Summarized in Department of Health and Human Services Centers for Disease Control and Prevention. 2005. Third National Report on Human Exposure to Environmental Chemicals, National Center for Environmental Health. NCEH publication no. 05-0570. Available online at: http://www.jhsph.edu/research/centers-and-institutes/center-for-excellence-in-environmental-health-tracking/Third_Report.pdf (accessed February 2, 2015).

Centers for Disease Control and Prevention (2005). Blood lead levels – United States, 1999–2002. MMWR Morb Mortal Wkly Rep 54: 513–516.

Chaffin D, Dinman B, Miller J et al. (1973). An evaluation of the effects of chronic mercury exposure on EMG and psychomotor functions. Document no. HSM-099-71-62. U.S. Department of Health and Human Services, National Institute of Occupational Safety and Health, Washington DC.

Chapman LJ, Sauter SL, Henning RA et al. (1991). Finger tremor after carbon disulfide-based pesticide exposures. Arch Neurol 48: 866–870.

Coon S, Stark A, Peterson E et al. (2006). Whole-body lifetime occupational lead exposure and risk of Parkinson's disease. Environ Health Perspect 114: 1872–1876.

Coulehan JL, Hirsch W, Brillman J et al. (1983). Gasoline sniffing and lead toxicity in Navajo adolescents. Pediatrics 71: 113–117.

Davidson PW, Myers GJ, Cox C et al. (1998). Effects of prenatal and postnatal methylmercury exposure from fish consumption on neurodevelopment: outcomes at 66 months of age in the Seychelles child development study. JAMA 280: 701–707.

Davis LE, Wands JR, Weiss SA et al. (1974). Central nervous system intoxication from mercurous chloride laxatives. Quantitative, histochemical, and ultrastructural studies. Arch Neurol 30: 428–431.

Deleu D, Hanssens Y (2000). Toluene induced postural tremor. J Neurol Neurosurg Psychiatry 68: 118.

Deuschl G, Wenzelburger R, Loffler K et al. (2000). Essential tremor and cerebellar dysfunction clinical and kinematic analysis of intention tremor. Brain 123 (Pt 8): 1568–1580.

Dick FD (2006). Solvent neurotoxicity. Occup Environ Med 63 (221–6): 179.

Dick F, Semple S, Chen R et al. (2000). Neurological deficits in solvent-exposed painters: a syndrome including impaired colour vision, cognitive defects, tremor and loss of vibration sensation. QJM 93: 655–661.

Dogu O, Sevim S, Camdeviren H et al. (2003). Prevalence of essential tremor: door-to-door neurologic exams in Mersin Province, Turkey. Neurology 61: 1804–1806.

Dogu O, Louis ED, Tamer L et al. (2007). Elevated blood lead concentrations in essential tremor: a case-control study in Mersin, Turkey. Environ Health Perspect 115: 1564–1568.

Dooyema CA, Neri A, Lo YC et al. (2012). Outbreak of fatal childhood lead poisoning related to artisanal gold mining in northwestern Nigeria, 2010. Environ Health Perspect 120: 601–607.

Eddleston M, Phillips MR (2004). Self poisoning with pesticides. BMJ 328: 42–44.

EPA (1993). Locating and Estimating Air Emissions from Sources of Mercury and Mercury Compounds, U.S. Environmental Protection Agency, Research Triangle Park, NC.

Eyer P (2003). The role of oximes in the management of organophosphorus pesticide poisoning. Toxicol Rev 22: 165–190.

Fawer RF, De Ribaupierre Y, Guillemin MP et al. (1983). Measurement of hand tremor induced by industrial exposure to metallic mercury. Br J Ind Med 40: 204–208.

Frumkin H (1998). Multiple system atrophy following chronic carbon disulfide exposure. Environ Health Perspect 106: 611–613.

Gironell A, Kulisevsky J, Lorenzo J et al. (2002). Transcranial magnetic stimulation of the cerebellum in essential tremor: a controlled study. Arch Neurol 59: 413–417.

Goldings AS, Stewart RM (1982). Organic lead encephalopathy: behavioral change and movement disorder following gasoline inhalation. J Clin Psychiatry 43: 70–72.

Goldman RH, White R, Kales SN et al. (1994). Lead poisoning from mobilization of bone stores during thyrotoxicosis. Am J Ind Med 25: 417–424.

Gorell JM, Johnson CC, Rybicki BA et al. (1997). Occupational exposures to metals as risk factors for Parkinson's disease. Neurology 48: 650–658.

Gorell JM, Peterson EL, Rybicki BA et al. (2004). Multiple risk factors for Parkinson's disease. J Neurol Sci 217: 169–174.

Grandjean P (1997). Mercurial uncertainties in environmental health. Ann N Y Acad Sci 837: 239–245.

Gulson BL, Mahaffey KR, Jameson CW et al. (1998). Mobilization of lead from the skeleton during the postnatal period is larger than during pregnancy. J Lab Clin Med 131: 324–329.

Gunnarsson LG, Bodin L, Soderfeldt B et al. (1992). A case-control study of motor neurone disease: its relation to heritability, and occupational exposures, particularly to solvents. Br J Ind Med 49: 791–798.

Hanninen H (1982). Behaviour effects of occupational exposure to mercury and lead. Acta Neurol Scand 66: 167–178.

Hryhorczuk DO, Rabinowitz MB, Hessl SM et al. (1985). Elimination kinetics of blood lead in workers with chronic lead intoxication. Am J Ind Med 8: 33–42.

HSE (1998). Health risks management: a guide to working with solvents. HSG188, Health and Safety Executive, Sudbury.

Hu H, Rabinowitz M, Smith D (1998). Bone lead as a biological marker in epidemiologic studies of chronic toxicity: conceptual paradigms. Environ Health Perspect 106: 1–8.

Hu H, Shih R, Rothenberg S et al. (2007). The epidemiology of lead toxicity in adults: measuring dose and consideration of other methodologic issues. Environ Health Perspect 115: 455–462.

Hursh JB, Cherian MG, Clarkson TW et al. (1976). Clearance of mercury (HG-197, HG-203) vapor inhaled by human subjects. Arch Environ Health 31: 302–309.

Iwata T, Mori H, Dakeishi M et al. (2005). Effects of mixed organic solvents on neuromotor functions among workers in Buddhist altar manufacturing factories. J Occup Health 47: 143–148.

Iwata T, Sakamoto M, Feng X et al. (2007). Effects of mercury vapor exposure on neuromotor function in Chinese miners and smelters. Int Arch Occup Environ Health 80: 381–387.

Jankovic J (2002). Essential tremor: a heterogeneous disorder. Mov Disord 17: 638–644.

Jett DA (2011). Neurotoxic pesticides and neurologic effects. Neurol Clin 29: 667–677.

Jimenez-Jimenez FJ, De Toledo-Heras M, Alonso-Navarro H et al. (2007). Environmental risk factors for essential tremor. Eur Neurol 58: 106–113.

Jimenez-Jimenez FJ, Alonso-Navarro H, Garcia-Martin E et al. (2013). Update on genetics of essential tremor. Acta Neurol Scand 128: 359–371.

Kaukiainen A, Hyvarinen HK, Akila R et al. (2009). Symptoms of chronic solvent encephalopathy: Euroquest questionnaire study. Neurotoxicology 30: 1187–1194.

Keski-Santti P, Mantyla R, Lamminen A et al. (2009). Magnetic resonance imaging in occupational chronic solvent encephalopathy. Int Arch Occup Environ Health 82: 595–602.

Knowles SO, Donaldson WE (1996). Dietary lead alters fatty acid composition and membrane peroxidation in chick liver microsomes. Poult Sci 75: 1498–1500.

Kosnett MJ, Wedeen RP, Rothenberg SJ et al. (2007). Recommendations for medical management of adult lead exposure. Environ Health Perspect 115: 463–471.

Kuhn W, Winkel R, Woitalla D et al. (1998). High prevalence of parkinsonism after occupational exposure to lead-sulfate batteries. Neurology 50: 1885–1886.

Kumar S, Jain S, Aggarwal CS et al. (1987). Encephalopathy due to inorganic lead exposure in an adult. Jpn J Med 26: 253–254.

Levine SP, Cavender GD, Langolf GD et al. (1982). Elemental mercury exposure: peripheral neurotoxicity. Br J Ind Med 39: 136–139.

Liew Z, Wang A, Bronstein J et al. (2014). Job exposure matrix (JEM)-derived estimates of lifetime occupational pesticide exposure and the risk of Parkinson's disease. Arch Environ Occup Health 69: 241–251.

Liu XL, Wang HB, Sun CW et al. (2011). The clinical analysis of mercury poisoning in 92 cases. Zhonghua Nei Ke Za Zhi 50: 687–689.

Louis ED (2005). Essential tremor. Lancet Neurol 4: 100–110.

Louis ED (2008). Environmental epidemiology of essential tremor. Neuroepidemiology 31: 139–149.

Louis ED, Ferreira JJ (2010). How common is the most common adult movement disorder? Update on the worldwide prevalence of essential tremor. Mov Disord 25: 534–541.

Louis ED, Ottman R, Hauser WA (1998). How common is the most common adult movement disorder? Estimates of the prevalence of essential tremor throughout the world. Mov Disord 13: 5–10.

Louis ED, Ford B, Frucht S et al. (2001). Risk of tremor and impairment from tremor in relatives of patients with essential tremor: a community-based family study. Ann Neurol 49: 761–769.

Louis ED, Shungu DC, Chan S et al. (2002). Metabolic abnormality in the cerebellum in patients with essential tremor: a proton magnetic resonance spectroscopic imaging study. Neurosci Lett 333: 17–20.

Louis ED, Jurewicz EC, Applegate L et al. (2003). Association between essential tremor and blood lead concentration. Environ Health Perspect 111: 1707–1711.

Louis ED, Applegate L, Graziano JH et al. (2005). Interaction between blood lead concentration and delta-aminolevulinic acid dehydratase gene polymorphisms increases the odds of essential tremor. Mov Disord 20: 1170–1177.

Louis ED, Factor-Litvak P, Parides M et al. (2006). Organochlorine pesticide exposure in essential tremor: a case-control study using biological and occupational exposure assessments. Neurotoxicology 27: 579–586.

Lucchini R, Cortesi I, Facco P et al. (2002). Neurotoxic effect of exposure to low doses of mercury. Med Lav 93: 202–214.

Manto M (2012). Toxic agents causing cerebellar ataxias. Handb Clin Neurol 103: 201–213.

Marshall VL, Reininger CB, Marquardt M et al. (2009). Parkinson's disease is overdiagnosed clinically at baseline in diagnostically uncertain cases: a 3-year European multicenter study with repeat [^{123}I]FP-CIT SPECT. Mov Disord 24: 500–508.

Mcconnell P, Berry M (1979). The effects of postnatal lead exposure on Purkinje cell dendritic development in the rat. Neuropathol Appl Neurobiol 5: 115–132.

Mcfarland RB, Reigel H (1978). Chronic mercury poisoning from a single brief exposure. J Occup Med 20: 532–534.

Miller J, Chaffin D, Smith R (1975). Subclinical psychomotor and neuromuscular changes to inorganic mercury. Am Ind Hyg Assoc J 36: 725–733.

National Research Council of the National Academies (2012). Potential Health Risks to DOD Firing-Range Personnel from Recurrent Lead Exposure, The National Academies Press, Washington, DC.

Newman EJ, Breen K, Patterson J et al. (2009). Accuracy of Parkinson's disease diagnosis in 610 general practice patients in the West of Scotland. Mov Disord 24: 2379–2385.

NINDS (2012). Tremor Fact Sheet. NIH publication no. 12-4734. Available online at: http://www.ninds.nih.gov/disorders/tremor/detail_tremor.htm (accessed February 2, 2015).

NIOSH (2013). Organic Solvents. NIOSH Pocket Guide to Chemical Hazards, Available online at: http://www.cdc.gov/niosh/npg/ (accessed February 2, 2015).

Nunes M, Tajara E (1998). Efeitos tardios dos praguicidas organoclorados no homem. (Delayed effects of organochlorine pesticides in man.). Rev Saude Publica 32: 372–382.

Orti-Pareja M, Jimenez-Jimenez FJ, Miquel J et al. (1996). Reversible myoclonus, tremor, and ataxia in a patient exposed to methyl ethyl ketone. Neurology 46: 272.

Paglia DE, Valentine WN, Dahlgren JG (1975). Effects of low-level lead exposure on pyrimidine 5'-nucleotidase and other erythrocyte enzymes. Possible role of pyrimidine 5'-nucleotidase in the pathogenesis of lead-induced anemia. J Clin Invest 56: 1164–1169.

Papageorgiou SG, Karantoni E, Pandis D et al. (2009). Severe dopaminergic pathways damage in a case of chronic toluene abuse. Clin Neurol Neurosurg 111: 864–867.

Patrick L (2006). Lead toxicity, a review of the literature. Part 1: Exposure, evaluation, and treatment. Altern Med Rev 11: 2–22.

Piikivi L, Hanninen H, Martelin T et al. (1984). Psychological performance and long-term exposure to mercury vapors. Scand J Work Environ Health 10: 35–41.

Plumb M, Bains P (2006). What is essential tremor? Essential Tremors: the facts, Oxford University Press, Oxford.

Rabinowitz MB (1991). Toxicokinetics of bone lead. Environ Health Perspect 91: 33–37.

Rafati-Rahimzadeh M, Rafati-Rahimzadeh M, Kazemi S et al. (2014). Current approaches of the management of mercury poisoning: need of the hour. Daru 22: 46.

Ridgway P, Nixon TE, Leach JP (2003). Occupational exposure to organic solvents and long-term nervous system damage detectable by brain imaging, neurophysiology or histopathology. Food Chem Toxicol 41: 153–187.

Riess ML, Halm JK (2007). Lead poisoning in an adult: lead mobilization by pregnancy? J Gen Intern Med 22: 1212–1215.

Rotenberg M, Shefi M, Dany S et al. (1995). Differentiation between organophosphate and carbamate poisoning. Clin Chim Acta 234: 11–21.

Sabbath EL, Glymour MM, Berr C et al. (2012). Occupational solvent exposure and cognition: does the association vary by level of education? Neurology 78: 1754–1760.

Sabbath EL, Gutierrez LA, Okechukwu CA et al. (2014). Time may not fully attenuate solvent-associated cognitive deficits in highly exposed workers. Neurology 82: 1716–1723.

Schulte PA, Burnett CA, Boeniger MF et al. (1996). Neurodegenerative diseases: occupational occurrence and potential risk factors, 1982 through 1991. Am J Public Health 86: 1281–1288.

Semple S (2004). Dermal exposure to chemicals in the workplace: just how important is skin absorption? Occup Environ Med 61: 376–382.

Seshia SS, Rjani KR, Boeckx RL et al. (1978). The neurological manifestations of chronic inhalation of leaded gasoline. Dev Med Child Neurol 20: 323–334.

Stolze H, Petersen G, Raethjen J et al. (2001). The gait disorder of advanced essential tremor. Brain 124: 2278–2286.

Takebayashi T, Omae K, Ishizuka C et al. (1998). Cross sectional observation of the effects of carbon disulphide on the nervous system, endocrine system, and subjective symptoms in rayon manufacturing workers. Occup Environ Med 55: 473–479.

Taylor JR (1982). Neurological manifestations in humans exposed to chlordecone and follow-up results. Neurotoxicology 3: 9–16.

Taylor JR, Selhorst JB, Houff SA et al. (1978). Chlordecone intoxication in man. I. Clinical observations. Neurology 28: 626–630.

Todd AC, Chettle DR (1994). In vivo X-ray fluorescence of lead in bone: review and current issues. Environ Health Perspect 102: 172–177.

Valpey R, Sumi SM, Copass MK et al. (1978). Acute and chronic progressive encephalopathy due to gasoline sniffing. Neurology 28: 507–510.

Van Der Hoek JA, Verberk MM, Van Der Laan G et al. (2001). Solvent-induced chronic encephalopathy; the 'solvent team' project. Ned Tijdschr Geneeskd 145: 256–260.

Van Der Laan G, Sainio M (2012). Chronic solvent induced encephalopathy: a step forward. Neurotoxicology 33: 897–901.

Van Valen E, Van Thriel C, Akila R et al. (2012). Chronic solvent-induced encephalopathy: European consensus of neuropsychological characteristics, assessment, and guidelines for diagnostics. Neurotoxicology 33: 710–726.

Verberk M, Salle H, Kemper C (1986). Tremor in workers with low exposure to metallic mercury. Am Ind Hyg Assoc J 47: 559–652.

Vroom FQ, Greer M (1972). Mercury vapour intoxication. Brain 95: 305–318.

Walsh TJ, Tilson HA (1984). Neurobehavioral toxicology of the organoleads. Neurotoxicology 5: 67–86.

Weisskopf MG, Weuve J, Nie H et al. (2010). Association of cumulative lead exposure with Parkinson's disease. Environ Health Perspect 118: 1609–1613.

Weuve J, Press DZ, Grodstein F et al. (2013). Cumulative exposure to lead and cognition in persons with Parkinson's disease. Mov Disord 28: 176–182.

WHO (1991). Inorganic mercury. Environmental Health Criteria, World Health Organization, Geneva.

Williamson AM, Teo R, Sanderson J (1982). Occupational mercury exposure and its consequences for behavior. Int Arch Occup Environ Health 50: 273–286.

Wills AJ, Jenkins IH, Thompson PD et al. (1995). A positron emission tomography study of cerebral activation associated with essential and writing tremor. Arch Neurol 52: 299–305.

Winegar DA, Levy BS, Andrews Jr JS et al. (1977). Chronic occupational exposure to lead: an evaluation of the health of smelter workers. J Occup Med 19: 603–606.

Winkel R, Kuhn W, Przuntek H (1995). Chronic intoxication with lead- and sulfur compounds may produce Parkinson's disease. J Neural Transm Suppl 46: 183–187.

World Health Organization International Agency for Research on Cancer (1997). IARC MONOGRAPHS 1997. Mercury and Mercury Compounds. Beryllium, Cadmium, Mercury, and Exposures in the Glass Manufacturing Industry. IARC, Lyon.

Yassin AS, Martonik JF, Davidson JL (2004). Blood lead levels in U.S. workers, 1988–1994. J Occup Environ Med 46: 720–728.

Young RS, Grzyb SE, Crismon L (1977). Recurrent cerebellar dysfunction as related to chronic gasoline sniffing in an adolescent girl. Lead poisoning from "leaded" gasoline as an attendent complication. Clin Pediatr (Phila) 16: 706–708.

Yucel M, Takagi M, Walterfang M et al. (2008). Toluene misuse and long-term harms: a systematic review of the neuropsychological and neuroimaging literature. Neurosci Biobehav Rev 32: 910–926.

Section 4

Toxic neuropathies

Handbook of Clinical Neurology, Vol. 131 (3rd series)
Occupational Neurology
M. Lotti and M.L. Bleecker, Editors

Chapter 16

Clinical description of toxic neuropathies

ANN A. LITTLE AND JAMES W. ALBERS*
Department of Neurology, University of Michigan Health System, Ann Arbor, MI, USA

INTRODUCTION

Toxic neuropathies, particularly those encountered in the workplace, represent relatively rare but important conditions. While toxic exposures from occupational, biologic, or environmental exposures are rare (with the exception of those related to medications) compared to a multitude of other causes, they are typically considered in any workup of neuropathy without clear cause. The term "toxic-metabolic" is often used indiscriminately in cases where no cause of neuropathy is apparent, with the implication that exposure to some toxin might be the driving force resulting in metabolic dysregulation leading to peripheral nerve damage. Some of the most common neuropathies are in fact toxic (e.g., alcoholic neuropathy) or metabolic (e.g., diabetic neuropathy). Other common causes include genetic, inflammatory, or systemic etiologies; only rarely is a toxic cause of neuropathy identified (Dyck et al., 1981).

There is increasing awareness and concern in the general public and among healthcare professionals of the adverse effects of chemical exposures, particularly in the workplace. New processes in manufacturing, food processing, and other industries have led to increased risk of exposures to potential neurotoxins. The relationship of these potential neurotoxins to a peripheral neuropathy is difficult to establish. While there is a great deal of published literature on the subject of various industrial toxins and their potential effects, much of the literature is flawed by imprecision of reporting (as in anecdotal reports) or, in the case of formal studies, inadequate study design with incomplete characterization of patients (symptoms, signs, medical comorbidities, and hereditary processes), insufficient concern for confounding factors related to other exposures (including use of alcohol and medications), and attribution of nonspecific "neuropathic" complaints to the toxic exposure. Careful epidemiologic studies with subjects matched for age, sex, level of education, and cultural background are necessary but rarely performed. Formal studies of exposures to a toxin can be particularly problematic. The level of exposure of individuals to a toxin is difficult to quantify: occupational exposures are often chronic and low level, may be episodic, and typically are not followed through air sampling or biological sampling programs. Patient complaints may be nonspecific.

The consideration of a toxic neuropathy through an occupational exposure may arise from known exposure to a toxic compound or may arise from consideration of causes of neuropathy when no other etiology is evident. There is unfortunately little about the characteristics of a toxic neuropathy that distinguish it from neuropathies of other broad etiologies. Further confounding the evaluation is the fact that up to 20–25% of peripheral neuropathies may remain idiopathic in nature, leaving questions in patients' and examiners' minds as to whether something "toxic" has been missed. Though toxic neuropathies are rare, particularly related to occupational exposures, the search for a toxic exposure is driven not only by a desire for explanation but also the possibility of halting or reversing a toxic neuropathy by removal from exposure and/or treatment. Removal from exposure to a toxin may preserve peripheral nerve function and even allow recovery (Sahenk, 1987; Schaumburg and Spencer, 1987). On a scientific level, toxic neuropathies are instructional in elucidating neuropathologic mechanisms of peripheral nerve dysfunction. Recognition of toxicity through review of clinical and laboratory manifestations in clearly affected individuals and experimentation with toxic exposures in animal models has enhanced our understanding of specific metabolic pathways that, when disrupted, cause peripheral neuropathy.

The purpose of this chapter is to discuss a standard approach to the clinical and electrodiagnostic evaluations

*Correspondence to: James W. Albers, M.D., Ph.D., University of Michigan Health System, F2647 UH South, 1500 East Medical Center Drive, Ann Arbor, MI 48109-5223, USA. Tel: +1-734-936-8586, Fax: +1-734-936-5185, E-mail: jwalbers@umich.edu

of a suspected peripheral neuropathy – an approach that may be applied to any neuropathy, toxic or otherwise. The methodology for establishing causality in the case of a known toxic exposure will be discussed. Common toxins encountered in occupational exposures will be reviewed to familiarize the reader with key clinical, laboratory, and electrodiagnostic features to aid in diagnosis. In particular, a classification scheme based on electrodiagnostic criteria will be introduced to aid in narrowing a differential diagnosis and suggesting alternative diagnoses to investigate.

EVALUATION

The evaluation of a toxic neuropathy proceeds as for any neuropathy of unknown cause. Cause is not required to proceed with evaluation and, in fact, a presumed cause may affect outcome of the evaluation, as alternative causes may not be included in a differential diagnosis. A first step is to verify that a patient actually has peripheral nerve involvement. The clinician uses history, clinical examination, and electrodiagnostic examination to localize a patient's symptoms and signs to the peripheral nervous system and then further characterize what part of the peripheral nervous system is involved: anterior horn cells, nerve roots, plexus, sensory or motor axons, muscle, or neuromuscular junction. The electrodiagnostic findings, considered with history and clinical examination, lead to a differential diagnosis, which may include a toxic neuropathy and which subsequently suggests further testing. Further testing may focus the differential diagnosis even further. Ideally, this iterative approach leads to a final diagnosis, which may at times include a disorder not considered initially. In the event that a toxic neuropathy is identified, early diagnosis may allow identification of other individuals at risk from the same toxic exposure. Early diagnosis can be important in final neurologic outcome. With removal of exposure to the offending agent, many toxic neuropathies resolve.

CLINICAL EXAMINATION

Evaluation begins with the clinical examination, focusing on careful history and physical examination. The history establishes time course and order of appearance of neuropathic symptoms or signs noticed by the patient. A careful history should include questions about sensory, motor, and autonomic complaints. Reported symptoms may be positive or negative in nature. Positive symptoms may include paresthesias or allodynia; negative symptoms may include numbness or weakness. Subtle sensory issues may be reported as loss of balance. Weakness may be reported as clumsiness or stiffness. Perceived weakness ("I drop things a lot") may actually represent sensory loss – the patient does not have appropriate sensory feedback to subtly adjust strength of grip. Fatigue may also be interpreted by the patient as weakness. Symptoms reported may be under- or overemphasized by the patient based on subjective experience. It is useful to ask how symptoms have impacted daily life: are there activities that can no longer be physically accomplished or which are more difficult or take longer?

Toxic neuropathies typically impact the largest and longest axons, the dysfunction of which results in numbness, paresthesias, or weakness in a length-dependent (stocking-glove) pattern (Schaumburg and Spencer, 1987). This is unfortunately a common, nonspecific pattern that can be seen in neuropathies from a variety of causes (the most common cause of which is diabetes). If a toxin is suspected as the etiology, directed questions should be asked about potential exposures in the workplace or at home (recreational or environmental exposures). Careful review of patients' medications, including nonprescription supplements (often overlooked by physicians) should be undertaken.

Further information that must be assessed within the clinical evaluation includes social history, family history, and a general review of systems. Particular important aspects of social history are personal use of alcohol, drugs, or recreational substances. These possibilities should be kept in mind even if denied, as patients can be reluctant to admit to them. Occupational and recreational history may reveal potential for toxic exposures. Location of home (rural or suburban), water supply (well or city), and travel experiences should be noted in case they provide clues to toxic or infectious exposures. Family history is always important, though not always well known by the patient. Family members with similar complaints to those of the patient raise the question of hereditary neuropathy. Finally, a comprehensive review of systems with the patient may identify clues to a systemic process (whether toxic, infectious, autoimmune, hereditary, or neoplastic). This could include skin changes, hair loss, fatigue, rash, vision changes, and autonomic complaints – essentially anything new or different noted, but possibly dismissed as irrelevant, by the patient. In particular, involvement of other organ systems is a feature that raises the question of a toxic exposure.

The neurologic examination provides more objective evidence of neuropathy. Sensation is often affected early in toxic neuropathies. The larger sensory axons subserve vibration, joint position, and light touch sensation; smaller sensory axons are responsible for pain and temperature sensation. Quantification of sensory deficits is difficult and confounded by the subjective nature of a patient's perception and reporting. Light touch is particularly difficult to quantify and is frequently abnormal and reported early in large-fiber neuropathies.

int position abnormalities typically appear later in rge-fiber neuropathies – often not until they are mod- ately severe. The assessment of joint position can be hanced by having the patient undergo Romberg test- g (standing still with feet together and eyes closed). omberg testing presupposes that vestibular disorders ave been excluded. Small-fiber dysfunction is best sted with pinprick, which activates pain pathways erved by those axons. Exclusively small-fiber neuropa- hies are rarely caused by toxins.

While the sensory portion of the neurologic examina- ion has a subjective component, there is valuable objec- ive information to be gained from testing of strength, eflexes, balance, and gait. Strength may be quantified, uch as with a grip dynamometer; however, this test is ubject to variation based on effort of the patient, either onsciously or subconsciously. Manual muscle testing y confrontation is the best determinant of weakness, as an experienced examiner is typically able to differen- tiate true weakness from "give-way" of poor or collaps- ing effort (Albers, 2003). When motor axons are involved, greatest weakness is typically seen in the lower extremities – related again to the greater involvement of longer axons in most neuropathies. Reflexes are not under voluntary control and their presence or absence is quite useful in establishing presence of peripheral neuropathy. Reflexes obtained with facilitation are not necessarily abnormal. Visual examination of balance and gait by an experienced examiner is also informative. A systematic neurologic examination is crucial in focusing subsequent laboratory and electrodiagnostic examination.

The general physical examination should not be overlooked, as some neuropathies are manifestations of a systemic process, which may affect other organ systems and give clues to diagnosis. Careful examination of skin, hair, eyes, and nails, for example, can give clues to an underlying systemic process. Structural abnormalities (high arches, hammertoes) can be clues to hereditary neuropathies. There are features that may suggest a toxic neuropathy (such as Mees' lines in arsenic toxicity); however, often these features appear long after neuropathy has developed, thus limiting usefulness in early diagnosis and removal of patient from exposure.

ELECTRODIAGNOSTIC EVALUATION

If the clinical history and examination are consistent with a peripheral neuropathy, the next step is electrodiagnostic evaluation with nerve conduction studies and needle electromyography (collectively referred to here as EMG). EMG is considered an extension and critical part of the clinical examination when evaluating neuropathy. It should never be performed in isolation. It allows not only documentation of neuropathy but localizes abnormalities of peripheral nerves anatomically, thus focusing and narrowing the differential diagnosis to some degree. A secondary role is identification of potential pathophysiology of nerve damage – this also focuses the differential diagnosis. EMG may additionally shed light on time course, severity, and activity of the underlying neuropathic process and additionally can identify pathophysiologic mechanisms (Albers, 1993). EMG is considered a "gold standard" for evaluation of peripheral neuropathies. The exception is in the evaluation of small axons and sensory receptors; however these are rarely involved in toxic neuropathies.

Nerve conduction studies are the most important part of the electrodiagnostic examination for neuropathy and include assessment of sensory nerves with sensory nerve action potentials (SNAPS) and motor nerves with compound motor unit action potentials (CMAPS). SNAPs and CMAPs for any given nerve have amplitudes that are direct (SNAPs) and indirect (CMAPs) measures of number of intact axons within a nerve. Loss of axons will result in responses of reduced amplitude, though it is notable that significant loss of axons has to occur before amplitude appreciably decreases. The needle examination is more sensitive for motor axonal loss early in a neuropathic process. Computed conduction velocities are typically measures of the integrity of the myelin sheath, the caveats being that dropout of the largest axons can cause mild slowing of conduction velocity and processes impacting ion channels (e.g., hyperglycemia or cold temperature) can also manifest as mild slowing of conduction velocity. The duration of an action potential reflects synchrony of nerve impulses passing under a recording electrode. Latency of response reflects time from nerve stimulation to recording of the initial impulse arriving at a distant site. In cases where defect in nerve transmission of impulse in a motor nerve is suspected, but distal-extremity nerve conduction studies are normal, the integrity of the proximal motor nerve can be assessed with F-wave response, in which a motor nerve is stimulated at a distal site and time of transmission to the spinal cord and back is measured. Since the pathway along the entire nerve and back is long, conduction abnormalities are accentuated. This technique is primarily used when proximal demyelination is suspected. Blink reflexes target peripheral and central reflex pathways that are activated when the trigeminal nerve is stimulated peripherally. Responses are recorded from facial muscles when the facial nerve is activated in this pathway. The pattern of response latency can localize a lesion within the peripheral or central nervous system, though evaluation of the peripheral system is limited to the trigeminal and facial nerves.

Motor nerves consist of large myelinated axons, whereas sensory nerves consist of both large and small axons. The sensory nerves tested with standard nerve conduction studies are large myelinated axons, which convey touch, vibratory perception, and joint perception. Nerve conduction studies are unable to assess the small sensory axons, which are diffusely distributed and convey pain and temperature or subserve the autonomic nervous system. Direct assessment of small-axon damage or loss is made by visual inspection of biopsied skin tissue for intraepidermal nerve fiber density. Within the autonomic system, sympathetic responses mediated by small sensory axons may be assessed using skin electrodes and measuring galvanic responses to stimuli such as startle or electric stimulation. The presence of a response argues against substantial abnormality of autonomic sudomotor function. More detailed testing of the autonomic nervous system can be accomplished with quantitative sudomotor axon reflex testing (QSART) and with tilt table blood pressure recordings.

It is occasionally necessary to assess for a defect in neuromuscular transmission as impaired neuromuscular transmission can result in weakness clinically and low-amplitude motor responses. This electrophysiologic assessment is done with repetitive stimulation studies. Motor nerves are stimulated at low (2–3 Hz) and at high (20–50 Hz) frequencies. Low-frequency stimulation of a motor nerve results in a decrement of amplitude of response if there is defective neuromuscular transmission. The repetitive nerve stimulation challenges the neuromuscular junction, where physiologically there is a decrease in availability of acetylcholine after a discharge, as mobilization of acetylcholine stores must occur. In normal neuromuscular transmission, there is a "safety factor," such that enough quanta of acetylcholine are available to fill an adequate percentage of acetylcholine receptors to result in all-or-none firing of all muscle fibers in continuity with the nerve in question. A presynaptic defect (inability to release acetylcholine in sufficient quantities or fast enough) or postsynaptic defect (reduced number of or, in the case of some toxins, blocked acetylcholine receptors) results in failure to reach all-or-none threshold for muscle fiber activation in a percentage of muscle fibers, resulting in a decremental response with sequential stimulation. Subsequent repetitive nerve stimulation at high frequencies can differentiate between a presynaptic or postsynaptic defect, as the high-frequency stimulation will overcome a presynaptic defect of acetylcholine release, resulting in facilitation of the response amplitude into normal range. No effect or further decrement will be seen in a postsynaptic defect. There are toxins encountered occupationally which may affect the neuromuscular junction, typically involving the postsynaptic neuromuscular junction. Acute organo-phosphate poisoning is a prime example of a toxin causing blockade of neuromuscular junction transmission.

Laboratory standards for nerve conduction studies for normal controls of various age groups are used to assess whether a response is normal or abnormal. In the case of abnormal responses, the nerve conduction studies can localize the disease process anatomically. The anatomical localization, in broad terms, defines the process as diffuse, focal, or multifocal. Involvement may be of sensory nerves, motor nerves, or a combination of sensory and motor nerves. The disease process may be further characterized as an axonal or demyelinating process, or in some cases, localize to muscle or the neuromuscular junction.

The needle examination of muscles is of secondary importance, but can provide further information of acuity or chronicity of nerve loss when motor nerves are involved. The needle examination can verify or refute a length-dependent neuropathic process. With the ability to examine proximal muscles, including paraspinal muscles, nerve root involvement can be assessed – this may be important in establishing localization of a disease process or ruling out radiculopathy as source of symptoms. The needle examination is sensitive for axonal loss and may reveal denervation of muscles early in a disease process before decline of amplitude of CMAPs on nerve conduction studies. Needle insertion into quiet muscle reveals abnormal spontaneous activity consisting of positive waves and fibrillation potentials. These may persist for weeks to months after initial appearance. The caveat is that, in an acute process, initial abnormalities of spontaneous activity on needle exam may not appear for 1–4 weeks, depending on the distance muscle being tested is from the site of nerve transection (Gilliatt and Taylor, 1959). The greater the distance, the longer is the time until fibrillation potentials are seen. This is related to the time it takes for Wallerian degeneration to be completed. Conduction along distal axons may remain essentially normal for days after a transection, delaying the development of the hypersensitivity that is the basis of fibrillation potentials. The ability of the distal portion of a transected nerve to conduct a stimulus resulting in muscle contraction for several days after nerve transection was demonstrated experimentally (Landau, 1953). The amplitudes of the elicited motor (or sensory) responses diminish over a number of days until lost. There is concomitant decline of conduction velocities during this process as largest, fastest-conducting axons are lost. At this point, fibrillation potentials may begin to appear. Fibrillation potentials result from hypersensitivity of muscle fibers to acetylcholine. With acute denervation – loss of contact of axons with their muscle fibers – there is proliferation and reorganization of acetylcholine receptors on the surface of muscle cells (Drachman, 1986). The reorganization typically occurs over the course of a week or more. The amplitude of the spontaneous activity declines over time as muscle fibers that have not been reinnervated undergo atrophy (Albers, 1993). In a chronic denervating process, needle EMG examination will reveal ongoing abnormal spontaneous activity of relatively small amplitude in quiet muscle and neurogenic voluntary motor units when muscle is activated. With ongoing denervation – an acute process on a chronic process – fibrillation potentials may be variable in their amplitude.

Neurogenic voluntary motor units are motor units of increased amplitude, prolonged duration, and decreased recruitment. Depending on whether the neurogenic insult is ongoing or when it occurred, there may be significant polyphasia as reinnervation occurs. With completed reinnervation, abnormalities of amplitude, recruitment, and duration persist, but morphology of the voluntary motor unit once again becomes triphasic (normal). The evolution of morphology and recruitment changes of voluntary motor units over time provide more insight into timing, severity, and extent of neuropathic involvement than do the nerve conduction studies (Albers, 1993).

The needle examination is also sensitive for muscle fiber necrosis. Muscle fiber necrosis cannot be differentiated from acute denervation on the basis of abnormalities of spontaneous activity, but the difference between a neuropathic and myopathic process is typically appreciated when voluntary activation of muscle is assessed, through characterization of motor unit amplitude, duration, phases, and recruitment. Single-fiber EMG is a special needle study used to assess for neuromuscular junction defect. It is, in fact, the most sensitive electrodiagnostic test for neuromuscular junction defect. Its application and importance in toxic disorders are limited, as repetitive nerve studies are typically sufficient to diagnose such a defect.

The initial EMG is done for the purpose of documenting the presence of, localizing, and characterizing neuropathy. It may be repeated to assess for interval progression or improvement of neuropathy. This can be quite helpful, as some neuropathies, including some toxic neuropathies (see later description of arsenic neuropathy), initially present in the acute phase with an electrodiagnostic picture consistent with demyelination, but are actually axonal in nature, as documented by later electrodiagnostic evaluation.

The degree and pattern of involvement of abnormality on EMG cannot be used to diagnose a specific neuropathy, but do focus a differential diagnosis and thus direct further workup. This applies not only to toxic neuropathies, but also to neuropathies of any etiology. Using electrodiagnostic abnormalities, neuropathies can be categorized into those with pure sensory, pure motor, or a combination of motor and sensory involvement. The type of pathophysiologic abnormalities of diseased nerves is limited and has been broadly classified into axonal degeneration, axonal stenosis, demyelination, and ion channel abnormalities (Aminoff and Albers, 1999).

Axonal degeneration leads to gradual decline of amplitude of nerve responses over time, with or without mild slowing of latency of responses. Concentric needle examination is remarkable for fibrillation potentials and neurogenic voluntary motor units. The most common pattern of axonal degeneration is a length-dependent one – an axonopathy – in which the longest axons degenerate as the nerve cell body is unable to sustain their metabolic requirements. This metabolic failure is postulated to be due to inability to transport one or more essential nutrients needed for health of the distal axon (Schaumburg et al., 1983). A length-dependent axonopathy is the most common pattern seen in most toxic neuropathies, though by no means is exclusive to toxic neuropathies.

Axonal stenosis describes greatly reduced caliber of distal-most axons, potentially reflecting atrophy of the distal axon prior to complete axonal degeneration. Velocity of conduction of nerve impulses is reduced consistent with reduction of the axonal diameter. As long as the axon continues to conduct a response, the amplitude of that response may remain normal but with prolonged latency. Muscle fibers remain in contact with their axons, preserving normal membrane mechanics and voluntary motor units demonstrate normal recruitment (Albers, 1993). This somewhat theoretical construct may explain why terminal latencies can be significantly prolonged with more proximal latencies and overall conduction velocity unaffected. Alternatively, some or all of the small-diameter distal-most axons could represent regenerating nerve fibers.

Demyelination (or dysmyelination) results in significant conduction slowing in axons. Disruption of the myelin sheath interferes with saltatory conduction of nerve impulses from node to node along the axon and produces areas of conduction block. Disruption of the myelin sheath may be through paranodal demyelination or segmental demyelination. Models of paranodal demyelination through focal compression suggest that under such conditions a myelin segment may "telescope" under another segment (Ochoa et al., 1972; Fowler and Ochoa, 1975) with subsequent blockage of current (ions) at the site, resulting in slowing or blocking of action potential (Ochoa, 1980). It is possible to conceive that some toxins might disrupt the tightly packed layers of myelin, allowing for compression and "telescoping" and subsequent conduction slowing or block. Alternatively, conduction slowing can occur separately from physical damage to the myelin sheath. Alternative reasons for conduction slowing include as benign a phenomenon as decreased temperature in which slowing of stimulus propagation results from the prolonged opening and closing of ion channels during depolarization of the axon. In the metabolic realm, transient hypoglycemia or hyperglycemia can produce conduction slowing (Troni et al., 1984), which resolves with correction of blood glucose levels. Inactivation of ionic channels through any means may result in conduction slowing without demyelination. There are a number of toxins that may affect ionic channels and lead to conduction slowing. Thus it is important to identify conduction slowing and carefully assess for a mechanism to explain it – primary demyelination, metabolic dysregulation, or channelopathy.

Conduction slowing is assessed on the basis of segmental conduction velocity, F-wave latency, and distal latency. Typically axonal neuropathies have little conduction slowing unless severe, in which case there is mild slowing secondary to dropout of the largest axons. Criteria for demyelination are reduced conduction velocity of motor nerves greater than 80% of the lower limit of normal or distal latencies and F-wave latencies exceeding 125% of the upper limit of normal (Cornblath et al., 1991). These abnormalities should be identified in several motor nerves. In general, slowing of conduction velocity of less than 70% of the lower limit of normal cannot be explained exclusively by axonal loss (Kelly, 1989). If temporal dispersion or conduction block is seen in the electrodiagnostic examination, this is evidence for acquired demyelination. However, it may not be present or be demonstrated on electrophysiologic testing.

Particularly in some toxic neuropathies, which by their nature are acquired conditions, there are atypical findings. Two explanations have been offered for the atypical findings in some toxic neuropathies in which conduction slowing is seen. One explanation is that early in the course of a rapidly progressive severe axonal neuropathy, before total axonal loss has occurred, nerve conduction studies may meet criteria for acquired demyelination. This is seen in severe vasculitic neuropathy, for example (Donofrio et al., 1987). Another explanation is that a toxin may primarily target the largest motor fibers. With dropout of largest and fastest-conducting axons, substantial conduction slowing may occur, mimicking hereditary dysmyelination, though the time course of progression may suggest otherwise.

Using a combination of the features discussed above – type of nerve involvement (motor and/or sensory), pattern of nerve involvement (diffuse, focal, multifocal), and presence of axonal degeneration versus conduction slowing – differential diagnoses can be developed which can be targeted for further testing. The differential diagnoses suggested by the above classifications are summarized in Tables 16.1–16.5. Note that several of the conditions, including some neurotoxicants, produce different forms of neuropathy, depending on the severity of the timing of the evaluation *vis-à-vis* onset.

A word of caution should be inserted here as to performance of EMG. It is of utmost importance that nerve conduction techniques are performed in a standardized manner and care is taken to avoid technical errors. There are a number of factors that can affect the results of nerve conduction studies and could potentially give false results that would confound an investigation. These factors include limb temperature, errors of measurement in approximating nerve length, stimulation parameters, technical issues with regard to machine settings and grounding, and patient factors of age, height, body habitus, and occupation. More detailed discussion of technical issues affecting quality and accuracy of nerve conduction studies can be reviewed elsewhere (Albers, 2005a).

ELECTRODIAGNOSTIC CLASSIFICATION OF NEUROPATHY

The following broad categories of electrodiagnostic classification of neuropathy may guide the resultant differential diagnosis. Clinical features of the patient including the temporal profile of progression of a neuropathy can further narrow diagnostic possibilities and suggest further testing. The differential diagnoses suggested by different patterns of the electrodiagnostic results are summarized in the following paragraphs.

Mixed sensory and motor neuropathy with no conduction slowing

A differential diagnosis for this type of neuropathy is given in Table 16.1. This sensorimotor neuropathy is generally nondescript, at least in terms of clinical and electrodiagnostic findings. Such findings are encountered frequently in clinical practice, arising from a broad range of etiologies and including a large number of neurotoxicants. For most, sensory symptoms and signs predominate and the motor component is comparatively mild. Although a common cause of this electrodiagnostic pattern is hereditary (axonal forms of Charcot–Marie–Tooth that include sensory loss), other causes include nutritional, autoimmune, metabolic, and nutritional disorders. In terms of neurotoxicants, a number of medications are included, as well as acrylamide, arsenic, and ethyl alcohol (all in the chronic stages of neuropathy). This type of neuropathy is the one most likely to remain idiopathic in origin after extensive testing.

Table 16.1

Mixed sensory and motor neuropathy with no conduction slowing (Donofrio and Albers, 1990; Albers, 2005c)

Hereditary/familial	**Toxic**
Hereditary motor and sensory neuropathy (HMSN II)	Acrylamide
Infectious	Amitriptyline
Human immunodeficiency virus (HIV)	Arsenic (chronic)
Leprosy	Carbon monoxide
Inflammatory (autoimmune)	Chloroquine
Churg–Strauss	Colchicine (neuromyopathy)
Connective tissue disorders	Ethambutol
Gout	Ethyl alcohol
Necrotizing angiopathy	Ethylene oxide
Periarteritis	Elemental mercury
Rheumatoid arthritis	Gold
Sarcoidosis	Hydralazine
Systemic lupus erythematosus	Isoniazid
Metabolic	Lithium
Acromegaly	Metronidazole
Hypothyroidism	Nitrofurantoin
Neoplastic (paraneoplastic)	Nitrous oxide (myeloneuropathy)
Amyloidosis	Paclitaxel
Carcinoma	Perhexiline
Lymphoma	Phenytoin
Monoclonal gammopathy of unknown significance (MGUS)	Vincristine
Multiple myeloma	
Nutritional	
Critical illness neuropathy	
Vitamin deficiency (e.g., folate, B_{12}, thiamine, postgastrectomy)	
Whipple disease	

Motor, motor greater than sensory, or mixed sensorimotor neuropathy with conduction slowing

As seen in Table 16.2, the differential diagnosis encompasses numerous diverse etiologies. It might appear that it would be easy to differentiate a toxic neuropathy from other causes producing the same electrodiagnostic pattern by clinical features. However, this is not always the case. For example, acute arsenic poisoning in its early stages can mimic Guillain–Barré syndrome (GBS), an acute inflammatory demyelinating polyneuropathy, or other neuropathies associated with immune dysregulation, such as monoclonal gammopathies, myeloma, macroglobulinemia, cryoglobulinemia, Castleman's disease, occult malignancy, or human immunodeficiency virus (HIV) infection. Over time, acute arsenic intoxication may result in clinical features of Mees' lines and dermatitis of hands and feet, and electrodiagnostic testing will

Table 16.2

Motor, motor more than sensory, or mixed sensorimotor neuropathy with conduction slowing (Donofrio and Albers, 1990; Albers, 2005c)

Hereditary/familial	**Toxic**
Friedreich's ataxia	Acrylamide
Giant axonal dystrophy	Arsenic (shortly after acute exposure)
Hereditary motor and sensory neuropathy (HMSN I, III)	Amiodarone
Hereditary tomaculous	Carbon disulfide
Myotonic dystrophy	Cytosine arabinoside (ara-C)
Infectious	Disulfiram
Human immunodeficiency virus (HIV)	Methyl *n*-butyl ketone
Inflammatory (autoimmune)	*n*-Hexane
Castleman's disease	Saxitoxin (sodium channel blocker)
Chronic inflammatory demyelinating	Suramin
Diphtheria	Tetrodotoxin (puffer fish)
Guillain–Barré syndrome	
Monoclonal gammopathy	
Porcine neural tissue	
Systemic lupus erythematosus	
Swine flu vaccine	
Vasculitis (confluent mononeuritis)	
Metabolic (or toxic-metabolic)	
Diabetes mellitus	
Uremic (end-stage renal disease)	
Neoplastic (paraneoplastic)	
Monoclonal gammopathy of unknown significance (MGUS)	
Multiple myeloma	
Osteosclerotic myeloma	

become characteristic of a severe length-dependent axonal neuropathy. The metabolic neuropathies associated with diabetes mellitus or uremia are among the conditions included in Table 16.2. The most common form of diabetic neuropathy is a distal symmetric polyneuropathy showing sensory greater than motor involvement. Nevertheless, it is included here because motor conduction slowing is a characteristic early and persistent finding associated with diabetic distal symmetric polyneuropathy. A similar EMG pattern may also be seen in patients with renal failure unrelated to diabetes (Dyck et al., 1975; Said et al., 1983). It is unclear if the conduction slowing in diabetes and uremia represents primary demyelination or axonal membrane dysfunction. In the early stages, it is important to consider the broader differential in order to perform additional testing to arrive at an earlier diagnosis. The conduction slowing associated with neuropathies in this category has several explanations, most often reflecting primary demyelination or, alternatively, loss of the largest myelinated axons.

A third mechanism of conduction slowing in this category is blockade of sodium channels (thus decreasing propagation of action potential), as is seen in ingestion of tetrodotoxin of the puffer fish and saxitoxin of shellfish contaminated by the "red tide" (Jacques et al., 1980; Lombet et al., 1988; Long et al., 1990).

Another toxin to be considered under this pattern of nerve conduction abnormalities is *n*-hexane. A common clinical presentation of acute toxicity is of progressive length-dependent sensory loss, weakness, and autonomic dysfunction. The conduction slowing seen on nerve conduction studies likely reflects myelin damage associated with giant axonal swellings due to accumulation of neurofilaments.

Motor or motor more than sensory neuropathy (or neuronopathy) with no conduction slowing

The differential diagnosis for this pattern of neuropathy is shown in Table 16.3. Hereditary causes for these motor predominant neuropathies (or neuronopathies) are common, and include some forms of porphyria. In this context, a variety of toxicants (including possibly lead) may act to cause porphyric attacks, which in turn result in porphyric neuropathy that is only indirectly related to the toxicant. Among the remaining toxicants causing this pattern of nerve conduction abnormalities, organophosphates are important to consider. Organophosphate esters are used as pesticides and in some "nerve gases" (Davis et al., 1985; Senanayake and Karalliedde, 1987a). Organophosphates inactivate acetylcholinesterase (AChE), with subsequent accumulation of acetylcholine in muscarinic and nicotinic cholinergic neurons (Lotti et al., 1984; Davis et al., 1985; Moretto and Lotti, 1998). In acute intoxication, muscarinic symptoms include miosis, increased secretions, sweating, gastric hyperactivity, and bradycardia. Nicotinic symptoms include fasciculations and skeletal muscle weakness. If exposure is high enough, there is inhibition of plasma butyrylcholinesterase (BuChE) and red blood cell cholinesterase levels. Neuropathy develops within a few weeks – after recovery from symptoms of acute intoxication. For this reason, it is described as organophosphate-induced delayed neurotoxicity (OPIDN) (Lotti et al., 1984; Senanayake and Karalliedde, 1987a).

Table 16.3

Motor or motor more than sensory neuropathy (or neuronopathy) with no conduction slowing (Donofrio and Albers, 1990; Albers, 2005c)

Hereditary/familial	**Toxic**
Hereditary and motor sensory neuropathy (HMSN type II, V)	Cimetidine
Hereditary neuropathy with liability to pressure palsy*	Dapsone*
Kennedy syndrome*	Doxorubicin
Motor neuron disease (MND)*	Lead (possibly)*
Neurotoxic esterase-associated MND	Nitrofurantoin (after initial sensory)
Porphyria (acute intermittent porphyria, coproporphyria, variegate)	Organophosphorus esters (organophosphate-induced delayed neuropathy)
Troyer syndrome	Tick paralysis
Infectious	Vincristine (after initial sensory)
Coxsackie virus*	
Botulism (presynaptic neuromuscular junction blockade)	
Poliomyelitis*	
West Nile virus*	
Inflammatory (autoimmune)	
Axonal Guillain–Barré syndrome	
Diabetic amyotrophy (radiculoplexus neuropathy)*	
Lewis–Sumner variant*	
Multifocal motor neuropathy (MMN)*	
Metabolic	
Critical illness neuropathy	
Hyperinsulin/hypoglycemia	
Neoplastic (paraneoplastic)	
Lymphoma	

*The weakness characterized by these conditions is usually asymmetric.

Sensory predominant neuropathy or neuronopathy with no conduction slowing

This combination of electrodiagnostic findings, in which sensory responses are absent or severely reduced in amplitude, either globally or in a length-dependent fashion, and motor conduction studies and the needle EMG examination are normal, is uncommon relative to the other forms of neuropathy. Typically patients complain of dysesthesias and numbness and develop gait ataxia and choreoathetoid movements. Clinical examination documents loss of reflexes and decreased vibratory and proprioceptive function with relatively preserved small-fiber function. The differential diagnosis based on this electrophysiologic pattern is presented in Table 16.4. Diagnostic consideration can be given to Sjögren's syndrome (perhaps the most common nontoxic cause of this uncommon pattern), Lyme disease, paraneoplastic processes, vitamin deficiencies (especially pyridoxine and tocopherol, whose effects are specific to sensory nerves), the Miller–Fisher variant of GBS, medications (many of which are chemotherapy agents), and alcohol (a common toxic etiology, which can have several presentations as to nerve involvement). Diabetes mellitus is included in this category because clinically evident weakness develops relatively late in the course of a diabetic distal symmetric polyneuropathy, and because some atypical forms of diabetic neuropathy

Table 16.4

Sensory predominant neuropathy or neuronopathy with no conduction slowing (Donofrio and Albers, 1990: Albers, 2005, 2012)

Hereditary/familial	**Toxic**
Amyloidosis*	Cisplatin
Fabry disease*	Ethyl alcohol*
Friedreich ataxia	Metronidazole
Kennedy syndrome	Pyridoxine
Sensory autonomic neuropathy*	Thalidomide
Tangier disease*	Thallium*
Infectious	Vacor
Human immunodeficiency virus (HIV)-associated*	Zinc (myeloneuropathy)
Leprosy	
Inflammatory (autoimmune)	
Fisher variant of Guillain–Barré syndrome	
Idiopathic sensory ganglionitis	
Insulin neuritis (treatment-induced neuritis)*	
Sjögren syndrome	
Metabolic	
Diabetes mellitus*	
Diabetic cachexia*	
Insulin neuritis (treatment-induced)*	
Metabolic syndrome*	
Neoplastic (paraneoplastic)	
Small-cell tumors	
Other malignancies	
Nutritional	
Vitamin deficiency	

*The sensory neuropathy produced by this condition is small-fiber predominant.

exist that are characterized by small-fiber sensory and autonomic involvement (e.g., diabetic cachexia and insulin neuritis). Thallium is unique among the neurotoxicants causing neuropathy in that it affects small-fiber sensory function more than large-fiber function.

Mononeuritis multiplex or asymmetric neuropathy (with or without conduction slowing)

These neuropathies (or neuronopathies) are usually asymmetric, making them unique. Etiologies are listed in Table 16.5. Though "mononeuritis" suggests inflammatory neuropathy, this category contains not only inflammatory neuropathies such as atypical forms of chronic inflammatory demyelinating polyneuropathy (CIDP) such as multifocal acquired demyelinating sensory and motor neuropathy, acute motor axonal neuropathy and multifocal motor neuropathy, it also includes hereditary neuropathies such as hereditary neuropathy with liability for pressure palsies and hereditary neuralgic amyotrophy. Infectious causes of asymmetric neuropathy include Lyme disease. Structural abnormalities (syrinx, for example) may produce asymmetric clinical symptoms of neuropathy (McGonagle et al., 1990).

Table 16.5

Mononeuritis multiplex or asymmetric neuropathy (with or without conduction slowing) (Albers and Teener, 2006)

Hereditary/familial	**Toxic**
Distal myopathy	Dapsone
Hereditary neuropathy with liability to pressure palsies (HNLPP)	Lead
Kennedy syndrome	Porphyria
Porphyria (acute intermittent porphyria, coproporphyria, variegate)	Trichloroethylene (cranial neuritis)
Progressive muscular atrophy	Toxic-oil syndrome
Idiopathic	L-Tryptophan
Atypical motor neuron disease (e.g., Aran–Duchenne)	
Inclusion body myositis	
Inflammatory/infectious	
Acute motor axonal neuropathy (AMAN)	
Diabetic amyotrophy (diabetic radiculoplexus neuropathy)	
Idiopathic brachial neuritis	
Lyme disease	
Multifocal acquired demyelinating sensory and motor (MADSAM)	
Multifocal motor neuropathy (MMN)	
Poliomyelitis	
Vasculitis	
West Nile virus	
Structural	
Cervical spondylosis	
Polyradiculopathy without neuropathy	
Syringomyelia	

Lead has been a toxin associated at times with an asymmetric motor-predominant neuropathy, and lead was considered in the preceding section as potentially precipitating porphyric neuropathy. Case reports of lead intoxication still appear on occasion, yet little pathophysiologic evidence has been presented to confirm its role in producing neuropathy. Reports of lead intoxication in the early 20th century may have been overrepresented due to lack of knowledge of other potential etiologies of asymmetric neuropathy.

Trichloroethylene is a solvent thought to produce an asymmetric neuropathy involving cranial nerves with a predilection for the trigeminal nerve (Feldman, 1970), perhaps due to reactivation of a latent viral infection. Lack of cranial nerve involvement should call the diagnosis of trichloroethylene toxicity into question (Fullerton, 1969b).

FURTHER TESTING

Once a peripheral neuropathy has been diagnosed on the basis of clinical history and examination and has been verified and characterized by EMG, further directed testing may occur. This undoubtedly will include laboratory assessment of organ function in general, but also may be targeted to the differential diagnosis generated by history, clinical examination, and electrodiagnostic testing to establish or rule out the varied diagnoses being entertained. Evaluation of other organ systems may be indicated, such as pulmonary function tests or cardiac evaluations.

Specific laboratory testing for toxic exposure, if suspected, may also be undertaken. This might involve blood or urine testing for acute exposures or hair, nail, skin, or bone evaluation if suspicion exists for certain chronic exposures. Care must be taken in interpretation of toxicity, as elevations of some toxins may be fleeting and inconsequential (as in elevated arsenic or organic mercury levels immediately after consumption of some fish) and underlying neuropathy may actually be due to other causes. Attribution of a peripheral neuropathy to a toxic cause without full consideration of other etiologies may delay diagnosis of other potentially treatable entities.

Further testing may be undertaken to better quantify sensory complaints. Quantitative sensory testing (QST) is one such measure. Its goal is to provide an objective measure of sensory discrimination of temperature, although a subjective component cannot be eliminated, as it requires the cooperation of a motivated patient. Specificity is limited, as the test does not distinguish between central and peripheral sensory disorders. The test, in a properly engaged patient, may provide objective measures of sensory loss; however, it remains complementary to standard nerve conduction studies, which establish peripheral nerve involvement. In a study of patients with diabetes and diabetic neuropathy, QST results were consistent with the presence of peripheral neuropathy; however, abnormalities of sural nerve response were the best predictor of mild neuropathy (Redmond et al., 1992). QST is not currently used on a routine basis in evaluation of peripheral neuropathy and specifically has not been generally undertaken in investigation of toxic neuropathy.

Small-fiber neuropathy may be suspected on the basis of clinical complaints of burning or lancinating pain, usually in distal extremities. Nerve conduction studies do not evaluate these small-diameter nerve fibers which are diffusely present in normal individuals throughout the epidermis and which also innervate sweat glands and are an important component of the autonomic nervous system. The most common cause of small-fiber neuropathy in the USA is pre-diabetes or diabetes, but other associations include autoimmune disorders such as Sjögren's disease, celiac disease, and metabolic dysregulation as seen in renal failure. Diagnosis of small-fiber neuropathy is made in the setting of typical clinical complaints of burning/lancinating pain, clinical examination (often demonstrating hyperalgesia in affected distal limbs), and normal or nearly normal nerve conduction studies, with confirmatory tests of small-nerve fiber dysfunction (QST, for example) (Lacomis, 2002). Pathologic diagnosis in the past relied on sural nerve biopsy with assessment of dropout of smallest unmyelinated axons; however more recently, skin biopsies have been utilized to examine the distal-most portions of small axons – those that terminate intraepidermally after passing through the basement membrane of the epidermis from the dermis. Abnormalities on biopsy are loss of fiber density (with intraepidermal fibers manually counted in tissue of prescribed thickness and length and then compared to findings in biopsies of normal control subjects of comparable age). More subtle early signs of small-fiber dysfunction may be swellings of intraepidermal axons, which precede axonal loss. Length-dependent reduction of intraepidermal nerve fiber density is the most common finding on skin biopsy, though nonlength-dependent patterns are described. One study (Holland et al., 1997) evaluated patients with complaints of burning feet without known etiology in the setting of normal clinical examination and normal EMG. Skin biopsies and sural nerve biopsies were compared. In most patients, the sural nerve biopsy and skin biopsy both had abnormalities; however there was a subset of patients with normal sural nerve biopsies and reduced intraepidermal nerve fiber density, suggesting that skin biopsy may be more sensitive than sural nerve biopsy in documenting the presence of a small-fiber neuropathy. The precise sensitivity, specificity, and reproducibility of findings of skin biopsy have not been defined; nonetheless, there is a place for it when diagnosis of a small-fiber neuropathy remains unclear. It is unfortunately unlikely to be of significant use in evaluation of toxic neuropathies, as few toxins cause a predominantly small-fiber neuropathy.

Sural nerve biopsy should be reserved for specific questions. In most axonal neuropathies, toxic or otherwise, the findings on sural nerve biopsy are those of nonspecific dropout of axons. Sural nerve biopsy is indicated when there is a question of vasculitis or infiltrative disease (i.e., sarcoid, amyloid, and lymphoma) or when confirmation of a demyelinating disorder is sought. Indiscriminate peripheral nerve biopsy is not useful in most cases of neuropathy, including toxic causes. Exceptions exist and include solvent-induced neuropathies (i.e., *n*-hexane intoxication) in which neurofilament aggregates form focal axonal swellings in peripheral nerves. Another example noted historically is result of nerve (and skin, fascia, and muscle) biopsy in patients with eosinophilia-myalgia syndrome associated with L-tryptophan intoxication in which there is perivascular inflammation with lymphocytes and eosinophils. Overall, nerve biopsy may be most useful in investigating neuropathic processes that are alternatives to a toxic neuropathy. Similarly, the utility of biopsy of other tissues is in establishing or ruling out competing diagnoses to that of a toxic neuropathy.

ESTABLISHING CAUSALITY

A major difficulty in dealing with or thinking about potentially toxic exposures is establishing causality. Toxic exposures are a rare cause of neuropathy when all causes are considered, but the term "toxic-metabolic" is commonly used when no evident cause of neuropathy exists, thus keeping the concern for a "missed" toxic exposure at the forefront. A patient may question whether potential exposure to toxins at work has caused neuropathy; alternatively consideration may be given to an exposure to a toxin in the environment derived from industrial waste or emissions, which may be unknown to the patient.

Occupational and environmental exposures to toxins are typically chronic and low level in nature. Acute high-level exposures to toxins occur primarily in industrial accidents. For any toxin to have adverse effect – on nerves or otherwise – a certain level of exposure to the respiratory system, mucous membranes, and/or skin has to occur. Exposure information is rarely available through routine monitoring of air or local environment. Meanwhile, the absolute time of exposure (hours per day – over weeks to months or longer) of a patient to a toxin is difficult to assess. Finally, individuals vary in their tissue susceptibility to toxic effect, based on genetic, physiologic, and acquired characteristics (e.g., nutritional status, obesity). Further muddling the evaluation is the fact that acute, subacute, and chronic exposures to the same toxin may manifest with different symptoms.

Timing of onset of neuropathy relative to potential exposure to toxin is of key importance. Clearly symptoms present prior to exposure to a toxin are not attributable to the toxin and alternative etiologies should be sought. Similarly, a neuropathy manifesting years after limited exposure to a toxin should not be attributed to exposure to that toxin. Agent Orange comes to mind, as a number of veterans developing neuropathy in their later years (60s or 70s typically) worry that their exposure to Agent Orange in the 1960s is the cause of their increasing nerve dysfunction. A toxin that causes neuropathy most always does so at the time of the exposure, and when exposure is ended, the progression of nerve damage stops; some recovery is possible in some instances. The exception to this rule is a phenomenon known as "coasting," in which neuropathy may develop during or late in exposure to a toxin and continues to progress for many weeks and occasionally longer after the toxic exposure is removed. This is a well-known phenomenon in cancer treatment, where many of the chemotherapy drugs may be neurotoxic to nerves to varying degrees.

Though timing may be correct for development of a toxic neuropathy and exposure may involve a toxin known to cause neuropathy, the exposure alone does not establish a toxic neuropathy related to that exposure. Patients who have occupational exposures may have systemic diseases known to be associated with neuropathy (i.e., diabetes, chronic renal failure), may take medications known to cause neuropathy (i.e., phenytoin, amiodarone), or may have personal habits known to predispose to neuropathy (i.e., alcohol consumption to excess), or may have relatives with similar symptoms (suggesting hereditary neuropathy). Prevalence of peripheral neuropathy in general makes it more likely that etiology is from other than toxic exposure in the workplace. Exposure to a toxin in the setting of an alternative etiology may be an association and not reflect causation. Alternatively, there may be substantial worsening of a preexisting neuropathy (e.g., diabetic, hereditary) following exposure to a known neurotoxic agent (e.g., chemotherapy) at dosages considered safe (Chaudhry et al., 2003). In formal consideration of a toxin as cause for neuropathy in an individual or group of individuals, one methodology is to apply the Bradford Hill criteria (Hill, 1965). The Bradford Hill criteria consist of a series of questions that address issues such as temporal association, plausibility, dose–response relationship, coherence, specificity and strength of association, consistency of symptomatology, and identification and elimination of alternative causes of neuropathy.

Temporal association

A temporal relationship to toxin exposure and development of neuropathy is essential. Neuropathy having symptoms developing prior to exposure to a toxin is clearly unrelated to subsequent exposure. Neuropathy developing long after (months or years) an acute exposure to a toxin is also inconsistent with toxic etiology, though this fact is not always appreciated by the public at large. Patients being evaluated for recent-onset neuropathy often cite potential toxic exposures that may have occurred decades prior to onset of current symptoms.

Plausibility

The clinical and electrodiagnostic characteristics of the neuropathy must present as expected based on the proposed biologic mechanism by which a toxin exerts its influence.

Dose–response

There should be a relationship between level of exposure and severity of neuropathy. Level of exposure can be difficult to quantify, although there are measurable markers in blood, urine, or hair for some toxins that may be compared to reference levels. Though a patient may be exposed to a toxin, the absorbed dose is the key factor in whether neuropathy or other systemic effects develop. A toxin cannot cause neuropathy unless present in the body in sufficient amount. Even after absorption of a toxin, there are a number of factors, such as half-life in different compartments, which impact the probability of developing a neuropathy. There are a number of methodologies to address dosage or absorption of a toxin which will not be described here, but an excellent review addressing issues of assessing toxin exposure was published by Ford (1999).

Coherence

Resolution or measurable improvement of neuropathy, in concert with normalization of biologic markers if available, with removal of exposure to toxin is compelling evidence for causation. Animal models are also helpful in studying effects of exposure of toxins and establishing proof of causality.

Specificity

Is the cause-and-effect relationship limited to exposed individuals? This specificity may be typified by cases of similar forms of neuropathy occurring in most or all individuals exposed to the same toxin in sufficient dose, with the neuropathy developing in a relatively uniform time interval.

Strength of association

Is there a consistently identified high risk of developing neuropathy with specific toxin exposure? This question is best answered by large epidemiologic studies (cohort or case–control studies) that can demonstrate a strong association between toxin and subsequent development of neuropathy. Isolated case reports or cross-sectional studies may suggest causality or raise interesting hypotheses but should not be accepted as definitive proof of causality. The process by which this is addressed is commonly referred to as "evidence-based medicine" or "evidence-based toxicology."

Consistency

Have there been repeated studies by different investigators that have drawn the same conclusions as to relationship of a toxin and development of specific neuropathy?

Differential diagnosis

Have other causes of neuropathy been considered and eliminated? This is perhaps the most difficult task in establishing causation and requires broad knowledge of systemic, inflammatory, infectious, hereditary, and nutritional causes of neuropathy, as well as other toxic causes that could present similarly.

Recognition of toxic exposures in the workplace or environment remains important in preserving peripheral nerve function. Increasing awareness of hazards associated with toxic exposures has led to more care in exposure – i.e., use of protective devices or institution of safety protocols in handling and disposing of toxic compounds. On the other hand, advances have resulted in an increased number of potential neurotoxicants that may cause neuropathy as well as have other potential systemic effects. Recognition and subsequent removal of a neurotoxicant or removal of the patient from toxic exposure may preserve or salvage nerve function. The remainder of this chapter will describe some important neurotoxicants that may be encountered in the workplace, with details of their mechanisms, clinical features, and electrophysiology.

EXAMPLES OF TOXIC NEUROPATHIES

Acrylamide

Acrylamide is an uncommon cause of clinically evident neuropathy. Essentially all reports of acrylamide neuropathy appeared shortly after acrylamide was commercially manufactured and involved occupational exposures (Gold and Schaumburg, 2000). The initial report was that of a 19-year-old laborer in a small factory who developed a severe neuropathy and dermatitis (Garland and Patterson, 1967). No potential exposures to known neurotoxicants were identified, but diligent investigation determined he had been working with acrylamide. Contact with other workers making acrylamide flocculators identified five with symptoms and signs of neuropathy, of which three also had dermatitis and four had hyperhidrosis (Garland and Patterson, 1967). The neurologic importance of acrylamide includes its contribution as an animal model of "dying-back" neuropathy.

The vinyl monomer, acrylamide, is neurotoxic, whereas polyacrylamide, the polymer, is not (Smith and Oehme, 1991; Mulloy, 1996). Acrylamide has commercial applications in soil grouting, water proofing, and as a flocculator in water treatment and municipal water systems, and it is a byproduct in some food products. Acrylamide is water soluble and absorbed after oral, dermal, or respiratory exposures. Intoxication produces skin irritation and hyperhydrosis in addition to neuropathy. The skin irritation may reflect a direct effect on the skin, as acrylamide produces aggregation of intermediate filaments in cultured human skin fibroblasts (Durham et al., 1983). The mechanism producing hyperhidrosis, a key feature of acrylamide exposure, is unclear but may have a neurologic basis, as mouse models of acrylamide neuropathy show degeneration of postganglionic sudomotor sympathetic efferent fibers and hypohidrosis (Navarro et al., 1993). Alternatively, hyperhidrosis associated with acrylamide exposure may represent sympathetic overactivity (Auld and Bedwell, 1967), whereas hypohidrosis due to dysautonomia ultimately results (Navarro et al., 1993), analogous to paresthesias or fasciculations prior to sensory loss or weakness in somatic neuropathy.

Acrylamide neuropathy is associated with axonal swellings due to accumulation of neurofilamentous axonal masses, a finding pathognomonic of the giant axonal neuropathies. Animal models of acrylamide neuropathy show axonal degeneration distal to neurofilamentous swellings (Gupta and Abou-Donia, 1997), with conduction failure in distal nerve terminals (Sumner and Asbury, 1975) and preferential involvement of the largest sensory neurons (Tandrup and Braendgaard, 1994; Ko et al., 1999). The mechanism whereby acrylamide produces axonal degeneration is unknown (Edwards et al., 1991), but the accumulation of neurofilaments is due to a cumulative dose–response impairment of retrograde transport (LeQuesne, 1985; Sayre et al., 1985; Moretto and Sabri, 1988; Gold et al., 1992; Harris et al., 1994; Gupta and Abou-Donia, 1997). Impaired transport is thought to involve a target-derived "trophic" signal to the neuronal perikaryon (Gold et al., 1992; Harris et al., 1994), possibly due to an increase in calmodulin-dependent protein phosphorylation (Reagan et al., 1995). Axonal sprouting and reinnervation of denervated muscle fibers also may be impaired (Edwards et al., 1991).

CASE PRESENTATION

The case has been repeated here, unmodified, as presented in the referenced publication (Davenport et al., 1976).

This 25-year-old man was admitted to the University of Washington Hospital because of loss of sensation and unsteady gait. He was the only worker in his factory who handled acrylamide and had been working with this chemical for 6 months. His work involved the mixing of raw acrylamide powder with other dry and liquid reagents in a sealed reactor vessel. He followed the company's safety directions, wearing coveralls, clear plastic gloves, and a filtering face mask. Shortly after beginning his job, intense irritation of his palms and soles developed, with moist, red, raised ulcerations. Over the next 3 months, he noted increasing fatigue, became anorectic, and lost 40 lb. Two weeks before admission, he awoke one morning with an unsteady gait; the day before he had been able to stand easily on a narrow ledge to paint a wall. He stopped work but had progressively worsening ataxia. A week later, he began to have tingling in his hands and found them clumsy for fine manipulations. Two days before consultation, he first noted slurred speech and vague dysphasia.

His medical history was unremarkable. He was taking no medication and had no allergies. He drank socially and had a 10-pack-year smoking history. No other toxins were identified in his environment.

On examination, his vital signs were normal. There was hyperhidrosis of the palms and soles with irregular areas of blistering. He had mild weakness of extensor muscles of the wrists and ankles; muscle tone was decreased. Terminal dysmetria and intention tremor of the upper extremities and a moderate truncal and gait ataxia with wide base were observed, and he was unable to walk in tandem. Appreciation of pinprick and temperature was impaired below the midcalf and midforearm, with raised threshold for light touch in the same areas. Position and vibration senses were absent below the knees and the Romberg test was positive. No deep tendon or plantar reflexes could be elicited. Eye movements were conjugate and full with fine sustained nystagmus on lateral gaze. There was a mild lingual dysarthria, with normal palatal and pharyngeal function.

Clinical manifestations

Symptoms of acrylamide neuropathy include distal numbness and weakness, clumsiness, and hyperhidrosis. Signs include stocking-glove sensory loss, unsteadiness, positive Romberg, distal weakness, and absent ankle reflexes (Garland and Patterson, 1967). Initial descriptions attributed unsteadiness to cerebellar involvement because gait ataxia seemed disproportionate to the sensory loss (Garland and Patterson, 1967), but later descriptions of marked vibration and joint position abnormalities with absent sensation below the ankles seem sufficient to explain an unsteady gait (Davenport et al., 1976).

Electrophysiology

Early descriptions of acrylamide neuropathy included borderline-reduced motor conduction velocities (~80–90% of the lower limit of normal) and needle EMG evidence of fibrillation potentials in distal leg muscles (Davenport et al., 1976). Other reports described low sensory and motor amplitudes and modest conduction slowing (Garland and Patterson, 1967; Fullerton, 1969a; Aminoff, 1985). Cross-sectional studies of asymptomatic acrylamide-exposed workers showed low sensory amplitudes and neurogenic changes on needle EMG, findings believed to precede clinically evident neuropathy (He et al., 1989). Most comparisons, including those involving tunnel workers previously exposed to acrylamide during grouting operations (2–10 years previously) and workers with no history of acrylamide exposure (Goffeng et al., 2008), have identified small group difference within the normal range. Specifically, the latter study described a significantly slower mean sural conduction velocity nerve in the previously exposed group compared to the control group, a finding interpreted to indicate a subclinical but persistent toxic effect. The small difference (44 vs 48 m/s, $p = 0.005$), however, made the interpretation suspect, as no significant group differences were found for any other nerve conduction study measures, including the sural amplitude recordings (6 vs 8 μV; $p = \text{NS}$) or any of the remaining median and ulnar sensory or fibular, tibial, median, and ulnar motor comparisons (Goffeng et al., 2008).

Differential diagnosis

A large number of substances and conditions cause a sensorimotor neuropathy, and the limited number of case descriptions of acrylamide neuropathy that include detailed electrophysiologic evaluations have produced disparate results in terms of the nerve fibers involved and the presence or absence of conduction slowing. Because acrylamide is one of the "giant axonal neuropathies," it might be anticipated that motor conduction slowing would be a prominent feature, but it is unclear whether conduction slowing is greater than can be explained by axonal degeneration of the largest motor fibers. For this reason, acrylamide is included among the sensorimotor neuropathies characterized by: (1) no conduction slowing (Table 16.1); and (2) conduction slowing (Table 16.2). This is not unique to acrylamide, as other neurotoxicants (e.g., arsenic) appear in both tables, possibly because the presence or absence of conduction slowing reflects the severity of the neuropathy or the timing of evaluation relative to onset. In this context, the association of neuropathy and a history of hyperhidrosis early in the course of acrylamide exposure are highly suggestive of a causal relationship, whereas later development is less specific.

Additional diagnostic evaluation

Sural nerve biopsy showing focal axonal swellings containing masses of neurofilaments is consistent with a diagnosis of acrylamide neuropathy. While not pathognomonic for this diagnosis, the relatively small number of neurotoxicants producing a giant neurofilamentous axonopathy (e.g., *n*-hexane, methyl *n*-butyl ketone, carbon disulfide) reduces the number of possible diagnostic considerations.

Treatment

Treatment of acrylamide neuropathy is removal from exposure.

Prognosis

The prognosis of recovery from acrylamide neuropathy is good. Removal from exposure before substantial neuropathy develops results in complete recovery (Gold and Schaumburg, 2000; Hagmar et al., 2001). In one study, all but two of 23 workers with neuropathy attributed to an acrylamide-containing grouting agent recovered within 18 months after the cessation of exposure (Gold and Schaumburg, 2000; Hagmar et al., 2001).

Arsenic

Arsenic is a metalloid found in both inorganic (toxic) and organic (nontoxic) forms. Occupational exposures to inorganic arsenic occur in numerous manufacturing processes, including smelting, mining, and tanning (Feldman et al., 1979), and environmental exposures occur from contaminated groundwater, combustion of contaminated fossil fuels and arsenate-treated wood, some pesticides and herbicides, and some medicine folk and herbal medicines (Tay and Seah, 1975; Kreiss et al., 1983; Mukherjee et al., 2003; Albers and Garabrant, 2004). Most individuals recognize arsenic as a poison, at least in part due to its well-documented homicidal applications (Chhuttani et al., 1967), a use romanticized in *Arsenic and Old Lace*, a dark comedy written by the American playwright Joseph Kesselring in the late 1930s. Although best known as a homicidal or suicidal agent, arsenic, like mercury, has a long history of medicinal use (supporting the concept that whether a substance is considered a medicine or poison depends on its intended use) (Chhuttani et al., 1967; Windebank, 1993; Gallagher, 1998). At present, arsenic is used to treat acute promyelocytic leukemia (Thery et al., 2008). Chronic exposure resulting from drinking arsenic-contaminated groundwater is a major environmental health hazard throughout the world (Mazumber and Dasgupta, 2011), and it is estimated that over 26 million people are so exposed in the state of West Bengal in India (Paul et al., 2013). Arsenic interferes with cellular energy metabolism and disrupts protein and enzyme activity (Albers and Garabrant, 2004). The neuropathology associated with arsenic-induced neuropathy is characterized by axonal degeneration with preferential involvement of large-diameter nerve fibers (Windebank, 1993).

CASE PRESENTATION

This patient was described previously in greater detail (Albers, 1999).

A 31-year-old man awoke experiencing nausea, vomiting, and diarrhea. These symptoms persisted for 1 week and were attributed to gastroenteritis. Shortly thereafter his symptoms attributed to gastroenteritis resolved, but he developed numbness and tingling of his feet that progressed over days to involve his legs and hands. He also noted unsteadiness of his gait, but did not seek medical evaluation until he developed rapidly progressive weakness of his extremities.

On physical examination, he showed mild respiratory distress and a weak cough. He had normal mental status and cranial nerve function, except for symmetric facial weakness. He had limb weakness most marked in distal muscles. Reflexes were hypoactive in the arms and absent in the legs. Pinprick sensation was normal, but he had impaired vibration and joint position sensations in a stocking-glove distribution.

His initial laboratory testing revealed pancytopenia with a white blood cell count of 1250 cells/μL and mildly abnormal liver function studies. Cerebrospinal fluid (CSF) protein was elevated (270 mg/dL) but cell count was normal (cytoalbumin dissociation), as was the CSF glucose – findings consistent with GBS. Sensory conduction studies showed absent sural responses and low-amplitude median sensory responses. Motor responses showed reduced amplitude with borderline-abnormal conduction velocity and distal latency. F waves were unobtainable. The findings were interpreted as consistent with a sensorimotor polyneuropathy of the demyelinating type, findings also consistent with GBS. Although rigid criteria for acquired demyelination were not fulfilled, partial conduction block in two motor nerves supported a diagnosis of acquired demyelination.

Clinical manifestations

Arsenic neuropathy appears about 5–10 days after a single massive exposure. Sensory symptoms herald the onset of neuropathy and include paresthesias and burning pain, distal numbness and tingling, and muscle aching. Muscle weakness follows, first in distal muscles and then more proximally. The neuropathy progresses over weeks despite no additional exposure ("coasting") and resembles GBS (Greenberg, 1996). When severe, the neuropathy results in flaccid paralysis with facial weakness, respiratory difficulty requiring ventilator support, sensory loss, and areflexia.

Arsenic neuropathy is but one part of a systemic illness. Other manifestations include cardiovascular (cardiomyopathy), gastrointestinal (nausea, vomiting, diarrhea, hepatitis, and hepatomegaly), hematologic (bone marrow failure, basophilic stippling), renal (acute tubular necrosis), and dermatologic (dermatitis, Mees' lines) involvement. Neuropathy was the most frequent manifestation of subacute arsenic poisoning among individuals who accidentally ingested excessive amounts of arsenic after a pipeline in a copper-smelting factory began to leak (Xu et al., 2008). Most experienced gastrointestinal symptoms (80%), with leucopenia (69%) and abnormal liver function (56%). Shortly thereafter, a few developed scattered papular eruptions and desquamation. Within 2–3 weeks, nearly 20% developed neuropathy with distal weakness, paresthesias, and stocking-glove distribution numbness. Features associated with chronic, lower-level environmental arsenic toxicity (arsenicosis) include bronchiectasis and obstructive pulmonary disease, atherosclerosis and hypertension, distal edema, and portal hypertension (Pimparkar and Bhave, 2010; Krishnan et al., 2012). Cutaneous manifestations include hyperpigmentation, hyperkeratosis, desquamation, and skin cancers (Sengupta et al., 2008). Following acute exposure, gastrointestinal symptoms antedate the onset of neuropathy and may be mistaken for an antecedent illness preceding GBS. Features such as Mees' lines appear only later, usually when the diagnosis is no longer in question (London and Albers, 2007). Chronic arsenic intoxication produces signs similar to those of acute intoxication but has a more insidious onset without Mees' lines. There are no convincing reports of symptomatic chronic arsenic neuropathy in the USA or Canada, despite inclusion of arsenic in the differential diagnosis of idiopathic neuropathy (Herkovitz et al., 2010).

Electrophysiology

Electrophysiologic evaluations performed soon after the onset of neuropathy are associated with motor conduction slowing and often suggest an acquired demyelinating neuropathy (Donofrio et al., 1987). Nerve conduction studies show low motor amplitudes with increased temporal dispersion or partial conduction block, slow conduction velocity, prolonged distal latencies, and absent F responses. Sensory studies may show low amplitudes, especially involving the sural nerve. The "normal sural – abnormal or absent median" sensory pattern frequently associated with GBS is not observed (Bromberg and Albers, 1993). The features resembling an acute demyelinating neuropathy appear prior to complete axonal degeneration and loss of recordable motor or sensory responses. Initial needle EMG abnormalities include reduced motor unit recruitment. Within weeks, EMG findings characteristic of a severe dying-back sensorimotor neuropathy develop, showing evidence of denervation (fibrillation potentials) and reinnervation (large-amplitude, polyphasic motor units).

Cross-sectional studies of neurologically normal arsenic-exposed workers involved in copper smelting showed few significant differences compared with controls, mostly involving decreased amplitude and conduction velocity (Sinczuk-Walczak et al., 2010). Measures important to the diagnosis of a generalized sensorimotor neuropathy (e.g., sural amplitude) showed no significant difference, calling into question the clinical importance of the observation. Nevertheless, similar results were identified associating chronic exposure to arsenic with sensory conduction velocity (higher cumulative exposure, slower conduction) among adolescents exposed to arsenic in their drinking water, implicating slowed conduction velocity as a potential early marker of chronic arsenic neuropathy (Tseng et al., 2006).

Differential diagnosis

A neuropathy characterized by prominent conduction slowing results in the differential diagnosis shown in Table 16.2. The gastrointestinal signs of acute arsenic intoxication frequently resolve before onset of neuropathy and may resemble a nonspecific antecedent illness associated with GBS. This confusion has resulted in accusations of failure to diagnose arsenic intoxication and thereby prevent subsequent exposure involving a patient who experienced episodes of gastroenteritis associated with neutropenia and weakness. His neuropathy was diagnosed as GBS prior to his death, but postmortem specimens showed elevated arsenic concentrations, and the wife of the decedent was charged with murder (Poklis and Saady, 1990). Although suggestive of GBS, the involvement of multiple systems, including the cardinal feature of a coexisting anemia, made arsenic intoxication an important and parsimonious consideration in the differential diagnosis. Arsenic neuropathy is supported by nerve conduction studies showing marked sensory abnormalities and absent sural sensory responses, a finding atypical of GBS, where the most specific presentation includes "normal sural-absent median" sensory responses.

Additional diagnostic evaluation

CSF cytoalbumin dissociation is sometimes considered a characteristic feature of GBS but is not specific for GBS or other inflammatory neuropathies, and patients with acute arsenic neuropathy may have elevated CSF protein levels. Arsenic is rapidly excreted in the urine, and increased urinary arsenic excretion is an important feature of recent exposure. A 24-hour urine collection can reliably establish excessive exposure to arsenic, but not necessarily intoxication. A conventional GBS evaluation includes screening the urine for arsenic. A source of confusion involves elevated arsenic levels resulting from ingestion of certain seafood such as shellfish and bottom-feeding fin fish that contain organic arsenic salts. Arsenic obtained from seafood is in the organic form and, unlike inorganic arsenic, is not neurotoxic. Although nontoxic, it can cause alarmingly high urine arsenic levels, because routine analyses do not distinguish organic arsenic from inorganic arsenic. Indiscriminant urine arsenic screening is not appropriate for patients with chronic neuropathy. Serum arsenic measurement is helpful only when collection is obtained within hours to a few days of acute ingestion because arsenic is rapidly cleared from the blood (Albers and Garabrant, 2004; Valenzuela et al., 2005).

Additional abnormalities reflecting systemic arsenic toxicity include abnormal liver function and renal studies, pancytopenia, and basophilic stippling of red blood cells, a nonspecific but important indicator of some toxic exposure. Persistent pancytopenia and progressive thrombocytopenia may result in examination of the bone marrow showing hypocellularity and megaloblastic changes. Sequential chest X-rays after arsenic poisoning may show progressive cardiomegaly and bilateral pleural effusions. When arsenic neuropathy is not considered as a diagnostic possibility until late in the course of the illness (e.g., such as when Mees' lines, pigmented dermatitis, or cardiomyopathy develop), the amount of arsenic present in hair or nails can be measured (Hindmarsh, 2002).

Treatment

The most important treatment involves supportive care and removal from additional exposure. Most cases of intentional arsenic poisoning involve a single, massive exposure, but homicidal poisonings include re-exposure of an individual who survived what had been presumed as GBS (Poklis and Saady, 1990). When arsenic neuropathy is misdiagnosed as presumed GBS, treatment has included therapeutic plasma exchange or intravenous immunoglobulins, neither of which has any demonstrated role in arsenic poisoning. Depending on when an acute arsenic poisoning is suspected, treatment may include gastric lavage, hemodialysis, and chelation. Despite use of hemodialysis and chelation for acute arsenic intoxication, there is little evidence that either influences neurologic progression or recovery, as irreversible damage likely occurs well before neurologic signs appear.

Prognosis

Patients who survive the systemic manifestations of acute arsenic intoxication experience substantial or complete neurologic recovery over months to several years, depending on the degree and extent of denervation, age, and the presence of underlying conditions.

Carbon disulfide

Carbon disulfide is an organic solvent that can be absorbed by inhalation of the fumes or by skin contact (Allen, 1979). Occupational exposure to carbon disulfide occurs in a variety of situations, including production of cellophane, plywood, vulcanized rubber, and some pesticides (Allen, 1979; Seppalainen and Haltia, 1980). Neurotoxicity most often has been reported following chronic low-level exposures associated with production of viscose rayon (Allen, 1979; Kim and Kang, 2010; Liu et al., 2012). Acute high-level exposures are said to result in acute psychosis, neurobehavioral abnormalities, parkinsonism, and polyneuropathy (Liu et al., 2012). The early descriptions included a perception that viscose rayon manufacturers with many years of carbon disulfide exposure had an increased frequency of parkinsonism (Seppalainen and Haltia, 1980). This association relied primarily on case reports and cross-sectional studies and remains unproven (Albers and Bromberg, 1995; Albin, 2000). Although uncommon, the best-established neurotoxicity is that of a sensorimotor polyneuropathy. Some of the initial investigations of chronic, low-level occupational exposure to carbon disulfide (generally below 20 ppm) found few signs of neuropathy, and symptoms were not consistently associated with exposure levels (Putz-Anderson et al., 1983). Small differences in nerve conduction study results were reported for asymptomatic workers having carbon disulfide levels up to 40 ppm and referent workers (discussed below). Many cross-sectional studies showing subclinical neuropathy involved small numbers of exposure workers who showed no improvement in conduction velocities after removal from exposure (Corsi et al., 1983), making the original diagnoses suspect and the findings likely explained by preexisting physiologic group differences, perhaps reflecting anthropometric or technical factors, not carbon disulfide exposure.

In contrast, cases of chronic carbon disulfide poisoning involving workers engaged in the viscose rayon industry reported in the 1980s appeared to support the association with neuropathy (Kim et al., 1989; Kim and Kang, 2010). The cases, most of whom had chronic exposure for more than 10 years, presented with gait disturbance, paresthesias, and weakness attributed to neuropathy, together with speech and visual disturbances attributed to central nervous system involvement. EMG studies confirmed the presence of a peripheral neuropathy in all cases. Cross-sectional comparisons of viscose rayon plant workers and referents found an increased frequency of unexplained neuropathy among exposed workers (Aaserud et al., 1990; Chu et al., 1995). In some cases, workers showed reduced conduction velocities compared to asymptomatic co-workers and other control subjects (Chu et al., 1995; Liu et al., 2012). Neuropathy was associated with exposure to high concentrations of carbon disulfide in some areas of the plant (e.g., fiber cutting), with peak exposures levels of 150–300 ppm and 8-hour time-weighted averages of 40–67 ppm (Chu et al., 1995). Sural nerve biopsy of a worker with neuropathy showed loss of large myelinated fibers and ongoing axonal degeneration, demyelination, and remyelination (Chu et al., 1996).

Although a causal association between occupational exposure to carbon disulfide and neuropathy remained uncertain, additional evidence identified carbon disulfide as a potential neurotoxicant. Disulfiram (Antabuse) is a medication used to promote abstinence from alcohol. Disulfiram is metabolized to acetaldehyde that produces nausea when combined with alcohol, forming the rationale for promoting alcohol abstinence. Disulfiram also is metabolized to carbon disulfide, and the medical literature includes numerous reports associating disulfiram with neuropathy, observations fairly compelling in support of peripheral neurotoxicity (Bradley and Hewer, 1966; Frisoni and Di Monda, 1989). Disulfiram neuropathy occurs after long use, usually exceeding 6 months. Importantly, the electrophysiology and neuropathology of disulfiram-induced neuropathy are relatively distinct and differ from alcohol or nutritional neuropathies. Most specifically, sural nerve biopsy of patients having suspected disulfiram neuropathy shows a neurofilamentous axonopathy with axons distended by neurofilaments (Ansbacher et al., 1982; Johnson et al., 1983), features resembling carbon disulfide neuropathy. Although the neuropathology is not pathognomonic for disulfiram or carbon disulfide, it is sufficiently distinct to exclude many other disorders associated with neuropathy (Ansbacher et al., 1982; Johnson et al., 1983).

The precise mechanism of carbon disulfide neurotoxicity is unknown. Laboratory animal models of carbon disulfide intoxication show a neuropathy characterized by axonal degeneration with paranodal and internodal swellings, features identical to those described in human sural nerve (Bilbao et al., 1984; Gottfried et al., 1985). These findings also are similar to those associated with other neurotoxicants (e.g., acrylamide and *n*-hexane) that produce giant axonal swellings reflecting neurofilament accumulations (Schaumburg and Spencer, 1984).

CASE PRESENTATION

(Details of the history have been modified to preserve the anonymity of the patient and the patient's family.)

This patient was previously described in greater detail elsewhere (Albers, 2005d).

A 38-year-old woman insidiously developed numbness and tingling in her feet and legs. Subsequently she noted numbness and clumsiness of hands, as well as weakness in distal lower extremities. Clinical examination several weeks after onset of symptoms revealed stocking-glove sensory loss, weakness of plantar- and dorsiflexion, and absent ankle reflexes. Her medication history included ongoing use of disulfiram (Antabuse), 250–500 mg daily, which she had taken for many years due to prior history of alcohol abuse. She denied current use of alcohol. She had no significant other past medical history or contributory family history. Disulfiram (her only medication) was suspected as etiology of neuropathy. Though disulfiram was discontinued, her symptoms progressed in severity and she developed severe pain in her feet. Four months from onset, her neurologic examination was significant for a wide-based slow, steppage gait. Romberg sign was positive. Strength was mild, reduced in hand intrinsic muscle, and was more prominently reduced in the lower extremities in all muscles below the knees, associated with atrophy in intrinsic foot muscles. Sensory deficits, in a stocking pattern, were to large- and small-fiber modalities of fine touch, pin pain, and vibration, with preservation of joint position and temperature.

An EMG examination showed a motor greater than sensory neuropathy, most likely of the axonal type. Slight slowing of motor conduction velocity was thought to reflect loss of the largest axons, not primary demyelination.

Clinical manifestations

The neuropathy attributed to carbon disulfide exposure is said to occur together with extrapyramidal symptoms (Vigliani, 1954; Seppalainen and Haltia, 1980; Peters et al., 1982, 1988), and combined neuropathy and parkinsonism in the appropriate setting should suggest possible carbon disulfide intoxication. The neuropathic symptoms are nonspecific and characterized by stocking-glove sensory loss and weakness. Signs are most prominent in the lower extremities, with stocking-glove sensory loss involving all sensory modalities (touch, vibration, joint position, pin pain, temperature), positive Romberg, distal weakness, and absent ankle reflexes (Schaumburg and Spencer, 1984; Albers and Wald, 2002). The neuropathy attributed to disulfiram can have subacute onset of weakness, thereby resembling GBS (Bradley and Hewer, 1966; Frisoni and Di Monda, 1989; Palliyath et al., 1990). Additional signs important to a diagnosis of carbon disulfide intoxication potentially include extrapyramidal features such as rest tremor, masked facies, bradykinesia, rigidity, and postural instability. Descriptions of reversible disulfiram-induced parkinsonism, sometimes developing after only days to weeks of use, suggest the basal ganglia are a target of disulfiram neurotoxicity (Laplane et al., 1992).

Electrophysiology

Occupational exposure to carbon disulfide is associated with slow conduction velocities among exposed compared to unexposed workers, a finding suggesting subclinical neuropathy (Seppalainen and Tolonen, 1974; Johnson et al., 1983). Nevertheless, the small differences are insufficient to establish a pathologic abnormality (Albers, 2005c). The nerve conduction study results show a dose–response relationship between conduction velocity and cumulative carbon disulfide exposure but are potentially confounded by an interaction with age-related decline in conduction velocity, as only older workers had an opportunity for long exposures (Albers, 2005c). Nevertheless, evaluations of exposed workers having clinically evident neuropathy include descriptions of reduced motor and sensory amplitudes, moderately slowed motor conduction velocities, and prolonged distal latencies, and needle EMG evidence of chronic, length-dependent denervation with decreased recruitment, large motor units, and fibrillation potentials (Corsi et al., 1983; Aaserud et al., 1990; Chu et al., 1996), findings comparable to those described in the case presentation. The mild conduction velocity slowing likely reflects degeneration of the largest myelinated axons.

Differential diagnosis

The clinical features of the neuropathy attributed to carbon disulfide are generally nonspecific. The combination of extrapyramidal signs and peripheral neuropathy, however, should result in consideration of carbon disulfide intoxication as part of the differential diagnosis. In the absence of extrapyramidal signs, recognizing the presence of motor greater than sensory involvement is helpful, because such neuropathies are not particularly common. Such recognition likely will result in an electrodiagnostic evaluation, and additional nerve conduction study findings of a sensorimotor or motor greater than sensory neuropathy with motor conduction slowing results in the differential diagnoses listed in Table 16.2. This list includes carbon disulfide, emphasizing the importance of the electrodiagnostic examination in focusing the differential diagnosis of an otherwise nonspecific neuropathy.

Additional diagnostic evaluation

Routine testing of blood and urine gives normal results, as does evaluation for heavy metals. In the unusual case of a rapid-onset carbon disulfide neuropathy resembling GBS, CSF examination presumably would be unremarkable. Most items included in the differential diagnosis list in Table 16.2 are readily excludable, although a late-onset familial neuropathy is always a consideration. Porphyric neuropathy is often asymmetric with an initial proximal greater than distal presentation (Bloomer and Bonkovsky, 1989; Sack, 1990). Laboratory testing can exclude (or confirm) the diagnosis of a hepatic porphyria. Sural nerve biopsy, as reported as part of several case reports, can confirm the presence of axonal swellings consistent with accumulation of neurofilaments, a finding important in establishing the diagnosis of carbon sulfide neuropathy.

Treatment

The only treatment for carbon disulfide neuropathy is removal from ongoing exposure.

Prognosis

Most toxic neuropathies improve or resolve after removal from exposure, the exceptions being some of the sensory neuronopathies. Recovery from carbon disulfide-associated neuropathy follows removal from exposure, although progression of symptoms and signs ("coasting") is typical and improvement may be slow and sometimes incomplete (Schaumburg and Spencer, 1984). This improvement includes workers with substantial impairments and sural nerve biopsy-confirmed neuropathy who show serial improvement in strength and nerve conduction study results over several years (Chu et al., 1996). In contrast, incomplete recovery was reported among some carbon disulfide-exposed workers after 3 years of follow-up, suggesting a persistent neuropathy (Huang et al., 2002). Similarly, EMG evaluation of workers diagnosed with mild carbon disulfide neuropathy before and 4 years after removal from exposure showed no improvement in conduction velocities, leading the authors to attribute this finding to permanent axonal damage (Corsi et al., 1983). This interpretation, however, is suspect because isolated conduction velocity is a poor surrogate for axonal degeneration or regeneration, and because preexisting group difference (e.g., including anthropometric factors) unrelated to carbon disulfide exposure likely explained the subclinical electrophysiologic findings.

Dioxin

Dioxin (tetrachlorodibenzo-*p*-dioxin) is accepted as a poison by most individuals, especially after the 2004 publicity surrounding Viktor Yushchenko, the former President of Ukraine, whose disfigurement by chloracne (a form of acne) was alleged to have been caused by ingestion of dioxin. Whether dioxin causes neuropathy, however, is a matter of disagreement. Support for the hypothesis comes mostly from case reports, cross-sectional studies, and a few longitudinal studies that are inconsistent in their conclusions. Exposure to dioxin occurs during the production of dioxin, as the results of industrial accidents, and from contact with dioxin-contaminated substances, including Agent Orange, a herbicide contaminated by dioxin and the source of great controversy in the USA (The Centers for Disease Control Vietnam Experience Study, 1988; Webb et al., 1989).

Descriptions of dioxin neuropathy

Most reports of dioxin neuropathy are limited by several factors, including use of nonspecific symptoms, few clinical signs, lack of electrophysiology, failure to consider competing explanations, and incomplete exposure assessment. Examples of the conflicting results follow.

Occupational exposure

Examples include the Czechoslovakia report of 17 dioxin-exposed herbicide manufacturers who developed an EMG-confirmed neuropathy (Jirasek et al., 1974: abstract translation only). Singer et al. (1982) reported that age- and temperature-adjusted nerve conduction study results obtained from 55 dioxin-exposed chemical manufacturing workers showed slowing of sural and median motor conduction velocities; response amplitudes were not reported, but the sural response had to be present to calculate a conduction velocity so any involvement had to be minimal. In contrast, workers exposed to dioxin during production and maintenance work and unexposed referents from the same plant showed no increased risk of peripheral nervous system abnormality (Suskind and Hertzberg, 1984). Similarly, clinical, quantitative sensory, and nerve conduction study comparisons of dioxin-exposed chemical workers and unexposed referents found no relationships between the frequency of peripheral neuropathy and exposure group or lipid-adjusted serum dioxin level (Sweeney et al., 1993). Thomke et al. (1999) evaluated 156 dioxin-exposed workers from a pesticide plant, including 42 with chloracne. After exclusions, workers with chloracne had significantly higher dioxin exposures (based on back-calculated lipid levels), more frequent clinically evident neuropathy (sensory and reflex signs), and more frequent nerve conduction study abnormalities. They concluded that a minority of severely exposed workers evidenced a mild length-dependent sensory neuropathy.

Industrial accidents

The most notorious industrial accident was the result of a chemical plant explosion in Seveos, Italy in the mid-1970s that contaminated the surrounding area with dioxin (Barbieri et al., 1988). Screening evaluations in the region having the highest soil concentrations suggested a higher prevalence of clinical and subclinical abnormalities, mostly reduced motor conduction velocities, than surrounding regions (Pocchiari et al., 1979). Contamination in some cases was associated with chloracne (e.g., comedones in a malar crescent or periauricular distribution), inferring dioxin exposure. A dose-related increased prevalence of neuropathy was reported among those who had experienced chloracne (Filippini et al., 1981), although different exclusion criteria were applied to the exposed and comparison groups. In another study of 152 subjects selected because they had experienced chloracne or elevated serum hepatic enzyme levels, no cases of overt neuropathy were found but this group also had an increased frequency of nerve conduction study abnormalities, including reduced sural amplitudes relative to control values (Barbieri et al., 1988). Among 13 nerve conduction study attributes among median (motor and sensory), peroneal motor, and sural nerve studies, only the sural amplitude differed significantly from control values. This amplitude was well within the normal range and the small 3 μV group difference was unlikely to be clinically important. A follow-up evaluation of 193 subjects who had chloracne identified only three thought to have a neuropathy, and no significant differences existed between the exposed and the unexposed groups for any of the neurologic tests, including peroneal and median motor and sural nerve conduction studies (Assennato et al., 1989).

A study of 41 dioxin-exposed Seveso residents included measurement of adipose tissue dioxin levels. This group included few with sensory signs and abnormal reflexes but no pattern of clinical abnormalities was related to dioxin levels (Webb et al., 1989). Serial evaluation of 35 technicians exposed to dioxin through handling samples from Seveso showed group differences on needle EMG examinations (Webb et al., 1989), but nerve conduction studies showed difference in motor and sensory conduction (velocity and distal latency measures) favoring the comparison group. Again, all mean values were within the normal range, and mean sensory amplitudes were comparable in both groups, limiting the importance of the observations.

In another incident involving workers who participated in the cleanup of a dioxin spill in Sturgeon, Missouri, 43 of 45 developed symptoms and signs of neuropathy (Klawans, 1987); nerve conduction studies were described only as being normal initially but later showing "mild generalized slowing." Finally, Urban et al. (2007) performed follow-up evaluations of 15 herbicide manufacturers accidently exposed to dioxin 35 years previously. Signs of neuropathy were found in nine and confirmed by nerve conduction studies in three (slowed sural conduction velocity), with secondary finding of a high frequency of lipid metabolism disorders.

Agent Orange

The US Centers for Disease Control and Prevention reported that Vietnam veterans had significantly more symptoms and signs of peripheral neuropathy relative to non-Vietnam veterans (The Centers for Disease Control Vietnam Experience Study, 1988), although Vietnam veterans with symptoms and signs of peripheral neuropathy were not overrepresented (odds ratio (OR) 1.4; 95% confidence interval (CI) 0.6–2.3). Michalek et al. (2001) subsequently reported the results of a longitudinal cohort study designed to address whether exposure to Agent Orange, and therefore to its contaminant dioxin, produced neuropathy. Participants included US Vietnam veterans involved in Operation Ranch Hand, the unit responsible for aerial herbicide spraying in Vietnam. Evaluations designed to identify health problems, including neuropathy, were performed at intervals for years after the last opportunity for dioxin exposure (1962–1971), categorized as "background," "low," and "high" using serum dioxin levels measured in 1987 or later, extrapolated back to estimate the dose at the end of exposure. The comparison group was matched by age, race, and military occupation. Neuropathy was based on clinical criteria, as nerve conduction studies were performed only early in the study and later repeated only for subjects having neuropathy. Examinations performed between 1982 and 1992 found no association between dioxin exposure and neuropathy. However, in 1997, a significantly increased risk of developing neuropathy was apparent for the high exposure group (OR = 1.7; 95% CI = 1.1–2.7) (Michalek et al., 2001). The results appeared robust, but in terms of a causal relationship, additional factors possibly explained the increased frequency of neuropathy.

These factors included concern of confounding by diabetes mellitus among the exposed subjects and the potential relationship between dioxin and diabetes (Longnecker and Michalek, 2000). Concern of a possible association between dioxin and diabetes mellitus originated in the 1980s (Pazderova-Vejlupkova et al., 1981; Bond et al., 1983; Moses et al., 1984), as a result of investigations of both persons residing near the dioxin-contaminated region in Seveso and US Vietnam veterans exposed to dioxin-contaminated herbicides. Although it is beyond the scope of this chapter to debate this association, suffice it to say that the relationship has resulted in controversy, with most being skeptical of a causal association between dioxin and diabetes. Despite this concern, Michalek et al. (2001) did not identify an interaction between diabetes and dioxin exposure in their analysis of neuropathy, nor was there evidence that diabetes modified the association between dioxin and peripheral neuropathy. Nevertheless, the results supporting a relationship between dioxin and neuropathy independent of diabetes involved complicated analyses and were limited by the relatively small numbers of subjects having diagnosable neuropathy but no diabetes. Further, 13 of the 14 subjects with neuropathy in the high-exposure category had diabetes or pre-diabetes (Michalek et al., 2001). The influence of diabetes on these data led to cautious interpretation of the results, something never fully resolved (Michalek et al., 2001).

Summary

At best, given these conflicting and seemingly weak results, it appears that most would conclude that the association between dioxin and neuropathy has not been established. Nevertheless, the US Department of Veterans Affairs in 2013 added acute and subacute peripheral neuropathy associated with exposure to certain herbicide agents to its list of presumptive service connection disorders.

Pseudocausation

In discussing dioxin and neuropathy, particularly in the context of the data involving Vietnam veterans, the potential confounding by diabetes is difficult to resolve. One possible explanation for this difficulty could be attributed to pseudocausation, whereby one of the study observations not initially thought to have causal qualities may be a cause of the main study outcomes. The diagnosis of diabetes mellitus is generally straightforward, although the threshold criteria for diagnosing diabetes appear less sensitive than those used to diagnose neuropathy, as evidenced by the frequent diagnosis of "diabetic" neuropathy based on nerve conduction study results before diabetes can be diagnosed. Likewise, elevated lipid levels, especially hypertriglyceridemia, may appear decades before diabetes is diagnosed. Consider the possibility that some factor such as the serum lipid level is also associated with the serum dioxin measurement, such that the dioxin levels recorded for some subjects could be influenced by this serum factor associated with pre-diabetes and at the same time result in a falsely elevated dioxin level, not the true body burden. In other words, such individuals with high serum lipid levels could have: (1) artificially high serum dioxin levels for a given exposure; (2) increased risk of developing diabetes; and (3) increased risk of developing diabetic neuropathy (possibly preceding diagnosable diabetes). Rather than causal associations between dioxin and diabetes and dioxin and neuropathy, the causal relationship could reflect pre-diabetes manifest by hyperlipidemia, in turn resulting in falsely elevated serum dioxin levels and eventually resulting in diabetic neuropathy and overt diabetes. This example of pseudocausation can influence any study design, including so-called "hypothesis-testing" longitudinal cohort studies. It is not clear that this example explains the apparent association between dioxin exposure and neuropathy, but it is one of the reasons that the association involving dioxin and neuropathy remains unanswered and requires cautious interpretation.

Elemental mercury

Mercury exists in elemental, mercurous, or mercuric forms depending on its oxidation state. It is ubiquitous in the environment, mostly as mercuric sulfide (cinnabar). Despite its known neuro- and nephrotoxic potential, elemental mercury has had medical use both in equipment (e.g., thermometers, sphygmomanometers) and in medications (Clarkson et al., 1988; Windebank, 1993). Medicinally, it has been used as a gastrointestinal purgative, an antiseptic (mercuric chloride), a diuretic, a preservative (e.g., in vaccines), and in dental amalgams (Clarkson et al., 1988; Windebank, 1993). In the 1900s, mercurial ointments were used to treat syphilitic skin lesions ("One night in the arms of Venus, the rest of your life with Mercury"). Mercury also has extensive industrial applications, including use in the chloralkali industry, metal plating, production of thermonuclear weapons, and tanning processes, where its use by hatters to cure felt was associated with the early description of occupational mercury poisoning. Mercury is still found in thermometers, barometers, electromechanical switches, fluorescent lamps, and a variety of electronic devices. Recent incidents involving elemental mercury intoxication include special manufacturing processes, such as extraction of silver from waste film (Kim and Kang, 2010).

Elemental mercury exposure occurs frequently, and mercury is widely recognized for its neurotoxic potential (Carman et al., 2013). Yet, reports of elemental mercury toxicity are relatively uncommon and few physicians have the opportunity to recognize a patient with mercury intoxication, despite the frequent requests for "heavy-metal screening" to detect abnormal excretion of lead, mercury, or arsenic. Signs of chronic mercury intoxication include postural tremor, peripheral neuropathy, and poorly defined behavioral changes, sometimes referred to, perhaps inappropriately, as "hatter's madness" (Hanninen, 1982; Bromberg, 2000). Elemental mercury is poorly absorbed through the skin or by the gastrointestinal tract, although some reports describe intoxication after topical use of traditional Chinese medicines containing mercury (Wu et al., 2013). On occasion, toxicity has resulted from subcutaneous injection (Malkani et al., 2011). In contrast, mercury vapor is readily absorbed through the lungs after inhalation, after which it is oxidized in blood and tissues and excreted in the bile and urine (Chang, 1980). Mercury combines with sulfhydryl groups in enzymes and structural proteins, but the specific mechanism by which mercury damages tissue is uncertain (Windebank, 1993).

CASE PRESENTATION (Albers, 2005b)

A 53-year-old male participated in a study of chloralkali workers who were potentially exposed to elemental mercury. The company's industrial hygiene program monitored mercury air levels and individual workers' urine mercury levels, collected monthly or after recognized mercury exposures associated with cleaning the "mercury cells" or after accidental spills (e.g., while removing a pipe that contained residual mercury). The industry threshold standard required that workers whose urine mercury level exceeded 500 µg/L were removed from additional exposure until their level fell below an acceptable threshold. The patient's study questionnaire inquired about current and past symptoms, as well as past medical, social, family, and occupational histories. He had no complaints referable to the nervous system, and his medical and family histories were unremarkable. He denied excessive alcohol use and had never used any medications for any chronic medical problems. Examination by a neurologist masked to the worker's job description, exposure history, and urine mercury levels identified a mild postural tremor of the outstretched hands, normal strength but mild atrophy of intrinsic foot muscles with high arches and mild hammertoe deformities, equivocally diminished vibration, pin pain, and touch sensations in a stocking distribution, excessive sway on Romberg testing, and absent ankle reflexes. He was unaware of his postural tremor and did not consider the appearance of his feet abnormal. The electrodiagnostic examination showed low-amplitude sensory responses and borderline-low motor responses having mildly prolonged latencies, most prominent in the legs. Needle EMG examination of an intrinsic foot muscle showed small-amplitude fibrillation potentials. The combined clinical and electrophysiologic findings were interpreted to represent a mild sensorimotor polyneuropathy of the axonal type.

Additional information included regular exposure to elemental mercury over nearly 8 years in the chloralkali plant. His urine mercury records for the 36 months prior to his neurologic evaluation included three samples exceeding 500 µg/L and three others exceeding 250 µg/L. In this plant, the workers' urine mercury levels corresponded to average air levels of approximately 100–200 µg/m^3. The urine mercury level used for removal from ongoing exposure was based on reports of tremor in some workers who had urine mercury levels of 300–600 µg/L (World Health Organization, 1976).

Clinical manifestations

Acute exposure to high-level mercury vapor produces an erosive bronchitis, interstitial pneumonitis with pulmonary edema, and renal insufficiency (Feldman, 1982; Berlin, 1986). Neurologic symptoms develop shortly thereafter in the form of increased excitability, postural tremor, and perioral numbness, an occasionally reported but unexplained symptom that may have reflected hyperventilation (Berlin, 1986; Bromberg, 2000). The neurologic effects associated with acute, high-level yet sublethal exposures are thought to resolve spontaneously (Albers and Bromberg, 1995), including cases with exposures sufficient to cause acute renal failure. Symptoms attributed to therapeutic mercury levels during medicinal use include gingivitis, excessive salivation, anorexia, and abdominal pain, whereas behavioral symptoms were more often attributed to toxicity (Waldron, 1983). Chronic occupational exposures to elemental mercury vapor are characterized by postural tremor, arguably the most sensitive clinical indicator of overexposure (Feldman, 1982; Albers et al., 1988). Other features include a large-fiber neuropathy associated with stocking-distribution sensory loss, positive Romberg sign, and absent ankle reflexes (Feldman, 1982; Albers et al., 1988). Motor involvement, when present, is mild, consisting mostly of atrophy of intrinsic foot muscles. The signs of mercury toxicity generally resolve after removal from exposure, although there may be residual tremor that usually is asymptomatic or insufficient to interfere with activities of daily living (Albers et al., 1988).

The association relating occupational elemental mercury and neuropathy is by no means robust. Evaluation of 138 chloralkali plant workers who had had substantial long-term mercury exposures (i.e., they were removed from ongoing exposure only when their urine mercury level exceeded 500 μg/L) identified only 18 workers with a neuropathy, all of whom were asymptomatic. These 18 workers had elevated urine mercury indexes compared with the remaining 120 subjects, associating their mercury exposure with a mild sensorimotor neuropathy, the degree of which correlated with the magnitude of exposure (Albers et al., 1982). In contrast, evaluation of 84 mercury-exposed thermometer-manufacturing workers whose urine mercury levels ranged from 1.3 to 344.5 μg/g creatinine showed that the exposed workers reported more symptoms and demonstrated a higher frequency of positive Romberg and difficulty with heel-to-toe gait than unexposed controls, but their symptoms or signs suggestive of neuropathy were not clearly associated with mercury exposure (Ehrenberg et al., 1991).

Electrophysiology

Nerve conduction study findings attributed to elemental mercury include low-amplitude sensory responses and normal or borderline-low amplitude motor responses. Distal latencies may be slightly prolonged, consistent with loss of large myelinated axons. Conduction velocities are normal. The EMG needle examination may show chronic neurogenic changes, most notable in the intrinsic foot muscles. Much of the evidence associating elemental mercury with a mild neuropathy reflects cross-sectional evaluations of exposed workers, not longitudinal studies. Workers frequently have markedly elevated urinary mercury levels, with some studies reporting levels exceeding 3000 μg/L (Albers et al., 1982, 1988).

The worker in the case presentation was a large male, 1.7 metre tall and weighing 117 kg (body mass index of 32 kg/m^2). Body size in terms of height and body mass index is of substantial importance to the clinical and electrophysiologic evaluations because height and body mass index are both associated with diminished distal sensation and reflexes, diminished sensory amplitudes and motor conduction velocities, and prolonged distal latencies, findings suggestive of neuropathy. Recognizing the influence of these anthropomorphic attributes, especially in terms of nerve conduction study normal values, is important in the assessment of individual workers.

Differential diagnosis

There is nothing distinctive about the mild sensorimotor neuropathy associated with elemental mercury exposure. The differential diagnosis of an axonal sensorimotor neuropathy is listed in Table 16.1, although the history and routine evaluations usually exclude many of the items from consideration. In this context, the absence of a coexisting postural tremor makes elemental mercury intoxication unlikely, whereas the combination of neuropathy, a postural tremor, and a characteristic maculopapular rash supports the possibility of elemental mercury intoxication, particularly in a worker having the opportunity for occupational exposure to mercury.

Additional diagnostic evaluation

The diagnosis of mercury intoxication requires confirmation by elevated urinary mercury excretion (Windebank, 1993). The preferred urine measurement is a 24-hour urine excretion or first morning void corrected for creatinine; spot samples are less reliable but often used for screening purposes or occupational studies. Because elemental mercury is nephrotoxic, laboratory testing may show impaired renal function. Despite reports attributing neuropathy to elemental mercury exposure on the basis of urine mercury levels of 30–40 μg/L, much of our understanding of elemental mercury toxicity derives from evaluations of mercury-exposed workers, many of whom have urinary mercury levels hundreds (and even thousands) of times higher than average levels observed in the general population, yet who have little or no evidence of neuropathy (Albers et al., 1988). In the early 1980s, such workers were removed from mercury vapor exposure only when routine sampling showed urine mercury levels exceeding 500 μg/L, a level in the range associated with abnormal postural tremor (World Health Organization, 1976). In contrast, the Environmental Protection Agency elemental mercury reference concentration for chronic inhalation exposure was 0.3 μg/m^3 as of 1995, based on an adjusted 8-hour time-weighted average lowest-observed adverse effect level. This value represents an estimate of the daily inhalation exposure unlikely to produce an appreciable risk during a lifetime of exposure. The American Conference of Governmental Industrial Hygienists time-weighted average threshold limit value of 25 μg/m^3 for daily occupational exposure to elemental mercury is a recommended threshold level below which adverse health effects are not expected (American Conference of Governmental Industrial Hygienists, 1999). Only rarely, however, are air-monitoring results available. The Environmental Protection Agency inhalation exposure level corresponds to a urine mercury level (excluding the order of magnitude uncertainty factor) in the range of 15 μg/L, using equivalent average concentrations in air and urine of 50 μg/m^3 and 150 μg/L, respectively (World Health Organization, 1976). This level is of similar magnitude to the biologic exposure index for mercury proposed for urine (20 μg/g creatinine) (Ellingsen et al., 1993; Lauwerys and Hoet, 1993; American Conference of Governmental Industrial Hygienists, 1999, 2013). In the context of the case presentation, the worker's urine-monitoring results suggested frequent exposures to mercury vapor exceeding 75 μg/m^3, with many above 150 μg/m^3, well above current recommended levels.

Serum mercury concentrations are a poor indicator of body burden, and hair is not a reliable tissue to document elemental mercury exposure because the measure probably reflects mercury absorption from the atmosphere, not the blood (Clarkson et al., 1988). Measurement of urine mercury levels after chelation (diagnostic chelation challenge) has been discredited as a useful biomarker of past mercury exposure (Frumkin et al., 2001).

Treatment

Treatment of elemental mercury intoxication involves removal from ongoing excessive exposure, recognizing that mercury is ubiquitous and some exposure is unavoidable. Chelation therapy using mercury-binding agents such as British anti-Lewisite, edetate calcium disodium, penicillamine, sodium 2,3-dimercaptopropane-1-sulfonate, and 2,3-dimercaptosuccinic acid is not generally considered necessary or effective, but is sometimes recommended after acute exposure (Oz et al., 2012), despite lack of proven efficacy.

Prognosis

The prognosis for resolution of acute symptoms and signs (including neuropathy) attributed to elemental mercury after removal from ongoing exposure is good, even among children (who are thought to be particularly susceptible) experiencing very high exposures, including some with urine levels exceeding 4500 μg/L (Carman et al., 2013). Case reports of elemental mercury neuropathy generally describe recovery or improvement after exposure is terminated. The exceptions include: (1) a case whose neuropathy resembled CIDP, an immune-mediated neuropathy unrelated to mercury exposure (Chu et al., 1998); (2) a case complicated by long-standing type 1 diabetes whose neuropathy developed in the setting of deteriorating glycemic control and improved substantially following cessation of exposure, chelation, and good control of the diabetes (Pelclova et al., 2002); and (3) a case showing improvement but having insufficient follow-up to document complete recovery (Shih and Gartner, 2001).

A similar overall favorable prognosis is reflected by the cross-sectional studies of workers having prior mercury exposure. Namely, examination of 502 subjects, 247 of whom had substantial elemental mercury exposures 20–35 years previously, found few significant differences between exposed and unexposed subjects (Albers et al., 1988). Subjects with clinically evident neuropathy had significantly higher urine mercury peak levels than normal subjects (850 vs 610 μg/L; $p=0.04$), and 28% of subjects with peak levels above 850 μg/L had neuropathy compared with 10% of remaining subjects ($p=0.005$). These results were confirmed by examination of 205 former workers, 104 of whom had been exposed to elemental mercury (Letz et al., 2000). In this study, peripheral nerve functions were significantly associated with prior cumulative mercury exposure, again suggesting that some long-term peripheral nervous system effects may be detectable decades after removal from exposure. Similarly, examination of 77 former chloralkali workers previously exposed to elemental mercury and 53 referents showed a higher prevalence of reduced distal sensation, postural tremor, and abnormal Romberg sign (Andersen et al., 1993). Although suggestive of a persistent neuropathy years after exposure cessation, the mild neurologic abnormalities could not be classified as disease in most cases, and in none of these studies was the residual neuropathy severe.

Hexacarbons

Peripheral neurotoxicity is often attributed to exposure to "organic solvents," possibly because of the association between ethyl alcohol and neuropathy, yet few solvents actually produce this effect. The hexacarbons, particularly *n*-hexane and methyl-*n*-butyl ketone (MBK), are the exceptions and among the small number of solvents capable of producing a clinically evident peripheral neuropathy (Korobkin et al., 1975; Scelsi et al., 1980; Ruff et al., 1981; Kuwabara et al., 1993). *n*-Hexane is a six-member carbon chain ring molecule found in industrial and household glues. MBK, a substituted hexacarbon also known as 2-hexanone, has been used in paint and paint thinners and to dissolve oils. Occupational exposure to *n*-hexane occurs in association with cabinet and shoe making, refinery work, and many construction jobs, where hexacarbon solvents were used extensively until the 1970s (Herskowitz et al., 1971; Spencer et al., 1975; Paulson and Waylonis, 1976; Scelsi et al., 1981; Iida, 1982). *n*-Hexane and MBK are absorbed across respiratory membranes and skin, and metabolized in the liver to 2,5-hexanedione, the likely neurotoxicant (Allen et al., 1975; DiVincenzo et al., 1980; O'Donoghue and Krasavage, 1980; Couri and Milks, 1985; Governa et al., 1987; Albers and Wald, 2002). Methyl ethyl ketone, a substituted hexacarbon similar to MBK, is not neurotoxic but may promote neuropathy due to other hexacarbons (Altenkirch et al., 1977).

The nerves of *n*-hexane-intoxicated animals show accumulations of neurofilaments (Scelsi et al., 1981; Cavanagh, 1985), presumably responsible for abnormalities of fast and slow axonal transport mechanisms

(Graham et al., 1985). At present, *n*-hexane neuropathy has been associated with use of degreasing solvents and cleansers (Chang et al., 1993; London and Albers, 2007), including use in the electronics and other light industries and press proofing (ink printing) (Kim and Kang, 2010; Liu et al., 2012). Most often, occupational exposures involve workers using glues without adequate ventilation (Sendur et al., 2009). Although the initial reports of hexacarbon neurotoxicity resulted from occupational exposures, most neurotoxic exposures at present are the result of intentional abuse in the form of recreational "glue sniffing" or "huffing," as illustrated in the following case presentation (Lalloo et al., 1981; Smith and Albers, 1997).

CASE PRESENTATION (From Smith and Albers, 1997 and Albers, 2005d)

A 25-year-old man experienced progressive weakness, sensory loss, and limb pain over several months, in association with a 25-lb weight loss over the previous 3 months despite reportedly good nutrition. He had a history of schizophrenia and auditory hallucinations, successfully treated with Thorazine and Prozac. He also reported previous polysubstance abuse, including crack cocaine, amphetamines, and occasional glue sniffing, but denied any drug use in the preceding year. All abnormalities involved his peripheral nervous system. He had distal weakness (graded 2–3+ on the Medical Resarch Council scale) and abnormal touch, vibration, joint position, pin, and temperature sensations. Arm reflexes were hypoactive, whereas leg reflexes were unobtainable. Initial laboratory testing was unremarkable, including liver function, HIV antibodies, phosphate, and glucose levels. The serum protein electrophoresis was normal, and no laboratory abnormalities suggested vasculitis. Twenty-four-hour urine measurements for arsenic, mercury, and lead were normal and he had a negative "toxicology screen." His CSF protein was slightly elevated (60 mg/dL) with normal glucose and cell count.

Electrodiagnostic evaluation about 1 month after onset of weakness showed slightly reduced sensory amplitudes. Motor responses were moderately reduced in amplitude, with abnormal temporal dispersion and/or partial conduction block, substantially reduced conduction velocities (e.g., median 44 m/s and ulnar 28 m/s), prolonged distal latencies (to an extent greater than could be explained by axonal loss), and absent F waves. The needle EMG examination showed profuse fibrillation potentials in distal muscles, decreased motor unit recruitment, but only slightly increased motor unit amplitude and duration.

His weakness progressed during the first 3 weeks of hospitalization, at which time he had less than antigravity function in his extremities, with markedly impaired activities of daily living. Despite his sensory loss, he complained of severe burning and electric shock-like pain in his extremities. Concern about an underlying vasculitis resulted in a sural nerve biopsy showing loss of large myelinated fibers but no evidence of vasculitis. The main finding was the presence of focal axonal swellings suggestive of *n*-hexane intoxication. Confronted with the nerve biopsy results, he admitted to a recent dramatic increase in glue sniffing, huffing rubber cement from a plastic bag, and using more than 40 4-oz bottles during the month preceding admission. Inspection of one of the bottles confirmed that the rubber cement contained *n*-hexane. About 1 year after discharge, and after receiving care for his substance abuse, he displayed only minimal distal weakness and sensory loss with normal reflexes. Serial electrodiagnostic evaluations showed progressive improvement in nerve conduction study measures, with increased sensory and motor amplitudes into the normal range.

Clinical manifestations

Substantial exposure to *n*-hexane results in central nervous system depression and narcosis; repeated exposures result in a length-dependent polyneuropathy with sensory loss, weakness, impaired reflexes, and autonomic dysfunction (Paulson and Waylonis, 1976; Ruff et al., 1981; London and Albers, 2007). The type and magnitude of clinical impairment are related to the potency, dose, and duration of exposure (Spencer and Schaumburg, 1985), and progression depends on the cumulative dose and whether exposure is ongoing. At times, the neuropathy progresses over weeks to a few months, resembling GBS, CIDP, or a confluent mononeuritis multiplex (Smith and Albers, 1997). In time, stocking-glove sensory loss involving all modalities and weakness develop, sometimes including foot drop or leg paralysis, progressing to involve the arms.

Electrophysiology

The electrophysiologic findings reflect the severity of neuropathy, its rate of progression, and the timing of the study relative to the onset of neuropathy. Nerve conduction evaluations of asymptomatic, hexacarbon-exposed workers suggest the presence of a subclinical neuropathy manifested by mildly slowed motor conduction velocities (Buiatti et al., 1978), sometimes with low sensory amplitudes (Paulson and Waylonis, 1976; Ruff et al., 1981; Chang et al., 1993). These findings correlate with urinary 2,5-hexanedione and are thought to indicate subclinical neuropathy (Neghab et al., 2012). *n*-Hexane-exposed printers with clinically evident neuropathy show a similar range of abnormalities, with low sensory and motor amplitudes, decreased conduction velocities, and needle EMG evidence of axonal degeneration among more severely involved workers (Chang et al., 1993).

More rapid-onset neuropathies develop after high-level *n*-hexane exposures and produce electrophysiologic findings more typical of those reported in the case presentation. These findings include evidence of substantial motor conduction slowing (often in the 20–40 m/s range, well below the lower limit of normal and in the range of a primary demyelinating disorder), with abnormal temporal dispersion, partial conduction block, prolonged motor distal latencies, and prolonged or absent F-response latencies (Yokoyama et al., 1990; Smith and Albers, 1997; Pastore et al., 2002). These features suggest nonuniform conduction abnormalities and distinguish the findings from those associated with hereditary neuropathies. Such findings were once thought to be near-pathognomonic of inflammatory nerve disease (Scelsi et al., 1980; Ruff et al., 1981; Kuwabara et al., 1993; Pastore et al., 2002), but it is now accepted that these conduction abnormalities are explained by secondary myelin damage associated with the axonal swellings demonstrated in peripheral and central nerve fibers (Korobkin et al., 1975; Oh and Kim, 1976; Griffin, 1981; Spencer and Schaumburg, 1985; Yokoyama et al., 1990; Chang et al., 1993; Kuwabara et al., 1993). Although sensory responses are usually abnormal, a sural response is often obtainable despite quite severe motor abnormalities. Some reports emphasize the prominent motor nerve findings, but a pure motor neuropathy would be incompatible with the sural nerve biopsy and electrophysiologic abnormalities (Oh and Kim, 1976; Lalloo et al., 1981; Tenenbein et al., 1984; Kuwabara et al., 1993). The needle EMG examination shows decreased recruitment and fibrillation potentials in distal-extremity muscles.

Differential diagnosis

The clinical and electrophysiologic features of *n*-hexane neuropathy, at least in the early stages of a severe intoxication, are those of a motor greater than sensory neuropathy with conduction slowing (Albers and Teener, 2006). The differential diagnosis for such a neuropathy, as shown in Table 16.2, includes several additional potential occupational toxicants, such as acrylamide, arsenic, carbon disulfide, and MBK. Far more common are the hereditary neuropathies, immune-mediated inflammatory neuropathies (e.g., GBS and CIDP), several closely related paraneoplastic immune-mediated neuropathies, and at least one infectious (HIV-associated) neuropathy. The differential diagnosis includes consideration of a confluent mononeuritis multiplex associated with systemic or isolated nervous system vasculitis, a condition that should be considered particularly when pain is prominent (Albers and Kelly, 1989). In general, medications are a common cause of toxic neuropathy, and amiodarone and disulfiram are included in the table, as both can cause a motor-predominant neuropathy similar to that of *n*-hexane.

Additional diagnostic evaluation

The CSF protein level may be slightly elevated in association with an *n*-hexane neuropathy, but not markedly, thereby reducing the likelihood of an inflammatory, immune-mediated neuropathy. A sural nerve biopsy showing axonal swellings supports a toxic etiology (Scelsi et al., 1980; Sayre et al., 1985; Yokoyama et al., 1990). *n*-Hexane and its metabolite 2,5-hexanedione can be demonstrated in the urine for a few days after acute exposure but are unlikely to be of diagnostic importance (Albers, 2012). Reports exist of brain magnetic resonance imaging (MRI) T2-weighted periventricular hyperintensities and proton MR spectroscopy abnormalities coexisting with *n*-hexane neuropathy owing to volitional abuse of *n*-hexane (Hashizume et al., 2011).

Treatment

Cessation of exposure is the primary means of treatment. Pain control may be required for painful dysesthesias.

Prognosis

The prognosis for recovery is favorable, as demonstrated in the case presentation and by reports of *n*-hexane-intoxicated workers (Misirli et al., 2008), including those with sensory and reflex loss, weakness, and muscle atrophy (Kuang et al., 2001: published in the Chinese literature with English abstract available). Despite cessation of exposure, weakness may progress for several weeks ("coasting"), typical of many other toxic neuropathies. Thereafter, the neuropathy stabilizes and improvement ensues. Even in cases of severe *n*-hexane neuropathy associated with quadriparesis, the peripheral abnormalities are reversible, but recovery may be protracted over several years. In parallel, electrophysiologic abnormalities progressively improve on serial evaluations and eventually resolve (Lalloo et al., 1981; Smith and Albers, 1997; Misirli et al., 2008). Abnormal spontaneous activity resolves in parallel with the improved nerve conduction study results, and reinnervation is documented by the presence of large-amplitude motor units on needle EMG (Iida, 1982). Serial evaluations have noted discordance between normal clinical examinations and residual EMG abnormalities (Sendur et al., 2009), an observation possibly due to insufficient duration of follow-up or permanent changes owing to reinnervation and remyelination of regenerating axons.

Lead

Lead is a nonferrous metal absorbed across respiratory and gastrointestinal membranes (Fischbein, 1983). Lead poisoning causes an encephalopathy in children but is associated with neuropathy in adults. Lead neuropathy is poorly characterized despite ancient recognition that lead was associated with illness (Albers and Berent, 2005). For example, between 1949 and 1989 fewer than 10 cases of lead neuropathy were identified among 3,500,000 new patient evaluations at the Mayo Clinic (Windebank, 1993), grouping lead neuropathy with the rarest of neurologic disorders. As of the mid-1980s, Wilbourn (1984) declared lead neuropathy almost extinct. There are frequent opportunities for lead exposure, however, as lead has widespread distribution, particularly since the development of metallurgy during the industrial revolution. Leaded fuel was a major source of environmental exposure in the USA but air levels have decreased in the last 30 years owing to the ban on leaded gasoline (Tavolato et al., 1980; Wu et al., 1995). Environmental exposure may still result, however, from several sources, including consumption of illicit whiskey made in lead-lined stills (a declining source of lead exposure confounded by simultaneous alcohol exposure) (Ellis and Lacy, 1998), or from unsuspected sources such as lead-containing candy, powdered food coloring, herbal medications, or food contaminated by glaze pigments in ceramic dishes (Anonymous, 1998; Phan et al., 1998; Ibrahim and Latif, 2002; Albers, 2005c, 2012).

Most cases of adult lead poisoning result from industrial exposures (Grandjean, 1978; Thomson and Parry, 2006) involving smelting, reprocessing of lead-containing products, demolition of lead-painted structures, burning lead-painted surfaces, manufacturing paint pigments, preparation of some plastic products, repair of automobile radiators, and manufacturing of lead-acid batteries, a process that accounts for 85% of the total lead usage in the USA (Conradi et al., 1982b; Thomson and Parry, 2006). In the Western world, improved industrial hygiene has reduced the risk of overt lead neuropathy (Thomson and Parry, 2006), but some countries, including Iran, continue to recognize occupational lead poisoning as an ongoing health problem in certain processes, such as the glazing of tiles (Balali-Mood et al., 2010). Nevertheless, few physicians will ever recognize a patient with lead neuropathy, despite its frequent inclusion in the differential diagnosis of many conditions.

Like many neurotoxicants, lead is a systemic poison and lead neuropathy never presents in the absence of hematologic, gastrointestinal, or renal consequences (Thomson and Parry, 2006). Experimental models of lead neuropathy contribute to the confusion, as lead produces a demyelinating neuropathy in some models relative to the axonal degeneration described in humans (Ehle, 1986; Thomson and Parry, 2006). The lesion responsible for lead neuropathy is unknown (Wilbourn, 1984; Thomson and Parry, 2006). There are parallels between lead and the hepatic porphyrias, including increased porphyrin excretion, interference with heme synthesis and resemblance of lead and porphyric neuropathies. This observation has led to speculation that the metabolic basis for lead neuropathy is interference with porphyrin metabolism (Thomson and Parry, 2006), and the possibility that lead neuropathy represents lead-induced porphyric neuropathy, not direct neurotoxicity (Donofrio et al., 1987; Rubens et al., 2001; Albers and Fink, 2004; Thomson and Parry, 2006). The wide variation in response to lead exposure may reflect genetic susceptibility toward lead, such that certain genotypes modify the relationship between lead and its peripheral neurotoxicity (Zheng et al., 2011).

Clinical manifestations

Acute lead intoxication results in abdominal cramping, constipation, and anorexia, possibly owing to decreased gastrointestinal motility (Thomson and Parry, 2006). Signs of chronic intoxication potentially include weight loss, constipation, renal insufficiency, anemia, and gingival lead lines (Feldman et al., 1977; Thomson and Parry, 2006; Kumar, 2010). In one study, gingival lead lines consisting of a dark blue pigmentation in gingival margin were found in 65% of lead-exposed workers, constituting the most common clinical sign (Balali-Mood et al., 2010).

The classic description of lead neuropathy as a pure motor neuropathy dates to the 19th century (Beritic, 1984), with subacute onset of asymmetric weakness and wasting of the wrist and finger extensors and with relative sparing of other muscles and sensation unless the neuropathy is severe (Aub et al., 1925; Beritic, 1984; Wilbourn, 1984; Ghosh, 2004). The extensors of the middle and ring fingers are involved first before spreading to other finger and thumb extensors and then wrist extensors (Thomson and Parry, 2006). Weakness was not length-dependent (Ehle, 1986) and it is unknown if the weakness was explained by predisposition to local trauma or pressure. More atypical patterns, as summarized by Wilbourn (1984) in his review, include weakness of proximal shoulder girdle, hand, and anterior leg muscles. Reports of lead neuropathy progressing to quadriplegia (Oh, 1975; Thomson and Parry, 2006) suggest a possible association between lead and progressive motor neuron diseases, including amyotrophic lateral sclerosis (Campbell et al., 1970; Roelofs-Iverson et al., 1984). This association remains speculative, as subsequent case–control studies have concluded that the limited nature of the association suggests a multifactorial etiologic mechanism (Armon et al., 1991). There also are descriptions associating lead with a distal sensorimotor neuropathy (Thomson and Parry, 2006; Kumar, 2010) or a mild sensory neuropathy (Rubens et al., 2001). Whether these different patterns reflect response to different exposure durations and concentration or even different diseases is unclear.

Electrophysiology

It is ironic that credible electrodiagnostic descriptions of lead neuropathy are lacking, as lead is thought to have intoxicated more individuals throughout the ages than all other metals (Wilbourn, 1984). One possible explanation is that most cases were clinically described before the modern electrophysiology era and prior to application of improved industrial hygiene. For the most part, the few electrophysiologic descriptions available to Wilbourn in the early 1980s are the same reports available today, 30 years later. A few descriptions of clinically evident lead neuropathy include nerve conduction study evidence of mixed axonal and demyelinating features with mild slowing of motor and sensory velocity or, in severe cases, EMG evidence of abnormal spontaneous activity and neurogenic motor units (Seppalainen et al., 1975; Ehle, 1986; Wu et al., 1995; Ghosh, 2004; Krishnan et al., 2012).

Most descriptions are the results of cross-sectional nerve conduction study comparisons of lead-exposed workers and referents, in the context of evaluating subclinical neurotoxicity. Such comparisons show puzzling combinations of no differences (Paulev et al., 1979) or slight group differences within the normal range, with exposed workers having slightly slowed motor conduction (Bordo et al., 1982; Yeh et al., 1995), slowed motor conduction with abnormal temporal dispersion/partial conduction block (Catton et al., 1970), slowed motor and sensory conduction (Feldman et al., 1977; Ashby, 1980; Murata et al., 1993), or reduced sensory amplitude (Rubens et al., 2001). These results are generally interpreted to denote a subclinical lead neuropathy, with little evidence that the subclinical differences ever progress to overt neuropathy (Thomson and Parry, 2006). Evidence that a subclinical neuropathy presents with dose-related abnormalities is included in the referenced studies (Seppalainen and Hernberg, 1972, 1980; Seppalainen et al., 1975; Landrigan et al., 1976; Singer et al., 1983; He et al., 1988). As a general observation, evidence of a subclinical neuropathy has not been found when serum lead levels are below about 60 μg/dL (Seppalainen et al., 1979; Nielsen et al., 1982; Baker et al., 1984; Triebig et al., 1984; Ehle, 1986; He et al., 1988). Although it is not clear if there is a threshold lead level below which evidence of a subclinical neuropathy does not exist, it is equally unclear if the relationships identified in these cross-sectional evaluations reflect exposure to lead or other unrecognized confounders. At best, the subclinical electrophysiologic associations remain unproven.

Differential diagnosis

Disorders associated with a motor or motor more than sensory neuropathy without substantial conduction slowing are listed in Table 16.3. Those conditions characterized by asymmetry are identified by an asterixis and include multifocal motor neuropathy, dapsone neuropathy, and most forms of motor neuron disease. Other conditions that mimic the asymmetric "motor neuropathy" ascribed to lead include inclusion body myositis and the mononeuropathies, although an absence of sensory involvement excludes most from consideration. When the classic description of lead neuropathy originated, a patient with lead exposure and asymmetric weakness could not have been diagnosed with many of the disorders listed in Table 16.3, for the simple reason that most were not yet described.

As an illustration that increased exposure to lead, even one resulting in a documented abnormal dose, does not establish causation, consider the report of a lead-exposed painter who developed painless right-arm weakness and anemia in the setting of elevated serum and urine lead levels (Albers and Bromberg, 1994). A diagnosis of lead neuropathy seemed secure until his brother developed similar findings despite normal lead levels, leading to the additional evaluation and, ultimately, a diagnosis of X-linked bulbospinomuscular atrophy (Kennedy's disease). Porphyric neuropathy is included in the differential diagnosis, and, in fact, may be directly linked to lead neuropathy. As discussed earlier, lead may be one of the numerous substances capable of inducing porphyric neuropathy, not a direct peripheral neurotoxicant.

Additional diagnostic evaluation

Confirmation of lead intoxication requires demonstration of elevated lead levels in blood (the primary screening test) or tissues, although what constitutes an acceptable amount of lead is subject of controversy. Systemic effects attributed to lead develop at levels that usually exceed 80 μg/dL and include a microcytic/hypochromic anemia, basophilic stippling of red blood cells, and abnormal excretion of heme precursors, delta-aminolevulinic acid and coproporphyrin (Baloh, 1974; Moore et al., 1987; Froom et al., 1999). Serum iron levels are normal, consistent with abnormal hemoglobin synthesis thought due to inhibition of D-aminolevulinic acid synthase. Lead has a long biologic half-life, reflecting storage in bone (Conradi et al., 1982a); lead can be identified on bone imaging or *in vitro* X-ray fluorescence (Landrigan and Todd, 1994). Nevertheless, excessive lead sequestration in bone does not necessarily translate into lead neurotoxicity (Thomson and Parry, 2006). Urine and hair lead levels can be measured but neither is widely used. Mobilization of lead after chelation challenge has limited utility (Whitaker et al., 1962). Laboratory evaluations directed at identifying a competing diagnosis, such as Kennedy syndrome, may be relevant.

Treatment

There are protocols that describe use of chelation to treat lead poisoning, but their effectiveness is undoubtedly limited by the large stores of lead in bone. Although the efficacy of any treatment for lead neuropathy is unknown, anecdotal case reports describe the successful use of chelation to treat symptomatic lead neuropathy following an acute exposure (Krishnan et al., 2012).

Prognosis

Prognosis for recovery is presumably good when exposure is terminated, within the context of lead's long biologic half-life. Nevertheless, there is little available information involving recovery from lead neuropathy.

Organophosphorus compounds

Organophosphorus (OP) compounds have widespread industrial and commercial applications, including use in insecticides, hydraulic fluids, lubricants, fuel additives, plasticizers, and flame retardants. The ready availability has resulted in its use in suicide attempts (Jayaratnam, 1990; Malik et al., 1998) and in acts of bioterrorism. OP insecticides produce cholinergic toxicity with muscarinic overactivity due to inhibition of AChE at cholinergic nerve endings, resulting in acute neuromuscular blockade (Kropp and Richardson, 2003). The inactivation of AChE is sometimes described as "irreversible" but is better designated as slowly reversible because AChE can be reactivated.

Whereas certain OP compounds produce dose-dependent cholinergic effects and neuromuscular blockade, massive exposure to some OPs is also associated with OPIDN, the result of inhibition and aging (molecular rearrangement) of neuropathy target esterase (NTE) (Lotti, 1991; Richardson, 1995). The initial mechanism in the development of OPIDN involves phosphorylation of NTE, followed by the aging reaction and a resultant loss of calcium homeostasis important to cytoskeletal proteins in the motor axon (Emerick et al., 2012). Some OP compounds that do not inhibit AChE also produce OPIND, indicating that the neuropathy is not related to the anticholinesterase effect (Dickoff et al., 1987). In the 1930s, tri-ortho-cresyl phosphate (TOCP) was associated with an outbreak of OPIDN. It was discovered that TOCP had been added to Jamaican ginger extract (Jake) as a plasticizer to circumvent prohibition regulations (Morgan and Penovich, 1978). Jake had been sold as a medicinal tonic but also consumed for its alcohol content. Shortly after TOCP was added to Jake, thousands of cases of OPIDN ("Jake paralysis") developed among those who consumed the product. In the 1970s, TOCP was the suspected offender in another outbreak of OPIDN, this time attributed to contaminated cooking oil that had been transported in containers previously used to store mineral oils (Senanayake and Jeyaratnam, 1981).

NTE may have implications beyond OPIDN, as mutations in the NTE gene are associated with an autosomal-recessive motor neuron disease (NTE-MND) (Rainier et al., 2011). NTE-MND is characterized by progressive, symmetric lower-extremity spastic weakness beginning in childhood with later development of atrophy of distal leg and intrinsic hand muscles, a combination of upper and lower motor neuron deficits similar to OPIDN. Motor neuron degeneration associated with NTE mutations supports the role of NTE and its biochemical cascade in the molecular pathogenesis of OPIDN and possibly other degenerative neurologic disorders (Rainier et al., 2011). Disease-specific NTE mutations support the importance of NTE in maintaining axonal integrity, raise the possibility that NTE pathway disturbances contribute to other MNDs, including amyotrophic lateral sclerosis, and support the role of NTE abnormalities in axonopathy produced by neuropathic OP compounds. The possibility that OP compounds contribute to motor neuron disease is supported by association of paraoxonase 1 (PON1) polymorphisms with amyotrophic lateral sclerosis and the selective degeneration of motor neurons in OPIDN, in which NTE is inhibited by organophosphorylation (Rainier et al., 2008).

In contrast to the established effects of massive OP exposure, our understanding of the consequences of long-term, low-dose OP exposures that do not produce acute cholinergic toxicity is less certain (Hernandez et al., 2006). At present, there is no evidence that prolonged, low-level OP insecticide exposures produce adverse peripheral nerve effects (Lotti, 2002). For example, long-term occupational exposure to chlorpyrifos (an OP pesticide) during the manufacturing process at levels sufficient to produce biologic effects on BuChE activity found no clinically evident or subclinical neurologic effects at baseline or after 1 year of additional exposure (Albers et al., 2004, 2007). Despite chlorpyrifos exposures to levels at which well-described physiologic effects on B-esterases exist, the frequency of symptoms or signs of neuropathy was similar among exposed and unexposed workers (Albers et al., 2007). In other occupational settings, however, adverse effects continue to be reported. For example, four workers exposed to sodium 3,5,6-trichloropyridin-2-olate used in the production of chlorpyrifos developed transient diffuse "chloracne-like" skin lesions and a poorly characterized neuropathy (Wu et al., 2012). Two of the cases had distal lower-extremity numbness and pain, weakness, and diminished ankle reflexes, whereas all four showed neurogenic EMG changes in the legs and prolonged mean tibial F response latencies (findings of uncertain importance in isolation).

PON1 is an enzyme found in liver and plasma that hydrolyzes a number of OP compounds. It has been proposed that polymorphisms of PON1 are potentially important in determining susceptibility to OP neurotoxicity because individuals having the PON1192Q allotype hydrolyze chlorpyrifos oxon less efficiently than the PON1192R allotype. This hypothesis was evaluated in studies of workers exposed to chlorpyrifos during the manufacturing process (Albers et al., 2010). Although BuChE activity was significantly inversely associated with measures of interim chlorpyrifos exposure, the biologic effects of chlorpyrifos exposure on BuChE activity were not modified by PON1 inferred genotype or chlorpyrifosoxonase activity.

CASE PRESENTATION

This patient was reported previously (Albers, 2005c) from an initial report presented in greater detail (Lotti et al., 1986).

This 42-year-old man attempted suicide by voluntarily ingesting an unknown amount of an OP pesticide. On arrival to the hospital, he was comatose and required intubation and ventilator support. He was given atropine to control cholinergic signs of sweating, tearing, and diarrhea. Laboratory testing identified markedly reduced plasma BuChE and red blood cell AChE levels, after which he was given pralidoxime to reactivate the phosphorylated cholinesterase. Following treatment, he stabilized and slowly recovered such that within 3 weeks he was asymptomatic and clinically normal. Nerve conduction studies, including tests of neuromuscular transmission, were performed prior to discharge and were normal. However, within 3 weeks of discharge, he noted onset of leg weakness and paresthesias, and a subsequent examination showed an impaired gait, diminished vibratory sensation in the legs, and loss of reflexes. Repeat nerve conduction studies showed low-amplitude motor and sensory responses and reduced conduction velocities. A needle EMG examination showed fibrillation potentials in distal lower-limb muscles.

Clinical manifestations

A massive OP exposure produces cholinergic stimulation of muscarinic and nicotinic receptors. Activation of muscarinic receptors produces miosis, increased secretions, sweating, gastric hyperactivity, and bradycardia, whereas activation of nicotinic receptors produces skeletal-muscle fasciculations and weakness. The effects of cholinergic stimulation may be severe and require intubation for airway protection and respiratory support. Some patients who recover from the initial cholinergic effects subsequently deteriorate and develop weakness of cranial, bulbar, and proximal-limb muscles. This "intermediate syndrome" develops 2–3 weeks after poisoning and is thought to reflect impaired neuromuscular transmission, similar to that of myasthenia gravis (Senanayake and Karalliedde, 1987b). A subset of patients requiring respiratory and intensive care unit support during the cholinergic phase develop OPIDN during or after recovery from the life-threatening cholinergic effects (Senanayake and Johnson, 1982). OPIDN is characterized by rapidly progressive distal weakness and reduced or absent ankle reflexes. Some patients experience dysesthesias at onset but show little or no sensory loss and no bladder or bowel involvement. The symmetric motor neuropathy is associated with normal or brisk reflexes proximally. This is a relevant observation, because as the neuropathy resolves, distal flaccid weakness is replaced by spasticity and hyperactive reflexes, signs of pyramidal tract motor involvement that was initially masked by the neuropathy. The resultant syndrome reflects the combination of upper and lower motor neuron involvement. OPIDN associated with TOCP presents similarly but without the initial cholinergic syndrome that characterizes exposure to most OPs.

Electrophysiology

During the cholinergic phase of an OP insecticide poisoning, motor responses show repetitive discharges after a single supramaximal depolarizing stimulus, presumably due to recurrent depolarization of the postsynaptic endplate (Gutmann and Besser, 1990), an effect identical to that seen in patients with myasthenia gravis receiving excessive amounts of cholinergic medication. If the cholinergic neuromuscular blockade is substantial, the motor response amplitude may be low and show a decrement to repetitive stimulation before the acute cholinergic effects resolve (over days to weeks). The decrement is said to be present at high stimulation rates (20 and 50 Hz) but not at low rates (e.g., 2 Hz) and displays no posttetanic facilitation, observations difficult to reconcile with current understanding of neuromuscular physiology (Senanayake and Karalliedde, 1987a; Wadia et al., 1987). Despite a few reports describing sensory involvement after an OP exposure (Kaplan et al., 1993), prospective study of workers having substantial OP insecticide exposure during the manufacturing process found no evidence of a clinically evident or subclinical sensory neuropathy (Albers et al., 2004). The abnormalities associated with OPIDN include low-amplitude motor responses without substantial conduction slowing, and needle EMG evidence of fibrillation potentials, decreased recruitment, and large-amplitude motor units in distal limb muscles (LeQuesne, 1978; Senanayake and Johnson, 1982; Wadia et al., 1987).

Differential diagnosis

OP intoxication results in a condition best classified as a motor neuropathy without substantial conduction slowing. Some conditions producing a motor or a motor greater than sensory neuropathy without substantial motor conduction slowing are listed in Table 16.3. Included are several hereditary or familial, infectious, and idiopathic forms of motor neuron disease, as well as disorders of the neuromuscular junction. Nevertheless, the initial cardinal features of cholinergic overactivity and diffuse paralysis with respiratory insufficiency suggest an OP intoxication. Absent the features of prominent cholinergic overactivity, the differential diagnosis for an acute quadriparesis with respiratory failure includes GBS or CIDP, acute intermittent porphyria, diphtheria, tick paralysis, botulinum intoxication, and acute inflammatory myositis. The onset of recurrent weakness within weeks of recovery from a life-threatening cholinergic syndrome requiring intensive support suggests the diagnosis of OPIDN (Moretto and Lotti, 1998). Diagnosing OPIDN resulting from an OP compound that does not produce cholinergic toxicity is more challenging, especially if the exposure was occult and not involving many individuals, but nonetheless would reflect a combined involvement of central and peripheral motor neurons.

Additional diagnostic evaluation

Depressed plasma BuChE and red blood cell AChE activity is used to document AChE inhibition due to an acute OP poisoning. In an occupational setting where the potential for an OP compound is known, baseline cholinesterase activity levels are generally available. Cholinergic symptoms generally are associated with at least a 70% or greater decrease from baseline cholinesterase activity (Osterloh et al., 1983). Metabolites of some OP insecticides can be measured in the urine to document the magnitude of exposure, such as 3,5,6 trichloro-2-pyridinol, a metabolite of chlorpyrifos. OPIDN is not characterized by any specific laboratory abnormalities.

Treatment

Gastric lavage of a recently ingested OP and supportive care, including intubation to protect the airway and provide respiratory support, are required in severe poisonings. Atropine reduces the muscarinic effects, and pralidoxime reactivates the phosphorylated cholinesterase at the neuromuscular junction, reducing weakness (Wills, 1970; Koller and Klawans, 1979; Sidell, 1994). No specific treatment is available to prevent the delayed upper motor neuron consequences of OPIDN.

Prognosis

Mortality is high after a severe OP poisoning, even with intensive care and respiratory support. Among patients who experience OPIDN, despite partial or full recovery from the lower motor neuron syndrome, prognosis remains poor due to the appearance of corticospinal tract abnormalities. A 13-year follow-up evaluation of 15 survivors of TOCP poisoning found that all had leg spasticity, ranging from mild to severe, with hyperreflexia (14 of 15), ankle clonus (11 of 15), Babinski signs (6 of 15), and Hoffmann signs (5 of 15) (Wang et al., 2009). Abnormal corticospinal tract signs included lower-limb hyperreflexia. All had normal sensation. Nerve conduction studies were normal in 13 of 15 cases.

Porcine neural tissue

In 2007, a cluster of swine abattoir (slaughterhouse) workers exposed to aerosolized porcine brain tissue developed symptoms attributed to a neuropathy (Holzbauer et al., 2010; Lachance et al., 2010). The initial group included 24 patients evaluated at the Mayo Clinic. All were employed in similar work in two pork-processing plants (in southern Minnesota and Indiana), an observation that immediately suggested a possible occupational exposure or a disease passed from pigs to humans (Deangelis and Shen, 2009). The Minnesota Department of Health investigated the facility and determined that all cases had exposure to aerosolized porcine brain tissue as the result of removing brains from the skull using compressed air at the "head table." All had a persistent (months) condition and most had features resembling an immune-mediated chronic CIDP. Three cases also had central nervous system involvement, characterized as transverse myelitis, meningoencephalitis, and aseptic meningitis (Tracy and Dyck, 2011). No systemic disorder or infectious agent was identified, and no family members became ill, further making an infectious agent unlikely. In contrast, sera from affected and many unaffected workers but no community controls showed a distinctive neural-reactive immunoglobulin G (IgG) that correlated with exposure risk. Specifically, the novel biomarker consisted of a unique IgG immunofluorescence pattern with binding to multiple neural elements, including cerebellar cortical white matter (consistent with myelin basic protein immunoreactivity), molecular-layer neuropil, Purkinje neuronal perikarya, cranial nerve fascicles, and autonomic neurons. This pattern was present in all 24 cases with neurologic findings, 34% of workers in the pork-processing plant, but no healthy controls. Dose–response included higher antibody titers in workers closer to the area of brain extraction relative to those farther away (Tracy and Dyck, 2011).

Although an occupational, immune-mediated polyradiculoneuropathy is distinctly unique, the pathophysiologic mechanism is consistent with other immune-mediated conditions resulting from exposure to neural tissue (Holzbauer et al., 2010). These include immune-mediated neuropathy (experimental allergic neuritis) and encephalopathy (experimental allergic encephalitis), as established in experimental animal models. Some immune-mediated neuropathies such as GBS develop within weeks of exposure to an antecedent respiratory, gastrointestinal illness (e.g., *Campylobacter jejuni*), or other infection (e.g., cytomegalovirus or Epstein–Barr virus). Similarly, the 1976 swine flu vaccination program resulted in an increased incidence of GBS. Influenza infections or vaccinations are not generally important trigger agents for GBS, but the association between the swine flu vaccine and GBS was well established in epidemiologic studies, showing a relative risk exceeding 7.0 during the 6 weeks after vaccination (Safranek et al., 1991). Although never fully established, the purported mechanism involved exposure to foreign protein antigens, contaminants associated with growing the virus in hen eggs. An immune-based mechanism also has been associated with penicillamine-induced myasthenia gravis after oral exposure to penicillamine used to treat several rheumatologic disorders (Albers et al., 1980). The combined evidence strongly supports the conclusion that the polyradiculoneuropathy observed among abattoir workers had an autoimmune origin related to occupational exposure to aerosolized porcine brain tissue antigens.

Clinical manifestations

None of the workers had symptoms of an antecedent illness prior to onset of neurologic symptoms (Lachance et al., 2010). The polyradiculoneuropathy was characterized by subacute onset of painful prickling/tingling paresthesias and sensory disturbance, usually associated with pain (legs, head, or neck) and fatigue as the predominant symptoms. Difficulty in walking or prolonged standing was common. Neurologic signs included symmetric distal sensory loss (19 of 24) and weakness (mild to moderate in 17 of 24), impaired tendon reflexes (17 of 24), and autonomic dysfunction (14 of 19 studied), developing at least 1 month after initial exposure (Lachance et al., 2010; Tracy and Dyck, 2011).

Electrophysiology

Electrophysiologic findings reported for the initial 24 cases included at least one nerve conduction study abnormality in all (Lachance et al., 2010; Tracy and Dyck, 2011). These included prolonged motor distal latencies, F-wave latencies, and trigeminal blink reflex latencies. Most patients showed normal conduction velocities along the intermediate segments of motor nerves (Tracy and Dyck, 2011), without evidence of conduction block or abnormal temporal dispersion, findings atypical of CIDP. Sensory responses showed low amplitudes, indicating involvement of the peripheral nerves (as opposed to only the dorsal nerve roots). The normal motor conduction velocities, in association with prolonged F wave, blink reflex, and motor distal latencies, localized the pathology to the extreme proximal and distal nerve segments where the blood–nerve barrier is most permeable (Lachance et al., 2010).

Differential diagnosis

The differential diagnosis for a sensorimotor neuropathy (or polyradiculoneuropathy) associated with some degree of motor conduction slowing is shown in Table 16.2. However, prior to recognition of this occupational disease among workers from two pork-processing plants, exposure to aerosolized porcine neural tissue would not have been included in the table. Regardless, physicians familiar with other forms of inflammatory neuropathies, such as GBS and CIDP, would (and did) recognize this as a likely immune-mediated condition. This recognition resulted in the epidemiologic studies that causally established the occupational association with exposure to aerosolized porcine brains (Adjemian et al., 2009; Deangelis and Shen, 2009; Lachance et al., 2010).

Additional diagnostic evaluation (Lachance et al., 2010; Tracy and Dyck, 2011)

CSF showed a normal cell count but an elevated protein concentration in 18 of 21 cases, averaging 134 mg/dL and supporting a diagnosis of an inflammatory polyradiculoneuropathy. Quantitative sweat testing showed a polyradicular pattern of sweat loss. MRI abnormalities were prominent and included enlarged, T2-enhancing cauda equina nerve roots (mostly ventral) or ganglia. Sural nerve biopsies showed findings of mild inflammatory demyelination, axonal degeneration, and foci of perivascular inflammation. The combined abnormalities are those of an immune-mediated inflammatory polyradiculoneuropathy.

Treatment

Treatment is based on experience with similar immune-mediated polyradiculoneuropathy and the observations from the initial cases. Among the available 23 cases, 6 improved without treatment. The remaining 17 cases received immunosuppressive or immune-modulating treatment with corticosteroids, intravenous immunoglobulin, or therapeutic plasma exchange, all conventional treatment for CIDP. One patient with a coexisting meningoencephalitis did not improve with intravenous methylprednisolone and plasma exchange but showed improvement after starting monthly intravenous cyclophosphamide (Lachance et al., 2010).

Prognosis

All cases of immune-mediated polyradiculoneuropathy, including those requiring immunosuppressive treatment, stabilized or improved over time (Deangelis and Shen, 2009). Most returned to work, although not in the processing area (Deangelis and Shen, 2009; Fitzgerald, 2009). Motor function improved in all treated patients, but pain and sensory dysfunction did not resolve completely in any of the patients (Lachance et al., 2010). The reported outcome is similar to existing experience in other forms of chronic, immune-mediated inflammatory neuropathies (aside from symptoms of prominent and persisting pain).

Since change in the occupational process and elimination of the compressed-air procedure for removing brains from the skull, no new cases have been reported. The prompt recognition of this unusual occupational disease emphasizes the importance of observation and following basic epidemiologic principles.

Thallium

Thallium is a ubiquitous heavy metal that, like arsenic, has been used in many insecticides and rodenticides (Bank, 1980; Moore et al., 1993). Like several other toxic metals or metalloids, thallium also has had medical applications, including use to treat infections (e.g., venereal diseases, tuberculosis, and ringworm) (Bank, 1980; Windebank, 1993), in the manufacturer of optical lenses, and in diagnostic imaging and scintillation counters (Moore et al., 1993; Saha, 2005). Poisoning occasionally occurs after accidental ingestion of rodent poison or in suicidal or homicidal poisonings (Meggs et al., 1994; Rusyniak et al., 2002), and the rationale of the 1972 US ban on thallium rodenticides was to reduce availability (Desenclos et al., 1992; London and Albers, 2007). Occupational exposure to thallium includes contact with thallium-containing materials or contaminated dust from pyrite burners, coal-fired power plants, smelting operations, cadmium production, and rodenticide manufacturing (Moore et al., 1993; Hirata et al., 1998; Saha, 2005; Liu et al., 2012). There are, however, relatively few reports of industrial or environmental thallium poisonings (Moore et al., 1993; Hirata et al., 1998). Thallium is readily absorbed across respiratory, gastrointestinal, and dermal membranes, and water-soluble thallium salts are distributed throughout the body prior to elimination in the stool and urine (Bank, 1980; Cavanagh, 1984). As a general systemic poison, thallium has gastrointestinal, hematologic, renal, and dermatologic involvement (including appearance of Mees' lines after acute exposures). The mechanism of thallium toxicity is unknown, but it enters cells through potassium channels and possibly competes with potassium in intracellular reactions and interferes with energy metabolism, oxidative phosphorylation, and glycolysis (Hoffman, 2003). Vacor is a rodenticide that is of historic importance because it was banned in the USA after it was found to destroy pancreatic beta cells and also cause a sensory more than motor neuropathy with profound dysautonomia independent of producing diabetes (LeWitt, 1980).

Clinical manifestations (Albers, 2012)

Reports of acute thallium toxicity usually emphasize painful stinging dysesthesias beginning in the feet and progressing up the legs (Bank, 1980; Wilbourn, 1984; Windebank, 1993; Kubis et al., 1997; Albers and Teener, 2006). Pain sometimes involves the joints, back, and chest, areas not affected by neuropathy (Wilbourn, 1984). Symptoms of autonomic dysfunction such as abdominal colic, nausea, vomiting, urinary retention, anhidrosis, and diarrhea may precede onset of neuropathy (Kalantri and Kurtz, 1988; Herrero et al., 1995). Stocking-glove sensory loss with prominent impairment of pin pain and temperature sensations suggests predominant involvement of small nerve fibers, an impression supported by signs of dysautonomia and preserved reflexes (at least initially) (Kalantri and Kurtz, 1988; Herrero et al., 1995; Nordentoft et al., 1998; Albers and Teener, 2006). In cases of severe poisoning, large-fiber sensory abnormalities and distal weakness may be present, although the degree of weakness is minor relative to the painful distal dysesthesias and hyperpathia (Aminoff and Albers, 2012). On occasion, weakness may progress to quadriparesis and respiratory failure, sometimes suggesting a diagnosis of GBS (Andersen, 1984; Cavanagh, 1984). Mild distal weakness may be overlooked because hyperpathia makes manual muscle testing difficult to interpret (Albers and Teener, 2006). A cardinal clinical feature that distinguishes thallium neuropathy from most other toxic neuropathies is the coexisting alopecia developing several weeks after acute intoxication (Bank, 1980; Windebank, 1993), as reported for all 49 patients with confirmed thallium poisoning in one study (Rangel-Guerra et al., 1990).

Electrophysiology (Albers, 2012)

Thallium produces a sensory greater than motor neuropathy characterized by prominent small-nerve fiber involvement, particularly early in the course of poisoning, when nerve conduction studies may be normal (London and Albers, 2007). In more severe cases, nerve conduction studies show more definite evidence of axonal loss (Dumitru and Kalantri, 1990; Kuo et al., 2005), with low-amplitude sensory and motor responses, and fibrillation potentials in the intrinsic foot muscles, sometimes apparent when the sural response is normal (Dumitru and Kalantri, 1990). Serial studies show progressive loss of response amplitudes but preserved conduction velocity, in the setting of profuse fibrillation potentials in distal leg muscles. Motor abnormalities may be present despite symptoms and signs suggesting only sensory involvement (London and Albers, 2007). Thallium neuropathy is of historic interest to electromyographers because Kaeser and Lambert used thallium to produce a "pure" axonal neuropathy in their pioneering descriptions of axonal loss lesions (Kaeser and Lambert, 1962; Wilbourn, 1984).

Differential diagnosis

Conditions associated with sensory predominant or exclusive neuropathies are listed in Table 16.4. Among the neurotoxicants, most produce large-fiber involvement, the exceptions being thallium, Vacor, and possibly ethyl alcohol. Thallium is among the painful neuropathies characterized by small-fiber involvement with abnormal pin pain and thermal sensations. The sensory abnormalities are sometimes present in the context of preserved ankle reflexes, a finding atypical for most neuropathies. The painful onset is similar to that associated with arsenic poisoning, but arsenic neuropathy is characterized by motor predominance and conduction slowing. In terms of the competing conditions listed in Table 8.4, none produces alopecia, a feature highly suggestive of thallium poisoning, albeit a feature usually not yet present when the diagnosis is in question.

Additional diagnostic evaluation

Laboratory evidence of anemia, nephropathy, and hepatotoxicity are nonspecific findings but nonetheless suggestive of systemic poisoning (Albers, 2012). Thallium is uniformly distributed throughout the body prior to elimination in the stool and urine (Bank, 1980; Cavanagh, 1984), and toxicologic evaluation is necessary to document exposure and estimate the dose. Thallium is excreted with an elimination half-time of up to 30 days (Moore et al., 1993; Albers, 2012). Thallium can be measured in serum, urine, and stool, but a 24-hour urine collection is the standard means of documenting excessive urinary excretion due to relatively recent exposures. More remote exposures can be documented as increased thallium levels in hair. Nerve biopsy results are nonspecific for thallium poisoning, showing loss of axons and degeneration of myelin sheaths along some axons into a series of ovoids (axon bodies) (Dumitru and Kalantri, 1990). Quantitative tests of autonomic function can confirm the presence of an autonomic neuropathy.

Treatment

Treatment of thallium poisoning includes removal from exposure, diuresis, and general medical support (Albers, 2005d). Prussian blue has been used as an antidote to counteract heavy-metal poisoning, and there are isolated reports of its use to increase the elimination of thallium in urine and feces (Pelclova et al., 2009).

Prognosis

Patients who survive the initial systemic effects of thallium poisoning have a good prognosis (Vergauwe et al., 1990). Even severe thallium neuropathy improves substantially over months and resolves over a few years (Dumitru and Kalantri, 1990). On the other hand, sensory neuropathies differ from motor neuropathies, in that they generally do not have the same excellent prognosis for full recovery. Consistent with this generalization, some reports of thallium neuropathy describe persisting residual clinical and electrophysiologic impairments during 2 years of follow-up (Pelclova et al., 2009). In this context, patients with Vacor-induced neuropathy displayed persistent dysautonomia and residual sensory loss, described as improved but not having returned to normal (LeWitt, 1980). Hair loss associated with thallium poisoning is fully reversible (Pelclova et al., 2009).

SUMMARY

Toxic neuropathies may be suspected in the presence of occupational (or other) exposures. As toxic-metabolic neuropathies are rare (with the exception of diabetes and certain medications), it can be a challenge to identify and confirm causality of a toxicant in a neuropathic process. A methodical approach begins with acquiring a detailed history and clinical examination and then proceeds with establishing that a neuropathy exists and characterizing its predominant pathophysiology through nerve conduction studies and needle electromyography. The electrodiagnostic results may focus the differential diagnosis and, when coupled with history and clinical examination, may suggest additional medical and

Table 16.6

Clinical pearls and key points (Albers, 2012)

- The diagnosis and characterization of a neuropathy proceed independently of knowing its cause, toxic or otherwise. Characterization may support suspicion of a specific toxic neuropathy and direct further evaluation
- Acrylamide monomer leads to accumulations of neurofilaments seen as axonal masses that impair axonal transport and result in a distal dying-back phenomenon (giant axonal neuropathy). Suggestive clinical features: dermatitis and hyperhidrosis progressing to anhidrosis
- Acute arsenic toxicity is a mimic of Guillain–Barré syndrome, electrophysiologically and clinically, in early stages. Electrophysiology evolves to reflect a severe axonal neuropathy in later stages. Suggestive clinical features: dermatitis, pancytopenia with basophilic stippling, Mees' lines
- Carbon disulfide produces a giant axonal neuropathy through accumulation of neurofilaments in axonal swellings. Electrophysiologically, this neuropathy is characterized as motor greater than sensory axonal loss. Suggestive clinical features: extrapyramidal (parkinsonian) signs
- Dioxin as a cause of neuropathy remains a controversial subject
- Elemental mercury produces a length-dependent sensory or sensorimotor axonal neuropathy. Suggestive clinical features: postural tremor, maculopapular rash, and possibly nonspecific behavioral changes
- *n*-Hexane toxicity results in accumulation of neurofilaments in axons, and as such, is another giant axonal neuropathy. Acute exposure results in neuropathy that resembles Guillain–Barré syndrome, though typically motor involvement is greater than sensory involvement. Suggestive clinical features: rapid onset, diffuse weakness, autonomic dysfunction, and increased cerebrospinal fluid protein
- Lead neuropathy is classically considered a pure motor disorder that is asymmetric and associated with wrist drop. It is remarkably rare in modern times. Suggestive clinical features: anemia, basophilic stippling of red blood cells, renal dysfunction, gastrointestinal dysfunction, gingival lead lines
- Organophosphate exposure leads to a delayed motor neuropathy after initial cholinergic symptoms caused by inhibition of acetylcholinesterase. As neuropathy resolves, upper motor neuron signs of spasticity and hyperactive reflexes appear. Suggestive clinical features: initial cholinergic symptoms of miosis, increased secretions, sweating, gastric hyperactivity, and bradycardia followed by fasciculations and weakness
- Porcine neural tissue in aerosolized state can cause autoimmune inflammatory neuropathy resembling chronic inflammatory demyelinating polyradiculoneuropathy. Suggestive clinical features: sensory disturbance, weakness, gait disturbance, autonomic dysfunction
- Thallium is one of the few toxic exposures that produces a sensory predominant small-fiber neuropathy associated with painful dysesthesias. Thallium is a systemic poison and abnormalities are seen in gastrointestinal, hematologic, renal, and dermatologic systems. Suggestive clinical features: alopecia, Mees' lines

laboratory investigations specific to a toxic or to alternative systemic disorders. In establishing causality of a toxin, it remains important to consider and rule out alternative diagnoses. A temporal relationship of toxin exposure (of sufficient dosage, acutely or chronically) and development of symptoms must be demonstrated. Familiarity with the electrodiagnostic pattern and systemic clinical features of a number of toxins may aid in final diagnosis. Early diagnosis is important to remove a patient from a toxic exposure, often allowing recovery of function to some degree. Selected "clinical pearls and key points" relevant to the topics discussed in this chapter are summarized in Table 16.6.

References

Aaserud O, Hommeren OJ, Tvedt B et al. (1990). Carbon disulfide exposure and neurotoxic sequelae among viscose rayon workers. Am J Ind Med 18: 25–37.

Adjemian JZ, Howell J, Holzbauer S et al. (2009). A clustering of immune-mediated polyradiculoneuropathy among swine abattoir workers exposed to aerosolized porcine brains, Indiana, United States. Int J Occup Environ Health 15: 331–338.

Albers JW (1993). Clinical neurophysiology of generalized polyneuropathy. J Clin Neurophysiol 10: 149–166.

Albers JW (1999). Toxic Neuropathies. In: ML Bleecker (Ed.), Continuum: Lifelong Learning in Neurology, Lippincott, Williams & Wilkins, Baltimore, pp. 27–50.

Albers JW (2003). Toxic neuropathies. In: MB Bromberg (Ed.), AAN 55th Annual Meeting Education Program Syllabus: Peripheral Neuropathy, American Academy of Neurology, Hawaii. pp. 7FC.006-149–7FC.006-168.

Albers JW (2005a). Clinical and electrodiagnostic evaluations of the peripheral nervous system. In: JW Albers, S Berent (Eds.), Neurobehavioral Toxicology. Neurological and Neuropsychological Perspectives. Vol. II. Peripheral Nervous system, Taylor & Francis, London, pp. 257–327.

Albers JW (2005b). Elemental mercury, in: Industrial and environmental agents. In: JW Albers, S Berent (Eds.), Neurobehavioral Toxicology. Neurological and Neuropsychological Perspectives. Vol. II. Peripheral Nervous System, Taylor & Francis, London, pp. 356–427.

Albers JW (2005c). Industrial and environmental agents. In: JW Albers, S Berent (Eds.), Neurobehavioral Toxicology. Neurological and Neuropsychological Perspectives. Vol. II. Peripheral Nervous System, Taylor & Francis, London, pp. 329–427.

Albers JW (2005d). Medications and substances of abuse. In: JW Albers, S Berent (Eds.), Neurobehavioral Toxicology. Neurological and Neuropsychological Perspectives. Vol. II. Peripheral Nervous System, Taylor & Francis, London, pp. 429–523.

Albers JW (2012). Occupational, biologic and environmental toxic neuropathies. In: PD Donofrio (Ed.), Textbook on Peripheral Neuropathy, Demos Medical, New York, pp. 87–105.

Albers JW, Berent S (2005). Neurobehavioral Toxicology. Neurological and Neuropsychological Perspectives, Volume II. Peripheral Nervous System, Taylor & Francis, London.

Albers JW, Bromberg MB (1994). X-linked bulbospinomuscular atrophy (Kennedy's disease) masquerading as lead neuropathy. Muscle Nerve 17: 419–423.

Albers JW, Bromberg MB (1995). Chemically induced toxic neuropathy. In: NL Rosenberg (Ed.), Occupational and Environmental Neurology. Butterworth-Heinemann, Boston, pp. 175–233.

Albers JW, Fink JK (2004). Porphyric neuropathy. Muscle Nerve 30: 410–422.

Albers JW, Garabrant DH (2004). Arsenic Poisoning. In: DJ Lynn, HB Newton (Eds.), The 5-Minute Neurology Consult, Lippincott Williams & Wilkins, Philadelphia, PA, pp. 70–71.

Albers JW, Kelly Jr JJ (1989). Acquired inflammatory demyelinating polyneuropathies; clinical and electrodiagnostic features. Muscle Nerve 12: 435–451.

Albers JW, Teener JW (2006). Toxic neuropathies. In: J Daube, F Maugwere (Eds.), Handbook of Clincial Neurophysiology, Elsevier, Philadelphia PA, pp. 669–694.

Albers JW, Wald JJ (2002). Industrial and environmental toxic neuropathy. In: WF Brown, CF Bolton, MJ Aminoff (Eds.), Clinical Neurophysiology and Neuromuscular Diseases, WB Saunders, Philadelphia, pp. 1143–1168.

Albers JW, Hodach RJ, Kimmel DW et al. (1980). Penicillamine-associated myasthenia gravis. Neurology 30: 1246–1250.

Albers JW, Cavender GF, Levine SP et al. (1982). Asymptomatic sensorimotor polyneuropathy in workers exposed to elemental mercury. Neurology 32: 1168–1174.

Albers JW, Kallenbach LR, Fine LJ et al. (1988). Neurological abnormalities associated with remote occupational elemental mercury exposure. Ann Neurol 24: 651–659.

Albers JW, Garabrant DH, Schweitzer SJ et al. (2004). The effects of occupational exposure to chlorpyrifos on the peripheral nervous system: a prospective cohort study. Occup Environ Med 61: 201–211.

Albers JW, Garabrant DH, Mattsson JL et al. (2007). Dose-effect analyses of occupational chlorpyrifos exposure and peripheral nerve electrophysiology. Toxicol Sci 97: 196–204.

Albers JW, Garabrant DH, Berent S et al. (2010). Paraoxonase (PON1) status and plasma butyrylcholinesterase activity in chlorpyrifos manufacturing workers. J Expo Sci Environ Epidemiol 20: 79–89.

Albin RL (2000). Basal ganglia neurotoxins. Neurol Clin 18: 665–680.

Allen N (1979). Solvents and other industrial organic compounds. In: PJ Vinken, GW Bruyn (Eds.), Handbook of Clinical Neurology. Elsevier/North Holland Biomedical Press, Amsterdam, pp. 361–389.

Allen N, Mendell JR, Billmaier DJ et al. (1975). Toxic polyneuropathy due to methyl n-butyl ketone: an industrial outbreak. Arch Neurol 32: 209–218.

Altenkirch R, Mager J, Stoltenburg G et al. (1977). Toxic neuropathies after sniffing a glue thinner. J Neurol 214: 137–152.

American Conference of Governmental Industrial Hygienists (1999). Mercury, Elemental and Inorganic. Documentation of the Threshold Limit Values and Biological Exposure Indices, American Congress of Governmental Industrial Hygienists, Cincinnati, OH, Vol I.

American Conference of Governmental Industrial Hygienists (ACGIH) (2013). TLVs and BEIs. Based on the Documentation of the Threshold Limit Values for Chemical Substances and Physical Agents & Biological Exposure Indices, American Conference of Governmental Industrial Hygienists, Cincinnati, OH.

Aminoff MJ (1985). Electrophysiologic recognition of certain occupation-related neurotoxic disorders. Neurol Clin 3: 687–697.

Aminoff MJ, Albers JW (1999). Electrophysiologic techniques in the evaluation of patients with suspected neurotoxic disorders. In: MJ Aminoff (Ed.), Electrodiagnosis in Clinical Neurology, Churchill Livingstone, New York, pp. 721–734.

Aminoff MJ, Albers JW (2012). Electrophysiologic techniques in the evaluation of patients with suspected neurotoxic disorders. In: MJ Aminoff (Ed.), Electrodiagnosis in Clinical Neurology, 6th ed. Elsevier Churchill Livingstone, Philadelphia.

Andersen O (1984). Clinical evidence and therapeutic indications in neurotoxicology, exemplified by thallotoxicosis. Acta Neurol Scand 70 (Suppl): 185–192.

Andersen A, Ellingsen DG, Morland T et al. (1993). A neurological and neurophysiological study of chloralkali workers previously exposed to mercury vapour. Acta Neurol Scand 88: 427–433.

Anonymous (1998). Lead poisoning associated with imported candy and powdered food coloring – California and Michigan. MMWR Morb Mortal Wkly Rep 47: 1041–1043.

Ansbacher LE, Bosch EP, Cancilla PA (1982). Disulfiram neuropathy: a neurofilamentous distal axonopathy. Neurology 32: 424–428.

Armon C, Kurland LT, Daube JR et al. (1991). Epidemiologic correlates of sporadic amyotrophic lateral sclerosis. Neurology 41: 1077–1084.

Ashby JS (1980). A neurological and biochemical study of early lead poisoning. Br Journal of Internal Medicine 37: 133–140.

Assennato G, Cervino D, Emmett EA et al. (1989). Follow-up of subjects who developed chloracne following TCDD exposure at Seveso. Am J Ind Med 16: 119–125.

Aub JC, Fairhall LT, Minot AS et al. (1925). Lead poisoning. Medicine (Baltimore) 4: 1–250.

Auld RB, Bedwell SF (1967). Peripheral neuropathy with sympathetic overactivity from industrial contact with acrylamide. Can Med Assoc J 96: 652–654.

Baker EL, Feldman RG, White RA et al. (1984). Occupational lead neurotoxicity: a behavioural and electrophysiological evaluation. Study design and year one results. Brit J Ind Med 41: 352–361.

Balali-Mood M, Shademanfar S, Rastegar MJ et al. (2010). Occupational lead poisoning in workers of traditional tile factories in Mashhad, Northeast of Iran. Int J Occup Environ Health 1: 29–38.

Baloh RW (1974). Laboratory diagnosis of increased lead absorption. Arch Environ Health 28: 198–208.

Bank WJ (1980). Thallium. In: PS Spencer, HH Schaumburg (Eds.), Experimental and Clinical Neurotoxicology, Williams and Wilkins, Baltimore, pp. 570–577.

Barbieri S, Pirovano C, Scarlato G et al. (1988). Long-term effects of 2,3,7,8-tetrachlorodibenzo-p-dioxin on the peripheral nervous system. Clinical and neurophysiological controlled study on subjects with chloracne from the Seveso area. Neuroepidemiology 7: 29–37.

Beritic T (1984). Lead neuropathy. Crit Rev Toxicol 12: 149–213.

Berlin M (1986). Mercury. In: L Friberg, GF Nordberg, VB Vouk (Eds.), Handbook on the Toxicology of Metal, Elsevier, New York, pp. 387–445.

Bilbao JM, Briggs SJ, Gray TA (1984). Filamentous axonopathy in disulfiram neuropathy. Ultrastruct Pathol 7: 295–300.

Bloomer JR, Bonkovsky HL (1989). The porphyrias. Dis Mon 35: 1–54.

Bond GG, Ott MG, Brenner FE et al. (1983). Medical and morbidity surveillance findings among employees potentially exposed to TCDD. Brit J Ind Med 40: 318–324.

Bordo BM, Filippini G, Massetto N et al. (1982). Electrophysiological study of subjects occupationally exposed to lead and with low levels of lead poisoning. Scand J Work Environ Health 8 (suppl 1): 142–147.

Bradley WG, Hewer RL (1966). Peripheral neuropathy due to disulfiram. Br Med J 2: 449.

Bromberg MB (2000). Peripheral neurotoxic disorders. Neurol Clin 18: 681–694.

Bromberg MB, Albers JW (1993). Patterns of sensory nerve conduction abnormalities in demyelinating and axonal peripheral nerve disorders. Muscle Nerve 16: 262–266.

Buiatti E, Cecchini S, Ronchi O et al. (1978). Relationship between clinical and electromyographic findings and exposure to solvents, in shoe and leather workers. Brit J Ind Med 35: 168–173.

Campbell AMG, Williams ER, Barltrop D (1970). Motor neurone disease and exposure to lead. J Neurol Neurosurg Psychiatry 33: 877–885.

Carman KB, Tutkun E, Yilmaz H et al. (2013). Acute mercury poisoning among children in two provinces of Turkey. Eur J Pediatr 172: 821–827.

Catton MJ, Harrison MJG, Fullerton PM et al. (1970). Subclinical neuropathy in lead workers. Br Med J 2: 80–82.

Cavanagh JB (1984). Neurotoxic effects of metal and their interaction. In: CL Galli, L Manzo, PS Spencer (Eds.), Recent Advances in Nervous System Toxicology, NATO ASI Series, Plenum Press, New York, pp. 177–202.

Cavanagh JB (1985). Solvent neurotoxicity. Brit J Ind Med 42: 433–434.

Chang LW (1980). Mercury. In: P Spencer, HH Schaumburg (Eds.), Experimental and Clinical Neurotoxicology, Williams & Wilkens, Baltimore, pp. 508–526.

Chang CM, Yu CW, Fong KY et al. (1993). n-Hexane neuropathy in offset printers. J Neurol Neurosurg Psychiatry 56: 538–542.

Chaudhry V, Chaudhry M, Crawford TO et al. (2003). Toxic neuropathy in patients with pre-existing neuropathy. Neurology 60: 337–340.

Chhuttani PN, Chawla LS, Sharma TD (1967). Arsenical neuropathy. Neurology 17: 269–274.

Chu CC, Huang CC, Chen RS et al. (1995). Polyneuropathy induced by carbon disulphide in viscose rayon workers. Occup Environ Med 52: 404–407.

Chu CC, Huang CC, Chu NS et al. (1996). Carbon disulfide induced polyneuropathy: sural nerve pathology, electrophysiology, and clinical correlation. Acta Neurol Scand 94: 258–263.

Chu CC, Huang CC, Ryu SJ et al. (1998). Chronic inorganic mercury induced peripheral neuropathy. Acta Neurol Scand 98: 461–465.

Clarkson TW, Friberg L, Nordberg GF et al. (1988). Mercury. In: Biological Monitoring of Toxic Metals, Plenum Press, New York, pp. 199–246.

Conradi S, Ronnevi L-O, Nise G et al. (1982a). Long-time penicillamine treatment in amyotrophic lateral sclerosis with parallel determination of lead in blood, plasma and urine. Acta Neurol Scand 65: 203–211.

Conradi S, Ronnevi LO, Norris FH (1982b). Motor neuron disease and toxic metals. Adv Neurol 36: 201–231.

Cornblath DR, Sumner AJ, Daube J et al. (1991). Conduction block in clinical practice. Muscle Nerve 14: 869–871.

Corsi G, Maestrelli GC, Picotti G et al. (1983). Chronic peripheral neuropathy in workers with previous exposure to carbon disulfide. Brit J Ind Med 40: 209–211.

Couri D, Milks MM (1985). Hexacarbon neuropathy: tracking a toxin. Neurotoxicology 6: 65–71.

Davenport JG, Farrell DF, Sumi M (1976). "Giant axonal neuropathy" caused by industrial chemicals: neurofilamentous axonal masses in man. Neurology 26: 919–923.

Davis CS, Johnson MK, Richardson DJ (1985). Organophosphorus compounds. In: JL O'Donoghue (Ed.), Neurotoxicity of Industrial and Commercial Chemicals. II. CRC Press, Inc., Boca Raton, pp. 1–24.

Deangelis TM, Shen L (2009). Outbreak of progressive inflammatory neuropathy following exposure to aerosolized porcine neural tissue. Mt Sinai J Med 76: 442–447.

Department of Veterans Affairs (2013). Disease associated with exposure to certain herbicide agents: peripheral neuropathy. Fed Regist 78: 54763–54766.

Desenclos JC, Wilder MH, Coppenger GW et al. (1992). Thallium poisoning: an outbreak in Florida, 1988. South Med J 85: 1203–1206.

Dickoff DJ, Gerber O, Turovsky Z (1987). Delayed neurotoxicity after ingestion of carbamate pesticide. Neurology 37: 1229–1231.

DiVincenzo GD, Hamilton ML, Kapalan CJ et al. (1980). Characterization of the metabolites of methyl n-butyl ketone. In: PS Spencer, HH Schaumburg (Eds.), Experimental and Clinical Neurotoxicology, Williams and Wilkins, Baltimore and London, pp. 846–855.

Donofrio PD, Albers JW (1990). Polyneuropathy: classification by nerve conduction studies and electromyography. Muscle Nerve 13: 889–903.

Donofrio PD, Wilbourn AJ, Albers JW et al. (1987). Acute arsenic intoxication presenting as Guillain-Barré-like syndrome. Muscle Nerve 10: 114–120.

Drachman DB (1986). Pathophysiology of the neuromuscular junction. In: AK Asbury, GM McKhann, WI McDonald (Eds.), Diseases of the Nervous System. Clinical Neurobiology, Saunders, Philadelphia, pp. 258–259.

Dumitru D, Kalantri A (1990). Electrophysiologic investigation of thallium poisoning. Muscle Nerve 13: 433–437.

Durham HD, Pena SD, Carpenter S (1983). The neurotoxins 2,5-hexanedione and acrylamide promote aggregation of intermediate filaments in cultured fibroblasts. Muscle Nerve 6: 631–637.

Dyck PJ, Johnson WJ, Nelson RA et al. (1975). Uremic neuropathy. III. Controlled study of restricted protein and fluid diet and infrequent hemodialysis versus conventional hemodialysis treatment. Mayo Clin Proc 50: 641–649.

Dyck PJ, Oviatt KF, Lambert EH (1981). Intensive evaluation of referred unclassified neuropathies yields improved diagnosis. Ann Neurol 10: 222–226.

Edwards PM, Sporel-Ozakat E, Gispen WH (1991). Neurotoxic acrylamide and neurotrophic melanocortin peptides – can contrasting actions provide clues about modes of action? Neuropathol Appl Neurobiol 17: 91–104.

Ehle AL (1986). Lead neuropathy and electrophysiological studies in low level lead exposure: a critical review. Neurotoxicology 7: 203–216.

Ehrenberg RL, Vogt RL, Smith AB et al. (1991). Effects of elemental mercury exposure at a thermometer plant. Am J Ind Med 19: 495–507.

Ellingsen DG, Thomassen Y, Langard S et al. (1993). Urinary mercury excretion in chloralkali workers after the cessation of exposure. Scand J Work Environ Health 19: 334–341.

Ellis T, Lacy R (1998). Illicit alcohol (moonshine) consumption in West Alabama revisited. South Med J 91: 858–860.

Emerick GL, DeOliveira GH, dos Santos AC et al. (2012). Mechanisms for consideration for intervention in the development of organophosphorus-induced delayed neuropathy. Chem Biol Interact 199: 177–184.

Feldman RG (1970). Facial nerve latency studies in man: effects of trichloroethylene intoxication. Electromyography Nr 1: 93–100.

Feldman RG (1982). Neurological manifestations of mercury intoxication. Acta Neurol Scand 66 (suppl 92): 201–209.

Feldman RG, Hayes MK, Younes R et al. (1977). Lead neuropathy in adults and children. Arch Neurol 34: 481–488.

Feldman RG, Niles CA, Kelly Hayes M et al. (1979). Peripheral neuropathy in arsenic smelter workers. Neurology 29: 939–944.

Filippini G, Bordo B, Crenna P et al. (1981). Relationship between clinical and electrophysiological findings and indicators of heavy exposure to 2,3,7,8-tetrachlorodibenzo-dioxin. Scand J Work Environ Health 7: 257–262.

Fischbein A (1983). Environmental and occupational lead exposure. Environ Occup Med 57: 433–447.

Fitzgerald S (2009). Condition of pork plant workers improving after developing neuropathy. Neurology Today 9: 18–20.

Ford DP (1999). Neurotoxicology: exposure assessment. Continuum: Lifelong Learning in Neurology 5: 9–25.

Fowler TJ, Ochoa J (1975). Unmyelinated fibers in normal and compressed peripheral nerves of the baboon. Neuropathol Appl Neurobiol 1: 247–255.

Frisoni GB, Di Monda V (1989). Disulfiram neuropathy: a review (1971–1988) and report of a case. Alcohol Alcohol 24: 429–437.

Froom P, Kristal-Boneh E, Benbassat J et al. (1999). Lead exposure in battery-factory workers is not associated with anemia. J Occup Environ Med 41: 120–123.

Frumkin H, Manning CC, Williams PL et al. (2001). Diagnostic chelation challenge with DMSA: a biomarker of long-term mercury exposure? Environ Health Perspect 109: 167–171.

Fullerton PM (1969a). Electrophysiological and histological observations on peripheral nerves in acrylamide poisoning in man. J Neurol Neurosurg Psychiatry 32: 186–192.

Fullerton PM (1969b). Toxic chemicals and peripheral neuropathy: clinical and epidemiological features. Proc R Soc Med 62: 201–204.

Gallagher RE (1998). Arsenic – new life for an old potion. N Engl J Med 339: 1389–1391.

Garland TO, Patterson MWH (1967). Six cases of acrylamide poisoning. Br Med J 4: 134–138.

Ghosh D (2004). Classic picture of lead neuropathy. Indian Pediatr 41: 1270–1271.

Gilliatt RW, Taylor JC (1959). Electrical changes following section of the facial nerve. Proc R Soc Med 52: 1080–1083.

Goffeng LO, Heier MS, Kjuus H et al. (2008). Nerve conduction, visual evoked responses and electroretinography in tunnel workers previously exposed to acrylamide and N-methylolacrylamide containing grouting agents. Neurotoxicol Teratol 30: 186–194.

Gold BG, Schaumburg HH (2000). Acrylamide. In: PS Spencer, HH Schaumburg (Eds.), Experimental and Clinical Neurotoxicology, 2nd edn. Oxford University Press, New York, pp. 124–132.

Gold BG, Griffin JW, Price DL (1992). Somatofugal axonal atrophy precedes development of axonal degeneration in acrylamide neuropathy. Arch Toxicol 66: 57–66.

Gottfried MR, Graham DG, Morgan M et al. (1985). The morphology of carbon disulfide neurotoxicity. Neurotoxicology 6: 89–96.

Governa M, Calisti R, Coppa G et al. (1987). Urinary excretion of 2,5-hexanedione and peripheral polyneuropathies workers exposed to hexane. J Toxicol Environ Health 20: 219–228.

Graham DG, Anthony DC, Szakal-Quin G et al. (1985). Covalent crosslinking of neurofilaments in the pathogenesis of n- hexane neuropathy. Neurotoxicology 6: 55–63.

Grandjean P (1978). Widening perspectives of lead toxicity. A review of health effects of lead exposure in adults. Environ Res 17: 303–321.

Greenberg SA (1996). Acute demyelinating polyneuropathy with arsenic ingestion. Muscle Nerve 19: 1611–1613.

Griffin JW (1981). Hexacarbon neurotoxicity. Neurobehav Toxicol Teratol 3: 437–444.

Gupta RP, Abou-Donia MB (1997). Acrylamide and carbon disulfide treatments increase the rate of rat brain tubulin polymerization. Molecular Chemistry & Neuropathology 30: 223–237.

Gutmann L, Besser R (1990). Organophosphate intoxication: pharmacologic, neurophysiologic, clinical, and therapeutic considerations. Semin Neurol 10: 46–51.

Hagmar L, Tornqvist M, Nordander C et al. (2001). Health effects of occupational exposure to acrylamide using hemoglobin adducts as biomarkers of internal dose. Scand J Work Environ Health 27: 219–226.

Hanninen H (1982). Behavioral effects of occupational exposure to mercury and lead. Acta Neurol Scand 66 (Suppl): 167–175.

Harris CH, Gulati AK, Friedman MA et al. (1994). Toxic neurofilamentous axonopathies and fast axonal transport. V. Reduced bidirectional vesicle transport in cultured neurons by acrylamide and glycidamide. J Toxicol Environ Health 42: 343–356.

Hashizume A, Koike H, Kawagashira Y et al. (2011). Central nervous system involvement in *n*-hexane polyneuropathy demonstrated by MRI and proton MR spectroscopy. Clin Neurol Neurosurg 113: 493–495.

He FS, Zhang SL, Li G et al. (1988). An electroneurographic assessment of subclinical lead neurotoxicity. Int Arch Occup Environ Health 61: 141–146.

He FS, Zhang SL, Wang HL et al. (1989). Neurological and electroneuromyographic assessment of the adverse effects of acrylamide on occupationally exposed workers. Scand J Work Environ Health 15: 125–129.

Herkovitz S, Scelsa SN, Schaumburg HH (2010). The toxic neuropathies: industrial, occupational, and environmental agents. In: Peripheral Neuropathies in Clinical Practice, Oxford University Press, London, pp. 301–310.

Hernandez AF, Amparo GM, Perez V et al. (2006). Influence of exposure to pesticides on serum components and enzyme activities of cytotoxicity among intensive agriculture farmers. Environ Res 102: 70–76.

Herrero F, Fernandez E, Gomez J et al. (1995). Thallium poisoning presenting with abdominal colic, paresthesia, and irritability. J Toxicol Sci 33: 261–264.

Herskowitz A, Ishii N, Schaumburg HH (1971). n-Hexane neuropathy: a syndrome occurring as a result of industrial exposure. N Engl J Med 285: 82–85.

Hill AB (1965). The environment and disease: association or causation? Proc R Soc Med 58: 295–300.

Hindmarsh JT (2002). Caveats in hair analysis in chronic arsenic poisoning. Clin Biochem 35: 1–11.

Hirata M, Taoda K, Ono-Ogasawara M et al. (1998). A probable case of chronic occupational thallium poisoning in a glass factory. Ind Health 36: 300–303.

Hoffman RS (2003). Thallium toxicity and the role of Prussian blue in therapy. Toxicol Rev 22: 29–40.

Holland NR, Stocks A, Hauer P et al. (1997). Intraepidermal nerve fiber density in patients with painful sensory neuropathy. Neurology 48: 708–711.

Holzbauer SM, Devries AS, Sejvar JJ et al. (2010). Epidemiologic investigation of immune-mediated polyradiculoneuropathy among abattoir workers exposed to porcine brain. PLoS One 5, e9782.

Huang CC, Chu CC, Wu TN et al. (2002). Clinical course in patients with chronic carbon disulfide polyneuropathy. Clin Neurol Neurosurg 104: 115–120.

Ibrahim AS, Latif AH (2002). Adult lead poisoning from a herbal medicine. Saudi Med J 23: 591–593.

Iida M (1982). Neurophysiological studies in n-hexane polyneuropathy in the sandal factory. Electroencephalogr Clin Neurophysiol 36 (Suppl): 671–681.

Jacques Y, Romey G, Cavey MT et al. (1980). Interaction of pyrethroids with the Na^+ channel in mammalian neuronal cells in culture. Biochim Biophys Acta 600: 882–897.

Jayaratnam J (1990). Acute pesticide poisoning: a major global health problem. World Health Stat Q 43: 139–144.

Jirasek L, Kalensky J, Kubec K et al. (1974). Acne chlorina, porphyria cutanea tarda and other manifestations of general poisoning during the manufacture of herbicides. II. Cesk Dermatol 49: 145–157.

Johnson BL, Boyd J, Burg JR et al. (1983). Effects on the peripheral nervous system of workers' exposure to carbon disulfide. Neurotoxicol 4: 53–66.

Kaeser HE, Lambert EH (1962). Nerve function studies in experimental polyneuritis (Proc 1st Int Congr Electromyography, Pavia). Electroencephalogr Clin Neurophysiol Suppl 22: 29–35.

Kalantri A, Kurtz E (1988). Electrodiagnosis in thallium toxicity: a case report (abstract). Muscle Nerve 11: 968A.

Kaplan JG, Rosenberg NL, Pack D et al. (1993). Sensory neuropathy associated with Dursban (chlorpyrifos) exposure. Neurology 43: 2193–2196.

Kelly JJJ (1989). Differential diagnosis of demyelinating polyneuropathies. In: AAN Course No. 146, Clinical EMG, American Academy of Neurology, St. Paul, MN, pp. 107–124.

Kim EA, Kang SK (2010). Occupational neurological disorders in Korea. J Korean Med Sci 25: S26–S35.

Kim SD, Song DB, Yum YT (1989). [Clinical analysis of 8 cases of chronic carbon disulfide poisoning in workers engaged in the viscose rayon industry.]. Korean J Occup Environ Med 1: 186–196.

Klawans HL (1987). Dystonia and tremor following exposure to 2,3,7,8-tetrachlorodibenzo-p-dioxin. Mov Disord 2: 255–261.

Ko MH, Chen WP, Lin-Shiau SY et al. (1999). Age-dependent acrylamide neurotoxicity in mice: morphology, physiology, and function. Exp Neurol 158: 37–46.

Koller WC, Klawans HL (1979). Organophosphorous intoxication. In: PJ Vinken, GW Bruyn (Eds.), Handbook of Clinical Neurology. Vol 37. North Holland Publishing, Amsterdam, pp. 541–562.

Korobkin R, Asbury AK, Sumner AJ et al. (1975). Glue-sniffing neuropathy. Arch Neurol 32: 158–162.

Kreiss K, Zack MM, Landrigan PJ et al. (1983). Neurologic evaluation of a population exposed to arsenic in Alaskan well water. Arch Environ Health 38: 116–121.

Krishnan AV, Park SB, Huynh W et al. (2012). Impaired energy-dependent processes underlie acute lead neuropathy. Muscle Nerve 46: 957–961.

Kropp TJ, Richardson RJ (2003). Relative inhibitory potencies of chlorpyrifos oxon, chlorpyrifos methyl oxon, and mipafox for acetylcholinesterase versus neuropathy target esterase. J Toxicol Environ Health 66: 1145–1157.

Kuang S, Huang H, Liu H et al. (2001). A clinical analysis of 102 cases of chronic n-hexane intoxication. Chung Hua Nei Ko Tsa Chih 40: 329–331.

Kubis N, Talamon C, Smadja D et al. (1997). Peripheral neuropathy caused by thallium poisoning. Rev Neurologique (Paris) 153: 599–601.

Kumar N (2010). Neurotoxic metals. In: HH Schaumburg (Ed.), American Academy of Neurology Education Program Syllabus: Neurotoxicology, American Academy of Neurology, Minneapolis, pp. 6–22.

Kuo HC, Huang CC, Tsai YT et al. (2005). Acute painful neuropathy in thallium poisoning. Neurology 65: 302–304.

Kuwabara S, Nakajima M, Tsuboi Y et al. (1993). Multifocal conduction block in n-hexane neuropathy. Muscle Nerve 16: 1416–1417.

Lachance DH, Lennon VA, Pittock SJ et al. (2010). An outbreak of neurological autoimmunity with polyradiculoneuropathy in workers exposed to aerosolised porcine neural tissue: a descriptive study. Lancet Neurol 9: 55–66.

Lacomis D (2002). Small-fiber neuropathy. Muscle Nerve 26: 173–188.

Lalloo M, Cosnett JE, Moosa A (1981). Benzine-sniffing neuropathy. S Afr Med J 59: 522–524.

Landau WM (1953). The duration of neuromuscular function after nerve section in man. J Neurosurg 10: 64–68.

Landrigan PJ, Todd AC (1994). Lead poisoning. Western J Med 161: 153–159.

Landrigan PJ, Baker Jr EL, Feldman RG et al. (1976). Increased lead absorption with anemia and slowed nerve conduction in children near a lead smelter. J Pediatr 89: 904–910.

Laplane D, Attal N, Sauron B et al. (1992). Lesions of basal ganglia due to disulfiram neurotoxicity. J Neurol Neurosurg Psychiatry 55: 925–929.

Lauwerys RR, Hoet P (1993). Mercury: metallic mercury and its organic salts. In: Industrial chemical Exposure. Guidelines for Biological Monitoring, Lewis Publishers, Boca Raton, pp. 75–83.

LeQuesne PM (1978). Neuropsychological investigations of subclinical and minimal toxic neuropathies. Muscle Nerve 1: 392–395.

LeQuesne PM (1985). Clinical and morphological findings in acrylamide toxicity. Neurotoxicology 6: 17–24.

Letz R, Gerr F, Cragle D et al. (2000). Residual neurologic deficits 30 years after occupational exposure to elemental mercury. Neurotoxicol 21: 459–474.

LeWitt PA (1980). The neurotoxicity of the rat poison vacor. A clinical study of 12 cases. N Engl J Med 302: 73–77.

Liu CH, Huang CY, Huang CC (2012). Occupational neurotoxic diseases in Taiwan. Safety and Health at Work 3: 257–267.

Lombet A, Mourre C, Lazdunski M (1988). Interaction of insecticides of the pyrethroid family with specific binding sites on the voltage-dependent sodium channel from mammalian brain. Brain Res Dev Brain Res 459: 44–53.

London Z, Albers JW (2007). Toxic neuropathies associated with pharmaceutical and industrial agents. Neurol Clin 25: 257–276.

Long RR, Sargent JC, Hammer K (1990). Paralytic shell fish poisoning: a case report and serial electrophysiologic observations. Neurology 40: 1310–1312.

Longnecker MP, Michalek JE (2000). Serum dioxin level in relation to diabetes mellitus among Air Force veterans with background levels of exposure. Epidemiology 11: 44–48.

Lotti M (1991). The pathogenesis of organophosphate polyneuropathy. Crit Rev Toxicol 21: 465–487.

Lotti M (2002). Low-level exposures to organophosphorus esters and peripheral nerve function. Muscle Nerve 25: 492–504.

Lotti M, Becker CE, Aminoff MJ (1984). Organophosphate polyneuropathy: pathogenesis and prevention. Neurology 34: 658–662.

Lotti M, Moretto A, Zoppellari R et al. (1986). Inhibition of lymphocytic neuropathy target esterase predicts the development of organophosphate-induced delayed polyneuropathy. Arch Toxicol 59: 176–179.

Malik GM, Mubarik M, Romshoo GJ (1998). Organophosphorus poisoning in the Kashmir Valley, 1994 to 1997 (letter). N Engl J Med 338: 1078.

Malkani R, Weinstein JM, Kumar N et al. (2011). Ataxia and cranial neuropathies from subcutaneously injected elemental mercury. Clin Toxicol 49: 334–336.

Mazumber DG, Dasgupta UB (2011). Chronic arsenic toxicity: studies in West Bengal, India. Kaohsiung J Med Sci 27: 360–370.

McGonagle TK, Levine SR, Donofrio PD et al. (1990). Spectrum of patients with EMG features of polyradiculopathy without neuropathy. Muscle Nerve 13: 63–69.

Meggs WJ, Hoffman RS, Shih RD et al. (1994). Thallium poisoning from maliciously contaminated food. J Toxicol Clin Toxicol 32: 723–730.

Michalek JE, Akhtar FZ, Arezzo JC et al. (2001). Serum dioxin and peripheral neuropathy in veterans of operation ranch hand. Neurotoxicol 22: 479–490.

Misirli H, Domac FM, Somay G et al. (2008). N-hexane induced polyneuropathy: a clinical and electrophysiological follow up. Electromyogr Clin Neurophysiol 48: 103–108.

Moore MR, McColl KEL, Rimington C et al. (1987). Disorders of porphyrin metabolism, Plenum, New York.

Moore D, House I, Dixon A (1993). Thallium poisoning. Diagnosis may be elusive but alopecia is the clue. Br Med J 306: 1527–1529.

Moretto A, Lotti M (1998). Poisoning by organophosphorus insecticides and sensory neuropathy. J Neurol Neurosurg Psychiatry 64: 463–468.

Moretto A, Sabri MI (1988). Progressive deficits in retrograde axon transport precede degeneration of motor axons in acrylamide neuropathy. Brain Res Dev Brain Res 440: 18–24.

Morgan JP, Penovich P (1978). Jamaica ginger paralysis. Arch Neurol 35: 530–532.

Moses M, Lilis R, Crow KD et al. (1984). Health status of workers with past exposure to 2,3,7,8-tetrachlorodibenzo-p-dioxin in the manufacture of 2,4,5-trichlorophenoxyacetic acid: comparison of findings with and without chloracne. Am J Ind Med 5: 161–182.

Mukherjee SC, Rahman MM, Chowdhury UK et al. (2003). Neuropathy in arsenic toxicity from groundwater arsenic contamination in West Bengal, India. J Environ Sci Health Part A 38: 165–183.

Mulloy KB (1996). Two case reports of neurological disease in coal mine preparation plant workers. Am J Ind Med 30: 56–61.

Murata K, Araki S, Yokoyama K et al. (1993). Assessment of central, peripheral, and autonomic nervous system functions in lead workers: neuroelectrophysiological studies. Environ Res 61: 323–336.

Navarro X, Verdu E, Guerrero J et al. (1993). Abnormalities of sympathetic sudomotor function in experimental acrylamide neuropathy. J Neurol Sci 114: 56–61.

Neghab M, Soleimani E, Khamoushian K (2012). Electrophysiological studies of shoemakers exposed to sub-TLV levels of n-hexane. J Occup Health 54: 376–382.

Nielsen CJ, Nielsen VK, Kirkby H et al. (1982). Absence of peripheral neuropathy in long-term lead-exposed subjects. Acta Neurol Scand 65: 241–247.

Nordentoft T, Andersen EB, Mogensen PH (1998). Initial sensorimotor and delayed autonomic neuropathy in acute thallium poisoning. Neurotoxicology 19: 421–426.

Ochoa J (1980). Nerve fiber pathology in acute and chronic compression. In: GE Omer, M Spinner (Eds.), Management of Peripheral Nerve Problems, Sanders, Philadelphia, pp. 487–501.

Ochoa J, Fowler TJ, Gilliatt RW (1972). Anatomical changes in peripheral nerves compressed by a pneumatic tourniquet. J Anat 113: 433–455.

O'Donoghue JL, Krasavage WJ (1980). Identification and characterization of methyl n-butyl ketone neurotoxicity in laboratory animals. In: PS Spencer, HH Schaumburg (Eds.), Experimental and Clinical Neurotoxicology, Williams and Wilkins, Baltimore, pp. 856–862.

Oh SJ (1975). Lead neuropathy: case report. Arch Phys Med Rehabil 56: 312–317.

Oh SJ, Kim JM (1976). Giant axonal swelling in "huffer's" neuropathy. Arch Neurol 33: 583–586.

Osterloh J, Lotti M, Pond SM (1983). Toxicologic studies in a fatal overdose of 2,4-D, MCPP, and chlorpyrifos. J Expo Anal Environ Epidemiol 7: 125–129.

Oz SG, Tozlu M, Yalcin SS et al. (2012). Mercury vapor inhalation and poisoning of a family. Inhal Toxicol 24: 652–658.

Palliyath SK, Schwartz BD, Gant L (1990). Peripheral nerve functions in chronic alcoholic patients on disulfiram: a six month follow up. J Neurol Neurosurg Psychiatry 53: 227–230.

Pastore C, Izura V, Marhuenda D et al. (2002). Partial conduction blocks in N-hexane neuropathy. Muscle Nerve 26: 132–135.

Paul S, Das N, Bhattacharjee P et al. (2013). Arsenic-induced toxicity and carcinogenicity: a two-wave cross-sectional study in arsenicosis individuals in West Bengal, India. J Exposure Sci & Environ Epidemiol 23: 156–162.

Paulev PE, Gry C, Dossing M (1979). Motor nerve conduction velocity in asymptomatic lead workers. Int Arch Occup Environ Health 43: 37–43.

Paulson GW, Waylonis GW (1976). Polyneuropathy due to n-hexane. Arch Int Med 136: 880–882.

Pazderova-Vejlupkova J, Lukas E, Nemcova M et al. (1981). The development and prognosis of chronic intoxication by tetrachlordibenzo-p-dioxin in men. Arch Environ Health 36: 5–11.

Pelclova D, Lukas E, Urban P et al. (2002). Mercury intoxication from skin ointment containing mercuric ammonium chloride. Int Arch Occup Environ Health 75 (Suppl-9).

Pelclova D, Urban P, Ridzon P et al. (2009). Two-year follow-up of two patients after severe thallium intoxication. Hum Exp Toxicol 28: 263–272.

Peters HA, Levine RL, Matthews CG et al. (1982). Carbon disulfide-induced neuropsychiatric changes in grain storage workers. Am J Ind Med 3: 373–391.

Peters HA, Levine RL, Matthews CG et al. (1988). Extrapyramidal and other neurologic manifestations associated with carbon disulfide fumigant exposure. Arch Neurol 45: 537–540.

Phan TG, Estell J, Duggin G et al. (1998). Lead poisoning from drinking Kombucha tea brewed in a ceramic pot. Med J Aust 169: 644–646.

Pimparkar BD, Bhave A (2010). Arsenicosis: review of recent advances. J Assoc Physicians India 58: 617–624.

Pocchiari F, Silano V, Zampieri A (1979). Human health effects from accidental release of tetrachlorodibenzo-p-dioxin (TCDD) at Seveso, Italy. Ann NY Acad Sci 320: 311–320.

Poklis A, Saady JJ (1990). Arsenic poisoning: acute or chronic? Suicide or murder? Am J Forensic Med Pathol 11: 226–232.

Putz-Anderson V, Albright BE, Lee ST et al. (1983). A behavioral examination of workers exposed to carbon disulfide. Neurotoxicol 4: 67–78.

Rainier S, Bui M, Mark E et al. (2008). Neuropathy target esterase gene mutations cause motor neuron disease. Am J Hum Genet 82: 780–785.

Rainier S, Albers JW, Dyck PJ et al. (2011). Motor neuron disease due to neuropathy target esterase gene mutation: clinical features of the index families. Muscle Nerve 43: 19–25.

Rangel-Guerra R, Martinez HR, Villarreal HJ (1990). Thallium poisoning. Experience with 50 patients. Gac Med Mex 126: 487–494.

Reagan KE, Wilmarth KR, Friedman MA et al. (1995). In vitro calcium and calmodulin-dependent kinase-mediated phosphorylation of rat brain and spinal cord neurofilament proteins is increased by glycidamide administration. Brain Res Dev Brain Res 671: 12–20.

Redmond JM, McKenna MJ, Feingold M et al. (1992). Sensory testing versus nerve conduction velocity in diabetic polyneuropathy. Muscle Nerve 15: 1334–1339.

Richardson RJ (1995). Assessment of the neurotoxic potential of chlorpyrifos relative to other organophosphorus compounds: a critical review of the literature. J Toxicol Environ Health 44: 135–165.

Roelofs-Iverson RA, Mulder DW, Elveback LR (1984). ALS and heavy metals: a pilot case-control study. Neurology 34: 393–395.

Rubens O, Logina I, Kravale I et al. (2001). Peripheral neuropathy in chronic occupational inorganic lead exposure: a clinical and electrophysiological study. J Neurol Neurosurg Psychiatry 71: 200–204.

Ruff RL, Petito CK, Acheson LS (1981). Neuropathy associated with chronic low level exposure to n-hexane. Clin Toxicology 18: 515–519.

Rusyniak DE, Furbee RB, Kirk MA (2002). Thallium and arsenic poisoning in a small midwestern town. Ann Emer Med 39: 307–311.

Sack GHJ (1990). Acute intermittent porphyria. JAMA 264: 1290–1293.

Safranek TJ, Lawrence DN, Kurland LT et al. (1991). Reassessment of the association between Guillain–Barré syndrome and receipt of swine influenza vaccine in 1976–1977: results of a two-state study. Expert Neurology Group. Am J Epidemiol 133: 940–951.

Saha A (2005). Thallium toxicity: a growing concern. IJOEM 9: 53–56.

Sahenk Z (1987). Toxic neuropathies. Sem Neurol 7: 9–17.

Said Jr GH, Boudier L, Selva J et al. (1983). Different patterns of uremic polyneuropathy. Neurology 33: 567–574.

Sayre LM, Autilio-Gambetti L, Gambetti P (1985). Pathogenesis of experimental giant neurofilamentous axonopathies: a unified hypothesis based on chemical modification of neurofilaments. Brain Res Dev Brain Res 357: 69–83.

Scelsi R, Poggi P, Fera L et al. (1980). Toxic polyneuropathy due to n-hexane. J Neurol Sci 47: 7–19.

Scelsi R, Poggi P, Fera L et al. (1981). Industrial neuropathy due to n-hexane: clinical and morphological findings in three cases. Clin Toxicol 18: 1387–1393.

Schaumburg HH, Spencer PS (1984). Human toxic neuropathy due to industrial agents. In: PJ Dyck, PK Thomas, EH Lambert et al. (Eds.), Peripheral Neuropathy, 2 edn. W.B. Saunders Company, Philadelphia, pp. 2115–2132.

Schaumburg HH, Spencer PS (1987). Recognizing neurotoxic disease. Neurology 37: 276–278.

Schaumburg HH, Spencer PS, Thomas PK (1983). Anatomical classification of PNS disorders, F.A. Davis Company, Philadelphia.

Senanayake N, Jeyaratnam J (1981). Toxic polyneuropathy due to gingili oil contaminated with tri-cresyl phosphate affecting adolescent girls in Sri Lanka. Lancet 1: 88–89.

Senanayake N, Johnson MK (1982). Acute polyneuropathy after poisoning by a new organophosphate insecticide. N Engl J Med 306: 155–157.

Senanayake N, Karalliedde L (1987a). Neurotoxic effects of organophosphorus insecticides. An intermediate syndrome. N Engl J Med 316: 761–763.

Senanayake N, Karalliedde L (1987b). Neurotoxic effects of organophosphorus insecticides. N Engl J Med 316: 761–763.

Sendur OF, Turan Y, Bal S et al. (2009). Toxic neuropathy due to N-hexane: report of three cases. Inhal Toxicol 21: 210–214.

Sengupta SR, Das NK, Datta PK (2008). Pathogenesis, clinical features and pathology of chronic arsenicosis. Indian J Derm, Venereology & Leprology 74: 559–570.

Seppalainen AM, Haltia M (1980). Carbon disulfide. In: PS Spencer, HH Schaumburg (Eds.), Experimental and Clinical Neurotoxicology, Williams & Wilkins, Baltimore, pp. 356–373.

Seppalainen AM, Hernberg S (1972). Sensitive technique for detecting subclinical lead neuropathy. Brit J Intern Med 29: 443–449.

Seppalainen AM, Hernberg S (1980). Subclinical lead neuropathy. Am J Ind Med 1: 413–420.

Seppalainen AM, Tolonen M (1974). Neurotoxicity of long term exposure to carbon disulfide in the viscose rayon industry. A neurophysiological study. Work & Environ Health 11: 145–153.

Seppalainen AM, Tola S, Hernberg S et al. (1975). Subclinical neuropathy at "safe" levels of lead exposure. Arch Environ Health 30: 180–183.

Seppalainen AM, Hernberg S, Kock B (1979). Relationship between lead levels and nerve conduction velocities. Neurotoxicol 1: 313–332.

Shih H, Gartner Jr JC (2001). Weight loss, hypertension, weakness, and limb pain in an 11-year-old boy. J Pediatr 138: 566–569.

Sidell FR (1994). Clinical effects of organophosphorus cholinesterase inhibitors. J Appl Toxicol 14: 111–113.

Sinczuk-Walczak H, Szymczak M, Halatek T (2010). Effects of occupational exposure to arsenic on the nervous system: clinical and neurophysiological studies. International

Journal of Occupational Medicine & Environmental Health 23: 347–355.

Singer R, Moses M, Valciukas J et al. (1982). Nerve conduction velocity studies of workers employed in the manufacture of phenoxy herbicides. Environ Res 29: 297–311.

Singer R, Valciukas JA, Lilis R (1983). Lead exposure and nerve conduction velocity: the differential time course of sensory and motor nerve effects. Neurotoxicol 4: 193–202.

Smith AG, Albers JW (1997). n-Hexane neuropathy due to rubber cement sniffing. Muscle Nerve 20: 1445–1450.

Smith EA, Oehme FW (1991). Acrylamide and polyacrylamide: a review of production, use, environmental fate and neurotoxicity. Rev Environ Health 9: 215–228.

Spencer PS, Schaumburg HH (1985). Organic solvent neurotoxicity. Facts and research needs. Scand J Work Environ Health 11 (Suppl 1): 53–60.

Spencer PS, Schaumburg HH, Raleigh RL et al. (1975). Nervous system degeneration produced by the industrial solvent methyl n-butyl ketone. Arch Neurol 32: 219–222.

Sumner AJ, Asbury AK (1975). Physiological studies of the dying-back phenomenon: muscle stretch afferents in acrylamide neuropathy. Brain 98: 91–100.

Suskind RR, Hertzberg VS (1984). Human health effects of 2,4,5-T and its toxic contaminants. JAMA 251: 2372–2380.

Sweeney MH, Fingerhut MA, Arezzo JC et al. (1993). Peripheral neuropathy after occupational exposure to 2,3,7,8-tetrachlorodibenzo-p-dioxin (TCDD). Am J Ind Med 23: 845–858.

Tandrup T, Braendgaard H (1994). Number and volume of rat dorsal root ganglion cells in acrylamide intoxication. J Neurocytol 23: 242–248.

Tavolato B, Licandro AC, Argentiero V (1980). Lead polyneuropathy of nonindustrial origin. Eur J Neurol 19: 273–276.

Tay CH, Seah CS (1975). Arsenic poisoning from antiasthmatic herbal preparations. Med J Australia 2: 424–428.

Tenenbein M, deGroot W, Rajani KR (1984). Peripheral neuropathy following intentional inhalation of naphtha fumes. Can Med Assoc J 131: 1077–1079.

The Centers for Disease Control Vietnam Experience Study (1988). Health status of Vietnam veterans. II. Physical Health. The Centers for Disease Control Vietnam Experience Study. JAMA 259: 2708–2714.

Thery JC, Jardin F, Massy N et al. (2008). Optical neuropathy possibly related to arsenic during acute promyelocytic leukemia treatment. Leuk Lymphoma 49: 168–170.

Thomke F, Jung D, Besser R et al. (1999). Increased risk of sensory neuropathy in workers with chloracne after exposure to 2,3,7,8-polychlorinated dioxins and furans. Acta Neurol Scand 100: 1–5.

Thomson RM, Parry GJ (2006). Neuropathies associated with excessive exposure to lead. Muscle Nerve 33: 732–741.

Tracy JA, Dyck PJ (2011). Auto-immune polyradiculoneuropathy and a novel IgG biomarker in workers exposed to aerosolized porcine brain. J Peripher Nerv Syst 16 (Suppl-7).

Triebig G, Weltle D, Valentin H (1984). Investigations on neurotoxicity of chemical substances at the workplace. V. Determination of the motor and sensory nerve conduction velocity in persons occupationally exposed to lead. Int Arch Occup Environ Health 53: 189–203.

Troni W, Carta Q, Cantello R et al. (1984). Peripheral nerve function and metabolic control in diabetes mellitus. Ann Neurol 16: 178–183.

Tseng HP, Wang YH, Wu MM et al. (2006). Association between chronic exposure to arsenic and slow nerve conduction velocity among adolescents in Taiwan. J Health, Population & Nutrition 24: 182–189.

Urban P, Pelclova D, Lukas E et al. (2007). Neurological and neurophysiological examinations on workers with chronic poisoning by 2,3,7,8-TCDD: follow-up 35 years after exposure. Eur J Neurol 14: 213–218.

Valenzuela OL, Borja-Aburto VH, Garcia-Vargas GG et al. (2005). Urinary trivalent methylated arsenic species in a population chronically exposed to inorganic arsenic. Environ Health Perspect 113: 250–254.

Vergauwe PL, Knockaert DC, Van Tittelboom TJ (1990). Near fatal subacute thallium poisoning necessitating prolonged mechanical ventilation. Am J Emer Med 8: 548–550.

Vigliani EC (1954). Carbon disulfide poisoning in viscose rayon factories. Brit J Ind Med 11: 235–244.

Wadia RS, Chitra S, Amin RB et al. (1987). Electrophysiological studies in acute organophosphate poisoning. J Neurol Neurosurg Psychiatry 50: 1442–1448.

Waldron HA (1983). Did the Mad Hatter have mercury poisoning? Br Med J 287: 1961.

Wang L, Liu YH, Xu Y et al. (2009). Thirteen-year follow-up of patients with tri-ortho-cresyl phosphate poisoning in northern suburbs of Xi'an in China. Neurotoxicol 30: 1084–1087.

Webb KB, Evans RG, Knutsen AP et al. (1989). Medical evaluation of subjects with known body levels of 2,3,7,8-tetrachlorodibenzo-p-dioxin. J Toxicol Environ Health 28: 183–193.

Whitaker JA, Austin W, Nelson JD (1962). Edathamil calcium disodium (versenate) diagnostic test for lead poisoning. Pediatrics 29 (82): 384–388.

Wilbourn AJ (1984). Metal neuropathies. In: AAEE 31st Annual Meeting, Course B1: Toxic Neuropathies, American Association of Electrodiagnostic Medicine, Kansas City, pp. 13–25.

Wills JH (1970). Toxicity of anticholinesterases and treatment of poisoning. In: AG Karczman (Ed.), Anticholinesterase Agents. International Encyclopedia of Pharmacology and Therapeutics, Section 13. I. Pergamon Press, Oxford, pp. 355–471.

Windebank AJ (1993). Metal neuropathy. In: PJ Dyck, PK Thomas, JW Griffin et al. (Eds.), Peripheral Neuropathy, 3rd edn. W.B. Saunders, Philadelphia, pp. 1549–1570.

World Health Organization (1976). International programme on chemical safety, Environmental Health Criteria 1: Mercury. World Health Organization, Geneva.

Wu PB, Kingery WS, Date ES (1995). An EMG case report of lead neuropathy 19 years after a shotgun injury. Muscle Nerve 18: 326–329.

Wu N, Hao F, Yu X (2012). Peripheral nerve and skin damage associated with working in a STCP factory: report of four cases. Clin Toxicol 50: 514–517.

Wu ML, Deng JF, Lin KP et al. (2013). Lead, mercury, and arsenic poisoning due to topical use of traditional Chinese medicines. Am J Med 126: 451–454.

Xu Y, Wang Y, Zheng Q et al. (2008). Clinical manifestations and arsenic methylation after a rare subacute arsenic poisoning accident. Toxicol Sci 103: 278–284.

Yeh JH, Chang YC, Wang JD (1995). Combined electroneurographic and electromyographic studies in lead workers. Occup Environ Med 52: 415–419.

Yokoyama K, Feldman RG, Sax DS et al. (1990). Relation of distribution of conduction velocities to nerve biopsy findings in n-hexane poisoning. Muscle Nerve 13: 314–320.

Zheng G, Tian L, Liang Y et al. (2011). Aminolevulinic acid dehydratase genotype predicts toxic effects of lead on workers' peripheral nervous system. Neurotoxicol 32: 374–382.

Section 5

Special sensory organ systems

Handbook of Clinical Neurology, Vol. 131 (3rd series)
Occupational Neurology
M. Lotti and M.L. Bleecker, Editors

Chapter 17

Neurotoxic exposure and impairment of the chemical senses of taste and smell

RICHARD L. DOTY*

Smell and Taste Center, Department of Otorhinolaryngology; Head and Neck Surgery, Perelman School of Medicine, University of Pennsylvania, Philadelphia, PA, USA

INTRODUCTION

The chemical senses of taste and smell are critical for quality of life, as they determine the flavor of foods and beverages, guide appropriate food intake, and provide a warning of such environmental hazards as spoiled or poisonous food, leaking natural gas, smoke, and airborne pollutants. Distortions or losses of the functioning of these senses can be devastating for those who experience them, particularly those dependent upon them for their safety or livelihood. This includes plumbers, fire fighters, cooks, perfumers, fragrance sales persons, wine merchants, coffee or tea tasters, food and beverage distributors, homemakers, and employees of numerous chemical, gas, and public works industries. Indeed, loss of smell function is a reason for dismissal from the armed services, including the reserves and the coast guard.

Because of the rather direct exposure of its receptors to the outside environment, the olfactory system is vulnerable to volatile neurotoxins from air pollution and industrial processes, as well as from viruses, bacteria, and other airborne agents. Damage to the olfactory epithelium can occur directly as an acute insult or indirectly via the induction of longer-term inflammatory processes. A number of xenobiotics, including metals, chemicals, and nanoparticles, have the potential to enter the brain via olfactory receptors or surrounding perineural spaces, thereby bypassing the blood–brain barrier. In some cases such agents damage central nervous system (CNS) processes, alter neurotransmitter function, and induce forms of central neuropathology similar to those seen in neurodegenerative diseases (Doty, 2008).

Although theoretically one might expect similar influences of neurotoxic agents on the taste system, there are surprisingly few empiric studies in support of this concept. A small number of scattered case reports suggest that dysgeusia (distorted taste sensation), such as the presence of a metallic taste, can arise in humans from exposure to some neurotoxic agents, including toxins from the dinoflagellate *Gambierdiscus toxicus* (Pearn, 2001) and pesticides (Grant, 1993). Conditioned aversions to saccharine and other food items by their pairing with pesticide-related illness have been induced in rodents (Schiffman and Nagle, 1992). Studies in fish have demonstrated that taste buds can be damaged from pollutants in water, including lead (Pb), mercury (Hg), cadmium (Cd), and so-called "hard" and "soft" detergents (Bardach et al., 1965; Pevzner et al., 1986; Borovyagin et al., 1989).

This chapter addresses the influences of neurotoxic exposures on human chemoreception and provides some information on the adverse influences of such exposures on rodent epithelia. Given the dearth of information on taste, the focus is on olfaction. Nonetheless, a brief review of the anatomy of the taste system is presented, along with practical means for its assessment, in order to inform the reader about specifics of this system and how it can be clinically evaluated. Hopefully this will help to encourage others to collect data on this topic.

ANATOMY AND PHYSIOLOGY

Olfaction

To activate the olfactory system, volatile chemicals must enter the region of the olfactory epithelium from either

*Correspondence to: Richard L. Doty, Ph.D., Professor and Director, Smell and Taste Center, Perelman School of Medicine, University of Pennsylvania, 5 Ravdin Pavilion, 3400 Spruce Street, Philadelphia, PA 19104, USA. Tel: +1-215-662-6580, E-mail: Richard.doty@uphs.upenn.edu

the front of the nose or from the oral cavity via the nasopharynx. This pseudostratified columnar epithelium contains the bipolar receptor cells responsive to odorants, as well as sustentacular (supporting) cells, microvillar cells, and several classes of basal cells from which the other cells are derived (Fig. 17.1A). It lines the upper regions of the nasal septum, cribriform plate, superior turbinate, and anterior sectors of the middle turbinate (Leopold et al., 2000). In humans, this epithelium loses its homogeneity over time. Indeed, within less than a month after birth, metaplastic islands of respiratory-like epithelia begin to appear within its borders, presumably reflecting insults from airborne toxins, viruses, and bacteria. These metaplastic islands accumulate during life, providing one basis for the well-established age-related decrement in smell function.

That being said, the olfactory epithelium is protected to a large degree from the outside environment by its location high in the nose and by mucus derived largely from specialized Bowman's glands. This mucus is comprised of water, mucopolysaccharides, enzymes, immune factors, and proteins, including odorant-binding proteins that aid in transporting hydrophobic odorants to the receptors. The receptor proteins proper are embedded within the cilia of the bipolar receptor cells (Fig. 17.1B). Unlike the respiratory cilia that move the mucus blanket across their surface, the olfactory cilia are relatively long and lack intrinsic motility, more or less wafting within the mucus. The 6–10 million olfactory receptor cells are insulated from one another by the surrounding sustentacular cells (Fig. 17.1A). Like the other cells within this epithelium, including the microvillar cells that may secrete nitric oxide to protect against microorganisms at the epithelial surface, the receptor cells germinate from stem cells located near its basement membrane (Moran et al., 1982).

In humans, the ciliary membranes of each receptor cell express one of nearly 400 different types of functional olfactory receptors. Most are members of the heptahelical G-protein-coupled receptor (GPCR) superfamily (Rouquier et al., 2000), although some may be members of the trace amine-associated receptor family (Liberles and Buck, 2006). While all but a few chromosomes have some olfactory receptor genes, the majority are found on chromosome 11. Others are located mainly on chromosomes 1, 6, and 9 (Glusman et al., 2001). When occupied, most odorant receptors activate a guanosine triphosphate-binding protein. Type III adenylyl cyclase is then released, catalyzing the production of 3',5'-cyclic monophosphate. This opens a cyclic nucleotide-gated channel, producing an influx of sodium and calcium ions and depolarization of the receptor neuron (Breer, 1994).

Despite the fact that only one type of receptor is expressed on a given receptor cell, most receptor cells are responsive to several odorants (Holley et al., 1974;

Fig. 17.1. **(A)** Low-power photomicrograph of cross-section of the human olfactory neuroepithelium showing the four major types of cells: bipolar receptor cells (arrows point to cilia at their dendritic knobs. bg, Bowman's gland; bs, basal cell undergoing mitosis; c, cell body of receptor cell; d, duct of Bowman's gland; lp, lamina propria; m, microvillar cell; n, collection of axons with an ensheathing cell; s, sustentacular (supporting) cell. (Courtesy of Dr. David Moran, Longmont, Colorado.) **(B)** Scanning electron micrograph showing the transition zone between the human olfactory epithelium (bottom) and the respiratory epithelium (top). Arrows signify two examples of olfactory receptor cilia dendrites with cilia that have been cut off. Bar = 5 μm. (Reproduced from Menco and Morrison (2003), with permission.)

Sicard and Holley, 1984). Hence, a given cell responds to a range of ligands that can overlap, to some degree, with ligands that activate other receptor cells. With rare exception, it is a combination of receptor cells that provides information to higher brain regions that ultimately results in the perception of an odor. The initial coding is associated with unique spatial maps induced by odorants within both the epithelium and the olfactory bulbs (Johnson and Leon, 2007).

Before projecting through the cribriform plate to the brain, the receptor cell axons coalesce within the lamina propria into bundles of about 200 axons. Each bundle is encapsulated by olfactory ensheathing cells and then by olfactory nerve fibroblasts which define the olfactory nerve (cranial nerve (CN) I) fila. As illustrated in Figure 17.2, a population of microglia and macrophages is found between the olfactory ensheathing cells that aid in thwarting the spread of viruses and other xenobiotics into the brain through the olfactory fascicles (Smithson and Kawaja, 2010). The direct projection of the receptor cells from the nasal cavity into the brain creates a passageway for xenobiotic invasion into the CNS that effectively bypasses the blood–brain barrier (Jackson et al., 1979; Doty, 2008).

The first synapses in the olfactory system are those between the olfactory receptor cells and mitral and tufted cells, the main projection neurons of the olfactory bulb. These synapses occur within spherical structures, termed glomeruli, located in a peripheral layer of the bulb (Fig. 17.3). Glutamate is the primary neurotransmitter released from the synaptic terminals of the receptor cells, activating both *N*-methyl-D-aspartate and α-amino-3-hydroxy-5-methyl-4-isoxazolepropionic acid receptors on the mitral and tufted cells. Glutamate release is modulated by juxtaglomerular cells via presynaptic D_1 and $GABA_B$ receptors on the olfactory receptor axons (O'Connor and Jacob, 2008). Similar modulation of the mitral and tufted cells occurs within the external plexiform layer of the bulb, mainly via GABAergic granule cells, the most prevalent cells in the bulb.

Central structures that receive projections from the mitral and tufted cell axons include the anterior olfactory nucleus, the piriform cortex, the rostral entorhinal cortex, and the corticobasal nuclei of the amygdala. The olfactory system is unique in that its initial projections go to cortical structures without first synapsing in the thalamus. Subsequent connections are formed with the thalamus, orbitofrontal cortex, hippocampus, hypothalamus, and cerebellum. The orbitofrontal cortex plays a vital role in flavor perception, combining input from several sensory modalities, including taste, texture, and smell (Rolls and Grabenhorst, 2008). Orbitofrontal cortex lesions in this area impair the identification of odors and flavors (Jones-Gotman and Zatorre, 1988).

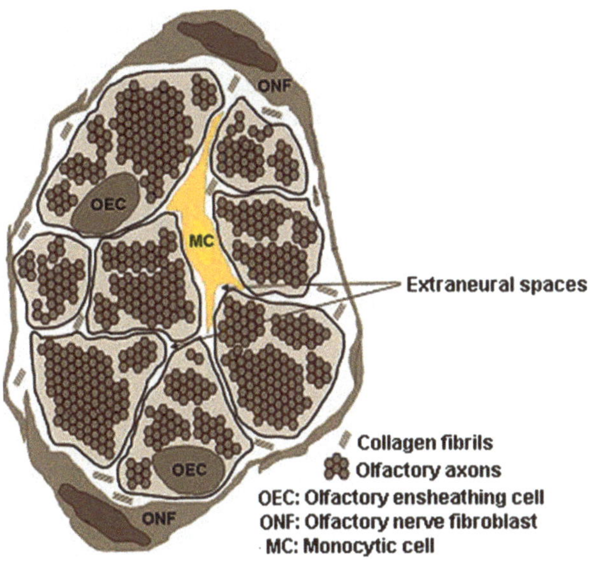

Fig. 17.2. Olfactory nerve fascicle. Olfactory axons are surrounded by bundles of olfactory ensheathing cells (OECs) which, in turn are surrounded by olfactory nerve fibroblasts (ONF). Monocytic cells (MC) and collagen fibrils are located within the extraneural spaces between the lamina bound units. (Modified from Hockman et al. (1971), with permission. © 2009 Wiley-Liss, Inc.)

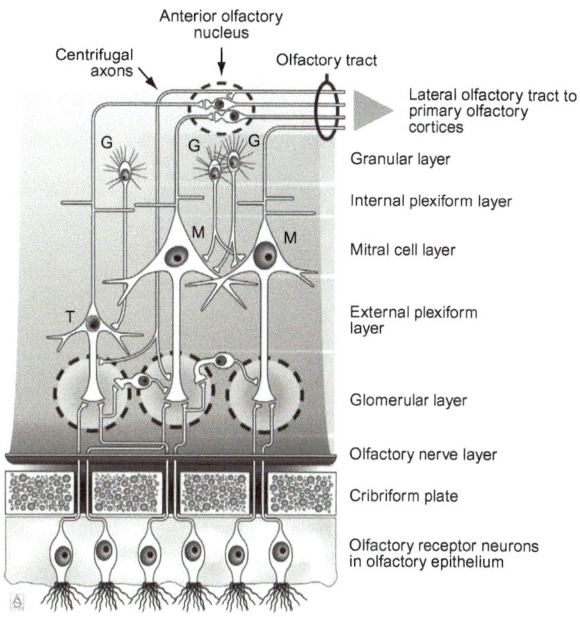

Fig. 17.3. Schematic of the major layers of the olfactory bulb and their interactions among cell types therein. G, granule cells; M, mitral cells; T, tufted cells. (Reprinted with permission from Duda, 2010. Copyright © 2010 Elsevier BV.)

Gustation

The receptors responsible for taste, i.e., sensations of sweet, bitter, salty, sour, fatty, and savory (umami), are found within the 5000–7500 taste buds located in the oral cavity. These flask-like structures are embedded in three of the four types of lingual papillae: the fungiform, foliate, and circumvallate (Fig. 17.4). Those on the fungiform papillae are supplied by the chorda tympani division of the facial nerve (CN VII), whereas those on the soft palate are supplied by the greater superficial petrosal division of this nerve. The buds on the posterior foliate papillae and on the circumvallate papillae are innervated by the glossopharyngeal nerve (CN IX). The vagus nerve (CN X) innervates the taste buds in the throat and nonoral taste receptors throughout the alimentary tract. Elements of the trigeminal nerve (CN V) project into the papillae and other oral mucosa surfaces and participate in the perception of flavor by signaling sensations of touch, pain, and temperature (e.g., the fizziness of carbonated drinks and the warmth of coffee). The front of the tongue is typically more sensitive than the back of the tongue to tastants, with the exception of bitter. Individuals differ considerably in terms of the number and distribution of their taste buds, which are most dense on the tongue tip and along its edges and back.

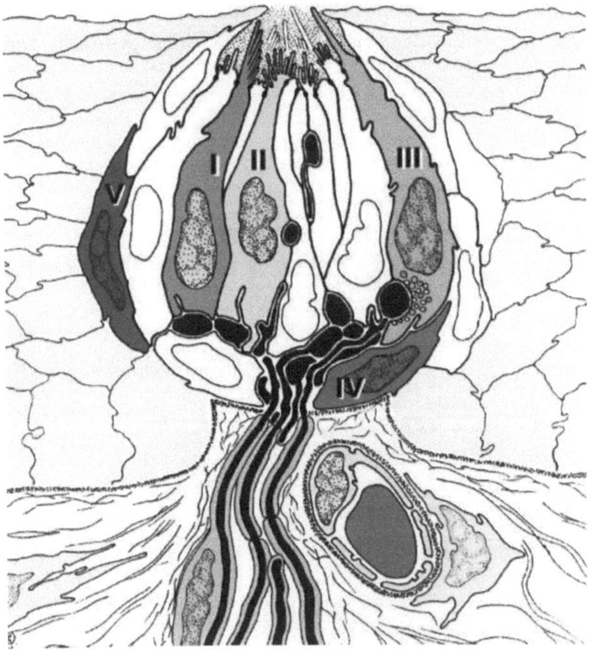

Fig. 17.4. Idealized drawing of longitudinal section of mammalian taste bud. Cells of type I, II, and III are elongated and form the sensory epithelium of the bud. These cells have different types of microvillae within the taste pit and may reach the taste pore. Type IV are basal cells and type V are marginal cells. Synapses are most apparent at the bases of type III cells. The connecting taste nerves have myelin sheaths. (Reproduced from Witt and Reutter (2015). Copyright © 2015 Richard L. Doty.)

Three types of taste-responsive cells within taste buds have been identified on the basis of their morphology (Murray, 1973). Type I cells mediate salty perception, with Na^+ ions entering specialized membrane channels such as the amiloride-sensitive Na^+ channel (Chaudhari and Roper, 2010). Type II cells are involved in the perception of sweet, bitter, and savory, with sweet and savory being mediated by a family of three GPCRs, termed the T1R1, T1R2, and T1R3 receptors. Bitter is mediated via a family of $\sim$30 GPCRs termed the T2R receptors (Chandrashekar et al., 2006; Meyerhof et al., 2010; Roudnitzky et al., 2011). Some fatty acids may also be detected via GPCRs located on such cells (Cartoni et al., 2010). Type III cells appear to be specialized for detecting sour tastes. H^+ ions pass through specialized proton channels of type III cells that do not involve GPCRs (Chang et al., 2010). Various ion channels are responsive to acids, including acid-sensing ion channels, potassium channels, and epithelial sodium channel-like channels (Shimada et al., 2006).

The nerves that innervate the taste buds (i.e., CN VII, IX, and X) send their projections from the oral cavity to the nucleus tractus solitarius of the brainstem. From there, projections are made via the medial lemniscus to the ventral posterior nucleus of the thalamus. Connections are then made to the amygdala and such brain regions as anterior insular cortex, the orbitofrontal cortex, and the primary somatosensory cortex. It is within these higher brain regions that intermodal responses are integrated, such as those related to taste, touch, and olfaction. Multimodal responses of cells in some of these areas develop as a result of conditioning of sensations from each of these modalities during deglutition. While hedonic responses to tastants can occur at the level of the brainstem, identification of a given tastant requires matching of the incoming information with memory stores of prior experiences with that tastant.

It is now evident that sweet and bitter taste-related receptor proteins are distributed not only in the oral cavity, but within the stomach, intestine, oropharynx, larynx, and the upper esophagus, where they serve diverse functions, including facilitation of digestion, bacterial defense, and chemical absorption (Breer et al., 2012; Reimann et al., 2012). One taste receptor (T2R38) is found in the epithelia of the human upper respiratory tract and releases nitric oxide in response to acyl-monoserine lactone, quorum-sensing molecules secreted by *Pseudomonas aeruginosa,* and other gram-negative bacteria. Interestingly, differences in T2R38 functionality, as related to the TAS2R38 genotype, are associated with susceptibility to upper respiratory infections (Lee et al., 2012).

QUANTITATIVE TESTING OF SMELL AND TASTE FUNCTION

Until being tested, many persons are unaware of less-than-total decrements in their senses and taste and smell. Hence, quantitative testing is necessary to identify such deficits. Chemosensory tests can be divided into three general classes: psychophysical, electrophysiological, and psychophysiological (Doty, 2007). Psychophysical tests require a conscious response on the part of the subject, and include tests of detection sensitivity, recognition, identification, discrimination, memory, hedonics, and suprathreshold intensity scaling. Electrophysiological tests measure stimulus-induced electric changes following sensory stimulation, such as event-related potentials measured from the scalp, whereas psychophysiologic tests assess odor-induced autonomic nervous system responses, such as changes in heart rate or respiration.

Psychophysical tests are the most popular tests for evaluating both taste and smell, reflecting considerations of reliability, practicality, and cost (Doty, 2007). In identification tests, which generally correlate with more time-consuming threshold tests, the patient is asked to identify the quality of each of a series of stimuli, usually from a list of names. In the most popular odor identification test, a series of microencapsulated (scratch and sniff) odorants is presented and the subject must choose, for each odorant, its name from a list of four response alternatives even if no stimulus is perceived or the sensation does not correspond to any of the alternatives (Fig. 17.5) (Doty et al., 1984). This and related tests can be self-administered and even sent to patients through the mail. The number of correct answers is determined and compared to norms based on thousands of normal subjects. This allows for an absolute (i.e., normosmia, mild microsmia, moderate microsmia, severe microsmia, anosmia) and relative (i.e., percentile rank relative to normal persons of the same age and sex) classification of function or dysfunction. Forced-choice testing focuses the subject's attention on each task and allows for detection of malingering, since malingerers often avoid the correct response beyond what would be expected by chance (Doty and Crastnopol, 2010).

Somewhat analogous taste tests are available using liquid tastants (Bartoshuk, 1978; Doty and Haxel, 2005; Stinton et al., 2010) or chemical tastants dried on pieces of filter paper (Linschoten and Kroeze, 1991; Zhao et al., 2003; Landis et al., 2009) or embedded in edible polymers (Smutzer et al., 2008; Desai et al., 2011). Electric taste testing is popular clinically, since there is no need to prepare stimuli or to rinse between trials, and stimuli can be accurately presented to a specific tongue region for a specified duration of time (Frank and Smith, 1991; Tomita and Ikeda, 2002). Electrogustometric thresholds correlate well with chemical thresholds (Berling et al., 2011) and the number of underlying papillae (Miller et al., 2002). Recently normative data have been published for electric thresholds determined using a staircase procedure on the anterior, posterior, and palate regions from 74 male and 82 female nonsmokers ranging in age from 10 to 80 years (Pavlidis et al., 2013).

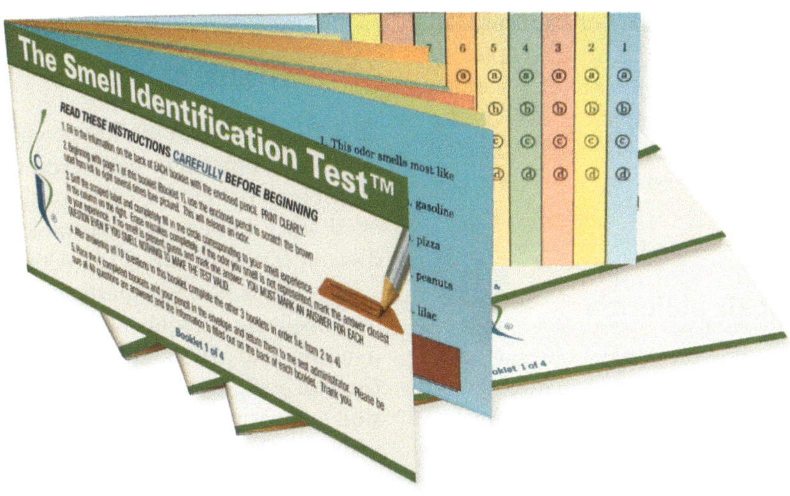

Fig. 17.5. The University of Pennsylvania Smell Identification Test (known commercially as the Smell Identification Test) (Doty et al., 1984). This test, developed in the early 1980s, is comprised of 40 microencapsulated odorants located next to forced-choice questions on each page of 10-page booklets. Over 90% of all olfactory studies of Parkinson's disease patients have used this test. (Copyright © 2004, Sensonics International, Haddon Heights, New Jersey.)

OLFACTORY DYSFUNCTION FROM ENVIRONMENTAL CHEMICAL EXPOSURES

In a pioneering review, Naus (1975) described the early and mostly European literature on the influences of airborne industrial toxins on smell function. A number of the reports were case studies or anecdotal. Nonforced-choice threshold testing, where there is potential for confounding the subject's response criterion with the measure of sensitivity, was common and in a number of studies basic controls for the effects of smoking and other factors were often absent. In these early studies, the reported exposure levels were generally rough estimates, as good industrial hygiene practices were not in force. Adverse effects on smell function were noted from a variety of industrial exposures, including those to spices, flour, cotton, Pb, coal, mercury, Cd, coke, grain, silicon dioxide, tobacco, paper, cement, ashes, and a range of industrial volatile chemicals (e.g., perfumes, menthol, carbonic disulfide, benzene, dichloromethane, pentachlorophenate, ammonia, bromine, trichloroethylene, nitrous gases, acids). In the second major review on this topic, Amoore (1986) provided a listing of compounds for which the strongest empirical evidence was available (Table 17.1).

Described below are relatively recent studies examining olfactory function in persons exposed to a number of environmental chemicals. The influences of general air pollution on such function are first presented, followed by those of specific metals, solvents, and gases. It should be kept in mind that, in many environments, including workplaces, exposures to multiple chemicals take place, either simultaneously or sequentially. This can confound inferences as to the specific agents involved. Rodent studies have shown that combinations of exposures to toxic agents can be more deleterious to the olfactory epithelium than exposures to single chemicals alone (Cassee et al., 1996) and that cumulative effects are possible even when exposures are short and separated from one another by long periods of time (Prediger et al., 2012).

Table 17.1

Substances or industrial processes that are reported by Amoore (1986) to alter smell function

Metallic compounds	**Organic compounds**
Cadmium compounds	Acetone
Cadmium oxide	Acetophenone
Chromate salts	Benzene
Manganese	Benzine
Nickel hydroxide	Butyl acetate
Zinc chromate	Chloromethanes
Metallurgic processes	Ethyl acetate
Chromium plating	Menthol
Lead smelting	Pentachlorophenol
Magnet production	Trichloroethylene
Mercury (chronic intoxication)	**Dusts**
Nickel plating	Cement
Nickel refining (electrolvtic)	Chemicals
Silver plating	Hardwoods
Steel production	Lime
Zinc production	Printing
Nonmetallic inorganic compounds	Silica
Ammonia	**Manufacturing processes**
Carbon disulfide	Acids
Carbon monoxide	Asphalt
Chlorine	Cutting oils
Fluorides	Fragrances
Hydrazine	Lead paints
Nitrogen dioxide	Paprika
Sulfur dioxide	Spices
	Tobacco
	Varnishes
	Wastewater (refinery)

Adapted and expanded from Amoore JE (1986) Effects of chemical exposure on olfaction in humans. In: Barrow CD (ed.) Toxicology of the Nasal Passages. Washington, DC: Hemisphere Publishing, pp. 155–190.

General air pollution

Air pollution, such as is common in major cities, is comprised of complex mixtures of particulate matter, gases, and organic compounds. Many persons living in highly polluted areas for extended periods of time exhibit measurable decrements in olfactory function (Hudson et al., 2006; Calderon-Garciduenas et al., 2010; Gudziol et al., 2010). Shorter-term exposure to extreme air pollution can also result in significant loss of smell. For example, following the World Trade Center collapse on September 11, 2001, more than 40 000 people were exposed for a number of months to high levels of a complex mixture of airborne pollutants. Many of these individuals developed smell deficits demonstrable even years after the disaster (Dalton et al., 2010; Altman et al., 2011).

Smell loss has been particularly well documented in persons living in highly polluted regions of Mexico City (Hudson et al., 2006; Calderon-Garciduenas et al., 2010; Gudziol et al., 2010). In the Calderon-Garciduenas et al. study, abnormal University of Pennsylvania Smell Identification Test (UPSIT) scores were present in 35.5% of the highly exposed subjects and in only 12% of controls of similar age and sex who lived in a much less polluted control city. Moreover, the influence of the apolipoprotein (APOE) epsilon 4 genotype – a risk factor for Alzheimer's disease – was related to the olfactory test scores of the Mexico City residents. Thus, highly exposed APOE epsilon 4 carriers failed, on average, 2.4 of the 10 UPSIT items identified in one study as being most strongly related to Alzheimer's disease (Tabert et al., 2005), while APOE 2/3 and 3/3 subjects failed only 1.36 such items ($p = 0.01$).

Autopsy studies performed on even young exposed Mexico City residents reveal marked olfactory bulb inflammation and the presence of nanoparticles – ultrafine particulate matter < 100 nm – in their olfactory bulbs (Calderon-Garciduenas et al., 2002, 2003, 2004, 2008, 2010, 2012). Such matter appears to induce innate immune responses within the brain, i.e., the induction of monocytes that produce and secrete tumor necrosis factor-α, interleukin-6, and interleukin-1β. These monocytes, in turn, recruit and increase the activity of other immune cells (Simard and Rivest, 2006; Calderon-Garciduenas et al., 2012). Thus, exposure to high levels of air pollution can activate endothelial cells, disrupt neurovascularization, induce innate immune system responses, and produce neuroinflammation and neurodegeneration (Calderon-Garciduenas et al., 2012). Such exposure may be responsible for the accumulation of Aβ and α-synuclein – defining features of Alzheimer's disease and Parkinson's disease – within the olfactory bulbs and other olfaction-related brain structures of residents of highly polluted cities (Calderon-Garciduenas et al., 2004).

Metals

As shown in Table 17.2, a number of metals are capable of being incorporated into the olfactory epithelium and transported into the olfactory bulb where, in some cases, they may cross synapses and enter into multiple brain regions. Thus, in addition to their potential to directly damage the olfactory system, per se, some forms of metals have the capability of entering the brain from the nose and damaging central brain structures.

Aluminum

Aluminum is the most abundant metal on the planet and, after oxygen and silicon, is the planet's most common element. Interest in the possibility that aluminum alters smell function comes from two sources: First, studies showing that forms of this metal can be taken up by the olfactory receptors and passed transsynaptically along olfactory pathways to a range of brain structures (Table 17.2). Second, evidence that Al accumulates in the brains of patients with Alzheimer's disease and some other neurodegenerative diseases accompanied by olfactory dysfunction (Perl and Good, 1991; Perl and Moalem, 2006). Perl and his associates used laser microprobe mass analysis to show that aluminum accumulates specifically in the neurofibrillary tangles of brains from patients with Alzheimer's disease (Good et al., 1992). Since aluminum is not intrinsic to the brain, any brain

Table 17.2

Examples of metals capable of incorporation into olfactory receptor cells from the nasal cavity and, in some cases, transported transneuronally to other brain regions

Metal	Species	Method of application	Receptor cell incorporation	Transneuronal transport	Example references
Aluminum lactate	Rabbit	Intranasal	Yes	Possible	Perl and Good (1987)
Aluminum silicate	Rabbit	Bedding	Yes	No	Hayek and Waite (1991)
Aluminum acetylacetonate	Rat	Inhalation	Yes	Yes	Zatta et al. (1993)
Cadmium	Rat	Intranasal	Yes	No	Tallkvist et al. (2002)
	Pike	Intranasal	Yes	No	Gottofrey and Tjalve (1991)
Cadmium chloride	Rat	Intranasal	Yes	No	Tjalve and Henriksson (1999)
Cadmium oxide	Rat	Aerosol	Yes	No	Hastings and Evans (1988)
Cobalt	Rat	Intranasal	Yes	Yes	Persson et al. (2003b)
	Salmon	Intranasal	Yes	Possible	Bazer et al. (1987)
Gold	Squirrel monkey	Intranasal	Yes	Yes	De Lorenzo (1970)
	Rabbit	Mucosal	Yes	No	Czerniawska (1970)
Iron oxide	Mouse	Inhalation	Yes	Yes	Wang et al. (2007)
Manganese	Pike	Intranasal	Yes	Yes	Tjalve et al. (1995)
	Rat	Intranasal	Yes	Yes	Henriksson et al. (1999)
Mercury	Rat and pike	Intranasal	Yes	No	Borg-Neczak and Tjalve (1996)
Nickel	Rat and pike	Intranasal	Yes	Yes	Tallkvist et al. (1998)
Zinc	Rat and pike	Intranasal	Yes	Yes	Persson et al. (2003a)

Modified from Baker and Genter (2003).

accumulation must come from outside sources. Nasally introduced aluminum can induce, in the olfactory bulb, piriform cortex, and hippocampus, granulomas containing large numbers of macrophages, lymphocytes, and plasma cells (Perl and Good, 1987).

Despite such observations, there have been no epidemiologic studies to assess whether olfaction is compromised in workers in aluminum foundries or machine shops where airborne exposures to aluminum dusts, aluminum oxide, or aluminum hydroxide occur.

Cadmium and nickel

Because many manufacturing processes co-involve Cd and nickel (Ni), these metals are discussed in this section together. Both of these metals are highly toxic and are employed in a number of manufacturing processes, including the production of batteries and the manufacturing of paints, plastics, and metal alloys. Hundreds of thousands of Americans are occupationally exposed to Cd and Ni fumes or dust each year. Municipal waste incineration, fossil fuel combustion, and sewage sludge and phosphate fertilizers contribute to the environmental levels of these metals and their related compounds.

Adverse effects of Cd and Ni on the human sense of smell have been known for years and animal studies have shown that the effect of Cd on the brain is region-specific and most pronounced in the olfactory bulb (Kumar et al., 1996). In 1950, Friberg reported that 44% of 43 workers employed in an alkaline battery factory complained of decreased ability to smell. These workers had been exposed to both Cd iron dust (5–15 mg/m^3 of air) and Ni graphite dust (10–150 mg/m^3 of air) for periods ranging from 9 to 34 years. About a third were said to be anosmic. Of 15 workers employed in the same areas of the factory for 4 years or less, only 10% complained of smell dysfunction. Edema, rhinitis, and general atrophy of the nasal epithelium were described in some workers with Cd poisoning by Baader (1952), who emphasized the potential for Cd to adversely influence olfaction. In 1961, Adams and Crabtree) administered an odor threshold test using phenyl to 106 Cd and Ni dust-exposed alkaline battery workers and 84 matched normal controls. The exposed workers performed significantly more poorly than the controls. Twenty-seven percent of those who were exposed were found to be anosmic, compared with 5% of the controls. Although the amount of exposure to Cd could only be roughly estimated, in some cases it was likely as high as 2.76 mg/m^3. Approximately half of the anosmic workers were unaware of their disorder. Proteinuria was positively correlated with the presence of anosmia, but no relationship was apparent between the degree of nasal irritation and the olfactory test measure. In a similar study, Potts (1965) reported that 64% of 70 battery factory workers who had been exposed to Cd dust (0.6–236 mg/m^3 of air) for 10–40 years were anosmic, although testing details were not presented.

Several more recent studies are worthy of note. In a 1985 study, 28% of workers who had worked 5 or more years in a Cd-refining plant were said to have anosmia (Yin-Zeng et al., 1985). The average concentration of airborne Cd was between 0.004 and 0.187 mg/m^3, a figure slightly above the current National Institute for Occupational Safety and Health permissible exposure limit of 0.005 mg/m^3. In 1992, Rose et al. administered a butanol odor detection threshold test and an odor discrimination test to 55 workers who had been exposed for an average of 12 years to Cd fumes (from brazing refrigerator coils). Airborne Cd, measured for the first time 13 years after production began, was 0.300 mg/m^3. Workers with the highest body burden of Cd, as measured by urinalysis, exhibited moderate to severe hyposmia, but not anosmia, on the detection threshold task, but performed normally on the odor discrimination task. These observations led to the suggestion that central elements of the olfactory pathway were not damaged.

Rydzewski and his associates assessed the olfactory function of 73 Polish workers involved in the production of Cd-Ni batteries and of 43 nonexposed age- and smoking-matched controls (Rydzewski et al., 1998; Sulkowski et al., 2000). Using Elsberg and Levy's (1935) blast injection olfactory test, in which odors of coffee, aniseed, lemon, and mint were employed, 26.0% of the exposed workers were said to be hyposmic and 17.8% parosmic. Only one case (1.4%) was reported as having anosmia. In most of the workers with olfactory dysfunction, hypertrophic changes in the nasal mucosa were identified that correlated with the duration of employment. Statistically significant associations between the olfactory impairment and Cd concentration in blood, urine, and the workplace air were observed. No relationship was found between the degree of olfactory dysfunction and the duration of employment. It is conceivable that the low number of anosmics observed in this study reflects the fact that the Elsberg and Levy procedure confounds olfactory and trigeminal responses (Jones, 1953), thereby likely underestimating the true number of anosmics. Despite this shortcoming, the results are in general accord with the concept that Cd exposure has significant influences on smell function.

To assess the influences of moderate exposures to Cd, i.e., exposures at levels close to or below American Conference of Governmental Industrial Hygienists (ACGIH) permissible exposure limits, Mascagni et al. (2003) administered odor identification and phenyl ethyl alcohol (PEA) detection thresholds to three groups of industrial workers: 33 male workers involved in Cd fusion, sintering and lamination of alloys, 39 nonexposed manual workers (drivers, warehousemen), and 23 iron/steel welders from a different facility. The thresholds of the Cd-exposed workers were about a half log unit higher than those of the iron/steel welders and approximately a full log unit higher than those of the manual workers. Odor identification tests were similar among the three groups. These data indicate that Cd can alter smell function even at relatively moderate levels of exposure.

Chromium

Chromium (Cr), like Cd, is employed in the manufacture of high-quality steel alloys. Compared to Cd and Ni, relatively little research on the influences of Cr on the sense of smell has been performed. Watanabe and Fukuchi (2000) assessed odor detection and recognition thresholds of 26 male and seven female employees of a chromate-producing factory who had a minimum of 7 years of exposure. Over half (51.4%) had perforated nasal septa and 54.5% exhibited elevated thresholds to the five target odorants (phenyl ethanol, cyclopentenolone, isovaleric acid, r-undecalactone, scatol). Two were anosmic. Duration of employment, but not the presence of septal perforations, was correlated with the degree of olfactory function.

Similar findings were noted in a study of 27 male chromate workers (mean age 32 years) in a factory in Korea (Kitamura et al., 2003). The target subjects exhibited nasal irritation but no septal perforations. Thresholds were correlated with exposure durations which ranged from 0.9 to 18.2 years (mean = 7.9 years). The environmental levels of both Cr^{3+} and Cr^{6+} were below the threshold limit values proposed by the ACGIH (2003); urinary levels of Cr of all subjects were within the biologic exposure indices of this organization. This study indicates that Cr can alter smell function even at exposure levels below those recommended by the ACGIH and that such exposures are dose-dependent.

Lead

It is well established that exposures to inorganic and organic Pb can produce transient or possibly permanent psychologic deficits, including problems related to attention, concentration, memory, visuoconstruction, and psychomotor speed and accuracy. Despite such effects, the occupational exposure to Pb appears to have only minor, if any, effects on the ability of adults to smell. Studies are limited, however, since exposed children have not been tested and children are at more risk than adults of Pb poisoning.

Two epidemiologic studies have addressed olfactory function relative to workplace exposure to Pb. In the first, the UPSIT and a battery of neuropsychologic tests were administered to 222 employees of a tetraethyl Pb-manufacturing plant (Schwartz et al., 1993). Estimates of exposure duration were obtained from personnel records and detailed occupational histories. Mean differences in the test scores were estimated by comparing the average scores of the moderate, high, and highest exposure groups to those of the low exposure (reference) group and by adjusting for the confounding influences of age, intellectual ability, race, and alcohol consumption. No influence of Pb on UPSIT scores was apparent, even though Pb exposure-related decrements were found for several of the neuropsychological measures. A subsequent study largely based on these data was performed by Bolla et al. (1995). In this study, UPSIT scores from 190 of the Pb-exposed subjects were compared to those from 144 solvent-exposed workers and 52 nonexposed comparison subjects. The UPSIT scores of Pb-exposed workers did not differ significantly from those of the controls (respective mean scores, 36.4 and 36.9). When the Pb-exposed workers were divided into exposure durations of less than 11 years, 11–17 years, and more than 17 years, significantly lower UPSIT scores were noted relative to the controls for the 11–17-year exposure group, but not for the other two groups.

Manganese

Manganese (Mn) is critical for human health. Mn ions are cofactors for numerous enzymes, including ones involved in the detoxification of supraoxide free radicals. However, too much exposure to environmental Mn can be neurotoxic, resulting in neurologic and psychologic deficits similar to ones seen in Alzheimer's and Parkinson's disease. Historically, such problems were seen in Mn ore miners and grinders, as well as in occupations where Mn was smelted or Mn dust was present.

Like some other metals, Mn is avidly taken up by rodent olfactory receptor cells and transported to CNS brain regions (Tjalve et al., 1996; Gianutsos et al., 1997; Henriksson et al., 1999; Leavens et al., 2007; Lehallier et al., 2012). Astrocytes are among the major targets of the Mn (Henriksson and Tjalve, 2000) and even systemic injections impact the olfactory bulbs, structures which are extremely sensitive to oxidative stress (Villalobos et al., 2009).

Mergler et al. (1994) administered a battery of sensory, neuromuscular, and psychologic tests to 74 workers who were employed in a large ferromanganese and silicomanganese alloy facility in Quebec. Seventy-four controls were also tested who had no history of occupational exposure to Mn. The controls were matched to the workers on the basis of age, sex, education, and family size. In addition to threshold sensitivity to the odorant phenyl ethyl methyl ethyl carbinol (PEMEC), visual acuity, contrast sensitivity, color vision, and vibrotactile perception were tested. The only sensory test that discriminated between the two groups was the olfactory threshold test. Interestingly, despite having higher blood levels of Mn at the time of testing, the thresholds of the Mn workers were significantly lower, not higher, than those of the controls. The investigators suggested that this unexpected hypersensitivity could reflect a dose-related excitatory component of the Mn intoxication.

A more recent study found smell dysfunction in 43 professional welders who had been exposed to airborne Mn-containing fumes while working in confined spaces during the reconstruction of the San Francisco Bay Bridge after the 1989 earthquake (Antunes et al., 2007). Most were not wearing respirators and all were engaged in either manual shielded metal arc welding or automated flux-corded arc welding. All had worked for 1–2 years on the bridge. At the time of testing, approximately half had elevated blood levels of Mn, but not elevated blood levels of iron, copper, or Pb. The UPSIT scores were significantly lower in the welders than in the controls (respective means (SEMS) = 29.51 (0.90) vs 36.55 (0.88)), with 88% of the welders scoring below their individually matched controls. As is typical in idiopathic Parkinson's disease, the welders' olfactory test scores were unrelated to a broad spectrum of neurologic and neuropsychologic test measures. Although time spent working on the bridge correlated with Mn blood levels, workers with the highest Mn blood levels at the time of testing had better olfactory function than those with the lowest Mn blood levels. Although, overall, the welders had the most olfactory dysfunction, this association with Mn was reminiscent of Mergler and Beauvais' observation that the Mn-exposed employees of the Quebec alloy facility, who had the highest blood levels of Mn, were actually more sensitive to the test odorant. A similar association with higher urine levels of Mn and better performance on a threshold test were also reported earlier by Lucchini et al. (1997).

Mercury

Liquid mercury is very toxic and, like Cr and a number of other heavy metals, can have a negative effect on smell function. Minamata disease (MD), a disease typically acquired from mercury-contaminated fish, has been strongly associated with smell loss. As reviewed by Furuta et al. (1994), several Japanese studies found smell dysfunction in patients with MD, including Nosaka et al. (1970), who noted smell disturbances in 16 of 29 (55%) MD cases. In one study, Furuta et al. (1994) administered the 12-item version of the UPSIT (the Brief Smell Identification Test) and a PEA detection threshold test to 19 MD patients ranging in age from 36 to 97 years (mean 78.8 years), 19 age- and sex-matched controls, and 16 patients with senile dementia. The mean score of the MD patients was 4.28 compared to 7.52 for the controls ($p = 0.007$), and did not differ from that of the dementia patients. The detection threshold was more than two log units higher in the MD patients than in the controls ($p < 0.003$) and was similar to that of the patients with dementia.

Solvents

Acute and chronic exposure to solvents occurs in many occupations, notably those associated with painting, chemical manufacturing, machining operations, equipment cleaning and degreasing, lumber and furniture fabrication, and floor and carpet laying. Evidence that exposure to solvents may influence smell ability comes from single case studies, case–control studies, and epidemiologic studies. Most solvents can cross the blood–brain barrier and mucosal surfaces. In extreme cases, such as chronic glue sniffers and chronically exposed painters, atrophy of central brain structures and evidence of encephalopathy have been reported (Rosenberg et al., 1988; Tvedt et al., 1989).

Listed below are common solvents for which there is some evidence suggestive of adverse influences on the human ability to smell. Although numerous animal studies suggest that many solvents, including benzene, benzyl acetate, and butyl acetate, adversely influence the olfactory epithelium, studies of specific influences on human olfactory function of acute or chronic exposure to such agents are lacking in the literature and therefore these agents are not discussed in this chapter. Unfortunately, too few studies have been performed on this topic, despite potential societal consequences and adverse influences of solvent exposures on workers.

Methyl isobutyl ketone

Methyl isobutyl ketone (MIBK) is a volatile chemical, closely related to acetone, which is used in manufacturing processes as a solvent for lacquers, nitrocellulose, and a number of polymers and resins. Gagnon et al. (1994) exposed four human male volunteers to MIKB in an inhalation chamber at concentrations of 20 ppm and 40 ppm for 7 hours. The 20 ppm and 40 ppm sessions were separated from one another by a 25-day interval. Detection thresholds for MIBK and for PEMEC were determined before exposure and at three successive time periods after exposure: immediately and 55 minutes and 95 minutes after exposure. Ratings of perceived odor intensity were ascertained on an hourly basis and, while initially high, diminished with time and stabilized after about 2 hours. Some subjects experienced headache and ear, nose, or throat irritation during the exposures. Thresholds for MIKB were always higher after exposures than before exposures, reflecting a ninefold concentration increase; recovery generally occurred and was dose-dependent, but was incomplete at the 95-minute postexposure test. No changes in thresholds were observed for PEMEC. The fact that the threshold deficit was specific to MIKB and was reversible suggested that the threshold changes reflected reversible adaptation processes. Whether long-term exposure to this agent has detrimental effects on smell function is unknown.

Sulfuryl fluoride

Sulfuryl fluoride is a colorless and odorless fumigant gas used mainly as a pesticide. It has largely replaced methyl bromide because of the latter's adverse influences on the ozone layer although it, too, is under ban consideration by the Environmental Protection Agency in light of its contribution of fluoride to the environment. Like methyl bromide, overexposures have been associated with rhinitis, pharyngitis, paresthesia, and conjunctivitis.

Calvert et al. (1998) assessed olfaction and a number of neurologic and neuropsychologic measures in 123 structural fumigation workers and 120 controls. The workers were exposed mostly to sulfuryl fluoride but also to methyl bromide for average durations of 2.85 and 1.20 years, respectively, and were, on average, 31 years of age. The respective adjusted mean UPSIT scores for the workers and controls were 33.6 and 34.5. Linear regression found that those workers with the highest sulfuryl fluoride exposure over the year before testing had the largest deficits in olfactory function. Several neurologic and neuropsychologic measures were also influenced by sulfuryl fluoride.

Toluene

A common solvent used in multiple industries is toluene. Mergler and Beauvais (1992) examined detection thresholds for toluene and PEMEC in five volunteers who were experimentally exposed for 7-hour periods to either toluene or xylene, or a mixture of these two agents in a counterbalanced order. A sixfold shift in the toluene detection threshold was found immediately after exposure to toluene, xylene, or their mixture. No differences were present in PEMEC thresholds after exposure to any of the compounds. Because the alterations in olfactory function were reversible, the shift in detection threshold is again most likely caused by odor adaptation, rather than by significant toxic insult to the olfactory receptor neurons.

Solvent mixtures

Many workers in paint-manufacturing and industrial fuel oil cleaning operations are exposed to multiple organic solvents. Such agents are highly lipophilic and are taken up by both the central and peripheral nervous systems, where they can influence neurotransmitter levels and sensory, motor, and cognitive function. In severe cases of solvent intoxication, as occurs in chronic glue sniffers, atrophy of the frontal lobes and cerebellum has been reported.

Numerous case reports exist regarding exposure to solvents and alterations in the loss of smell which are not described in this review (see reviews mentioned earlier in the chapter). Ahlstrom et al. (1986) compared the smell thresholds of 20 industrial fuel oil tank cleaners who had been exposed to relatively high concentrations of hydrocarbon solvents to those of sex-, age-, and smoking habit-matched controls. The tank cleaners exhibited higher *n*-butanol and fuel oil vapor thresholds, with the largest deficits occurring for fuel oil, but this effect did not last for more than 2 days. In both cases, the deficits still fell within what would be considered the normal range. Elevated thresholds were not seen for the odors of pyridine and dimethyl disulfide. These findings accord well with the concept of long-term adaptation ("industrial anosmia"), where there is a reduction in sensitivity confined to the odors to which the subjects are exposed and normal function returns within a week of removal from the exposure.

Sandmark et al. (1989) administered the UPSIT to 54 painters who had been exposed on a near-daily basis to organic solvents and to 42 unexposed controls. On average the subjects were in their mid-30s. Although the painters exhibited somewhat lower UPSIT scores than the controls, when multiple regression analysis was used to control for the effects of age and smoking on the test scores, no significant differences in test scores between the painters and the controls were found, suggesting that the effect, if present, was very weak.

In the first of a series of more recent studies, Schwartz et al. (1990) assessed the effects of low-level organic solvent exposure (2–40% of the existing threshold limit values) on odor identification in 187 workers in two paint-manufacturing plants. Only nonsmokers exhibited deficits in test scores – scores that were related to the degree of dose-related solvent exposure. Those in the highest exposure category had UPSIT scores falling below the fifth percentile for their age. These findings suggested to the authors that the olfactory system may be a critical target organ for the neurotoxic effects of even low levels of solvents.

In a second study on this same subject population, correlations were determined between the UPSIT scores and scores on eight neuropsychologic tests that tapped learning and memory (Schwartz et al., 1991). Although positive correlations were found between such measures, separate linear regression analyses on each of eight psychologic measures with UPSIT score, age, vocabulary score, and cumulative solvent exposures as predictor variables found no strong evidence that UPSIT scores were predictive of cognitive function when these other variables were controlled. Interestingly, scrutiny of their regression findings also suggested, using the same criteria, that solvent exposure was similarly nonpredictive in most of these analyses.

In a third study of this subject population (which was mentioned above under Pb exposure), Bolla et al. (1995) compared UPSIT scores from 144 of the aforementioned solvent-exposed workers to those from 52 nonexposed comparison subjects. To minimize the differences among groups in age and verbal intelligence, only those workers over the age of 30 years and with vocabulary scores between 25 and 68 were included in the study group. Although odor identification was clearly and significantly lower in the exposed than in the nonexposed workers, the average UPSIT scores of even the solvent-exposed workers were, on average, still in the low normal range (respective UPSIT (SD) scores = 34.7 (5.4) and 36.9 (2.2); $p = 0.005$). The UPSIT scores of the solvent-exposed workers differed significantly from those of the Pb-exposed workers ($p < 0.001$), whose UPSIT scores (36.4 (3.8)) were essentially the same as those of the nonexposed workers. Unlike the earlier studies, smoking behavior was not included as a variable in this study.

Irritant gases and pollutants

Numerous irritant gases, including ones that make up both indoor and outdoor pollution (e.g., ozone, formaldehyde, acetaldehyde, and sulfur dioxide), have been associated with olfactory dysfunction. Although it has not been empirically assessed, it would seem likely that the more irritating an agent, the higher its potential to damage epithelial tissue such as the olfactory epithelium. A large rodent literature supports the idea that chronic exposure to a number of irritants can damage the olfactory epithelium if at high enough concentrations and duration, although the degree to which such data are directly applicable to humans is debatable, as noted later in the chapter.

Acetaldehyde

Acetaldehyde, a chemical similar in both chemistry and toxicity to formaldehyde, is a ubiquitous air pollutant and an important industrial chemical used in synthetic chemistry. It is involved in the manufacture of acetic acid, flavorings, aniline dyes, plastics, and synthetic rubber. It is also produced by the combustion of numerous substances, including tobacco, wood, and paper. It is typically metabolized and detoxified by aldehyde dehydrogenase. Rodent studies indicate that chronic exposures to this agent induce, over a wide dose range, epithelial degeneration, metaplasia, and loss of olfactory receptor neurons and Bowman's glands (Table 17.2). Its metabolism to acetic acid is critical for its olfactory toxicity (Morris, 1997).

Long-term chronic exposure studies to acetaldehyde are not available for humans. In the sole acute human study, 24 subjects between the ages of 20 and 35 years were tested following 4-hour exposures to acetyaldehyde (50 ppm) and air on separate days in an exposure chamber (Muttray et al., 2009). *n*-Butanol threshold testing was administered and the threshold values did not differ significantly between these two conditions. The authors suggested that the exposure conditions were probably not long enough to see an effect and that more extensive olfactory testing may be needed.

Acrylic acid and its esters

Acrylic acid, when reacted with an alcohol, forms esters which are collectively termed acrylates. Acrylic acid and its associated acrylates play an important role in the manufacturing of plastics, adhesives, floor polishes, paints, and coatings and adhesives of various types. However, they are very corrosive and can damage the skin, nasal passages, and respiratory tract. In rodents, exposures to acrylic acid, ethyl acrylate, and methyl methacrylate (5–300 ppm) can damage the olfactory epithelium, producing, for example, inflammation, degeneration, loss of receptor cells, focal necrosis, and hyperplasia of submucosal glandular elements (Miller et al., 1981).

Schwartz et al. (1989) administered the UPSIT to 731 workers exposed to acrylic acid and a variety of acrylates and methacrylates in various job positions at a chemical manufacturing plant. The levels of exposure were typically below their threshold limit values. A nested case–control study demonstrated that the olfactory test scores decreased with indices of increased cumulative exposure, as determined from exposure histories related to job titles and the length of service within each job position. As with the case of solvent mixtures, the highest relative risk of olfactory dysfunction occurred in workers who had never smoked, suggesting they were particularly vulnerable to the exposures. Evidence was presented that the smell loss was likely reversible by removal from the exposure situation.

Chlorine

Chlorine is widely used in industrial processes, such as plastics manufacturing, and as a disinfectant in water systems and swimming pools, usually in the form of sodium hypochlorite, which is formed from chlorine and sodium hydroxide. When it comes into contact with mucosal surfaces, chlorine can form hydrochloric acid, and accidental exposures are known to be capable of producing permanent anosmia (Benjamin and Pickles, 1997). Airway disturbances in swimmers have been attributed to chlorinated swimming-pool water (Bougault and Boulet, 2012) and the potential impact of chlorine gas accidents is enormous. In fact, from 2007 to 2010, at least nine terrorist attacks on chlorine transport and/or storage facilities were recorded in Iraq in attempts to cause mass casualties (White and Martin, 2010).

Aside from the aforementioned case report of anosmia by Benjamin and Pickles, no published quantitative studies of olfactory function have been performed in patients who have been acutely exposed to toxic levels of chlorine. At our center, at least one patient has presented with smell loss secondary to accidental poisoning while chlorinating a home pool, and newspaper accounts report that some persons have suffered from the aftermath of chlorine release in transportation accidents, with difficulty in smelling (Kay, 2011).

Formaldehyde

Formadehyde, a ubiquitous and highly reactive gas, is used in a variety of manufacturing processes and in situations where preserving or fixing of tissues is required (e.g., biology classes, hospital and university histology departments, mortuaries). Although exposure to formaldehyde is often said to decrease olfactory acuity (Spealman, 1954), relatively few data are available on this point. In one study, workers exposed to formaldehyde (0.05–0.50 mg/m^3) or formaldehyde (0.2–0.3 mg/m^3) plus wood dust (1–2 mg/ m^3) for years exhibited a significant elevation in detection thresholds for pyridine when compared to controls (14.2 vs 15.6 binary threshold steps; $p < 0.05$ for each comparison) (Holstrom and Wilhelmsson, 1988). In another study, an assessment of bromine olfactory thresholds was made in 41 medical students (Hisamitsu et al., 2011) prior to and immediately after and 6 months after a cadaver course. Thirty-two percent of the students exhibited smell dysfunction immediately after the course – dysfunction that disappeared within the next 6 months.

Hydrogen sulfide

Hydrogen sulfide (H_2S) produces a "rotten egg-like" smell that arises from bacterial and decomposition of proteins, such as occurs in animal containment facilities and sewers. It is a major component of bad breath (halitosis). Its toxicity is dose-related and concentrations greater than 1000 ppm can result in respiratory paralysis and sudden death (Brenneman et al., 2000). Nonfatal exposures commonly produce nausea, pulmonary edema, respiratory distress, and temporary loss of consciousness. Persistent anosmia has been reported in several cases of survivors of near-fatal high-level exposures to H_2S (Tvedt et al., 1991).

Styrene

Styrene, a clear colorless liquid, is a derivative of benzene used extensively in the development of a range of polymers and copolymers important in the manufacture of plastic, rubber, insulation, fiberglass, carpet backings, and related products. It is widely used in the fiberglass boat industry. The major route of occupational exposure is inhalation. Exposure to airborne concentrations of styrene ranging from 20 to 50 ppm induce histopathologic changes within the olfactory neuroepithelium of rodents (Cruzan et al., 1997, 1998).

Several studies have examined the influences of styrene exposure on human olfactory function. In the first, Dalton et al. (2003) tested the odor identification performance and detection threshold sensitivity to styrene and PEA of 62 workers who had been exposed to styrene in the reinforced-plastics industry. Sixty-seven age- and sex-matched unexposed controls were also tested. The subjects came from two facilities that employed an open-mold process to manufacture fiberglass-reinforced shower and tub enclosures, one facility that used a closed-mold process to manufacture a number of reinforced-plastic products, including sinks, truck parts, and fan blades, and one facility that manufactured reinforced-fiberglass paneling. All exposed workers were tested during their regular 8- or 10-hour work shift in a styrene-free environment (<1 ppm styrene). The styrene thresholds were markedly elevated in the exposed subjects relative to the nonexposed subjects, a difference that was magnified in the older workers. No exposure effects were evident for the PEA thresholds or the other tests, all of which were within normal limits, leading the authors to conclude that the elevated styrene thresholds most likely reflected long-term adaptation specific to styrene.

A subsequent study by these investigators similarly evaluated olfactory function in 30 workers in a reinforced-plastics boat-manufacturing facility (Dalton et al., 2007). Fifteen of the workers had job positions where relatively high exposures to styrene occurred, whereas 15 had job positions where much less exposure occurred. As before, styrene thresholds, but not PEA thresholds, were elevated in the exposed workers. However, scores on an odor identification test did differ between the two study groups, with the more exposed workers performing more poorly than the less exposed workers. Since no associations were evident between the identification test scores and (a) exposure levels on the day before and the day of the olfactory testing and (b) lifetime exposure profiles of the workers, the investigators suggested that these effects were likely due to factors independent of current or cumulative styrene exposures.

A similar conclusion was reached in a study performed in Taiwan (Cheng et al., 2004). This study assessed odor identification and 1-butanol thresholds in 52 workers in injection molding operations who had been exposed to the vapors of decomposition products of acrylonitrile-butadiene-styrene for a minimum of 6 months. Seventy-two workers who were not similarly exposed were tested over the same time periods. Testing occurred before and after their work shifts. The detection thresholds were significantly higher in the exposed workers than in the nonexposed ones after their work shifts, but not before their work shifts. No influence on the odor identification test was observed. Since the pre-shift testing found no differences between the two groups of subjects, these data support the concept that the olfactory thresholds reflect adaptation rather than sustained neurologic damage to the olfactory epithelium or other elements of the olfactory system.

Sulfur dioxide

Sulfur dioxide is a common component of urban air pollution and has been associated with significant nasal irritation and inflammation. Harada et al. (1983) administered a single-series ascending detection threshold test for five odorants to eight chemical plant workers who had been exposed to sulfur dioxide and ammonia for several years, seven of whom had been exposed only to ammonia, and to six controls. The mean thresholds for all odorants were reportedly higher in the group exposed to both sulfur dioxide and ammonia than in the other two groups, although an examination of the data presented in a figure showing this effect suggests this effect was not statistically significant.

THE MULTIPLE CHEMICAL SENSITIVITY SYNDROME

There is a group of persons who report being hypersensitive to environmental odors, including those of plastics, household cleaners, drugs, organic solvents, combustion engine exhausts, soaps, insecticides, preservatives, cigarette smoke, fragrances, perfumes, and colognes. They also report that exposures to such chemicals induce a wide variety of somatic symptoms, including anxiety, depression, fatigue, general malaise, mental confusion, lightheadedness, headache, insomnia, myalgia, loss of appetite, and numbness. However, no documentable signs of medical disease are apparent (Ross et al., 1999) and no single widely accepted test of physiologic function has been found to correlate with the symptoms (Miller and Mitzel, 1995).

In the 1960s, Randolph (1962) named this apparent hypersensitivity and multiorgan complex of unorthodox chemically induced symptoms "environmental illness" or "chemical reactivity." The condition is now more commonly termed "multiple chemical sensitivities" (MCS) (Cullen, 1987), although numerous other terms have been used, including ecologic illness, multiple chemical intolerance, idiopathic environmental intolerance, universal allergy, and toxicant-induced loss of tolerance (Miller, 2003). MCS has been associated with so-called sick building syndrome, chronic fatigue syndrome, and fibromyalgia, among others. Whatever its name, numerous attempts have been made to operationally define the disorder. Based on Cullen's original set of definitional criteria and a survey of 89 clinicians and scientists familiar with this symptom complex but of differing opinions as to its origins, Bartha et al. (1999) provided a list of six "consensus criteria" for the diagnosis of MCS:

1. chronicity
2. reliably recurring symptoms
3. induction by low levels of chemical exposure
4. involvement of multiple chemicals
5. resolution after removal of incitants
6. symptoms present in multiple organ systems.

Although not all patients with the putative MCS syndrome report heightened smell sensitivity, enough have done so to catalyze a number of empiric olfactory studies on this topic. Two studies have examined olfactory thresholds in an attempt to verify the olfactory hypersensitivity. In the first, PEA and methyl ethyl ketone detection thresholds were measured in 18 patients with putative MCS and 18 age- and sex-matched controls (Doty et al., 1988). Although the thresholds did not differ significantly from one another, the MCS patients exhibited higher nasal resistances to airflow, respiration rates, and Beck Depression Inventory scores (Doty et al., 1988). In the second, PEA thresholds of 23 MCS subjects and of 23 controls also were found to not differ from one another (Hummel et al., 1996). Paradoxically, the MCS subjects of this study had more difficulty than controls in identifying and discriminating among suprathreshold odors and exhibited smaller odor-induced event-related potentials, suggesting that the suprathreshold olfactory function of MCS patients may have been decreased, not increased.

Ojima and colleagues (2002) administered both the UPSIT and the 12-item B-SIT (formally known as the Cross-Cultural Smell Identification Test) to 25 MCS subjects and 50 age- and sex-matched controls. No differences in the test scores of the MCS and controls were present for either test (indeed, the means were exactly the same in both cases). Relative to the controls, the MCS patients did report a larger number of the odorants as unpleasant. The data of this study support the earlier conclusion that olfactory sensitivity is not heightened in MCS patients, although they are in disaccord with the concept that such patients may be less sensitive to suprathreshold odors.

A number of cases of mild cognitive impairment likely have a psychogenic basis. Zucco et al. (2008) tested a 36-year-old-woman who met a strict set of criteria for MCS. The UPSIT and other quantitative tests found no evidence of olfactory dysfunction. A series of odorants were then presented along with some blanks following two sets of instructions. One set suggested the presented odors were harmful and the other set that they were benign. The patient's task was to rate the intensity of the odors and to indicate which ones induced the mild cognitive impairment symptoms. When the odorants were presented as being harmless, they were judged significantly less intense, and triggered fewer symptoms, than when they were presented as being harmful. In turn, when they were presented as potentially harmful, the patient rated them more intense and as triggering more symptoms than when they were presented as being harmless. The authors suggest this simple test may useful in identifying patients whose MCS is psychogenic.

MECHANISMS OF TOXIN-INDUCED OLFACTORY LOSS

Animal studies suggest that the primary means by which environmental xenobiotics alter the olfactory system is by direct interaction with the olfactory epithelium. Some agents are very specific as to the types of cells they damage. Thus, the neurotoxin beta,beta'-iminodipropionitrile damages sensory cells while sparing sustentacular cells (Genter et al., 1992), whereas sustentacular cells are targets for methyl bromide and dibasic esters. In the latter case, receptor cell degeneration and necrosis follow soon thereafter (Hurtt et al., 1988; Trela et al., 1992). Some chemicals, such as dimethylamine and 3-methylfuran, damage the epithelium indirectly by producing metabolic byproducts that, in turn, damage the mucosa (Brandt et al., 1990). Other toxins induce acute or chronic inflammation that can destroy the olfactory epithelium and, in turn, depress local immunoglobulin A production, thereby promoting bacterial colonization of the olfactory mucosa that can adversely affect olfaction (Harkema et al., 2006). In rare cases, volatile toxins can increase mucosal cellular proliferation, resulting in neoplastic processes that also alter the ability to smell. Increased prevalence of nasal and sinus carcinomas has been noted in workers exposed to Ni dust, ionizing radiation, wood dust, formaldehyde, and leather dust, among others (Leopold, 1994). Importantly, a number of factors dictate the degree to which chemicals produce their toxicity, including age, weakened defense mechanisms, and simultaneous exposures to several toxins that, individually, would not produce the same degree of damage (Reuzel et al., 1990).

As shown in Table 17.3, there is a large literature on the types of chemicals known to damage the olfactory epithelium of rodents. However, the degree to which such findings generalize to humans is unclear. Although biopsies can be obtained from human olfactory epithelial tissue, sampling issues are involved and relatively few human histologic studies based on toxin-induced olfactory dysfunction are available. Importantly, rodents have much more complex noses than humans and devote proportionally more of their nasal epithelia to smell. Thus, $\sim$ 50% of the nasal cavity surface area in F344 rats is lined by olfactory neuroepithelium (Gross et al., 1982), in contrast to humans, where only 1–3% of the nasal cavity is lined by this epithelium (Escada et al., 2009). In addition to being obligate nose breathers, rodents possess maxillary turbinates and an additional row of ethmoidal turbinates, resulting in a more complex absorption distribution of chemicals than occurs in humans (Zhao et al., 2006). During resting breathing, a larger proportion of inspired air passes through olfactory-lined regions of the rat than the human nose (Kimbell, 2006).

Based on these factors and differences in enzymatic activity within the olfactory mucosa, some toxins may have more of an adverse effect on rodent than on human olfactory epithelia, and vice versa. For example, one study exposed adult male rats for 20 weeks to relatively high levels of airborne Cd (0.5 mg/m^3) (Sun et al., 1996). No evidence of pathologic alterations of either the olfactory or nasal respiratory epithelia was present and behavioral tests were unable to document any Cd-related adverse effect on detection threshold sensitivity, unlike the situation in humans exposed for long periods to Cd. This is in spite of the fact that lung and cardiac toxicity were found and other rat studies have noted that high exposures to Cd increase markers of oxidative stress in the olfactory bulb and other CNS structures (Kumar et al., 1996; Erikson et al., 2005). Although similar studies (using Ni and Cr) have found Ni-related damage to the rodent olfactory epithelium, behavioral deficits were not observed (Evans et al., 1995).

It should be noted that a number of reflex or reflex-like responses serve to protect the nasal and olfactory mucosae from damage from inhaled toxic chemicals. For example, upon encountering an irritating or extremely bad-smelling chemical, reflex-like motor

Table 17.3

Examples of effects of toxic agents on the olfactory mucosa of common laboratory rodents (mice, rats and hamsters)

Compound	Duration	Effects on histology and function	Example references
Acetaldehyde (400–5000 ppm)	6 hours/day, 5 days/week, 1–28 months	Dose-related degeneration, metaplasia, loss of Bowman's glands and olfactory neurons and nerve bundles, adenomas, and squamous cell carcinomas	Appelman et al. (1982, 1986); Dorman et al. (2008)
Acrolein (1.7 ppm)	6 hours/day, 5 days/week	Hypertrophy, hyperplasia, erosion, ulceration, necrosis inflammation	Buckley et al. (1985)
Acrylic acid (5–75 ppm)	6 hours/day, 5 days/week, 13 weeks	Degeneration, replacement with respiratory epithelium, inflammation, hyperplasia of Bowman's glands	Miller et al. (1981)
Allyl glycidyl ether (7.1 ppm)	4 days	Erosion of the olfactory epithelium	Gagnaire et al. (1987)
Benomyl (50–200 mg/m^3)	5 minutes/day, 3 days	Degeneration	Warheit et al. (1989)
Bis(chloromethyl) ether (0.1 ppm)	6 hours/day, 5 days/week, up to 80 days	Replacement of olfactory epithelium with cuboidal to columnar cells; neuroepitheliomas	Kuschner et al. (1975)
Bromobenzene (25 mol/kg ip)*	5 days/week, 13 weeks	Degeneration of olfactory epithelium and Bowman's glands	Brittebo et al. (1990)
Cadmium (250–500 g/m^3)	5 hours/day, 5 days/week, 20 weeks	Little change	Sun et al. (1996)
Chlorine gas (0.4–11 ppm)	6 hours/day, 5 days/week, 16 weeks	Degeneration, septal perforations, intracellular deposits of eosinophilic material, mucus cell hypertrophy	Jiang et al. (1983); Buckley et al. (1984); Wolf et al. (1995)
Chloroform (300 ppm; 34–400 mg/kg/day)*	6 hours/day, 7 days; stomach gavage – 5 days/week, 3 weeks	Degeneration of Bowman's glands, receptor cells, cell proliferation in periosteum and bone	Mery et al. (1994); Dorman et al. (1997)
Chloropicrin (8 ppm)	6 hours/day, 5 days	Hypertrophy, hyperplasia, ulceration, necrosis, inflammation	Buckley et al. (1984)
Coumarin (50 mg/kg ip)	48 hours	Necrosis, cell loss, and basal cell metaplasia in the olfactory mucosa	Zhuo et al. (1999)
Chlorthiamid (6–50 mg/kg ip)*	8 hours, 7 days	Degeneration of olfactory epithelium and Bowman's glands, replacement with respiratory epithelium, fibrosis in lamina propria	Brittebo et al. (1991)
2,4-Decadienal*	Stomach gavage; 5 days/week, 14 weeks	Lesions of olfactory epithelium	Chan (2011)
Dibasic esters (20–900 mg/m^3)	4 hours/day, 7–13 weeks	Degeneration, sustentacular cells injured initially, cell proliferation	Keenan et al. (1990); Bogdanffy and Frame (1994)
1,2 dibromo-3-chloropropane (5–60 ppm)	6 hours/day, 5 days/week, 13 weeks	Degeneration, metaplasia, hyperplasia	Reznik et al. (1980)
1,2 dibromo ethane (3–75 ppm)	6 hours/ day 5 days/week, 13 weeks	Degeneration, metaplasia, hyperplasia	Reznik et al. (1980)

Table 17.3

Continued

Compound	Duration	Effects on histology and function	Example references
Dichlobenil (12–50 mg/kg ip)	8 hours, 7 days	Degeneration of olfactory epithelium and Bowman's glands	Brandt et al. (1990)
1,3-dichloro propene (30–150 ppm)	6 hours/day, 5 days/week, 6–24 months	Degeneration and metaplasia	Stott et al. (1988); Lomax et al. (1989)
Dimethylamine (10–511 ppm)	6 hours/day, 5 days/week, 6–12 months	Degeneration, loss of nerve bundles, hypertrophy of Bowman's glands	Buckley et al. (1984, 1985)
Dimethylethylamine (300 ppm)	10 days	Desquamation of epithelium with only basal and sustentacular cells remaining	Gaskell (1990)
1,4 dithiane (105–420 mg/kg ip)	90 days until sacrifice	Anisotrophic crystals in giant cells (undetermined chemical composition)	Schieferstein et al. (1988)
Epichlorohydrin (687 ppm)	6 hours/day, 5 days	Ulceration, necrosis	Buckley et al. (1984)
Ethyl acrylate (25–75 ppm)	6 hours/day, 5 days/week, 27 months	Degeneration and metaplasia	Miller et al. (1985)
Ferrocene (3–30 mg/m^3)	6 hours/day, 5 days/week, 13 weeks	Iron accumulation, necrotizing inflammation, metaplasia	Nikula et al. (1993)
Formaldehyde (0.25–15 ppm)	6 hours/day, 5 days/week, 4 months 8 hours/day, 5 days/week, 13 weeks	Decrease number of bipolar cells, increase number of basal cells, degeneration of nerve bundles, reduced odor discrimination; in repeated dosing at marginally cytotoxic concentrations, total dose less important than concentration	Wilmer et al. (1989); Apfelbach (1991); Apfelbach et al. (1992)
Furfural (250–400 ppm)	7 hours/day, 5 days/week, 52 weeks	Disorientation of sensory cells, degeneration of Bowman's glands, cyst-like structures in lamina propria	Feron and Kruysse (1978)
Furfural alcohol (2–250 ppm)	13 weeks	Squamous and respiratory metaplasia of olfactory epithelium, inflammation, hyaline droplets, squamous metaplasia of ducts	Irwin et al. (1997)
5-(Hyrdroxoymethyl)-2-furfural*	Stomach gavage, daily for 3 months	Incidences of degeneration, metaplasia	National Toxicology Program (2010b)
Hexamethylene di-isocyanate (0.005–175 ppm)	6 hours/day, 5 days/week, 12 months	Degeneration, mucus hyperplasia	Reed and De (1989)
Hydrazine (75–750 ppm)	1 hours/day, 1–10 days	Degeneration	Latendresse et al. (1995)
Beta-myrcene*	Stomach gavage, 5 days/week, 14 weeks	Degeneration of epithelium in some cases at 2 g/kg	National Toxicology Program (2010a)
β,β'-minodi-propionitrile (200–400 mg / kg ip)	6 hours – 56 days until sacrifice	Degeneration of axon bundles, increase of glial fibrillary acidic protein	Genter et al. (1992)

Continued

Table 17.3

Continued

Compound	Duration	Effects on histology and function	Example references
Methyl bromide (200 ppm)	4 hours/day, 4 days/week	Degeneration, decreased carnosine, behavioral deficits	Hastings and Evans (1991)
3-methylfuran (148-322 mol/l)	1 hour	Degeneration, more severe in rats than hamsters	Morse et al. (1984)
3-methylindole (100–400 mg/kg ip)	7–90 until sacrifice	Degeneration, fibrous, adhesions, osseous remodeling, Bowman's gland hypertrophy, behavioral deficits	Turk et al. (1987); Papi (1990)
Methyl isocyanate (10, 30 ppm)	2 hours	Degeneration	Uraih et al. (1987)
Naphthalene (400–1600 mg/kg ip)	24 hours and 5 days	Cytotoxicity (mice and hamsters), necrosis (rats); degeneration	Plopper et al. (1992); Dodd et al. (2010)
Nickel subsulfide (0.11–1.8 mg/m^3	6 hours/day, 5 days/week, 13 weeks	Atrophy	Dunnick et al. (1989)
Nickel sulfate (3.5–635 mg/m^3)	6 hours/day, 12–16 consecutive days	Atrophy, degeneration, decrease in carnosine	Evans et al. (1995)
N-nitroso-dimethylamine (20–80 mg/ kg ip)	6 hours – 30 days until sacrifice	Degeneration of olfactory epithelium and Bowman's glands	Rangga-Tabbu and Sleight (1992)
N-nitroso-pyrrolidine (30–100 mg/kg ip)	6 hours – 30 days until sacrifice	Degeneration of olfactory epithelium and Bowman's glands	Rangga-Tabbu and Sleight (1992)
Propylene oxide (10–525 ppm)	4 weeks	Degeneration	Eldridge et al. (1995)
Pyridine (5–444 ppm)	6 hours – 4 days	Degeneration of olfactory epithelium	Nikula and Lewis (1994)
RP 73401 (1 mg/ kg/ day)	1 hour – 5 days	Degeneration of olfactory epithelium and Bowman's glands	Pino et al. (1999)
Sulfur dioxide (10–117 ppm)	72 hours or 6 hours/day, 5 days	Necrosis, edema, destruction, hyperplasia, hypertrophy	Buckley et al. (1984); Imaizumi et al. (1998)
Tetralin (120 ppm)	6 hours/day, 5 days/week, 12–105 weeks	Degeneration, glandular hypertrophy	National Toxicology Program (2011)
2,4-toluene di-isocyanate (0.4 ppm)	6 hours/day, 5 days	Ulceration, necrosis, inflammation, degeneration	Buckley et al. (1984)
3-trifluoromethyl 1-pyridine (0.1–329 ppm)	6 hours – 90 days until sacrifice	Degeneration, reduced Bowman's activity	Gaskell et al. (1988, 1990)
Urethane polymeric dispersion (flurotelometer-based)*	Daily gavage; 90 days	Degeneration, necrosis	Stadler et al. (2008)
Vinyl chloride (5000 ppm)	7 hours/day, 5 days/week, 40 weeks	Hyperplasia and neoplastic alterations; proliferation of atypical basal and Bowman gland cells	Feron and Kroes (1979)

*Exposures other than air or nasal instillation.

and secretory rejection responses come into play (e.g., sneezing, halting of inhalation, increased mucus secretion, engorgement of intranasal erectile tissue) (Tourbier and Doty, 2007). Longer-term changes in the epithelium, such as regeneration of cells that have been invaded by toxins or viruses, also provide a means to thwart the penetration of viruses and other xenobiotics into the CNS via the olfactory epithelium. The time required for restoration after the sloughing off of the epithelium depends on the involved chemical and its concentration. In some cases, restoration occurs in a matter of weeks, whereas in other cases it can take months. The full complement of cells only rarely returns following such damage. The degree of reconstitution is dependent upon the degree of insult; when the lamina propria containing the stem cells is damaged, poor reconstitution is expected (Uraih et al., 1987).

In addition to the aforementioned protections, the olfactory mucosa, like the nasal mucosa, possesses important chemical defense mechanisms. These include the secretion of antibodies and antimicrobial proteins, such as lactoferrin and lysozyme, to deal with inhaled pathogens (Mellert et al., 1992). Phase I and phase II enzymes (e.g., cytochrome P-450s, glutathione, and related enzymes) are found in the sustentacular cells and Bowman gland secretions, often at higher concentrations on a per gram basis than in the liver or lungs (Reed, 1993). In some cases, chronic exposure to low doses of toxic compounds induces an enzymatic metabolic protective status so that subsequent higher exposures produce little or no toxicity, conceivably explaining why cigarette smokers may be protected to some degree against smell loss induced by some industrial chemicals (Schwartz et al., 1989, 1990). There is also evidence that some hormones, for example estrogen, may mitigate olfactory epithelial damage produced by neurotoxins such as 3-methyl indole (Dhong et al., 1999).

Strategies to ameliorate the dysfunction resulting from exposure to toxic compounds are limited. In some instances, inhibition of enzymatic pathways that produce toxins can eliminate the damaging effects of such chemicals as dimethylamine and 3-methylfuran (Reed, 1993). Topical steroids, which are routinely used for alleviation of rhinitis and polyposis, may do more than to simply increase airway patency by shrinking inflamed tissue. Thus, they may hasten the reconstitution of the olfactory epithelium after insult (Kimura et al., 1991), although large doses may have an adverse effect (Kratskin et al., 1999). New therapeutic avenues will undoubtedly develop in the future as we come to better understand factors that regulate neurogenesis and apoptosis in the olfactory epithelium as well as those involved in the guidance of axonal propagation through the cribriform plate.

CONCLUSIONS

It is apparent from the studies reviewed in this chapter that the olfactory system can be compromised by a number of environmental toxins, either by accident or by exposures to agents in the workplace or in the general environment (e.g., air pollution). Such exposures can lead to permanent or temporary alterations, depending upon the duration of exposure and the nature of the chemical involved. While the weight of the evidence is that most toxic chemicals produce olfactory damage at the level of the neuroepithelium, there is a dearth of information regarding the potential influences of such agents on central nervous system structures related to chemosensation. Clearly, elements of air pollution can produce neuropathologic changes within the olfactory bulb and elsewhere in the brain. Although most studies suggest that industrial chemicals do not damage the olfactory membrane at concentrations below current threshold limit values, this is not always the case and more data are clearly needed for a variety of chemicals to determine cases where this is not true. In such cases, revision of threshold value limits may be in order.

The question arises as to how generalizable rodent data are to the human condition. It is also not known whether the human olfactory system is, in fact, cumulatively damaged by intermittent exposure to environmental toxic chemicals. More research is needed on this point, as it is clear that studies of the toxic effects of airborne agents on the human olfactory system are still in their infancy. In contrast to olfaction, there is little information regarding the effects of neurotoxic chemical exposures to the sense of taste, i.e., sweet, sour, bitter, salty, and savory sensations. Taste receptors are exposed to the oral environment, much like olfactory receptors are exposed to the outside environment, and future testing is needed to determine whether the taste system is also vulnerable to similar toxic exposures.

References

Adams RG, Crabtree N (1961). Anosmia in alkaline battery workers. Br J Ind Med 18: 216–221.

Ahlstrom R, Berglund B, Berglund U et al. (1986). Impaired odor perception in tank cleaners. Scand J Work Environ Health 12: 574–581.

Altman KW, Desai SC, Moline J et al. (2011). Odor identification ability and self-reported upper respiratory symptoms in workers at the post-9/11 World Trade Center site. Int Arch Occup Environ Health 84: 131–137.

American Conference of Governmental Industrial Hygienists (2003). TVLs and BEIs. ACGHI, Cincinnati.

Amoore JE (1986). Effects of chemical exposure on olfaction in humans. In: CS Barrows (Ed.), Toxicology of the Nasal Passages. Hemisphere Publishing Corp, Washington, D.C., pp. 155–190.

Antunes MB, Bowler R, Doty RL (2007). San-Francisco/Oakland Bay Bridge welder study: Olfactory function. Neurology 69: 1278–1284.

Apfelbach R (1991). Sensitivity to odors in Wistar rats is reduced after low-level formaldehyde-gas exposure. Naturwissenschaften 78: 221–223.

Apfelbach R, Reibenspies M, Schmidt R et al. (1992). Behavioral and structural modifications in the olfactory epithelium after low level formaldehyde-gas exposure. In: Methylation and Demethylation Processes, Proceedings of the 3rd International Conference on Role of Formaldehyde in Biological Systems, Hungarian Biochemical Society, Sopron, Hungary.

Appelman LM, Woutersen RA, Feron VJ (1982). Inhalation toxicity of acetaldehyde in rats. I. Acute and subacute studies. Toxicology 23: 293–307.

Appelman LM, Woutersen RA, Feron VJ et al. (1986). Effect of variable versus fixed exposure levels on the toxicity of acetaldehyde in rats. J Appl Toxicol 6: 331–336.

Baader EW (1952). Chronic cadmium poisoning. IMS Ind Med Surg 21: 427–430.

Baker H, Genter MB (2003). The olfactory system and the nasal mucosa as portals of entry of viruses, drugs, and other exogenous agents into the brain. In: RL Doty (Ed.), Handbook of Olfaction and Gustation. Marcel Dekker, New York, pp. 549–573.

Bardach JE, Fujaya M, Holl A (1965). Detergents: Effects on the chemical senses of the fish *Ictalurus natalis* (Le Sueur). Science 148: 1605–1607.

Bartha L, Baumzweiger W, Buscher DS et al. (1999). Multiple chemical sensitivity: a 1999 consensus. Arch Environ Health 54: 147–149.

Bartoshuk LM (1978). The psychophysics of taste. Am J Clin Nutr 31: 1068–1077.

Bazer GT, Ebbesson SOE, Reynolds JB et al. (1987). A cobalt-lysine study of primary olfactory projections in king salmon fry (Oncorhynchus-Tshawytscha Walbaum). Cell Tissue Res 248: 499–503.

Benjamin E, Pickles J (1997). Chlorine-induced anosmia. A case presentation. J Laryngol Otol 111: 1075–1076.

Berling K, Knutsson J, Rosenblad A et al. (2011). Evaluation of electrogustometry and the filter paper disc method for taste assessment. Acta Otolaryngol 131: 488–493.

Bogdanffy MS, Frame SR (1994). Olfactory mucosal toxicity. Integration of morphological and biochemical data in mechanistic studies: dibasic esters as an example. Inhal Toxicol 6: 205–219.

Bolla KI, Schwartz BS, Stewart W et al. (1995). Comparison of neurobehavioral function in workers exposed to a mixture of organic and inorganic lead and in workers exposed to solvents. Am J Ind Med 27: 231–246.

Borg-Neczak K, Tjalve H (1996). Uptake of 203Hg2+ in the olfactory system in pike. Toxicol Lett 84: 107–112.

Borovyagin V, Hernadi L, Salanki J (1989). Mercury and cadmium induced structural alterations in the taste buds of the fish *Alburnus alburnus*. Acta Biol Hung 40: 237–254.

Bougault V, Boulet LP (2012). Airway dysfunction in swimmers. Br J Sports Med 46: 402–406.

Brandt I, Brittebo EB, Feil VJ et al. (1990). Irreversible binding and toxicity of the herbicide dichlobenil (2,6-dichlorobenzonitrile) in the olfactory mucosa of mice. Toxicol Appl Pharmacol 103: 491–501.

Breer H (1994). Odor recognition and second messenger signaling in olfactory receptor neurons. Semin Cell Biol 5: 25–32, 1994 Feb.

Breer H, Eberle J, Frick C et al. (2012). Gastrointestinal chemosensation: chemosensory cells in the alimentary tract. Histochem Cell Biol 138: 13–24.

Brenneman KA, James RA, Gross EA et al. (2000). Olfactory neuron loss in adult male CD rats following subchronic inhalation exposure to hydrogen sulfide. Toxicol Pathol 28: 326–333.

Brittebo EB, Eriksson C, Brandt I (1990). Activation and toxicity of bromobenzene in nasal tissues in mice. Arch Toxicol 64: 54–60.

Brittebo EB, Eriksson C, Feil V et al. (1991). Toxicity of 2,6-dichlorothiobenzamide (chlorthiamid) and 2,6-dichlorobenzamide in the olfactory nasal mucosa of mice. Fundam Appl Toxicol 17: 92–102.

Buckley LA, Jiang XZ, James RA et al. (1984). Respiratory tract lesions induced by sensory irritants at the RD50 concentration. Toxicol Appl Pharmacol 74: 417–429.

Buckley LA, Morgan KT, Swenberg JA et al. (1985). The toxicity of dimethylamine in F-344 rats and B6C3F1 mice following a 1-year inhalation exposure. Fundam Appl Toxicol 5: 341–352.

Calderon-Garciduenas L, Azzarelli B, Acuna H et al. (2002). Air pollution and brain damage. Toxicol Pathol 30: 373–389.

Calderon-Garciduenas L, Mora-Tiscareno A, Fordham LA et al. (2003). Respiratory damage in children exposed to urban pollution. Pediatr Pulmonol 36: 148–161.

Calderon-Garciduenas L, Reed W, Maronpot RR et al. (2004). Brain inflammation and Alzheimer's-like pathology in individuals exposed to severe air pollution. Toxicol Pathol 32: 650–658.

Calderon-Garciduenas L, Solt AC, Henriquez-Roldan C et al. (2008). Long-term air pollution exposure is associated with neuroinflammation, an altered innate immune response, disruption of the blood–brain barrier, ultrafine particulate deposition, and accumulation of amyloid beta-42 and alpha-synuclein in children and young adults. Toxicol Pathol 36: 289–310.

Calderon-Garciduenas L, Franco-Lira M, Henriquez-Roldan C et al. (2010). Urban air pollution: influences on olfactory function and pathology in exposed children and young adults. Experimental and Toxicological Pathology 62: 91–102.

Calderon-Garciduenas L, Kavanaugh M, Block M et al. (2012). Neuroinflammation, hyperphosphorylated tau, diffuse amyloid plaques, and down-regulation of the cellular prion protein in air pollution exposed children and young adults. J Alzheimers Dis 28: 93–107.

Calvert GM, Mueller CA, Fajen JM et al. (1998). Health effects associated with sulfuryl fluoride and methyl bromide exposure among structural fumigation workers. Am J Public Health 88: 1774–1780.

Cartoni C, Yasumatsu K, Ohkuri T et al. (2010). Taste preference for fatty acids is mediated by GPR40 and GPR120. J Neurosci 30: 8376–8382.

Cassee FR, Arts JH, Groten JP et al. (1996). Sensory irritation to mixtures of formaldehyde, acrolein, and acetaldehyde in rats. Arch Toxicol 70: 329–337.

Chan PC (2011). NTP toxicity studies of toxicity studies of 2,4-decadienal (CAS No. 25152-84-5) administered by gavage to F344/N Rats and B6C3F1 mice. Toxic Rep Ser 1–94.

Chandrashekar J, Hoon MA, Ryba NJ et al. (2006). The receptors and cells for mammalian taste. Nature 444: 288–294.

Chang RB, Waters H, Liman ER (2010). A proton current drives action potentials in genetically identified sour taste cells. Proc Natl Acad Sci U S A 107: 22320–22325.

Chaudhari N, Roper SD (2010). The cell biology of taste. J Cell Biol 190: 285–296.

Cheng SF, Chen ML, Hung PC et al. (2004). Olfactory loss in poly(acrylonitrile-butadiene-styrene) plastic injection-moulding workers. Occup Med 54: 469–474.

Cruzan G, Cushman JR, Andrews LS et al. (1997). Subchronic inhalation studies of styrene in CD rats and CD-1 mice. Fundam Appl Toxicol 35: 152–165.

Cruzan G, Cushman JR, Andrews LS et al. (1998). Chronic toxicity/oncogenicity study of styrene in CD rats by inhalation exposure for 104 weeks. Toxicol Sci 46: 266–281.

Cullen MR (1987). Workers with multiple chemical hypersensitivities. State of the Art Reviews of Occupational Medicine 2: 655–806.

Czerniawska A (1970). Experimental investigations on penetration of Au-198 from nasal mucous membrane into cerebrospinal fluid. Acta Otolaryngol 70: 58.

Dalton P, Cowart B, Dilks D et al. (2003). Olfactory function in workers exposed to styrene in the reinforced-plastics industry. Am J Ind Med 44: 1–11.

Dalton P, Lees PS, Gould M et al. (2007). Evaluation of long-term occupational exposure to styrene vapor on olfactory function. Chem Senses 32: 739–747.

Dalton PH, Opiekun RE, Gould M et al. (2010). Chemosensory loss: functional consequences of the World Trade Center disaster. Environ Health Perspect 118: 1251–1256.

De Lorenzo AJD (1970). The olfactory neuron and the blood brain barrier. In: GEW Wolstenholme, J Knight (Eds.), Taste and Smell In Vertebrates. Churchill Livingstone, London, pp. 151–171.

Desai H, Smutzer G, Coldwell SE et al. (2011). Validation of edible taste strips for identifying PROP taste recognition thresholds. Laryngoscope 121: 1177–1183.

Dhong HJ, Chung SK, Doty RL (1999). Estrogen protects against 3-methylindole-induced olfactory loss. Brain Res 824: 312–315.

Dodd DE, Gross EA, Miller RA et al. (2010). Nasal olfactory epithelial lesions in F344 and SD rats following 1- and 5-day inhalation exposure to naphthalene vapor. Int J Toxicol 29: 175–184.

Dorman DC, Miller KL, D'Antonio A et al. (1997). Chloroform-induced olfactory mucosal degeneration and osseous ethmoid hyperplasia are not associated with olfactory deficits in Fischer 344 rats. Toxicology 122: 39–50.

Dorman DC, Struve MF, Wong BA et al. (2008). Derivation of an inhalation reference concentration based upon olfactory neuronal loss in male rats following subchronic acetaldehyde inhalation. Inhal Toxicol 20: 245–256.

Doty RL (2007). Office procedures for quantitative assessment of olfactory function. Am J Rhinol 21: 460–473.

Doty RL (2008). The olfactory vector hypothesis of neurodegenerative disease: is it viable? Ann Neurol 63: 7–15.

Doty RL, Crastnopol B (2010). Correlates of chemosensory malingering. Laryngoscope 120: 707–711.

Doty RL, Haxel BR (2005). Objective assessment of terbinafine-induced taste loss. Laryngoscope 115: 2035–2037.

Doty RL, Shaman P, Dann M (1984). Development of the University of Pennsylvania Smell Identification Test: a standardized microencapsulated test of olfactory function. Physiol Behav 32: 489–502.

Doty RL, Deems DA, Frye RE et al. (1988). Olfactory sensitivity, nasal resistance, and autonomic function in patients with multiple chemical sensitivities. Arch Otolaryngol Head Neck Surg 114: 1422–1427.

Duda JE (2010). Olfactory system pathology as a model of Lewy neurodegenerative disease. J Neurol Sci 289: 49–54.

Dunnick JK, Elwell MR, Benson JM et al. (1989). Lung toxicity after 13-week inhalation exposure to nickel oxide, nickel subsulfide, or nickel sulfate hexahydrate in F344/N rats and B6C3F1 mice. Fundam Appl Toxicol 12: 584–594.

Eldridge SR, Bogdanffy MS, Jokinen MP et al. (1995). Effects of propylene oxide on nasal epithelial cell proliferation in F344 rats. Fundam Appl Toxicol 27: 25–32.

Elsberg CA, Levy I (1935). The sense of smell: I. A new and simple method of quantitative olfactometry. Bull Neurol Inst NY 4: 5–19.

Erikson KM, Dorman DC, Lash LH et al. (2005). Persistent alterations in biomarkers of oxidative stress resulting from combined in utero and neonatal manganese inhalation. Biol Trace Elem Res 104: 151–163.

Escada PA, Lima C, da Silva JM (2009). The human olfactory mucosa. Eur Arch Otorhinolaryngol 266: 1675–1680.

Evans JE, Miller ML, Andringa A et al. (1995). Behavioral, histological, and neurochemical effects of nickel (II) on the rat olfactory system. Toxicol Appl Pharmacol 130: 209–220.

Feron VJ, Kroes R (1979). One-year time-sequence inhalation toxicity study of vinyl chloride in rats. II. Morphological changes in the respiratory tract, ceruminous glands, brain, kidneys, heart and spleen. Toxicology 13: 131–141.

Feron VJ, Kruysse A (1978). Effects of exposure to furfural vapour in hamsters simultaneously treated with benzo [alpha] pyrene or diethylnitrosamine. Toxicology 11: 127–144.

Frank ME, Smith DV (1991). Electrogustometry: a simple way to test taste. In: TV Getchell, RL Doty, LM Bartoshuk et al. (Eds.), Smell and Taste in Health and Disease. Raven Press, New York, pp. 503–514.

Friberg L (1950). Health hazards in the manufacture of akaline accumulators with special reference to chronic cadmium poisoning. Acta Med Scand 138: 1–124.

Furuta S, Nishimoto K, Egawa M et al. (1994). Olfactory dysfunction in patients with Minamata disease. Am J Rhinol 8: 259–263.

Gagnaire F, Zissu D, Bonnet P et al. (1987). Nasal and pulmonary toxicity of allyl glycidyl ether in mice. Toxicol Lett 39: 139–145.

Gagnon P, Mergler D, Lapare S (1994). Olfactory adaptation, threshold shift and recovery at low levels of exposure to methyl isobutyl ketone (MIBK). Neurotoxicology 15: 637–642.

Gaskell BA (1990). Nonneoplastic changes in the olfactory epithelium – experimental studies. Environ Health Perspect 85: 275–289.

Gaskell BA, Hext PM, Pigott GH et al. (1988). Olfactory and hepatic changes following inhalation of 3-trifluoromethyl pyridine in rats. Toxicology 50: 57–68.

Gaskell BA, Hext PM, Pigott GH et al. (1990). Olfactory and hepatic changes following a single inhalation exposure of 3-trifluoromethyl pyridine in rats: concentration and temporal aspects. Toxicology 62: 35–51.

Genter MB, Llorens J, O'Callaghan JP et al. (1992). Olfactory toxicity of beta, beta'-iminodipropionitrile in the rat. J Pharmacol Exp Ther 263: 1432–1439.

Gianutsos G, Morrow GR, Morris JB (1997). Accumulation of manganese in rat brain following intranasal administration. Fundam Appl Toxicol 37: 102–105.

Glusman G, Yanai I, Rubin I et al. (2001). The complete human olfactory subgenome. Genome Res 11: 685–702.

Good PF, Perl DP, Bierer LM et al. (1992). Selective accumulation of aluminum and iron in the neurofibrillary tangles of Alzheimer's disease: a laser microprobe (LAMMA) study. Ann Neurol 31: 286–292.

Gottofrey J, Tjalve H (1991). Axonal transport of cadmium in the olfactory nerve of the pike. Pharmacol Toxicol 69: 242–252.

Grant SM (1993). An unusual cause of burning mouth sensation. Br Dent J 175: 378–380.

Gross EA, Swenberg JA, Fields S et al. (1982). Comparative morphometry of the nasal cavity in rats and mice. J Anat 135: 83–88.

Gudziol V, Pietsch J, Witt M et al. (2010). Theophylline induces changes in the electro-olfactogram of the mouse. Eur Arch Otorhinolaryngol 267: 239–243.

Harada N, Fujii M, Dodo H (1983). Olfactory disorders in chemical plant workers exposed to SO_2 and/or NH_3. Journal of Science and Labour 59: 17–23.

Harkema JR, Carey SA, Wagner JG (2006). The nose revisited: a brief review of the comparative structure, function, and toxicologic pathology of the nasal epithelium. Toxicol Pathol 34: 252–269.

Hastings L, Evans J (1988). Transaxonal transport of cadmium in the olfactory system. Chem Senses 13: 696.

Hastings L, Evans JE (1991). Olfactory primary neurons as a route of entry for toxic agents into the CNS. Neurotoxicology 12: 707–714.

Hayek R, Waite PME (1991). The olfactory pathway as a possible route for aluminum entry to the brain. J Neurochem 57: S113.

Henriksson J, Tjalve H (2000). Manganese taken up into the CNS via the olfactory pathway in rats affects astrocytes. Toxicol Sci 55: 392–398.

Henriksson J, Tallkvist J, Tjalve H (1999). Transport of manganese via the olfactory pathway in rats: dosage dependency of the uptake and subcellular distribution of the metal in the olfactory epithelium and the brain. Toxicol Appl Pharmacol 156: 119–128.

Hisamitsu M, Okamoto Y, Chazono H et al. (2011). The influence of environmental exposure to formaldehyde in nasal mucosa of medical students during cadaver dissection. Allergol Int 60: 373–379.

Hockman CH, Lloyd KG, Farley IJ et al. (1971). Experimental midbrain lesions: neurochemical comparison between the animal model and Parkinson's disease. Brain Res 35: 613–618.

Holley A, Duchamp A, Revial MF et al. (1974). Qualitative and quantitative discrimination in the frog olfactory receptors: analysis from electrophysiological data. Ann N Y Acad Sci 237: 102–128.

Holstrom M, Wilhelmsson B (1988). Respiratory symptoms and pathophysiological effects of occupational exposure to formaldehyde and wood dust. Scand J Work Environ Health 14: 306–311.

Hudson R, Arriola A, Martinez-Gomez M et al. (2006). Effect of air pollution on olfactory function in residents of Mexico City. Chem Senses 31: 79–85.

Hummel T, Roscher S, Jaumann MP et al. (1996). Intranasal chemoreception in patients with multiple chemical sensitivities: a double-blind investigation. Regul Toxicol Pharmacol 24: S79–S86.

Hurtt ME, Thomas DA, Working PK et al. (1988). Degeneration and regeneration of the olfactory epithelium following inhalation exposure to methyl bromide: pathology, cell kinetics, and olfactory function. Toxicol Appl Pharmacol 94: 311–328.

Imaizumi S, Mori K, Kiritani S et al. (1998). Task-dependent laterality for cue decoding during spoken language processing. Neuroreport 9: 899–903.

Irwin RD, Chou BJ, Mellick PW et al. (1997). Toxicity of furfuryl alcohol to F344 rats and B6C3F1 mice exposed by inhalation. J Appl Toxicol 17: 159–169.

Jackson RT, Tigges J, Arnold W (1979). Subarachnoid space of the CNS, nasal mucosa, and lymphatic system. Arch Otolaryngol 105: 180–184.

Jiang XZ, Buckley LA, Morgan KT (1983). Pathology of toxic responses to the RD50 concentration of chlorine gas in the nasal passages of rats and mice. Toxicol Appl Pharmacol 71: 225–236.

Johnson BA, Leon M (2007). Chemotopic odorant coding in a mammalian olfactory system. J Comp Neurol 503: 1–34.

Jones FN (1953). A test of the validity of the Elsberg method of olfactometry. Am J Psychol 66: 401–409.

Jones-Gotman M, Zatorre RJ (1988). Olfactory identification deficits in patients with focal cerebral excision. Neuropsychologia 26: 387–400.

Kay J (2011). Chlorine accidents take a big human toll. Available online at: http:/www.scientificamerican.com/article.cfm?id=chlorine-accidents-take-big-human-toll.

Keenan CM, Kelly DP, Bogdanffy MS (1990). Degeneration and recovery of rat olfactory epithelium following inhalation of dibasic esters. Fundam Appl Toxicol 15: 381–393.

Kimbell JS (2006). Nasal dosimetry of inhaled gases and particles: where do inhaled agents go in the nose? Toxicol Pathol 34: 270–273.

Kimura Y, Miwa T, Furukawa M et al. (1991). Effects of topical application of steroids on olfactory disturbance in mice. Chem Senses 297–302.

Kitamura F, Yokoyama K, Araki S et al. (2003). Increase of olfactory threshold in plating factory workers exposed to chromium in Korea. Ind Health 41: 279–285.

Kratskin IL, Kimura Y, Hastings L et al. (1999). Chronic dexamethasone treatment potentiates insult to olfactory receptor cells produced by 3-methylindole. Brain Res 847: 240–246.

Kumar R, Agarwal AK, Seth PK (1996). Oxidative stress-mediated neurotoxicity of cadmium. Toxicol Lett 89: 65–69.

Kuschner M, Laskin S, Drew RT et al. (1975). Inhalation carcinogenicity of alpha halo ethers. III. Lifetime and limited period inhalation studies with bis(chloromethyl)ether at 0.1 ppm. Arch Environ Health 30: 73–77.

Landis BN, Welge-Luessen A, Bramerson A et al. (2009). "Taste Strips" – a rapid, lateralized, gustatory bedside identification test based on impregnated filter papers. J Neurol 256: 242–248.

Latendresse JR, Marit GB, Vernot EH et al. (1995). Oncogenic potential of inhaled hydrazine in the nose of rats and hamsters after 1 or 10 1-hr exposures. Fundam Appl Toxicol 27: 33–48.

Leavens TL, Rao D, Andersen ME et al. (2007). Evaluating transport of manganese from olfactory mucosa to striatum by pharmacokinetic modeling. Toxicol Sci 97: 265–278.

Lee RJ, Xiong G, Kofonow JM et al. (2012). T2R38 taste receptor polymorphisms underlie susceptibility to upper respiratory infection. J Clin Invest 122: 4145–4159.

Lehallier B, Coureaud G, Maurin Y et al. (2012). Effects of manganese injected into rat nostrils: implications for in vivo functional study of olfaction using MEMRI. Magn Reson Imaging 30: 62–69.

Leopold DA (1994). Nasal toxicity: end points of concern in humans. Inhal Toxicol 6 (suppl): 23–39.

Leopold DA, Hummel T, Schwob JE et al. (2000). Anterior distribution of human olfactory epithelium. Laryngoscope 110: 417–421.

Liberles SD, Buck LB (2006). A second class of chemosensory receptors in the olfactory epithelium. Nature 442: 645–650.

Linschoten MR, Kroeze JHA (1991). Spatial summation in taste: NaCl thresholds and stimulated area on the anterior human tongue. Chem Senses 16: 219–224.

Lomax LG, Stott WT, Johnson KA et al. (1989). The chronic toxicity and oncogenicity of inhaled technical-grade 1,3-dichloropropene in rats and mice. Fundam Appl Toxicol 12: 418–431.

Lucchini R, Bergamaschi E, Smargiassi A et al. (1997). Motor function, olfactory threshold, and hematological indices in manganese-exposed ferroalloy workers. [erratum appears in Environ Res 1997 Nov;75(2):187]. Environ Res 73: 175–180.

Mascagni P, Consonni D, Bregante G et al. (2003). Olfactory function in workers exposed to moderate airborne cadmium levels. Neurotoxicology 24: 717–724.

Mellert TK, Getchell ML, Sparks L et al. (1992). Characterization of the immune barrier in human olfactory mucosa. Otolaryngol Head Neck Surg 106: 181–188.

Menco BPM, Morrison EE (2003). Morphology of the mammalian olfactory epithelium: form, fine structure, function, and pathology. In: RL Doty (Ed.), Handbook of Olfaction and Gustation, 2nd edn Marcel Dekker, New York, pp. 17–50.

Mergler D, Beauvais B (1992). Olfactory threshold shift following controlled 7-hour exposure to toluene and/or xylene. Neurotoxicology 13: 211–215.

Mergler D, Huel G, Bowler R et al. (1994). Nervous system dysfunction among workers with long-term exposure to manganese. Environ Res 64: 151–180.

Mery S, Larson JL, Butterworth BE et al. (1994). Nasal toxicity of chloroform in male F-344 rats and female B6C3F1 mice following a 1-week inhalation exposure. Toxicol Appl Pharmacol 125: 214–227.

Meyerhof W, Batram C, Kuhn C et al. (2010). The molecular receptive ranges of human TAS2R bitter taste receptors. Chem Senses 35: 157–170.

Miller CS (2003). Multiple chemical intolerance. In: RL Doty (Ed.), Handbook of Olfaction and Gustation. Marcel Dekker, New York, pp. 525–547.

Miller CS, Mitzel HC (1995). Chemical sensitivity attributed to pesticide exposure versus remodeling. Arch Environ Health 50: 119–129.

Miller RR, Ayres JA, Jersey GC et al. (1981). Inhalation toxicity of acrylic acid. Fundam Appl Toxicol 1: 271–277.

Miller RR, Young JT, Kociba RJ et al. (1985). Chronic toxicity and oncogenicity bioassay of inhaled ethyl acrylate in Fischer 344 rats and B6C3F1 mice. Drug Chem Toxicol 8: 1–42.

Miller SL, Mirza N, Doty RL (2002). Electrogustometric thresholds: relationship to anterior tongue locus, area of stimulation, and number of fungiform papillae. Physiol Behav 75: 753–757.

Moran DT, Rowley III JC, Jafek BW et al. (1982). The fine structure of the olfactory mucosa in man. J Neurocytol 11: 721–746.

Morris JB (1997). Dosimetry, toxicity and carcinogenicity of inspired acetaldehyde in the rat. Mutat Res 380: 113–124.

Morse CC, Boyd MR, Witschi H (1984). The effect of 3-methylfuran inhalation exposure on the rat nasal cavity. Toxicology 30: 195–204.

Murray RG (1973). The ultrastructure of taste buds. In: I Friedmann (Ed.), The Ultrastructure of Sensory Organs. North Holland Publishing Company, Amsterdam, pp. 1–81.

Muttray A, Gosepath J, Brieger J et al. (2009). No acute effects of an exposure to 50 ppm acetaldehyde on the upper airways. Int Arch Occup Environ Health 82: 481–488.

National Toxicology Program (2010a). NTP technical report on the toxicology and carcinogenesis studies of beta-myrcene (CAS No. 123-35-3) in F344/N rats and B6C3F1 mice (gavage studies). Natl Toxicol Program Tech Rep Ser 1–163.

National Toxicology Program (2010b). NTP toxicology and carcinogenesis studies of 5-(hydroxymethyl)-2-furfural (CAS No. 67-47-0) in F344/N rats and B6C3F1 mice (gavage studies). Natl Toxicol Program Tech Rep Ser 7–9, 21.

National Toxicology Program (2011). Toxicology and carcinogenesis studies of tetralin (CAS No. 119-64-2) in F344/N rats and B6C3F1 mice (inhalation studies). Natl Toxicol Program Tech Rep Ser 1–198.

Naus A (1975). Olphactoric properties of industrial matters. Charles University, Prague.

Nikula KJ, Lewis JL (1994). Olfactory mucosal lesions in F344 rats following inhalation exposure to pyridine at threshold limit value concentrations. Fundam Appl Toxicol 23: 510–517.

Nikula KJ, Sun JD, Barr EB et al. (1993). Thirteen-week, repeated inhalation exposure of F344/N rats and B6C3F1 mice to ferrocene. Fundam Appl Toxicol 21: 127–139.

Nosaka Y, Sadanaga M, Shiga A et al. (1970). Development of disturbance of hearing acuity, vestibular function, sense of taste and speech in Minamata disease. Nihon Jibiinkoka Gakkai Kaiho 73 (Suppl-7).

O'Connor S, Jacob TJ (2008). Neuropharmacology of the olfactory bulb. Curr Mol Pharmacol 1: 181–190.

Ojima M, Tonori H, Sato T et al. (2002). Odor perception in patients with multiple chemical sensitivity. Tohoku J Exp Med 198: 163–173.

Papi F (1990). Olfactory navigation in birds. Experientia 46: 351–363.

Pavlidis P, Gouveris H, Antonia A et al. (2013). Age-related changes in electrogustometry thresholds, tongue tip vascularization, density, and form of the fungiform papillae in humans. Chem Senses 38: 35–43.

Pearn J (2001). Neurology of ciguatera. J Neurol Neurosurg Psychiatry 70: 4–8.

Perl DP, Good PF (1987). Uptake of aluminium into central nervous system along nasal-olfactory pathways. Lancet 1: 1028.

Perl DP, Good PF (1991). Aluminum, Alzheimer's disease, and the olfactory system. Ann N Y Acad Sci 640: 8–13.

Perl DP, Moalem S (2006). Aluminum and Alzheimer's disease, a personal perspective after 25 years. J Alzheimers Dis 9: 291–300.

Persson E, Henriksson J, Tallkvist J et al. (2003a). Transport and subcellular distribution of intranasally administered zinc in the olfactory system of rats and pikes. Toxicology 191: 97–108.

Persson E, Henriksson J, Tjalve H (2003b). Uptake of cobalt from the nasal mucosa into the brain via olfactory pathways in rats. Toxicol Lett 145: 19–27.

Pevzner RA, Hernadi L, Salanki J (1986). Effect of mercury on the fish (*Alburnus alburnus*) chemoreceptor taste buds. A scanning electron microscopic study. Acta Biol Hung 37: 159–167.

Pino MV, Valerio MG, Miller GK et al. (1999). Toxicologic and carcinogenic effects of the type IV phosphodiesterase inhibitor RP 73401 on the nasal olfactory tissue in rats. Toxicol Pathol 27: 383–394.

Plopper CG, Suverkropp C, Morin D et al. (1992). Relationship of cytochrome P-450 activity to Clara cell cytotoxicity. I. Histopathologic comparison of the respiratory tract of mice, rats and hamsters after parenteral administration of naphthalene. J Pharmacol Exp Ther 261: 353–363.

Potts CL (1965). Cadmium proteinuria – the health of battery workers exposed to cadmium oxide dust. Annuals of Occupational Hygiene 8: 55–61.

Prediger RD, Aguiar Jr AS, Matheus FC et al. (2012). Intranasal administration of neurotoxicants in animals: support for the olfactory vector hypothesis of Parkinson's disease. Neurotox Res 21: 80–116.

Randolph TG (1962). Human ecology and susceptibility to the chemical environment. Charles C. Thomas, Springfield, IL.

Rangga-Tabbu C, Sleight SD (1992). Sequential study in rats of nasal and hepatic lesions induced by N-nitrosodimethylamine and N-nitrosopyrrolidine. Fundam Appl Toxicol 19: 147–156.

Reed CJ (1993). Drug metabolism in the nasal cavity: relevance to toxicology. Drug Metab Rev 25: 173–205.

Reed CJ, De MF (1989). Cumene hydroperoxide-dependent oxidation of NNN'N'-tetramethyl-p-phenylenediamine and 7-ethoxycoumarin by cytochrome P-450. Comparison between the haemoproteins from liver and olfactory tissue. Biochem J 261: 793–800.

Reimann F, Tolhurst G, Gribble FM (2012). G-protein-coupled receptors in intestinal chemosensation. Cell Metab 15: 421–431.

Reuzel PG, Wilmer JW, Woutersen RA et al. (1990). Interactive effects of ozone and formaldehyde on the nasal respiratory lining epithelium in rats. J Toxicol Environ Health 29: 279–292.

Reznik G, Stinson SF, Ward JM (1980). Respiratory pathology in rats and mice after inhalation of 1,2-dibromo-3-chloropropane or 1,2 dibromoethane for 13 weeks. Arch Toxicol 46: 233–240.

Rolls ET, Grabenhorst F (2008). The orbitofrontal cortex and beyond: from affect to decision-making. Prog Neurobiol 86: 216–244.

Rose CS, Heywood PG, Costanzo RM (1992). Olfactory impairment after chronic occupational cadmium exposure. J Occup Med 34: 600–605.

Rosenberg NL, Kleinschmidt-DeMasters BK, Davis KA et al. (1988). Toluene abuse causes diffuse central nervous system white matter changes. Ann Neurol 23: 611–614.

Ross PM, Whysner J, Covello VT et al. (1999). Olfaction and symptoms in the multiple chemical sensitivities syndrome. Prev Med 28: 467–480.

Roudnitzky N, Bufe B, Thalmann S et al. (2011). Genomic, genetic and functional dissection of bitter taste responses to artificial sweeteners. Hum Mol Genet 20: 3437–3449.

Rouquier S, Blancher A, Giorgi D (2000). The olfactory receptor gene repertoire in primates and mouse: evidence for reduction of the functional fraction in primates. Proc Natl Acad Sci U S A 97: 2870–2874.

Rydzewski B, Sulkowski W, Miarzynska M (1998). Olfactory disorders induced by cadmium exposure: a clinical study. Int J Occup Med Enviro Health 11: 235–245.

Sandmark B, Broms I, Lofgren L et al. (1989). Olfactory function in painters exposed to organic solvents. Scand J Work Environ Health 15: 60–63.

Schieferstein GJ, Sheldon WG, Cantrell SA et al. (1988). Subchronic toxicity study of 1,4-dithiane in the rat. Fundam Appl Toxicol 11: 703–714.

Schiffman SS, Nagle HT (1992). Effect of environmental pollutants on taste and smell. Otolaryngol Head Neck Surg 106: 693–700.

Schwartz BS, Doty RL, Monroe C et al. (1989). Olfactory function in chemical workers exposed to acrylate and methacrylate vapors. Am J Public Health 79: 613–618.

Schwartz BS, Ford DP, Bolla KI et al. (1990). Solvent-associated decrements in olfactory function in paint manufacturing workers. Am J Ind Med 18: 697–706.

Schwartz BS, Ford DP, Bolla KI et al. (1991). Solvent-associated olfactory dysfunction: not a predictor of deficits in learning and memory. Am J Psychiatry 148: 751–756.

Schwartz BS, Bolla KI, Stewart W et al. (1993). Decrements in neurobehavioral performance associated with mixed exposure to organic and inorganic lead. Am J Epidemiol 137: 1006–1021.

Shimada S, Ueda T, Ishida Y et al. (2006). Acid-sensing ion channels in taste buds. Arch Histol Cytol 69: 227–231.

Sicard G, Holley A (1984). Receptor cell responses to odorants: similarities and differences among odorants. Brain Res 292: 283–296.

Simard AR, Rivest S (2006). Neuroprotective properties of the innate immune system and bone marrow stem cells in Alzheimer's disease. Mol Psychiatry 11: 327–335.

Smithson LJ, Kawaja MD (2010). Microglial/macrophage cells in mammalian olfactory nerve fascicles. J Neurosci Res 88: 858–865.

Smutzer G, Lam S, Hastings L et al. (2008). A test for measuring gustatory function. Laryngoscope 118: 1411–1416.

Spealman CR (1954). Odors, odorants, and deodorants in aviation. Ann N Y Acad Sci 58: 40–43.

Stadler JC, Delker DA, Malley LA et al. (2008). Subchronic, reproductive, and developmental toxicity of a fluorotelomer-based urethane polymeric product. Drug Chem Toxicol 31: 317–337.

Stinton N, Atif MA, Barkat N et al. (2010). Influence of smell loss on taste function. Behav Neurosci 124: 256–264.

Stott WT, Young JT, Calhoun LL et al. (1988). Subchronic toxicity of inhaled technical grade 1,3-dichloropropene in rats and mice. Fundam Appl Toxicol 11: 207–220.

Sulkowski WJ, Rydzewski B, Miarzynska M (2000). Smell impairment in workers occupationally exposed to cadmium. Acta Otolaryngol 120: 316–318.

Sun TJ, Miller ML, Hastings L (1996). Effects of inhalation of cadmium on the rat olfactory system: behavior and morphology. Neurotoxicol Teratol 18: 89–98.

Tabert MH, Liu X, Doty RL et al. (2005). A 10-item smell identification scale related to risk for Alzheimer's disease. Ann Neurol 58: 155–160.

Tallkvist J, Henriksson J, d'Argy R et al. (1998). Transport and subcellular distribution of nickel in the olfactory system of pikes and rats. Toxicol Sci 43: 196–203.

Tallkvist J, Persson E, Henriksson J et al. (2002). Cadmium-metallothionein interactions in the olfactory pathways of rats and pikes. Toxicol Sci 67: 108–113.

Tjalve H, Henriksson J (1999). Uptake of metals in the brain via olfactory pathways. Neurotoxicology 20: 181–195.

Tjalve H, Mejare C, Borg-Neczak K (1995). Uptake and transport of manganese in primary and secondary olfactory neurones in pike. Pharmacol Toxicol 77: 23–31.

Tjalve H, Henriksson J, Tallkvist J et al. (1996). Uptake of manganese and cadmium from the nasal mucosa into the central nervous system via olfactory pathways in rats. Pharmacol Toxicol 79: 347–356.

Tomita H, Ikeda M (2002). Clinical use of electrogustometry: strengths and limitations. Acta Otolaryngol 547 (Suppl): 27–38.

Tourbier IA, Doty RL (2007). Sniff magnitude test: Relationship to odor identification, detection, and memory tests in a clinic population. Chem Senses 32: 515–523.

Trela BA, Frame SR, Bogdanffy MS (1992). A microscopic and ultrastructural evaluation of dibasic esters (DBE) toxicity in rat nasal explants. Exp Mol Pathol 56: 208–218.

Turk MA, Henk WG, Flory W (1987). 3-Methylindole-induced nasal mucosal damage in mice. Vet Pathol 24: 400–403.

Tvedt B, Skyberg K, Berstad J (1989). Chronic toxic encephalopathy among house painters with disability pension. J Oslo City Hosp 39: 74–80.

Tvedt B, Skyberg K, Aaserud O et al. (1991). Brain damage caused by hydrogen sulfide: a follow-up study of six patients. Am J Ind Med 20: 91–101.

Uraih LC, Talley FA, Mitsumori K et al. (1987). Ultrastructural changes in the nasal mucosa of Fischer 344 rats and B6C3F1 mice following an acute exposure to methyl isocyanate. Environ Health Perspect 72: 77–88.

Villalobos V, Bonilla E, Castellano A et al. (2009). Ultrastructural changes of the olfactory bulb in manganese-treated mice. Biocell 33: 187–197.

Wang B, Feng WY, Wang M et al. (2007). Transport of intranasally instilled fine Fe2O3 particles into the brain: Microdistribution, chemical states, and histopathological observation. Biol Trace Elem Res 118: 233–243.

Warheit DB, Kelly DP, Carakostas MC et al. (1989). A 90-day inhalation toxicity study with benomyl in rats. Fundam Appl Toxicol 12: 333–345.

Watanabe S, Fukuchi Y (2000). Occupational impairment of the olfactory sense of chromate producing workers. Sangyo Igaku 23: 606–611.

White CW, Martin JG (2010). Chlorine gas inhalation: human clinical evidence of toxicity and experience in animal models. Proc Am Thorac Soc 7: 257–263.

Wilmer JW, Woutersen RA, Appelman LM et al. (1989). Subchronic (13-week) inhalation toxicity study of formaldehyde in male rats: 8-hour intermittent versus 8-hour continuous exposures. Toxicol Lett 47: 287–293.

Witt M, Reutter K (2015). Anatomy of the tongue and taste buds. In: RL Doty (Ed.), Handbook of Olfaction and Gustation, 3rd edition. Wiley-Liss, New York, pp. 639–665.

Wolf DC, Morgan KT, Gross EA et al. (1995). Two-year inhalation exposure of female and male B6C3F1 mice and F344 rats to chlorine gas induces lesions confined to the nose. Fundam Appl Toxicol 24: 111–131.

Yin-Zeng L, Jin-Xiang H, Cheng-Mo L et al. (1985). Effects of cadmium on cadmium smelter workers. Scand J Work Environ Health 11: 29–32.

Zatta P, Favarato M, Nicolini M (1993). Deposition of aluminum in brain-tissues of rats exposed to inhalation of aluminum acetylacetonate. Neuroreport 4: 1119–1122.

Zhao L, Kirkmeyer SV, Tepper BJ (2003). A paper screening test to assess genetic taste sensitivity to 6-n-propylthiouracil. Physiol Behav 78: 625–633.

Zhao K, Pribitkin EA, Cowart BJ et al. (2006). Numerical modeling of nasal obstruction and endoscopic surgical intervention: outcome to airflow and olfaction. Am J Rhinol 20: 308–316.

Zhuo X, Gu J, Zhang QY, Spink DC et al. (1999). Biotransformation of coumarin by rodent and human cytochromes P-450: metabolic basis of tissue-selective toxicity in olfactory mucosa of rats and mice. J Pharmacol Exp Ther 288: 463–471.

Zucco GM, Militello C, Doty RL (2008). Discriminating between organic and psychological determinants of multiple chemical sensitivity: a case study. Neurocase 14: 485–493.

Handbook of Clinical Neurology, Vol. 131 (3rd series)
Occupational Neurology
M. Lotti and M.L. Bleecker, Editors

Chapter 18

Retinal and visual system: occupational and environmental toxicology

DONALD A. FOX*

Departments of Vision Sciences, Biology and Biochemistry, Pharmacology, and Health and Human Performance, University of Houston, Houston, TX, USA

INTRODUCTION

Approximately 3000 chemicals are toxic to the eye and visual system (Grant and Schuman, 1993). Occupational exposure to chemicals often results in structural and functional alterations in sensory systems, such that ~50% of all neurotoxic chemicals affect some aspect of sensory function (Anger and Johnson, 1985; Crofton and Sheets, 1989; Fox and Boyes, 2013). In many cases following chemical exposure, disturbances in sensory system function are the first clinical symptoms noted or reported (Damstra, 1978; Hanninen et al., 1978; Baker et al., 1984; Mergler et al., 1987; Iregren et al., 2002). Importantly, these alterations are often present in the absence of any other clinical signs or symptoms of toxicity (Damstra, 1978; Hanninen et al., 1978; Baker et al., 1984; Anger and Johnson, 1985; Crofton and Sheets, 1989). These findings and observations indicate that sensory systems are especially vulnerable to toxic insult. Overall, the most frequently reported sensory system alterations occur in the visual system (Anger and Johnson, 1985; Crofton and Sheets, 1989; Fox and Boyes, 2013). Moreover, alterations in the structure and/or function of the eye or central visual system are among the criteria utilized for setting permissible occupational or environmental exposure levels for many different chemicals in the USA (Anger and Johnson, 1985; Environmental Protection Agency, 2014; National Institute for Occupational Safety and Health, 2014). Subtle alterations in visual processing of information (e.g., visual perceptual, visual motor, and visual spatial) can have profound immediate, long-term, and delayed effects on the mental, social, and physical health and performance of an individual. Finally, sensory system impairments, especially auditory and visual, can lead to increased occupational injuries.

The aims of this chapter are fourfold. First, to provide easily available and excellent references on basic retinal and central visual pathways/system structure and function as well as clinically available techniques used to assess alterations in these tissues. Second, to introduce the features that make the retina especially vulnerable to toxic chemicals. Third, to review the clinical and then, where appropriate, the corresponding experimental data regarding the structural and functional retinal and visual pathways/system deficits produced by selected neurotoxic chemicals commonly found in workplace environments. These are organic solvents (e.g., carbon disulfide, chlorinated organic solvents trichloroethylene (TCE) and tetrachloroethylene (PERC), styrene, toluene, mixtures) and metals (inorganic lead, methyl mercury, and mercury vapor). The retinotoxic effects of pesticides are not discussed herein as they were reviewed recently and no reports on exposed workers have been published since 2006 (Fox and Boyes, 2013). Fourth, to discuss occupational toxicants as risk factors for retinal diseases.

Many of these toxicants adversely affect retinal photoreceptors – either cone-mediated color vision or rod-mediated (scotopic) night vision. The effects of drugs and toxicants on color vision often follow Kollner's rule, which generalized that the adverse effects of drugs on the retina produced blue–yellow deficits whereas those on the central visual pathways/system including the optic nerve/tract caused red–green deficits (Kollner, 1912). In addition, neurotoxicant-induced alterations in the central visual pathways/system can result in reduced peripheral vision, spatial or temporal contrast sensitivity, and

*Correspondence to: Donald A. Fox, PhD, University of Houston, College of Optometry, 4901 Calhoun Road, Houston, TX 77204-2020, USA. Tel: +1-713-743-1964, E-mail: dafox@uh.edu

visual search performance. Moreover, since different populations of intrinsically photosensitive retinal ganglion cells control the pupillary light reflex and circadian rhythm (Chen et al., 2012), selective damage to these cells could selectively alter these essential functions. Except for the recently updated chapter in *Casarett & Doull's Essentials of Toxicology* entitled "Toxic responses of the ocular and visual system" (Fox and Boyes, 2013), the most recent chapters on occupational retinal and visual system toxicology were published 15–20 years ago (Mergler, 1995; Fox, 1998). Critical reviews on color vision deficits in occupational workers exposed to solvents and inorganic mercury were published over a decade ago (Geller and Hudnell, 1997; Iregren et al., 2002; Gobba and Cavalleri, 2003; Lomax et al., 2004).

An excellent comprehensive, peer-reviewed, and up-to-date website and book that present the anatomy, biochemistry, cell/molecular biology, and electrophysiology of the human, monkey, and rodent retina and central visual pathways/system by recognized experts are http://www.webvision.med.utah.edu/ and *Adler's Physiology of the Eye* (Levin et al., 2011). Fox and Boyes (2013) addressed ocular pharmacodynamics and pharmacokinetics, evaluation of ocular toxicity and visual function, and target sites and mechanisms of action of numerous drugs and neurotoxicants. There are several excellent reviews and discussion of the critical issues in the use and analysis of color vision deficits resulting from neurotoxicant exposure (Geller and Hudnell, 1997; Iregren et al., 2002; Gobba and Cavalleri, 2003; Lomax et al., 2004). Acquired color vision deficits, as quantitatively assessed by the color confusion index (CCI), were usually determined with the Farnsworth D15, Farnsworth 100-hue, Lanthony D-15, or Lanthony D-15d (desaturated). The Farnsworth D15 and Lanthony D-15d tests, which contain a reference disk and 15 numbered disks that make up an incomplete color circle, are used to separate medium-strong and mild-strong color vision deficits, respectively, from normal color vision. These tests determine blue–yellow and red–green color vision deficits. Ishihara pseudo-isochromatic plates also were used to detect red–green deficiencies.

The Lanthony D-15d is the most sensitive test for determining subclinical color vision deficits (Geller and Hudnell, 1997; Gobba and Cavalleri, 2003). As discussed in the reviews and individual original papers (see below), these acquired color vision deficits using sex- and age-matched controls were usually along the blue–yellow color axis (termed blue–yellow, tritan axis or type III color deficit) or blue–yellow and red–green color axes (termed mixed deutan and protan axes or type II color deficit). In the normal population, blue–yellow deficits or tritanomalies are rare, as mutations of the short-wavelength (S cone) gene located on chromosome 7 occur in less than 0.01% of males and females (http://retina.umh.es/webvision/KallColor.html). However, transient tritan color vision axis deficits occur at high altitude under hypoxic conditions (Tekavcic-Pompe and Tekavcic, 2008; Willmann et al., 2010).

Penha et al. (2010) review the use of animal models and toxicity assays for retinal and ocular toxicity testing. Rosolen et al. (2008) present the applications and limits of different electrophysiology techniques for assessing retinal toxicity (e.g., flash, pattern reversal, and focal electroretinograms (ERGs), and visual evoked potentials (VEPs)). Guidelines and procedures for different types of VEP techniques are presented by the American Clinical Neurophysiology Society (2008). The importance of multimodal clinical, color, spatial and temporal contrast sensitivity, electrophysiologic, spectral domain-optical coherence tomography (SD-OCT), visual field and automated perimetry, eye movement, visual search, and visuomotor testing to assess the extent of structural and functional visual damage and loss has been noted in several studies (Trobe et al., 1996; Ojanpää et al., 2006; Wilensky and Hawkins, 2001; Näsänen et al., 2005; Costa et al., 2012; Lacerda et al., 2012; Ayton et al., 2014).

The human, monkey, and rodent retina share several common anatomic, biologic/biochemical and functional features. Several of these inherent features make their retinas especially sensitive and vulnerable to the effects of chemicals and drugs. First, these retinas have a dual circulatory supply: choroidal and retinal vasculature systems in the outer and inner retina, respectively. The choroidal vasculature is characterized by a dense, one-layered network of fenestrated highly permeable vessels formed by the short posterior ciliary arteries located next to the retinal pigment epithelium (RPE). The retinal vessels have tight junctions similar to those that form the blood–brain barrier in the cerebral capillaries and under normal physiologic conditions they are relatively impermeable. However, at the level of the optic disk, the blood–retinal barrier lacks tight junctions and thus hydrophilic molecules can enter the optic nerve head by diffusion from the extravascular space (Flammer and Mozaffarieh, 2008; Nickla and Wallman, 2010). Second, the retina, and especially the photoreceptors, has one of the highest rates of oxygen consumption of any tissue in the body (Graymore, 1969; Medrano and Fox, 1995; Shulman and Fox, 1996). Together, the rod and cone photoreceptor inner segments and synaptic terminals contain ~75% of the retinal mitochondria and the mitochondria in each of these four compartments have different substructure and vulnerability to chemicals (He et al., 2003; Perkins et al., 2003, 2012; Johnson et al., 2007). Similarly, the visual cortex has a high rate of basal oxygen consumption that is increased during visual stimulation (Vafaee et al., 1999; Lin et al., 2010). Third, melanin in the RPE has a high binding affinity for polycyclic aromatic hydrocarbons, electrophiles, calcium, and heavy metals such as aluminum, iron, lead, and mercury (Potts and Au, 1976; Ulshafer et al., 1990; Eichenbaum and Zheng, 2000; Erie et al., 2005, 2009a). Although this may initially have a protective role, it can result in the excessive accumulation, long-term storage, and slow release of drugs and chemicals from melanin into the outer retina. For example, in donated retinas from nonoccupationally exposed subjects the retina and RPE lead concentration was higher than in other ocular tissues (Eichenbaum and Zheng, 2000; Erie et al., 2005, 2009a). Fourth, the retina/choroid has a unique profile of phase 1 and phase 2 xenobiotic enzymes (Fox and Boyes, 2013). For example, the major P450 drug-metabolizing enzymes CYP3A4, CYP2C8/9, and CYP2C19 likely are not in the retina/choroid, although CYP1A2 and CYP2D6 are present. Although the mRNA profile of retinal drug efflux and uptake transporters is similar to that other tissues (Zhang et al., 2008), the retinal endothelial cells of the inner blood–retinal barrier express and utilize a variety of unique transporters (Hosoya and Tachikawa, 2009). Finally, the retina is a "window into the brain" (London et al., 2013; Jindal, 2014). It is an embryologic extension of the central nervous system that shares anatomic, functional, and physiologic similarities to the brain. An increased awareness of retinal injury and altered visual function, understanding of retinal pathophysiology and progression, and focus on retinal therapeutics should provide insight into inherited and acquired neurodegenerative disorders.

ORGANIC SOLVENTS

Occupational exposure to different organic solvents (i.e., carbon disulfide, chlorinated organic solvents, TCE and PERC, styrene, toluene) and/or mixtures produce color vision deficits, visual field loss, reduced contrast sensitivity at different spatial frequencies, and slowed visual search speed (Muttray et al., 1997; Iregren et al., 2002; Gobba and Cavalleri, 2003; Paramei et al., 2004; Näsänen et al., 2005; Ojanpää et al., 2006; Attarchi et al., 2010; Costa et al., 2012; Lacerda et al., 2012). Comprehensive reviews and a meta-analysis report that the color vision deficits are usually blue–yellow and dependent upon dose and duration of exposure (Iregren et al., 2002; Gobba and Cavalleri, 2003; Lomax et al., 2004; Paramei et al., 2004).

Carbon disulfide

Human studies

Carbon disulfide is an important industrial solvent used since the 1850s. Occupational exposure to this chemical is a potential health hazard worldwide, most notably in the viscose rayon industry. It is also used in the production of carbon tetrachloride and cellophane; as a solvent for rubber, sulfur, oils, resins and waxes; and in agriculture as a pesticide and fumigant. Carbon disulfide has been found in at least 210 of the 1430 National Priorities List sites identified by the Environmental Protection Agency (ATSDR, 1996a). Over 25 000 workers in the USA are exposed daily to this industrial chemical (ATSDR, 1996b). Carbon disulfide is an irritant and metabolic toxicant whose neurologic alterations include central and peripheral distal axonopathies, as well as vascular alterations in the retina, kidney, and cardiovascular system. Retinal alterations include retrobulbar neuritis, pre-senile arteriosclerotic changes in the fundus, microaneurysms and small hemorrhages, delayed peripapillary filling, and irregularities in the caliber of the retinal arterioles and veins (Raitta et al., 1974; Seppalainen and Haltia, 1980; De Rouck et al., 1986).

Alterations in human visual function following carbon disulfide exposure were described over 75 years ago (McDonald, 1938; Gordy and Trumper, 1940). Clinical reports described a severe loss of central vision (central scotoma), depressed sensitivity in the peripheral visual field (rod scotoma), and accompanying optic atrophy, blurred vision, impaired color vision, and pupillary changes (McDonald, 1938; Gordy and Trumper, 1940; Savic, 1967; Raitta et al., 1981). Electrophysiologic studies revealed a decreased light/dark ratio of the electrooculogram (EOG), amplitude of the dark-adapted (rod or scotopic) ERG, and amplitude of the light-adapted (cone or photopic) ERGs (Palacz et al., 1980; Raitta et al., 1981; De Rouck et al., 1986). EOG and ERG alterations were still observable in a 4-year follow-up study (De Rouck et al., 1986). More recent studies describe blue–yellow color vision deficits (Iregren et al., 2002; Gobba and Cavalleri, 2003).

Animal studies

Mice, rabbits, rats, and/or monkeys exposed to carbon disulfide have retinal ganglion cell loss and neurofilamentous axonopathy in the optic nerve and tract (Pappolla et al., 1987; Eskin et al., 1988). Monkeys examined 1 year following exposure still had marked central retinal ganglion cell degeneration, decreased anterograde axonal transport from eye to the lateral geniculate nucleus and visual cortex, and decreased cytochrome oxidase activity (mitochondrial-encoded terminal electron transport enzyme) in these central visual centers (Eskin et al., 1988). These findings are consistent with the clinical reports of optic atrophy, central scotoma, altered color vision, and altered photopic ERG (McDonald, 1938; Gordy and Trumper, 1940; Savic, 1967; Palacz et al., 1980; Raitta et al., 1981; De Rouck et al., 1986; Iregren et al., 2002; Gobba and Cavalleri, 2003). Acute and chronic exposure of rats to carbon disulfide decreased the amplitude and increased the latency of flash VEPs, and decreased the amplitude of pattern reversal VEPs, consistent with a central visual pathway/system site of action (Herr et al., 1992). Moreover, monkeys exposed to carbon disulfide exhibited severe and permanent reductions in visual acuity and contrast sensitivity (Merigan et al., 1988), consistent with clinical reports of decreased visual sensitivity and impaired color vision (Gordy and Trumper, 1940; Savic, 1967; Raitta et al., 1981; Iregren et al., 2002; Gobba and Cavalleri, 2003).

Chlorinated unsaturated organic solvents: TCE and PERC

Human studies

In the USA, approximately 400 000 and 700 000 workers have been exposed to TCE and PERC, respectively (ATSDR, 1997a, b). EPA's Toxic Release Inventory database revealed that TCE and PERC release levels in the USA were ~4 and ~2.5 million lb in 2012. (http://www2.epa.gov/toxics-release-inventory-tri-program/2012-tri-national-analysis). Occupational exposure to these chemicals is a potential worldwide health hazard. Potential TCE exposure occurs for dry cleaners, mechanics, oil processors, printers, resin workers, rubber cementers, shoe makers, textile and fabric cleaners, tobacco denicotinizers, and varnish workers. Potential PERC exposure occurs for dry cleaners, degreasing and metal cleaners, plastic extruders, electronic assemblers, and product manufacturers.

Animal studies

Bale et al. (2011) review and compare the neurologic, neurochemical, pathologic, and visual system alterations in animals exposed to TCE or PERC as well as their effects on ion channels and receptors from *in vitro* cellular systems. They hypothesized that, since the central visual system changes in TCE- and PERC-exposed animals were relatively similar, they may share one of two common mechanisms of action: increased glutamate release and activation of *N*-methyl-D-aspartate receptors, and/or direct inhibition of cholinergic nicotinic receptors (Bale et al., 2011).

Trichloroethylene

Human studies

Short-term exposure to high concentrations of TCE is associated with cranial nerve palsies: especially of the trigeminal (V) and, to a lesser extent, the optic (II), facial (VII), and auditory (VIII) nerves (Feldman, 1979). Damage to the optic nerve may result in loss of vision due to optic-nerve atrophy or retrobulbar neuritis (Tabbacchi et al., 1966). Although the cellular mechanisms responsible for these alterations are unknown, they are consistent with demyelinating neuropathy (Feldman, 1979). Consistent with this hypothesis, workers acutely or chronically exposed to TCE exhibit altered trigeminal evoked potentials and VEPs (Barret et al., 1982; Winneke, 1982). Clinical observations suggest that chronic TCE exposure produces effects similar to cerebellar dysfunction (Feldman, 1979). This is consistent with experimental findings in human volunteers showing vestibulo-oculomotor reflex (VOR), but not flicker fusion or form perception, abnormalities during and following 1–2-hour exposures to TCE above the threshold limit values (Vernon and Ferguson, 1969; Larsby et al., 1986).

Animal studies

Consistent with the human clinical and experimental findings, animals exposed to TCE also exhibited vestibulo-oculomotor alterations. Rabbits and rats exhibited increased nystagmus, either static (positional) or that induced by accelerated rotation, after acute exposure to moderate TCE levels above the threshold limit value (Tham et al., 1979, 1984; Niklasson et al., 1990). Pharmacologic experiments in rats suggest that TCE enhanced the VOR by activating γ-aminobutyric acid (GABA) B receptors in the GABAergic cerebellar VOR pathway (Niklasson et al., 1994). Four hours following the injection of TCE, the ERG b-waves, but not a-waves, of light-adapted rabbits were decreased (Blain et al., 1990). Experimental flash VEP studies following acute or subchronic exposure to TCE were equivocal (Rebert et al., 1991; Blain et al., 1992). Light-adapted rabbits chronically exposed to 350 or 700 ppm TCE via inhalation for 12 weeks exhibited decreased and increased N1 and P3 flash VEP amplitudes, respectively, which recovered 6 weeks following exposure (Blain et al., 1992). Light-adapted rats chronically exposed to 1600 or 3200 ppm TCE via inhalation for 12 weeks exhibited no flash VEP or somatosensory EP changes, but instead exhibited a high-frequency hearing loss, as assessed by the brainstem auditory evoked potential (Rebert et al., 1991). In contrast, acute TCE reduced the amplitude of the pattern reversal VEP second harmonic component, suggesting a central visual pathway/system effect (Boyes et al., 2003, 2005).

Tetrachloroethylene

Human studies

Long-term low-level occupational and environmental PERC exposure produces adverse retinal and central visual system/pathway effects. Dry cleaners exposed to low-level PERC for 8–10 years had blue–yellow deficits (Cavalleri et al., 1994). A 2-year follow-up study showed that these deficits increased in workers who were exposed to higher concentrations of PERC for 2 additional years (1.7–4.3 ppm), yet remained the same when the concentration decreased from 2.9 to 0.7 ppm during the same time period (Cavalleri et al., 1994; Gobba et al., 1998). The blue–yellow color vision deficits are similar to those from long-term occupational and environmental exposure to mixed organic solvents, although workers exposed to mixtures often have red–green deficits too (Muttray et al., 1997; Päällysaho et al., 2007; Attarchi et al., 2010; Lacerda et al., 2012).

Decreased spatial contrast sensitivity across all frequencies, decreased visual memory, and increased simple reaction time were observed in long-term residents (6–11 years) of apartment buildings with co-located dry cleaners and neighborhoods containing dry cleaners (Altmann et al., 1995; Schreiber et al., 2002). No changes in visual acuity or flash VEP latencies were seen (Altmann et al., 1995; Schreiber et al., 2002). A more recent study found decreased contrast sensitivity at intermediate spatial frequencies in adult and child residents in apartment buildings with or without dry cleaners using PERC (Storm et al., 2011). The pathophysiologic site and mechanism underlying spatial contrast sensitivity deficits are not well understood. Results from these studies suggest that the PERC-induced alterations might have occurred in the retinal ganglion cells and/or the central visual pathways/system. A detailed meta-analysis of the Altmann et al. (1995) and Schreiber et al. (2002) studies has been reported (Benignus et al., 2009). Interestingly, workers with mixed solvent exposure have reduced spatial contrast sensitivity at all frequencies (Näsänen et al., 2005; Costa et al., 2012) or at intermediate frequencies (Frenette et al., 1991; Mergler et al., 1991; Donoghue et al., 1995; Gong et al., 2003).

A recent population-based retrospective cohort study found an association between prenatal and early childhood PERC-contaminated drinking water and blue–yellow color and spatial contrast sensitivity deficits at high spatial frequencies in 30-year-old developmentally exposed adults without retinal disease (Getz et al., 2012). No effects on visual acuity were observed (Getz et al., 2012). These findings reveal persistent adverse subclinical effects on the retina three decades after PERC exposure. Follow-up structural magnetic resonance imaging studies found no alterations in white- or gray-matter volumes or white-matter hypointensities (Janulewicz et al., 2013), suggesting that the observed color vision and contrast sensitivity alterations are due to retinal changes (i.e., photoreceptors and parvocellular or P-type retinal ganglion cells). At first glance, the spatial contrast sensitivity and color vision deficits appear similar to those from children prenatally exposed to TCE (Till et al., 2005). However, the children with prenatal TCE exposure had reductions in contrast sensitivity in the low- and intermediate-frequency range and abnormal red–green, but not blue–yellow, vision (Till et al., 2005).

Human subjects exposed to PERC (4 hours for 4 days) had decreased contrast sensitivity at low and intermediate spatial frequencies and increased flash VEP latencies (Altmann et al., 1995). Acute PERC inhalation exposure (250–4000 ppm for 2 hours) produced a dose-dependent reduction in steady-state pattern VEP second harmonic (F2 component) amplitudes in rats (Boyes et al., 2009). Subchronic PERC inhalation exposure (800 ppm, 6 hours/day for 13 weeks) decreased the amplitude of the long-latency N3 component of the flash VEP (Mattsson et al., 1998).

Styrene

Human studies

Styrene is used in the manufacture of glass-fiber reinforced plastics and resins. Styrene has been found in at least 251 of 1699 National Priorities List sites identified by the Environmental Protection Agency (ATSDR, 2012). Numerous studies report that healthy workers exposed to mean atmospheric concentrations of styrene ranging from 4 to 70 ppm for 5–17 years exhibited concentration-, but not exposure duration-dependent alterations in color vision, as evidenced by an increased CCI (Gobba et al., 1991; Fallas et al., 1992; Chia et al., 1994; Campagna et al., 1995; Eguchi et al., 1995; Mergler et al., 1996; Kishi et al., 2001; Iregren et al., 2002), although two of the studies did not have no-exposure control groups (Campagna et al., 1995; Mergler et al., 1996). Almost all of the color vision deficits were type III blue–yellow. These CCI findings, by different groups of investigators in different countries, argue convincingly for the reproducibility and validity of these styrene-induced CCI deficits and use of the Lanthony D-15d as a sensitive and reliable test for detecting color vision deficits. One study reports no effect of styrene exposure for 6 years (40 ppm) or 15 years (27 ppm) on CCI or spatial contrast sensitivity (Seeber et al., 2009). These authors note two potential problems with this study: there was no control group and the CCI had a large variance that was due to language translation.

Data analysis from different studies suggests that the threshold styrene concentration for the elevated CCI is $\leq$4 ppm (Campagna et al., 1996) or 10–20 ppm (Kishi et al., 2001). This is at or below the threshold limit value time-weighted average value for any country (range 20–50 ppm). The urinary metabolite of styrene, mandelic acid, can be used as a sensitive and reliable biomarker of exposure (Eguchi et al., 1995; Kishi et al., 2001; Iregren et al., 2002). A meta-analysis of the CCI data was reported (Benignus et al., 2005). The reversibility of these CCI impairments is unknown; however, in one study no recovery was seen following a month of no exposure (Gobba et al., 1991). Overall, the above findings reveal the sensitivity and vulnerability of the retina to styrene exposure.

Animal studies

Although the site and mechanism of action are unknown, an experimental study using retinas isolated from rats exposed to 300 ppm styrene 6 hours/day for 12 weeks examined the structure and function of dopaminergic amacrine cells (Vettori et al., 2000). They found a 30% loss of tyrosine hydroxylase immunoreactive amacrine cells, a decrease in retinal dopamine and 3,4-dihydroxyphenylacetic acid (main dopamine metabolite) content, and a decrease in reduced glutathione. Finally, and similar to results following TCE and toluene exposure, rats exposed to styrene exhibited increased nystagmus (Tham et al., 1984).

Toluene

Human studies

Toluene is widely used as a solvent, degreaser, constituent of products such as paints and glues, and component of gasoline and other fuels. Acute toluene exposure, even at high levels, to print shop workers had no effect on color vision (Muttray et al., 1995, 1999). In contrast, prolonged toluene exposure (15–20 years) increased the CCI, as assessed by the Lanthony D-15, and produced concentration-dependent blue–yellow type II deficits and blue–yellow and red–green type III deficits in different groups of print shop workers (Zavalic et al., 1998; Zavalić et al., 1998; Campagna et al., 2001) and rubber plant workers (Cavalleri et al., 2000).

Animal studies

In rats, acute toluene exposure enhanced the VOR function, which was blocked by a $GABA_B$ receptor antagonist (Niklasson et al., 1995). Acute or chronic exposure to high concentrations of toluene slowed and reduced the amplitude of different flash- or pattern-elicited VEP components (Rebert et al., 1989; Mattsson et al., 1990; Boyes et al., 2007).

Solvent mixtures

Human studies

The neurotoxicity of organic solvents is well established. Loss of color vision (Iregren et al., 2002; Gobba and Cavalleri, 2003; Lomax et al., 2004; Parmei et al., 2004; Päällysaho et al., 2007; Attarchi et al., 2010; Costa et al., 2012; Lacerda et al., 2012) and decreased spatial contrast sensitivity (Frenette et al., 1991; Mergler et al., 1991; Donoghue et al., 1995; Gong et al., 2003; Näsänen et al., 2005; Costa et al., 2012; Lacerda et al., 2012) have been reported in workers exposed to organic solvent mixtures at gas stations, microelectronic plants, print shops, paint-manufacturing facilities, and as painters. Their occupational exposure varied from 1 to 38 years. However, quantitative dosimetry of the mixtures causing adverse health effects on the retinal and visual system has not been conducted. Most, but not all, studies conducted comprehensive ophthalmologic exams prior to testing. These studies excluded subjects who did not have normal or corrected visual acuity, clear ocular media, normal pupil responses, normal central visual fields, and in one case normal spatial contrast sensitivity. Almost all studies, regardless of the ocular exclusions, found increased CCIs (Iregren et al., 2002; Gobba and Cavalleri, 2003; Lomax et al., 2004; Päällysaho et al., 2007; Attarchi et al., 2010; Costa et al., 2012; Lacerda et al., 2012). The color vision losses were blue–yellow or blue–yellow and red–green deficits, which often did not correlate with amount of exposure, duration of exposure, or age of subject. However, Costa et al. (2012) found a significant correlation between working years and loss of color vision. Furthermore, Mergler et al. (1987) documented a concentration-dependent occurrence of blue–yellow color vision deficits in workers at a paint-manufacturing plant, which was accompanied by a red–green deficit at higher exposure. Moreover, 35% of the subjects had a diminished macular light reflex (Mergler et al., 1987), reminiscent of the macular changes and decreased foveal reflex seen in *n*-hexane-exposed workers (Raitta et al., 1978) and 2,5-hexanedione-exposed rats with rod and cone photoreceptor degeneration (Backstrom and Collins, 1992; Backstrom et al., 1993).

Many of the occupationally exposed workers also exhibited decreased spatial contrast sensitivity. Most contrast sensitivity function decreases occurred at intermediate spatial frequencies (Frenette et al., 1991; Mergler et al., 1991; Donoghue et al., 1995; Gong et al., 2003); however, other studies found decreases at high spatial frequency (Lacerda et al., 2012) or all spatial frequencies (Näsänen et al., 2005; Costa et al., 2012). Lacerda et al. (2012) also reported visual field losses at 10–30° eccentricity and no change in temporal contrast sensitivity.

The pathophysiologic site and mechanism underlying spatial contrast sensitivity deficits are not well understood. Decreases in contrast sensitivity and visual field loss are seen in patients with glaucoma (Wilensky and Hawkins, 2001) and optic neuritis (Trobe et al., 1996), suggesting that solvent exposure could be a risk factor for these retinal disorders. Thus, results from the contrast sensitivity studies with mixed solvent exposure, like those with PERC exposure, suggest that organic solvents target retinal ganglion cells and/or the central visual pathways/system.

Using transient and sweep VEP techniques, the effect of prenatal occupational exposure to mixed organic solvents was assessed in children (Till et al., 2005). Transient VEP testing showed that over 25% of the 21 infants from solvent-exposed mothers exhibited abnormal red–green, but not yellow–blue, color vision compared to none of the age-matched controls. Sweep VEPs revealed a decrease in spatial contrast sensitivity at low and intermediate frequencies. Moreover, children in the high-exposed solvent group compared to controls and the low-exposed group exhibited decreased grating visual acuity. These results (Till et al., 2005) differ substantially from those in children from mothers exposed to PERC during pregnancy (Getz et al., 2012).

In summary, these studies differed in the amount and duration of exposures, exposure history, age, and sex of subject. Despite these differences, the overall evidence indicates that organic solvents produce color vision and spatial contrast sensitivity deficits in occupationally exposed workers and their offspring. This suggests that simple, but reliable, tools such as the Lanthony D-15d color vision test, Humphrey visual field analyzer, and Pelli–Robson spatial contrast sensitivity chart (Trobe et al., 1996) combined with a thorough ophthalmologic exam should be employed as visual system screening tools for occupationally exposed workers.

Inorganic lead

Human studies

Inorganic lead is probably the oldest and best-studied environmental neurotoxicant. Occupational lead exposure is still a worldwide problem that can affect auto repairers, battery plant workers, bridge reconstruction workers, lead product casting workers (e.g., pipes, plates, sinkers), e-waste and scrap metal recyclers, firing-range instructors and cleaners, foundry workers, jewelry and stained-glass makers, painters, plumbers, pottery glazers, and home remodelers and refinishers (New York State Department of Health, 2009). Children of these lead-exposed workers are at increased risk for paraoccupational lead exposure (Baker et al., 1977; MMWR, 1989; Roscoe et al., 1999) and visual system disorders such as strabismus (Hakim et al., 1991) as lead accumulates in the eye (Eichenbaum and Zheng, 2000; Erie et al., 2005).

Frank lead poisoning (mean blood lead concentrations [BPb] $\geq$80 μg/dL) and moderate ([BPb] 40–80 μg/dL) produce visual symptoms as well as retinal and visual system pathology (Otto and Fox, 1993; Fox, 1998; Fox and Boyes, 2013). As discussed below, clinical, behavior, and electrophysiologic studies report optic neuritis or atrophy, peripheral and central scotomas, oculomotor deficits, and scotopic vision deficits in occupationally exposed factory workers. Retrobulbar optic neuritis and optic-nerve atrophy occurred in workers either chronically exposed to moderate-level lead (>5 years) or acute high-level lead (Baghdassarian, 1968; Baloh et al., 1979; Karai et al., 1982). These cases usually presented with peripheral or paracentral scotomas, and the earliest observable scotomas were most frequently detected under scotopic or mesopic viewing conditions.

These ophthalmologic findings correlate with the ERG data observed in similarly exposed lead workers and animals. In workers with mean [BPb] of 35–47 μg/dL, consistent decreases in critical flicker fusion threshold (i.e., temporal resolution) were detected under scotopic conditions or when green lights were used, but not under photopic conditions or when red lights were used (Cavelleri et al., 1982; Betta et al., 1983; Signorino et al., 1983; Campara et al., 1984). Furthermore, lead workers with mean [BPb] of $\geq$49 μg/dL exhibited decreases in the critical flicker fusion threshold when tested under mesopic to photopic conditions (Jeyaratnam et al., 1986; Williamson and Teo, 1986). In occupationally lead-exposed workers, the sensitivity and amplitude of the dark-adapted scotopic ERG a-wave and/or b-wave were decreased (Scholl and Zrenner, 2000). These results are consistent with *in vitro* extracellular and intracellular electrophysiologic studies showing that nM to μM Pb^{2+} selectively depressed the amplitude and absolute sensitivity of the rod, but not cone, photoreceptor potential (Fox and Sillman, 1979; Tessier-Lavigne et al., 1985). Consistent with clinical observations of optic nerve damage, lead-exposed workers (mean [BPb] ~43 μg/dL) had increased latency of the N145 component for pattern reversal VEPs that positively correlated with duration (1–29 years) of exposure (Hirata and Kosaka, 1993). Similarly, increased latency of the P300 wave of checkerboard pattern-stimulated event-related potentials was observed in chronically lead-exposed workers (Hirata et al., 2004).

A recent quantitative retinal structural assessment compared three groups of young (mean 27–32 years) lead-exposed battery workers (exposed ~4 years) with mean [BPb] of 16.9, 29.3, 46.2 μg/dL lead to an age-matched control group of office workers with a mean [BPb] of 2.85 μg/dL lead (Ekinci et al., 2014). All subjects had full ophthalmologic exams and were corrected for best visual acuity. SD-OCT, a noninvasive imaging technology, was used to determine retinal nerve fiber layer thickness, macular thickness, and choroidal thickness. Retinal nerve fiber layer thickness exhibited concentration-dependent decreases compared to controls, although the two lower groups were not different from each other. Macular thickness and choroidal thickness exhibited concentration-dependent decreases for each lead-exposed group compared to controls.

Oculomotor deficits also occur in chronically lead-exposed workers who have no observable ophthalmologic abnormalities. The mean accuracy of saccadic eye movements was lower, saccade maximum velocity was decreased, and number of overshoots were increased in lead-exposed workers (Baloh et al., 1979; Spivey et al., 1980; Specchio et al., 1981; Glickman et al., 1984). Moreover, abnormal smooth-pursuit eye movements were observed (Specchio et al., 1981). Although the site and mechanism of action underlying these alterations are unknown, they likely result from centrally mediated alterations. In summary, these results show that occupational lead exposure produces concentration- and time-dependent alterations in the retina and central visual pathway/system. Lower lead exposure primarily affects rod photoreceptors, whereas higher lead exposure adversely affects rod and cone photoreceptors, the optic nerve, and central visual system.

Using pattern reversal VEP techniques, conducted under mesopic-to-photopic luminance levels, the effect of prenatal environmental exposure to lead was assessed in 10-year-old children (Ethier et al., 2012). Cord blood lead from 0.83 to 19.48 μg/dL was associated with a concentration-dependent delay in the latency of the N150 wave at low and high contrast levels. The authors suggest that the alterations in their prospective study were due to visual pathway and/or visual cortical alterations (Ethier et al., 2012). An earlier prospective epidemiologic ERG study in 7–10-year-old children found that gestational lead exposure was associated with increased amplitude of the rod and bipolar cell-mediated scotopic ERG a- and b-waves as well as increased b-wave sensitivity, termed ERG supernormality (Rothenberg et al., 2002). This novel scotopic ERG supernormality also was observed in a 10-year-old patient with early developmental lead exposure (Nagpal and Brodie, 2009). A rat animal model, using a human equivalent gestational lead exposure model, confirmed the ERG supernormality (Fox et al., 2008). Moreover, rat and mouse studies with the same model revealed a lead-induced selective increase in the number of rod photoreceptors and bipolar cells that likely underlies these scotopic ERG alterations (Fox et al., 2008; Giddabasappa et al., 2011).

Animal studies

Consistent with these clinical findings, rod-selective deficits in sensitivity and temporal resolution were observed in both *in vivo* and *ex vivo* animal studies. Lead exposure to adult animals produced retinal damage and functional deficits. The degree and extent of these alterations depended upon the dose and duration of lead exposure. High-level lead exposure to adult rabbits for 60–300 days (Hass et al., 1964; Brown, 1974; Hughes and Coogan, 1974) resulted in focal necrosis of the rod photoreceptor outer and inner segments, and inner nuclear layer (Hass et al., 1964; Brown, 1974; Hughes and Coogan, 1974). Six weeks of moderate-level lead exposure to adult rats produced rod- and bipolar cell-selective apoptotic cell death (Fox et al., 1997). Consistent with these findings, brief (15 minutes) exposure of isolated adult rat retinas to nM to μM Pb^{2+} produced rod-selective apoptosis (He et al., 2000). This resulted from a direct Pb^{2+}-induced inhibition of rod outer-segment cGMP-phosphodiesterase, a resultant elevation of rod [cGMP] that gates the nonselective cation channel, and an increased mitochondrial [Ca^{2+}] and [Pb^{2+}], which initiated Bax translocation to the mitochondria and subsequently activation of the cyctochrome c-caspase cascade of apoptosis (Srivastava et al., 1995; Fox et al., 1997; He et al., 2003).

Methylmercury

Human studies

The retina and gray, but not white, matter of the cerebral and visual cortex are major targets of methylmercury exposure (Merigan et al., 1983; Iregren et al., 2002; Gobba and Cavalleri, 2003; Fox and Boyes, 2013). Methylmercury exposure mainly occurs by eating fish contaminated with mercury from coal-fired power stations, cement-manufacturing plants, incineration of medical and municipal waste, amalgam burning during gold-mining operations, and residential heating systems (WHO, 2003; UNEP, 2008). Methylmercury produces immediate and persistent scotopic vision deficits, type II and III color vision deficits, visual field constriction, and reduced contrast sensitivity in fish-eating Amazonian populations with hair levels of mercury below the clinical threshold of 50 μg/g (Lebel et al., 1996; Rodrigues et al., 2007; Fillion et al., 2011, 2013; Fox and Boyes, 2013) and in nonhuman primates (Berlin et al., 1975; Evans et al., 1975; Merigan et al., 1983). Methylmercury, in contrast to inorganic mercury, accumulates in the brain (i.e., cerebral and visual cortex, cerebellar Purkinje cells, and basal nuclei), outer and inner retinal layers, RPE, and retinal capillaries, where it is retained for over 10 years (Berlin et al., 1975; DuVal et al., 1987; WHO, 2003; Mela et al., 2010, 2012; Fox and Boyes, 2013).

Animal studies

Monkeys and dogs exposed to methylmercury showed greater damage in the calcarine fissure/sulcus (anterior and posterior parts are location of peripheral and central visual fields, respectively), associated with higher regional concentrations of protein-bound mercury, than in other brain regions (Yoshino et al., 1966; Berlin et al., 1975). Using pattern reversal VEP techniques, Ethier et al. (2012) found that cord blood mercury from 1.9 to 92.4 μg/dL was associated with a concentration-dependent decrease in N75 amplitude at high contrast and delay in the N75 latency at medium contrast in 10-year-old children. They suggested an alteration in the parvocellular pathway, consistent with mercury-induced alterations in visual acuity and color vision (Lebel et al., 1996; Cavalleri and Gobba, 1998; Ventura et al., 2004; Fillion et al., 2011).

Acute exposure of isolated frog or axolotl retinas to mercury produces selective decreases in the rod light response (Fox and Sillman, 1979; Tessier-Lavigne et al., 1985). Rats injected with methylmercury for 7 days had disrupted photoreceptors and altered RPE, which only partially recovered following exposure (Braekevelt, 1982). A single injection of methylmercury dose- and time-dependently decreased the fish horizontal cell responses (Tanan et al., 2006), number of protein kinase C-immunoreactive ON-bipolar cells and number of parvalbumin-immunoreactive amacrine cells (Bonci et al., 2006). Consistent with the above clinical and experimental findings, subacute dosing with methylmercury (30 days) produced dose- and time-dependent decreases in the photopic and then scotopic ERG of rats (Goto et al., 2001). Thus, methylmercury is both a direct retinal as well as visual cortical structural and functional neurotoxicant.

Mercury vapor

Human studies

Elemental or metallic mercury (Hg^0) is used worldwide by manufacturers of barometers, batteries, electric switches and relays, fluorescent and mercury lamps, solar cells, solders and thermometers; for electroplating and chloroalkali production; and as an amalgam by dentists. Mercury vapor exposure also occurs with e-waste and scrap metal recyclers. Occupational and environmental exposure to elemental mercury, occurs through inhaling mercury vapors. The average atmospheric mercury level is currently three to six times higher than that estimated for preindustrial ambient air (WHO, 2003; UNEP, 2008). The American Conference of Governmental Industrial Hygienists (ACGIH) recommends a threshold limit of $\leq$35 μg of urinary Hg/g creatinine as the safety level (ACGIH, 2001).

Similar to methylmercury, the retina and visual cortex are major target organs of mercury vapor exposure. To date, however, there are no comprehensive reviews of this subject. Comprehensive clinical, electrophysiologic, and psychophysical testing reveals that mercury vapor produces concentration- and duration-dependent color vision deficits, decreases in rod- and cone-mediated ERGs, central and peripheral visual field deficits, and reduced contrast sensitivity in occupationally exposed industry workers (Cavalleri et al., 1995; Cavalleri and Gobba, 1998; Urban et al., 2003; Ventura et al., 2004, 2005; Feitosa-Santana et al., 2007, 2008, 2010; Barboni et al., 2008, 2009; Costa et al., 2008) and dentists (Canto-Pereira et al., 2005). These color and visual system alterations occur in chronically exposed workers (>1.5 years) with 20–22 μg of urinary Hg/g creatinine and no neuropsychologic dysfunction (Urban et al., 2003; Barboni et al., 2009) as well as long-time dentists with 2 μg of urinary Hg/g creatinine (Canto-Pereira et al., 2005). These adverse functional changes occur below the ACGIH safe limit levels.

The deficits are irreversible if the exposure is greater than 1 year (Cavalleri and Gobba, 1998), as they persist or worsen even 6–7.5 years following removal from the exposure resulting in patients with <3 μg of urinary Hg/g creatinine (Ventura et al., 2005; Feitosa-Santana et al., 2007, 2008, 2010; Barboni et al., 2008; Costa et al., 2008). The color vision alterations are primarily type III tritan deficits, although longer exposures produce alterations in the red–green axis (Ventura et al., 2004, 2005; Feitosa-Santana et al., 2007, 2008, 2010). Contrast sensitivity is reduced at all spatial frequencies, but especially at low and intermediate frequencies (Ventura et al., 2005; Costa et al., 2008; Barboni et al., 2009). The visual field deficits are initially mid-peripheral and peripheral and eventually central (Ventura et al., 2004; Barboni et al., 2008, 2009).

Taken together, these studies reveal several important aspects about the retinal and visual toxicology of mercury vapor exposure. They are generally dependent on the duration and concentration of exposure, irreversible and occur below the current ACGIH safe level for occupational workers. The findings suggest direct retinal and visual cortical sites of action as well as the parvocellular koniocellular and magnocellular visual system pathways. Consistent with this idea, following exposure to mercury vapor, an accumulation of mercury is found in the brain, retina, optic disk, RPE, and retinal capillaries of rats and monkeys (Khayat and Dencker, 1984; Warfvinge and Bruun, 1996).

TOXICANTS AS RISK FACTORS FOR LATE-ONSET RETINAL DISEASES

The World Health Organization estimates that three billion workers are exposed to hazardous risks such as chemicals, noise, particulates, and stressors at their workplaces. These risks can produce a wide variety of adverse health outcomes (WHO, 2004). The interaction of chemical/toxicant exposure with underlying genetics is an area of increased concern and study by the World Health Organization, National Institutes of Health, and investigators (WHO, 2004; NIEHS, 2012), although relatively little data exist. Several recent studies suggest that occupational and environmental exposure to neurotoxicants may increase the risk for a number of retinal diseases. Three studies ($n=4$ subjects) suggest that occupational exposure to organic solvents could trigger Leber's hereditary optic neuropathy: a maternally inherited loss of central vision due to one of three mitochondrial DNA (mtDNA) point mutations (Johns et al., 1992; Rufa et al., 2005; Carelli et al., 2007). Two subjects had the 3460 mtDNA mutation (Johns et al., 1992) and the other two had 11 778 mtDNA mutations (Rufa et al., 2005; Carelli et al., 2007). Increased lead accumulation, as measured by hair lead levels, was positively associated with increased primary open-angle glaucoma in 58 females who were $\sim$60 years old, but not in 40 males who were $\sim$53 years old (Yuki et al., 2009). Moreover, this association was significant in females with low-, but not high-tension glaucoma (Yuki et al., 2009).

A recent study revealed the association between the incidence of early age-related macular degeneration and elevated nitrate-nitrogen in the drinking water in a rural population due to herbicides such as atrazine (Klein et al., 2013). Although this is a multifactorial disease, *N*-nitrosoamine/*N*-nitrosoasmide-induced oxidative stress and inflammation were suggested as possible contributors (Klein et al., 2013). Elevated neural retinal lead is positively associated with an increased incidence of age-related macular degeneration (AMD) in 50 donor eyes compared to 72 control eyes (Erie et al., 2009a). Reduced levels of RPE copper and zinc are found in donor retinas with defined AMD relative to controls (Erie et al., 2009b). Lead accumulates in the retina and RPE of nonoccupationally affected humans (Eichenbaum and Zheng, 2000; Erie et al., 2005), avidly binds to melanin like copper and zinc (Ulshafer et al., 1990), and can displace copper and zinc from several proteins (Kirberger and Yang, 2008). Thus, lead may increase the risk of AMD via divalent cation displacement. Alternatively, lead-induced increases in oxidative stress and/or inflammatory cytokine production (Stohs and Bagchi, 1995) are potential mechanisms contributing to AMD. In summary, these studies reveal the importance of exploring and uncovering the role of toxicant–disease (gene) interactions in late-onset retinal degenerative disorders.

CONCLUSIONS

Retinal and visual system toxicology are fields of immense importance and relevance to the occupationally exposed workers and their paraoccupationally exposed family members. The latter is especially important in vulnerable populations like the developing fetus, young children, and aged adults as delayed and/or late-onset neurotoxic effects to low levels of environmental and occupational toxicants are increasingly documented (Williams and Ross, 2007; Cohen Hubal et al., 2008; Grandjean and Landrigan, 2014). These complex visual system tissues have unique structural and functional biology/biochemistry make them exquisitely sensitive and vulnerable to neurotoxicant insult, are early indicators of broader and impending central nervous system toxicity, and are often ignored during worker monitoring. However, subtle alterations in retinal and/or visual processing as well as circadian rhythm can have profound immediate, long-term, and maybe even delayed effects on the mental, social, and physical health and performance of an individual. In addition, such alterations can markedly affect the quality of life. Noninvasive behavioral, clinical, electrophysiologic, and imaging tools such as the Lanthony D-15d color test, automated perimetry, scotopic and photopic ERGs, pattern reversal ERGs and VEPs, and SD-OCT are needed to routinely screen and monitor toxicant-exposed workers for retinal and visual system alterations. Most of these behavioral and clinical tests are available in relatively inexpensive, easily administrable and inexpensive formats. If a visual alteration or deficit is observed with these initial tests, more sophisticated and expensive electrophysiologic and imaging techniques can be employed to determine the sites of action, degree of injury, as well as potential recovery during follow-up exams. Several important and standardizable methodologic variables such as lighting sources, test–retest protocols, appropriate statistical analysis, and use of proper and valid criteria need to be followed to reliably and validly identify these deficits.

Experimental studies investigating dose–, duration–, and time–response relationships as well as determining the cellular and molecular mechanisms underlying neurotoxicant-induced retinal and visual system disorders are nascent. The experimental data cited above indicate that *in vivo* and *ex vivo* animal models can faithfully and reliably predict and reproduce most of the structural, functional, and/or behavioral deficits in the retina and visual system produced by neurotoxicant chemicals. Reliable and informative alternative *in vitro* test systems are not yet available to assess neurotoxicant-induced structural and functional alterations in the adult retina and/or visual system. Such assays, although not likely high-throughput, would advance worker safety and retinal/visual system risk assessment.

References

ACGIH (2001). Threshold Limit Value for Chemical Substances and Biological Exposure Indices. American Conference of Governmental Industrial Hygienists, Cincinnati.

Altmann L, Neuhann HF, Kramer U et al. (1995). Neurobehavioral and neurophysiological outcomes of chronic low-level tetrachloroethene exposure measured in neighborhoods of dry cleaning shops. Environ Res 69: 83–89.

American Clinical Neurophysiology Society (2008). Guideline 9B: Guidelines on visual evoked potentials. Available online at: http://www.acns.org/pdf/guidelines/Guideline-9B.pdf.

Anger WK, Johnson BL (1985). Chemicals affecting behavior. In: JL O'Donoghue (Ed.), Neurotoxicity of Industrial and Commercial Chemicals. CRC Press, Boca Raton, FL, pp. 51–148.

ATSDR (1996a). Toxicological profile for carbon disulphide. Agency for Toxic Substances and Disease Registry. US Department of Health and Human Services, Atlanta, GA.

ATSDR (1996b). Public health statement: carbon disulfide. Available online at: http://www.atsdr.cdc.gov/ToxProfiles/tp82-c1-b.pdf.

ATSDR (1997a). Toxicological profile for tetrachloroethylene. Agency for Toxic Substances and Disease Registry. US Department of Health and Human Services, Atlanta, GA.

ATSDR (1997b). Toxicological profile for trichloroethylene. Agency for Toxic Substances and Disease Registry. US Department of Health and Human Services, Atlanta, GA.

ATSDR (2012). Styrene – ToxFAQs. Available online at: http://www.atsdr.cdc.gov/toxfaqs/tfacts53.pdf.

Attarchi MS, Labbafinejad Y, Mohammadi S (2010). Occupational exposure to different levels of mixed organic solvents and colour vision impairment. Neurotoxicol Teratol 32: 558–562.

Ayton LN, Apollo NV, Varsamidis M et al. (2014). Assessing residual visual function in severe vision loss. Invest Ophthalmol Vis Sci 55: 1332–1338.

Backstrom B, Collins VP (1992). The effects of 2,5-hexanedione on rods and cones of the retina of albino rats. Neurotoxicology 13: 199–202.

Backstrom B, Nylen P, Hagman M et al. (1993). Effect of exposure to 2,5-hexanediol in light or darkness on the retina of albino and pigmented rats. I. Morphology. Arch Toxicol 67: 277–283.

Baghdassarian SA (1968). Optic neuropathy due to lead poisoning. Arch Ophthalmol 80: 721–723.

Baker EL, Folland DS, Taylor TA et al. (1977). Lead poisoning in children of lead workers: home contamination with industrial dust. N Engl J Med 296: 260–261.

Baker EL, Feldman RG, White RA et al. (1984). Occupational lead neurotoxicity: a behavioral and electrophysiological evaluation. Br J Ind Med 41: 352–361.

Bale AS, Barone Jr S, Scott CS et al. (2011). A review of potential neurotoxic mechanisms among three chlorinated organic solvents. Toxicol Appl Pharmacol 255: 113–126.

Baloh RW, Spivey GH, Brown CP et al. (1979). Subclinical effects of chronic increased lead absorption – A prospective study. II. Results of baseline neurologic testing. J Occup Med 21: 490–496.

Barboni MT, da Costa MF, Moura AL et al. (2008). Visual field losses in workers exposed to mercury vapor. Environ Res 107: 124–131.

Barboni MT, Feitosa-Santana C, Zachi EC et al. (2009). Preliminary findings on the effects of occupational exposure to mercury vapor below safety levels on visual and neuropsychological functions. J Occup Environ Med 51: 1403–1412.

Barret L, Arsac P, Vincent M et al. (1982). Evoked trigeminal potential in chronic trichloroethylene exposure. J Toxicol Clin Toxicol 19: 419–423.

Benignus VA, Geller AM, Boyes WK et al. (2005). Human neurobehavioral effects of long-term exposure to styrene: a meta-analysis. Environ Health Perspect 113: 532–538.

Benignus VA, Boyes WK, Geller AM et al. (2009). Long-term perchloroethylene exposure: a meta-analysis of neurobehavioral deficits in occupationally and residentially exposed groups. J Toxicol Environ Health A 72: 824–831.

Berlin M, Grant CA, Hellberg J et al. (1975). Neurotoxicity of methyl mercury in squirrel monkeys. Cerebral cortical pathology, interference with scotopic vision, and changes in operant behavior. Arch Environ Health 30: 340–348.

Betta A, De Santa A, Savonitto C et al. (1983). Flicker fusion test and occupational toxicology performance evaluation in workers exposed to lead and solvents. Hum Toxicol 2: 83–90.

Blain L, Lachapelle P, Molotchnikoff S (1990). The effect of acute trichloroethylene exposure on electroretinogram components. Neurotoxicol Teratol 12: 633–636.

Blain L, Lachapelle P, Molotchnikoff S (1992). Evoked potentials are modified by long-term exposure to trichloroethylene. Neurotoxicology 13: 203–206.

Bonci DM, Lima SM, Grötzner SR et al. (2006). Losses of immunoreactive parvalbumin amacrine and immunoreactive alpha protein kinase C bipolar cells caused by methylmercury chloride intoxication in the retina of the tropical fish *Hoplias malabaricus*. Braz J Med Biol Res 39: 405–410.

Boyes WK, Bercegeay M, Ali JS et al. (2003). Dose-based duration adjustments for the effects of inhaled trichloroethylene on rat visual function. Toxicol Sci 76: 121–130.

Boyes WK, Bercegeay M, Krantz T et al. (2005). Momentary brain concentration of trichloroethylene predicts the effects on rat visual function. Toxicol Sci 87: 187–196.

Boyes WK, Bercegeay M, Krantz QT et al. (2007). Acute toluene exposure and rat visual function in proportion to momentary brain concentration. Toxicol Sci 99: 572–581.

Boyes WK, Bercegeay M, Oshiro WM et al. (2009). Acute perchloroethylene exposure alters rat visual-evoked potentials in relation to brain concentrations. Toxicol Sci 108: 159–172.

Braekevelt CR (1982). Morphological changes in rat retinal photoreceptors with acute methyl mercury intoxication. In: JG Hollyfield (Ed.), The Structure of the Eye. Elsevier, New York, pp. 123–131.

Brown DVL (1974). Reaction of the rabbit retinal pigment epithelium to systemic lead poisoning. Trans Am Ophthalmol Soc 72: 404–447.

Campagna D, Gobba F, Mergler D et al. (1996). Color vision loss among styrene-exposed workers: neurotoxicological threshold assessment. Neurotoxicology 17: 367–373.

Campagna D, Mergler D, Huel G et al. (1995). Visual dysfunction among styrene-exposed workers. Scand J Work Environ Health 21: 382–390.

Campagna D, Stengel B, Mergler D et al. (2001). Color vision and occupational toluene exposure. Neurotoxicol Teratol 23: 473–480.

Campara P, D'Andrea F, Micciolo R et al. (1984). Psychological performance of workers with blood-lead concentration below the current threshold limit value. Int Arch Occup Environ Health 53: 233–246.

Canto-Pereira LH, Lago M, Costa MF et al. (2005). Visual impairment on dentists related to occupational mercury exposure. Environ Toxicol Pharmacol 19: 517–522.

Carelli V, Franceschini F, Venturi S et al. (2007). Grand rounds: could occupational exposure to n-hexane and other solvents precipitate visual failure in Leber hereditary optic neuropathy? Environ Health Perspect 115: 113–115.

Cavalleri A, Gobba F (1998). Reversible color vision loss in occupational exposure to metallic mercury. Environ Res 77: 173–177.

Cavalleri A, Trimarchi F, Gelmi C et al. (1982). Effects of lead on the visual system of occupationally exposed subjects. Scand J Work Environ Health 8 (suppl 1): 148–151.

Cavalleri A, Gobba F, Paltrinieri M et al. (1994). Perchloroethylene exposure can induce colour vision loss. Neurosci Lett 179: 162–166.

Cavalleri A, Belotti L, Gobba F et al. (1995). Colour vision loss in workers exposed to elemental mercury vapour. Toxicol Lett 77: 351–356.

Cavalleri A, Gobba F, Nicali E et al. (2000). Dose-related color vision impairment in toluene-exposed workers. Arch Environ Health 55: 399–404.

Chen SK, Badea TC, Hattar S (2012). Photoentrainment and pupillary light reflex are mediated by distinct populations of ipRGCs. Nature 476: 92–95.

Chia SE, Jeyaratnam J, Ong CN et al. (1994). Impairment of color vision among workers exposed to low concentrations of styrene. Am J Ind Med 26: 481–488.

Cohen Hubal EA, Moya J, Selevan SG (2008). A lifestage approach to assessing children's exposure. Birth Defects Res B Dev Reprod Toxicol 83: 522–529.

Costa MF, Tomaz S, de Souza JM et al. (2008). Electrophysiological evidence for impairment of contrast sensitivity in mercury vapor occupational intoxication. Environ Res 107: 132–138.

Costa TL, Barboni MT, Moura AL et al. (2012). Long-term occupational exposure to organic solvents affects color vision, contrast sensitivity and visual fields. PLoS One 7: e42961.

Crofton KM, Sheets LP (1989). Evaluation of sensory system function using reflex modification of the startle response. J Am Coll Toxicol 8: 199–211.

Damstra T (1978). Environmental chemicals and nervous system dysfunction. Yale J Biol Med 51: 457–468.

De Rouck A, De Laey JJ, Van Hoorne M et al. (1986). Chronic carbon disulfide poisoning: a 4 year follow-up study of the ophthalmological signs. Int Ophthalmol 9: 17–27.

Donoghue AM, Dryson EW, Wynn-Williams G (1995). Contrast sensitivity in organic-solvent-induced chronic toxic encephalopathy. J Occup Environ Med 37: 1357–1363.

DuVal G, Grubb BR, Bentley PJ (1987). Mercury accumulation in the eye following administration of methyl mercury. Exp Eye Res 44: 161–164.

Eguchi T, Kishi R, Harabuchi I et al. (1995). Impaired colour discrimination among workers exposed to styrene: relevance of a urinary metabolite. Occup Environ Med 52: 534–538.

Eichenbaum JW, Zheng W (2000). Distribution of lead and transthyretin in human eyes. Clin Toxicol 38: 371–381.

Ekinci M, Ceylan E, Cağatay HH et al. (2014). Occupational exposure to lead decreases macular, choroidal, and retinal nerve fiber layer thickness in industrial battery workers. Curr Eye Res 39: 853–858.

Environmental Protection Agency (2014). Integrated Risk Information System (IRIS). Available online at: http://www.epa.gov/iris/index.html.

Erie JC, Butz JA, Good JA et al. (2005). Heavy metal concentrations in human eyes. Am J Ophthalmol 139: 888–893.

Erie JC, Good JA, Butz JA (2009a). Excess lead in the neural retina in age-related macular degeneration. Am J Ophthalmol 148: 890–894.

Erie JC, Good JA, Butz JA et al. (2009b). Reduced zinc and copper in the retinal pigment epithelium and choroid in age-related macular degeneration. Am J Ophthalmol 147: 276–282.

Eskin TA, Merigan WH, Wood RW (1988). Carbon disulfide effects on the visual system. II. Retinogeniculate degeneration. Invest Ophthalmol Vis Sci 29: 519–527.

Ethier AA, Muckle G, Bastien C et al. (2012). Effects of environmental contaminant exposure on visual brain development: a prospective electrophysiological study in school-aged children. Neurotoxicology 33: 1075–1085.

Evans HL, Laties VG, Weiss B (1975). Behavioral effects of mercury and methylmercury. Fed Proc 34: 1858–1865.

Fallas C, Fallas J, Maslard P et al. (1992). Subclinical impairment of colour vision among workers exposed to styrene. Br J Ind Med 49: 679–682.

Feitosa-Santana C, Costa MF, Lago M et al. (2007). Long-term loss of color vision after exposure to mercury vapor. Braz J Med Biol Res 40: 409–414.

Feitosa-Santana C, Barboni MT, Oiwa NN et al. (2008). Irreversible color vision losses in patients with chronic mercury vapor intoxication. Vis Neurosci 25: 487–491.

Feitosa-Santana C, Bimler DL, Paramei GV et al. (2010). Color-space distortions following long-term occupational exposure to mercury vapor. Ophthalmic Physiol Opt 30: 724–730.

Feldman RG (1979). Trichloroethylene. In: PJ Vinken, GW Bruyn (Eds.), In: Handbook of Clinical Neurology, Intoxications of the Nervous System, 36. North-Holland Publishing Company, Amsterdam, pp. 457–464, Part I.

Fillion M, Philibert A, Mertens F et al. (2011). Neurotoxic sequelae of mercury exposure: an intervention and follow-up study in the Brazilian Amazon. Ecohealth 8: 210–222.

Fillion M, Lemire M, Philibert A et al. (2013). Toxic risks and nutritional benefits of traditional diet on near visual contrast sensitivity and color vision in the Brazilian Amazon. Neurotoxicology 37: 173–181.

Flammer J, Mozaffarieh M (2008). Autoregulation: A balancing act between supply and demand. Can J Ophthalmol 43: 317–321.

Fox DA (1998). Sensory system alterations following occupational exposure to chemicals. In: L Manzo, LG Costa (Eds.), Occupational Neurotoxicology. CRC Press, Boca Raton, pp. 169–184.

Fox DA, Boyes WK (2013). Toxic responses of the ocular and visual system. In: CD Klaassen, JB Watkins (Eds.), Casarett & Doull's Essentials of Toxicology, 8th Edition, McGraw-Hill Press, New York, pp. 767–798.

Fox DA, Sillman AJ (1979). Heavy metals affect rod, but not cone, photoreceptors. Science 206: 78–80.

Fox DA, Campbell ML, Blocker YS (1997). Functional alterations and apoptotic cell death in the retina following developmental or adult lead exposure. Neurotoxicology 18: 645–665.

Fox DA, Kala SV, Hamilton WR et al. (2008). Low-level human equivalent gestational lead exposure produces supernormal scotopic electroretinograms, increased retinal neurogenesis and decreased dopamine utilization in rats. Environ Health Perspect 116: 618–625.

Frenette B, Mergler D, Bowler R (1991). Contrast-sensitivity loss in a group of former microelectronics workers with normal visual acuity. Optom Vis Sci 68: 556–560.

Geller AM, Hudnell HK (1997). Critical issues in the use and analysis of the Lanthony desaturate color vision test. Neurotoxicol Teratol 19: 455–465.

Getz KD, Janulewicz PA, Rowe S et al. (2012). Prenatal and early childhood exposure to tetrachloroethylene and adult vision. Environ Health Perspect 120: 1327–1332.

Giddabasappa A, Hamilton WR, Chaney S et al. (2011). Gestational lead exposure selectively increases the proliferation and number of late-born rod photoreceptors and bipolar cells. Environ Health Perspect 119: 71–77.

Glickman L, Valciukas JA, Lilis R et al. (1984). Occupational lead exposure. Effects on saccadic eye movements. Int Arch Occup Environ Health 54: 115–125.

Gobba F, Cavalleri A (2003). Color vision impairment in workers exposed to neurotoxic chemicals. Neurotoxicology 24: 693–702.

Gobba F, Galassi C, Imbriani M et al. (1991). Acquired dyschromatopsia among styrene-exposed workers. J Occup Med 33: 761–765.

Gobba F, Righi E, Fantuzzi G et al. (1998). Two-year evolution of perchloroethylene-induced color-vision loss. Arch Environ Health 53: 196–198.

Gong Y, Kishi R, Kasai S et al. (2003). Visual dysfunction in workers exposed to a mixture of organic solvents. Neurotoxicology 24: 703–710.

Gordy ST, Trumper M (1940). Carbon disulfide poisoning. Ind Med 9: 231–235.

Goto Y, Shigematsu J, Tobimatsu S et al. (2001). Different vulnerability of rat retinal cells to methylmercury exposure. Curr Eye Res 23: 171–178.

Grandjean P, Landrigan PJ (2014). Neurobehavioural effects of developmental toxicity. Lancet Neurol 13: 330–338.

Grant WM, Schuman JS (1993). Toxicology of the Eye, 4th ed. CC Thomas, Springfield, IL.

Graymore CN (1969). General aspects of metabolism in the retina. In: H Davson (Ed.), 2nd Ed. In: The Eye. 1. Academic Press, New York, pp. 601–645.

Hakim RB, Stewart WF, Canner JK et al. (1991). Occupational lead exposure and strabismus in offspring: a case-control study. Am J Epidemiol 133: 351–356.

Hanninen H, Nurminen M, Tolonen M et al. (1978). Psychological tests as indicators of excessive exposure to carbon disulfide. Scand J Work Environ Health 19: 163–174.

Hass GM, Brown DVL, Eisenstein R et al. (1964). Relation between lead poisoning in rabbit and man. Am J Pathol 45: 691–727.

He L, Poblenz AT, Medrano CJ et al. (2000). Lead and calcium produce photoreceptor cell apoptosis by opening the mitochondrial permeability transition pore. J Biol Chem 275: 12175–12184.

He L, Perkins GA, Poblenz AT et al. (2003). Bcl-xL overexpression blocks bax-mediated mitochondrial contact site formation and apoptosis in rod photoreceptors of lead-exposed mice. Proc Natl Acad Sci U. S. A. 100: 1022–1027.

Herr DW, Boyes WK, Dyer RS (1992). Alterations in rat flash and pattern reversal evoked potentials after acute or repeated administration of carbon disulfide (CS2). Fundam Appl Toxicol 18: 328–342.

Hirata M, Kosaka H (1993). Effects of lead exposure on neurophysiological parameters. Environ Res 63: 60–69.

Hirata M, Kosaka H, Yoshida T (2004). A study on the effect of lead on event-related potentials among lead-exposed workers. Ind Health 42: 431–434.

Hosoya K, Tachikawa M (2009). Inner blood-retinal barrier transporters: role of retinal drug delivery. Pharm Res 26: 2055–2065.

Hughes WF, Coogan PS (1974). Pathology of the pigment epithelium and retina in rabbits poisoned with lead. Am J Pathol 77: 237–254.

Iregren A, Andersson M, Nylén P (2002). Color vision and occupational chemical exposures: I. An overview of tests and effects. Neurotoxicology 23: 719–733.

Janulewicz PA, Killiany RJ, White RF et al. (2013). Structural magnetic resonance imaging in an adult cohort following prenatal and early postnatal exposure to tetrachloroethylene (PCE)-contaminated drinking water. Neurotoxicol Teratol 38: 13–20.

Jeyaratnam J, Boey KW, Ong CN et al. (1986). Neuropsychological studies on lead workers in Singapore. Br J Ind Med 43: 626–629.

Jindal V (2014). Interconnection between brain and retinal neurodegeneration. Mol Neurobiol May 15. [Epub ahead of print] PMID: 24826919.

Johns DR, Smith KH, Miller NR (1992). Leber's hereditary optic neuropathy. Clinical manifestations of the 3460 mutation. Arch Ophthalmol 110: 1577–15781.

Johnson JE, Perkins GA, Giddabasappa A et al. (2007). Spatiotemporal regulation of ATP and Ca^{2+} dynamics in vertebrate rod and cone photoreceptor ribbon synapses. Mol Vis 13: 887–919.

Karai I, Horiguchi SH, Nishikawa N (1982). Optic atrophy with visual field defect in workers occupationally exposed to lead for 30 years. J Toxicol Clin Toxicol 19: 409–418.

Khayat A, Dencker L (1984). Organ and cellular distribution of inhaled metallic mercury in the rat and marmoset monkey (*Callithrix jacchus*): influence of ethyl alcohol pretreatment. Acta Pharmacol Toxicol (Copenh) 55: 145–152.

Kirberger M, Yang JJ (2008). Structural differences between Pb^{2+}- and Ca^{2+}-binding sites in proteins: implications with respect to toxicity. J Inorg Biochem 102: 1901–1909.

Kishi R, Eguchi T, Yuasa J et al. (2001). Effects of low-level occupational exposure to styrene on color vision: dose relation with a urinary metabolite. Environ Res 85: 25–30.

Klein BE, McElroy JA, Klein R et al. (2013). Nitrate-nitrogen levels in rural drinking water: is there an association with age-related macular degeneration? J Environ Sci Health A Tox Hazard Subst Environ Eng 48: 1757–1763.

Kollner H (1912). Die Storungen Des Farbensinnes, ihre Klinische Bedeutung und ihre Diagnose. S Karger, Berlin.

Lacerda EM, Lima MG, Rodrigues AR et al. (2012). Psychophysical evaluation of achromatic and chromatic vision of workers chronically exposed to organic solvents. J Environ Public Health 2012: 1–7.

Larsby B, Tham R, Eriksson B et al. (1986). Effects of trichloroethylene on the human vestibulo-oculomotor system. Acta Otolaryngol 101: 193–199.

Lebel J, Mergler D, Lucotte M et al. (1996). Evidence of early nervous system dysfunction in Amazonian populations exposed to low-levels of methylmercury. Neurotoxicology 17: 157–167.

Levin LA, Nilsson SFE, Ver Hoeve J et al. (2011). Adler's Physiology of the Eye. 11th Ed. Elsevier, New York.

Lin AL, Fox PT, Hardies J et al. (2010). Nonlinear coupling between cerebral blood flow, oxygen consumption, and ATP production in human visual cortex. Proc Natl Acad Sci U. S. A. 107: 8446–8451.

Lomax RB, Ridgway P, Meldrum M (2004). Does occupational exposure to organic solvents affect colour discrimination? Toxicol Rev 23: 91–121.

London A, Benhar I, Schwartz M (2013). The retina as a window to the brain-from eye research to CNS disorders. Nat Rev Neurol 9: 44–53.

Mattsson JL, Gorzinski SJ, Albee RR et al. (1990). Evoked potential changes from 13 weeks of simulated toluene abuse in rats. Pharmacol Biochem Behav 36: 683–689.

Mattsson JL, Albee RR, Yano BL et al. (1998). Neurotoxicologic examination of rats exposed to 1,1,2,2-tetrachloroethylene (perchloroethylene) vapor for 13 weeks. Neurotoxicol Teratol 20: 83–98.

McDonald RW (1938). Carbon disulfide poisoning. Arch Ophthalmol 20: 839–843.

Medrano CJ, Fox DA (1995). Oxygen consumption in rat outer and inner retina: Light- and pharmacologically induced inhibition. Exp Eye Res 61: 273–284.

Mela M, Cambier S, Mesmer-Dudons N et al. (2010). Methylmercury localization in *Danio rerio* retina after trophic and subchronic exposure: a basis for neurotoxicology. Neurotoxicology 31: 448–453.

Mela M, Grötzner SR, Legeay A et al. (2012). Morphological evidence of neurotoxicity in retina after methylmercury exposure. Neurotoxicology 33: 407–415.

Mergler D (1995). Behavioral neurophysiology: Quantitative measures of sensory toxicity. In: LW Chang, W Slikker (Eds.), Neurotoxicology: Approaches and Methods. Academic Press, New York, pp. 727–736.

Mergler D, Blain L, Lagace JP (1987). Solvent related colour vision loss: an indicator of damage? Int Arch Occup Environ Health 59: 313–321.

Mergler D, Huel G, Bowler R et al. (1991). Visual dysfunction among former microelectronics assembly workers. Arch Environ Health 46: 326–334.

Mergler D, Huel G, Bélanger S et al. (1996). Surveillance of early neurotoxic dysfunction. Neurotoxicology 17: 803–812.

Merigan WH, Maurrisen JPJ, Weiss B et al. (1983). Neurotoxic actions of methylmercury on the primate visual system. Neurobehav Toxicol Teratol 5: 649–658.

Merigan WH, Wood RW, Zehl D et al. (1988). Carbon disulfide effects on the visual system: I. Visual thresholds and ophthalmoscopy. Invest Ophthalmol Vis Sci 29: 512–518.

MMWR Morb Mortal Wkly Rep (1989). Occupational and paraoccupational exposure to lead – Colorado. MMWR Morb Mortal Wkly Rep 38: 338–340.

Muttray A, Wolters V, Mayer-Popken O et al. (1995). Effect of subacute occupational exposure to toluene on color vision. Int J Occup Med Environ Health 8: 339–345.

Muttray A, Wolff U, Jung D et al. (1997). Blue-yellow deficiency in workers exposed to low concentrations of organic solvents. Int Arch Occup Environ Health 70: 407–412.

Muttray A, Wolters V, Jung D et al. (1999). Effects of high doses of toluene on color vision. Neurotoxicol Teratol 21: 41–45.

Nagpal AG, Brodie SE (2009). Supranormal electroretinogram in a 10-year-old girl with lead toxicity. Doc Ophthalmol 118: 163–166.

Näsänen R, Kaukiainen A, Hero V et al. (2005). Effects of long-term occupational solvent exposure on contrast sensitivity and performance in visual search. Environ Toxicol Pharmacol 19: 497–504.

National Institute for Occupational Safety and Health (2014). NIOSH pocket guide to chemical hazards. Available online at: http://www.cdc.gov/niosh/npg/.

New York State Department of Health (2009). Lead exposure in adults – a guide for health care providers. Available online at: http://www.health.ny.gov/publications/2584/.

Nickla DL, Wallman J (2010). The multifunctional choroid. Prog Retin Eye Res 29: 144–168.

NIEHS (2012). Advancing Science, Improving Health: A Plan for Environmental Health Research. Strategic Plan 2012-2017. National Institute of Environmental Health Sciences, NIH Pub. No. 12-7935.

Niklasson M, Tham R, Larsby B et al. (1990). The influence of visual and somatosensory input on the vestibulo-oculomotor reflex of pigmented rats. J Vestib Res 1: 251–262.

Niklasson M, Tham R, Larsby B et al. (1994). Effects of GABAB activation and inhibition on vestibulo-ocular and optokinetic responses in the pigmented rat. Brain Res 649: 151–158.

Niklasson M, Stengard K, Tham R (1995). Are the effects of toluene on the vestibulo and opto-ocular motor system inhibited by the action of GABAB antagonist CGP 35348? Neurotoxicol Teratol 17: 351–357.

Ojanpää H, Näsänen R, Päällysaho J et al. (2006). Visual search and eye movements in patients with chronic solvent-induced toxic encephalopathy. Neurotoxicology 27: 1013–1023.

Otto DA, Fox DA (1993). Auditory and visual dysfunction following lead exposure. Neurotoxicology 14: 191–208.

Päällysaho J, Näsänen R, Mäntyjärvi M et al. (2007). Colour vision defects in occupational chronic solvent encephalopathy. Hum Exp Toxicol 26: 375–383.

Palacz O, Szymanska K, Czepita D (1980). Studies on the ERG in subjects with chronic exposure to carbon disulfide: I. Assessment of the condition of the visual system taking into account ERG findings depending on the duration of exposure to CS2. Klin Oczna 82: 65–68.

Pappolla M, Penton R, Weiss HS et al. (1987). Carbon disulfide axonopathy. Another experimental model characterized by acceleration of neurofilament transport and distinct changes of axonal size. Brain Res 424: 272–280.

Paramei GV, Meyer-Baron M, Seeber A (2004). Impairments of colour vision induced by organic solvents: a meta-analysis study. Neurotoxicology 25: 803–816.

Penha FM, Rodrigues EB, Maia M et al. (2010). Retinal and ocular toxicity in ocular application of drugs and chemicals – Part I: Animal models and toxicity assays. Ophthalmic Res 44: 82–104.

Perkins GA, Ellisman MH, Fox DA (2003). Three-dimensional analysis of mouse rod and cone mitochondrial cristae architecture: comparisons and bioenergetic implications. Mol Vis 9: 60–73.

Perkins GA, Scott R, Perez A et al. (2012). Bcl-xL-mediated remodeling of rod and cone synaptic mitochondria after postnatal lead exposure: electron microscopy, tomography and oxygen consumption. Mol Vis 18: 3029–3048.

Potts AM, Au PC (1976). The affinity of melanin for inorganic ions. Exp Eye Res 22: 487–491.

Raitta C, Tolonen M, Nurminen M (1974). Microcirculation of ocular fundus in viscose rayon workers exposed to carbon disulfide. Albrecht Von Graefes Arch Klin Exp Ophthalmol 191: 151–164.

Raitta C, Seppalainen AN, Huuskonen MS (1978). N-hexane maculopathy in industrial workers. Albrecht Von Graefes Arch Klin Exp Ophthalmol 209: 99–110.

Raitta C, Teir H, Tolonen M et al. (1981). Impaired color discrimination among viscose rayon workers exposed to carbon disulfide. J Occup Med 23: 189–192.

Rebert CS, Matteucci MJ, Pryor GT (1989). Acute electrophysiologic effects of inhaled toluene on adult male Long-Evans rats. Pharmacol Biochem Behav 33: 157–165.

Rebert CS, Day VL, Matteucci MJ et al. (1991). Sensory-evoked potentials in rats chronically exposed to trichloroethylene: Predominant auditory discrimination. Neurotoxicol Teratol 13: 83–90.

Rodrigues AR, Souza CR, Braga AM et al. (2007). Mercury toxicity in the Amazon: contrast sensitivity and color discrimination of subjects exposed to mercury. Braz J Med Biol Res 40: 415–424.

Roscoe RJ, Gittleman JL, Deddens JA et al. (1999). Blood lead levels among children of lead-exposed workers: a meta-analysis. Am J Ind Med 36: 475–481.

Rosolen SG, Kolomiets B, Varela O et al. (2008). Retinal electrophysiology for toxicology studies: applications and limits of ERG in animals and ex vivo recordings. Exp Toxicol Pathol 60: 17–32.

Rothenberg SJ, Schnaas L, Salgado-Valladares M et al. (2002). Increased ERG a- and b-wave amplitudes in 7–10 year old children resulting from prenatal lead exposure. Invest Ophthalmol Vis Sci 43: 2036–2044.

Rufa A, Dotti MT, Cardaioli E et al. (2005). Leber hereditary optic neuropathy in 2 of 4 siblings with 11778 mtDNA mutation: clinical variability or effect of toxic environmental exposure? Eur Neurol 53: 32–34.

Savic SM (1967). Influence of carbon disulfide on the eye. Arch Environ Health 14: 325–326.

Scholl HPN, Zrenner E (2000). Electrophysiology in the investigation of acquired retinal disorders. Surv Ophthalmol 45: 29–47.

Schreiber JS, Hudnell HK, Geller AM et al. (2002). Apartment residents' and day care workers' exposures to tetrachloroethylene and deficits in visual contrast sensitivity. Environ Health Perspect 110: 655–664.

Seeber A, Bruckner T, Triebig G (2009). Occupational styrene exposure, colour vision and contrast sensitivity: a cohort study with repeated measurements. Int Arch Occup Environ Health 82: 757–770.

Seppalainen AM, Haltia M (1980). Carbon disulfide. In: PS Spencer, HH HH Schaumberg (Eds.), Experimental and Clinical Neurotoxicology. William & Wilkins, Baltimore, pp. 356–373.

Shulman LM, Fox DA (1996). Dopamine inhibits mammalian photoreceptor Na+, K+-ATPase activity via a selective effect on the α3 isozyme. Proc Natl Acad Sci U. S. A. 93: 8034–8039.

Signorino M, Scarpino O, Provincialli L et al. (1983). Modification of the electroretinogram and of different components of the visual evoked potentials in workers exposed to lead. Ital Electroencephalogr J 10, 51P.

Specchio LM, Bellomo R, Dicuonzo F et al. (1981). Smooth pursuit eye movements among storage battery workers. Clin Toxicol 18: 1269–1276.

Spivey GH, Baloh RW, Brown CP et al. (1980). Subclinical effects of chronic increased lead absorption – a prospective study: III. Neurologic findings at follow-up examination. J Occup Med 22: 607–612.

Srivastava D, Hurwitz RL, Fox DA (1995). Lead- and calcium-mediated inhibition of bovine rod cGMP phosphodiesterase: Interactions with magnesium. Toxicol Appl Pharmacol 134: 43–52.

Stohs SJ, Bagchi D (1995). Oxidative mechanisms in the toxicity of metal ions. Free Radic Biol Med 18: 321–336.

Storm JE, Mazor KA, Aldous KM et al. (2011). Visual contrast sensitivity in children exposed to tetrachloroethylene. Arch Environ Occup Health 66: 166–177.

Tabbacchi G, Corsico R, Gallinelli R (1966). Retrobulbar neuritis caused by suspected chronic trichloroethylene poisoning. Ann Ottalmol Clin Ocul 92: 787–792.

Tanan CL, Ventura DF, de Souza JM et al. (2006). Effects of mercury intoxication on the response of horizontal cells of the retina of thraira fish (*Hoplias malabaricus*). Braz J Med Biol Res 39: 987–995.

Tekavcic-Pompe M, Tekavcic I (2008). Color vision in the tritan axis is predominantly affected at high altitude. High Alt Med Biol 9: 38–42.

Tessier-Lavigne M, Mobbs P, Attwell D (1985). Lead and mercury toxicity and the rod light response. Invest Ophthalmol Vis Sci 26: 1117–1123.

Tham R, Larsby B, Odkvist LM et al. (1979). The influence of trichloroethylene and related drugs on the vestibular system. Acta Pharmacol Toxicol (Copenh) 44: 336–342.

Tham R, Bunnfors I, Eriksson B et al. (1984). Vestibulo-ocular disturbances in rats exposed to organic solvents. Acta Pharmacol Toxicol 54: 58–63.

Till C, Westall CA, Koren G et al. (2005). Vision abnormalities in young children exposed prenatally to organic solvents. Neurotoxicology 26: 599–613.

Trobe JD, Beck RW, Moke PS et al. (1996). Contrast sensitivity and other vision tests in the optic neuritis treatment trial. Am J Ophthalmol 121: 547–553.

Ulshafer RJ, Allen CB, Rubin ML (1990). Distributions of elements in the human retinal pigment epithelium. Arch Ophthalmol 108: 113–117.

UNEP (2008). The global atmospheric mercury assessment: Sources, emissions and transport. United Nations Environment Programme Chemicals Branch, Switzerland, Geneva.

Urban P, Gobba F, Nerudová J et al. (2003). Color discrimination impairment in workers exposed to mercury vapor. Neurotoxicology 24: 711–716.

Vafaee MS, Meyer E, Marrett S et al. (1999). Frequency-dependent changes in cerebral metabolic rate of oxygen during activation of human visual cortex. J Cereb Blood Flow Metab 19: 272–277.

Ventura DF, Costa MT, Costa MF et al. (2004). Multifocal and full-field electroretinogram changes associated with color-vision loss in mercury vapor exposure. Vis Neurosci 21: 421–429.

Ventura DF, Simões AL, Tomaz S et al. (2005). Colour vision and contrast sensitivity losses of mercury intoxicated industry workers in Brazil. Environ Toxicol Pharmacol 19: 523–529.

Vernon RJ, Ferguson K (1969). Effects of trichloroethylene on visual-motor performance. Arch Environ Health 18: 894–900.

Vettori MV, Corradi D, Coccini T et al. (2000). Styrene-induced changes in amacrine retinal cells: an experimental study in the rat. Neurotoxicology 21: 607–614.

Warfvinge K, Bruun A (1996). Mercury accumulation in the squirrel monkey eye after mercury vapour exposure. Toxicology 107: 189–200.

WHO (2003). Elemental mercury and inorganic mercury compounds: Human health effects. Concise International Chemical Assessment Document 50, World Health Organization, Geneva, Switzerland.

WHO (2004). Selected Occupational Risk Factors, Chapter 21. In: M Ezzati, AD Lopez, A Rodgers et al. (Eds.), Comparative Quantification of Health Risks: Global and Regional Burden of Diseases Attributable to Selected Major Risk Factors, World Health Organization. 1. Geneva, Switzerland, pp. 1651–1802.

Wilensky JT, Hawkins A (2001). Comparison of contrast sensitivity, visual acuity, and Humphrey visual field testing in patients with glaucoma. Trans Am Ophthalmol Soc 99: 213–217.

Williams JH, Ross L (2007). Consequences of prenatal toxin exposure for mental health in children and adolescents: a systematic review. Eur Child Adolesc Psychiatry 16: 243–253.

Williamson AM, Teo RKC (1986). Neurobehavioral effects of occupational exposure to lead. Br J Ind Med 43: 374–380.

Willmann G, Ivanov IV, Fischer MD et al. (2010). Effects on colour discrimination during long term exposure to high altitudes on Mt Everest. Br J Ophthalmol 94: 1393–1397.

Winneke G (1982). Acute behavioral effects of exposure to some organic solvents. Acta Neurobiol Exp 66: 117–129.

Yoshino Y, Mozai T, Nako K (1966). Distribution of mercury in the brain and its subcellular units in experimental organic mercury poisonings. J Neurochem 13: 397–406.

Yuki K, Dogru M, Imamura Y et al. (2009). Lead accumulation as possible risk factor for primary open-angle glaucoma. Biol Trace Elem Res 132: 1–8.

Zavalic M, Mandic Z, Turk R et al. (1998). Assessment of colour vision impairment in male workers exposed to toluene generally above occupational exposure limits. Occup Med (Lond) 48: 175–180.

Zavalić M, Mandić Z, Turk R et al. (1998). Qualitative color vision impairment in toluene-exposed workers. Int Arch Occup Environ Health 71: 194–200.

Zhang T, Xiang CD, Gale D et al. (2008). Drug transporter and cytochrome P450 mRNA expression in human ocular barriers: Implications for ocular drug disposition. Drug Metab Dispos 36: 1300–1307.

Handbook of Clinical Neurology, Vol. 131 (3rd series)
Occupational Neurology
M. Lotti and M.L. Bleecker, Editors

Chapter 19

Hearing

MARIOLA SLIWINSKA-KOWALSKA*
Department of Audiology and Phoniatrics, Nofer Institute of Occupational Medicine, Lodz, Poland

INTRODUCTION

According to the World Health Organization (WHO), approximately 5% of people in the world (360 million) have disabling hearing impairment, which is defined as an average hearing loss higher than 40 dB hearing level (HL) at speech frequencies (0.5, 1, 2, and 4 kHz), as measured in the better ear. Since the prevalence of hearing loss doubles with every decade of life, it may concern over 33% of individuals aged 65 or above (www.who.int/healthinfo/statistics/bod_hearingloss.pdf).

Overexposure to noise is one of the main causes of progressive prevalence of hearing loss in the aging population. It has been estimated that, worldwide, 16% of disabling hearing loss in adults is attributed to occupational noise, and this number rises to 21% in industrialized countries (Nelson et al., 2005). In the USA about 30 million individuals are exposed to high levels of noise, and the same number of people are overexposed to noise in Europe. Taking into account the entire lifespan, as many as 500 million people can be overexposed to noise worldwide at a certain period of their life.

The risk of developing occupational noise-induced hearing loss (NIHL) depends primarily on the type of industry. The highest noise exposures are present in heavy industries, such as mining, metallurgy, textiles, shipbuilding, but also in the construction industry, agriculture, and entertainment industry. The entertainment industry includes, among others, professional musicians and night-club and bar employees.

Exposure to noise can lead to auditory and nonauditory health effects. Well-recognized nonauditory effects include sleep disturbance, irritability, cognitive impairment, and mental health problems, as well as increased risk of cardiovascular disease and myocardial infarction. Nonauditory effects can be observed even at moderate levels of prolonged exposure to noise (such as transportation noise), which do not exceed occupational limits for hearing.

Simultaneous exposure to noise and chemicals is very common in industry. In the USA, among 30 million workers exposed to noise, about 10 million are co-exposed to ototoxic chemicals. The substances known to be possibly dangerous for hearing are: organic solvents (e.g., styrene, toluene, *n*-hexane, mixtures of solvents), heavy metals (e.g., lead, mercury, arsenic, and cadmium), asphyxiants, and pesticides. Combined exposure to noise and chemicals can have additive or even supra-additive effects on hearing. This fact is not taken into account when determining the risk of occupational NIHL.

This chapter describes current knowledge on pathophysiology, diagnostics, and clinical aspects of hearing loss related to occupational exposure to noise and chemicals, and provides information on selected methods of hearing loss assessment and current European and international noise regulation standards.

ACOUSTICS AND PSYCHOACOUSTICS

Noise is defined as all sounds of different acoustic characteristics, which are unwanted in a given situation and/or for a given person. A more general definition describes noise as all sounds that are undesirable, unpleasant, annoying, or harmful acoustic air vibrations perceived by the ear or other parts of a human body.

The two most important acoustic parameters of noise are level and frequency. They are measured as sound pressure level (SPL) in decibels (dB), and frequency in hertz (Hz), respectively.

The decibel is a logarithmic unit, defined as ratio between the value of the acoustic pressure level of a given sound, measured in pascals (Pa), to the value of 20 μPa, which is the perception threshold (i.e., the lowest

*Correspondence to: Mariola Sliwinska-Kowalska, MD, PhD, Department of Audiology and Phoniatrics, Nofer Institute of Occupational Medicine, Lodz, Poland, 8, St. Teresa Str, 91-348 Lodz. Tel: 48-42-63-14520, E-mail: marsliw@imp.lodz.pl

sound pressure perceived by human ear). The formula used to convert a linear scale (in Pa) to the logarithmic scale (in dB) is given below:

$$L = 10 \log (p/p_o)^2 = 20 \log (p/p_o)$$

where L is the SPL in relative units decibels; p is the sound pressure in pascals; $p_o = 20$ μPa is the reference sound pressure, i.e., perception threshold.

In a logarithmic dB unit scale, the simple addition is not possible. Usually the formula of 3 dB exchange rate is used, which means that doubling in sound energy is equal to a 3 dB increase of SPL expressed in dB (i.e., 100 dB + 100 dB = 103 dB).

To interpret the effects of noise on hearing, the psychoacoustics of human ear must be understood. Noise parameters that are physically measured as the SPL and frequency are perceived by humans as the loudness of sound (in phons) and pitch of sound (in Hz), respectively.

Human ear is sensitive to the sounds from 20 Hz to 20 kHz, thus the audible noise refers to this range of frequencies (Fig. 19.1A). Sounds below 20 Hz are defined as infrasound noise, and sounds above 20 kHz as ultrasound noise. As for the loudness of noise, the highest level of noise which can be tolerated by the human ear is considered to be 120 dB (threshold of pain) (Fig. 19.1A and Table 19.1).

The human auditory system is not equally sensitive to sounds by frequency. This means that the perception thresholds of sounds (as well as their thresholds of pain) vary depending on the sound frequency. Thus, two sounds of different frequencies at the same level will not in general have the same loudness (in other words, the same SPL in dB corresponds to different loudness levels, expressed in phons at different frequencies). This unique property of human hearing is characterized by equal loudness contours (Fig. 19.1B) (ISO 226: 2003, 2003). As can be seen in Figure 19.1B, the frequencies 3–4 kHz are the most sensitive within the sound frequency range from 20 Hz to 20 kHz that can be heard by the human ear, and the distance between sound perception hearing threshold and pain threshold for these frequencies is the largest. For lower than 3–4 kHz and higher sound frequencies, the ear becomes less sensitive.

To refer the physical sound measurement to the psychoacoustics of the human ear, an A-weighted SPL is measured and expressed in dBA. The letter "A" indicates that the measurement has been performed using a sound level meter incorporating an A-weighting filter, whose frequency characteristics are modeled on the equal loudness contours of 40 phons. For low-frequency noise, the C-weighted dB units are also used. The C-weighting characteristics reflect the equal loudness contour of 100 phons.

Usually, the SPL varies over time. Thus, in order to evaluate nonsteady (fluctuating) noise, the so-called A-weighted equivalent continuous SPL ($L_{Aeq,T}$) is applied. $L_{Aeq,T}$ is the imaginary constant SPL that would produce the same energy as the fluctuating noise level over a given time interval. $L_{Aeq,T}$ normalized to a nominal 8-hour working day (or 40-hour working week) is called the daily (or weekly) noise exposure level $L_{EX,8h}$ ($L_{EX,w}$) and it is applied in the assessment of workers' exposure to noise in an occupational environment. Below the value $L_{EX,8h}$ of 85 dB, the risk of NIHL is small, and no hearing loss due to exposure to occupational noise is anticipated at values below 80 dB.

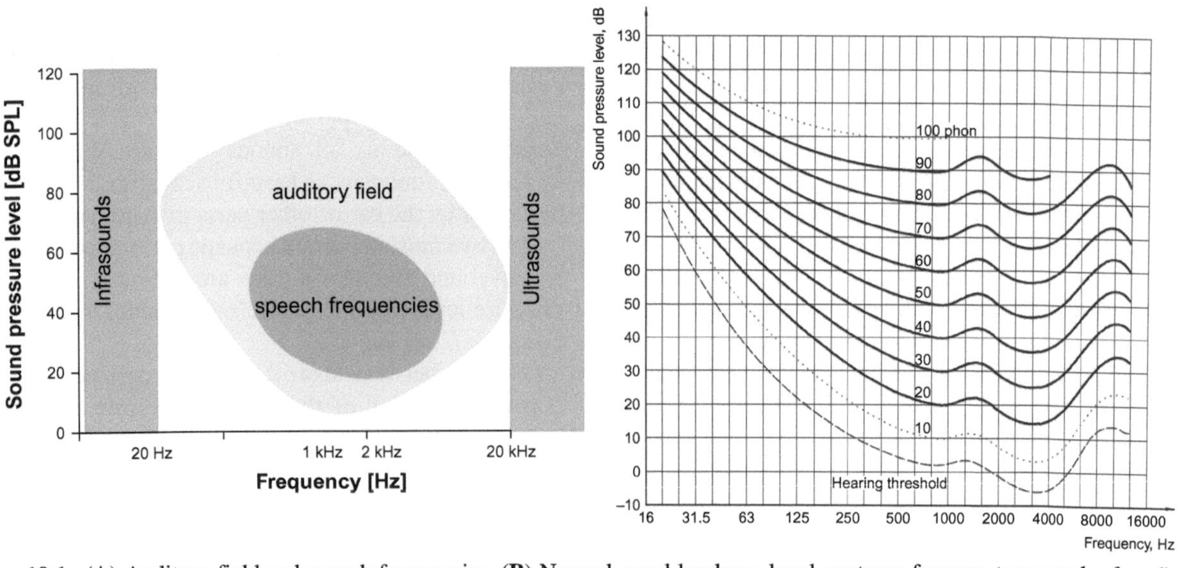

Fig. 19.1. (**A**) Auditory field and speech frequencies. (**B**) Normal equal loudness level contours for pure tones under free-field listening conditions according to ISO 226:2003. (Reproduced with permission from the British Standards Institution (BSI)).

Table 19.1

Examples of pressure levels of sounds perceived by the human ear

Type of sound	Sound pressure level (dB)
Perception threshold	0
Sound of falling leaves	20
Normal speech	60
Motor vehicles	60–100
Jet plane during take-off	120
Pain threshold	120–140

PHYSIOLOGY OF HEARING

Hearing is a process comprising (Fig. 19.2):

- transmission of acoustic waves through the external and middle ear
- reception and conversion of acoustic waves into electric impulses of the auditory nerve fibers in the inner ear
- auditory processing of the electric impulses in the central parts of the auditory pathway (auditory nerve, cochlear nuclei, midbrain, subcortical and cortical auditory centers).

The peripheral part of the human auditory organ is composed of three compounds: external ear, middle ear, and inner ear. Acoustic waves are gathered by the concha, transmitted through the external ear canal, and amplified in the middle ear by the movement of three ossicles (malleus, incus, and stapes). The vibration of the stapes is converted to inner-ear fluid movement in its anterior-lower part, named the cochlea.

When presented with a high-intensity sound stimulus, usually exceeding 80 dB SPL, the stapedius muscle contracts, increasing the stiffness of the ossicular chain and decreasing the acoustic energy reaching the inner ear by about 10 dB. The strength of the stapedius contraction increases as the stimulus increases, and the latency of this reflex decreases from typical threshold responses of around 100 ms to 25 ms for stimuli of very high intensity. This mechanism plays a crucial role in protecting the inner ear against acoustic trauma and is most effective for sound stimuli below the frequency of 2 kHz.

The cochlea is in the shape of a snail shell, with 2.5 turns. The interior space of the cochlea is divided in the middle by basilar membrane along its entire length (Fig. 19.3A). The membrane is narrowest in the basilar turn of the cochlea, and widest in the apical turn. On the basilar membrane lies the organ of Corti, a receptive organ responsible for converting acoustic stimuli into electric nerve stimuli. Basilar membrane is organized tonotopically, which means that every sound frequency is analyzed in a strictly defined place. The highest audible

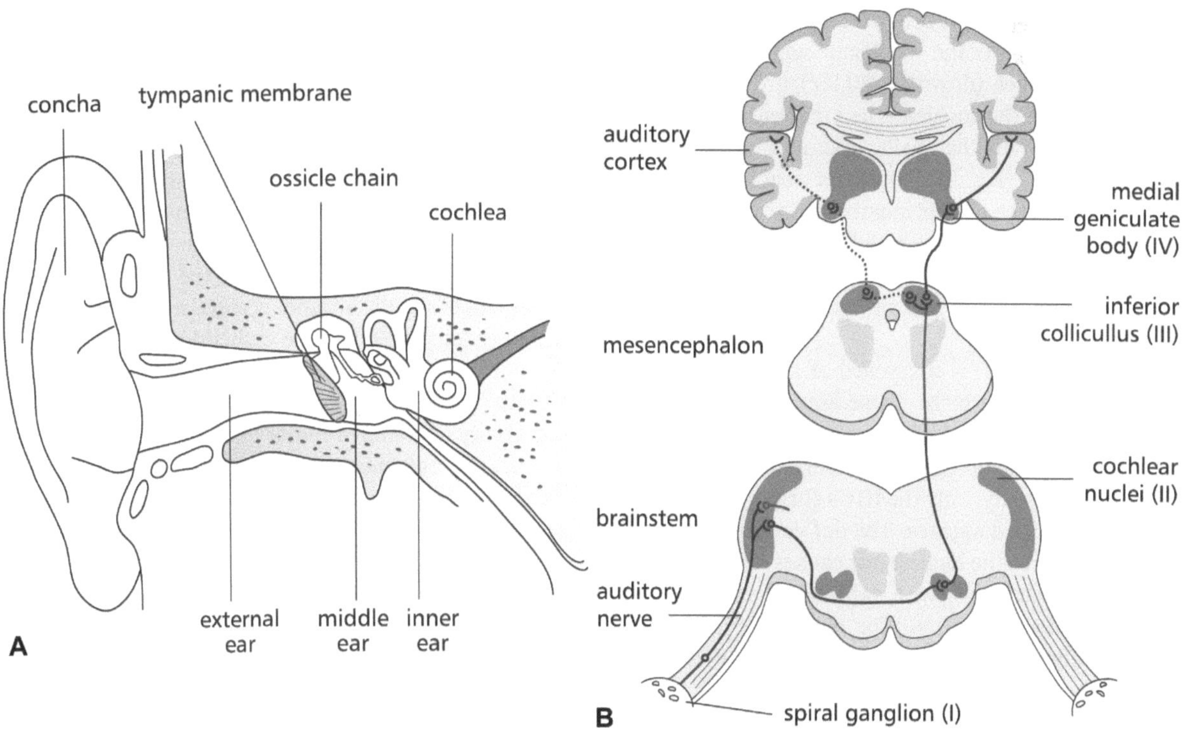

Fig. 19.2. (**A**) Peripheral and (**B**) central parts of the auditory pathway.

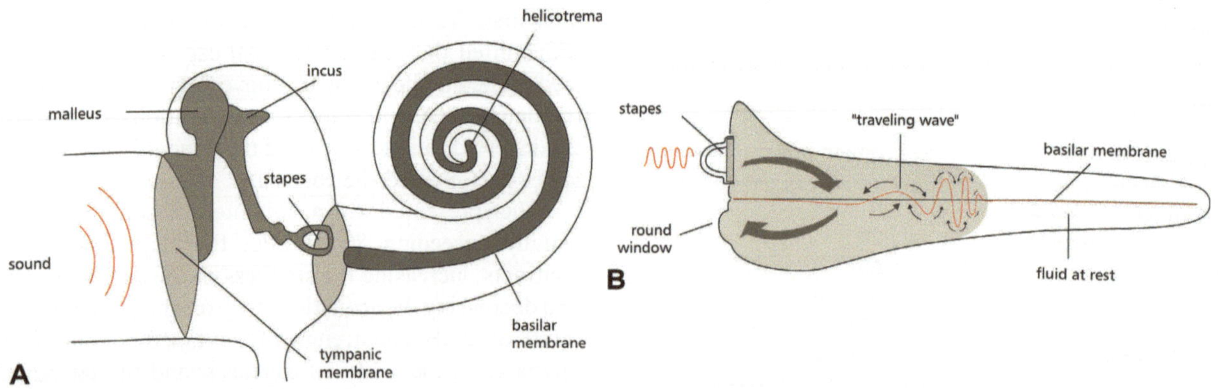

Fig. 19.3. Schematic of the basilar membrane. (**A**) Basilar membrane divides the cochlear space into two compartments. (**B**) "Traveling wave."

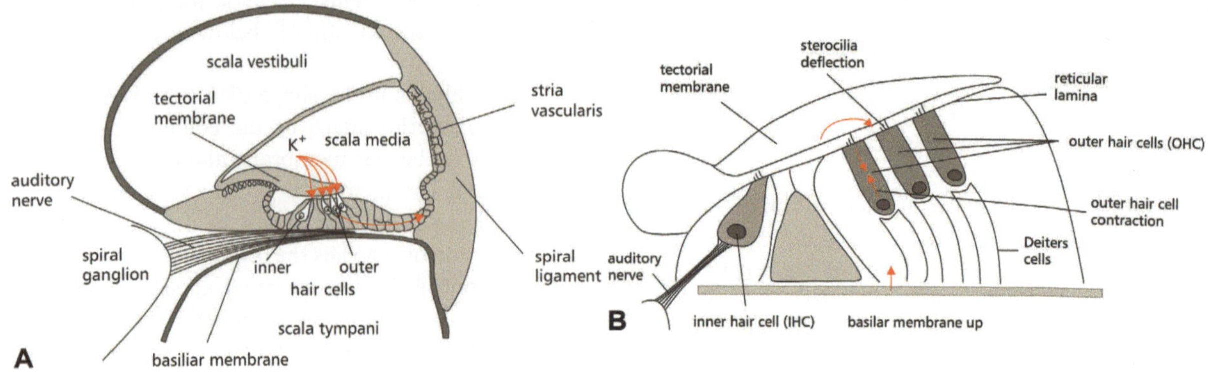

Fig. 19.4. Schematic of the organ of Corti. (**A**) Structure; (**B**) cochlear micromechanics.

frequencies of sounds (up to 20 kHz) are analyzed in the basilar turn of the cochlea, while the lowest audible frequencies (down to 20 Hz) are analyzed in the apical part of the cochlea. The inner-ear fluid movement sets in motion the basilar membrane, creating a "traveling wave" (Fig. 19.3B). After reaching a maximum deflection, the basilar membrane "traveling wave" is instantly suppressed. The maximum deflection place of basilar membrane corresponds to the place of reception of the specific frequency sound in the organ of Corti.

The organ of Corti is composed of two kinds of sensory cells, called hair cells. These are organized in one row of inner hair cells (IHCs: around 3500cells), and three rows of outer hair cells (OHCs: around 12 000 cells) (Fig. 19.4). Practically, only the IHCs form synapses with the auditory nerve dendrites. The deflection of hair cell stereocilia towards the highest one causes the opening of potassium ion channels, with subsequent influx of K^+ and hair cell depolarization. The receptor current triggers a cascade of processes inside the cell, eventually resulting in a release of the neurotransmitter (glutamate) into the synapse, and activation of the auditory nerve fibers.

The function of OHCs is different than the function of IHCs. Although there are almost three times as many OHCs as IHCs, the former practically do not form afferent synapses. In contrast, they remain under the control of the efferent system of innervation. The OHCs have unique capabilities of constant and fast (measured in μs) contraction after their depolarization. The major somatic motor unit responsible for OHC shrinkage is prestin. The OHCs, together with the basilar membrane, on which these cells lie, and tectorial membrane, in which the stereocilia of OHC are embedded, form a so-called "cochlear amplifier." The cochlear amplifier is responsible for the auditory organ's very high sensitivity and specificity to sound stimulation. This process is possible due to the enhancing movement of the endocochlear fluid towards the IHC, which makes these cells more responsive to very low sound stimuli. On the other hand, the activation of efferent system feedback due to exposure to high levels of sound leads to slow contraction of the OHC (measured in ms), which may play a protective role in acoustic trauma. Motile constant activities of OHC require high oxygen and energy supply. Any deficit related to, for example, oxidative stress after exposure to noise may lead to OHC damage.

PATHOPHYSIOLOGY OF NOISE-INDUCED HEARING LOSS

Mechanisms

There are several mechanisms responsible for developing hearing loss after prolonged exposure to noise. The most important one is based on the processes of oxidative stress (Henderson et al., 2006). Excessive acoustic stimulation of the ear increases the metabolism of hair cells, primarily OHCs, leading to overproduction of reactive oxygen and nitrogen species (so-called "free radicals"). The disturbance in the balance between the production of free radicals (along with lipid peroxidation) and antioxidant defenses is defined as an oxidative stress. Prolonged oxidative stress may lead to damage of all components of the cell, including proteins, lipids, and DNA, and eventually to damage of the hair cell, which dies through a process of apoptosis.

The other mechanism of OHC loss involves a mechanical microdestruction which can be caused by relatively high levels of acoustic stimuli (in animal studies, levels exceeding 120 dB SPL) (Borg et al., 1995). The most fragile structures are stereocilia that can become disorganized, broken, or completely destroyed. If the destruction to sterocilia is irreversible, the hair cell dies. Some role in OHC damage may also be caused by blood supply disturbances (Shi, 2011). The contraction of capillaries of stria vascularis may lead to cell hypoxia, while subsequent vasodilatation may promote the sudden release of toxic metabolites accumulated in the hair cell and its destruction.

Additionally, very high levels of noise can lead to an excessive release of glutamate neurotransmitter to the afferent synapses formed between IHCs and the dendrites of the auditory nerve, causing so-called "excitotoxicity" (Puel et al., 2002). This process underlies the pathogenesis of acute acoustic trauma and is to some extent reversible. Extremely high exposure to noise or explosion can also mechanically damage the structures of the external and middle ear (tympanic membrane rupture, disruption of chain of ossicles).

Morphologic changes

In the early stages of NIHL, morphologic changes refer mostly to OHC and hair cell stereocilia. The stereocilia become bent and disorganized and some can be broken or missing. The OHCs become damaged in a strictly defined order. The most sensitive to noise is the first row of OHCs, followed by the third row of OHCs. The most resistant to noise is the second row of OHCs. IHCs stay unchanged as they are highly resistant to moderate levels of noise (Borg et al., 1995).

Chronic exposure to noise causes also the damage of other structures of the inner ear, like seam of spiral lamina, spiral ligament, and stria vascularis. In addition to damaging OHC, prolonged exposure to noise can produce degeneration of auditory nerve fibers, and some ganglion cells (the first neuron of the auditory pathway) located in the spiral ganglion in the cochlea. The latest animal studies show that progressive auditory nerve fiber degeneration along with a slow loss of ganglion cells may be triggered by acute exposure to very high levels of noise, and can persist despite full recovery of hearing thresholds. This phenomenon is called "hidden hearing loss" (Lin et al., 2011).

Functional consequences

The loss of OHCs significantly decreases the sensitivity and selectivity of hearing. The tuning curves registered from the auditory neurons lose their sharpness and become wider. The primary functional symptom of OHC damage is elevation of hearing thresholds. The loss of at least 30% of hair cells results in hearing threshold shift at frequencies related to the place of damage of the organ of Corti. The maximum loss of hearing acuity due to the loss of all OHCs is limited to 50–70 dB, as long as IHCs remain undamaged by noise.

It has been shown in animal studies that overexposure to narrowband noise results in hearing loss at the frequencies corresponding to the place of damage of the organ of Corti, as well as to the frequency of half an octave higher than the center frequency of noise (Borg et al., 1995). Exposure to wideband noise, typically present in workplaces, results in hearing loss at high frequencies in animals. The most sensitive region of the cochlea in humans is the place corresponding to the frequencies in the range 3–6 kHz. Thus, the first sign of NIHL is an audiometric notch at high frequencies (typically 4 or 6 kHz).

Damage to the peripheral part of the auditory organ may have central physiologic consequences. One of the hallmarks of damage to the OHC is loudness recruitment. Recruitment refers to the perceptual phenomenon of sounds becoming rapidly louder with increasing sound level. In consequence, people with cochlear hearing loss request others to speak louder, but a rise in the volume of a voice causes patient discomfort, as it is too loud. The mechanism of recruitment phenomenon resulting from damage to OHCs is still not fully understood. It can be related to loss of cochlear compressive nonlinearity, widening of tuning curves of the nerve fibers, as well as decrease in the spread of thresholds.

Another consequence of noise-induced damage to the cochlea is impairment of speech intelligibility in noisy environments. It has been hypothesized that this is due to a selective loss of high-threshold auditory nerve fibers related to prolonged exposure to noise or to exposure to noise at very high levels.

The loss of IHC ribbons (deafferentation) and increased central neural excitability following acoustic trauma may lead to tinnitus. Tinnitus is defined as the perception of sound in the absence of an acoustic stimulus. In humans, tinnitus is accompanied by increased gamma activity in the right cortex after exposure to noise.

Susceptibility to NIHL

Susceptibility to the damaging effects of noise differs among individuals, meaning that after an identical exposure to noise there can be a large difference in resulting hearing loss. Another definition of increased vulnerability states that in susceptible subjects hearing threshold shift can be developed after exposures to noise that in principle do not result in hearing damage in most subjects in the population. This interindividual variability depends on complex interactions between environmental and individual factors, including genetics. Among environmental factors the most important are: co-exposures to noise and chemicals, ototoxic drug intake, heat, and co-exposure to noise and vibration (paleness of fingers in cold conditions).

Individual factors associated with increased susceptibility for NIHL include gender (males are more susceptible than females), smoking, elevated blood pressure, high level of lipids, skin pigmentation (melanin has a protective effect). Genetics seems to play a significant role, and several gene polymorphisms were shown to increase the susceptibility to noise. These include potassium ion recycling channel genes (*KCNQ1, KCNQ4, KCNE1*), myosin 14 gene (*MYO14*), heat shock protein gene (*HSP*), protocadherin 15 gene (*PCDH15*), and cadherin 23 gene (*CDH23*). The association was also found between noise and catalase (Sliwinska-Kowalska and Pawelczyk, 2013).

Temporary threshold shift and permanent threshold shift

Insufficient hearing protection for a short exposure to industrial noise (e.g., over a work shift time) can lead to temporary threshold shift (TTS) that recovers over several hours of rest from noise. Although there is no simple translation of TTS to the permanent threshold shift (PTS) which occurs after years of exposure to industrial noise, recent data show that TTS can have a high sensitivity for predicting the development of NIHL (Moshammer et al., 2014).

PATHOPHYSIOLOGY OF ORGANIC SOLVENT-INDUCED HEARING LOSS

The ototoxicity of organic solvents was first recognized in research performed at the beginning of the 1980s, when high concentrations of toluene were given to rats. Subsequently, such properties were shown for styrene, ethylbenzene, trichloroethylene, *n*-hexane, and *p*-xylene.

Prolonged exposure to organic solvents at high concentration leads to damage of the supporting cells (Hensen cells and Deiter's cells), as well as OHCs of the organ of Corti. This fact was proven mainly for toluene and styrene (Campo et al., 1999). After inhalation, these chemicals are carried by blood, and penetrate through the stria vascularis and supporting cells of the organ of Corti, reaching the OHCs. The most sensitive to damage is the third row of OHCs, followed by the second and the first rows of OHCs, and this differentiate the damage related to solvents from noise (Campo et al., 2001). Similarly to noise exposure, the IHCs are immune to damage by solvents. Organic solvents, which are lipophilic toxic substances, damage cell membranes, causing cell death through apoptosis. As in the case of NIHL, oxidative stress plays a major role in the pathogenesis of organic solvent-induced hearing loss (Fig. 19.5).

The morphologic changes to the cochlea result in hearing loss. In rats toluene and styrene cause a hearing threshold shift in the middle (16–20 kHz) and low (4 kHz) frequencies (the range of frequencies perceived by the inner ear of a rat is 1–50 kHz) (Campo et al., 1997). This audiometric characteristic allows hearing loss caused by organic solvents to be differentiated from high-frequency hearing loss caused by noise. Concentrations of styrene greater than 600 ppm show threshold shifts directly related to styrene concentration.

The ototoxic properties of organic solvents seem to depend on the number of carbon atoms in their chemical structure. The most toxic is styrene, followed by toluene and other organic solvents. The ototoxic effect of these

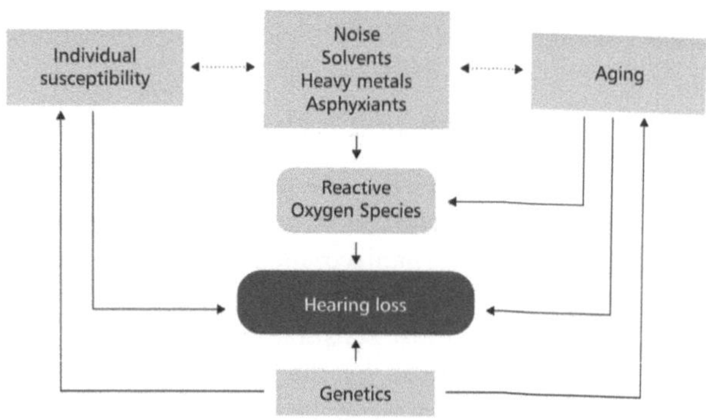

Fig. 19.5. The mechanisms of hearing loss due to exposure to noise and chemicals.

chemicals depends also on the animal, as hearing damage shown in rats was not confirmed in guinea pigs.

Where there was co-exposure to noise and organic solvents such as styrene and toluene, a synergistic effect on hearing threshold shift was observed in rats (Lataye et al., 2000; Makitie et al., 2003). It seems that the chemicals modify the structure of the OHC membrane, making the cells more sensitive to noise damage. Thus, the same acoustic energy reaching the cochlea may exert more profound damage to the organ of Corti. In turn, the ototoxicity of organic solvents seems to depend on the level of noise. The consequences of combined exposure to noise and organic solvents primarily depend on the level of sound pressure, although at low levels of sound, co-exposure to organic solvents may lead to substantial hearing loss that could not be attributed to noise exposure alone (Campo and Maguin, 2007).

Apart from destructive effects to the inner ear, organic solvents may cause damage to the central part of the auditory system. Such consequences were shown in rats using auditory brainstem responses (ABR) and contralateral acoustic reflex suppression (Rumeau et al., 2011). Research in which rats were exposed to *n*-hexane showed abnormalities in central components of the brainstem response, which were reduced by toluene (Pryor and Rebert, 1992). In humans, exposure to styrene was shown to be related to central auditory processing disorders (Zamyslowska-Smytke et al., 2009).

Although several studies suggested peripheral, as well as central auditory, dysfunction induced by occupational exposure to organic solvents, the pathophysiology of the damage in humans is still not clear (Vyskocil et al., 2012). Also, the dose–response relationship for developing hearing loss due to exposure to these substances in occupational settings is not established. Several factors stand in the way of proving the results of the research in humans to be conclusive. First, the concentration of organic solvents in the workplace is much lower than used in animal studies. Second, employees are generally exposed to mixtures of solvents, thus it is difficult to single out the effect of a particular solvent. Finally, workers are very often co-exposed to noise at different levels and organic solvents of different concentrations, which makes human research even more challenging.

Figure 19.5 summarizes possible pathomechanisms of developing hearing loss due to exposure to noise and chemicals in association with susceptibility factors and aging.

METHODS OF HEARING ASSESSMENT

The type of hearing loss or hearing impairment depends on where in the auditory pathway the damage occurs (Table 19.2). While conductive hearing loss is a consequence of pathology in the external and middle ear, sensorineural hearing loss relates predominantly to OHC damage in the cochlea, and retrocochlear hearing loss refers to the IHC dendrites, auditory nerve, and brainstem pathology. The damage to more central structures of the auditory pathway is a cause of central auditory processing disorders, with no or little effect on hearing acuity (hearing threshold). Auditory processing disorders can be described as deficits in the effectiveness by which the central nervous system utilizes auditory information, including interhemispheric communication.

Table 19.2

Type of hearing loss/impairment in relation to the site of the lesion in the auditory pathway

Type of hearing loss/ impairment	Site of the lesion (cause)
Conductive hearing loss	• External ear (e.g., excess of earwax) • Middle ear (e.g., otitis media, destruction of the ossicular chain)
Sensorineural hearing loss	• Inner ear (loss of outer hair cells, e.g., due to exposure to noise or chemicals)
Retrocochlear hearing loss	• Inner ear (inner hair cell dendrite loss, ganglion cell loss) • Auditory nerve (e.g., acoustic neuroma) • Cochlear nuclei (e.g., brainstem tumors)
Auditory processing disorder	• Central structures of auditory pathway (e.g., due to ototoxic chemicals)

Audiometric examination should be preceded by otoscopy. Its role is to check for excessive ear wax, pathologic discharge from the ear, and pathologic changes in the tympanic membrane. The basic test battery of hearing evaluation includes examination by whisper, tuning-fork tests, pure-tone audiometry (a "gold" standard for audiometric tests), impedance audiometry, and otoacoustic emissions (OAEs). More advanced tests are auditory evoked potentials, speech audiometry, and behavioral auditory processing tests.

Examination by whisper

Examination by whisper is the simplest and most natural way to assess hearing acuity. It should be done in a quiet room, separately for each ear. During the examination, the patient stands sideways, with the tested ear towards the examiner. The contralateral ear is drowned out during the test by rhythmic finger pressure on the earlobe. In addition, the eye on the tested side can be obscured by an assistant, so that the patient cannot lip read the words.

Commonly known words should be whispered, first from the furthest position – approximately 4–6 meters. If the patient is not able to understand, the examiner repeats the words while approaching the patient.

Understanding whispered words from a distance of ≥4 m is generally an acceptable norm. Not understanding the words whispered "ad concham" indicates a profound hearing loss. However, the sensitivity of the test in the diagnostics of the degree of hearing loss is rather poor. As speech frequencies are limited to the range of 0.5–4 kHz, higher-frequency threshold shifts may remain undiscovered.

Tuning-fork tests

Commonly used tuning-fork tests are the Weber test and the Rinne test. The Weber test allows for very simple and fast comparison of differences in hearing acuity between the ears. The Rinne test assesses air conduction and bone conduction in the same ear, allowing for quick differential diagnostics of conductive hearing loss from sensorineural hearing loss.

Weber test

After setting the tuning fork (usually $C_2 = 512$ Hz) in vibration, its stem is placed at the top of the skull, exactly in midline. The role of the patient is to answer whether the tone is heard in the center of head, in the right ear, or in the left ear. In a normal-hearing subject, or in a subject with side-symmetric sensorineural hearing loss (as in NIHL), there is no lateralization of the tone to either ear. In a patient with one-sided hearing impairment, the result of the test depends on the type of hearing loss. Where there is conductive hearing loss, the tone strays to the impaired ear, while with sensorineural hearing loss the tone strays to the normal-hearing ear.

Rinne test

After setting the tuning fork (usually $C_2 = 512$ Hz) in vibration, its stem is placed on the mastoid process. The patient is asked to report when s/he can no longer hear the tone. Then, the fork is placed close to the concha on the same side, while ensuring that the prongs are parallel to the axis of the ear canal. The patient is asked whether the tone can still be heard. If so, the result of the test is "positive," indicating that either the patient's hearing on this side is normal, or that the patient has a sensorineural hearing loss. In the latter case both bone conduction and air conduction times are shortened, and the result of the test is described as "small positive." If the patient does not hear the tone, the test should be repeated in reverse order – first, the tuning fork should be placed close to the concha, and once the patient can no longer hear the tone, its position has to be changed to the mastoid process on the same side. If the time of hearing the pure tone emitted by the tuning fork through the bone is longer than through the air, the result is "negative." This indicates a conductive hearing loss in the ear on the tested side. The same procedure is repeated for the opposite ear.

Pure-tone audiometry

Pure-tone audiometry is a "gold" standard test of audiologic examination. Its role is to assess whether hearing acuity is normal or impaired. Air conduction hearing thresholds are measured for tonal stimuli at the range of frequencies from 0.125 kHz to 8 kHz with the use of headphones. Then, bone conduction hearing thresholds are measured for tonal stimuli at the range of frequencies from 0.25 to 4 kHz, with the use of a headband with oscillator. The graph, plotted by connecting all hearing threshold values for all tested frequencies, is called "audiogram," "pure-tone audiogram," or "tonal audiogram." Hearing thresholds are measured in dB HL units, which are calibrated on normal-hearing young populations so that the bone conduction curve lies slightly above the air conduction curve.

In normal-hearing ears both curves are placed on the audiogram plot within the range of values that do not exceed 20 dB HL (Fig. 19.6A). In conductive hearing loss, air conduction thresholds worsen, so the air conduction curve is shifted down, by up to 50 dB HL, while bone conduction thresholds remain unchanged (Fig. 19.6B). The difference between air conduction curve and bone conduction curve is called an air–bone gap, and indicates the possible improvement in hearing if the cause of impaired sound transmission was removed. In sensorineural hearing loss both air conduction and bone conduction curves worsen, and no air–bone gap is present (Fig. 19.6C). The shape of the audiogram can differ depending on the disease. High-frequency hearing loss is the most frequent type of sensorineural hearing loss, since it is typical for both age-related and noise-induced hearing damage. Low-frequency hearing loss is seen in Ménière's disease, and a "U"-shape audiogram is typical for some hereditary hearing loss. Where there is both middle- and inner-ear pathology, an audiogram indicates a mixed type of hearing loss. Both air conduction and bone conduction curves worsen, but air–bone gap is also present (Fig. 19.6D).

Pure-tone audiometry should be performed in a soundproof booth (parameters are defined in International Organization for Standardization (ISO) norm (ISO 11957:1996, 1996). Calibration of the test equipment and the stimuli is needed before testing according to ISO standards. A rest period of at least 16 hours after last

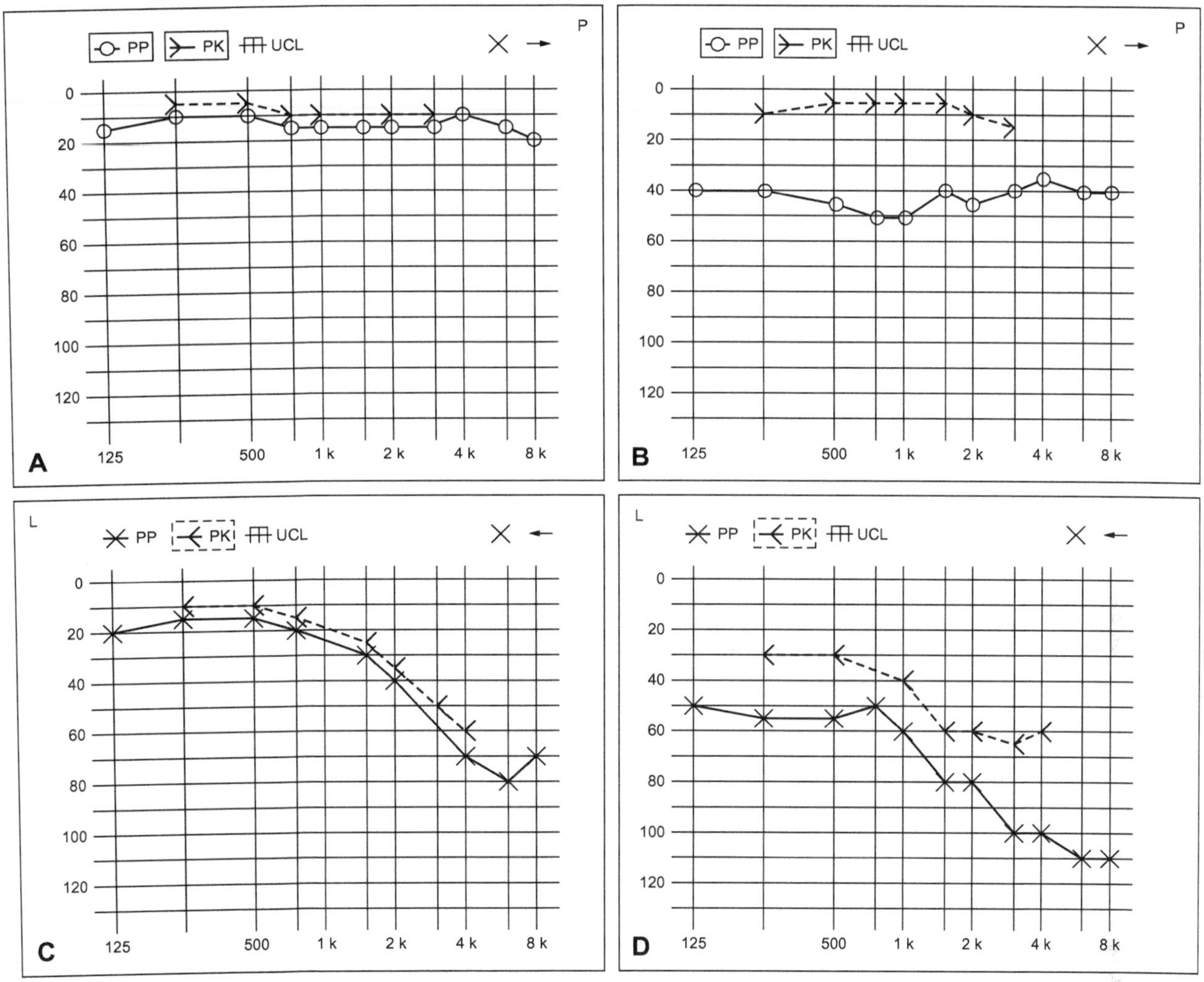

Fig. 19.6. Types of hearing loss assessed with pure-tone audiometry. (**A**) Normal hearing; (**B**) conductive hearing loss; (**C**) sensorineural hearing loss; (**D**) mixed hearing loss.

exposure to noise is advised, in order to avoid TTS influence on the audiometric test result.

Cochlear recruitment

The presence of cochlear recruitment (a perceptual phenomenon of faster increase of sound loudness in relation to the increase of its intensity) is typically assessed with suprathreshold audiometric tests, such as Fowler test, Reger test or Short Increment Sensitivity Index (SISI), among others.

In the SISI test, the intensity of a pure tone is amplified by 1 dB 20 times, starting from the intensity of 20 dB SL (sensation level, i.e., level above audiometric hearing threshold). The result is presented as a percent of observed increases in the intensity of sound reported by the patient. The normal-hearing ear does not differentiate such small changes in sound intensity. In contrast, in the case of sensorineural hearing loss due to OHC damage, the patient recognizes all or almost all increases in sound intensity. Values above 80% (according to some authors, above 60%) indicate the presence of cochlear recruitment. The test is calibrated for 1 kHz, but in practice it is used also for higher frequencies (2 and 4 kHz).

Impedance audiometry

Impedance or immitance audiometry is an objective assessment method of the function of the middle ear. By increasing acoustic pressure in the external ear canal, impedance audiometry measures the sum of resistance which a sound wave encounters on its way through the middle ear to the cochlear receptor. Middle-ear resistance depends on the physical properties of its elements, such as the mass of ossicles, thickness and flexibility of tympanic membrane, and the stiffness of the middle-ear tendons and muscles. The measurement is simple and can be performed in a quiet room without or with very limited cooperation on the patient's side.

The resistance of the tympanic membrane and the middle ear is measured in cubic centimeters. Normal values range from 0.3 to 1.3 cm^3, with abnormal values indicating pathology of the middle ear. Changes in resistance of the conductive system due to increasing acoustic pressure in the external ear canal are graphically presented as a tympanogram. The shape of the tympanogram, the peak of its amplitude, and value of acoustic pressure at which the curve reaches the maximum allow us to draw conclusions about the type of pathology in the middle ear. The type "A" tympanogram is typical for normal middle ears. The peak of the tympanometric curve is reached at 0 mmH_2O, and its amplitude is normal. Where there is increased resistance of the middle ear, as is observed in a presence of fluid in the middle ear, the tympanogram is flat (type "B"). Finally, where there is decreased air pressure in the middle ear, as is common in Eustachian tube dysfunction, the tympanogram is of type "C." The shape of the tympanogram C is similar to tympanogram A, but the peak of the curve is shifted towards lower values of sound pressure, i.e. below −100 mmH_2O. In NIHL, the tympanogram is of type A, since no changes in the middle ear are present.

The secondary purpose of impedance audiometry is to register the acoustic reflex from the stapedius muscle, as well as to evaluate the function of structures involved in reflex arc pathway, i.e., the cochlea, auditory nerve, ventral cochlear nucleus, trapezoid body, medial superior olive, facial nerve and its motor nucleus, and stapedius muscle. Stapedius muscle contraction occurs bilaterally in response to intense sounds with either unilateral (ipsi- or contralateral) or bilateral stimulation. Acoustic reflex thresholds are generally determined in response to the stimuli of 0.5, 1, 2, and 4 kHz. The thresholds for acoustic reflexes in young individuals with normal hearing are in the range of 70–100 dB HL (on average 80 dB) above audiometric threshold.

Impedance audiometry cannot be used to directly assess auditory sensitivity, but its results can be interpreted in conjunction with other threshold measures. For mild to moderate sensorineural hearing loss with recruitment, acoustic reflex thresholds remain within the normal range. Thus, the distance between audiometric thresholds and acoustic reflex thresholds decreases to 60 dB or less. This symptom, known as the Metz symptom, is used to confirm cochlear recruitment.

Otoacoustic emissions

OAEs are weak acoustic signals emitted from the cochlea spontaneously or in response to sound stimulation. OAEs are the byproducts of the OHC contraction and production of a reverse acoustic wave emitted back through the middle ear to the external ear canal. When testing the OAE, a small microphone is installed in the external ear canal to register the signals.

Transiently evoked OAEs (TEOAEs) and distortion-product OAEs (DPOAEs) are the tests commonly used in clinics. TEOAE is generated after stimulation of the cochlea with a click (a wideband stimulus) and OAE is assessed as a whole response or a response by frequency using Fourier transform function. DPOAE is generated by two pure tones, slightly differing in frequency (usually $f_1/f_2 = 1.20–1.25$), and the OAE signal is measured at the third frequency $f_3 = 2f_1 - f_2$, where $f_2 > f_1$. OAE tests can be performed in a quiet room. The measurement is simple, fast, and objective.

Both TEOAEs and DPOAEs are very sensitive to the loss of OHC, and the changes in OAEs are believed to precede the audiometric threshold shifts. Thus, OAEs are widely used in monitoring NIHL. The advantage of DPOAE over TEOAE is higher frequency specificity, larger spectrum of analyzed frequencies in the high-frequency region, and wider dynamic range. TEOAE is usually absent when hearing loss exceeds the level of 30–40 dB HL, while DPOAE is absent when hearing loss exceeds 45–50 dB HL. Both tests are characterized by large intersubject variability.

The normal results of TEOAE and DPOAE are shown in Figure 19.7.

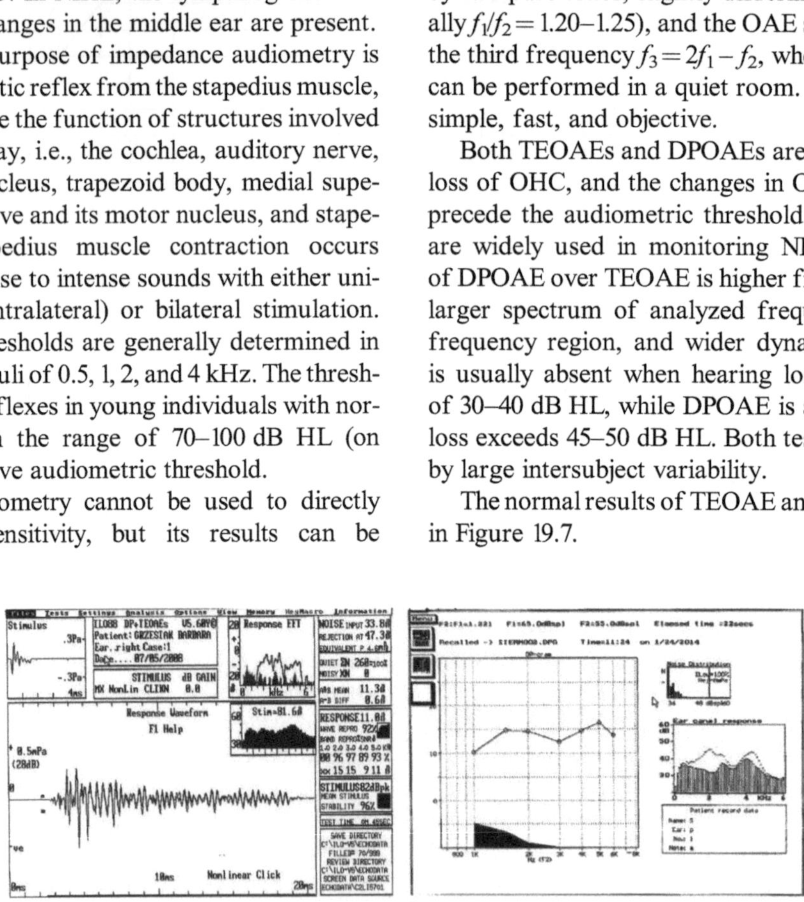

A B

Fig. 19.7. Normal results of (**A**) transiently evoked otoacoustic emission (TEOAE) and (**B**) distortion-product otoacoustic emission (DPOAE).

Auditory-evoked potentials

Auditory evoked potentials measure the bioelectric function of the auditory pathway in response to sound (or electric) stimuli. The most common test is the auditory brainstem response (ABR).

ABR assesses postreceptor function of the cochlea, and functionality of auditory nerve and brainstem. Three electrodes are placed on the patient's head: one on each mastoid process and the third one on the forehead in the midline below the hairline. Click or tone burst acoustic stimuli are presented through the earphones. An average response of the auditory pathway to an acoustic stimulus is plotted for each ear. The plot contains five waveforms: I, II, III, IV, and V. Wave V is of highest amplitude and its presence determines the threshold of the response. The waveforms represent the electric activity derived in specific anatomic sites, i.e., the cochlear nerve and nuclei (waves I and II), superior olivary nucleus (wave III), lateral lemniscus (wave IV), and inferior colliculi (wave V).

In differential diagnostics of NIHL, the ABR allows for differentiation between cochlear and retrocochlear site of lesion:

- normal hearing – morphology of the waveforms and wave and interwave latencies are normal
- conductive hearing loss – the latencies of all waveforms can be increased, but interwave latencies are normal
- sensorineural hearing loss (with recruitment) – the latency of wave I can be increased, but the interwave latencies are normal
- retrocochlear hearing loss – latency of wave I is normal; there is elongation of interwave I–V and III–V latencies (central conduction time delay); the morphology of waves may be abnormal.

In uncooperative subjects the test is used to reconstruct hearing thresholds. More information about the measurement technique can be found in audiologic handbooks.

Behavioral auditory processing tests

According to the American Speech–Language–Hearing Association (ASHA), central auditory processing refers to the perceptual processing of auditory information in the central nervous system and the neurobiologic activity which underlies this processing (www.asha.org/policy). Auditory processing disorders can be described as deficits in the effectiveness of the central nervous system to utilize auditory information, including interhemispheric communication. Auditory processing is related to the following auditory abilities or skills:

- auditory discrimination
- temporal aspects of audition (sound sequencing, ordering, and resolution)
- sound localization and lateralization
- auditory performance in competing acoustic signals
- auditory performance with degraded acoustic signals.

Behavioral tests of central auditory processing are used to assess auditory processing disorders. They are categorized as verbal and nonverbal tests. Among verbal tests which gained widespread acceptance are the Dichotic Digit (DD) test and the Hearing in Noise Test (HINT). Nonverbal tests commonly used in clinics are the Duration Pattern Test (DPT), the Frequency Pattern Test (FPT) and the Gaps in Noise (GIN) test, among others.

Dichotic Digit test

The DD test is a verbal test which assesses auditory discrimination of two pairs of digits presented simultaneously to the right and left ear, with one pair of digits applied to each side. The duration of each digit-pair presentation is 500 ms, with a silent interval of 500 ms. The test is administered at 50 dB above the subject's audiometric threshold at 1 kHz. The binaural integration task requires patients to repeat what was heard in both ears. The results are presented as the percentage of correct answers.

Hearing in Noise Test

HINT is a verbal test assessing auditory performance in competing acoustic signals (Soli and Wong, 2008). Lists of 10 sentences are presented bilaterally in four test conditions, which are: (1) sentences with no competing noise; (2) sentences with competing noise of steady loudness 65 dB(A) presented in front of the patient; (3) sentences with competing noise presented to the right ear; and (4) sentences with competing noise presented to the left ear. The subject is required to listen to the sentence and repeat it. All words in the sentence must be repeated correctly. The results are presented as the smaller signal-to-noise ratio, indicating how loud the sentences need to be turned up (above 65 dB(A) noise floor) in order for the patient to be able to correctly repeat 50% of sentence material. The higher the signal-to-noise ratio, the more difficulty the patient has with hearing.

The Duration Pattern Test

The DPT is a nonverbal test assessing temporal sequencing and ordering of two or more auditory stimuli of different duration. The test performance is based on presenting sets of 30 sequences in two tones differing in duration (250 and 500 ms), arranged in three-tone tokens (e.g. "short-long-long" or "long-short-long") (Musiek, 1994). The tones are presented monaurally to each ear at 50 dB above the subject's audiometric threshold at 1000 Hz. The tone frequency is 1000 Hz, with a rise/fall time of 10 ms, and with an intertone interval of 300 ms. The task of the subject is to verbalize the duration of each tone in the right sequence (e.g., "short-long-long"). The answers are registered on the score sheet and the percentage of correct answers is calculated. The 10th percentile is usually a cut-off score to distinguish between "normal" and "abnormal" test results.

The Frequency Pattern Test

The FPT is a nonverbal test assessing temporal sequencing and ordering of two or more auditory stimuli of different frequencies. The test performance is usually based on presenting sets of 30 sequences in two tones differing in pitch (880 and 1122 Hz), arranged in three-tone tokens (e.g., "low-high-high" or "high-low-high") (Musiek, 1994). The tones are presented monaurally to each ear at 50 dB above the subject's audiometric threshold at 1000 Hz. The tone duration is 150 ms with a rise/fall time of 10 ms, and with an intertone interval of 200 ms. The task of the subject is to verbalize the pitch of each tone in the right sequence (e.g., "low-high-high"). The answers are registered on the score sheet and the percentage of correct answers is calculated. The 10th percentile is usually a cut-off score to distinguish between "normal" and "abnormal" test results.

The Gaps in Noise Test

The GIN is a nonverbal test assessing temporal resolution by referring to the shortest duration of time in which subjects can discriminate between two auditory signals (Musiek et al., 2005). Noise stimuli are presented monaurally at 50 dB HL above the subject's audiometric threshold at 1000 Hz. This test consists of a series of 36 different 6-second segments of white noise, each containing 0–3 gaps. The interstimulus interval between noise segments is 5 seconds. The gap durations in this test are: 2, 3, 4, 5, 6, 8, 10, 12, and 20 ms, with each silence gap duration occurring six times. GIN contains a total of 60 gaps per list. While undergoing the test, the subject needs to listen for any silence gap that may or may not occur within each noise segment. As soon as the noise segment is finished the subject has to state how many gaps (0–3) took place in the noise segment presented. The approximate threshold is used for analysis by considering the following two criteria: (1) at least four out of six gaps are correctly identified; (2) performance for longer gap durations is not worse than for four out of six gaps correctly identified.

CLINICAL ASPECTS OF NOISE-INDUCED HEARING LOSS

Chronic exposure to noise leads to the development of NIHL. NIHL is defined as binaural symmetric (or almost symmetric) sensorineural hearing loss, pertaining predominantly to high audiometric frequencies and slowly progressing over the years of exposure. In the early stages of NIHL a notch at the frequencies 3–6 kHz is typical.

Acute exposure to noise at very high levels can also lead to acoustic trauma, i.e., suddenly developed sensorineural hearing loss. In contrast to NIHL, acoustic trauma may be unilateral. Both NIHL and acoustic trauma are often accompanied by tinnitus. From a medicolegal point of view, compensation can be claimed for NIHL as an occupational disease related to occupational exposure. On the other hand, acoustic trauma related to acute exposure to noise in the workplace is categorized as an accident at work. The following parts of this chapter will refer only to occupational NIHL.

Diagnostics of NIHL

Diagnostics of NIHL in a cooperative patient is relatively easy. In principle it is based on medical history (anamnesis), otoscopy, and audiometric evaluation.

Anamnesis

Anamnesis should include detailed information on age, time of employment, occupation, occupational hazards, such as exposure to noise (current and in previous workplaces), exposure and co-exposure to chemicals, exposure and co-exposure to local vibration, nonoccupational exposure to aforementioned hazards, signs and symptoms of the disease, as well as other risk factors promoting the development of hearing loss. The most important data that should be gathered through medical history is listed in Table 19.3.

The age, time of employment, and actual occupation of the patient is key information. For instance, a substantial progression of hearing loss in a retiring person is more probably due to the aging of the hearing organ than to exposure to noise in the past. The probability of developing NIHL related to occupational exposure is very high for some professions (e.g., in heavy-metal industry workers), but negligible or none in other (e.g., teachers or call center operators). The type of noise in the workplace also plays a role. It has been proven that exposure to impulse noise is more risky to hearing than exposure to steady-state noise. This is because the sound stimuli related to impulse noise grow too rapidly for the stapedius muscle to contract and have a protective effect.

Table 19.3

Medical history

Exposure
• Time of employment and occupation
• Characteristics of noise at workplace (high/moderate levels of noise; steady-state/impulse noise; continuous/intermittent exposure; the use of hearing protectors)
• Exposure to chemicals (type of chemicals; route of exposure: airways/skin; protection)
• Exposure/co-exposure to vibration
Signs and symptoms
• Hearing loss (sudden/chronic onset; profoundness, symmetry, speech intelligibility in noisy environment)
• Presence of tinnitus
• Presence of vertigo or dizziness
Risk factors for hearing loss
• Head trauma in the past
• Middle-ear inflammation and surgery in the past
• Smoking
• Hypertension
• Other diseases (diabetes, kidney disease, meningitis)
• Ototoxic drugs (now and in the past)
• Acoustic trauma and barotrauma in the past
• Family history of hearing loss

Co-exposure to chemicals and vibration can have an additive or supra-additive impact on the development of NIHL. Chronic occupational exposure to local vibration (through the hands) can lead to the development of vascular and nerve signs and symptoms, called hand–arm vibration syndrome (HAVS). The most characteristic sign is paleness of the fingers (fading fingers) in the cold. In a group of workers with HAVS exposed additionally to noise in the workplace, permanent hearing thresholds were about 10 dB higher than in workers exposed to noise only, at the comparable level.

Hearing loss due to prolonged exposure to occupational noise develops very slowly, over years of exposure. In the early stage of NIHL, hearing problems can remain unnoticed, because hearing threshold shifts concern only the high audiometric frequencies, i.e., the region that is not important for speech understanding. Problems with hearing are more often first noticed by family members than by the subject. However, over time hearing acuity worsens, the patient asks others to talk louder, and can have increasing difficulties with understanding speech, especially in noisy environments. The rate of increase in hearing loss is highest in the first 10–15 years of exposure to noise. Even though hearing loss can constantly progress over the time of overexposure to noise, its degree never reaches a profound level, i.e., requiring shouting in order to communicate. If this is the case, another cause of hearing loss should be suspected.

Subjective tinnitus is a symptom which very often accompanies both NIHL and acoustic trauma. While in the general population the prevalence of tinnitus is estimated at 5–10%, in subjects with NIHL this symptom can be present in up to 50–70% of people. Tinnitus is perceived as different sounds (squeaks, rustles, rumbles). There is no association between the presence of tinnitus and the degree of hearing loss.

Due to the close proximity of sensory receptors of the auditory and vestibular systems, several diseases can involve both hearing loss and vertigo. However, this is not the case with NIHL. Although extremely high exposures may cause transient vertigo (so-called Tulio symptom) and some studies show saccular degeneration after prolonged noise exposure, noise does not induce vertigo. The presence of spinning vertigo with accompanying nausea or vomiting indicates inner-ear disease other than NIHL. Dizziness defined as nonspinning vertigo, without accompanying nausea or vomiting, is of central origin, and does not contraindicate the recognition of NIHL.

Medical history should also include any information about other major risk factors of hearing loss. In some cases these could pose a more significant threat to hearing than noise. Some examples include severe head trauma, barotrauma, middle-ear surgery due to chronic inflammation, some systemic and metabolic diseases (like diabetes), and intake of ototoxic drugs. Another very important question is family history, as it is believed that the same genetic polymorphisms may be responsible for increased susceptibility for age-related hearing loss and NIHL.

Otoscopic examination

Otoscopy is the most important part of a physical examination. If a patient is suspected of suffering only NIHL, the appearance of the tympanic membrane should be normal. If this is not the case, the patient needs to undergo a detailed otorhinolaryngologic evaluation. The same is true for patients complaining of chronic nasal congestion.

Audiometric evaluation

The standard first-line test in the diagnostics of NIHL is pure-tone audiometry. Suprathreshold audiometric tests (like SISI) are required to confirm cochlear recruitment. The most typical picture of NIHL at an early stage is shown in Figure 19.8. Both ears exhibit sensorineural (perceptive) hearing loss with typical notch at the frequencies of 3–6 kHz (Fig. 19.8A). Hearing thresholds are slightly worse in the left ear than in the right ear. This coincides with epidemiologic studies, which show slightly higher susceptibility of the left ear to develop NIHL.

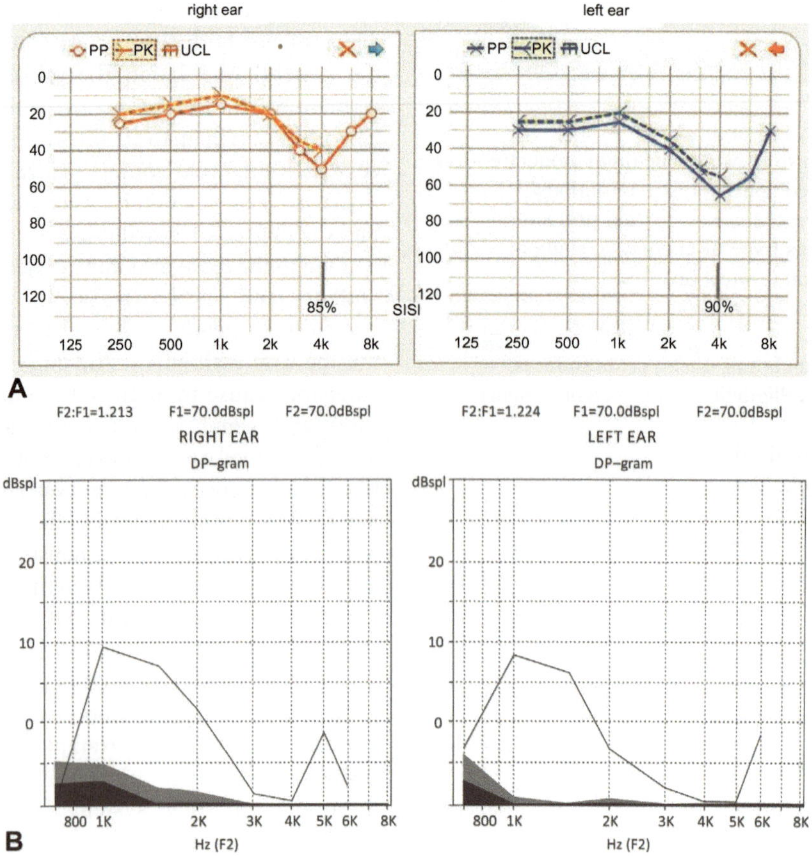

Fig. 19.8. Hearing test results in a 40-year-old man exposed to noise on a farm. (**A**) Pure-tone audiometry; (**B**) distortion-product otoacoustic emission (DPOAE).

The results of SISI are 85% and 90% at 4 kHz in the right and left ear respectively, indicating cochlear origin of hearing loss. The loss of OHCs is further confirmed by DPOAE. A lack of DPOAE signals is seen at 4 kHz in the right ear and at 4–5 kHz in the left ear, corresponding to the frequencies of audiometric threshold shifts (Fig. 19.8B). Along with the 20-year history of the subject's exposure to noise on a farm, audiometric test results clearly indicate the presence of NIHL.

Diagnosing NIHL becomes more difficult with increasing years of exposure to noise and degeneration of the auditory nerve fibers (and some ganglion cells), which is secondary to the loss of OHCs. The audiometric curve may lose its typical notch at high frequencies, although hearing threshold shifts are always most profound at high frequencies, and only minor effects of noise exposure can be seen in the middle and low frequencies (Fig. 19.9A). Audiometric suprathreshold tests, like SISI, may not be pathognomonic for cochlear recruitment. The cochlear site of lesion may be confirmed by OAEs, which are usually absent in the damaged high-frequency region (Fig. 19.9B). Also the differences between audiometric thresholds and the thresholds of acoustic stapedius reflexes are smaller than 60 dB, indicating a cochlear recruitment phenomenon (Metz symptom) (Fig. 19.9A).

An example of audiometric test results in a 65-year-old man, exposed to occupational noise for over 35 years, is presented in Figure 19.9.

In every case of NIHL, speech audiometry is impaired. Understanding speech in a noisy environment becomes most challenging, thus speech in noise tests like HINT are highly recommended for diagnostics. A summary of audiometric test results characteristic for NIHL is given in Table 19.4.

Differential diagnostics

Occupational NIHL should be differentiated from other pathologies involving the inner ear and central parts of the auditory pathway, including:

- age-related hearing loss (presbycusis)
- sensorineural hearing loss due to causes other than exposure to noise
- retrocochlear hearing loss

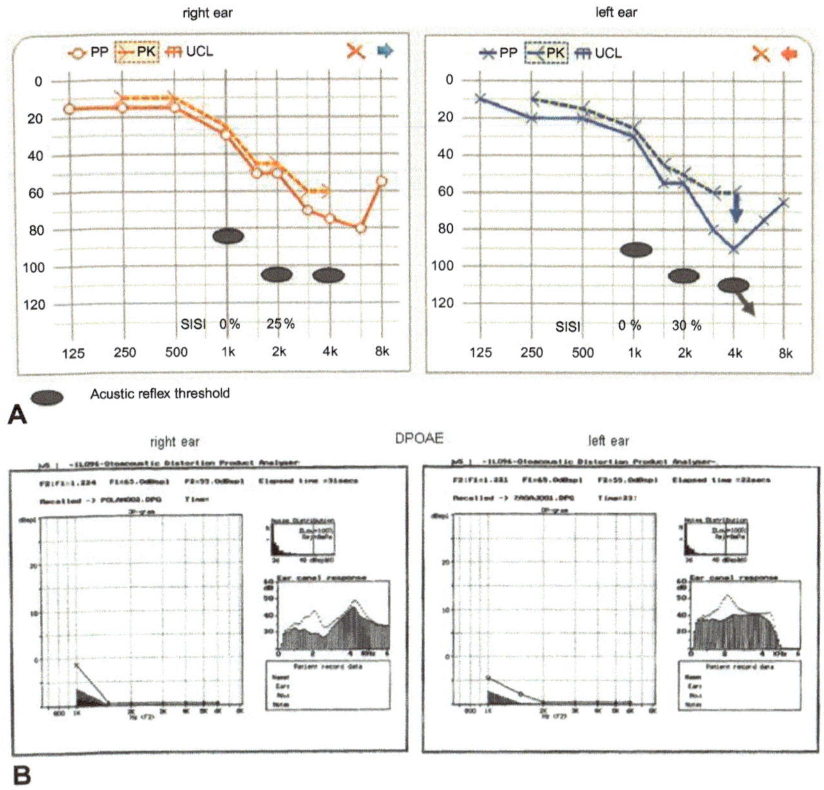

Fig. 19.9. Hearing test results in a 65-year-old man exposed to noise at the workplace for 35 years. (**A**) Pure-tone audiograms, Short Increment Sensitivity Index (SISI), and acoustic reflex thresholds; (**B**) distortion-product otoacoustic emission (DPOAE).

- central auditory pathologies
- malingering of hearing loss (in medicolegal cases).

Age-related hearing loss is the most frequent type of hearing impairment in the adult population. As in NIHL, hearing loss is of sensorineural type, and concerns mainly high audiometric frequencies, although no typical notch at 3–6 kHz can be seen. Usually this kind of hearing loss is not observed in subjects below 55 years of age. Differential diagnostics is based on the growth rate of hearing loss over the lifetime. In age-related hearing loss this rate is fastest in elderly subjects, while for NIHL it is fastest in the first 10–15 years of exposure to noise.

Apart from NIHL and aging, several other diseases can affect the inner ear. These include sudden deafness, autoimmune inner-ear disease, and Ménière's disease, among other. Sudden deafness develops within hours or days, autoimmune inner disease progresses over months, and hearing loss fluctuates, while in NIHL hearing loss develops over years and is irreversible. The other symptom that is important for diagnostics is vertigo, which is present in several inner-ear pathologies, like Ménière's disease, but absent in NIHL. If hearing loss is unilateral, it also contraindicates NIHL.

Some retrocochlear and central auditory diseases can result in sensorineural hearing loss, if the blood supply to the cochlea is impaired. The most common cases relate to acoustic neuromas, which can be accompanied by unilateral hearing loss. Tumors of the central nervous system can also result in bilateral symmetric high-frequency hearing loss with a distinctive, sharp slope of the audiometric curve and deep hearing threshold shifts at highest audiometric frequencies. Auditory evoked potentials and brain imaging determine the diagnosis.

The diagnostics of occupational NIHL can be difficult in subjects who aggravate hearing loss, which is commonly related to medicolegal cases. If this occurs, the profoundness of hearing loss should be proven on the basis of reconstruction of the audiogram by means of ABR.

Management

NIHL is an irreversible disease, thus maximum effort should be exerted to prevent hearing damage. Hearing conservation programs in occupational settings are based on attenuating the noise from its sources, using hearing protectors, monitoring hearing, and training workers on the risks of hearing loss. It should be

Table 19.4

Summary of audiometric changes related to noise-induced hearing loss (NIHL)

Audiometric test	Test results
Pure-tone audiometry	• Hearing threshold shifts mainly at high frequencies • Characteristic notch at 3–6 kHz, deepening and widening over the years of exposure to noise • Hearing loss binaural, symmetric, or almost symmetric • The highest rate of growth of hearing loss in the first 10–15 years of exposure to noise
Recruitment	• Present • Suprathreshold audiometric tests, like Short Increment Sensitivity Index, pathognomonic in the early stages of NIHL • Acoustic reflex threshold assessment for Metz symptom recommended in the late stages of NIHL
Otoacoustic emission	• Attenuation or loss of evoked otoacoustic emissions • Changes in frequencies corresponding to audiometric hearing threshold shifts
Speech audiometry	• Impairment of speech intelligibility in noisy environment (Hearing in Noise Test (HINT) recommended)

remembered that hearing protection is effective only if it is used throughout whole workday. TTS evaluation with the use of audiometry or OAEs can be helpful in real-life verification of the effectiveness of hearing protectors.

Pharmacologic intervention is applied in acoustic trauma. The spectrum of applied drugs is vast, and includes corticosteroids, antioxidative drugs, and vitamins, among other.

AUDITORY AND OTONEUROLOGIC EFFECTS OF ORGANIC SOLVENTS

There is growing clinical evidence that organic solvents may cause dysfunction of both peripheral and central parts of the auditory organ, as well as vestibular organ and balance system. Recently a critical systematic review on the ototoxic potential of industrial chemicals was published (Vyskocil et al., 2012). Results of human studies and evidence from animal studies show that three organic solvents, styrene, toluene, and trichloroethylene, should be regarded as ototoxic agents. There is also convincing evidence that, after combined exposure to toluene and noise, an ototoxic interaction can induce more severe hearing loss. It is possible that similar interactions exist for styrene and trichloroethylene, but further studies are necessary to draw a final conclusion. In addition, given the current evidence from animal studies, three other organic solvents, *n*-hexane, ethyl benzene, and *p*-xylene (and consequently a mixture of xylene isomers), should be considered as possible ototoxic agents. No conclusive data exist on the ototoxicity of *n*-heptane. There is no evidence of ototoxicity of hexachlorobenzene, *p*-tert-buthyltoluene, or methylene chloride, or of their interaction with noise.

Solvent mixtures are the most common kind of exposure to organic solvents in industry. Less common are isolated exposure to styrene, toluene, and *n*-hexane. Studies on exposure to mixtures of solvents included painters, paint and lacquer manufacturing workers (Morata et al., 1993; Sliwinska-Kowalska et al., 2001), dockyard (Sliwinska-Kowalska et al., 2004) and petrochemical industry workers (Morata et al., 1997a), coating factory workers (Fuente et al., 2009), and employees exposed to JP-8 (jet fuel) (Kaufman et al., 2005; Hughes and Hunting, 2013), among other. Mixtures were usually composed of xylenes, toluene, benzene, ethyl benzene, methyl-ethyl-ketone, methyl-isobuthyl-ketone, trichloroethylene, and *n*-hexane. Studies on styrene-only exposure included employees of a glass fibre-reinforced plastic products industry (yachts, lavatory pans, and wash basins) (Sliwinska-Kowalska et al., 2003; Hoffmann et al., 2006; Johnson et al., 2006). Finally studies on toluene-only exposure included rotogravure printing plant workers (Morata et al., 1997b; Schäper et al., 2008).

The next paragraphs summarize the results of studies aimed at assessing the ototoxicity of organic solvents used in industry. All studies were cross-sectional, not free of methodologic flaws and bias related to the selection of study subjects and confounding factors. Thus, their results must be interpreted with caution.

Auditory effects

Risk of hearing loss

Occupational exposure to organic solvents over a long period of time may increase the risk of hearing loss. Depending on the audiometric cut-off criterion (usually it is hearing threshold above 25 dB HL for at least one audiometric frequency in the range from 0.125 to 8 kHz), the incidence of hearing loss in workers exposed to mixtures of solvents, including with toluene as one of the main compounds, ranged between 18% and 57% (Morata et al., 1993; Sliwinska-Kowalska et al., 2001), and in workers exposed to styrene reached a value of 63% (Sliwinska-Kowalska et al., 2003). Combined exposure to noise and organic solvents significantly increased these numbers, up to 61% for mixtures (Sliwinska-Kowalska et al., 2004; Kim et al., 2005), and 80 and 86% for co-exposure to noise and styrene or toluene, respectively (Sliwinska-Kowalska et al., 2003; Chang et al., 2006). In reference groups, the incidence of hearing loss was in the range of 17–45% in noise-only exposed workers, and in the range of 5–8% in control non-exposed groups.

The risk of hearing loss due to occupational exposure to organic solvents depends primarily on time of employment, the composition of the mixture of organic solvents, and co-exposure to noise. Once the risk of hearing loss due to sole exposure to noise is roughly 2–4 times higher as compared with the nonexposed population, and similar numbers are reported for sole exposures to organic solvents, the probability of hearing threshold shift significantly increases in case of co-exposure. The highest risk (odds ratio 21.5; confidence interval 5.1–90.0) of hearing damage was found in workers co-exposed to noise and a mixture of styrene, toluene, and *n*-hexane (Sliwinska-Kowalska et al., 2005). Low-moderate solvent concentration and exposure lasting less than 5 years were not shown to impair hearing (Schäper et al., 2008; Hughes and Hunting, 2013).

Pure-tone audiometry

Exposure to organic solvents may cause sensorineural hearing loss related to OHC damage that is manifested by audiometric hearing threshold shifts. Studies show that hearing threshold shift of a small to moderate degree can result from isolated exposure to styrene. In such a case, hearing loss can potentially be shown in all standard audiometric frequencies, as well as extended high frequencies (Muijser et al., 1988; Morioka et al., 2000; Sliwinska-Kowalska et al., 2005; Hoffmann et al., 2006). Since the basal turn of the cochlea in humans is the most sensitive to ototoxic chemicals, it is believed that extended high-frequency hearing threshold shift can be a first sign of hearing damage. Unfortunately, high-frequency audiometry is not a standardized test used in clinics, and no normative data for this test is accessible.

In case of co-exposure to noise and the mixture of solvents, noise-induced hearing threshold shift seems to be dominant, although some additional effect of solvents can be also seen. It pertains predominantly to high audiometric frequencies (Sliwinska-Kowalska et al., 2005).

Central auditory lesion

Occupational exposures to organic solvents can have damaging effects on the central part of the auditory pathway (Juárez-Pérez et al., 2014). Thus, the dysfunction can be manifested not only as a sensorineural hearing loss related to cochlear damage, but also as an auditory neuropathy or central auditory processing disturbance (Fuente et al., 2011).

Examples of central auditory dysfunction that may be related to occupational exposure to organic solvents are well documented (Fuente et al., 2011). The dysfunction due to the mixture of solvents, as well as toluene and styrene, was shown using large batteries of tests, including ABR (Abbate et al., 1993; Prasher et al., 2005), cortical evoked responses (Vrca et al., 1996; Johnson et al., 2006), interrupted speech and speech in noise tests (Johnson et al., 2006), behavioral auditory processing tests, like the DD test (Fuente et al., 2009), FPT, Latency Pattern test, and GIN test (Zamyslowska-Szmytke et al., 2009), HINT, and Random Gap Detection (Fuente et al., 2013). Since auditory processing disorders can be present despite a lack of audiometric threshold shift, it is recommended that workers exposed to organic solvents are routinely examined for central processing disorders. The most widely used tests in this area are the DD test and HINT.

The possible sites of damage of the auditory pathway due to exposure to organic solvents are listed in Table 19.5.

Relationship between concentration of solvents and hearing loss

Several studies examined the correlation between solvent concentration and risk of hearing loss as well as the degree of hearing threshold shift at certain

Table 19.5

Summary of audiometric test results related to occupational exposures to organic solvents by site of lesion in the auditory pathway

Site of lesion	Test results
Damage to the cochlea	• Increased hearing thresholds in pure-tone audiometry (0.5–8 kHz) • Increased hearing thresholds in extended high frequencies (12 and 16 kHz) • Decrease in otoacoustic emission
Damage to the auditory nerve	• Elongation of wave I–V interval • Increased latency of wave V
Damage to the central auditory structures	• Pathologic results of contralateral acoustic reflex • Abnormalities in cortical evoked responses, cognitive event-related potentials • Abnormalities in auditory processing assessed with Dichotic Digits (DD) test, Random Gap Detection (RGD) test, Gaps in Noise (GIN) test, Distorted Speech (DS), Interrupted Speech (IS), Filtered Speech (FS), and Hearing in Noise Test (HINT)

frequencies. The exposures were evaluated according to years of exposure, concentration of a solvent in the air, and concentration of solvent metabolites in urine. Where there was exposure to a mixture of organic solvents, a whole working-life exposure index was calculated for all mixture compounds. A correlation was found between the years/concentration of exposure and the risk of hearing loss in standard frequencies (mixtures of solvents), the degree of hearing threshold shift at standard audiometric frequencies, and extended high frequencies (for styrene). However, so far no maximum allowable exposure values for organic solvents have been determined in respect to their ototoxic effects.

Otoneurologic effects

Vestibular organ and balance system dysfunction related to occupational exposures to organic solvents is less documented than auditory effects. The most evident abnormalities were found in patients with chronic toxic encephalopathy (CTE), which was developed as a consequence of exposure to solvents. Two major tests are commonly used in otoneurologic diagnostics. These are posturography (static and dynamic), and electro(video) nystagmography (E(V)NG). Abnormalities in kinetic visual suppression test were the most common pathology in E(V)NG, followed by an increased number of square waves, increased latency of saccades, and increased vestibular reactivity (Ödkvist et al., 1987; Niklasson et al., 1997). Posturography test results were abnormal for opened and closed eyes on foam platform; swing scope of the center of gravity was larger (Niklasson et al., 1997). Moreover, incorrect results for motor coordination were reported in psychomotor tests. These abnormalities predominantly indicate a damage of the central part of the vestibular and balance systems.

Several vestibular and balance abnormalities were also found in subjects exposed to organic solvents who did not develop the symptoms of CTE. Again, E(V)NG mostly indicated central damage of the vestibular system. Abnormalities were found in optokinetic test, pendulum tracking test, and saccades (Prasher et al., 2005; Zamyslowska-Szmytke and Sliwinska-Kowalska, 2011; Zamyslowska-Szmytke et al., 2011). More rarely, an asymmetric hyperreactivity or hyporeactivity of labyrinths was found (Calabrese, 1996). Less consistent changes were present in posturographic examination. No clear correlation between the concentration of solvents and E(V)NG and posturography test parameters was found.

Few studies which were designed to localize the damage caused by organic solvents in the vestibular system suggest that the main sensitive central structures are the paramedian pontine reticular formation and/or the complex of vestibular nuclei. Abnormal test results of eye fixation in patients with CTE can be related to the pathology of rostral pole of inferior colliculus, located in the mesencephalon.

HEAVY METALS

Heavy metals are defined as metals with density above 4.5 g/cm^3. These substances access human body in water and food, and can be inhaled with air through the lungs or absorbed through the skin. Heavy metals cannot be metabolized. Once accumulated in the soft tissue and bones, heavy metals exert several toxic effects. The possible ototoxic chemicals are: lead, mercury, cadmium, and arsenic; however their effects for concentrations present in industrial settings are still not fully recognized (Prasher, 2009). A summary of available research on ototoxicity of heavy metals is given in Table 19.6.

Lead

The annual world lead production is estimated at 2.5 million tons. Most of this metal is used for batteries. The remainder is used for cable coverings, plumbing, ammunition, and fuel additives, among others.

Occupational exposure to lead can cause several neurobehavioral and endocrine alterations, like hyperactivity, neuropathy, headaches, and cognitive disabilities. Given the current evidence from human studies, lead is also

Table 19.6

Summary of study results related to the ototoxicity of heavy metals

Metal	Auditory effect
Lead	• Inconsistent findings on lead-induced hearing loss, but delayed wave I of auditory brainstem responses (ABR) implies cochlear dysfunction • Increased risk of hearing loss and association between blood concentration of lead and hearing thresholds in occupationally exposed populations • Not clear whether lead and noise exacerbate hearing loss
Mercury	• Central conduction time delay (ABR I–V, III–V) • Cochlear function may be unaffected
Cadmium	• Dose-dependent hearing loss in rats • Delayed wave I implies cochlear dysfunction • Zinc-enriched diet reduces ototoxic effect • Cadmium and noise show synergistic effect at 4 and 6 kHz
Arsenic	• Low-frequency hearing loss • Balance disturbances

recommended to be considered as an ototoxic agent (Vyskocil et al., 2012).

Among several metals (manganese, copper, zinc, arsenic, cadmium, and lead), only elevated concentration of lead over 7 mg/L in blood was significantly associated with hearing loss (Hwang et al., 2009). Mainly high audiometric frequencies, from 3 to 8 kHz, were affected. A case-control epidemiologic study shows a positive association between the concentration of lead and hearing threshold, while an inverse association was found for selenium blood concentration. It may indicate that selenium is an antagonist to lead ototoxicity (Chuang et al., 2007).

It is not clear whether co-exposure to lead and noise exacerbates hearing loss. Raised auditory thresholds from 3 to 8 kHz due to combined lead and noise exposure were suggested in ceramic glazing industry workers exposed for long periods of time (Counter and Buchanan, 2002). A significant correlation between high long-term lead exposure (duration of employment and ambient lead concentration) with hearing threshold at 4 kHz was also found in workers of a lead-battery manufacturing factory (Wu et al., 2000). Environmental exposure is among the major confounding factors in occupationally lead-exposed adults.

Mercury

Mercury is widely used in dental amalgams, thermometers, vaccine preservatives, topical antiseptics, aquatic sediments, mercury-based skin creams, fungicides, gold mining, and the chloralkali industry. Environmental intoxication is related to consumption of fish (e.g., shark, swordfish, tuna). In 1953, a severe neurologic disorder was recognized among persons living in the vicinity of Minimata, Japan, where mercury-containing effluent flowing from a chemical manufacturing plant into the local bay contaminated shellfish.

Mercury can cause several neurologic and psychologic symptoms, like tremor, personality change, restlessness, anxiety, depression, sleep disturbance, and paresthesia. Intoxication with this metal may also cause hearing loss in humans and animals. Deterioration in hearing and deafness were reported among other neurologic symptoms (Prasher, 2009). However, given the level of current evidence, it is not possible to draw any conclusions regarding ototoxicity of mercury's compounds and vapors, as well as their interaction with noise (Vyskocil et al., 2012).

Human studies show that environmental exposure to mercury can cause damage of central parts of the auditory pathway, while cochlear function may be unaffected. Intoxication with this metal in children leads to blood concentration-dependent elongation of the interwave I–V and III–V latencies of ABR (Counter et al., 1998; Murata et al., 1999, 2002). Wave III of ABR delay can be used as a biomarker of prenatal methyl mercury toxicity from contaminated seafood. Ototoxicity of mercury depends on the chemical structure of the substance, as methyl mercury was shown to be more toxic than mercuric chloride (Prasher, 2009). However, no relationship was found between biomarkers of mercury poisoning and the degree of lesion in the auditory system.

Cadmium

Cadmium is used in nickel-cadmium batteries, PVC plastics, and paint pigments. It can be found in soil, due to the fact that insecticides, fungicides, sludge, and commercial fertilizers use cadmium in agriculture. Lesser-known sources of exposure are dental alloys, electroplating, motor oil, and exhaust fumes.

Inhalation accounts for 15–50% of absorption; 2–7% of ingested cadmium is absorbed in the gastrointestinal system. Cadmium is considered to be a carcinogenic occupational factor for lung cancer.

Cadmium can cause hearing loss. In rats, a dose-dependent effect on auditory system was found (Prasher, 2009). Increased blood and renal cortical cadmium levels were associated with high cadmium accumulation in ear ossicles and labyrinth in rats exposed to cadmium. The ABR changes indicate hair cell damage (elongation of wave I latency without changes in other waves and interwave latencies) (Ozcaglar et al., 2001).

A zinc-enriched diet reduces cadmium accumulation in the kidneys and prevents nephrotoxicity in rats. The same mechanism can refer to cadmium-related ototoxicity. Zinc addition to the diet without altering cadmium content in ear ossicles may help to prevent cadmium-induced hearing loss (Agirdir et al., 2002). It has been suggested that hair cells are more sensitive to cadmium than kidney tubule cells and that the cochlear component of hearing is more vulnerable to cadmium toxicity than other parts of the auditory system.

Combined exposure to cadmium fumes and noise was shown to have additive effects at the frequencies of 4–6 kHz (de Abreu and Suzuki, 2002).

Arsenic

Arsenic is released into the environment by the smelting process of copper, zinc, and lead, as well as by the manufacture of chemicals and glass. Arsine gas is a common byproduct produced by the manufacture of pesticides that contain arsenic. Arsenic may also be found in water supplies worldwide, leading to exposure of shellfish, cod, and haddock. Other sources are paint, rat poison, fungicides, and wood preservatives.

Environmental overexposure of teenagers to arsenic was shown to result in disorders in the organ of Corti beginning at the apex. Hearing losses were greater in the lower frequencies at 125 and 250 kHz, and high frequency at 8 kHz (Bencko and Symon, 1977; Prasher, 2009). Balance problems were also noted (Prasher, 2009).

OTHER CHEMICALS

Asphyxiants

Limited study results are available on potentially ototoxic asphyxiants such as carbon monoxide (CO), hydrogen cyanide, and carbon disulfide (CS_2).

No studies on hearing loss in humans exposed to CO are available, and no ototoxic effect of such exposure was shown in animals. However, the actual animal data suggest that CO can possibly potentiate NIHL at moderate noise exposures (Fechter et al., 2000).

Similarly, no human studies are available on the ototoxicity of hydrogen cyanide, but limited research on animals shows harmful effect of this chemical on hearing and its synergistic effect with noise (Fechter et al., 2002).

Human and animal studies on the ototoxic effect of CS_2 from occupational exposure as well as human studies on its interaction with noise are not conclusive.

Pesticides

The literature on ototoxicity of pesticides is very limited, and it is not possible to draw any conclusions about the association between pesticide exposure and hearing loss (Gatto et al., 2014).

NOISE REGULATION

The current European legislation regulating protection of workers from the risk of NIHL is Directive 2003/10/EC of the European Parliament and the Council of 6 February 2003 on the minimum health and safety requirements regarding the exposure of workers to the risks arising from physical agents (noise) (Directive 2003/10/EC of European Parliament and of the Council of 6 February 2003). Based on the ISO 1999:1990 standard (ISO 1999:1990, 1990) the Directive sets the limits of exposure depending on equivalent noise level for an 8-hour working day and defines the obligation of the employer to take suitable steps if the limits are exceeded.

The Noise at Work Regulation (Directive 2003/10/EC) recommends three action levels for occupational settings:

- the first lower action level, set at $L_{EX,8h}=80$ dB(A); and $p_{peak}=112$ Pa
- the second upper action level, set at $L_{EX,8h}=85$ dB (A); and $p_{peak}=140$ Pa
- the maximum exposure limit, set at $L_{EX,8h}=87$ dB (A); and $p_{peak}=200$ Pa.

The limit values are to be converted using the time–intensity trade-off of 3 dB increase for halving the time, so, for example, exposure to noise at 97 dB(A) for 30 minutes a day would equate to the second action level, under the assumption of long-term exposure (Table 19.7).

The first (minimum) action level of 80 dB(A) is the most conservative value, as it is assumed that below this level no consequences of exposure to occupational noise can be expected. If the noise level exceeds this limit, the employer should train employees about the risk of hearing loss related to exposure to noise in the workplace, and provide workers with hearing protection, which they can freely use.

If the exposure to noise exceeds the second (upper) action level of 85 dB(A), in addition to the above-mentioned actions, the employer shall oblige workers to wear hearing protectors, as well as provide them with hearing monitoring. Personal hearing protectors

Table 19.7

Examples of equivalent time–intensity levels relating to the action levels according to Directive 2003/10/EC

Action level	$L_{EX,8h}$	Equivalent time–intensity levels
First action level (minimum): provide protection	80 dB(A)	83 dB(A): 4 hours; 86 dB(A): 2 hours 89 dB(A): 1 hour; 92 dB(A): 30 minutes; 95 dB(A): 15 minutes 98 dB(A): 8 minutes; 101 dB(A): 4 minutes 104 dB(A): 2 minutes; 107 dB(A): 1 minute
Second action level: mandatory protection	85 dB(A)	88 dB(A): 4 hours; 91 dB(A): 2 hours 94 dB(A): 1 hour; 97 dB(A): 30 minutes; 100 dB(A): 15 minutes 105 dB(A): 5 minutes; 111 dB(A): 1 minute
Maximum exposure limit value	87 dB(A)	90 dB(A): 4 hours; 93 dB(A): 2 hours 96 dB(A): 1 hour; 99 dB(A): 30 minutes; 102 dB(A): 15 minutes 107 dB(A): 5 minutes; 113 dB(A): 1 minute

Table 19.8

Duration (in hours) of allowable exposures based on Occupational Safety and Health Administration (OSHA) and National Institute for Occupational Safety and Health (NIOSH) criteria

Level, in dB(A)	85	88	90	92	94	95	100	105	110	115
OSHA permissible exposure limit	16		8			4	2	1	0.5	0.25
NIOSH recommended exposure limit	8	4			1		0.25			

must be properly fitting, in accordance with Directive 89/656/EEC (Directive 89/656/EEC).

The maximum exposure limit of 87 dB(A) should never be exceeded for safety requirements. Noise measurement should take into account the attenuation provided by personal hearing protectors worn by workers.

The Noise at Work Regulation (Directive 2003/10/EC) defines that the employer shall assess and, if necessary, measure the levels of exposure to noise to which workers are exposed. This has to be done in accordance with the obligations laid down in the Framework Directive 89/391/EEC of 12 June 1989 (Directive 89/391/EEC). Carrying out the risk assessment, the employer must pay particular attention to level, type, and duration of exposure, exposure limit/action values, individual workers' susceptibility to noise, interactions with other hazards, in particular organic solvents and vibration, exposure to noise beyond normal working hours under the employer's responsibility, and noise caused by warning signals at work.

The US regulation of noise control is based on either recommended exposure limits according to the National Institute for Occupational Safety and Health standard, or permissible exposure levels according to the Occupational Safety and Health Administration standard. Noise exposure levels/times exceeding the values shown in Table 19.8 require the use of hearing protection.

As for exposure to organic solvents and other chemicals, neither no observed adverse effect level (NOAEL) nor lowest observed adverse effect level (LOAEL) due to ototoxicity has been yet established (Sliwinska-Kowalska et al., 2007). It is also not clear at what level of noise preventive action should be taken in case of co-exposure to noise and chemicals. Since occupational chemicals can exacerbate NIHL, it would be reasonable to implement hearing monitoring for all workers exposed to noise exceeding the first minimum action level of 80 dB(A). In addition to pure-tone audiometry, workers should be also screened for central auditory processing disorders.

References

Abbate C, Giorgianni C, Munaò F et al. (1993). Neurotoxicity induced by exposure to toluene. An electrophysiologic study. Int Arch Occup Environ Health 64 (6): 389–392.

Agirdir BV, Bilgen I, Dinc O et al. (2002). Effect of zinc ion on cadmium-induced auditory changes. Biol Trace Elem Res 88 (2): 153–163.

Bencko V, Symon K (1977). Test of environmental exposure to arsenic and hearing changes in exposed children. Environ Health Perspect 19: 95–101.

Borg E, Canlon B, Engstrom B (1995). Noise-induced hearing loss. Literature review and experiments in rabbits. Morphological and electrophysiological features, exposure parameters and temporal factors, variability and interactions. Scand Audiol Suppl 40: 1–147.

Calabrese G (1996). Otoneurological study in workers exposed to styrene in the fiberglass industry. Int Arch Occup Environ Health 68 (4): 219–223.

Campo P, Maguin K (2007). Solvent-induced hearing loss: mechanisms and prevention strategy. Int J Occup Med Environ Health 20 (3): 265–270.

Campo P, Lataye R, Cossec B et al. (1997). Toluene-induced hearing loss: a mid-frequency location of the cochlear lesions. Neurotoxicol Teratol 19 (2): 129–140.

Campo P, Loquet G, Blachère V et al. (1999). Toluene and styrene intoxication route in the rat cochlea. Neurotoxicol Teratol 21 (4): 427–434.

Campo P, Lataye R, Loquet G et al. (2001). Styrene-induced hearing loss: a membrane insult. Hear Res 154 (1–2): 170–180.

Chang SJ, Chen CJ, Lien CH et al. (2006). Hearing loss in workers exposed to toluene and noise. Environ Health Perspect 114 (8): 1283–1286.

Chuang HY, Kuo CH, Chiu YW et al. (2007). A case-control study on the relationship of hearing function and blood concentrations of lead, manganese, arsenic, and selenium. Sci Total Environ 387 (1–3): 79–85.

Counter SA, Buchanan LH (2002). Neuro-ototoxicity in Andean adults with chronic lead and noise exposure. J Occup Environ Med 44 (1): 30–38.

Counter SA, Buchanan LH, Laurell G et al. (1998). Blood mercury and auditory neuro-sensory responses in children and adults in the Nambija gold mining area of Ecuador. Neurotoxicology 19 (2): 185–196.

de Abreu MT, Suzuki FA (2002). Audiometric evaluation of noise and cadmium occupationally exposed workers. Braz J Otorhinolaryngol 68 (6): 488–494.

Directive 2003/10/EC of European Parliament and of the Council of 6 February 2003 on the minimum health and safety requirements regarding the exposure of workers to the risks arising from physical agents (noise) (17th individual Directive within the meaning of Article 16(1) of

Directive 89/391/EEC). Official Journal of the European Communities. No L42/38, 15.2.2003, 2000.

Directive 89/391/EEC of 12 June 1989 on the introduction of measures to encourage improvements in the safety and health of workers at work. Official Journal L 183: 1–8.

Directive 89/656/EEC Use of personal protective equipment of 30 November 1989 on the minimum health and safety requirements for the use by workers of personal protective equipment at the workplace (third individual directive within the meaning of Article 16 (1) of Directive 89/391/EEC).

Fechter LD, Cheng GD, Rao D et al. (2000). Predicting exposure conditions that facilitate the potentiation of noise-induced hearing loss by carbon monoxide. Toxicol Sci 58 (2): 315–323, Dec.

Fechter LD, Chen GD, Johnson DL (2002). Potentiation of noise-induced hearing loss by low concentrations of hydrogen cyanide in rats. Toxicol Sci 66 (1): 131–138.

Fuente A, Slade MD, Taylor T et al. (2009). Peripheral and central auditory dysfunction induced by occupational exposure to organic solvents. J Occup Environ Med 51 (10): 1202–1211.

Fuente A, McPherson B, Hickson L (2011). Central auditory dysfunction associated with exposure to a mixture of solvents. Int J Audiol 50 (12): 857–865.

Fuente A, McPherson B, Hickson L (2013). Auditory dysfunction associated with solvent exposure. BMC Public Health 13: 39–50.

Gatto MP, Fioretti M, Fabrizi G et al. (2014). Effects of potential neurotoxic pesticides on hearing loss: a review. Neurotoxicology 42: 24–32.

Henderson D, Bielehed EC, Harris KC et al. (2006). The role of oxidative stress in noise-induced hearing loss. Ear Hear 27: 1–19.

Hoffmann J, Ihrig A, Hoth S et al. (2006). Field study to explore possible effects of styrene on auditory function in exposed workers. Ind Health 44 (2): 283–286.

Hughes H, Hunting KL (2013). Evaluation of the effects of exposure to organic solvents and hazardous noise among US Air Force Reserve personnel. Noise Health 15 (67): 379–387.

Hwang YH, Chiang HY, Yen-Jean MC et al. (2009). The association between low levels of lead in blood and occupational noise-induced hearing loss in steel workers. Sci Total Environ 408 (1): 43–49.

ISO 11957:1996 (1996). Acoustics. Determination of sound insulation performance of cabins. Laboratory and in situ measurements. International Committee for Standardisation, Geneva.

ISO 1999:1990 (1990). Acoustics. Determination of occupational noise exposure and estimation of noiseinduced hearing impairment. International Committee for Standardisation, Geneva.

ISO 226:2003 (2003). Acoustics. Normal equal-loudness-level contours, Edition: 2. International Committee for Standardisation, Geneva.

Johnson AC, Morata TC, Lindblad AC et al. (2006). Audiological findings in workers exposed to styrene alone or in concert with noise. Noise Health 8 (30): 45–57.

Juárez-Pérez CA, Torres-Valenzuela A, Haro-García LC et al. (2014). Ototoxicity effects of low exposure to solvent mixture among paint manufacturing workers. Int J Audiol 53 (6): 370–376.

Kaufman LR, LeMasters GK, Olsen DM et al. (2005). Effects of concurrent noise and jet fuel exposure on hearing loss. J Occup Environ Med 47 (3): 212–218.

Kim J, Park H, Ha E et al. (2005). Combined effects of noise and mixed solvents exposure on the hearing function among workers in the aviation industry. Ind Health 43 (3): 567–573.

Lataye R, Campo P, Loquet G (2000). Combined effects of noise and styrene exposure on hearing function in the rat. Hear Res 139 (1-2): 86–96.

Lin HW, Furman AC, Kujawa SG et al. (2011). Primary neural degeneration in the guinea pig cochlea after reversible noise-induced threshold shift. J Assoc Res Otolaryngol 12 (5): 605–616.

Makitie AA, Pirvola U, Pyykko I et al. (2003). The ototoxic interaction of styrene and noise. Hear Res 179 (1-2): 9–20.

Morata TC, Dunn DE, Kretschmer LW et al. (1993). Effects of occupational exposure to organic solvents and noise on hearing. Scand J Work Environ Health 19 (4): 245–254.

Morata TC, Engel T, Durão A et al. (1997a). Hearing loss from combined exposures among petroleum refinery workers. Scand Audiol 26 (3): 141–149.

Morata TC, Fiorini AC, Fischer FM et al. (1997b). Toluene-induced hearing loss among rotogravure printing workers. Scand J Work Environ Health 23 (4): 289–298.

Morioka I, Miyai N, Yamamoto H et al. (2000). Evaluation of combined effect of organic solvents and noise by the upper limit of hearing. Ind Health 38 (2): 252–257.

Moshammer H, Kundi M, Wallner P et al. (2014). Early predictors of noise-induced hearing loss. Occup Environ Med. http://dx.doi.org/10.1136/oemed-2014-102200.

Muijser H, Hoogendijk EM, Hooisma J (1988). The effects of occupational exposure to styrenee on high-frequency hearing thresholds. Toxicology 49 (2-3): 331–340.

Murata K, Weihe P, Araki S et al. (1999). Evoked potentials in Faroese children prenatally exposed to methylmercury. Neurotoxicol Teratol 21: 471–472.

Murata K, Budtz-Jørgensen E, Grandjean P (2002). Benchmark dose calculations for methylmercury-associated delays on evoked potential latencies in two cohorts of children. Risk Anal 22 (3): 465–474.

Musiek FE (1994). Frequency (pitch) and duration pattern tests. J Am Acad Audiol 5 (4): 265–268.

Musiek FE, Shinn JB, Jirsa R et al. (2005). GIN (Gaps-In-Noise) test performance in subjects with confirmed central auditory nervous system involvement. Ear Hear 26 (6): 608–618.

Nelson DI, Nelson RY, Concha-Barrientos M et al. (2005). The global burden of occupational noiseinduced hearing loss. Am J Ind Med 48 (6): 446–458.

Niklasson M, Möller C, Ödkvist LM et al. (1997). Are deficits in the equilibrium system relevant to the clinical investigation of solvent-induced neurotoxicity? Scand J Work Environ Health 23 (3): 206–213.

Ödkvist LM, Arlinger SD, Edling C et al. (1987). Audiological and vestibulo-oculomotor findings in workers exposed to solvents and jet fuel. Scand Audiol 16: 75–81.

Ozcaglar HU, Agirdir B, Dinc O et al. (2001). Effects of cadmium on the hearing system. Acta Otolaryngol 121 (3): 393–397.

Prasher D (2009). Heavy metals and noise exposure: health effects. Noise Health 11 (44): 141–144.

Prasher D, Al-Hajjaj H, Aylott S et al. (2005). Effect of exposure to a mixture of solvents and noise on hearing and balance in aircraft maintenance workers. Noise Health 7 (29): 31–39.

Pryor GT, Rebert CS (1992). Interactive effects of toluene and hexane on behavior and neurophysiologic responses in Fischer-344 rats. Neurotoxicology 13 (1): 225–234.

Puel JL, Ruel J, Guitton M et al. (2002). The inner hair cell afferent/efferent synapses revisited: a basis for new therapeutic strategies. Adv Otorhinolaryngol 59: 124–130.

Rumeau C, Campo P, Venet T et al. (2011). Toluene effect on the olivocochlear reflex. Toxicol Sci 121 (1): 140–145.

Schäper M, Seeber A, van Thriel C (2008). The effects of toluene plus noise on hearing thresholds: an evaluation based on repeated measurements in the German printing industry. Int J Occup Med Environ Health 21 (3): 191–200.

Shi X (2011). Physiopathology of the cochlear microcirculation. Hear Res 282 (1–2): 10–24.

Sliwinska-Kowalska M, Pawelczyk M (2013). Contribution of genetic factors to noise-induced hearing loss: a human studies review. Mutation Research/Reviews in Mutation Research 752 (1): 61–65.

Sliwinska-Kowalska M, Zamyslowska-Szmytke E, Szymczak W et al. (2001). Hearing loss among workers exposed to moderate concentrations of solvents. Scand J Work Environ Health 27 (5): 335–342.

Sliwinska-Kowalska M, Zamyslowska-Szmytke E, Szymczak W et al. (2003). Ototoxic effects of occupational exposure to styrene and co-exposure to styrene and noise. J Occup Environ Med 45 (1): 15–24.

Sliwinska-Kowalska M, Zamyslowska-Szmytke E, Szymczak W et al. (2004). Effects of coexposure to noise and mixture of organic solvents on hearing in dockyard workers. J Occup Environ Med 46 (1): 30–38.

Sliwinska-Kowalska M, Zamyslowska-Szmytke E, Szymczak W et al. (2005). Exacerbation of noise-induced hearing loss by co-exposure to workplace chemicals. Environ Toxicol Pharmacol 19: 547–553.

Sliwinska-Kowalska M, Prasher D, Rodrigues CA et al. (2007). Ototoxicity of organic solvents – from scientific evidence to health policy. Int J Occup Med Environ Health 20 (2): 215–222.

Soli S, Wong L (2008). Assessment of speech intelligibility in noise with the Hearing in Noise Test. Int J Audiol 47 (6): 356–361.

Vrca A, Karacić V, Bozicević D et al. (1996). Brainstem auditory evoked potentials in individuals exposed to long-term low concentrations of toluene. Am J Ind Med 30 (1): 62–66.

Vyskocil A, Truchon G, Leroux T et al. (2012). A weight of evidence approach for the assessment of the ototoxic potential of industrial chemicals. Toxicol Ind Health 28 (9): 796–819.

Wu TN, Shen CY, Lai JS et al. (2000). Effects of lead and noise exposures on hearing ability. Arch Environ Health 55 (2): 109–114.

Zamyslowska-Szmytke E, Sliwinska-Kowalska M (2011). Otoneurological assessment in workers exposed to styrene and dichloromethane. Int J Audiol 50 (11): 815–822.

Zamyslowska-Szmytke E, Fuente A, Niebudek-Bogusz E et al. (2009). Temporal processing disorder associated with styrene exposure. Audiol Neurootol 14 (5): 296–302.

Zamyslowska-Szmytke E, Politanski P, Sliwinska-Kowalska M (2011). Balance system assessment in workers exposed to organic solvent mixture. J Occup Environ Med 53 (4): 441–447.

Section 6

Musculoskeletal disorders

Handbook of Clinical Neurology, Vol. 131 (3rd series)
Occupational Neurology
M. Lotti and M.L. Bleecker, Editors

Chapter 20

Mechanical stress

STEPHEN BAO*

SHARP Program, Washington State Department of Labor and Industries, Olympia, WA, USA

INTRODUCTION

Work-related musculoskeletal disorders (MSDs) constitute a significant proportion of occupational morbidity, lost work days, and costs (Silverstein et al., 2009). According to the US Bureau of Labor Statistics, there were 388 060 MSD cases in 2012, accounting for 34% of all injury and illness cases. Both the incidence rate and case count remained statistically unchanged in 2012 from the previous year; however, the median days away from work increased by 1 day to a median of 12 days. Laborer and freight, stock, and material movers had the highest number of MSD cases and an incidence rate of 164 per 10 000 full-time workers – up from 140 in 2011 (Bureau of Labor Statistics, 2013).

Neck, back, and upper-extremity work-related MSDs represent 27% of all workers' compensation claims in Washington State – 36% of the claims result in four or more lost work days and more than 43% of all costs (Silverstein and Adams, 2007). Knee disorders occurred often among lower-extremity disorders, accounting for 65% of lower-extremity MSDs and 5% of total body MSDs (Reid et al., 2010).

MSD prevalence seems to vary by industries and occupations. For example, among shipyard employees, 37%, 22%, and 15% of employees reported complaints of low-back, shoulder/neck, and hand/wrist during the past 12 months, respectively (Alexopoulos et al., 2006). Among ironworkers, the prevalence of doctor-diagnosed MSD was 19% for tendonitis, 18% of ruptured disc in the back, 15% for bursitis in the shoulder, and 12% for carpal tunnel syndrome (CTS) (Forde et al., 2005). Based on a New Zealand survey, 88% of nurses, postal workers, and office workers had at least one MSD lasting longer than a day and 72% reported an MSD present for at least 7 days over a 12-month period (Harcombe et al., 2009). Postal workers had the highest prevalence of elbow and wrist/hand pain, and nurses of knee pain. A review conducted by Walker-Bone and Palmer (2002) concluded that there is strong evidence relating to osteoarthritis of the hip among farm workers, and suggestive evidence that farmers more often have knee osteoarthritis and low-back pain than workers in occupations with fewer physical demands.

It is generally agreed among MSD researchers that certain risk factors caused by the various work activities in different occupations are associated with the development of work-related MSD. Some of these risk factors can be labeled as mechanical stressors (or physical exposures) apart from certain psychosocial and individual factors (Armstrong et al., 1993). These risk factors include high forceful exertions, awkward postures, repetitive motions, whole-body and hand–arm vibrations, and contact stresses (NIOSH, 1997). In discussions of work-related MSD, these mechanical stressors are also often referred to as physical risk factors (or workload) or biomechanical risk factors.

In order to facilitate our discussions on relationships between mechanical stressors and work-related MSD, measurements of mechanical stressor, and workplace intervention for prevention of MSD, it may be necessary to have a discussion on a conceptual model of the exposure–response relationship for work-related MSD. Several conceptual models for work-related MSD have been proposed by researchers over the years (Armstrong et al., 1993; Winkel and Mathiassen, 1994). Most of these conceptual models include key elements such as external exposures which are work requirements from jobs, and various responses (e.g., fatigue, and MSD symptoms, or physical capacity improvements) as a result of these external exposures (Armstrong et al., 1993; Winkel and Mathiassen, 1994). For the purpose of our discussions, the conceptual model shown in Figure 20.1 will be used.

*Correspondence to: Stephen Bao, Ph.D., CPE, CCPE, SHARP Program, Washington State Department of Labor and Industries, Olympia WA 98504 4330, USA. E-mail: baos235@lni.wa.gov

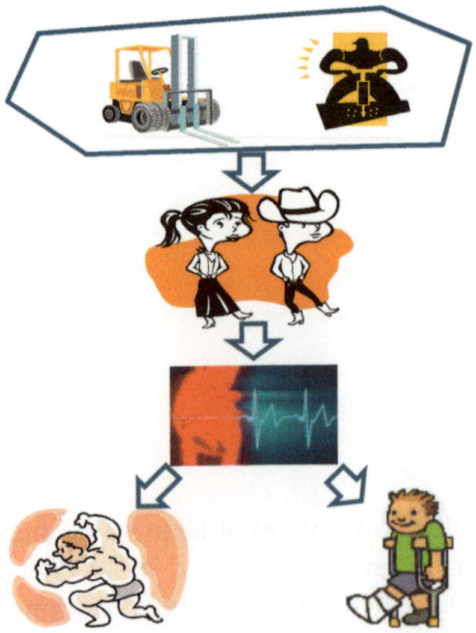

Fig. 20.1. A conceptual model of occupational mechanical stressor on worker health.

There are various mechanical stressors in many jobs. Some examples include manually handling heavy objects, such as lifting a heavy jack hammer in a construction worksite or stocking milk containers in a grocery store, and being exposed to vibrations for workers driving a forklift truck, or using a powerful vibrating tool. These mechanical stressors are requirements of the job, and are determined by the job contents (e.g., machine design, assembly methods, and tools to be used) and how the work is organized (e.g., by one worker, or multiple workers). Their existence is not determined by workers, i.e., worker A is exposed to these job exposures on one shift. If worker B comes to work on the same job next day, he will be exposed to the same exposures. Some researchers call these workplace or job mechanical stressors external physical exposures (Winkel and Mathiassen, 1994).

These mechanical stressors may impact workers in many different ways. They often cause different physiologic responses in workers (e.g., increased muscle activities, awkward joint postures, higher joint torques and increased heart rate). The same amount of external exposures applied to workers with different physical capacities may impact them very differently. For example, it may not be as risky for a young male worker with greater lifting strength (capacity) to install 15-kg wheels asfor an older female worker with chronic low-back pain. Another example is that a shorter worker may have to reach over her head level in order to reach objects on a shelf of 140 cm, compared to a tall worker who may not have to raise his arm above 90°. This results in very different postural exposures for these two workers. Work techniques may also impact the actual load applied on the workers. For instance, an experienced technician with neutral hand/wrist and shoulder postures applying the same amount of 20 newton-meter torque using a wrench may put less stresses on the shoulder and wrist joints compared to a newly hired worker assuming awkward hand/wrist and shoulder postures. Workers' responses (to external physical exposures) of the musculoskeletal systems are often referred to as internal physical exposures (Winkel and Mathiassen, 1994).

It is these internal physical exposures that may produce wanted or unwanted health outcomes. For example, the proper amount of physical activity may improve individuals' muscle strength and cardiovascular capacity, and reduce obesity in many workplaces (Bridger et al., 2013). However, many mechanical stressors can be excessive and can cause injuries to workers' musculoskeletal systems (Hagberg et al., 1995; NIOSH, 1997). It is the latter type of mechanical stressors that we will focus on in our discussions.

WORK-RELATED MUSCULOSKELETAL DISORDERS

Work-related MSDs are defined differently in different studies (NIOSH, 1997). Some studies may have used workers' compensation data, or company injury statistics, and others may have used specific diagnostic criteria specific to the study. Therefore, the absolute numbers in prevalence and incidence data from various epidemiologic studies sometimes may not be directly comparable. However, based on a number of studies, the body parts most often injured include neck, shoulders, back, elbows, hands/wrists, and knees. It is also generally agreed that jobs with certain mechanical stressors tend to cause higher prevalence or incidences in certain industries. According to Washington State workers' compensation claims data (Silverstein and Adams, 2007), the top 12 industries for compensable work-related MSD claims were:

1. couriers with a rate ratio (RR) of 4.3 (meaning that the rate of claims for this industry is 4.3 times the overall state rate)
2. scheduled air transportation, RR = 3.5
3. foundation, structure, andand exterior building contractors, RR = 2.5
4. nursing care facilities, RR = 2.6
5. general freight trucking, RR = 2.5
6. building finishing contractors, RR = 2.3
7. community care facilities for elderly, RR = 2.4
8. residential building construction, RR = 2.0
9. general medical surgical hospitals, RR = 1.9
10. other wood products manufacturing, RR = 2.3
11. grocery stores, RR = 1.8
12. waste collection, RR = 3.4.

Table 20.1

Examples of high-risk jobs in the top 12 industries for compensable work-related musculoskeletal disorder claims (unpublished data)

Industry sector	Job examples	
Couriers	Truck drivers, light or delivery services Laborers and freight, stock, and material movers, hand Couriers and messengers	Truck drivers, heavy and tractor-trailer Production workers, all other
Scheduled air transportation	Flight attendants Laborers and freight, stocking, and material movers, hand Transportation workers, all others	Reservation and transportation ticket agents and travel clerks Customer service represent
Foundation, structure and exterior building contractors	Carpenters Construction craft laborers Roofers	First-line supervisors/managers of construction trades and extraction workers Cement masons, concrete finishers, and terrazzo workers
Nursing care facilities	Nursing aides, orderlies, and attendants Registered nurses Licensed practical and licensed vocational nurses	Maids and housekeeping cleaners Healthcare support workers
General freight trucking	Truck drivers, heavy and tractor-trailer Truck drivers, light or delivery services Laborers and freight, stock, and material movers, hand	Bus and truck mechanics and diesel engine specialists Production workers, all other
Building finishing contractors	Drywall and ceiling tile installers Painters, construction and maintenance Carpenters	First-line supervisors/managers of construction trades and extraction workers Construction craft laborer
Community care facilities for elderly	Nursing aides, orderlies, and attendants Personal and home care aides Home health aides	Maids and housekeeping cleaners Licensed practical and licensed vocational nurses
Residential building construction	Carpenters Construction craft laborer First-line supervisors/managers of construction	Construction and related workers, all other Roofers
General medical surgical hospitals	Registered nurses Nursing aides, orderlies, and attendants Maids and housekeeping cleaners	Healthcare support workers Licensed practical and licensed vocational nurses
Other wood products manufacturing	Production workers, all other Laborers and freight, stock, and material movers, hand Sawing machine setters, operators, and tenders, wood	Woodworking machine setters, operators, and tenders, except sawing Carpenters
Grocery stores	Cashiers Laborers and freight, stock, and material movers, hand Stock clerks and order fillers	First-line supervisors/managers of retail sales workers Combined food preparation
Waste collection	Refuse and recyclable material collectors Truck drivers, heavy and tractor-trailer Truck drivers, light or delivery services	Laborers and freight, stock, and material movers, hand Bus and truck mechanics and diesel engine specialists

Table 20.1 shows examples of high-risk jobs in these top 12 industries for compensable work-related MSD claims.

According to a study of the State of Maryland workers' compensation data, back, knee, and shoulder injuries were the top three single-injury contributors to total-injury costs across all claims in each year when taking into account both frequency and cost (Mroz et al., 2014). Again, using workers' compensation data in

Washington State, it was found that nontraumatic MSDs of the neck, back, and upper extremities are the most frequent injuries across all sectors and occupations in the trucking industry (Smith and Williams, 2014).

Some MSDs have no specific diagnoses (e.g., joint pain, and joint stiffness), although many MSD claims do have specific clinical diagnoses. Table 20.2 lists specific diagnoses for MSDs for the neck, shoulder, back, elbow, hand/wrist, and knee.

Low-back problems

The claim incidence rate for sciatica was 5.7 with a severity rate (number of lost days per 10 000 full-time employees (FTEs)) of 1954 days per 10 000 FTEs (Silverstein and Adams, 2007). On average, each compensable sciatica claim costs $69 237, and 77.5% of claims were compensable with an average time loss of 554 days. Table 20.3 shows the variation in incidence and severity rates of compensable back claims by different industries based on unpublished data from Washington State.

Work-related low-back disorders have been one of the most studied injuries in workplaces. While low-back pain is common in the general population (Andersson, 1981), it is clear that back disorder rates vary substantially by industry, occupation, and by job within given industries (Bigos et al., 1986; Riihimaki et al., 1989; Skovron et al., 1994; Schibye et al., 1995; Forde et al., 2005; Wilkins and Mackenzie, 2007; Lipscomb et al., 2009; Bell and Steele, 2012; Maguire and Smith, 2013; Smith and Williams, 2014). Certain mechanical stresssors may have caused the high incidence of back disorders in many physically demanding jobs.

Among the many mechanical stressors existing in workplaces, heavy physical work, lifting and forceful movements, awkward postures in bending and twisting, whole-body vibration, and static work postures are often studied in relation to work-related back disorders (NIOSH, 1997). Heavy physical work usually has been defined as work that has high energy demands or requires some measure of physical strength (NIOSH, 1997). Based on reviews of many epidemiologic studies, it has been found that low-back disorders are associated with heavy physical work (NIOSH, 1997). For example, Bergenudd and Nilsson (1988) did a survey among a sample of 575 55-year-old residents and found that back pain was more common among individuals who worked at physically more strenuous jobs.

There is strong evidence that low-back disorders are associated with jobs requiring lifting and forceful movements (NIOSH, 1997). A case-referent study conducted in an automobile assembly plant (Punnett et al., 1991) found that lifting was associated with back disorder, with an odds ratio of 2.16 (95% confidence interval (CI) 1.0–4.7). When the subset with physical medical findings was examined, associations were more pronounced. In a retrospective case-control study from Sweden, it was found that heavy physical work involving lifting and carrying loads exceeding 15 kg and other risk factors were associated with excess risk of low-back pain (Thorbjornsson et al., 2000; Vingard et al., 2000). Heavy lifting and carrying were also identified with increased risk of low-back pain among professional drivers (Lyons, 2002). Several studies were also able to find dose–response relationships between low-back disorders and objective measures of lifting (Chaffin and Park, 1973; Punnett et al., 1991; Marras et al., 1995). Some of these relationships were later used in mechanical stressor evaluation tools.

It is also evident that jobs requiring workers to assume awkward postures are associated with low-back disorders. In a cross-sectional study involving three different groups of workers in a tank terminal company, it was found that the observed work time in trunk flexion and rotation was associated with the occurrence of back pain (Burdorf et al., 1997). Prolonged sitting and awkward postures were found to be associated with increased risk for low-back pain and injury (Lyons, 2002). A case-referent study with 95 cases and 124 referents in an automobile assembly plant (Punnett et al., 1991) found that time in nonneutral postures was strongly associated with back disorders, with an odds ratio of 8.09 (95% CI 1.4–44). In a study consisting of 769 workers in nursing homes and homes for the elderly, a dose–response relationship between trunk flexion and low-back pain was found (Jansen and Burdorf, 2003). Some researchers used an index method to quantify the awkwardness of postures and found a statistically significant association between the posture index and back symptoms, with an odds ratio of 1.23 (Burdorf et al., 1991). Stooping and kneeling postures were found to be statistically significantly associated with low-back pain, particularly with severe disorders, with odds ratios of 2.6 and 3.5 respectively (Holmstrom et al., 1992).

A strong association between whole-body vibration and low-back disorders has been found in many epidemiologic studies. In a cross-sectional study comparing a population of 1155 male tractor drivers exposed to whole-body vibration and postural stress with a control group of 220 office workers, Bovenzi and Betta (1994) found that tractor drivers demonstrated an odds ratio of 3.22 (95% CI 2.1–5.2 for lifetime low-back problem). In a study on the role of exposure to driving and other covariates in reports of back and other MSDs, cohorts in Sweden and the USA were compared (Magnusson et al., 1996). It was found that the most common risk factors (as measured by corresponding odds ratios) for back

Table 20.2

Disorders with specific diagnoses by body parts

Neck	Shoulder	Back	Elbow	Hand/wrist	Knee
Spondylosis and allied disorders	Pain in joint involving shoulder region	Spinal enthesopathy	Pain in joint involving forearm	Mononeuritis of upper limb or multiple regions	Pain in joint involving lower leg
Cervical spondylosis without myelopathy	Stiffness of joint not elsewhere classified involving shoulder region	Sacroiliitis not elsewhere classified	Stiffness of joint not elsewhere classified involving forearm	Carpal tunnel syndrome	Stiffness of joint not elsewhere classified involving lower leg
Cervical spondylosis with myelopathy	Adhesive capsulitis of shoulder	Lumbosacral spondylosis without myelopathy	Enthesopathy of elbow	Contracture of hand joint	Enthesopathy of knee
Intervertebral disc disorders	Rotator cuff syndrome	Spondylosis with myelopathy lumbar region	Medial epicondylitis	Stiffness of joint not elsewhere classified	Unspecified enthesopathy of knee
Displacement of cervical intervertebral disc without myelopathy	Disorders of bursae and tendons in shoulder region unspecified	Displacement of thoracic/lumbar intervertebral disc without myelopathy	Lateral epicondylitis	Enthesopathy of wrist and carpus	Pes anserinus tendinitis or bursitis
Degeneration of cervical intervertebral disc	Calcifying tendinitis of shoulder	Displacement of intervertebral disc site unspecified without myelopathy	Olecranon bursitis	Trigger finger	Tibial collateral ligament bursitis
Intervertebral disc disorder with myelopathy cervical region	Bicipital tenosynovitis	Degeneration of lumbar or lumbosacral intervertebral disc	Other enthesopathy of elbow region	Radial styloid tenosynovitis	Fibular collateral ligament bursitis
Other and unspecified disc disorder of cervical region	Other specified disorders of bursae and tendons in shoulder region	Degeneration of intervertebral disc site unspecified	Sprain of other specified sites of elbow and forearm	Other tenosynovitis of hand and wrist	Patellar tendinitis
Cervicalgia	Complete rupture of rotator cuff	Intervertebral disc disorder with myelopathy	Sprain of unspecified site of elbow and forearm	Nontraumatic rupture of extensor tendons of han/wrist	Prepatellar bursitis
Cervicobrachial syndrome	Infraspinatus sprain and strain	Intervertebral disc disorder with myelopathy lumbar region		Nontraumatic rupture of flexor tendons of hand/wrist	Other enthesopathy of knee
Brachial neuritis or radiculitis not otherwise specified	Rotator cuff sprain and strain	Other and unspecified disc disorders			Synovial cyst of popliteal space
Unspecified musculoskeletal disorders and symptoms referable to neck	Subscapularis sprain and strain	Spinal stenosis, lumbar region without neurogenic claudication			Nontraumatic rupture of patellar tendon
Nonallopathic lesions of cervical region not elsewhere classified	Supraspinatus sprain and strain	Spinal stenosis of other region			Plica syndrome
Neck sprain		Lumbago			
		Sciatica			
		Thoracic/lumbosacral neuritis/radiculitis unspecified			
		Backache unspecified			
		Nonallopathic lesions of lumbar region not elsewhere classified			
		Nonallopathic lesions of sacral region not elsewhere classified			
		Lumbar sprain			

Table 20.3

Differences in incidence rates and severity rates of the top 25 Washington State all compensable back claims by industries (between 2002 and 2010 according to unpublished Washington Workers' Compensation data)

Industry sector	Incidence rate*	Severity rate†
Nursing care facilities	131.9	22 576.5
Foundation, structure, and building exterior contractors	117.7	36 101.0
Scheduled air transportation	137.4	3564.8
Couriers and express delivery services	126.2	2851.5
Building finishing contractors	98.0	31 295.4
General medical and surgical hospitals	79.0	483.8
Services to buildings and dwellings	87.3	18 176.0
General freight trucking	99.0	16 748.2
Residential building construction	93.3	28 392.5
Community care facilities for the elderly	103.3	21 658.8
Other ambulatory healthcare services	135.6	6110.1
Specialized freight trucking	122.7	24 658.9
Department stores	71.7	700.4
Waste collection	123.3	11 467.9
Other specialty trade contractors	86.5	27 518.8
Residential mental retardation, mental health	103.8	19 969.7
Grocery and related product merchant wholesale	71.7	8358.1
Grocery stores	66.8	4167.3
Building material and supplies dealers	73.4	8591.3
Executive legislative, and other general government	62.2	1597.0
Building equipment contractors	63.2	18 328.7
Other residential care facilities	85.4	19 187.0
Vocational rehabilitation services	77.2	11 532.6
Other general merchandise stores	65.9	3709.7
Beer, wine, and distilled alcoholic beverage	94.0	10,202.6

*Incidence rate: number of claims per 10 000 full-time employees (FTEs).
†Severity rate: time loss in days per 10 000 FTE.

and neck pain were long-term vibration exposure, heavy lifting, and frequent lifting. A combination of long-term vibration exposure and frequent lifting carried the highest risk of low-back pain.

It has been shown that multiple risk factors occurring at the same job may increase the risk of low-back problems. In an epidemiologic case-control study in Connecticut, USA, during 1979 and 1981, it was found that the highest risk for prolapsed lumbar disc was associated with jobs involving lifting objects of more than 11.3 kg with the body usually twisted and the knees not bent (Kelsey et al., 1984). In a cross-sectional study, it was found that back injuries were associated with spinal loading during lifting, which included simultaneous exposures to weight lifted, lifting frequency, trunk lateral movement velocity, trunk twisting velocity, and trunk flexion angle (Marras et al., 1993, 1995).

Neck/shoulder problems

The claim incidence rate for rotator cuff syndrome was 17.3, with a severity rate of 2344 days per 10 000 FTEs (Silverstein and Adams, 2007). On average, each compensable rotator cuff syndrome claim costs $32 169, and 63% were compensable, with an average time loss of 334 days.

While low-back problem has been historically the most common MSD, neck and shoulder problems are increasing. This may be partially due to the technology changes in workplaces where more and more workers become sedentary when working in front of computers. Many epidemiologic studies have examined mechanical stressors in the workplace and their relationship with neck and shoulder MSD. Strong evidence was found that workers on jobs with high levels of static contraction, prolonged static loading, or extreme working postures are at increased risk for neck and shoulder MSD (NIOSH, 1997). There are also evidences showing that forceful exertion and highly repetitive work are associated with neck and shoulder MSD (NIOSH, 1997).

A prospective epidemiologic study that followed 364 new trailer assemblers for 45 months found that the risk for male workers having neck/shoulder or upper-limb disorders in strenuous tasks was about threefold (hazard ratio (HR) 3.2 with a 95% CI of 1.1–9.4 for upper-limb disorders and HR 2.7 with a 95% CI of 0.9–8.1 for neck/shoulder disorders) that of nonstrenuous tasks (Hakkanen et al., 2001).

Many epidemiologic studies found that jobs requiring repetitive work or repetitive neck/shoulder movements are significantly associated with neck and shoulder disorders. A cross-sectional study among female industrial workers exposed to repetitive work tasks found statistically significant associations between exposure to repetitive work and neck and shoulders diagnoses, with a prevalence odds ratio of 4.6 (Ohlsson et al., 1995).

A prospective study on 1832 male machine operators showed that machine operators exposed to static loading had an increased risk of developing severe neck pains (Viikari-Juntura et al., 1994). The same research group did another study among 5180 Finnish forest industry workers and found that duration of work with a hand above shoulder level is associated with radiating neck pain (Viikari-Juntura et al., 2001). An ergonomics intervention study carried out in a Norwegian electronics assembly workplace found that reduced static loading on the neck/shoulder region by improved workstation design (hence improved work postures) was able to reduce sick leave due to MSD from 5.3% to 3.1%, and turnover rate from 30.1% to 7.6% (Aarås, 1994).

Hand/wrist disorders

According to the Washington State workers' compensation data, the claim incidence rate for CTS was 20.4, with a severity rate of 2478 days per 10 000 FTEs (Silverstein and Adams, 2007). On average, each compensable CTS costs $22 686, and 65% were compensable, with an average time loss of 258 days. The claim incidence rate for hand/wrist tendonitis was 15.3, with a severity of 1199 per 10 000 FTEs. On average, each compensable hand/wrist tendonitis costs $14 045, and 41% were compensable, with an average time loss of 264 days.

Jobs involving high levels of mechanical stressors are associated with hand/wrist disorders. These mechanical stressors include forceful exertions, high repetition, awkward hand/wrist postures, and hand/arm vibration (NIOSH, 1997; Mani and Gerr, 2000). High hand force has been considered the major risk factor responsible for hand/wrist disorders. Moore and Garg (1994) evaluated 32 jobs in a pork-processing plant together with reviews of past Occupational Safety and Health Administration (OSHA) 200 logs and plant medical records for CTS cases and found that workers at jobs requiring high hand forces had higher risk of developing CTS.

Awkward hand/wrist postures have often been reported to be a risk factor associated with hand/wrist problems (NIOSH, 1997; Mani and Gerr, 2000). However, it has been reported that there is insufficient evidence to link awkward hand/wrist posture alone with hand/wrist MSDs, as often awkward hand/wrist postures occur at the same time as other risk factors, such as high hand forces and high repetitions (NIOSH, 1997).

In order to study the prevalence of vascular, neurologic, and MSDs, a group of female workers performing mechanical or hand sanding jobs in the furniture industry was studied and compared with a control group of female office workers (Bovenzi et al., 2005). The study showed that the occurrence of MSD of the upper limb increased significantly with the increase of both daily vibration exposure and the Strain Index score (Moore and Garg, 1995), which quantifies exposure to hand force, repetition, and awkward hand/wrist postures.

Jobs involving high force and high repetition were found to be 15 times more likely to develop CTS than jobs with low force and low repetition (Silverstein et al., 1987).

Elbow disorders

According to Washington State workers' compensation data, the claim incidence rate for epicondylitis was 11.1, with a severity rate of 546.6 days per 10 000 FTEs (Silverstein and Adams, 2007). On average, each compensable epicondylitis claim costs $12 006, and 41.8% were compensable, with an average time loss of 271 days.

Jobs involving forceful exertions have been reported to contribute to epicondylitis (Fan et al., 2009). High incidences of epicondylitis were found among workers performing jobs involving high hand force and repetitive motions. For instance, in a 31-month follow-up study among 377 workers in strenuous manual jobs and 338 employees in manually nonstrenuous work in a large meat-processing factory, the annual incidence of epicondylitis was about 1% for employees in nonstrenuous jobs, 11.3% for female sausage markers, 7.0% for female packers, and 6.4% for male meat cutters respectively (Kurppa et al., 1991).

There is strong evidence for a relationship between exposure to a combination of risk factors such as high hand force and high repetition, or high hand force and awkward postures (NIOSH, 1997). A Finnish study among a sample of population aged 30–64 years found an interaction between repetitive movements of the arms and forceful activities for developing lateral epicondylitis (Shiri et al., 2006). The odds ratio for jobs involving both repetitive movements and forceful activities compared to jobs without such activities was 5.6, with a 95% CI of 1.9 and 16.5. This study also found that repetitive movements and forceful activities independently of each other showed significant associations with medial epicondylitis.

Reviews have shown that there is insufficient evidence to draw conclusions about the relationship of postural factors alone and epicondylitis, or repetitive work alone and epicondylitis (NIOSH, 1997). Studies found that awkward postures or repetitive movements as risk factors usually have other risk factors present in the jobs. For example, a study among 733 active workers in 12 different worksites showed that forearm supination $\geq$45° for $\geq$5% work time with high lifting force was a significant risk factor for lateral epicondylitis (Fan et al., 2009). The same study also found that high frequency of forceful exertions is a risk factor for lateral epicondylitis.

Some work-related elbow problems are related to ulnar neuropathy of the elbow. According to unpublished Washington State workers' compensation data, there were 363 accepted cases with specific diagnosis of lesion of ulnar nerve between 2000 and 2013. Many of the cases occurred in the construction industry.

Knee problems

Again, according to Washington State workers' compensation data, the claim incidence rate for all work-related knee claims was 18.2 per 10 000 FTEs (Spector et al., 2011). The average severity rate for the top 10 industries for compensable work-related knee MSDs was 7250 days per 10 000 FTEs. On average, each compensable work-related knee MSD claim costs $39 901, and 48.9% were compensable, with an average time loss of 213 days.

A cross-sectional study examined associations of occupational tasks with radiographic and symptomatic osteoarthritis in a community-based sample, and found that several occupational tasks were associated with increased odds of symptomatic knee problems (Allen et al., 2010). These tasks included lifting more than 10 lb (4.5 kg) weights, crawling, and doing heavy work while standing. More occupational walking and standing and less sitting were also associated with symptomatic knee osteoarthritis.

Job tasks requiring prolonged kneeling or squatting, walking >2 miles/day (3 km/day) and regularly lifting weights of at least 25 kg were found to have increased odds (odds ratios 1.9, 1.9, and 1.7 respectively) of developing knee osteoarthritis (Coggon et al., 2000). The risk associated with kneeling and squatting was higher in workers who also reported occupational lifting and appeared to interact multiplicatively with the risk conferred by obesity. Workers with a body mass index (BMI) of $\geq$30 kg/m^2 whose work had entailed prolonged kneeling or squatting had an odds ratio of 14.7 (95 CI 7.2–30.2), compared with workers with a BMI <25 kg/m^2 who were not exposed to occupational kneeling or squatting (Coggon et al., 2000). Kneeling and crawling are believed to cause high knee injuries among floor layers (Kivimaki et al., 1992; Jensen and Kofoed, 2002; Jensen et al., 2010, 2012; Reid et al., 2010). It has also been shown that carpet layers have a high prevalence of occupational knee morbidity (Tanaka et al., 1985, 1989). One of the main causes is that they need to frequently "kick" the bumper on the rear end of the knee kicker with one knee when laying a carpet (Huang and Wu, 2012). The repetitive contact stress is believed to be one of the risk factors causing knee injuries.

A study examined the relationship between occupational exposures to frequent squatting/kneeling and/or heavy lifting with cartilage morphology based on magnetic resonance imaging at the tibiofemoral and patellofemoral joints among male workers (Amin et al., 2008). It found that jobs involving squatting and kneeling, and heavy lifting have shown a modest increased risk for worse cartilage morphology scores at the patellofemoral joint with an odds ratio of 1.8 (95% CI 1.1–3.2) and medial tibiofemoral joint with an odds ratio of 1.6 (95% CI 0.9–3.0) respectively. Kneeling in many of the construction trades also creates elevated contact stress on the skin of the knees. A cross-sectional survey found that a history of past skin infection in the knee area was about four times more prevalent among floor layers who spent a significant amount of time in kneeling than among brick layers and millwrights who do not spent that much time in kneeling (Tanaka et al., 1985). Unilateral or bilateral patellar skin thickening and erythema were found to be more prevalent among those (floor layers and tile layers) who kneel.

PHYSICAL RISK FACTORS

According to many of the epidemiologic studies on work-related MSD of the various body parts, it is evident that certain physical risk factors or mechanical stressors are significantly associated with these disorders. It is necessary to examine these mechanical stressors systematically so that methods to quantify the level of risks due to these mechanical stressors and strategies to improve work conditions and protect workers can be developed. For the purpose of work-related MSD discussions, mechanical stressors or physical risk factors can be categorized into force, posture, repetition, contact stress, and vibration. These mechanical stressors are associated with the specific job demands in workplaces, as shown in Figure 20.1. They are applied to individual workers performing these jobs. Recognizing these risk factors is important in performing proper risk assessments or job evaluations and developing relevant solutions.

Force

Manual forces generated by human muscles have been, and are still being, used to accomplish many industrial tasks in today's workplaces, in spite of increasing levels of mechanization in recent decades. In the field of occupational ergonomics, the forces found in workplaces are usually divided into two major groups: (1) forces related to manual material handling, and (2) forces related to intensive, repetitive, or forceful hand activities.

Manual material handling tasks often involve lifting/lowering, pushing/pulling, and carrying (Fig. 20.2).

Lifting is the operation that brings or moves something from a lower level to a higher one and lowering moves objects in the opposite direction. Some examples of lifting and lowering include stocking milk jars on or from a pallet in a grocery store, and lifting and lowering tires by a mechanic in a car repair shop.

Pushing and pulling is the operation of using push or pull forces to move something from one point to another (when the object is usually in contact with the ground). Some examples of pushing and pulling include pushing a dolly loaded with fence boards in a lumber mill and pulling a loaded cart in a manufacturing facility. It should be noted that, in the scenario where a worker "pushes" an object from his assembly station to the next station without moving his feet, it is not considered "pushing" in the category of manual material handling in the field of occupational ergonomics. This scenario will be discussed later, in the discussion on other hand forceful exertions.

Carrying is the operation of moving something from one point to another with the hands (when the object is usually off the ground). Some examples of manual carrying include carrying a basket of coffee beans by a coffee harvester and carrying a tool box by a millwright. It should be noted that carrying often occurs together with lifting/lowering operations. Carrying involves not only upper-body strength to hold the object, but also the lower extremities when the worker walks from one point to another.

Gripping is one form of intensive or repetitive hand activity (Fig. 20.3). Often it is also called power gripping. It happens often in operations that require the worker to forcefully grip an object. Some examples include gripping a knife in meat processing, and gripping a control in machine operation.

Pinching is another form of intensive or repetitive hand activity. It happens often in operations that require the worker to pinch an object with fingers. Some examples include pinching a piece of glass and tightening a small screw with fingers. A major difference between gripping and pinching is that the fingers can be wrapped around the object that the worker is gripping, compared to when the fingers are not able to get around the object so that they are usually held straight. Pinching requires more muscular effort compared to gripping when the external force demand is the same; in other words, pinching strength is generally lower than gripping strength for most people.

There are many other types of hand forceful exertions in different occupational operations. Some examples include using fingers to press (finger press), such as in operations of inserting rubber seals, forceful twisting, as in turning knobs by machine operators, and pushing or pulling with one or two hands (without moving the feet) in some industrial operations.

High forces can be applied repetitively, or held for long period of time. The former is usually called repetitive exertions and the latter static loading. The pattern of loading can have significant physiologic effect on the human body (Byström et al., 1991a, b; Mathiassen, 1993). For example, Mathiassen (1993) did a study on the influence of an exercise–rest schedule on the physiologic response to isometric shoulder–neck exercise, and found that endurance times are significantly related to both exertion cycle time (or frequency) and duty cycle (percent time in exertion). The exertion duty cycles influenced arterial blood pressure, heart rate, electromyography (EMG) amplitude, and perceived fatigue (Mathiassen, 1993). Therefore, it is not only important to quantify the level of a forceful exertion, but also to

Fig. 20.2. Manual material handling tasks involve lifting/lowering, pushing/pulling, and carrying.

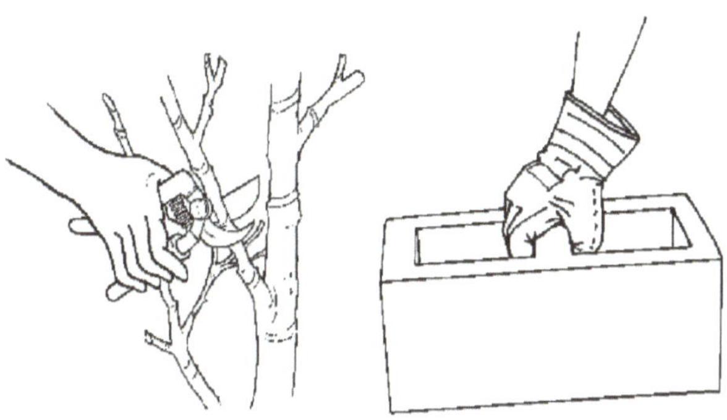

Fig. 20.3. Gripping and pinching often occur in hand-intensive activities.

quantify its frequency and duty cycle. The total duration of exertion should also be quantified in forceful exertion job evaluations, as the health effect of physical exposures is often accumulative.

Posture

Workers assume various postures in order to perform their jobs. Some of the postures are comfortable and others may cause discomfort due to their awkwardness. Awkward postures are often risk factors associated with many MSDs in workplaces. Awkward postures are defined as body parts away from their neutral positions, and may cause increased strain to body parts (Fig. 20.4). Examples of awkward postures in workplaces include working overhead for long periods of time when performing assembly work, bent wrists using a power tool in maintenance work, and lifting with back flexed about 90° in many construction jobs.

In quantifying the risk of awkward postures, it is important to evaluate the awkwardness (or level) of the posture (i.e., how much the body part deviates from its neutral position, for example, back bending of more than 45°), duration (time spent in the awkward posture, for example, neck bending of 30° for 3 hours in a shift), and frequency (i.e., how frequently the body part moves into the awkward position, for example, upper arm raised overhead 10 time in an hour). Holding a body part in an awkward posture for long periods of time, such as in some welding jobs, is often called static postures, and is considered to be one of the major risk factors among many computer users as well.

Repetition

Repetition often occurs in short cyclic tasks when the same body parts are used over and over again. As discussed above, repetition often is believed to be a risk factor associated with many types of MSD in the workplace. Some repetitive exertion examples include tasks in meat-processing factories where manual cutting requires repetitive forceful hand exertions, and apple-packing facilities, where operators repetitively use their fingers to pick up apples, place them into trays, and use their shoulder muscles in the movements. Repetitive tasks may require the use of repetitive muscle exertions of many different body parts at the same time (Fig. 20.5). Different muscles may have different capacities of performing repetitive tasks. In general, smaller muscles (e.g., muscles controlling eye blinks) can move more frequently or faster than larger muscles (e.g., trunk muscles).

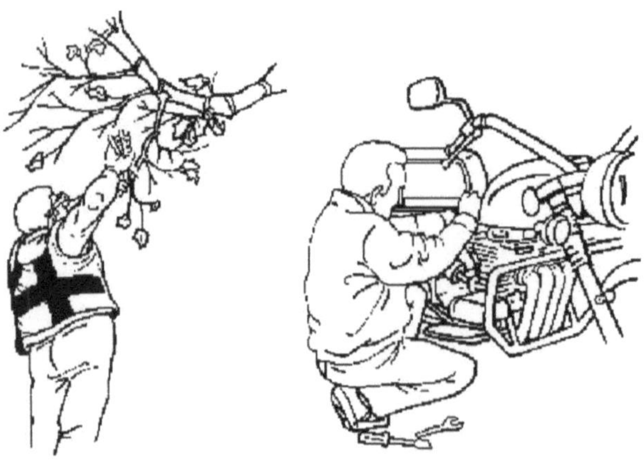

Fig. 20.4. Awkward postures are defined as body parts away from their neutral positions.

Contact stress

When workers have to place their bodies (such as knees and forearms) on a hard surface in order to support themselves, pressures on the localized small contact areas increase (Fig. 20.6). This is so-called contact stress, which is considered a risk factor associated with some MSD in workplaces. Some examples of contact stress include leaning on the sharp edge of a workstation in

Fig. 20.5. Repetition often occurs in short cyclic tasks when the same body parts are used over and over again.

Fig. 20.6. Contact stress causes pressures on localized small contact areas to increase.

Fig. 20.7. The energy of hand–arm vibration is transmitted to the upper arm(s) from the operator's hand(s) in contact with a vibrating tool.

order to reach work objects which are too far from the operator (contact stress on the thighs) and resting the forearms on a sharp edge of a desk due to placement of keyboard which is too close to the edge of the desk (contact stress on the forearms), and holding a pair of pliers with short handles in bare hands (contact stress on the palms).

Vibration

Many power tools and machinery are used in today's workplaces. Without doubt, they often increase work productivity and make impossible jobs possible. However, they sometimes also bring some unwanted negative health problems to operators. There are two types of vibration that are considered risk factors associated with MSD. One is labeled as whole-body vibrations, in which vibration energy is transmitted to the whole bodies of operators through contact with the buttocks (when operators sit on vibrating equipment) and/or feet (when operators stand on vibrating equipment). Examples of whole-body vibration include driving a forklift truck on an uneven log yard, or standing on a vibrating platform handling heavy boards in a sawmill.

Another type of human vibration is called hand–arm vibration, in which vibration energy is transmitted to the upper arm(s) from the operator's hand(s) (Fig. 20.7). Examples of hand–arm vibration include using a power chain saw or operating a hand-guided lawn mower. Using power tools is usually the main source of hand–arm vibration exposure.

MECHANICAL STRESSOR QUANTIFICATION

Quantification of mechanical stressors is important in research into the relationships between risk factors and MSDs, developing threshold limit values (TLVs), performing job evaluations to determine risk levels, and guiding intervention efforts. Effective evaluations of job improvement efforts also need the quantification of mechanical stressors.

Depending on the purposes of mechanical stressor quantification and the resources available, there are many options. In order to understand these methods, it is important to discuss the terminologies and concepts.

Risk factors

Whenever we talk about risk factors here, we are referring to those mechanical stressors in workplaces and jobs (Fig. 20.1) that are believed to be associated with the development of work-related MSDs. The presence of risk factors alone does not warrant the development of an injury or MSD. The development of MSD is determined by the dose of the exposure to the risk factor(s) and the individual workers. The dose of the exposure is related to the total duration that a worker is exposed to the specific exposures.

Dimension of exposure

Any workplace exposure can be characterized in terms of its level, frequency, and duration. The level of exposure (mechanical stressor), sometimes referred to as amplitude of exposure, determines how high or how awkward an exposure is. For example, it is the magnitude of forces in lifting, pushing and pulling, carrying, gripping and pinch gripping. It is the awkwardness of work postures (i.e., how far away a body part is from its neutral position). It is the level of vibration level that a power tool generates. In terms of repetitive movements, it is the number of times a body part moves back and forth.

The frequency of exposure is the number of times that a risk factor (mechanical stressor) occurs. For example, the frequency for lifting may be 10 lifts per hour; for forceful gripping in a poultry-processing job it may be 30 times per minute.

The duration of exposure usually refers to the percent time that a particular risk factor (mechanical stressor) occurs. This is also often referred as duty cycle or percent time (% time) in order to differentiate it from task duration (number of hours performing a task in a workshift). For a cyclic task where forceful exertion is applied, the duty cycle (% time in exertion) is calculated by the actual exertion time divided by the cycle time.

Hazard level

Hazard levels are used to determine the likelihood that exposure to a risk factor may cause MSD. It is a threshold level of exposures to risk factors that are associated with work-related MSDs. Often hazard levels are determined by considerations of epidemiologic study results, biomechanical analysis, and physiologic studies (Waters et al., 1994; Moore and Garg, 1995; ACGIH, 2001; David et al., 2008). While hazard level is a continuous variable, many guidelines and evaluation methods use categoric limit values in order to allow practitioners to make decisions easily. Three hazard-level categories (safe, caution, and hazard) are most commonly used.

Self-reports

Self-reports are methods that gather mechanical stressor information through the workers themselves who are exposed to these mechanical stressors. This is the type of methods that many epidemiologists use in order to gather mechanical stressor information among a large population group. Occupational safety and health practitioners also use this type of method to quickly and economically quantify job mechanical stressors in order to perform quick job evaluations.

Self-reports may be used to ask workers about their hand force levels when they are performing hand-intensive tasks. Often a scale such as the Borg scale (Fig. 20.8) (Borg, 1982), or visual analog scale (VAS: Werner et al., 2005) is used. Self-report methods have also been used to gather information on repetitive movements (Latko et al., 1997; Spielholz et al., 2001; Bao et al., 2006) and awkward postures (Wiktorin et al., 1993; Jensen et al., 2000; Spielholz et al., 2001; Balogh et al., 2004).

The benefits of using self-reports are obvious in that they can gather information among a large group of population quickly and economically. No specific equipment is required. However, as expected, this method may not be very accurate as it depends on the subjective judgment of the individuals who provide the estimates. People are different when you perceive mechanical stressors. Therefore, the interrater reliability of self-reports may not be good. As it is subjective, information collected using self-report methods may not be very specific either. For example, when asked about perceived hand exertion with the same amount of hand grip forces with different load on the upper arms, people may report different levels of force (Bao and Silverstein, 2005). This may be because of perceptions of loading on other upper-arm muscles rather than actual hand grip forces.

Rated by the Worker Using Borg Scale

Rating of Perceived Exertion (RPE) – Borg Scale (1982)	10	Maximum
	9	Very, very hard
	8	Very hard
	7	Hard
	6	
	5	Somewhat hard
	4	Moderately hard
	3	
	2	Light
	1	Very light
	0.5	Very, very light

Fig. 20.8. A Borg scale can be used to collect exertion level using a self-report method (Borg, 1982).

Instrumentation

In order to obtain more specific and accurate mechanical stressor estimates, special mechanical stressor measurement instruments are often used. This approach is also often called direct measurement. For example, in order to measure hand forces when operating power tools, instrumented tool handles can be used in experiments where force gauges are built into the tool handles (McGorry, 2001; Seo and Armstrong, 2008; Seo et al., 2008).

Specific muscle activities during task performances can be quantified using surface EMG techniques (Winkel et al., 1983; Aarås and Westgaard, 1987; Radwin et al., 1989; Bao et al., 1996, 2001; De Luca, 1997). Muscle activities are responses of individual muscles to the workload requirements of jobs. Researchers and practitioners are often interested to know how much muscle loading is required when performing a particular task. This is considered the amplitude domain analysis of EMG signals.

In the field of occupational health and safety, and ergonomics, EMGs are usually measured with surface electrodes which are attached to the skins of muscles. Since surface electrodes are used, only muscles that are directly under the skin can be measured. The muscle activities measured by these electrodes are very specific or localized. For example, electrodes attached to the skin of the descending portion of the right trapezius muscle quantify the muscle activities of that particular part of the muscle, instead of loading on the general "shoulder region," as estimated by a self-report method.

The same workload requirements of jobs (the external exposure) may cause different EMG responses (internal exposure) depending on individual workers' characteristics and work techniques (so that different muscles may be engaged in performing the jobs). Variations of electrode attachments may also influence EMG signals measured (Woods and Bigland, 1983). Therefore, clear descriptions of electrode locations are required when performing EMG measurements. Standardized electrode placement locations have also been suggested (NIOSH, 1992; Perotto, 1994).

While EMG activities are proportionally related to external job workload requirements, the individual variations require a calibration procedure before vocational

EMG measurements. Two EMG calibration procedures (also known as normalization) are commonly used (Mathiassen et al., 1995). The first one is to reference work EMG activities to the EMG amplitude of that particular muscle during a maximal voluntary contraction: the work EMG is expressed as a percentage of the EMG level during the maximal voluntary contraction. This approach relates the work EMG activities to the maximal muscle capacity of individuals. The second method is to reference work EMG activities to the EMG amplitude when the individual is applying a known load on a particular muscle (this is usually done by asking the individual to apply a known force in a standardized posture). The work EMG is then expressed as a percentage of the reference voluntary contraction. Of course, if the objective of muscle activity measurements is to compare two conditions (e.g., before and after an ergonomics intervention), the electric potential levels of the muscle activities in μV from the two work conditions can be used directly and compared.

Surface EMGs are also frequently used to quantify muscle fatigue objectively (Cook et al., 1998; Christensen et al., 2000; Bosch et al., 2007; Ferguson et al., 2013). This differs from the subjective fatigue measurement using self-report method by asking workers what level of fatigue they feel. Estimation of muscle fatigue relies on changes in the spectral characteristics of EMG signals. Objective measures of EMG spectral characteristics can be in the amplitude domain and frequency domain. In general, EMG of a fatigued muscle has increased amplitude and decreased frequencies (the measures of the median or mean frequency of the power spectrum of the EMG signal are often used).

Another example of instrumentation methods is the measurement of joint postures using electrogoniometers. An electrogoniometer is a protractor with electric circuits inside that gives electric readouts about an angle. Typical electrogoniometers have two reference arms. These two arms are attached to the skins of the long axes of the adjoining body segments with the center of the protractor over the estimated center of the joint. After proper calibration, the electric readout indicates the posture angle of a particular body joint. Using proper kinematics calculations, one can obtain not only the posture angles of the body joint, but also displacement (or changes of postures), angular velocity, and acceleration of the body joint. Electrogoniometers have been used in many postural mechanical stressor evaluations of different occupations (Balogh et al., 1999; Stal et al., 1999; Wahlstrom et al., 2000; Jones and Kumar, 2008; Fethke et al., 2012).

As we can see, instrumentation methods provide more accurate data compared to self-report methods. However, accuracy depends on the proper use of the instruments. This includes proper sensor positioning (e.g., proper EMG electrode attachment, and correct electrogoniometer mounting), monitoring electric interference (e.g., electric signal from the power line), recognizing artifact signals, standardized calibration procedures, and expert result interpretations. Mechanical stressor data collected by instrumentation methods are usually more objective (in contrast to the subjective nature of self-report methods). They are also very specific (e.g., loading from a specific muscle and movement of a specific body joint). However, instrumentation methods also require investment in sophisticated equipment; tedious procedures in attaching sensors; conducting equipment calibration and performing data collection. Enormous expertise is required to operate the instruments, perform calculations, and interpret results. Therefore, it is very difficult to use instrumentation methods in quantifying mechanical stressors among large populations in workplaces, but only among a limited number of subjects. This type of method is more frequently used by researchers than ergonomics practitioners.

Observational methods

A good compromise between the self-report and instrumentation methods is the observational methods used by many researchers and practitioners in collecting mechanical stressor data. Observational methods probably present the best options for individual exposure assessment to mechanical stressors in large-scale epidemiologic studies as well as in job assessments by practicing ergonomists. Observations are either done on site, when analysts complete their estimations of interested mechanical stressors directly through observing workers performing their tasks, or off site, when video recordings of workers' task performances are reviewed and mechanical stressors are identified and estimated.

Observational methods often use categoric scoring of body positions, movement frequencies, types of hand grip, and levels of forces based on analysts' estimations during observations (Corlett and Bishop, 1976; Karhu et al., 1977; Armstrong et al., 1982; Kemmlert and Kilbom, 1986; Keyserling, 1986). The method used by Armstrong et al. (1982), for example, classified wrist postures into five categories: (1) neutral; (2) flexion; (3) extreme flexion; (4) extension; and (5) extreme extension. In general, video-based analysis may be the most appropriate observational method for mechanical stressor estimations and identification of work activities in large-scale epidemiologic studies as well as applied job assessments because it allows the analysts sufficient time to estimate the mechanical stressors and provides the possibility to reanalyze the data for quality control purposes or communicate with others if needed.

Several new computer programs (e.g., Multimedia Video Task Analysis (MVTA) from the University of Wisconsin, and The Observer XT, from Noldus) have been marketed to make observation of mechanical stressors easier. For example, the MVTA system (Yen and Radwin, 1995) is able to set accurate time codes for various user-defined events on video recordings and perform time studies on various time-based events (e.g., tasks, and wrist postures). A drawback of this type of video analysis is that the analyst has no control over the angles of observation, and has to depend on the quality of the video recordings. Therefore, it is important to have good-quality video recordings in order to obtain reliable and adequate mechanical stressor estimates based on video recording observations. Another disadvantage with video recording-based observation analysis is that one cannot obtain direct measurements of force parameters (e.g., object weights and hand forces) easily from video recordings. In contrast, when observations are done on site, the analyst can most often communicate or interact with the workers to obtain information about those force parameters. To overcome this shortfall, it is often recommended to obtain force information using direct measurement methods on site in conjunction with video observation methods for other posture- and task time-dependent parameters.

Similar to self-report methods, observational methods may also be biased by the perception of the analyst (instead of the workers, as in self-report methods). After all, observational methods are based on the perception of analysts through their observations and interpretations of workers performing their tasks. Therefore, some variations between and within analysts are unavoidable. Measures to reduce such variations should be taken through proper training, clear instructions and guidelines, standardized procedures, and providing references that may help standardize analysts' judgment.

JOB ASSESSMENT

One of the ultimate objectives of quantifying mechanical stressors is to measure risk levels of a job and protect workers' safety. Based on quantified mechanical stressor information and injury MSD statistics, researchers have developed various job evaluation tools to assess job risk levels in workplaces. Many of these job evaluation tools include one or more risk factors (e.g., hand forces, hand/wrist postures and repetitions) in the model, and consider all three dimensions of mechanical stressors (level, frequency and duration). Following are introductions to several published and widely used job assessment methods that address various job situations, such as lifting/lowering, pushing/pulling, carrying, intensive hand activities, and exposure to vibrations.

NIOSH Lifting Equation

In 1985, the US National Institute for Occupational Safety and Health (NIOSH) convened an *ad hoc* committee of experts who reviewed the current literature on lifting, recommended criteria for defining lifting capacity, and in 1991 developed a guideline for evaluating lifting and lowering tasks, the revised NIOSH lifting equation (Waters et al., 1993; NIOSH, 1994). This job evaluation method collects job-related mechanical stressor variables, such as: (1) object weights; (2) horizontal location of the origin from where a lifting task starts; (3) vertical location of the origin from where a lifting task starts; (4) vertical travel distance of the lifting; (5) trunk twisting; and (6) hand coupling together with the frequency and duration of the lifting task. Using this information, the method calculates the recommended weight limit (RWL) of the lifting job that would allow most workers (99% of male workers and 75% of female workers, or 90% of the work population) to be likely to perform this lifting job safely. Using the RWL, one can calculate a lifting index (LI), which is the ratio of the object weight lifted to the RWL.

The RWL equation and LI are based on the concept that the risk of lifting-related low-back pain increases as the demands of the lifting task increase. According to NIOSH, as the magnitude of the LI increases, the level of the risk for a given worker would be increased and a greater percentage of the workforce is likely to be at risk for developing lifting-related low-back pain. The shape of the risk function, however, is not known. Without additional data showing the relationship between low-back pain and the LI, it is impossible to predict the magnitude of the risk for a given individual or the exact percent of the work population who would be at elevated risk for low-back pain. NIOSH suggests that, when LI is ≤ 1, the lifting job is considered safe; otherwise, it is at increased risk. In practice, many users have been setting two limit values for the LI (the job is safe when $LI \leq 1$, job has increased risk and improvement may be needed, if feasible, when $1 < LI \leq 3$, job has high risk and intervention is definitely needed when $LI > 3$). This is also the position of NIOSH and many experts of work-related MSD that nearly all workers will be at an increased risk of a work-related injury when performing highly stressful lifting tasks (i.e., lifting tasks that would exceed an LI of 3.0).

The NIOSH lifting equation is one of the most widely used evaluation tools for lifting and lowering jobs. Electronic NIOSH lifting equation calculators are available both commercially and as freeware found on the internet.

ACGIH lifting threshold limit values

The American Conference of Governmental Industrial Hygienists (ACGIH) publishes TLVs for various chemical and physical agents commonly encountered in today's workplaces. ACGIH developed a set of TLVs that recommend workplace lifting conditions under which it is believed nearly all workers may be repeatedly exposed, day after day, without developing work-related low-back and shoulder disorders associated with repetitive lifting tasks (ACGIH, 2005).

The ACGIH lifting TLVs are based on biomechanical, psychophysical, and epidemiologic studies, which together demonstrate a causal association between lifting activities and increased risk of low-back disorders characterized by pain and temporary or prolonged inability to perform normal occupational and nonoccupational activities. The ACGIH lifting TLVs are built on the foundation of the Washington State's lifting calculator (Washington State Department of Labor and Industries, 2014). The original Washington State's lifting calculator (which will be introduced later) was based on the revised NIOSH lifting equation (NIOSH, 1994) and set a bar at an LI of approximately 1.9. The experts in the ACGIH committee suggested that the weight limits used in the Washington State's lifting calculator did not provide sufficient reduction in risk for low-back disorders associated with lifting tasks at shoulder height and above and below knuckle height after reviewing the scientific evidence (ACGIH, 2005). Therefore, a more protective set of lifting TLVs is adopted in the ACGIH lifting TLVs.

The ACGIH lifting TLVs consist of three tables with weight limits for two-handed, mono-lifting tasks within 30° of the sagittal plane (the neutral standing position). The three tables are used for conditions of lifting tasks according to their duration (hours per day) and frequency (number of lifts per hour). These conditions are: (1) ≤ 2 hours per day with ≤ 60 lifts per hour or > 2 hours per day with ≤ 12 lifts per hour; (2) > 2 hours per day with > 12 and ≤ 30 lifts per hour or ≤ 2 hours per day with > 60 and ≤ 360 lifts per hour; and (3) > 2 hours per day with > 30 and ≤ 360 lifts per hour, respectively.

For each of the TLV tables, the users need to enter the location of objects to be lifted in the horizontal and vertical zones. Each of the zones is divided into three to four sections: (1) close; (2) intermediate; and (3) extended for the horizontal zone, and: (1) reach limit or 30 cm above shoulder to 8 cm below shoulder height; (2) knuckle height to below shoulder; (3) middle shin to knuckle height; and (4) floor to middle shin height for the vertical zone respectively.

Under each of the lifting task conditions based on lifting duration, frequency, and an object location zone, a TLV for the lifting task is determined. For instance, when lifting an object located close to the body (horizontal zone) and from floor level (vertical zone), the weight limit is set at 14 kg if the task is performed less frequently (≤ 60 lifts per hour) for ≤ 2 hours per day; however, the weight limits are only 9 kg if the task is performed more frequently (>60 but ≤ 360 lifts per hour) for ≤ 2 hours per day. Under the same task condition but with longer task duration (>2 hours per day), the experts in the ACGIH lifting TLV committee did not provide any weight limit values as they believe that there is no known safe limit for such repetitive lifting (ACGIH, 2005).

Liberty Mutual tables for push/pull and carry jobs

While either the revised NIOSH lifting equation or the ACGIH lifting TLV can be used for lifting and lowering jobs, the Liberty Mutual tables can be used for pushing, pulling, and carrying jobs (Snook and Ciriello, 1991). Although there is also a set of tables that Liberty Mutual developed for lifting and lowering job evaluations, most practitioners prefer to use the revised NIOSH lifting equation or the ACGIH lifting TLVs for lifting job evaluations.

Liberty Mutual Group Loss Prevention field organization has been analyzing and evaluating lifting, lowering, pushing, pulling, and carrying tasks using the so-called "psychophysical tables," commonly known as the Liberty Mutual tables or Snook tables. These tables are based on research work carried out by Drs. Stover Snook and Vincent Ciriello at the Liberty Mutual Research Institute for Safety (Snook and Ciriello, 1991). The tables were developed with the goal of controlling costs associated with manual material handling operations. These costs can be attributed to high low-back disability costs, reduced productivity and quality due to poor job design. These tables provide users with an objective risk assessment of a problem manual material handling job and the foundation on which to build a solution.

These tables were obtained through a series of experiments carried out in the laboratory of the institute (Snook and Ciriello, 1991). Researchers made the tasks of these experiments as realistic as possible. For instance, pushing tasks were dynamic pushes through a given horizontal distance, and not just a maximum isometric push. Test sessions lasted approximately 4 hours. Experimental subjects were workers from local industry. These subjects were instructed to work on an incentive basis, working as hard as they could without straining themselves, or without becoming unusually tired, weakened, overheated, or out of breath. The experiments used a psychophysical methodology with measurements of oxygen consumption, heart rate, and anthropometric characteristics. Task frequency, distance, and hand positions were independent variables. The nature of the psychophysical methodology is that the subjects were able to adjust the amount of force required to push/pull the carts and the object weights that they carried until they felt they were the maximal that they could handle for the whole work day. The forces and weights are the maximal allowed limits for the workers.

The tables are arranged by women and men separately in different percentiles of the population (10%, 25%, 50%, 75%, and 90%). Since women usually have lower manual material handling capacities, the design goals for pushing, pulling, and carrying tasks that are considered acceptable for 75% of women are commonly used. Using these tables to evaluate a manual material handling job, one usually needs to collect a set of input task variables.

For pushing and pulling tasks, the following task variables need to be obtained:

- Is the task pulling or pushing?
- hand position during the task performance (high point or at or above shoulder height, middle point at or about the elbow height, and low point at or below knuckle height)
- distances that a cart is pushed or pulled (in feet or meters)
- frequency of pushing/pulling tasks
- the actual pushing or pulling forces required when the push/pull starts (initial force) and during the push/pull (sustained force); these forces can be measured using push/pull force gauges.

With the input pushing/pulling task variables, one can find the maximal acceptable initial and sustained forces. These are compared to the measured initial and sustained forces. If the measured force(s) is (are) greater than those obtained from the table, the job is considered not acceptable and need to be improved.

For carrying tasks, the following task variables need to be collected:

- hand position during carrying (elbow bent/about waist height or elbow straight/at about knuckle height)
- distance that the object is carried (in feet or meters)
- frequency of the carrying task
- the actual object weight that is carried.

With the input carrying task variables, one can find the maximal acceptable weight of the carrying task. This weight value is compared to the actual object weight carried. If the actual weight is greater than the value obtained from the table, the job is considered not acceptable and needs to be improved.

The use of the tables is somewhat cumbersome. Thanks to advances in computer technology, many electronic versions can be found either online as freeware or as standalone commercial programs.

ACGIH hand activity level TLVs

Another mechanical stressor that the ACGIH has suggested for TLVs is intensive hand activities, such as those found in assembly and meat processing jobs. This TLV is focused on mechanical stressors on the hand, wrist, and forearm (ACGIH, 2001). This TLV deals with hand activities that include mechanical stressors such as forceful hand exertions, repetitions, and awkward hand/wrist postures. It is based on epidemiologic, psychophysical, and biomechanical studies and is intended for "mono-task" jobs performed for 4 hours or more per day. The ACGIH defines a mono-task job as a job involving performing a similar set of motions or exertions repeatedly, such as working on an assembly line or using a keyboard and a mouse. It is believed that nearly all workers may be repeatedly exposed to the conditions that this TLV believes to be safe without adverse health effects.

Two variables need to be determined in order to evaluate a hand-intensive job using this TLV. The first one is the hand activity level or HAL. It is based on the frequency of hand exertions and the duty cycle (distribution of work and recovery periods). HAL can be determined with a VAS by a trained observer or calculated using information on exertion frequency and duty cycle (work/recovery ratio), as described by ACGIH (2001).

The second variable is the peak hand force, which is normalized on a scale of 0–10, which corresponds to 0–100% of the applicable population reference hand grip strength. ACGIH gives users several options to obtain peak hand force. One is to use a VAS rating scale by trained observer, or to be rated by workers themselves using a Borg scale (Borg, 1982), or measured using instrumentation such as a force gauge, or EMG. ACGIH also suggests that, in some cases, the peak hand force can be calculated using biomechanical methods (ACGIH, 2001). Peak hand force is normalized by dividing the job force requirement (external exposure) by the strength capability of the work population for that job activity.

While the hand/wrist posture is not explicitly used in the intensive hand activity job evaluation, it is considered through the normalized peak hand force. The strength capability of the target work population, which is used in the normalization calculation, is dependent on hand/wrist posture. When the job is performed in an awkward hand/wrist posture, the strength capability is reduced. Therefore the resulting normalized hand peak force is increased, as is the risk level of this hand-intensive job.

After obtaining these two variables, the risk level of the hand-intensive job is determined by using the chart shown in Figure 20.9. The HAL is on the horizontal axis, and the normalized peak hand force on the vertical axis. If the interaction of the HAL and the normalized peak

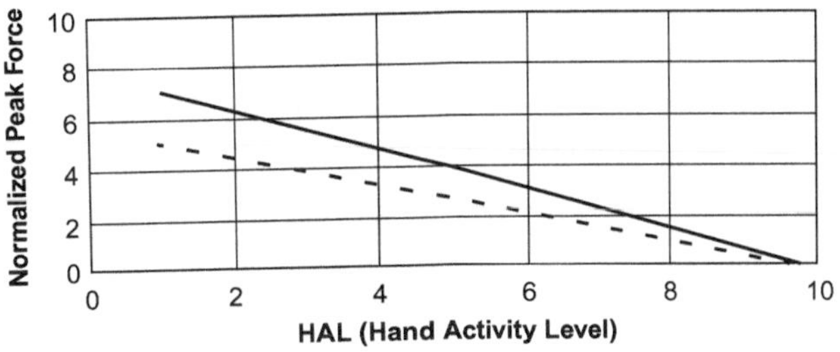

Fig. 20.9. Risk level determination based on hand activity level and normalized peak hand force. The solid line depicts the threshold limit value and dashed line is an action limit, for which general controls are recommended.

hand force lands above the solid line, the risk level is considered high and the job needs to be improved. If the interaction is below the dashed line, the job is considered safe. The dashed line is considered an action limit. If the interaction is above the action limit but under the TLV, the job has increased risk and general controls are recommended (ACGIH, 2001). If the interaction is at or above the TLV, the risk is high and control to lower the risk is needed.

The Strain Index

Another widely used hand-intensive job evaluation tool is the strain index method, developed by Moore and Garg (1995). Again, this method was developed based on existing knowledge and theory of the physiology, biomechanics, and epidemiology of distal upper-extremity disorders. In order to use this method, one needs to collect six task variables. These are: (1) intensity of exertion; (2) duration of exertion per cycle (or the duty cycle); (3) efforts per minute (frequency) of exertion; (4) speed of exertion; (5) hand/wrist posture; and (6) duration of task per day.

1. Intensity of exertion is an estimate of the hand force requirement of a task in relation to the individual who performs the task. It is defined as the percentage of maximum strength of the person who performs the task. The proposed methodology of obtaining this variable involves estimating the intensity of hand exertion using verbal descriptors (light, somewhat hard, hard, very hard, and near maximal) through job observation. A rating corresponding to the verbal descriptor is then assigned using a five-point rating scale (from 1 to 5, with 1 the lowest and 5 the highest intensity of exertion). In the strain index method, this variable is the most critical variable, and has the highest weight in the final risk determination.
2. Duration of exertion relates to how long an exertion is maintained. It is characterized as the percentage of time an exertion is applied in a task cycle or an exertion cycle. This is also often called the duty cycle. To obtain this variable, observation on site with the use of a stop watch or performing video time study on recorded task performance can be used. Average cycle time as well as average exertion time will be recorded and used to calculate the duration of exertion variable. Duration of exertion is divided into five categories: $<10\%$, 10–30%, 30–50%, 50–80%, and $\geq 80\%$. A five-point rating scale linking to the five categories of duration of exertion (1 for $<10\%$ and 5 for $\geq 80\%$) is then used.
3. Efforts per minute measures the repetitiveness of a task. It is the number of exertions per minute. It is also referred to as frequency of exertion. This variable can be obtained through on-site observation by counting the number of exertions in a representative period of task performance, or making video recordings of the task performance. Efforts per minute is also divided into five categories: <4, 4–9, 9–15, 15–20, and ≥ 20 efforts/minute. A five-point rating scale linking to this five categories of efforts per minute (1 for <4 efforts/minute, and 5 for ≥ 20 efforts/minute) is used.
4. Speed of work estimates perceived pace of the task. It is a subjective estimate using a verbal anchor to quantify its value (very slow, slow, fair, fast, and very fast). The five categoric quantities correspond to a five-point scale ranging from 1 (very slow) to 5 (very fast). Speed of work is different from the variable of efforts per minute. It is a measure of the pace of performing a task, rather than the rate (or repetitiveness) of movements (efforts/minute). In the "very fast" category, the workers may either not or barely manage to keep up with the required pace of the job. These workers often report being overly rushed and intently focused on the task. In reality, the variables of speed of work and efforts per minute often correlate to each other to some extent.
5. Hand/wrist posture refers to the anatomic position of hand and wrist relative to its neutral position. Since the hand/wrist postures have an effect on hand grip strength, this variable modifies the physical capacity of the hand/wrist by considering the hand/wrist postures. A system divided into five descriptive categories (very good, good, fair, bad, and very bad) is used for this variable. Analysts can use their professional judgment or perception to quantify the hand/wrist posture variable. Alternatively, the authors of the strain index method provided a list of categories (in degrees of angles) for wrist extension, flexion, and ulnar deviation to facilitate the quantification of this variable.

6. Duration of task per day reflects the total time that a task is performed per day. This variable adds the dimension of duration into the job evaluation. Depending on the number of hours spent on tasks, this variable uses a five-category system (≤1, 1–2, 2–4, 4–8, >8 hours) together with a corresponding five-point scale (1, 2, 3, 4, and 5).

After each of the six strain index variables receives its rating in a job evaluation, a corresponding multiplier is assigned for each variable. This multiplier system adds different weights to the six different variables as well as their different ratings in order to consider their physiologic importance in the development of hand and arm disorders. For example, the intensity of exertion is considered the most important factor in terms of work-related hand–arm disorders, hence it receives a larger multiplier compared to that of the other variables. This variable increases exponentially with the increase in its rating. Some variables (e.g., hand/wrist posture and speed of work) may have a constant weight until the rating reaches 3 or 4. For some variables (e.g., duration of exertion, efforts per minute, and duration per day), it is believed that lower exposures may be protective. Therefore a weight with reduction effect (multiplier < 1) is used for lower exposure categories.

Once the multipliers for the exposure ratings of the six different variables are determined, the final job evaluation is determined by the strain index score, which is the product of all six multipliers. The higher the strain index score, the higher the risk of the job. Based on the strain index score, job risk levels can be determined. A three-risk-level system is proposed: ≤3, safe; 3–7, action level; and ≥7, hazard level.

The originally proposed strain index method (Moore and Garg, 1995) only applies to mono-task jobs, i.e., when the task is cyclic with one type of exertion. In the real world, many jobs may involve multiple tasks (e.g., job rotation) and many different task activities (e.g., using different tools for different components). Recently, more complicated methods (Drinkaus et al., 2005; Bao et al., 2009) have been developed in order to consider more complicated job scenarios.

RULA/REBA

The rapid upper-limb assessment (RULA) method was originally developed for use in ergonomics investigations of workplaces where work-related upper-extremity disorders are reported (McAtamney and Corlett, 1993). This is a method that is often used when postural risk factors are of concern. A slightly different version of the RULA method, rapid entire-body assessment (REBA), was later developed (Hignett and McAtamney, 2000), in order to fill a perceived need for a practitioner's field tool. REBA is specifically designed to be sensitive to the type of unpredictable working postures found in healthcare and other service industries.

The RULA method requires working postures to be recorded. The body is divided into various segments, which form two groups (A and B). Group A includes the upper arm, forearm, and wrist. Group B includes the neck, trunk, and lower extremities. The posture for each body part is divided into a number of categories. For example, for the upper arms, there are five postural categories: ≤20° of extension to ≤20° of flexion; >20° extension; 20–45° of flexion; 45–90° of flexion; and >90° of flexion. Based on the category of the posture, a score is assigned (e.g., score ranging from 1 to 4 for the upper arms). The higher the score, the higher the risk.

After the posture scoring, a modifying score is also awarded to the corresponding body part if some conditions exist. For example, the total upper-body score needs to add 1 if the shoulder is raised, and/or add 1 if the upper arm is abducted, and/or subtract 1 if leaning or supporting the weight of the arm.

After the scorings of the postures of all body parts, they are combined for groups A and B for all postures of the upper extremities and neck, trunk, and lower extremities respectively, resulting in score A and score B. Muscle use and force/load scores, if present, are then added to each group. If posture is mainly static, or if action repeatedly occurs ≥4 times/minute, one needs to add a muscle use score of 1 to the total posture score to that corresponding group. If load is < 2 kg and exertion is intermittent, there will be no force/load score to be added to the total posture score for that group. However, if load is 2–10 kg and the exertion is intermittent, or load is 2–10 kg and the exertion is static or repeated, a force/load score of 1 or 2 will be added to the total posture score of that group. If load is > 10 kg and repeated in the form of shocks, a force/load score of 3 will be added to the total posture score of that group.

After adding the muscle use and force/load scores to score A and score B, score C and D are obtained respectively. A matrix is then constructed to obtain the so-called grand score, which is the indicator of the estimated risk level of the job (Fig. 20.10). The higher the grand score, the higher the job risk. The authors proposed four job risk levels (labeled as action levels) based on the grand scores (McAtamney and Corlett, 1993):

- Action level 1: A score of 1 or 2 indicates that posture is acceptable if it is not maintained or repeated for long periods of time.
- Action level 2: A score of 3 or 4 indicates that further investigation is needed and changes may be required.

C Score (upper extremities)	D Score (neck, trunk, lower extremities) 1	2	3	4	5	6	7+
1	1	2	3	3	4	5	5
2	2	2	3	4	4	5	5
3	3	3	3	4	4	5	6
4	3	3	3	4	5	6	6
5	4	4	4	5	6	7	7
6	4	4	5	6	6	7	7
7	5	5	6	6	7	7	7
8	5	5	6	7	7	7	7

Fig. 20.10. A matrix to obtain the grand score based on score C and D. (After McAtamney and Corlett, 1993.)

- Action level 3: A score of 5 or 6 indicates that investigation and changes are required soon.
- Action level 4: A score of 7 indicates that investigation and changes are required immediately.

The REBA method (Hignett and McAtamney, 2000) has fundamentally the same conceptual structure as the RULA method. A main difference of REBA as compared to RULA is that it adds the consideration of coupling (a descriptive measure to quantify how the load is supported) to the total score of the upper arm, forearm, and hand/wrist group in the handling of loads and recognizes that coupling may not always be through the hands. The other difference is that it only includes load/force modification to the group of the trunk, neck, and lower extremity. The ranges (1–12) of total scores A and B are different from that of the RULA method. The job risk level classification is different as well. For the REBA method, the following action levels are defined according to final REBA scores:

- Action level 0: A REBA score of 1: risk level is considered negligible, and no action is necessary.
- Action level 1: A score of 2–3: low risk and some action may be necessary.
- Action level 2: A score of 4–7: moderate risk and action is necessary.
- Action level 3: A score of 8–10: high risk and action is necessary soon.
- Action level 4: A score of 11–15: very high risk and action is necessary now.

The Michigan 3-D Static Strength Prediction Program (3DSSPP)

The 3DSSPP is a computer program developed by the Center for Ergonomics at the University of Michigan, USA, to predict static strength requirements for tasks such as lifting, pressing, and pulling based on occupational biomechanical modeling and human static strength capability data (The University of Michigan, 2008). This is a versatile method to help ergonomists evaluate task risk levels and with prevention of work-related MSDs both proactively and reactively. This method is based on biomechanical modeling calculations of external loads and work postures when performing tasks, and using the calculated loading on the various body joints and low back to compare with population strength data of corresponding body parts.

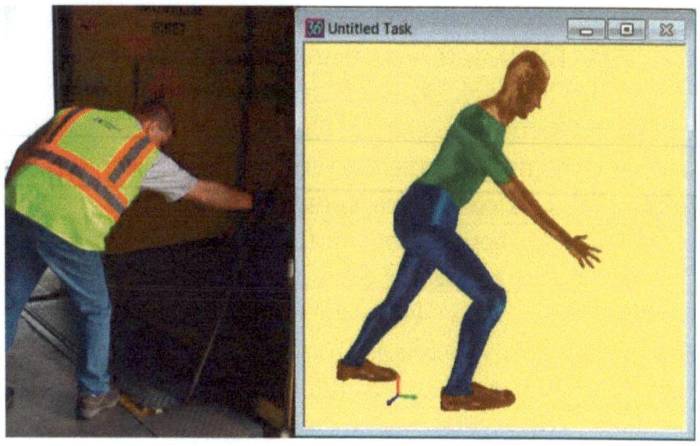

Fig. 20.11. A human graphic figure mimicking the task performance posture of the worker.

We will use a task evaluation example to introduce this method. In a trucking company, one of the tasks for truck drivers is to set a metal plate to bridge warehouse dock and a trailer. In order to do this, the driver uses a lever to operate the plate. An on-site visit documented the work posture when performing the task, and obtained a measurement of a one-hand pulling force of 50 lb (23 kg) using a force gauge. The various posture angles of the work posture and the pulling force are the input variables for this job evaluation.

In addition to these input variables, the strength of the evaluation population needs to be determined. There are several options in the program from which users can choose: data for female and male populations, data at the 5th, 50th, and 95th percentiles, or data corresponding to specific body height and weight values. This can be determined by the nature of the evaluation. In our example, we entered the value of the volunteer driver's data (height: 69 inches (175 cm), weight: 178 lb (81 kg)).

After entering the postural data, one can check the correctness of the work posture by comparing the posture of the human graphic figure (Fig. 20.11) and the actual work posture of the image. Running the analysis, a screen provides users with a list of summary results of the analysis. These include low-back compression force at L4/L5 of the lumbar spine, and percentages of workers having sufficient strengths at the respective joints capable of performing this particular job (strength percent capable). The black lines indicate the calculated values of the respective measures. The program uses a color coding scheme to indicate risk levels: green depicts acceptable and red is unacceptable. The yellow means moderate risk between the safe and hazard zones. In the case of our particular example (Fig. 20.12), the evaluation results show that the low-back compression is 653 lb (296 kg), which is in the safe zone, and many of the calculated torques at the various joints impose increased risks (yellow zone), although none is considered hazardous.

Using the 3DSSPP, one can also perform various evaluations in the design stage of a product, a workstation, or a job. It may also be incorporated into computer-aided design software to allow engineers to perform job risk evaluations together with their product engineering design. To solve existing problems in workplaces, one can not only use this method to evaluate the risk level, but also test various improvement ideas by running the program with different input variable values.

Washington State checklists

Washington State adopted an ergonomics regulation (Ergonomics Rule) in 2000 that focused on primary prevention of work-related MSDs (Foley et al., 2009). In November 2003 the regulation was repealed by an industry-funded voter initiative arguing that the implementation of this rule was unnecessary because the number of ergonomic-related injuries is declining and the complexity and cost of the rule would contribute to job losses and damage Washington's business climate (Guppy, 2003; Foley et al., 2009). While the attempt to adopt the ergonomics rule in Washington State was repealed (Ballotpedia.org, 2003), the checklists accompanying the regulation have been widely used by practitioners. There are two checklists: the Caution Zone and Hazard Zone checklists (LNI, 2000a, b). Since these checklists were developed in the line of the repealed ergonomics rule and were intended to have regulatory effects on employers, they impose minimal requirements on employers. Some experts believe that some of the limit values may be less protective (ACGIH, 2005). One should keep this in mind when using these checklists.

The Caution Zone checklist answers the question: Does the job have any MSD risk factors that need to be addressed? It is a screening tool. Jobs reach the thresholds of the caution zone needed to be further evaluated to determine whether they are in the hazard level. Caution zone jobs may not need to be fixed. Only the analysis results from the Hazard Zone checklist will make the determination. The Hazard Zone checklist is the job evaluation tool.

The Caution Zone checklist screens some of the risk factors in workplaces that are believed to be associated with MSDs. This checklist covers:

- awkward posture
- high hand force
- highly repetitive motion
- repeated impact
- heavy, frequent, or awkward lifting
- moderate to high hand–arm vibration.

Under each category, there are a number of questions addressing various aspects of these risk factors. For example, one of the checklist questions for the awkward

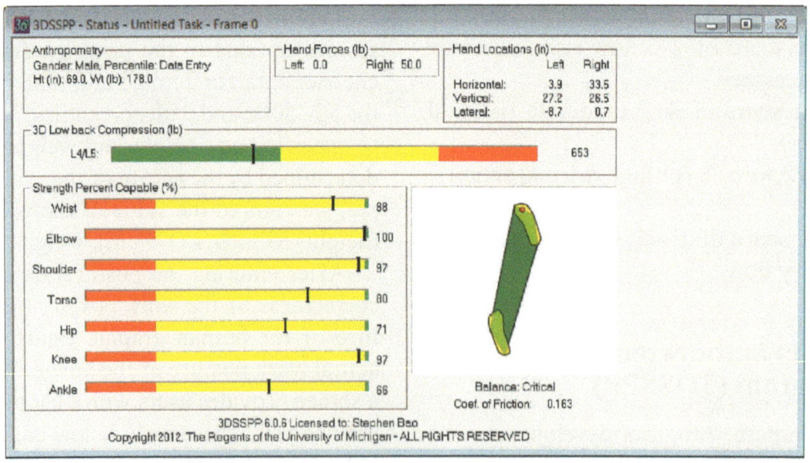

Fig. 20.12. A screen of the summary results using the 3-D Static Strength Prediction Program (3DSSPP).

posture category is (whether the job requires the worker) "squatting more than 2 hours total per day?" Another example for the repeated-impact category is (whether the job requires) "using the hand (heel/base of palm) or knee as a hammer more than 10 times per hour, more than 2 hours total per day?" If the condition described in the checklist question exists, the box next to the question is checked. Whenever there is a confirmed condition on the Caution Zone checklist, the Hazard Zone checklist is then required to be used to evaluate the job. Table 20.4 shows the questions for the Caution Zone checklist.

The Hazard Zone checklist (Table 20.5) covers the same job risk categories as the Caution Zone checklist. It uses a "calculator" available online to estimate heavy, frequent, or awkward lifting, and another "calculator" to determine hand–arm vibration risk level. The Hazard Zone checklist also differs from the Caution Zone checklist on the consideration of more risk combinations, as well as longer exposure time. For example, exposure to the risk factor of squatting for more than 4 hours total per day will put the job in the hazard zone (meaning the job has high risk level and improvement measures need to be developed). For the risk factor of pinching

Table 20.4

Washington State Caution Zone checklist (modified after LNI 2000a)

Caution Zone checklist (use one sheet for each position evaluated)	
Movements or postures that are a regular and foreseeable part of the job, occurring more than one day per week, and more frequently than one week per year	If done in this job position ✓ the box
Awkward postures	
1. Working with the hand(s) above the head, or the elbow(s) above the shoulders more than 2 hours total per day	☐
2. Working with the neck or back bent more than 30° (without support and without the ability to vary posture) more than 2 hours total per day	☐
3. Squatting more than 2 hours total per day	☐
4. Kneeling more than 2 hours total per day	☐
High hand force	☐
5. Pinching an unsupported object(s) weighing 2 or more pounds per hand, or pinching with a force of 4 or more pounds per hand, more than 2 hours total per day (comparable to pinching half a ream of paper)	☐
6. Gripping an unsupported objects(s) weighing 10 or more pounds per hand, or gripping with a force of 10 or more pounds per hand, more than 2 hours total per day (comparable to clamping light-duty automotive jumper cables on to a battery)	☐
Highly repetitive motion	
7. Repeating the same motion with the neck, shoulders, elbows, wrists, or hands (excluding keying activities) with little or no variation every few seconds, more than 2 hours total per day	☐
8. Performing intensive keying more than 4 hours total per day	☐
Repeated impact	☐
9. Using the hand (heel/base of palm) or knee as a hammer more than 10 times per hour, more than 2 hours total per day	☐
Heavy, frequent, or awkward lifting (a simple scale can be used to determine the weight of materials)	
10. Lifting object weighing more than 75 pounds once per day or more than 55 pounds more than 10 times per day	☐
11. Lifting objects weighing more than 10 pounds if done more than twice per minute, more than 2 hours total per day	☐
12. Lifting objects weighing more than 25 pounds above the shoulders, below the knees or at arms' length more than 25 times per day	☐
Moderate to high hand–arm vibration (closely estimate or obtain the vibration value of the tool in use)	
13. Using impact wrenches, carpet strippers, chain saws, percussive tools (jack hammers, scalers, riveting or chipping hammers) or other tools that typically have high vibration levels, more than 30 minutes total per day	☐
14. Using grinders, sanders, jigsaws or other hand tools that typically have moderate vibration levels more than 2 hours total per day	☐

Table 20.5

Washington State Hazard Zone checklist (modified after LNI 2000b)

Hazard Zone Job checklist		
(For each "caution zone job" find any physical risk factors that apply. If a hazard exists, it must be reduced below the hazard level or to the degree technologically and economically feasible)		
Movements or postures that are a regular and forseeable part of the job, occurring more than 1 day per week, and more frequently than 1 week per year		Hazard exists ✓
Awkward posture		
1. Working with the hand(s) above the head, or the elbows above the shoulders	More than 4 hours total per day	☐
2. Repeatedly raising the hand(s) above the head, or the elbow(s) above the shoulder(s) more than once per minute	More than 4 hours total per day	☐
3. Working with the neck bent more than 45° (without support or the ability to vary posture)	More than 4 hours total per day	☐
4. Working with the back bent forward more than 30° (without support or the ability to vary posture)	More than 4 hours total per day	☐
5. Working with the back bent forward more than 45° (without support or the ability to vary posture)	More than 2 hours total per day	☐
6. Squatting	More than 4 hours total per day	☐
7. Kneeling	More than 4 hours total per day	☐
High hand force		
Pinching an unsupported object(s) weighing 2 lb or more per hand, or pinching with a force of 4 lb or more per hand (comparable to pinching a half a ream of paper)		
+ highly repetitive motion	More than 3 hours total per day	☐
+ wrist flexion $\geq$30°, or wrist extension$\geq$45° or wrist ulnar deviation $\geq$30°	More than 3 hours total per day	☐
No other risk factors	More than 4 hours total per day	☐
Gripping an unsupported object(s) weighing 10 lb or more per hand, or gripping with a force of 10 lb or more per hand (comparable to clamping light-duty automotive jumper cables on to a battery)		
+ highly repetitive motion	More than 3 hours total per day	☐
+ wrist flexion $\geq$30°, or wrist extension$\geq$45° or wrist ulnar deviation $\geq$30°	More than 3 hours total per day	☐
No other risk factors	More than 4 hours total per day	☐
Highly repetitive motion		
Using the same motion with little or no variation every few seconds (excluding keying activities)		
+ wrist flexion $\geq$30°, or wrist extension$\geq$45° or wrist ulnar deviation $\geq$30° + high, forceful exertions with the hand(s)	More than 2 hours total per day	☐
No other risk factors	More than 6 hours total per day	☐
Intensive keying		
+ wrist flexion $\geq$30°, or wrist extension$\geq$45° or wrist ulnar deviation $\geq$30°	More than 4 hours total per day	☐
No other risk factors	More than 7 hours total per day	☐
Repeated impact		
Using the hand (heel/base of palm) as a hammer more than once per minute	More than 2 hours total per day	☐
Using the knee as a hammer more than once per minute	More than 2 hours total per day	☐

force, the force is considered independently from other risk factors in the Caution Zone checklist. However, this risk factor is considered together with highly repetitive motions or awkward hand/wrist postures with different exposure duration limits in the Hazard Zone checklist.

The lifting calculator considers object weight to be lifted, vertical and horizontal hand location where the lift occurs, lifting frequency and duration, as well as whether trunk twisting occurs during the lift in order to determine the lifting risk level. Considering the critics were less protective of the original lifting calculator, an updated lifting calculator was recently published on the website of the Washington State Department of Labor and Industries (2014). This calculator is more in-line with the ACGIH's lifting TLV (as it

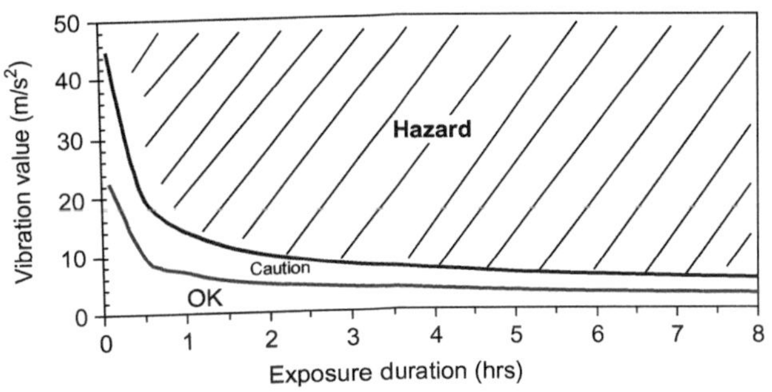

Fig. 20.13. Determining of hand–arm vibration risk level using exposure duration and vibration value of a power tool (LNI, 2000b).

is more protective than the previous one) since the current lifting calculator is no longer tied to a regulatory requirement.

An innovative hand–arm vibration exposure evaluation is included in the Hazard Zone checklist. While a more accurate hand–arm vibration exposure evaluation using sophisticated direct measurement systems (e.g., accelerometer, vibration signal recording, and signal-processing devices) is always preferred, most practitioners have neither the resources nor the expertise to perform this type of evaluation. The Hazard Zone checklist proposes a simpler method to roughly estimate the hand–arm vibration exposure when vibrating power tools are used. This method starts with obtaining the so-called vibration value(s) for the tool(s) that a worker uses in his/her job. This can usually be done by finding the declared vibration value from the tool manufacturer's catalogue (vibration values of power tools need to be declared when they are sold in the European market), or from a few websites (e.g., NIOSH and University of Umea, Sweden) where vibration values of thousands of popular tools of various brands are listed. It is then required to determine the number of hours that a tool is (or tools are) used in a work shift.

Using the tool vibration value and exposure duration, one can use a chart to determine the hand–arm vibration risk level. When the calculated vibration exposure falls in the "OK" zone, the job is considered safe. If the calculated vibration exposure falls in the "hazard" zone, the job is considered unsafe and improvement measures are required to reduce exposure. If the exposure is between the "OK" and "hazard" zones, moderate vibration exposure is determined and, if possible, improvement measures should be explored. The lines dividing the "caution" and "hazard" zones correspond to the action value of 2.5 m/s^2 and the limit value of 5 m/s^2 standardized to an 8-hour reference period. They are the permissible values for hand–arm vibration in the standard of the European Union (2002). When multiple vibrating tools are used in a workday, the total daily vibration exposure can be calculated through the partial daily vibration exposure of each of the tools (Fig. 20.13).

The Quick Exposure Check (QEC)

The QEC is an observational tool developed for occupational safety and health practitioners in the UK to assess exposure to risk for work-related MSDs and provide a basis for ergonomic interventions (David et al., 2008). This is a task-based assessment tool with quantitative scores indicating risk levels. With this scoring system, the QEC method can best be used in intervention studies where before and after scores can be compared. The QEC method is designed to be used by occupational safety and health practitioners with the participation of workers who actually perform the tasks. Both occupational safety and health practitioners and workers provide input information to complete the QEC task assessment. There are eight questions for occupational safety and health practitioners, and another eight questions for workers. In general, most task assessments can be completed in 10 minutes.

The QEC method assesses four main body parts: back, shoulder/arm, wrist/hand, and neck. Occupational safety and health practitioners (the observers) collect data related to postures and movement frequencies (repetition). The workers provide information related to force/weight, duration, as well as task-related questions such as task visual demand, driving, power tool use, work pace and their stress levels. Table 20.6 shows the questions that are used by both observers and workers to collect the information for the four body parts as well as a few other questions related to driving, vibration, work pace, and job stress.

After the completion of the QEC assessment form, a QEC scoring form (Table 20.7) is used to provide numeric indications of risk levels for the four body parts. The scoring system incorporates multiple combination risk factors simultaneously. For the consideration of back posture and weight handled (Table 20.7), the scoring system uses information from the observer's assessment (A in Table 20.6) and worker's assessment information (H in Table 20.6) to determine the risk level of the combined risk factors. The scores provide an estimation of risk levels (the higher the score, the higher the risk). Based on the scores for the different body parts, it also suggests a system of priority levels for intervention (Table 20.8).

Table 20.6

General structure of the Quick Exposure Check data collection form (modified after David et al., 2008)

Observer's assessment	Worker's assessment
Back	**H Is the maximum weight handled manually by you in this task**
A When performing the task, is the back (select worst-case situation)	H1 Light (5 kg or less)?
A1 Almost neutral?	H2 Moderate (6–10 kg)?
A2 Moderately flexed or twisted or side bent?	H3 Heavy (11–20 kg)?
A3 Excessively flexed or twisted or side bent?	H4 Very heavy (more than 20 kg)?
B Select only one of the two following task options: either	J On average, how much time do you spend per day on this task?
For seated or standing stationary tasks: Does the back remain in a static position most of the time?	J1 Less than 2 hours
B1 No	J2 2–4 hours
B2 Yes	J3 More than 4 hours
or For lifting, pushing/pulling and carrying tasks (i.e. moving a load): Is the movement of the back	K When performing this task, is the maximum force level exerted by one hand
B3 Infrequent (around 3 times per minute or less)?	K1 Low (e.g. less than 1 kg)?
B4 Frequent (around 8 times per minute)?	K2 Medium (e.g. 1–4 kg)?
B5 Very frequent (around 12 times per minute or more)?	K3 High (e.g. more than 4 kg)?
Shoulder/arm	**L Is the visual demand of this task**
C When the task is performed, are the hands (select worst-case situation)	L1 Low (almost no need to view fine details)?
C1 At or below waist height?	*L2 High (need to view some fine details)?
C2 At about chest height?	*If high, please give details in the box below
C3 At or above shoulder height?	M At work do you drive a vehicle for
D Is the shoulder/arm movement	M1 Less than 1 hour per day or never?
D1 Infrequent (some intermittent movement)?	M2 Between 1 and 4 hours per day?
D2 Frequent (regular movement with some pauses)?	M3 More than 4 hours per day?
D3 Very frequent (almost continuous movement)?	N At work do you use vibrating tools for
Wrist/hand	**N1 Less than 1 hour per day or never?**
E Is the task performed with (select worst-case situation)	N2 Between 1 and 4 hours per day?
E1 An almost straight wrist?	N3 More than 4 hours per day?
E2 A deviated or bent wrist?	P Do you have difficulty keeping up with this work?
F Are similar motion patterns repeated	P1 Never
F1 10 times per minute or less?	P2 Sometimes
F2 11–20 times per minute?	*P3 Often
F3 More than 20 times per minute?	*If often, please give details in the box below
Neck	**Q In general, how do you find this job?**
G When performing the task, is the head/neck bent or twisted?	Q1 Not at all stressful
G1 No	Q2 Mildly stressful
G2 Yes, occasionally	*Q3 Moderately stressful
G3 Yes, continuously	*Q4 Very stressful
	*If moderately or very, please give details in the box below
Additional details for L, P and Q if appropriate	
*L	
*P	
*Q	

Table 20.7

Quick Exposure Check exposure scoring system (modified after David et al., 2008)

Back	Shoulder/arm	Wrist/hand	Neck
Back posture (A) and weight (H)	Height (C) and weight (H)	Repeated motion (F) and force (K)	Neck posture (G) and duration (J)

Back posture (A) and weight (H)

	A1	A2	A3
H1	2	4	6
H2	4	6	8
H3	6	8	10
H4	8	10	12

Score 1: _____

Height (C) and weight (H)

	C1	C2	C3
H1	2	4	6
H2	4	6	8
H3	6	8	10
H4	8	10	12

Score 1: _____

Repeated motion (F) and force (K)

	F1	F2	F3
K1	2	4	6
K2	4	6	8
K3	6	8	10

Score 1: _____

Neck posture (G) and duration (J)

	G1	G2	G3
J1	2	4	6
J2	4	6	8
J3	6	8	10

Score 1: _____

Back posture (A) and duration (J)

	A1	A2	A3
J1	2	4	6
J2	4	6	8
J3	6	8	10

Score 2: _____

Height (C) and duration (J)

	C1	C2	C3
J1	2	4	6
J2	4	6	8
J3	6	8	10

Score 2: _____

Repeated motion (F) and duration (J)

	F1	F2	F3
J1	2	4	6
J2	4	6	8
J3	6	8	10

Score 2: _____

Visual demand (L) and duration (J)

	L1	L2
J1	2	4
J2	4	6
J3	6	8

Score 2: _____

Duration (J) and weight (H)

	J1	J2	J3
H1	2	4	6
H2	4	6	8
H3	6	8	10
H4	8	10	12

Score 3: _____

Duration (J) and weight (H)

	J1	J2	J3
H1	2	4	6
H2	4	6	8
H3	6	8	10
H4	8	10	12

Score 3: _____

Duration (J) and force (K)

	J1	J2	J3
K1	2	4	6
K2	4	6	8
K3	6	8	10

Score 3: _____

Total score for neck
Sum of scores 1 to 2
Neck score: ______

Now do only 4 if static
or 5 and 6 if manual handling

Static posture (B) and duration (J)

	B1	B2
J1	2	4
J2	4	6
J3	6	8

Score 4: _____

Frequency (D) and weight (H)

	D1	D2	D3
H1	2	4	6
H2	4	6	8
H3	6	8	10
H4	8	10	12

Score 4: _____

Wrist posture (E) and force (K)

	E1	E2
K1	2	4
K2	4	6
K3	6	8

Score 4: _____

Driving (M)

M1	M2	M3
1	4	9

Total for driving: _____

Frequency (B) and weight (H)

	B3	B4	B5
H1	2	4	6
H2	4	6	8
H3	6	8	10
H4	8	10	12

Score 5: _____

Frequency (D) and duration (J)

	D1	D2	D3
J1	2	4	6
J2	4	6	8
J3	6	8	10

Score 5: _____

Wrist posture (E) and duration (J)

	E1	E2
J1	2	4
J2	4	6
J3	6	8

Score 5: _____

Vibration (N)

N1	N2	N3
1	4	9

Total for vibration: ____

Continued

Table 20.7

Continued

Frequency (B) and duration (J)

	B3	B4	B5
J1	2	4	6
J2	4	6	8
J3	6	8	10

Score6: _____

Total score for back	Total score for shoulder/arm	Total score for wrist/hand
Sum of scores 1 to 4 or scores 1 to 3 plus 5 and 6	Sum of scores 1 to 5	Sum of scores 1 to 5
Back score:_______	Shoulder/arm score: _____	Wrist/hand score: ____

Work pace (P)

P1	P2	P3
1	4	9

Total for work pace: ___

Stress (Q)

Q1	Q2	Q3	Q4
1	4	9	16

Total for stress: _____

Table 20.8

Suggested exposure level for prioritization of intervention (modified after David et al., 2008)

	Exposure level			
Quick Exposure Check score	Low	Moderate	High	Very high
Back (static)	8–15	16–22	23–29	29–40
Back (moving)	10–20	21–30	31–40	41–56
Shoulder/arm	10–20	21–30	31–40	41–56
Wrist/hand	10–20	21–30	31–40	41–46
Neck	4–6	8–10	12–14	16–18
Driving	1	4	9	–
Vibration	1	4	9	–
Work pace	1	4	9	–
Stress	1	4	9	16

References

Aarås A (1994). The impact of ergonomic intervention on individual health and corporate prosperity in a telecommunications environment. Ergonomics 37: 1679–1696.

Aarås A, Westgaard RH (1987). Further studies of postural load and musculo-skeletal injuries of workers at an electro-mechanical assembly plant. Appl Ergon 18: 211–219.

ACGIH (2001). Documentation of the Threshold Limit Values for Physical Agents, 7th edition. American Conference of Governmental Industrial Hygienists, Cincinnati, Ohio.

ACGIH (2005). Lifting Documentation, American Conference of Governmental Industrial Hygienists, Cincinnati, Ohio.

Alexopoulos EC, Tanagra D, Konstantinou E et al. (2006). Musculoskeletal disorders in shipyard industry: prevalence, health care use, and absenteeism. BMC Musculoskelet Disord 7: 88.

Allen KD, Chen JC, Callahan LF et al. (2010). Associations of occupational tasks with knee and hip osteoarthritis: the Johnston County Osteoarthritis Project. J Rheumatol 37: 842–850.

Amin S, Goggins J, Niu J et al. (2008). Occupation-related squatting, kneeling, and heavy lifting and the knee joint: a magnetic resonance imaging-based study in men. J Rheumatol 35: 1645–1649.

Andersson GBJ (1981). Epidemiologic aspects on low-back pain in industry. Spine 6: 53–60.

Armstrong TJ, Foulke JA, Joseph BS et al. (1982). Investigation of cumulative trauma disorders in a poultry processing plant. Am Ind Hyg Assoc J 43: 103–116.

Armstrong TJ, Buckle P, Fine LJ et al. (1993). A conceptual model for work-related neck and upper-limb musculo-skeletal disorders. Scand J Work Environ Health 19: 73–84.

Ballotpedia.org (2003). Washington Repeal of Ergonomic Regulatons, Initiative 841 [Online]. Ballotpedia.org.

Balogh I, Hansson GA, Ohlsson K et al. (1999). Interindividual variation of physical load in a work task. Scand J Work Environ Health 25: 57–66.

Balogh I, Orbaek P, Ohlsson K et al. (2004). Self-assessed and directly measured occupational physical activities – influence of musculoskeletal complaints, age and gender. Appl Ergon 35: 49–56.

Bao S, Silverstein B (2005). Estimation of hand force in ergonomic job evaluations. Ergonomics 48: 288–301.

Bao S, Mathiassen SE, Winkel J (1996). Ergonomic effects of a management-based rationalization in assembly work – a case study. Appl Ergon 27: 89–99.

Bao S, Silverstein B, Cohen M (2001). An electromyography study in three high risk poultry processing jobs. Int J Ind Ergon 27: 375–385.

Bao S, Spielholz P, Howard N et al. (2006). Quantifying repetitive hand activity for epidemiological research on musculoskeletal disorders – part I: individual exposure assessment. Ergonomics 49: 361–380.

Bao S, Spieholz P, Howard N et al. (2009). Application of the Strain Index in multiple task jobs. Appl Ergon 40: 56–68.

Bell AF, Steele JR (2012). Risk of musculoskeletal injury among cleaners during vacuuming. Ergonomics 55: 237–247.

Bergenudd H, Nilsson B (1988). Back pain in middle age; occupational workload and psychologic factors: an epidemiologic survey. Spine (Phila Pa 1976) 13: 58–60.

Bigos SJ, Spengler DM, Martin NA et al. (1986). Back injuries in industry: a retrospective study. II. Injury factors. Spine (Phila Pa 1976) 11: 246–251.

Borg GAV (1982). Psychophysical bases of perceived exertion. Med Sci Sport Exer 14: 377–381.

Bosch T, de Looze MP, van Dieen JH (2007). Development of fatigue and discomfort in the upper trapezius muscle during light manual work. Ergonomics 50: 161–177.

Bovenzi M, Betta A (1994). Low-back disorders in agricultural tractor drivers exposed to whole-body vibration and postural stress. Appl Ergon 25: 231–241.

Bovenzi M, Della Vedova A, Nataletti P et al. (2005). Work-related disorders of the upper limb in female workers using orbital sanders. Int Arch Occup Environ Health 78: 303–310.

Bridger RS, Brasher K, Bennett A (2013). Sustaining person-environment fit with a changing workforce. Ergonomics 56: 565–577.

Burdorf A, Govaert G, Elders L (1991). Postural load and back pain of workers in the manufacturing of prefabricated concrete elements. Ergonomics 34: 909–918.

Burdorf A, van Riel M, Brand T (1997). Physical load as risk factor for musculoskeletal complaints among tank terminal workers. Am Ind Hyg Assoc J 58: 489–497.

Bureau of Labor Statistics (2013). Nonfatal occupational injuries and illnesses requiring days away from work, 2012, U.S. Department of Labor, Washington, DC.

Byström S, Mathiassen SE, Fransson C (1991a). Isometric handgrip exercise and micropauses. In: Y Quéinnec, F Daniellon (Eds.), Taylor and Francis, Paris, pp. 182–184.

Byström SEG, Mathiassen SE, Fransson-Hall C (1991b). Physiological effects of micropauses in isometric handgrip exercise. Eur J Appl Physiol 63: 405–411.

Chaffin DB, Park KS (1973). A longitudinal study of low-back pain as associated with occupational weight lifting factors. Am Ind Hyg Assoc J 34: 513–525.

Christensen H, Sogaard K, Pilegaard M et al. (2000). The importance of the work/rest pattern as a risk factor in repetitive monotonous work. Int J Ind Ergon 25: 367–373.

Coggon D, Croft P, Kellingray S et al. (2000). Occupational physical activities and osteoarthritis of the knee. Arthritis Rheum 43: 1443–1449.

Cook T, Rosecrance J, Zimmermann C et al. (1998). Electromyographic analysis of a repetitive hand gripping task. Int J Occup Saf Ergon 4: 185–200.

Corlett EN, Bishop RP (1976). A technique for assessing postural discomfort. Ergon 19: 175–182.

David G, Woods V, Li G et al. (2008). The development of the Quick Exposure Check (QEC) for assessing exposure to risk factors for work-related musculoskeletal disorders. Appl Ergon 39: 57–69.

De Luca C (1997). The use of surface electromyography in biomechanics. J Appl Biomech 13: 135–163.

Drinkaus P, Bloswick DS, Sesek R et al. (2005). Job level risk assessment using task level strain index scores: a pilot study. Int J Occup Saf Ergon 11: 141–152.

European Union (2002). Directive of the European Parliament and of the Council on the minimum health and safety requirements regarding the exposure of workers to the risks arising from physical agents (vibration) (sixteenth individual Directive within the meaning of Article 16(1) of Directive 89/391/EEC), European Parliament, Brussels.

Fan ZJ, Silverstein BA, Bao S et al. (2009). Quantitative exposure–response relations between physical workload and prevalence of lateral epicondylitis in a working population. Am J Ind Med 52: 479–490.

Ferguson SA, Allread WG, Le P et al. (2013). Shoulder muscle fatigue during repetitive tasks as measured by electromyography and near-infrared spectroscopy. Hum Factors 55: 1077–1087.

Fethke NB, Gerr F, Anton D et al. (2012). Variability in muscle activity and wrist motion measurements among workers performing non-cyclic work. J Occup Environ Hyg 9: 25–35.

Foley M, Silverstein B, Polissar N et al. (2009). Impact of implementing the Washington State ergonomics rule on employer reported risk factors and hazard reduction activity. Am J Ind Med 52: 1–16.

Forde MS, Punnett L, Wegman DH (2005). Prevalence of musculoskeletal disorders in union ironworkers. J Occup Environ Hyg 2: 203–212.

Guppy P (2003). An Overview of Initiative 841 - Repeal of State Ergonomics Regulations [Online]. Washington Policy Center, Seattle. Available: http://www.washingtonpolicy.org/sites/default/files/ErgonomicsText_0.pdf (accessed 22.12.14.).

Hagberg M, Silverstein B, Wells R et al. (Eds.), (1995). Work related musculoskeletal disorders (WMSDs): A reference book for prevention, Taylor and Francis, London.

Hakkanen M, Viikari-Juntura E, Martikainen R (2001). Incidence of musculoskeletal disorders among newly employed manufacturing workers. Scand J Work Environ Health 27: 381–387.

Harcombe H, McBride D, Derrett S et al. (2009). Prevalence and impact of musculoskeletal disorders in New Zealand nurses, postal workers and office workers. Aust N Z J Public Health 33: 437–441.

Hignett S, McAtamney L (2000). Rapid entire body assessment (REBA). Appl Ergon 31: 201–205.

Holmstrom EB, Lindell J, Mortiz U (1992). Low back and neck/shoulder pain in construction workers: occupational workload and psychosocial risk factors. Part 1: relationship to low back pain. Spine 17: 663–671.

Huang WF, Wu CF (2012). Evaluation of a knee-kicker bumper design for reducing knee morbidity among carpet layers. Appl Ergon 43: 850–858.

Jansen JP, Burdorf A (2003). Effects of measurement strategy and statistical analysis on dose–response relations between physical workload and low back pain. Occup Environ Med 60: 942–947.

Jensen LK, Kofoed LB (2002). Musculoskeletal disorders among floor layers: is prevention possible? Appl Occup Environ Hyg 17: 797–806.

Jensen LK, Eenberg W, Mikkelsen S (2000). Validity of self-reporting and video-recording for measuring knee-straining work postures. Ergonomics 43: 310–316.

Jensen LK, Rytter S, Bonde JP (2010). Exposure assessment of kneeling work activities among floor layers. Appl Ergon 41: 319–325.

Jensen LK, Rytter S, Bonde JP (2012). Symptomatic knee disorders in floor layers and graphic designers. A cross-sectional study. BMC Musculoskelet Disord 13: 188.

Jones T, Kumar S (2008). Comparison of ergonomic risk assessment output in a repetitive sawmill occupation: Trim-saw operator. Work 31: 367–376.

Karhu O, Kansi P, Kuorinka I (1977). Correcting working postures in industry: a practical method for analysis. Appl Ergon 8: 199–201.

Kelsey JL, Githens PB, White 3rd AA et al. (1984). An epidemiologic study of lifting and twisting on the job and risk for acute prolapsed lumbar intervertebral disc. J Orthop Res 2: 61–66.

Kemmlert K, Kilbom A (1986). Method for the identification of musculo-skeletal stress factors which may have injurious effects-PLIBEL, National Board of Occupational Safety and Health, Research Department, Work Physiology Unit, Solna, Sweden.

Keyserling WM (1986). postural analysis of the trunk and shoulders in simulated real time. Ergon 29: 569–583.

Kivimaki J, Riihimaki H, Hanninen K (1992). Knee disorders in carpet and floor layers and painters. Scand J Work Environ Health 18: 310–316.

Kurppa K, Viikari JE, Kuosma E et al. (1991). Incidence of tenosynovitis or peritendinitis and epicondylitis in a meat-processing factory. Scand J Work Environ Health 17: 32–37.

Latko WA, Armstrong TJ, Foulke JA et al. (1997). Development and evaluation of an observation method for assessing repetition in hand tasks. Am Ind Hyg Assoc J 58: 278–285.

Lipscomb HJ, Dement JM, Silverstein B et al. (2009). Health care utilization for musculoskeletal back disorders, Washington State union carpenters, 1989–2003. J Occup Environ Med 51: 604–611.

LNI (2000a). Caution Zone Checklist [Online]. Washington State Department of Labor and Industries, Olympia, WA. Available: http://www.lni.wa.gov/safety/SprainsStrains/evaltools/CautionZones2.pdf [Accessed September 2, 2014].

LNI (2000b). Hazard Zone Jobs Checklist [Online], Washington State Department of Labor and Industries, Olympia, WA. Available: http://www.lni.wa.gov/safety/SprainsStrains/evaltools/HazardZoneChecklist.PDF [Accessed September 2, 2014].

Lyons J (2002). Factors contributing to low back pain among professional drivers: A review of current literature and possible ergonomic controls. Work 19: 95–102.

Magnusson ML, Pope MH, Wilder DG et al. (1996). Are occupational drivers at an increased risk for developing musculoskeletal disorders? Spine 21: 710–717.

Maguire BJ, Smith S (2013). Injuries and fatalities among emergency medical technicians and paramedics in the United States. Prehosp Disaster Med 28: 376–382.

Mani L, Gerr F (2000). Work-related upper extremity musculoskeletal disorders. Prim Care 27: 845–864.

Marras WS, Lavender SA, Leurgans SE et al. (1993). The role of dynamic three-dimensional trunk motion in occupationally-related low back disorders. The effects of workplace factors, trunk position, and trunk motion characteristics on risk of injury. Spine (Phila Pa 1976) 18: 617–628.

Marras WS, Lavender SA, Leurgans SE et al. (1995). Biomechanical risk factors for occupationally related low back disorders. Ergonomics 38: 377–410.

Mathiassen SE (1993). The influence of exercise/rest schedule on the physiological and psychophysical response to isometric shoulder-neck exercise. Eur J Appl Physiol Occup Physiol 67: 528–539.

Mathiassen SE, Winkel J, Hagg GM (1995). Normalization of surface EMG amplitude from the upper trapezius muscle in ergonomic studies – a review. J Electromyogr Kinesiol 5: 197–226.

McAtamney L, Corlett EN (1993). RULA: a survey method for the investigation of work-related upper limb disorders. Appl Ergon 24: 91–99.

McGorry RW (2001). A system for the measurement of grip forces and applied moments during hand tool use. Appl Ergon 32: 271–279.

Moore JS, Garg A (1994). Upper extremity disorders in a pork processing plant: relationships between job risk factors and morbidity. Am Ind Hyg Assoc J 55: 703–715.

Moore JS, Garg A (1995). The Strain Index: a proposed method to analyze jobs for risk of distal upper extremity disorders. Am Ind Hyg Assoc J 56: 443–458.

Mroz M, Carlini AR, Archer KR et al. (2014). Frequency and cost of claims by injury type from a State Workers' Compensation Fund, 1998–2008. Arch Phys Med Rehabil 95: 1048–1054.

NIOSH (1992). Selected Topics in Surface Electromyography for Use in the Occupational Setting - Expert Perspectives, U.S. Department of Health and Human Services, Cincinnati, OH.

NIOSH KR (Ed.), (1994). Applications Manual for the Revised NIOSH Lifting Equation, National Institute for Occupational Safety and Health, Springfield, VA.

NIOSH (1997). Musculoskeletal Disorders and Workplace Factors: A Critical Review of Epidemiologic Evidence for Work-Related Musculoskeletal Disorders of the Neck, Upper Extremity, and Low Back, National Institute for Occupational Safety and Health, DHHS, Cincinnati, OH.

Ohlsson K, Attewell RG, Palsson B et al. (1995). Repetitive industrial work and neck and upper limb disorders in females. Am J Ind Med 27: 731–747.

Perotto AO (Ed.), (1994). Anatomy for the Electromyographer, Charles C. Thomas, Springfield, IL.

Punnett L, Fine LJ, Keyserling WM et al. (1991). Back disorders and nonneutral trunk postures of automobile assembly workers. Scand J Work Environ Health 17: 337–346.

Radwin RG, VanBergeijk E, Armstrong TJ (1989). Muscle response to pneumatic hand tool torque reaction forces. Ergonomics 32: 655–674.

Reid CR, Bush PM, Cummings NH et al. (2010). A review of occupational knee disorders. J Occup Rehabil 20: 489–501.

Riihimaki H, Tola S, Videman T et al. (1989). Low-back pain and occupation. A cross-sectional questionnaire study of men in machine operating, dynamic physical work, and sedentary work. Spine (Phila Pa 1976) 14: 204–209.

Schibye B, Skov T, Ekner D et al. (1995). Musculoskeletal symptoms among sewing machine operators. Scand J Work Environ Health 21: 427–434.

Seo NJ, Armstrong TJ (2008). Investigation of grip force, normal force, contact area, hand size, and handle size for cylindrical handles. Hum Factors 50: 734–744.

Seo NJ, Armstrong TJ, Chaffin DB et al. (2008). The effect of handle friction and inward or outward torque on maximum axial push force. Hum Factors 50: 227–236.

Shiri R, Viikari-Juntura E, Varonen H et al. (2006). Prevalence and determinants of lateral and medial epicondylitis: a population study. Am J Epidemiol 164: 1065–1074.

Silverstein B, Adams D (2007). Work-related musculoskeletal disorders of the neck, back and upper extremity in Washington State. Washington State Department of Labor and Industries, Olympia, WA.

Silverstein BA, Fine LJ, Armstrong TJ (1987). Occupational factors and carpal tunnel syndrome. Am J Ind Med 11: 343–358.

Silverstein B, Fan ZJ, Smith CK et al. (2009). Gender adjustment or stratification in discerning upper extremity musculoskeletal disorder risk? Scand J Work Environ Health 35: 113–126.

Skovron ML, Szpalski M, Nordin M et al. (1994). Sociocultural factors and back pain. A population-based study in Belgian adults. Spine (Phila Pa 1976) 19: 129–137.

Smith CK, Williams J (2014). Work related injuries in Washington State's Trucking Industry, by industry sector and occupation. Accid Anal Prev 65: 63–71.

Snook SH, Ciriello VM (1991). The design of manual handling tasks: revised tables of maximum acceptable weights and forces. Ergon 34: 1197–1213.

Spector JT, Adams D, Silverstein B (2011). Burden of work-related knee disorders in Washington State, 1999 to 2007. J Occup Environ Med 53: 537–547.

Spielholz P, Silverstein B, Morgan M et al. (2001). Comparison of self-report, video observation and direct measurement methods for upper extremity musculoskeletal disorder physical risk factors. Ergonomics 44: 588–613.

Stal M, Hansson GA, Moritz U (1999). Wrist positions and movements as possible risk factors during machine milking. Appl Ergon 30: 527–533.

Tanaka S, Halperin WE, Smith AB et al. (1985). Skin effects of occupational kneeling. Am J Ind Med 8: 341–349.

Tanaka S, Lee ST, Halperin WE et al. (1989). Reducing knee morbidity among carpetlayers. Am J Public Health 79: 334–335.

The University of Michigan (2008). 3D static strength prediction program (Version 6.0.0). The University of Michigan, Center for Ergonomics, Ann Arbor, MI.

Thorbjornsson CB, Alfredsson L, Fredriksson K et al. (2000). Physical and psychosocial factors related to low back pain during a 24-year period. A nested case-control analysis. Spine (Phila Pa 1976) 25: 369–374, discussion 375.

Viikari-Juntura E, Riihimaki H, Tola S et al. (1994). Neck trouble in machine operating, dynamic physical work and sedentary work: a prospective study on occupational and individual risk factors. J Clin Epidemiol 47: 1411–1422.

Viikari-Juntura E, Martikainen R, Luukkonen R et al. (2001). Longitudinal study on work related and individual risk factors affecting radiating neck pain. Occup Environ Med 58: 345–352.

Vingard E, Alfredsson L, Hagberg M et al. (2000). To what extent do current and past physical and psychosocial occupational factors explain care-seeking for low back pain in a working population? Results from the Musculoskeletal Intervention Center-Norrtalje Study. Spine (Phila Pa 1976) 25: 493–500.

Wahlstrom J, Svensson J, Hagberg M et al. (2000). Differences between work methods and gender in computer use. Scand J Work Environ Health 26: 390–397.

Walker-Bone K, Palmer KT (2002). Musculoskeletal disorders in farmers and farm workers. Occup Med (Lond) 52: 441–450.

Washington State Department of Labor and Industries (2014). How much can you safely lift [Online]. Washington State Department of Labor and Industries, Olympia, WA. Available: http://www.orosha.org/interactive/lifting/lift_safety.html [Accessed August 14, 2014].

Waters TR, Putz-Anderson V, Garg A et al. (1993). Revised NIOSH equation for the design and evaluation of manual lifting tasks. Ergonomics 36: 749–776.

Waters TR, Putz-Anderson V, Garg A (Eds.), (1994). Applications manual for the revised NIOSH lifting equation. Cincinnati, OH: U.S. Department of Health and Human Services Publishing Public Health Service, Cincinnati, OH.

Werner RA, Franzblau A, Gell N et al. (2005). Predictors of upper extremity discomfort: a longitudinal study of industrial and clerical workers. J Occup Rehabil 15: 27–35.

Wiktorin C, Karlqvist L, Winkel J et al. (1993). Validity of self-reported exposures to work postures and manual materials handling. Scand J Work Environ Health 19: 208–214.

Wilkins K, Mackenzie SG (2007). Work injuries. Health Rep 18: 25–42.

Winkel J, Mathiassen SE (1994). Assessment of physical work load in epidemiologic studies: concepts, issues and operational considerations. Ergonomics 37: 979–988.

Winkel J, Ekblom B, Hagberg M et al. (1983). The working environment of cleaners. Evaluation of physical strain in mopping and swabbing as a basis for job redesign, Butterworths, London.

Woods JJ, Bigland RB (1983). Linear and non-linear surface EMG/force relationships in human muscles – an anatomical/functional argument for the existence of both. Am J Phys Med 62: 287–299.

Yen TY, Radwin RG (1995). A video-based system for acquiring biomechanical data synchronized with arbitrary events and activities. IEEE-Transactions-on-Biomedical-Engineering 42: 944–948.

Handbook of Clinical Neurology, Vol. 131 (3rd series)
Occupational Neurology
M. Lotti and M.L. Bleecker, Editors

Chapter 21

Low-back pain

FRANCESCO S. VIOLANTE*, STEFANO MATTIOLI, AND ROBERTA BONFIGLIOLI
Occupational Health Unit, Department of Medical and Surgical Sciences, University of Bologna, Bologna, Italy

INTRODUCTION

Low-back pain is one of the most common painful conditions experienced by human bs throughout their life.

Lifetime prevalence has been reported to be as high as 84% (Balagué et al., 2012), depending on the case definition used, and no age group is spared, even children (Calvo-Muñoz et al., 2013).

Low-back pain is also an occupational disorder and, probably, the most common among occupational disorders worldwide (Punnett et al., 2005).

Although low-back pain is not a lethal condition, it was estimated at the third rank among all diseases by disability-adjusted life-years in 2010 in the USA (US Burden of Disease Collaborators, 2013), after ischemic heart disease and chronic obstructive pulmonary disease, and at the first rank by years lived with disability. It also ranked high (13th) globally for the same year, in disability-adjusted life-years (Murray et al., 2012).

This chapter will provide an overview of modern concepts of low-back pain (in general) and will outline some distinctive features of work-related low-back pain.

DEFINITION AND ATTRIBUTES OF LOW-BACK PAIN

It has been written that "low back pain is neither a disease nor a diagnostic entity of any sort" (Ehrlich, 2003): nevertheless, this condition is a leading cause of disability and it consumes a great amount of health resources (Chou et al., 2007).

Low-back pain is defined as "pain, muscle tension or stiffness localized below the costal margin and above the inferior gluteal folds, with or without referred leg pain" (Airaksinen et al., 2006; van Tulder et al., 2006; Chou, 2011).

Low-back pain also stimulates a number of scientific studies: while this chapter was being written, a simple PubMed search with the string "low-back pain" retrieved more than 24 000 articles (up from 176 in the year 1983 to 1723 in 2013 – a 10-fold increase in 30 years).

One of the first successful efforts at categorizing work-related low-back pain was made by the Quebec Task Force on Spinal Disorders (1987), which classified 11 categories of this disorder, based on symptoms or status.

Presently, clinical guidelines (Hegmann, 2011) usually call for distinguishing two main categories of low-back pain:

1. nonspecific low-back pain, which is defined as a condition attributable to no recognizable known specific pathology (including low-back pain deemed to be of mechanical origin)
2. specific low-back pain, which is defined as a condition attributable to a recognizable known specific pathology (e.g., infection, tumor, fracture, inflammatory process, radicular syndrome).

Nonspecific low-back pain accounts for the vast majority (85%) of cases of this condition (van Tulder et al., 2006): most of these cases will recover spontaneously within a couple of weeks (Quebec Task Force on Spinal Disorders, 1987).

However, in some cases pain and disability have a longer duration, so low-back pain is currently classified, with regard to duration of symptoms, as:

- acute: an episode of low-back pain lasting less than 6 weeks

*Correspondence to: Francesco S. Violante, M.D., Occupational Health Unit, University of Bologna – Sant'Orsola Malpighi Hospital, via Palagi 9, 40138 Bologna, Italy. Tel: +39-051-636-2611, E-mail: francesco.violante@unibo.it

- subacute: duration between 6 and 12 weeks
- chronic: duration more than 12 weeks.

The distinction between nonspecific/specific and acute/subacute/chronic low-back pain is not only useful for epidemiologic studies, but also (mainly) for choosing the appropriate strategy for the diagnosis and treatment of the disorder.

The most distinctive feature of low-back pain is its recurrence: 24–87% of individuals who have an episode of low-back pain will suffer a recurrence within 1 year (Stanton et al., 2009). The variability of recurrence estimates is probably due to the lack of a common definition of an episode of low-back pain, namely, when an episode is finished and when a new one has commenced (de Vet et al., 2002).

Based on a systematic review of published studies (Stanton et al., 2009), it has been proposed that a new episode of low-back pain is observed if:

- the patient has been painfree for at least 30 days and
- has been experiencing pain for at least 24 hours and
- the pain has an intensity at least equal to a defined minimal clinically important change on a chosen pain intensity scale and
- the pain is associated with a functional limitation at least equal to a defined minimal clinically important change on a chosen functional limitation scale.

Some practitioners still label low-back pain cases according to the presumed cause of the pain (discogenic, facetogenic, sacroiliac), but in most cases this is an unsupported assumption (see section on pathophysiology, below).

In order to better standardize the diagnosis of low-back pain for epidemiologic studies, some researchers have proposed "consensus definitions" based on the location of the symptoms, and their frequency, duration, and severity (Dionne et al., 2008).

In addition, a number of instruments have been proposed for grading the severity of low-back pain (Ostelo et al., 2008) and the associated disability (Smeets et al., 2011).

Based on the presently available knowledge (Axén and Leboeuf-Yde, 2013; Dunn et al., 2013), it is possible to conceive low-back pain as a condition (or, maybe, a set of different conditions) which can assume a diverse clinical picture along a continuum which has on one extreme the individuals who will (almost) never experience an episode of low-back pain throughout their life, and at the other extreme the subjects who will suffer from continuous low-back pain for most days in their life and, in between, subjects who will suffer from repeated episodes of low-back pain, variable for frequency, duration, and severity (who, however, will be painfree between those episodes).

EPIDEMIOLOGY OF LOW-BACK PAIN

While this chapter was being written, a simple PubMed search with the string "low back pain/epidemiology" (Medical Subject Headings: MeSH) retrieved more than 1700 articles: more than 100 were systematic reviews about different aspects of this multifaceted condition. Before providing an overview of occupational low-back pain, however, it is useful to summarize some of the available epidemiologic evidence about this condition in the general population.

It is worth noting that what is presently known about the epidemiology of low-back pain is based on studies which reflect the interpretation of this condition in the past. Traditionally, low-back pain has been viewed as a result of wear and tear of the spine throughout the life (hence the emphasis on spine pathology, and especially disc disease), or as a consequence of an injury, especially at the workplace. As a result, a great number of studies have been performed on low-back pain in adults, focusing on prevalence and incidence of the condition, personal risk factors, and antecedents of low-back pain episodes.

The traditional view of low-back pain, however, has been recently challenged in different ways.

The view that low-back pain results from wear and tear of the spine and, mainly, from disc disease has been challenged by several lines of evidence:

- Even young children suffer from low-back pain, in the absence of any underlying spine pathology.
- Intervertebral discs in women are usually in better condition than in men of the same age group, but, on average, women report low-back pain more frequently than men.
- Disc disease is very frequent in the totally asymptomatic adult population.
- Studies on identical twins suggest that genetic factors seem to influence disc degeneration as much as (or more than) age and biomechanical overload.

However, as most of the studies so far published are based on the traditional view of low-back pain, we will first summarize the information on low-back pain and disc disease prevalence and incidence in the general population, and the relevant risk factors. After that, we will summarize the information on work-related low-back pain and disc disease.

Epidemiology of low-back pain in the general population

According to a systematic review of studies about the prevalence of low-back pain in adults (Hoy et al., 2012), the point prevalence of low-back pain (subjects who report low-back pain at a point in time), based on

243 estimates, has a mean of 18.3%, with a standard deviation of 11.7. One-month prevalence (145 estimates) has a mean of 30.8% and a standard deviation of 12.7, 1-year prevalence (271 estimates) has a mean of 38% and a standard deviation of 19.4, and lifetime prevalence (133 estimates) has a mean of 38.9% and a standard deviation of 24.3. The large variation in these estimates reflects the heterogeneity of the studies and the lack of use of a uniform case definition for an episode of low-back pain.

According to a systematic review of studies on the first-time incidence of low-back pain in adults (Taylor et al., 2014), the summary pooled estimate for community-based populations was 26%, with 24–28% as 95% confidence interval. A similar value (27%) was obtained in studies that investigated only subjects who were painfree at baseline.

Low-back pain seems to be common also among young people: according to a meta-analysis of studies on the prevalence of low-back pain (Calvo-Muñoz et al., 2013), the mean point prevalence obtained from 10 studies was 12% (95% confidence interval: 9–15.9), the mean period prevalence at 1 week obtained from six studies was 17.7% (95% confidence interval: 12.4–24.7), and the mean period prevalence at 1 year obtained from 13 studies was 33.6% (95% confidence interval: 26.9–41), whereas the mean lifetime prevalence obtained from 30 studies was 39.9% (95% confidence interval: 34.2–45.9). According to the authors, lifetime prevalence exhibited a positive, statistically significant relationship with the mean age of the participants in the samples and with the publication year of the studies. The most recent studies showed higher prevalence rates than the oldest ones, and studies with better methodology exhibited higher lifetime prevalence rates than studies that were methodologically poor.

Incidence of low-back pain in children has also been studied. According to a study (Nissinen et al., 1994), the incidence of low-back pain in a group of 859 children (mean age at the start and end of the follow-up 11.8–13.8 years) was 18.4% in girls and 16.9% in boys. Another study (Jones et al., 2003), based on an original cohort of 1046 schoolchildren, aged 11–14 years at baseline, identified as being free of low-back pain, reported new onset of low-back pain in 12.5% of children at age 11, 16% at age 12, 23.8% at age 13, and 24.1% at age 14 years. The proportion of children reporting low-back pain was 20.1% among schoolgirls and 16.9% among schoolboys.

Prevalence of low-back pain in children does not show striking differences from corresponding values in the adult population and, as reported earlier, challenges the idea that low-back pain is mostly an adult-acquired condition due to wear and tear of the spine. On the other hand, the incidence of low-back pain in children seems to be lower than in adults, up to the age of 12 years.

Factors associated with low-back pain in the general population

A number of factors have been studied in association with low-back pain in the general population (Hoy et al., 2012):

- Female prevalence of low-back pain is higher than among males across all age groups.
- Prevalence increases with age until middle age, decreasing afterwards.
- Prevalence in high-income countries seems to be higher than in middle- and low-income countries.

Obesity and smoking have been found to be associated with low-back pain in twin studies (Ferreira et al., 2013). However, the magnitude of the pooled odds ratios from cohort studies for both smoking (Shiri et al., 2010a) and obesity (Shiri et al., 2010b) was between 1.3 and 1.5.

Nonoccupational strenuous physical activity has been found to be associated with an increased risk of low-back pain, whereas leisure time activity has been found to be protective (Heneweer et al., 2011).

Although their role is still uncertain, a variety of psychosocial factors seem to be associated with low-back pain, possibly with a predictive value greater than the results of nuclear magnetic resonance (Pincus et al., 2002; Carragee et al., 2005). More than "causing" low-back pain, psychosocial factors could be implicated in the transition to a chronic pain status (Ramond et al., 2011).

A number of factors have been studied in association with low-back pain also in children: in a systematic review of the studies examining risk factors for first episode of low-back pain in children (Hill and Keating, 2010), the authors reported that 13 of the 47 risk factors studied were significantly associated with a first episode of low-back pain. However no risk factor was found to be associated with future low-back pain in children in more than one study and inconsistency in definitions of low-back pain, predefined recall periods, and methods used to collect and analyze data limit conclusions that can be drawn about factors that identify children at risk of developing low-back pain.

Epidemiology of lumbar disc disease in the general population

Intervertebral disc degeneration is a concept for which in medicine, up to the present time, there is no widely agreed definition. Conceptually, disc degeneration may be defined as the product of the process of deterioration of intervertebral discs (spanning the lifetime) associated with remodeling of the same discs and adjoining vertebrae, including adaptation of disc structure to variations of physical load or the occasional lesion.

Operationally, disc degeneration is defined by the diagnostic method used, which at the present time is mostly nuclear magnetic resonance (Battié and Videman, 2006).

Since 2012, "intervertebral disc degeneration" is a MeSH term, defined as "degenerative changes in the intervertebral disc due to aging or structural damage, especially to the vertebral end-plates" (http:/www.ncbi.nlm.nih.gov/mesh/68055959).

Lately, lumbar disc degeneration has been studied in the general population by means of nuclear magnetic resonance: unfortunately, the absence of a generally agreed standard makes it difficult to compare the results of different studies and probably explains the wide range of reported results. As an example, lumbar disc bulging prevalence in asymptomatic subjects has been reported to be between 10% and 80% (and higher), and the prevalence of annular tears has been reported to range between 6% and 56% (Battié et al., 2004).

The prevalence of disc degeneration is already elevated in young adults: a study in subjects between 20 and 22 years of age has reported that almost half had at least one degenerated disc and a quarter had a bulging disc (Takatalo et al., 2009).

Epidemiology of lumbar disc herniation suffers from the same limitations of disc degeneration study: incidence and prevalence vary according to the case definition (surgery, magnetic resonance imaging) and they seem to have the same distribution of disc degeneration in relation to age and gender, although lumbar disc herniation seems to be much rarer. Among adults, prevalence has been estimated to range between 1% and 3% (Andersson, 1997), whereas incidence has been estimated in adult males at around 1.7% (Jørgensen et al., 2013).

Lumbar disc herniation is well known (although rare) even in infancy (Martínez-Lage et al., 2003; Haidar et al., 2010). This is not a new finding, as the presence of lumbar disc degeneration in children has been known for more than a century (Beneke, 1897): specimens from autopsies or surgical procedures have demonstrated the presence of vertebral cartilage pathology or annular tears of the discs in children between 3 and 10 years of age.

Sciatica is the symptom more often associated with lumbar disc herniation: lifetime prevalence has been estimated at 12.2–26%, annual prevalence has been estimated at 2.2–19.5%, and point prevalence has been estimated at 1.6–4.8% (Konstantinou and Dunn, 2008). Incidence of sciatica seems to be variable as well, with reported values in the range 0.65–37% (Cook et al., 2014).

Factors associated with lumbar disc disease in the general population

Recent studies point to a changing view of the determinants of lumbar disc degeneration (Battié et al., 2009).

The prevalent view of lumbar disc disease determinants at the beginning of the 1990s (Frymoyer, 1992) was: "Among the factors associated with its occurrence are age, gender, occupation, cigarette smoking, and exposure to vehicular vibration. The contribution of other factors such as height, weight, and genetics is less certain."

Ten years later (Ala-Kokko, 2002), the view had dramatically changed:

> *Even though several environmental and constitutional risk factors have been implicated in this disease, their effects are relatively minor, and recent family and twin studies have suggested that sciatica, disc herniation and disc degeneration may be explained to a large degree by genetic factors.*

Although the relative contribution of genetic and environmental factors is not precisely known, the current evidence about determinants of lumbar disc degeneration can be summarized as follows.

Genetic factors play an important role in lumbar disc degeneration and this role seems to be more important for the lumbar discs between L1 and L4 (Battié et al., 1995).

Age is another important factor: frequency of lumbar disc degeneration increases with increasing age (Miller et al., 1988).

Female gender seemed to be protective for discs (Miller et al., 1988): recent evidence suggests that the actual progression rate of disc degeneration with age among women and men is converging (Lee et al., 2012).

Body weight, smoking, and height seem also to be associated with lumbar disc degeneration (Wahlström et al., 2012).

Epidemiology of low-back pain in the occupational population

The association between manual material handling and low-back pain has been known since ancient times: Bernardino Ramazzini (1700), in his book on workers' diseases, describes the different diseases attributed to the porter's occupation.

The scientific evidence of a relation between low-back pain and occupational factors was reviewed by a US National Institute for Occupational Safety and Health (NIOSH) expert group (Bernard, 1997), which concluded that there was strong evidence of work-relatedness:

> *A causal relationship is shown to be very likely between intense or long-duration exposure to the specific risk factor(s) and musculoskeletal disorder when the epidemiologic criteria of causality are used. A positive relationship has been observed between exposure to the specific risk factor and musculoskeletal disorder in studies in which chance, bias, and confounding factors could be ruled out with reasonable confidence in at least several studies*

linking "lifting/forceful movement" and "whole body vibration" to back disorders. Evidence of work-relatedness, e.g.,

> *Some convincing epidemiologic evidence shows a causal relationship when the epidemiologic criteria of causality for intense or long-duration exposure to the specific risk factor(s) and musculoskeletal disorder are used. A positive relationship has been observed between exposure to the specific risk factor and musculoskeletal disorder in studies in which chance, bias, and confounding factors are not the likely explanation*

was found between "awkward posture" and "heavy physical work" and back disorders. No evidence of work-relatedness was found between "static work posture" and back disorders.

Since the substantial NIOSH review, a number of studies have been published concerning occupational low-back pain epidemiology, including systematic revisions of these studies.

A systematic review (Hoogendoorn et al., 1999) including 28 cohort and three case-referent studies (no cross-sectional study was included) concluded for the existence of strong evidence for manual materials handling, bending and twisting, and whole-body vibration as risk factors for back pain. The evidence was moderate for patient handling and heavy physical work, and no evidence was found for standing or walking, sitting, sports, and total leisure-time physical activity.

Another systematic review (Bakker et al., 2009), which used stricter inclusion criteria and chose only 18 cohort studies, found strong evidence that leisure-time sport or exercises, sitting, and prolonged standing/walking were not associated with low-back pain. Evidence for associations in leisure-time activities (e.g., do-it-yourself home repair, gardening), whole-body vibration, nursing tasks, heavy physical work, and working with one's trunk in a bent and/or twisted position and low-back pain was conflicting. However, of the 12 studies concerning "heavy physical work," five reported a statistically significant increase in the risk of low-back pain and seven reported some increase in low-back pain risk as well, although not statistically significant.

A systematic review (Ribeiro et al., 2012) did not find a clear dose–response relationship for work posture exposures and low-back pain. Limited evidence was found for trunk range of motion and duration of sustained flexed posture as risk factors for low-back pain. No evidence was found for frequency of trunk flexion as a risk factor for low-back pain.

A meta-analysis of mechanical workplace risk factors and low-back pain using individual participant data (Griffith et al., 2012) found small to moderate odds ratios for the association of mechanical exposures and low-back pain, although the relationships were complex. The odds ratios for posture exposures ranged from 1.1 to 2.0, whereas for force exposure they ranged from 1.4 to 2.1. The magnitudes of the odds ratios differed according to the definition of low-back pain, and heterogeneity was associated with both study-level and individual-level characteristics.

Low-back pain has been particularly studied in some occupational groups: among occupations in which women are the majority, an increased risk of low-back pain has been consistently reported in nurses (Yassi and Lockhart, 2013), with risk estimates in the range 1.2–5.5. Among occupations in which men are the majority, an increased risk of low-back pain has been consistently reported in construction workers (Boschman et al., 2011), with odds ratios in the range 2.3–3.0. Similar evidence has been reported for farmers (Osborne et al., 2012).

Factors associated with low-back pain in the occupational population

Manual material handling has been traditionally associated with low-back pain. However, manual material handling is not a single risk factor but, rather, a common task which can result in an overload of the spine (which is, ultimately, the risk factor for low-back pain). Lifting heavy loads, especially in an awkward posture or with bending and twisting of the trunk, may result in an overload of the spine: as manual lifting is a very common task, it is not surprising that sometimes the terms manual material handling and manual lifting are used interchangeably. An overload of the spine may also be the result of pushing and pulling heavy loads: all these issues are thoroughly reviewed in an excellent manual (Chaffin et al., 2006).

The relation between low-back pain prevalence and lifting has been studied using the NIOSH lifting index (see section on management of work-related low-back pain, below) as an approximate measure of the exposure.

In a cross-sectional study (Waters et al., 1999), the odds ratio of low-back pain was 1.04 for workers engaged in tasks in which the lifting index value was below 1, 1.96 for lifting index values between 1 and 2, 2.20 for lifting index values between 2 and 3, and 1.09 for lifting index above 3 (the highest category of exposure). The authors concluded that more data were needed for rigorous examination of the correlation between the lifting index and risk of low-back pain and that more research was needed to investigate the possible effects of psychosocial and personal factors on the reports of low-back pain that had been shown in other studies.

The same group of authors published a similar study 12 years later (Waters et al., 2011) obtaining, more or less, the same results (odds ratio of low-back pain was 1.06 for workers engaged in tasks in which the lifting index value was below 1, 1.22 for lifting index values between 1 and 2, 1.88 for lifting index values between 2 and 3, and 1.36 for lifting index above 3). Their conclusion was that, as found in the first data collection, the risk in the highest exposure group (lifting index above 3) was less than in the group with exposure lifting index values between 2 and 3. As they noted previously, it is possible that this may be due to problems with the predictive power of the equation: however, they concluded that this was not likely the case, but probably due to a combination of "worker selection" and "survivor" effects. Nonetheless, they also warned that, as the lifting index increases, workers with low-back pain are more likely to believe that their disorder is due to the repeated activity that they do on the job.

The reduction in self-reported low-back pain in highest estimated exposure groups seems to be confirmed also in later longitudinal studies (Garg et al., 2014a, b).

Exposure to whole-body vibrations (mainly as a result of driving heavy equipment vehicles) has been studied as a risk factor for the development of low-back pain and there are several reviews on the subject: a meta-analysis of 18 observational studies reported a meta-relative risk of 2.21 in drivers of heavy equipment vehicles as compared to those not exposed (Waters et al., 2008). However, the authors stated that the methodologic qualities of the published studies ranged from marginal to average and that prospective cohort studies were urgently needed to confirm the results and that the biologic plausibility should be further explored.

Psychosocial factors at work seem to be also associated with musculoskeletal pain in general and also with low-back pain: a longitudinal study of a large working population (Sterud and Tynes, 2013) found that odds ratios for low-back pain and high job demands were 1.41 (95% confidence interval: 1.16–1.72) and for low job control were 1.26 (95% confidence interval: 1.01–1.57). The same magnitude results had been reported in a systematic review and meta-analysis on the same issue (Lang et al., 2012).

Contrary to popular opinion, the epidemiologic literature does not support the notion that sitting while at work is associated with low-back pain (Hartvigsen et al., 2000).

Epidemiology of lumbar disc disease in the occupational population

There are to date no systematic studies on lumbar disc disease in the occupational population, mainly due to the fact that, as noted above, intervertebral disc degeneration is a concept for which in medicine, up to the present time, there is no widely agreed definition. So far, most of the available evidence about the epidemiology of lumbar disc disease in the occupational population comes from studies which have examined the occupational risk factors for this outcome (see below).

Factors associated with lumbar disc disease in the occupational population

Manual material handling and heavy work (which usually includes manual material handling) have been found to be associated with lumbar disc disease in several studies.

A large cohort study of Swedish construction workers (Wahlström et al., 2012) found an increased risk for hospitalization due to lumbar disc disease for several occupational groups compared with white-collar workers and foremen. Occupational groups with high biomechanical loads had the highest risks: for example, the relative risk (RR) for concrete workers was 1.55 (95% confidence interval: 1.29–1.87). A taller stature was consistently associated with increased risk. The RR for a man of 190–199-cm height was 1.55 (95% confidence interval: 1.30–1.86) compared with a man of 170–179-cm height. Body weight and smoking were also risk factors, but weaker than height. Workers in the age span of 30–39 years had the highest RR (1.87; 95% confidence interval: 1.58–2.23) compared with those aged 20–29 years, whereas men aged 60–65 years had a lower RR (0.86; 95% confidence interval: 0.68–1.09).

A cohort study of Danish male workers followed for 33 years (Sørensen et al., 2011) found that the strongest predictor of hospitalization for herniated lumbar disc disease was frequent strenuous physical activity at work: compared with unexposed subjects, the hazard ratio was 3.9 (95% confidence interval: 1.82–8.38). Also, body height was a significant predictor, whereas body weight was only insignificantly associated with herniated lumbar disc disease.

Exposure to whole-body vibration has been linked to lumbar disc degeneration, however, a systematic review (Bible et al., 2012) which included five retrospective cohort and two cross-sectional studies which had used imaging modalities for the assessment of lumbar disc disease found that, although mixed results and conclusions were found, the majority of studies did not identify an association between whole-body vibration exposure and an abnormal spinal imaging finding indicating damage of the spine. The authors also stressed that each included study had limitations secondary to quantifying whole-body vibration exposure accurately, both as a single encounter and as a total exposure over years.

A major concern about the interpretation of the available studies about occupational factors and lumbar disc disease is that, so far, no study actually examined the relation with the putative occupational factor and the specific level of disc disease. Apparently, the outcome studied was "any disc disease at any lumbar level," whereas biomechanics provide evidence that lumbar load during manual material handling, for example, is very different for different discs in the lumbar spine, even adjacent ones.

Based on measures of compressive forces on intervertebral discs in healthy volunteers, available since the 1960s (Nachemson and Morris, 1963), biomechanical models have been developed to estimate loads on lumbar spine during manual lifting: according to these models, "the L5–S1 disc incurs the greatest moment in lifting activities because it most often has the largest moment arm relative to any load in the hands" (Chaffin et al., 2006, p. 131).

According to these models, also shear forces acting on lumbar disc during lifting varies with the disc level (Arjmand et al., 2011): for example, it has been estimated that, whereas axial compression forces acting on L4–L5 and L5–S1 discs during manual lifting are somewhat comparable, disc shear reactive forces are much greater on L5–S1 disc (range 129–1437 N) than on L4–L5 disc (range 3–571 N).

As biomechanical knowledge suggests that "exposure" of lumbar discs to axial and shear forces during manual material handling (and, possibly, to whole-body vibration) is not the same for each level, future studies should be designed to provide evidence linking manual material handling (and whole-body vibration) with disc disease at specific anatomic levels.

Some evidence about specific patterns of lumbar disc disease seems to emerge (Cheung et al., 2010).

PATHOPHYSIOLOGY OF WORK-RELATED LOW-BACK PAIN AND DISC DISEASE

It is not the scope of this chapter to provide a comprehensive review of low-back pain: however, to introduce the diagnosis and management of work-related low-back pain, it is useful to summarize some concepts about pathophysiology of low-back pain in general.

As stated previously, low-back pain is a symptom (similarly to pain in any particular body region) and not a particular disease: that is, the pain is a consequence of different pathologic processes, which result in this common symptom.

The human spine is a complex structure which maintains our erect position and has the ability to move in different directions in space: it is also subject to different forces acting on our body while we stand, move, lift, carry, or push/pull loads.

Many structures of the body located in the region where low-back pain is felt are capable of originating pain (Vora et al., 2010). In a given case, however, it may be difficult to trace the pain to one (or more) specific structure(s), without the use of invasive tests.

Intervertebral discs are composed of a central nucleus pulposus surrounded by a tough annulus fibrosus, and they are avascular. Nutrition of the disc is believed to be provided by terminal vessels to the vertebral endplate which supply the outer surface of the annulus fibrosus: the inner portion of the disc depends on diffusion for nutrition and metabolite accumulation or removal. Intervertebral discs are innervated and capable of producing pain, possibly by different mechanisms (mechanical or biochemical), and pain may be acute, as demonstrated by pain associated with discography, or chronic, as in disc degeneration (Bogduk, 2012). Why degenerated discs do not produce pain in a number of subjects remains a matter of research.

Disc degeneration has been traditionally viewed as a preceding condition for disc herniation: however, recent experimental evidence seems not to support this view (Lama et al., 2013), as earlier investigations demonstrated that disc herniation could not be produced by compression alone, but that it was necessary to first remove the posterior part of the intervertebral motion segment and then load the disc in compression and flexion (Chaffin et al., 2006, p. 32).

Each lumbar vertebra has two posterolateral zygapophysial joints, commonly referred to as facet joints: these joints articulate with the corresponding facets of the superior and inferior vertebrae. Work-related low-back pain could theoretically result from degenerative changes in the facet joints of the lumbar vertebrae due to repeated excessive load (as a consequence of manual material handling). However, there are no studies which have examined this specific hypothesis and facet osteoarthritis seems to be more common at the L4–L5 level (Kalichman et al., 2009) and not at L5–S1 level, which is the spinal segment subjected to the highest load during manual lifting (Chaffin et al., 2006).

Other structures which may be a source of pain in the low back are the sacroiliac joint–ligament complex, the ligaments of the lumbosacral spine, the muscles of the lumbosacral spine, the thoracolumbar fascia (Vora et al., 2010), and the peridural membrane of the spinal canal (Ansari et al., 2012).

As well as low-back pain, sciatica is not a diagnosis but a symptom: the traditional view was that it was a result of compression on a nerve root due to an herniated disc, but current opinion is that inflammation is also necessary (Valat et al., 2010).

DIAGNOSIS OF WORK-RELATED LOW-BACK PAIN

Diagnosis of work-related low-back pain is extensively covered by a number of guidelines (Hegmann, 2011): in this chapter we will only review some specific issues which are peculiar to the diagnosis of work-related low-back pain. In order to do this, we will first summarize current recommendations on the diagnosis of low-back pain in general.

Diagnosis of low-back pain in general

The presentation of a case of low-back pain to the physician is usually "acute": the patient will report the onset of a pain in the back (extending to a leg, in some cases) which developed in a short time (sometimes abruptly, in other cases over a period of a few hours).

Although in most cases the physician will face a case of nonspecific low-back pain (which in most cases is benign and self-limiting), it is currently recommended that the physician always search for signs of a possible severe underlying condition (red flags). History may reveal a major trauma and pain which reduce spontaneously in the supine position (vertebral fracture), whereas a first episode of pain after 50 years of age, a history of cancer, weight loss, symptoms such as fever, chills, a recent bacterial infection, and immunosuppression, will raise the suspicion of cancer or infection. Severe or progressive sensory alteration or weakness, bladder or bowel dysfunction, or evidence of neurologic deficit in the legs or perineum will prompt investigation for a more serious neurologic condition.

Physical examination is considered a sufficient diagnostic workup in a case of acute low-back pain (and sciatica) without red flags: early magnetic resonance imaging use in acute, work-related, disabling low-back pain is considered to be associated with a strong iatrogenic effect, regardless of radiculopathy status (Webster et al., 2013).

Pharmacologic treatment of low-back pain should be based on the use of analgesic drugs, with acetaminophen as a first-line treatment: if control of pain is not satisfactorily achieved, nonsteroidal anti-inflammatory drugs are used and, in some cases, opioids.

Most cases of acute low-back pain will resolve in less than 2 weeks (Quebec Task Force on Spinal Disorders, 1987): if the case does not resolve within 4–6 weeks, different options should be considered and diagnostic imaging is indicated (Hegmann, 2011).

Yellow flags have been developed as warning signals for the physician for the possible development of chronic pain: in most cases, this evolution seems to be linked to psychosocial factors (Chou and Shekelle, 2010), although biologic issues should not be overlooked (Hancock et al., 2011). Yellow flags (Nicholas et al., 2011) have been categorized as:

- beliefs, appraisals, and judgments: unhelpful beliefs about pain, such as indication of injury as uncontrollable or likely to worsen, expectations of poor treatment outcome, delayed return to work
- emotional responses: distress not meeting criteria for diagnosis of mental disorder, worry, fears, anxiety
- pain behavior (including pain coping strategies): avoidance of activities due to expectations of pain and possible reinjury, overreliance on passive treatments (hot packs, cold packs, analgesics).

Diagnosis of work-related low-back pain

By definition, work-related low-back pain is theoretically a case in which occupational activities have been the cause of the episode or have significantly contributed to the episode. In this sense, the diagnosis of work-related low-back pain is essentially an etiologic diagnosis, which can be summarized as: is this case of nonspecific low-back pain due to the patient's job?

It's a difficult question to answer, so we will review some issues which can be considered in order to provide a rational basis to the diagnosis (for simplicity, we will discuss the issue of acute nonspecific low-back pain, but the concepts used are pertinent also to the other outcomes, previously cited).

In some cases workers will present with a history of an inciting event that produced immediate low-back pain, such as lifting and/or twisting the trunk while holding a heavy object: in this case, the work-relatedness of the (*bona fide*) episode is easy to ascertain.

In other cases the worker will present without a history of an inciting event that produced immediate low-back pain, but it is known that the worker belongs to a category in which epidemiology has shown an increased risk of low-back pain (nurses, construction workers): in such cases, we can resort to epidemiology to evaluate the *a priori* probability of facing a case of work-related low-back pain.

If we know, for example, that the category to which the worker belongs has an RR for the outcome greater than 2 (compared to unexposed workers), than the attributable fraction (i.e., the proportion of cases caused by the exposure) is greater than 50%, according to the formula RR – 1/RR. This *a priori* probability, however, is attributable to every case, including the cases caused by the natural background. The higher the RR within the group of workers to which the case belongs, the higher the *a priori* probability that a random case belongs to the group of cases caused by exposure (for example, if the RR for the outcome in a population of exposed workers equals 5, 80% of the cases in the that population are produced by the exposure and only 20% by the natural background).

In the case of low-back pain, sciatica, and disc disease, as we have previously discussed, the outcome is common in the unexposed population and the RR is not particularly elevated among exposed subjects (below 2, in most cases of manual lifting, for example) so that an exclusive reliance on general epidemiology alone would probably bring one to an exclusion of work-relatedness in most cases.

In order to achieve a sounder base for the diagnosis of work-relatedness of an individual case of low-back pain, it becomes crucial to have an epidemiologic picture of the specific population to which that case belongs (ideally, the specific workplace, provided that it includes a sufficient number of subjects).

Careful consideration should also be given to the nonwork-related risk factors for the outcome in question, bearing in mind that, as a rule of thumb, RR for personal factors is usually not lower than that of occupational factors.

MANAGEMENT OF WORK-RELATED LOW-BACK PAIN

The management of work-related low-back pain integrates the clinical treatment with some specific issues relating to work: thus, we will review some concepts about the treatment of the condition, the assessment of fitness for work, and the prevention of low-back pain.

Treatment of work-related low-back pain

Medical treatment of work-related low-back pain does not differ from the general treatment of the condition (analgesics to control the pain, advice to stay active to avoid deconditioning the musculature).

However, in the case of work-related low-back pain, additional issues should be considered in order to:

- avoid the transition of the subject to a chronic case
- facilitate the return to work as soon as possible.

To reach these goals the occupational physician should consider not only the yellow flags, already mentioned, but also the so-called blue flags and black flags (Main and Burton, 2000).

Blue flags include perceptions about the relationship between work and health such as belief that work is too onerous and likely to cause further injury and/or belief that workplace supervisor and workmates are unsupportive, whereas black flags were termed the more observable characteristics of the workplace and nature of the work, as well the insurance and compensation system under which workplace injuries are managed, such as legislation restricting options for return to work, conflict with insurance staff over injury claim, overly solicitous family and healthcare providers, or heavy work, with little opportunity to modify duties.

A recent review on the "flags" subject concluded that:

> *Consistent evidence has been found to support the role of various psychological factors in prognosis, although questions remain about which factors are the most important, both individually and in combination, and how they affect outcomes. Published early interventions have reported mixed results, but, overall, the evidence suggests that targeting yellow flags, particularly when they are at high levels, does seem to lead to more consistently positive results than either ignoring them or providing omnibus interventions to people regardless of psychological risk factors. Psychological risk factors for poor prognosis can be identified clinically and addressed within interventions, but questions remain in relation to issues such as timing, necessary skills, content of treatments, and context. In addition, there is still a need to elucidate mechanisms of change and better integrate this understanding into the broader context of secondary prevention of chronic pain and disability (Nicholas et al., 2011).*

As the prognosis of low-back pain is related to the clinical picture of the case, the individual and psychologic characteristics, as well as the work and social environment (Hayden et al., 2010), it is important that the necessary attention is paid to all these issues to achieve the best possible outcome for the individual and society, especially in the worker who has still problems 6–12 weeks after the onset of the episode of low-back pain.

Assessment of fitness for work

Whether or not a single case of low-back pain is deemed to be work-related, the issue of fitness for work will be crucial.

The advice to stay active implies that workers affected by low-back pain remain in their job or return to it very soon, even if this may result in some low-back pain. In some cases, however, the job may require exceptional levels of fitness (e.g., emergency rescue services) or unavoidable manual loading of heavy loads: in these (infrequent) cases, return to work may require the temporary modification of the duties of the worker, to facilitate return to work. As any prolonged period off work raises the chance that symptoms will became chronic, it is preferable to devise temporary "light" duties or a pattern of work to encourage uninterrupted employment or early return to work (Palmer et al., 2013).

The issue of work restrictions for low-back pain is a debated one. Although is common practice to issue work restrictions for low-back pain, there is not much evidence about the effectiveness of this practice. A study on the outcome of work restriction for low-back pain in a utility company (Hiebert et al., 2003) found that restrictions were given to 43% of workers, the median duration of restricted duty was 32.5 days and, for 22% of workers, restricted duty was never lifted. Sickness absence duration did not differ between those who had received restrictions and those who had not (adjusted hazard ratio, 1.12); recurrence appeared less likely to occur among those who had work restrictions in their initial episode, but, the difference was not statistically significant. The authors concluded that no evidence of an association between a prescription of work restriction and early return to work was found.

The utility of working restriction in subjects who are engaged in manual lifting, even after spine surgery, had already been questioned (Carragee et al., 1999), suggesting that no postoperative activity restrictions after limited discectomy allowed shortened time to return to work and that complication rates appeared comparable to those reported in the literature for patients under postoperative restrictions.

Prevention of work-related low-back pain

Historically, the risk associated with manual lifting has been addressed, initially setting limits on the load which could be lifted manually, without consideration of the specific conditions in which the load was lifted. For example, in 1943 the Bureau of Labor Standards of the US Department of Labor published the *Bulletin No. 11 ± A Guide to the Prevention of Weight Lifting Injuries*: the bulletin recommended a maximum lifting weight of 50 lb (23 kg) for men and 25 lb (11 kg) for women. Up to the late 1970s, many national legislations included maximum weights for manual lifting for male and female workers.

A major shift occurred in 1981 with the publication of the NIOSH *Work Practices Guide for Manual Lifting*, which used the available biomechanical evidence from three different approaches (biomechanical, physiologic, and psychophysical) "for recommending load limits given different task factors."

The NIOSH equation was updated 10 years later (Waters et al., 1993) and published in 1994 (Waters et al., 1994). The criteria used to design the new equation were:

- biomechanical: maintain L5–S1 compression forces below 3400 N
- psychophysical: loads are acceptable "to 75% of females and about 99% of males"
- physiologic: limit energy expenditure to values ranging from 2.2 to 4.7 kcal/min depending on the duration and vertical displacement of the lifts.

Subsequent evaluation, however, has found that the multiplicative nature of the equation actually produces recommended weight limits which are even lower than those based on the more protective criteria on which the equation is based (Potvin, 2014). The revised NIOSH equation is probably, to date, the most well-known approach to the design of safe manual lifting tasks (although it is often used as a risk evaluation tool).

Although lifting is a principal component of manual material handling, manual pushing/pulling of a load is also common (carrying is less frequent in most jobs).

In order to avoid work-related low-back pain (and disc disease and sciatica), guidelines can be followed to design jobs in such a way that all the components of manual material handling are within safe limits.

The revised NIOSH equation, as shown above, provides a very conservative guidance for the design of manual lifting jobs: according to recent evidence (Potvin, 2014), the revised NIOSH lifting equation-recommended weight limit was found to be much more conservative than would be expected if it strictly adhered to its stated biomechanical, psychophysical, and physiologic criteria. The recommended weight limit was generally consistent with the biomechanical criteria (3400 N of lumbar compression force) for lifts below knuckle height, but resulted in compression forces that were much lower than the criterion at higher lifting heights. In moderate lifting frequencies, the average NIOSH-recommended weight limit is acceptable to more than 95% of female workers according to the psychophysical criterion. At the highest frequencies, the recommended weight limit was found to be an average of 63% of the values based on either the psychophysical or physiologic criteria.

For pushing/pulling tasks, psychophysically based tables have been compiled and updated based on studies published in 1978 and 1991 (Snook, 1978; Snook and Ciriello, 1991). Recently, prediction equations of acceptable pushing/pulling forces (initial and sustained) have been published, to be used on a wider range of frequency of exertions, distance, and handle height (Garg et al., 2014c). It should be remembered, however, that psychophysically based guidance is not based on health risks, but on acceptability of the task by subjects.

References

Airaksinen O, Brox JI, Cedraschi C et al. (2006). COST B13 Working Group on Guidelines for Chronic Low Back Pain. Chapter 4. European guidelines for the management of chronic nonspecific low back pain. Eur Spine J (Suppl 2): S192–S300.

Ala-Kokko L (2002). Genetic risk factors for lumbar disc disease. Ann Med 34 (1): 42–47.

Andersson G (1997). Epidemiology of spinal disorders. In: JW Frymoyer, TB Ducker, NM Hadler et al. (Eds.), The adult spine: principles and practice. Raven Press, New York, pp. 93–141.

Ansari S, Heavner JE, McConnell DJ et al. (2012). The peridural membrane of the spinal canal: a critical review. Pain Pract 12 (4): 315–325.

Arjmand N, Plamondon A, Shirazi-Adl A et al. (2011). Predictive equations to estimate spinal loads in symmetric lifting tasks. J Biomech 44 (1): 84–91.

Axén I, Leboeuf-Yde C (2013). Trajectories of low back pain. Best Pract Res Clin Rheumatol 27 (5): 601–612.

Bakker EW, Verhagen AP, van Trijffel E et al. (2009). Spinal mechanical load as a risk factor for low back pain: a systematic review of prospective cohort studies. Spine (Phila Pa 1976) 34 (8): E281–E293.

Balagué F, Mannion AF, Pellisé F et al. (2012). Non-specific low back pain. Lancet 379 (9814): 482–491.

Battié MC, Videman T (2006). Lumbar disc degeneration: epidemiology and genetics. J Bone Joint Surg Am 88 (Suppl 2): 3–9.

Battié MC, Videman T, Gibbons LE et al. (1995). 1995 Volvo Award in clinical sciences. Determinants of lumbar disc degeneration. A study relating lifetime exposures and magnetic resonance imaging findings in identical twins. Spine (Phila Pa 1976) 20 (24): 2601–2612.

Battié MC, Videman T, Parent E (2004). Lumbar disc degeneration: epidemiology and genetic influences. Spine (Phila Pa 1976) 29 (23): 2679–2690.

Battié MC, Videman T, Kaprio J et al. (2009). The Twin Spine Study: contributions to a changing view of disc degeneration. Spine J 9 (1): 47–59.

Beneke R (1897). Zur Lehre von der Spondylitis deformans. Festschrift zur 69 Versammlung Deutscher Naturforscher und ÄrzteBraunschweig, Germany.

Bernard BP (1997). Musculoskeletal Disorders and Workplace Factors. A Critical Review of Epidemiologic Evidence for Work-Related Musculoskeletal Disorders of the Neck, Upper Extremity, and Low Back. DHHS (NIOSH) Publication No. 97B141.

Bible JE, Choemprayong S, O'Neill KR et al. (2012). Whole-body vibration: is there a causal relationship to specific imaging findings of the spine? Spine (Phila Pa 1976) 37 (21): E1348–E1355.

Bogduk N (2012). Degenerative joint disease of the spine. Radiol Clin North Am 50 (4): 613–628.

Boschman JS, van der Molen HF, Sluiter JK et al. (2011). Occupational demands and health effects for bricklayers and construction supervisors: a systematic review. Am J Ind Med 54 (1): 55–77.

Calvo-Muñoz I, Gómez-Conesa A, Sánchez-Meca J (2013). Prevalence of low back pain in children and adolescents: a meta-analysis. BMC Pediatr 13: 14.

Carragee EJ, Han MY, Yang B et al. (1999). Activity restrictions after posterior lumbar discectomy. A prospective study of outcomes in 152 cases with no postoperative restrictions. Spine (Phila Pa 1976) 24 (22): 2346–2351.

Carragee EJ, Alamin TF, Miller JL et al. (2005). Discographic, MRI and psychosocial determinants of low back pain disability and remission: a prospective study in subjects with benign persistent back pain. Spine J 5 (1): 24–35.

Chaffin DB, Andersson GBJ, Martin BJ (2006). Occupational Biomechanics. Wiley-Interscience, Hoboken, New Jersey, USA.

Cheung KM, Samartzis D, Karppinen J et al. (2010). Intervertebral disc degeneration: new insights based on "skipped" level disc pathology. Arthritis Rheum 62 (8): 2392–2400.

Chou R (2011). Low back pain (Chronic). Clin Evid Handbook 403–445.

Chou R, Shekelle P (2010). Will this patient develop persistent disabling low back pain? JAMA 303 (13): 1295–1302.

Chou R, Qaseem A, Snow V et al. (2007). Clinical Efficacy Assessment Subcommittee of the American College of Physicians; American College of Physicians; American Pain Society Low Back Pain Guidelines Panel. Diagnosis and treatment of low back pain: a joint clinical practice guideline from the American College of Physicians and the American Pain Society. Ann Intern Med 147 (7): 478–491.

Cook CE, Taylor J, Wright A et al. (2014). Risk factors for first time incidence sciatica: a systematic review. Physiother Res Int 19 (2): 65–78.

de Vet HC, Heymans MW, Dunn KM et al. (2002). Episodes of low back pain: a proposal for uniform definitions to be used in research. Spine (Phila Pa 1976) 27 (21): 2409–2416.

Dionne CE, Dunn KM, Croft PR et al. (2008). A consensus approach toward the standardization of back pain definitions for use in prevalence studies. Spine (Phila Pa 1976) 33 (1): 95–103.

Dunn KM, Hestbaek L, Cassidy JD (2013). Low back pain across the life course. Best Pract Res Clin Rheumatol 27 (5): 591–600.

Ehrlich GE (2003). Low back pain. Bull World Health Organ 81 (9): 671.

Ferreira PH, Beckenkamp P, Maher CG et al. (2013). Nature or nurture in low back pain? Results of a systematic review of studies based on twin samples. Eur J Pain 17 (7): 957–971.

Frymoyer JW (1992). Lumbar disk disease: epidemiology. Instr Course Lect 41: 217–223.

Garg A, Boda S, Hegmann KT et al. (2014a). The NIOSH lifting equation and low-back pain, Part 1: Association with low-back pain in the backworks prospective cohort study. Hum Factors 56 (1): 6–28.

Garg A, Kapellusch JM, Hegmann KT et al. (2014b). The NIOSH lifting equation and low-back pain, Part 2: Association with seeking care in the backworks prospective cohort study. Hum Factors 56 (1): 44–57.

Garg A, Waters T, Kapellush J et al. (2014c). Psychophysical basis for maximum pushing and pulling forces: a review and recommendations. Int J Ind Ergon 44 (2): 281–291.

Griffith LE, Shannon HS, Wells RP et al. (2012). Individual participant data meta-analysis of mechanical workplace risk factors and low back pain. Am J Public Health 102 (2): 309–318.

Haidar R, Ghanem I, Saad S et al. (2010). Lumbar disc herniation in young children. Acta Paediatr 99 (1): 19–23.

Hancock MJ, Maher CG, Laslett M et al. (2011). Discussion paper: what happened to the 'bio' in the bio-psychosocial model of low back pain? Eur Spine J 20 (12): 2105–2110.

Hartvigsen J, Leboeuf-Yde C, Lings S et al. (2000). Is sitting-while-at-work associated with low back pain? A systematic, critical literature review. Scand J Public Health 28 (3): 230–239.

Hayden JA, Dunn KM, van der Windt DA et al. (2010). What is the prognosis of back pain? Best Pract Res Clin Rheumatol 24 (2): 167–179.

Hegmann KT (2011). Occupational medicine practice guidelines. In: Evaluation and management of common health problems and functional recovery in workers, 3rd ed. American College of Occupational and Environmental Medicine (ACOEM), Elk Grove Village (IL).

Heneweer H, Staes F, Aufdemkampe G et al. (2011). Physical activity and low back pain: a systematic review of recent literature. Eur Spine J 20 (6): 826–845.

Hiebert R, Skovron ML, Nordin M et al. (2003). Work restrictions and outcome of nonspecific low back pain. Spine (Phila Pa 1976) 28 (7): 722–728.

Hill JJ, Keating JL (2010). Risk factors for the first episode of low back pain in children are infrequently validated across samples and conditions: a systematic review. J Physiother 56 (4): 237–244.

Hoogendoorn WE, van Poppel MN, Bongers PM et al. (1999). Physical load during work and leisure time as risk factors for back pain. Scand J Work Environ Health 25 (5): 387–403.

Hoy D, Bain C, Williams G et al. (2012). A systematic review of the global prevalence of low back pain. Arthritis Rheum 64 (6): 2028–2037.

Jones GT, Watson KD, Silman AJ et al. (2003). Predictors of lowback pain in British schoolchildren: a population-based prospective cohort study. Pediatrics 111 (4 Pt 1): 822–828.

Jørgensen MB, Holtermann A, Gyntelberg F et al. (2013). Physical fitness as a predictor of herniated lumbar disc disease – a 33-year follow-up in the Copenhagen male study. BMC Musculoskelet Disord 14: 86.

Kalichman L, Suri P, Guermazi A, Li L et al. (2009). Facet orientation and tropism: associations with facet joint osteoarthritis and degenerative spondylolisthesis. Spine (Phila Pa 1976) 34 (16): E579–E585.

Konstantinou K, Dunn KM (2008). Sciatica: review of epidemiological studies and prevalence estimates. Spine (Phila Pa 1976) 33 (22): 2464–2472.

Lama P, Le Maitre CL, Dolan P et al. (2013). Do intervertebral discs degenerate before they herniate, or after? Bone Joint J 95-B (8): 1127–1133.

Lang J, Ochsmann E, Kraus T et al. (2012). Psychosocial work stressors as antecedents of musculoskeletal problems: a systematic review and meta-analysis of stability-adjusted longitudinal studies. Soc Sci Med 75 (7): 1163–1174.

Lee MJ, Dettori JR, Standaert CJ et al. (2012). The natural history of degeneration of the lumbar and cervical spines: a systematic review. Spine (Phila Pa 1976) 37 (22 Suppl): S18–S30.

Main CJ, Burton AK (2000). Economic and occupational influences on pain and disability. In: CJ Main, CC Spanswick (Eds.), Pain Management: An Interdisciplinary Approach, Churchill Livingstone, Edinburgh, Scotland, pp. 63–87.

Martínez-Lage JF, Fernández Cornejo V, Lopez F et al. (2003). Lumbar disc herniation in early childhood: case report and literature review. Childs Nerv Syst 19 (4): 258–260.

Miller JA, Schmatz C, Schultz AB (1988). Lumbar disc degeneration: correlation with age, sex, and spine level in 600 autopsy specimens. Spine (Phila Pa 1976) 13 (2): 173–178.

Murray CJ, Vos T, Lozano R et al. (2012). Disability-adjusted life years (DALYs) for 291 diseases and injuries in 21 regions, 1990–2010: a systematic analysis for the Global Burden of Disease Study 2010. Lancet 380 (9859): 2197–2223.

Nachemson A, Morris J (1963). Lumbar discometry. Lumbar intradiscal pressure measurements in vivo. Lancet 1 (7291): 1140–1142.

Nicholas MK, Linton SJ, Watson PJ et al. (2011). Early identification and management of psychological risk factors ("yellow flags") in patients with low back pain: a reappraisal. Phys Ther 91 (5): 737–753.

NIOSH (1981). Work Practices Guide for Manual Lifting. U.S. Department of Health and Human Services, Cincinnati, Ohio, (NIOSH) Publication 81-122.

Nissinen M, Heliövaara M, Seitsamo J et al. (1994). Anthropometric measurements and the incidence of low back pain in a cohort of pubertal children. Spine (Phila Pa 1976) 19 (12): 1367–1370.

Osborne A, Blake C, Fullen BM et al. (2012). Risk factors for musculoskeletal disorders among farm owners and farm workers: a systematic review. Am J Ind Med 55 (4): 376–389.

Ostelo RW, Deyo RA, Stratford P et al. (2008). Interpreting change scores for pain and functional status in low back pain: towards international consensus regarding minimal important change. Spine (Phila Pa 1976) 33 (1): 90–94.

Palmer KT, Brown I, Hobson J (2013). Fitness for Work: The Medical Aspects. Oxford University Press, Oxford, UK.

Pincus T, Burton AK, Vogel S et al. (2002). A systematic review of psychological factors as predictors of chronicity/disability in prospective cohorts of low back pain. Spine (Phila Pa 1976) 27 (5): E109–E120.

Potvin J (2014). Comparing the revised NIOSH lifting equation to the psychophysical, biomechanical and physiological criteria used in its development. Int J Ind Ergon 44 (2): 246–252.

Punnett L, Prüss-Utün A, Nelson DI et al. (2005). Estimating the global burden of low back pain attributable to combined occupational exposures. Am J Ind Med 48 (6): 459–469.

Quebec Task Force on Spinal Disorders (1987). Scientific approach to the assessment and management of activity-related spinal disorders. A monograph for clinicians. Spine (Phila Pa 1976) 12 (7 Suppl): S1–59.

Ramazzini B (1700). De Morbis Artificum Diatriba, Capponi, Padua, Italy.

Ramond A, Bouton C, Richard I et al. (2011). Psychosocial risk factors for chronic low back pain in primary care – a systematic review. Fam Pract 28 (1): 12–21.

Ribeiro DC, Aldabe D, Abbott JH et al. (2012). Dose–response relationship between work-related cumulative postural exposure and low back pain: a systematic review. Ann Occup Hyg 56 (6): 684–696.

Shiri R, Karppinen J, Leino-Arjas P et al. (2010a). The association between smoking and low back pain: a meta-analysis. Am J Med 123 (1): 87.e7–87.e35.

Shiri R, Karppinen J, Leino-Arjas P et al. (2010b). The association between obesity and low back pain: a meta-analysis. Am J Epidemiol 17 (2): 135–154.

Smeets R, Köke A, Lin CW et al. (2011). Measures of function in low back pain/disorders: Low Back Pain Rating Scale (LBPRS), Oswestry Disability Index (ODI), Progressive Isoinertial Lifting Evaluation (PILE), Quebec Back Pain Disability Scale (QBPDS), and Roland-Morris Disability Questionnaire (RDQ). Arthritis Care Res (Hoboken) 63 (Suppl 11): S158–S173.

Snook SH (1978). The design of manual handling tasks. Ergonomics 21 (12): 963–985.

Snook SH, Ciriello VM (1991). The design of manual handling tasks: revised tables of maximum acceptable weights and forces. Ergonomics 34 (9): 1197–1213.

Sørensen IG, Jacobsen P, Gyntelberg F et al. (2011). Occupational and other predictors of herniated lumbar disc disease – a 33-year follow-up in the Copenhagen male study. Spine (Phila Pa 1976) 36 (19): 1541–1546.

Stanton TR, Latimer J, Maher CG et al. (2009). Definitions of recurrence of an episode of low back pain: a systematic review. Spine (Phila Pa 1976) 34 (9): E316–E322.

Sterud T, Tynes T (2013). Work-related psychosocial and mechanical risk factors for low back pain: a 3-year follow-up study of the general working population in Norway. Occup Environ Med 70 (5): 296–302.

Takatalo J, Karppinen J, Niinimäki J et al. (2009). Prevalence of degenerative imaging findings in lumbar magnetic resonance imaging among young adults. Spine (Phila Pa 1976) 34 (16): 1716–1721.

Taylor JB, Goode AP, George SZ et al. (2014). Incidence and risk factors for first-time incident low back pain: a systematic review and meta-analysis. Spine J 14 (10): 2299–2319.

U.S. Department of Labor Bureau of Labor Standards (1943). Bulletin No. 11 – A Guide to the Prevention of Weight Lifting Injuries. Washington, USA.

US Burden of Disease Collaborators (2013). The state of US health, 1990–2010: burden of diseases, injuries, and risk factors. JAMA 310 (6): 591–608.

Valat JP, Genevay S, Marty M et al. (2010). Sciatica. Best Pract Res Clin Rheumatol 24 (2): 241–252.

van Tulder M, Becker A, Bekkering T et al. (2006). COST B13 Working Group on Guidelines for the Management of Acute Low Back Pain in Primary Care. Chapter 3. European guidelines for the management of acute nonspecific low back pain in primary care. Eur Spine J 15 (Suppl 2): S169–S191.

Vora AJ, Doerr KD, Wolfer LR (2010). Functional anatomy and pathophysiology of axial low back pain: disc, posterior elements, sacroiliac joint, and associated pain generators. Phys Med Rehabil Clin N Am 21 (4): 679–709.

Wahlström J, Burström L, Nilsson T et al. (2012). Risk factors for hospitalization due to lumbar disc disease. Spine (Phila Pa 1976) 37 (15): 1334–1339.

Waters TR, Putz-Anderson V, Garg A et al. (1993). Revised NIOSH equation for the design and evaluation of manual lifting tasks. Ergonomics 36 (7): 749–776.

Waters TR, Putz-Anderson V, Garg A (1994). Application Manual for the Revised NIOSH Lifting Equation. U.S. Department of Health and Human Services, Cincinnati, Ohio, (NIOSH) Publication 94-110.

Waters TR, Baron SL, Piacitelli LA et al. (1999). Evaluation of the revised NIOSH lifting equation. A cross-sectional epidemiologic study. Spine (Phila Pa 1976) 24 (4): 386–394.

Waters TR, Genaidy A, Barriera Viruet H et al. (2008). The impact of operating heavy equipment vehicles on lower back disorders. Ergonomics 51 (5): 602–636.

Waters TR, Lu ML, Piacitelli LA et al. (2011). Efficacy of the revised NIOSH lifting equation to predict risk of low back pain due to manual lifting: expanded cross-sectional analysis. J Occup Environ Med 53 (9): 1061–1067.

Webster BS, Bauer AZ, Choi Y et al. (2013). Iatrogenic consequences of early magnetic resonance imaging in acute, work-related, disabling low back pain. Spine (Phila Pa 1976) 38 (22): 1939–1946.

Yassi A, Lockhart K (2013). Work-relatedness of low back pain in nursing personnel: a systematic review. Int J Occup Environ Health 19 (3): 223–244.

Handbook of Clinical Neurology, Vol. 131 (3rd series)
Occupational Neurology
M. Lotti and M.L. Bleecker, Editors

Chapter 22

Occupational mononeuropathies in industry

ROBERTA BONFIGLIOLI*, STEFANO MATTIOLI, AND FRANCESCO S. VIOLANTE
Occupational Health Unit, Department of Medical and Surgical Sciences, University of Bologna, Bologna, Italy

INTRODUCTION

Entrapment and compression neuropathies of the upper and lower extremities are a heterogeneous group of focal neuropathy syndromes related to peripheral nerve compression injuries; these conditions have the potential to cause significant disability and can be commonly associated with recreational (e.g., sports) and occupational activities. Specific history features are important, such as the type of activity that aggravates symptoms and the temporal relation of symptoms to activity.

Entrapment neuropathies can be acute or chronic and are characterized by sensory and motor symptoms such as numbness, paresthesia, weakness, and pain in a specific anatomic area supplied by the injured nerve.

Recovery of nerve function is more likely with a mild injury and a shorter duration of compression. Recovery is faster if the repetitive activities that exacerbate the injury can be decreased or ceased (Neal and Fields, 2010).

Acute nerve injuries related to violent trauma (crush injuries, transection) cause axonal injury followed by Wallerian degeneration. Although the myelin changes after chronic nerve injury were previously thought to be a mild form of Wallerian degeneration, modern evidence points to a distinct pathophysiology involving Schwann cell mechanosensitivity. When repeated mechanical trauma occurs due to external forces, such as when maintaining a body part in the same position or holding a tool in a certain manner, or when the nerve is compressed between two muscle edges during repetitive motions, evidence suggests that chronic nerve compressive injury is a Schwann cell-mediated disease. In the early stages of the disorder, changes in Schwann cell proliferation and apoptosis develop, followed by myelin protein downregulation, local demyelination and remyelination in the region of injury, and subsequent axonal sprouting (Pham and Gupta, 2009; Tapadia et al., 2010).

A contribution of reduced blood flow can be expected, such as mechanical compression impact on microvascular circulation. Physical examination shows sensory and motor impairment in the affected region, while nerve conduction slowing is recorded when electrodiagnostic tests are performed.

A strong interaction between manual work and nerve function exists. Upper-limb nerves warrant fine movements of the hand, together with strength. The median nerve supplies the muscles involved in precision grip, the ulnar nerve controls power grip, the radial nerve contributes to providing fixation of the wrist. Other movements, like forearm pronation, supination, and flexion, are controlled by muscles supplied through the contribution of several nerves. Holding tools or performing manual forceful work tasks impose mechanical stress to musculoskeletal tissues and peripheral nerves passing through, especially in particular anatomic narrow spaces like carpal tunnel, thus placing workers at increased risk for peripheral compressive nerve disorders. At the same time nerve injury impairs hand function and reduces work ability.

Age, sex, anatomic variations, and systemic diseases may predispose some individuals. The same is true for rare genetic conditions such as "hereditary neuropathy with liability to pressure palsies," which consists of recurrent focal compression neuropathies associated with autosomal-dominant inheritance.

Although acute or chronic compression-related injury may occur in essentially any peripheral nerve, anatomic considerations may predispose certain nerves to intrinsic

*Correspondence to: Roberta Bonfiglioli, Occupational Health Unit, University of Bologna Sant'Orsola Malpighi Hospital, via P. Palagi 9, 40138 Bologna, Italy. Tel: +39-051-214-2611, E-mail: roberta.bonfiglioli@unibo.it

or extrinsic compression-related injury (Arnold and Elsheikh, 2013).

Work-relatedness of entrapment compression neuropathies should be recognized if the working history demonstrates exposure to risk factors for which scientific evidence exists of their role in causing chronic nerve injury and when individual risk factors alone do not justify development of the disorder (Feldman et al., 1983).

This chapter will present a narrative review of the existing evidence of the association between peripheral compressive nerve disorders and work-related risk factors. Search strings proposed by Mattioli et al. (2010, 2013) to study occupational etiology for diseases that are not generally thought to be work-related have been used to support a literature search.

UPPER-LIMB ENTRAPMENT NEUROPATHIES

Thoracic outlet syndrome

Thoracic outlet syndrome is a controversial cause of neck and shoulder pain due to complex mechanisms involving muscular dysfunction and nerve compression. It is caused by a large number of etiologies, which can affect the elements of the neurovascular bundle in combination or separately, resulting in compression of the brachial plexus as well as the subclavian or axillary artery. The neural symptoms include local and referred pain secondary to pressure, paresthesia in the hand and forearm along the medial aspect, and weakness of the intrinsic hand muscles. Pain and sensory changes are usually in the ulnar innervated fingers (Roos, 1976, 1980).

Various forms of thoracic outlet syndrome are distinguished: vascular forms (arterial or venous), which raise few diagnostic problems, and "neurologic" forms, which are by far the most frequent, as they represent more than 95% of all cases (Roos, 1976).

The main causes are congenital abnormalities (often reported and associated with traumatic or functional causes) like bone anomalies (cervical rib, prolonged transverse process), fibrous anomalies (e.g., transversocostal, costocostal), or muscular anomalies (e.g., scalenus anticus muscle, sickle-shaped scalenus medius). Other causes are traumatic or "functional" acquired causes, probably the most frequent, due to overuse. According to the literature, workers at increased risk perform repetitive movements and use the arms in elevation (e.g., hairdressers, switchboard operators, workers on assembly lines), in antepulsion or abduction, carry heavy loads on the shoulders or with the arm outstretched. Postural factors incriminated in sedentary jobs are mainly prolonged cervical flexion (protraction of the head), and abduction and internal rotation of the shoulders (Laulan et al., 2011).

Prevalence varies across studies, mainly due to the case definition adopted. Sallstrom and Schmidt (1984) found, among 191 industry and service workers with symptoms from the cervicobrachial region, that 18% had thoracic outlet symptoms (27% of women and 11% of men). Pascarelli and Hsu (2001) found a 70% prevalence in a population with upper-limb complaints of computer users and musicians (70% and 28% respectively). Thoracic outlet syndrome was one of the most common disorders diagnosed among 56 patients with upper-extremity symptoms associated with occupational keyboard/mouse use (Bleecker et al., 2011). Laulan et al. (2011) reviewed the definitions, diagnosis, and management of thoracic outlet syndrome and the links with the workplace in the pathogenesis and course of this disease. Overall, studies presented many methodologic limitations and no conclusion can be drawn concerning a significant association between thoracic outlet syndrome and occupational exposure risk factors; despite the low level of evidence, when the role of occupation is alleged, a multidisciplinary approach including work rehabilitation can allow early diagnosis and management of these patients.

Standard radiographs or computed tomography scans can demonstrate a cervical rib or a prolonged C7 transverse process; however the presence of this anatomic feature is not sufficient to confirm the diagnosis. Some series of thoracic outlet syndrome report the presence of a cervical rib in only 10% of cases. Imaging is also useful where there is suspicion of mass lesions.

Electrodiagnostic study helps localize and quantify a lesion in the brachial plexus. It is also important to rule out other more frequent segmental or systemic neuropathies (Tolson, 2004). The most sensitive electrodiagnostic measure for thoracic outlet syndrome is the amplitude of the medial antebrachial cutaneous nerve action potential and also side-to-side comparisons (Nishida et al., 1993; Seror, 2002). Tsao et al. (2014) found electrodiagnostic evidence of a chronic axon loss process that affected the lower portion of the brachial plexus (involving the T1 more than the C8 sensory and motor fibers) in a group of 32 patients with surgically verified true neurogenic thoracic outlet syndrome; 89% of these patients showed abnormal medial antebrachial cutaneous sensory nerve conduction.

Suprascapular nerve neuropathy

The suprascapular nerve derives from the upper trunk of the brachial plexus (C5 and C6 roots). It reaches the upper border of the scapula behind the brachial plexus to enter the suprascapular notch and innervates the supraspinatus and the infraspinatus muscles.

A common site of entrapment is at the level of the suprascapular notch beneath the transverse scapular ligament. Clinical findings include shoulder pain due to involvement of the deep sensory fibers for the shoulder joint, weakness in shoulder external rotation, and abduction if both the infraspinatus and supraspinatus muscles are affected; when the site of compression is in a more distal site, at the spinoglenoidal notch, only the infraspinatus shows changes. Differential diagnoses of suprascapular neuropathy comprise shoulder joint disorders and rotator cuff tears and may require magnetic resonance (Jacobson et al., 2010).

The etiology is traction or compression of the suprascapular nerve. This can result from a space-occupying lesion at the suprascapular or spinoglenoid notch, such as a ganglion cyst, or a traction injury as a result of repetitive overhead activities (Boykin et al., 2010; Freehill et al., 2012).

Suprascapular nerve injury is more frequently associated with repetitive overhead loading as a consequence of repetitive movement of the scapula that involves shoulder abduction and protraction, causing traction on the nerve (Shapiro and Preston, 2009).

Weight lifters, professional volleyball or tennis players, as well as dancers are prone to suprascapular neuropathy (Fansa and Schneider, 2003). Only sporadic work-related cases have been reported (Arboleya and García, 1993; Araujo and Fragoso, 2004). Newsreel camera operators, carrying a heavy mobile camera on the shoulder, have experienced suprascapular nerve entrapment (Karataş and Göğüş, 2003).

A systematic review carried out to provide a quantitative assessment of the exposure–response relationships between work-related physical and psychosocial factors and the occurrence of specific shoulder disorders in occupational populations concluded that there was an absence of any articles reporting associations between work-related factors and the occurrence of suprascapular nerve compression (van Rijn et al., 2010).

Electrodiagnostic studies are useful to determine if abnormalities are restricted to the suprascapular nerve or involve cervical roots or brachial plexus. Imaging (computed tomography or magnetic resonance) of the shoulder region should be done to exclude mass lesions.

Median nerve neuropathies

The median nerve arises from the lateral and medial cord of the brachial plexus. It is a mixed nerve originating from the C6–T1 roots. It gives sensation to the skin over the lateral aspect of the palm and the dorsal surfaces of the terminal phalanges, along with the volar surfaces of the thumb, the index and middle fingers, and the lateral aspect of the fourth digit. The median nerve innervates most flexor muscles in the forearm and muscles of the thenar eminence. Along the course in the upper arm the median nerve innervates no muscles. The median nerve enters the forearm beyond the elbow joint by passing between the two heads of the pronator teres and gives supply to the pronator teres, flexor carpi radialis, and flexor digitorum superficialis. In the forearm it gives rise to a pure motor branch, the anterior interosseous nerve, which innervates the flexor pollicis longus, pronator quadratus, and flexor digitorum profundus I and II. The main branch descends distally and, at the end of the forearm, it passes through the carpal tunnel. It supplies lumbricalis I and II after giving off the recurrent thenar nerve at the distal edge of the carpal ligament. The motor branch to the thenar eminence innervates the abductor pollicis brevis, the lateral half of the flexor pollicis brevis, and the opponens pollicis.

Entrapment of the median nerve at the wrist causes so-called carpal tunnel syndrome (CTS), which is the most frequent peripheral compressive neuropathy. Other less common sites of injury are the elbow, where the median nerve can be compressed between the two heads of the pronator teres, and more distally, the forearm (anterior interosseous syndrome). The median nerve can be compressed at the elbow in the presence of a variant bone (the supracondylar process) beneath a fibrous band called the ligament of Struthers (Jacobson et al., 2010).

Median nerve neuropathies at the forearm: pronator teres syndrome and anterior interosseous nerve syndrome

Pronator teres syndrome is caused by the entrapment of the median nerve in the proximal forearm between the ulnar and humeral heads of the pronator teres. It is a rare condition that is more common in women than in men. Clinical symptoms, pain, and numbness of the volar elbow, forearm, and wrist, are exacerbated by repeated pronation and supination. No muscle weakness is detectable.

Weakness of deep flexor muscles to the thumb and index fingers characterize the clinical presentation of anterior interosseous syndrome.

Very few case reports suggest an association between median nerve compression at the forearm and certain occupations.

Morris and Peters (1976) report seven men with pronator syndrome who have occupations (mechanic, barber, woodworker) which require considerable muscular activity of the forearm and hand, especially forceful pronation accompanying finger flexion.

Stål et al. (1998) described the pronator teres syndrome in a group of female machine milkers suffering from pain in the volar part of the forearm, numbness, tingling, and decreased muscle strength in their hands (mostly in the hands, which usually were statically loaded with heavy equipment). The authors hypothesized that the external exposure of the arm during the application of the milking cluster probably causes muscle and fascial tensions that induce compression of the nerve.

A single case report of a 57-year-old woman who suffered from weakness in the thumb was reported by Wertsch (1992). The woman presented difficulties while gardening; she described difficulty pulling weeds out because of an inability to get a firm grip when using the thumb.

Carpal tunnel syndrome

CTS is one of the most common entrapment upper-limb neuropathies; it results from compression of the median nerve at the wrist level where it crosses the carpal canal. CTS is characterized by numbness, tingling, and pain in the typical distribution of the median nerve, and symptoms are usually worse at night; in more severe cases hand weakness and thenar atrophy may occur. Local and regional causes may play a role in the etiology of CTS, such as inflammation (i.e., tenosynovitis), trauma, tumors, anatomic abnormalities, amyloidosis, rheumatoid arthritis, and gout. Systemic causes (diabetes, hypothyroidism, renal diseases, and hemodialysis) can predispose to CTS. Females are at increased risk, especially during pregnancy and menopause; higher rates are also related to hormonal factors due to contraception. Rarely, a sudden increase of pressure inside the carpal tunnel (bone fracture, local infection, or swelling of tendon sheath following overuse) is responsible for acute CTS (Aroori and Spence, 2008).

The incidence and prevalence of CTS vary according to the case definition adopted and is higher among women and in working groups exposed to repetitive manual work.

In the general population CTS prevalence is between 5% and 16% depending on the criteria used for diagnosis (de Krom et al., 1992; Tanaka et al., 1994; Atroshi et al., 1999).

In Sweden Atroshi et al. (1999) found a prevalence in women of 17.3% based on symptoms, of 4.6% according to symptoms and physical examination, and of only 3% when abnormal nerve conduction was required for case definition; among men the corresponding figures were 10.4%, 2.8%, and 2.1%.

The crude overall incidence of CTS in the population of Rochester, Minnesota, USA, from 1961 through 1980 was estimated as 0.99 cases per 1000 person-years (age-adjusted rates were 0.52 for men, 1.49 for women, and 1.05 for both sexes combined) (Stevens et al., 1988). The incidence of CTS in the same population was updated through 2005: the overall age- and sex-adjusted incidence of CTS was 3.76 per 1000 person-years, greater in women (4.91 per 1000 person-years) than men (2.58 per 1000 person-years) (Gelfman et al., 2009).

According to Atroshi et al. (2011), in Sweden the annual incidence was 4.28 in women and 1.82 in men per 1000 adults; females have a two- to threefold incidence compared to males depending on age. Corresponding figures for surgery-treated CTS were 2.20 and 0.85 per 1000 adults.

In the Korean general adult population the incidence of clinically and electrophysiologically diagnosed CTS was 4.96 and 0.98 per 1000 person-years, respectively. The incidence of surgically treated CTS was 0.29 per 1000 person-years. The age-adjusted female to male incidence ratio of diagnosed CTS was 2.58 (95% confidence interval (CI), 2.56–2.59) and that of surgically treated CTS was 5.82 (95% CI, 5.64–6.00). Women had the highest incidence, especially if they were aged 50–59 years (Roh et al., 2010).

In the Italian population the overall age-standardized incidence rate of in-hospital cases of CTS was 1.06 per 1000 person-years; the age-standardized incidence rate for women was 3.8-fold that for men and showed a peak in the 50–54-year age group (Mattioli et al., 2008). Two- to threefold higher rates were recorded for clinically/electrodiagnostically diagnosed CTS in the general population: the calculated crude incidence was 3.29 per 1000 person-years while the age-standardized rate corresponded to 2.76 per 1000 person-years (Mondelli et al., 2002).

Figures in the working population are quite different. In the early 1990s Hagberg et al. (1992) reported the results of a literature survey showing a highly variable prevalence of CTS in different occupational groups, ranging from 0.6% to 61%; the highest prevalence was noted for grinders, butchers, grocery store workers, frozen-food factory workers, plating workers, and workers with high-force, high-repetitive manual movements. The attributable fraction in the diagnosis of CTS in manual workers exposed to high force, high repetition, and vibration was estimated as 80% or higher.

In the USA the overall estimated prevalence of self-reported clinician-diagnosed CTS among current/recent workers was 6.7%, while the 12-month prevalence was 3.1% (Luckhaupt et al., 2013).

High attributable risk fractions of clinically/electrodiagnostically diagnosed CTS among workers were observed in lower-grade white-collar occupations for women (43–67%) and blue-collar occupations for men (60–74%) and women (48–88%) enrolled in a French prospective study. The highest fractions were recorded in manufacturing (42–93% for both sexes), construction

(66% for men), and personal service industries (66% for women), and in the trade and commerce sectors (49% for women) (Roquelaure et al., 2008).

The results of a population-based multicenter case-control study carried out in Italy showed that the broad socio-occupational class of blue-collar workers and housewives appeared to have a sevenfold (point estimate) risk of surgical treatment for CTS, as compared with white-collar workers. Blue-collar work as well as the combination of biomechanical risk factors that includes repetition and sustained force are risk factors for surgically treated CTS (Mattioli et al., 2009).

CTS is the most common peripheral neuropathy and also the most studied occupational disease in working populations employed in manual repetitive and forceful activities. The work-relatedness of CTS is essentially based on epidemiologic evidence and the results of experimental studies showed the capability of repetitive wrist extreme postures, associated with hand–wrist forceful exertions, to increase the pressure inside the carpal tunnel and to compress the median nerve.

During the past decades convincing evidence of the association between CTS and manual work has grown up. Physical risk factors, such as repetition, force, extreme wrist postures, and hand–arm vibration, especially if combined while performing the job, have been found to be associated with an increased risk of CTS (National Institute of Occupational Safety and Health, 1997; National Research Council and the Institute of Medicine, 2001).

Experimental studies in humans and animal models have been used to examine the effect of compression on the peripheral nerve; mechanic and ischemic effects depend on both magnitude and duration of compression. Pressures, which are approximately 30 mmHg or more, are able to interfere with blood flow and axonal transport (Lundborg et al., 1982). Acute median nerve conduction impairment occurred experimentally when carpal canal pressure increased between 40 and 50 mmHg; nerve dysfunction, caused by nonneutral hand postures, grip forces and external pressure, has been demonstrated (Lundborg et al., 1982; Gelberman et al., 1983; Keir and Rempel, 2005). The acute effect of hand activity on the median nerve has also been observed using ultrasonography: in response to manual exercise, an acute increase in the cross-sectional area of the median nerve at the pisiform level has been detected before and after activity or keyboarding (Massy-Westropp et al., 2001; Toosi et al., 2011).

Joint motions impose longitudinal and transverse movements of the median nerve and may cause friction and strain; excursions during fingers and wrist flexion or extension can reach 10 mm in proximal or distal direction (Wright et al., 1996).

Human experimental studies evaluating mechanical properties of the carpal tunnel usually involve small sample sizes and seldom discuss gender differences. Li (2005) found different mechanical properties of the transverse carpal ligament between males and females and hypothesized that lower carpal tunnel compliance among females could play a role as a predisposing factor for median nerve compression at wrist level. However the results are not conclusive. Given that the stiffness of the transverse carpal ligament is affected by the size of the contact area when applying force and that women tend to have smaller wrists than men, the relationship of wrist size to applied load size may partially explain some of Li's findings (Keir and Rempel, 2005; Holmes et al., 2011).

The effect of cumulative mechanical trauma on nerve function has also been shown experimentally in animals. In a rat model, animals trained to perform a high-repetition, high-force task, grasping for food pellets for 12 weeks, were tested for median nerve function and showed a 15–16% reduction of conduction velocity (Clark et al., 2004).

Repetitive or continuous compression induced a reduction of compound nerve action potential amplitudes and a decrease of nerve blood flow in rabbits, followed by a slower recovery after repetitive compression (Yoshii et al., 2010).

According to previous experience, carpal tunnel pressure can increase while performing manual activities, including pinch grip, or finger pressure, or wrist flexion/extension or deviation (Rempel et al., 1997; Keir et al., 1998; Rempel and Diao, 2004). High pressures have also been recorded in experimental studies during forearm full supination and 90° metacarpophalangeal joint flexion in humans (Rempel et al., 1998a).

Comparing studies, Keir and Rempel (2005) summarized the effects of posture, stating that lowest mean hydrostatic carpal tunnel pressure is found in a neutral or slightly flexed wrist posture and that it increases with deviation from this posture in flexion and extension as well as in lateral deviation; also pinching and gripping are able to increase pressure. In CTS patients the effects of posture are heightened.

Epidemiologic evidence displays that jobs with a high occurrence of CTS are the meat- and fish-processing industry, forestry work with chain saws, and electronic assembly work. Physical risk factors associated with CTS are: hand force (>4 kg), repetitiveness (cycle time <10 seconds or >50% of a cycle time spent performing the same movements), and hand–arm vibration (van Rijn et al., 2009a).

However the vast majority of the studies performed to evaluate the relationship between biomechanical risk factors and CTS are cross-sectional or case–control.

Few longitudinal studies have investigated the impact of manual work on CTS. Nathan et al. (2005) followed a cohort of industrial workers for 17 years but almost two-thirds of the original sample were lost at last follow-up; obesity and gender resulted as consistent predictors of CTS, while workplace demands appear to bear an uncertain relationship to CTS.

Among a large Italian cohort of more than 4000 industrial and service "blue-collar" and "white-collar" workers, a dose–response relationship between the incidence of CTS and the biomechanical load associated with increased hand activity level (HAL) and force was observed. Two case definitions for CTS were explored: one based on classic hand symptoms and the other confirmed by nerve conduction studies; according to the latter, after 2 years, the calculated CTS incidence was 9.6 per 1000 person-years among men and 30.5 per 1000 person-years among female workers. Job characteristics were quite similar by gender; however the exposure to force was more heterogeneous among men. CTS was more sensitive to repetitiveness than force (Violante et al., 2007; Bonfiglioli et al., 2013).

In the USA Burt et al. (2013) examined a cohort of healthcare and manufacturing workers; after 2 years the incidence of CTS was 51.1 per 1000 person-years and the percentage of time spent working with forceful exertion contributed (with individual risk factors as obesity) to the risk for CTS.

Even if CTS has been widely studied in working populations, the small number of longitudinal studies, the variability, and the heterogeneous quality of exposure assessment hamper the possibility of reaching definitive conclusions in terms of causality for several risk factors and intensity of exposures.

Exposure assessment is a challenging issue; several methods have been developed to assess biomechanical exposures – the vast majority are based on observation and subjective judgment, but only some of these tools have been tested in a systematic manner for validity and repeatability. Thus no single tool appears to have a clear advantage over any other (Takala et al., 2010). Observational tools are time consuming as they require long periods of observation on the field or video analysis; thus, job title or self-reporting is often used as a proxy of exposure, reducing the capability to obtain detailed information about the exposure–response relationship or about which factor is actually responsible for the increased risk.

Furthermore, workplace factors are transitory and difficult to assess with precision, especially over time if compared to most individual factors, such as gender, weight, or medical conditions, which tend to be stable and are easier to monitor (Nathan et al., 2005).

Another issue that should be taken into account is the case definition adopted. Ideally case definitions based on a combination of clinical history and instrumental findings would be preferred for both surveillance and epidemiologic purposes. Discordances are known to exist between the presence of clinical symptoms of CTS and the presence of nerve conduction abnormalities both in the general population (Redmond and Rinver, 1988) and in at-risk groups of workers (Homan et al., 1999; Violante et al., 2004).

Furthermore, when interpreting the results of electrodiagnostic studies, population-based normative data may not be appropriate for manual workers. The literature reports a large proportion of industrial asymptomatic workers with median nerve conduction slowing at wrist level (Franzblau et al., 1993); in a group of meat industry workers highly significant differences between symptomatic and asymptomatic hands were found for median wrist sensory and motor parameters, with mean values recorded in the asymptomatic hands falling between mean values recorded in the general population and in the symptomatic hands (Violante et al., 2004). A possible explanation is that occupational overuse could expose the median nerve to mechanical trauma at the wrist level, consequently modifying median nerve conduction. Manual asymptomatic workers may thus represent the reference population. Werner and Andary (2011) stated that caution should be used in defining an abnormality when evaluating an active worker with moderate slowing of the median nerve relative to the ulnar and that firm guidelines for interpretation of nerve conduction studies using exact numbers in the setting of peripheral neuropathy and other diseases are not available.

According to scientific recommendations, median nerve conduction studies can confirm a clinical diagnosis of CTS with a high degree of sensitivity and specificity (Jablecki et al., 2002).

The same criteria are also valid for epidemiologic research in the occupational setting (Rempel et al., 1998b).

There is probably no "best" case definition for CTS. The goal of the study together with its context (clinical or epidemiologic) should guide the selection of criteria for diagnosis; when applying different case definitions for CTS in a working population a fair degree of agreement has been found, suggesting the possibility of pooling data from different studies (Descatha et al., 2011).

In conclusion, CTS is a common disorder in working populations; many occupations characterized by repeated and prolonged exposure to repetitive and forceful exertions (e.g., assembly line workers, food and fish processing and packing) and jobs involving exposure to vibrating tools are known to increase the risk for CTS (Palmer, 2011).

When analyzing risk factors, a meta-analysis showed a strong relationship between CTS (symptoms combined with nerve conduction abnormality) and the use of vibratory tools, with a fivefold increased risk, as well as with hand force (fourfold increase); repetition and a combination of repetition and force accounted for a double risk of CTS (Barcenilla et al., 2012). The results of more recent cohort studies are in line with these findings. Burt et al. (2013) found that time spent in forceful exertion can be a greater risk for CTS: the risk increased significantly when forceful exertion $\geq$20% of the time and was higher if forceful exertions $\geq$60% of the time vs <20% (hazard ratios (HRs) 19.57 (5.96, 64.24)). Harris-Adamson et al. (2015) found an increased risk of incident-dominant hand CTS associated with percent time spent in forceful hand exertions (HR = 2.05; 95% CI 1.34–3.15) in a large cohort of 2474 incident-eligible workers followed for 6.5 years.

As reported from previous cross-sectional studies also, repetition plays an important role (Latko et al., 1999); when exposure has been assessed using the combination of HAL and normalized peak force, CTS incidence was more sensitive to repetition in a cohort of industrial manual workers followed for 3 years (Bonfiglioli et al., 2013).

Based on the results of these studies, preventive measures should focus on recognizing and minimizing exposure to the above-mentioned risk factors. Scientific evidence on prevention of work-related CTS is still lacking. The American Conference of Governmental Industrial Hygienists (ACGIH) proposed threshold limit values (TLVs) based on a specific combination of HAL and normalized peak force (ACGIH, 2000). Moore and Garg (1995) proposed a method (the strain index) to analyze jobs at risk of distal upper-extremity disorders. A dose–response relationship between ACGIH TLV classification and risk of CTS has been found (Bonfiglioli et al., 2013). TLVs for HAL and the strain index have been judged useful metrics to assess exposure to biomechanical stressors and to predict risk of CTS in a cohort study (Garg et al., 2012a).

Even if the level of evidence is not conclusive, some recommendations for the management of CTS in the occupational setting are included in the American College of Occupational and Environmental Medicine (ACOEM) guidelines; for patients with CTS, it is recommended that their work be restricted to tasks that do not involve high-force, stereotypical hand gripping or pinching or the use of high-acceleration vibrating hand-held tools. Some physicians place work restrictions on patients with CTS; others do not, as there is no quality evidence to suggest that restrictions are required (ACOEM, 2010).

Radial nerve neuropathies

The radial nerve is a continuation of the posterior cord of the brachial plexus and derives its axons from C5 to T1 roots. The radial nerve courses distally in the posterior aspect of the upper arm; it innervates the triceps and the anconeus, then enters the spiral groove winding around the humerus posteriorly from the medial to the lateral side, where it gives supply to the brachioradialis muscle and the extensor carpi radialis longus. At the level of the lateral epicondyle the radial nerve is located in the anterolateral aspect of the lower arm between the brachioradialis and the brachialis, where it divides into superficial and deep branches. The deep branch is a motor branch and is also referred to as the posterior interosseous nerve. The posterior interosseous nerve innervates the supinator muscle traveling between its two heads, all the extensor muscles in the forearm (extensor carpi radialis brevis, extensor carpi ulnaris, extensor digitorum, extensor digiti minimi, extensor pollicis longus and brevis, extensor indicis) and the abductor pollicis longus. The sensory branch courses distal in the anterolateral forearm and more superficial in the distal third of the forearm, where it provides sensory innervation to the dorsoradial aspect of the hand, of the thumb, index, and middle finger.

The radial nerve is predisposed to entrapment and compression injury at the axilla, in the spiral groove of the humerus and at the elbow, where the deep branch can be entrapped between the two heads of the supinator muscle (posterior interosseous nerve syndrome, supinator syndrome). The superficial branch is subjected to injury over the first dorsal wrist compartment (Wartenberg's syndrome) (Jacobson et al., 2010).

Radial nerve compression at the axilla and spiral groove syndrome

Radial nerve compression at the axilla and at the spiral groove is usually related to arm positioning during sleep (Saturday-night palsy). Spiral groove syndrome is also common after humeral diaphysis fracture. The clinical presentation is characterized by sensory and motor loss with typical wrist drop. When the injury is at the spiral groove both elbow extension and triceps reflex remain normal as the triceps is regularly supplied (Shapiro and Preston, 2009).

No occupational conditions seem to predispose radial nerve compressive neuropathies at a proximal site.

Posterior interosseous nerve syndrome

Posterior interosseous nerve syndrome is also called supinator syndrome since its origin is the result of an entrapment of the deep branch of the radial nerve at the level of the supinator muscle. The superficial head of the supinator muscle forms a fibrous arch, the arcade of Frohse, which is the most common site of compression of the posterior interosseous nerve.

Other causes include prior trauma (radius fracture) or compression from adjacent mass or joint process. Clinical features are painless lack of finger and thumb extension without wrist drop due to the spread of the extensor carpi radialis which comes off prior to the posterior interosseous nerve. This syndrome has been related to supination and pronation overuse, such as in sports (swimming, tennis) or playing the violin (Jacobson et al., 2010). A more controversial painful form presents as forearm pain without muscle weakness (referred to as radial tunnel syndrome); differential diagnosis with persistent lateral epicondylitis is sometimes challenging. It was hypothesized that the chronic tennis-elbow syndrome was caused by an entrapment neuropathy of the radial nerve (Roles and Maudsley, 1972); detailed examination revealed that this syndrome couldn't be explained by an entrapment neuropathy (van Rossum et al., 1978). Rosenbaum (1999) recommended reserving the term "radial tunnel syndrome" for the truly neurogenic cases identifiable by focal motor weakness in the distribution of the posterior interosseous nerve.

Different elements could affect the deep branch of the radial nerve in the radial tunnel and cause an entrapment syndrome: a capsule–tendon–aponeurotic complex on the anterior aspect of both the humeroradial joint and the radial head, the vascular arcade formed by the radial recurrent artery, and its branches, the arcade formed by the medial edge of extensor carpi radialis brevis muscle, and the superior and inferior arcades of the superficial layer of the supinator muscle (Riffaud et al., 1999).

Even if the posterior interosseous nerve syndrome has been associated with repetitive pronation and supination of the forearm, few data exist concerning the relationship between this syndrome and occupational manual activities.

Single cases have been reported in manual workers (Lee et al., 2006; Krishnan et al., 2009).

Furthermore, one case-referent study assessed occupational risk factors for radial tunnel syndrome in 21 surgically treated manual workers (mainly employed in assembling and sewing duties) compared to 21 controls. The regular use of a force of at least 1 kg (odds ratio (OR) = 9.1 (1.4–56.9)), static work (static load applied to the hand during work) (OR = 5.9 (1.2–29.9)) as well as work with the elbow constantly extended was associated with an increased risk of radial tunnel syndrome (OR = 4.9 (1.0–25.0)) (Roquelaure et al., 2000).

Interestingly, the analysis of functional outcome after surgical treatment for posterior interosseous nerve syndrome highlighted a predominance of men manual workers (Quignon et al., 2012). A less successful outcome after surgical treatment was observed if there were coexisting additional nerve compression syndromes or lateral epicondylitis, if the patient was receiving workers' compensation, or in patients who had work-related injuries (Ritts et al., 1987; Eaton and Lister, 1992; Lee et al., 2008; Naam and Nemani, 2012).

Ulnar nerve neuropathies

The ulnar nerve is the terminal branch of the medial cord of the brachial plexus and contains fibers from the C8 and T1 spinal nerve roots; it descends along the arm, reaches the elbow region where it becomes superficial behind the medial epicondyle, then passes through the cubital tunnel. In the forearm the ulnar nerve goes across the two heads of the flexor carpi ulnaris muscle, passes along the medial aspect of the wrist, and enters the hand through the Guyon canal. The ulnar nerve supplies the flexor carpi ulnaris, partially the flexor digitorum profundus and, along its course from the hypothenar to the thenar region of the hand, it innervates the majority of the hand muscles.

Common sites of entrapment of the ulnar nerve are the elbow, the Guyon canal, and the heel of the hand where its deep motor branch may be subjected to external trauma.

Experimental studies have shown significant excursion of the ulnar nerve during upper-extremity motion, ranging from an average of 5 mm to more than 20 mm; ulnar nerve strain of 15% or greater was experienced in cadaver specimens at the elbow with elbow flexion and at the wrist with wrist extension and radial deviation. Any factor that limits excursion at these sites could result in repetitive traction of the nerve and possibly play a role in the pathophysiology of cubital tunnel syndrome or ulnar neuropathy at Guyon's canal (Wright et al., 2001).

Ulnar nerve neuropathy at the elbow level: cubital tunnel syndrome

At elbow level the ulnar nerve can be compressed or stretched either at the ulnar groove or more distally at the cubital tunnel; the two forms result in analogous clinical presentation and some use the term cubital tunnel syndrome to refer to all lesions of the ulnar nerve around the elbow (Shapiro and Preston, 2009).

Cubital tunnel syndrome is the second most common peripheral nerve entrapment syndrome in the human body. Posttraumatic deformities and recurrent subluxation can cause ulnar nerve injury at elbow level, usually referred to as "tardy ulnar palsy." Ulnar nerve injury commonly occurs at the elbow due to repeated trauma; the nerve is superficial in this region; furthermore, elbow flexion and extension require the ulnar nerve to stretch and slide through the cubital tunnel. Elbow flexion narrows the cubital tunnel causing a significant increase in intraneural pressure, especially with simultaneous wrist extension and shoulder abduction (Cutts, 2007).

Variations in the anatomy of the cubital tunnel retinaculum documented in cadavers may explain certain types of ulnar neuropathy (O'Driscoll et al., 1991).

The appearance of bilateral ulnar neuropathy in a large number of patients suggests a congenital predisposition to this syndrome (Miller, 1979). A tight tunnel may predispose to friction and compression during repetitive movements of the upper limb.

Ulnar neuropathy at the elbow is more common in men; gender differences in the anatomy of the elbow region represent a possible explanation as men have less subcutaneous fat and larger tubercles of the coronoid process, which predispose to compression-induced ischemia (Contreras et al., 1998).

Few data are available about the incidence of ulnar nerve neuropathy at the elbow in the general population. Mondelli et al. (2005) found a standardized incidence of 20.9 cases per 100 000 person-years in the province of Siena (Italy). Cases were surgically treated or confirmed by electrodiagnostic testing; sex-specific incidences were 32.7 for men and 17.2 for women, higher in male residents of a health subdistrict where manual work was dominant (57.1 cases per 100 000 person-years).

A Dutch hospital-based case–control study, performed including 110 patients surgically treated for ulnar nerve compression at the elbow and 192 controls, revealed that lower level of education predisposes to an increased risk of ulnar nerve compression at the elbow. Heavy work was also a risk factor, even if less important than education; repetitive movement of the arm during work was not a risk factor; however data on repetitiveness were not collected precisely to draw definitive conclusions. Surprisingly, body mass index, gender, and fractures were not found to be potential risk factors, while smoking was a predisposing factor (Bartels and Verbeek, 2007).

Cubital tunnel syndrome may develop because of various factors, including repetitive elbow motion, prolonged elbow flexion, or direct compression (Bozentka, 1998). Direct mechanical compression, impairment of axonal transportation or of intrinsic blood supply to the nerve due to compressive forces (>30 mmHg) can be responsible for nerve damage.

Chronic mechanical external compression and repeated trauma (such as habitual leaning on the elbow) are common causes of ulnar neuropathy at the groove. Patients whose work requires prolonged periods of elbow flexion with the elbow pressed against a hard surface are at increased risk of cubital tunnel syndrome.

Descatha et al. (2004) found a relationship between ulnar nerve entrapment at the elbow and occupational risk factors like "holding a tool in position" (OR 3.53) in a cohort of almost 600 workers followed for 3 years. Workers performing repetitive tasks belonged to one of the following activity sectors: assembly line (packaging excluded), clothing and shoes industry (packaging excluded), food industry (packaging excluded), packaging, supermarket cashiers. Annual incidence was estimated at 0.8 per 100 person-years; clinically diagnosed cubital tunnel syndrome was also associated with personal factors (obesity), and with the presence of other upper-limb disorders, especially medial epicondylitis and other nerve entrapments. However, cases were not confirmed by nerve conduction studies and postures and biomechanical constraints were self-assessed; another limitation was the small number of cases (Descatha et al., 2004). No other occupational risk factors arose from a systematic literature review carried out to assess the association between work-related factors and specific disorders at the elbow (van Rijn et al., 2009b).

Subsequent Danish studies showed an association between ulnar neuropathy and forceful work with an exposure–response pattern reaching an OR of 3.85 (95% CI 2.04–7.24); nonneutral postures strengthened the effects of forceful work. No relation was observed with repetitive movements (Svendsen et al., 2012).

Furthermore, in a cohort study, poor prognosis with respect to symptoms and disability among patients with ulnar nerve neuropathy was demonstrated when abnormal nerve conduction studies were recorded and high-force-requirement jobs were performed, suggesting a negative impact of continuous and unchanged occupational biomechanical exposure among these patients (Svendsen et al., 2013).

Cross-sectional studies have recorded the prevalence of ulnar neuropathy at the elbow in specific jobs such as floor cleaners (6.8% in a group of 103 female cleaners) (Mondelli et al., 2006) and in computer keyboard operators, where it was notably high (74.5% in a group of more than 800 subjects). However, authors recommended caution in interpreting these results because electrodiagnostic examination was performed by an examiner not blinded for symptoms (Nainzadeh et al., 2011).

Case reports have described the association between entrapment ulnar neuropathy at the elbow and particular jobs like handcrafting household fittings and toys using bamboo (Bonfiglioli et al., 1996), or performing repetitive manual handling activities in poultry industry (Svendsen and Juhl, 2008).

Cubital tunnel syndrome was also observed in 42.5% of 167 Hungarian workers exposed to hand–arm vibration (Kákosy, 1994).

Ulnar nerve neuropathies at the hand–wrist level

Common causes of compression of the distal part of the ulnar nerve are ganglion, other space-occupying lesions, such as tumors, abnormal or anomalous muscles or carpal bones, and a thrombosed ulnar artery (Waugh and Pellegrini, 2007).

Compression at Guyon's canal (Guyon's canal syndrome) has been associated with activities and occupations that involve repetitive movements or force against the ulnar aspect of the wrist, such as manual laborers who use tools that cause repetitive pressure against the hypothenar eminence (Shapiro and Preston, 2009). Bilateral median and ulnar neuropathy at the wrist has been reported in a 43-year-old male parquet floorer who used his hands to hit and press the wood blocks, holding his hands in an awkward posture (hand dorsiflexion) (dell'Omo et al., 1995). Distal ulnar neuropathy at the wrist has also been described in the dominant hand of a professional chef (Krishnan et al., 2009) and in meat packers (Streib and Sun, 1984).

Hypothenar hammer syndrome is a rarely diagnosed condition that occurs when the heel of the hand and/or the hypothenar region are used as a tool to hammer, push, or squeeze hard objects. It is due to vascular damage to the distal part of the ulnar artery, probably caused by acute or repetitive blunt trauma to the hypothenar region. The incidence of hypothenar hammer syndrome in the general population is likely less than 1%, even if activities in which the hands are used as a hammer are quite common. Possible explanations include anatomic variability, subclinical diseases, or abnormal arterial histology (Swanson et al., 2012).

Hypothenar hammer syndrome has been associated with sports activities and manual work; however, mainly case reports have been published, while epidemiologic data are almost absent. A case–control study confirmed that work-related acute or repetitive blunt trauma to the hypothenar region can damage the distal part of the ulnar artery: risk factors are using the hand as a hammer on a daily basis (adjusted odds ratio (aOR) 17.04, 95% CI 5.51–52.67) and daily pressure to the palm of the hand (aOR 4.96, 95% CI 1.39–17.71) (Scharnbacher et al., 2013).

Workers considered at risk are auto mechanics, metal workers, bricklayers, carpenters, butchers, and bakers (Carter et al., 2013).

Hypothenar hammer syndrome results primarily from direct trauma to the ulnar artery (compressed against the hook of the hamate), thrombosis and occlusion, ischemia, and consequent nerve damage. The distal branch of the ulnar nerve can also be impaired by the trauma itself; repeated or prolonged dorsiflexion of the hand may increase the compression and independently cause nerve injury.

A systematic review was carried out to determine whether ulnar hammer syndrome has an occupational association; however, the heterogeneity in study design and outcome measures limited definitive conclusions about occupational association (Vartija et al., 2013).

Early clinical features in patients with ulnar nerve compression at any level include numbness, tingling, and sensory loss over the fifth digit and the ulnar side of the ring finger; when the nerve is compressed at the Guyon canal or distally, only palmar sensation is altered, since the dorsal cutaneous branch is spared.

Weakness of the small muscles of the hand and the ulnar-sided muscles of the forearm occurs as the condition progresses.

Motor and sensory symptoms may vary according to the anatomic site of compression.

Involvement of the palmar deep branch at wrist level results exclusively in motor impairment. Hypothenar muscles may also be spared.

Electrodiagnostic testing can confirm the diagnosis and severity of injury to the nerve. Imaging (X-ray, ultrasound, magnetic resonance) could be useful, especially if there is a history of trauma or anatomic lesions are suspected. Surgical intervention is indicated when nonoperative treatment does not relieve the symptoms.

LOWER-LIMB ENTRAPMENT NEUROPATHIES

The spinal cord ends at the L1–L2 intervertebral space level; the cauda equina is formed by the lumbar and sacral roots which assume a downward direction from the conus medullaris towards their respective exit foramina. The main peripheral entrapment syndromes involving nerves derived from the lumbar plexus (L1–L4) and the sacral plexus (L5–S2) are described and possible work-related associations discussed.

Lateral femoral cutaneous nerve

The lateral femoral cutaneous nerve is the first sensory branch of the lumbar plexus; the nerve reaches the upper thigh after going to a tunnel formed by the lateral attachment of the inguinal ligament and the anterior superior iliac spine, then divides into two branches: the anterior supplies the skin over the lateral and anterior surface of the thigh, the other innervates the posterior and lateral aspect of the thigh.

The term "meralgia paresthetica" is used to identify the disorder of the lateral femoral cutaneous nerve that results in symptoms of anterolateral thigh paresthesias and dysesthesias without associated loss of reflexes or motor weakness. Most cases are idiopathic or consequent to local compression from mass lesions. Occupational causes have been described in workers who wear heavy tool belts. Case reports of meralgia paresthetica have been published concerning police officers and soldiers who wear protective and lifesaving body armor (Korkmaz and Ozçakar, 2004; Fargo and Konitzer, 2007).

Common peroneal nerve

The common peroneal nerve is one of the two terminal branches of the sciatic nerve; it derives from the L4, L5, and S1 roots and originates at the popliteal fossa. Soon after its origin the nerve becomes superficial and winds around the neck of the fibula laterally, entering the lateral compartment of the leg where it passes inside the fibular tunnel (between the two heads of the peroneus longus). At this location, usually considered a common site of entrapment, it gives off a small recurrent nerve that supplies sensation to the patella and then it divides into its two terminal branches: the superficial peroneal nerve and the deep peroneal nerve (also called anterior tibialis nerve). The superficial peroneal nerve supplies the peroneus longus and brevis, which plantarflex and evert the foot, and with its sensory branches supplies the anterolateral aspect of the lower leg, the dorsum of the foot and toes. The deep peroneal nerve innervates the muscles that dorsiflex and evert the foot and supplies the skin over a small area between the first and the second toes.

Common peroneal neuropathy is one of the most common acute focal compression mononeuropathies of the lower limb; it often occurs when the common peroneal nerve is compressed at the fibular neck, giving rise to a combination of deep and superficial peroneal neuropathy (weakness of the toe and ankle dorsiflexion, ankle eversion, and sensory disturbance over the dorsum of the foot and the lateral calf below the knee). Also known as foot drop and strawberry picker's palsy, this palsy occurs with bed positioning, squatting, casting, and tight boots.

Common peroneal nerve neuropathy at the fibular neck

Prolonged and repetitive maintenance of the crossed-leg posture and repetitive stretch from squatting have been associated with peroneal nerve neuropathy at the fibular neck where the nerve is superficial ("dropfoot" palsy). Severe weight loss and acute direct trauma have also been associated with peroneal mononeuropathy at this level (Shapiro and Preston, 2009).

Since the mid- 1970s agricultural activities requiring prolonged squatting have been associated with common peroneal nerve entrapment at the fibular neck. Several case reports can be found in the literature covering a large spectrum of tasks performed for harvesting crops (Sandhu and Sandhey, 1976; Piza-Katzer and Pilz, 1997). Peroneal mononeuropathy has been described in strawberry pickers (Seppalainen et al., 1977), hazelnut gatherers (Laterza and Nappo, 1977), and olive harvesters (Bautista et al., 1986).

Certain practices, commonly used in India, such as hoeing in a sitting position, have been associated with common peroneal nerve compression among farm workers (Sangwan et al., 2004).

Garg et al. (2012b) examined a group of 50 males affected by footdrop and employed as farmers in India. Nerve conduction studies showed peroneal neuropathy of the affected limb associated with abnormal peroneal nerve conduction parameters in the opposite side, suggesting a common cause of mechanical trauma. The authors hypothesized that the nerve damage could be due to repeated trauma to the peroneal nerve at the fibular head or to a particular sitting position that the patient assumes. During harvesting or weeding, the farmer squats, bends the dominant knee completely and the other knee partially, and then makes to-and-fro movements of the dominant hand to cut the plants or weeds. Whether it is the complete flexion of the knee or the repeated trauma to the fibular head, made by the forearm, the injuries to the peroneal nerve are not fully understood.

Common peroneal nerve injury at knee level has also been related to repeated overstretch and contusion during military training or secondary to performing flight duties in the cockpit due to the practice of bracing the "collective" with the knee (McGrail, 1999; Solmaz et al., 2011).

Cases of peroneal palsy due to a prolonged sitting position have been occasionally recorded in a painter's model, among workers at archeologic digs and in a student of biology who remained in a squatting position for several hours in order to process water samples (Mendel et al., 1987; Rivera Redondo and Sáenz Lope, 1988; Schröter et al., 1990).

Electrodiagnostic studies are particularly helpful in diagnosis as they can be useful to localize the site of the lesion and to differentiate between demyelinating and axonal lesions; needle electromyography assesses the severity of damage and guides treatment and prognosis.

Tibial nerve neuropathies

The tibial or posterior tibial nerve is the extension of the medial popliteal nerve that arises at the bifurcation of the sciatic nerve in the popliteal fossa. It supplies the gastrocnemius and soleus, the tibialis posterior, flexor digitorum longus, and flexor halluces longus; it enters the foot passing between the medial malleolus and the flexor retinaculum, then it divides into medial and lateral plantar nerves after giving off a calcaneal branch.

Tarsal tunnel syndrome

Tarsal tunnel syndrome is an entrapment neuropathy of the posterior tibial nerve or its branches within its fibro-osseous tunnel beneath the flexor retinaculum on the medial side of the ankle, leading to a range of symptoms affecting the plantar aspect of the foot. The role of electrodiagnostic techniques in its diagnosis is still controversial, with inability to predict which cases respond to decompression (Ahmad et al., 2012).

The syndrome has been described in reviews and case reports. Most cases are idiopathic; local trauma, degenerative bone disease, or rheumatic disorders may predispose to tarsal tunnel syndrome. Lam (1967) looked for common risk factors among a series of 10 patients with tarsal tunnel syndrome but found that none shared a common occupation.

A literature review on the etiology of foot and ankle disorders (hallux valgus, interdigital neuroma, tarsal tunnel syndrome, lesser-toe deformity, heel pain, adult acquired flatfoot, and foot and ankle osteoarthritis) concluded that there was an absence of evidence to consider cumulative industrial trauma as a clear etiology of these disorders of the adult foot and ankle. The authors hypothesized that the evolutionary adaptation of the human foot to prolonged ambulation and the absence of industrial demands that significantly differ from this task likely account for this reduced vulnerability of the foot to industrial repetitive motion disorders compared to the upper extremity (Guyton et al., 2000).

Among the possible causes that have been suggested for tarsal tunnel syndrome are included local diseases (such as lipomas, ganglion cysts, fractures) as well as tenosynovitis and synovial hypertrophy. Even if the latter two are considered as causal factors for CTS in workers exposed to upper-limb biomechanical overload, insufficient evidence exists for such a relation between occupation and tarsal tunnel syndrome. A case was reported of a woman working as a seamstress sat at an electric sewing machine; she operated using her right foot to press on a pedal 40 hours a week, depressing the foot pedal several hundred times a day (Forst and Hryhorczuk, 1988). Occupations or activities that include repetitive weight bearing, like jogging, increase the risk (Shapiro and Preston, 2009).

Diagnosis is mainly based on symptoms and signs; nerve conduction studies can be quite difficult, partly because of the absence of evidence-based studies on their validity; however they can be used to exclude more proximal lesions (Shapiro and Preston, 2009).

CONCLUDING REMARKS

The association between peripheral compressive nerve disorders and occupational activities has been reported. Scientific literature encompasses several contributions showing that a significant proportion of workers employed in manual forceful jobs are at risk of upper-limb musculoskeletal disorders and peripheral nerve diseases at the wrist or elbow level. Prolonged awkward upper-limb postures as well as repetitive hand–wrist forceful exertions can overstretch peripheral nerves in certain anatomic regions, where they can also be particularly exposed to local compression. Lower-limb neuropathies are instead less frequently reported and often described as single cases.

The assembly industry, food processing and packaging, the use of hand–arm vibrating tools and generally all jobs involving high-repetition, high-force tasks place workers at risk for CTS, the most frequently diagnosed entrapment peripheral neuropathy both in the general population and among manual workers. Less strong evidence exists of the association between ulnar elbow neuropathy and tasks including repetitive elbow motion, prolonged elbow flexion, or direct compression. Repetitive stretch on squatting, as happens for workers whose job is carried out on the floor, are at increased risk for peroneal nerve neuropathy at the fibular head. Very few reports are available about the association between occupation and other compressive peripheral nerve injuries.

Although abundant epidemiologic evidence exists for median nerve injury at the wrist, the majority of the studies are cross-sectional and the body of evidence supporting a causal relationship between ergonomic risk factors is still growing also for CTS. Several occupational risk factors have shown the capability to produce mechanical injury to the median nerve as it passes the carpal tunnel. Unquestionably, repetitive hand movements, extreme wrist postures, frequent forceful exertions, and the use of vibrating tools are able to produce a mechanical load over the nerve; furthermore the results of epidemiologic studies received experimental and laboratory support in animal models and human studies and are coherent with the pathophysiology of chronic compressive nerve injury. More recent evidence from cohort studies is shedding light on the relationship between CTS incidence and the exposure to risk factors and on the contribution of each of them or their combination; the debate on the relative impact of repetition and force is still ongoing. Deeper knowledge on this will improve our capability to set up preventive measures and to define health-based occupational exposure limits.

References

Ahmad M, Tsang K, Mackenney PJ et al. (2012). Tarsal tunnel syndrome: a literature review. Foot Ankle Surg 18 (3): 149–152, Sep.

American College of Occupational and Environmental Medicine (ACOEM) (2010). Occupational medicine practice guidelines: evaluation and management of common health problems and functional recovery in workers. Third Edition,ACOEM, Elk Grove Village, IL. Chapter 11.

American Conference of Governmental Industrial Hygienists (ACGIH) (2000). TLVs and BEIs. American Conference of Governmental Industrial Hygienists, Cincinnati.

Araujo MG, Fragoso YD (2004). Occupational lesion of the suprascapular nerve leading to marked atrophy of the supra- and infraspinatus muscles. J Clin Neuromuscul Dis 6 (1): 22–23, Sep.

Arboleya L, García A (1993). Suprascapular nerve entrapment of occupational etiology: clinical and electrophysiological characteristics. Clin Exp Rheumatol 11 (6): 665–668.

Arnold WD, Elsheikh BH (2013). Entrapment neuropathies. Neurol Clin 31 (2): 405–424, May.

Aroori S, Spence RA (2008). Carpal tunnel syndrome. Ulster Med J 77 (1): 6–17, Jan.

Atroshi I, Gummesson C, Johnsson R et al. (1999). Prevalence of carpal tunnel syndrome in a general population. JAMA 282 (2): 153–158.

Atroshi I, Englund M, Turkiewicz A et al. (2011). Incidence of physician-diagnosed carpal tunnel syndrome in the general population. Arch Intern Med 171 (10): 943–944.

Barcenilla A, March LM, Chen JS et al. (2012). Carpal tunnel syndrome and its relationship to occupation: a meta-analysis. Rheumatology (Oxford) 51 (2): 250–261.

Bartels RH, Verbeek AL (2007). Risk factors for ulnar nerve compression at the elbow: a case control study. Acta Neurochir (Wien) 149 (7): 669–674.

Bautista J, Sureda B, Gil Néciga E et al. (1986). Occupational neuropathy during olive harvesting. Neurologia 1 (6): 263–265.

Bleecker ML, Celio MA, Barnes SK (2011). A medical-ergonomic program for symptomatic keyboard/mouse users. J Occup Environ Med 53 (5): 562–568.

Bonfiglioli R, Lodi V, Tabanelli S et al. (1996). [Entrapment of the ulnar nerve at the elbow caused by repetitive movements: description of a clinical case.] Intrappolamento del nervo ulnare al gomito da movimenti ripetitivi: descrizione di un caso clinico. Med Lav 87 (2): 147–151.

Bonfiglioli R, Mattioli S, Armstrong TJ et al. (2013). Validation of the ACGIH TLV for hand activity level in the OCTOPUS cohort: a two-year longitudinal study of carpal tunnel syndrome. Scand J Work Environ Health 39 (2): 155–163.

Boykin RE, Friedman DJ, Higgins LD et al. (2010). Suprascapular neuropathy. J Bone Joint Surg Am 92 (13): 2348–2364.

Bozentka DJ (1998). Cubital tunnel syndrome pathophysiology. Clin Orthop Relat Res (351): 90–94.

Burt S, Deddens JA, Crombie K et al. (2013). A prospective study of carpal tunnel syndrome: workplace and individual risk factors. Occup Environ Med 70 (8): 568–574.

Carter PM, Hollinshead PA, Desmond JS (2013). Hypothenar hammer syndrome: case report and review. J Emerg Med 45 (1): 22–25.

Clark BD, Al-Shatti TA, Barr AE et al. (2004). Performance of a high-repetition, high-force task induces carpal tunnel syndrome in rats. J Orthop Sports Phys Ther 34 (5): 244–253.

Contreras MG, Warner MA, Charboneau WJ et al. (1998). Anatomy of the ulnar nerve at the elbow: potential relationship of acute ulnar neuropathy to gender differences. Clin Anat 11 (6): 372–378.

Cutts S (2007). Cubital tunnel syndrome. Postgrad Med J 83: 28–31.

de Krom MC, Knipschild PG, Kester AD et al. (1992). Carpal tunnel syndrome: prevalence in the general population. J Clin Epidemiol 45 (4): 373–376.

dell'Omo M, Muzi G, Cantisani TA et al. (1995). Bilateral median and ulnar neuropathy at the wrist in a parquet floorer. Occup Environ Med 52 (3): 211–213.

Descatha A, Leclerc A, Chastang JF et al. (2004). Study Group on Repetitive Work. Incidence of ulnar nerve entrapment at the elbow in repetitive work. Scand J Work Environ Health 30 (3): 234–240.

Descatha A, Dale AM, Franzblau A et al. (2011). Comparison of research case definitions for carpal tunnel syndrome. Scand J Work Environ Health 37 (4): 298–306.

Eaton CJ, Lister GD (1992). Radial nerve compression. Hand Clin 8 (2): 345–357.

Fansa H, Schneider W (2003). [Suprascapular nerve entrapment.] Die isolierte Kompression des N. suprascapularis. Handchir Mikrochir Plast Chir 35 (2): 122–126.

Fargo MV, Konitzer LN (2007). Meralgia paresthetica due to body armor wear in U.S. soldiers serving in Iraq: a case report and review of the literature. Mil Med 172 (6): 663–665.

Feldman RG, Goldman R, Keyserling WM (1983). Classical syndromes in occupational medicine. Peripheral nerve entrapment syndromes and ergonomic factors. Am J Ind Med 4 (5): 661–681.

Forst L, Hryhorczuk D (1988). Occupational tarsal tunnel syndrome. Br J Ind Med 45 (4): 277–278.

Franzblau A, Werner RA, Valle J et al. (1993). Workplace surveillance for carpal tunnel syndrome: a comparison of methods. J Occup Rehabil 3: 1–14.

Freehill MT, Shi LL, Tompson JD et al. (2012). Suprascapular neuropathy: diagnosis and management. Phys Sportsmed 40 (1): 72–83.

Garg A, Kapellusch J, Hegmann K et al. (2012a). Hand Study Research Team. The Strain Index (SI) and Threshold Limit Value (TLV) for Hand Activity Level (HAL): risk of carpal tunnel syndrome (CTS) in a prospective cohort. Ergonomics 55 (4): 396–414.

Garg R, Singh Arora K, Bansal N et al. (2012b). Footdrop in the farmers in Punjab: a retrospective electrodiagnostic study. Clin Diagn Res 6 (10): 1653–1657.

Gelberman RH, Szabo RM, Williamson RV et al. (1983). Tissue pressure threshold for peripheral nerve viability. Clin Orthop Relat Res 178: 285–291.

Gelfman R, Melton 3rd LJ, Yawn BP et al. (2009). Long-term trends in carpal tunnel syndrome. Neurology 72 (1): 33–41.

Guyton GP, Mann RA, Kreiger LE et al. (2000). Cumulative industrial trauma as an etiology of seven common disorders in the foot and ankle: what is the evidence? Foot Ankle Int 21 (12): 1047–1056.

Hagberg M, Morgenstern H, Kelsh M (1992). Impact of occupations and job tasks on the prevalence of carpal tunnel syndrome. Scand J Work Environ Health 18 (6): 337–345.

Harris-Adamson C, Eisen EA, Kapellusch J et al. (2015). Biomechanical risk factors for carpal tunnel syndrome: a pooled study of 2474 workers. Occup Environ Med 72 (1): 33–41.

Holmes MW, Howarth SJ, Callaghan JP et al. (2011). Carpal tunnel and transverse carpal ligament stiffness with changes in wrist posture and indenter size. J Orthop Res 29 (11): 1682–1687.

Homan MM, Franzblau A, Werner RA et al. (1999). Agreement between symptom surveys, physical examination procedures and electrodiagnostic findings for the carpal tunnel syndrome. Scand J Work Environ Health 25 (2): 115–124.

Jablecki CK, Andary MT, Floeter MK et al. (2002). Practice parameter: Electrodiagnostic studies in carpal tunnel syndrome. Report of the American Association of Electrodiagnostic Medicine, American Academy of Neurology, and the American Academy of Physical Medicine and Rehabilitation. Neurology 58 (11): 1589–1592.

Jacobson JA, Fessell DP, Lobo Lda G et al. (2010). Entrapment neuropathies I: upper limb (carpal tunnel excluded). Semin Musculoskelet Radiol 14 (5): 473–486.

Kákosy T (1994). Tunnel syndromes of the upper extremities in workers using hand-operated vibrating tools. Med Lav 85 (6): 474–480.

Karataş GK, Göğüş F (2003). Suprascapular nerve entrapment in newsreel cameramen. Am J Phys Med Rehabil 82 (3): 192–196.

Keir PJ, Rempel DM (2005). Pathomechanics of peripheral nerve loading. Evidence in carpal tunnel syndrome. J Hand Ther 18 (2): 259–269.

Keir PJ, Bach JM, Rempel DM (1998). Fingertip loading and carpal tunnel pressure: differences between a pinching and a pressing task. J Orthop Res 16 (1): 112–115.

Korkmaz N, Ozçakar L (2004). Meralgia paresthetica in a policeman: the belt or the gun. Plast Reconstr Surg 114 (4): 1012–1013.

Krishnan AV, Fulham MJ, Kiernan MC (2009). Another cause of occupational entrapment neuropathy: la main du cuisinier (the chef's hand). J Clin Neurophysiol 26 (2): 129–131.

Lam SJS (1967). Tarsal tunnel syndrome. J Bone Joint Surg 49B: 87–92.

Laterza A, Nappo A (1977). Paralysis of the peroneal nerve in hazelnut gatherers. Acta Neurol (Napoli) 32 (5): 606–612.

Latko WA, Armstrong TJ, Franzblau A et al. (1999). Cross-sectional study of the relationship between repetitive work and the prevalence of upper limb musculoskeletal disorders. Am J Ind Med 36 (2): 248–259.

Laulan J, Fouquet B, Rodaix C et al. (2011). Thoracic outlet syndrome: definition, aetiological factors, diagnosis, management and occupational impact. J Occup Rehabil 21 (3): 366–373.

Lee YK, Kim YI, Choy WS (2006). Radial nerve compression between the brachialis and brachioradialis muscles in a manual worker: a case report. J Hand Surg Am 31 (5): 744–746.

Lee JT, Azari K, Jones NF (2008). Long term results of radial tunnel release – the effect of co-existing tennis elbow, multiple compression syndromes and workers' compensation. J Plast Reconstr Aesthet Surg 61 (9): 1095–1099.

Li ZM (2005). Gender difference in carpal tunnel compliance. J Musculoske Res 09: 153.

Luckhaupt SE, Dahlhamer JM, Ward BW et al. (2013). Prevalence and work-relatedness of carpal tunnel syndrome in the working population, United States, 2010 National Health Interview Survey. Am J Ind Med 56 (6): 615–624.

Lundborg G, Gelberman RH, Minteer-Convery M et al. (1982). Median nerve compression in the carpal tunnel – functional response to experimentally induced controlled pressure. J Hand Surg Am 7 (3): 252–259.

Massy-Westropp N, Grimmer K, Bain G et al. (2001). The effect of a standard activity on the size of the median nerve as determined by ultrasound visualization. J Hand Surg Am 26 (4): 649–654.

Mattioli S, Baldasseroni A, Curti S et al. (2008). Incidence rates of in-hospital carpal tunnel syndrome in the general population and possible associations with marital status. BMC Public Health 28 (8): 374, Oct.

Mattioli S, Baldasseroni A, Bovenzi M et al. (2009). Risk factors for operated carpal tunnel syndrome: a multicenterpopulation-based case-control study. BMC Public Health 9: 343.

Mattioli S, Zanardi F, Baldasseroni A et al. (2010). Search strings for the study of putative occupational determinants of disease. Occup Environ Med 67 (7): 436–443.

Mattioli S, Gori D, Di Gregori V et al. (2013). PubMed search strings for the study of agricultural workers' diseases. Am J Ind Med 56 (12): 1473–1481.

McGrail MA (1999). Common peroneal nerve palsy in a UH-60 aviator. Mil Med 164 (6): 446–447.

Mendel F, Gordin A, Herness D et al. (1987). Transient bilateral peroneal palsy in a painter's model. J Occup Med 29 (2): 109–111.

Miller RG (1979). The cubital tunnel syndrome: diagnosis and precise localization. Ann Neurol 6 (1): 56–59.

Mondelli M, Giannini F, Giacchi M (2002). Carpal tunnel syndrome incidence in a general population. Neurology 58 (2): 289–294.

Mondelli M, Giannini F, Ballerini M et al. (2005). Incidence of ulnar neuropathy at the elbow in the province of Siena (Italy). J Neurol Sci 234 (1–2): 5–10.

Mondelli M, Grippo A, Mariani M et al. (2006). Carpal tunnel syndrome and ulnar neuropathy at the elbow in floor cleaners. Neurophysiol Clin 36 (4): 245–253.

Moore JS, Garg A (1995). The Strain Index: a proposed method to analyze jobs for risk of distal upper extremity disorders. Am Ind Hyg Assoc J 56 (5): 443–458.

Morris HH, Peters BH (1976). Pronator syndrome: clinical and electrophysiological features in seven cases. J Neurol Neurosurg Psychiatry 39: 461–464.

Naam NH, Nemani S (2012). Radial tunnel syndrome. Orthop Clin North Am 43 (4): 529–536.

Nainzadeh NK, Ilizarov S, Piligian G et al. (2011). Ulnar neuropathy at the elbow in computer keyboard operators. Work 39 (2): 93–101.

Nathan PA, Istvan JA, Meadows KD (2005). A longitudinal study of predictors of research-defined carpal tunnel syndrome in industrial workers: findings at 17 years. J Hand Surg Br 30: 593–598.

National Institute of Occupational Safety and Health (1997). Musculoskeletal Disorders and workplace factors. Publication N° 97–141 1997. Available online at: www.cdc.gov/niosh/ergoscil.html.

National Research Council and the Institute of Medicine (2001). Musculoskeletal Disorders and the Workplace: Low Back and Upper Extremities. Panel on Musculoskeletal Disorders and the Workplace. Commission on Behavioral and Social Sciences and Education. National Academy Press, Washington, DC.

Neal S, Fields KB (2010). Peripheral nerve entrapment and injury in the upper extremity. Am Fam Physician 81 (2): 147–155.

Nishida T, Price SJ, Minieka MM (1993). Medial antebrachial cutaneous nerve conduction in true neurogenic thoracic outlet syndrome. Electromyogr Clin Neurophysiol 33 (5): 285–288.

O'Driscoll SW, Horii E, Carmichael SW et al. (1991). The cubital tunnel and ulnar neuropathy. J Bone Joint Surg Br 73 (4): 613–617.

Palmer KT (2011). Carpal tunnel syndrome: the role of occupational factors. Best Pract Res Clin Rheumatol 25 (1): 15–29.

Pascarelli EF, Hsu YP (2001). Understanding work-related upper extremity disorders: clinical findings in 485 computers users, musicians, and others. J Occup Rehabil 11: 1–21.

Pham K, Gupta R (2009). Understanding the mechanisms of entrapment neuropathies. Review article Neurosurg Focus 26 (2).

Piza-Katzer H, Pilz E (1997). Compression syndrome of the superficial fibular nerve. Case report. Handchir Mikrochir Plast Chir 29 (3): 124–126.

Quignon R, Marteau E, Penaud A et al. (2012). Posterior interosseous nerve palsy. A series of 18 cases and literature review. Chir Main 31 (1): 18–23.

Redmond MD, Rinver MH (1988). False positive electrodiagnostic tests in carpal tunnel syndrome. Muscle Nerve 11: 511–518.

Rempel DM, Diao E (2004). Entrapment neuropathies: pathophysiology and pathogenesis. J Electromyogr Kinesiol 14 (1): 71–75.

Rempel D, Keir PJ, Smutz WP, Hargens A (1997). Effects of static fingertip loading on carpal tunnel pressure. J Orthop Res 15 (3): 422–426.

Rempel D, Bach JM, Gordon L et al. (1998a). Effects of forearm pronation/supination on carpal tunnel pressure. J Hand Surg 23A: 38–42.

Rempel D, Evanoff B, Amadio PC et al. (1998b). Consensus criteria for the classification of carpal tunnel syndrome in epidemiologic studies. Am J Public Health 88 (10): 1447–1451.

Riffaud L, Morandi X, Godey B et al. (1999). Anatomic bases for the compression and neurolysis of the deep branch of the radial nerve in the radial tunnel. Surg Radiol Anat 21 (4): 229–233.

Ritts GD, Wood MB, Linscheid RL (1987). Radial tunnel syndrome. A ten-year surgical experience. Clin Orthop Relat Res (219): 201–205.

Rivera Redondo J, Sáenz Lope E (1988). Postural paralysis among workers at archeological digs. Neurologia 3 (2): 84.

Roh YH, Chung MS, Baek GH et al. (2010). Incidence of clinically diagnosed and surgically treated carpal tunnel syndrome in Korea. J Hand Surg Am 35 (9): 1410–1417.

Roles NC, Maudsley RH (1972). Radial tunnel syndrome: resistant tennis elbow as a nerve entrapment. J Bone Joint Surg Br 54 (3): 499–508.

Roos DB (1976). Congenital anomalies associated with thoracic outlet syndrome. Am J Surg 132: 771–778.

Roos DB (1980). Pathophysiology of congenital anomalies in thoracic outlet syndrome. Acta Chir Belg 79 (5): 353–361.

Roquelaure Y, Raimbeau G, Dano C et al. (2000). Occupational risk factors for radial tunnel syndrome in industrial workers. Scand J Work Environ Health 26 (6): 507–513.

Roquelaure Y, Ha C, Nicolas G et al. (2008). Attributable risk of carpal tunnel syndrome according to industry and occupation in a general population. Arthritis Rheum 59 (9): 1341–1348.

Rosenbaum R (1999). Disputed radial tunnel syndrome. Muscle Nerve 22 (7): 960–967.

Sallstrom J, Schmidt H (1984). Cervicobrachial disorders in certain occupations, with special reference to compression in the thoracic outlet. Am J Ind Med 6: 45–52.

Sandhu HS, Sandhey BS (1976). Occupational compression of the common peroneal nerve at the neck of the fibula. Aust N Z J Surg 46 (2): 160–163.

Sangwan SS, Marya KM, Kundu ZS et al. (2004). Compressive peroneal neuropathy during harvesting season in Indian farmers. Trop Doct 34 (4): 244–246.

Scharnbacher J, Claus M, Reichert J et al. (2013). Hypothenar hammer syndrome: a multicenter case-control study. Am J Ind Med 56 (11): 1352–1358.

Schröter C, Braune HJ, Huffmann G (1990). The so-called turnip-harvester palsy – a rare differential diagnosis today. Fortschr Neurol Psychiatr 58 (9): 351–353.

Seppalainen AM, Aho K, Uusitupa M (1977). Strawberry pickers' peroneal palsy. Electromyogr Clin Neurophysiol 17 (5): 375–384.

Seror P (2002). The medial antebrachial cutaneous nerve: antidromic and orthodromic conduction studies. Muscle Nerve 26 (3): 421–423.

Shapiro BE, Preston DC (2009). Entrapment and compressive neuropathies. Med Clin North Am 93 (2): 285–315.

Solmaz I, Cetinalp EN, Göçmez C et al. (2011). Management outcome of peroneal nerve injury at knee level: experience of a single military institution. Neurol Neurochir Pol 45 (5): 461–466.

Stål M, Hagert CG, Moritz U (1998). Upper extremity nerve involvement in Swedish female machine milkers. Am J Ind Med 33 (6): 551–559.

Stevens JC, Sun S, Beard CM et al. (1988). Carpal tunnel syndrome in Rochester, Minnesota, 1961 to 1980. Neurology 38 (1): 134–138.

Streib EW, Sun SF (1984). Distal ulnar neuropathy in meat packers. An occupational disease? J Occup Med 26 (11): 842–843.

Svendsen SW, Juhl AH (2008). Ulnar neuropathy in a poultry worker. Ugeskr Laeger 170 (40): 3131–3132.

Svendsen SW, Johnsen B, Fuglsang-Frederiksen A et al. (2012). Ulnar neuropathy and ulnar neuropathy-like symptoms in relation to biomechanical exposures assessed by a job exposure matrix: a triple case-referent study. Occup Environ Med 69 (11): 773–780.

Svendsen SW, Johnsen B, Fuglsang-Frederiksen A et al. (2013). Prognosis of ulnar neuropathy and ulnar neuropathy-like symptoms in relation to occupational biomechanical exposures and lifestyle. Scand J Work Environ Health 39 (5): 506–514.

Swanson KE, Bartholomew JR, Paulson R (2012). Hypothenar hammer syndrome: a case and brief review. Vasc Med 17 (2): 108–115.

Takala EP, Pehkonen I, Forsman M et al. (2010). Systematic evaluation of observational methods assessing biomechanical exposures at work. Scand J Work Environ Health 36 (1): 3–24.

Tanaka S, Wild DK, Seligman PJ et al. (1994). The US prevalence of self-reported carpal tunnel syndrome: 1988 National Health Interview Survey data. Am J Public Health 84 (11): 1846–1848.

Tapadia M, Mozaffar T, Gupta R (2010). Compressive neuropathies of the upper extremity: update on pathophysiology, classification, and electrodiagnostic findings. J Hand Surg Am 35 (4): 668–677.

Tolson TD (2004). "EMG" for thoracic outlet syndrome. Hand Clin 20 (1): 37–42.

Toosi KK, Impink BG, Baker NA et al. (2011). Effects of computer keyboarding on ultrasonographic measures of the median nerve. Am J Ind Med 54 (11): 826–833.

Tsao BE, Ferrante MA, Wilbourn AJ et al. (2014). Electrodiagnostic features of true neurogenic thoracic outlet syndrome. Muscle Nerve 49 (5): 724–727.

van Rijn RM, Huisstede BM, Koes BW et al. (2009a). Associations between work-related factors and the carpal tunnel syndrome – a systematic review. Scand J Work Environ Health 35 (1): 19–36.

van Rijn RM, Huisstede BM, Koes BW et al. (2009b). Associations between work-related factors and specific disorders at the elbow: a systematic literature review. Rheumatology (Oxford) 48 (5): 528–529.

van Rijn RM, Huisstede BM, Koes BW et al. (2010). Associations between work-related factors and specific disorders of the shoulder – a systematic review of the literature. Scand J Work Environ Health 36 (3): 185–187.

van Rossum J, Buruma OJ, Kamphuisen HA et al. (1978). Tennis elbow – a radial tunnel syndrome? J Bone Joint Surg Br 60-B (2): 197–198.

Vartija L, Cheung K, Kaur M et al. (2013). Ulnar hammer syndrome: a systematic review of the literature. Plast Reconstr Surg 132 (5): 1181–1191.

Violante FS, Bonfiglioli R, Isolani L et al. (2004). Levels of agreement of nerve conduction studies and symptoms in workers at risk of carpal tunnel syndrome. Int Arch Occup Environ Health 77 (8): 552–558.

Violante FS, Armstrong TJ, Fiorentini C et al. (2007). Carpal tunnel syndrome and manual work: a longitudinal study. J Occup Environ Med 49 (11): 1189–1196.

Waugh RP, Pellegrini Jr VD (2007). Ulnar tunnel syndrome. Hand Clin 23 (3): 301–310, v.

Werner RA, Andary M (2011). Electrodiagnostic evaluation of carpal tunnel syndrome. Muscle Nerve 44 (4): 597–607.

Wertsch JJ (1992). AAEM case report #25: anterior interosseous nerve syndrome. Muscle Nerve 15 (9): 977–983.

Wright TW, Glowczewskie Jr F, Wheeler D et al. (1996). Excursion and strain of the median nerve. J Bone Joint Surg Am 78: 1897–1903.

Wright TW, Glowczewskie Jr F, Cowin D et al. (2001). Ulnar nerve excursion and strain at the elbow and wrist associated with upper extremity motion. J Hand Surg Am 26 (4): 655–662.

Yoshii Y, Nishiura Y, Terui N et al. (2010). The effects of repetitive compression on nerve conduction and blood flow in the rabbit sciatic nerve. J Hand Surg Eur 35 (4): 269–278.

Handbook of Clinical Neurology, Vol. 131 (3rd series)
Occupational Neurology
M. Lotti and M.L. Bleecker, Editors

Chapter 23

Upper-extremity and neck disorders associated with keyboard and mouse use

STEFANO MATTIOLI*, FRANCESCO S. VIOLANTE, AND ROBERTA BONFIGLIOLI
Occupational Health Unit, Department of Medical and Surgical Sciences, University of Bologna, Bologna, Italy

INTRODUCTION

The use of personal computers in offices and factories, in the last two decades of the 20th century, determined among employees and scientists the perception of a risk due to occupational exposure to video display units, considered as a source of radiation (including ionizing radiation) and as a possible cause of eye damage. A great deal of research and years of studies have concluded that the use of computers is not a causal factor in serious eye diseases or cancers; nevertheless, the research underlined the possibility that computer users may suffer because of musculoskeletal disorders of the spine and of the upper limbs.

The aim of this chapter is to review the scientific literature regarding a possible association between the use of computers (and peripheral devices, such as keyboard or mouse) at work, and workers' neck or upper-extremity disorders.

METHODS

We decided to explore the literature, in a scoping review, searching on MEDLINE, through PubMed, to retrieve already published systematic reviews and informative, analytical studies (randomized controlled trials, cohort studies, and case-control studies) published later on. For each disorder/disease studies we describe the information obtained from systematic reviews and informative studies (not included in systematic reviews). Where we were unable to find at least one systematic review, we included narrative reviews and informative studies.

Different literature searches were carried out for individual disorders: cervical pain, shoulder pain, tennis elbow, ulnar neuropathy, carpal tunnel syndrome, de Quervain's syndrome, and trigger finger. Moreover a literature search was carried out with regard to unspecified upper-limb disorders.

Every search strategy contained: (1) the "more sensitive" search string proposed by Mattioli et al. (2010) to filter papers on occupational etiology of diseases; (2) a search string to filter papers on computers and their use; and (3) a search string to filter papers on the disorder studied.

The search strings were built with relevant PubMed Medical Subject Heading (MeSH) terms and relevant synonyms used as free terms.

RESULTS

Neck disorders

With regard to neck disorders, the search strategy produced 157 citations, of which 25 were relevant: four systematic reviews, 11 analytical studies (10 cohort studies and one case-control study) and 10 cross-sectional studies.

The systematic review by Paksaichol et al. (2012) focused on office work risk factors for nonspecific neck pain. The authors searched in the literature until 2011 and included in their review seven cohort studies, five of which were considered high-quality papers. The authors, analyzing the results of the papers included, concluded that only female gender and a previous history of neck complaints were considered predictors of neck pain.

The systematic review by Waersted et al. (2010) studied computer use and cervical pain and other upper-extremity disorders, with the exception of carpal tunnel syndrome. The review, updated at the beginning of 2010, has the declared limit of language restriction, since it included only papers in English. The authors included

*Correspondence to: Stefano Mattioli, Occupational Health Unit, University of Bologna Sant'Orsola Malpighi Hospital, via Palagi 9, 40138 Bologna, Italy. E-mail: s.mattioli@unibo.it

papers in which the association between the health problem (evaluated after a physical examination, and at times after imaging exams) and computer work was studied. The literature search was expanded to five different databases, but the strings used were not reported. On the other hand, the quality assessment was detailed and the strength of evidence – obtained by the studies – was classified into five categories, according to categories suggested by the Scientific Committee of the Danish Society of Occupational and Environmental Medicine (Jensen, 2008). On the basis of highly informative studies (as well as cross-sectional studies), the authors concluded that, concerning neck pain, there is limited evidence for a causal relationship with computer work in general and with hours of mouse use, but not with hours of keyboard use. The authors emphasized that there are indications about the relevance of the personal way the computer is used and of the (dis-)ergonomic layout of the workstation.

A systematic review by Côté et al. (2009), with data from 2007, published in 2009, was part of a larger project of systematic reviews exploring different aspects of cervical pain. In this particular case, the authors investigated the possible association of cervical pain with work exposure, considering only informative studies and their quality.

For the vast majority of cases, the authors retrieved papers about computer work. This gave them the opportunity to make distinctions between different aspects of computer work. They found evidence of an increase in risk of neck pain in the following cases: prolonged periods of sedentary work, repetitive or precision work, working with the cervical spine in flexion for prolonged periods, poor physical environment (lighting, temperature, air quality, room size, and acoustics), inadequate keyboard position (close to the desk edge), and use of telephone shoulder rests. They also found inconsistent evidence that mouse position, screen height, or arm rests may increase the risk of neck pain.

A systematic review by Ijmker et al. (2007) was based on nine informative studies. With regard to neck and shoulder disorders, the authors concluded that there was insufficient evidence of an association with duration of mouse or keyboard use, since inconsistent results were found. Moreover, the authors did not find an association between neck and shoulder disorders and duration of total computer use.

We also retrieved 11 analytical studies: six of them were not included in the four reviews mentioned above. They are briefly summarized below.

As a part of the international collaborative project Cultural and Psychosocial Influences on Disability (CUPID: Coggon et al., 2012), an Iranian cohort study by Sadeghian et al. (2013) explored the incidence and persistence of neck and shoulder pain at work using the standardized CUPID questionnaire. Analysis was based on the 218 nurses (89%) and 165 office workers (91%) who completed the follow-up. The authors concluded that incidence and persistence of neck and shoulder pain were higher in office workers than in nurses, which would be compatible with hazard from prolonged computer use.

The cohort study by Huysmans et al. (2012) was a prospective study conducted among almost 2000 office workers, with a 2-year follow-up, with the aim of studying the risk factors for neck and shoulder pain and for arm, wrist, and hand pain (Prospective Research On Musculoskeletal disorders in Office workers study: PROMO). With regard to neck and shoulder symptoms and computer use, they distinguished a significant, even if weak, increased risk for: duration of mouse use >4 hours/day, supporting the arms during keyboard use (most of the time, either on the desk or chair arm supports), and low task variation. Instead, unsurprisingly, they found that workers who had suffered disabling neck and shoulder symptoms 1 year before had a risk more than fivefold higher than individuals who recently suffered the same disorders.

A paper published by Richter et al. (2012) showed other results from the same PROMO cohort study. In this case, the authors aimed to relate periods of high-peak computer exposure to the occurrence of neck and upper-extremity symptoms: they defined high-peak exposure as days or weeks with extensive computer use (per day) or with high-input device use (mouse frequency $\geq$20 clicks per minute; keyboard frequency $\geq$160 strokes per minute). The results showed that there was no indication that periods of high-peak exposure to computer usage were related to neck and shoulder symptoms, nor to arm, wrist, and hand symptoms.

The paper by Eltayeb et al. (2009) presented the results of another cohort study, conducted among 264 Dutch computer users with a 2-year follow-up. The authors' aim was to study if body posture and workplace design were associated with the occurrence of symptoms in the neck, shoulder, and forearms or hands. Moreover, they also studied some psychosocial aspects of computer work, with the same purpose.

The results showed a weak, even if statistically significant, association between neck complaints and "irregular" head and body posture, task difficulty, and number of working hours with the computer. In contrast, a previous history of neck complaints was the main risk factor, with an adjusted odds ratio of 7.2.

In 2009 a cohort study conducted in Sweden by Tornqvist et al. was published. This study involved, for 10 months, 1283 workers and explored possible associations between some aspects of computer use and the incidence of neck disorders. The results of the multivariate analysis showed a risk ratio that was modestly increased for high job strain and a weak increase in risk ratio where the computer work environment was not very comfortable.

Finally, a cohort study by Grimby-Ekman et al. (2009) was carried out on university students. These authors were able to show, using multivariate analysis, that at least two 4-hour periods of computer use, without a break, showed an increase in risk for neck pain.

In conclusion, with regard to neck pain, the evidence of an association with hours of computer or mouse use is limited and inconsistent, while there is limited and inconsistent evidence of no association with keyboard use.

Shoulder disorders

With regard to shoulder disorders, our search strategy produced 342 citations, of which 26 were relevant: two systematic reviews, one narrative review, eight cohort studies, and 15 cross-sectional studies.

The systematic review by Waersted et al. (2010) also reported information on shoulder disorders and use of computers in the workplace. The authors found insufficient evidence for a causal relationship between shoulder complaints and not only computer work per se, but also mouse and keyboard time use per day.

The results of the systematic review by Ijmker et al. (2007) have already been cited in the neck disorders section, above: there was insufficient evidence for an association between neck and shoulder disorders and duration of mouse and keyboard use, and there was no association between neck and shoulder disorders and duration of total computer use.

We also retrieved eight cohort studies: four were published after the two systematic reviews had been updated, and they are briefly summarized below.

The cohort study by Huysmans et al. (2012), described above in the neck disorders section, explored neck and shoulder symptoms and computer use. The authors reported a weak increased risk for duration of mouse use >4 hours/day, for supporting the arms during keyboard use (most of the time, either on the desk or on chair arm supports), and for low task variation. Moreover, workers who had suffered disabling neck and shoulder symptoms 1 year before showed a fivefold higher risk for new episodes of these complaints.

The results of the cohort study by Richter et al. (2012), already cited above in the neck disorders section, showed that there was no indication that periods of high-peak exposure in computer usage were related to neck and shoulder symptoms.

The paper by Eltayeb et al. (2009), described in the neck disorders section above, is a cohort study whose results, with regard to shoulder complaints and computer use, are similar to those obtained concerning neck disorders: weak associations with "irregular" head and body posture and average number of hours working with the computer per day. Also in this case, the main risk factor was a previous history of shoulder complaints (odds ratio (OR) 19.2; 95% confidence interval (CI) 9.8–28.6).

The cohort study by Tornqvist et al. (2009), also described in the neck disorders section, showed, for shoulder symptoms, a significant association with uncomfortable computer work environment. Inconsistent results were obtained regarding the duration of mouse use (hours per day).

In conclusion, regarding shoulder pain, the evidence of an association with hours of computer use is limited and inconsistent, while there is limited evidence of association with hours of mouse or keyboard use.

Forearm, wrist, and hand disorders

From the results produced by the different search strategies, we were also able to retrieve studies concerning computer work and unspecified forearm, wrist, and hand disorders: one systematic review and six cohort studies.

The systematic review by Waersted et al. (2010), already cited, also reported information on forearm pain and computer use. The authors concluded that there was limited evidence for a causal effect of mouse time on forearm pain diagnoses, while there is insufficient evidence of a causal link between forearm pain and computer work per se or keyboard time.

Recently, Mikkelsen et al. (2012) published the results of a cohort study in which they examined the association between objectively recorded computer work activity data and the development of acute, prolonged, and chronic elbow, forearm, and wrist/hand pain, in a follow-up period of 1 year of a cohort of 9480 technical assistants and machine technicians. The authors recorded occupational exposure to duration of mouse and keyboard work, duration of sustained-activity periods, speed, and number of micropauses.

The results of this study showed that acute pain had a weak, even if significant, increase in all three regions with increasing mouse time use in the preceding week. The authors emphasized that there was no indication of an exposure–response threshold. At the same time, keyboard time had no significant effect on acute pain, while elbow pain decreased significantly with increasing number of keyboard micropauses. Finally, the results did not show any relation between computer work and prolonged pain or chronic pain. The authors concluded that, on the basis of objective computer activity recordings, their findings did not support any serious musculoskeletal health effect from computer work.

The cohort study by Huysmans et al. (2012), already described in the neck pain section, also explored the onset of arm, wrist, and hand symptoms. The authors found a small, weak increase in risk for aspects of computer use: often/always using the computer and the phone at the same time, inability to move the mouse freely on the desk due to space restrictions, at least 4 hours per day of (self-reported) computer use during leisure time, low task variation, high level of commitment, female gender, age (49–68 years). The strength of the associations was generally low, with the exception of previously disabling arm, wrist, and hand symptoms in the past year, identified as the most important risk factor for the onset of arm, wrist, and hand symptoms among office workers.

The results of the cohort study by Richter et al. (2012) have already been cited in the neck disorders section. They showed that, as in the case of neck and shoulder symptoms, there was no indication that periods of high-peak exposure to computer usage were related to arm, wrist, and hand symptoms.

The paper by Eltayeb et al. (2009), described in the neck disorders section, above, is a Dutch cohort study whose results, regarding shoulder complaints and computer use, are similar to those concerning neck disorders: weak associations with "irregular" head and body posture and average number of hours working with the computer per day. Also in this case, the main risk factor was a previous history of forearm/hand complaints. Similar results have been obtained by Eltayeb et al. (2011) from a cohort study executed in Sudan. Also in Sudan, as well as in The Netherlands, the higher risk of neck, shoulder, forearm, and hand pain for computer workers was related to a previous history of symptoms.

The cohort study by Tornqvist et al. (2009), described in the neck disorders section above, showed, as in the case of shoulder symptoms, a significant association between hand or arm pain and uncomfortable computer work environment. On the other hand, in contrast to neck and shoulder pain, in the case of the hand andarm there was a weak but significant association with duration of daily mouse use.

In conclusion, regarding forearm/wrist pain, the evidence of an association with hours of computer use or keyboard use is insufficient, while there is limited evidence of association with hours of mouse use.

Carpal tunnel syndrome

The search strategy regarding carpal tunnel syndrome produced 106 citations, of which 36 were relevant: four systematic reviews (one with meta-analysis), one scoping review, three narrative reviews, six analytical studies (two cohort studies and four case-control studies), one laboratory study, and 21 cross-sectional studies.

The recent systematic review and meta-analysis published by Mediouni et al. (2014) studied the possible association between carpal tunnel syndrome and computer work. The authors explored four databases, and included only recent studies (from 1992 to June 2012), without any language restriction. At the end of their search, they retrieved 18 papers, of which six were included in the review (two cohort studies and four cross-sectional studies). From these six studies, the authors obtained sufficient information to perform a meta-analysis that showed no association between carpal tunnel syndrome and computer use or mouse and keyboard use.

Some years before, Andersen et al. (2011) published an overview of existing reviews on computer use and carpal tunnel syndrome (and other upper-extremity musculoskeletal disorders). With regard to carpal tunnel syndrome, they included three reviews – the same reviews we also retrieved from our search (Palmer et al., 2007; Thomsen et al., 2008; van Rijn et al., 2009. Considering the results of these three reviews, Andersen and co-workers concluded that there was not sufficient epidemiologic evidence for an association between computer use and carpal tunnel syndrome. In contrast, Andersen et al. (2003) published a 1-year large follow-up study, and showed that the risk for carpal tunnel syndrome was increased by mouse use of more than 20 hours per week.

Seror and Seror (2012) published a case-control study which compared severe and nonsevere cases of carpal tunnel syndrome diagnosed after median nerve conduction studies. They classified the subjects into three categories: (1) nonworkers (retirees, housewives, unemployed people); (2) white-collar workers using computers (daily use of a computer, mouse, and keyboard for at least 1 hour); and (3) blue-collar workers (patients with jobs that required heavy physical hand workload but did not require computer use). On multivariate analysis, blue-collar or manual workers and nonworkers did not differ with regard to risk of severe carpal tunnel syndrome; instead, this risk was lower for white-collar workers using computers than for nonworkers (OR = 0.41; 95% CI 0.20–0.83) and blue-collar workers (OR = 0.43; 95% CI 0.21–0.71). These results show that computer use allows individuals to avoid performing more strenuous manual activities, such as the duties of blue-collar workers and nonworkers.

In conclusion, regarding carpal tunnel syndrome, there is sufficient evidence of no association with computer, mouse, or keyboard use.

Other diseases of elbow, forearm, wrist, and hand

Tennis elbow

With regard to tennis elbow, the search strategy produced 11 citations, of which two were relevant: one narrative review and one cross-sectional study.

The review by Pedersen and Jensen (1999) was conducted to study the possible occupational origin of tennis elbow and concluded that there was no controlled study showing an association between the use of a computer (and of a mouse) at work and epicondylitis.

Ulnar neuropathy

With regard to ulnar neuropathy, the search strategy produced three citations, of which two were relevant: one case-control study and one case series.

The case-control study was performed in Denmark by Andersen et al. (2012). These authors examined some potential occupational risk factors of cases (546 patients with ulnar neuropathy, 633 patients with ulnar neuropathy-like symptoms) and three matched community referents per case. Especially in the ulnar neuropathy patients, the results showed a significant trend of progressive protection offered by length of computer use (OR 0.30 for >4 hours a day). Conversely, workers leaning on an elbow for more than 2 hours per workday were at double risk.

De Quervain's syndrome

With regard to de Quervain's syndrome, the search strategy produced three citations, of which two were relevant: one narrative review and one cross-sectional study.

The review conducted by Gigante et al. (2012) showed the same results that we obtained with our search: there are no informative papers regarding the putative association between computer use and de Quervain's syndrome.

Trigger finger

The search strategy produced, regarding trigger finger and other tendonitis, 40 citations, of which four were relevant: one case-control study and three cross-sectional studies.

The case-control study was performed in Spain in 2006 and included 250 operated cases of trigger finger (De la Parra-Márquez et al., 2008). The results showed that the disease can be more frequent among secretaries who probably use a computer at work, but also among housewives and seamstresses.

CONCLUDING REMARKS

The reviews included in this overview, as well as the more recent informative studies, provide limited/insufficient (and, in some cases, inconsistent) evidence indicating that neck, shoulder, and forearm complaints have an association with computer use. In their review of reviews, Andersen et al. (2011) included those reviews we retrieved and concluded that there is no evidence of an association between computer use and specific diseases or chronic pain development. For the authors there is some evidence of acute pain after intensive mouse use, but a causal relation is still uncertain. In their review they also emphasize that there is not sufficient evidence regarding the effectiveness of interventions carried out to prevent upper-limb disorders among computer workers. This may due to the fact that there is no serious effect on workers' health because of the use of computers and peripheral devices (Andersen et al., 2011).

It must be underlined that the vast majority of retrieved papers concerned subjective symptoms (pain or discomfort of neck, shoulder, upper limb) instead of properly objectifiable diseases, apart from the case of carpal tunnel syndrome: we were not able to find studies on, for example, cervical disc hernia or rotator cuff tears.

In the case of carpal tunnel syndrome, there is no evidence of an association between computer work and disease. Moreover, office work seems to provide, conversely, protection against median compression due to the biomechanical overload of many manual activities. The same increasing protective trend seems to be offered by hours of computer use against the development of ulnar neuropathy.

Furthermore, we were not able to find studies regarding tennis elbow and de Quervain's syndrome and their putative association with the use of computers at work.

Considered together, this means that no specific indication can be provided to limit neck, shoulder, and upper-limb disorders among computer workers. At the same time, the application of ergonomic principles appears to be the best strategy to limit job strain and discomfort, and to maintain office workers' well-being.

Finally, occupational physicians have to share their knowledge regarding this matter, because, as already emphasized (Scangas et al., 2008), some upper-limb disorders are still attributed, in parascientific environments, to the use of a keyboard and mouse.

APPENDIX 1 SEARCH STRINGS

#1

(occupational diseases [MH] OR occupational exposure [MH] OR occupational exposure* [TW] OR "occupational health" OR "occupational medicine" OR work-related OR working environment [TW] OR at work [TW] OR work environment [TW] OR occupations [MH] OR work [MH] OR workplace* [TW] OR workload OR occupation* OR worke* OR work place* [TW] OR work site* [TW] OR job* [TW] OR occupational groups [MH] OR employment OR worksite* OR industry)

#2

(video display terminal* OR computer terminal work* OR vdt operator* OR vdt work* OR computer use* OR computer operator* OR computer work* OR data entry operator* OR data entry clerk* OR keyboard clerk* OR office clerk* OR office work* OR secretar* OR keyboard operator* OR keyboard OR keyboard operating clerk* OR typist* OR stenographer* OR word processor operator* OR word-processor operator* OR word processing operator* OR receptionist* OR clerk* OR video display unit operator* OR video display unit work* OR vdu operator* OR vdu worker* OR "data entry" OR (mouse NOT (mice* OR mice [MH])))

#3

For cervical pain

(neck pain [MH] OR Neck/physiopathology [MH] OR Neck/pathology [MH] OR "radicular pain syndrome" OR "somatic pain syndrome" OR "cervical myofascial pain syndrome" OR "cervical pain syndrome" OR cervicalgi* OR neck disorder* OR neckach* OR neck ache* OR trapezius pain*)

For shoulder disorders

(shoulder pain [MH] OR rotator cuff [MH] OR shoulder impingement syndrome [MH] OR shoulder [MH] OR "shoulder pain syndrome" OR tendinitis of biceps tendon* OR rotator cuff tear* OR suprascapular nerve compression* OR infraspinatus OR supraspinatus OR subscapularis OR shoulder)

For tennis elbow

(tennis elbow [MH] OR elbow/physiopathology [MH] OR tennis elbow* or lateral epicondyl* OR lateral humeral epicondyl* OR humeral epicondyl* OR "lateral elbow tendinopathy" OR "lateral epicondylalgia")

For carpal tunnel syndrome

(carpal tunnel syndrom* OR carpal tunnel syndrome [MH])

For de Quervain's syndrome

(de quervain disease [MH] OR "de quervains disease" OR "de quervain's disease" OR "de quervain disease" OR "blackberry thumb" OR "gamer's thumb" OR "washerwoman's sprain" OR "radial styloid tenosynovitis" OR "de quervain's tenosynovitis" OR "de quervains tenosynovitis" OR "de quervain tenosynovitis" OR "de Quervain's stenosing tenosynovitis" OR "mother's wrist" OR "mommy thumb" OR "de quervain syndrome" OR de quervain*)

For tendonitis/trigger finger

(trigger finger disorder [MH] OR trigger finger* OR "trigger digit" OR "stenosing tenosynovit*" OR "trigger thumb" OR "trigger digit" OR trigger finger disorder* OR tenosynovitis OR tendinopathy [MH] OR tendin* OR tendo*)

For upper extremity in general

(upper limb* OR upper extremit* OR Upper Extremity/physiopathology [MH] OR WRUL* OR UEMSD* OR (arm AND (symptom* OR pain OR disorder)))

The search strategies were composed as follows: #1 AND #2 AND #3

References

Andersen JH, Thomsen JF, Overgaard E et al. (2003). Computer use and carpal tunnel syndrome: a 1-year follow-up study. JAMA 289: 2963–2969.

Andersen JH, Fallentin N, Thomsen JF et al. (2011). Risk factors for neck and upper extremity disorders among computers users and the effect of interventions: an overview of systematic reviews. PLoS One 6, e19691.

Andersen JH, Frost P, Fuglsang-Frederiksen A et al. (2012). Computer use and ulnar neuropathy: results from a case-referent study. Work 41 (Suppl 1): 2434–2437.

Coggon D, Ntani G, Palmer KT et al. (2012). The CUPID (Cultural and Psychosocial Influences on Disability) study: methods of data collection and characteristics of study sample. PLoS One 7, e39820.

Côté P, van der Velde G, Cassidy JD et al. (2009). The burden and determinants of neck pain in workers: results of the Bone and Joint Decade 2000–2010 Task Force on Neck Pain and Its Associated Disorders. J Manipulative Physiol Ther 32 (Suppl 2): S70–86.

De la Parra-Márquez ML, Tamez-Cavazos R, Zertuche-Cedillo L et al. (2008). Factores de riesgo asociados a tenosinovitis estenosante. Estudio de casos y controles [Risk factors associated with trigger finger. Case-control study]. Cir Cir 76: 323–327.

Eltayeb S, Staal JB, Hassan A et al. (2009). Work related risk factors for neck, shoulder and arms complaints: a cohort study among Dutch computer office workers. J Occup Rehabil 19: 315–322.

Eltayeb SM, Staal JB, Khamis AH et al. (2011). Symptoms of neck, shoulder, forearms, and hands: a cohort study among computer office workers in Sudan. Clin J Pain 27: 275–281.

Gigante MR, Martinotti I, Cirla PE (2012). Utilizzo di videoterminale e sindrome di De Quervain: approccio basato sull'evidenza [Computer work and De Quervain's tenosynovitis: an evidence based approach]. G Ital Med Lav Ergon 34 (Suppl 3): 116–118.

Grimby-Ekman A, Andersson EM, Hagberg M (2009). Analyzing musculoskeletal neck pain, measured as present pain and periods of pain, with three different regression models: a cohort study. BMC Musculoskelet Disord 10: 73.

Huysmans MA, Ijmker S, Blatter BM et al. (2012). The relative contribution of work exposure, leisure time exposure, and individual characteristics in the onset of arm-wrist-hand and neck-shoulder symptoms among office workers. Int Arch Occup Environ Health 85: 651–666.

Ijmker S, Huysmans MA, Blatter BM et al. (2007). Should office workers spend fewer hours at their computer? A systematic review of the literature. Occup Environ Med 64: 211–222.

Jensen LK (2008). Knee osteoarthritis: influence of work involving heavy lifting, kneeling, climbing stairs or ladders, or kneeling/squatting combined with heavy lifting. Occup Environ Med 65: 72–89.

Mattioli S, Zanardi F, Baldasseroni A et al. (2010). Search strings for the study of putative occupational determinants of disease. Occup Environ Med 67: 436–443.

Mediouni Z, de Roquemaurel A, Dumontier C et al. (2014). Is carpal tunnel syndrome related to computer exposure at work? A review and meta-analysis. J Occup Environ Med 56: 204–208.

Mikkelsen S, Lassen CF, Vilstrup I et al. (2012). Does computer use affect the incidence of distal arm pain? A one-year prospective study using objective measures of computer use. Int Arch Occup Environ Health 85: 139–152.

Paksaichol A, Janwantanakul P, Purepong N et al. (2012). Office workers' risk factors for the development of non-specific neck pain: a systematic review of prospective cohort studies. Occup Environ Med 69: 610–618.

Palmer KT, Harris EC, Coggon D (2007). Carpal tunnel syndrome and its relation to occupation: a systematic literature review. Occup Med (Lond) 57: 57–66.

Pedersen LK, Jensen LK (1999). Sammenhaeng mellem erhverv og albuesmerter, epicondylitis. [Relationship between occupation and elbow pain, epicondylitis]. Ugeskr Laeger 161: 4751–4755.

Richter JM, van den Heuvel SG, Huysmans MA et al. (2012). Is peak exposure to computer use a risk factor for neck and upper-extremity symptoms? Scand J Work Environ Health 38: 155–162.

Sadeghian F, Raei M, Ntani G et al. (2013). Predictors of incident and persistent neck/shoulder pain in Iranian workers: a cohort study. PLoS One 8, e57544.

Scangas G, Lozano-Calderón S, Ring D (2008). Disparity between popular (Internet) and scientific illness concepts of carpal tunnel syndrome causation. J Hand Surg Am 33: 1076–1080.

Seror P, Seror R (2012). Hand workload, computer use and risk of severe median nerve lesions at the wrist. Rheumatology (Oxford) 51: 362–367.

Thomsen JF, Gerr F, Atroshi I (2008). Carpal tunnel syndrome and the use of computer mouse and keyboard: a systematic review. BMC Musculoskelet Disord 9: 134.

Tornqvist EW, Hagberg M, Hagman M et al. (2009). The influence of working conditions and individual factors on the incidence of neck and upper limb symptoms among professional computer users. Int Arch Occup Environ Health 82: 689–702.

van Rijn RM, Huisstede BM, Koes BW et al. (2009). Associations between work-related factors and the carpal tunnel syndrome – a systematic review. Scand J Work Environ Health 35: 19–36.

Waersted M, Hanvold TN, Veiersted KB (2010). Computer work and musculoskeletal disorders of the neck and upper extremity: a systematic review. BMC Musculoskelet Disord 11: 79.

Section 7

Other neurologic disorders and work restriction/accommodation

Handbook of Clinical Neurology, Vol. 131 (3rd series)
Occupational Neurology
M. Lotti and M.L. Bleecker, Editors

Chapter 24

Sleep deprivation due to shift work

GIOVANNI COSTA*

Department of Clinical Sciences and Community Health, University of Milan and IRCCS "Ca' Granda Ospedale Maggiore Policlinico" Foundation, Milan, Italy

INTRODUCTION

Shift work, night work, irregular and flexible working hours, together with new technologies, are the milestones of the transition to the "24-hour society." The borders between working and social times are no longer fixed and rigidly determined by the normal diurnal working day: working hours are extended to evening and night hours, and to weekend days, and hours of duty have become more and more variable. Shift work includes any arrangement of daily working hours differing from standard day work: it is aimed at extending the organization operational time from 8 hours up to 24 hours per day, by a succession of different teams of work.

It enables round-the-clock activities not only in relation to rigid technologic conditions (e.g., chemical and steel industry, power plants) and necessary social services (e.g., hospitals, transport, electricity, telecommunications), but also to support productive and economic choices (e.g., textile, paper, food, mechanical industry, banking), as well as a wider use of leisure time (e.g., entertainment).

In Europe, only 27% of employed and 8% of self-employed people are working the classic 5-day working week (i.e., 7–8 a.m. to 5–6 p.m., Monday to Friday) (Costa et al., 2004): overall, 21.9% of men and 10.7% of women work on shifts, including night work. In the USA, estimates suggest that 16.7% of men and 12.4% of women who are full-time salaried usually work on shifts, including nights. According to the USA 2012 Bedroom Poll (National Sleep Foundation, 2013), 59% of respondents slept 6 hours 44 minutes on work days, on average, as compared to 7 hours 35 minutes on nonwork days. Less sleep than needed was reported by 37% on work days or week days and by 17% on nonwork days or weekends; 47% slept less than 7 hours, and 14% less than 6 hours per day: among these, 27% complained of sleep disorders, 26% of daytime sleepiness, and 32% used sleeping pills regularly.

Shift and night work are well recognized risk factors for health and well-being as they interfere with four main spheres of human life, namely: (1) basic biologic functions: due to disturbance of psychophysiologic circadian rhythms, beginning with the sleep/wake cycle; (2) performance and work ability: due to fluctuations in work performance and efficiency over the 24-hour span, with consequent errors and accidents; (3) social relations: dealing with difficulties in maintaining usual relationships both at family and social levels, with negative influences on marital relations, care of children, and social contacts; and (4) health: due to deterioration of psychophysical conditions, that can be manifested in the short term by sleeping troubles, digestive disturbances, anxiety, irritability, and perturbation of women's hormonal and reproductive function. In the long term, this may result in more severe disorders mostly of gastrointestinal, neuropsychologic, or cardiovascular function and, probably, cancer.

CHRONOBIOLOGIC ASPECTS

Humans are daylight creatures; in the course of their evolutionary adaptation, they have associated wakefulness and activity with the day/light period, and sleep and rest with the night/dark period. Although in modern society artificial lighting makes it possible to have light for the whole 24-hour span, body functions (e.g., hormonal, metabolic, digestive, cardiovascular, mental) are still mainly influenced by the natural light/dark cycle, showing periodic oscillations that have, in general, peaks (acrophases) during the daytime and troughs (nadirs) at night.

*Correspondence to: Prof. Giovanni Costa, Dipartimento di Scienze Cliniche e di Comunità, Clinica del Lavoro, Via S. Barnaba, 8, 20122 Milano, Italy. Tel: +39-02-50320151, Fax: +39-02-55035304, E-mail: giovanni.costa@unimi.it

The perturbation of the sleep/wake cycle, associated with the modified activity/rest pattern, is a significant stress for the endogenous regulation of the "circadian" (about 24 hours) rhythms of biologic functions, which are driven by the biologic master clock located in the suprachiasmatic nuclei (SCN) of the encephalon. They are synchronized by environmental cues, the light/dark cycle in particular, through nonvision-related photic stimuli from melanopsin-containing retinal ganglion cells, via the retinohypothalamic tract. The SCN, via the superior cervical ganglion, regulates melatonin production by the pineal gland: secretion is inhibited by light exposure, while it increases during the dark period, indicating biologic night (Reppert and Weaver, 2002; Roenneberg et al., 2007).

With shift work, including night work in particular, the mismatch of the normal oscillation of biologic functions is documented by a flattening of amplitude and a shift of acrophase of circadian rhythms, which can be more or less pronounced according to the number of successive night shifts worked, as well as to the clockwise (i.e., morning–afternoon–night) or counterclockwise (i.e., afternoon–morning–night) rotation of the shift periods.

Workers involved in rotating shift work (the large majority) are subjected to a continuous stress to adjust as quickly as possible to the variable duty periods, which is partial and invariably frustrated by the continuous changeovers. In contrast, permanent night workers may adjust almost completely, provided that they keep maintaining their inverted sleep/wake cycle also on their days off (Folkard, 2008). As a consequence, a desynchronization is observed among the partial adjustments of the different functions, that is also due to the "masking" influence of the rest/activity pattern. It has more impact on functions having a prevalent exogenous component (e.g., heart rate, catecholamines) than those with a strong endogenous control (e.g., body temperature, cortisol). Hence, workers suffer to a greater or a lesser extent from a syndrome similar to jet lag, characterized by feeling of fatigue, sleepiness, insomnia, digestive troubles, irritability, impaired mental agility, and reduced performance efficiency (Comperatore and Krueger, 1990).

SLEEP DEPRIVATION AND PSYCHONEUROTIC TROUBLES

Sleep loss is the main complaint of shift and night workers since they have to change sleep times and strategies according to the duty periods; consequently, both sleep length and quality can be considerably affected according to the variable start and finish times of the different shifts. After a night shift, people have to sleep during the normal rising phase of biologic rhythms, which sustains wakefulness. This makes it difficult to fall asleep and sleep longer, because the environmental conditions (lighting and noise in particular) are often unfavorable. Consequently, sleep is reduced by 2–4 hours, more frequently or prematurely interrupted, and has poorer nonrapid-eye-movement (NREM) stage 2 and rapid-eye-movement (REM) sleep. Also in the early-morning shift (starting at 6 a.m. or earlier), sleep can be notably reduced (particularly REM sleep) and disturbed due to the advanced waking time, not usually compensated for by a corresponding advancing of bedtime the night before, due to family constraints and social habits: this causes excessive daytime sleepiness during the waking period (Åkerstedt, 2003; Åkerstedt et al., 2010).

According to several laboratory and field studies on sleep deprivation, having less than 5 hours' sleep in the preceding 24 hours, as well as less than 12 hours' sleep in the preceding 48 hours, increases the risk of significant performance impairment (Dawson and McCulloch, 2005).

A comparative analysis of more than 18 000 shift workers from 11 countries showed that sleep troubles were reported by 10–30% of day workers, 5–30% of shift workers without night shifts, 10–95% of shift workers with night shifts, and 35–55% of permanent night workers, with 15% of former shift workers (Knauth, 1983).

Also in the case of jet lag due to long transmeridian flights, people have to cope with a shift of waking/working hours in a changed environmental context, where a time shift has occurred as well. These conflicts are similar to those of current shift work, but are often aggravated by fatigue and sleep deficit due to extended duty periods, as well as by loss of the usual external time cues. On account of circadian desynchronization, and in association with sleep deprivation and increased daytime sleepiness and fatigue, performance on many psychomotor activities (e.g., reaction time, hand–eye coordination, logical reasoning, vigilance) also shows an acute 8–10% decrement that may last for 3–5 days (in the case of six or more time zones crossed). The speed of readjustment may differ among individuals (i.e., age) and variables ("internal dissociation"), being generally more rapid in westbound (about 60 minutes per day) than eastbound (about 90 minutes per day) flights, due to a progressive phase delay of circadian rhythms in the former case, whereas in the latter there is a phase advance due to the compressed day (directional asymmetry). Also sleep times and patterns appear to be much more variable and disrupted after eastward than westward flights across an equivalent number of time zones, in particular as concerns a reduction of REM sleep and increase of slow-wave sleep phases (Gundel and Wegmann, 1987; Suvanto et al., 1990; Minors et al., 1994).

According to the International Classification of Sleep Disorders, version 2 (ICSD-2), the "shift work sleep disorder" (307.45-1) "consists of symptoms of insomnia or excessive sleepiness that occur as transient phenomena in relation to work schedules" (American Academy of Sleep Medicinc, 2005). The disorder may be diagnosed by history and better qualified by polysomnography and the Multiple Sleep Latency Test (MSLT), which can also help in the differential diagnosis with other medical or mental sleep disorders, in particular with narcolepsy, sleep apnea syndrome, primitive circadian rhythm sleep disorders, REM and non-REM parasomnias, traumatic brain syndromes, chronic anxiety and depression, epilepsy, migraine, chronic fatigue, posttraumatic stress disorder, burnout, or drug- and alcohol-dependency sleep disorders. It may by defined as "acute" (less than 7 days), "subacute" (less than 3 months), or "chronic" (more than 3 months).

About 10% of night and rotating shift workers, aged between 18 and 65 years, have been estimated to have a diagnosable "shift work sleep disorder" (Drake et al., 2004).

The key point of the clinical evaluation concerns the ability to differentiate tolerable and transitory troubles (compatible with short-lived perturbation of the sleep/wake cycle) from more severe or pathologic disorders, asking for prompt interventions at work (i.e., transfer to day work) and at a personal level (therapy, rehabilitation).

For the latter, the occupational health physician needs the constant help and support of sleep experts for a careful diagnostic process, considering all possible intervening and confounding factors, the most appropriate therapy, and the forensic implications connected with the diagnosis of shift work sleep disorder as an occupational or job-related disease.

In the long run, frequent sleep perturbations and deprivations may cause persistent and severe disturbances of sleep itself, chronic fatigue, and psychoneurotic syndromes (such as chronic anxiety or depression), often needing treatment by psychotropic drugs (sedatives and hypnotics). As for seasonal affective disorders, some depressive states reported by shift workers could be associated with perturbed circadian rhythms and lower exposure to bright light (Cole et al., 1990).

These symptoms may act as a risk or aggravating factor for other psychosomatic complaints or diseases, including gastrointestinal and cardiovascular health. It has been also evidenced that insomnia was, among 37 cofactors, the most predictable factor for absenteeism at work (Leigh, 1991).

SLEEPINESS, ERRORS, AND ACCIDENTS

Alertness can be substantially reduced by irregular rest/activity patterns. In general, performance efficiency appears to parallel the circadian rhythm in body temperature, but it can peak earlier or later in the day depending on task demands (e.g., physical, cognitive, or memory-loaded), on the length of time that has elapsed since the individual's last proper sleep, and on the person's general level of arousal and motivation (Cajochen et al., 2014).

Hence, both homeostatic (time elapsed since prior sleep termination) and circadian (sleep/wake cycle) components interact in determining the extent of the reduction in alertness and psychophysical performance over the waking day, and even more so at night.

This may be further aggravated by features of the work schedule, and in particular the number of night shifts in succession, the start time of morning shifts, and shorter rest periods between shifts (i.e., "quick returns" or double shift in the same day) (Pallesen et al., 2007; Vittasalo et al., 2008; Ohayon et al., 2010).

The "postlunch dip" is only partially dependent on the meal itself, but it also reflects 12-hour or shorter, ultradian rhythms in alertness and wakefulness superimposed on the 24-hour circadian cycle (Rutenfranz and Colquhoun, 1979).

In addition, excessive daytime sleepiness, due to the truncation of sleep by an early start of the morning shift, has been proved to cause a higher frequency of errors and accidents during the day, while increased sleepiness and changes in electroencephalogram ultradian rhythms (bursts of alpha and theta power density) have been recorded on the night shift, indicating a high propensity for the workers concerned to fall asleep "on the job" (Åkerstedt, 1996).

Sleepiness, sleep disturbances, chronic fatigue, and oscillatory fluctuations of alertness and vigilance associated with shift and night work are key factors in causing human errors, and consequent work accidents and injuries, by interacting with organizational factors, such as environmental conditions, job content, workload, and time pressure. Some well-controlled studies, both in industry and transport sectors, show a common trend of accident risk, which appears to parallel the mean trend in sleep propensity over the 24-hour day. It is higher in the central hours of the night (0200–0400 hours), it shows a second minor peak in the early afternoon (1400–1600 hours) corresponding to the "postlunch dip," and it is lower in the late morning (1000–1200 hours) and late afternoon (1800–2000 hours) (Folkard, 1997).

Although it is worth noting that working conditions and risk may change between day and night in relation to fluctuation of work pacing, number of workers on duty, type of task, and supervision, some epidemiologic studies, estimating the relative risk of industrial accidents in the 3 × 8-hour shift systems, under comparable working conditions and after controlling for other potential confounders, showed a 18% increased risk in the afternoon shift, and 30% in the night shift, as compared with the morning shift. Moreover, other studies reported that the risk increases over successive shifts, being about 6% higher in the second night, 17% higher in the third night, and 36% higher in the fourth night (Folkard and Tucker, 2003).

Also the length of hours on duty is a key factor for fatigue-related industrial accidents, as reported by studies which examined trends of national accident statistics in some European countries. According to these studies it is possible to estimate a double risk of accident when working in 12-hour shifts as compared with 8-hour shifts (Mitchell and Williamson, 2000; Folkard and Tucker, 2003). Also, a recent survey of more than 75 000 US workers over a 4-year period confirmed a higher risk of injury strictly related to a progressive increase of working hours and reduction of sleep duration (Lombardi et al., 2010).

Beside industry, the transport sector is particularly sensitive to such problems. Many studies carried out in the past decades have evidenced how drowsiness and fatigue are key risk factors in road and railway accidents of professional drivers (Philip and Åkerstedt, 2006).

Accidents, caused by driver fatigue or lapses of attention due to sleep deprivation, are often quite severe as the drowsy and/or fatigued driver may not take the appropriate actions to avoid an accident. It was also reported that sleepy drivers often do not perceive their risky condition, and frequently drive with closed eyes for 5–50-second periods (microsleeps) (Kecklund and Åkerstedt, 1993; Horne and Reyer, 1995; Sagberg, 1999; Herrmann et al., 2010).

It was found that being sleepy while driving increases by two- to eightfold the risk of a serious injury crash (Connor et al., 2002; Howard et al., 2004). Moreover, several studies showed that "single-vehicle" accidents in roads have the greatest probability of occurring at night or in the early morning, also once traffic density has been controlled (Pack et al., 1995; Åkerstedt and Kecklung, 2001; Åkerstedt et al., 2001). In 1999, the US National Transportation Safety Board estimated that 52% of single-vehicle accidents involving heavy trucks were fatigue-related, and 17.6% of drivers admitted to having fallen asleep.

Similar findings come from studies on train drivers (Ingre et al., 2000; Lamond et al., 2005) and seafarers (Philips, 2000). In a Finnish study on train drivers and railway traffic controllers (Härmä et al., 2002), the risk of severe sleepiness was 6–14 times higher in the night shift and about twice as high in the morning shift as compared with the day shift.

Also in the healthcare sector several studies have reported a higher injury rate in evening and night shifts as compared to day shifts (Parks et al., 2000; Horwitz and McCall, 2004; Ayas et al., 2006). As to clinical risk, most studies on nurses and physicians documented that the primary predictor of error at work was sleepiness with relation to prolonged duty periods, shift and night work (Gaba and Howard, 2002; Landrigan et al., 2004; Scott et al., 2006; Tanaka et al., 2010).

Hence there are severe consequences not only for workers' health, but also for production/service quality and quantity, company and social costs (e.g., for damages and insurance), as well as higher risk of injuries and damages to the general population. It has been estimated that 52% of work accidents in the USA are potentially related to sleepiness, with an associated cost of $24.7 billion (Léger, 1994). According to Moore-Ede (1993), the total costs of human failure in the 24-hour society in the year 1991 were over 77 billion dollars in the USA and over 377 billion dollars in the world, due to work catastrophes and accidents, decreased productivity and quality, increased healthcare for sleep and chronic illnesses, as well as negative consequences for family life and public security.

It is clear that careful attention has to be paid to the potential accident risk when considering work schedules and other organizational factors. Indeed, it is noteworthy that the nuclear incidents at Three Mile Island and Chernobyl, the Bhopal disaster, and many air accidents, including the Challenger space shuttle, all occurred at night and, in each case, shift scheduling and fatigue due to sustained operation were cited as important contributory factors (Price and Holley, 1990). In recent years several biomathematical models have been developed, aimed at predicting times of reduced alertness and performance impairment, due to cumulative effects of "time of day" and "time on duty" period, as well as establishing times that are more suitable for restful recovery sleep and napping, and developing ergonomic shift schedules that are both safe and productive (Mallis et al., 2004; Van Dongen, 2004; Dawson et al., 2011).

OTHER EFFECTS ON HEALTH

An increasing amount of epidemiologic studies carried out in the latest decades show that shift and night work may be a significant risk factor for several health disorders, with consequent high economic and social costs for both the individual and the society.

Beside the above-mentioned sleep and psychoneurologic disorders, they relate in particular to gastrointestinal, metabolic, and cardiovascular diseases, cancer, toxicologic risk, women's reproductive function, and social life.

After sleep, digestive troubles are the most frequently reported by shift workers (20–75% vs 10–25% of day workers). According to epidemiologic studies, shift and night work increases (two to five times) the risk for gastrointestinal diseases, such as gastroduodenitis, peptic ulcer, irritable-bowel syndrome, as well as for metabolic disorders (dyslipidemic syndrome, type 2 diabetes) (Ha and Park, 2005; Knutsson and Boggild, 2010).

Ischemic heart disease and hypertension are reported to be 1.3–3 times higher in shift workers according to age groups. This association is partially due to the combination of the stress connected with inverted sleep/wake cycle and related circadian disruption with disturbed cardiac autonomic control, sleep deprivation, work/family conflicts, and lifestyle changes (e.g., smoking and obesity are more prevalent among shift workers) (Frost et al., 2009; Puttonen et al., 2010).

The International Agency on Research on Cancer has recently classified "shift work that involves circadian disruption" as "probably carcinogenic to humans" (group 2A), on the basis of "limited evidence in humans for the carcinogenicity of shift-work that involves night work," and "sufficient evidence in experimental animals for the carcinogenicity of light during the daily dark period (biological night)." This has been related to breast cancer in women in particular, but there are some other sporadic indications for cancer of the endometrium, prostate, colon, rectum, and non-Hodgkin lymphoma. The three plausible mechanisms are: (1) the multilevel endocrine changes caused by circadian disruption with melatonin suppression through light by night; (2) the epigenetic changes of clock genes; and (3) chronic sleep deprivation with consequent impairment of immunologic surveillance (Straif et al., 2007; Costa et al., 2010). The relevance of a robust and stable phase locking between the central clock and the peripheral cell cycles is becoming more and more evident according to recent studies, pointing out the significant implications for the pathogenesis of many diseases, including cancer, of a disordered circadian function (Feillet et al., 2014).

Shift workers may be particularly susceptible to xenobiotics due to fluctuations in the circadian metabolism and excretion of toxic substances, as well as variations in susceptibility after changes to the light/dark regimen. Thus the balance between "biokinetics" of a substance and "chronoesthesia" of the biologic system is more likely to be unfavorable at night when metabolic function is slowing down, as was dramatically demonstrated by the Bhopal disaster (Smolenski and Reinberg, 1990).

A higher incidence of altered menstrual cycle, premenstrual syndrome, and menstrual pains has been reported in many groups of female shift workers, such as nurses, air crews, and blue-collar workers. Some studies also reported a higher incidence of miscarriage and impaired fetal development, including preterm birth and low birth weight (Nurminen, 1998; Zhu et al., 2003).

At a social level, people engaged in irregular working hours are frequently out of phase with society and may encounter greater difficulties in their social relationships. This can lead to some social "marginalization" due to the interference between the workers' time budgets (working hours, commuting and leisure times) and the complex organization of social activities, particularly when these refer to groups of persons and require regular contact. Irregular working hours impose heavier burdens on the organization of family activities and timetables that can be more or less demanding according to family composition (e.g., number and age of children, cohabiting persons), personal duties (e.g., school, housework), availability of social support services (e.g., shop hours and transport). "Stress," "time pressure," and related complaints are constant conditions among those who have a high family burden, and this may have negative consequences on partners' relationships, parental roles, and children's education (Pisarki et al., 2008; Shechter et al., 2008).

The scientific literature offers a quite extensive mass of knowledge on all these aspects, but it also emphasizes the difficulty of unanimous agreement, as the problem is multifaceted and changing quite rapidly, in particular as concerns working time organization and workload, on the one hand, and individual psychophysical and social conditions on the other.

Consequently, the outcomes can vary greatly according to different working and living conditions of the groups and individuals examined, as well as the various approaches and methods used. Moreover, changes in historic and epidemiologic relevance of the disorders, modifications of health perception, extension of medical surveillance, as well as combined effects with other individual and social risk factors, make the problem quite complex and sometimes difficult to interpret and manage properly.

HEALTH SURVEILLANCE AND ASSESSMENT OF FITNESS TO WORK

There are many pathologic conditions, either directly associated with shift and night work, as seen above, or independent from it, that may be potential contraindications for shift and/or night work. They must be carefully evaluated both in terms of severity and possibility of appropriate therapy, in the process of assessment of fitness to work, with or without limitations and/or prescriptions, on a temporary or permanent basis (Costa, 1998).

That is the case, in particular, of persistent sleep disorders (i.e., chronic insomnia, obstructive sleep apnea syndrome, parasomnias), severe gastrointestinal disorders (i.e., peptic ulcer, chronic hepatitis or pancreatitis, Crohn's disease), cardiovascular diseases (i.e., ischemic heart disease, hyperkinetic syndromes, and severe hypertension), neurologic disorders (i.e., brain injuries with sequelae, epilepsy, spasmophilia), psychic syndromes (i.e., chronic anxiety and/or depression, seasonal affective disorder, alcoholism, and other drug addictions), metabolic (i.e., insulin-dependent diabetes) and hormonal (i.e., thyroid and suprarenal) diseases, chronic renal impairment, immunologic disorders, cancer, and pregnancy, particularly if there is a known risk of miscarriage.

Moreover, the assessment of psychophysical impairment and therapeutic action and fitness to (shift) work has to take into account that shift schedules may be quite different (i.e., continuous/discontinuous shift rosters, slow/fast and clockwise/counterclockwise rotation, prolonged duty periods, rest days between shifts), and the consequent effects on health and well-being may also be strongly influenced by the coexistence of other occupational risk factors (i.e., work load, environmental conditions, type of task).

It also has to take into consideration that many psychosomatic troubles and diseases, ascribable to or aggravated by shift work, have a multifactorial etiology (i.e., genetic inheritance, psychologic characteristics, lifestyle, socioeconomic conditions, intervening illnesses), a chronic trend, and their chronic degenerative trend makes their manifestation more likely to occur after long-term exposure and with increasing age. In this context, shift work may act as a further stress factor or a trigger as it combines conflicts between endogenous rhythms and social synchronizers with demanding working conditions and interference with family and social life.

Often these triggers may develop in a light form, not significantly affecting the worker's psychophysical work ability, and recent advances in pharmacology and rehabilitation allow the worker to recover or, at least, to significantly limit the severity and consequences (i.e., in the case of peptic ulcer, hypertension, or ischemic heart disease). On the other hand, shift work may hamper the effectiveness of their pharmacologic control, particularly when this requires a precisely timed administration and/or a stable life regimen, as is the case for diabetes, hypertension, asthma, hormonal pathologies, epilepsy, and depression. Also, individuals affected by seasonal affective disorder may encounter more difficulty in coping with night work, due to the lower exposure to bright light.

Consequently, the impact on health of shift and night work is the result of the interaction among several risk factors dealing with different domains, which can have a different weight and relevance both in terms of severity and timing of manifestation during the working life.

Obviously such disorders are more critical in the case of night work, whereas they may be compatible with day shifts, except for interference with sleep (such as in the case of early-morning shifts), diet (in metabolic and digestive disorders), and physical work load (in cardiovascular disease), and current pharmacologic therapy (in terms of timing and therapeutic effectiveness, as well as interference to vigilance and sleep). Regular health checks at the work site by occupational health physicians should be aimed at detecting early signs of intolerance to night work which may require prompt organizational (e.g., change of shift schedule) and/or individual (e.g., improved coping strategies or transfer to day work) intervention. Such checks should focus primarily on sleeping times and troubles, eating and digestive problems, psychosomatic complaints, drug consumption, housing and commuting problems, work load, and out-of-job activities, preferably using standardized questionnaires (Barton et al., 1995) or checklists in order to facilitate a comparison of the worker's health over the years.

These periodic checks may be usefully complemented with laboratory and instrumental tests, and specialized clinical visits, by neurologists, sleep experts, cardiologists, gastroenterologists, or endocrinologists, according to the specific troubles suffered. As concerns sleep troubles and disorders, in particular, subjective rating scales for sleepiness (i.e., Epworth, Karolinska and Stanford Sleepiness Scales), sleep propensity tests (MSLT, Maintenance of Wakefulness Test), actigraphy, polysonography, and other neurophysiologic tests, can be used to better define type and severity of the condition, to distinguish a transitory "masking" effect due to the altered sleep/wake cycle from a more persistent effect due to pathologic dysfunction, and to differentiate primary or secondary sleep disorders.

INTERINDIVIDUAL DIFFERENCES IN TOLERANCE

The impact of shift and night work on health shows a high interindividual variability, due to the interaction between individual characteristics, social and working conditions (Costa, 2003; Saksvik-Lehouillier et al., 2012).

According to some epidemiologic studies, 15–20% of workers suffer from serious troubles that force them to quit night work in the first 1–2 years, mainly due to circadian disruption and severe sleep problems. On the other hand, 10% do no complain, while the majority show various levels of maladaptation, that may be more or less manifested at different times and with different severity, concerning both short-term circadian adjustments and long-term tolerance.

Some individual characteristics have been associated with either adverse health effects or a better tolerance – in particular, age, gender, circadian structure and type, neuroticism, rigidity/flexibility of sleeping habits, hardiness, and stability of circadian rhythms.

Aging people can face increasing difficulties in coping with irregular work and living patterns due to reduced psychophysical fitness, decreased restorative properties of sleep, higher proneness to internal desynchronization of circadian rhythms, increased mental rigidity, and resistance to stressors. This can favor age discrimination or unequal treatment in the workplace of older workers and increase social problems in a progressively aging society (Costa and Di Milia, 2008).

Also people having a small amplitude of circadian rhythms show a less stable circadian structure and may be more prone to internal desynchronization when working on irregular shift schedules (Reinberg and Ashkenazi, 2008).

As mentioned above, women can face more specific adverse problems as regards their hormonal, family, and social conditions. In addition to the already mentioned disorders of menstrual cycle and reproductive function, women shift workers (those married with small children, in particular) can have more difficulties in combining their irregular working schedules with their additional domestic duties, thus suffering more sleep troubles and chronic fatigue.

Morning types ("larks") cope quite well with early-morning shifts, thanks to their advanced phase of biologic rhythms, but suffer greatly from night work. On the other hand, evening types ("owls") tolerate night work better, having a delayed circadian phase, but face greater problems in waking up early in the morning, and may suffer from an even greater truncation of their night sleep taken before early-morning shifts (Folkard and Hunt, 2000).

Also the characteristics of rigidity/flexibility of sleeping habits and languidity/vigorousness to overcome drowsiness may certainly hinder/help the worker's adaptation to shift and night work. High levels of neuroticism may be an unfavorable factor for tolerance, although it is not yet clear whether neuroticism is a predictor or an effect of shift work intolerance (Folkard et al., 1979; Van Dongen, 2006).

Moreover, unfavorable living and social conditions, often reported by surveys in developing countries, and usually associated with poor working conditions and long working hours, may aggravate the impact of shift work on health (Ong and Kogi, 1990; Fischer, 2001).

On the other hand, good physical fitness increases tolerance by lessening fatigue and improving recovery mechanisms, as well as a high motivation and a strong commitment to shift work, since this is associated with more stable sleep timings and other circadian behaviours (Härmä, 1996; Smith et al., 2001).

All these factors may also influence the self-selection process ("healthy worker's effect") that often occurs among shift workers and may make it difficult to compare different groups, both among shift workers and between shift and day workers, also because many studies are cross-sectional.

PREVENTIVE AND CORRECTIVE ACTIONS

It is quite obvious that a careful health surveillance has to proceed in parallel with corrective and preventive actions on working time organization, in particular shift scheduling.

There are thousands of different shift schedules which may have quite a different impact on workers' sleep, health, and safety, in particular with regard to amount of night work, timing and duration of shifts, length of shift cycle, speed and direction of shift rotation, position and length of rest days, and continuous or discontinuous shift systems (Driscoll et al., 2007; Bambra et al., 2008; Sallinen and Kecklund, 2010).

Hence, particular attention has to be given to the organization of the shift schedules, in order to take into account not only economic reasons, but primarily the physiology of the human body and psychologic and social well-being.

Shift schedules should be designed according to some ergonomic criteria, suitable to limit the above-mentioned adverse effects, by avoiding or minimizing circadian disruption, accumulation of sleep deficit and fatigue, as well as marked interference with family and social life, so that almost any worker can cope with shift and night work without significant perturbations of their health and well-being (Kogi 1996, 2001; Knauth and Hornberger, 2003).

These criteria relate to: (1) reducing night work as much as possible; (2) avoiding many consecutive night shifts by planning quickly rotating (every 1–3 days) shift rosters in order to limit the interference on circadian rhythms and minimize the cumulative sleep deficit; (3) preferring clockwise rotation (morning/afternoon/night) to counterclockwise (afternoon/morning/night) rotation, since it parallels the endogenous circadian rhythms (which show a periodicity slightly longer than 24 hours in "free-running" conditions); (4) avoiding quick changeovers (e.g., double shift in the same day) and allowing longer rest periods for immediate recovery from fatigue and sleep loss; (5) tailoring length of shifts according to psychophysical demand and personal characteristics; (6) avoiding too-early morning shifts in order to reduce the truncation of night sleep (REM phase in particular) and consequent daytime sleepiness; (7) setting an adequate number of rest days between shifts, particularly after nights; (8) keeping the shift roster as regular as possible; and (9) allowing flexible working time arrangements.

Further useful countermeasures, aimed at reducing or eliminating inconvenience, relate to additional rest breaks for meals and naps, supplementary rest days or holidays to improve recovery, better canteen facilities and social services (i.e., transport, school, and shop hours), training and rehabilitation courses, financial support for improving housing conditions (i.e., to soundproof bedrooms), periodic transfer to day work and/or progressive reduction of night work with increasing age.

Moreover, counseling and training on how to cope best with irregular working hours have to be delivered at individual and group levels through educational programs, dealing with improving self-care coping strategies, in particular as concerns sleep hygiene (tight scheduling of sleeping hours, use of naps, arrangements to avoid disturbances), abuse of medicaments or caffeinated drinks, diet, stress management, off-job activities, and exposure to bright light (Costa et al., 2006; Knauth et al., 2006; Pallesen et al., 2010).

The use of artificial bright light at night has been proposed to counteract the drop in circadian rhythms and vigilance (Czeisler et al., 1990; Boivin and James, 2005). The role of light is complex, since if bright enough it can promote phase adjustment of the body clock but, at a lower intensity, it has a more general positive effect, possibly through some forms of "general activation" of the central nervous system. However, the use and timing of bright light during night work need to be carefully tailored according to whether adjustment of circadian rhythms is recommended or not, and to sustain alertness without disturbing work activity (Bjorvatn and Pallesen, 2009).

Napping is a good countermeasure for main sleep deprivation, improving alertness and alleviating fatigue. More than length, its right timing in relation to the wake/duty period appears to be the key factor, having a prophylactic, strategic, or compensative function if taken before, during, or after the duty period, respectively (Gillberg et al., 1996; Garbarino et al., 2004).

The administration of melatonin is another tool proposed to help adjustment of biologic circadian rhythms and sleep/wake cycles. As with bright light, the effect may be different depending on dose and intention of timing, related either to the intention to speed up the phase adjustment of circadian rhythms or simply to favor sleep recovery (Arendt, 2010).

As for stimulant substances, caffeine is most commonly used to counteract sleepiness, having a quick onset of action and a sufficiently long effect. However, its use does not address the underlying sleep deprivation, and thus it is not a viable long-term solution, beside considering eventual side effects (Roehrs and Roth, 2008). Other medicaments (i.e., modafinil and armodafinil) seem to be able to reduce sleepiness during the night shift (Czeisler et al., 2009), but they must be used temporarily and caution is needed for permanent use as long-term side effects are not well known as yet.

References

Åkerstedt T (1996). Wide awake at odd hours. Swedish Council for Work Life Research, Stockholm, pp. 1–116.

Åkerstedt T (2003). Shift work and disturbed sleep/wakefulness. Occup Med 53: 89–94.

Åkerstedt T, Kecklund G (2001). Age, gender and early morning highway accidents. J Sleep Res 10: 105–110.

Åkerstedt T, Kecklund G, Horte L-G (2001). Night driving, season, and the risk of highway accidents. Sleep 24: 401–406.

Åkerstedt T, Keklund G, Selén J (2010). Early morning work – prevalence and relation to sleep/wake problems: a national representative survey. Chronobiol Int 27: 975–986.

American Academy of Sleep Medicine (2005). The International Classification of Sleep Disorders: Diagnostic & Coding manual, 2nd ed. American Academy of Sleep Medicine, Westchester, IL.

Arendt J (2010). Shift work: coping with the biological clock. Occup Med (Lond) 60: 10–20.

Ayas Nt, Barger LK, Cade BE et al. (2006). Extended work duration and the risk of self-reported percutaneous injuries in interns. JAMA 296: 1055–1062.

Bambra CL, Whithead MM, Sowden AJ et al. (2008). Shifting schedules: the health effects of reorganizing shift work. Am J Prev Med 34 (5): 427–434.

Barton J, Spelten E, Totterdell P et al. (1995). The Standard Shiftwork Index: a battery of questionnaires for assessing shiftwork related problems. Work Stress 9: 4–30.

Bjorvatn B, Pallesen S (2009). A practical approach to circadian rhythm sleep disorders. Sleep Med Rev 13: 47–60.

Boivin DB, James FO (2005). Light treatment and circadian adaptation to shift work. Ind Health 43: 34–48.

Cajochen C, Chellappa SL, Schmidt C (2014). Circadian and light effects on human sleepiness-alertness. In: S Garbarino, L Nobili, G Costa (Eds.), Sleepiness and Human Impact Assessment. Springer-Verlag, Milan, pp. 2–22.

Cole RJ, Loving RT, Kripke DF (1990). Psychiatric aspects of shiftwork. Occup Med 5: 301–314.

Comperatore CA, Krueger GP (1990). Circadian rhythm desynchronosis, jet-lag, shift lag, and coping strategies. Occup Med 5: 323–341.

Connor J, Norton R, Ameratunga S et al. (2002). Driver sleepiness and risk of serious injury to car occupants: population based case control study. Br Med J 324 (7346): 1125.

Costa G (1998). Guidelines for the medical surveillance of shiftworkers. Scand J Work Environ Health 24 (Suppl 3): 151–155.

Costa G (2003). Factors influencing health of workers and tolerance to shift work. Theor Issues Ergon Sci 4: 263–288.

Costa G, Di Milia L (2008). Ageing and shiftwork: a complex problem to face. Chronobiol Int 25: 165–181.

Costa G, Åkerstedt T, Nachreiner F et al. (2004). Flexible working hours, health, and well-being in Europe: some considerations from a SALTSA project. Chronobiol Int 21: 831–844.

Costa G, Sartori S, Åkerstedt T (2006). Influence of flexibility and variability of working hours on health and well-being. Chronobiol Int 23: 1–13.

Costa G, Haus E, Stevens R (2010). Shiftwork and cancer: considerations on rationale, mechanisms, and epidemiology. Scand J Work Environ Health 36: 163–179.

Czeisler CA, Johnson MP, Duffy JF et al. (1990). Exposure to bright light and darkness to treat physiologic maladaptation to night work. N Engl J Med 322: 1253–1259.

Czeisler CA, Walsh JK, Wesnes KA et al. (2009). Armodafinil for treatment of excessive sleepiness associated with shift work disorder: a randomized controlled study. Mayo Clin Proc 84: 958–972.

Dawson D, McCulloch K (2005). Managing fatigue: it's about sleep. Sleep Med Rev 9 (5): 365–380.

Dawson D, Noy YI, Härmä M et al. (2011). Fatigue modeling: practices and principles in real world settings. Acc Anal Prev 43: 549–564.

Drake CL, Roehrs T, Richardson G et al. (2004). Shift work sleep disorder: prevalence and consequences beyond that of symptomatic day workers. Sleep 27: 1453–1462.

Driscoll TR, Grunstein RR, Rogers NL (2007). A systematic review of the neurobehavioural and physiological effects of shiftwork systems. Sleep Med Rev 11: 179–194.

Feillet C, Kruscheb P, Tamanini F et al. (2014). Phase locking and multiple oscillating attractors for the coupled mammalian clock and cell cycle. Proc Natl Acad Sci U S A 111: 9828–9833.

Fischer FM (2001). Shiftworkers in developing countries: health and well-being and support measures. J Hum Ergol 30: 155–160.

Folkard S (1997). Black times: temporal determinants of transport safety. Accident Anal Prev 29 (4): 417–430.

Folkard S (2008). Do permanent night workers show circadian adjustment? A review based on the endogenous melatonin rhythm. Chronobiol Int 25: 215–224.

Folkard S, Hunt LJ (2000). Morningness-eveningness and long-term shiftwork tolerance. In: S Hornberger, P Knauth, G Costa et al. (Eds.), Shiftwork in the 21st Century. Peter Lang, Frankfurt am Main, pp. 311–316.

Folkard S, Tucker P (2003). Shiftwork, safety and productivity. Occup Med 53: 95–101.

Folkard S, Monk TH, Lobban MC (1979). Towards a predictive test of adjustment to shift work. Ergonomics 22: 79–91.

Frost P, Kolstad HA, Bonde JP (2009). Shift work and the risk of ischemic heart disease – a systematic review of the epidemiologic evidence. Scand J Work Environ Health 35: 163–179.

Gaba DM, Howard SK (2002). Fatigue among clinicians and the safety of patients. N Engl J Med 347: 1249–1255.

Garbarino S, Mascialino B, Penco MA et al. (2004). Professional shift-work drivers who adopt prophylactic naps can reduce the risk of car accidents during night work. Sleep 27 (7): 1295–1302.

Gillberg M, Kecklund G, Axelsson J et al. (1996). The effects of a short daytime nap after restricted night sleep. Sleep 19: 570–575.

Gundel A, Wegmann H (1987). Resynchronisation of the circadian system following a 9-hr advance or a delay zeitgeber shift: real flights and simulations by a Van-der-Pol oscillator. Prog Clin Biol Res 227B: 391–401.

Ha M, Park J (2005). Shiftwork and metabolic risk factors of cardiovascular disease. J Occup Health 47: 89–95.

Härmä M (1996). Ageing, physical fitness and shiftwork tolerance. Appl Ergon 27: 25–29.

Härmä M, Sallinen M, Ranta R et al. (2002). The effect of an irregular shift system on sleepiness at work in train drivers and railway traffic controllers. J Sleep Res 11: 141–151.

Herrmann Us, Hess CW, Guggisberg AG et al. (2010). Sleepiness is not always perceived before falling asleep in healthy, sleep-deprived subjects. Sleep Med 11 (8): 747–751.

Horne JA, Reyner LA (1995). Sleep related vehicle accidents. Br Med J 310: 565–567.

Horwitz IB, McCall BP (2004). The impact of shift work on the risk and severity of injuries for hospital employees: an analysis using Oregon workers' compensation data. Occup Med 54: 556–563.

Howard ME, Desai AV, Grunstein RR et al. (2004). Sleepiness, sleep-disordered breathing, and accidents in commercial vehicle bus drivers. Am J Respir Crit Care Med 170 (9): 1014–1021.

Ingre M, Soderstrom M, Kecklund G et al. (2000). Train Drivers Work Situation. Working Hours, Sleep, Stress and Safety, Stress Research Report no. 292, Swedish Institute for Psychosocial Medicine, Stockholm.

Kecklund G, Åkerstedt T (1993). Sleepiness in long distance truck driving: an ambulatory EEG study of night driving. Ergonomics 36: 1007–1017.

Knauth P (1983). Ergonomische Beitrage zu Sicherheitsaspekten der Arbeitszeitorganisation. Fortschr.-Ber. Riehe 17, No 8. VDI-Verlag, Dusseldorf.

Knauth P, Hornberger S (2003). Preventive and compensatory measures for shift workers. Occup Med 53: 109–116.

Knauth P, Jung D, Bopp W et al. (2006). Compensation for unfavorable characteristics of irregular individual shift rotas. Chronobiol Int 23: 1277–1284.

Knutsson A, Boggild H (2010). Gastrointestinal disorders among shift workers. Scand J Work Environ Health 36: 85–95.

Kogi K (1996). Improving shift workers' health and tolerance to shiftwork: recent advances. Appl Ergon 27: 5–8.

Kogi K (2001). Healthy shiftwork, healthy shiftworkers. J Hum Ergol 30: 3–8.

Lamond N, Darwent S, Dawson D (2005). Train drivers' sleep and alertness during short relay operations. Appl Ergon 36: 313–318.

Landrigan CP, Rothschild JM, Cronin JW et al. (2004). Effect of reducing interns' work hours on serious medical errors in intensive-care units. N Engl J Med 351: 1838–1848.

Léger D (1994). The cost of sleep-related accidents: a report for the National Commission on Sleep Disorders Research. Sleep 17: 84–93.

Leigh JP (1991). Employee and job attributes as predictors of absenteeism in a national sample of workers: the importance of health and dangerous working conditions. Soc Sci Med 33 (2): 127–137.

Lombardi DA, Folkard S, Willetts JL et al. (2010). Smith GS. Daily sleep, weekly working hours, and risk of work-related injury: US national health interview survey (2004–2008). Chronobiol Int 27: 1013–1030.

Mallis MM, Mejdal S, Nguyen TT et al. (2004). Summary of the key features of seven biomathematical models of human fatigue and performance. Aviat Space Environ Med 75: A4–A14.

Minors D, Akerstedt T, Waterhouse J (1994). The adjustment of the circadian rhythm of body temperature to simulated time-zone transitions: a comparison of the effect of using raw versus unmasked data. Chronobiol Int 11: 356–366.

Mitchell RJ, Williamson AM (2000). Evaluation of an 8-h versus a 12-h shift roster on employees at a power station. Appl Ergon 31: 83–93.

Moore-Ede M (1993). The twenty-four hour society: understanding human limits in a world that never stops. Addison-Wesley Publishing Company, Reading, MA.

National Sleep Foundation (2013). 2012 Bedroom Poll. http://www.sleepfoundation.org/sites/default/files/bedroom2012.pdf (accessed 05.05.2014).

National Transportation Safety Board (1999). Factors that affect fatigue in heavy truck accidents. Washington, D.C. PB95-917001,NTSB/SS-95/01;1995.

Nurminen T (1998). Shift work and reproductive health. Scand J Work Environ Health 24 (suppl 3): 28–34.

Ohayon MM, Smolensky MH, Roth T (2010). Consequences of shiftworking on sleep duration, sleepiness, and sleep attacks. Chronobiol Int 27: 575–589.

Ong CN, Kogi K (1990). Shiftwork in developing countries: current issues and trends. Occup Med 5: 417–428.

Pack AI, Pack AM, Rodgman E et al. (1995). Characteristics of crashes attributed to the driver having fallen asleep. Accident Anal Prev 27: 769–775.

Pallesen S, Nordhus IH, Omvik S et al. (2007). Prevalence and risk factors of subjective sleepiness in the general adult population. Sleep 30: 619–624.

Pallesen S, Bjorvatn B, Mageroy N et al. (2010). Scand J Work Environ Health 36: 109–120.

Parks DK, Yetman RJ, McNeese MC et al. (2000). Day–night pattern in accidental exposures to blood-borne pathogens among medical students and residents. Chronobiol Int 17: 61–70.

Philip P, Åkerstedt T (2006). Transport and industrial safety, how are they affected by sleepiness and sleep restriction? Sleep Med Rev 10: 347–356.

Philips R (2000). Sleep, watchkeeping and accidents: a content analysis of incident at sea reports. Transp Res 3: 229–240.

Pisarki A, Lawrence SA, Bohle P et al. (2008). Organizational influences on the work life conflicts and health of shiftworkers. Appl Ergon 39: 580–588.

Price WJ, Holley DC (1990). Shiftwork and safety in aviation. Occup Med 5: 343–377.

Puttonen S, Härmä M, Hublin C (2010). Shift work and cardiovascular disease – pathways from circadian stress to morbidity. Scand J Work Environ Health 36: 96–108.

Reinberg A, Ashkenazi I (2008). Internal desynchronization of circadian rhythms and tolerance to shiftwork. Chronobiol Int 25: 625–643.

Reppert SM, Weaver DR (2002). Coordination of circadian timing in mammals. Nature 418 (6901): 935–941.

Roehrs T, Roth T (2008). Caffeine: sleep and daytime sleepiness. Sleep Med Rev 12: 153–162.

Roenneberg T, Kumar CJ, Merrow M (2007). The human circadian clock entrains to sun time. Curr Biol 17: R44–R45.

Rutenfranz J, Colquhoun P (1979). Circadian rhythms in human performance. Scand J Work Environ Health 5: 167–177.

Sagberg F (1999). Road accidents caused by drivers falling asleep. Accident Anal Prev 31: 639–649.

Saksvik-Lehouillier I, Bjorvatn B, Hetland H et al. (2012). Personality factors predicting changes in shift work tolerance: a longitudinal study among nurses working rotating shifts. Work Stress 26: 143–160.

Sallinen M, Kecklund G (2010). Shift work, sleep and sleepiness – differences between shift schedules and systems. Scand J Work Environ Health 36: 121–133.

Scott LD, Rogers AE, Hwang W-T et al. (2006). Effects of critical care nurses' work hours on vigilance and patients' safety. Am J Crit Care 15: 30–37.

Shechter A, James FO, Boivin DB (2008). Circadian rhythms and shift working women. Sleep Med 3: 13–24.

Smith L, Norman P, Folkard S (2001). Predicting shiftwork-related outcomes: shiftwork locus of control and circadian type. J Hum Ergol 30: 59–64.

Smolenski MH, Reinberg A (1990). Clinical chronobiology: relevance and applications to the practice of occupational medicine. Occup Med 5: 239–272.

Straif K, Baan R, Grosse Y et al. (2007). Carcinogenicity of shift-work, painting, and fire-fighting. Lancet Oncol 8: 1065–1066.

Suvanto S, Partinen M, Härmä M et al. (1990). Flight attendant's desynchronosis after rapid time zone changes. Aviat Space Environ Med 61: 543–547.

Tanaka K, Takahashi M, Hiro H et al. (2010). Differences in medical error risk among nurses working two- and three-shift systems at teaching hospitals: a six-month prospective study. Ind Health 48: 357–364.

Van Dongen HPA (2004). Comparison of mathematical model predictions to experimental data of fatigue and performance. Aviat Space Environ Med 75: A15–A36.

Van Dongen HPA (2006). Shift work and inter-individual differences in sleep and sleepiness. Chronobiol Int 23: 1139–1148.

Vittasalo K, Kuosma E, Laitinen J et al. (2008). Effects of shift rotation and the flexibility of a shift system on daytime alertness and cardiovascular risk factors. Scand J Work Environ Health 34: 198–205.

Zhu JL, Hjollund NH, Boggild H, Olsen J et al. (2003). Shift work and subfecundity: a causal link or an artefact? Occup Environ Med 60, e12.

Handbook of Clinical Neurology, Vol. 131 (3rd series)
Occupational Neurology
M. Lotti and M.L. Bleecker, Editors

Chapter 25

Headache and migraine: a leading cause of absenteeism

KHOSROW BAIGI[1] AND WALTER FRANCIS STEWART[2]*
[1]*Montefiore Headache Center, Bronx, NY, USA*
[2]*Sutter Health, Walnut Creek, CA, USA*

INTRODUCTION

Headache is a common problem in working-age adults, a dominant cause of lost work time, and the second most common pain problem that leads to lost work time. Headache causes a significant amount of individual and societal burden. Primary headache disorders – headache without underlying cause – account for most headache episodes that occur in the general population. These include tension-type headache (TTH), migraine headache, cluster headaches (and other trigeminal-autonomic cephalgias), and other less common primary headache disorder (Headache Classification Subcommittee of International Headache Society, 2004). While the individual effects of migraine headaches are the most disabling, tension-type and other nonmigrainous headaches are far more common and as a consequence have an equivalent population impact on lost work time and related costs (i.e., otherwise known as "indirect costs") (Schwartz et al., 1997).

While the pain from headache is a dominant concern for individual sufferers, the disabling effects of headaches result in individual impact on work productivity and substantial employer and societal burden from direct medical costs, lost work time, and underemployment, and, in more severe persistent headache, unemployment (Stewart et al., 1996a). The lost work time costs greatly exceed medical care costs. Disability encompasses limitations to activities at work and home, and also in social roles. Work-related disability is the most important determinant of the economic impact of illness (Stewart et al., 1996a), because lost productive time can be translated into labor costs in the form of absenteeism and presenteeism (Stewart et al., 2011). Absenteeism is defined as absence from work due to disability and presenteeism is defined as loss of productive capacity (disability) that occurs when a sick individual goes to work (Serrano et al., 2013). The burden from headaches also results in societal costs from underemployment and unemployment among those of working age who suffer from migraine or TTH (Stang et al., 1998; Stewart et al., 2011).

In this chapter we review the epidemiology, burden, and comorbidities of migraine and TTH; next we consider estimating employer burden and strategies for management, and we close with different treatment options for both migraine and TTH.

MIGRAINE: PREVALENCE AND INCIDENCE

Prevalence of migraine

The work impact of migraine on employers is directly related to the prevalence of active migraine in a workforce and number of days with headache. These two factors vary substantially in the population by demographics and the burden that migraine poses to an employer can be determined, in part, by workforce demographics. Migraine prevalence varies by age, gender, and other sociodemographic variables, and headache frequency also varies by some of these factors.

Estimates of the 1-year prevalence of migraine have remained remarkably stable across different populations and time. Three large-scale population studies in the USA (American Migraine Study I, II, and III) conducted over a 15-year period demonstrate this consistency with prevalence estimates of 12% overall – 18% in females, and 6% in males. Migraine is significantly more common in females but the female-to-male prevalence ratio varies with age (Stewart and Lipton, 1996; Lipton et al., 2001, 2007). The prevalence of migraine

*Correspondence to: Walter "Buzz" Stewart, Ph.D., MPH, VP, Chief Research Officer, Sutter Health, 2121 N. California Blvd, Suite 310, Walnut Creek, CA 94596, USA. Tel: +1-925-287-4012, E-mail: StewarWF@sutterhealth.org

peaks between 25 and 55 years for both men and women (Lipton and Bigal, 2005). The gender ratio varies considerably by age, ranging from 2.0 to 3.3. The highest gender ratio is between the ages of 30 and 45 years (Stewart et al., 1992). A recent analysis from the American Migraine Prevalence and Prevention (AMPP) study found that the adjusted female-to-male prevalence ratios for migraine range from 1.5:1 for those 12–17 years old and 3.3:1 for those 18–29 years old (Buse et al., 2013).

North American studies consistently find an inverse relationship between the prevalence of migraine and socioeconomic status, as measured by income or education. In the American Migraine Study (I, II, and III), migraine prevalence increased directly in relation to decreasing household income (Stewart et al., 1996b; Lipton et al., 2001; Buse et al., 2012). These findings were confirmed by a study conducted in the UK using methods that were the same as methods used in a US study (Stewart et al., 1996b). In contrast, studies from the Netherlands, Denmark, and Canada do not support this inverse relationship (Rasmussen, 1992; O'Brien et al., 1994; Launer et al., 1999), and it has been suggested that differences among western countries may be explained by differences in the level of social and economic support systems.

The higher prevalence in lower-income groups may be explained by a higher incidence of new cases or a lower remission rate (i.e., longer duration of time over which cases suffer from attacks). Three hypotheses have been posited to explain the variation in socioeconomic status (Stewart et al., 1992). The social causation hypothesis suggests that poor diet, stress, and other factors associated with low income may precipitate migraine attacks. Alternatively, access to good healthcare may decrease the duration of the illness and therefore the prevalence among higher-income groups. The social selection hypothesis states that in some individuals migraine may cause low income. That is, headache-related disability may seriously disrupt function at work or school. The phenomenon of "downward" socioeconomic drift has been described in other disabling conditions (Eaton et al., 1988; Liberatos et al., 1988). In turn, occupational and social disruption may exacerbate the migraine condition and create a downward spiral (Stewart et al., 1992). Recent evidence indicates that a higher incidence of migraine in lower-income groups accounts for the difference in prevalence by income, as remission rates do not appear to vary by household income in the USA (Stewart et al., 2013).

After adjusting for socioeconomic status, migraine prevalence varies by race in the USA. The prevalence is highest in white females and males (20.4% and 8.6%, respectively), intermediate in African American females and males (16.2% and 7.2%, respectively), and lowest in Asian American females and males (9.2%, 4.2%, respectively) (Stewart et al., 1995; Robbins and Lipton, 2010). The overall prevalence of migraine is higher in North America (13%) and Europe (15%) and lower in Asia (9%), Central/South America (9%), and Africa (5%) (Stovner et al., 2007). Differences in the prevalence of migraine between African Americans and Africans, or Asian Americans and Asians may indicate the significance of environmental stressors in triggering onset of migraine, a notion that is consistent with recent evidence on incidence and remission of migraine (Stewart et al., 2013).

Incidence of migraine

Several studies have estimated the sex- and age-specific incidence of migraine. A Copenhagen group of investigators conducted a 12-year longitudinal population study of 673 eligible adults, aged 25–64 years. Of them, 549 (81.6%) participants completed follow-up. The annual incidence rate of migraine was 8.1 per 1000 person-years and the female-to-male ratio was 6:1. No significant differences in age or gender were found when comparing migraine with aura and migraine without aura. In both genders, the incidence decreased markedly with increasing age. The highest incidence rates were found in the age group 25–34 years, with annual incidence rates of 6.5 per 1000 person-years for males and 22.8 per 1000 person-years for females. In the 55–64-year age group, the incidence of primary headache was less than 5 per 1000 person-years.

Stewart et al. (2008) used data from the the AMPP study to measure cumulative lifetime incidence of migraine in women and men. Migraine incidence peaked between 20 and 24 years of age for women, at 18.2 cases per 1000 person-years and between 15 and 19 for men, at 6.2 cases per 1000-years. Broadly, these findings agree with the results of the Copenhagen study and other incidence studies (Fig. 25.1). The median age of onset was 25 years in females and 24 years in males, and the cumulative lifetime incidence was 43% in females and 18% in males. More than 75% of new-onset cases in males and more than 85% of new-onset cases in females occur after 14 years of age. For both men and women, approximately 75% of new-onset cases occur before age 35 years (Stewart et al., 2008).

Breslau et al. (2003) estimated an annual incidence of 19.5/1000 person-years among a predominantly female (79%) cohort with a mean age of 42 years. This estimate, which is a weighted composite of the rate for those with and without depression, is more than twice as high as the incidence rate reported by Stewart et al. (2008) for the same age group. In another study, investigators estimated the annual incidence of migraine in an

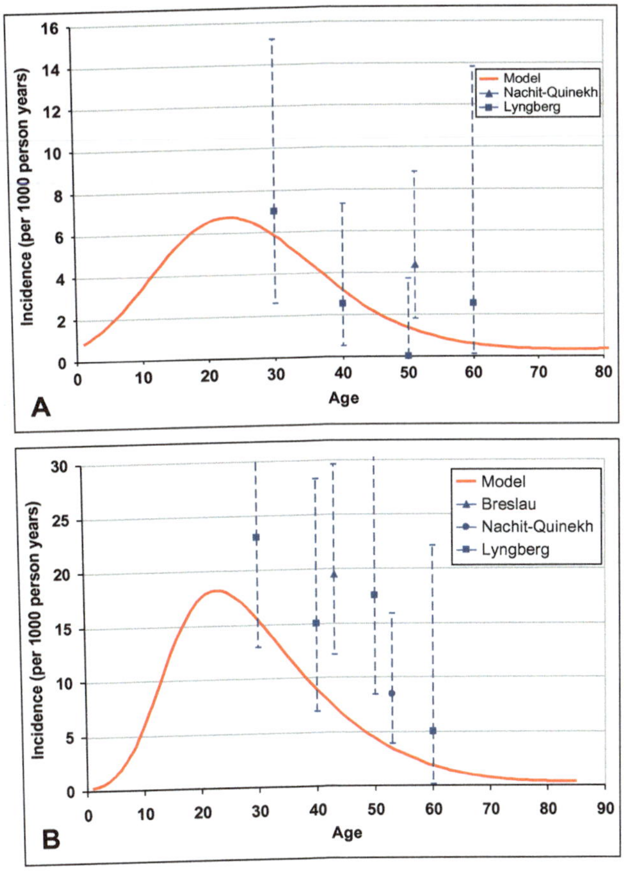

Fig. 25.1. Model-based (**A**) male and (**B**) female 5-year age-specific migraine incidence compared to estimates from longitudinal studies. From Stewart et al. (2008).

occupational cohort of men (n = 180) and women (n = 60) with a median age of 46 years at baseline and an average age of 51 years between the two surveys 10 years apart (Bigal et al., 2008a). The annual incidence of migraine was approximately 8.8 per 1000 person-years for women and 3.8 per 1000 person-years for men. Both of these estimates are substantially higher than the incidence rate reported by Stewart et al. and the Copenhagen group. In total these estimates show a declining incidence after age 30 years, a pattern consistent in all studies (Stewart et al., 2008).

Risk factors associated with increased incidence of migraine include poor living conditions, stress, low educational level at baseline, workload, and frequent TTH (Lyngberg et al., 2005a).

CHRONIC MIGRAINE: PREVALENCE AND INCIDENCE

Chronic migraine is defined by headaches of more than 4 hours' duration occurring on 15 or more days per month for at least 3 months. Chronic migraine is relatively common in the general population, but because of its significant disabling impact on the work role, it is less common in the workforce.

Prevalence of chronic migraine

Estimates of chronic migraine prevalence in the population vary with case definition. Natoli et al. (2010) performed a systematic review and identified 12 unique studies. The Silberstein and Lipton diagnostic criteria were most commonly used in these studies (Silberstein et al., 1994). Depending on the case definition, the prevalence of chronic migraine ranged from 0.9% to 5.1% in the general population, with estimates typically in the range of 1.4–2.2%. Katsarava et al. (2004) conducted a study to estimate the 1-year prevalence of both migraine and TTH in the general population of the Republic of Georgia, and reported chronic migraine prevalence as 1.4%. In a population-based, cross-sectional study conducted in a rural area of Brazil, 1-year prevalence of chronic migraine was reported as 1.36%.

In a recent analysis of US population data from the AMPP study, the overall prevalence of chronic migraine was 0.91% (1.29% of females and 0.48% of males) (Buse et al., 2012). Relative to 12–17-year-olds, the age- and sex-specific prevalence for chronic migraine peaked in the 40s, at 1.89% for females and 0.79% for males (prevalence ratio 3.35). The gender prevalence ratio was (4.57) and the prevalence of chronic migraine was inversely related to household income. Chronic migraine represented 7.7% of migraine cases overall, and the proportion generally increased with age.

Incidence of chronic migraine

Few data are available on the incidence of chronic migraine in the population. Based on results from the AMPP study, among persons with episodic migraine, the incidence of chronic migraine in a subsequent year was 2.5%. It is not known, however, if this estimate represents a composite of recurrence and new onset, as evidence indicates that individuals with chronic migraine can remit to episodic migraine, recur, or fully remit.

The transition from episodic to chronic migraine represents a form of progression that could be temporary or persistent. Potential modifiable risk factors for progression of episodic to chronic migraine include obesity (Scher et al., 2003a), medication overuse (Bigal et al., 2008b), sleep apnea and snoring (Scher et al., 2003b), caffeine intake (Scher et al., 2004), comorbid pain syndromes (Scher et al., 2005a), allodynia (Lipton et al., 2008), and major life changes (Scher et al., 2008). Other risk factors, such as younger age, low education and socioeconomic status, and white race (Scher et al., 2003a), are not modifiable and are surrogates for a host of other endogenous and exogenous factors. Comorbid illnesses and headache frequency are potentially modifiable risk factors.

PREVALENCE AND INCIDENCE OF TTH

Lifetime prevalence of TTH is estimated to be 78% (Rasmussen et al., 1991), but the true estimate may be higher given that recall of TTH among remitted cases probably declines with time since last headache. This factor and, more importantly, the methods of asking about lifetime history and defining TTH, explain the wide variation in estimates from different studies. Prevalence varies by age, sex, and continent (Rasmussen et al., 1991; Schwartz et al., 1998). In general, TTH is more common in Europe, with various studies reporting a 1-year prevalence as high as 80% to lower estimates of 20–30% in Asia and 38% in the USA (38%). The prevalence of TTH is slightly higher in women in all age, race, and education groups, with a female-to-male ratio in various studies ranging from 1.16:1 (Lavados and Tenhamm, 1998; Schwartz et al., 1998).

In the USA, the 1-year prevalence from one of the largest population-based studies, reported by Schwartz et al. (1998), was 38.3% in adults. Women had an average 16% higher prevalence than men in all age, race, and educational-level subgroups. Prevalence peaked in the 30–39-year-old group in men (42.3%) and women (46.9%), and then declined thereafter with increasing age. Prevalence was 18% higher among Caucasians than African Americans. Finally, after adjusting a number of demographic factors, prevalence of episodic TTH (ETTH) increased with increasing level of education in both sexes, reaching a peak in individuals with graduate-school education of 49% for both sexes (Schwartz et al., 1998).

The incidence of TTH has only been estimated in a single study of a random sample of Danish adults. Individuals were surveyed at two different time points, with an intervening 12-year period. The overall incidence of frequent TTH, designated as 15 or more headache days per year, was estimated at 14.2 per 1000 person-years, with a female-to-male ratio of 3:1. Incidence of frequent TTH was highest between the ages of 25 and 34 years for both men and women, declining in older age groups. Risk factors for incidence of frequent TTH included poor self-related health and sleep, and difficulty with relaxation after work (Lyngberg et al., 2005a).

Chronic TTH (CTTH) is defined as headache occurring on $\geq$15 days/month for 3 months or more, usually bilateral and nonpulsating and occurring with no more than one of photophobia, phonophobia, or mild nausea. Prevalence of CTTH is higher in Europe and the USA compared with Asia and South America. Prevalence of CTTH is also generally higher in females than males. Schwartz et al. (1998) reported a 1-year prevalence of CTTH of 2.8% for women and 1.4% for men. African Americans had a lower prevalence than did Caucasians. Prevalence of CTTH declined with increasing education and this pattern was more apparent in women.

HEADACHE PROGRESSION AND REMISSION

Headache progression is generally defined by the increasing frequency of headache attacks or days with headache and by the persistence of having frequent headache. Chronic daily headache (CDH) represents a widely accepted stage of progression. CDH is defined as a group of primary headaches that occur on 15 days or more per month for at least 3 months. CDH has a number of variants. It can occur for those with migraine, defined as chronic migraine, for those with those with episodic tension headache, defined as CTTH, and is expressed in other forms, including new daily persistent headache, and hemicrania continua. A dominant hypothesis for progression is based on dynamic process, where recurrent severe headaches lead to central sensitization with the generation of free radicals and anatomic changes in areas of the central nervous system that control pain. Central sensitization decreases the threshold to pain and increases the risk of more frequent pain episodes. Repeated and more frequent attacks of headache may decrease the threshold for activation of primary and modulating pain pathways, leading to a pain-prone state marked by cutaneous allodynia and sensitization. It is notable that increases in number of headache days is directly related to an increase in the likelihood that other symptoms (e.g., nausea, photophobia, phonophobia) co-occur with migraine.

Scher et al. (2003a) used data from the Frequent Headache Epidemiology Study to identify factors that predict the onset or remission of CDH in a population sample of 1134 individuals with CDH and with episodic headache. Over the course of 1 year, 3% of individuals with episodic headache (headache frequency 2–104 days per year) progressed to CDH. Risk factors measured at baseline that were statistically significant predictors of the 1-year progression to CDH included a higher number of headache days, obesity, habitual snoring independent of obesity, and history of neck injury. Other risk factors for incidence onset of CDH included high daily caffeine consumption, particularly in women younger than 40 years, and major social life events (e.g., divorce, marital separation, death of relative or friend, and occupational change) in individuals older than 40 years of age (Scher et al., 2003a).

A study of 8219 individuals with episodic migraine from the 2005 AMPP database found that baseline frequency of headache days predicted onset of chronic migraine, even when controlling for monthly use of acute pain medication. The risk of chronic migraine increased exponentially with higher baseline headache frequency (days/month). It is noteworthy that treatment with opioids and barbiturates, medications that are often prescribed for migraine, was associated with progression of episodic migraine to chronic migraine (Scher et al., 2003a).

One important goal in the treatment of headache is to reduce attack frequency so that chronic migraine remits to episodic migraine (Silberstein et al., 2008). Relatively few studies have examined remission rates for chronic migraine. Manack et al. (2011) utilized the longitudinal AMPP study dataset to estimate remission rates in a population-based sample of persons with chronic migraine, to identify potential predictors of chronic migraine remission. Over a 2-year period, approximately one-fourth of those with chronic migraine remitted to episodic migraine or other headache types. In multivariate models, headache frequency (average days/month) and absence of allodynia predicted remission (Manack et al., 2011). Stewart et al. (2013) used the same data to examine whether higher prevalence of migraine in individuals with lower household income is explained by a higher incidence rate or a lower remission rate. Age-specific incidence was significantly higher among those in the lower-household- income groups for both males and females; however, age-specific remission rate did not differ by household income in either females or males.

Remission from CTTH to fewer than 180 headaches per year is common. Estimates range from 33% to 67% over a year. However, a transition from CTTH to a "normal" frequency of less than one headache per week is uncommon. A population-based study was conducted among 146 patients with frequent TTH and 15 patients with CTTH at baseline. After 12 years' follow-up, 45% of these individuals had infrequent or no TTH, 39% had frequent TTH, and 16% had CTTH. Baseline factors that predicted persistence of CTTH included baseline CTTH, coexisting migraine, not being married, and having problems with sleep (Lyngberg et al., 2005b).

THE IMPACT OF HEADACHE

Burden of migraine and associated symptoms

According to *The World Health Organization Report 2001*, migraine is the 19th leading cause of years lived with disability among both males and females of all ages and the 12th leading cause of years lived with disability among females of all ages (WHO, 2001).

Migraine headache pain and associated symptoms cause substantial individual, employer, and societal burdens that are mediated by high prevalence among working-age individuals. But individuals vary by attack frequency and duration, as well as other factors that influence or mediate headache-related disability (Mennini et al., 2008).

Approximately 18% of females and 6% of males have migraines, have one or more attacks of migraine in a year, and most have one or more migraines per month. The age of highest prevalence of migraine occurs during the ages of peak productive years – between 25 and 55 years of age. The prevalence is higher in women than men at all postpubertal ages, but the sex ratio varies with age.

Attack-related impact of migraine is mediated by pain severity, attack duration, associated features, comorbidities, and psychologic factors, including fear of attacks. Measures of severity may be useful as predictors of disability and healthcare use. Moreover, such measures may also be useful in assessing the need for patient care and treatment and help target those who are more disabled by migraine. About one-third of migraine sufferers experience severe disability or the need for bed rest following attacks; another 50% report mild to moderate disability (Peres et al., 2007).

The burden of migraine occurs both during (ictal) and between (interictal) attacks (Rasmussen et al., 1991; Buse et al., 2007). The ictal burden is related to head pain, nausea, vomiting, and sensitivity to environmental stimuli, all of which limit function during attacks. Approximately half of those with migraine in the 2009 AMPP study reported having nausea half the time or more often with their attacks (Lipton et al., 2013). Nausea was more common in females than males. Frequent nausea with migraine was associated with more frequent headache symptoms, in general, and also resulted in more severe headache-related impact, as measured by the Headache Impact Test-6. High-frequency nausea was associated with greater work-related disability, greater risk of medical leave, and a higher cost from headache medications or from the adverse effects of these medications on paid work, school work, household work, and family or leisure activities (Lipton et al., 2013).

The interictal burden from migraine encompasses challenges to planning activities or restrictions to activities, in part because of fear of the next attack. Anxiety in anticipation of the next migraine attack (i.e., interictal anxiety) may lead migraineurs to use acute pain medications before any symptoms of an attack occur, ultimately resulting in the overuse of these medications. Avoidance of activities because of fear of headache (i.e., cephalalgiaphobia) is an important interictal burden from migraine. Cephalalgiaphobia (fear of the next headache) may also decrease the threshold for initiating the consumption of analgesics, leading to medication overuse (Peres et al., 2007).

Migraine headache is associated with substantial impact on Health-Related Quality of Life (HRQoL). For example, investigators conducted a telephone survey with 200 randomly selected patients with migraine and 200 matched nonmigraine controls among residents of Greater London. The migraineurs scored significantly lower than controls on HRQoL, which was found to be inversely related to headache-related disability (Lipton et al., 2003).

The overall costs attributable to the migraine are difficult to measure. However, most of the cost impact is associated with disability, decreased ability to function, and the consequent impact on the work role and cost to employers (Hu et al., 1999). The annual US cost of migraine from lost work time in 1998 was estimated to be $13 billion. This estimate does not include the additional societal costs of underemployment and unemployment. Stewart and colleagues (2003) conducted a 2-week survey of a random sample of 28 902 working adults across the USA. Overall, 12.7% of the workforce lost productive time in a 2-week period due to a common pain condition. Headache was the most common pain condition (followed by back pain and arthritis), resulting in an average of 3.1 ± 0.1 hours of lost productive time per week (Stewart et al., 2003). Hawkins and colleagues (2007) compared the cost to employers of workers with migraine to workers without migraine headache in the workplace. On average, the costs from absenteeism, short-term disability, and workers' compensation claims for those with migraine were $2800 higher than these same costs among those without migraine.

Population-based studies have demonstrated that episodic migraine has a substantially greater impact than chronic migraine on headache-related disability, of lost productive time (Stang et al., 1998), use of healthcare resource utilization, and decrements in HRQoL (Meletiche et al., 2001). In contrast, the individual impact of chronic migraine is greater than that of episodic migraine. In the International Burden of Migraine Study I, the costs associated with lost productive time over 3 months for chronic migraine participants were about four times those of episodic migraine participants in full-time workers and about eight times in part-time workers. In addition, a majority of those with chronic migraine had a Migraine Disability Assessment Test grade of IVa or IVb, indicating severe or very severe disability (chronic migraine 79.8% vs episodic migraine 21.6%). Another analysis of 24 000 individuals with "severe" headache who responded to the 2005 AMPP survey identified 655 respondents with chronic migraine and 11 249 respondents with episodic migraine. Respondents with chronic migraine (relative to episodic migraine) had statistically significantly lower levels of household income, were less likely to be employed full-time, and were more likely to be occupationally disabled (Stovner et al., 2007). Among persons with chronic migraine, 25% had very severe headache-related disability compared with 3% of those in the episodic migraine group and, after adjusting for sociodemographic covariates, the odds of headache-related disability among persons with chronic migraine were nearly four times greater than they were for those with episodic migraine (odds ratio (OR) = 3.90; 95% confidence interval (CI) 3.5–4.3, $p < 0.001$) (Buse et al., 2012).

Comorbidities of migraine

Migraine co-occurs with a number of other health problem more often than expected by chance (i.e., the proportion of individuals with migraine who have another condition is higher than what would be predicted given the prevalence of the other condition among those without migraine). Comorbidities usually occur because they have a cause in common with migraine, because the other condition and migraine are indirectly related to another condition, or because the comorbidity causes migraine, or vice versa. Comorbidities can be defined as other chronic episodic disorders like asthma (i.e., acute episodes of signs and symptoms that can come and go for years) or chronic progressive conditions like cardiovascular disease (CVD). Chronic episodic conditions can be further stratified as psychiatric, neurologic, and other. Migraine comorbidities are common and diverse and include asthma, myocardial infarction, obstructive sleep apnea, systemic lupus erythematosus, Raynaud's phenomenon, and nonheadache pain disorders, including temporomandibular disorder (TMD) and fibromyalgia (Scher et al., 2005a; Rhode et al., 2007; De Tommaso et al., 2008). Neurologic comorbidities include stroke, restless-legs syndrome, multiple sclerosis, and epilepsy. Psychiatric comorbidities include depression, anxiety, and bipolar disorder. A number of studies indicate that migraine co-occurs with depression, anxiety disorders, bipolar disorder, posttraumatic stress disorder, personality disorders, and suicide attempts, among other psychiatric conditions. Among these, depression and anxiety have received the most attention in population and clinical studies. In a large health plan-based study, the prevalence of major depression was three times higher in migraineurs when compared to a nonmigraine control group (Patel et al., 2004). A large Norwegian population study assessed depression and anxiety disorders in more than 50 000 persons with migraine, other nonmigraine headache, and headache-free controls. After adjustment for other psychiatric comorbidities (Zwart et al., 2003), depression and anxiety were 1.9 (95% CI 1.6–2.3) and 2.7 (95% CI 2.4–3.1) times more common, respectively, among those with migraine when compared to headache-free controls (Zwart et al., 2003).

Three longitudinal population-based studies reported a bidirectional association between migraine and depression. In these studies, the risk of new-onset migraine in people with depression ranged from 2.8 to 3.5. Conversely, the risk of new onset of depression in people with migraine ranged from 2.4 to 5.8. Ashina et al. (2012) used data from the AMPP study to assess the role of depression as a predictor of new onset of chronic migraine in individuals with episodic migraine in the preceding year. In fully adjusted models, depression was a significant predictor of chronic migraine onset (OR = 1.7; 95% CI 1.1–2.5) and risk increased with the severity of depression, suggesting a dose–response effect.

Epilepsy is also a well-studied comorbidity of migraine. The Epilepsy Family Study of Columbia University revealed that the incidence of epilepsy is two times higher among those with migraine and that the incidence of migraine is two times higher among those with epilepsy (Ottman and Lipton, 1994).

Several chronic pain disorders are known to be associated with migraine and these associations are consistent with a central nervous system sensitization hypothesis (Woolf, 2011). Von Korff et al. (2005) found that patients with self-reported chronic spine pain had a 5.2 OR for migraine. Fibromyalgia was very common in patients with migraine, where the prevalence of diagnosis was between 22% and 40% (Peres et al., 2001; Ifergane et al., 2006; De Tommaso et al., 2009). The development of fibromyalgia was highly associated with migraine frequency. Patients who suffered from both migraine and fibromyalgia reported a higher prevalence of insomnia, lower quality of life, and more mental stress (Becker et al., 2008; De Tommaso et al., 2009).

Regarding asthma, one large epidemiologic study (Head-HUNT study) performed in Norway reported that both migraine and nonmigrainous headaches were related to asthma and chronic bronchitis (Aamodt et al., 2007). But, in another study, the incidence of newly diagnosed asthma was not different between those with and without migraine (OR 1.17) (Becker et al., 2008). The difference between these studies of comorbid occurrence and incidence of onset may be explained by those with migraine having asthma for a longer duration of time (i.e., a lower remission rate).

The association between migraine and ischemic stroke has been well studied. Stang et al. (2005) used data from the US Atherosclerosis Risk in Communities Study and showed that migraine with aura was associated with stroke symptoms (OR = 5.46), transient ischemic attack symptoms (OR = 4.28), and verified ischemic stroke events (OR = 2.81). Scher et al. (2005b) demonstrated that migraineurs (especially those with migraine with aura) have a higher CVD risk profile than people without migraine. In the Women's Health Study of nearly 40 000 female healthcare professionals aged 45 years or older, Kurth et al. (2005) studied migraine with aura, defined as a "warning" of impending (3610 females) attacks (Kurth et al., 2006). Females with active migraine with aura had a twofold increase in risk for major CVD, ischemic stroke, nonfatal myocardial infarction, coronary revascularization, angina, and death. In the Physicians' Health Study of males aged 40 years or older without a history of CVD, cancer, or other major illnesses (Kurth et al., 2007), those with migraine were at increased risk of major CVD and myocardial infarction. This study did not differentiate between migraine with or without aura. A meta-analysis of the migraine stroke data showed that the risk for ischemic stroke was two times higher for those with migraine than for those without migraine. However, the higher risk was confined to those with migraine with aura and not found among those who only had migraine without aura (Schurks et al., 2009).

Migraine with aura is a risk factor for brain lesions. These lesions include deep white-matter lesions, subtentorial white-matter lesions, and stroke-like posterior circulation areas. A large population study of persons with migraine with aura , migraine without aura , and controls showed that posterior circulation stroke-like lesions were more likely in the migraine with aura group and more common in those with more attacks of migraine with aura (Kruit et al., 2004). The adjusted OR for migraine with aura was 13.7, with an OR of 15.8 for those with migraine with aura attacks occurring more than once per month.

Both clinic- and population-based studies have demonstrated higher rates of comorbid medical and psychiatric conditions in persons with chronic migraine (relative to episodic migraine) (Buse et al., 2010; Chen et al., 2012). Those with chronic migraine are approximately twice as likely as those with episodic migraine to have depression, anxiety, and chronic pain. Respiratory disorders, including asthma, bronchitis, and chronic obstructive pulmonary disease, and cardiac risk factors, including hypertension, diabetes, high cholesterol level, and obesity, were also significantly more likely to be reported by those with chronic migraine (Stokes et al., 2011). The International Burden of Illness study also revealed a similar pattern (Stokes et al., 2011).

Burden of tension-type headache

TTH is less severe than migraine. However, as previously noted, TTH is very common, affecting as many as 78% of the general population in the USA. In both sexes the prevalence of TTH peaks between 30 and 39 years of age at 42.3% in men and 46.9% in women. In those with TTH, frequency of attacks rather than severity of headaches is more often correlated with disability and HRQoL (Lenaerts, 2006).

Female gender, low education level, sleeping problems, analgesic overuse, depression, and anxiety are risk factors for progression from ETTH to the chronic form of the disease (Lipton et al., 2013). The 2–3% of the population that suffers from the chronic form of TTH do not have specific treatment options. This is in contrast to ETTH, for which there are a diversity of treatment options. As a consequence, those with CTTH may have a higher risk of medication overuse, more disability and comorbidity. In the study by Holroyd et al. (2000), CTTH had a profound negative effect on the emotional life of affected patients. They were seven times more likely than control subjects to be classified as impaired on all subscales of the applied quality of life instruments (Jensen, 2003).

The work impact of TTH is substantial. In a prospective population-based study by Schwartz et al. (1998), 8.3% of participants with ETTH reported absenteeism and 43.6% reported presenteeism during 1-year follow-up. The means of absenteeism and presenteeism were 8.9 and 5.0 days per year respectively; 12% of respondents with CTTH reported absenteeism and 47% reported presenteeism during the same period of time. However both presenteeism and absenteeism were three times higher among those with CTTH. These findings demonstrate the greater individual impact of CTTH; however CTTH is relatively uncommon and has a far smaller impact on the population than ETTH (Lyngberg et al., 2005a).

Comorbidities of TTH

TTH, and particularly CTTH, are associated with certain medical and psychiatric conditions. Psychiatric disorders such as anxiety and depression are common among those with TTH. In a population-based study in Brazil, generalized anxiety was found to be the most frequent comorbidity (38.5%), followed by major depression (32.7%) in those with TTH (da Silva et al., 2010). Psychiatric disorders occur more frequently in CTTH than ETTH and they are more likely to contribute to an increased burden of accompanying headache symptoms (Juang et al., 2000).

An association between TMD and TTH has been reported by several studies. Prevalence of headache in those with TMD is significantly higher compared to those without TMD (73% vs 38% respectively) (Ciancaglini and Radaelli, 2001). TMD may play an important role in the progression from ETTH to CTTH through pericranial and central nociceptors which converge in the trigeminal nucleus caudalis (Ciancaglini and Radaelli, 2001).

It is important to identify and treat psychiatric and physical comorbidities associated with TTH as they confer a worse prognosis and increase healthcare costs and burden of TTH at both individual and societal level. Treatment should address both headache and associated comorbidities in order to be effective.

Epidemiologic studies provide evidence that the magnitude of headache burden varies by type of headache, frequency of attacks, and number of days with pain, severity, symptom burden, and comorbidities. Societal and employer impact depend on prevalence and demographic distribution, lost productive time, cost of lost productive time, and associated comorbidities. Cost also extends more to governmental agencies when burden of headache includes underemployment and unemployment. Migraine has a considerable impact on functional capacity, resulting in disrupted work and social activities: many migraineurs do not seek medical attention because they have not been accurately diagnosed by a physician or do not use prescribed medication. As a conclusion, the identification of patients who need more care and their treatment should reduce the impact of migraine on the individual and the burden of migraine on society.

TREATMENT

Pharmacologic treatment of migraine

Treatment of migraine headache can be acute or preventive. The goals of acute treatment are timely alleviation of pain and associated symptoms without recurrence, the restoring of ability to function, minimizing the use of back-up and rescue medications with minimal adverse events, and providing the best cost-effective management. An effective treatment is defined as complete headache relief within 2 hours or reducing pain intensity from severe to mild or moderate to no pain within 2 hours (Matchar et al., 2000).

The goals of migraine-preventive medications are decreasing future attack frequency, severity, and duration; improving responsiveness to acute treatments, and improving overall function and decreasing disability. Preventing analgesic overuse and "transformation" to

CDH is another goal of preventive treatment. It has been shown that preventive medications improve quality of life and health outcomes (D'Amico et al., 2006), and reduce disability and healthcare costs (Brown et al., 2005).

Preventive medications should be considered in migraine patients "reporting either 6 or more headache days per month, 4 or more headache days with at least some impairment, or 3 or more headache days with severe impairment or requiring bed rest," and should be considered for patients "with 4 or 5 migraine days per month with normal functioning, 3 migraine days with some impairment, or 2 migraine days with severe impairment" (Lipton et al., 2007).

Acute treatment

Acute treatments can be specific to migraine or nonspecific. Migraine-specific medications include triptans and ergot alkaloids. Nonspecific treatments include the nonsteroidal anti-inflammatory drugs (NSAIDs), opioid analgesics, neuroleptics, and combination analgesics, among others The first choice for mild to moderate migraine attacks is NSAIDs such as naproxen, ibuprofen, and diclofenac. Acetaminophen is usually used in combination with aspirin and caffeine (Lipton et al., 1998). Ibuprofen and diclofenac potassium have very rapid absorption from the gastrointestinal tract, and therefore the potential for a rapid onset of action. Naproxen sodium, on the other hand, has slower absorption but a much longer half-life. There is a soluble formulation of diclofenac potassium with a buffer that is dissolved in water and ingested (Cambia); it works faster and on more patients than the same 50 mg of diclofenac potassium as a tablet (Diener et al., 2006). It is the only NSAID that is approved by the US Food and Drug Administration (FDA) for migraine.

It has been shown that a combination of acetaminophen, aspirin, and caffeine has greater efficacy than any one of its components in comparable doses alone (Diener et al., 2005). Combination analgesics are associated with a higher risk of rebound headache and medication overuse headache and should not be used more than 10 days per month (Headache Classification Committee of the International Headache Society (HIS), 2013). NSAIDs are contraindicated if the patient has a history of gastrointestinal ulcers and hemorrhagic diathesis. Acetaminophen is generally safe; however it is contraindicated in the presence of hepatic or renal insufficiency.

Triptans are agonists at 5-HT1B/D receptors. Stimulation of presynaptic 5-HT1D receptors inhibits release of calcitonin gene-related peptide, substance P, and neurokinin A, and therefore prevents dural vasodilation and plasma extravasation (Moskowitz, 1992; Shepheard et al., 1997). The 5HT1D receptors are primarily located in the trigeminal nuclei and nerves and vascular smooth muscle. Since these receptors are also localized on noncranial vessels, some coronary-artery vasoconstriction is unavoidable (Humphrey and Feniuk, 1991; Ferrari and Saxena, 1993).

Seven derivatives of the triptan family are currently available in North America, along with their various formulations. In general, two of them, frovatriptan and naratriptan, have relatively long plasma half-lives along with slow onset of pain relief and a low headache recurrence rate in the following 24 hours if relief occurs. Oral rizatriptan and eletriptan have a fast onset of action. Tablet formulation can play a significant role, and a systematic review found that the onset of action of the fast-dissolving sumatriptan tablet was comparable to rizatriptan, one of the oral triptans that have fast onset of action (Pascual et al., 2007). Among the triptans, subcutaneous sumatriptan (6 mg) has shown the highest headache response rates in patient groups (Silberstein et al., 2008), and it should be considered in patients with severe migraine attacks, particularly if they have not had a satisfactory response to oral triptans. Intranasal zolmitriptan (5 mg) can also be a helpful option for these patients (Charlesworth et al., 2003). It has been shown that absorption in migraineurs is compromised both during a migraine attack and between attacks (Aurora et al., 2007). When this is the case, a nasal spray, such as zolmitriptan, or a subcutaneous injection of sumatriptan would be a better choice.

Investigators conducted a meta-analysis of 53 clinical trials, involving 24 089 patients. Compared with 100 mg sumatriptan, 10 mg rizatriptan and 80 mg eletriptan showed higher response rate (improvement from moderate or severe to mild or no pain in 2 hours) and painfree rate (improvement to no pain in 2 hours). Almotriptan also showed higher painfree rate and 12.5 mg almotriptan, 2.5 mg naratriptan showed better tolerability (Ferrari, 2003). In another meta-analysis among triptans, eletriptan provided the highest 24-hour sustained free rate with no recurrence of headache for 24 hours after treatment (Thorlund et al., 2014).

Triptans are effective during the prodrome and ictal phases of attack; however they are more effective if taken at the beginning of the ictal phase of an attack when the pain is still mild (Burstein et al., 2004; Dowson et al., 2004). The differences in response to one particular triptan among patients are considered to be greater than the efficacy differences between the oral triptans, and side effect profiles and headache recurrence rates are additional factors to be considered when deciding which triptan to be used. A meta-analysis found that triptans also reduce the severity of, or eliminate, migraine-associated symptoms such as nausea, photophobia, and phonophobia (Ferrari, 2003).

The response rate of triptans is about 60–80% and on average 40% of patients fail to respond to a particular triptan. Nonresponder is defined if the same triptan is attempted three times in three different attacks but fails to abort the attacks. A Nonresponder to one particular triptan may have benefit from another triptan (Ferrari, 2003).

There is some evidence that combining sumatriptan with naproxen provides greater efficacy than using either drug alone. The combination provides a higher 2-hour headache relief rate than either drug alone, with a 65% rate for a combination of sumatriptan 85 mg/naproxen sodium 500 mg as compared to 55% for sumatriptan monotherapy ($p=0.009$) in one study, and 57% vs 50% in another ($p=0.02$) (Brandes et al., 2007). The combination also has been shown to provide a higher 24-hour sustained pain response (no more than mild pain and no rescue medication at 2 hours and for 24 hours postdose) than either drug alone (Smith et al., 2005).

Triptans with a longer half-life have a lower rate of recurrence, and a combination of triptan and NSAIDs might help to avoid recurrence of headache, decrease lost productive time in the workplace, and improve nonworkplace activity (Brandes et al., 2007). Two double-blind single-dose randomized trials compared a combination of sumatriptan/naproxen sodium with sumatriptan only, naproxen sodium only, or placebo. The means of total lost productive time and nonwork activity time were significantly lower in the sumatriptan/naproxen sodium group compared with any other alternative treatment groups (Landy et al., 2007).

Combination analgesics with opioids including codeine should not be used routinely in migraine because there is not enough evidence for efficacy beyond that provided by NSAIDs and there is a risk of overuse, with resultant medication overuse headache. When their use is necessary because of unresponsiveness or contraindications to other medications, combination analgesics with tramadol may be the preferred option (Silberstein et al., 2005). Administration of barbiturate-containing analgesics and opioids like morphine and butorphanol should be limited to exceptional cases because of the risk of addiction and medication overuse headache (Silberstein, 2000).

Ergot alkaloids are also effective in the treatment of acute migraine attack. Compared with triptans, they are less effective at reducing the intensity of headache from severe or moderate to mild or none, but have a longer duration of action. One of the side effects of ergot alkaloids as well as ergot combinations (ergot–caffeine–pentobarbital or ergotamine–metoclopramide) is nausea and vomiting and it should be distinguished from the nausea and vomiting associated with migraine attack. Ergots have the same contraindication profile as triptans and also carry a high risk for inducing medication overuse headache (Wammes-van der Heijden et al., 2006).

To avoid medication overuse headache, use of migraine-specific drugs should be limited to 10 intake days per month and migraine-nonspecific drugs can be used on up to 15 days a month (Headache Classification Committee of the International Headache Society (IHS), 2013).

Preventive treatment

Acute treatment is effective for the control of acute onset of migraine but it is not sufficient to prevent the occurrence of future attacks or to address the needs of the 20% of patients who do not respond to abortive therapies, including injectable sumatriptan (Farmer et al., 2001). Lack of effective pain control may lead to medication overuse and rebound headaches (Rapoport, 1998). Preventive medications offer an alternative both for patients who do not respond to abortive therapy and to those with frequent attacks.

An "expert panel" of migraine providers arrived at a consensus statement that preventive medications should be offered to migraine patients "reporting either 6 or more headache days per month, 4 or more headache days with at least some impairment, or 3 or more headache days with severe impairment or requiring bed rest," and should be considered for patients "with 4 or 5 migraine days per month with normal functioning, 3 migraine days with some impairment, or 2 migraine days with severe impairment." (Lipton et al., 2007) They also agreed that preventive medications are not indicated for patients "with less than 4 headache days per month and no impairment, or subjects with no more than 1 headache day per month regardless of the impairment." By these criteria, approximately 38% of patients with migraine would likely benefit from preventive medications (Lipton et al., 2007).

Most preventive treatments are generic medications initially developed for other problems, such as anticonvulsants, antidepressants, and antihypertensive medications. Only four drugs (propranolol, timolol, divalproex sodium, and topiramate) are currently approved by the US FDA for episodic migraine prevention and only one drug (onabotulinumtoxinA) for chronic migraine prevention. A 50% reduction in frequency in headache days is often considered successful therapy and a trial of at least 2 months at the appropriate dosage is typically required before effectiveness can be adequately assessed. Polypharmacy with more than one preventive medication may improve overall effectiveness in some patients (Goslin et al., 1999; Matchar et al., 2000).

Among the anticonvulsants, efficacy evidence is strongest for valproic acid and topiramate. Topiramate (i.e., 100 mg/day) was evaluated among 325 employed migraineurs in two double-blind randomized trials for a period of 6 month. Topiramate significantly reduced the lost productive work time, expressed as the average absenteeism and presenteeism per person-month at 3.9 and 5.6 days (i.e., 9.5 total days/month) in the treatment group, compared to 2.3 and 5.2 days (i.e., 7.5 total days) in the placebo group (Lofland et al., 2007).

Antidepressants that are widely used for migraine prevention include tricyclic antidepressants such as amitriptyline, doxepin, desipramine, and protriptyline. Protriptyline is a convenient choice for patients who are obese or concerned about weight gain. Protriptyline offers the advantage of no sedative effect but the risk and severity of dry mouth and constipation are higher than with other antidepressants. Venlafaxine, fluoxetine, and monoamine oxidase inhibitors (MAOIs) such as phenelzine are also effective in preventing migraine attacks. All patient taking MAOIs must be on a tyramine-restricted diet and sympathomimetic drugs must be vigorously avoided (Silberstein and Marmura, 2010).

Antihypertensives include beta-blockers such as propranolol, timolol, and metoprolol used for migraine prevention medications. Metoprolol has less bronchoconstrictive effect and is an option for patients with mild asthma. Beta-blockers may rarely produce depression; however, they are useful in controlling symptoms of anxiety which often co-occurs with migraine. Angiotensin-converting enzyme inhibitors such as lisinopril and angiotensin receptor blockers such as candesartan and telmisartan are effective in migraine prevention and have been studied in placebo-controlled trials. Persistent cough that does not improve with continued use is an adverse effect that occurs in about 1 in 4 patients who use lisinopril and that leads to discontinuation (Silberstein and Marmura, 2010).

Botulinum toxin-A is among the recent proven treatments for migraine preventions. The PREEMPT (Phase III Research Evaluating Migraine Prophylaxis Therapy) study has shown that treatment of chronic migraine with onabotulinumtoxinA results in a significant and clinically meaningful reduction in headache impact and improvement in HRQoL (Lipton et al., 2011).

A number of acute analgesic NSAID medications (e.g., fenoprofen, ibuprofen, ketoprofen, naproxen) may be effective as preventive medications when taken on a daily basis; however caution should be exercised due to potential concerns for induction of medication overuse headache or transformed migraine, depending on the frequency of baseline headaches (Lipton, 2008).

The American Headache Society and the American Academy of Neurology issued updated evidence-based practice guidelines for use of medications and complementary therapies in prevention of episodic migraine in adults (Holland et al., 2012; Silberstein et al., 2012). Recommendations from these guidelines are based solely on efficacy results, rather than side effects or costs, and are summarized in Table 25.1.

Natural preventives

Petasites is a purified extract of the butterbur plant. Two placebo-controlled trials have shown that petasites at a dose of 75 mg twice daily is effective in preventing migraine headache. The only common side effect is burping (Diener et al., 2004; Lipton et al., 2004). The effectiveness of the other natural products, including riboflavin (Schoenen et al., 1998), magnesium (Peikert et al., 1996), coenzyme-Q (Sándor et al., 2005), and feverfew (Murphy et al., 1988), has been shown in randomized clinical trials.

Nonpharmacologic approach

Goals of behavioral and physical treatments for migraine include reduction of frequency and severity of headache and headache-related disability; reduction of reliance on poorly tolerated pharmacotherapies, headache-related distress, and psychologic symptoms (Campbell et al., 2000a).

Non-pharmacologic interventions are particularly preferred in patients with poor tolerance or contraindications for specific pharmacologic treatments, insufficient or no response to pharmacologic treatment; during pregnancy, planned pregnancy, or nursing. It is also preferred to prevent excessive use of analgesics or acute medications or in the patient with significant stress or deficient stress-coping skills (Campbell et al., 2000a). Several nonpharmacologic modalities are supported by strong evidence (grade A). These include relaxation training, biofeedback, and cognitive-behavioral therapy (CBT) (Campbell et al., 2000a). CBT strategies focus on identifying and challenging maladaptive and dysfunctional thoughts, beliefs, and response to stress (Beck et al., 1979). For migraine management, the cognitive foci of CBT include enhancing self-efficacy, encouraging adoption of internal locus of control, and reducing or eliminating catastrophizing thoughts (Heath et al., 2008). Biofeedback is used to make patients aware of physiologic functions associated with migraine and stress, and offer a means for patients to learn to control their physiologic response (Holroyd and Penzien, 1994). Some relaxation skills, such as diaphragmatic breathing or visualization to induce "relaxation response," are often incorporated into biofeedback training. Through relaxation training, patients learn to minimize physiologic responses to stress and decrease sympathetic arousal (Rime and Andrasik, 2011). There are several

Table 25.1

American Headache Society – American Academy of Neurology guidelines for preventive drugs for episodic migraine

Drug	Examples of studied doses
Level A: established as effective	
Should be offered to patients requiring migraine prophylaxis	
Divalproex/sodium valproate*	400–1000 mg/day
Metoprolol	47.5–200 mg/day
Petasites (butterbur)	50–75 mg bid
Propranolol*	120–240 mg/day
Timolol	10–15 mg bid
Topiramate*	25–200 mg/day
Level B: probably effective	
Should be considered for patients requiring migraine prophylaxis	
Amitriptyline	25–150 mg/day
Fenoprofen	200–600 mg tid
Feverfew	50–300 mg bid; 2.08–18.75 mg tid for MIG-99
Histamine	1–10 ng subcutaneously twice a week
Ibuprofen	200 mg bid
Ketoprofen	50 mg tid
Magnesium	600 mg tri-magnesium dicitrate qd
Naproxen/naproxen sodium	500–1100 mg/day for naproxen; 550 mg bid for naproxen sodium
Riboflavin	400 mg/day
Venlafaxine	150 mg extended-release/ day
Atenolol	100 mg/day
Level C: possibly effective	
May be considered for patients requiring migraine prophylaxis	
Candesartan	16 mg/day
Carbamazepine	600 mg/day
Clonidine	0.75–0.15 mg/day; patch formulations also studied
Guanfacine	0.5–1 mg/day
Lisinopril	10–20 mg/day
Nebivolol	5 mg/day
Pindolol	10 mg/day
Flurbiprofen	200 mg/day
Mefenamic acid	500 mg tid
Coenzyme Q10	100 mg tid
Cyproheptadine	4 mg/day

*Food and Drug Administration-approved.

Table derived from Loder et al. (2012), with permission. Dosages listed are those tested and not intended specifically as a recommendation for treatment. All drugs given orally unless otherwise noted.

bid, twice a day; MIG-99, carbon dioxide extracted feverfew; qd, daily; tid, three times a day.

techniques, such as meditation, yoga, and guided relaxation, that have successfully been used to achieve the benefits from relaxation.

Patient education regarding genetic predisposition, hormonal factors, stress response, and relevant cognitive, emotional, and behavioral factors can improve patients' understanding of the how to use pharmacologic treatment and how they work, as it does for behavioral treatment. Review of patient diary at medical visits helps both patient and provider to understand potential factors involved in developing headache in patients (Packard, 1979; Berliner et al., 1999).

Physical treatment such as acupuncture, chiropractic, physical and occupational therapies and massage may have some benefit for patients for whom various soft-tissue, neurogenic, or osseous structures of the head, neck, or upper body contribute to triggering migraine attacks.

Treatment of TTH

Pharmacologic treatment of ETTH

NSAIDs have remained the mainstay of treatment for ETTH. Acetaminophen and aspirin are effective compared to placebo. However, acetaminophen is associated with fewer gastrointestinal side effects. Combination medications such as naproxen sodium or aspirin and acetaminophen with caffeine are also effective in treatment of ETTH. Caffeine increases the efficacy of these other treatments, although the addition of opioids to these combinations does not improve efficacy (Bendtsen et al., 2010).

Pharmacologic treatment of CTTH

Treatment of CTTH with acute treatment medications should be limited to avoid the development of medication overuse headache. The goal of preventive medications for headache is a reduction of 50% or more in attack frequency, and primary outcome should be measured by the number of headache days or area under the headache curve (AUC). AUC is also called headache index and calculated by multiplying the duration and intensity of the headache (Lenaerts, 2009).

Tricyclic antidepressants such as amitriptyline, nortriptyline, and clomipramine are effective as preventive medications for the treatment of CTTH. The effectiveness of tricyclic antidepressants increases over time and amitriptyline is the drug of choice for preventive treatment of CTTH (Bendtsen et al., 1996). Mirtazapine increases noradrenergic and serotoninergic transmission and, in a placebo-controlled, double-blind trial, a dose of 30 mg/day reduced the AUC by 34% compared with placebo. Mirtazapine is recommended as a second-line treatment for patients who cannot tolerate or fail to respond to amitriptyline (Bendtsen and Jensen, 2004). Venlafaxine, a serotonin and norepinephrine reuptake inhibitor, reduced average headache per month but did not improve the headache intensity (Zissis et al., 2007). There is not enough evidence that selective serotonin reuptake inhibitors are effective in the treatment of CTTH.

Nonpharmacologic treatments of CTTH

Trigger identification and avoidance are important steps in the treatment of TTH. The most common triggers are physical and mental stress, dehydration, inadequate and inappropriate food intake, too much or too little sleep, and hormonal changes (Bendtsen et al., 2010). The US Headache Consortium 2000 recommended relaxation training, thermal biofeedback combined with relaxation training, electromyogram biofeedback and CBT as grade A evidence-based strategies for the treatment of CTTH (Campbell et al., 2000b). In a randomized, placebo-controlled trial, patients treated with stress management and antidepressants were more likely to experience a 50% or greater reduction in headache index scores than those receiving placebo or either treatment alone (Verhagen et al., 2009).

Despite several randomized trials, the effectiveness of spinal manipulation, physical therapy, and acupuncture is controversial in the treatment of CTTH.

References

Aamodt AH, Stovner LJ, Langhammer A et al. (2007 Feb). Is headache related to asthma, hay fever, and chronic bronchitis? The Head-HUNT Study. Headache 47 (2): 204–212.

Ashina S, Serrano D, Lipton RB et al. (2012). Depression and risk of transformation of episodic to chronic migraine. J Headache Pain 13 (8): 615–624.

Aurora S, Kori S, Barrodale P et al. (2007). Gastric stasis occurs in spontaneous, visually induced, and interictal migraine. Headache 47: 1443–1446.

Beck AT, Shaw BF, Emery G (1979). Cognitive therapy of depression, Guilford Press, New York.

Becker C, Brobert GP, Almqvist PM et al. (2008 Apr). The risk of newly diagnosed asthma in migraineurs with or without previous triptan prescriptions. Headache 48 (4): 606–610.

Bendtsen L, Jensen R (2004). Mirtazapine is effective in the prophylactic treatment of chronic tension-type headache. Neurology 62 (10): 1706–1711.

Bendtsen L, Jensen R, Olesen J (1996). A non-selective (amitriptyline), but not a selective (citalopram), serotonin reuptake inhibitor is effective in the prophylactic treatment of chronic tension-type headache. J Neurol Neurosurg Psychiatry 61 (3): 285–290.

Bendtsen L, Evers S, Linde M et al. (2010 Nov). EFNS guideline on the treatment of tension-type headache - report of an EFNS task force. Eur J Neurol 17 (11): 1318–1325.

Berliner R, Solomon S, Newman LC et al. (1999). Migraine: clinical features and diagnosis. Compr Ther 25: 397–402.

Bigal ME, Serrano D, Buse D et al. (2008a). Acute migraine medications and evolution from episodic to chronic migraine: a longitudinal population-based study. Headache 48: 1157–1168.

Bigal ME, Serrano D, Reed M et al. (2008 Aug 19b). Chronic migraine in the population: burden, diagnosis, and satisfaction with treatment. Neurology 71 (8): 559–566.

Brandes JL, Kudrow D, Stark SR et al. (2007). Sumatriptannaproxen for acute treatment of migraine: a randomized trial. JAMA 297: 1443–1454.

Breslau N, Lipton RB, Stewart WF et al. (2003). Comorbidity of migraine and depression: investigating potential etiology and prognosis. Neurology 22: 1308–1312.

Brown JS, Papadopoulos G, Neumann PJ et al. (2005). Cost-effectiveness of topiramate in migraine prevention: results from a pharmacoeconomic model of topiramate treatment. Headache 45: 1012–1022.

Burstein R, Collins B, Jakubowski M (2004 Jan). Defeating migraine pain with triptans: a race against the development of cutaneous allodynia. Ann Neurol 55 (1): 19–26.

Buse DC, Bigal MB, Rupnow M et al. (2007). Development and validation of the Migraine Interictal Burden Scale (MIBS): a self-administered instrument for measuring the burden of migraine between attacks. Neurology 68 (Suppl. 1): A89.

Buse DC, Manack A, Serrano D et al. (2010). Sociodemographic and comorbidity profiles of chronic migraine and episodic migraine sufferers. J Neurol Neurosurg Psychiatry 81: 428–432.

Buse DC, Manack AN, Fanning K et al. (2012). Chronic migraine prevalence, disability, and sociodemographic factors: results from the American Migraine Prevalence and Prevention Study. Headache 52: 1456–1470.

Buse DC, Loder E, Gorman JA et al. (2013 Sep). Sex differences in prevalence, symptoms, and other features of migraine, probable migraine and other severe headache: results of the American Migraine Prevalence and Prevention Study. Headache 53 (8): 1278–1299.

Campbell JK, Freitag FG, Frishberg B et al. (2000a). Evidence-Based Guidelines For Migraine Headache: Behavioral and Physical Treatments. US Headache Consortium. http://www.researchgate.net/profile/Donald_Penzien/publication/254479333_Evidenced-Based_Guidelines_For_Migraine_Headache_Behavioral_and_Physical_Treatments/links/546c3cbd0cf20dedafd540f5.pdf.

Campbell JK, Penzien DB, Wall EM (2000b). Evidence-Based guideline for Migraine treatment: Behavioral and Physical Treatments, Available online at: http:/www.aan.com/professionals/practice/pdfs/gl0089.pdf.

Charlesworth BR, Dowson AJ, Purdy A et al. (2003). Speed of onset and efficacy of zolmitriptan nasal spray in the acute treatment of migraine: a randomised, doubleblind, placebo-controlled, dose-ranging study vs zolmitriptan tablet. CNS Drugs 17: 653–667.

Chen YC, Tang CH, Ng K et al. (2012). Comorbidity profiles of chronic migraine sufferers in a national database in Taiwan. J Headache Pain 13: 311–319.

Ciancaglini R, Radaelli G (2001 Feb). The relationship between headache and symptoms of temporomandibular disorder in the general population. J Dent 29 (2): 93–98.

D'Amico D, Solari A, Usai S et al. (2006). Improvement in quality of life and activity limitations in migraine patients after prophylaxis. A prospective longitudinal multicentre study. Cephalalgia 26: 691–696.

da Silva A, Costa Jr EC, Gomes JB et al. (2010 Sep). Chronic headache and comorbidities: a two-phase, population-based, cross-sectional study. Headache 50 (8): 1306–1312.

De Tommaso M, Sardaro M, Serpino C et al. (2008). Fibromyalgia comorbidity in primary headaches. Cephalalgia 29: 453–464.

de Tommaso M, Sardaro M, Serpino C et al. (2009). Fibromyalgia comorbidity in primary headaches. Cephalalgia 29 (4): 453–464.

Diener HC, Rahlfs VW, Danesch U (2004). The first placebo-controlled trial of a special butterbur root extract for the prevention of migraine: reanalysis of efficacy criteria. Eur Neurol 51 (2): 89–97.

Diener HC, Pfaffenrath V, Pageler L et al. (2005). The fixed combination of acetylsalicylic acid, paracetamol and caffeine is more effective than single substances and dual combination for the treatment of headache: a multicentre, randomized, double-blind, single-dose, placebo-controlled parallel group study. Cephalalgia 25: 776–787.

Diener H-C, Montagna P, Gacs G et al. (2006). Efficacy and tolerability of diclofenac potassium sachets in migraine: a randomized, double-blind, cross-over study in comparison with diclofenac potassium tablets and placebo. Cephalalgia 26: 537–547.

Dowson AJ, Massiou H, Lainez JM et al. (2004). Almotriptan improves response rates when treatment is within 1 hour of migraine onset. Headache 44 (4): 318–322.

Eaton WW, Day R, Kramer M (1988). The use of epidemiology for risk factor research in schizophrenia: an overview in methodologic critique. In: MT Tsuang, JC Simpson (Eds.), Handbook of schizophrenia-nosology, epidemiology and genetics. Elsevier Science, New York, NY, pp. 169–204.

Farmer K, Cady R, Bleiberg J et al. (2001). Sumatriptan nasal spray and cognitive function during migraine: results of an open-label study. Headache 41 (4): 377–384.

Ferrari MD (2003). Current perspectives on effective migraine treatments: are small clinical differences important for patients? Drugs Today (Barc) 39(Suppl D): 37–41.

Ferrari MD, Saxena PR (1993). Clinical and experimental effects of sumatriptan in humans. Trends Pharmacol Sci 14: 129–134.

Goslin RE, Gray RN, McCrory DC, et al. Behavioral and Physical Treatments for Migraine Headache. Technical Review 2.2, February 1999. Prepared for the Agency for Health Care Policy and Research under Contract No. 290-94-2025. Available from the National Technical Information Service; NTIS Accession No. 127946.

Hawkins K, Wang S, Rupnow MF (2007 Apr). Indirect cost burden of migraine in the United States. J Occup Environ Med 49 (4): 368–374.

Headache Classification Committee of the International Headache Society (IHS) (2013). The International Classification of Headache Disorders, 3rd edition (beta version). Cephalalgia 33: 629–808.

Headache Classification Subcommittee of International Headache Society (2004). The International Headache Disorders, 2nd edition. Cephalalgia 24 (suppl 1): 9–160.

Heath RL, Saliba M, Mahmassani O et al. (2008 Oct). Locus of control moderates the relationship between headache pain and depression. J Headache Pain 9 (5): 301–308.

Holland S, Silberstein SD, Freitag F et al. (2012). Evidence-based guideline update: NSAIDs and other complementary treatments for episodic migraine prevention in adults: report of the Quality Standards Subcommittee of the American Academy of Neurology and the American Headache Society. Neurology 78: 1346–1353.

Holroyd KA, Penzien DB (1994). Psychosocial interventions in the management of recurrent headache disorders. 1: Overview and effectiveness. Behav Med. Summer: 20 (2): 53–63.

Holroyd KA, Stensland M, Lipchik GL et al. (2000). Psychosocial correlates and impact of chronic tension-type headache. Headache 40: 3–16.

u XH, Markson LE, Lipton RB et al. (1999 Apr 26). Burden of migraine in the United States: disability and economic costs. Arch Intern Med 159 (8): 813–818.

umphrey PPA, Feniuk W (1991). Mode of action of the anti-migraine drug sumatriptan. Trends Pharmacol Sci 12: 444–446.

ergane G, Buskila D, Simiseshvely N et al. (2006 Apr). Prevalence of fibromyalgia syndrome in migraine patients. Cephalalgia 26 (4): 451–456.

ensen R (2003 Dec). Diagnosis, epidemiology, and impact of tension-type headache. Curr Pain Headache Rep 7 (6): 455–459.

uang KD, Wang SJ, Fuh JL et al. (2000). Comorbidity of depressive and anxiety disorders in chronic daily headache and its subtypes. Headache 40 (10): 818–823.

Katsarava Z, Schneeweiss S, Kurth T et al. (2004). Incidence and predictors for chronicity of headache in patients with episodic migraine. Neurology 62: 788–790.

Kruit MC, van Buchem MA, Hofman PAM et al. (2004). Migraine as a risk factor for subclinical brain lesions. JAMA 291: 427–434.

Kurth T, Slomke MA, Kase CS et al. (2005). Migraine, headache, and the risk of stroke in women: a prospective study. Neurology 64: 1020–1026.

Kurth T, Gaziano JM, Cook NR et al. (2006). Migraine and the risk of cardiovascular disease in women. JAMA 296: 283–291.

Kurth T, Gaziano JM, Cook NR et al. (2007). Migraine and risk of cardiovascular disease in men. Arch Intern Med 167: 795–801.

Landy S, DeRossett SE, Rapoport A et al. (2007). Two double-blind, multicenter, randomized, placebo-controlled, single-dose studies of sumatriptan/naproxen sodium in the acute treatment of migraine: function, productivity, and satisfaction outcomes. Med Gen Med 9 (2): 53.

Launer L, Terwindt GM, Ferrari MD (1999). The prevalence and characteristics of migraine in a population-based cohort: the GEM study. Neurology 53: 537–542.

Lavados PM, Tenhamm E (1998). Epidemiology of tension-type headache in Santiago, Chile: a prevalence study. Cephalalgia 18: 552–558.

Lenaerts ME (2006 Dec). Burden of tension-type headache. Curr Pain Headache Rep 10 (6): 459–462.

Lenaerts ME (2009 Jun). Pharmacotherapy of tension-type headache (TTH). Expert Opin Pharmacother 10 (8): 1261–1271.

Liberatos P, Link BG, Kelsey JL (1988). The measurement of social class in epidemiology. Epidemiol Rev 10: 87–121.

Lipton RB (2008). Acute migraine medications and evolution from episodic to chronic migraine: a longitudinal population-based study. Headache 48: 1157.

Lipton RB, Bigal ME (2005 Apr). Migraine: epidemiology, impact, and risk factors for progression. Headache 45 (Suppl 1): S3–S13.

Lipton RB, Stewart WF, Ryan Jr RE et al. (1998 Feb). Efficacy and safety of acetaminophen, aspirin, and caffeine in alleviating migraine headache pain: three double-blind, randomized, placebo-controlled trials. Arch Neurol 55 (2): 210–217.

Lipton RB, Stewart WF, Diamond S et al. (2001). Prevalence and burden of migraine in the United States: data from the American Migraine Study II. Headache 41: 646–657.

Lipton RB, Liberman JN, Kolodner KB et al. (2003). Migraine headache disability and health-related quality-of-life: a population-based case-control study from England. Cephalalgia 23: 441–450.

Lipton RB, Göbel H, Einhäupl KM et al. (2004 Dec 28). *Petasites hybridus* root (butterbur) is an effective preventive treatment for migraine. Neurology 63 (12): 2240–2244.

Lipton RB, Bigal ME, Diamond M et al. (2007). Migraine prevalence, disease burden, and the need for preventive therapy. Neurology 68: 343–349.

Lipton RB, Bigal ME, Ashina S et al. (2008). Cutaneous allodynia in the migraine population. Ann Neurol 63: 148–158.

Lipton RB, Varon SF, Grosberg B et al. (2011 Oct 11). OnabotulinumtoxinA improves quality of life and reduces impact of chronic migraine. Neurology 77 (15): 1465–1472.

Lipton RB, Buse DC, Saiers J et al. (2013). Frequency and burden of headache-related nausea: results from the American Migraine Prevalence and Prevention (AMPP) study. Headache 53 (1): 93–103.

Loder E, Burch R, Rizzoli P (2012). The 2012 AHS/AAN guidelines for prevention of episodic migraine: a summary and comparison with other recent clinical practice guidelines. Headache 52: 930–945.

Lofland JH, Gagne JJ, Pizzi LT et al. (2007 Mar). Impact of topiramate migraine prophylaxis on workplace productivity: results from two US randomized, double-blind, placebo-controlled, multicenter trials. J Occup Environ Med 49 (3): 252–257.

Lyngberg AC, Rasmussen BK, Jørgensen T et al. (2005 Aug 23a). Prognosis of migraine and tension-type headache: a population-based follow-up study. Neurology 65 (4): 580–585.

Lyngberg AC, Rasmussen BK, Jørgensen T et al. (2005 Jun 1b). Incidence of primary headache: a Danish epidemiologic follow-up study. Am J Epidemiol 161 (11): 1066–1073.

Manack A, Buse DC, Serrano D et al. (2011 Feb 22). Rates, predictors, and consequences of remission from chronic migraine to episodic migraine. Neurology 76 (8): 711–718.

Matchar DB, Young WB, Rosenberg JH et al. (2000). Evidence-based guidelines for migraine headache in the primary care setting: pharmacological management of acute attacks. Available at http:/tools.aan.com/professionals/practice/pdfs/gl0087.pdf.

Meletiche DM, Lofland JH, Young WB (2001). Quality of life differences between patients with episodic and transformed migraine. Headache 41: 573–578.

Mennini FS, Gitto L, Martelletti P (2008 Aug). Improving care through health economics analyses: cost of illness and headache. J Headache Pain 9 (4): 199–206.

Moskowitz MA (1992). Neurogenic versus vascular mechanisms of sumatriptan and ergot alkaloids in migraine. Trends Pharmacol Sci 13: 307–312.

Murphy JJ, Heptinstall S, Mitchell JR (1988 Jul 23). Randomised double-blind placebo-controlled trial of feverfew in migraine prevention. Lancet 2 (8604): 189–192.

Natoli JL, Manack A, Dean B et al. (2010 May). Global prevalence of chronic migraine: a systematic review. Cephalalgia 30 (5): 599–609.

O'Brien B, Goeree R, Streiner D (1994). Prevalence of migraine headache in Canada: a population-based survey. Int J Epidemiol 23: 1020–1026.

Ottman R, Lipton RB (1994). Comorbidity of migraine and epilepsy. Neurology 44: 2105–2110.

Packard RC (1979 Nov). What does the headache patient want? Headache 19 (7): 370–374.

Pascual J, Mateos V, Roig C et al. (2007). Marketed oral triptans in the acute treatment of migraine: a systematic review on efficacy and tolerability. Headache 47: 1152–1168.

Patel NV, Bigal ME, Kolodner KB et al. (2004). Prevalence and impact of migraine and probable migraine in a health plan. Neurology 63: 1432–1438.

Peikert A, Wilimzig C, Köhne-Volland R (1996 Jun). Prophylaxis of migraine with oral magnesium: results from a prospective, multi-center, placebo-controlled and double-blind randomized study. Cephalalgia 16 (4): 257–263.

Peres MF, Young WB, Kaup AO et al. (2001 Oct 9). Fibromyalgia is common in patients with transformed migraine. Neurology 57 (7): 1326–1328.

Peres MF, Mercante JP, Guendler VZ et al. (2007). Cephalalgiaphobia: a possible specific phobia of illness. J Headache Pain 8: 56–59.

Rapoport AM (1998). Pharmacological prevention of migraine. Clin Neurosci 5 (1): 55–59.

Rasmussen BK (1992). Migraine and tension-type headache in a general population: psychosocial factors. Int J Epidemiol 21: 1138–1143.

Rasmussen BK, Jensen R, Schroll M et al. (1991). Epidemiology of headache in a general population: a prevalence study. J Clin Epidemiol 44: 1147–1157.

Rhode AM, Hosing VG, Happe S et al. (2007). Comorbidity of migraine and restless legs syndrome – a case-control study. Cephalalgia 27: 1255–1260.

Rime C, Andrasik F (2011). Relaxation techniques and guide imagery. In: SD Waldman (Ed.), Pain Management. Vol. 2. Saunders/Elsevier, Philadelphia, PA, pp. 967–977.

Robbins MS, Lipton RB (2010 Apr). The epidemiology of primary headache disorders. Semin Neurol 30 (2): 107–119.

Sándor PS, Di Clemente L, Coppola G et al. (2005 Feb 22). Efficacy of coenzyme Q10 in migraine prophylaxis: a randomized controlled trial. Neurology 64 (4): 713–715.

Scher AI, Lipton RB, Stewart WF (2003a). Habitual snoring as a risk factor for chronic daily headache. Neurology 60: 1366–1368.

Scher AI, Stewart WF, Ricci JA et al. (2003b). Factors associated with the onset and remission of chronic daily headache in a population-based study. Pain 106: 81–89.

Scher AI, Stewart WF, Lipton RB (2004). Caffeine as a risk factor for chronic daily headache. Neurology 63: 2022–2027.

Scher AI, Bigal ME, Lipton RB (2005a). Comorbidity of migraine. Curr Opin Neurol 18: 305–310.

Scher AI, Terwindt GM, Picavet HSJ et al. (2005b). Cardiovascular risk factors and migraine: the GEM population-based study. Neurology 64: 614–620.

Scher AI, Stewart WF, Buse D et al. (2008). Major life changes before and after the onset of chronic daily headache: a population-based study. Cephalalgia 28: 868–876.

Schoenen J, Jacquy J, Lenaerts M (1998 Feb). Effectiveness of high-dose riboflavin in migraine prophylaxis. A randomized controlled trial. Neurology 50 (2): 466–470.

Schurks M, Rist PM, Bigal ME et al. (2009). Migraine and cardiovascular disease: systematic review and meta-analysis. BMJ 339: 1–11.

Schwartz BS, Stewart WF, Lipton RB (1997 Apr). Lost workdays and decreased work effectiveness associated with headache in the workplace. J Occup Environ Med 39 (4): 320–327.

Schwartz BS, Stewart WF, Simon D et al. (1998). Epidemiology of tension-type headache. JAMA 279: 381–383.

Serrano D, Manack AN, Reed ML et al. (2013 Jan-Feb). Cost and predictors of lost productive time in chronic migraine and episodic migraine: results from the American Migraine Prevalence and Prevention (AMPP) Study. Value Health 16 (1): 31–38.

Shepheard SL, Williamson DJ, Beer MS et al. (1997). Differential effects of 5-HT1B-1D receptor agonists on neurogenic dural plasma extravasation and vasodilation in anaesthetized rats. Neuropharmacology 36: 525–533.

Silberstein SD (2000). Practice parameter: Evidence-based guidelines for migraine headache (an evidence based review): report of the Quality Standards Subcommittee of the American Academy of Neurology. Neurology 55: 754–762.

Silberstein SD, Marmura MJ (2010). Essential Neuropharmacology: The Prescriber's Guide. Cambridge University Press, New York.

Silberstein SD, Lipton RB, Solomon S et al. (1994). Classification of daily and near-daily headaches: proposed revisions to the IHS criteria. Headache 34: 1–7.

Silberstein SD, Freitag FG, Rozen TD et al. (2005). Tramadol/acetaminophen for the treatment of acute migraine pain: Findings of a randomized, placebo-controlled trial. Headache 45: 1317–1327.

Silberstein S, Tfelt-Hansen P, Dodick DW et al. (2008). Guidelines for controlled trials of prophylactic treatment of chronic migraine in adults. Cephalalgia 28: 484–495.

Silberstein SD, Holland S, Freitag F et al. (2012). Evidence-based guideline update: Pharmacologic treatment for episodic migraine prevention in adults: report of the Quality Standards Subcommittee of the American Academy of Neurology and the American Headache Society. Neurology 78: 1337–1345.

Smith TR, Sunshine A, Stark SR et al. (2005). Sumatriptan and naproxen sodium for the acute treatment of migraine. Headache 45: 983–991.

Stang P, Von Korff M, Galer BS (1998 May). Reduced labor force participation among primary care patients with headache. J Gen Intern Med 13 (5): 296–302.

Stang PE, Carson AP, Rose KM et al. (2005 May 10). Headache, cerebrovascular symptoms, and stroke: the Atherosclerosis Risk in Communities Study. Neurology 64 (9): 1573–1577.

Stewart WF, Lipton RB (1996). Work-related disability: results from the American Migraine Study. Cephalalgia 16: 231–238.
Stewart WF, Lipton RB, Celentano DD et al. (1992 Jan 1). Prevalence of migraine headache in the United States. Relation to age, income, race, and other sociodemographic factors. JAMA 267 (1): 64–69.
Stewart WF, Lipton RB, Liberman J (1996a). Variation in migraine prevalence by race. Neurology 47: 52–59.
Stewart WF, Lipton RB, Simon D (1996b). Work-related disability: results from the American migraine study. Cephalalgia 16 (4): 231–238, discussion 215.
Stewart WF, Ricci JA, Chee E et al. (2003 Nov 12). Lost productive time and cost due to common pain conditions in the US workforce. JAMA 290 (18): 2443–2454.
Stewart WF, Wood C, Reed ML et al. (2008 Nov). Cumulative lifetime migraine incidence in women and men. Cephalalgia 28 (11): 1170–1178.
Stewart WF, Wood GC, Bruce C et al. (2011 May). Longitudinal change in migraine headache-days and indirect cost consequences. J Occup Environ Med 53 (5): 478–487.
Stewart WF, Roy J, Lipton RB (2013 Sep 10). Migraine prevalence, socioeconomic status, and social causation. Neurology 81 (11): 948–955.
Stokes MS, Werner JB, Lipton RB et al. (2011). Cost of health care among patients with chronic migraine and episodic migraine in Canada and the USA: results from the International Burden of Migraine Study (IBMS). Headache 51: 1058–1077.
Stovner Lj, Hagen K, Jensen R et al. (2007 Mar). The global burden of headache: a documentation of headache prevalence and disability worldwide. Cephalalgia 27 (3): 193–210.
Thorlund K, Mills EJ, Wu P et al. (2014). Comparative efficacy of triptans for the abortive treatment of migraine: a multiple treatment comparison metaanalysis. Cephalalgia 34: 258–267.
Verhagen AP, Damen L, Berger MY et al. (2009 Summer). Behavioral treatments of chronic tension-type headache in adults: are they beneficial? CNS Neurosci Ther 15 (2): 183–205.
Von Korff M, Crane P, Lane M et al. (2005 Feb). Chronic spinal pain and physical-mental comorbidity in the United States: results from the national comorbidity survey replication. Pain 113 (3): 331–339.
Wammes-van der Heijden EA, Rahimtoola H, Leufkens HG et al. (2006 Oct 10). Risk of ischemic complications related to the intensity of triptan and ergotamine use. Neurology 67 (7): 1128–1134.
WHO (2001). The World Health Organization Report 2001: Mental health, new understanding new hope, World Health Organization, Geneva, Switzerland. Available online at: http:/www.who.int/whr/2001/main/en/overview/outline.htm.
Woolf CJ (2011 Mar). Central sensitization: implications for the diagnosis and treatment of pain. Pain 152 (3 Suppl): S2–S15.
Zissis NP, Harmoussi S, Vlaikidis N et al. (2007 Apr). A randomized, double-blind, placebo-controlled study of venlafaxine XR in out-patients with tension-type headache. Cephalalgia 27 (4): 315–324.
Zwart JA, Dyb G, Hagen K et al. (2003). Depression and anxiety disorders associated with headache frequency. The Nord-Trondelag Health Study. Eur J Neurol 10: 147–152.

Handbook of Clinical Neurology, Vol. 131 (3rd series)
Occupational Neurology
M. Lotti and M.L. Bleecker, Editors

Chapter 26

Considerations for return to work following traumatic brain injury

DEBORAH M. LITTLE[1,2]*, ANDREW J. COOK[2,3], SANDRA B. MORISSETTE[2,3], AND JOHN W. KLOCEK[4]
[1]*Baylor Scott and White Healthcare, Temple, TX, USA*
[2]*Neuroscience Institute, Texas A&M Health Science Center College of Medicine, Temple, TX, USA*
[3]*Central Texas Veterans Healthcare System, Temple, TX, USA*
[4]*Department of Psychology and Neuroscience, Baylor University, Waco, TX, USA*

Traumatic brain injury (TBI) is a significant public health concern, with annual costs at nearly $17 billion each year in healthcare costs and lost productivity in the USA (Finkelstein et al., 2006). In the USA alone, almost 1.7 million new cases of TBI present to emergency departments or require hospitalization each year (Langlois et al., 2006). The gross majority (~75%) of these TBI cases are classified as mild in severity, with infants and young children and individuals in late adolescence and early adulthood as the most prevalent group (Langlois et al., 2005). Within the range of mild TBI, it is reported that the gross majority will fully recover (Bigler and Maxwell, 2012; Eierud et al., 2014), while 10–20% will not have symptom resolution following a single event (Bazarian and Atabaki, 2001). The risk of incomplete recovery (as evidenced by sustained symptoms) may be increased by genetic loading (McAllister, 2010), affected by gender (Niemeier et al., 2014), and influenced by acute management and adherence to safety regulations for prevention and return-to-play and return-to-work standards following the injury (for example, Harmon et al., 2013). Importantly, TBI severity, which is a classification made based on acute TBI variables, is only grossly predictive of outcome (e.g., those who sustain milder injuries tend to show better recovery and less long-term impairment than those with severe injuries) and not by itself directly predictive of chronic outcomes within each severity grade (Dikmen and Levin, 1993; Hoofien et al., 2001). Impairment, in this chapter, is used to describe sustained alterations in cognition, behavior, and mood, as all of these variables contribute to lifetime disability, degree of independence, and return to the workforce. In addition, because TBI is a traumatic insult to the central nervous system that can and does co-occur with other physical trauma, other factors including development of chronic pain and chronic headache must also be considered.

Further complicating the multifactorial relationship between TBI and return to work is an age-dependent increase in risk for TBI. Across the lifespan, those at greatest risk for TBI are children from birth to age 4 (2193 per 100000), who present most commonly with TBI due to falls and abuse (Centers for Disease Control and Prevention (CDC), The National Institutes of Health (NIH), The Department of Defense (DoD), and The Department of Veterans Affairs (VA) Leadership Panel, 2013). The second most prevalent age group to suffer from TBI are those approaching the legal age of motor vehicle operation (982 per 100 000) (Centers for Disease Control and Prevention (CDC) et al., 2013). In these populations one must consider both direct impacts on independence and the indirect relationship between age of injury, scholastic performance, and the effects of scholastic performance on highest degree of formal education later in life, as all have been associated with occupational attainment. Multiple longitudinal studies are ongoing with a focus on further defining these relationships.

In addition to age-related increases in TBI presentation, one must also consider groups at higher risk for TBI due to occupational exposure. It is this group that is the focus of this chapter, as the occupation is not only

*Correspondence to: Deborah M. Little, PhD, Neuroscience Institute, Baylor Scott and White Healthcare, 5701 Airport Road, MS AR D120, Temple TX 76508, USA. E-mail: deborahmlittle@gmail.com

the cause for TBI exposure but also requires an almost immediate consideration of the effect of that injury on return to work, demands efforts to optimize the workplace for the injured party, and raises disability benefit considerations. The occupations at greatest risk for TBI include professional drivers (taxi cab drivers, truck drivers), construction workers, first responders, professional athletes, and those in the military. For both first responders and veterans, extreme stress at the time of injury and in the period following injury must also be considered.

In this chapter we provide first a definition of TBI and brief introduction to the pathophysiology of TBI as well as considerations for return to work and risk of premature return. We then provide a detailed review of the most commonly impaired domains, methods to assess these domains, and discussion of the comorbid and co-occurring conditions that affect recovery and a successful return to work. We also provide a brief review of the current state of the literature on long-term health outcomes for which TBI is a risk factor.

TRAUMATIC BRAIN INJURY

Although multiple definitions and criteria exist, we, for the purposes of this chapter, use the definition issued by the US Assistant Secretary of Defense for Health Affairs (2007), which is:

> *a traumatically induced structural injury and/or physiological disruption of brain function as the result of an external force that is indicated by new onset or worsening of at least one of the following clinical signs, immediately following the event: any period of loss of or decreased level of consciousness; any loss of memory for events immediately before or after the injury; neurologic deficits; intracranial lesion.*

Within the range of TBI injuries, we further subspecify as a function of injury severity based upon acute injury variables, including duration of loss of consciousness (LOC), Glasgow Coma Score (GCS), and duration of posttraumatic amnesia (PTA) (Levin et al., 1979). Mild TBI, which requires a LOC of less than 30 minutes, a GCS of 13–15, and period of PTA not to exceed 24 hours, is the most common severity, with a recent World Health Organization taskforce reporting that 70–90% of all treated TBI fell into this category (Holm et al., 2005). Those who exceed the thresholds for mild TBI are then graded as either moderate or severe based upon a cutoff of 24 hours of LOC and 7 days of PTA. These standards are consistent with the National Institute of Neurological Disorders and Stroke common data elements, the American Congress on Rehabilitation Medicine (1993) the American Academy of Neurology practice guidelines on concussion grading (1997) and the report on definition of TBI and TBI severity from the US Assistant Secretary of Defense for Health Affairs. Positive neuroimaging is also commonly included but, with the exception of the more severe injuries, not predictive of outcome.

MECHANISMS OF TBI

Common civilian mechanisms of injury include, in order of frequency, falls, being struck by or against an object (excluding motor vehicle accidents), motor vehicle accidents, and assaults. In terms of specific at-risk populations, sports-related concussions are believed to represent anywhere from 1.6 to 3.8 million new cases per year in the USA alone. The final mechanisms that require identification are TBIs due to military service and are associated with a blast.

Of these mechanisms of injury, recovery patterns are generally similar. Those who suffer injuries secondary to assaults, those with injuries that co-occur with other major trauma, and those sustained in the combat theater of operations appear to have, based upon the limited information available, different degrees of recovery and impairment. Sports-related concussions are also commonly identified in the media but do not appear unique in outcome data. Patients with sports-related concussions, as with occupational risk, are more likely to report multiple lifetime TBIs and are at a greater risk for repeated injuries. It is for this reason, and not based upon the pathophysiology, that these mechanisms are commonly separated from discussion.

With the exception of blast-related injuries, most TBIs are acceleration–deceleration injuries. The primary mechanism of injury in acceleration–deceleration injury is from the movement of brain within the skull with or without focal injuries, including contusions. Of course, many acceleration–deceleration injuries also involve rotational components. This rotational component may explain damage to the optic nerve and self-reports following head trauma of visual defects (Gutierrez et al., 2001). Severe rotational injury also can result in damage to the cranial nerves, damage to cerebral vasculature resulting in bleeds, and damage to the hippocampus (Gutierrez et al., 2001).

The specific deformations of the brain during acceleration and deceleration were investigated *in vivo* in humans using magnetic resonance imaging (MRI) techniques. Four volunteers had MRI data acquired during acceleration and deceleration impact (Bayly et al., 2005). First, the head was lifted off the table by a small amount. The subject then initiated a release of the head on to the table. Using optic triggers, MRI data collection was initiated as the head was released. The data showed a shortening or compression of tissues in the anterior and frontal tissue and a lengthening in the posterior and inferior tissue (Bayly et al., 2005). How well these data describe the actual acceleration–deceleration forces needed to cause alteration in consciousness is unknown, but the study does demonstrate that the shear strain on tissue is not homogeneous. Confirmatory evidence for an acceleration–deceleration model more selective of diffuse axonal injury (DAI, which is more common in certain regions) has been reported as part of retrospective chart and literature review. This damage is likely worse near the contrecoup injury because of the greater deformation observed away from the injury (Drew and Drew, 2004). It is this selective finding of a common location that distinguishes human from animal models in TBI and calls the validity of these models into question, because many animal models of DAI do not show this same pattern of damage (Maxwell et al., 1997). However, differences in skull shape may account for such differences. Indeed, the best model appears to be with primates but, due to cost and other factors, these studies are uncommon. One of the few primate models of DAI not only demonstrates a similar preference for these selective DAI locations observed in human studies, but also provides a timeline for components of DAI (Gennarelli et al., 1982).

In contrast to acceleration–deceleration injury, blast-related TBI offers a number of additional concerns, including that the blast itself can affect central nervous system function (Moore et al., 2009) and can result in alterations in intracranial pressure as well as direct shear and strain injury due to mechanical forces (Chafi et al., 2010). Similar to acceleration–deceleration TBI, blast TBI is associated with damage to axons, cerebral edema, and small hemorrhages in white matter, the cerebellum, and brainstem (Dennis and Kochanek, 2007). The specific pathophysiology of damage due to blast is uncertain, but at least three main mechanisms have been proposed. The first is that the blast waves travel directly through the brain, rapidly increasing and decreasing intracranial pressure and shearing tissue (DePalma et al., 2005; Taber et al., 2006; Chavko et al., 2007; Dennis and Kochanek, 2007). Second, blast-wave overpressure forces the skull away from the blast, leaving the brain to lag behind and then impact the side of the skull (Levi et al., 1990). This is then followed by negative pressure, which pulls the skull back toward the direction of blast again, with the brain lagging behind. The initial blast effects are proposed to be similar to acceleration–deceleration injury. However, how these effects interact with the second force of returning the skull toward the blast is unknown. Finally, it has also been proposed that blast wave energy in the torso and neck forces organs to constrict and relax, thus producing a surge of blood into the brain (Cernak et al., 1999). Although the specific mechanism is unknown, it is likely that brain injury secondary to blast wave involves at least two of these components, if not all.

It has also been suggested that, in addition to primary brain injury from blast, the biochemical cascade that follows acute injury impairs the ability of the cerebral vascular system to compensate, making it more susceptible to additional damage such as from the effects of acceleration–deceleration injury resulting from the blast (DeWitt and Prough, 2009).

As with civilian injury, brain injury due to blast includes significant shear and strain injury, DAI, hematomas, and hemorrhage (Hicks et al., 2010). Primary blast-related injury (barotrauma) is commonly followed by an acceleration–deceleration injury due to objects striking the individual or due to the individual striking an object. Blast-related TBI shares significant pathophysiology with nonblast TBI, including DAI, oxidative stress, and excitotoxic cell damage. The acute sequelae, including retrograde amnesia, compromised executive function, headache, confusion, amnesia, difficulty concentrating, mood disturbance, alterations in sleep patterns, and anxiety, also appear similar to civilian TBI (Ling et al., 2009). In chronic TBI, early investigations also report similar patterns of cognitive function between civilian and veteran populations (Belanger et al., 2009).

PATHOPHYSIOLOGY OF TBI

Certain aspects of the pathophysiology of TBI may be especially pertinent to the risk for neurobehavioral sequelae. There are several important components that can contribute to neurobehavioral outcome, including the location and severity of the injury, diffuse effects, and secondary biochemical mechanisms of injury. Primary neurologic injury due to TBI can be direct and/or indirect. Contusions are common following TBI, and can directly disrupt function in a specific area. Certain areas of the brain may be more vulnerable to contusion following trauma, such as the frontal and anterior temporal cortices, due to their position within the skull. (Adams et al., 1980; Levin et al., 1992). Importantly, disruption of function can also result from direct and indirect damage to white-matter fiber tracts, the links in neural networks. The primary pathophysiology in head injury is DAI and appears to be the most common in parasagittal white matter, the corpus callosum, brainstem, and any gray-matter–white-matter junction (Gentry et al., 1988; Meythaler et al., 2001). We define DAI in this chapter to include both increased microglia as a marker of degeneration of white matter and Wallerian-type axonal degeneration. This damage is not necessarily shearing of the axon itself, but instead is a marker of a pathophysiologic process that damages the integrity of the axon. Maxwell and colleagues (1993) demonstrated that there is shearing to axons acutely following injury, but that axonal membranes seal themselves within the first hour. Following this, axons also demonstrate swelling, which results in disconnection of the axon. It is likely this secondary process that leads to the effects we observe in humans. The degree of damage and/or depth of damage in the brain following acceleration–deceleration injury may serve as an independent marker of severity of acceleration–deceleration forces (Levin et al., 1997).

Secondary processes, including inflammation and free radical formation, also contribute to white-matter damage (Bandak, 1995). DAI can disrupt critical cortical-subcortical pathways (Gennarelli et al., 1982; Povlishok, 1992), as these white-matter fiber tracts are particularly susceptible to the shearing forces that often occur with TBI, especially at gray–white-matter junctions like the thalamus (Graham et al., 2002). Brain edema and shift can also compromise blood supply and lead to secondary infarction in the corpus callosum and deep gray matter, and elevated intracranial pressure can cause damage to the brainstem and optic nerve in TBI (Graham et al., 1987). Although the diagnosis of DAI can only be definitively confirmed by microscopic examination, it may be inferred from specific neuroimaging findings, such as hemorrhages in the corpus callosum or areas of rostral brainstem (Geddes, 1997; Geddes et al., 1997). DAI may be the only significant pathology found in certain cases of TBI, and has been identified via direct pathologic studies as well as on both standard clinical neuroimaging in severe cases and in high-resolution imaging of the white matter in mild to moderate injury (Povlishock et al., 1983; Graham et al., 1989; Blumbergs et al., 1994, 1995; Goodman, 1994; Mittl et al., 1994; Aihara et al., 1995; Gennarelli, 1996; Inglese et al., 2005b).

Changes in cerebral white matter have been shown, even in mild TBI (Inglese et al., 2005a, b), and in both acute and chronic phases of recovery. Regions reported to have been affected include the corpus callosum, internal capsule, and centrum semiovale (Inglese et al., 2005b). Another issue is the specificity of the types of lesion found in mild TBI. Kurca et al. (2006) reviewed MRI results on 30 patients with reported mild TBI within 96 hours postinjury, and 30 matched controls. They divided MRI findings into traumatic and nonspecific, but were not only looking at white matter. They found that the group with the defined traumatic lesions showed significantly greater impairment on neuropsychologic evaluations and subjective reports of postconcussion syndrome-type symptoms. However, it is not entirely clear that the lesion types that the authors define as nontraumatic are actually nontraumatic. In addition, this is also another study on an acute, not chronic population.

In moderate to severe injuries, white-matter abnormalities have been more clearly demonstrated, as would be expected. Mathias et al. (2004) studied a group of 25 moderate and severe TBI patients and 25 controls, all receiving the same neuropsychologic test battery. The average time out from injury for the group was 212.9 days (SD = 86.6). Of the TBI group, 18 underwent MRI on a 1.5 T scanner. They acquired 5-mm-thick axial proton density and T2-weighted turbo spin echo, fluid-attenuated inversion recovery (FLAIR), and T2*-weighted gradient echo images. Quantitative MRI analysis for whole-brain white-matter volume, hippocampal volume, and corpus callosum area was performed. The primary outcome was the level of corpus callosum atrophy in the TBI group. The corpus callosum measurements for the TBI group were found to be smaller than the comparison group, except for the splenium. They also found that reaction time data from the neuropsychologic tasks they chose correlated negatively with the total white-matter volume, but the correlations did not reach significance. The controls were not imaged; rather, they used normative data previously collected from a different healthy control group. It is not clear how the authors selected or even matched these groups.

NEUROBEHAVIORAL OUTCOMES FOLLOWING TBI

Neurobehavioral sequelae of TBI often underlie functional disability and include disorders of cognition, mood, and behavior (CDC, 2003). The most common of these behavioral and psychologic concerns are presented in Table 26.1. Cognitive deficits after TBI are common and well documented across severities. Disorders of mood are also common following TBI, and results of epidemiologic studies have typically demonstrated high prevalence rates for major depression, bipolar disorder, and generalized anxiety in patients with a coexisting history of TBI (van Reekum et al., 1996). Of these, apathy and depressive symptoms are the most prevalent following TBI, with estimates extending to greater than 50% (Dikmen et al., 2004).

These deficits are logical given the nature of the neuropathology following TBI (Levin et al., 1992; Levin and Kraus, 1994), and they are often the primary source of disability that prevents the patient from successful social and occupational reintegration (Levin et al., 1992; Levin and Kraus, 1994) The observed cognitive changes that can follow TBI include impairments in attention and

Table 26.1

Most common behavioral diagnoses following traumatic brain injury

Diagnosis	Frequency
Apathy	60%[1]
Depression	10–50%[2]
Agitation	25%[3]
Posttraumatic stress disorder	11–18%[4]
Psychosis	7–10%[5]

[1]Kant et al. (1998).
[2]McAllister (1992); National Institute of Mental Health (2001); Holsinger et al. (2002); Dikmen et al. (2004); Guskiewicz et al. (2007).
[3]Nott et al. (2006).
[4]Fedoroff et al. (1992); Jorge et al. (1993, 2004); Jorge and Robinson (2003).
[5]Davison and Bagley (1969).

executive functions, such as decreased mental flexibility, and trouble shifting sets (Levin and Kraus, 1994; Miller, 2000; Godefroy, 2003). Patients also demonstrate additional executive functioning deficits, including poor planning, lack of organization, problems with sequencing, impaired judgment, deficits in verbal fluency, problems with working memory, as well as impulsivity errors (Levin and Kraus, 1994; Miller, 2000; Godefroy, 2003). Mood and behavioral disturbances are also common, including various symptoms of depression, irritability, impulsivity, apathy, and amotivation (Levin et al., 1992, 2005; McAllister, 1992; van Reekum et al., 1996; Hoofien et al., 2001; Dikmen et al., 2004; Jorge et al., 2004; Moldover et al., 2004). Neurobehavioral deficits often persist and can even result in long-term problems for patients with mild TBI (Gronwall and Wrightson, 1974; Rimel et al., 1981; Bohnen et al., 1994, 1995; Ruff et al., 1994; Alexander, 1995; Roberts et al., 1995; Binder et al., 1997; Ashman et al., 2006; Marsh and Kersel, 2006).

Importantly, while an older CDC report (CDC, 2003) emphasizes that there are numerous studies demonstrating neuropsychologic deficits in mild TBI, it underscores that there is controversy as to how long these problems persist. It is this interindividual variability in longer-term outcomes that needs to be studied further, especially to discern the differences between those with good recovery and those with persistent disability.

In a combat-injured sample of TBI patients treated at Walter Reed Medical Center between 2003 and 2005, approximately 90% reported postconcussive symptoms, including over 40% with reported problems in memory, attention and concentration, and irritability, as well as other behavioral disturbances (Warden, 2006). Even when variables such as age, litigation, substance abuse, and prior psychiatric history are controlled for, there is still unexplained variability in mild TBI, as well as in more severe injuries.

Depression is often poorly conceptualized following TBI, and may be underrecognized and undertreated in acute care and rehabilitation models and must be considered in a return-to-work setting (National Institute of Mental Health, 2001). There are likely multiple etiologic pathways that converge to a common clinical presentation of disturbed mood with a myriad of features (McAllister, 1992). A 2007 study of professional athletes in the National Football League showed that those players who had had at least three concussions were three times as likely to develop depression (Guskiewicz et al., 2007). Those employees for whom the TBI is not the first lifetime injury should be monitored more closely for depression in the years following a TBI. Additionally, it is reasonable to posit that those with depression prior to the injury may have poorer outcomes.

Dikmen and colleagues (2004) conducted a 3–5-year prospective study examining the rates and risk factors associated with depression in moderate to severe TBI at a Level I trauma hospital. The Center for Epidemiological Studies Depression (CES-D) scale was used as the primary outcome measure. The authors found that rates of moderate to severe depression ranged from 31% at 1 month to 17% at 3–5 years, suggesting a time-dependent decline in depression rates from the time of injury. The rates of depression did not appear dependent upon the severity of injury (Dikmen et al., 2004). Premorbid depression and psychosocial factors were predictive of worsening depression following TBI, which suggests a strong need to screen this vulnerable group after head trauma (Dikmen et al., 2004). Beyond the impact of depression on quality of life, it has also been well documented to impede the achievement of optimal functional outcome in both acute and chronic stages (Rapoport et al., 2003).

As emphasized previously, TBI commonly results in disturbances in mood and in changes in cognition and behavior. Additionally, other issues, such as chronic pain (e.g., headaches), which are relatively common in TBI, can play a significant role in the extent of neurobehavioral symptoms. As such, these other factors must also be characterized and addressed. All of these features may interact with each other. Outside of the neurobehavioral symptoms described above, there are three additional comorbid and co-occurring conditions that affect return to the workplace. These are potential issues not only with successful return to work but as conditions that may limit the patient from return to prior duties. These include chronic posttraumatic headache (PTH) and chronic pain.

CHRONIC POSTTRAUMATIC HEADACHE

Headache is the most common symptom after TBI, often one of a constellation of symptoms of postconcussive syndrome. There is an inverse relationship between the severity of TBI and prevalence of PTH, with the highest rates found for mild TBI (Branca, 2006). The incidence of chronic PTH (CPTH) after mild head injury is between 30% and 90%, and varies between countries (Ramadan and Keidel, 2000; Lane and Arciniegas, 2002). Headache is the most common residual impairment 5 years following TBI (Hillier et al., 1997). The most common form of CPTH by a large margin is tension-type headache, followed by migraine, occipital neuralgia, and other headache presentations (e.g., mixed, cluster) (Chibnall and Duckro, 1994; Seifert and Evans, 2010). The pathophysiology of PTH associated with mild TBI is thought to involve DAI and neurometabolic processes (e.g., excitatory neurotransmitters such as aspartate, glutamate, and acetylcholine) (Branca, 2006; Seifert and Evans, 2010). Adaptive gray-matter changes in pain-processing centers have been identified in patients with CPTH (Obermann et al., 2009). CPTH is a common complaint among returning military service members due to the prevalence of mild TBI. More than 90% of those reporting blast injuries have headaches, mostly of the migraine type (Seifert and Evans, 2010).

In the original 1988 edition of the International Classification of Headache Disorders (ICHD: Headache Classification Committee of the International Headache Society, 1988), persistence of PTH beyond 8 weeks was classified as chronic. In the second ([The International Classification of Headache Disorders: 2nd edition], 2006) and third (World Health Organization, 2013) editions of ICHD, CPTH is characterized by persistence beyond 3 months, and resulting from either mild or moderate to severe head injury (World Health Organization, 2013). These classifications are presented in Table 26.2.

Due to the episodic nature of headache and variations in its course, both within and between individuals, measurement of functional disability from chronic headache is difficult (Andrasik et al., 2011). Many of the validated self-report headache instruments use a 1–3-month reporting period to provide a representative assessment. There are many self-administered instruments for assessing headache-related functional impairment and disability (Borkum, 2005; Andrasik et al., 2011), including the Headache Impact Test (Kosinski et al., 2003), Headache Impact Questionnaire (Stewart et al., 1998), and the Henry Ford Hospital Headache Disability Inventory (Jacobson et al., 1994). Though no standardized criteria exist for impairment ratings in CPTH, a systematic ratings system has been proposed (Packard and Ham, 1993).

Table 26.2

ICHD-III classification of headache. (From the International Classification of Headache Disorders, 3rd edition, 2013, with permission from Sage Publications Ltd.)

5.1 Acute headache attributed to traumatic injury to the head
5.1.1 Acute headache attributed to moderate or severe traumatic injury to the head
5.1.2 Acute headache attributed to mild traumatic injury to the head
5.2 Persistent headache attributed to traumatic injury to the head
5.2.1 Persistent headache attributed to moderate or severe traumatic injury to the head
5.2.2 Persistent headache attributed to mild traumatic injury to the head
5.3 Acute headache attributed to whiplash
5.4 Persistent headache attributed to whiplash
5.5 Acute headache attributed to craniotomy
5.6 Persistent headache attributed to craniotomy

Important information related to functioning and treatment is also obtained by assessment of cognitive status and psychosocial constructs such as stress, locus of control (i.e., the degree to which an individual perceives his or her headaches to be under personal or external control), self-efficacy (i.e., perceived ability to achieve a desired action or outcome), and catastrophizing (i.e., a pattern of thinking characterized by hopelessness and helplessness) (Andrasik et al., 2011); these are common assessments done in patients with depression as well.

Though most patients experiencing PTH recover within a few months, chronic problems are experienced by a significant minority. CPTH affects almost 25% of patients at 4 years (Seifert and Evans, 2010). Factors hypothesized to contribute to poor recovery include insufficient treatment, analgesic rebound headache due to excessive use of analgesics, and/or psychosocial comorbidities such as anxiety, depression, insomnia, or substance abuse (Lane and Arciniegas, 2002; Lipton et al., 2003; Walker et al., 2006). Posttraumatic stress disorder (PTSD) is a common comorbidity in some CPTH subpopulations due to co-experienced psychologic and physical trauma (Chibnall and Duckro, 1994; Borkum, 2005). Though there has been active debate about the association between accident-related CPTH and presence of litigation and/or potential for financial gain (Sheftell et al., 2007; Obermann et al., 2010; Seifert and Evans, 2010), there is evidence for its persistence beyond time of legal settlements (Packard, 1992) and some have cautioned against premature oversimplification of potentially complex etiologic factors (Obermann et al., 2010).

Of greatest relevance to the present discussion, chronic headache produces a high level of occupational sability. For chronic tension-type headache, 12% of ıfferers report missed work days, with a very significant average annual absenteeism of 27 days (Borkum, 005). Traumatic onset is known to be associated with igher levels of reported disability in clinical samples, elieved to be due to the commonly comorbid symptoms uch as postconcussive syndrome, PTSD, and depression Borkum, 2005). CPTH is associated with greater reducions in activity level, more complete disability, and ɔwer physical function than nontraumatic chronic headache (Marcus, 2003). Though a clear understanding of redictors of vocational rehabilitation is lacking, identiied factors from past research include age at injury, everity and duration of posttraumatic amnesia, types of persisting impairments, and premorbid occupation and characteristics.

The often slow and challenging course of treatment can be demoralizing to patients and their families, who can be inadequately prepared with strategies for long-term coping (Barnat, 1986). Self-awareness and acceptance of limitations are important to vocational rehabilitation (Branca, 2006). Cognitive symptoms (e.g., difficulties with memory, concentration, and/or information processing) are common among CPTH sufferers, especially females (Packard et al., 1993), with high potential for impact on occupational function. Age and performance on cognitive tests have been found to predict work status in a majority of military service members with mild TBI (Branca, 2006). The affective component of CPTH should also be considered in occupational rehabilitation for workers with CPTH. Depression predicts functional disability in CPTH, partially mediated by anger (both suppressed and expressed) (Duckro et al., 1995). When CPTH sufferers return to work, they may experience improvements in psychosocial domains, such as less depression and reduced likelihood of substance abuse (Branca, 2006).

Making the return to work and treatment more complex is the absence of randomized, placebo-controlled studies to demonstrate efficacy for CPTH treatments. As such, treatment recommendations are based on expert opinion and evidence-based guidelines for specific primary headache phenotypes, such as migraine and tension-type (Vargas and Dodick, 2012). As with rehabilitation for other chronic pain disorders, there is general consensus that multimodal and interdisciplinary, or at least multidisciplinary, treatment is most effective for CPTH, especially when early resolution is not achieved. Complex comorbid conditions, such as chronic pain, TBI, and PTSD, also warrant an interdisciplinary approach (Otis et al., 2011). Medical management (prophylactic, abortive, physical) (Bell et al., 1999) and cognitive-behavioral therapy (Gurr and Coetzer, 2005) can be effective components of multimodal treatment.

Management of stress in the workplace is critical to effective employee functioning (von Onciul, 1996) and can play an important role in successful return to work for workers with CPTH. Neuropsychologic assessment can provide valuable input on current cognitive status as well as premorbid level of functioning (Branca, 2006). Behavioral treatment that is anchored in the biopsychosocial model, that is, multimodal treatment, has been shown to be effective for treatment of CPTH following occupational trauma (Duckro et al., 1985).

Successful reintegration for workers with CPTH is supported by the presence of health insurance, effective tailoring of work duties, supported employment (SE), and a socially inclusive work environment (Branca, 2006). It has been suggested that the costs of occupational rehabilitation may be influenced more by the relative decline from an individual's premorbid functioning than by the absolute level of neuropsychologic impairment (Branca, 2006). General modalities and principles of TBI rehabilitation and chronic headache management (e.g., Whyte et al., 2010) typically provide a foundation for successful treatment of CPTH and work reintegration.

NONHEADACHE-RELATED CHRONIC PAIN

As with TBI, it is critical to start any discussion on the topic of chronic pain with a definition. In this chapter we use the definition provided by the International Association for the Study of Pain as "an unpleasant sensory and emotional experience associated with actual or potential tissue damage, or described in terms of such damage" (Merskey and Bogduk, 1994). The taxonomy of chronic pain conditions is extensive and beyond the scope of this chapter. Reference here will be limited to the broad classifications of chronic pain utilized by the various studies assessing the presence and role of nonheadache-related chronic pain following TBI. Further complicating the relationship between TBI and nonheadache-related chronic pain is the lack of an obvious etiologic connection between the brain injury and the painful region (except in the case of TBI with concurrent other physical trauma). The literature investigating the prevalence and role of nonheadache-related chronic pain following TBI is limited, as the majority of studies are inclusive of both headache and nonheadache-related chronic pain. Nevertheless, an emerging body of research indicates that nonheadache-related chronic pain is a significant challenge to individuals following TBI. While not as prevalent as CPTH, nonheadache-related chronic pain is common following TBI (Nampiaparampil, 2008), with estimates ranging from 24% among individuals with severe brain injury

(Bryant et al., 1999) to 75% among individuals with mild TBI (Alfano et al., 2000). Lahz and Bryant (1996) found similar rates of chronic neck and shoulder pain among individuals with moderate to severe TBI ($n=79$; 24%) and mild TBI ($n=53$; 28%). They also found similar rates of chronic low-back pain among individuals with moderate to severe TBI ($n=79$; 16%) and mild TBI ($n=53$; 19%). Pain can complicate recovery from TBI. In a sample of 67 adults studied, 9% were found to have pain as a significant barrier to recovery following a mild TBI (Mooney et al., 2005). Of note, they also found that 49% showed mild TBI, pain, and a psychiatric diagnosis as the factors which substantially interfered with recovery.

As is true with chronic headache, nonheadache-related chronic pain also results in significant impairment in occupational function and increased healthcare costs (Pizzi et al., 2005). Return to work in non-CTPH-related chronic pain has been found to be impacted by over 100 distinct factors (Krause et al., 2001). These include factors that are both subject to modification with treatment and those which are not subject to change through treatment. For example, demographic variables such as older age, lower education, lower income, and blue-collar or labor-related profession were negatively associated with return to work (Krause et al., 2001). However, other authors have noted that variables such as an individual's prediction of when s/he may be able to return to work, ability to change professions or jobs, and active participation in a return-to-work treatment program can be modified – and have successfully been the target of interventions (Hamer et al., 2013).

LONG-TERM CONSIDERATIONS FOLLOWING TBI

What is known is that the best predictor of future concussion is past concussion. As discussed, TBI and repeated TBI have been associated with increased risk of depression (O'Connor et al., 2012; Rapoport, 2012; Reeves and Laizer, 2012; Scheibel et al., 2012; Vasterling et al., 2012; Bryan et al., 2013), increased impairments in memory (Vasterling et al., 2012), increased postconcussive symptoms (van der Horn et al., 2013), increased sleep disturbance (Bryan, 2013), and decreased quality of life (Williamson et al., 2013). Repeated concussion risk can however be partially mitigated, as is the case with robust and comprehensive concussion identification. The challenge is that there are few validated measures to allow an accurate lifetime history to be obtained. This is critical in an occupational setting where risk for TBI exposure is part of an employee's job duties. At present, one of the best measures is the Ohio State University TBI Identification tool (Corrigan and Bogner, 2007), which was initially validated in part on a prison population (Bogner and Corrigan, 2009). However, this is a lengthy assessment tool that might exceed time available outside of a research setting.

Beyond these concerns that present in the acute to chronic stages following TBI, repeated concussions are also a risk factor for long-term health concerns, specifically, neurodegenerative diseases. Epidemiology data show increased risk for chronic traumatic encephalopathy, Alzheimer's disease, Parkinson's disease, and amyotrophic lateral sclerosis. However, unlike for depression, pain, and cognitive impairment, there is little convincing evidence that a history positive for a single uncomplicated concussion increases risk for these neurodegenerative diseases. However, although there is public concern for the risk of neurodegenerative disease, it is critical to remember that the bounding conditions for this increased risk are not known in any detail beyond presence of multiple injuries. For example, the threshold for the minimum number of lifetime TBIs, effects of time between TBIs for increased risk, or factors such as age at first or last TBI are still unknown. Multiple large-scale studies supported by federal grants are ongoing on these topics.

The reasonable person judgment should be followed when considering risk and potential disability due to repeated injuries. In this case, a reasonable person (or employer) should put forth effort to ensure the employee has recovered and is safe to return to work after a period of rest. This rest period is believed to be critical in facilitating recovery and should help limit lost productivity and costs associated with disability. Although no data on rest and repeated concussion exist to inform long-term risk for neurodegenerative disease, it seems prudent to ensure return to work occurs only after that period of rest. Informal reports from the combat theater also support the positive effects of rest as it relates to full recovery. Early position papers from the field of sports medicine also support this approach. One unique challenge for occupational safety is that, unlike sports and the military, baseline testing is generally not used but may be of benefit for those occupations at the greatest risk for head injury, such as firefighters. Wider use of baseline testing would provide critical tools for those assessing return to work. Factors that influence successful return to work and available interventions to assist with return to work following injury are discussed below.

RE-ENTRY TO THE WORKFORCE

The *International Classification of Functioning, Disability and Health* (ICF), created by the World Health Organization (2002), recognizes that return to work is a central component to successful reintegration. Those who are employed report greater sense of wellbeing, health status, and quality of life (Corrigan et al., 2001; Steadman-Pare et al., 2001). Reciprocally, failure to return to work can create tremendous psychosocial and financial strain for both the person with TBI and his/her family. Although a significant proportion of individuals with severe TBI successfully return to work, rates vary widely (12–70%) and are usually contingent upon a comprehensive medical and psychosocial rehabilitation approach (Shames et al., 2007). In a review of 49 studies, Van Velzen and colleagues (2009a) found that 40% of those with traumatic or nontraumatic brain injury returned to work after 1–2 years. Evaluation of successful return, however, is often complicated by how success is defined (i.e., return to previous job/premorbid levels of functioning, degree of job modification, full- vs part-time).

Myriad factors influence functional recovery and return to work following brain injury, including severity and area of the injury, educational and occupational background prior to the injury, and age of injury (Wehman et al., 2005). Data on injury severity are mixed, likely due to varying definitions, with some studies indicating that patients with more severe injuries have difficulty returning to work (Brooks et al., 1987) and others finding no effect (Van Velzen et al., 2009b). Further, a significant proportion of those with mild TBI experience lingering postconcussive symptoms that can interfere with work functioning, such as dizziness (Chamalian and Feinstein, 2004) and tiredness (van Velzen et al., 2011). In turn, persistent postconcussive symptoms are further complicated by commonly co-occurring mental health disorders, such as PTSD and depression (Morissette et al., 2011), which also can negatively impact work functioning (Zivin et al., 2011). Problematically, deficits related to mild TBI can be hard to detect during formal neuropsychologic testing, and testing situations do not reflect challenges related to real-world work settings in which environments are less structured, have distractions, and require longer performance times (Sbordone, 2001).

In terms of prior education and work history, skilled professionals and those with higher education are more likely than manual workers to return to work (Greenspan et al., 1996; Gollaher et al., 1998; Walker et al., 2006). Moreover, age over 40 negatively predicts return to work, which could be related to physiologic factors associated with neurologic recovery along with cultural influences regarding age and re-employment (Ponsford et al., 1995; Keyser-Marcus et al., 2002). Finally, a particularly important factor influencing return to work is self-awareness, impairments in which may curtail motivation and ability to set reasonable goals (Ben-Yishay and Lakin, 1989; Sherer et al., 1998). For this reason, self-awareness is often a key component of rehabilitation treatment.

With respect to rehabilitation treatment programs for work re-entry, approaches are generally cognitive or cognitive-behavioral in nature, although content, length, and intensity vary considerably based on funding and availability of trained professionals.

INTERVENTIONS FOR RETURN TO WORK

One of the most widely studied interventions is called supported employment or SE, which aims to increase competitive employment rates among those with TBI (Wehman et al., 1993). SE involves job placement that is supported by job coaches who advocate for the patient and facilitate training and work accommodations. In addition to making physical adaptations, SE includes behavioral and skills training, social adjustment, and development of cognitive strategies. Coordination between rehabilitation specialists and employers is a critical component. Wehman and colleagues demonstrated job retention rates of more than 70%. Further, the SE approach has been found to be effective in obtaining and maintaining employment for individuals with a wide variety of mental health conditions (e.g., Bond et al., 2001; Rogers et al., 2006; Davis et al., 2012). Adjunctive cognitive rehabilitation treatment has also been successfully added to SE to address cognitive impairments that might limit effectiveness (McGurk et al., 2007).

Finally, approximately 15–30% of military personnel serving in the current wars in Iraq and Afghanistan report TBI (Hoge et al., 2007, 2008), making this a key concern to the Veterans Health Administration, US Department of Defense, and public health initiatives. Ability to return to work following head injury can be a critical piece of the reintegration process for service members who are trying to reset for future missions or find their next "mission" as civilians. To that end, in a randomized controlled trial of two acute inpatient rehabilitation approaches (integrated cognitive-didactic versus functional-experiential rehabilitation therapy) for TBI in active duty and veterans, Vanderploeg and colleagues (2008) did not find group differences in return to work. However, within exploratory subgroup analyses, younger participants who engaged in the cognitive-didactic therapy returned to work/school at higher rates than same-aged participants receiving functional-experiential therapy. In contrast, those who were over 30 years of age and had higher educational level had greater rates of independent living at 1-year follow-up when receiving functional therapy compared to those in the cognitive treatment. These findings are particularly important in the context of military demographics, in which 42.7% of all active duty service members are aged 25 years or younger, whereas 43.8% of officers are 36 years or older. Clearly, future research is needed to determine whether different strategies may be needed for differing military demographics to assist them with recovery.

CONCLUSIONS

TBI presents many challenges, whether the injury is a function of the job duties or if a person with a previous injury is returning to or entering the workplace. In the case of occupational injury, return to work, after a period of rest, should be considered relative to a resolution of symptoms. In the case of milder injuries, this may be sufficient for optimizing the return and for reducing the risk of long-term disability. In the case of complicated mild injuries, mild injuries that result in sustained impairments, or more severe injuries, then an assessment of those impairments and limitations, relative to job duties, should be considered proactively. The literature is clear that factors such as frustration related to challenges in performing previous duties that had been easy result in poorer long-term outcomes. Similarly, one should be aware of areas of greatest challenge, including multitasking and learning new skills. A prospective approach and involvement of the injured employee in this process will be beneficial to all parties. In the cases where cognitive impairment is present, options for interventions should be considered as early in the return-to-work process as possible. For occupations where work is associated with risk of repeated injury, additional concern is warranted not only to optimize acute recovery but for reduction in long-term risk.

Patient and staff education is also helpful at optimizing return to work. TBI is associated with behavioral changes, including frustration, anger, impulsivity, anxiety, and depression. These can have a negative impact on the workplace and should be monitored early in the process. Employee health should also re-examine medication use if warranted by the job duties, as many medications can impact employee safety. These are of greatest concern in populations that exhibit the most common comorbid diagnoses, including chronic pain and chronic headache.

Special consideration should be given to those who must seek other employment when brain injury makes return to the same occupation unsafe. These cases are most common among military veterans and first responders. For these populations, consideration must go beyond ratings of disability and include plans for identification of new careers and training to support these careers.

ACKNOWLEDGMENTS

We would like to thank Dr. Suzy B. Gulliver and Dr. Sara L. Dolan for their contribution to this chapter.

REFERENCES

Adams HJ, Graham DI, Scott G et al. (1980). Brain damage in fatal non-missile head injury. J Clin Pathol 33 (12): 1132–1145.

Aihara N, Hall J, Pitts L et al. (1995). Altered immunoexpression of microglia and macrophages after mild head injury. J Neurotrauma 12: 53–63.

Alexander M (1995). Mild traumatic brain injury: pathophysiology, natural history, and clinical management. Neurology 45: 1253–1260.

Alfano D, Asmundson G, Larsen D (2000). MTBI and chronic pain: preliminary findings. Arch Clin Neuropsychol 15: 831–832.

American Academy of Neurology (1997). Practice Parameter: The management of concussion in sports (summary statement). Report of the Quality Standards Subcommittee. Neurology 48: 581–585.

American Congress on Rehabilitation Medicine (1993). Definition of mild traumatic brain injury. J Head Trauma Rehabil 8: 86–87.

Andrasik F, Buse D, Lettich A (2011). Assessment of headaches (D Turk, R Melzack Eds.), Guildford Press, New York.

Ashman T, Gordon W, Cantor J et al. (2006). Neurobehavioral consequences of traumatic brain injury. Mt Sinai J Med 73 (7): 999–1005.

Assistant Secretary of Defense for Health Affairs (2007). Health Affairs Memorandum. Traumatic Brain Injury: Definition and Reporting.

Bandak FA (1995). On the mechanics of impact neurotrauma – a review and critical synthesis. J Neurotrauma 12 (4): 635–649.

Barnat MR (1986). Post-traumatic headache patients I: Demographics, injuries, headache and health status. Headache 26 (6): 271–277.

Bayly P, Cohen T, Leister E et al. (2005). Deformation of the human brain induced by mild acceleration. J Neurotrauma 22 (8): 845–856.

Bazarian JJ, Atabaki S (2001). Predicting postconcussion syndrome after minor traumatic brain injury. Acad Emerg Med 8 (8): 788–795.

Belanger HG, Kretzmer T, Yoash-Gantz R et al. (2009). Cognitive sequelae of blast-related versus other mechanisms of brain trauma. J Int Neuropsychol Soc 15 (1): 1–8.

Bell KR, Kraus EE, Zasler ND (1999). Medical management of posttraumatic headaches: pharmacological and physical treatment. J Head Trauma Rehabil 14 (1): 34–48.

Ben-Yishay Y, Lakin P (1989). Structured group treatment for brain injury survivors, Kluwer Academic, Boston.

Bigler ED, Maxwell WL (2012). Neuropathology of mild traumatic brain injury: relationship to neuroimaging findings. Brain Imaging Behav 6 (2): 108–136.

inder L, Rohling M, Larrabee J (1997). A review of mild head trauma, part I: meta-analytic review of neuropsychological studies. J Clin Exp Neuropsychol 19: 421–431.

lumbergs P, Scott G, Manavis J et al. (1994). Staining of amyloid precursor protein to study axonal damage in mild head injury. Lancet 344: 1055–1056.

lumbergs P, Scott G, Manavis J et al. (1995). Topography of axonal injury as defined by amyloid precursor protein and the sector scoring method in mild and severe closed head injury. J Neurotrauma 12: 565–572.

ogner J, Corrigan JD (2009). Reliability and predictive validity of the Ohio State University TBI identification method with prisoners. J Head Trauma Rehabil 24 (4): 279–291.

Bohnen NI, Van Zutphen W, Twinjstra A (1994). Late outcome of mild head injury: results from a controlled postal survey. Brain Inj 8 (8): 701–708.

Bohnen NJ, Wijnen G, Twijnstra A et al. (1995). The constellation of late posttraumatic symptoms of mild head-injury patients. J Neurol Rehabil 9 (1): 33–39.

Bond GR, Becker DR, Drake RE et al. (2001). Implementing supported employment as an evidence-based practice. Psychiatr Serv 52 (3): 313–322.

Borkum J (2005). Predicting disability from headache. In: I Schultz, R Gatchel (Eds.), Handbook of complex occupational disability claims: early risk identification, intervention and prevention. Springer, New York.

Branca B (2006). Neuropsychologic aspects of post-traumatic headache and chronic daily headache. Curr Pain Headache Rep 10 (1): 54–66.

Brooks N, McKinlay W, Symington C et al. (1987). Return to work within the first seven years of severe head injury. Brain Inj 1 (1): 5–19.

Bryan CJ (2013). Repetitive traumatic brain injury (or concussion) increases severity of sleep disturbance among deployed military personnel. Sleep 36 (6): 941–946.

Bryan CJ, Clemans TA, Hernandez AM et al. (2013). Loss of consciousness, depression, posttraumatic stress disorder, and suicide risk among deployed military personnel with mild traumatic brain injury. J Head Trauma Rehabil 28 (1): 13–20.

Bryant RA, Marosszeky JE, Crooks J et al. (1999). Interaction of posttraumatic stress disorder and chronic pain following traumatic brain injury. J Head Trauma Rehabil 14 (6): 588–594.

CDC (2003). Report to Congress on Mild Traumatic Brain Injury in the United States: Steps to Prevent a Serious Public Health Problem. Centers for Disease Control, Atlanta, GA.

Centers for Disease Control and Prevention (CDC), The National Institutes of Health (NIH), The Department of Defense (DoD), The Department of Veterans Affairs (VA) Leadership Panel (2013). Report to Congress on Traumatic Brain Injuries in the United States: Understanding the Public Health Problem among Current and Former Military Personnel. Available online at: http://www.cdc.gov/traumaticbraininjury/pdf/Report_to_Congress_on_Traumatic_Brain_Injury_2013-a.pdf.

Cernak I, Savic J, Ignjatovic D et al. (1999). Blast injury from explosive munitions. J Trauma 47: 96–104.

Chafi M, Karami G, Ziejewski M (2010). Biomechanical assessment of brain dynamic responses due to blast pressure waves. Ann Biomed Eng 38: 490–504.

Chamalian I, Feinstein A (2004). Outcome after mild to moderate TBI: the role of dizziness. Arch Phys Med Rehabil 85: 1662–1666.

Chavko M, Koller W, Prusaczyk W et al. (2007). Measurement of blast wave by a miniature fiber optic pressure transducer in the rat brain. J Neurosci Meth 159 (2): 277–281.

Chibnall JT, Duckro PN (1994). Post-traumatic stress disorder in chronic post-traumatic headache patients. Headache 34 (6): 357–361.

Corrigan JD, Bogner J (2007). Initial reliability and validity of the Ohio State University TBI Identification Method. J Head Trauma Rehabil 22 (6): 318–329.

Corrigan JD, Bogner JA, Mysiw WJ et al. (2001). Life satisfaction after traumatic brain injury. J Head Trauma Rehabil 16 (6): 543–555.

Davis LL, Leon AC, Toscano R et al. (2012). A randomized controlled trial of supported employment among veterans with posttraumatic stress disorder. Psychiatr Serv 63 (5): 464–470.

Davison K, Bagley C (1969). Schizophrenia-like psychosis associated with organic disorder of the CNS. Br J Psychiatry 4: 113–184.

Dennis A, Kochanek P (2007). Pathobiology of blast injury. Intensive Care Med 21.

DePalma R, Burris D, Champion H et al. (2005). Blast injuries. N Engl J Med 352: 1335–1342.

DeWitt D, Prough D (2009). Blast-induced brain injury and post-traumatic hypotension and hypoxemia. J Neurotrauma 26: 1–11.

Dikmen S, Levin H (1993). Methodological issues in the study of mild head injury. J Head Trauma Rehabil 8 (3): 30–37.

Dikmen SS, Bombardier CH, Machamer JE et al. (2004). Natural history of depression in traumatic brain injury. Arch Phys Med Rehabil 85 (9): 1457–1464.

Drew L, Drew W (2004). The contrecoup-coup phenomenon: a new understanding of the mechanism of closed head injury. Neurocrit Care 3: 385–390.

Duckro PN, Tait R, Margolis RB et al. (1985). Behavioral treatment of headache following occupational trauma. Headache 25 (6): 328–331.

Duckro PN, Chibnall JT, Tomazic TJ (1995). Anger, depression, and disability: a path analysis of relationships in a sample of chronic posttraumatic headache patients. Headache 35 (1): 7–9.

Eierud C, Craddock RC, Fletcher S et al. (2014). Neuroimaging after mild traumatic brain injury: review and meta-analysis. Neuroimage Clin 4: 283–294.

Fedoroff JP, Starkstein S, Forrester AW et al. (1992). Depression in patients with acute traumatic brain injury. Am J Psychiatr 149: 918–923.

Finkelstein E, Corso P, Miller T (2006). The Incidence and Economic Burden of Injuries in the United States. Oxford University Press, New York.

Geddes J (1997). What's new in the diagnosis of head injury? J Clin Pathol 50 (4).

Geddes J, Vowles G, Beer T et al. (1997). The diagnosis of diffuse axonal injury: implications for forensic practice. Neuropathol Appl Neurobiol 23 (4): 339–347.

Gennarelli T (1996). The spectrum of traumatic axonal injury. Neuropathol Appl Neurobiol 22: 509–513.

Gennarelli TA, Thibault LE, Adams JH et al. (1982). Diffuse axonal injury and traumatic coma in the primate. Ann Neurol 12 (6): 564–574.

Gentry L, Godersky J, Thompson B (1988). MR imaging of head trauma: review of the distribution and radiopathologic features of traumatic lesions. AJR Am J Roentgenol 150: 663–672.

Godefroy O (2003). Frontal syndrome and disorders of executive functions. J Neurol 250 (1): 1–6.

Gollaher K, High W, Sherer M et al. (1998). Prediction of employment outcome one to three years following traumatic brain injury (TBI). Brain Inj 12 (4): 255–263.

Goodman J (1994). Pathologic changes in mild head injury. Semin Neurol 14: 19–24.

Graham D, Lawrence A, Adams J et al. (1987). Brain damage in non-missile head injury secondary to high intracranial pressure. Neuropathol Appl Neurobiol 13: 209–217.

Graham D, Lawrence A, Adams J et al. (1989). Pathology of mild head injury. In: Mild to Moderate Head Injury, Blackwell Scientific, Boston, pp. 63–75.

Graham D, Gennarelli T, McIntosh T (2002). Trauma. In: D Graham, P Lantos (Eds.), Greenfield's Neurpathology. Vol. 2. Arnold, Hodder Headline Group, London.

Greenspan AI, Wrigley JM, Kresnow M et al. (1996). Factors influencing failure to return to work due to traumatic brain injury. Brain Inj 10 (3): 207–218.

Gronwall D, Wrightson P (1974). Delayed recovery of intellectual function after minor head injury. Lancet 2 (7881): 605–609.

Gurr B, Coetzer BR (2005). The effectiveness of cognitive-behavioural therapy for post-traumatic headaches. Brain Inj 19 (7): 481–491.

Guskiewicz K, Marshall S, Bailes J et al. (2007). Recurrent concussion and risk of depression in retired professional football players. Med Sci Sports Exerc 39 (6): 903–909.

Gutierrez E, Huang YC, Haglid K et al. (2001). A new model for diffuse brain injury by rotational acceleration: I model, gross appearance, and astrocytosis. J Neurotrauma 18 (3): 247–257.

Hamer H, Gandhi R, Wong S et al. (2013). Predicting return to work following treatment of chronic pain disorder. Occup Med (Lond) 63 (4): 253–259.

Harmon KG, Drezner J, Gammons M et al. (2013). American Medical Society for Sports Medicine position statement: concussion in sport. Clin J Sport Med 23 (1): 1–18.

Headache Classification Committee of the International Headache Society (1988). Classification and diagnostic criteria for headache disorders, cranial neuralgias and facial pain. Cephalalgia (8 Suppl 7): 1–96.

Hicks R, Fertig S, Desrocher R et al. (2010). Neurologic effects of blast injury. J Trauma 68: 1257–1263.

Hillier SL, Sharpe MH, Metzer J (1997). Outcomes 5 years post-traumatic brain injury (with further reference to neurophysical impairment and disability). Brain Inj 11 (9): 661–675.

Hoge CW, Terhakopian A, Castro CA et al. (2007). Association of posttraumatic stress disorder with somatic symptoms, health care visits, and absenteeism among Iraq war veterans. Am J Psychiatr 164: 150–153.

Hoge CW, McGurk D, Thomas JL et al. (2008). Mild traumatic brain injury in US soldiers returning from Iraq. N Engl J Med 358: 453–463.

Holm L, Cassidy JD, Carroll LJ et al. (2005). Neurotrauma task force on mild traumatic brain injury of the WHO Collaborating Centre. Summary of the WHO Collaborating Centre for Neurotrauma Task Force on Mild Traumatic Brain Injury. J Rehabil Med 37 (3): 137–141.

Holsinger T, Steffens DC, Phillips C et al. (2002). Head injury in early adulthood and the lifetime risk of depression. Arch Gen Psychiatry 59 (1): 17–22.

Hoofien D, Gilboa A, Vakil El et al. (2001). Traumatic brain injury (TBI) 10–20 years later: a comprehensive study of psychiatric symptomatology, cognitive abilities and psychosocial functioning. Brain Inj 15 (3): 189–209.

Inglese M, Bomsztyk E, Gonen O et al. (2005a). Dilated perivascular spaces: hallmarks of mild traumatic brain injury. Am J Neuroradiol 26 (4): 719–724.

Inglese M, Makani S, Johnson G et al. (2005b). Diffuse axonal injury in mild traumatic brain injury: a diffusion tensor imaging study. J Neurosurg 103 (2): 298–303.

Jacobson GP, Ramadan NM, Aggarwal SK et al. (1994). The Henry Ford Hospital Headache Disability Inventory (HDI). Neurology 44 (5): 837–842.

Jorge R, Robinson RG (2003). Mood disorders following traumatic brain injury. Int Rev Psychiatry 15 (4): 317–327.

Jorge R, Robinson RG, Arndt SV et al. (1993). Comparison between acute-and delayed-onset depression following traumatic brain injury. J Neuropsychiatr 5 (1): 43–49.

Jorge RE, Robinson RG, Moser D et al. (2004). Major depression following traumatic brain injury. Arch Gen Psychiatry 61 (1): 42–50.

Kant R, Duffy J, Pivovarnik A (1998). Prevalence of apathy following head injury. Brain Inj 12: 87–92.

Keyser-Marcus LA, Bricout JC, Wehman P et al. (2002). Acute predictors of return to employment after traumatic brain injury: a longitudinal follow-up. Arch Phys Med Rehabil 83 (5): 635–641.

Kosinski M, Bayliss MS, Bjorner JB et al. (2003). A six-item short-form survey for measuring headache impact: the HIT-6. Qual Life Res 12 (8): 963–974.

Krause N, Frank J, Dasinger L et al. (2001). Determinants of duration of disability and return-to-work after work-related injury and illness: challenges for future research. Am J Ind Med 40: 464–484.

Kurca E, Sivak S, Kucera P (2006). Impaired cognitive functions in mild traumatic brain injury patients with normal and pathologic magnetic resonance imaging. Neuroradiology 48 (9).

Lahz S, Bryant RA (1996). Incidence of chronic pain following traumatic brain injury. Arch Phys Med Rehabil 77 (9): 889–891.

Lane JC, Arciniegas DB (2002). Post-traumatic headache. Curr Treat Options Neurol 4 (1): 89–104.

Langlois JA, Rutland-Brown W, Thomas KE (2005). The incidence of traumatic brain injury among children in the United States: differences by race. J Head Trauma Rehabil 20 (3): 229–238.

Langlois JA, Rutland-Brown W, Wald MM (2006). The epidemiology and impact of traumatic brain injury: a brief overview. J Head Trauma Rehabil 21 (5): 375–378.

Levi L, Borovich B, Guilburd J (1990). Wartime neurosurgical experience in Lebanon, 1982–1985. Israel J Med Sci 26: 555–578.

Levin H, Kraus M (1994). The frontal lobes and traumatic brain injury. J Neuropsychiatry Clin Neurosci 6: 443–454.

Levin HS, O'Donnell VM, Grossman RG (1979). The Galveston Orientation and Amnesia Test: a practical scale to assess cognition after head injury. J Nerv Ment Dis 167 (11): 675–684.

Levin HS, Williams DH, Eisenberg HM et al. (1992). Serial magnetic resonance imaging and neurobehavioral findings after mild to moderate closed head injury. J Neurol Neurosurg Psychiatry 55 (4): 255–262.

Levin H, Mendelsohn D, Lilly M et al. (1997). Magnetic resonance imaging in relation to functional outcome of pediatric closed head injury: a test of the Ommaya-Gennarelli model. Neurosurgery 40: 432–441.

Levin HS, McCauley SR, Josic CP et al. (2005). Predicting depression following mild traumatic brain injury. Arch Gen Psychiatry 62 (5): 523–528.

Ling G, Bandak F, Armonda R et al. (2009). Explosive blast neurotrauma. J Neurotrauma 26: 1–11.

Lipton RB, Silberstein SD, Saper JR et al. (2003). Why headache treatment fails. Neurology 60 (7): 1064–1070.

Marcus DA (2003). Disability and chronic posttraumatic headache. Headache 43 (2): 117–121.

Marsh N, Kersel D (2006). Frequency of behavioural problems at one year following traumatic brain injury: correspondence between patient and caregiver reports. Neuropsychol Rehabil 16 (6): 684–694.

Mathias J, Bigler E, Jones N et al. (2004). Neuropsychological and Informational processing performance and its relationship to white matter changes following moderate and severe traumatic brain injury: a preliminary study. Appl Neuropsychol 11 (3): 134–152.

Maxwell WL, Watt C, Graham DI et al. (1993). Ultrastructural evidence of axonal shearing as a result of lateral acceleration of the head in nonhuman-primates. Acta Neuropathol 86 (2): 136–144.

Maxwell W, Povlishock J, Graham D (1997). A mechanistic analysis of nondisruptive axonal injury: a review. J Neurotrauma 14: 419–440.

McAllister TW (1992). Neuropsychiatric sequelae of head injuries. Psychiatr Clin North Am 15 (2): 395–413.

McAllister TW (2010). Genetic factors modulating outcome after neurotrauma. PM R 2 (12 Suppl 2): S241–S252.

McGurk SR, Mueser KT, Feldman K et al. (2007). Cognitive training for supported employment: 2–3 year outcomes of a randomized controlled trial. Am J Psychiatry 164: 437–441.

Merskey H, Bogduk N (1994). Part III: Pain Terms, A Current List with Definitions and Notes on Usage. In: Classification of Chronic Pain, 2nd edn. In: IASP Task Force on Taxonomy. IASP Press, Seattle, pp. 209–214.

Meythaler J, Peduzzi J, Eleftherious E et al. (2001). Current concepts: diffuse axonal injury associated traumatic brain injury. Arch Phys Med Rehabil 82: 1461–1471.

Miller E (2000). The prefrontal cortex and cognitive control. Nat Rev Neurosci 1 (1): 59–65.

Mittl R, Grossman R, Hiehle J et al. (1994). Prevalence of MR evidence of diffuse axonal injury in patients with mild head injury and normal head CT findings. AJNR Am J Neuroradiol 15: 1583–1589.

Moldover JE, Goldberg KB, Prout MF (2004). Depression after traumatic brain injury: a review of evidence for clinical heterogeneity. Neuropsychol Rev 14 (3): 143–154.

Mooney G, Speed J, Sheppard S (2005). Factors related to recovery after mild traumatic brain injury. Brain Inj 19 (12): 975–987.

Moore D, Jerusalem A, Nyein M et al. (2009). Computational biology – modeling of primary blast effects on the central nervous system. Neuroimage 47 (Suppl 2): T10–T20.

Morissette SB, Woodward M, Kimbrel NA et al. (2011). Deployment-related TBI, persistent postconcussive symptoms, PTSD, and depression in OEF/OIF veterans. Rehabil Psychol 56 (4): 340–350.

Nampiaparampil DE (2008). Prevalence of chronic pain after traumatic brain injury: a systematic review. JAMA 300 (6): 711–719.

National Institute of Mental Health (2001). The numbers count: mental disorders in America: a summary of statistics describing the prevalence of mental disorders in America. Retrieved from, http://www.nimh.nih.gov/publicat/numbers.cfm.

Niemeier JP, Perrin PB, Holcomb MG et al. (2014). Gender differences in awareness and outcomes during acute traumatic brain injury recovery. J Womens Health (Larchmt) 23 (7): 573–580.

Nott M, Chapparo C, Baguley I (2006). Agitation following traumatic brain injury: an Australian sample. Brain Inj 20: 1175–1182.

Obermann M, Nebel K, Schumann C et al. (2009). Gray matter changes related to chronic posttraumatic headache. Neurology 73 (12): 978–983.

Obermann M, Keidel M, Diener HC (2010). Post-traumatic headache: is it for real? Crossfire debates on headache: pro. Headache 50 (4): 710–715.

O'Connor SS, Zatzick DF, Wang J et al. (2012). Association between posttraumatic stress, depression, and functional impairments in adolescents 24 months after traumatic brain injury. J Trauma Stress 25 (3): 264–271.

Otis JD, McGlinchey R, Vasterling JJ et al. (2011). Complicating factors associated with mild traumatic brain injury: impact on pain and posttraumatic stress disorder treatment. J Clin Psychol Med Settings 18 (2): 145–154.

Packard RC (1992). Posttraumatic headache: permanency and relationship to legal settlement. Headache 31 (10): 496–500.

Packard RC, Ham LP (1993). Impairment ratings for posttraumatic headache. Headache 33 (7): 359–364.

Packard RC, Weaver R, Ham LP (1993). Cognitive symptoms in patients with posttraumatic headache. Headache 33 (7): 365–368.

Pizzi L, Carter C, Howell J et al. (2005). Work loss, healthcare utilization, and costs among US employees with chronic pain. Dis Manag Health Outcome (13): 201–208.

Ponsford JL, Olver JH, Curran C et al. (1995). Prediction of employment status 2 years after traumatic brain injury. Brain Inj 9 (1): 11–20.

Povlishock J, Becker D, Cheng C et al. (1983). Axonal change in minor head injury. J Neuropathol Exper Neurol 42: 225–242.

Povlishok JT (1992). Traumatically induced axonal injury: pathogenesis and pathobiological implications. Brain Pathol 2 (1): 1–12.

Ramadan N, Keidel M (2000). Chronic posttraumatic headache. In: J Olesen, P Tfelt-Hansen, K Welch (Eds.), The Headaches, 2nd edn. Lippincott Williams and Wilkins, Philadelphia, pp. 771–780.

Rapoport MJ (2012). Depression following traumatic brain injury: epidemiology, risk factors and management. CNS Drugs 26 (2): 111–121.

Rapoport MJ, McCullagh S, Streiner D et al. (2003). The clinical significance of major depression following mild traumatic brain injury. Psychosomatics 44 (1): 31–37.

Reeves RR, Laizer JT (2012). Traumatic brain injury and suicide. J Psychosoc Nurs Ment Health Serv 50 (3): 32–38.

Rimel RW, Giordani B, Barth JT et al. (1981). Disability caused by minor head injury. Neurosurgery 9 (3): 221–228.

Roberts MA, Manshadi FF, Bushnell DL et al. (1995). Neurobehavioural dysfunction following mild traumatic brain injury in childhood: a case report with positive findings on positron emission tomography (PET). Brain Inj 9 (5): 427–436.

Rogers E, Anthony W, Lyass A et al. (2006). A randomized clinical trial of vocational rehabilitation for people with psychiatric disabilities. Rehabil Counsel Bull 49: 143–156.

Ruff RM, Crouch JA, Troster AI et al. (1994). Selected cases of poor outcome following a minor brain trauma: comparing neuropsychological and positron emission tomography assessment. Brain Inj 8 (4): 297–308.

Sbordone RJ (2001). Limitations of neuropsychological testing to predict the cognitive and behavioral functioning of persons with brain injury in real-world settings. Neurorehabilitation 16 (4): 199–201.

Scheibel RS, Newsome MR, Troyanskaya M et al. (2012). Altered brain activation in military personnel with one or more traumatic brain injuries following blast. J Int Neuropsychol Soc 18 (1): 89–100.

Seifert TD, Evans RW (2010). Posttraumatic headache: a review. Curr Pain Headache Rep 14 (4): 292–298.

Shames J, Treger I, Ring H et al. (2007). Return to work following traumatic brain injury: trends and challenges. Disabil Rehabil 29 (17): 1387–1395.

Sheftell FD, Tepper SJ, Lay CL et al. (2007). Post-traumatic headache: emphasis on chronic types following mild closed head injury. Neurol Sci 28 (Suppl 2): S203–S207.

Sherer M, Bergloff P, Levin E et al. (1998). Impaired awareness and employment outcome after traumatic brain injury. J Head Trauma Rehabil 13 (5): 52–61.

Steadman-Pare D, Colantonio A, Ratcliff G et al. (2001). Factors associated with perceived quality of life many years after traumatic brain injury. J Head Trauma Rehabil 16 (4): 330–342.

Stewart WF, Lipton RB, Simon D et al. (1998). Reliability of an illness severity measure for headache in a population sample of migraine sufferers. Cephalalgia 18 (1): 44–51.

Taber KH, Warden DL, Hurley RA (2006). Blast-related traumatic brain injury: What is known? J Neuropsychiatry Clin Neurosci 18 (2): 141–145.

[The International Classification of Headache Disorders: 2nd edition] (2006). Neurol Neurochir Pol 40(2 Suppl 1): S7–S41.

van der Horn HJ, Spikman JM, Jacobs B et al. (2013). Postconcussive complaints, anxiety, and depression related to vocational outcome in minor to severe traumatic brain injury. Arch Phys Med Rehabil 94 (5): 867–874.

van Reekum R, Bolago I, Finlayson MA et al. (1996). Psychiatric disorders after traumatic brain injury. Brain Inj 10 (5): 319–327.

Van Velzen J, Van Bennekom C, Edelaar M et al. (2009a). How many people return to work after acquired brain injury? A systematic review. Brain Inj 23: 473–488.

Van Velzen J, Van Bennekom C, Edelaar M et al. (2009b). Prognostic factors of return to work after acquired brain injury: a systematic review. Brain Inj 23: 385–395.

van Velzen J, van Bennekom C, van Dormolen M et al. (2011). Factors influencing return to work experienced by people with acquired brain injury: a qualitative research study. Disabil Rehabil 33: 23–24.

Vanderploeg RD, Schwab K, Walker WC et al. (2008). Rehabilitation of traumatic brain injury in active duty military personnel and veterans: defense and Veterans Brain Injury Center randomized controlled trial of two rehabilitation approaches. Arch Phys Med Rehabil 89: 2227–2238.

Vargas BB, Dodick DW (2012). Posttraumatic headache. Curr Opin Neurol 25 (3): 284–289.

Vasterling JJ, Brailey K, Proctor SP et al. (2012). Neuropsychological outcomes of mild traumatic brain injury, post-traumatic stress disorder and depression in Iraq-deployed US Army soldiers. Br J Psychiatry 201 (3): 186–192.

von Onciul J (1996). ABC of work related disorders. Stress at work. Br Med J 313 (7059): 745–748.

Walker WC, Marwitz JH, Kreutzer JS et al. (2006). Occupational categories and return to work after traumatic brain injury: a multicenter study. Arch Phys Med Rehabil 87 (12): 1576–1582.

Warden D (2006). Military TBI during the Iraq and Afghanistan wars. J Head Trauma Rehabil 21 (5): 398–402.

Wehman P, Kregel J, Sherron P et al. (1993). Critical factors associated with the successful supported employment placement of patients with severe traumatic brain injury. Brain Inj 7 (1): 31–44.

Wehman P, Targett P, West M et al. (2005). Productive work and employment for persons with traumatic brain injury: what have we learned after 20 years? J Head Trauma Rehabil 20 (2): 115–127.

Whyte J, Ponsford J, Watanabe T et al. (2010). Traumatic brain injury. In: W Frontera (Ed.), Delisa's Physical Medicine and Rehabilitation: Principles and Practice, 5th edn. Lippincott, Williams & Wilkins, Philadelphia, PA.

Williamson ML, Elliott TR, Berry JW et al. (2013). Predictors of health-related quality-of-life following traumatic brain injury. Brain Inj 27 (9): 992–999.

World Health Organization (2002). International classification of functioning, disability and health. ICF: World Health Organization, Geneva.

World Health Organization (2013). The International Classification of Headache Disorders, 3rd edition (beta version). Cephalalgia 33 (9): 629–808.

Zivin K, Bohnert AS, Mezuk B et al. (2011). Employment status of patients in the VA health system: implications for mental health services. Psychiatr Serv 62 (1): 35–38.

Handbook of Clinical Neurology, Vol. 131 (3rd series)
Occupational Neurology
M. Lotti and M.L. Bleecker, Editors

Chapter 27

The complex regional pain syndrome

ANNE LOUISE OAKLANDER[1,2]* AND STEVEN H. HOROWITZ[1,3]

[1]*Department of Neurology, Massachusetts General Hospital, Boston, MA, USA*

[2]*Department of Pathology (Neuropathology), Massachusetts General Hospital, Boston, MA, USA*

[3]*Department of Neurology, School of Medicine, Tufts University, Boston, MA, USA*

INTRODUCTION AND NOMENCLATURE

The term complex regional pain syndrome (CRPS) refers to chronic pain of neuropathic origin in the region of or distal to, an injured area, despite healing of visible tissues. CRPS most commonly follows injury to a limb (sprains/strains, fractures, surgery) or limb immobilization, with up to 10% of cases developing without known trauma (possibly reflecting internal neural damage, e.g., nerve infarction, entrapment) (de Rooij et al., 2010).

Although described earlier in fragmentary form, this particular endophenotype was first codified by American neurologist S. Weir Mitchell in soldiers with meticulously characterized major injuries to their peripheral nervous system during the American Civil War (Mitchell, 1864; Mitchell et al., 1864). Mitchell named this condition "causalgia" from the Greek, *causos* = heat; *algia* = pain. In 1946, Evans described similar symptoms of chronic distal-limb pain plus microvascular dysregulation in a case series of patients without major traumas or known nerve injuries. However, the patients had been only cursorily examined by today's standards, and in many if not most, the underlying injury or diagnosis was unknown (e.g., "flat feet"). He termed this condition "reflex sympathetic dystrophy" (RSD), believing that the rubor, pallor, sweating, edema, skin, and bone atrophy were secondary to reflexive involvement of the sympathetic nervous system that was responsive to sympathetic blockade (Evans, 1946). Of note, his patients were not examined by nerve specialists, nor did they undergo electrodiagnostic or other studies capable of detecting underlying nerve injuries. RSD thus became a "wastebasket" term applied to patients with prolonged or seemingly excess pain that lingered after limb traumas of varying degrees of severity and diagnostic certainty.

In 1994 the International Association for the Study of Pain (IASP) created the name "complex regional pain syndrome" and developed formal, albeit loose definitions for CRPS-I and CRPS-II (Table 27.1) (Merskey and Bogduk, 1994). CRPS-I replaced the term "reflex sympathetic dystrophy" for patients without known nerve injuries, and CRPS-II replaced "causalgia" for patients with identified nerve injuries. However, the exact means by which nerve injuries should be diagnosed was never specified, thus the distinction remained dubious. Algodystrophy is the common name in Europe, and "shoulder–hand syndrome" and "Sudeck's atrophy" are historic terms (Sudeck, 1901).

The term "CRPS type II" lacks anatomic precision, and we and others have seen patients deprived of potentially curative treatments because of its use (Birch, 2009), plus it deters medical providers. Patients' interests are better served by labels that describe the anatomic localization of their nerve injuries. Even in CRPS-I, where underlying lesions may not have been precisely localized or confirmed, each patient's most likely cause should be stated. In the USA at least, a patient labeled as "probable peroneal nerve injury from cast compression at the fibular head" will receive better care than one labeled "CRPS-I after leg fracture." A recent review (Borchers and Gershwin, 2014) questions the similarity between CRPS-I and II, discusses potential subgroups, and concludes that CRPS "is a poorly defined term used to describe a variety of disorders characterized by pain disproportional to the inciting event."

Recent modifications proposed for research and clinical use (Table 27.2), informally known as "the 2003

*Correspondence to: A.L. Oaklander, Massachusetts General Hospital, 275 Charles St./Warren 310, Boston MA 02114, USA. Tel: +1-617-726-0260, Fax: +1-617-726-0473, E-mail: aloaklander@mgh.harvard.edu

Table 27.1

The 1994 International Association for the Study of Pain (IASP) diagnostic criteria for complex regional pain syndrome (CRPS)-I (reflex sympathetic dystrophy)

1. The presence of an initiating noxious event or a cause of immobilization
2. Continuing pain, allodynia, or hyperalgesia, the pain of which is disproportionate to any inciting event
3. Evidence at some time of edema, changes in skin blood flow, or abnormal sudomotor activity in the region of pain
4. This diagnosis is excluded by the existence of conditions that would otherwise account for the degree of pain and dysfunction.

Note: Criteria 2–4 must be satisfied

IASP diagnostic criteria for CRPS-II (causalgia)

1. The presence of continuing pain, allodynia, or hyperalgesia after a nerve injury, not necessarily limited to the distribution of the injured nerve
2. Evidence at some time of edema, changes in skin blood flow, or abnormal sudomotor activity in the region of pain
3. This diagnosis is excluded by the existence of conditions that would otherwise account for the degree of pain and dysfunction.

Note: All three criteria must be satisfied

Adapted from Merskey and Bogduk (1994).

Budapest criteria" (Harden et al., 2007), were validated and endorsed by the IASP (Harden et al., 2010) but are not yet widely used in the USA. These include more restrictive research criteria and less restrictive clinical criteria – of 975 patients who satisfied the 1994 IASP criteria, only 71.5% met the Budapest clinical criteria and 45.8% met the research criteria, partly because of its requirement for motor symptoms and signs. Patients meeting the Budapest criteria had lower quality-of-life scores, consistent with worse disease (van Velzen et al., 2014).

CLINICAL SYMPTOMS

CRPS symptoms can range from mild to severe, but a Dutch hospital-based survey of 975 patients found quality of life was lower for CRPS patients than for those with other chronic-pain diagnoses (van Velzen et al., 2014). Community-based surveys from mid-western USA and the Netherlands estimated the incidence of CRPS-I to be 5.46 and 26.2 per 100 000 person-years with median/mean ages of onset of 46 years or 52.7 years, respectively (Sandroni et al., 2003; de Mos et al., 2007). The reason for the disparate prevalences is unclear, although cultural/regional and sampling differences (Sumitani et al., 2010), and geographic, genetic, and environmental differences might contribute (Herlyn et al., 2010).

Table 27.2

The 2003 Budapest clinical diagnostic criteria for complex regional pain syndrome (CRPS)*

General definition

CRPS describes an array of painful conditions characterized by continuing (spontaneous and/or evoked) regional pain seemingly disproportionate in time or degree to the usual course of any known trauma or other lesion. The pain is regional (not in a specific nerve territory or dermatome) and usually has a distal predominance of abnormal sensory, motor, sudomotor, vasomotor, and/or trophic findings. The syndrome shows variable progression over time

To make the clinical diagnosis the following criteria must be met:

1. Continuing pain, which is disproportionate to any inciting event
2. Must report at least one symptom in three of the four following categories:
   - Sensory: Reports of hyperesthesia and/or allodynia
   - Vasomotor: Reports of temperature asymmetry and/or skin color changes and/or skin color asymmetry
   - Sudomotor/edema: Reports of edema and/or sweating changes and/or sweating asymmetry
   - Motor/trophic: Reports of decreased range of motion and/or motor dysfunction (weakness, tremor, dystonia) and/or trophic changes (hair, nail, skin)
3. Must display at least one sign at time of evaluation in two or more of the following categories:
   - Sensory: Evidence of hyperalgesia (to pinprick) and/or allodynia (to light touch and/or temperature sensation and/or deep somatic pressure and/or joint movement)
   - Vasomotor: Evidence of temperature asymmetry (>1°C) and/or skin color changes and/or asymmetry
   - Sudomotor/edema: Evidence of edema and/or sweating changes and/or sweating asymmetry
   - Motor/trophic: Evidence of decreased range of motion and/or motor dysfunction (weakness, tremor, dystonia) and/or trophic changes (hair, nail, skin)
4. There is no other diagnosis that better explains the signs and symptoms

For research purposes: Must report at least one symptom in all four symptom categories and at least one sign (observed at evaluation) in two or more sign categories.

*Current distinctions between CRPS-I and CRPS-II are retained and a third subtype, CRPS-NOS (not otherwise specified) is recommended. From Harden, Bruehl and Perez et al. (2010).

The role of occupational causes has not been formally studied. An Ovid literature search on May 9, 2014 using the terms "occupational diseases" and "CRPS" yielded only two relevant references out of 38 983 citations since 1996. In a study of 485 patients with upper-extremity pain in the workplace only 3 (0.6%) were labeled with CRPS (Pascarelli and Hsu, 2001). One limitation is this syndrome's multiple names; another is that patients' locations at the time of their trauma is not usually reported (Sandroni et al., 2003; de Mos et al., 2007). However, since the most common causal injuries are fractures and sprains, both of which can be occupational, some CRPS is occupational.

CRPS also has a more specific link to occupational injury. Although first associated with battlefield injury, iatrogenic medical injury is now a more common cause, thus occupational injuries treated with medical procedures that cause CRPS are also occupationally caused. CRPS can be triggered by any type of invasive medical procedure, small or large. The substrate appears to be traumatic injury to sensory axons, although immobility from pain, disuse, or casting can contribute (Ota et al., 2014). For instance, a 56-year-old member of a hotel cleaning staff developed lower-leg CRPS due to superficial peroneal nerve injury in the context of a fracture caused by falling while cleaning a hotel bathtub. Her occupational CRPS-causal injury could have been caused by her cast or subsequent surgery, not only by the fall.

Perhaps the best-characterized iatrogenic cause of CRPS is routine phlebotomy at the antecubital fossa (Newman and Waxman, 1996), even in uncomplicated blood draws. The pathologic substrate is penetration of the medial or lateral cutaneous nerves of the forearm, which encircle the basilic and cephalic veins respectively (Horowitz, 2000). This provided the pathologic basis for one of the rodent models of CRPS based on a single needle puncture through rats' tibial nerve (Siegel et al., 2007; Klein et al., 2012). Regarding surgical causes, systematic review of neuropathic pain triggered by surgery found that the prevalence of pain differed in various types of surgery, probably depending on the likelihood of surgical iatrogenic nerve injury (Haroutiunian et al., 2013). CRPS has been reported in 2–5% of patients with carpal tunnel syndrome following surgical repair (Li et al., 2010). Whatever the cause, in one study 68% of CRPS patients returned to work within 15 months on average (Dumas et al., 2011).

Chronic pain and other sensory abnormalities

Several large case series show that CRPS-associated pain is disproportionately severe or prolonged compared to the inciting injury, and this is its defining characteristic. Common descriptors of CRPS pain include "burning," "shooting," "lancinating," "electric," "tearing," and "aching," often deep within the limb (Veldman et al., 1993; Birklein et al., 2000). Spontaneous (stimulus-independent) pain was reported in 74.6% of CRPS-I and 91.3% of CRPS-II patients with acute or chronic disease (Birklein et al., 2000). In chronic upper-limb CRPS, spontaneous ongoing hand pain was present in 65.3%, and exaggerated pain with physical activity in 80.5% (Huge et al., 2011). Distal pain with proximal radiation was common. Mechanical allodynia (pain from light contact) is common and often disabling as it may prevent patients from wearing clothes or using the affected limb. External stimuli, particularly moving the limb, and cold temperature can worsen pain (Veldman et al., 1993). Usually, CRPS pain begins immediately or shortly after the inciting injury, but it usually improves and changes over time. Initial spontaneous, constant, intense, "burning," "electric" qualities can evolve into "dull," "boring," "aching," or "tingling," with intermittent exacerbations, often after mechanical stimulation (Horowitz, 2001).

A close but incomplete relationship exists between spontaneous pain and stimulus-provoked hyperesthesias (allodynia or hyperalgesia) in CRPS. One group found that 93.4% of CRPS-I patients and 95.7% of CRPS-II patients had both (Birklein et al., 2000) and another found both in 75% of CRPS-I patients (Veldman et al., 1993). However, a few patients lacked stimulus-provoked hyperesthesias despite having spontaneous pain. Importantly, some patients also have positive or negative sensory phenomena (hypoesthesia) in glove or stocking patterns or even more diffusely, indicating an associated polyneuropathy. Occasional patients fulfill all other diagnostic criteria for CRPS-I (6.6–7.0%) or CRPS-II (4.3%), despite having no pain or pain caused only by dystonia or edema (Veldman et al., 1993; Birklein et al., 2000; Huge et al., 2011). These diverse findings by reputable investigators illustrate the difficulties with symptom-based diagnoses. For some patients, neither the 1994 nor the 2003 diagnostic CRPS criteria apply. Differences likely reflect different study samples, different interpretations of diagnostic criteria, and different patient and examiner interpretations of hyperesthesias and other diagnostic features.

The symptoms and signs of CRPS typically spread beyond classic territories of individual nerves. Although this is cited as a reason why CRPS-I could not be caused by nerve lesions as CRPS-II is, it reflects misunderstanding of peripheral-nerve anatomy and function. Indeed, when CRPS patients are asked to outline the epicenter, or most abnormal area, it often is in the distribution of a specific nerve (Fig. 27.1; Table 27.3), even though the pain extends further (van Rijn et al., 2011). The tendency of pain to spread is also documented in post-herpetic neuralgia, another very focal lesion that almost always affects only a single dorsal root ganglion (Petersen and Rowbotham, 2010).

This misimpression may have arisen because the nerve maps in textbooks depict the autonomous territories of myelinated motor and sensory axons, but CRPS primarily involves the thinly myelinated (A-delta) and unmyelinated C-fibers. Microneurographic study demonstrated that these have far larger receptive fields and innervate bone and blood vessels outside the traditional nerve-map areas (Schmelz et al., 1997). Also, small-fiber axon terminals can be electrically coupled to arbors of adjacent neurons (Meyer et al., 1985), permitting lateral communication between neurons with adjacent or overlapping receptive fields. Additionally, injured or degenerating axons can profoundly influence their uninjured neighbors within nerve trunks, roots, the spinal cord, and the brain. Small-fiber axonal injuries thus have multiple sites of spread and thus quite typically cause pain and dysfunction that extend beyond conventional nerve territories in all types of painful neuropathic pain conditions, not only in CRPS.

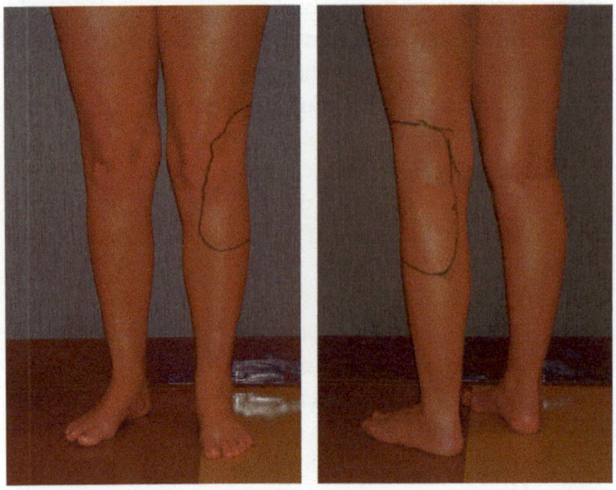

Fig. 27.1. This woman had chronic, refractory, posttraumatic left lateral knee and lower-leg pain and edema, diagnosed as complex regional pain syndrome type I (CRPS-I). Mechanical allodynia precluded wearing tight pants or skirts below the knee. When asked to identify the borders of her most affected area, she outlined the territory of the lateral cutaneous nerve of the calf (LCNC), a small sensory-only branch of the common peroneal nerve (Stewart, 2000). Targeted nerve conduction study confirmed a left-only LCNC lesion (the LCNC is not examined in standard nerve conduction studies). Small nerve injuries are common in CRPS-I. Many remain undiagnosed because few patients are examined by nerve specialists or with customized diagnostic tests. (Reproduced with permission from Oaklander and Ochoa, 2012.)

Another pattern of pain spread is to another limb, reported in one study to affect as many as 45% of patients with single-limb CRPS (van Rijn et al., 2011). This occurred most often ipsilaterally or mirror image contralaterally. It could occur spontaneously or after a separate trauma to the second limb. Spread was associated with younger age of onset and more severe illness. The authors postulated that spread is due to spinal or cortically mediated mechanisms, but contralateral "mirror image" is a fundamental property of focal nerve injuries in experimental animals as well as humans, as reviewed in Koltzenburg et al. (1999), and independently of cause (Oaklander and Belzberg, 1997; Oaklander et al., 1998; Oaklander and Brown, 2004; Fink and Oaklander, 2006). Thus, it is not specific for CRPS.

Vasomotor (microvascular) and sudomotor (sweating) abnormalities

CRPS is definitionally distinguished from posttraumatic neuralgia by its additional visible signs, specifically "evidence at some time of edema, changes in skin blood

Table 27.3

Well-characterized iatrogenic nerve injuries that can cause complex regional pain syndrome

Medical procedure	Nerve damaged	Epicenter of symptoms
Carpal tunnel release	Palmar cutaneous branch of median	Thenar eminence
Venepuncture at antecubital fossa of cephalic/basilic veins	Medial or lateral antebrachial cutaneous nerves	Medial or lateral inner forearm
Venepuncture on back of hand	Radial nerve/dorsal digital nerves	Back of hand
Femoral artery catheterization	Femoral nerve	Anterior thigh
Arthroscopic or open-knee surgery	Infrapatellar branch of saphenous nerve	Lower anterior knee, knee joint, medial lower leg (descending branch of saphenous)
Arthroscopic or open-knee surgery	Lateral cutaneous nerve of calf	Lateral knee and proximal lower leg (Fig. 27.2)
Casting or compression below knee at fibular head	Peroneal nerve	Outer lower leg, dorsum of foot
Saphenous vein stripping or harvesting for bypass graft	Descending branch of saphenous nerve	Medial lower leg, arch of foot (variable)

flow, or abnormal sudomotor activity in the region of the pain" (Merskey and Bogduk, 1994). Vasomotor instability (microvasculopathy) is the primary cause of edema, abnormal skin color, and temperature differences, and it can contribute to sweating abnormalities. In most cases, these phenomena are due to malfunction or degeneration of the small-fiber innervation of blood vessels (Fig. 27.2) (Albrecht et al., 2006). An analysis of 145 cases found edema in 92% of CRPS patients within the first 5 weeks, but only in 55% of patients with chronic CRPS-I (range 25–468 weeks) (Birklein et al., 2000). Temperature changes, hyperhidrosis, and abnormal hair growth also tend to improve over time (Huge et al., 2011). As CRPS resolves, which three-quarters of cases do even without treatment (Sandroni et al., 2003), the microvascular signs distinguishing CRPS from posttraumatic neuralgia (Fig. 27.3) usually resolve before the pain disappears; thus most CRPS patients pass through a stage where they no longer meet full diagnostic criteria for CRPS but only for post-traumatic neuralgia, suggesting fundamental unity of these conditions (Table 27.3).

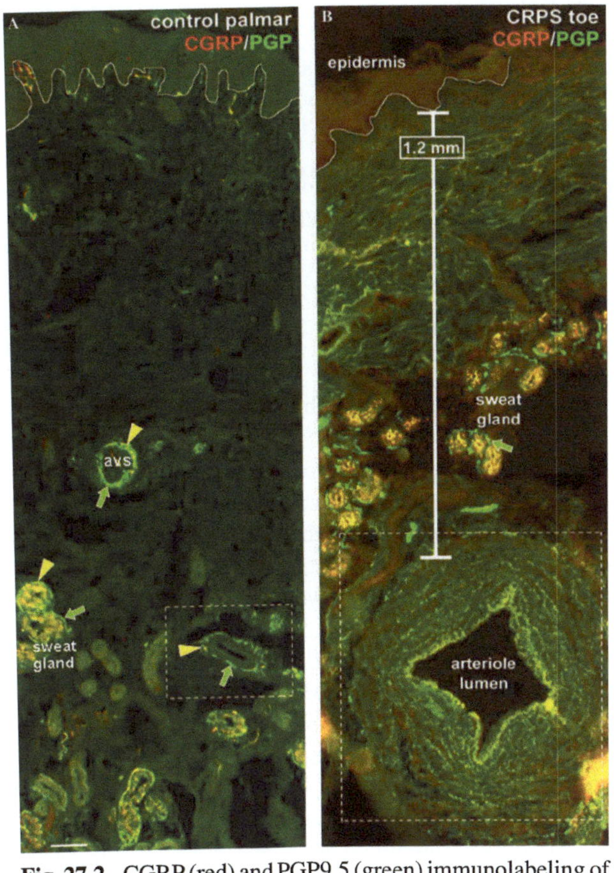

Fig. 27.2. CGRP (red) and PGP9.5 (green) immunolabeling of cutaneous innervation. (**A**) Control palmar skin demonstrating normal abundant innervation of sweat glands, arteries (box), and arteriovenous shunts (avs) located within 2 mm of the epidermis (basement membrane delineated by the white line). All contain abundant PGP^+ innervation (arrows) and some coexpress calcitonin gene-related peptide (CGRP: arrowheads). (**B**) In contrast, complex regional pain syndrome-affected skin contained sweat glands with reduced PGP^+ innervation (arrow), no coexpression of CGRP, and denervated arteries (box) that are extremely hypertrophied. Scale bar = 100 μm. (Reproduced from Albrecht et al., 2006, with permission.)

Motor abnormalities

Historic descriptions of CRPS focused on the chronic pain and microvascular symptoms, but there has been recent recognition that most patients also have impaired movement and motor function, such as weakness, bradykinesia (slow movement), reduced dexterity and range of motion, dystonia, myoclonus, tremor, tics, and motor neglect (Galer et al., 1998; Veldman et al., 1993; Birklein et al., 2000; van Rijn et al., 2007). They were not requisite in the 1994 IASP CRPS criteria (Table 27.1), but are given equal weight to other criteria in the 2003 Budapest revisions (Table 27.2). They may have originally been discounted due to erroneous attribution to either secondary defensive withdrawal of a painful limb or psychogenic origins.

Fixed limb dystonia is a dramatic and common movement disorder specifically associated with CRPS (van Rijn et al., 2007). It usually begins distally in the affected limb (Fig. 27.3A) but sometimes spreads to other limbs or the torso. Some movement-disorder experts attribute CRPS/dystonia to psychogenic causes (Munts and Koehler, 2010), but others report neurophysiologic and pharmacotherapeutic evidence of non-psychogenic contributors to posttraumatic dystonia as well (Schrag et al., 2004; van Rooijen et al., 2011). A systematic review of 133 publications presenting findings on 713 patients with peripherally induced movement disorders found that more than one-third had CRPS and that in many cases the peripheral trauma was associated with nerve injuries that triggered a central process that caused the movement disorder (van Rooijen et al., 2011).

There is limited information about the muscles, motor axons, and spinal motor systems in CRPS. Much of the literature has focused on the primary (M1) motor cortex, where changes in the cortical representation of CRPS-affected limbs have been reported. A meta-analysis of 18 studies of motor cortex function reviewed changes in M1 cortical excitability, spatial representation, reactivity, and glucose metabolism (Di et al., 2013). The authors found no conclusive abnormalities, except for limited evidence from two studies of bilateral M1 disinhibition in CRPS of the upper limb (Schwenkreis et al., 2003; Eisenberg et al., 2005).

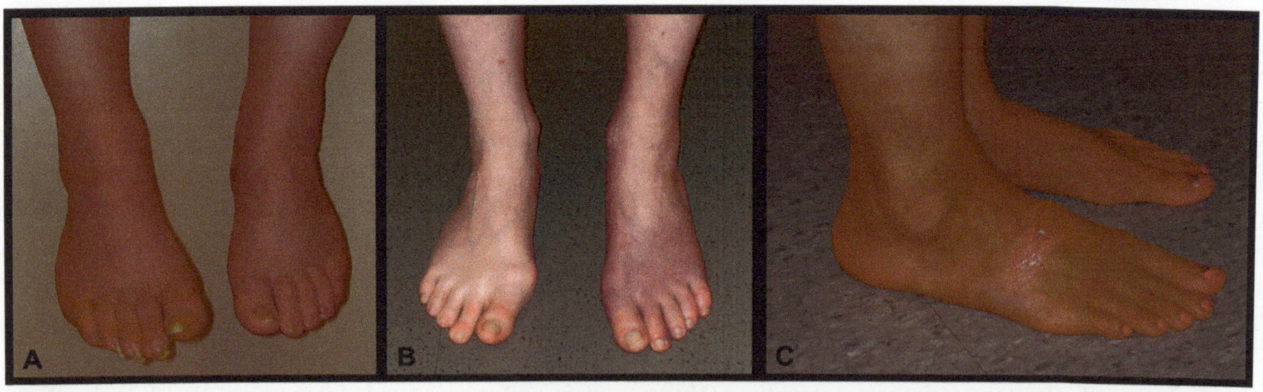

Fig. 27.3. Photographs depicting the microvascular dysregulation that distinguishes complex regional pain syndrome from standard posttraumatic neuralgia. These additional signs often gradually abate during recovery before complete pain resolution.

(**A**) A laborer 4 years after crush of his right foot by a steel beam. Severe ongoing and touch-induced pain in his great toe spread to his entire foot, which he guarded from contact even with bath water. Poor hygiene and mild edema, color change, and toe dystonia are visible. His history and a persistent Tinel's sign at his right fibular head suggested peroneal-nerve injury, but he declined electromyogram/nerve conduction study.

(**B**) A researcher in her 30s with chronic pain, edema, and vasodysregulation in her left foot and lower leg following bunionectomy years ago. She also developed hypertrophic ossification of her proximal left tibia. She had a history of autoimmune thyroid disease and asthma. When conventional medical therapy was inadequate she found some relief from hyperbaric oxygen therapy, and sustained benefit from intravenous immunoglobulin.

(**C**). An otherwise healthy woman in her 40s several years after healed second-degree burn from hot-coffee spill on her right foot. Afterwards, any contact with the dorsum of her foot lasting longer than a few minutes trigged severe pain and blistering (shown) and the resulting ulcer took weeks to heal. Inability to wear a shoe precluded employment. (Reproduced from Oaklander and Fields, 2009, with permission.)

Disorders of higher brain function

Secondary brain dysfunction can develop in CRPS as in other chronic pain disorders. Specific problems associated with CRPS include centrally mediated motor difficulties, impaired cognition, and neglect of CRPS-affected limbs, which were identified in 84% of 224 CRPS patients (Galer et al., 1995). "Neglect" refers to the phenomenon of impaired self-perception of a part of the body that is innervated by damaged neural structures. In mild form, it can distort perception or reduce awareness of the affected body part (Moseley, 2005, 2006; Forderreuther et al., 2004). Motor consequences of neglect include impaired control and use of the limb (Moseley, 2004b). In extreme form, some patients deny that their affected limb belongs to, or is attached to them. Neglect phenomena were previously thought to affect only patients after prior strokes and other structural brain injuries, but their identification in CRPS demonstrates the potential of even the most peripheral injuries to have transsynaptic effects on the brain (Legrain et al., 2012).

A study of upper-limb neglect in CRPS patients, normal controls, and patients with non-CRPS pain found similar neglect in CRPS and non-CRPS pain patients; however this differed clinically from the classic neglect seen in parietal-lobe stroke patients (Kolb et al., 2012). CRPS patients may be more likely to develop limb neglect, and it may be more severe than in patients with other causes of chronic limb pain (Frettlöh et al., 2006). Early on, neglect of a CRPS-affected limb was considered a self-defensive protective strategy or evidence of psychopathology, but functional imaging and neuropsychologic testing now implicate maladaptive reorganization of central sensorimotor systems, including the posterior parietal cortex, where important nociceptive and multimodal processing occurs (Legrain et al., 2012).

Other CRPS-associated cortical reorganization includes changes in contralateral primary (S1) and secondary (S2) somatosensory cortices – increased cortical inhibition, shrinkage of cortical hand representation, increased two-point discrimination thresholds. These can be detected with somatosensory evoked potential and functional magnetic resonance imaging (fMRI) blood oxygen level-dependent signal studies (Maihöfner et al., 2003; Pleger et al., 2004, 2006). Abnormalities correlated with pain intensity and reverted toward normal as CRPS resolved (Maihöfner et al., 2004). In CRPS patients, pinprick hyperalgesia and mechanical allodynia evoked widespread cortical activation – contralateral S1, motor and parietal association cortices, bilateral S2, insular, frontal, and cingulated cortices – whereas cortical activation by the unaffected limb was more restricted (Maihöfner et al., 2006).

Depression is another common brain complication of CRPS and other painful and disabling conditions. Other brain sequelae of CRPS include deficits in executive functions, working memory, and verbal fluency (Robinson et al., 2011), and in recognition of the emotional states of others (Shin et al., 2013). fMRI, magnetoencephalographic imaging, and transcranial magnetic stimulation (TMS) have demonstrated altered processing in multiple locations and pathways, with strong evidence of deficient inhibition of the motor cortex, particularly in patients with CRPS dystonia (Juottonen et al., 2002; Eisenberg et al., 2005; Walton et al., 2010). Structural MRI of 15 right-handed women with right-handed CRPS revealed reduced volume of gray matter in several pain-related regions of interest including the dorsal insula, left orbitofrontal cortex, and cingulate cortex (Barad et al., 2014). The dorsal putamen and right hypothalamus had an increase in gray matter.

Brain imaging changes are not specific for CRPS, nor are they primary. The brain is now recognized to be far more plastic than appreciated by earlier structural imaging techniques. Potential contributors include reduced use and function of the affected limb, depression, and medication use (Hsu et al., 2009). Fortunately, the limited evidence so far suggests that these brain changes reverse during recovery (Maihöfner et al., 2004; Pleger et al., 2005).

Additional clinical features

CRPS can affect the bones and joints to cause areas of bone marrow edema, demineralization, and, at worst, resorption of bones and joints (Kozin et al., 1976). Pathologically, these changes primarily reflect the effects of small-fiber axonopathy on the affected sclerotome (Basle et al., 1983), although disuse, lack of weight bearing, and impaired microcirculation presumably also contribute. Osteoprotegerin appears to be a molecular mediator, with elevated levels found in CRPS patients (Krämer et al., 2014).

Even less is known about the effects of CRPS on the spinal cord. Clearly, involvement of both the peripheral nervous system and the brain implicates the spinal cord as well, but this has not been adequately studied. Neuropathologic study of the spinal cord from a single patient with 7 years of CRPS triggered by a bicycle injury to her left calf showed loss of dorsal-horn (sensory) neurons and activation of microglia and astrocytes, not only at the level of the CRPS, but throughout cervical, thoracic, and lumber spinal cord levels (Del Valle et al., 2009). Similar spread of pathologic findings in the spinal cord has been described in a rodent model of CRPS triggered by a needlestick through the tibial nerve of rats (Lee et al., 2009).

In the past, CRPS was often attributed to psychologic illness or malingering, like many pain syndromes of uncertain causality, despite the fact that CRPS patients have visible signs in addition to subjective symptoms. In the American Medical Association *Guides to the Evaluation of Permanent Impairment*, the sections on CRPS state that the differential diagnosis includes "somatoform disorders, factitious disorder, and malingering," that the diagnostic workup may include psychologic evaluation and psychologic testing, and in one of the patient examples there was a comprehensive psychologic evaluation and review of the examinee's psychologic records from his entire life (Rondinelli, 2008). Thus, outdated views persist in influential places.

Recent studies rigorously examine the hypothesis of psychologic causality. A large population-based epidemiologic study in the Netherlands analyzed the earlier medical records of patients who would later develop CRPS and found no association with psychologic or psychiatric illnesses (de Mos et al., 2008). Systematic review of the published studies on adult CRPS and psychologic and psychiatric illnesses (most of low quality) also found no association (Beerthuizen et al., 2009). So, although patients with CRPS, like those with other causes of chronic pain and disability, have increased depression and anxiety, the evidence suggests these are secondary rather than preexisting risk factors.

CRPS is not a homogeneous disorder; variations in clinical picture are typical. While earlier reviews postulated three definite sequential stages (acute, dystrophic, atrophic), a later study using cluster analysis found three subgroups: (1) a relatively limited syndrome, mostly with vasomotor signs; (2) a relatively limited syndrome with neuropathic pain and/or sensory abnormalities; and (3) a florid syndrome with high levels of symptoms and signs in all categories (Bruehl et al., 2002).

DIAGNOSTIC TESTING

There are no objective diagnostic tests for CRPS. Since its pathophysiologic basis is unknown, the diagnosis is clinical, based on a history of limb trauma (usually) plus persistent pain and added features. In a small study, levels of serum osteoprotegrin had sensitivity of 0.74, specificity of 0.80, positive predictive value of 68%, and negative predictive value of 84%, while receiver operating characteristic curve analysis showed an area under the curve of 0.80 (Krämer et al., 2014), so further study is warranted. The following tests can help support the clinical diagnosis or rule out other entities.

Neurophysiological studies

Nerve conduction study (NCS) and electromyography (EMG) are the most common tests for documenting nerve damage, but they have only limited utility for CRPS. Standard NCS using surface electrodes only captures the activity of large (rapidly conducting) myelinated A-alpha or A-beta fibers in mixed nerves. It is insensitive to activity of the small A delta and C fibers that appear to be the major axon type affected in CRPS. With EMG, abnormal "neurogenic" patterns only identify axonal injuries that affect muscles. Therefore, EMG and NCS are useful when they are abnormal and identify damage of large-diameter sensory axons and/or motor axons (e.g., CRPS-II), but normal EMG and NCS have no negative predictive value. They cannot rule out the presence of small-fiber nerve damage, or of subthreshold injuries, or injuries to nonstudied nerves (Fig. 27.1). Negative studies have low "negative predictive value," meaning they cannot accurately exclude the presence of a nerve injury, although they are routinely used this way in CRPS. Additional limitations of EMG/NCS are that routine studies may not sample the actual nerve injured in that specific patient, or the injury may be to a nerve branch too small to study, or the proportion of injured axons may be too small to detect.

Quantitative sensory testing

Quantitative sensory testing (QST) is a more formal and sensitive method of measuring sensory thresholds (most commonly, thermal, mechanical, and vibratory) than clinical examination. As with clinical examination, the findings are subjective and depend on patient motivation and alertness (Freeman et al., 2003), and there are no CRPS-specific patterns. Maier and colleagues (2010) studied 1236 patients with various neuropathic pain syndromes with QST and found thermal and mechanical hyperalgesias to be most frequent in CRPS. QST is mainly a research tool at present.

Radiologic studies

Three-phase bone scintigraphy (TPBS), and, to a lesser extent, X-ray and bone densitometry have been used, as they can reveal focal subchondral or subperiosteal osteoporosis. One study found TPBS diagnostically helpful (Wuppenhorst et al., 2010), whereas another did not (Moon et al., 2012), likely in part because not all CRPS-causing nerve injuries involve nerves that also innervate bone (e.g., Fig. 27.1). TPBS can help identify when extensive bone remodeling may be contributing to pain, as this can be treated with nasal calcitonin or bisphosphonates (see below). Computed axial tomography and MRI are less often useful but can identify neurogenic edema of affected bones, joints, and deep tissues. MR neurography and ultrasound are emerging techniques for visualizing nerves, especially in entrapment neuropathies (Pham et al., 2014).

Neurodiagnostic skin biopsy

This involves removing a small skin punch under local anesthesia, vertically sectioning and immunolabeling it with a panaxonal marker (i.e., PGP 9.5) to permit visualization and quantitation of axons in the dermis and epidermis. Skin biopsy, currently the best objective diagnostic test for small-fiber polyneuropathy (SFPN) (England et al., 2009a), provided research evidence of focal small-fiber nerve injury in CRPS-I-affected skin, essentially obviating the distinction between CRPS-I and CRPS II (Oaklander et al., 2006). Because normative densities differ at different body locations and there is considerable interindividual variability, skin biopsies are not useful for CRPS diagnosis. They can rarely be used to compare innervation within a patient's CRPS-affected and matching unaffected areas to confirm focal axonopathy in the CRPS-affected area.

In research use, skin biopsies, mostly from the thigh and distal leg, identified reductions to less than the fifth centile of published norms in 20% among a cohort of 43 patients with lower-limb or "generalized disease affecting all extremities, face, and torso," consistent with the hypothesis that a minority of CRPS patients have small-fiber polyneuropathy (those with body-wide symptoms) (Kharkar et al., 2012; Birley and Goebel, 2014).

Sympathetic nerve blocks

Following Evans' enthusiasm, for decades these blocks were used not only for treatment, but a positive response was required to diagnose reflex sympathetic dystrophy (Evans, 1946). Later, patients with temporary postblock pain reduction were labeled as having sympathetically maintained pain and failure to respond was termed "sympathetically independent pain." This distinction did not prove useful either diagnostically or therapeutically and again, experimental evidence does not support a major role for sympathetic overactivity in CRPS (Campero et al., 2010).

DIFFERENTIAL DIAGNOSIS

Since the CRPS diagnosis is clinical, it is by nature imprecise, and alternate diagnoses should always be considered. Foremost is SFPN, which causes CRPS-like signs and symptoms, including spontaneous and/or stimulus-evoked pain (allodynia and hyperalgesia), and sweating and vasomotor changes (Amato and Oaklander, 2004). Indeed, it is the similarity to SFPN that inspired recognition that small-fiber axonopathy contributes to CRPS as well. If large (A alpha beta) nerve fibers are affected by polyneuropathy, patients can have reduced reflexes, weakness, and vibratory and proprioceptive deficits. However, in SFPN, the signs and symptoms are usually widespread and bilateral, usually begin in the lower legs and feet, and they can spread to affect the upper extremities and sometimes the torso and head as well. Hence the importance of thorough neurologic examination of CRPS patients.

The second important difference is that SFPN appears spontaneously and is not caused by trauma. Third, the tempo is different, since CRPS tends to appear rapidly in full-blown form within days of the causal trauma and then slowly improve, whereas SFPN usually develops insidiously and worsens gradually. It is important not to miss a diagnosis of SFPN since this should prompt testing for its many identifiable and treatable causes, including diabetes mellitus, alcoholism, chronic infections, vitamin deficiencies (B_{12}) or excesses (B_6), and hematologic dyscrasias (England et al., 2009b). Each of these can be markedly improved or cured by appropriate treatment. Patients labeled as "total-body CRPS," particularly when there is no history of focal causal trauma, usually have undiagnosed SFPN or another cause of widespread pain such as a rheumatologic disorder, since CRPS is by definition a focal or regional pain syndrome.

However, CRPS and SFPN may not be entirely unrelated. It seems likely, although not tested, that underlying SFPN might increase risk of developing or failing to recover from CRPS. Such a relationship is clearly established for diabetic polyneuropathy, a proven risk factor for developing focal neuropathies, including ulnar and median-nerve lesions (carpal tunnel syndrome) (Gulabi et al., 2014; Rota et al., 2014). It is to be expected that nerves that are metabolically compromised and perhaps swollen would be more likely to become damaged, entrapped, and/or less likely to regenerate and recover from trauma. Indeed, some 10% of CRPS patients also report chronic widespread pain (Birley and Goebel, 2014), and another study reported that up to 37% of patients with painful diabetic neuropathy met the clinical criteria of CRPS (Galer et al., 1998).

There are other causes of focal chronic pain. Nontraumatic entrapment neuropathies such as carpal tunnel, cubital tunnel, or thoracic outlet syndromes often progress gradually, and they may not meet full CRPS criteria since they are not started by trauma. However, they may underlie the approximately 10% of CRPS cases that lack a history of external trauma. Unilateral vascular abnormalities affecting a limb such as deep-vein thrombosis or thrombophlebitis, or arterial insufficiency can cause distal-limb swelling, pain, temperature changes, and discoloration, thus they should also be considered. Vascular ultrasound can help diagnose them. Lymphedema caused by surgery, radiation, or infection develops insidiously without trauma, and the pain is usually deep and aching. It can be diagnosed with lymphoscintigraphy. Focal joint inflammatory diseases such as gout, pseudogout, tenosynovitis due to arthropathy, focal vasculitis, and regional scleroderma must also be considered in the differential diagnosis, as these also require specific treatment for cure (Ramchandren et al., 2008). Focal lesions of the spinal cord or diseases affecting the nerve plexi can also produce regional pain syndromes involving the limbs. In short, we recommend detailed investigation of "nontraumatic" CRPS because of the possibility of an internal cause that can be medically or surgically cured.

PATHOPHYSIOLOGIC BASIS OF SYMPTOMS

CRPS is a complex disorder involving multiple pathophysiologies and symptoms. It is currently uncertain which of the described changes are primary and which are secondary. It is not clear that every patient with the CRPS endophenotype has the same underlying cause and why some patients develop the full syndrome whereas others have only portions. The microvascular symptoms were previously attributed to excessive activity of injured sympathetic small-fiber axons, hence the name "reflex sympathetic dystrophy." More recent studies fail to support this hypothesis, with some finding sympathetic activity to be, if anything, reduced in CRPS-affected limbs. Microneurographic recordings of electrophysiologic interactions between small numbers of sympathetic and nociceptive axons in 13 patients with CRPS-I and II patients showed no ability of sympathetic activation to augment or influence the firing of nociceptive pain fibers (Campero et al., 2010), although the CRPS-II subjects did not meet current diagnostic standards.

Furthermore, although the vast majority of CRPS cases are temporally and anatomically linked to prior limb trauma, in multiple series a few patients have no external trauma (de Rooij et al., 2010). They provide important clues to pathogenesis. In some, detailed investigation reveals an internal focal lesion causing nerve entrapment, compression, or irritation (Thimineur and Saberski, 1996). There are more than 15 published cases of CRPS triggered by shingles (herpes zoster) affecting the arms or legs (Berry et al., 2004). A recent well-studied patient with shingles affecting the hand (Förster et al., 2012) was informative. That patient met all criteria for CRPS, including bone-scan abnormalities. These cases of CRPS associated with postherpetic neuralgia and other nontraumatic causes suggest that lesion location may be a key important determinant of whether CRPS develops after a focal problem rather than trauma as the specific cause. The CRPS endophenotype develops only after injuries to the limbs, but not to the torso or head. The lack of collateral circulation and need to overcome gravity during return of blood and lymph from the limbs to the torso are likely contributors (Oaklander, 2012).

The genetic contributors to CRPS remain largely unknown since the lack of a robust case definition and diagnostic test makes it difficult to assemble large homogeneous populations for study. There is increasing identification of genetic contributions to both perception of pain in normal people and to development of chronic-pain conditions in both rodent models and people (Mogil et al., 1999; Diatchenko et al., 2005). There is additional evidence of genetic risk of neuropathic pain after traumatic injury; including data from postsurgical patients (Devor, 2004). De Rooij and colleagues (2009) studied the risk of CRPS in siblings of CRPS patients, and found a slight but significant increase in siblings younger than 50. The most robust genetic association has been with the human leukocyte antigen (HLA) system. CRPS with or without dystonia has been linked to HLA-DQ8, whereas HLA-B62 has been associated with CRPS with dystonia but not with CRPS without dystonia (van Rooijen et al., 2012). We proposed an association between CRPS and Ehlers–Danlos, a family of inherited disorders of collagen that causes joint hyperflexibility and predisposes patients to injury and surgery (Stoler and Oaklander, 2006).

Nerve injury

Persistent injury to peripheral axons that do not recover completely from a causal trauma is central in CRPS. CRPS-II is the term used for patients with identified nerve injuries. The problem is with the more-common, similar-appearing patients, labeled as "CRPS-I" because they have no diagnosed nerve injuries. However, the ability to detect subtle nerve injuries depends strongly on the examiner's experience and on the tests used. Partial injuries that predominantly affect distal sensory axons are very difficult to detect and localize (Oaklander and Fields, 2009).

There is considerable evidence for distal axonal injuries in CRPS-I as well as in CRPS-II, obviating meaningful distinctions between CRPS-I and CRPS-II. Careful neurologic examination of CRPS-I patients by nerve specialists, or pathologic study of neural tissues identifies traumatic mononeuropathies in many patients (e.g., Table 27.3). These can be subtle and remain undetected for several reasons. Some lesions involve terminal branches of sensory-only nerves without motor axonopathy. Partial lesions may be undetectable by routine electrodiagnostic study (Fig. 27.1). In CRPS-I, relative sparing of motor and large-fiber sensory axons can preserve many functions such as muscle strength, whose loss normally helps to diagnose nerve injury.

Pathologic examination of limbs amputated from CRPS-I patients directly demonstrates nerve injuries. The seminal study by van der Laan et al. (1998a) of sural nerves and muscles from limbs amputated from 8 patients with CRPS-I identified ultrastructural evidence of degeneration of small fibers in 4/8 sural nerves. A more recent study found a loss of larger myelinated fibers (>12um) in sural nerves in CRPS-I patients (Geertzen et al., 2015). Since the sural is but one of the three major nerves innervating the lower leg, these findings are significant. Another pathologic study of muscles from the amputated limbs from CRPS-I patients identified nerve injury in 5/8 (Hulsman et al., 2009).

Skin biopsies from living patients have been used in larger studies with congruent findings. A study of skin from limbs amputated from two patients with CRPS-I (Fig. 27.2) showed reduced small-fiber innervation and new neuropeptide profiles in the axons innervating cutaneous microvessels, sweat glands, and hair follicles (Albrecht et al., 2006). These findings offer a biologically plausible explanation for many of the previously mysterious symptoms of CRPS. Our group compared immunolabeled skin biopsies from affected limbs of CRPS-I patients to contralateral control skin samples and found fewer axons at the CRPS-affected sites of 17/18 subjects, with reductions averaging 29% (Oaklander et al., 2006). Together, these pathology studies demonstrate that nerve damage is present in CRPS-I as well as CRPS-II.

Rat models suggest that small nerve traumas such as single needlestick penetration can cause endoneurial inflammation and damage to nearby uninjured nerves, including the uninjured contralateral "mirror-image" nerve (Oaklander and Brown, 2004). Perhaps this contributes to the regional and contralateral spread of pain and other symptoms. Furthermore, even small distal nerve injuries can have profound effects on the spinal cord and brain (Lee et al., 2009).

The authors' current interpretation is that CRPS is likely to be an extreme "pain-plus" endophenotype of posttraumatic neuralgia that is characterized by prominent microvascular, and often sweating and motor abnormalities on top of persistent neuropathic pain (Fig. 27.3). Although many cases of posttraumatic neuralgia lack the CRPS-specific features, continued distinction between the two may be without justification.

Microvascular dysfunction, tissue hypoxia, and ischemia

Microvasculopathy, hypoxia, and inflammation are additional major causes of pain, disability, weakness, and dysfunction in CRPS, and CRPS is definitionally distinguished from posttraumatic neuralgia by "evidence at some time of edema, changes in skin blood flow, or abnormal sudomotor activity in the region of the pain" (Merskey and Bogduk, 1994). Vasomotor instability (microvasculopathy) is the primary cause of edema, abnormal skin color, and temperature differences and it can contribute to sweating abnormalities. The distal limbs are vulnerable to microvascular dysregulation because blood flow is limited by the relative paucity of collateral circulation and by the requirement for blood and lymph to overcome gravity to return to the heart. Extravascular edema caused by microvascular dysregulation further impedes circulation and promotes inflammation. Abnormal perfusion causes chronic edema and changes in skin color and temperature, and can leave tissues devoid of adequate oxygen and nutrient exchange. Furthermore, hypoxia is a potent trigger of inflammation (Eltzschig and Carmeliet, 2011). Of course, the distal axons are exquisitely sensitive to any disruption in their energy supply, and healing of any axonal injuries will be slowed, and at worst, new axonal degeneration will be triggered.

The red, swollen skin most often seen during acute severe CRPS has been convincingly linked to denervation of cutaneous arteriovenous shunts (Fig. 27.2), which permits them to remain tonically open and allows blood to bypass the capillary beds where gas and nutrient exchange normally occurs. This arteriovenous shunting can leave underlying tissues ischemic or hypoxic and acidotic, and impede muscle function, causing fatigue and disability and impeding healing. In most cases these phenomena are due to malfunction or degeneration of the small-fiber innervation of blood vessels (Fig. 27.2) (Albrecht et al., 2006).

Microvasculopathy often improves during the course of CRPS; Birklein et al. (2000) found edema in 92% of CRPS patients within the first 5 weeks, but only in 55% of patients with chronic CRPS-I (range 25–468 weeks). As CRPS resolves, the microvascular signs distinguishing CRPS from posttraumatic neuralgia (Fig. 27.3) disappear before the pain, so most patients eventually no longer meet full diagnostic criteria for CRPS, but only for posttraumatic neuralgia.

A rat model of CRPS-I was developed based on causing distal-limb ischemia by transiently occluding blood flow at the ankle and then removing it to allow reperfusion of the ischemic tissues (Coderre et al., 2004). However, another group attempting to test this proposed mechanism by ligating the femoral artery of rats for 3 hours. When this failed to model any findings of CRPS, they concluded that ischemia and reperfusion may not play a major role in CRPS-I (Ludwig et al., 2007).

Inflammation

Inflammation is a pathogenetic contributor to CRPS and to postinjury neuropathic pain in general (Iwatsuki et al., 2013). The axonal and microvascular abnormalities reviewed above are both proinflammatory (Eltzschig and Carmeliet, 2011). Small fibers have direct and paracrine effects on immunocytes that permit them to influence inflammatory responses to injury, a phenomenon known as "neurogenic inflammation" (Riol-Blanco et al., 2014), and physiologic study has found this to be enhanced in CRPS patients (Weber et al., 2001).

Molecular markers of enhanced inflammation have been identified in CRPS patients. Systematic review of 22 such studies documents enhanced proinflammatory markers in blood, cerebrospinal fluid (CSF), and tissue fluids, although many of the studies did not meet all quality criteria (Parkitny et al., 2013). Different inflammatory profiles were present in acute vs chronic cases. In acute CRPS, concentrations of interleukin (IL)-8 and soluble tumor necrosis factor receptors I (sTNF-RI) and II (sTNF-RII) were significantly increased in blood.

This review concluded that, in chronic CRPS, there were increases in TNF-α, bradykinin, sIL-1RI, IL-1Ra, IL-2, sIL-2Ra, IL-4, IL-7, interferon-gamma, monocyte chemoattractant protein-1 (MCP-1), and sRAGE (soluble receptor for advanced glycation end products) in blood. CRPS-affected tissue fluids had high levels of IL-1Ra, MCP-1, macrophage inflammatory protein-1β (MIP-1b) and IL-6; CSF had excess IL-1β and IL-6. Chronic CRPS was also associated with decreased substance P, sE-selectin, sL-selectin, sP-selectin, and sGP130 in blood; and soluble intercellular adhesion molecule-1 (sICAM-1) in CSF (Parkitny et al., 2013). One study of acute CRPS found activated keratinocytes, epidermal thickening, mast cell accumulation, and upregulation of TNF-α and IL-6 expression, whereas in chronic CRPS there was reduced keratinocyte proliferation and epidermal thinning with no increases in mast cells (Birklein et al., 2014).

Similarly, a recent study found that after 6 months, a proinflammatory cytokine profile of early CRPS resolved, and did not predict clinical outcome (Lenz et al., 2013). Other caveats are that levels of inflammatory mediators are not generally higher in CRPS patients' affected vs unaffected limbs, and few studies included non-CRPS chronic-pain controls, so these biomarkers may not be CRPS-specific. Indeed, there is evidence of similar changes nonspecifically in postinjury neuropathic pain (Iwatsuki et al., 2013), and levels of inflammatory mediators do not necessarily correlate with the clinical symptoms of CRPS (Wesseldijk et al., 2007). Overabundance of inflammatory cytokines has been reported in non-CRPS neuropathic pain syndromes, perhaps additional evidence that CRPS is fundamentally a neuropathic syndrome (Üçeyler et al., 2007). Unfortunately, clinical experience suggests that CRPS symptoms are not improved by treatment with conventional anti-inflammatory medications, although most have not been formally studied. Furthermore, a large clinical trial of lenalidomide, an inhibitor of TNF-α, did not find efficacy (see treatment section, below).

Disordered immunity

A number of older studies found corticosteroids to be effective for acute CRPS, and three more recent lines of evidence support the hypothesis that autoimmunity may be an underlying pathogenic cause of CRPS. German investigators have demonstrated that serum from CRPS patients contains autoantibodies that react against antigenic markers associated with small-fiber axons (Kohr et al., 2009, 2011).

Additionally, Mayo Clinic investigators provided strong pathologic and physiologic evidence that routine surgeries (which breach the blood–nerve barrier) trigger autoimmune polyneuropathies in some patients (Staff et al., 2010). Although CRPS was not explicitly discussed, surgeries are a common trigger for CRPS, and it must be inferred that nonsurgical traumas are similarly capable of eliciting antinerve immune attacks. Additional support comes from a small trial of immunoglobulins for treatment of long-standing CRPS with encouraging results (see treatment section, below) (Goebel et al., 2010). Goebel and Blaes (2013) have hypothesized that underlying neuroinflammation, at least in some cases of CRPS, is a unique regional autoimmune response to the presence of neoantigens, initiated or activated by trauma. Several lines of experimental evidence support the presence of autoimmunity with roles for B cells and IgM (Li et al., 2014) and IgG antibodies (Tekus et al., 2014).

EVIDENCE-BASED TREATMENT

The vast majority of CRPS patients improve markedly within the first year (Bean et al., 2014), so the prognosis is less grim than commonly perceived. Several recent systematic reviews of treatment options for CRPS (Perez et al., 2010; Cossins et al., 2013; Harden et al., 2013; Borchers and Gershwin, 2014; Gierthmuhlen et al., 2014) can be consulted for completeness. An analysis of all randomized controlled trials published between 1966 and February 2012 identified 29 that provided evidence for efficacy of 10 treatments (Cossins et al., 2013). There was strong evidence for three treatments (bisphosphonates, repetitive TMS, and graded motor imagery – all discussed below), moderate evidence for one, and limited evidence for six. Importantly, this analysis also found evidence of the inefficacy of 15 treatments (Cossins et al., 2013), and the largest controlled, blinded clinical trial in 184 patients with chronic CRPS using the Budapest research criteria recently found lenalidomide, a TNF-α, to be ineffective (Manning et al., 2014). The heterogeneity of studied interventions and the low quality of many trials prevented the authors from drawing conclusions about the utility of most CRPS treatments. We know of only one study that focused on prevention. Zollinger and colleagues (2007) reported that 50 days of treatment with oral vitamin C (a free-radical scavenger) reduced the risk of CRPS after wrist fracture.

Since three-quarters of cases of acute CRPS resolve even without treatment (Sandroni et al., 2003), all treatments are more effective when used early. Mild cases may require nothing more than prompt rehabilitation. Several small clinical trials conducted for several drug treatments have documented efficacy for early CRPS. These include small trials reporting efficacy of corticosteroids (Christensen et al., 1982; Braus et al., 1994). Calcitonin and the bisphosphonates have also been evaluated and a systematic review of four randomized controlled trials found efficacy of the bisphosphonates in reducing pain, swelling, and disability (Brunner et al., 2009). A high quality multicenter, pivotal trial of a bisphosphonate, neridronic acid, is underway in the U.S. and in the European Union. It is intended to support application to regulatory agencies for a CRPS-I indication. Based on the evidence of exaggerated inflammation in acute CRPS, anti-inflammatory agents such as antioxidants and free-radical scavengers, including mannitol, dimethyl sulfoxide, and *N*-acetylcysteine are also widely used for early CRPS in the Netherlands and have been formally studied, demonstrated. These approaches have not been adequately studied in chronic CRPS.

In long-lasting and more severe cases, the prognosis becomes guarded. As with any illness, the longer it has lasted and the more treatments have been ineffective, the less likely it is that a new treatment will be effective (Zyluk and Puchalski, 2013). Use of refractory patients for clinical trials may obscure efficacy for the larger group of less severely affected patients (Manning et al., 2014).

Regarding psychotherapy, CRPS, like other painful and disabling conditions, can cause profound psychologic difficulties for affected individuals and their families. Patients can develop depression, anxiety, or posttraumatic stress disorder, all of which heighten the perception of pain and interfere with rehabilitation. It is important to treat these secondary problems so that patients can recover more effectively.

Rehabilitation modalities, including physiotherapy

These are the cornerstone of treatment, and may be all that is required for patients with acute or mild CRPS. The lower risk and cost of these interventions add to their appeal. Their primary goal is augmenting limb activity and restoring function. This is particularly important in CRPS because immobility and poor circulation can worsen tissue damage and symptoms. In addition to treating primary CRPS symptoms, exercise-based regimens help prevent secondary effects including deconditioning, depression, contractures, osteoporosis, and tissue atrophy. Thus there is much to recommend them for established as well as acute CRPS. Children in particular do well with these treatments (Sherry et al., 1999). Although not formally studied, methods that improve circulation, tissue nutrition, and oxygenation, including regular aerobic exercise, and treating risks for peripheral vascular disease (including smoking cessation) likely have general as well as CRPS-specific benefit at low risk and cost.

Unfortunately, despite widespread clinical use of various rehabilitation-based therapies, the quality of evidence is overall poor, with only four studies rated as high-quality (Cossins et al., 2013). Among them, two studies by the same group evaluated graded motor imagery for long-standing CRPS after wrist fracture. This attempts to eliminate the mismatch between motor intention and sensory feedback that may contribute to neuropathic pain in CRPS. It decreased pain significantly compared with control or another active treatment (Moseley, 2004a, 2006). Mirror therapy is a related physiotherapy intervention where mirror reflections of the healthy limb are used to provide (false) visual feedback that the affected limb appears and functions well. Mirror therapy decreased pain compared with placebo in early CRPS after stroke in one high-quality trial (Cacchio et al., 2009a). In a low-quality trial, it reduced pain more than mental imagery for long-standing CRPS after stroke (Cacchio et al., 2009b). Small prospective randomized controlled trials support the efficacy of physical and, to a lesser extent, occupational therapy for CRPS in adults and children (Oerlemans et al., 1999; Lee et al., 2002). Systematic review reveals no evidentiary support for stress loading, scrubbing, or other stimulation-based therapies that try to "desensitize" patients by repeated exposure to rough mechanical contact (Daly and Bialocerkowski, 2009). These are often painful and poorly tolerated. One premise behind these therapies is that changes in the spinal cord and brain occur secondary to peripheral nerve injury and these changes affect pain perception and limb function. It is hoped that the aforementioned abnormal cortical plasticity will be limited or reversed with one or more of these techniques. Desensitization strategies attempt to reduce the allodynia and hyperalgesia found in CRPS.

Pharmacologic treatments

Pharmacotherapy is indicated for patients who continue to have pain and/or disability despite aggressive rehabilitation. No medication has a US Food and Drug Administration indication for CRPS or posttraumatic neuralgia. Nine studies performed between 2000 and 2012 met inclusion criteria in a recent systematic analysis of treatments for CRPS (Cossins et al., 2013) and are discussed below, along with a few other studies that report efficacy of treatments to consider for patients who fail primary and secondary treatment options.

Treatments for neuropathic pain

Because these treatments have been available and in use for a long time, there are few studies that meet modern methodologic standards. Based on randomized controlled trials for other neuropathic conditions, such as diabetic neuropathy and postherpetic neuralgia (Collins et al., 2000; Hempenstall et al., 2005), rational first-line therapy for pain CRPS includes second-generation tricyclics (especially nortriptyline), the gabapentinoids (gabapentin and pregabalin), serotonin-norepinephrine reuptake inhibitors (e.g., duloxetine, venlafaxine), and opioids. In CRPS, gabapentin was effective in an open-label uncontrolled study of nine adult patients given 600 mg/day (a lower dose than usually used now) (Mellick and Mellick, 1997). One study comparing morphine with memantine to morphine alone found the combination more effective for long-standing CRPS restricted to the upper limb (Gustin et al., 2010).

Topical application of local anesthetics is worth considering, since these do not penetrate systemically to trigger distant averse effects. A randomized trial of 40 patients with focal neuropathic pain including CRPS found the 5% lidocaine patch efficacious (Meier et al., 2003), as did an uncontrolled open-label add-on study containing six CRPS patients (Devers and Galer, 2000). Systemic local anesthetics block sodium channels and reduce conduction in unmyelinated pain C-fibers at lower doses than those required to block motor and sensory conduction in myelinated axons. Systematic review also found systemic administration of local anesthetics effective for neuropathic pain (Challapalli et al., 2005). Several small randomized trials in CRPS-I and -II document their efficacy and safety (Linchitz and Raheb, 1999; Wallace et al., 2000; Schwartzman et al., 2009b). In practice, patients should try oral mexiletine before trying continuous systemic lidocaine administered using an insulin pump (Ferrini and Paice, 2004), as efficacy and safety of long-term or out-of-hospital use are not documented in the literature.

Treatments for pain from bone remodeling

The bones and joints are affected in most cases; indeed, bone fracture is a leading cause of CRPS. Excess resorption and remodeling of bones and joints in the affected limb can contribute to pain in CRPS, since osteoclasts secrete protons to dissolve hydroxyapatite crystals, and CRPS patients have high levels of osteoprotegerin, a marker of bone remodeling (Krämer et al., 2014). The microenvironment may become acidic enough to activate acid-sensing ion channels and transient receptor potential vanilloid type 1 channels expressed on the small-fiber nerve endings that densely innervate compact bone and periosteum. Calcitonin was the first agent studied, and efficacy is supported by systematic analysis (Perez et al., 2001). The antinociceptive mechanism of calcitonin is unclear, since, in addition to regulating calcium metabolism in bone, this polypeptide hormone from the thyroid also binds in the brain and may have antinociceptive effects via other mechanisms (Braga, 1994; Walker et al., 2002). Given that it can be administered nasally, and has a low rate of adverse effects, calcitonin should be considered more often.

The efficacy of calcitonin led to consideration of bisphosphonates for CRPS, and these later trials are better designed. A systematic review of trials of bisphosphonates for CRPS-I that screened publications up to 2007 identified four trials of moderate quality that reported significant efficacy against pain and a trend towards improvement in quality of life (Brunner et al., 2009). As with calcitonin, it is uncertain whether the pain relief is mediated exclusively by reducing osteoclast activity. Bisphosphonates inhibit prostaglandin E_2, proteolytic enzymes, lactic acid, and proinflammatory cytokines. Long-term administration of bisphosphonates may increase bone fragility and risk of atypical hip fractures, and intravenous use has been associated with jaw osteonecrosis. If given intravenously, bisphosphonates must be administered slowly to prevent kidney damage. Less serious adverse effects include transient responses to intravenous infusions and esophagitis after oral alendronate, the least expensive treatment option. Again a multicenter pivotal trial of neridronate is currently in process.

Ketamine

Ketamine, an *N*-methyl D-aspartate (NMDA) antagonist with anesthetic properties, is increasingly used as a tertiary treatment for refractory CRPS because of the important role of increased glutamate release and NMDA receptor function in spinal dorsal-horn physiology. Five days of anesthetic doses were efficacious in an open-label phase II study of 20 patients, although some had serious complications (Kiefer et al., 2008). Lesser doses (up to 100 mg for 4 hours daily for 10 days) were efficacious for CRPS pain in a randomized double-blind placebo-controlled study (Schwartzman et al., 2009a). Three cases of ketamine-induced liver injury have been reported in CRPS patients (Noppers et al., 2011).

Treatments for neuroinflammation

There is no evidence that conventional nonsteroidal anti-inflammatory drugs such as aspirin, ibuprofen, or naproxen are effective, but this has not been adequately studied. Corticosteroids were trialed in acute CRPS, with early reports of efficacy, but a recent trial of a single intrathecal dose of methyprednisolone (60 mg) was stopped prematurely because no effect of treatment was evident at 6 weeks (Munts et al., 2010).

The finding of local overproduction of the proinflammatory cytokine TNF-α in blister fluids and skin biopsies from CRPS patients (Huygen et al., 2002; Wesseldijk et al., 2007, 2008; Krämer et al., 2011) prompted consideration of therapies to reduce or block TNF-α. Adalimumab, a recombinant human monoclonal antibody that blocks TNF-α from activating its receptor was tested in an uncontrolled trial of 10 CRPS-I patients, of whom six responded partially or well (Eisenberg et al., 2013). Several case reports document improvement after treatment with thalidomide, a TNF-α inhibitor, including one study in 42 patients (Schwartzman et al., 2003), but use is limited by concerns about teratogenicity, particularly since three-quarters of CRPS patients are female. A class I, controlled, blinded, clinical trial evaluated efficacy of oral lenalidomide, a TNF-α inhibitor that is a safer derivative of thalidomide in a phase-II randomized controlled trial of 184 carefully selected subjects with CRPS-I. Efficacy was no greater for lenalidomide than placebo, but this study demonstrated the feasibility of conducting large class I trials in CRPS-I and provided considerations for designing future CRPS trials (Manning et al., 2014).

Local injections of botulinum toxin are used to selectively weaken muscles in movement disorders, spasticity, and for cosmetic use. These effects are based on its ability to block release of acetylcholine from motor axons at neuromuscular synapses. Botulinum toxin also inhibits release of noncholinergic neurotransmitters (e.g., glutamate) and neuropeptides (substance P and calcitonin gene-related peptide) from sensory nerve terminals by blocking transient receptor potential vanilloid 1 exocytosis. These effects offer the possibility of mediating the neurogenic inflammation that these paracrine effectors cause and several studies document the efficacy of botulinum toxin for neuropathic pain. Regional intradermal injections of botulinum toxin improved spontaneous pain, brush allodynia, and cold pain thresholds at the painful sites of 25 patients with focal painful neuropathies (Ranoux et al., 2008) and, when used in conjunction with sympathetic blockade with bupivacaine, botulinum toxin extended the duration of analgesia in a subset of CRPS patients (Carroll et al., 2009). However, botulinum toxin was not found to be effective and was poorly tolerated in 14 CRPS patients with allodynia (Safarpour et al., 2010). Given the fact that botulinum toxin is expensive and requires reinjection at multiple sites several times yearly it is unlikely to become a primary treatment option.

Treatments for poor tissue perfusion

Whether primary or secondary, inadequate tissue perfusion, nutrition, and oxygenation contribute to pain, disability, and limb dysfunction in CRPS. In severe cases they lead to ulceration, infection, gangrene, and amputation (van der Laan et al., 1998b). Limb edema further worsens the situation. Tissue perfusion should be assessed in all CRPS patients and treated aggressively if necessary. Limited trials of vasoactive therapies such as calcium-channel blockers (Prough et al., 1985; Muizelaar et al., 1997) and alpha-adrenergic agonists (Ghostine et al., 1984; Muizelaar et al., 1997) do not demonstrate good efficacy.

Nonpharmacologic treatments for poor tissue perfusion are far more useful, including smoking cessation and control of other vascular risk factors, such as hyperlipidemia, regular exercise, limb elevation, and compressive garments if tolerated. A few studies investigated hyperbaric oxygen therapy (HBOT) for CRPS. In HBOT, patients lie in a tank of pressurized oxygen, which dramatically increases tissue oxygenation. Two double-blinded randomized controlled trials (Tuter et al., 1997; Kiralp et al., 2004) and two case reports (Peach, 1995; Williams et al., 2009) support the efficacy of HBOT in reducing pain and swelling, and improving range of motion.

Treatments for dysimmunity

Publications from Blaes and colleagues report that sera from CRPS patients harbor autoantibodies that cross-react with autonomic neurons (Kohr et al., 2009, 2011). IgG isolated from these sera have reproduced some features of CRPS in mice (Tekus et al., 2014). Although not yet reproduced, these findings prompted consideration of immunomodulatory therapy for established CRPS. Intravenous immunoglobulin (IVIg) was studied because efficacy is established for treating chronic inflammatory demyelinating polyneuropathy. To date, a single small, well-designed, randomized clinical trial reported modest efficacy of IVIg (1 g/kg) for established refractory CRPS (Goebel et al., 2010).

Interventional treatments, including nerve blocks

Sympathetic nerve blocks using local anesthetics with or without corticosteroids are controversial. For CRPS-affected upper limbs, the stellate ganglion is blocked, and the lumbar sympathetic chain is targeted for lower-limb CRPS. Few, if any, early studies of sympathetic nerve blocks are considered adequate today since they were unblinded and often uncontrolled; and the importance of the placebo response, which is strong for interventions, was unrecognized. Patients with temporary postblock pain reduction were labeled as having sympathetically maintained pain and failure to respond was termed "sympathetically independent pain." This distinction did not prove useful diagnostically or therapeutically. As with other invasive procedures, these blocks can injure the neural structures targeted and rarely, have contributed to or exacerbated CRPS. Given their technical and interpretational limitations, and the evidence that somatic axons co-innervate most structures traditionally thought only autonomically innervated, use has finally waned. A revised systematic review of the efficacy of sympathetic blockade by local anesthetics for CRPS analyzed 10 studies, to again conclude that the data do not demonstrate even short-term efficacy for such blocks in reducing CRPS pain (Stanton et al., 2013). Systematic review of RCTs for CRPS found lack of efficacy for most interventional treatments studied, including intravenous regional blockage using methylprednisolone, lidocaine, stellate ganglion blockade, and intravenous mannitol (Munts et al., 2010; Cossins et al., 2013). Since these treatments are expensive and have rare serious adverse effects, their general use is not indicated.

Intrathecal pumps deliver pain-relieving medications directly into the CSF that bathes the spinal cord. Typical agents include opioids, local anesthetic agents, clonidine, or baclofen, sometimes in combination. The advantage is that pain-signaling targets in the spinal cord can be attained using doses far lower than those required for oral administration, which decreases side effects in other organs and increases drug effectiveness. There are no studies that show benefit specifically for CRPS. A trial of single-bolus intrathecal methylprednisolone found it ineffective for established CRPS (Munts et al., 2010).

Surgical treatment

Neurosurgical options deserve consideration for patients with nonresolving or treatment-resistant CRPS. Unlike pharmacotherapy, surgery is a disease-modifying treatment that offers the possibility of major improvement or cure if the right procedure is performed in patients with appropriate, localized lesions, even in cases lasting more than a decade (Watson et al., 2014). Although concern is often raised that surgery will trigger new CRPS or worsen existing symptoms, this is not reported by experienced practitioners. Clearly when contemplating surgery, the location of the causal lesion needs to be carefully established in advance, whether by history (Table 27.3), examination (Fig. 27.1), electrophysiologic study, or imaging.

Birch (2009) summarized this well:

> *Clinicians faced with treating severe neuropathic pain ought to be alert to the possibility of a continuing active provocative agent at work on the nerve. When that pain follows wounding of a nerve, fracture, foreign body, expanding hematoma, or other arterial injury, or results from damage inflicted during an operation, then that possibility becomes a near certainty. In such cases, exploration should be undertaken urgently.*

Since nerve lesions are highly individual, it is difficult to assemble enough patients with similar lesions to conduct clinical trials, and blinding is impractical for surgery, but considerable numbers of case reports and expert neurosurgical opinion establish that CRPS is sometimes caused or maintained by nerve entrapment, compression, or neuroma formation, that requires surgery to treat definitively (Thimineur and Saberski, 1996). In the approximately 10% of CRPS patients without known external trauma (de Rooij et al., 2010), some have internal focal lesions such as infections (e.g., leprosy), tumors, or vascular malformations that benefit from surgical exploration for diagnosis and initiation of treatment. New imaging techniques such as ultrasound and magnetic resonance neurography may help better identify and localize such lesions (Baumer et al., 2014).

In contrast, ablative neurosurgery – cutting the nerves or nerve roots that are proximal to and transmitting pain sensation from painful areas – has proven ineffective despite its simplistic appeal, and it can compromise neurologic function and lead to untreatable deafferentation pain (anesthesia dolorosa). Limb amputation was studied in 28 CRPS patients after 34 amputations in 31 limbs. This had been performed for untenable pain (5), recurrent infection (14), or to improve residual function (15). Only two patients were relieved of pain by amputation, 10/14 were cured of infection, and 9/15 had improved residual function. In 28 amputations, CRPS recurred in the limb stump, especially after amputation through symptomatic tissues. Because of these stump problems, only two patients could wear a prosthesis (Dielissen et al., 1995). Amputation is only very rarely indicated for serious complications such as gangrene and, if performed, arrangements should be made in advance for an academic center to perform detailed pathologic study to identify the causal lesions in that patient and to advance understanding of the causes of CRPS (van der Laan et al., 1998a; Geertzen et al., 2015).

Surgical sympathectomy has been perceived as effective and was widely performed for CRPS with some evidentiary support (Hassantash et al., 2003; Singh et al., 2003). Of note, the hypothesis that pain is initiated or maintained by sympathetic overactivity in CRPS is contradicted by the bulk of more-recent studies (Goldstein et al., 2000). Compensatory hyperhidrosis and pathologic sweating during eating are adverse effects of sympathectomy.

Invasive and noninvasive neuromodulation therapies

Another common procedural treatment for severe established CRPS is electric stimulation of neurons (neuromodulation), usually with implanted bipolar electrodes. Stimulation can be delivered at various locations along the pain pathways, including near injured nerves (peripheral nerve stimulators), to the spinal cord, outside the membranes around the brain (motor cortex stimulation with dural electrodes), and within the parts of the brain that control pain (deep-brain stimulation). The evidence shows that significant pain relief and functional gains are possible for properly selected patients.

Stimulation of the dorsal column of the spinal cord is common, because the leads can be placed through an epidural cannula. Dorsal-column stimulation is thought to increase activity of inhibitory interneurons and circuits by using nonnociceptive neurons to activate inhibitory spinal-cord circuits. The patient perceives this as nonpainful paresthesias. At least one well-designed study (Kemler et al., 2000) and several reviews and meta-analyses (Stanton-Hicks, 1999; Turner et al., 2004) demonstrate efficacy of dorsal-column stimulation in about half of implanted CRPS patients. Limitations include expense, possibility for infections, and frequent need for surgical revisions. Spinal-cord stimulation may not be maximally effective within the first year of CRPS and efficacy can wane due to lead migration or other complications. Whereas spinal-cord stimulation has been reserved for patients with continued symptomatology after maximal medical and rehabilitation treatment, a recent meta-analysis concluded that it can be clinically beneficial and cost-effective if used earlier, especially before long-term opioid administration (Poree et al., 2013).

For patients with CRPS caused by damage to a single identified nerve, surgical implantation of leads on the injured nerve proximal to the lesion offers even better chances of significant pain relief – about 80% in some series (Campbell and Long, 1976). In a series of 30 - consecutive implanted patients with CRPS-I, 63% obtained good pain relief (Hassenbusch et al., 1996).

The dorsal-root sensory ganglion innervating a CRPS-affected lower limb is a more recent target of stimulation (Liem et al., 2013). In a small, well-designed trial (van Buyten et al., 2015), 11 subjects with uni- or bilateral lower-extremity CRPS had quadripolar epidural leads percutaneously placed near lumbar dorsal root ganglia and all eight with successful trials had permanent implantation and programming optimization. Some had pain in their feet or other areas that were hard to treat with spinal-cord stimulation. All eight reported some pain relief and improved function. Self-reported pain scores averaged 62% of baseline values at 1 month postimplantation, and pain relief persisted through 12 months in most subjects. In some patients, CRPS-associated edema and trophic skin changes and function also improved (van Buyten et al., 2015).

Brain stimulation is another effective treatment. Stimulation of deep-brain structures is being replaced by stimulation of the motor cortex with more superficial epidural electrodes, which effectively relieve chronic neuropathic pain (Tsubokawa, 1995) and relieved pain in a small study of five CRPS patients (Velasco et al., 2009). Although craniotomy is required, the electrode array is sutured to the outside of the dura, at minimal brain risk.

The success of epidural motor cortex stimulation for relieving pain led to development of entirely noninvasive (extracranial) forms of stimulation. The most developed at present is TMS. Here, electromagnetic induction is used to trigger action potentials in immediately underlying cortical neurons. The advantage is that no surgery is needed; the disadvantage is that repeated treatment sessions are needed to maintain benefit. Multiple meta-analyses show that repeated application of thousands of pulses on consecutive days effectively relieves several types of neuropathic pain (O'Connell et al., 2014). Safety guidelines have been formulated, and patients with metal or microelectronic devices in their heads or upper bodies are not appropriate candidates due to the risk of overheating or malfunction (Rossi et al., 2009). The only significant adverse effect is the possibility of inducing a single seizure. TMS has mostly been used in CRPS to investigate brain function, but one study documents benefit for pain in a class I randomized controlled trial of 23 patients with upper-limb CRPS (Picarelli et al., 2010). Repetitive TMS was compared to sham-TMS using a cross-over design. The cortical representation of the hand on the cortical convexity is easier to access then the representation of the leg and foot, which is parafalcine. However, it is not clear whether or not the exact cortical location corresponding to the painful area needs to be stimulated (Lefaucheur et al., 2006).

CONCLUSION

New developments in this historic syndrome include placing CRPS on the spectrum of post-traumatic neuralgias, improving the diagnostic criteria to enhance sensitivity and specificity, appreciating its effects on brain plasticity and "network patterns," and delineating underlying axonal, inflammatory, and microvascular contributors. These demonstrate that treatment should be multifaceted. Unfortunately, while prognosis is good for early and mild cases, that is not true for the most severely affected, chronic patients. They require thoughtful reevaluation to identify all possible systemic or local factors that may be impeding recovery, as these can still be correctable, even after years. Disease-modifying treatments are more useful than mere symptom palliation. Preliminary findings that the brain abnormalities associated with CRPS and other chronic pain disorders are nonspecific and reversible (Maihöfner et al., 2004) give hope that even the most chronically affected patients can be helped. It is never too late.

In the occupational setting, the incidence of CRPS appears to be low; nonetheless, recognition that CRPS is primarily a neurologic disorder and not a psychiatric one is extremely important for accurate diagnosis and effective treatment.

ACKNOWLEDGMENTS

This study was supported in part by the Public Health Service (NINDS K24-NS059892) and several nonprofit foundations. The authors have no commercial or corporate affiliations or conflicts of interest.

References

Albrecht PJ, Hines S, Eisenberg E et al. (2006). Pathologic alterations of cutaneous innervation and vasculature in affected limbs from patients with complex regional pain syndrome. Pain 120: 244–266.

Amato AA, Oaklander AL (2004). Case records of the Massachusetts General Hospital. Weekly clinicopathological exercises. Case 16-2004. A 76-year-old woman with numbness and pain in the feet and legs. N Engl J Med 350: 2181–2189.

Barad MJ, Ueno T, Younger J et al. (2014). Complex regional pain syndrome is associated with structural abnormalities in pain-related regions of the human brain. J Pain 15: 197–203.

Basle MF, Rebel A, Renier JC (1983). Bone tissue in reflex sympathetic dystrophy syndrome – Sudeck's atrophy: structural and ultrastructural studies. Metab Bone Dis Relat Res 4: 305–311.

Baumer P, Reimann M, Decker C et al. (2014). Peripheral nerve perfusion by dynamic contrast-enhanced magnetic resonance imaging: demonstration of feasibility. Invest Radiol 49: 518–523.

Bean DJ, Johnson MH, Kydd RR (2014). The outcome of complex regional pain syndrome type 1: a systematic review. J Pain 15: 677–690.

Beerthuizen A, Spijker A, Huygen FJPM et al. (2009). Is there an association between psychological factors and the complex regional pain syndrome type 1 (CRPS1) in adults? A systematic review. Pain 145: 52–59.

Berry JD, Rowbotham MC, Petersen KL (2004). Complex regional pain syndrome-like symptoms during herpes zoster. Pain 110: 8–9.

Birch R (2009). Causalgia: a restatement. Neurosurgery 65: A222–A228.

Birklein F, Riedl B, Sieweke N et al. (2000). Neurological findings in complex regional pain syndromes – analysis of 145 cases. Acta Neurol Scand 101: 262–269.

Birklein F, Drummond PD, Li W et al. (2014). Activation of cutaneous immune responses in complex regional pain syndrome. J Pain 15: 485–495.

Birley T, Goebel A (2014). Widespread pain in patients with complex regional pain syndrome. Pain Pract 14: 526–531.

Borchers AT, Gershwin ME (2014). Complex regional pain syndrome: a comprehensive and critical review. Autoimmun Rev 13: 242–265.

Braga PC (1994). Calcitonin and its antinociceptive activity: animal and human investigations 1975–1992. Agents Actions 41: 121–131.

Braus DF, Krauss JK, Strobel J (1994). The shoulder–hand syndrome after stroke: a prospective clinical trial. Ann Neurol 36: 728–733.

Bruehl S, Harden RN, Galer BS et al. (2002). Complex regional pain syndrome: are there distinct subtypes and sequential stages of the syndrome? Pain 95: 119–124.

Brunner F, Schmid A, Kissling R et al. (2009). Biphosphonates for the therapy of complex regional pain syndrome I – systematic review. Eur J Pain 13: 17–21.

Cacchio A, De Blasis E, De Blasis V et al. (2009a). Mirror therapy in complex regional pain syndrome type 1 of the upper limb in stroke patients. Neurorehabil Neural Repair 23: 792–799.

Cacchio A, De Blasis E, Necozione S et al. (2009b). Mirror therapy for chronic complex regional pain syndrome type 1 and stroke. N Engl J Med 361: 634–636.

Campbell JN, Long DM (1976). Peripheral nerve stimulation in the treatment of intractable pain. J Neurosurg 45: 692–699.

Campero M, Bostock H, Baumann TK et al. (2010). A search for activation of C nociceptors by sympathetic fibers in complex regional pain syndrome. Clin Neurophysiol 121: 1072–1079.

Carroll I, Clark JD, Mackey S (2009). Sympathetic block with botulinum toxin to treat complex regional pain syndrome. Ann Neurol 65: 348–351.

Challapalli V, Tremont-Lukats IW, McNicol ED et al. (2005). Systemic administration of local anesthetic agents to relieve neuropathic pain. Cochrane Database Syst Rev CD003345.

Christensen K, Jensen EM, Noer I (1982). The reflex dystrophy syndrome response to treatment with systemic corticosteroids. Acta Chir Scand 148: 653–655.

Coderre TJ, Xanthos DN, Francis L et al. (2004). Chronic post-ischemia pain (CPIP): a novel animal model of complex regional pain syndrome-type I (CRPS-I; reflex sympathetic dystrophy) produced by prolonged hindpaw ischemia and reperfusion in the rat. Pain 112: 94–105.

Collins SL, Moore RA, McQuay HJ et al. (2000). Antidepressants and anticonvulsants for diabetic neuropathy and postherpetic neuralgia: a quantitative systematic review. J Pain Symptom Manage 20: 449–458.

Cossins L, Okell RW, Cameron H et al. (2013). Treatment of complex regional pain syndrome in adults: a systematic review of randomized controlled trials published from June 2000 to February 2012. Eur J Pain 17: 158–173.

Daly AE, Bialocerkowski AE (2009). Does evidence support physiotherapy management of adult complex regional pain syndrome type One? A systematic review. Eur J Pain 13: 339–353.

Del Valle L, Schwartzman RJ, Alexander G (2009). Spinal cord histopathological alterations in a patient with long-standing complex regional pain syndrome. Brain Behav Immun 23: 85–91.

de Mos M, de Bruijn AGJ, Huygen FJPM et al. (2007). The incidence of complex regional pain syndrome: A population-based study. Pain 129: 12–20.

de Mos M, Huygen FJ, Dieleman JP et al. (2008). Medical history and the onset of complex regional pain syndrome (CRPS). Pain 139: 458–466.

de Rooij AM, de MM, van Hilten JJ et al. (2009). Increased risk of complex regional pain syndrome in siblings of patients? J Pain 10: 1250–1255.

de Rooij AM, Perez RSGM, Huygen FJ et al. (2010). Spontaneous onset of complex regional pain syndrome. Eur J Pain 14: 510–513.

Devers A, Galer BS (2000). Topical lidocaine patch relieves a variety of neuropathic pain conditions: an open-label study. Clin J Pain 16: 205–208.

Devor M (2004). Evidence for heritability of pain in patients with traumatic neuropathy. Pain 108: 200–201.

Di PF, McAuley JH, Parkitny L et al. (2013). Primary motor cortex function in complex regional pain syndrome: a systematic review and meta-analysis. J Pain 14: 1270–1288.

Diatchenko L, Slade GD, Nackley AG et al. (2005). Genetic basis for individual variations in pain perception and the development of a chronic pain condition. Hum Mol Genet 14: 135–143.

Dielissen PW, Claassen AT, Veldman PH et al. (1995). Amputation for reflex sympathetic dystrophy. Journal of Bone and Joint Surgery - British Volume 77-B: 270–273.

Dumas S, Pichon B, Dapolito AC et al. (2011). Work prognosis of complex regional pain syndrome type I: multicenter retrospective study on the determinants and time to return to work. J Occup Environ Med 53: 1354–1356.

Eisenberg E, Chistyakov AV, Yudashkin M et al. (2005). Evidence for cortical hyperexcitability of the affected limb representation area in CRPS: a psychophysical and transcranial magnetic stimulation study. Pain 113: 99–105.

Eisenberg E, Sandler I, Treister R et al. (2013). Anti tumor necrosis factor-alpha adalimumab for complex regional pain syndrome type 1 (CRPS-I): a case series. Pain Pract 13: 649–656.

Eltzschig HK, Carmeliet P (2011). Hypoxia and inflammation. New Engl J Med 364: 656–665.

England JD, Gronseth GS, Franklin G et al. (2009a). Practice Parameter: Evaluation of distal symmetric polyneuropathy: role of autonomic testing, nerve biopsy, and skin biopsy (an evidence-based review). Report of the AAN, AANEM, and AAPMR. Neurology 72: 177–184.

England JD, Gronseth GS, Franklin G et al. (2009b). Practice Parameter: Evaluation of distal symmetric polyneuropathy: role of laboratory and genetic testing (an evidence-based review). Report of the AAN, AANEM, and AAPMR. Neurology 72: 185–192.

Evans JA (1946). Reflex sympathetic dystrophy. Surg Gynecol Obstet 82: 36–44.

Ferrini R, Paice J (2004). How to initiate and monitor infusional lidocaine for severe and/or neuropathic pain. J Support Oncol 2: 90–94.

Fink E, Oaklander AL (2006). Contralesional "mirror image" neuralgia caused by unilateral sural nerve biopsy. J Neuropathic Pain 1: 25–28.

Forderreuther S, Sailer U, Straube A (2004). Impaired self-perception of the hand in complex regional pain syndrome (CRPS). Pain 110: 756–761.

Förster M, Umnus A, Siebrecht D et al. (2012). A case of pain, motor impairment, and swelling of the arm after acute herpes zoster infection. Pain 153: 2478–2481.

Freeman R, Chase KP, Risk MR (2003). Quantitative sensory testing cannot differentiate simulated sensory loss from sensory neuropathy. Neurology 60: 465–470.

Frettlöh J, Hüppe M, Maier C (2006). Severity and specificity of neglect-like symptoms in patients with complex regional pain syndrome (CRPS) compared to chronic limb pain of other origins. Pain 124: 184–189.

Galer BS, Butler S, Jensen MP (1995). Case reports and hypothesis: a neglect-like syndrome may be responsible for the motor disturbance in reflex sympathetic dystrophy (complex regional pain syndrome-1). J Pain Symptom Manage 10: 385–391.

Galer BS, Bruehl S, Harden RN (1998). IASP diagnostic criteria for complex regional pain syndrome: a preliminary empirical validation study. Clin J Pain 14: 48–54.

Geertzen JH, Bodde MI, van den Dungen JJ et al. (2015). Peripheral nerve pathology in patients with severely affected complex regional pain syndrome type I. Int J Rehabil Res 38: 121–130.

Ghostine SY, Comair YG, Turner DM et al. (1984). Phenoxybenzamine in the treatment of causalgia. Report of 40 cases. J Neurosurg 60: 1263–1268.

Gierthmuhlen J, Binder A, Baron R (2014). Mechanism-based treatment in complex regional pain syndromes. Nat Rev Neurol 10: 518–528.

Goebel A, Blaes F (2013). Complex regional pain syndrome, prototype of a novel kind of autoimmune disease. Autoimmun Rev 12: 682–686.

Goebel A, Baranowski A, Maurer K et al. (2010). Intravenous immunoglobulin treatment of the complex regional pain syndrome: a randomized trial. Ann Intern Med 152: 152–158.

Goldstein DS, Tack C, Li ST (2000). Sympathetic innervation and function in reflex sympathetic dystrophy. Ann Neurol 48: 49–59.

Gulabi D, Cecen G, Guclu B et al. (2014). Carpal tunnel release in patients with diabetes result in poorer outcome in long-term study. Eur J Orthop Surg Traumatol 24: 1181–1184.

Gustin SM, Schwarz A, Birbaumer N et al. (2010). NMDA-receptor antagonist and morphine decrease CRPS-pain and cerebral pain representation. Pain 151: 69–76.

Harden RN, Bruehl S, Stanton-Hicks M et al. (2007). Proposed new diagnostic criteria for complex regional pain syndrome. Pain Med 8: 326–331.

Harden RN, Bruehl S, Perez RSGM et al. (2010). Validation of proposed diagnostic criteria (the Budapest Criteria) for complex regional pain syndrome. Pain 150: 268–274.

Harden RN, Oaklander AL, Burton AW et al. (2013). Complex regional pain syndrome: practical diagnostic and treatment guidelines, 4th edition. Pain Med 14: 180–229.

Haroutiunian S, Nikolajsen L, Finnerup NB et al. (2013). The neuropathic component in persistent postsurgical pain: a systematic literature review. Pain 154: 95–102.

Hassantash SA, Afrakhteh M, Maier RV (2003). Causalgia: a meta-analysis of the literature. Arch Surg 138: 1226–1231.

Hassenbusch SJ, Schoppa D, Walsh JG et al. (1996). Long-term results of peripheral nerve stimulation for reflex sympathetic dystrophy. J Neurosurg 84: 415–423.

Hempenstall K, Nurmikko TJ, Johnson RW et al. (2005). Analgesic therapy in postherpetic neuralgia: a quantitative systematic review. PLoS Med 2: e164.

Herlyn P, Muller-Hilke B, Wendt M et al. (2010). Frequencies of polymorphisms in cytokines, neurotransmitters and adrenergic receptors in patients with complex regional pain syndrome type I after distal radial fracture. Clin J Pain 26: 175–181.

Horowitz SH (2000). Venipuncture-induced causalgia: anatomic relations of upper extremity superficial veins and nerves, and clinical considerations. Transfusion 40: 1036–1040.

Horowitz SH (2001). Venipuncture-induced neuropathic pain: the clinical syndrome, with comparisons to experimental nerve injury models. Pain 94: 225–229.

Hsu MC, Harris RE, Sundgren PC et al. (2009). No consistent difference in gray matter volume between individuals with fibromyalgia and age-matched healthy subjects when controlling for affective disorder. Pain 143: 262–267.

Huge V, Lauchart M, Magerl W et al. (2011). Complex interaction of sensory and motor signs and symptoms in chronic CRPS. PLoS One 6: e18775.

Hulsman NM, Geertzen JH, Dijkstra PU et al. (2009). Myopathy in CRPS-I: disuse or neurogenic? Eur J Pain 13: 731–736.

Huygen FJ, De Bruijn AG, De Bruin MT et al. (2002). Evidence for local inflammation in complex regional pain syndrome type 1. Mediators Inflamm 11: 47–51.

Iwatsuki K, Arai T, Ota H et al. (2013). Targeting anti-inflammatory treatment can ameliorate injury-induced neuropathic pain. PLoS One 8e57721.

Juottonen K, Gockel M, Silen T et al. (2002). Altered central sensorimotor processing in patients with complex regional pain syndrome. Pain 98: 315–323.

Kemler MA, Barendse GA, van Kleef M et al. (2000). Spinal cord stimulation in patients with chronic reflex sympathetic dystrophy. N Engl J Med 343: 618–624.

Kharkar S, Venkatesh YS, Grothusen JR et al. (2012). Skin biopsy in complex regional pain syndrome: case series and literature review. Pain Physician 15: 255–266.

Kiefer RT, Rohr P, Ploppa A et al. (2008). Efficacy of ketamine in anesthetic dosage for the treatment of refractory complex regional pain syndrome: an open-label phase II study. Pain Med 9: 1173–1201.

Kiralp MZ, Yildiz S, Vural D et al. (2004). Effectiveness of hyperbaric oxygen therapy in the treatment of complex regional pain syndrome. J Int Med Res 32: 258–262.

Klein MM, Lee JW, Siegel SM et al. (2012). Endoneurial pathology of the needlestick-nerve-injury model of complex regional pain syndrome, including rats with and without pain behaviors. Eur J Pain 16: 28–37.

Kohr D, Tschernatsch M, Schmitz K et al. (2009). Autoantibodies in complex regional pain syndrome bind to a differentiation-dependent neuronal surface autoantigen. Pain 143: 246–251.

Kohr D, Singh P, Tschernatsch M et al. (2011). Autoimmunity against the beta(2) adrenergic receptor and muscarinic-2 receptor in complex regional pain syndrome. Pain 152: 2690–2700.

Kolb L, Lang C, Seifert F et al. (2012). Cognitive correlates of "neglect-like syndrome" in patients with complex regional pain syndrome. Pain 153: 1063–1073.

Koltzenburg M, Wall PD, McMahon SB (1999). Does the right side know what the left is doing? Trends Neurosci 22: 122–127.

Kozin F, McCarty DJ, Sims J et al. (1976). The reflex sympathetic dystrophy syndrome. I. Clinical and histologic studies: evidence for bilaterality, response to corticosteroids and articular involvement. Am J Med 60: 321–331.

Krämer HH, Eberle T, Üçeyler N et al. (2011). TNF-alpha in CRPS and 'normal' trauma – significant differences between tissue and serum. Pain 152: 285–290.

Krämer HH, Hofbauer LC, Szalay G et al. (2014). Osteoprotegerin: a new biomarker for impaired bone metabolism in complex regional pain syndrome? Pain 155: 889–895.

Lee BH, Scharff L, Sethna NF et al. (2002). Physical therapy and cognitive-behavioral treatment for complex regional pain syndromes. J Pediatr 141: 135–140.

Lee JW, Siegel SM, Oaklander AL (2009). Effects of distal nerve injuries on dorsal-horn neurons and glia: relationships between lesion size and mechanical hyperalgesia. Neuroscience 158: 904–914.

Lefaucheur JP, Hatem S, Nineb A et al. (2006). Somatotopic organization of the analgesic effects of motor cortex rTMS in neuropathic pain. Neurology 67: 1998–2004.

Legrain V, Bultitude JH, De Paepe AL et al. (2012). Pain, body, and space: what do patients with complex regional pain syndrome really neglect? Pain 153: 948–951.

Lenz M, Üçeyler N, Frettloh J et al. (2013). Local cytokine changes in complex regional pain syndrome type I (CRPS I) resolve after 6 months. Pain 154: 2142–2149.

Li Z, Smith BP, Tuohy C et al. (2010). Complex regional pain syndrome after hand surgery. Hand Clin 26: 281–289.

Li WW, Guo TZ, Shi X et al. (2014). Autoimmunity contributes to nociceptive sensitization in a mouse model of complex regional pain syndrome. Pain 155: 2377–2389.

Liem L, Russo M, Huygen FJ et al. (2013). A multicenter, prospective trial to assess the safety and performance of the spinal modulation dorsal root ganglion neurostimulator system in the treatment of chronic pain. Neuromodulation 16: 471–482.

Linchitz RM, Raheb JC (1999). Subcutaneous infusion of lidocaine provides effective pain relief for CRPS patients. Clin J Pain 15: 67–72.

Ludwig J, Gorodetskaya N, Schattschneider J et al. (2007). Behavioral and sensory changes after direct ischemia-reperfusion injury in rats. Eur J Pain 11: 677–684.

Maier C, Baron R, Tolle TR et al. (2010). Quantitative sensory testing in the German Research Network on Neuropathic Pain (DFNS): somatosensory abnormalities in 1236 patients with different neuropathic pain syndromes. Pain 150: 439–450.

Maihöfner C, Handwerker HO, Neundorfer B et al. (2003). Cortical reorganization during recovery from complex regional pain syndrome. Neurology 61: 1707–1715.

Maihöfner C, Handwerker HO, Neundörfer B et al. (2004). Cortical reorganization during recovery from complex regional pain syndrome. Neurology 63: 693–701.

Maihöfner C, Handwerker HO, Birklein F (2006). Functional imaging of allodynia in complex regional pain syndrome. Neurology 66: 711–717.

Manning DC, Alexander G, Arezzo JC et al. (2014). Lenalidomide for complex regional pain syndrome type 1: lack of efficacy in a phase II randomized study. J Pain 15: 1366–1376.

Meier T, Wasner G, Faust M et al. (2003). Efficacy of lidocaine patch 5% in the treatment of focal peripheral neuropathic pain syndromes: a randomized, double-blind, placebo-controlled study. Pain 106: 151–158.

Mellick GA, Mellick LB (1997). Reflex sympathetic dystrophy treated with gabapentin. Arch Phys Med Rehabil 78: 98–105.

Merskey H, Bogduk N (1994). Classification of Chronic Pain: Descriptions of Chronic Pain Syndromes and Definitions of Pain Terms. Second Edition, IASP Press, Seattle.

Meyer RA, Raja SN, Campbell JN (1985). Coupling of action potential activity between unmyelinated fibers in the peripheral nerve of monkey. Science 227: 184–187.

Mitchell SW (1864). Injuries of Nerves and their Consequences (vols. 6), Dover Publications, New York, American Academy of Neurology Reprint Series (1965 edition).

Mitchell SW, Morehouse GR, Keen WM (1864). Gunshot Wounds and Other Injuries of Nerves. J.B. Lippincott, Philadelphia.

Mogil JS, Wilson SG, Bon K et al. (1999). Heritability of nociception I: responses of 11 inbred mouse strains on 12 measures of nociception. Pain 80: 67–82.

Moon JY, Park SY, Kim YC et al. (2012). Analysis of patterns of three-phase bone scintigraphy for patients with complex regional pain syndrome diagnosed using the proposed research criteria (the 'Budapest Criteria'). Br J Anaesth 108: 655–661.

Moseley GL (2004a). Graded motor imagery is effective for long-standing complex regional pain syndrome: a randomised controlled trial. Pain 108: 192–198.

Moseley GL (2004b). Why do people with complex regional pain syndrome take longer to recognize their affected hand? Neurology 62: 2182–2186.

Moseley GL (2005). Distorted body image in complex regional pain syndrome. Neurology 65: 773.

Moseley GL (2006). Graded motor imagery for pathologic pain: a randomized controlled trial. Neurology 67: 2129–2134.

Muizelaar JP, Kleyer M, Hertogs IA et al. (1997). Complex regional pain syndrome (reflex sympathetic dystrophy and causalgia): management with the calcium channel blocker nifedipine and/or the alpha-sympathetic blocker phenoxybenzamine in 59 patients. Clin Neurol Neurosurg 99: 26–30.

Munts AG, Koehler PJ (2010). How psychogenic is dystonia? Views from past to present. Brain 133: 1552–1564.

Munts AG, van der Plas AA, Ferrari MD et al. (2010). Efficacy and safety of a single intrathecal methylprednisolone bolus in chronic complex regional pain syndrome. Eur J Pain 14: 523–528.

Newman BH, Waxman DA (1996). Blood donation-related neurologic needle injury: evaluation of 2 years' worth of data from a large blood center. Transfusion 36: 213–215.

Noppers IM, Niesters M, Aarts LPHJ et al. (2011). Drug-induced liver injury following a repeated course of ketamine treatment for chronic pain in CRPS type 1 patients: a report of 3 cases. Pain 152: 2173–2178.

Oaklander AL (2012). Development of CRPS after shingles: It's all about location. Pain 153: 2309–2310.

Oaklander AL, Belzberg AJ (1997). Unilateral nerve injury down-regulates mRNA for Na^+ channel SCN10A bilaterally in rat dorsal root ganglia. Mol Brain Res 52: 162–165.

Oaklander AL, Brown JM (2004). Unilateral nerve injury produces bilateral loss of distal innervation. Ann Neurol 55: 639–644.

Oaklander AL, Fields HL (2009). Is reflex sympathetic dystrophy/complex regional pain syndrome type I a small-fiber neuropathy? Ann Neurol 65: 629–638.

Oaklander AL, Ochoa JL (2012). Mononeuropathy of the lateral cutaneous nerve of the calf (LCNC) in patients previously labeled with "complex regional pain syndrome-I". Ann Neurol 72: S16.

Oaklander AL, Romans K, Horasek S et al. (1998). Unilateral postherpetic neuralgia is associated with bilateral sensory neuron damage. Ann Neurol 44: 789–795.

Oaklander AL, Rissmiller JG, Gelman LB et al. (2006). Evidence of focal small-fiber axonal degeneration in complex regional pain syndrome-I (reflex sympathetic dystrophy). Pain 120: 235–243.

O'Connell NE, Wand BM, Marston L et al. (2014). Non-invasive brain stimulation techniques for chronic pain. Cochrane Database Syst Rev 4: CD008208.

Oerlemans HM, Oostendorp RA, de Boo T et al. (1999). Pain and reduced mobility in complex regional pain syndrome I: outcome of a prospective randomised controlled clinical trial of adjuvant physical therapy versus occupational therapy. Pain 83: 77–83.

Ota H, Arai T, Iwatsuki K et al. (2014). Pathological mechanism of musculoskeletal manifestations associated with CRPS type II: an animal study. Pain 155: 1976–1985.

Parkitny L, McAuley JH, Di Pietro F et al. (2013). Inflammation in complex regional pain syndrome: a systematic review and meta-analysis. Neurology 80: 106–117.

Pascarelli EF, Hsu YP (2001). Understanding work-related upper extremity disorders: clinical findings in 485 computer users, musicians, and others. J Occup Rehabil 11: 1–21.

Peach G (1995). Hyperbaric oxygen and the reflex sympathetic dystrophy syndrome: a case report. Undersea Hyperb Med 22: 407–408.

Perez RS, Kwakkel G, Zuurmond WW et al. (2001). Treatment of reflex sympathetic dystrophy (CRPS type 1): a research synthesis of 21 randomized clinical trials. J Pain Symptom Manage 21: 511–526.

Perez R, Zollinger P, Dijkstra P et al. (2010). Evidence based guidelines for complex regional pain syndrome type 1. BMC Neurol 10: 20.

Petersen KL, Rowbotham MC (2010). Natural history of sensory function after herpes zoster. Pain 150: 83–92.

Pham M, Baumer P, Meinck HM et al. (2014). Anterior interosseous nerve syndrome: fascicular motor lesions of median nerve trunk. Neurology 82: 598–606.

Picarelli H, Teixeira MJ, de Andrade DC et al. (2010). Repetitive transcranial magnetic stimulation is efficacious as an add-on to pharmacological therapy in complex regional pain syndrome (CRPS) type I. J Pain 11: 1203–1210.

Pleger B, Tegenthoff M, Schwenkries P et al. (2004). Mean sustained pain levels are linked to hemispherical side-to-side differences of primary somatosensory cortex in the complex regional pain syndrome I. Exp Brain Res 155: 115–119.

Pleger B, Tegenthoff M, Ragert P et al. (2005). Sensorimotor returning in complex regional pain syndrome parallels pain reduction. Ann Neurol 57: 425–429.

Pleger B, Ragert P, Schwenkries P et al. (2006). Patterns of cortical reorganization parallel impaired tactile discrimination and pain intensity in complex regional pain syndrome. Neuroimage 31: 503–510.

Poree L, Krames E, Pope J et al. (2013). Spinal cord stimulation as treatment for complex regional pain syndrome should be considered earlier than last resort therapy. Neuromodulation 16: 125–141.

Prough DS, McLeskey CH, Poehling GG et al. (1985). Efficacy of oral nifedipine in the treatment of reflex sympathetic dystrophy. Anesthesiology 62: 796–799.

Ramchandren S, Chaudhry V, Hoke A et al. (2008). Peripheral nerve vasculitis presenting as complex regional pain syndrome. J Clin Neuromuscul Dis 10: 61–64.

Ranoux D, Attal N, Morain F et al. (2008). Botulinum toxin type A induces direct analgesic effects in chronic neuropathic pain. Ann Neurol 64: 274–283.

Riol-Blanco L, Ordovas-Montanes J, Perro M et al. (2014). Nociceptive sensory neurons drive interleukin-23-mediated psoriasiform skin inflammation. Nature 510: 157–161.

Robinson G, Cohen H, Goebel A (2011). A case of complex regional pain syndrome with agnosia for object orientation. Pain 152: 1674–1681.

Rondinelli RD (2008). Complex regional pain syndrome impairment. In: Robert D Rondinelli (Ed.), AMA Guides to the Evaluation of Permanent Impairment, 6th ed. American Medical Association, Chicago, IL, pp. 450–454.

Rossi S, Hallett M, Rossini PM et al. (2009). Safety, ethical considerations, and application guidelines for the use of transcranial magnetic stimulation in clinical practice and research. Clin Neurophysiol 120: 2008–2039.

Rota E, Zavaroni D, Parietti L et al. (2014). Ulnar entrapment neuropathy in patients with type 2 diabetes mellitus: an electrodiagnostic study. Diabetes Res Clin Pract 104: 73–78.

Safarpour D, Salardini A, Richardson D et al. (2010). Botulinum toxin A for treatment of allodynia of complex regional pain syndrome: a pilot study. Pain Med 11: 1411–1414.

Sandroni P, Benrud-Larson LM, McClelland RL et al. (2003). Complex regional pain syndrome type I: incidence and prevalence in Olmsted county, a population-based study. Pain 103: 199–207.

Schmelz M, Schmidt R, Bickel A et al. (1997). Specific C-receptors for itch in human skin. J Neurosci 17: 8003–8008.

Schrag A, Trimble M, Quinn N et al. (2004). The syndrome of fixed dystonia: an evaluation of 103 patients. Brain 127: 2360–2372.

Schwartzman RJ, Chevlen E, Bengtson K (2003). Thalidomide has activity in treating complex regional pain syndrome. Arch Intern Med 163: 1487–1488.

Schwartzman RJ, Alexander GM, Grothusen JR et al. (2009a). Outpatient intravenous ketamine for the treatment of complex regional pain syndrome: a double-blind placebo controlled study. Pain 147: 107–115.

Schwartzman RJ, Patel M, Grothusen JR et al. (2009b). Efficacy of 5-day continuous lidocaine infusion for the treatment of refractory complex regional pain syndrome. Pain Med 10: 401–412.

Schwenkreis P, Janssen F, Rommel O et al. (2003). Bilateral motor cortex disinhibition in complex regional pain syndrome (CRPS) type I of the hand. Neurology 61: 515–519.

Sherry DD, Wallace CA, Kelley C et al. (1999). Short- and long-term outcomes of children with complex regional pain syndrome type I treated with exercise therapy. Clin J Pain 15: 218–223.

Shin NY, Kang DH, Jang JH et al. (2013). Impaired recognition of social emotion in patients with complex regional pain syndrome. J Pain 14: 1304–1309.

Siegel SM, Lee JW, Oaklander AL (2007). Needlestick distal nerve injury in rats models symptoms of complex regional pain syndrome. Anesth Analg 105: 1820–1829.

Singh B, Moodley J, Shaik AS et al. (2003). Sympathectomy for complex regional pain syndrome. J Vasc Surg 37: 508–511.

Staff NP, Engelstad J, Klein CJ et al. (2010). Post-surgical inflammatory neuropathy. Brain 133: 2866–2880.

Stanton TR, Wand BM, Carr DB et al. (2013). Local anaesthetic sympathetic blockade for complex regional pain syndrome. Cochrane Database Syst Rev 8: CD004598.

Stanton-Hicks M (1999). Spinal cord stimulation for the management of complex regional pain syndromes. Neuromodulation 2: 193–201.

Stewart JD (2000). Focal Peripheral Neuropathies, 3rd ed. Lippincott Williams & Wilkins, Philadelphia.

Stoler JM, Oaklander AL (2006). Patients with Ehlers Danlos syndrome and CRPS: a possible association? Pain 123: 204–209.

Sudeck PHM (1901). Über die akute (reflektorische) Knochenatrophie nach Entzündungen und Verletzungen an den Extremitäten und ihre klinischen Erscheinungen. Fortschr Geb Rontgenstr 5: 277.

Sumitani M, Shibata M, Sakaue G et al. (2010). Development of comprehensive diagnostic criteria for complex regional pain syndrome in the Japanese population. Pain 150: 243–249.

Tekus V, Hajna Z, Borbely E et al. (2014). A CRPS-IgG-transfer-trauma model reproducing inflammatory and positive sensory signs associated with complex regional pain syndrome. Pain 155: 299–308.

Thimineur MA, Saberski L (1996). Complex regional pain syndrome type I (RSD) or peripheral mononeuropathy: a discussion of three cases. Clin J Pain 12: 145–150.

Tsubokawa T (1995). Motor cortex stimulation for deafferentation pain relief in various clinical syndromes and its possible mechanism. In: JM Besson, G Guilbaud, H Ollat (Eds.), Forebrain Areas Involved in Pain Processing. John Libbey Eurotext, Paris, pp. 261–276.

Turner JA, Loeser JD, Deyo RA et al. (2004). Spinal cord stimulation for patients with failed back surgery syndrome or complex regional pain syndrome: a systematic review of effectiveness and complications. Pain 108: 137–147.

Tuter NV, Danilov AB, Poliakova LV (1997). [The treatment of a complex regional pain syndrome.] Zh. Nevrol. Psikhiatr. Im S.S. Korsakova 97: 33–35.

Üçeyler N, Rogausch JP, Toyka KV et al. (2007). Differential expression of cytokines in painful and painless neuropathies. Neurology 69: 42–49.

van Buyten JP, Smet I, Liem L et al. (2015). Stimulation of dorsal root ganglia for the management of complex regional pain syndrome: a prospective case series. Pain Pract 15: 208–216.

van der Laan L, ter Laak HJ, Gabreels-Festen A et al. (1998a). Complex regional pain syndrome type I (RSD): pathology of skeletal muscle and peripheral nerve. Neurology 51: 20–25.

van der Laan L, Veldman PH, Goris RJ (1998b). Severe complications of reflex sympathetic dystrophy: infection, ulcers, chronic edema, dystonia, and myoclonus. Arch Phys Med Rehabil 79: 424–429.

van Rijn MA, Marinus J, Putter H et al. (2007). Onset and progression of dystonia in complex regional pain syndrome. Pain 130: 287–293.

van Rijn MA, Marinus J, Putter H et al. (2011). Spreading of complex regional pain syndrome: not a random process. J Neural Transm 118: 1301–1309.

van Rooijen DE, Geraedts EJ, Marinus J et al. (2011). Peripheral trauma and movement disorders: a systematic review of reported cases. J Neurol Neurosurg Psychiatry 82: 892–898.

van Rooijen DE, Roelen DL, Verduijn W et al. (2012). Genetic HLA associations in complex regional pain syndrome with and without dystonia. J Pain 13: 784–789.

van Velzen GAJ, Perez RSGM, van Gestel MA et al. (2014). Health-related quality of life in 975 patients with complex regional pain syndrome type 1. Pain 155: 629–634.

Velasco F, Carrillo-Ruiz JD, Castro G et al. (2009). Motor cortex electrical stimulation applied to patients with complex regional pain syndrome. Pain 147: 91–98.

Veldman PH, Reynen HM, Arntz IE et al. (1993). Signs and symptoms of reflex sympathetic dystrophy: prospective study of 829 patients. Lancet 342: 1012–1016.

Walker K, Medhurst SJ, Kidd BL et al. (2002). Disease modifying and anti-nociceptive effects of the bisphosphonate, zoledronic acid in a model of bone cancer pain. Pain 100: 219–229.

Wallace MS, Ridgeway BM, Leung AY et al. (2000). Concentration-effect relationship of intravenous lidocaine on the allodynia of complex regional pain syndrome types I and II. Anesthesiology 92: 75–83.

Walton KD, Dubois M, Llinàs RR (2010). Abnormal thalamocortical activity in patients with complex regional pain syndrome (CRPS) Type I. Pain 150: 41–51.

Watson CPN, Mackinnon SE, Dostrovsky JO et al. (2014). Nerve resection, crush and re-location relieve complex regional pain syndrome type II: A case report. Pain 155: 1168–1173.

Weber M, Birklein F, Neundörfer B et al. (2001). Facilitated neurogenic inflammation in complex regional pain syndrome. Pain 91: 251–257.

Wesseldijk F, Huygen FJ, Heijmans-Antonissen C et al. (2007). Tumor necrosis factor-alpha and interleukin-6 are not correlated with the characteristics of complex regional pain syndrome type 1 in 66 patients. Eur J Pain 12: 716–721.

Wesseldijk F, Huygen FJ, Heijmans-Antonissen C et al. (2008). Six years follow-up of the levels of TNF-alpha and IL-6 in patients with complex regional pain syndrome type 1. Mediators Inflamm 2008: 469439.

Williams ST, Davies A, Bryson P (2009). Chronic regional pain syndrome after subtalar arthrodesis is not prevented by early hyperbaric oxygen. Pain Physician 12: E335–E339.

Wuppenhorst N, Maier C, Frettloh J et al. (2010). Sensitivity and specificity of 3-phase bone scintigraphy in the diagnosis of complex regional pain syndrome of the upper extremity. Clin J Pain 26: 182–189.

Zollinger PE, Tuinebreijer WE, Breederveld RS et al. (2007). Can vitamin C prevent complex regional pain syndrome in patients with wrist fractures? A randomized, controlled, multicenter dose-response study. J Bone Joint Surg Am 89: 1424–1431.

Zyluk A, Puchalski P (2013). Complex regional pain syndrome: observations on diagnosis, treatment and definition of a new subgroup. J Hand Surg Eur 38: 599–606.

Handbook of Clinical Neurology, Vol. 131 (3rd series)
Occupational Neurology
M. Lotti and M.L. Bleecker, Editors

Chapter 28

Accommodation of workers with chronic neurologic disorders

MARGIT L. BLEECKER* AND SHERYL K. BARNES
Center for Occupational and Environmental Neurology, Baltimore, MD, USA

INTRODUCTION

This chapter focuses on neurologic disorders that develop in the workforce or may be present when a worker is hired, and how the associated disabilities may be accommodated. In the second part the key points of the Americans with Disabilities Act (ADA) and the Americans with Disabilities Act Amendment Act (ADAAA) are summarized to clarify the term "disability" under the law and what constitutes reasonable accommodation, along with the responsibilities of the employer and employee under this law. Table 28.1 provides a list of some of the more common neurologic disorders present in the workforce. Aging nervous system is included, as the workforce over the age of 65 years is steadily increasing, according to the Bureau of Labor Statistics, which predicts by 2022 a third of Americans aged between 64 and 75 years will still be working (Toossi 2013). Therefore healthcare providers must be made aware of the needs of this aging workforce, owing to the significant changes with age in the nervous system that may require accommodations in the workplace to maintain productivity and safety. Just as older workers maintain the ability to work for more years, whether full-time or part-time, improvements in treatments for workers with chronic neurologic disorders will allow them to maintain employment for many more years. The healthcare provider must realize that, for many with disabilities, the ability to work is important and may require accommodations.

SIGNS AND SYMPTOMS

Signs and symptoms common to many chronic neurologic disorders that may require accommodation to maintain employment are listed in Table 28.2.

Fatigue

Fatigue is a common symptom of workers with chronic neurologic conditions, particularly multiple sclerosis (MS), Parkinson disease's (PD), myasthenia gravis (MG), traumatic brain injury (TBI), and stroke (Kluger et al., 2013). With chronic neurologic illnesses some believe fatigue is an intrinsic part of the underlying disease (Kluger et al., 2013), while others believe it is part of sleep disturbances and depression, which are strong predictors of fatigue (Strober and Arnett, 2005).

FATIGUE IN MULTIPLE SCLEROSIS

MS symptoms often disrupt sleep, leading to the need for daytime naps (Stanton et al., 2006). The underlying MS pathology significantly associated with fatigue includes reduced glucose metabolism in the frontal cortex and basal ganglia (Roelcke et al., 1997; Bakshi et al., 1998), widespread axonal dysfunction (Tartaglia et al., 2004), and brain atrophy (Marrie et al., 2005). Fatigue with MS often determines if the individual is employed full-time, had to reduce work hours, or had to leave the workforce (Johnson et al., 2004a; Smith and Arnett, 2005). Other factors related to employment status are the Expanded Disability Status Scale, which assesses information-processing speed and the personality characteristic of persistence (Strober et al., 2012).

Forty percent of those unemployed due to MS would like to be employed (Johnson et al., 2004a). Many patients with MS will continue to work for years after their physician has recommended they stop because of fatigue, as work is therapeutic. However it is important to be able to organize the work day around the issue of fatigue, as it can be profound and variable during the day and not visible to co-workers (Johnson et al., 2004b). Fatigue is also reflected in cognitive changes, such as

*Correspondence to: Margit L. Bleecker, MD, PhD, Director, Center for Occupational and Environmental Neurology, 2 Hamill Road, Suite 225, Baltimore, MD, USA. E-mail: coen@msn.com

Table 28.1

Neurologic disorders present in the workplace

Aging nervous system
Cerebrovascular disease
Cervical and lumbar spine disorders
Dementia
Epilepsy
Essential tremor
Fibromyalgia
Migraine
Multiple sclerosis
Muscle disease
Myasthenia gravis
Parkinson's disease
Peripheral neuropathy
Postpolio syndrome
Traumatic brain injury

Table 28.2

Signs and symptoms common to many chronic neurologic disorders that require accommodations to maintain employment

Fatigue
Poor endurance
Decreased fine motor abilities
Tremor
Muscle weakness and cramps
Pain
Impaired gait
Impaired balance
Slowed movement
Mild cognitive impairment
Slowed thinking
Decreased concentration

difficulty learning new material, slowed thinking, inability to multitask, and even talking, which may require more effort, as otherwise words are slurred (Johnson et al., 2004a).

Fatigue in Parkinson's disease

Fatigue in PD is common, and sometimes the most troubling nonmotor symptom present in 68% of PD patients, who can differentiate "fatigue-related" from normal tiredness (Martinez-Martin et al., 2006; Friedman et al., 2007). However fatigue was not even considered in the original Unified Parkinson's Disease Rating Scale (UPDRS) (Fahn et al., 1987). Since up to 50% of PD patients have depression, it was difficult to separate the fatigue associated with depression from that associated with PD. Exercise may not improve fatigue in PD but may have positive effects on endurance, self-confidence, and depression, and therefore benefit the patient (Friedman et al., 2007). However it was the presence of depression and anxiety, and not motor symptoms, that increased the likelihood of PD patients leaving the workforce (Armstrong et al., 2014). In another study, 32% of PD workers reported that slowness, fatigue, and tremor were their greatest difficulties, causing them to leave work, and they believed that changes in work schedule, type of work, and having a longer time to complete tasks would have allowed them to remain employed (Murphy et al., 2013).

Fatigue in myasthenia gravis

MG patients' fatigue is due to muscle weakness and physical tiredness from impaired peripheral neurotransmission. Even though MG causes peripheral muscle fatigue and weakness, there is fatigue associated with cognitive and social function (Paul et al., 2000). This is why fatigue in MG is proposed to have both peripheral and central mechanisms. As with MS and PD, fatigue is associated with depression (Kittiwatanapaisan et al., 2003); however, when this variable is controlled in the analysis, the best predictor of fatigue in MG is activity restriction. This was reflected in the validation study for Myasthenia Gravis Quality of Life 15-item (MG-QoL15) questionnaire (Burns et al., 2011, 2012), where items most significantly scored abnormal were "I am frustrated by my MG," "My MG limits my ability to enjoy hobbies and fun activities," and "I have to make plans around my MG." It is useful to follow scores on the MG-QoL15 for successful maintenance of MG patients' symptoms in the workplace.

Poststroke fatigue

Poststroke fatigue (PSF) is defined as "a state specifically caused by focal brain dysfunction and characterized by a long-standing lack of effectiveness while executing any motor and mental tasks, independent of poststroke depression, significant cognitive or neurologic sequelae or other clinical dysfunction" (Radman et al., 2012). Even with minor ischemic strokes, those with no neurologic deficits, PSF is correlated with attentional and executive impairments whose neural circuits are distributed throughout the brain. Following a minor stroke a reduction in work activities is reported by 55%, and may improve with attention to PSF and cognitive deficits (Kutlubaev and Mead, 2012; Radman et al., 2012).

Fatigue in postpolio syndrome

Fatigue in postpolio syndrome is severe and persistent and is associated with physical functioning, pain, sleep disorders, mental health, and task-oriented coping (Tersteeg et al., 2011).

Fatigue in traumatic brain injury

Fatigue in TBI is the most common sequela present in 50–80% of TBI individuals (Bushnik et al., 2008) and can persist in 73% 5 years postinjury (Olver et al., 1996). Even with mild TBI, 18% tire more easily at 3 months postinjury (Englander et al., 1992) and 15% are still symptomatic at 1 year. This is important, as the incidence of mild TBI is 27.4/100 000 and usually affects young men. The most prevalent concomitant problem is excessive daytime sleepiness that responds to modafinil (Kaiser et al., 2010), followed by depression and pain (Bushnik et al., 2008).

Allowing patients in the workforce with chronic neurologic disease and significant fatigue to work from home removes the stress of commuting and allows the patient to conserve energy for work. This makes it easier for energy management and allows for a brief nap to increase productivity, especially with MG, MS, and PD. At the workplace, permitting patients to have control over arranging their daily duties allows them to start early, when complex material can be learned, and later focus on tasks that require less effort and then leave early. If the patient is allowed to set up routine schedules, then the day is predictable with rest breaks. This helps with fatigue and with a routine the cognitive stress is less.

COGNITIVE IMPAIRMENTS

Cognitive impairment in multiple sclerosis

Cognitive impairment in MS may determine employment status that was not influenced by physical disability or disease duration (Rao et al., 1991). Cognitively impaired MS patients have the same degree of fatigue, anxiety, and depression as cognitively preserved MS patients but the fractional anisotropy values on diffusion tensor imaging studies were more reduced in the cognitively impaired group for the corpus callosum, the superior and inferior longitudinal fasciculus, the corticospinal tracts, the forceps minor, the entire cingulum and fornices (Hulst et al., 2013). Memory impairment in MS is due to inefficient new learning that responds to training with a modified story memory technique addressing skills for imagery and context. These skills translated to improved measures of everyday memory (Chiaravalloti et al., 2013). Use of electronic memory aids such as computer, palm or pocket PC, or cell phone can serve as cognitive aids for monitoring medication, remembering daily routine at work, and navigating in the community (Johnson et al., 2009). Monitoring stress and fatigue is important as both can enhance cognitive difficulties.

Cognitive impairment in Parkinson's disease

The role of fatigue in PD-related cognitive difficulties is unknown, but virtually all PD patients have deficits in visuospatial abilities, memory retrieval, set shifting, and attention early in the course of the disease (Dubois and Pillon, 1997). The Wisconsin Card Sort task, a measure of executive function, is associated with reduced frontal-lobe perfusion in PD (Abe et al., 2000). When attentional demands are increased in PD, motor slowing occurs; this is believed to be secondary to the fatigue and stress caused by the mental challenges (Rochester et al., 2004).

Cognitive impairment in traumatic brain injury

Very mild TBI is associated with prefrontal axonal injury involving the dorsolateral prefrontal cortex found on diffusion tensor imaging. This results in impaired executive function that continues to develop over time following the injury (Lipton et al., 2009). Executive function impairment demonstrated increased errors of omission on continuous performance task and slower times to complete a maze. Use of methylphenidate helped the attentional component and memory in these patients (Cernich et al., 2010). Individuals with persistent neuropsychologic deficits following minor head injury did not differ according to the neurologic deficit immediately following the injury – namely, loss of consciousness, disorientation, or confusion (Leininger et al., 1990). Functional recovery to being disability-free at 1 year from moderate/severe TBI was significantly influenced by educational attainment, a measure of cognitive reserve (Schneider et al., 2014). Response to cognitive training for attention and executive functions after brain injury is dependent upon functional brain network modularity (Arnemann et al., 2015).

Cognitive impairment in brain tumor

Brain tumor survivors at work had difficulty following the flow of events, remembering their train of thought while speaking, putting together materials for a task, shifting between tasks, and following written instructions (Collins et al., 2013). They reported more deficits at work with tasks that require working memory, executive function, and attention.

Cognitive impairment in elderly people

In the elderly worker with mild cognitive impairment or early dementia, intervention with exercise that improved balance and strength improved cognitive performance and behavior (Fundaro and Casale, 2012). This exercise rehabilitation program is believed to increase neuroplasticity and thereby maintain the ability to work in the elderly population with early memory problems. The role of cognitive training for mild dementia has not shown significant benefits (Bahar-Fuchs et al., 2013).

MOTOR IMPAIRMENT

Motor impairment in Parkinson's disease

In PD, male sex was the only predictor of worse motor symptoms that were levodopa-responsive, a process related to degeneration of the nigrostriatal dopaminergic neurons (Velseboer et al., 2013). Faster progression of disability occurred with the onset of levodopa-nonresponsive motor symptoms and cognitive dysfunction (Velseboer et al., 2013). Increased risk for falls in PD is well known with impaired postural stability and shuffling gait. Studies show that coupling steps to music or the beat of a metronome leads to lasting motor improvement by increasing motor speed and stride length (Bella et al., 2015). Moving to music, whether ballroom dancing (Shanahan et al., 2015), walking (Benoit et al., 2014), or tai chi (Gao et al., 2013; Ni et al., 2014), has helped PD patients with balance and a more stable gait.

Motor impairment with increasing age

Just age alone increases fall risk: one-third of the population older than 65 years is at risk for at least one or more fall a year. When there is a reduction in stride length variability through music-based multitask training on gait, balance is improved (Kressig et al., 2005; Trombetti et al., 2011). This program involved walking to music and responding to changes in music rhythm patterns and challenged balance with multidirectional weight shifting.

Motor impairment in neuropathy

The presence of neuropathies further increases fall risk due to impaired proprioception and reduced lower-limb strength (Richardson et al., 2014).

Motor impairment in postpolio syndrome

Weakness in postpolio syndrome is progressive but has responded to interventions with isotonic strengthening and isokinetic and isometric strengthening. However muscle overuse should be avoided (Trojan and Cashman, 2005).

PAIN

Common pain conditions in the workplace (arthritis, back, migraine, tension-type headaches, and general musculoskeletal conditions) resulted in reduced work performance but did not increase lost time due to absenteeism (Stewart et al., 2003). Productive time lost due to reduced performance from pain conditions was four times more than time lost from absenteeism. For patients with work-related neck and low-back disorders using an interdisciplinary functional restoration program provided a high rate of return to work and work retention (Hartzell et al., 2014). The most feasible accommodation for back pain was avoiding awkward postures by change in physical workload, work environment, schedule, or finding alternate work and arranging for assistance (Shaw et al., 2014). In workers at risk for neck and upper-extremity musculoskeletal disorders, intervention with a program that improved upper-extremity strength resulted in reduction of pain symptoms and disability (Rasotto et al., 2015).

Diabetic polyneuropathy can be very painful but usually responds to pharmacologic and nonpharmacologic interventions. Studies examining quality of life in patients with painful diabetic polyneuropathy found worse scores for energy, sleep, pain, physical mobility, and emotional reactions (Burns et al., 2012). Poor proprioception in this disorder leads to poor balance and increased risk for falls that can become an issue in workplace safety.

TREATMENT

It is imperative that the healthcare provider maximize the medical management of any neurologic condition in the workplace to ensure that workplace accommodations are successful. This includes treating comorbidities such as depression, sleep disruption, and pain and referring patients in the workforce at increased risk for falls to participate in programs to improve balance and lower-extremity strength. See Table 28.3 for validated questionnaires that will help with following symptom response to intervention by treatment or accommodation.

Table 28.3

Validated symptoms questionnaires

Fatigue Impact Scale (D-FIS)[1]
Fatigue Severity Scale (FSS)[2]
McGill Pain Questionnaire (MPQ)[3]
Myasthenia Gravis Quality of Life Questionnaire (MGQoL15)[4]
Norfolk Quality of Life Questionnaire – Diabetic Neuropathy (QoL-DN)[5]
Parkinson Disease Quality of Life Questionnaire (PDQ-8)[6]

[1]Fisk and Doble (2002); Benito-Leon et al. (2007).
[2]Valko et al. (2008).
[3]Byrne et al. (1982).
[4]Burns et al. (2011).
[5]Vinik et al. (2005).
[6]Jenkinson et al. (1997).

The next section summarizes the key points of the law covering individuals with disabilities in the workplace and the role of the healthcare provider along with accommodations in the workplace.

SUMMARY OF THE ADA AND THE ADAAA OF 2008 (http://www.eeoc.gov/laws/regulations/ada.qa.final.rule.cfm)

Under the ADA and the ADAAA, the role of physician as gatekeeper and authority in terms of patient and work is clearly defined.

The gatekeeper

The gatekeeper determines when a patient's diagnosis is becoming a component of daily functioning and performing work tasks. The physician needs to broach the subject, and allow for questions, emotions, fears, and facts to be clarified.

Authority

The physician acts as an additional team member (beyond the patient) to endorse, request, require, and document the changes needed by the patient and when.

The ADA and its follow-up, the ADAAA, are current and evolving federal legislation that holds many options for patients as they work through the many facets of their neurologic conditions.

The US Equal Employment Opportunity Commission (EEOC), which is separate from Congress, is charged by Congress to clarify and define "disability" under the act. The ADA was first implemented in 1999. Many lauded this ground-breaking position by the federal government to offer the idea that people with disabilities were to be given an equal footing on which to participate in work that they were qualified for. The original ADA did not offer much more than that important idea. But the ADAAA of 2008 provided clarity as to what were the responsibilities of the parties concerned.

The following highlights items covered under ADAAA 2008:

- It is ot retroactive prior to January 1, 2009.
- Disability includes many types of impairment, such as progressive diseases "such as epilepsy, diabetes, multiple sclerosis, major depression, and bipolar disorder."
- ADAAA increases the number of agencies and organizations that are now required to comply with basically every federal, state, local, county, and municipality formal organization.
- ADAAA broadened the term "disability" to include a three-pronged approach: "A physical or mental impairment that substantially limits one or more major life activity, or a record of a physical or mental impairment that substantially limited a major life activity, or when a covered entity takes an action prohibited by the ADA because of an actual or perceived impairment that is not both transitory and minor."
- ADAAA reinforced the "reasonable accommodation" method for a person to broach the issue of his or her need and therefore automatically be allowed to request what needs to be done to continue to work.
- ADAAA decreased the burden of "providing a record" as the only vehicle to document needs.

Of course, there are many other important issues that were clarified under the ADAAA that should not be overlooked. The bottom line now is that any aspect of disability can be brought up as a legitimate need by any person, including patients themselves.

Congress charged the EEOC with further defining "disability" and revisions were made that were implemented in 2011. Now most private companies and federal programs fall under the clearer ADAAA. Small enterprises of 15 employees or less may not fall under the EEOC jurisdiction but the Job Accommodation Network reminds us that now state and local laws reflect the purpose of the ADAAA. So every employer falls under the expectations and guidelines set up by the EEOC.

The ADAAA clarifies what a disability is under the law with the important wording: "Major life activities," including: caring for oneself, performing manual tasks, seeing, hearing, eating, sleeping, walking, standing, sitting, reaching, lifting, bending, speaking, breathing, learning, reading, concentrating, thinking, communicating, interacting with others, and working. The final regulations also state that major life activities include the operation of "major bodily functions," including: functions of the immune system, special sensory organs and skin, normal cell growth, digestive, genitourinary, bowel, bladder, neurologic, brain, respiratory, circulatory, cardiovascular, endocrine, hemic, lymphatic, musculoskeletal, and reproductive functions. Although not specifically stated in the Notice of Proposed Rulemaking, the final regulations state that major bodily functions include the operation of an individual organ within a body system (e.g., the operation of the kidney, liver, or pancreas).

The ADA and ADAAA are links between employee and employer, and may act as a model for problem solving in the workplace. This allows for a workable, malleable, measurable dialog or form of communication that employers and employees use to dial in the needs of the relationship. There is no federal accommodation form. In essence, the ADA and ADAAA equip employees with a method of requesting a change to their employment agreement because of a disability. This communication is called an accommodation. Under the ADAAA, the request for the accommodation has been streamlined and is now a very simple process that can be implemented by the employee, family, friend, caregiver, or doctor. The request need not be in writing and documentation of the reason should not be lengthy. The employee can then go on doing his or her job and fulfilling the role for which s/he was hired. The employee can "enjoy the Benefits and Privileges of Employment, and all the good things that come with that." Concurrently, the ADA gives employers an adaptable, measurable response to that specific employee need or accommodation. Then the employer can benefit from the skills of the employee, knowing the core qualifications of the employee are enriching the employer's concerns and the job is being performed by a qualified individual.

BENEFITS AND PRIVILEGES OF EMPLOYMENT

The ADA ensures that the entire scope of working is privy to accommodations. It "Covers the whole process from application, testing, and training, meeting job performance standards, taking part in services available from the job like the Credit Union or job parties." However the employee or prospective employee must have the necessary qualifications for the job. The diagnosis is not required on the job application, unless accommodations are requested for pre-employment testing. The employer cannot charge for these pre-employment testing accommodations, that may include:

- More time, allowance for breaks, multiple day testing.
- A scribe, a reader, providing a proctor, using a clock or watch, or calculator, allowing test taker to use written task list.
- Alternate communication format, bolding, coloring, enlarging print, using a dictionary or thesaurus, allowing for verbal response.
- Different space, reducing/eliminating distractions.
- Using symbols instead of words, allowing assistance of caregiver or care animal during the testing time, allowing test administrators a plan of action should symptoms arise.
- Allowing sunglasses or baseball cap, glare guard.

REQUEST FOR REASONABLE ACCOMMODATIONS BY EMPLOYEE

- Be specific, be very specific.
- Ask for what you want.
- Ask for the length of time you will need.
- Give examples of what is hard/difficult to do now that will be made better by having this accommodation.

The ADAAA encourages a natural back-and-forth communication about what solutions have been tried and which ones have been effective, allowing for a number of possible fixes for the specific needs to be broached. So, therefore, there should be some confidence in asking for an accommodation or working with a supervisor on past solutions that have been effective for others. Additionally, if needs for accommodation arise because of any disability or side-effects of treatments, either known before employment or discovered during employment, these are the employee's needs covered by the ADA. However, illegal drug use of any kind (prescription or nonprescription) is not covered under ADA.

EMPLOYER RESPONSIBILITY

- Dialog about accommodations cannot be initiated by the employer.
- Employer may request documentation but may not request documentation of the disability if it is very plain to see or has been previously documented.
- Employer may require a "nexus" between reasonable accommodations and employee limitations. For example, the employee cannot demand to work from home because gas is too expensive, as this does not represent a link between the disability and accommodation.
- Employer may try cheaper, easier accommodation, but the accommodation must be effective.
- Employer may cite undue hardship if the reasonable accommodation is too expensive for the scope of the job, the employer, or the fundamental way that business has been done. "Undue hardship means significant difficulty or expense and focuses on the resources and circumstances of the particular employer in relationship to the cost or difficulty in providing a specific accommodation. Undue hardship not only refers to financial difficulty, but to reasonable accommodation requests that are unduly extensive, substantial, and disruptive or those that would fundamentally alter the nature or operation of the business."

Table 28.4 is a summary of accommodations by neurologic disorder. These accommodations are further expanded below.

Table 28.4

Accommodations by neurologic disorder

		Conditions							
	Accommodations	Aging	Back pain	Brain injury	Chronic pain	CRPS/RSD	Multiple sclerosis	Myasthenia gravis	Parkinson's disease
	Activities of daily living, physical limitations, allow for treatment/therapy visits	*	*	*	*	*	*	*	*
	Fatigue/weakness	*			*	*	*	*	*
	Physical discomfort	*	*	*	*	*			
	Fine and gross motor	*	*	*	*	*	*	*	*
	Mood issues (depression/anxiety)	*	*	*	*	*	*	*	*
Sense organs	Hearing	*							
	Speech			*			*	*	*
	Vision	*					*	*	
Special considerations	Respiratory	*							
	Stress management		*		*	*			
	Cognitive		*	*		*			
	Concentration		*	*	*	*			
	Symptom fluctuation		*			*	*		
	Meeting deadlines			*					
	Getting along with co-workers			*	*				
	Temperature sensitivity	*				*	*		
	Stamina							*	

Data compiled from: Job Accommodation Network (JAN) Accommodations and Compliance Series from The Office of Disability Employment Policy (ODEP), an agency within the US Department of Labor. http:/askjan.org/media/atoz.htm.
CRPS, chronic regional pain syndrome; RSD, reflex sympathetic dystrophy.

Activities of daily living

- Allow use of a personal attendant at work.
- Allow use of a service animal at work.
- Make sure the facility is accessible.
- Move workstation closer to the restroom.
- Allow longer breaks or more frequent, shorter breaks.

Physical limitations

- Install ramps and handrails, and provide handicap parking spaces.
- Install level-style door handles.
- Clear pathways of travel of any unnecessary equipment or furniture.

Allow for treatment/therapy visits

- Provide flexible schedules.
- Provide flexible leave.
- Provide part-time work schedules.
- Provide flexible leave for health problems.

Fatigue/weakness

- Reduce or eliminate physical exertion and workplace stress.
- Schedule periodic rest breaks away from the workstation.
- Allow a flexible work schedule and flexible use of leave time.
- Allow a self-paced workload.
- Allow the employee to work from home.
- Provide a scooter or other mobility aid if walking cannot be reduced.
- Provide parking close to the work site and an accessible entrance.
- Install automatic door openers.
- Provide an accessible route of travel to other work areas used by the employee.
- Move workstation close to other work areas, office equipment, and break rooms.

Physical discomfort

- Implement ergonomic workstation design, i.e., ergonomic chair.
- Use a workstation that alternates between being able to sit or stand.
- Reduce repetitive tasks or interrupt the tasks with other duties.
- Provide carts and lifting aids.
- Modify work-site temperature and/or dress code.
- Use fan/air conditioner or heater at the workstation.
- Allow the employee to work from home during extremely hot or cold weather.

Fine motor impairment

- Implement ergonomic workstation design.
- Provide alternative computer access.
- Provide alternative telephone access.
- Provide arm supports.
- Provide writing and grip aids.
- Provide a page turner and a book holder.
- Provide a note taker.

Gross motor impairment

General

- Modify the work site to make it accessible.
- Provide parking close to the work site.
- Provide an accessible entrance.
- Install automatic door openers.
- Provide an accessible restroom and break room.
- Provide an accessible route of travel to other work areas used by the employee.
- Modify the workstation to make it accessible.
- Make sure materials and equipment are within reach range.
- Move workstation close to other work areas and break rooms.

Industrial

- Provide overhead structure for lifting devices.
- Modify the work area to make it adjustable.
- Place frequently used tools and supplies at or near waist height.
- Provide low-task chairs for work that cannot be brought to waist height.
- Provide stand/lean stools and antifatigue mats for standing work.
- Make wheelchairs, scooters, industrial tricycles, or golf carts available if walking long distances is required.
- Provide compact lifting devices to push and pull supplies and tools from storage.

Office settings

- Provide a height-adjustable desk and ergonomic chair.
- Move workstation close to commonly used office equipment.
- Provide low-task chair and rolling safety ladder to access high and low file drawers and supplies.
- Provide a cart to move files, mail, and supplies.

- Provide a lazy Susan carousel or desktop organizers to access frequently used materials.

Service settings

- Provide antifatigue mats and stand/lean stools for functions requiring long periods of standing.
- Provide a height-adjustable desk and ergonomic chair.
- Move workstation close to commonly used office equipment.
- Provide compact lifting devices to push and pull stock and supplies from shelves,
- Provide carts to move supplies and stock.

Medical settings

- Provide a spring-bottomed linen cart.
- Make patient-lifting and transfer devices available.
- Make wheelchairs, scooters, industrial tricycles, or golf carts available if walking long distances is required.
- Train employees on proper lifting techniques and on proper use of patient-lifting and transfer devices.
- Provide powered beds for transporting patients.
- Provide adjustable exam tables.
- Provide a height-adjustable desk and ergonomic task chairs.

Mood issues (depression/anxiety)

- Reduce distractions in work environment.
- Provide to-do lists and written instructions.
- Remind employee of important deadlines and meetings.
- Allow time off for counseling.
- Provide clear expectations of responsibilities and consequences.
- Provide sensitivity training to co-workers.
- Allow breaks to use stress management techniques.
- Develop strategies to deal with work problems before they arise.
- Allow telephone calls during work hours to doctors and others for support.
- Provide information on counseling and employee assistance programs.

SENSORY ORGANS

Hearing limitations

- Provide visual or tactile alerting device.
- Implement a buddy system.
- Provide an assistive listening device.
- Provide communication access real-time translation.
- Provide computer-assisted note taking.
- Address environmental factors, i.e., background noise, lighting, and positioning.
- Provide an interpreter.
- Use web-based meeting software or video conferencing.
- Provide standard note taking or other text information.
- Allow tape recording of meetings.
- Provide speech recognition software.

Speech impairment

- Provide speech amplification, speech enhancement, or other communication device.
- Use written communication, such as e-mail or fax.
- Transfer to a position that does not require a lot of communication.
- Allow periodic rest breaks.

Vision impairment

- Magnify written material using hand/stand/optical magnifiers.
- Provide large-print material or screen-reading software.
- Control glare by adding a glare screen to the computer.
- Install proper office lighting.
- Allow frequent rest breaks.

SPECIAL CONSIDERATIONS

Respiratory difficulties

- Provide adjustable ventilation.
- Keep work environment free from dust, smoke, odor, and fumes.
- Implement a "fragrance-free" workplace policy and a "smoke-free" building policy.
- Avoid temperature extremes.
- Use fan/air conditioner or heater at the workstation.
- Redirect air-conditioning and heating vents.

Stress management

- Allow longer or more frequent work breaks as needed.
- Provide backup coverage for when the employee needs to take breaks.
- Provide additional time to learn new responsibilities.

- Restructure job to include only essential functions during times of stress.
- Assign a supervisor, manager, or mentor to answer the employee's questions.
- Allow for time off for counseling.

Cognitive

- Provide written job instructions when possible with more structure.
- Prioritize job assignments.
- Allow flexible work hours.
- Allow periodic rest breaks to reorient.
- Provide memory aids, such as schedulers or organizers.
- Minimize distractions.
- Allow a self-paced workload.
- Reduce job stress.

Concentration

- Reduce distractions in the work area.
- Provide space enclosures or a private office.
- Allow for use of white noise or environmental sound machines.
- Allow the employee to play soothing music using a mobile device and headset.
- Increase natural lighting or provide full-spectrum lighting.
- Reduce clutter in the employee's work environment.
- Plan for uninterrupted work time.
- Divide large assignments into smaller tasks and steps.
- Restructure job to include only essential functions.

Symptom fluctuation

- Allow for open channel of communication between employee and supervisor.
- Plan out projects in advance.
- Use a wall calendar to emphasize due dates.
- Use a handheld or electronic calendar to tick off days until a project is due.
- Provide a checklist of assignments.
- Plan and structure times for projects to begin, be evaluated, and be finalized.
- Allow additional time for new tasks, and allow relinquishment of routine tasks as new tasks are assigned.
- Color-code tasks according to priority.
- Use a job coach or shared job to reinforce completion of a project.
- Assign a mentor or friend on the job.
- Automate paperwork by creating electronic files wherever possible.
- Provide templates and boilerplate solutions to routine forms.

Meeting deadlines and staying organized

- Make daily to-do lists and check items off as they are completed.
- Use several calendars to mark meetings and deadlines.
- Remind employee of important deadlines via memos or e-mail or weekly supervision.
- Use a watch or App with timer capability.
- Use electronic organizers.
- Divide large assignments into smaller tasks and steps.
- Assign a mentor to assist employee in determining goals and provide daily guidance.
- Schedule weekly meetings with supervisor, manager, or mentor to determine if goals are being met.

Getting along with co-workers

- Provide positive praise and reinforcement.
- Provide written job instructions.
- Write clear expectations of responsibilities and the consequences of not meeting them.
- Allow for open communication with managers and supervisors.
- Establish written long-term and short-term goals.
- Develop strategies to deal with problems before they arise.
- Provide written work agreements.
- Develop a procedure to evaluate the effectiveness of the accommodation.

Temperature sensitivity

- Reduce work-site temperature.
- Use cool vest or other cooling clothing.
- Use fan/air conditioner at the workstation.
- Allow flexible scheduling and flexible use of leave time.
- Allow the employee to work from home during hot weather.

Stamina

- Permit flexible scheduling; allow longer or more frequent work breaks.
- Provide additional time to learn new responsibilities.

- Provide self-paced workload.
- Provide backup coverage for when the employee needs to take breaks.
- Allow time off for counseling.
- Allow for use of supportive employment and job coaches.
- Allow employee to work from home during part of the day.
- Provide for job-sharing opportunities.
- Allow part-time work schedules.

Employers reported benefits from accommodating workers because they retained a qualified employee, increased worker productivity, eliminated the costs of training a new employee, improved interactions with co-workers, and increased overall company morale (Solovieva et al., 2011).

References

Abe K, Takanashi M, Yanagihara T et al. (2000). Fatigue in patients with Parkinson's disease. Behav Neurol 12: 103–106.

Armstrong MJ, Gruber-Baldini AL, Reich SG et al. (2014). Which features of Parkinson's disease predict earlier exit from the workforce? Parkinsonism Rel Disord 20: 1257–1259.

Arnemann KL, Chen AJW, Novakovic-Agopian T et al. (2015). Functional brain network modularity predicts response to cognitive training after brain injury. Neurology 84: 1568–1574.

Bahar-Fuchs A, Clare L, Woods B (2013). Cognitive training and cognitive rehabilitation for persons with mild to moderate dementia of the Alzheimer's or vascular type; a review. Alzheimer's Res Ther 5: 35.

Bakshi R, Miletich RD, Kinkel PR et al. (1998). High resolution fluorodeoxyglucose positron emission tomography shows both global and regional cerebral hypometabolism in multiple sclerosis. J Neuroimaging 8: 228–234.

Bella SD, Benoit CE, Farrugia N et al. (2015). Effects of musically cued gait training in Parkinson's disease: beyond a motor benefit. Ann N Y Acad Sci 1337: 77–85.

Benito-Leon J, Martinez-Martin P, Frades B et al. (2007). Impact of fatigue in multiple sclerosis: the Fatigue Impact Scale for Daily Use (D-FIS). Mult Scler 13 (5): 645–651.

Benoit CE, Bella SD, Farrugia N et al. (2014). Musically cued gait-training improves both perceptual and motor timing in Parkinson's disease. Front Neuro Sci 8: 494.

Burns TM, Grouse CK, Conaway MR et al. (2011). The MG-QOL15 for following the health- related quality of life for patients with myasthenia gravis. Muscle Nerve 43: 14–18.

Burns TM, Graham CD, Rose MR et al. (2012). Quality of life and measures of quality of life in patients with neuromuscular disorders. Muscle Nerve 46: 9–25.

Bushnik T, Englander J, Wright J (2008). The experience of fatigue in the first 2 years after moderate-to-severe traumatic brain injury: a preliminary report. J Head Trauma Rehabil 23: 17–24.

Byrne M, Troy A, Bradley LA et al. (1982). Cross-validation of the factor structure of the McGill Pain Scale Questionnaire. Pain 13: 193–201.

Cernich AN, Kurtz SM, Mordecai KL et al. (2010). Cognitive rehabilitation in traumatic brain injury. Curr Treat Options Neurol 12: 412–423.

Chiaravalloti ND, Moore NB, Nikelsphur OM et al. (2013). An RTC to treat learning impairment in multiple sclerosis. The MEMREHHA trial. Neurology 81: 2006–2072.

Collins C, Gehrke A, Feuerstein M (2013). Cognitive tasks challenging brain tumor survivors at work. J Occup Env Med 55: 1426–1430.

Dubois B, Pillon B (1997). Cognitive deficits in Parkinson's disease. J Neurol 244: 2–8.

Englander J, Hall K, Stimpson T et al. (1992). Mild traumatic brain injury in an insured population: subjective complaints and return to employment. Brain Inj 6: 161–166.

Fahn S, Elton RLthe UPDRS Development Committee (1987). Unified Parkinson's Disease Rating Scale. In: S Fahn, CD Mardsen, D Calne et al. (Eds.), Recent developments in Parkinson's Disease, Vol. 2. Macmillian Healthcare Information, Florham Park, pp. 153–163.

Fisk JD, Doble SE (2002). Construction and validation of a fatigue impact scale for daily administration (D-FIS). Qual Life Res 11: 263–272.

Friedman JH, Brown RG, Comella C et al. (2007). Fatigue in Parkinson's disease: a review. Mov Disord 22: 297–308.

Fundaro C, Casale R (2012). Rehabilitation and dementia in the era of elderly workers. G Ital Med Lav Erg 34 (2): 197–201.

Gao Q, Leung A, Yang Y et al. (2013). Effects of Tai Chi on balance and fall prevention in Parkinson's disease: a randomized controlled trial. Clin Rehabil 28: 748–753.

Hartzell MM, Mayer TG, Asih S et al. (2014). Evaluation of functional restoration outcomes for chronic disabling occupational cervical disorders. J Occup Env Med 56: 959–964.

Hulst HE, Steenwijk MD, Versteeg A et al. (2013). Cognitive impairment in MS. Neurology 80: 1025–1032.

Jenkinson C, Fitzpatrick R, Peto V et al. (1997). The PDQ-8: Development and validation of a short-form Parkinson's disease questionnaire. Psychol Health 12: 805–814.

Johnson KL, Yorkston KM, Klasner ER et al. (2004a). The cost and benefit of employment: a qualitative study of experiences of persons with multiple sclerosis. Arch Phys Med Rehabil 85: 201–209.

Johnson KL, Amtmann D, Yorkston KM et al. (2004b). Medical, psychological, social, and programmatic barriers to employment for people with multiple sclerosis. J Rehabil 70: 38–49.

Johnson KL, Bamer AM, Yorkston KM et al. (2009). Use of cognitive aids and other assistive technology by individuals with multiple sclerosis. Disabil Rehabil Assist Technol 4: 1–8.

Kaiser PR, Valko PO, Werth E et al. (2010). Modafinil ameliorates excessive daytime sleepiness after traumatic brain injury. Neurology 75: 1780–1785.

Kittiwatanapaisan W, Gauthier D, Williams A et al. (2003). Fatigue in myasthenia gravis patients. J Neurosci Nurs 35: 87–93.

Kluger BM, Krupp LB, Enoka RM (2013). Fatigue and fatigability in neurologic illnesses: a proposal for a unified taxonomy. Am Acad Neuro 80: 409–416.

Kressig RW, Allali G, Beauchert O (2005). Long-term practice of Jacques-Dalcroze eurhythmics prevent age-related increase of gait variability under a dual task. J Am Geriatr Soc 53 (4): 728–729.

Kutlubaev MS, Mead GE (2012). One step closer to understanding poststroke fatigue. Neurology 79: 1414–1415.

Leininger BE, Gramling SE, Farrell AD et al. (1990). Neuropsychological deficits in symptomatic minor head injury patients after concussion and mild concussion. J Neurol Neurosurg Psychiatry 53: 293–296.

Lipton ML, Gulko E, Zimmerman ME et al. (2009). Diffusion-tensor imaging implicates prefrontal axonal injury in executive function impairment following very mild traumatic brain injury. Radiology 242: 816–824.

Marrie RA, Fisher E, Miller DM et al. (2005). Association of fatigue and brain atrophy in multiple sclerosis. J Neuro Sci 228: 161–166.

Martinez-Martin P, Catalan M, Benito-Leon J et al. (2006). Impact of fatigue in Parkinson's disease: the Fatigue Impact Scale for Daily Use (D-FIS). Qual Life Res 15: 597–606.

Murphy R, Tubridy N, Kevelighan H et al. (2013). Parkinson's disease: how is employment affected. Ir J Med Sci 182: 415–419.

Ni X, Liu S, Lu F et al. (2014). Efficacy and safety of tai Chi for Parkinson's disease: a systemic review and meta-analysis of randomized controlled trials. PLoS One 9: e99377.

Olver JH, Ponsford JL, Curran CA (1996). Outcome following TBI: a comparison between 2 and 5 years post injury. Brain Inj 10: 841–848.

Paul RH, Cohen RA, Goldstein JM et al. (2000). Fatigue and its impact on patients with myasthenia gravis. Muscle Nerve 23: 1402–1406.

Radman N, Staub F, Aboulafia-Brakha T et al. (2012). Poststroke fatigue following minor infarcts: a prospective study. Neurology 79: 1422–1427.

Rao SM, Leo GJ, Ellington L et al. (1991). Cognitive dysfunction in multiple sclerosis: II: impact on employment and social functioning. Neurology 41: 692–696.

Rasotto C, Bergamin M, Sieverdes JC et al. (2015). A tailored exercise program for women at risk for neck and upper limb musculoskeletal disorders. A randomized controlled trial. J Occup Env Med 57: 178–183.

Richardson JK, Demott T, Allet L et al. (2014). Hip strength: ankle proprioceptive threshold ratio predicts falls and injury in diabetic neuropathy. Muscle Nerve 50: 437–442.

Rochester L, Hetherington V, Jones D et al. (2004). Attending to the task: interference effects of functional tasks on walking in Parkinson's disease and the roles of cognition, depression, fatigue and balance. Arch Phys Med Rehabil 85: 1578–1585.

Roelcke U, Kappes L, Lechner-Scott J et al. (1997). Reduced glucose metabolism in the frontal cortex and basal ganglia of multiple sclerosis patients with fatigue: a 18F-fluorodeoxyglucose positron emission tomography study. Neurology 48: 1566–1571.

Schneider EB, Sur S, Raymont V et al. (2014). Functional recovery after moderate/severe traumatic brain injury. A role for cognitive reserve? Neurology 82: 1636–1642.

Shanahan J, Morris ME, Bhriain ON et al. (2015). Dance for people with Parkinson disease: what is the evidence telling us? Am Congenit Rehabil Med 96: 141–163.

Shaw WS, Kristman VL, Williams-Whitt K et al. (2014). The Job Accommodation Scale (JAS): psychometric evaluation of a new measure of employer support for temporary job modifications. J Occup Rehabil 24: 755–765.

Smith MM, Arnett PA (2005). Factors related to employment status changes in individuals with multiple sclerosis. Mult Scler 11: 602–609.

Solovieva TI, Dowler DL, Walls RT (2011). Employer benefits from making workplace accommodations. Disabil Health J 4: 39–45.

Stanton BR, Barnes F, Silber E (2006). Sleep and fatigue in multiple sclerosis. Mult Scler 12: 481–486.

Stewart WF, Ricci JA, Chee E et al. (2003). Lost productive time and costs due to common pain conditions in the US Workforce. JAMA 290: 2443–2454.

Strober LB, Arnett PA (2005). An examination of four models predicting fatigue in multiple sclerosis. Arch Clin Neuropsychol 20: 631–646.

Strober LB, Christodoulou C, Benedict RH et al. (2012). Unemployment in multiple sclerosis: the contribution of personality and disease. Mult Scler 18: 647–653.

Tartaglia MC, Narayanan S, Francis SJ et al. (2004). The relationship between diffuse axonal damage and fatigue in multiple sclerosis. Arch Neurol 61: 201–207.

Tersteeg IM, Koopman FS, Stolwijk-Swuste JM et al. (2011). A 5-year longitudinal study of fatigue in patients with late-onset sequelae of poliomyelitis. Arch Phys Med Rehabil 92: 899–904.

Toossi M (2013). Labor force projections to 2022: The labor force participation rate continues to fall. US Bureau of Labor Statistics, Monthly Labor Review. December. Available online at: http:/www.bls.gov/opub/mlr/2013/article/pdf/labor-force-projections-to-2022-the-labor-force-participation-rate-continues-to-fall.pdf.

Trojan DA, Cashman NR (2005). Post-poliomyelitis syndrome. Muscle Nerve 31: 6–19.

Trombetti A, Hars M, Herrmann FR et al. (2011). Effect of music-based multitask training on gait, balance, and fall risk in elderly people. Arch Intern Med 171: 525–533.

Valko PO, Bassetti CL, Bloch KE et al. (2008). Validation of the Fatigue Severity Scale in a Swiss cohort. Sleep 31 (11): 1601–1607.

Velseboer DC, Broeders M, Post B et al. (2013). Prognostic factors of motor impairment, disability, and quality of life in newly diagnosed PD. Neurology 80: 627–633.

Vinik EJ, Hayes RP, Oglesby A et al. (2005). The development and validation of the Norfolk QoL-DN, a new measure of patients' perception of the effects of diabetes and diabetic neuropathy. Diabetes Technol Ther 7: 497–508.

Index

NB: Page numbers in *italics* refer to boxes, figures or tables.

D

N

U

V

W

X

Y

Z

FSC
www.fsc.org
MIX
Papier | Fördert gute Waldnutzung
FSC® C083411

Kolibri 360 GmbH
Ferdinand-Jühlke-Straße 7
99095 Erfurt, Deutschland
produktsicherheit@kolibri360.de